Conversion Factors for U.S. Customary and SI Units[a]

QUANTITY	U.S. CUSTOMARY TO SI[b]	SI TO U.S. CUSTOMARY[b]
length	**1 in = 2.540 000 E + 01 mm**	**1 mm = 3.937 008 E − 02 in**
	1 in = 2.540 000 E − 02 m	**1 m = 3.937 008 E + 01 in**
	1 ft = 3.048 000 E − 01 m	**1 m = 3.280 840 E + 00 ft**
	1 yd = 9.144 000 E − 01 m	**1 m = 1.093 613 E + 00 yd**
	1 mi (intl.) = 1.609 344 E + 00 km	**1 km = 6.213 712 E − 01 mi (intl.)**
	1 mi (naut.) = 1.609 347 E + 00 km	1 km = 6.213 700 E − 01 mi (naut.)
force	1 lb = 4.448 222 E + 00 N	1 N = 2.248 089 E − 01 lb
	1 oz = 2.780 139 E − 01 N	1 N = 3.596 942 E + 00 oz
	1 kip = 4.448 222 E + 03 N	1 kN = 2.248 089 E − 01 kip
mass	1 slug = 1.459 390 E + 01 kg	1 kg = 6.852 178 E − 02 slug
	1 lbm[c] = 4.535 924 E − 01 kg	1 kg = 2.204 622 E + 00 lbm
area	**1 in^2 = 6.451 600 E − 04 m^2**	**1 m^2 = 1.550 003 E + 03 in^2**
	1 ft^2 = 9.290 304 E − 02 m^2	**1 m^2 = 1.076 391 E + 01 ft^2**
volume, solids	1 in^3 = 1.638 706 E − 05 m^3	1 m^3 = 6.102 376 E + 04 in^3
	1 ft^3 = 2.831 685 E − 02 m^3	1 m^3 = 3.531 466 E + 01 ft^3
volume, liquids	1 gal (U.S.) = 3.785 412 E − 03 m^3	1 m^3 = 2.641 720 E + 02 gal (U.S.)
	1 gal (U.S.) = 3.785 412 E + 00 L	1 L = 2.641 720 E − 01 gal (U.S.)
	1 qt (U.S.) = 9.463 529 E − 01 L	1 L = 1.056 688 E + 00 qt (U.S.)
velocity	**1 in/s = 2.540 000 E − 02 m/s**	**1 m/s = 3.937 008 E + 01 in/s**
	1 ft/s = 3.048 000 E − 01 m/s	**1 m/s = 3.280 840 E + 00 ft/s**
	1 mi/h (intl.) = 4.470 400 E − 01 m/s	**1 m/s = 2.236 936 E + 00 mi/h (intl.)**
	1 mi/h (intl.) = 1.609 344 E + 00 km/h	**1 km/h = 6.213 712 E − 01 mi/h (intl.)**
acceleration	**1 in/s^2 = 2.540 000 E − 02 m/s^2**	**1 m/s^2 = 3.937 008 E + 01 in/s^2**
	1 ft/s^2 = 3.048 000 E − 01 m/s^2	**1 m/s^2 = 3.280 840 E + 00 ft/s^2**
pressure/stress	1 lb/in^2 = 6.894 757 E + 03 N/m^2	1 N/m^2 = 1.450 377 E − 04 lb/in^2
	1 lb/ft^2 = 4.788 026 E + 01 N/m^2	1 N/m^2 = 2.088 544 E − 02 lb/ft^2
energy/work	1 in·lb = 1.129 848 E − 01 N·m	1 N·m = 8.850 748 E + 00 in·lb
	1 ft·lb = 1.355 818 E + 00 N·m	1 N·m = 7.375 621 E − 01 ft·lb
power	1 in·lb/s = 1.129 848 E − 01 N·m/s	1 N·m/s = 8.850 748 E + 00 in·lb/s
	1 ft·lb/s = 1.355 818 E + 00 N·m/s	1 N·m/s = 7.375 621 E − 01 ft·lb/s
momentum/impulse	1 lb·s = 4.448 222 E + 00 N·s	1 N·s = 2.248 089 E − 01 lb·s
moment of force	1 lb·in = 1.129 848 E − 01 N·m	1 N·m = 8.850 748 E + 00 lb·in
	1 lb·ft = 1.355 818 E + 00 N·m	1 N·m = 7.375 621 E − 01 lb·ft
moment of inertia of area	1 in^4 = 4.162 314 E − 07 m^4	1 m^4 = 2.402 510 E + 06 in^4
	1 in^4 = 4.162 314 E + 05 mm^4	1 mm^4 = 2.402 510 E − 06 in^4
moment of inertia of mass	1 lb·in·s^2 = 1.129 848 E − 01 N·m·s^2	1 N·m·s^2 = 8.850 748 E + 00 lb·in·s^2
	1 lb·ft·s^2 = 1.355 818 E + 00 N·m·s^2	1 N·m·s^2 = 7.375 621 E − 01 lb·ft·s^2

a. Each conversion factor is written as it would appear on a computer readout or in electronic data transmission: as a number equal to or greater than 1 and less than 10, with six or fewer decimal places, followed by the letter E (to denote exponent of 10), a plus or minus sign, and two digits that indicate the power of 10 by which the preceding number must be multiplied. For example, the factor 9.463 529 E − 04 is, in exponential notation, $9.463\,529 \times 10^{-4}$, or 0.000 946 3529.

b. Factors printed in boldface are exact.

c. The abbreviation lbm stands for "pound-mass." Although the pound is interpreted as a unit of force in this book, you should be aware that it is occasionally used as a unit of mass.

Dedicated to our children:
Jennifer, Annette, and Nancy
Marshall, Charles, and Catherine

CONTENTS

TO THE INSTRUCTOR

Our intention in writing this book is to provide a thorough, rigorous presentation of mechanics, augmented with proven learning techniques for the benefit of instructor and student. Our first objective is to present the topics thoroughly and directly, allowing fundamental principles to emerge through applications to real-world problems. We emphasize concepts, derivations, and interpretations of the general principles, and we discuss the applicability and limitations of each principle. We illustrate that general rules frequently have exceptions; however, we do not dwell on these to the extent that they become a distraction. This book presents, in the clearest form possible, more theorems, more proofs, and more explanation than you will find in most introductory engineering mechanics texts.

Our second objective is unique. We have attempted to integrate learning principles and teaching techniques that improve students' ability to grasp and absorb concepts. In general, texts in engineering mechanics focus exclusively on the technical principles, with no structure to increase student comprehension. The integrated use of learning aids in this book is based on our experience that students can be taught efffective study habits while they learn mechanics.

Each chapter in this book is organized around the SQ3R study-reading method.[1] This structured approach to reading directs the student to:

- develop a global view of the course material one section or chapter at a time,
- organize the material into manageable pieces and read each piece for content and comprehension, and
- review the material as a coherent whole.

Details of the SQ3R method are described in *To the Student*, which also offers tips on note taking and presents a formal strategy for problem solving. The SQ3R method offers an integrated approach to studying mechanics that requires no additional preparation or follow-up by the instructor.

FEATURES AND DESIGN

To maximize students' success, instructors must capture their interest and attention. The instructor's task is to cultivate students' interest, while educating them in the fundamentals and broad applications of mechanics. To facilitate that task, we attempt to present mechanics in the most exciting and relevant context possible. Since the study of mechanics is not limited to engineering practice, we include examples and illustrations of mechanics at work in many different applications. For a wider perspective and a sense of tradition, we include historical references to balance the technical presentation.

Much of engineering mechanics involves the development of mathematical models of the physical world. In particular, visualization of three-dimensional physical systems and development of appropriate models are of paramount importance. We facilitate this process throughout the text by including pertinent photographs and figures emphasizing a wide range of applications from simple to complex.

1. SQ3R is an acronym for Survey, Question, Read, Recite, and Review (Robinson, 1970).

Throughout the text, we integrate *Problem-Solving Techniques.* These are not step-by-step solution methods to be used mechanically by the students, but rather summaries of the key concepts and issues that must be considered to solve particular types of problems.

A large number of example and homework problems provide students with ample opportunities to apply concepts and principles to familiar, meaningful situations. The homework problems include relatively simple introductory problems that test students' basic understanding, as well as more challenging problems requiring a deeper grasp of concepts and techniques. They also include problems that are best solved with the aid of computer-based tools (spreadsheets, equation solvers, and plotting tools) and problems that are design-oriented. The design problems are intended to introduce students to the kinds of decisions that will be expected of them throughout their engineering studies and in professional practice and to give them an appreciation for the importance of engineering mechanics in the design process. These problems do not cover all aspects of design, but are intended to introduce students to the challenge of creating a part or a system that will perform a particular function. Icons identify these three types of homework problems:

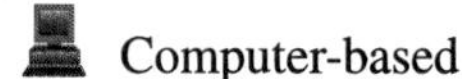

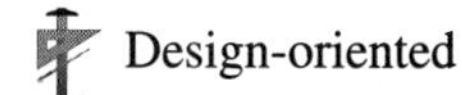

Solving numerical problems alone is not sufficient to allow students to learn the fundamental concepts that underlie the equations and solution techniques used in mechanics. So, we have included in each chapter *Survey Questions, Checkpoints,* and *Review Questions* that challenge students' understanding of fundamental concepts. These short-answer questions, appropriate for homework assignments or examinations, help students develop a balance between understanding concepts and applying them to problem solving.

TEXT ORGANIZATION

Our primary objective is the application of the concepts and principles of mechanics to real-world problems. We have organized this book so that students can solve practical problems as early as possible.

- Chapter 1 serves as a review of the basic concepts of mechanics as taught in physics courses for engineers. This chapter may be used as a reference throughout the entire course.
- Chapter 2 introduces the concepts of force and other vector quantities.
- Chapter 3, on particle equilibrium, gives the students the background and opportunity to solve real-world problems. The critical importance of free-body diagrams is emphasized.
- Chapter 4 expands on the principles developed in the first few chapters by focusing on forces acting on rigid bodies, moments, and rigid-body equilibrium.
- The general theory of three-dimensional statics of a rigid body is presented in Chapter 5.
- Chapter 6 discusses statics of planar and space trusses.
- Chapter 7 examines several topics related to structures and machines.
- Distributed force resultants, centroids and body forces, fluid statics, and friction are presented in Chapters 8, 9, and 10 and are incorporated into the already familiar concept of rigid-body equilibrium.

- Chapter 11 focuses on the analysis of beams and shafts under concentrated and distributed forces and couples.
- Chapter 12 introduces work principles as an alternative method for solving statics problems and as a preparation for dynamics.
- Short appendixes provide reference material on simultaneous algebraic equations and determinants; geometric and trigonometric relations; centroids of lines, areas, and volumes; and area moments of inertia. (Some regard the study of moments of inertia as more appropriate for courses in mechanics of materials. However, we include a brief discussion here for those instructors who incorporate this topic in their statics courses.)

HOMEWORK PROBLEMS AND SOLUTIONS

Each homework problem in this book was solved by both the authors and students. We used student problem solvers to help us test the clarity of the problem statements, identify the relative difficulty of the problems, and obtain approximate solution times. As a result of this process, we edited or rewrote some problems. Each section of the Problems at the end of each chapter is arranged approximately in order of increasing time required to obtain the solution. The solution times are given in the *Solution Manual.*

We developed the solutions presented in the *Solution Manual* to be used by students as teaching aids after they have attempted to solve the problems. For the most part, the solutions are fairly detailed and similar in format to those presented in the text for the Examples. This approach allowed us to outline the problem-solving process, as well as give correct answers. The Preface to the *Solution Manual* describes more completely the process used to develop the problems and their solutions.

We believe that we have developed homework problems that are student-tried and student-tested, with solutions that are correct and complete. Nevertheless, we are solely responsible for any errors that exist. So, we welcome any feedback that you may wish to offer. Send corrections, comments, or suggestions for improvements to:

Arthur P. Boresi or Richard J. Schmidt
Department of Civil and Architectural Engineering
University of Wyoming
PO Box 3295
Laramie WY 82071-3295

ACKNOWLEDGMENTS

We wish to thank the following reviewers of the text:

Leon Bahar, Drexel University
William Bickford, Arizona State University
John B. Brunski, Rensselaer Polytechnic Institute
Susan L. Gerth, Kansas State University
Tribikram Kundu, University of Arizona
Vernon C. Matsen, North Carolina State University
George H. Staab, Ohio State University
William H. Walston, Jr., University of Maryland at College Park
Jonathan Wickert, Carnegie Mellon University

Special recognition goes to the following students for their exceptional contributions to the evaluation of the text and the production of the *Solution Manual:*

Angie Eicke
Joe D. Hall
Paul M. Schuman
Teresa Williams

The following students contributed to the text evaluation and to the problem solutions and time evaluations:

Robert Berland
Kathryn Call
Kathie Cook
Michelle Cummings
Angie Eicke
Joe D. Hall
David Hooper
Doug Jonart
Joe Kolnik
Scrinivas Kotha
John Levar
Jane Mari
Matt Ostrander
Matthew D. Peavy
Charles E. Prior
Damian Ray
Jennifer M. Reusser
Paul M. Schuman
David Stierwalt
Derek Swanson
Dong Qing Wang
Teresa Williams

Dr. Thomas R. Parish and Dr. Sally Steadman taught a beta version of this text to classes at the University of Wyoming.

TO THE STUDENT

The purpose of this book is to start you on the road to becoming a good engineer. Although fundamentals of physics and mathematics are important in this endeavor, we emphasize applications of physical and mathematical principles to engineering. While physicists are interested primarily in understanding the principles that govern the natural world and mathematicians focus on developing mathematical models to describe natural phenomena, engineers seek to create what does not exist in nature and to enrich people's lives by solving problems that confront modern society. Indeed, an engineer is a problem solver. To be an effective engineer, you must develop a thorough understanding of physical and mathematical principles and their application to the world around you.

Statics and dynamics form the basis for a large part of your study in engineering. The importance of these fields cannot be overstated. For many of you, the study of statics and dynamics will be your first opportunity to formally apply your skills in algebra, trigonometry, and calculus and to expand your understanding of how the laws of physics apply to real-world problems.

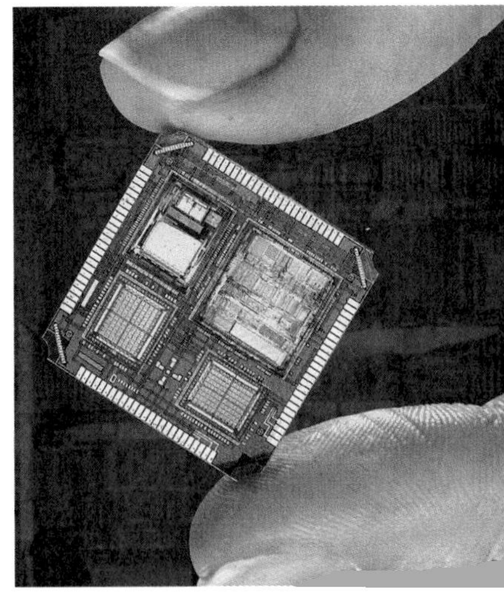

The designs of all engineering systems—from massive space vehicles to tiny microprocessor chips—are based on the fundamental laws of mechanics.

A systematic approach facilitates the solution of problems in mechanics. Knowledge of a set of mathematical tools and techniques for solving problems is not sufficient: You must also have motivation, ingenuity, and an organized strategy for attacking new problems. Our first objective in writing this book on engineering mechanics is to present the topics in a thorough and systematic way so that fundamental principles become evident through application to real-world problems, including examples from everyday engineering practice. We emphasize concepts, derivations, and interpretations of the general principles. Our goal is to give you a clear grasp of major concepts by presenting the simplest and most elementary derivations that are consistent with the demands of rigor. Then, we apply those concepts and derivations in example problems to illustrate their use.

Our second objective is to help you develop your study skills and problem-solving abilities. These cannot always be developed intuitively, but the use of problem-solving strategies can accelerate their development. We present such strategies, which you can apply to become a more efficient problem solver.

STUDYING AND LEARNING TOOLS

Engineering is a challenging field, both as a student and as a professional. To succeed, you must develop effective study habits, adding structure and discipline to your routine. We offer the following tools designed to help you accomplish this goal.

Reading—The SQ3R Method The process of reading a textbook for understanding is not the same as that of reading a novel for entertainment. One does not ordinarily skim over a novel to get the high points, jump to the last pages to learn the ending, and then go back to fill in the gaps. This approach would defeat the purpose of reading a novel. However, it is just what you should do to maximize your understanding and retention when reading a textbook. Developed by Francis Robinson (1970), the SQ3R method is a formal procedure that has proven effective in improving textbook reading comprehension. The name SQ3R (or SQRRR) stands for Survey, Question, Read, Recite, and Review.

- *Survey.* The best way to begin a new chapter is to skim over it.[2] Read the section headings, figure captions, highlighted items, chapter summary, and other emphasized material that indicates key ideas. This gives you a general idea of what will be discussed, how it is organized, and how the topics relate to one another and to what you already know.
- *Question.* Before reading each section, ask yourself what the section will cover and what you should learn from it. Use the section heading and subheadings, figure captions, and highlighted items as guides to ask yourself questions about the material. Be as specific as possible. Jot down your questions in outline form, leaving room for the answers. The questions need not be particularly insightful or complex; focus on the basics. This step allows you to create a mental framework for remembering the detailed information you need to learn while you read.
- *Read.* Once you have prepared yourself to absorb the new information, read the text and think about what you are reading. Are the questions that you raised being answered? Do you understand the relationships among the topics? Read a small amount of material at a time. If the material seems difficult or confusing, don't go

2. Here we use the word "chapter." However, in engineering courses, you will generally be given reading assignments that involve less than a whole chapter.

on. Go back and reread the section, maybe just a paragraph or two at a time, until it makes sense.

- *Recite.* At the end of each section, stop. In your own words, recite the major ideas. Answer the questions on your outline. If your initial set of questions is incomplete, ask—and answer—further questions. Don't go on to the next section until you have stated the main ideas in your own words. If you cannot answer some of your questions, take them to class and discuss them with your instructor.
- *Review.* At the end of the chapter, review all of its material and your notes; then ask and answer your questions again. You should be able to see the topical relationships within each section and among the sections. Once you have grasped the main concepts, individual facts become easier to remember. It may be helpful to work with a classmate and quiz each other. Finally, combine your SQ3R notes with your lecture notes to create a coherent, well-organized package for the chapter. Then, take a break and relax. Approach each chapter with a fresh outlook.

Note Taking Taking notes during a classroom lecture is far different from being a court stenographer. A stenographer records every word spoken, but the point of note taking in class is capture the main points accurately and in an organized way so that you can use your notes later as a study aid. Students who develop a mental outline of a lecture's main ideas and concentrate on capturing those in a few words typically remember more than students who try to write down every detail.

Here are some simple hints for effective note taking:

- Come to class prepared. Do the assigned reading before class and bring your SQ3R notes. Ask your questions when the instructor is discussing the topic that confuses you. Jot down his or her answer on your outline. Use the lecture time to fill in the gaps in your own understanding, not to provide you with your first exposure to the topic. If you prepare for the lecture ahead of time, you will already have your general outline, and you will recognize important points.
- Every day, imagine that you are taking notes for your best friend who could not attend class that day. Your objective is to give your friend an organized set of notes that record the *key points* of the lecture. Your lecture notes should be a supplement to your SQ3R notes.
- During the lecture, listen for key words that signal main ideas ("The three primary reasons that . . ."), an alternative explanation ("In other words, . . ."), a conclusion or summary statement ("Finally, we can say that . . ."), or a change of direction ("On the other hand . . .").
- Do not assume that the instructor will write on the board everything that you are expected to know. Read everything that is written, and listen to everything that is said; but be selective in what you write down in your notes. Note taking takes concentration. Take brief notes so that you are not distracted from the lecture for long periods of time.
- After class and before the next class, combine your SQ3R notes and your lecture notes into a single, organized unit. This is another opportunity to review the material and solidify your understanding. If you missed a point or two during the lecture, leave space in your rewritten notes for additional detail. Then talk to your instructor to fill in the gaps. When the time comes for an exam, your rewritten notes will be well-organized and complete—perfect for exam preparation.

Developing Problem-Solving Skills To be an effective problem solver, you must follow an organized strategy.

- *Define the problem.* Read the given information thoroughly and carefully before working the problem. Unnecessary, incorrect, or misleading information could distract you from the real problem.
- *Think about it.* Be sure you have an accurate understanding of the problem. Don't jump into the solution process without identifying the facts and features of the problem situation. Summarize and organize what is given. Do you have enough information to solve the problem? Have you solved a similar problem before?
- *Plan your approach.* Make a flowchart outlining the solution process. Decide where to start and what sequence of steps to follow to reach the solution. Break the problem down into manageable pieces that can be solved one at a time. Think about alternate plans in case your first plan does not succeed.
- *Carry out your plan.* Persevere, and do not lose confidence in your plan.
- *Look back and evaluate.* Make sure you solved the problem that was posed. Determine whether your solution is reasonable. Does it agree with what your intuition tells you? Check all calculations for math errors, correct signs, decimal place locations, transposed digits, proper units, and so on. Be certain that your solution and the answer are presented in a readable, logical form that will not be misinterpreted.[3]

CHAPTER ORGANIZATION

We have organized this book to facilitate use of the SQ3R method. Features of each chapter include the following:

- *A Look Forward* provides a general perspective on the chapter. The main topics are introduced in relatively nontechnical terms so that you can get a sense of what is important before you begin detailed study.
- *Survey Questions* are included to start you on the SQ3R method. These questions are fairly simple and are based on an overview of the contents of the chapter. You are encouraged to perform your own survey and develop additional questions.
- *Key Terms* are printed in the margins of the text as signposts to recognize during reading. On completing a chapter, you should be able to define each of its key terms.
- *Key Concepts* are highlighted statements of the chapter's main ideas that appear throughout the text. They are relatively general so that you can grasp the big picture, then fill in the details from the text discussion.
- *Learning Objectives* are specific, performance-related statements at the beginning of each section. They help you assess your learning by telling you specifically what you should know or understand after you have studied a section.

3. In Section 1.5, we expand on these ideas and apply them explicitly to problem solving in engineering mechanics. Throughout this book, we include step-by-step strategies, called *Problem-Solving Techniques*, to help you develop your own methods of problem solving.

- *Checkpoints* at the ends of sections present true/false or short-answer questions that remind you to recite a section's main points. As you answer these questions, work on your chapter outline and check your survey questions. Answers are given below each Checkpoint box, but try to answer the questions before looking at them.
- *Chapter Highlights* are provided at the end of each chapter for your survey and review.
- *Problems* are grouped at the end of each chapter rather than immediately after the text and examples to which they relate. This organization preserves the continuity and flow of the text material and emphasizes that problem solving should follow thorough study of the chapter. However, we have divided the problems by section and have arranged the problems for each section in increasing order of difficulty. Icons next to problem numbers identify three special types of problems:

    Problems that are more challenging

    Problems that are best solved using computer-based tools

    Problems that are design-oriented

- *Review Questions* at the end of each chapter allow you to explain the concepts in your own words. Your ability to paraphrase key concepts will help you apply them during problem solving.

FINAL WORDS

An old saying goes: "What I hear, I forget; what I see, I may remember; what I experience, I know for life." Many engineers are visual, sensory learners, relying on examples, demonstrations, and illustrations to help them generalize and understand fundamental, often abstract, concepts. For this reason, this book is illustrated with photographs and drawings to enliven and enrich your study of mechanics. Apply what you learn to your own life, and know it for life. We have done all that we can to get you off on the right foot. The rest is up to you. Now start your study of mechanics with our best wishes for your success.

Arthur P. Boresi
Richard J. Schmidt

Chapter 1 REVIEW OF CONCEPTS IN MECHANICS

A Look Forward

IN THIS CHAPTER, we will review some history of mechanics and certain fundamental concepts that form the foundation of modern engineering mechanics. We also outline problem-solving methods that will serve you well throughout your engineering career. Included in the discussion is the proper use of numerical calculations to ensure accuracy in engineering mechanics.

The application of mechanics principles requires the use of trigonometry, algebra, and calculus. Therefore, we advise you to refer to your mathematics textbooks to review those topics as needed. The material in this chapter is intended as reference material for use throughout the entire book. You should return to it if you have questions regarding definitions and conversions of units, accuracy of calculations, and dimensions.

Finally, you should use this book as a reference in your future courses in engineering. Statics and dynamics form the basis for much of your later coursework (and engineering practice), so you will often need to refer back to these topics to refresh your understanding.

Renaissance scientist Galileo Galilei (1564–1642) was the first person to develop the modern quantitative description of the motion of a projectile in a vacuum.

Survey Questions

- What is mechanics? Are there subfields within mechanics?
- How old is the field of mechanics?
- Have the principles of mechanics changed substantially since the time of Newton?
- Isn't a particle just a really small body?
- What role do units of measure play in the study of mechanics?

1.1 BRIEF HISTORY OF MECHANICS

After studying this section, you should be able to:

- Define the science of mechanics and briefly outline its history.
- Identify and distinguish between systems composed of particles and those composed of rigid bodies.

Key Concepts

Mechanics is the science that considers the motion of bodies under the action of forces.

Statics is the study of motionless systems or systems that move with constant velocity.

mechanics *Mechanics* is the science that considers the motion of bodies and the effects of forces on that motion. Mechanics includes *statics,* which deals with the special case of a body at rest **statics** (one that does not move) or a body that moves with constant velocity. A body at rest or moving with constant velocity is said to be in *equilibrium.* **equilibrium**

The origins of the science of mechanics are lost in antiquity. However, many historians associate the birth of mechanics with the research of the Greek mathematician Archimedes (287–212 B.C.), who developed principles for the analysis of parallel forces and applied these principles to the statics of simple levers, systems of pulleys, floating bodies, and centers of gravity of bodies.

The successful analysis of nonparallel forces was not accomplished until nearly 2000 years after the death of Archimedes, when the Flemish mathematician and inventor Simon Stevin (1548–1620) solved the inclined plane problem (which involves nonparallel forces). Stevin also used directed line segments to represent forces and included an arrowhead on a line segment to indicate the sense of the force along the line (Fig. 1.1). He **resultant** showed how to add two forces to obtain their *resultant* by constructing a force parallelogram with the forces (arrows) as the sides. The diagonal of the parallelogram then represents the sum, or resultant, of the two forces (Fig. 1.2; see also Sec. 2.3). Quantities that **vectors** add like forces are called *vectors.*[1] The French scientist René Descartes (1596–1650) de-

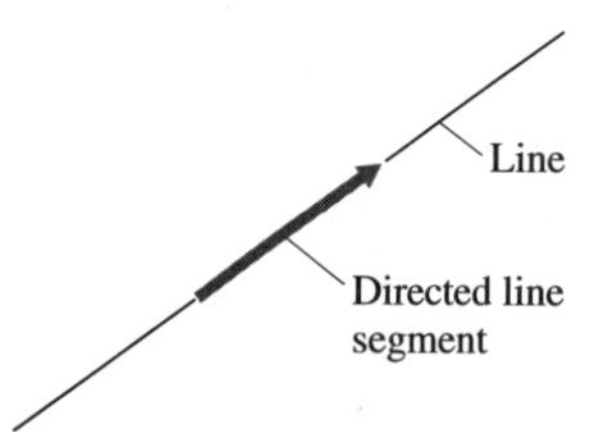

Figure 1.1 Directed line segment.

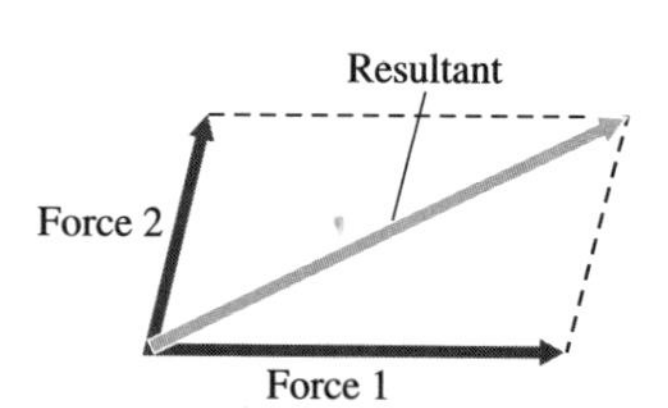

Figure 1.2 Force parallelogram.

1. According to Lamb (1924), the term "vector" (or "carrier") for quantities such as force that possess both magnitude and direction was introduced by Sir W. R. Hamilton. Lamb also credits Hamilton with the first use of the term "scalar."

coordinate axes veloped the idea of resolving vectors into projections parallel to *coordinate axes*.[2] Complementing Stevin's parallelogram law, this idea greatly facilitates computations in both two and three dimensions in terms of projections of vectors (Sec. 2.5). Vector quantities **scalar** are distinguished from *scalars* (e.g., temperature) in that scalars possess only magnitude.

Stevin also had an understanding of the *principle of virtual work*. This principle led to an alternative theory of equilibrium—the *virtual work method* (see Sec. 12.5).

Early Greek scholars, particularly Aristotle (384–322 B.C.), attempted to explain the motion of bodies acted on by forces. However, because of an inability to measure distance or time accurately and a false belief that force was necessary to maintain motion, these attempts led to erroneous conclusions. One example of these errors is the Aristotelian theory that heavy bodies fall faster in a gravity field than light bodies do.

Following these early studies, the field of mechanics gradually evolved into three main divisions: statics, kinematics, and kinetics. Statics is concerned with motionless systems or systems that move with constant speed in a straight line and with forces that act to establish these states of motion. For instance, a book resting on your desk or your car traveling along a straight highway at a constant speed are subject to and analyzed with the laws of **kinematics** statics. *Kinematics* is concerned with rates of change of geometrical quantities in a moving system; it does not involve the concept of force. *Kinetics* treats the causes and the na- **kinetics** ture of motion that results from specified forces. Kinematics and kinetics together form the field of *dynamics*. The relationships among position, velocity, and acceleration of a mov- **dynamics** ing body (e.g., a thrown ball) are defined by kinematics. The relationship between forces that act on the body (e.g., wind and gravity) and the motion of the body involves kinetics.

It was not until the 17th century that Galileo (1564–1642) laid the foundation for the science of dynamics by careful experiments and analysis. Galileo contributed to the theory of equilibrium (statics) and to kinematics. However, he is known principally for his contributions to kinetics. Galileo's work was the first successful effort to dispel the false dynamical doctrines of Aristotle, which had been taught without serious question or confirmation for nearly 2000 years. Galileo understood the law of inertia, as evidenced by his statement that a body in motion and free from external forces will keep moving at a constant speed and in a straight line. He realized that the acceleration (the rate of change of velocity) of a body is determined by external forces and that acceleration therefore depends on the forces applied to the body and the inertia (mass) of the body.

Key Concept

A particle is a body whose size does not influence its response to the forces acting on it.

The study of mechanics may also be classified according to the kind of physical sys- **particle** tem under consideration. The simplest mechanical system is a single particle. A *particle* is defined as a body whose size, in a given physical situation, is not significant in the analysis of its response to the forces that act on it. In other words, the body may be modeled as a point of concentrated mass, and the rotational motion of the body can be ignored. In a physical situation in which the size or rotation of a body cannot be ignored, the body cannot be treated as a particle.

2. The concept of Cartesian (rectangular) coordinates was first developed by Descartes.

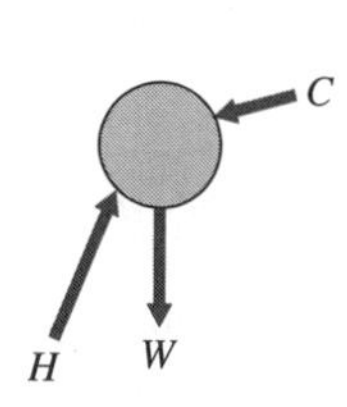

Dan O'Brien, Olympic champion in the decathlon, is poised to put the shot. The shot can be modeled as a particle, subject to three forces: *H* from O'Brien's hand, *C* from his chin, and *W* (the weight of the shot) due to gravity.

A more complicated system is that of two or more particles that interact (exert forces on each other). The particles may also be subjected to forces from bodies outside of the system. This type of model was used by Newton to study the motion of the planets.

Key Concept

A rigid body is a body that does not deform under the action of forces. Generally, the size of a rigid body influences its response to forces.

In some physical situations, the deformation of a body has a significant effect on the motion (or equilibrium) of the body due to forces acting on it; such deformation cannot be ignored. Alternatively, if the deformation of the body is very small and has little effect on its motion or equilibrium, that deformation can be ignored when considering the body's motion or equilibrium. The body can then be assumed to be rigid and is referred to as a *rigid body.* Each point in a rigid body is always at a constant distance from any other point in the body.

rigid body

In a crash-worthiness test, a GM car collides with a wall. Large deformations of the car occur. Subsequent motion is affected by these deformations.

A dozen locomotives were used to test this Northern Pacific Railroad bridge in Bismarck, North Dakota in 1882. Such a bridge can be analyzed as if it were rigid—that is, as if it does not deform.

Depending on the physical situation, a given body may be treated as a particle, as a rigid body, or as a deformable body. For example, in the computation of the motion of the earth about the sun, the earth may be accurately modeled as a particle, since the rotation of the earth has a very small effect on its motion about the sun. Similarly, with respect to the rotation of the earth about its own axis, the earth can be modeled as a rigid body, since the deformation of the earth has a very small effect on its rotation. However, in the determination of the shape of the earth, the earth's elastic and plastic deformations must be considered; that is, the earth can no longer be assumed to be rigid.

classical mechanics

Collectively, the study of statics and dynamics is called *classical mechanics.* Classical mechanics treats the motion of bodies of "ordinary" size that move at speeds that are small compared to the speed of light. It includes the special case of statics (equilibrium) in which

Depending on the purpose of the analysis, the earth might be modeled as a particle or as a rigid body.

a body is at rest or moves in a straight line with constant speed, as well as the special case in which a body is sufficiently rigid or stiff that its deformation has no effect on its motion.

Sir Isaac Newton (1642–1727), who was born the year Galileo died, summarized, clarified, and extended Galileo's work. In addition, he formulated the law of universal gravitation and the mathematics of calculus. Newton introduced and clarified the concepts of force and mass. He also formulated the three laws of motion that are the basis of engineering applications of mechanics. Although these three laws had been discovered experimentally by Galileo about four years before Newton was born, Newton was the first to systematize them. The great importance of Newton's contributions to mechanics is evidenced by the fact that classical mechanics is often called *Newtonian mechanics.*

Newtonian mechanics

The theories of relativity and quantum mechanics show that Newtonian mechanics is inexact when speeds approach the speed of light or when the motion of subatomic particles is considered. However, Newtonian mechanics is extremely precise for all other cases. Consequently, Newtonian mechanics forms the basis for the analysis of the vast majority of modern engineering problems.

1.2 NEWTONIAN MECHANICS, ENGINEERING MECHANICS

After studying this section, you should be able to:

- State and discuss the laws that govern Newtonian mechanics.

The fundamental principles of Newtonian mechanics may be used to study the conditions of rest or motion of rigid bodies of ordinary size that move at ordinary speeds. These principles are also applicable, in part, to the study of fluids and the study of deformable solids. The field of *engineering mechanics,* which may be considered to be modern Newtonian mechanics, includes the mechanics of rigid bodies, the mechanics of fluids, and the mechanics of deformable solids.

NEWTON'S LAWS OF MOTION

Sir Isaac Newton, in his treatise *Principia,*[3] announced three famous laws that describe the motion of a particle. Newton established these laws of motion from his study of the motion of planets. Since the sizes of the planets are extremely small compared to the distances involved, the motion of a planet is accurately predicted by considering it to be a particle. Thus, Newton's laws apply directly to a particle—that is, a body that may be treated as a point mass. Newton's laws may be stated in terms of a particle as follows:

First Law: In the absence of applied forces, a particle originally at rest or moving with constant speed in a straight line will remain at rest or continue to move with constant speed in a straight line.

Second Law: If a particle is subjected to a force, the particle will be accelerated (i.e., its velocity will change). The acceleration of the particle will be in the direction of the force, and the magnitude of the acceleration will be proportional to the magnitude of the force and inversely proportional to the mass of the particle.

Third Law: For every action, there is an equal and opposite reaction. Or, the mutual forces exerted by particles on each other are always equal and oppositely directed.

3. Newton's *Principia,* as translated by Andrew Motte and revised by Florian Cajori, is reproduced in *Great Books,* Vol. 34, pp. 1–372, Encyclopaedia Britannica, Inc., Chicago.

> **Key Concept** The acceleration of a particle is proportional to the net force that acts on the particle and inversely proportional to the mass of the particle.

Newton's first law is a special case of the second law. It implies the conditions under which a particle is in a static equilibrium state. It also implies that a particle resists changes in its motion; that is, it possesses *mass* (also referred to as *inertia*). Newton's second law is a quantitative law that is expressed mathematically in the form

$$\mathbf{F} = m\mathbf{a} \qquad \text{or} \qquad \mathbf{a} = \frac{\mathbf{F}}{m} \tag{1.1}$$

where **F** denotes force, m mass, and **a** acceleration.[4]

Equation (1.1) indicates that acceleration is proportional to force and inversely proportional to mass (inertia). Hence, the larger the mass of the particle, the larger the force required to produce a given acceleration.

> **Key Concept** Every force (action) is accompanied by an equal and opposite force (reaction).

action

reaction

The second law assumes that a single (resultant) force acts on a single particle. The third law, however, notes that a single force, or *action,* does not exist alone; rather, it is always opposed by a *reaction.* The significance of Newton's third law is that it allows his second law, which applies to a single particle, to be extended to a system of two or more particles acted on by a system of forces.

CHECKPOINT 

1. Arrange the following names in chronological order: Archimedes, Aristotle, Newton, Galileo.
2. ***True or False:*** Galileo refuted Archimedes' theory that heavy bodies fall faster than light bodies do.
3. ***True or False:*** Kinetics is the study of rates of change of geometrical quantities in a moving system.
4. ***True or False:*** Newton's first law is a special case of his third law.
5. ***True or False:*** A particle is defined as a body whose size, mass, and rotation are of no significance in the analysis of its response to the forces that act on it.
6. ***True or False:*** A body that can be modeled as a particle in one situation can be modeled as a particle in any situation.

Answers: 1. Aristotle, Archimedes, Galileo, Newton; 2. False; 3. False; 4. False; 5. False; 6. False

4. The force **F** and the acceleration **a** are vectors (see Chapter 2). Boldface letters are used to denote vectors. Scalar variables, such as m for mass, are italic letters.

1.3 UNITS OF MEASURE AND PHYSICAL DIMENSIONS

After studying this section, you should be able to:

- Express numerical values in proper SI or U.S. Customary units.
- Manipulate physical dimensions and units as algebraic quantities.
- Convert numerical values from one set of units to another.
- Determine whether an algebraic expression is dimensionally homogeneous.

units of measure

In engineering computations, *units of measure* are used to describe physical quantities such as forces. Currently, Le Système International d'Unités (the International System of Units, abbreviated SI) is used by nearly all industrialized nations. A modern version of the metric system, SI was adopted in 1960 by the delegates of the Eleventh General Conference on Weights and Measures. Although the United States participated in this conference, it has not yet fully adopted SI. The U.S. Customary System of units is still widely used in the United States. However, the Metric Conversion Act of 1975, as amended by the Omnibus Trade and Competitiveness Act of 1988, established SI as the preferred measurement system in the United States. This act required that, to the extent feasible, SI be used in federal procurement, grants, and business-related activities by September 30, 1992. However, the deadline was extended to January 1, 1994, because of resistance from Congress and other groups. This deadline was not met either.

Technician Max Peltz examines a precast concrete beam being tested for strength and ductility at the NIST tridirectional test facility.

What does all this mean to you? It means that you will have to be proficient not only in SI but also in the U.S. Customary System. Therefore, for your benefit, both systems of units are used in this book.

The field of engineering mechanics relies heavily on the study of physical phenomena and experimental results. Engineers use numbers to quantify physical phenomena and experimental results. These numbers are called *physical quantities.* To each physical quantity is attached a unit. A measurement of a quantity is actually a comparison to a reference stan-

physical quantities

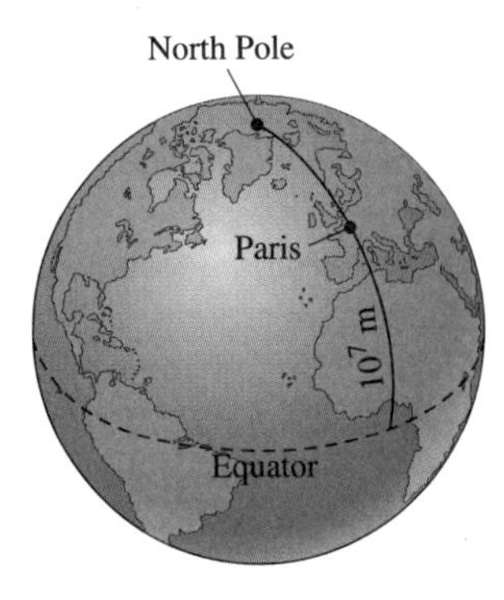

Figure 1.3 Original definition of the meter.

dard. When you refer to a 400 meter race in track and field, for example, you mean that the race distance is 400 times as long as a meter, the international standard unit of length.

To ensure precise measurements, engineers need standard units of measurement that do not change and that can be duplicated by experimenters. When the metric system of measurement was established by the Paris Academy of Sciences in 1791, the meter (m) was defined as the distance (at zero degrees Celsius) between two marks on a platinum-iridium bar that is preserved at the International Bureau of Weights and Measures in Sèvres, France (near Paris). This distance was equivalent to 1 ten-millionth (1×10^{-7}) of the distance from the equator to the North Pole measured along the meridian through Paris (Fig. 1.3). Since 1791, several other definitions of the meter have been used. The current definition, established in November 1983, defines the meter to be the distance that light travels in a vacuum in 1/(299 792 458) second.[5] This definition has the effect of defining the speed of light to be precisely 299 792 458 meters per second. One second (s) is defined as the time required for 9 192 631 770 cycles of radiation to be emitted by the cesium atom as it makes the transition between its two lowest energy states (Young, 1992). These definitions of the meter and the second are useless in a machine shop, but they make the base units of length and time accessible to physicists throughout the world without recourse to, say, a standard meter bar.

Key Concept

Weight is a force, not a mass. The kilogram is a unit of mass, not a unit of weight.

UNITS OF MASS AND FORCE

balance-type scale

The mass of an object may be measured by an instrument called the *balance-type scale,* or simply the *balance.* The basic balance-type scale has a symmetrical pivoting beam with two identical pans suspended from its ends (Fig. 1.4). In scientific terminology, *a balance-type*

Figure 1.4 Balance-type scale.

5. In SI, a space, rather than a comma, is used to separate digits into groups of three. See note a of Table 1.1 for further explanation.

scale measures mass in a gravitational field. A body whose mass is to be measured is placed in one of the pans. The gravitational force of the earth acting on the mass of the body is balanced by the gravitational force of the earth acting on known masses placed in the other pan. Under these conditions, the mass of the body is equal to that of the known masses.

standard kilogram

The standard unit of mass is the kilogram (kg). The kilogram was initially intended to be equal to the mass of 1000 cubic centimeters (1 liter) of water, a centimeter (cm) being 0.01 meter. Presently, the kilogram is the mass of a platinum cylinder, 39 millimeters long and 39 millimeters in diameter, a millimeter (mm) being 0.001 meter. The *standard kilogram* is stored at the International Bureau of Weights and Measures. The mass of an object is 1 kilogram if the object is exactly balanced by the standard kilogram on a balance-type scale with the fulcrum at the center of the balance beam.

force of gravity

Balance-type scales are used in commerce to measure mass because they are not affected by differences in the *force of gravity* (the attraction of the earth on a body) in different geographical locations. At a given location, the gravitational attraction of the earth acts equally on the body whose mass is to be measured and on the known masses in the other pan.

As Newton pointed out, all bodies attract each other.[6] However, since the attractions between small bodies are very small, they can be detected only with highly sensitive instruments. You can, nevertheless, sense the force of attraction between an ordinary-size body and the earth (a large body) when you lift the body. This force is so familiar that it has been given a special name—the *weight* of the body.

A balance does not measure weight, since the weight of a body depends on the location of the body with respect to the earth. For example, if a spaceship were to travel away from the earth at constant velocity, the weights of objects in the ship would decrease steadily. Travelers in the ship would feel themselves growing lighter, and they would be able to lift large objects that they could not move on earth. A balance in equilibrium on the spaceship would not detect this phenomenon at all. It would remain in equilibrium, since the masses in the two pans would remain equal to each other.[7] For example, a kilogram of sugar, measured on a balance-type scale, is the same amount of sugar (i.e., the same number of molecules of sugar) whether the measurement is made on the earth or on the moon.

spring scale

Unlike the balance, which measures mass, the *spring scale* is a device that measures force. The experiments of Robert Hooke (1635–1703) showed that the extension of an elastic spring is proportional to the force that is applied to the spring. The spring scale operates on this principle. On a spring scale, an object of fixed mass will have a variable weight, depending on the gravitational field it is in. For instance, as shown in Fig. 1.5, a 1 kilogram mass will have no weight in deep space, will weigh about 1.6 newtons on the moon, and will weigh about 9.8 newtons on earth.

6. Newton's law of universal gravitation, expressed by the formula $F = Gmm'/r^2$, allows computation of the force of attraction between bodies (see Sec. 14.1 for a discussion of this law).

7. This statement is true in a gravitational field, since a balance compares weights of two bodies in such a field with great precision (to 1 part in 10^6 parts). Hence, it compares masses, since the weight W of a body in a gravitational field is related to the mass m of the body by the equation $W = mg$, where g is the acceleration due to gravity. In a zero-gravity field, Newton's second law must be used to determine the mass of a body by applying a known force of magnitude F to the body, measuring the magnitude a of the body's acceleration, and computing the body's mass m as the ratio F/a. This inertial method, or a variation of it, is used to measure the masses of objects in orbiting space stations. In scientific work, the inertial method of measuring masses is preferred to the gravitational balance method. However, gravitational masses (masses measured by the gravitational balance method) match inertial masses (masses measured by the inertial method) very closely.

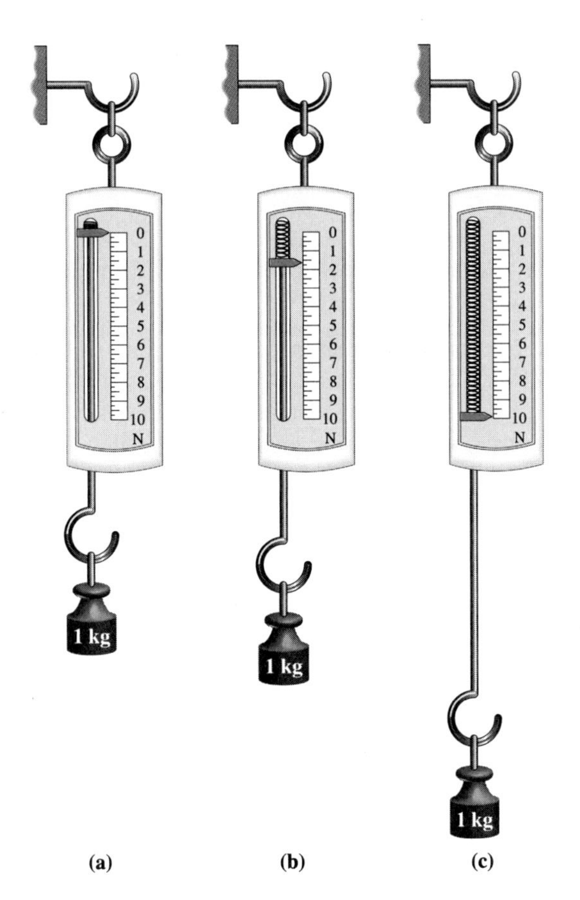

Figure 1.5 Spring scale: (**a**) in a zero-gravity field; (**b**) on the moon; (**c**) on earth.

Figure 1.6 A force of 1 N acting on a mass of 1 kg produces an acceleration of 1 m/s².

newton

The SI unit of force is the *newton* (N). By definition, a newton is the force that will impart an acceleration of 1 meter per second squared (m/s²) to the standard kilogram (Fig. 1.6). The reason for this convention is discussed in the study of kinetics (however, see footnote 7).

The standard kilogram weighs 9.806 65 N at a point on the earth where the acceleration of gravity has the standard value, $g = 9.806\ 65$ m/s² (Fig. 1.7). This follows from Newton's second law, Eq. (1.1). If the mass is 1 kg and the acceleration due to gravity on the surface of the earth is 9.806 65 m/s², Eq. (1.1) yields

$$F = ma = 1 \text{ kg} \times 9.806\ 65 \text{ m/s}^2 = 9.806\ 65 \text{ kg}\cdot\text{m/s}^2 = 9.806\ 65 \text{ N}$$

where 1 kg·m/s² is defined to be a newton. Thus, in SI, the unit of force, the newton (N), is derived from Newton's second law in terms of the SI base units kilogram (kg), meter (m), and second (s).[8]

The weight of a body is the force with which the earth attracts the body. The force of attraction is called the force of gravity. Since weight is a force, it is expressed in new-

Figure 1.7 The weight of a unit mass equals g units of force.

8. In the U.S. Customary System of units (discussed later), the pound force (lb) is the unit of force. It is approximately the weight of 0.4536 kg on earth. Consequently, 1 N = 0.2248 lb and 1 lb = 4.4482 N, approximately. A force equal to 1000 lb, called a *kip,* is commonly used in engineering.

tons. By Eq. (1.1), with $a = g$, the weight W of a body of mass 1 kg at a location where $g = 9.806\ 65\ \text{m/s}^2$ is

$$W = mg = 1\ \text{kg} \times 9.806\ 65\ \text{m/s}^2 = 9.806\ 65\ \text{N} \tag{1.2}$$

More generally, if, in consistent units, we set $m = 1$ in the equation $W = mg$, we obtain $W = g$. This result has the following meaning:

> For consistency with the equation $\mathbf{F} = m\mathbf{a}$, the weight of a unit of mass must be exactly g units of force.

This condition is independent of the local acceleration of gravity, for if g varies, the weight of an object varies proportionally.

An incredible amount of confusion has arisen because the concepts and units of weight and mass have been intermingled in everyday usage, even in scientific and engineering circles. The adoption of SI is a supreme effort to end this confusion. The success of SI is due, in part, to its unequivocal support of this statement:

> Weight is a force, never a mass; the kilogram is a unit of mass, never a unit of force.[9]

THE INTERNATIONAL SYSTEM OF UNITS (SI)

Key Concept

SI is an absolute system of units. The U.S. Customary System is a gravitational system of units.

kinetic units

This section has introduced four physical quantities: length, time, mass, and force. The units associated with these quantities are sometimes referred to as *kinetic units,* since they

Canadian sprinter Donavan Bailey celebrates his world-record time in the men's 100 m race at the 1996 Olympic Games in Atlanta, Georgia.

9. In the U.S. Customary System, the pound is used as both a unit of force and a unit of mass. However, the pound unit should be interpreted consistently. The predominant view has been that the pound is a unit of force, and it is so interpreted in this book.

form a consistent set of units that satisfy Eq. (1.1). In SI, these units are meter, second, kilogram, and newton, respectively. The first three of these units are SI *base units,* or *fundamental units;* that is, these units are defined independently of any others. The fourth unit, the newton, is a *derived unit;* that is, it is defined in terms of the base units. Hence, of the four kinetic units, only meter, second, and kilogram are independent. The choice of base units for a system is arbitrary.

base units

In SI, the base units (meter, second, and kilogram) are defined such that they do not depend on location. For example, a kilogram of mass on the earth is the same amount of mass on any other planet or on the moon; that is, the kilogram is independent of the force due to gravity. For this reason, SI is said to be an *absolute system.*

absolute system

In the U.S. Customary System, the base units chosen for length, time, and force are foot (ft), second (s), and pound (lb), respectively. The unit of mass, called the *slug,* is a derived unit. In this system, the definition of the pound depends on location and, therefore, on the force of gravity at that location. For this reason, the U.S. Customary System is said to be a *gravitational system.*

gravitational system

In addition to being an absolute system of units, SI is a *decimal system;* that is, the units of any physical quantity can be expressed in terms of SI base units or their multiples expressed in powers of 10 (e.g., 10^3 or 10^{-3}). For instance, a millimeter is 10^{-3} meter. Thus, in SI, prefixes are added to the names of base units to form multiples of those base units (e.g., MN represents meganewtons, or 10^6 newtons). The kilogram is the exception; it is the base unit of mass, and its name includes the multiplier prefix *kilo-* (see Table 1.1).

decimal system

The simplicity of SI is enhanced by the fact that there is only one base unit for each physical quantity; there are no conversion factors or additional constants to remember.[10] For example, in SI, the meter (with its decimal multiples) is the single unit of length; whereas in the U.S. Customary System, the mil, inch, foot, yard, fathom, rod, chain, furlong, and mile are used as units of length. The small number of base units and the use of decimal arithmetic make SI a logical and convenient measurement system (see ASTM, 1992).

Another decimal system that is still being used is the cgs system, based on the centimeter, gram, and second. The centimeter (cm) is defined as 0.01 m. The gram (g) is defined as 0.001 kg. In general, the cgs system has been replaced by SI.

SI units are divided into three classes: base units, supplementary units, and derived units. As previously noted, three SI base units are used in the study of kinetics. In other scientific and engineering fields, four additional base units are regarded as dimensionally independent. They are the ampere (A), the kelvin (K), the mole (mol), and the candela (cd); see Table 1.2. Thus, SI is based on seven well-defined units that, by convention, are regarded as dimensionally independent. The class of *supplementary units* contains only two units, the radian (rad) and the steradian (sr); see Table 1.3. However, since the radian is defined as the ratio of two lengths and the steradian is defined as a ratio between an area and the square of a length, these units may also be regarded as dimensionless derived units (ASTM, 1992; see also the footnote to Table 1.3).

supplementary units

Derived units are obtained from base and supplementary units through algebraic expressions that relate the associated physical quantities [e.g., Eq. (1.1)]. That is, the units of derived quantities are obtained from the mathematical relationships that define the quantities. For example, the velocity v of a particle that travels along a straight line is defined by the time derivative of its position x along the line: $v = dx/dt$ (see Sec. 13.2). Since the

derived units

10. Because of these facts, the SI units of measurement are said to be coherent (see ASTM, 1992, Appendix X1.9).

Table 1.1
SI prefixes and symbols

MULTIPLICATION FACTOR[a]	PREFIX[b]	SYMBOL
1 000 000 000 000 000 000 = 10^{18}	exa	E
1 000 000 000 000 000 = 10^{15}	peta	P
1 000 000 000 000 = 10^{12}	tera	T
1 000 000 000 = 10^{9}	giga	G
1 000 000 = 10^{6}	mega	M
1 000 = 10^{3}	kilo	k
100 = 10^{2}	hecto[c]	h
10 = 10^{1}	deka[c]	da
0.1 = 10^{-1}	deci[c]	d
0.01 = 10^{-2}	centi[c]	c
0.001 = 10^{-3}	milli	m
0.000 001 = 10^{-6}	micro	μ
0.000 000 001 = 10^{-9}	nano	n
0.000 000 000 001 = 10^{-12}	pico	p
0.000 000 000 000 001 = 10^{-15}	femto	f
0.000 000 000 000 000 001 = 10^{-18}	atto	a

a. With SI, except in certain specialized applications, such as engineering drawings, a space for a separator is recommended to separate digits into groups of three, counting from the decimal point toward the left and toward the right. This practice is intended to prevent confusion arising from the comma's use as a decimal marker, a common practice outside the United States. In the United States, a comma is generally used to separate digits into groups of three (e.g., the number 45,983 in the United States might be interpreted as 45.983 outside the United States). In numbers with four digits on either side of the decimal point, the space is not usually necessary, except for uniformity in tables. Examples of the use of this convention are 45 980, 4598, 0.4598, and 4.598 070. Note that this convention does not apply to expressions for amounts of money.
b. In the United States, preferred pronunciation accents the first syllable of every prefix to assure that the prefix retains its identity. For example, giga is pronounced *jig-a* (*i* as in *jig, a* as in *about*), and kilometer is therefore pronounced *kill-oh-meter.* (See also ASTM, 1992, Table 8.)
c. These prefixes should be avoided where practical, except for particular applications—for example, centimeters are used for body measurements and clothing sizes.

Table 1.2
Base units in SI

QUANTITY	UNIT	SYMBOL
length[a]	meter/metre	m
mass[a]	kilogram	kg
time[a]	second	s
electric current	ampere	A
thermodynamic temperature[b]	kelvin	K
amount of matter	mole	mol
luminous intensity	candela	cd

a. These base units are used in statics and dynamics.
b. The 1 degree interval (°C) of the Celsius temperature scale (formerly the centigrade scale) is equal to the kelvin (K) of thermodynamic temperature (TT). There is a shift of 273.15 K between the scales, with the zero of the Celsius scale occurring at 273.15 K. Thus, °C = TT − 273.15 K. Water freezes at 273.15 K (0°C) and boils at 373.15 K (100°C).

Table 1.3
Supplementary units in SI

QUANTITY	UNIT	SYMBOL
plane angle	radian[a]	rad
solid angle	steradian[a]	sr

a. These units are dimensionless.

dimension of x is length, the units of v in terms of the base units meter and second are m/s, or $\mathrm{m \cdot s^{-1}}$. A list of some derived units used in engineering mechanics is given in Table 1.4. [Note that it is frequently simpler to express derived units in terms of other derived units that have special names; e.g., the watt (W) may be expressed in terms of the joule (J) and the second (s) as J/s, rather than as N·m/s.] Along with the base units the meter, kilogram, and second (Table 1.2), the derived units watt and joule are commonly used in mechanics.

The kilogram is the only base unit whose name contains a prefix, *kilo-*. Therefore, names of decimal multiples of the unit of mass are obtained by attaching prefixes to the word "gram" (g)—for example, megagram (Mg). It is recommended (ASTM, 1992) that only one prefix be used in forming a multiple of a compound unit. Normally, the prefix should be attached to a unit in the numerator; the exception is when kilogram occurs in the denominator. For example, use kN/m (not N/mm) and MJ/kg (not kJ/g).

Prefixes may be applied to the meter and to other SI units. In general, prefixes are used to represent multiplication or division of a base unit by multiples of 10^3. For example, 1 millimeter = 1 mm = 0.001 m = 10^{-3} m, and 1 MN = 1000 kN = 1 000 000 N = 10^6 N.

Table 1.4
Derived units commonly used in statics and dynamics

QUANTITY	UNIT	SYMBOL	EQUIVALENT
acceleration	meter per second squared	—	$\mathrm{m/s^2}$
angle	radian	rad	—
angular acceleration	radian per second squared	—	$\mathrm{rad/s^2}$
angular velocity	radian per second	—	rad/s
area	meter squared	—	$\mathrm{m^2}$
density, mass	kilogram per meter cubed	—	$\mathrm{kg/m^3}$
energy	joule	J	N·m
force	newton	N	$\mathrm{kg \cdot m/s^2}$
frequency	hertz	Hz	1/s (or 1 $\mathrm{s^{-1}}$)
impulse	newton-second	—	N·s (or kg·m/s)
moment of force	newton-meter	—	N·m (or $\mathrm{kg \cdot m^2/s^2}$)
power	watt	W	J/s (or N·m/s)
pressure	pascal	Pa	$\mathrm{N/m^2}$ [or $\mathrm{kg/(m \cdot s^2)}$]
stress	pascal	Pa	$\mathrm{N/m^2}$ [or $\mathrm{kg/(m \cdot s^2)}$]
velocity	meter per second	—	m/s
volume, solid	cubic meter	—	$\mathrm{m^3}$
volume, liquid	liter	L	10^{-3} $\mathrm{m^3}$
work	joule	J	N·m

U.S. CUSTOMARY SYSTEM OF UNITS

The U.S. Customary System of units is used principally in the United States and is gradually being replaced by SI. The U.S. Customary System uses dimensions of length, force, and time. These U.S. Customary System base units are officially defined in terms of SI units.

Length: 1 inch = 2.54 cm = 0.0254 m (exact)

Force: 1 pound = 4.448 221 611 526 N (exact)

Time: 1 second (same as the SI unit)

Table 1.5 lists conversion factors between U.S. Customary System units and SI units for quantities commonly used in statics and dynamics.

standard pound

In the U.S. Customary System, the definition of the pound is given in terms of the *weight* of a platinum bar, called the *standard pound.* This is a bar of mass 0.453 592 43 kg that is preserved at the National Institute of Standards and Technology (formerly called the National Bureau of Standards) in Gaithersburg, Maryland. Since the weight of a body is due to the gravitational attraction of the earth, which depends on location, included in the definition of the pound is the specification that it is the weight of the standard pound at sea level and 45° north latitude. On the moon, the standard pound would weigh about $\frac{1}{6}$ of its weight on earth, because the gravitational attraction of the moon is about $\frac{1}{6}$ that of the earth.

Comparisons between U.S. Customary units and equivalent SI units are useful for obtaining a feel for their relative magnitudes. For example, 1 millimeter is about $\frac{1}{25}$ inch, or slightly less than the thickness of a U.S. dime. One meter is about 10% longer than a yard. One inch is just a fraction longer than 25 millimeters. One foot is slightly longer than 300 millimeters. Therefore, engineers often use approximations such as 300 mm $\approx$ 1 ft and 1 in $\approx$ 25 mm to obtain approximate estimates of a quantity. (The symbol $\approx$ stands for "is approximately equal to." You should make a list of approximate equivalents for comparison purposes. For example, what object has a mass of approximately 1 gram? 1 kilogram? How much does a baseball weigh on the earth's surface in newtons and in pounds?)

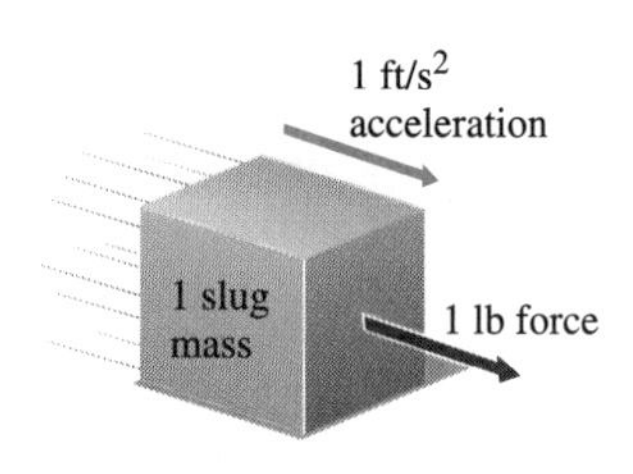

Figure 1.8 A force of 1 lb acting on a mass of 1 slug produces an acceleration of 1 ft/s^2.

In the U.S. Customary System, the unit of mass, the slug, is a derived quantity. By Eq. (1.1), a slug is defined to be the amount of mass that receives an acceleration of 1 ft/s^2 when acted on by a force of 1 lb (Fig. 1.8):

$$1 \text{ slug} = \frac{1 \text{ lb}}{1 \text{ ft/s}^2} = 1 \text{ lb}\cdot\text{s}^2\text{/ft} \tag{1.3}$$

CHECKPOINT

1. ***True or False:*** The laws of mechanics are dependent on the system of units used.
2. ***True or False:*** The definition of the kilogram is based on an arbitrary standard, not a physical constant.
3. ***True or False:*** A gravitational system of units includes the acceleration of gravity as a base unit.
4. The SI base units used in mechanics are _______, _______, and _______.
5. The U.S. Customary System base units used in mechanics are _______, _______, and _______.
6. ***True or False:*** The U.S. Customary System base units are defined in terms of SI base units.

Answers: 1. False; 2. True; 3. False; 4. meter, kilogram, second; 5. foot, pound, second; 6. True.

Table 1.5 *Conversion factors for U.S. Customary and SI units*[a]

QUANTITY	U.S. CUSTOMARY TO SI[b]	SI TO U.S. CUSTOMARY[b]
length	**1 in = 2.540 000 E + 01 mm**	**1 mm = 3.937 008 E − 02 in**
	1 in = 2.540 000 E − 02 m	**1 m = 3.937 008 E + 01 in**
	1 ft = 3.048 000 E − 01 m	**1 m = 3.280 840 E + 00 ft**
	1 yd = 9.144 000 E − 01 m	**1 m = 1.093 613 E + 00 yd**
	1 mi (intl.) = 1.609 344 E + 00 km	**1 km = 6.213 712 E − 01 mi (intl.)**
	1 mi (naut.) = 1.609 347 E + 00 km	1 km = 6.213 700 E − 01 mi (naut.)
force	1 lb = 4.448 222 E + 00 N	1 N = 2.248 089 E − 01 lb
	1 oz = 2.780 139 E − 01 N	1 N = 3.596 942 E + 00 oz
	1 kip = 4.448 222 E + 03 N	1 kN = 2.248 089 E − 01 kip
mass	1 slug = 1.459 390 E + 01 kg	1 kg = 6.852 178 E − 02 slug
	1 lbm[c] = 4.535 924 E − 01 kg	1 kg = 2.204 622 E + 00 lbm
area	**1 in^2 = 6.451 600 E − 04 m^2**	**1 m^2 = 1.550 003 E + 03 in^2**
	1 ft^2 = 9.290 304 E − 02 m^2	**1 m^2 = 1.076 391 E + 01 ft^2**
volume, solids	1 in^3 = 1.638 706 E − 05 m^3	1 m^3 = 6.102 376 E + 04 in^3
	1 ft^3 = 2.831 685 E − 02 m^3	1 m^3 = 3.531 466 E + 01 ft^3
volume, liquids	1 gal (U.S.) = 3.785 412 E − 03 m^3	1 m^3 = 2.641 720 E + 02 gal (U.S.)
	1 gal (U.S.) = 3.785 412 E + 00 L	1 L = 2.641 720 E − 01 gal (U.S.)
	1 qt (U.S.) = 9.463 529 E − 01 L	1 L = 1.056 688 E + 00 qt (U.S.)
velocity	**1 in/s = 2.540 000 E − 02 m/s**	**1 m/s = 3.937 008 E + 01 in/s**
	1 ft/s = 3.048 000 E − 01 m/s	**1 m/s = 3.280 840 E + 00 ft/s**
	1 mi/h (intl.) = 4.470 400 E − 01 m/s	**1 m/s = 2.236 936 E + 00 mi/h (intl.)**
	1 mi/h (intl.) = 1.609 344 E + 00 km/h	**1 km/h = 6.213 712 E − 01 mi/h (intl.)**
acceleration	**1 in/s^2 = 2.540 000 E − 02 m/s^2**	**1 m/s^2 = 3.937 008 E + 01 in/s^2**
	1 ft/s^2 = 3.048 000 E − 01 m/s^2	**1 m/s^2 = 3.280 840 E + 00 ft/s^2**
pressure/stress	1 lb/in^2 = 6.894 757 E + 03 N/m^2	1 N/m^2 = 1.450 377 E − 04 lb/in^2
	1 lb/ft^2 = 4.788 026 E + 01 N/m^2	1 N/m^2 = 2.088 544 E − 02 lb/ft^2
energy/work	1 in·lb = 1.129 848 E − 01 N·m	1 N·m = 8.850 748 E + 00 in·lb
	1 ft·lb = 1.355 818 E + 00 N·m	1 N·m = 7.375 621 E − 01 ft·lb
power	1 in·lb/s = 1.129 848 E − 01 N·m/s	1 N·m/s = 8.850 748 E + 00 in·lb/s
	1 ft·lb/s = 1.355 818 E + 00 N·m/s	1 N·m/s = 7.375 621 E − 01 ft·lb/s
momentum/impulse	1 lb·s = 4.448 222 E + 00 N·s	1 N·s = 2.248 089 E − 01 lb·s
moment of force	1 lb·in = 1.129 848 E − 01 N·m	1 N·m = 8.850 748 E + 00 lb·in
	1 lb·ft = 1.355 818 E + 00 N·m	1 N·m = 7.375 621 E − 01 lb·ft
moment of inertia of area	1 in^4 = 4.162 314 E − 07 m^4	1 m^4 = 2.402 510 E + 06 in^4
	1 in^4 = 4.162 314 E + 05 mm^4	1 mm^4 = 2.402 510 E − 06 in^4
moment of inertia of mass	1 lb·in·s^2 = 1.129 848 E − 01 N·m·s^2	1 N·m·s^2 = 8.850 748 E + 00 lb·in·s^2
	1 lb·ft·s^2 = 1.355 818 E + 00 N·m·s^2	1 N·m·s^2 = 7.375 621 E − 01 lb·ft·s^2

a. Each conversion factor is written as it would appear on a computer readout or in electronic data transmission: as a number equal to or greater than 1 and less than 10, with six or fewer decimal places, followed by the letter E (to denote exponent of 10), a plus or minus sign, and two digits that indicate the power of 10 by which the preceding number must be multiplied. For example, the factor 9.463 529 E − 04 is, in exponential notation, $9.463\,529 \times 10^{-4}$, or 0.000 946 3529.

b. Factors printed in boldface are exact.

c. The abbreviation lbm stands for "pound-mass." Although the pound is interpreted as a unit of force in this book, you should be aware that it is occasionally used as a unit of mass.

PHYSICAL DIMENSIONS AND THEIR USE

As you have seen, the physical quantities usually used in mechanics are mass, length, time, force, velocity, acceleration, and angle. In other branches of physics, other physical quantities, such as electric current, temperature, amount of matter, and luminous intensity, are used (see Table 1.2).

Key Concept Units and dimensions can be manipulated like algebraic quantities; that is, they can be added, multiplied, divided, combined, or canceled.

According to SI (an *absolute* system of units), the fundamental quantities of mechanics are mass, length, and time. According to the U.S. Customary System (a *gravitational* system), however, the fundamental quantities of mechanics are force, length, and time. The fundamental quantities of SI are designated by the symbols M, L, and T, the first letters of the words "mass," "length," and "time," enclosed in brackets: $[M]$, $[L]$, and $[T]$. Similarly, the fundamental quantities of the U.S. Customary System are designated by the symbols $[F]$ for "force," $[L]$ for "length," and $[T]$ for "time." These quantities, and products of them, are called *dimensions.*[11]

dimensions

Dimensions were used by the French mathematician J. Fourier (1768–1830) as a means of clarifying units of physical quantities. The dimensions of physical quantities follow from definitions or from physical laws. For example, the velocity v of a particle that moves on the x axis is defined to be the derivative of distance along the x axis with respect to time t, or $v = dx/dt$. Since dx is an increment of length and dt is an increment of time, the dimension of velocity is $[L/T]$, or $[LT^{-1}]$. Similarly, since acceleration is defined by the derivative dv/dt, in which dv is an increment of velocity, the dimension of acceleration is $[L/T^2]$, or $[LT^{-2}]$. These dimensions show that velocity may be expressed in units of meters per second (m/s), feet per second (ft/s), kilometers per hour (km/h), miles per hour (mi/h), and so on. Likewise, acceleration may be expressed in meters per second squared (m/s^2), feet per second squared (ft/s^2), kilometers per hour squared (km/h^2), miles per hour squared (mi/h^2), and so on.

In the U.S. Customary System, mass is a derived quantity. Since force and acceleration have the respective dimensions $[F]$ and $[LT^{-2}]$, Newton's law, $\mathbf{F} = m\mathbf{a}$, may be used to determine the dimension of mass:

$$[\text{force}] = [\text{mass} \times \text{acceleration}]$$

Thus, the dimension of mass is

$$[\text{mass}] = [M] = [\text{force/acceleration}] = [FL^{-1}T^2]$$

Similarly, in SI, force is a derived quantity that has the dimension

$$[\text{force}] = [F] = [\text{mass} \times \text{acceleration}] = [MLT^{-2}]$$

11. The notation $[Q]$, where Q denotes a physical quantity, is read "the dimension of Q." Thus, the notation [force] is read as "the dimension of force." Symbolically, $[F]$ = [force].

Therefore, to convert dimensions from the U.S. Customary System to SI, you substitute the dimensional expression $[MLT^{-2}]$ for $[F]$.[12] Likewise, to convert from SI to the U.S. Customary System, you substitute the dimensional expression $[FL^{-1}T^2]$ for $[M]$. For example, in SI, the dimension of mass density (mass per unit volume) is $[ML^{-3}]$, and in the U.S. Customary System, it is

$$[(FL^{-1}T^2)(L^{-3})] = [FL^{-4}T^2]$$

Thus, the dimension of a quantity varies with the system of units. For example, consider the quantity pressure. Pressure is defined as force per unit area (see Tables 1.4 and 1.5). In the U.S. Customary System, the dimension of pressure is [pressure] $= [FL^{-2}]$; in SI, it is [pressure] $= [ML^{-1}T^{-2}]$. Note also that dimensions may be treated as algebraic quantities; that is, their exponents may be combined by rules of algebra.

The above discussion illustrates the fact noted by the physicist Max Planck: "To inquire into the 'real' dimension of a quantity has no more meaning than to inquire into the 'real' name of an object."

If a quantity Q is dimensionless—for example, a radian (Table 1.3)—then $[Q] = [M^0L^0T^0]$ in SI units, or $[Q] = [F^0L^0T^0]$ in U.S. Customary units. Since a number raised to the zero power is unity, these relations are more conveniently expressed as $[Q] = [1]$.

In the following subsection, you will often need to convert the units of a quantity from one system of measurement to another system. The concept of dimensions facilitates such conversions; that is, *dimensions serve as a code telling you how the numerical value of a quantity changes when the base units of measurement are changed.* The examples in the following subsection illustrate this fact. Where appropriate, the examples use approximate values of conversion factors.

CONVERSION OF UNITS

In some situations, it may be desirable to change from one set of units to another. You may wish to express quantities in terms of U.S. Customary units instead of SI units, or you may want to use the unit millimeter instead of meter or inch instead of foot. For instance, consider the acceleration of gravity. As noted earlier, the acceleration of gravity in SI units is $g = 9.806\,65$ m/s^2. Rounding g to two decimal places gives $g = 9.81$ m/s^2, which is usually sufficiently accurate for most engineering calculations.

In SI, conversions are quite straightforward, since they usually involve multiplication or division by powers of 10. For example, to convert the units of g from m/s^2 to mm/s^2, we substitute 1000 millimeters for 1 meter, eliminating the meter unit:

$$g = (9.81\text{ m/s}^2)\left(\frac{1000\text{ mm}}{1\text{ m}}\right) = 9810\text{ mm/s}^2 \tag{1.4}$$

We may also convert g to U.S. Customary units by expressing the meter in terms of equivalent units of foot. Thus, using Table 1.5, we have

$$g = (9.81\text{ m/s}^2)\left(\frac{1\text{ ft}}{0.3048\text{ m}}\right) = 32.18\text{ ft/s}^2 \tag{1.5}$$

12. Langhaar (1980) lists dimension symbols in the order *M, L, T* for the dimension of a physical quantity in SI. This book follows this convention as well as the order *F, L, T* for dimensions in the U.S. Customary System.

Consequently, in U.S. Customary units, the weight W of a unit mass (1 slug) is, by Eq. (1.1) with $W = F$ and $a = g$,

$$
\begin{aligned}
W &= mg \\
&= (1 \text{ slug})(32.18 \text{ ft/s}^2) \qquad (1.6) \\
&= (1 \text{ lb}\cdot\text{s}^2/\text{ft})(32.18 \text{ ft/s}^2) = 32.18 \text{ lb}
\end{aligned}
$$

Note that this result is consistent with the fact that the weight of a unit mass must be exactly g units of force [see also Eq. (1.2)].

Conversion from one unit to another in the U.S. Customary System is not as direct as in SI, since the conversion factors (e.g., 1 ft = 12 in) are not simple multiples of 10. As an example, consider an object with mass equal to 1 slug. We wish to convert the weight of the object on earth (where g is 32.18 ft/s²) from pounds to ounces:

$$W = (32.18 \text{ lb})\left(\frac{16 \text{ oz}}{1 \text{ lb}}\right) = 54.88 \text{ oz} \qquad (1.7)$$

Example 1.1

Conversion of Speed from mi/h to ft/s

Problem Statement Since speed has the dimension $[L/T]$, it can be expressed in any combination of units of length and time. For example, if the units mile and hour are associated with L and T, respectively, then the speed of a car may be expressed as 60 mi/h. Convert the units of mile and hour to foot and second, respectively. That is, express the speed of the car in units of ft/s.

Solution The conversion is carried out as follows:

$$60 \frac{\text{mi}}{\text{h}} = \left(\frac{60 \text{ mi}}{1 \text{ h}}\right)\left(\frac{1 \text{ h}}{3600 \text{ s}}\right)\left(\frac{5280 \text{ ft}}{1 \text{ mi}}\right) = 88 \frac{\text{ft}}{\text{s}}$$

■

Example 1.2

Conversion of Acceleration from mm/s² to mi/h²

Problem Statement Since acceleration has the dimension $[LT^{-2}]$, it can be expressed in any combination of units of length and time. For example, assign the units millimeter and second to length and time, respectively. Then, the acceleration of gravity is $g = 9810$ mm/s². Convert g into units of mile and hour.

Solution First, we convert millimeters to miles. From Table 1.5, 1 km = 10^3 m = 10^6 mm = 0.621 37 mi, or 1 mm = 6.2137×10^{-7} mi. Likewise, we convert seconds to hours by noting that 1 h = 3600 s, or 1 s = 2.778×10^{-4} h. The conversion proceeds as follows:

$$9810 \frac{\text{mm}}{\text{s}^2} = \left(\frac{9810 \text{ mm}}{1 \text{ s}^2}\right)\left(\frac{6.2137 \times 10^{-7} \text{ mi}}{1 \text{ mm}}\right)\left(\frac{1 \text{ s}}{2.7778 \times 10^{-4} \text{ h}}\right)^2 = 78{,}970 \frac{\text{mi}}{\text{h}^2}$$

■

These examples show that units (or, more generally, dimensions) may be treated as algebraic quantities in conversion problems. They may be canceled or combined. The method illustrated by these examples is perfectly general.

DIMENSIONAL HOMOGENEITY

The French mathematician Fourier noted that the laws of nature are independent of human systems of measurement. Therefore, the equations that represent natural phenomena must be independent of units; that is, these equations must be valid for SI, for the U.S. Customary System, or for any other system of measurement. This property of an equation is called *dimensional homogeneity.* For example, the scalar form of Newton's second law, $F = ma$, is valid whether the units of its quantities are meter, kilogram, and second or feet, slug, and second. Therefore, $F = ma$ is said to be dimensionally homogeneous.

dimensional homogeneity

dimensional analysis

The concept of dimensional homogeneity leads to an extensive theory called *dimensional analysis* (Langhaar, 1980). This theory has proved useful for many types of physical studies. For example, one theorem of dimensional analysis states that an equation of the type $x = a + b + c$ is dimensionally homogeneous (valid for all measurement systems) if, and only if, the variables x, a, b, and c all have the same dimensions. This theorem is useful for checking algebraic equations. If a derived equation contains a sum or difference of two or more terms that have different dimensions, a mistake has been made. In common terms, you cannot add apples to oranges and get apples as a sum. For example, consider a problem for which you have derived the equation $F - mv^2 = 0$, where F denotes force, m mass, and v velocity. If you check the dimensions of the terms, you discover that the equation is not dimensionally homogeneous, since the expression mv^2 does not have the dimension of force. In fact, you see that

$$[mv^2] = [FL^{-1}T^2][L^2T^{-2}] = [FL] \neq [F]$$

Consequently, the relation $F - mv^2 = 0$ cannot represent a general physical law.

Warning Dimensional homogeneity is a necessary, but not sufficient, condition for an equation to correctly describe a physical phenomenon. An equation can have the same dimensions in each term and still not be physically meaningful. For example, the equation $FL = mv^2$ is dimensionally homogeneous, but it does not represent a physical phenomenon.

Example 1.3

An Equation of Kinematics

Problem Statement In kinematics, for *constant acceleration* of a particle,

$$x = \frac{1}{2}at^2 + v_0t$$

where x denotes the distance (dimension of length, $[L]$) traveled by the particle, a is acceleration, t is time, v_0 is velocity, and $\frac{1}{2}$ is a dimensionless constant.[13] This equation represents the displacement of a particle that has constant acceleration a and velocity v_0 at time $t = 0$. Show that this equation is dimensionally homogeneous.

Solution This equation is dimensionally homogeneous because each term has the dimension $[L]$:

$$[x] = \left[\frac{1}{2}\right][a][t^2] + [v_0][t]$$

or

$$[L] = [1][LT^{-2}][T^2] + [LT^{-1}][T]$$
$$[L] = [L] + [L]$$
$$[L] = [L]$$

where $\left[\frac{1}{2}\right] = [M^0L^0T^0] = [1]$.

13. For a dimensionless constant C, $[C] = [M^0L^0T^0] = [1]$.

CHECKPOINT 

1. ***True or False:*** The dimension of force is symbolized as $[F]$.
2. ***True or False:*** The dimension of force can be written as $[MLT^{-2}]$.
3. ***True or False:*** The dimensions of physical quantities follow from definitions.
4. ***True or False:*** The dimensions of physical quantities follow from physical laws.
5. ***True or False:*** The dimension of a dimensionless quantity Q is expressed as $[Q] = [0]$.
6. ***True or False:*** An equation that represents a natural phenomenon must be dimensionally homogeneous.
7. ***True or False:*** If an equation is dimensionally homogeneous, it must represent a physical phenomenon.

Answers: 1. True; 2. True; 3. True; 4. True; 5. False; 6. True; 7. False

1.4 NUMERICAL COMPUTATIONS IN ENGINEERING

After studying this section, you should be able to:

- Distinguish between a physical system and a model of the system.
- Identify the sources of error in the solution of a problem.
- Use the appropriate number of significant figures to reflect the accuracy of a numerical value.

MODELING

model

A realistic and successful solution to an engineering problem usually begins with an accurate model of the problem and an understanding of the assumptions used. Physicists and engineers define a *model* to be an approximation of a system that is sufficiently accurate for the purpose of physical analysis. For example, suppose that you are given the problem of computing the motion of a baseball thrown or batted into the air with a given velocity (see Young, 1992, for an excellent introduction to this problem). The ball is usually modeled as a particle—that is, as a point mass. Is this model of a baseball a good one? To answer this question, we must consider a real baseball: Is it a point of concentrated mass? No! Is this fact important? It could be, if rotation of the ball is important! A slow-motion replay of a videotape of the ball in flight would show that it spins (rotates). Recall that a particle does not spin—that is, its rotation has an insignificant effect on its motion. However, as every baseball fan knows, the spin imparted to a baseball by a pitcher has a significant effect on the motion of the ball. The velocity of the ball, its spin, and its interaction with the air all affect its motion. (Because of these complications, physicists are still puzzling over why some pitchers throw better curve balls than others.) Another factor that must be considered is the effect of gravity. Is the force of gravity that acts on the ball constant? No, the force of gravity changes with the altitude of the ball. Is this change significant for this problem? Probably not. Does the rotation of the earth affect the motion of the ball? Yes! Do you think this effect is important for this problem? Are there other factors that affect the ball's motion? Can you think of any?

In brief, if we try to include all these effects in our analysis, the problem becomes extremely complex. So, what do we do? We simplify the problem to allow ourselves to

In his major league baseball career, Nolan Ryan pitched 7 no-hit games and struck out 5308 batters—both league records. Although he depended mainly on his 90+ mi/h fast ball, he used his excellent curve ball to set up many of his strike-out victims.

obtain a sufficiently accurate approximation to the solution of the real problem. First, we ignore the size and shape of the ball; that is, we consider it to be a particle. Then we neglect air resistance and wind effects and assume that the ball travels in a vacuum. Finally, we neglect changes in the force of gravity and the effect of the earth's rotation. Now we have a model of the problem: a particle in a constant-gravity field moving in a vacuum with an initial velocity. With this simplified model, the problem is manageable. The solution is that the ball moves along a parabolic arc. But how accurately does this model solution represent the real solution? We will examine this question in Chapter 14.

The search for answers to such questions is the art of engineering. A significant component of good engineering is the art of good modeling. In general, the objective of good modeling is to keep the most important effects in a problem and discard the less important effects. Of course, you must not throw out too much. For example, if, in addition to the simplifications discussed above, you neglected the force of gravity entirely (rather than keeping it constant), the solution to the baseball problem would be that the ball would travel at a constant speed in a straight line, since no force is acting on it (Newton's first law). It would continue to travel in a straight line into outer space, never to return to earth! Obviously, this is a poor solution to the real problem.

ACCURACY OF DATA

As noted above, good modeling involves an approximation of a real system by an acceptable idealization (a model). Approximations affect the accuracy of solutions. In engineering mechanics, physical models such as the particle and the rigid body are used extensively. Other types of approximations that affect the accuracy of solutions are also common in engineering. For example, engineers are constantly dealing with numbers that are obtained more or less from experimental measurements. Experimental observations always have inaccuracies. Therefore, wise use of numbers based on experimental measurements requires engineers to account for these inaccuracies.

Inaccuracies may occur because of allowable manufacturing tolerances. For example, in standard mill production of a steel beam, variations in cross-sectional area and weight of ±2.5% from specified values are allowed (AISC, 1994). Hence, in conversion of such areas and weights from one set of units to another, it would be nonsense to use conversion factors accurate to, say, nine digits. For example, if a specification requires that the cross-sectional area of a beam be 10 in^2, a steel mill may produce beams with cross-sectional areas from 9.75 to 10.25 in^2—that is, in the range 10±0.25 in^2. Since engineering computations will use the specified area of 10 in^2, a conversion factor accurate to three digits is sufficient. In other words, the number of digits used to report a numerical value should not misrepresent the level of accuracy associated with the value.

systematic errors

Measurement errors may be systematic or random. *Systematic errors* are errors associated with a particular instrument or measurement technique and, as such, are repeatable. For example, suppose you have a pencil that is 180 mm long. You measure it by laying a ruler along it, with the zero end of the ruler aligned with one end of the pencil. If the zero end of the ruler started at 2 mm instead of 0 mm, because 2 mm of the ruler had been broken off, the length of the pencil might mistakenly be recorded as 182 mm. Other pencils of the same length measured with this ruler would also be recorded as being 182 mm long. This error in measurement is a systematic error. Systematic errors can also be introduced in the measurement of temperatures if an improperly calibrated thermometer is used.

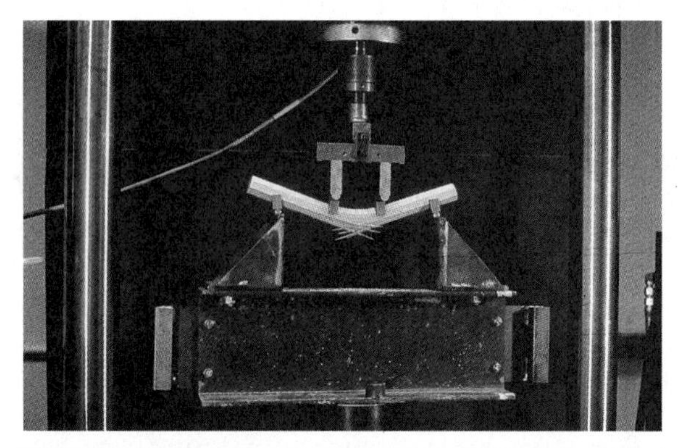

The bending strength of wooden pegs used in timber frame construction is tested experimentally. Inaccuracies in such measurements are due to errors in calibration of the test equipment, limits to the precision of the data-acquisition equipment, and natural variability of the materials being tested.

random errors

Random errors are produced by a number of unpredictable and unknown differences in experimental measurements, such as fluctuations in temperature, voltage, light intensity, and so on. Random errors are frequently distributed according to a simple law, and they may be analyzed by statistical methods. These methods are treated in courses on statistics.

Scientists and engineers often indicate the accuracy of a measurement or number by writing a number followed by the symbol ± and a second number indicating the likely uncertainty. For example, if a series of measurements of the diameter of a steel rod is reported as $3\frac{1}{2} \pm \frac{1}{16}$ in or, in shorthand notation, as $3\frac{1}{2}(\frac{1}{16})$ in, this means that the true diameter of the rod is unlikely to be greater than $3\frac{9}{16}$ in or less than $3\frac{7}{16}$ in. Accuracy can also be indicated in terms of the maximum likely percentage of uncertainty—for example, 5.6±2%.

When you use an uncertain number in a computation with other numbers, the products or sums are also uncertain. For instance, if you multiply an uncertain number by an exact number, the product is uncertain. Thus, the product of 20.0±1.0 (or 20.0±5%) and the exact number 2.0 is 40.0±2.0 (or 40.0±5%).

The *accuracy of a solution* to a problem depends on:

1. the accuracy of the data
2. the accuracy of the analysis (including modeling and the use of a proper theory)
3. the accuracy of the numerical calculations

The accuracy of a solution is no greater than the accuracy of these three items. Since the accuracy of data can be expressed in amounts of likely uncertainties, any computation utilizing the data is subject to these uncertainties. By taking great care with the analysis and numerical computations, you can reduce the uncertainties in these items to acceptable levels.

SIGNIFICANT FIGURES

The phrase "number of significant figures" means the number of accurately known digits in a number.[14] For example, the cross-sectional area A of a steel rod with circular cross section is given by the formula $A = \pi d^2/4$, where d is the diameter of the rod and $\pi = 3.141\,592\,654$ to ten significant figures. Suppose you measure the diameter of the rod with a micrometer and find it to be 18.12 mm, rounded off to four significant figures. In this case, there is no benefit in using π to ten significant figures. It is sufficiently accurate to round π to four significant figures, $\pi = 3.142$. You use your calculator to obtain $A = (3.142)(18.12\text{ mm})^2/4 = 2.579 \times 10^2\text{ mm}^2$, rounded off to four digits. Since you have rounded off, the actual value of A lies in the range $2.5785 \times 10^2 \leq A \leq 2.5795 \times 10^2$. The absolute error in A is between $-5 \times 10^{-2}\text{ mm}^2$ and $5 \times 10^{-2}\text{ mm}^2$. The use of powers of 10 allows you to show significant figures directly.

scientific notation

The use of powers of 10 to display numbers is called *scientific notation.* The usual practice when using scientific notation is to express a quantity as a number between 1 and 10 with the appropriate number of digits, multiplied by a power of 10. For example, the speed of light is about 300 000 000 m/s. When it is written in this manner, there is no indication of the number of significant figures, and ordinarily all nine figures would not be considered significant. However, when the speed of light is written as 3.00×10^8 m/s, it is clear that there are three significant figures.

In computations with approximate numbers, some figures to the right of the decimal may be of doubtful accuracy. They should therefore be rounded off. The following rules usually give the largest number of significant figures that it is reasonable to retain.

Addition and Subtraction The last significant figure in a sum or a difference of numbers is the figure in the last full column. For example, the following sum has four significant figures:

$$\begin{array}{r} 3.546 \\ 1.234\,5 \\ +\ 0.846\,75 \\ \hline 5.627 \end{array}$$

14. An accurately known digit, other than zeros that locate the decimal point, is called a *significant figure.* For example, the number 3.60 has three significant figures, and 3.609 has four. The number 0.0036 0 has three significant figures, since the first three zeros merely locate the decimal point.

Multiplication and Division Ordinarily, you should keep no more significant figures in the result than the fewest in any number involved in the calculation. For example, if both numbers are rounded, $3.2345 \times 1.20 = 3.88$. However, if 1.20 is an exact number, then five significant digits may be retained in the product. In a series of multiplications or divisions of approximate (rounded) numbers, keep one additional figure until the final result is obtained, then round off appropriately. In cases where there is a larger difference in the relative accuracy of the numbers—that is, in the number of significant figures—round off all the numbers to one more significant figure than in the least accurate number. This procedure will probably introduce a small error in the last figure.

CHECKPOINT 

1. ***True or False:*** Physical systems are usually too complex to analyze exactly.
2. ***True or False:*** Models based on simplifying assumptions are frequently used to obtain approximate solutions of physical problems.
3. ***True or False:*** There are formal procedures for constructing models of physical systems and for determining simplifying assumptions that are realistic.
4. ***True or False:*** To express π to seven significant figures, you should write $\pi = 3.141\,592\,7$.

Answers: 1. True; 2. True; 3. False; 4. False

1.5 SUGGESTIONS FOR PROBLEM SOLVING

After studying this section, you should be able to:

- Develop a strategy for effective and efficient problem solving.

At the beginning of this book, "To the Student" outlines a five-step strategy for problem solving that can be applied to a wide variety of problems. While this generic strategy has proven effective in improving the problem-solving skills of students, this section offers suggestions as to how it can be tailored to problems of mechanics. The merit of these techniques may not be obvious at this time; however, as you continue your studies and are faced with more and more complex problems, the value of these techniques should become apparent.

Define the Problem With all problems, you must *carefully read the problem statement* and determine what data are given and what results are required. You may often be confronted with a lot of data, some of which may not be significant. Do not be misled by extraneous, erroneous, or insignificant data.

Think about the Problem and Identify Its Basic Features Identify the appropriate concepts or principles that apply to the problem. Do you need to convert units? Does the problem involve statics, kinematics, or kinetics? Does it require a combination of these branches of mechanics? Can you model the system with particles or with rigid bodies?

Plan Your Analysis In the application of Newton's laws, always use an appropriate model (say, a particle or a rigid body) that represents the system with sufficient accuracy for your purposes. Sketch an accurate free-body diagram of the system, showing all the forces (magnitudes, directions, and senses) acting on the system.[15] Select a convenient coordinate system on which to represent forces in the free-body diagram. Draw a diagram of the coordinate system, showing the origin, the directions of the axes, and their positive senses. Label the axes accordingly. If you know the direction and positive sense of an important quantity (such as the force on a particle), orient your coordinate axes to correspond to this direction and sense.

Determine the known and unknown quantities, and assign an identifying symbol (letter) to each unknown quantity. If you know the direction but not the sense of a force, assume the sense. If your solution shows the magnitude of this force as positive, your assumed sense is correct. If the magnitude of the force turns out to be negative, the sense of the force is opposite to what you assumed.

Apply mechanics principles to the body in your free-body diagram, and derive independent equations, equal in number to the number of unknown quantities. These equations must include one or more of the unknowns. In some cases, you may need to draw free-body diagrams of different parts of the system in order to obtain enough equations to solve for the unknowns. Pay careful attention to dimensions to ensure that the equations are dimensionally homogeneous (Sec. 1.3). Organize the equations in an order that will simplify their algebraic solution. Check limiting cases. For example, determine what happens as an angle goes to 0°, 90°, and 180° or as friction goes to zero. This type of check may reveal errors in the equations.

Carry Out Your Plan Solve the equations for the unknowns.

Check signs to be sure they are consistent. Once you define the direction of an axis and its positive sense, then displacement, velocity, acceleration, and force projections in the direction of the positive sense of that axis are also positive. Otherwise, they are negative.

Represent your numerical answers with the appropriate number of significant figures and in consistent units. You should realize, however, that it is impractical to write every number in a textbook's examples and problems with the correct number of significant figures. You will find that most examples and problems in this book give data to three or four significant figures. Frequently, though, a length is given as 6 m or a mass as 10 kg, for example, rather than as 6.00 m or 10.0 kg. In general, unless stated otherwise, you may assume that all data given in this book are accurate to three significant figures.

Look Back and Appraise Your Results Did you solve the problem that was posed? Examine your results. Are they logical? Do the numbers (magnitude and sign) make sense? If not, did you make errors in arithmetic? In other parts of the mathematics? Check for errors in signs, decimal point locations, transposed digits, and so forth. Are the units correct? Is your solution physically reasonable? Compare the magnitudes of your answers to the magnitudes of the given data. If your numerical results differ by several orders of magnitude from the given data, you may have made an error.

Present your results in a clear, logical, and readable format.

15. Construction of free-body diagrams is discussed in Chapter 3 for particles and in Chapter 4 for rigid bodies.

Chapter Highlights

- Force is a quantitative measure of the interaction between two particles.
- Newton's first law implies that if no force acts on a particle, the particle is said to be in equilibrium. If the particle is initially at rest, it remains at rest; if it is initially in motion, it continues to move with constant velocity.
- The inertial properties of a particle are characterized by its mass. The acceleration **a** of a particle under the action of several forces is directly proportional to its resultant force **F** and inversely proportional to its mass m; that is, $\mathbf{a} = \mathbf{F}/m$ [Eq. (1.1)].
- The unit of force is defined in terms of the units of mass and acceleration. In SI, the unit of force is the newton (N), equal to 1 kg·m/s^2. In the U.S. Customary Systcm, thc unit of force is the pound (lb), equal to 1 slug·ft/s^2.
- The weight of a body on earth is the gravitational force exerted on it by the earth. Weight is a force. The magnitude W of the weight of a body at a specific location on earth is equal to the product of the body's mass m and the magnitude g of the acceleration due to the earth's gravity at that location; that is, $W = mg$. On another planet—say, Mars—the weight of the body is the gravitational force exerted on it by that planet. Therefore, the weight of a body depends on its location. However, the mass of the body is constant, independent of location.
- Newton's third law states that action equals reaction. Therefore, two particles exert forces on each other that are equal in magnitude but opposite in sense.
- The base physical quantities in SI are mass, length, and time, with units of kilogram (kg), meter (m), and second (s), respectively. Multiples of SI units are obtained by adding prefixes (representing powers of 10) to the base units—for example, *kilo*meter. Derived units are defined in terms of the SI base units.
- SI is an absolute decimal system because the fundamental physical quantities are defined to be independent of location, and all derived units are decimal multiples of the base units.
- Base physical quantities in the U.S. Customary System are force, length, and time, with units of pound (lb), foot (ft), and second (s), respectively.
- The U.S. Customary System of units is a gravitational system because the definition of the pound depends on the local force of gravity.
- Equations that describe physical phenomena must be dimensionally homogeneous. Terms in these equations must have the same units before they can be manipulated algebraically. In particular, the two sides of an equation must have the same units. When one set of units is converted to another, the units can be treated like algebraic quantities.
- The uncertainty of a number is indicated by the number of significant figures or stated explicitly (often as a percentage).

Problems

1.2 *Newtonian Mechanics, Engineering Mechanics*

1.1 Look up or estimate the maximum speed of each of the following objects:

a. a satellite that orbits the earth
b. a racing car
c. a military fighter plane
d. a baseball thrown by a major league pitcher
e. an automobile piston

Express these speeds in both SI and U.S. Customary units. All these objects may be analyzed using Newtonian mechanics.

1.2 Albert Einstein, in his theory of relativity, related the mass m of a particle to its speed v by the formula $m = m_0(1 - v^2/c^2)^{-1/2}$, where $c = 3 \times 10^8$ m/s is the speed of light and m_0 is the mass of the particle at rest.

a. By what percentage does the mass of a particle change from its rest mass if its speed is (i) 3×10^4 m/s, (ii) 3×10^6 m/s, or (iii) 3×10^7 m/s?
b. What do you think of the validity of using Newtonian mechanics (which assumes that m_0 remains constant) for ordinary-size bodies that travel at ordinary speeds?
c. What does Einstein's theory predict would be the mass of the particle if it could travel at the speed of light?

1.3 In an experiment to determine the acceleration of a particle, a student records the (x, y, z) components of the force **F** that acts on the particle. The student also measures the (x, y, z) components of the acceleration **a**. The measurements are (8, 10, 30) for **F** (units of N) and (2.5, 3.125, 10) for **a** (units of m/s²). What do you think of the quality of this data? Explain your answer.

1.4 The student of Problem 1.3 finds that the accelerometer is accurate and that the z component of **F** (30 N) is correct. What are the correct x and y components of **F**?

1.3 *Units of Measure and Physical Dimensions*

1.5 Standing on top of Mt. McKinley, Alaska (elevation 20,320 ft), you discover a small meteorite. You have a spring scale for measuring the weight of fish, and you decide to "weigh" the meteorite. Discuss the significance of your measurement.

1.6 The U.S. national debt is approximately $5 trillion.

a. Express this amount using a prefix from Table 1.1.

b. If you could lay that many dollar bills end to end, how many times would they wrap around the earth at the equator? The circumference of the earth at the equator is approximately 25,000 mi.

1.7 A computer server operating with Microsoft's Windows NT™ and Internet Information Server™ software is advertised as being able to process 1200 hits per second. Using the prefixes in Table 1.1, determine how many transactions can be processed in:

a. 1 day

b. 1 year

1.8 What are the kinetic units in SI?

1.9 In SI, is MN·m a base unit or a derived unit? Explain.

1.10 In SI, is N·m/s a derived unit or a base unit? Explain.

1.11 What is your height in inches, in feet, in yards, in meters, and in millimeters?

1.12 What is your body mass in slugs and in kilograms?

1.13 Use the prefixes in Table 1.1 to express each of the following quantities:

a. 10^{-6} phone

b. 10^{6} lopolis

c. 2×10^{1} cards

d. 10^{-6} scope

e. 10^{6} bucks

Get the idea? Can you come up with a few similar expressions?

1.14 a. If runner A runs a 100 m dash in the same time as runner B runs a 100 yd dash, which runner is faster? How much faster than the slower runner, as a percentage?

b. Runner A runs 10 km, and runner B runs 6 mi. Which runner runs the longer distance? What percentage longer is that distance relative to the shorter distance?

1.15 Noting that 1 in = 25.4 mm, determine the number of miles in 1.00 km.

1.16 The mass density of water is 10^3 kg/m³.

a. What is the density of water in slug/ft³?

b. What is the density of water in g/cm³?

1.17 A Navy jet plane can fly at 1450 mi/h (about twice the speed of sound—that is, at Mach 2).

a. What is its speed in mi/s?

b. What is its speed in km/min?

1.18 The piston displacement of a General Motors V-8 engine is 350 in³. What is the displacement in liters? (Note that 1 L = 1000 cm³, 1 cm = 10 mm, and 1 in = 25.4 mm.)

1.19 The fuel consumption of a truck is 6 mi/gal. Determine the fuel consumption of the truck in km/L.

1.20 Determine the conversion factors between each pair of units:

a. a square mile and a square kilometer

b. a cubic meter and a cubic yard

c. a square inch and a square yard

1.21 The mass of the earth is approximately 4.08×10^{23} lb·s²/ft.

a. Express the mass of the earth in the units kip·yr²/mi (1 kip = 1000 lb; 1 yr = 365.25 days; 1 day = 86 400 s).

b. Express the mass of the earth in the units N·s²/m.

1.22 A real estate investor purchased a rectangular tract of land that is $\frac{1}{8}$ mi wide and $\frac{1}{2}$ mi long for $32,000. How much did the investor pay per square mile, per acre (1 mi² = 640 acres), per square meter, and per square foot?

1.23 A mass is given as 1×10^{-5} kip·h²/mi. What is the magnitude of the mass expressed in lb·s²/in, in grams, and in slugs? What is the weight, in pounds, of the mass under standard gravitational attraction?

1.24 The position y of a particle that moves along the y axis is given by the equation $y = at^2 + bt^4$, where t denotes time.

a. What dimensions must a and b have?

b. If y is measured in feet and t in seconds, what units must a and b have?

c. If y is in miles and t is in hours, what units must a and b have?

d. If y is in kilometers and t is in seconds, what units must a and b have?

1.25 The period T of a satellite that orbits the earth is given by the equation $T = ca^{3/2}$, where c is a constant of proportionality and a is the length of the major semiaxis of the satellite's elliptic orbit. When T is in hours and a is in miles, $c = 5.65 \times 10^{-6}$ h/mi$^{3/2}$. Determine the magnitude and the units of c when T and a are measured in minutes and meters, respectively.

1.26 Determine whether or not the following equations are dimensionally homogeneous. In these equations, F denotes force $[F]$, m mass $[M]$, s distance $[L]$, v speed $[LT^{-1}]$, and t time $[T]$.

a. $Fs = \frac{1}{2}mv^2$

b. $Fs = \dfrac{dA}{dt}$, $[A] = [mvs]$

c. $Ft = mv^2$

1.27 A simple electric circuit (Fig. P1.27) with inductance, capacitance, and resistance is described by the differential equation

$$\frac{d^2v}{dt^2} + a\frac{dv}{dt} + bv = 0$$

where t denotes time in seconds and v denotes electric potential with units N·m·s^{-1}·A^{-1}. For this equation to be dimensionally homogeneous, what are the units of the coefficients a and b?

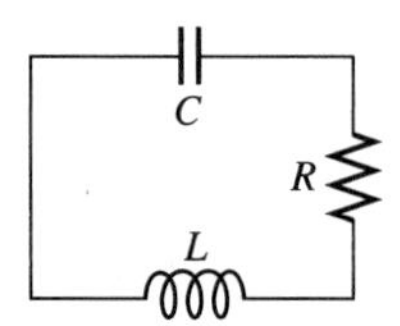

Figure P1.27

1.28 A cantilever beam of length L is subjected to a force P at its free end (Fig. P1.28). The deflection d of the beam's neutral axis (shown as a dashed line in the figure) at a distance x from the free end is given by the equation

$$d = \frac{P}{6EI}(2L^3 - 3L^2x + x^3)$$

What are the dimensions of the product EI if this equation is dimensionally homogeneous?

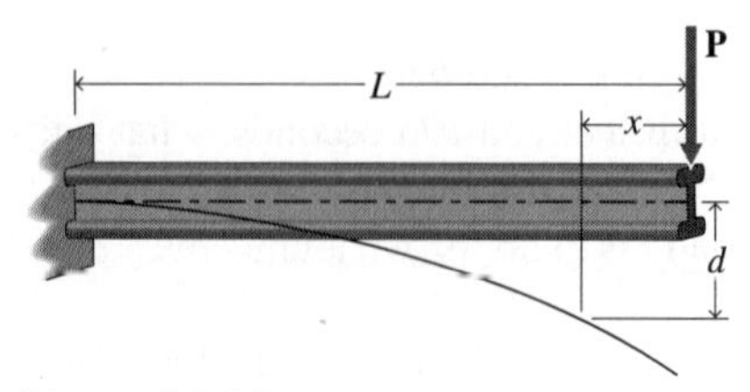

Figure P1.28

1.29 The dimension of a quantity is $[F^{1/2}L^{-3}T^{-2}]$. By what factor is the numerical value of the quantity changed if the unit of force is changed from pound to newton, the unit of length from foot to meter, and the unit of time from minute to second?

1.30 The dimension of a quantity is $[M^3L^{-1}T^2]$. By what factor is the numerical value of the quantity changed if the unit of mass is changed from slug to kilogram and the unit of length from foot to meter, while the unit of time remains second?

1.4 *Numerical Computations in Engineering*

1.31 The numerical value of π is approximately 3.141 592 654. Determine the percent error in the following approximate values of π by comparing them to that value:

a. $\frac{22}{7}$

b. $\frac{355}{113}$

1.32 According to elasticity theory, the stress σ at the center of a solid uniform rotating disk is given by the dimensionally homogeneous formula

$$\sigma = \frac{1}{8}(3 + \nu)\rho\omega^2r^2$$

where ν is a dimensionless constant of the material (Poisson's ratio), ω is the angular velocity of the disk, r is the radius of the disk, and ρ is the mass density (mass per unit volume). Using this formula, compute the stress (in lb/in^2) at the center of a steel disk for which $\nu = 0.30$, $r = 12$ in, and $\omega = 200$ rad/s. Steel weighs 0.28 lb/in^3 on earth.

1.33 The mass m and the radius r of the earth are approximately 4.08×10^{23} slugs and 20.9×10^6 ft, respectively.

a. Calculate the average mass density (mass per unit volume) of the earth in slug/ft^3. Use powers-of-10 notation and the correct number of significant figures. The volume V of a sphere of radius r is $V = 4\pi r^3/3$.

b. Calculate the average mass density of the earth in kg/m^3. Again, use powers-of-10 notation and the correct number of significant figures.

REVIEW QUESTIONS

1.34 Who was Archimedes? What contributions did he make to mechanics?

1.35 Who was Aristotle? What contributions did he make to mechanics?

1.36 Who was Simon Stevin? What contributions did he make to mechanics?

1.37 Who was René Descartes? What contributions did he make to mechanics?

1.38 Who was Galileo? What contributions did he make to mechanics?

1.39 Who was Isaac Newton? What contributions did he make to mechanics?

1.40 Newton has been quoted as saying that he was able to develop his three laws by standing on the shoulders of giants. Explain what he meant.

1.41 Who is credited with first analyzing nonparallel forces?

1.42 Who is credited with the concept of the force parallelogram?

1.43 What name is given to quantities that add like forces?

1.44 Who developed the idea of resolving forces into projections parallel to coordinate axes?

1.45 Who is credited with the introduction of the terms *vector* and *scalar*?

1.46 Define statics, kinematics, kinetics, and dynamics.

1.47 What are Newton's three laws?

1.48 How is Newton's first law related to his second law?

1.49 What is the name given to a point mass?

1.50 What is the relationship between the acceleration of a particle, the force that acts on the particle, and the mass of the particle?

1.51 What property of a particle is mass a measure of?

1.52 What is a physical quantity? What is a unit?

1.53 Why do engineers need standard units of measurement?

1.54 What is the current standard unit of mass? Of time? Of length?

1.55 Why is the balance-type scale unsatisfactory as a universal device for measuring force?

1.56 Why is the standard unit of force not defined using the properties of a standard spring?

1.57 What is meant by the *weight* of a body? Is weight a force? Is weight a mass?

1.58 What are kinetic units?

1.59 What is meant by *base (or fundamental) units*, by *derived units*, and by *supplementary units*?

1.60 What does SI stand for? What is SI?

1.61 What is meant by *absolute system of units*, *gravitational system of units*, and *decimal system of units*?

1.62 What is the standard pound?

1.63 What is meant by *conversion of units*?

1.64 What are dimensions?

1.65 What is meant by *dimensional homogeneity*?

1.66 What is meant by *modeling*, by *data accuracy*, and by *significant figures*?

1.67 What is meant by *equilibrium of a particle*?

Chapter 2 FORCES AND OTHER VECTOR QUANTITIES

By Newton's third law of motion, the force (action) of the boy on the 576 pound sumo wrestler is opposite to the force (reaction) of the sumo wrestler on the boy.

A Look Forward

IN THIS CHAPTER, we explore the characteristics of forces and discuss the rules of force addition, subtraction, and projection. The rules for manipulating forces also apply to other physical quantities called *vector quantities,* or simply *vectors.* Since an understanding of the characteristics of forces and the rules by which they combine is essential to the study of their effects on physical systems, this chapter lays a foundation for the discussions of equilibrium in Chapters 3, 4, and 5.

The topics in this chapter are presented from three points of view. First, we discuss force resultants and force components in terms of graphical representations. We consider the parallelogram law and the polygon construction for composition (addition) and resolution of forces. This graphical approach demonstrates that operations on forces are independent of a coordinate system. With this approach, the concepts are easily visualized and hence understood. Next, we explore the concepts of axes and projections of forces on rectangular Cartesian coordinate axes. This second approach is more generally applicable than the graphical approach, particularly for three-dimensional problems. Finally, we consider the concept of unit vectors and their use in vector operations. This third approach is beneficial for considering equilibrium of rigid bodies.

After you have studied this chapter, you should have a clear understanding of the characteristics of forces and how forces combine. You should know what is meant by a force resultant, the components and projections of a force, the parallelogram law and the polygon construction, the composition and resolution of forces, axes, right-handed and left-handed coordinate systems (in a plane and in space), and unit vectors. You should also know how to apply your knowledge to specific situations.

Survey Questions

- What are the characteristics of a force? What kinds of forces are there?
- Is Newton's third law useful in classifying forces?

- How are forces combined? Can a single force be broken into two or more components?
- How do a force and a vector differ?
- What are the characteristics of a vector? Do the rules for operating on forces apply to vectors?
- How are unit vectors used?

2.1 CHARACTERISTICS OF FORCES

After studying this section, you should be able to:

- List and describe the characteristics of a force.
- Identify different types of forces, including concentrated, distributed, and body forces.
- Recognize when forces are concurrent, collinear, and/or coplanar.

Key Concepts

A force is a physical vector quantity created by the interaction between two bodies.

Forces occur in pairs. One force of the pair is called the action; the other is called the reaction.

force A *force* is created by the interaction between two bodies. As was mentioned in Chapter 1, the vector character of a force is emphasized by denoting the force using a symbol in boldface type. Thus, **F** represents a force completely; that is, it represents both the magnitude and the direction of the force. The symbol F (in lightface italic type) denotes only the magnitude of the force. In handwritten work, the notation $\vec{F}$ or $\underline{F}$ is often used to symbolize a force.

When two bodies interact, one of the bodies is assumed to exert the force and the other to resist it. For example, the force of gravity acts between an apple and the earth (see Fig. 2.1). The normal assumption is that the earth exerts the force $\mathbf{F}_{AE}$ on the apple. However, the apple exerts an equal force $\mathbf{F}_{EA}$ on the earth. Newton's third law states that for every **action** force (*action*), there is an equal and oppositely directed force (*reaction*). Whether you consider a force to be an action or a reaction depends on your point of view. For example, in **reaction** Fig. 2.1, if you consider a system consisting of the apple alone, the force $\mathbf{F}_{AE}$ that the earth exerts on the apple is an action (on the apple). Then, the force $\mathbf{F}_{EA}$ that the apple exerts on the earth is the reaction to the force $\mathbf{F}_{AE}$. Alternatively, if you consider a system consisting of the earth alone, the force $\mathbf{F}_{EA}$ that the apple exerts on the earth is an action (on the earth), and the force $\mathbf{F}_{AE}$ is the reaction to $\mathbf{F}_{EA}$ (see Sec. 2.2 for additional discussion of Newton's third law).

The force of gravity is usually expressed in terms of a single characteristic—the weight of a body. However, a force, including the force of gravity, is not characterized by magnitude alone. The direction of the force is also an essential property. For example, the attraction due to earth's gravity is directed toward the center of the earth. Hence, the force of gravity is directed toward the center of the earth (Fig. 2.1). Since this fact is well known, the direction of weight is generally not specified. It is understood to be directed toward the center of the earth. Generally, any force must be specified in terms of all its characteristics—namely, its magnitude, direction line, and sense.

Figure 2.1 Action and reaction: $\mathbf{F}_{AE}$ = force exerted on the apple by the earth; $\mathbf{F}_{EA}$ = force exerted on the earth by the apple.

The essential characteristics of a force (its magnitude and direction) may be graphically illustrated with a directed arrow (see Fig. 2.2). Note in Fig. 2.2a that the end of the arrow shaft with the arrowhead is called the *tip* and the other end is called the *tail.* The line on which the force lies is along the shaft of the arrow and is called the *direction line* of the force. The length of the arrow from tip to tail represents the *magnitude* of the force, with respect to an arbitrarily chosen scale (Fig. 2.2b). The direction line along which the force acts is also referred to as the *line of action* of the force. The line of action of the force may be located relative to a given reference line in terms of an angle θ (Fig. 2.2c).

direction line

line of action

The arrowhead indicates which way along its line of action the force acts. In Figs. 2.2a, 2.2b, and 2.2c, the forces **F** act up to the right. If the arrow is reversed in sense, as in Fig. 2.2d, the force $-\mathbf{F}$ (opposite to **F**) acts down to the left. Finally, a point of application of a force may be required to correctly determine the physical effects the force may produce on a body on which it acts (Fig. 2.2c).

A summary of the characteristics of a force is given in Fig. 2.3. Many writers use the word "direction" to include both the direction line and the sense of the force. Then, a force is fully defined when its magnitude and direction are prescribed, with the understanding that *the given direction includes both the sense and the line of action of the force.* This

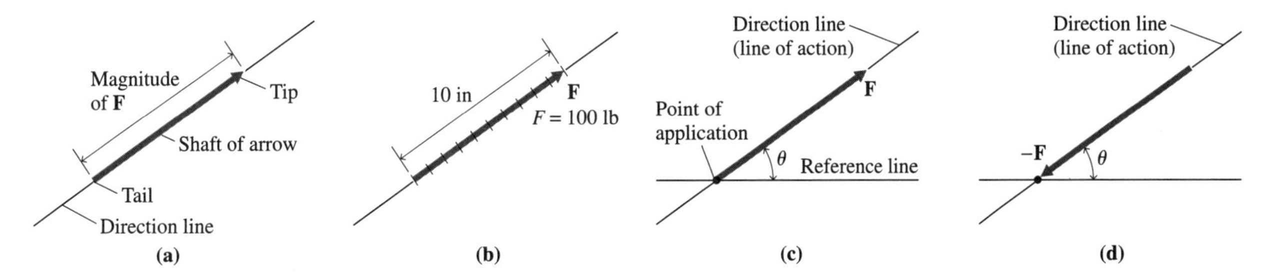

Figure 2.2 **(a)** Representation of a force **F** by an arrow. **(b)** Magnitude of **F** (1 in = 10 lb). **(c)** Line of action and point of application of **F**. **(d)** Negative force vector $-\mathbf{F}$.

Tension in the cables and straps applies forces to the pulley. These forces can vary in both magnitude and direction.

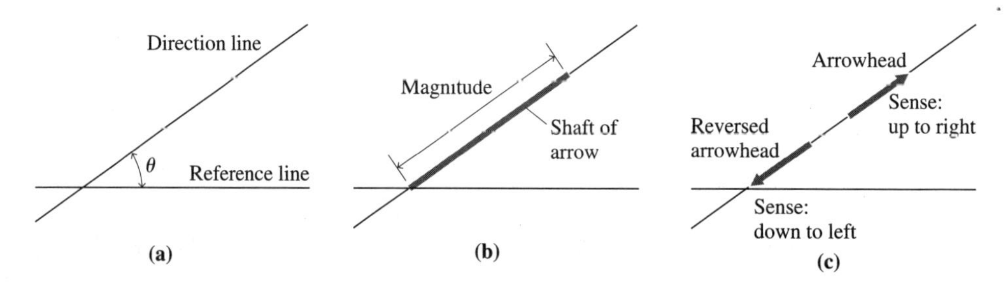

Figure 2.3 Graphical illustration of force characteristics: **(a)** direction line of a force (at angle θ); **(b)** magnitude of force (proportional to length of shaft); **(c)** sense of force (designated by arrowhead).

book will follow the customary practice and use the word "direction" to mean *both* the sense and the direction line (line of action) of the force. Thus, two forces $\mathbf{F}_1$ and $\mathbf{F}_2$ have *opposite directions* if they lie along two parallel direction lines or along the same direction line and have opposite senses (Fig. 2.4). If there is a need for further clarity, however, the sense and the direction line of a force will be given separately. Also, the point of application of the force will be specified when it is necessary to do so.

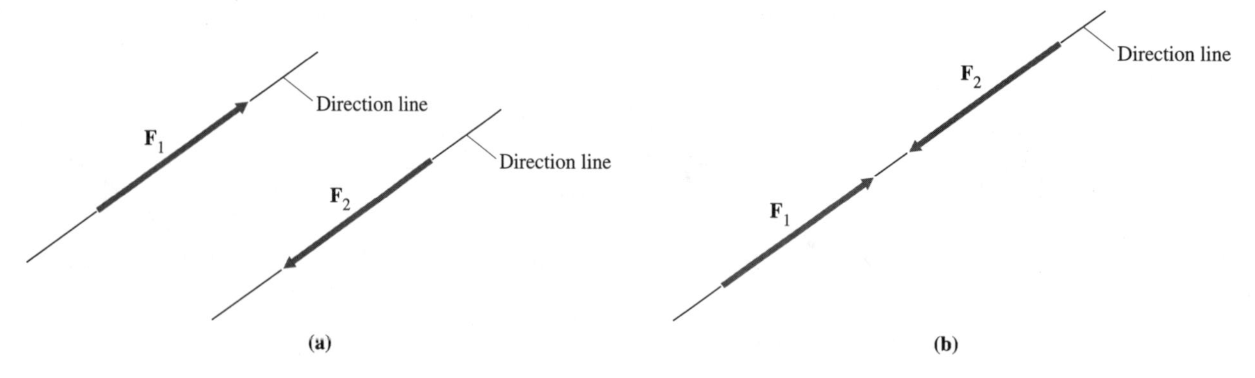

Figure 2.4 Forces $\mathbf{F}_1$ and $\mathbf{F}_2$ with opposite directions: **(a)** parallel direction lines; **(b)** coincident direction lines.

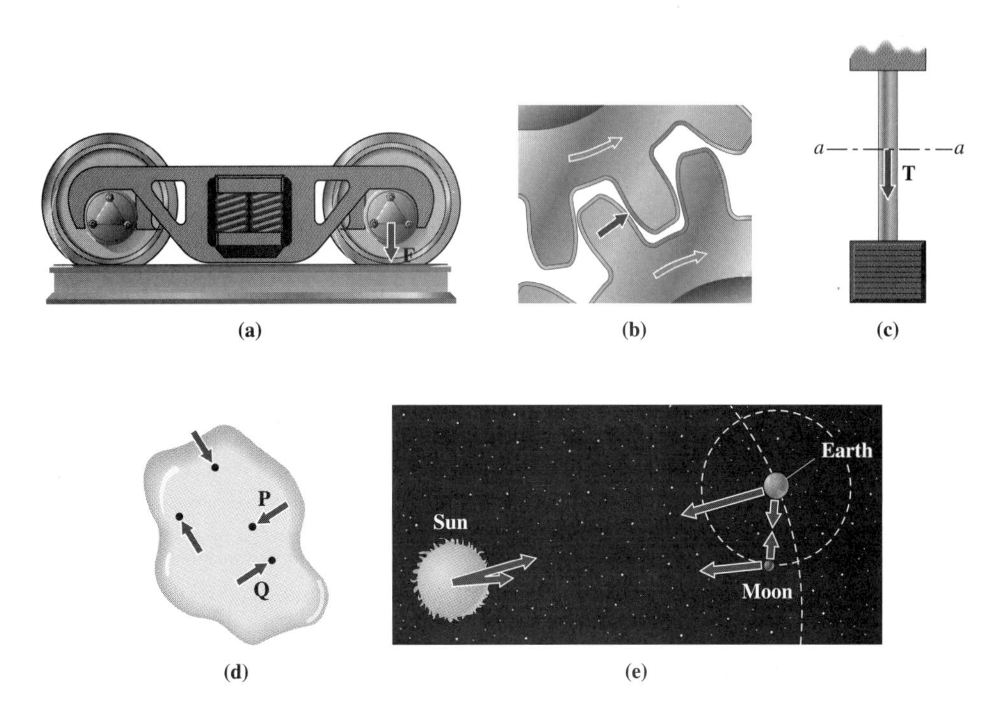

Figure 2.5 Representation of forces by means of arrows: (**a**) wheel on a rail; (**b**) two gear teeth; (**c**) weight on a bar; (**d**) forces acting on an arbitrary body; (**e**) gravitational forces among sun, earth, and moon.

Examples of several types of forces that occur in various systems are shown in Fig. 2.5. In Fig. 2.5a, the vertical vector represents the force that a railroad car wheel exerts on the rail. Figure 2.5b illustrates the force that one gear tooth exerts on another. Figure 2.5c shows the force that the part of a bar below the plane *a–a* exerts on the part above the plane. Figure 2.5d represents several forces acting on an arbitrary body, where forces **P** and **Q** have the same magnitude but opposite directions. Figure 2.5e represents the gravitational forces that act among the earth, moon, and sun.

concentrated force

In Fig. 2.5, the forces are shown acting at specific points of the bodies; that is, they are considered to be *concentrated forces,* or *point forces.* However, these situations are idealized. In the real world, there are no concentrated forces. For example, both the railroad car wheel and the rail flatten slightly where they come into contact. Consequently, the force that the wheel transmits to the rail is distributed over a small contact area on the surface of the rail. Hence, the force due to the wheel could be considered a *distributed force.* A similar distribution of force occurs on the contact area of two gear teeth. Forces that result from direct physical contact between bodies are called *contact forces.* The gravitational force that the sun exerts on the earth is distributed throughout the earth. Such a force is called a *body force,* since it is distributed throughout the volume of the body on which it acts. In general, body forces arise because of gravitational action between bodies or because of other physical phenomena such as electromagnetic or inertial effects.

distributed force

contact force

body force

If a force *could* be concentrated at a point on a body, it would pierce the body. In many cases, however, the areas on which forces are distributed are relatively small and thus may be approximated as points. Also, the theory of point forces is an essential preliminary to the theory of distributed forces (Chapter 8). So let's first consider point forces, recognizing that they are approximations of real forces.

CONCURRENT FORCES

concurrent forces

collinear forces

Two or more forces that act at the same point are called *concurrent forces.* Concurrent forces need not have the same direction. They simply act at the same point. If they do have the same direction, they are *collinear forces* (see also Sec. 3.5). However, two collinear forces need not be concurrent; they may have different points of application along the same line.

COPLANAR FORCES

coplanar forces

Two or more forces whose directed arrows lie in the same plane are called *coplanar forces.* Since two concurrent forces always lie in a common plane, they are always coplanar. Three or more concurrent forces are not necessarily coplanar. Conversely, coplanar forces are not necessarily concurrent. However, forces are often both coplanar and concurrent.

Example 2.1

Types of Forces

Problem Statement Determine the type of each of the forces R shown in Figs. E2.1a through E2.1d.

Solution In Fig. E2.1a, R is a body force. It is the attraction of the sun acting on the earth and is the resultant of the gravitational forces acting on each of the mass elements in the earth. In Fig. E2.1b, R is a distributed contact force between the tire and the pavement. In Fig E2.1c, R is the resultant of the resisting contact forces between the board and the nail, the resistance of the board to the penetration force of the nail as it is driven by the hammer. The forces R in Fig. E2.1d are contact forces of the small steel ball on the two larger ones.

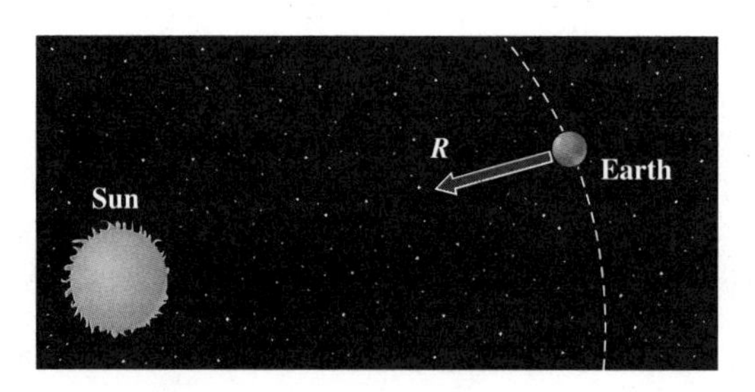

Figure E2.1a

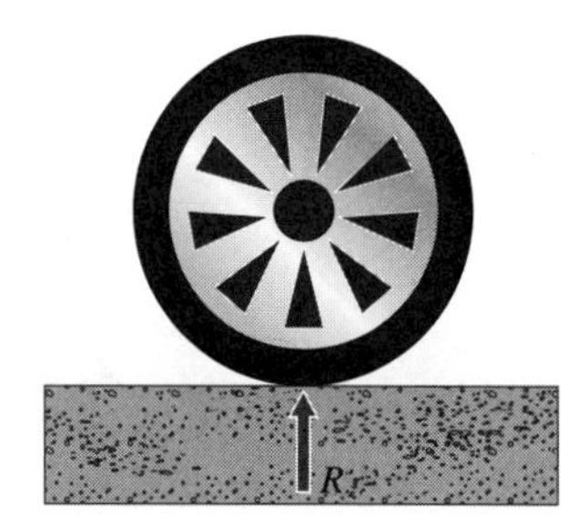

Figure E2.1b

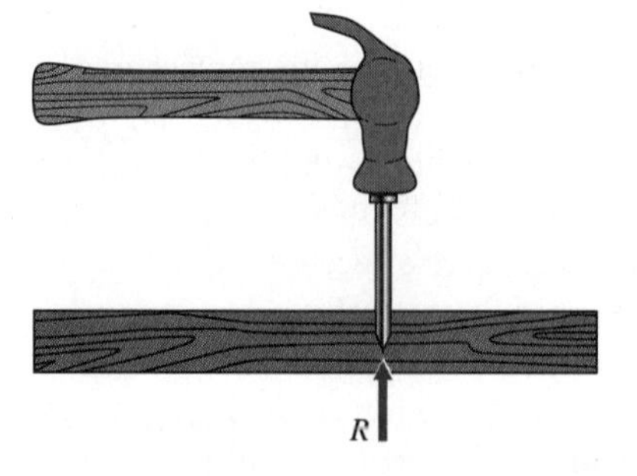

Figure E2.1c

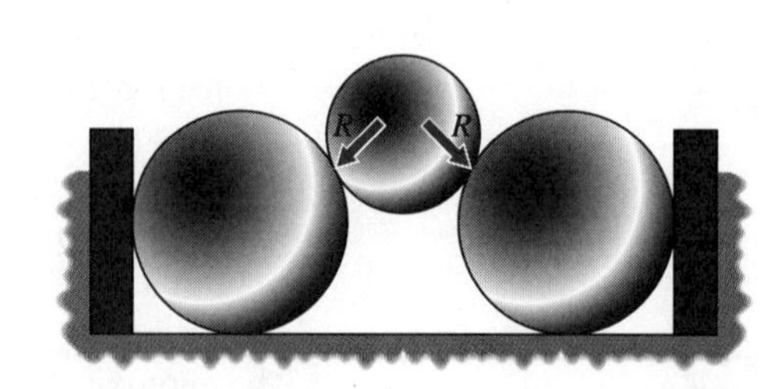

Figure E2.1.d

Example 2.2

Characteristics of a Force System

Problem Statement Determine the type of force system for each of the systems shown in Figs. E2.2a through E2.2d.

Solution In Fig. E2.2a, the force system is concurrent and coplanar. The force system in Fig. E2.2b is collinear and, hence, coplanar. In Fig. E2.2c, the force system is coplanar. The force system in Fig. E2.2d is concurrent but not coplanar. It is noncoplanar, or three-dimensional.

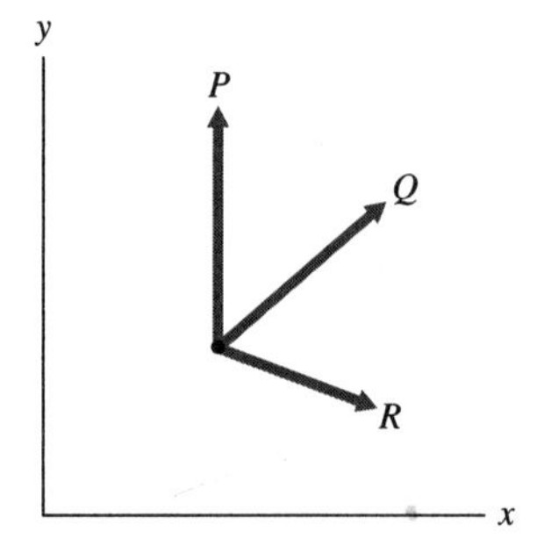

Figure E2.2a

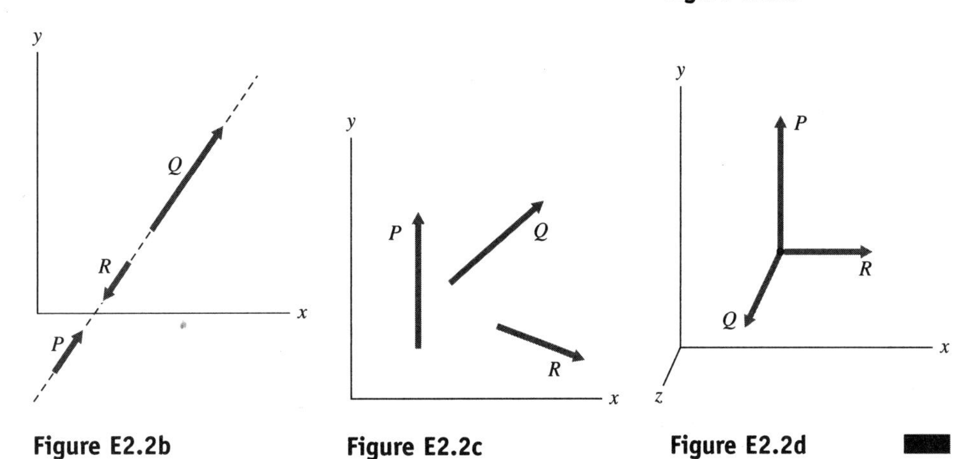

Figure E2.2b **Figure E2.2c** **Figure E2.2d**

CHECKPOINT ✓

1. ***True or False:*** There are two essential properties of a force—magnitude and sense.
2. ***True or False:*** A point of application is an essential characteristic of a force.
3. ***True or False:*** A point force is an idealization of a real force that is distributed over a small area of application.
4. ***True or False:*** Concurrent forces must be coplanar.
5. ***True or False:*** Coplanar forces must be concurrent.
6. ***True or False:*** Collinear forces are concurrent and coplanar.

Answers: 1. False; 2. False; 3. True; 4. False; 5. False; 6. False

2.2 NEWTON'S THIRD LAW

After studying this section, you should be able to:

- Use the concept of action-reaction to classify a force as an internal or external force for a given system.

law of action-reaction

As was noted in Chapter 1, Newton's third law (the *law of action-reaction*) can be expressed as follows:

> For every action, there is an equal and opposite reaction. Or, the mutual forces of two bodies (particles) on each other are always equal and oppositely directed.

To further explain this law, Newton wrote: "Whatever draws or presses another is as much drawn or pressed by that other. If you press a stone with your finger, the finger is also pressed by the stone. If a horse draws a stone tied by a rope, the horse (if I may so say) will be equally drawn back toward the stone."

Although Newton did not use the word "force," he clearly meant that if body A exerts a force on body B, then body B exerts a force of the same magnitude, but opposite direction, on body A. This law signifies that forces are mated in pairs—action and reaction. The reaction to a given force $\mathbf{F}$ is understood to act on the body that causes or exerts the force $\mathbf{F}$. For example, in Fig. 2.5a, the downward force $\mathbf{F}$ that the wheel exerts on the rail is reacted to by an upward force of magnitude F that the rail exerts on the wheel. Similarly, in Fig. 2.5c, not only does the lower part of the bar exert a downward force $\mathbf{T}$ of magnitude T on the upper part, but the upper part also exerts an upward force $\mathbf{T}'$ of equal magnitude on the lower part. Both of these forces acting at plane a–a through the bar are indicated in Fig. 2.6. In Fig. 2.5c, the gravitational forces among the earth, moon, and sun are seen to occur in opposing pairs of equal magnitude.

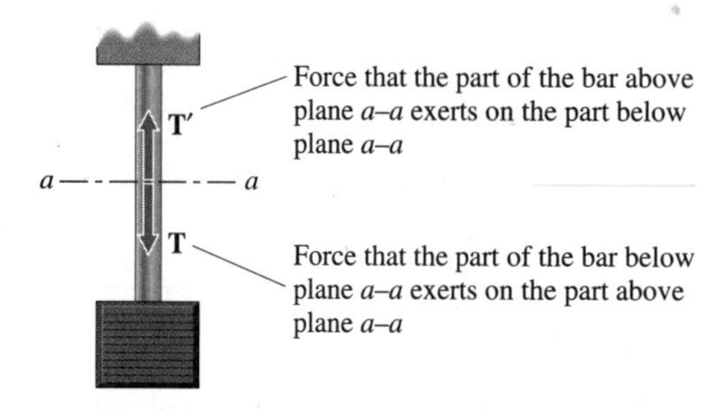

Figure 2.6 Force and reaction at a cross section of a bar.

internal force

external force

If a force $\mathbf{F}$ acts on a system, its reaction acts on another part of the same system or on a body outside the system. In the first case, the force is called an *internal force;* in the second case, it is called an *external force.* For example, if the system under consideration is the whole bar shown in Fig. 2.6, the force $\mathbf{T}$ that the lower part of the bar exerts on the upper part is an internal force, since its reaction (i.e., the force that the upper part exerts on the lower part) also acts on a part of the system. On the other hand, if in Fig. 2.5a the system consists of a wheel alone, the force that the rail exerts on the wheel is an external force. This is because the reaction of that force acts on the rail, which is not a part of the wheel. In Fig. 2.5e, the forces that the sun and moon exert on the earth are external forces if the earth alone is the designated system, but they are internal forces if the earth, moon, and sun together constitute the system.

The analysis of structures or machines is often concerned with the force that a certain part of a member exerts on another part. For example, consider a member AB, with forces of equal magnitude $F_A = F_B$ acting on its ends (see Fig. 2.7a). Imagine that the member is separated into two parts, AC and DB (see Fig. 2.7b). Then the two bodies AC and DB, be-

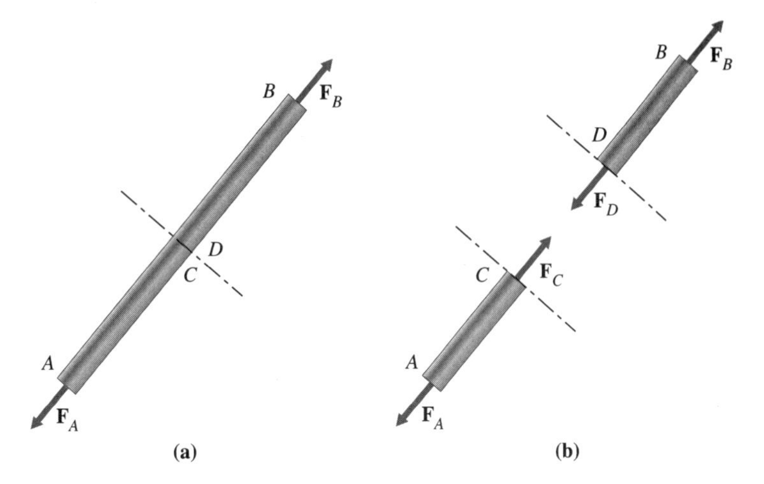

Figure 2.7 A member in equilibrium.

tween which the two forces $\mathbf{F}_C$ and $\mathbf{F}_D$ at C and D act, are actually two parts of the same member. In this context, with respect to member AB, $\mathbf{F}_A$ and $\mathbf{F}_B$ are *external* forces, and $\mathbf{F}_C$ and $\mathbf{F}_D$ are *internal* forces.

Example 2.3

Action and Reaction

Problem Statement Consider the bar shown in Fig. E2.3a. It is acted on by a force P and rests on two rollers A and B, which in turn rest on the floor. The rollers exert forces Q and R on the bar (Fig. E2.3b). Assume that the bar and rollers are weightless.

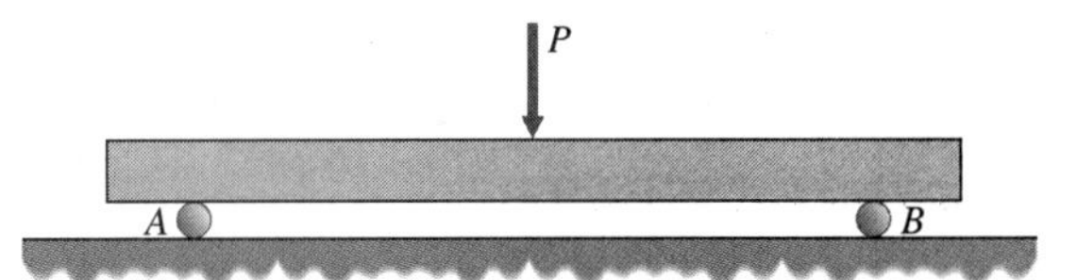

Figure E2.3a

a. Determine the reactions of forces Q and R.

b. Determine the forces exerted by the floor on the rollers.

c. Determine the reactions of the forces exerted by the floor on the rollers.

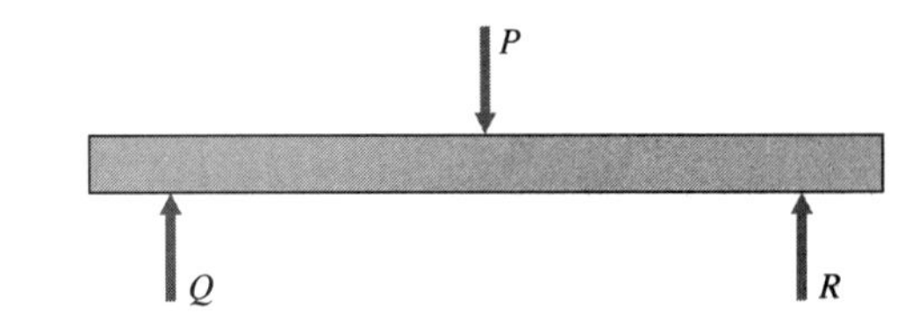

Figure E2.3b

Solution

a. Since rollers A and B exert forces Q and R, respectively, on the bar, by the law of action-reaction (Newton's third law), the reactions S and T act on the rollers (Fig. E2.3c). They have the same magnitude as Q and R but are opposite in sense.

Figure E2.3c

b. The forces that act on the rollers are shown in Fig. E2.3d, where U and V are forces exerted by the floor on the rollers. By Newton's third law, U and V are the reactions to forces S and T, respectively. Hence, $U = S$ and $V = T$.

c. Consider U and V to be actions (forces the floor exerts on the rollers). Then, from Fig. E2.3d, the reactions to U and V are S and T, respectively.

Figure E2.3d

2.3 FORCE RESULTANTS AND FORCE COMPONENTS

After studying this section, you should be able to:

- Determine the resultant of a system of two or more forces using either graphical or trigonometric methods.
- Resolve a force into its components along a specified set of axes.

The investigations of Stevin, Descartes, Newton, and other scientists of the 17th century led to the conclusion that *the effects of several forces that act at a common point P of a body invariably may be produced by a single force that acts at point P.* The single force that is equivalent to the several forces is called the *resultant force,* or simply the *resultant.*

resultant

THE PARALLELOGRAM LAW AND COMPOSITION OF FORCES

Key Concept

Forces add by the parallelogram law. The addition of forces is called *composition of forces.*

Two forces **F** and **G** that act at a point P of a body are represented by two arrows drawn outward from point P (see Fig. 2.8). These arrows form two sides of a parallelogram. The diagonal **R** of the parallelogram, drawn from point P, represents the resultant of the forces **F** and **G**. In all respects, the single force **R** is equivalent to the resultant of the two forces **F** and **G**. This statement expresses the *parallelogram law* of force, a principle that

parallelogram law

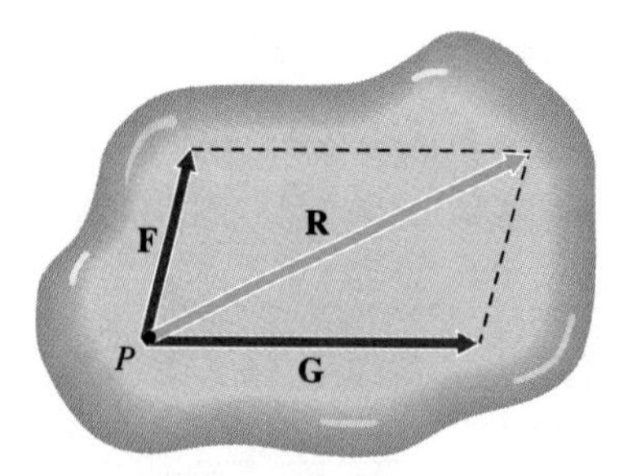

Figure 2.8 Parallelogram construction of the resultant of two concurrent forces.

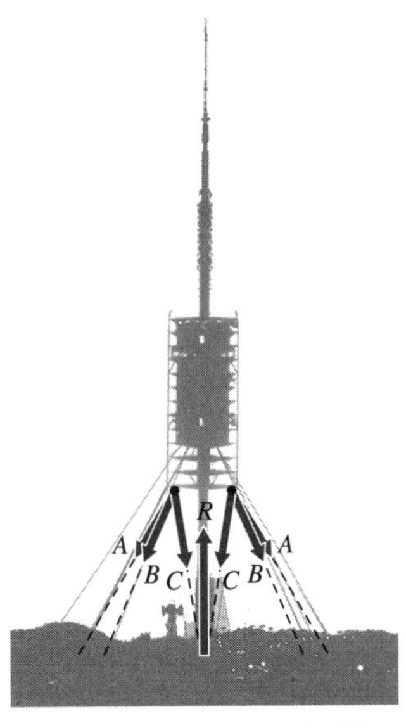

The Collserola communication tower in Barcelona, Spain is stabilized with guy lines, each of which exerts a force on the tower. The resultant of these forces is aligned with the vertical axis of the tower.

has been verified by direct experiments. The process of combining two or more concurrent forces into a single resultant force is called *composition of forces.*

composition of forces

THE PARALLELOGRAM LAW AND RESOLUTION OF FORCES

> ***Key Concept*** Resolution of a force is the process of breaking it down into component parts.

resolution of a force

The reverse of composition of forces is called *resolution of a force.* By this process, a given force **R** is resolved into two forces **S** and **T**. These two forces are known as components of **R**; that is, **S** and **T** can be recombined to obtain the resultant **R**. To perform the resolution operation, we lay off a line segment *PA,* representing the force **R** (see Fig. 2.9). We let *PM* and *PN* be two straight lines through *P* such that lines *PM, PN,* and *PA* all lie in a plane, and we construct lines *AB* and *AC* parallel to lines *PN* and *PM,* respectively. Then, by the parallelogram law, the two line segments *PB* and *PC* represent forces **S** and **T** whose resultant is **R**. That is, the given force **R** has been *resolved into the components* **S** and **T**. In all respects, the two forces **S** and **T** together are equivalent to the given force **R**. The forces **R**, **S**, and **T** are coplanar and concurrent.

You are free to choose the directions of lines *PM* and *PN,* provided that lines *PM* and *PN* and the direction line of force **R** all lie in the same plane. So, the resolution of **R** into two components **S** and **T** can be performed in an infinite number of ways. Since their selection is arbitrary, it is often most convenient to choose the lines *PM* and *PN* (see Fig. 2.9) to lie along rectangular coordinate axes that are well suited for solving a particular problem. Then it may be advantageous to resolve all forces under consideration into components that have the directions of the coordinate axes. Details of this approach are presented in Sec. 2.5.

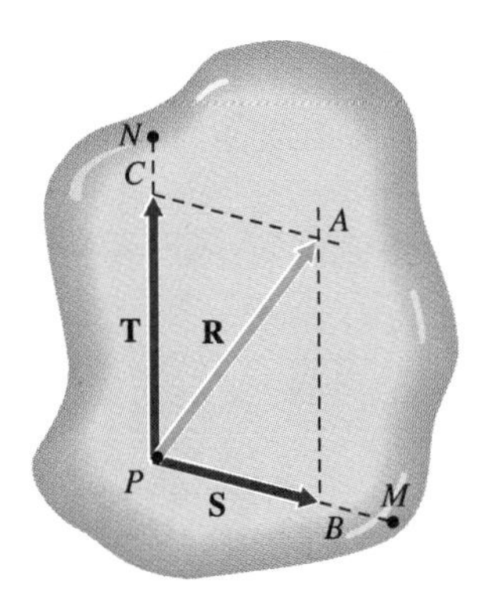

Figure 2.9 Resolution of a force into two components.

POLYGON LAW OF FORCES

The parallelogram construction of the resultant **R** of two concurrent forces **F** and **G** is illustrated in Fig. 2.10a. Since the sides *AB* and *PC* are equal and parallel, we may simplify the construction by drawing only the triangle *PBA* (Fig. 2.10b). The side *BA* represents the force **F**, although *it does not designate the actual line of action of* **F**. For the graphical construction of the resultant **R**, we displace the line segment representing force **F**. You must remember, however, that all forces act at point *P*. Alternatively, the resultant **R** may be obtained by displacing the line segment *PB* that represents force **G** to coincide with *CA* (Fig. 2.10c). This method is called the *triangle construction* of the resultant force.

triangle construction

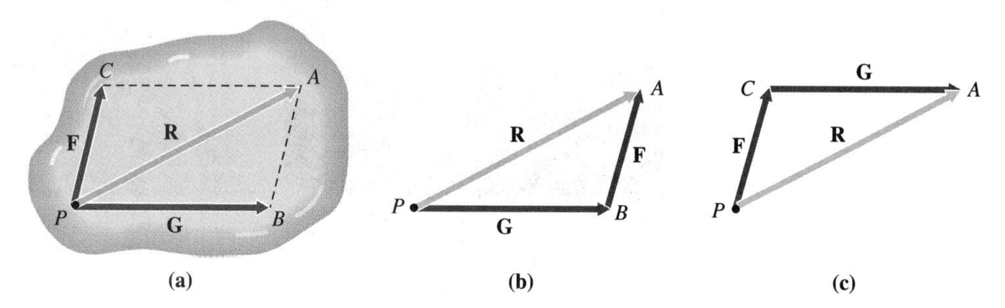

Figure 2.10 Triangle construction of the resultant of two concurrent forces: **(a)** concurrent force system; **(b)** construction of triangle *PBA* by displacing **F**; **(c)** construction of triangle *PCA* by displacing **G**.

The triangle construction is easily generalized to the case where there are more than two concurrent forces. For example, suppose there are three forces **F**, **G**, and **H** acting at a point *P* (Fig. 2.11a). The triangle construction first yields the resultant **R′** (dashed in Fig. 2.11b) of the two forces **F** and **G**. Application of the triangle construction again with the forces **R′** and **H** gives the resultant **R** of the three forces **F**, **G**, and **H**. The intermediate resultant **R′** is shown only for illustrative purposes. To obtain the resultant **R**, you can simply arrange the directed arrows representing the forces **F**, **G**, and **H** in a chain, maintaining their proper directions. Then the arrow from the initial point *P* to the terminal point of the chain represents the resultant force. This construction is not limited to three forces; it applies for any number of concurrent forces. Figure 2.12 illustrates the

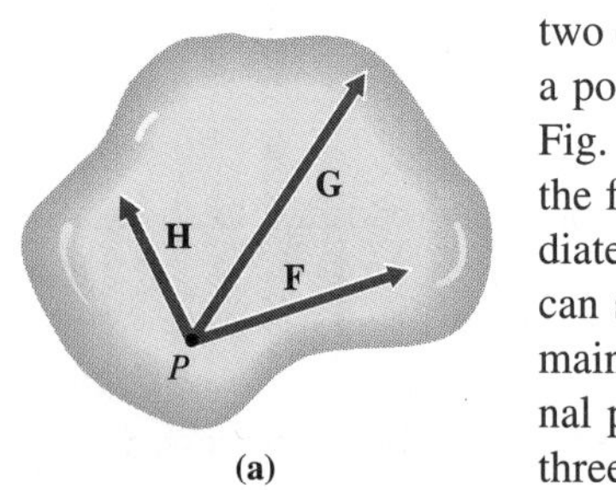

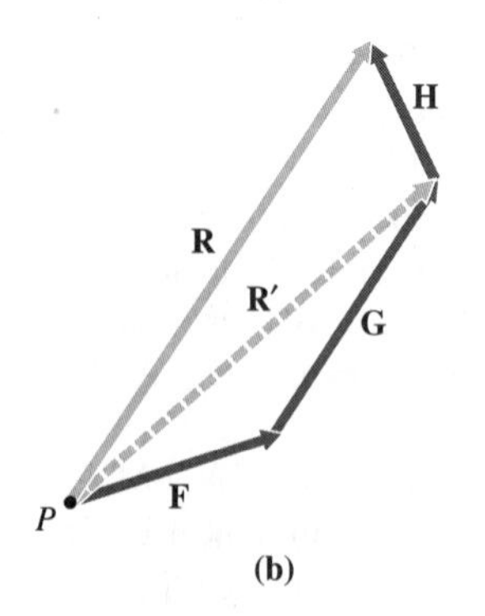

Figure 2.11 Polygon construction of the resultant of three concurrent forces.

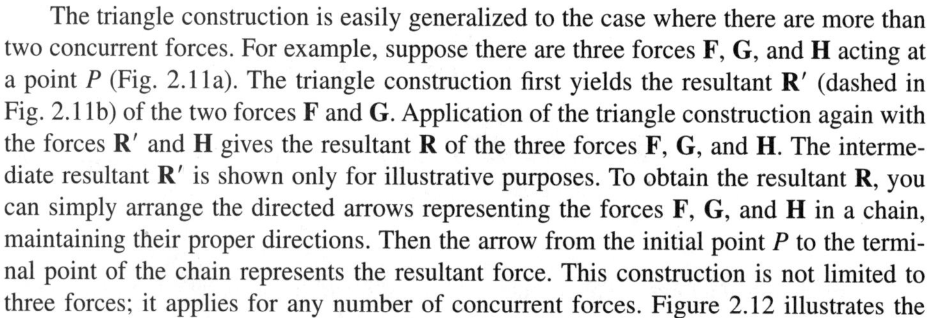

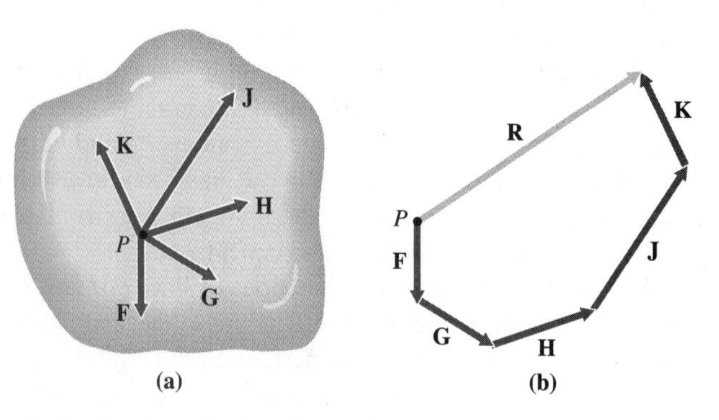

Figure 2.12 Polygon construction of the resultant of five concurrent forces: **(a)** system of five forces; **(b)** Force polygon.

polygon construction

construction of the resultant of five forces that act at a point *P*. The method illustrated by Figs. 2.11 and 2.12 is called the *polygon construction* of the resultant of several concurrent forces. As with triangle construction, the resultant force that is obtained by the polygon construction does not depend on the order in which the forces are incorporated in the polygon, but proof of this statement is deferred until later. The polygon construction is not limited to cases in which the forces are coplanar. However, as a practical graphical technique, it has obvious shortcomings if the forces are not coplanar—that is, if the forces are acting in three-dimensions. The concurrent forces **F**, **G**, **H**, . . . , whose resultant is the force **R**, are called components of **R**. Again, remember that all the forces act at a common point.

PROBLEM-SOLVING TECHNIQUE

Resultant of Two Concurrent Forces

To find the resultant of a pair of concurrent forces:

1. Draw the two concurrent forces, to a convenient scale, from a common point of application, with direction angles measured from a common reference line.
2. Construct a parallelogram with the given forces as sides.
3. Draw the resultant as the diagonal of the parallelogram, with the tail of the arrow at the point of application of the two given forces.
4. Measure the magnitude of the resultant with a scaled ruler; measure the direction angle of the resultant, relative to the common reference line, with a protractor. Alternatively, find the magnitude and direction of the resultant by trigonometry.

Example 2.4

Resultant of Two Concurrent Forces by Parallelogram Construction

Problem Statement Two forces **A** and **B** with magnitudes 80 N and 40 N, respectively, act at point *O* of a body. Force **A** acts horizontally, and force **B** acts at an angle of 60° counterclockwise from **A** (see Fig. E2.4). Determine the resultant force **R**.

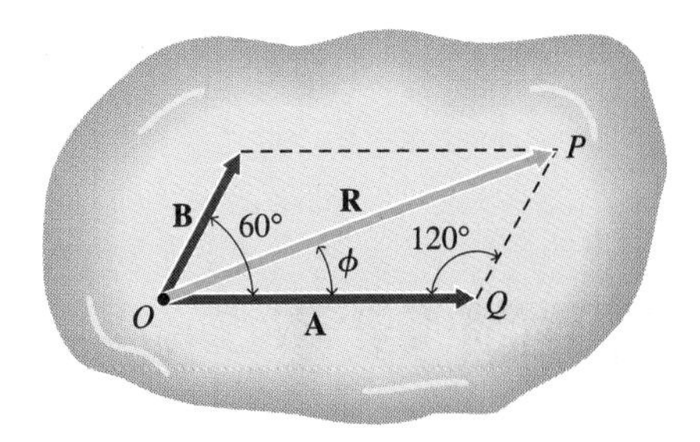

Figure E2.4

Solution To determine the resultant force, we construct the force parallelogram to some convenient scale. In the parallelogram, line *QP* has the same length and orientation as **B**. The resultant **R** is then the diagonal arrow from *O* to *P*. The magnitude *R* of the resultant and the angle ϕ that defines its direction may be determined by direct measurements or by trigonometry.

To compute the magnitude and the direction of the resultant by trigonometry, note that the angle *OQP* is 120°. The sides *OQ* and *QP* of the triangle *OQP* are 80 units and 40 units in length, respectively. Consequently, by the law of cosines,

$$R^2 = 80^2 + 40^2 - 2(80)(40)(\cos 120°)$$

Solving for R, we obtain $R = 105.83$ N.

The law of sines now gives

$$\frac{\sin\phi}{40} = \frac{\sin 120°}{R}$$

Therefore, $\phi = 19.11°$. Thus, the resultant **R** has magnitude $R = 105.83$ N and is oriented at 19.11° counterclockwise from the line of action of force **A**.

Example 2.5

Resolution of a Force into Components

Problem Statement A resultant force **R** has magnitude $R = 30$ N and components **P** and **Q**. The magnitude of **P** is $P = 50$ N. Component **P** forms a 60° angle with **R** (see Fig. E2.5). Determine the magnitude and direction of component **Q**.

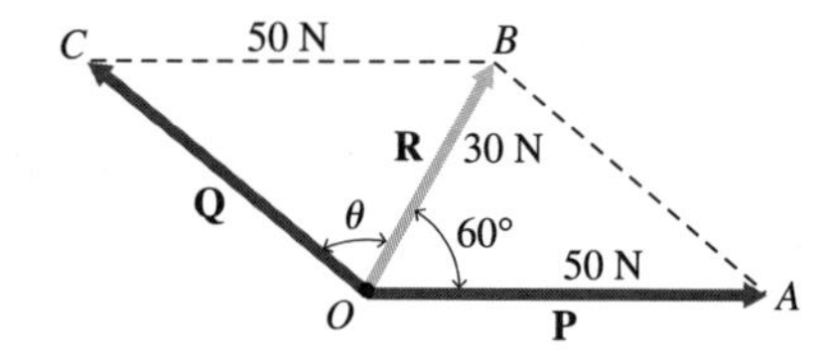

Figure E2.5

Solution In Fig. E2.5, the given force **P** and resultant **R** are laid off to a convenient scale and include the given angle of 60°. The force parallelogram is completed by laying off the lines OC and BC parallel to the lines AB and OA, respectively. The side OC represents component Q. The magnitude of **Q** and the angle θ may be determined by measuring the parallelogram or by trigonometry.

To compute the magnitude of component **Q** by trigonometry, we apply the law of cosines to determine side AB, which is equal in magnitude to **Q**. Thus,

$$Q^2 = 30^2 + 50^2 - 2(30)(50)(\cos 60°)$$

Therefore, $Q = 43.59$ N.

Applying the law of cosines to the triangle OBC and solving for θ, we obtain

$$50^2 = Q^2 + 30^2 - 2(30)(Q)(\cos\theta)$$

$$\cos\theta = 0.11471$$

$$\theta = 83.4°$$

Example 2.6

Polygon Construction of the Resultant of Four Concurrent Forces

Problem Statement Four coplanar, concurrent forces act at point O of a body (see Fig. E2.6a). Use the polygon construction and trigonometry to determine the magnitude (in newtons) and direction of their resultant **R**.

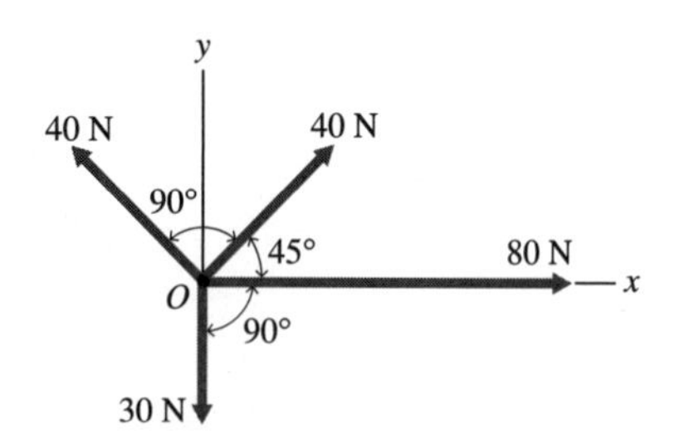

Figure E2.6a

Solution In the polygon construction of the resultant **R**, the forces may be arranged in various orders. Three different force polygons are shown in Figs. E2.6b, E2.6c, and E2.6d. Note that these different constructions all provide the same resultant **R**.

To compute the resultant, we set up rectangular coordinate axes, as illustrated in Fig. E2.6b. The abscissa of point Q is $x = 80$ N. The ordinate of point Q is

$$y = 40 \cos 45° + 40 \cos 45° - 30 = 26.57 \text{ N}$$

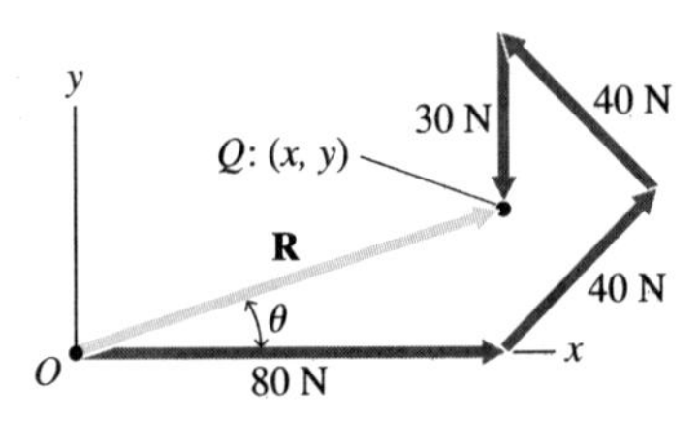

Figure E2.6b

By the Pythagorean theorem,

$$R^2 = x^2 + y^2 = 80^2 + 26.57^2 = 7106$$

Hence, $R = 84.3$ N. The angle θ is determined by $\tan \theta = y/x$:

$$\tan \theta = \frac{26.57}{80} = 0.3321$$

Therefore, $\theta = 18.37°$.

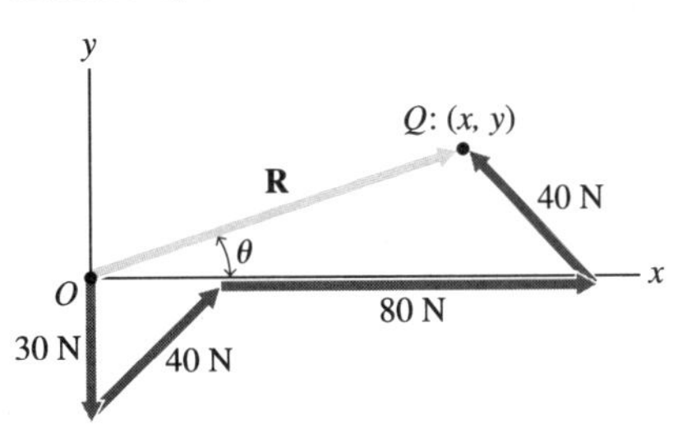

Figure E2.6c

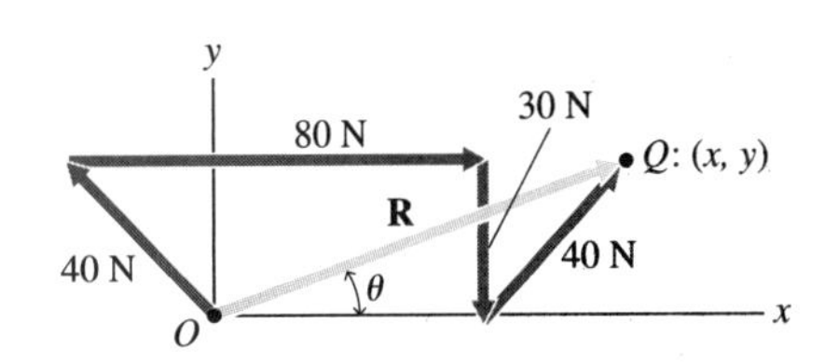

Figure E2.6d

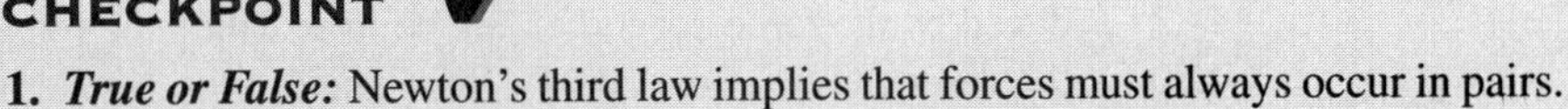

CHECKPOINT

1. ***True or False:*** Newton's third law implies that forces must always occur in pairs.
2. ***True or False:*** An internal force can become an external force if the definition of the system is changed.
3. ***True or False:*** The parallelogram law is the basis of a graphical technique for composition of forces.
4. ***True or False:*** The parallelogram construction is used for composition of forces but not for resolving a force into components.
5. ***True or False:*** The polygon law of forces is derived from Newton's third law.
6. ***True or False:*** The polygon construction of the resultant of forces can be used only for two-dimensional problems.

Answers: 1. True; 2. True; 3. True; 4. False; 5. False; 6. False

2.4 VECTORS AND VECTOR QUANTITIES

After studying this section, you should be able to:

- Distinguish between physical quantities that can be classified as vectors and those that cannot.
- Identify vectors as fixed, slide, or free.
- Add two vectors, subtract one vector from another, and multiply a vector by a scalar.

Key Concept

Physical quantities that obey the same rules of composition and resolution as forces are called vectors.

Forces are not the only physical quantities that can be represented by arrows and combined by the parallelogram construction or the polygon construction. In other words, forces are not the only *vector quantities* (see Sec. 1.1). Many physical quantities can be represented as vectors. For example, displacements, velocities, and accelerations of particles are all vector quantities. The arrow that represents a vector quantity is called a *vector.* The term "vectors" is often used to refer to vector quantities. In contrast to vector quantities, which have both magnitude and direction, a physical quantity that has magnitude alone is called a *scalar quantity.* For instance, temperature is a scalar quantity. The number that represents a scalar quantity is called a *scalar.*

vector quantity

vector

scalar quantity

scalar

VECTOR TYPES

As was mentioned in Chapter 1, it is customary to distinguish vectors from scalars by printing vectors in boldface type. For example, the symbol **R** can be used to denote a force. It stands for the magnitude and direction of the force, although it does not signify the point at which the force acts. Ordinarily, the point of action of a force is designated by an explicit statement or a diagram. The symbol *R* (in italics) represents only the magnitude of the vector **R**. Consequently, *R* is a nonnegative number—that is, a number that is positive or zero. For handwritten work, the notation $\vec{R}$ or $\underline{R}$ is often used to symbolize a vector **R**.

If the point of application of a vector cannot be changed without changing the physical significance or the physical effect of the vector, the vector is a *fixed vector.* A force that acts on a body is represented by a fixed vector when the deformation and the stresses produced by the force depend on the point at which the force acts. For instance, the weight of the diver can be regarded as a fixed vector with respect to the diving board in Fig. 2.13a, since the deformation of the diving board depends on the position of the diver.

fixed vector

If a vector has no point of action or if it may be considered to act at any point, it is a *free vector.* For example, a vector that represents the velocity of a body, such as that of the car in Fig. 2.13b, is a free vector.

free vector

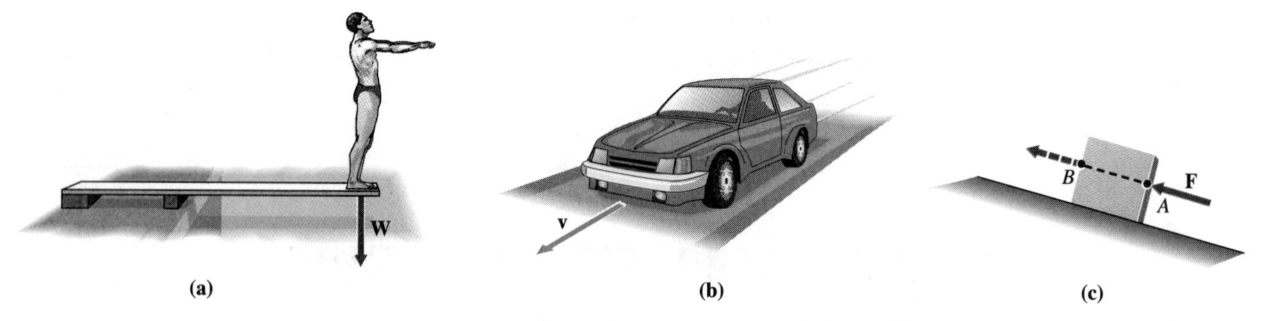

Figure 2.13 Types of vectors: **(a)** fixed vector; **(b)** free vector; **(c)** slide vector.

You will also encounter vectors that can be displaced arbitrarily along the straight lines on which they lie, but cannot be displaced from those lines without changing their physical effects or their physical meanings. These are called *slide vectors,* since they may

slide vector

"slide" along their lines of action without changing any associated physical effects. An example of a slide vector is the force **F** acting on the rigid block in Fig. 2.13c. The point of application of the force can slide between points A and B along its line of action and not change the interaction between the block and the ramp.

NONVECTOR QUANTITIES

Not all physical quantities that can be represented by arrows are vectors (or vector quantities). For example, a finite rotation of a rigid body about an axis (a line with a given orientation and sense) may be represented by a directed line segment (an arrow) that lies on the axis of rotation and that has magnitude equal to the angular displacement. The sense of rotation may be defined by the *right-hand screw rule* (see Fig. 2.14). That is, the body is rotated in the sense that would cause a right-hand screw to advance in the direction of the arrow. If you imagine that you are grasping an arrow in your right hand with your thumb pointing in the direction of the arrowhead, your fingers curl around the arrow in the positive sense of rotation.

right-hand screw rule

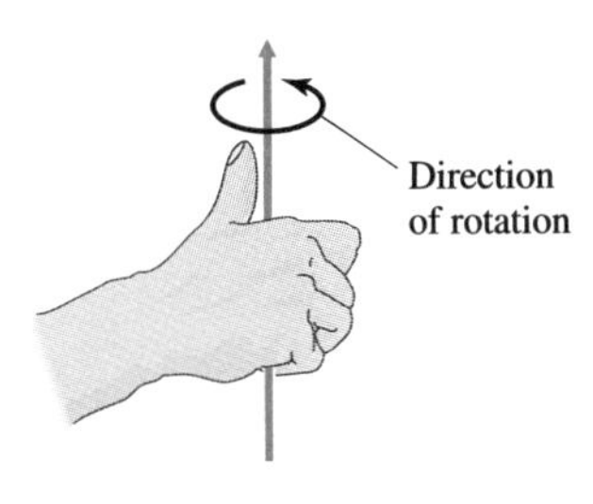

Figure 2.14 Right-hand screw rule.

You might expect a finite rotation of a rigid body to be a vector quantity. Determining whether this is true involves determining whether or not rotations combine by the polygon construction. A simple experiment shows that finite angular displacements do not combine in the same way vectors do. In Fig. 2.15a, two 90° angular displacements of a die are indicated by the arrows A and B, which lie parallel to edges of the die. These line segments are considered to be fixed in space, even as the die is displaced. Figure 2.15b shows the position of the die after the rotations have been performed in the order A, B. Figure 2.15c shows the position of the die after the rotations have been performed in the order B, A. Clearly, the final position of the die depends on the order in which the two rotations are performed. This behavior is inconsistent with the fact that the resultant vector obtained by the triangle construction is independent of the order of the vectors in the triangle, as illustrated by Figs. 2.10b and 2.10c. Thus, a finite angular displacement of a rigid body is not a vector quantity, even though it may be represented by an arrow. You must always look to physical principles to determine whether a quantity that can be represented by an arrow is a vector quantity.

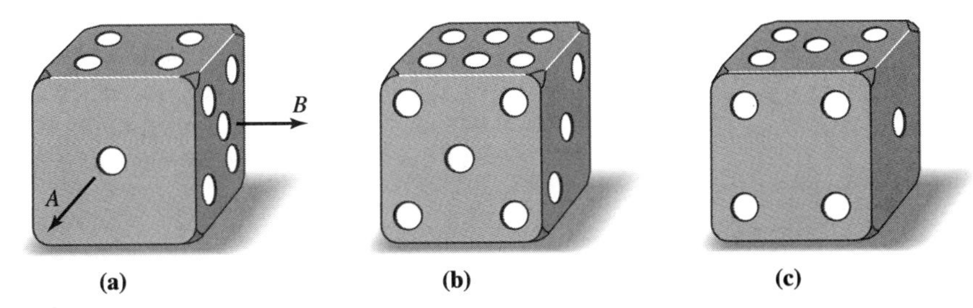

Figure 2.15 Rotations of a playing die represented by arrows: **(a)** arrows A and B lie parallel to edges of the die; **(b)** 90° rotation about A, then about B; **(c)** 90° rotation about B, then about A.

VECTOR ARITHMETIC

Equality of Vectors The vector equation $\mathbf{A} = \mathbf{B}$ means that the vectors **A** and **B** have the same magnitude and the same direction, but not necessarily the same point of application or the same line of action. The equation $\mathbf{A} = \mathbf{B}$ implies that $A = B$.

Negative Vector The vector $-\mathbf{F}$ is defined to have the same magnitude as vector **F** but the opposite sense. If **F** denotes a force that acts on a body, its reaction is $-\mathbf{F}$.

Vector Sum, Vector Addition The resultant **R** of several vectors, $\mathbf{F}_1$, $\mathbf{F}_2$, . . . , is called the *sum* of the vectors, represented symbolically as $\mathbf{R} = \mathbf{F}_1 + \mathbf{F}_2 + \cdots$. The process of obtaining the sum by the polygon construction or other means is called *vector addition.* The expressions "vector addition" and "composition of vectors" have the same meaning. It may seem strange that the symbol + is used for an operation that is different from scalar addition. As we shall see, however, vector addition has certain features in common with scalar addition, and the use of the symbol + serves to emphasize those commonalities. There are also distinct differences. For example, the relation $\mathbf{R} = \mathbf{F} + \mathbf{G}$ does not imply that $R = F + G$. The expression $F + G$ denotes the scalar sum of the magnitudes F and G, but, in general, $F + G$ is greater than R, since the three vectors **R**, **F**, and **G** form the sides of a triangle (Fig. 2.10b).

vector addition

Subtraction of Vectors Subtraction of a vector is defined as addition of the negative vector: $\mathbf{A} - \mathbf{B} = \mathbf{A} + (-\mathbf{B})$. That is, to subtract a vector **B** from a vector **A**, you add $-\mathbf{B}$ to **A**. You may subtract the same vector from both sides of a vector equation. For example, if $\mathbf{A} + \mathbf{B} = \mathbf{F}$, then, analogous to scalar subtraction, $\mathbf{A} + \mathbf{B} - \mathbf{B} = \mathbf{F} - \mathbf{B}$, from which $\mathbf{A} = \mathbf{F} - \mathbf{B}$. These relations are illustrated in Fig. 2.16.

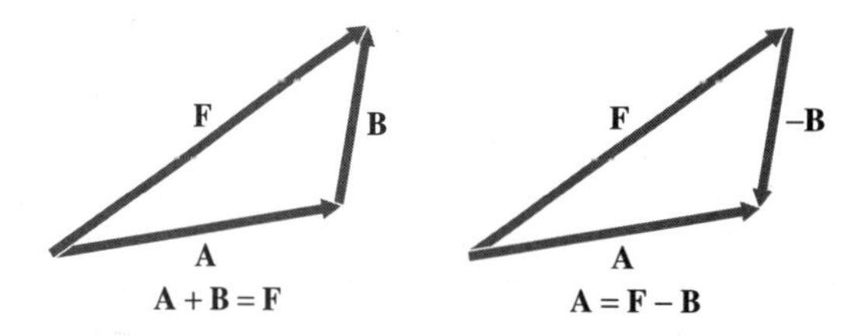

Figure 2.16 Subtraction of a vector.

Product of a Scalar and a Vector Analogous to scalar addition, it is natural to write $\mathbf{F} + \mathbf{F} = 2\mathbf{F}$, $2\mathbf{F} + \mathbf{F} = 3\mathbf{F}$, and so forth. Consequently, if k is a nonnegative number, $k\mathbf{F}$, the *product of a scalar and a vector*, is defined to be a vector with the direction of **F** and magnitude kF. In view of the definition of a negative vector, the product of a negative number k and a vector **F** is defined to be a vector with magnitude $|k|F$ and sense opposite to that of vector **F**. The notation $|k|$ designates the absolute value of k. The effect of multiplying a vector by a scalar is to change the magnitude or the sense, or both, of the vector, while leaving its direction line unchanged. For example, multiplication of a vector by the scalar -1 reverses the sense of the vector, leaving its magnitude unchanged. These effects are illustrated in Fig. 2.17.

product of a scalar and a vector

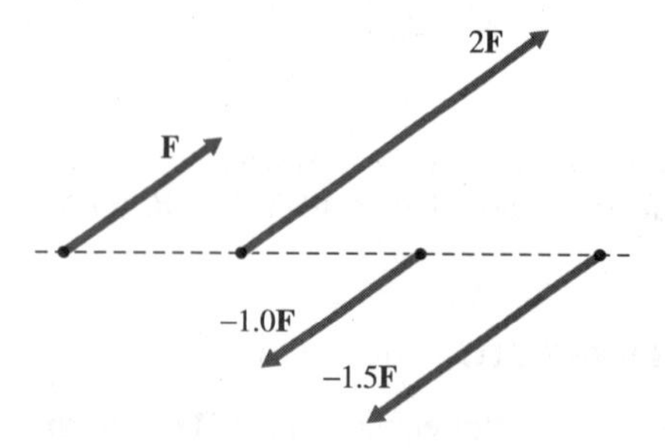

Figure 2.17 Product of a scalar and a vector.

The foregoing vector symbolism and further vector notations to be discussed later (in Sec. 2.6) were introduced by the American physicist J. Willard Gibbs (1839–1903) and by the English physicist Oliver Heaviside (1850–1925). They are known as *Gibbs's vector notation.*[1]

Example 2.7

Vector Arithmetic

Problem Statement Three concurrent, coplanar vectors **A**, **B**, and **C** have the following magnitudes and directions relative to the x axis (see Fig. E2.7a):

A: 100 N, 0°

B: 200 N, 90°

C: 150 N, 30°

a. Find the vector **D** given by $\mathbf{D} = \mathbf{A} + 0.5\mathbf{B} - \mathbf{C}$.

b. Find the scalar multipliers k_1 and k_2 such that $k_1\mathbf{A} + k_2\mathbf{B} + \mathbf{C} = 0$.

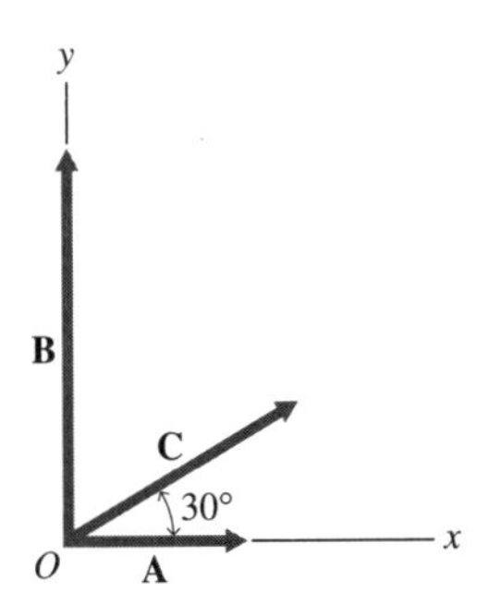

Figure E2.7a

Solution

a. We construct the force polygon (Fig. E2.7b) by displacing **B** laterally from O to point P, at the tip of **A**. Then **C** is displaced laterally from O to Q, at the tip of 0.5**B**, and the sense of **C** is also reversed since we are subtracting it from **A** + 0.5**B**. The vector **D** is the vector from the tail of **A** (point O) to the tip of −**C** (point R). By the geometry of the force polygon, R has (x, y) coordinates (−29.90, 25.0). Hence, the vector **D** has magnitude $D = 38.98$ N and is oriented at an angle of 140.1° counterclockwise from the x axis.

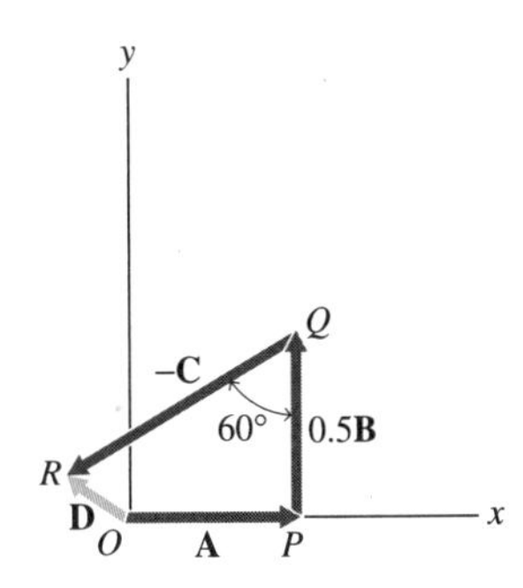

Figure E2.7b

b. By inspection of the three vectors **A**, **B**, and **C** in Fig. E2.7a, we see that, for the sum $k_1\mathbf{A} + k_2\mathbf{B} + \mathbf{C}$ to be zero, both k_1 and k_2 must be negative. In effect, k_1 and k_2 are scale factors on the horizontal and vertical legs of the force triangle OST (Fig. E2.7c). By geometry of the triangle,

$$-k_1A = C\cos 30° \qquad \text{(a)}$$

$$-k_2B = C\sin 30° \qquad \text{(b)}$$

From Eqs. (a) and (b), we find $k_1 = -1.299$ and $k_2 = -0.375$.

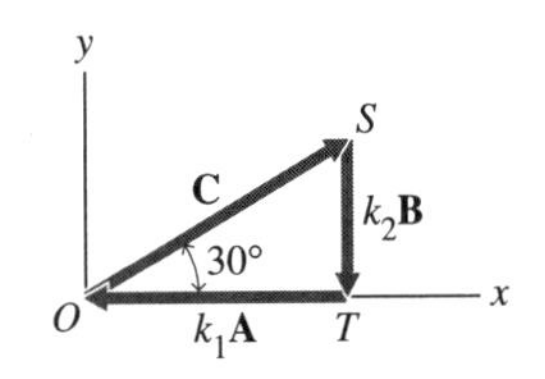

Figure E2.7c

1. Aristotle and Archimedes used directed line segments to describe the effects of forces, Stevin used vectors in his composition of forces by the parallelogram law, and Descartes introduced the idea of resolving vectors into geometric components parallel to coordinate axes (see Sec. 1.1). The algebra of vectors was developed simultaneously and independently in the 1870s by Josiah Willard Gibbs, a mathematical physicist at Yale University, and by the English mathematical physicist Oliver Heaviside. The works of Gibbs and Heaviside grew out of mathematical theories developed some years earlier by the Irish mathematician Sir William Hamilton (1805–1865) and by the German linguist, physicist, and geometer Hermann Grassman (1809–1877).

Example 2.8

Vector Addition

Problem Statement A support ring is subjected to the forces **A**, **B**, and **C** (Fig. E2.8a) having magnitudes

$$A = 100 \text{ lb} \qquad B = 75 \text{ lb} \qquad C = 100 \text{ lb} \tag{a}$$

a. Find the force **D** = **A** + **C**.

b. Find the resultant force **R** = **A** + **B** + **C**.

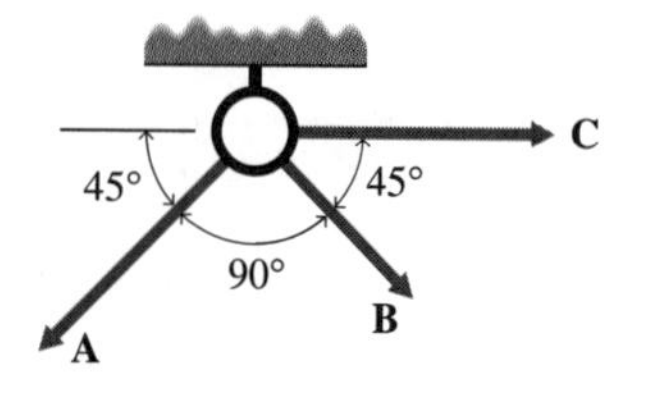

Figure E2.8a

Solution **a.** From Fig. E2.8a and Eqs. (a),

D = **A** + **C** = 45° 100 lb + 100 lb (b)

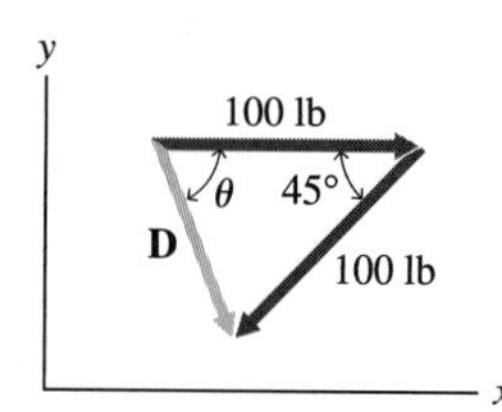

Figure E2.8b

The force polygon for Eq. (b) is shown in Fig. E2.8b. By Fig. E2.8b and the law of cosines,

$$D^2 = 100^2 + 100^2 - 2(100)(100)(\cos 45°)$$

or

$$D = 76.5 \text{ lb} \tag{c}$$

Also, by Fig. E2.8b and the law of sines,

$$\frac{100}{\sin\theta} = \frac{D}{\sin 45°}$$

From Eqs. (c) and (d),

$$\theta = 67.5° \tag{d}$$

Therefore, **D** = 76.5 lb at $\theta = 67.5°$.

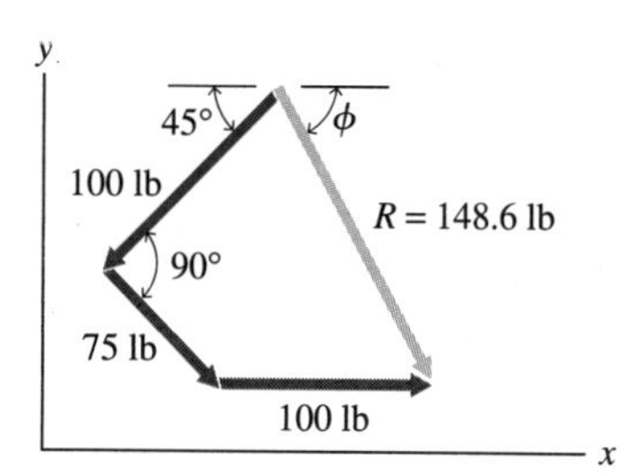

Figure E2.8c

b. By graphical construction, the resultant **R** = **A** + **B** + **C** is obtained as shown in Fig. E2.8c. Measurement yields

$$R = 148.6 \text{ lb} \qquad \phi = 56.4°$$

Example 2.9

Vector Subtraction

Problem Statement Given the forces that act on the ring of Example 2.8, determine the following forces:

a. **F** = **A** − **B** + **C**

b. **G** = **A** − **B** − **C**

Solution **a.** The polygon construction of **F** = **A** − **B** + **C** is shown in Fig. E2.9a. Since, from Example 2.8, $A = 100$ lb, $B = 75$ lb, and $C = 100$ lb, measurement of Fig. E2.9a yields

$$F = 30 \text{ lb} \qquad \theta = 143°$$

b. The polygon construction of **G** = **A** − **B** − **C** is shown in Fig. E2.9b. By measurement of Fig. E2.9b, we obtain

$$G = 224 \text{ lb} \qquad \phi = 175°$$

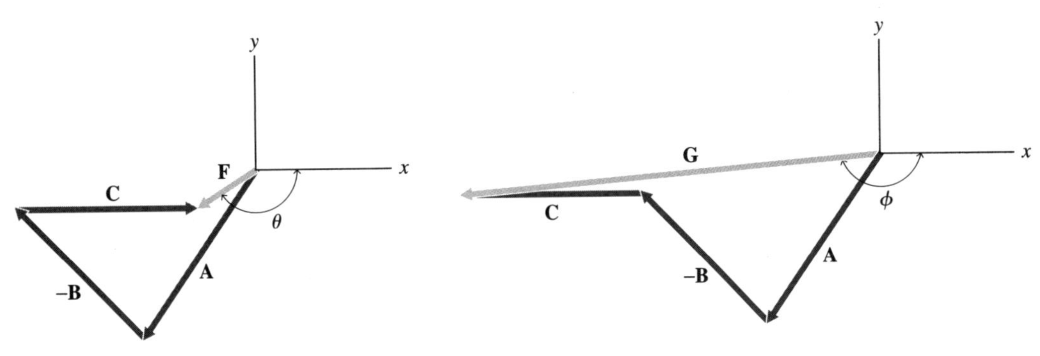

Figure E2.9a **Figure E2.9b**

CHECKPOINT

1. ***True or False:*** Vectors have the same essential characteristics as forces.
2. The essential characteristics of a vector are ________ and ________.
3. A scalar differs from a vector in that it lacks the property of ________.
4. ***True or False:*** The sum of two vectors is a vector.
5. ***True or False:*** The product of a scalar and a vector is a scalar.

Answers: 1. True; 2. magnitude, direction; 3. direction; 4. True; 5. False

2.5 RECTANGULAR COMPONENTS OF A VECTOR

After studying this section, you should be able to:

- Determine the components of a vector with respect to an arbitrary rectangular Cartesian coordinate system.
- Manipulate vectors using their rectangular Cartesian projections.
- Represent and manipulate vectors using their direction cosines.

AXES

axis An *axis* is a straight line with a given orientation in space and a given sense (Fig. 2.18). Axes are used to set directions with respect to some fixed system. For most engineering purposes, directions are set with respect to some fixed position on the earth. The sense of an axis is indicated by an arrowhead on the straight line. The *positive sense* of the axis corresponds to the direction in which the arrowhead points; the *negative sense* is the opposite direction. An axis is not a vector since it has no magnitude, only direction. For simplicity in diagrams, the arrowhead is sometimes omitted when the positive sense of the axis cannot be misinterpreted.

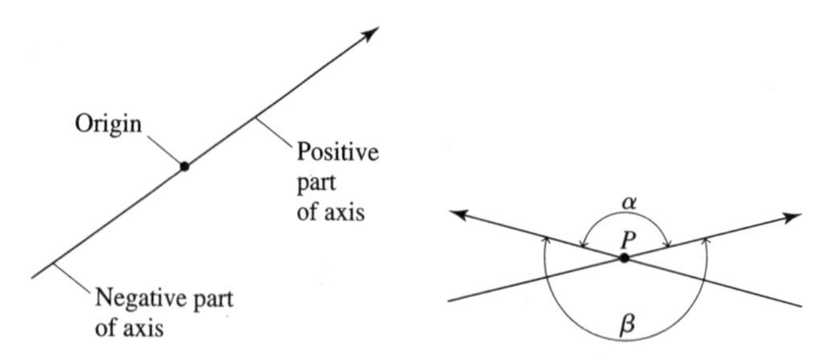

Figure 2.18 An axis.

Figure 2.19 Angle between two axes.

A given point on an axis, known as the *origin,* separates the axis into two parts, the *positive part* and the *negative part.* These parts lie in the positive sense and the negative sense from the origin, respectively.

To define the angle between two axes, let the axes be translated (i.e., displaced without change of direction), if necessary, so that they intersect at point *P.* The angle between the axes is the angle between their positive parts. In three-dimensional problems, the angle is understood to be no greater than 180°. For instance, the angle between the pair of axes shown in Fig. 2.19 is denoted by α. In two-dimensional problems, however, the angle between two axes is sometimes considered to be a *reflex angle* (an angle greater than 180°), as indicated by the angle β in Fig. 2.19. In any case, a meaning is ascribed to the angle between two axes, even though the axes might not intersect. The angle between two vectors or the angle between a vector and an axis is defined in the same way as the angle between two axes.

RECTANGULAR CARTESIAN COORDINATE SYSTEMS

Cartesian coordinate system

origin

A rectangular *Cartesian coordinate system* in space is composed of three mutually perpendicular axes that intersect at a common point, called the *origin.* The axes are designated by three different symbols, commonly x, y, and z. There are two distinct types of rectangular Cartesian coordinate systems: right-handed systems and left-handed systems. They are illustrated in Fig. 2.20. These systems differ in that they cannot be brought into congruence with each other in such a way that the axes with the same letters have the same sense. With the right-handed system, if you grasp the z axis in your right hand with your thumb pointing in the positive sense of the axis, your fingers curl around the z axis from the x axis to the y axis (Fig. 2.20a). The left hand bears the same relationship to a left-handed coordinate system (Fig. 2.20b).

The distinction between right-handed and left-handed coordinate systems also exists in two dimensions. If the coordinate systems shown in Fig. 2.21 remain in the plane of the paper, they cannot be brought into congruence with each other in such a way that the axes with the same symbol have the same sense. If you imagine that there is a z axis perpendicular to the plane of the paper, pointing toward yourself, the system shown in Fig. 2.21a is right-handed, and that shown in Fig. 2.21b is left-handed. To avoid confusion, this book usually uses right-handed coordinates for both two-dimensional and three-dimensional systems.

ORTHOGONAL PROJECTION OF A VECTOR ON AN AXIS

orthogonal projection of a vector

Some of the most important properties of vectors are concerned with orthogonal projections of vectors on fixed axes (Fig. 2.22). An *orthogonal projection of a vector* **A** on an

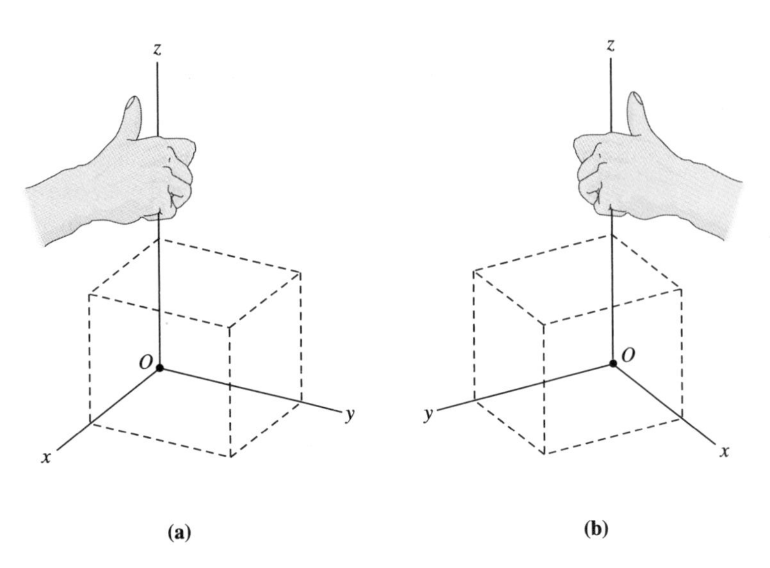

Figure 2.20 Cartesian coordinate systems in space: **(a)** right-handed system; **(b)** left-handed system.

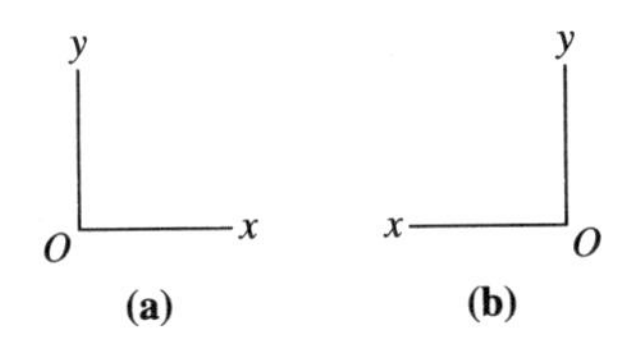

Figure 2.21 Planar Cartesian coordinate systems: **(a)** right-handed system; **(b)** left-handed system.

axis L is constructed by drawing projection lines (shown as dashed lines in Fig. 2.22) from the axis to the tail and tip of the vector; these projection lines are perpendicular (orthogonal) to the axis. The projection has a magnitude defined by the distance between the two projection lines. The projection has a sign determined by the orientation of the vector **A** with respect to the axis L. The sign of the projection is positive if the component of **A** parallel to L has the same sense as the axis; otherwise, the sign of the projection is negative (see Fig. 2.22). If the vector is perpendicular to the axis, its projection on the axis is zero. If a vector of magnitude A forms an angle θ with an axis, the projection of the vector on the axis is represented in magnitude and in sign by $A \cos \theta$. This is true whether θ is interpreted to be the smaller angle or the reflex angle that the vector forms with the axis (Fig. 2.19). Therefore, *the projection of a vector on an axis is regarded as a scalar since it is characterized as a signed numerical value.* Consequently, the projection of a vector has *no point of application.* In this book, a reference to the *projection* of a vector on an axis should be understood to imply an orthogonal projection. When nonorthogonal projections are considered, they will be explicitly identified as such.

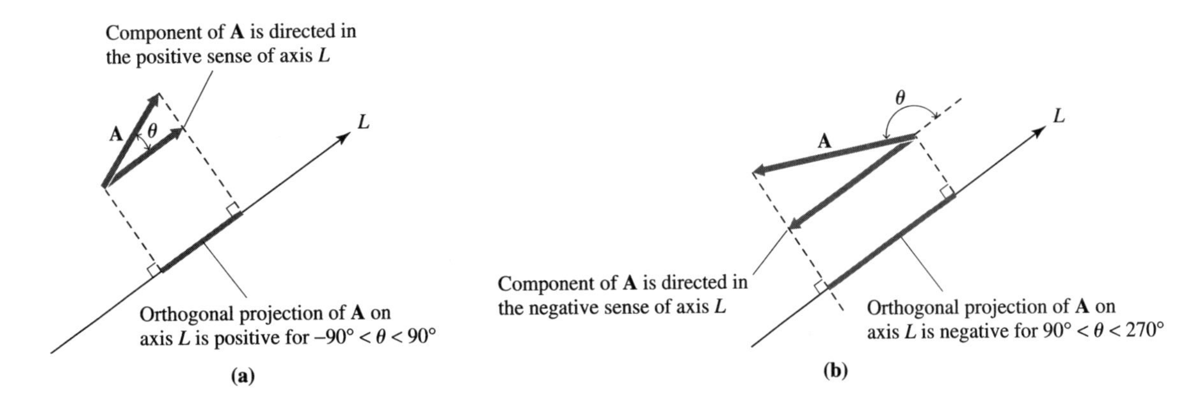

Figure 2.22 Orthogonal projection of a vector on an axis.

RECTANGULAR CARTESIAN PROJECTIONS OF A VECTOR

The projections of a vector on rectangular Cartesian *xyz* axes in space are denoted by appending the subscripts *x*, *y*, and *z* to the symbol that denotes the magnitude of the vector. For example, the projections of a vector **F** on the *xyz* axes are denoted by (F_x, F_y, F_z); see Fig. 2.23. The terms F_x, F_y, and F_z are called the rectangular Cartesian projections of the vector **F**.

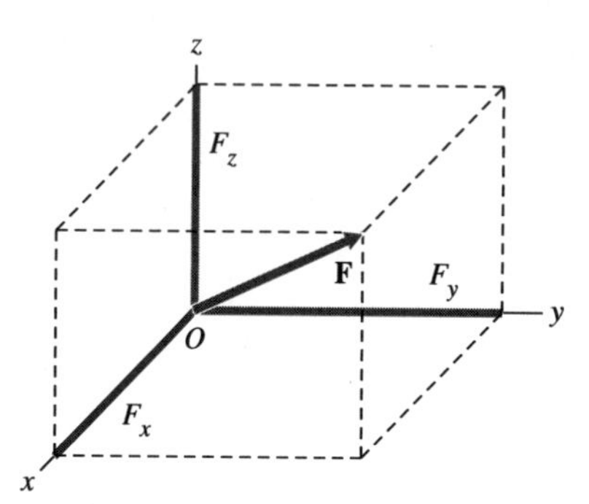

Figure 2.23 Rectangular Cartesian projections of a vector.

direction angle

The angle between a vector and a Cartesian coordinate axis is called the *direction angle* of the vector with respect to that axis. The three direction angles θ_x, θ_y, and θ_z of a vector with reference to the three rectangular Cartesian axes determine the direction of the vector (see Fig. 2.24). The direction angles of a vector are specified to lie in the range from 0 to 180°, inclusive. Consequently, a direction angle can be determined by its cosine. If the cosine is negative, the angle is greater than 90°. The cosines of the direction angles of a vector are called the *direction cosines* of the vector.

direction cosines

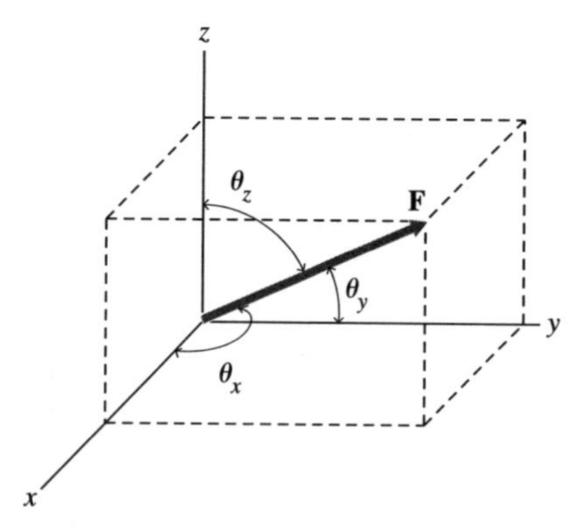

Figure 2.24 Direction angles of a vector.

If θ_x, θ_y, and θ_z are the direction angles of vector **F** and if F denotes the magnitude of **F**, then

$$F_x = F\cos\theta_x \qquad F_y = F\cos\theta_y \qquad F_z = F\cos\theta_z \tag{2.1}$$

In terms of the rectangular Cartesian projections, the magnitude of any vector **F** is determined by the equation

$$F^2 = F_x^2 + F_y^2 + F_z^2 \tag{2.2}$$

To derive this equation, we refer to Fig. 2.25. The Pythagorean theorem yields $OP^2 = OB^2 + BP^2$, since OBP is a right triangle. Also, since OAB is a right triangle, $OB^2 = AB^2 + OA^2$. Using this equation to eliminate OB from the preceding equation, we obtain $OP^2 = AB^2 + OA^2 + BP^2$. Since $OP = F$, $AB = F_x$, $OA = F_y$, and $BP = F_z$, Eq. (2.2) results.

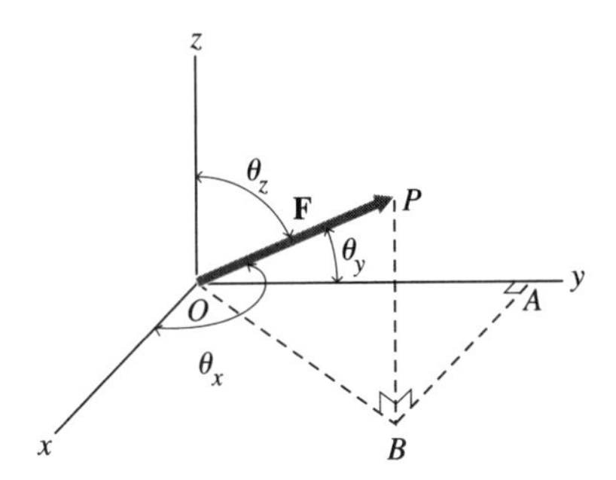

Figure 2.25 Vector in space.

Also, substituting Eqs. (2.1) into Eq. (2.2) and canceling F^2, we get

$$\cos^2\theta_x + \cos^2\theta_y + \cos^2\theta_z = 1 \tag{2.3}$$

Equation (2.3) shows that the three direction cosines of a vector in space are not independent of each other; any one can be determined, except for sign, by the other two.

In two-dimensional problems, you need consider only x and y axes, since then $\theta_z = 90°$ and therefore $F_z = 0$. Analysis of coplanar forces ordinarily employs plane Cartesian coordinates (x, y) in a right-handed system. The direction of a vector **F** can then be specified by the single angle θ that the vector forms with the x axis (see Fig. 2.26). Usually, θ ranges from 0 to 360° and is measured in the counterclockwise sense.[2]

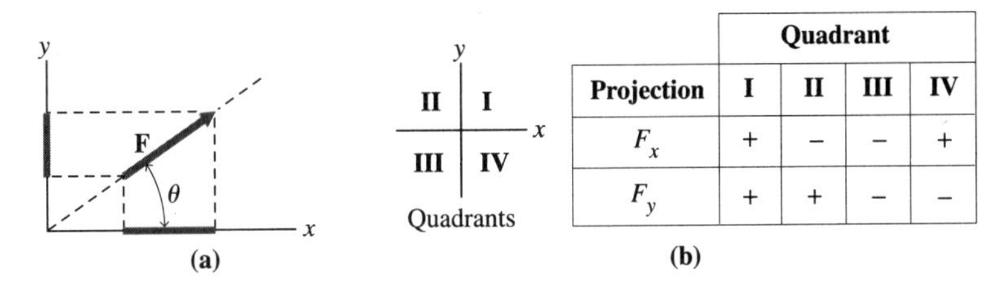

Projection	Quadrant I	II	III	IV
F_x	+	–	–	+
F_y	+	+	–	–

Figure 2.26 **(a)** Direction of a vector **F** in a plane. **(b)** Sign of projection of **F** in each quadrant of the xy plane.

The projections of a vector **F** on the x and y axes are

$$F_x = F\cos\theta \qquad F_y = F\sin\theta \tag{2.4}$$

where, as usual, F denotes the magnitude of vector **F**. The signs of the projections (F_x, F_y) of **F** in each quadrant of the xy plane are given in Fig. 2.26b. You can also see, by Eq. (2.4), that

$$F^2 = F_x^2 + F_y^2 \tag{2.5}$$

2. The range $-180° < \theta \leq 180°$ is often used in mathematical texts.

ADDITION OF VECTORS USING RECTANGULAR CARTESIAN PROJECTIONS

The polygon construction of the resultant of several vectors is shown in Fig. 2.27. You can see from Fig. 2.27 that the projection AE of the resultant **R** on the x axis is equal to the algebraic sum of the projections AB, BC, CD, and DE of the vectors **F**, **G**, **H**, and **J** on that axis.[3] In other words, $R_x = F_x + G_x + H_x + J_x$, where F_x, G_x, H_x, and J_x are the projections of vectors **F**, **G**, **H**, and **J** on the x axis, respectively. In the case of rectangular Cartesian xyz axes, the expression for R_x may be supplemented by similar relations R_y and R_z for projections on the y and z axes. The complete set of relations is

$$\begin{aligned} R_x &= F_x + G_x + H_x + J_x \\ R_y &= F_y + G_y + H_y + J_y \\ R_z &= F_z + G_z + H_z + J_z \end{aligned} \tag{2.6}$$

or, in more concise vector notation,

$$\mathbf{R} = \mathbf{F} + \mathbf{G} + \mathbf{H} + \mathbf{J} \tag{2.7}$$

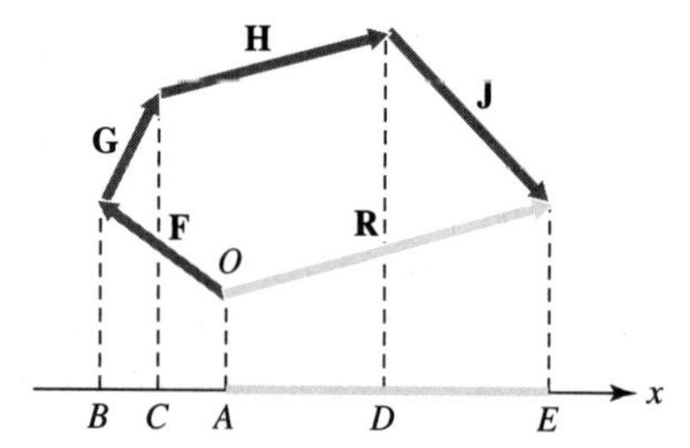

Figure 2.27 Projection of resultant vector, represented as the algebraic sum of projections of the components.

Thus, the projections of a resultant vector on xyz coordinate axes are equal to the sums of the projections of the individual vectors on the same xyz axes. This leads to the following theorem.[4]

Theorem 2.1

The resultant of several vectors is a vector whose projection on any axis is the algebraic sum of the projections of the original vectors on the axis.

This theorem tells you that you can obtain the resultant of several vectors by projecting each of the vectors onto rectangular xyz axes and, for each axis, adding the corresponding projections to obtain the (x, y, z) projections (R_x, R_y, R_z) of the resultant **R**. By the Pythagorean theorem [Eq. (2.2)] with (R_x, R_y, R_z) you can obtain the magnitude R of the resultant **R**. Then, by Eq. (2.1), you can compute the direction cosines ($\cos\theta_x$, $\cos\theta_y$,

3. When used in this sense, the word "algebraic" means that the signs of the terms are taken into account. For example, the projection AB is negative, and the projections BC, CD, and DE are positive. Thus, the projection AE is equal to the sum $BC + CD + DE - AB = BE - AB$.

4. In general, the word "theorem" denotes a fact that has been established as a principle or law.

$\cos\theta_z$) and, hence, the angles (θ_x, θ_y, θ_z) that **R** forms with the *xyz* axes. In view of Eq. (2.7), **R** is called the *vector sum* of **F**, **G**, **H**, and **J**. The expressions "vector sum" and "resultant" are used interchangeably. Equation (2.7) suggests use of the polygon construction to find the resultant vector **R**. However, the projection equations [Eqs. (2.6)] and the vector equation [Eq. (2.7)] are equivalent. The polygon construction is useful in visual (graphical) work, but the Cartesian projections are often more convenient for analytical (algebraic) work.

vector sum

ASSOCIATIVE LAW OF VECTOR ADDITION

The sum of several vectors can be determined by the polygon construction. If the order in which the vectors are combined is varied, the resultant vector is not changed. To show that this is true, we must show that

$$\mathbf{F} + \mathbf{G} + (\mathbf{H} + \mathbf{J}) = (\mathbf{F} + \mathbf{G}) + \mathbf{H} + \mathbf{J} = \mathbf{F} + (\mathbf{G} + \mathbf{H}) + \mathbf{J} \tag{2.8}$$

and so on. The proof of Eq. (2.8) follows immediately from Eqs. (2.6), since the projections (scalars) add algebraically. Thus,

$$\begin{aligned} F_x + G_x + (H_x + J_x) &= (F_x + G_x) + H_x + J_x = F_x + (G_x + H_x) + J_x \\ F_y + G_y + (H_y + J_y) &= (F_y + G_y) + H_y + J_y = F_y + (G_y + H_y) + J_y \\ F_z + G_z + (H_z + J_z) &= (F_z + G_z) + H_z + J_z = F_z + (G_z + H_z) + J_z \end{aligned} \tag{2.9}$$

and so on. Equation (2.8) expresses the *associative law of vector addition.* In view of the associative law, parentheses are unnecessary in vector addition. Consequently, the vector sum of **F**, **G**, **H**, and **J** is represented by **F** + **G** + **H** + **J**, as in Eq. (2.7). This conclusion is expressed by the statement that *vector addition is associative.*

COMMUTATIVE LAW OF VECTOR ADDITION

Equations (2.6) also serve to verify that

$$\mathbf{F} + \mathbf{G} + \mathbf{H} + \mathbf{J} = \mathbf{F} + \mathbf{H} + \mathbf{G} + \mathbf{J} = \mathbf{H} + \mathbf{G} + \mathbf{J} + \mathbf{F} \tag{2.10}$$

and so on. This may be shown immediately from the fact that the Cartesian projections [Eqs. (2.6)] of vectors obey the commutative law:

$$\begin{aligned} F_x + G_x + H_x + J_x &= F_x + H_x + G_x + J_x = H_x + G_x + J_x + F_x \\ F_y + G_y + H_y + J_y &= F_y + H_y + G_y + J_y = H_y + G_y + J_y + F_y \\ F_z + G_z + H_z + J_z &= F_z + H_z + G_z + J_z = H_z + G_z + J_z + F_z \end{aligned} \tag{2.11}$$

Therefore, changing the order of vectors in the polygon in Fig. 2.27 to **G**, **J**, **H**, **F** gives **R** = **G** + **J** + **H** + **F**, which is equivalent to Eq. (2.7). Equation (2.10) expresses the *commutative law of vector addition.* Consequently, the order of vectors in the polygon construction does not affect the resultant that is obtained. This conclusion is expressed by the statement that *vector addition is commutative.*

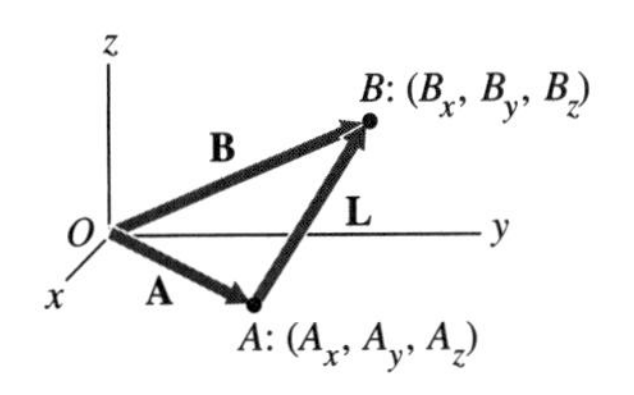

Figure 2.28 General vector **L** in space.

GENERAL VECTOR IN SPACE

Consider two vectors **A** and **B** that extend from the origin to points A: (A_x, A_y, A_z) and B: (B_x, B_y, B_z), respectively. Let vector **L** extend from A to B (see Fig. 2.28). Hence, **L** = **B** − **A**. The magnitude (or length) L is determined by Eq. (2.2):

$$L = [L_x^2 + L_y^2 + L_z^2]^{1/2} \tag{2.12}$$

where $L_x = B_x - A_x$, $L_y = B_y - A_y$, and $L_z = B_z - A_z$ are the projections of vector **L** on the x, y, and z axes. By Eqs. (2.1), the direction cosines of **L** are

$$\cos\theta_x = \frac{L_x}{L} = \frac{B_x - A_x}{L}$$

$$\cos\theta_y = \frac{L_y}{L} = \frac{B_y - A_y}{L} \tag{2.13}$$

$$\cos\theta_z = \frac{L_z}{L} = \frac{B_z - A_z}{L}$$

PROBLEM-SOLVING TECHNIQUE

Resultant of Forces by Projections

A. Three-dimensional (noncoplanar) force systems To find the resultant **R** of a set of concurrent, noncoplanar forces $\mathbf{F}_i$, $i = 1, \ldots, n$, using direction cosines:

1. Determine the direction angles θ_{xi}, θ_{yi}, and θ_{zi} for each force $\mathbf{F}_i$. Recall that the direction angles for noncoplanar forces are taken to lie in the range 0 to 180°, inclusive. Compute the (x, y, z) projections of each force $\mathbf{F}_i$ by the formulas

$$F_{xi} = F_i \cos\theta_{xi} \qquad F_{yi} = F_i \cos\theta_{yi} \qquad F_{zi} = F_i \cos\theta_{zi}$$

   Depending on the values of $\cos\theta_{xi}$, $\cos\theta_{yi}$, and $\cos\theta_{zi}$, some projections may be positive, and some may be negative.

2. Add the individual projections (F_{xi}, F_{yi}, F_{zi}) algebraically to find R_x, R_y, and R_z, the (x, y, z) projections of the resultant **R**.
3. Determine the magnitude R and the direction cosines (and, hence, the direction angles) using the formulas

$$R = (R_x^2 + R_y^2 + R_z^2)^{1/2}$$

$$\cos\theta_x = \frac{R_x}{R}$$

$$\cos\theta_y = \frac{R_y}{R}$$

$$\cos\theta_z = \frac{R_z}{R}$$

4. Verify that $\cos^2\theta_x + \cos^2\theta_y + \cos^2\theta_z = 1$.

B. Two-dimensional (coplanar) force systems To find the resultant **R** of a set of concurrent, coplanar forces $\mathbf{F}_i$, $i = 1, \ldots, n$, using direction cosines:

1. Determine the (x, y) projections of each force. If a force is described by its magnitude F_i and its direction angle θ_i measured counterclockwise from the positive x axis, then the projections are given by $F_{xi} = F_i \cos\theta_i$ and $F_{yi} = F_i \sin\theta_i$. In general, some projections may be positive, and some may be neg-

ative. Use Fig. 2.26b to check signs. If the force is given in some other way—say, using a direction angle relative to some reference line other than the positive x axis—be careful with signs. Make a table showing the projections for each force.

2. Add the individual projections (F_{xi}, F_{yi}) algebraically to find R_x and R_y, the x and y projections, respectively, of the resultant **R**.
3. Determine the magnitude R and the direction angle θ of the resultant **R** from the formulas

$$R = (R_x^2 + R_y^2)^{1/2}$$

$$\theta = \tan^{-1}\frac{R_y}{R_x}$$

Remember that $R \geq 0$ and, for coplanar forces, $0 \leq \theta < 360°$.

Example 2.10 Characteristics of Forces by Projection

Problem Statement Determine the magnitude and direction, relative to the x axis, of each of the forces **R** shown in Figs. E2.10a through E2.10d.

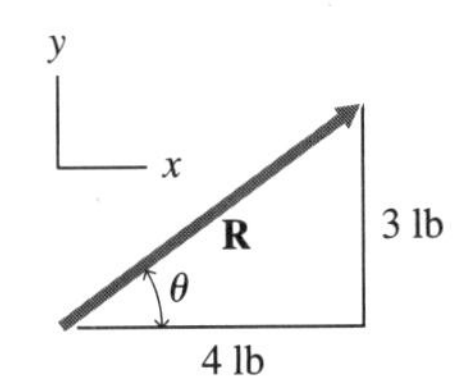

Figure E2.10a

Solution In Fig. E2.10a, the magnitude of **R** is

$$R = \sqrt{3^2 + 4^2} = 5 \text{ lb}$$

The direction of **R** is given by

$$\tan\theta = \tfrac{3}{4} \quad \text{or} \quad \theta = 36.87° \text{ counterclockwise}$$

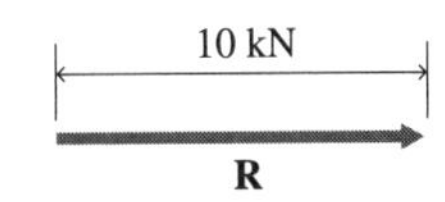

Figure E2.10b

In Fig E2.10b, the magnitude of **R** is

$$R = 10 \text{ kN}$$

The direction of **R** is given by

$$\theta = 0$$

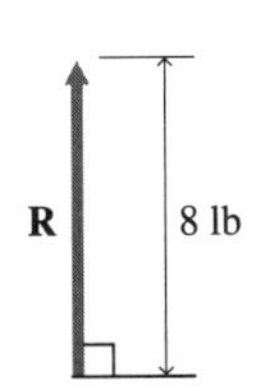

Figure E2.10c

The magnitude of **R** in Fig. E2.10c is

$$R = 8 \text{ lb}$$

The direction of **R** is given by

$$\theta = 90° \text{ counterclockwise}$$

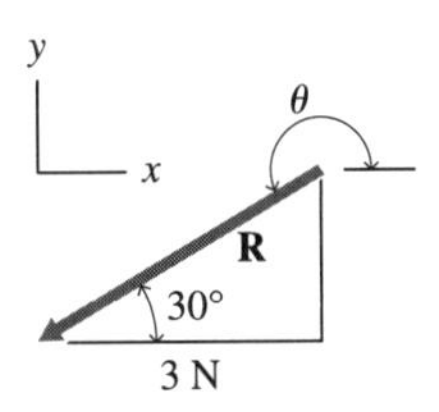

Figure E2.10d

The magnitude of **R** in Fig. E2.10d is

$$R = \frac{3}{\cos 30°} = 3.46 \text{ N}$$

The direction of **R** is given by

$$\theta = 180° + 30° = 210°$$

Example 2.11

Resultant of Forces by Projections on Coordinate Axes

Problem Statement Determine the resultant **R** of three coplanar forces **A**, **B**, and **C** that act at point O of a body. The magnitudes of **A**, **B**, and **C** are 20 N, 40 N, and 60 N, respectively. The directions of these forces are shown in Fig. E2.11.

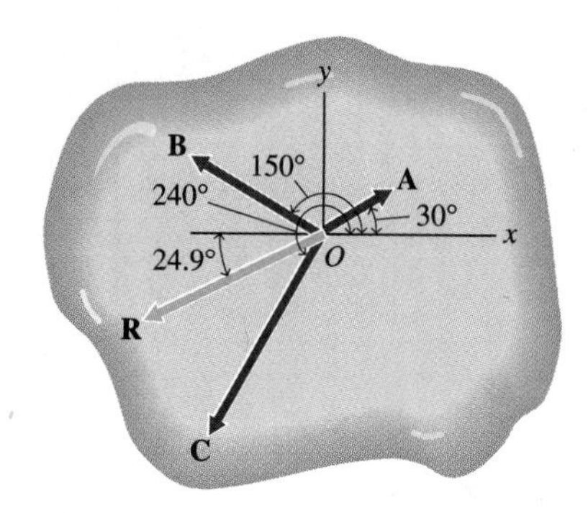

Figure E2.11

Solution The projections of forces **A**, **B**, and **C** and their sums are found using the given force magnitudes and orientations (see Table E2.11). Since the signs of the projections $R_x = -47.32$ N and $R_y = -21.96$ N are negative, the (x, y) components of the resultant **R** have the negative senses of the coordinate axes. Therefore, **R** lies in the third quadrant. By trigonometry, the angle θ that **R** forms with the x axis is determined by $\tan\theta = -21.96/-47.32 = 0.4641$. Thus, $\theta = 204.9°$. The magnitude R of the resultant **R** is, by Eq. (2.5),

$$R = \sqrt{R_x^2 + R_y^2} = 52.17 \text{ N}$$

The resultant **R** is shown in Fig. E2.11. It is located $204.9° - 180° = 24.9°$ counterclockwise from the negative x axis.

Table E2.11

FORCE	PROJECTION ON x AXIS	PROJECTION ON y AXIS
A	$20 \cos 30° = 17.32$ N	$20 \sin 30° = 10.00$ N
B	$40 \cos 150° = -34.64$ N	$40 \sin 150° = 20.00$ N
C	$60 \cos 240° = -30.00$ N	$60 \sin 240° = -51.96$ N
R	-47.32 N	-21.96 N

Example 2.12

Resultant of Concurrent Noncoplanar Forces

Problem Statement Determine the resultant **R** of three forces **P**, **S**, and **T**, which are concurrent at point O of the cube shown in Fig. E2.12. Forces **P**, **S**, and **T** are directed toward the corners F, G, and D of the cube, respectively, and have magnitudes $P = 20$ N, $S = 30$ N, and $T = 50$ N.

Solution For a unit cube, the lengths of lines OF, OG, and OD are $\sqrt{2}$, $\sqrt{3}$, and $\sqrt{2}$, respectively. So, the direction cosines $(\cos\theta_x, \cos\theta_y, \cos\theta_z)$ are $(1/\sqrt{2}, 0, 1/\sqrt{2})$ for **P**, $(1/\sqrt{3}, 1/\sqrt{3}, 1/\sqrt{3})$ for **S**, and $(0, 1/\sqrt{2}, 1/\sqrt{2})$ for **T**. The projections of the forces and their sums are given in Table E2.12. Hence, by Eq. (2.2),

$$R = \sqrt{R_x^2 + R_y^2 + R_z^2} = \sqrt{31.46^2 + 52.68^2 + 66.82^2} = 90.72 \text{ N}$$

By Eqs. (2.1), the direction cosines of **R** are

$$\cos\theta_x = \frac{R_x}{R} = 0.3468$$

$$\cos\theta_y = \frac{R_y}{R} = 0.5806$$

$$\cos\theta_z = \frac{R_z}{R} = 0.7366$$

Therefore, $\theta_x = 69.70°$, $\theta_y = 54.50°$, and $\theta_z = 42.56°$. As a check, you can verify that the direction cosines satisfy Eq. (2.3): $0.3468^2 + 0.5806^2 + 0.7366^2 = 1.00$.

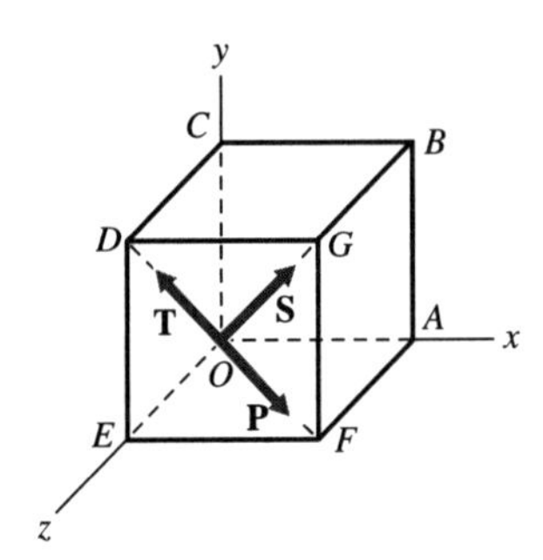

Figure E2.12

Table E2.12

FORCE	PROJECTION ON *x* AXIS	PROJECTION ON *y* AXIS	PROJECTION ON *z* AXIS
P	$20/\sqrt{2} = 14.14$ N	0 N	$20/\sqrt{2} = 14.14$ N
S	$30/\sqrt{3} = 17.32$ N	$30/\sqrt{3} = 17.32$ N	$30/\sqrt{3} = 17.32$ N
T	0 N	$50/\sqrt{2} = 35.36$ N	$50/\sqrt{2} = 35.36$ N
R	31.46 N	52.68 N	66.82 N

CHECKPOINT ✓

1. ***True or False:*** An axis is a vector.
2. ***True or False:*** A right-handed coordinate system is congruent with a left-handed coordinate system in two dimensions, but not in three.
3. ***True or False:*** The projection of a vector on an axis is a vector.
4. How many independent direction cosines does a three-dimensional vector have?
5. ***True or False:*** The direction cosines for a vector can be found from the rectangular Cartesian projections of the vector.

Answers: 1. False; 2. False; 3. False; 4. Two; 5. True

2.6 UNIT VECTORS

After studying this section, you should be able to:

- Represent and manipulate a force vector using unit vectors.

Key Concept Use of unit vectors systematizes the composition and resolution of forces.

unit vector A *unit vector* **n** is a vector that has a magnitude $n = 1$. The magnitude n is dimensionless; it has no physical units. The main purpose of a unit vector is to denote a direction in space. As we shall see, unit vectors are a convenient tool for expressing composition of vectors in terms of projections.

THREE-DIMENSIONAL VECTORS

For three-dimensional (noncoplanar) problems, we can define unit vectors **i**, **j**, and **k** that point in the positive directions of rectangular Cartesian xyz axes (see Fig. 2.29a). Then, if we express a vector **A** in terms of its projections (A_x, A_y, A_z) (Fig. 2.29a) and also in terms of its Cartesian components $(\mathbf{A}_x, \mathbf{A}_y, \mathbf{A}_z)$ (Fig. 2.29b), we can relate the components to the projections as follows:

$$\mathbf{A}_x = A_x\mathbf{i} \qquad \mathbf{A}_y = A_y\mathbf{j} \qquad \mathbf{A}_z = A_z\mathbf{k} \tag{2.14}$$

The vector **A** can then be expressed in the form

$$\mathbf{A} = A_x\mathbf{i} + A_y\mathbf{j} + A_z\mathbf{k} \tag{2.15}$$

In this form, the projections (A_x, A_y, A_z) may be thought of as *scale factors* that multiply the unit vectors **i**, **j**, and **k**, respectively. The given vector **A** is then described as the sum (resultant) of the scaled unit vectors [Eq. (2.15)].

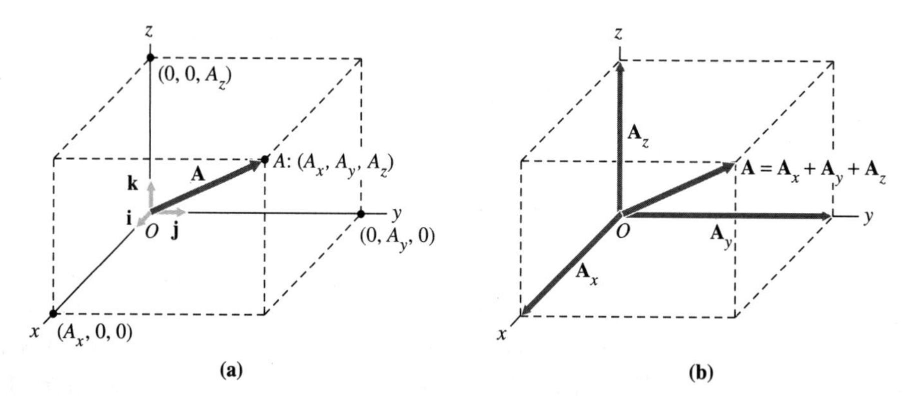

Figure 2.29 Unit vectors along coordinate axes: **(a)** projections of vector **A**; **(b)** components of vector **A**.

The direction of an arbitrary line or vector in space may also be defined by a unit vector **n** (Fig. 2.30). For example, since $n = 1$, by Eqs. (2.1), the unit vector **n** can be expressed as

$$\mathbf{n} = (\cos\theta_x)\mathbf{i} + (\cos\theta_y)\mathbf{j} + (\cos\theta_z)\mathbf{k} \tag{2.16}$$

Thus, the projections (n_x, n_y, n_z) of the unit vector **n** on the xyz coordinate axes are

$$n_x = \cos\theta_x \qquad n_y = \cos\theta_y \qquad n_z = \cos\theta_z \tag{2.17}$$

That is, the projections (n_x, n_y, n_z) are identical to the direction cosines of **n**.

Since the magnitude of **n** is $n = 1$, we have, by Eqs. (2.16) and (2.17),

$$n^2 = n_x^2 + n_y^2 + n_z^2 = 1 \tag{2.18}$$

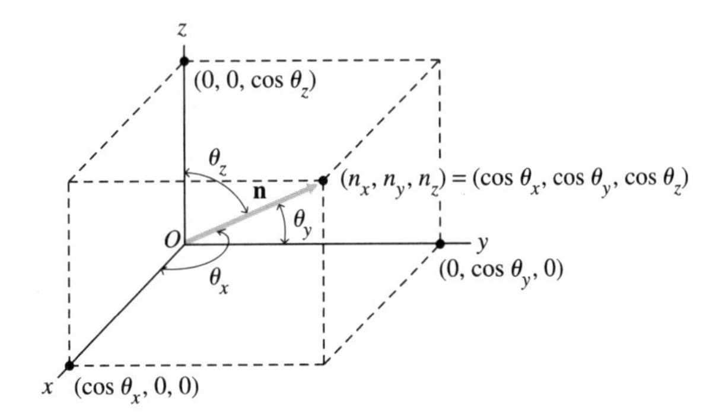

Figure 2.30 Direction in space defined by a unit vector **n**.

or

$$\cos^2 \theta_x + \cos^2 \theta_y + \cos^2 \theta_z = 1 \tag{2.19}$$

Equation (2.19) agrees with the result obtained earlier; see Eq. (2.3).

If the direction of the unit vector **n** coincides with that of the vector **F**, then **F** can be represented as the product of its magnitude F and **n**. That is, **F** is a vector with magnitude F and direction corresponding to that of **n** (Fig. 2.31). Thus, we may write

$$\mathbf{F} = F\mathbf{n} \tag{2.20}$$

Substitution of Eq. (2.16) into Eq. (2.20) gives

$$\mathbf{F} = (F\cos\theta_x)\mathbf{i} + (F\cos\theta_y)\mathbf{j} + (F\cos\theta_z)\mathbf{k} \tag{2.21}$$

where, by Eqs. (2.1),

$$F\cos\theta_x = F_x \qquad F\cos\theta_y = F_y \qquad F\cos\theta_z = F_z \tag{2.22}$$

The scalar factors ($F\cos\theta_x$, $F\cos\theta_y$, $F\cos\theta_z$) are the projections (F_x, F_y, F_z) of **F** on the xyz coordinate axes. Consequently, by Eqs. (2.22), the direction cosines of **F** are given by the relations

$$\cos\theta_x = \frac{F_x}{F} \qquad \cos\theta_y = \frac{F_y}{F} \qquad \cos\theta_z = \frac{F_z}{F} \tag{2.23}$$

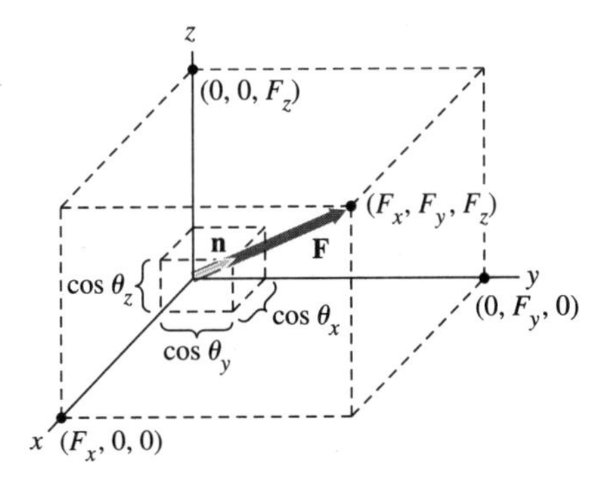

Figure 2.31 Unit vector **n** in direction of vector **F**.

By Eqs. (2.21) and (2.22), the unit vector **n** in the direction of **F** is

$$\mathbf{n} = \frac{\mathbf{F}}{F} = \frac{F_x}{F}\mathbf{i} + \frac{F_y}{F}\mathbf{j} + \frac{F_z}{F}\mathbf{k} \tag{2.24}$$

Various arithmetic operations that are performed on vectors (Sec. 2.4) can be expressed with unit vectors using Eq. (2.21). Consider two vectors **A** and **B** with magnitudes A and B and direction cosines ($\cos\alpha_x$, $\cos\alpha_y$, $\cos\alpha_z$) and ($\cos\beta_x$, $\cos\beta_y$, $\cos\beta_z$), respectively. Then, by Eq. (2.21), **A** and **B** can be represented by

$$\mathbf{A} = (A\cos\alpha_x)\mathbf{i} + (A\cos\alpha_y)\mathbf{j} + (A\cos\alpha_z)\mathbf{k} \tag{2.25a}$$

$$\mathbf{B} = (B\cos\beta_x)\mathbf{i} + (B\cos\beta_y)\mathbf{j} + (B\cos\beta_z)\mathbf{k} \tag{2.25b}$$

or, in terms of projections,

$$\begin{aligned} \mathbf{A} &= A_x\mathbf{i} + A_y\mathbf{j} + A_z\mathbf{k} \\ \mathbf{B} &= B_x\mathbf{i} + B_y\mathbf{j} + B_z\mathbf{k} \end{aligned} \tag{2.26}$$

where

$$\begin{aligned} A_x &= A\cos\alpha_x \qquad A_y = A\cos\alpha_y \qquad A_z = A\cos\alpha_z \\ B_x &= B\cos\beta_x \qquad B_y = B\cos\beta_y \qquad B_z = B\cos\beta_z \end{aligned} \tag{2.27}$$

Analogous to the discussion in Sec. 2.4, equality of **A** and **B** requires that

$$\begin{aligned} A\cos\alpha_x &= B\cos\beta_x \\ A\cos\alpha_y &= B\cos\beta_y \\ A\cos\alpha_z &= B\cos\beta_z \end{aligned} \tag{2.28}$$

or

$$A_x = B_x \qquad A_y = B_y \qquad A_z = B_z \tag{2.29}$$

Likewise, from Eq. (2.25a), the negative of a vector **A** is

$$-\mathbf{A} = -(A\cos\alpha_x)\mathbf{i} - (A\cos\alpha_y)\mathbf{j} - (A\cos\alpha_z)\mathbf{k} \tag{2.30}$$

or

$$-\mathbf{A} = -A_x\mathbf{i} - A_y\mathbf{j} - A_z\mathbf{k} \tag{2.31}$$

The resultant **R** of vectors **A** and **B** is, with Eqs. (2.25),

$$\mathbf{R} = (A\cos\alpha_x + B\cos\beta_x)\mathbf{i} + (A\cos\alpha_y + B\cos\beta_y)\mathbf{j} + (A\cos\alpha_z + B\cos\beta_z)\mathbf{k} \tag{2.32}$$

or

$$\mathbf{R} = (A_x + B_x)\mathbf{i} + (A_y + B_y)\mathbf{j} + (A_z + B_z)\mathbf{k} \tag{2.33}$$

Consequently, the projections (R_x, R_y, R_z) of **R** are

$$\begin{aligned} R_x &= A\cos\alpha_x + B\cos\beta_x = A_x + B_x \\ R_y &= A\cos\alpha_y + B\cos\beta_y = A_y + B_y \\ R_z &= A\cos\alpha_z + B\cos\beta_z = A_z + B_z \end{aligned} \tag{2.34}$$

TWO-DIMENSIONAL VECTORS

For two-dimensional (coplanar) vectors in the xy plane, the z projection of any vector **A** is zero—that is, $A_z = 0$. Then, the vector components $(\mathbf{A}_x, \mathbf{A}_y)$ of **A** are, in terms of unit vectors **i** and **j** [see Eqs. (2.14)]

$$\mathbf{A}_x = A_x\mathbf{i} \qquad \mathbf{A}_y = A_y\mathbf{j} \tag{2.35}$$

where (A_x, A_y) are the (x, y) projections of **A**.

Analogous to the case of noncoplanar vectors, a unit vector **n**, with direction angle θ relative to the positive x axis, is (see Fig. 2.32)

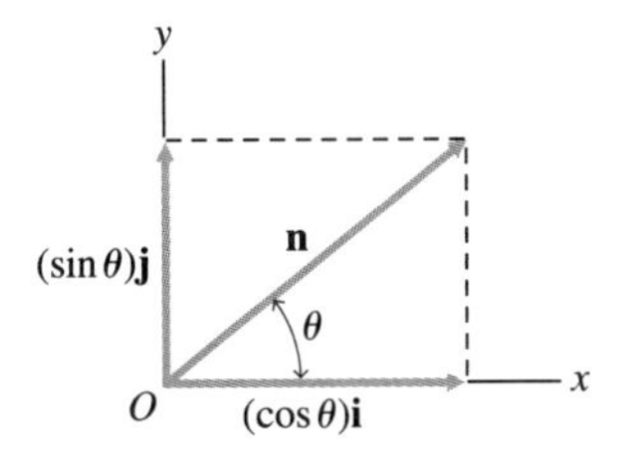

Figure 2.32 Direction in a plane defined by a unit vector **n**.

$$\mathbf{n} = (\cos\theta)\,\mathbf{i} + (\sin\theta)\,\mathbf{j} \tag{2.36}$$

Therefore, the (x, y) projections of **n** are

$$n_x = \cos\theta \qquad n_y = \sin\theta \tag{2.37}$$

Consequently, since the magnitude of **n** is $n = 1$, by Eqs. (2.37),

$$n^2 = n_x^2 + n_y^2 = \cos^2\theta + \sin^2\theta = 1 \tag{2.38}$$

If the direction and sense of the unit vector **n** coincide with those of vector **A**, then [see Eq. (2.20)]

$$\mathbf{A} = A\mathbf{n} \tag{2.39}$$

Equations (2.36) and (2.39) yield

$$\mathbf{A} = (A\cos\theta)\,\mathbf{i} + (A\sin\theta)\,\mathbf{j} \tag{2.40}$$

or

$$\mathbf{A} = A_x\mathbf{i} + A_y\mathbf{j} \tag{2.41}$$

where the scale factors of the unit vectors **i** and **j**, namely,

$$A_x = A\cos\theta \qquad A_y = A\sin\theta \tag{2.42}$$

are the (x, y) projections of **A**. Consequently, by Eqs. (2.42), the direction cosines of **A** are

$$\begin{aligned} \cos\theta_x &= \cos\theta = \frac{A_x}{A} \\ \cos\theta_y &= \sin\theta = \frac{A_y}{A} \end{aligned} \tag{2.43}$$

By Eqs. (2.39) and (2.41), the unit vector **n** in the direction of **A** is

$$\mathbf{n} = \frac{\mathbf{A}}{A} = \frac{A_x}{A}\mathbf{i} + \frac{A_y}{A}\mathbf{j} \tag{2.44}$$

Analogous to the case of noncoplanar vectors [see Eq. (2.33)], the resultant **R** of two vectors **A** and **B** is

$$\mathbf{R} = (A_x + B_x)\,\mathbf{i} + (A_y + B_y)\,\mathbf{j} \tag{2.45}$$

and the projections (R_x, R_y) of **R** are

$$R_x = A_x + B_x \qquad R_y = A_y + B_y \tag{2.46}$$

The general properties of forces (and other vector quantities) discussed in this chapter are summarized in Tables 2.1 and 2.2.

Table 2.1
Summary of properties of forces (vectors)

PROPERTY	EXPLANATION	GRAPHICAL REPRESENTATION	REPRESENTATION BY PROJECTIONS ON RECTANGULAR COORDINATE *xyz* AXES
equality	$\mathbf{A} = \mathbf{B}$ if $A = B$ and the vectors have the same directions and senses	A, B	$A_x = B_x$ $A_y = B_y$ $A_z = B_z$
addition (resultant)	$\mathbf{R} = \mathbf{A} + \mathbf{B} = \mathbf{B} + \mathbf{A}$	B, R, A	$R_x = A_x + B_x$ $R_y = A_y + B_y$ $R_z = A_z + B_z$
negative of a force	$\mathbf{A} = -\mathbf{B}$ if $\lvert A \rvert = \lvert B \rvert$ and the vectors have the same directions but opposite senses	A, B	$A_x = -B_x$ $A_y = -B_y$ $A_z = -B_z$
subtraction	$\mathbf{R} = \mathbf{A} - \mathbf{B}$	B, A, –B, R	$R_x = A_x - B_x$ $R_y = A_y - B_y$ $R_z = A_z - B_z$
multiplication by a scalar	$\mathbf{B} = s\mathbf{A}$ if $B = sA$, where s is a scalar and the directions of B and A are the same	B = *s*A, A	$B_x = sA_x$ $B_y = sA_y$ $B_z = sA_z$

PROBLEM-SOLVING TECHNIQUE

Composition of Vectors Using Unit Vectors

To find the resultant **R** of a set of vectors **A**, **B**, . . . , using unit vectors:

1. Define the coordinate system (axes). It is probably best to use a right-handed *xyz* coordinate system. Show the location of the origin and the positive direction of each axis.
2. Assign unit vectors (**i**, **j**, **k**) with positive directions the same as those of the *xyz* axes.
3. Resolve each vector, **A**, **B**, . . . , into (x, y, z) components ($\mathbf{A}_x$, $\mathbf{A}_y$, $\mathbf{A}_z$), ($\mathbf{B}_x$, $\mathbf{B}_y$, $\mathbf{B}_z$), ...

(continued)

Table 2.2
Summary of unit vectors and direction cosines

QUANTITY	SYMBOLIC REPRESENTATION	GRAPHICAL REPRESENTATION
unit vector	$\mathbf{n}$	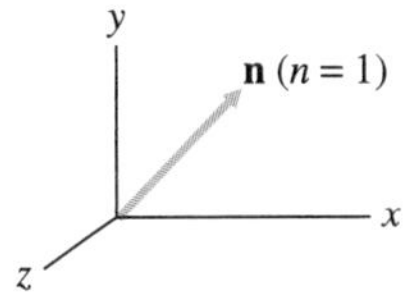
(x, y, z) projections of unit vector $\mathbf{n}$	(n_x, n_y, n_z)	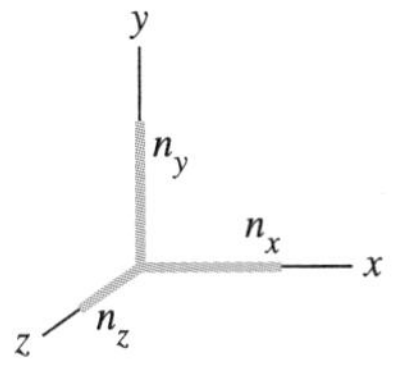
(x, y, z) direction angles of unit vector $\mathbf{n}$	$(\theta_x, \theta_y, \theta_z)$	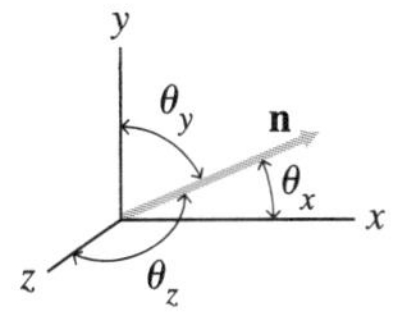
unit vectors directed along xyz coordinate axes	$\mathbf{i}, \mathbf{j}, \mathbf{k}$	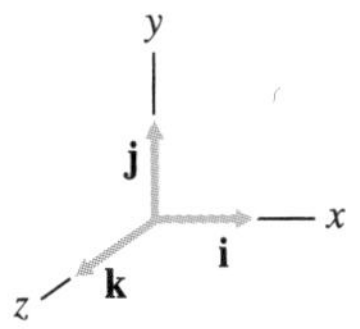
unit vector $\mathbf{n}$ in terms of projections and unit vectors $\mathbf{i}, \mathbf{j}, \mathbf{k}$	$\mathbf{n} = n_x\mathbf{i} + n_y\mathbf{j} + n_z\mathbf{k}$	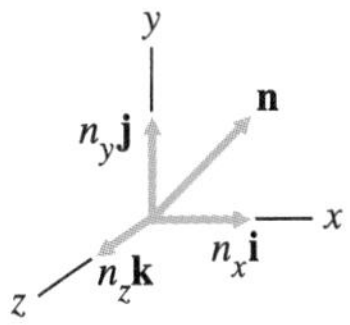
unit vector $\mathbf{n}$ in terms of direction cosines and unit vectors $\mathbf{i}, \mathbf{j}, \mathbf{k}$	$\mathbf{n} = (\cos\theta_x)\mathbf{i} + (\cos\theta_y)\mathbf{j} + (\cos\theta_z)\mathbf{k}$	
unit vector projections and direction cosines	$n_x = \cos\theta_x,\ n_y = \cos\theta_y,\ n_z = \cos\theta_z$	
direction cosine rule	$\cos^2\theta_x + \cos^2\theta_y + \cos^2\theta_z = 1$	
vector $\mathbf{A}$, of magnitude A, in direction of unit vector $\mathbf{n}$	$\mathbf{A} = A\mathbf{n} = A_x\mathbf{i} + A_y\mathbf{j} + A_z\mathbf{k}$ $A_x = A\cos\theta_x,\ A_y = A\cos\theta_y,\ A_z = A\cos\theta_z$	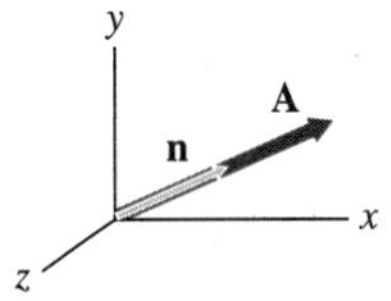

PROBLEM-SOLVING TECHNIQUE *(CONT.)*

4. Determine the projections associated with each vector component. Be careful of signs. Once you have defined an axis—say, the x axis—and its positive direction, the sign of an x projection is positive if the corresponding vector component points in the positive x direction; otherwise, it is negative.
5. Sum the projections on each of the axes, x, y, and z, and multiply the x projections by $\mathbf{i}$, the y projections by $\mathbf{j}$, and the z projections by $\mathbf{k}$. Add the vectors thus formed to obtain the resultant vector

$$\mathbf{R} = (A_x + B_x + \cdots)\mathbf{i} + (A_y + B_y + \cdots)\mathbf{j} + (A_z + B_z + \cdots)\mathbf{k}$$

or, in brief, $\mathbf{R} = R_x\mathbf{i} + R_y\mathbf{j} + R_z\mathbf{k}$.

6. If required, determine the magnitude R and the direction angles θ_x, θ_y, and θ_z using these formulas:

$$R = (R_x^2 + R_y^2 + R_z^2)^{1/2}$$

$$\cos\theta_x = \frac{R_x}{R} \qquad \cos\theta_y = \frac{R_y}{R} \qquad \cos\theta_z = \frac{R_z}{R}$$

Example 2.13

Sum of Two Vectors

Problem Statement The sum of two vectors **A** and **B** is given by

$$\mathbf{S} = \mathbf{A} + \mathbf{B} \tag{a}$$

where

$$\mathbf{A} = 40\mathbf{i} + 30\mathbf{j} \qquad \mathbf{B} = 30\mathbf{i} - 40\mathbf{j} \tag{b}$$

and **i** and **j** are unit vectors directed along the x and y axes, resepctively.

a. Determine the vector **S**.

b. Determine the angle **S** makes with the x axis.

Solution **a.** From Eqs. (a) and (b),

$$\mathbf{S} = (40\mathbf{i} + 30\mathbf{j}) + (30\mathbf{i} - 40\mathbf{j})$$

or

$$\mathbf{S} = 70\mathbf{i} - 10\mathbf{j} \tag{c}$$

b. By Eq. (c), the unit vector in the direction of **S** is

$$\mathbf{n} = \frac{\mathbf{S}}{S} = \frac{70\mathbf{i} - 10\mathbf{j}}{\sqrt{70^2 + 10^2}} = 0.9899\mathbf{i} - 0.1414\mathbf{j}$$

Hence, the direction cosines of **S** are

$$\cos\alpha = 0.9899 \qquad \cos\beta = -0.1414 \tag{d}$$

So, the angle **S** makes with the x axis is

$$\alpha = \cos^{-1}(0.9899) = 8.13°$$

■

Example 2.14

Direction Cosines, Components, and Projections of a Vector

Problem Statement A vector $\mathbf{A} = 6\mathbf{i} + 5\mathbf{j} - 4\mathbf{k}$ has a line of action that passes through the point (3, 2, 1), where **i**, **j**, and **k** comprise an orthogonal triad of unit vectors directed along xyz coordinate axes.

a. Determine the direction cosines of **A**.

b. Determine the (x, y, z) components of **A**.

c. Determine the (x, y, z) projections of **A** on the coordinate axes.

Solution **a.** The magnitude of **A** is

$$A = \sqrt{6^2 + 5^2 + 4^2} = 8.775$$

Hence, the unit vector in the direction of **A** is

$$\mathbf{n} = \frac{\mathbf{A}}{A} = 0.6838\mathbf{i} + 0.5698\mathbf{j} - 0.4558\mathbf{k}$$

The direction cosines of **A** are

$$\cos\alpha = 0.6838 \qquad \cos\beta = 0.5698 \qquad \cos\gamma = -0.4558$$

b. The (x, y, z) components of **A** are

$$\mathbf{A}_x = 6\mathbf{i} \qquad \mathbf{A}_y = 5\mathbf{j} \qquad \mathbf{A}_z = -4\mathbf{k}$$

c. The (x, y, z) projections of **A** are

$$A_x = 6 \qquad A_y = 5 \qquad A_z = -4$$

Example 2.15

Position of a Point in Space

Problem Statement **a.** Use the unit vectors **i**, **j**, and **k**, parallel to xyz coordinate axes, to express the position vectors $\mathbf{r}_A$ and $\mathbf{r}_B$ for the points A: (4, −3, 5) and B: (−4, 3, 3), relative to the origin.

b. Express the vector $\mathbf{r}_{B/A}$ from point A to point B in terms of the unit vectors **i**, **j**, and **k**.

Solution **a.** Since the vector $\mathbf{r}_A$ extends from the origin of the xyz axes to point A, its projections on the xyz axes are 4, −3, and 5. Hence,

$$\mathbf{r}_A = 4\mathbf{i} - 3\mathbf{j} + 5\mathbf{k}$$

Similarly,

$$\mathbf{r}_B = -4\mathbf{i} + 3\mathbf{j} + 3\mathbf{k}$$

b. Since the vector $\mathbf{r}_A$ extends from A to B, it is given by $\mathbf{r}_{B/A} = \mathbf{r}_B - \mathbf{r}_A$, or

$$\mathbf{r}_{B/A} = -8\mathbf{i} + 6\mathbf{j} - 2\mathbf{k}$$

Example 2.16 Vector Arithmetic with Unit Vectors

Problem Statement Three concurrent forces **A**, **B**, and **C** act at point O, as shown in Fig. E2.16, where each tic mark along the coordinate axes indicates an increment of 1 kN.

a. Write an expression for each force in terms of unit vectors **i**, **j**, and **k**.

b. Determine the direction cosines for each force.

c. Determine the resultant **R** of the three forces.

d. Find the component of **R** that lies in the xy plane.

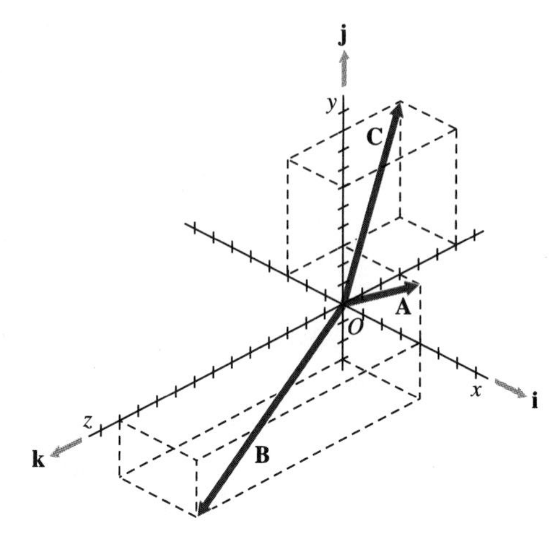

Figure E2.16

Solution

a. The projections of each force onto the coordinate axes are easily seen, and the resulting vector expressions are

$$\begin{aligned} \mathbf{A} &= 4\mathbf{i} + 3\mathbf{j} \\ \mathbf{B} &= 4\mathbf{i} - 3\mathbf{j} + 12\mathbf{k} \\ \mathbf{C} &= -3\mathbf{i} + 6\mathbf{j} - 6\mathbf{k} \end{aligned} \qquad \text{(a)}$$

b. The magnitude of each force is determined by Eq. (2.2), where the force projections are taken from Eqs. (a).

$$\begin{aligned} A &= \sqrt{4^2 + 3^2} = 5 \text{ kN} \\ B &= \sqrt{4^2 + (-3)^2 + 12^2} = 13 \text{ kN} \\ C &= \sqrt{(-3)^2 + 6^2 + (-6)^2} = 9 \text{ kN} \end{aligned} \qquad \text{(b)}$$

The direction cosines follow directly from the values for the force projections on each of the coordinate axes and the force magnitudes in Eqs. (b). Therefore, by Eqs. (2.23),

$$\cos\alpha_x = \tfrac{4}{5} = 0.8 \qquad \cos\alpha_y = \tfrac{3}{5} = 0.6 \qquad \cos\alpha_z = \tfrac{0}{5} = 0$$

$$\cos\beta_x = \tfrac{4}{13} = 0.308 \qquad \cos\beta_y = -\tfrac{3}{13} = -0.231 \qquad \cos\beta_z = \tfrac{12}{13} = 0.923$$

$$\cos\gamma_x = -\tfrac{3}{9} = -0.333 \qquad \cos\gamma_y = \tfrac{6}{9} = 0.667 \qquad \cos\gamma_z = -\tfrac{6}{9} = -0.667$$

where $(\alpha_x, \alpha_y, \alpha_z)$, $(\beta_x, \beta_y, \beta_z)$, and $(\gamma_x, \gamma_y, \gamma_z)$ are the direction angles for forces **A**, **B**, and **C**, respectively.

c. With the three forces in the form of Eqs. (a), the resultant **R** is obtained simply by summing the individual force projections for each coordinate axis:

$$\mathbf{R} = 5\mathbf{i} + 6\mathbf{j} + 6\mathbf{k} \qquad \text{(c)}$$

From Eq. (c), the magnitude of the resultant is $R = \sqrt{5^2 + 6^2 + 6^2} = 9.85$ kN.

d. The expression for **R** in terms of force projections and unit vectors is a convenient form from which to determine the component that lies in the xy plane. In this case, we simply discard the z component of the resultant. Hence, the component of **R** in the xy plane is

$$\mathbf{R}_{xy} = 5\mathbf{i} + 6\mathbf{j}$$

Remark You can see in this example that expressing a vector in terms of projections and unit vectors has an advantage over describing a vector in terms of magnitude and direction. Other advantages to this technique for writing vectors will become apparent in Chapter 5, when we discuss equilibrium for bodies in three dimensions.

CHECKPOINT 

1. ***True or False:*** Unit vectors are dimensionless.
2. ***True or False:*** Any vector can be expressed as the sum of three unit vectors, each with an appropriate scalar multiplier.
3. ***True or False:*** In the expression $\mathbf{A} = A_x\mathbf{i} + A_y\mathbf{j} + A_z\mathbf{k}$, the scalars A_x, A_y, and A_z are (x, y, z) components of $\mathbf{A}$.
4. ***True or False:*** The (x, y, z) direction cosines of a vector are the cosines of the angles that the vector forms with the x, y, and z axes.
5. ***True or False:*** The (x, y, z) projections of a resultant vector $\mathbf{R}$ of several vectors, $\mathbf{A}$, $\mathbf{B}$, $\mathbf{C}$, . . . , are the sums of the (x, y, z) projections, respectively, of $\mathbf{A}$, $\mathbf{B}$, $\mathbf{C}$, ...
6. ***True or False:*** A unit vector times a positive scalar gives a resultant vector that has the same direction as the unit vector but magnitude greater than unity.

Answers: 1. True; 2. True; 3. False; 4. True; 5. True; 6. False

Chapter Highlights

- Force is a vector quantity. It is a quantitative measure of the interaction between two bodies.
- When two or more concurrent forces act on a body, the net effect of the forces is equivalent to the action of a single force, known as the resultant. The resultant is equal to the vector sum of the forces.
- The weight of a body is the gravitational force exerted on it by the earth. Weight is a vector quantity that is directed toward the center of the earth. In diagrams, it is usually directed downward.
- Newton's third law states that forces occur in pairs; the two forces in such an action-reaction pair are collinear, equal in magnitude, and opposite in direction. The action and the reaction always act on two different bodies; they never act on the same body.

Problems

2.1 *Characteristics of Forces*

2.1 List the magnitude and the direction relative to the x axis of each of the coplanar forces **A**, **B**, **C**, and **D** shown in Fig. P2.1.

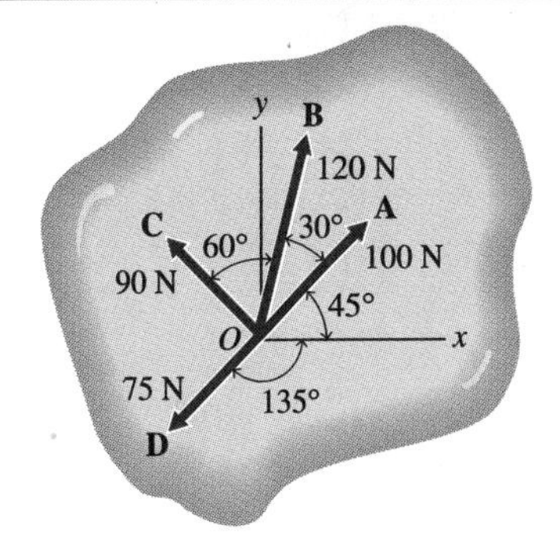

Figure P2.1

2.2 Describe the characteristics of the two forces **A** and **B** such that

a. $\mathbf{A} + \mathbf{B} = \mathbf{C}$ and $A + B = C$

b. $\mathbf{A} + \mathbf{B} = \mathbf{A} - \mathbf{B}$

c. $\mathbf{A} + \mathbf{B} = \mathbf{C}$ and $A^2 + B^2 = C^2$

2.3 Classify the four force systems shown in Fig. P2.3.

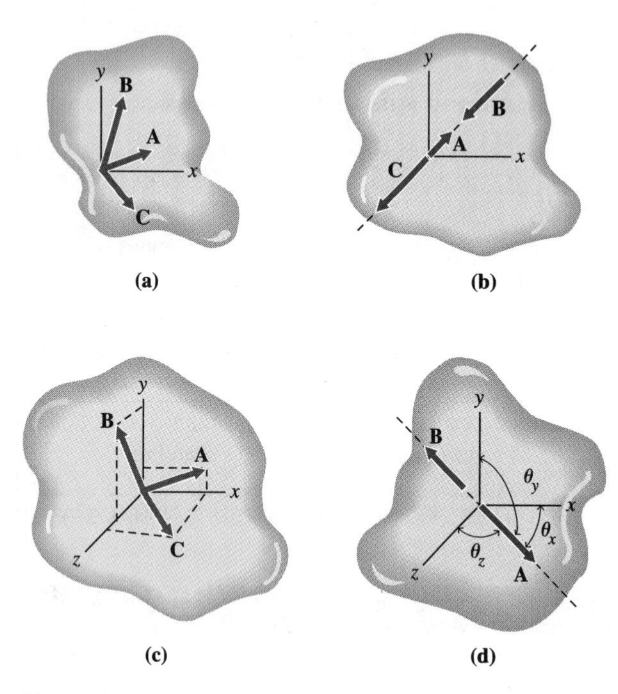

Figure P2.3

2.4 Three forces that act on a frame are shown in Fig. P2.4. For each force, write a formal statement describing

a. its direction line

b. its magnitude

c. its sense

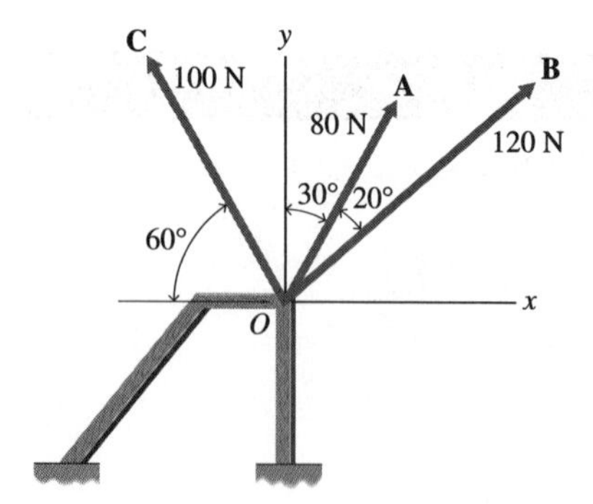

Figure P2.4

2.5 For each of the three forces shown in Fig. P2.5, write a formal description of

a. its direction line

b. its magnitude

c. its sense

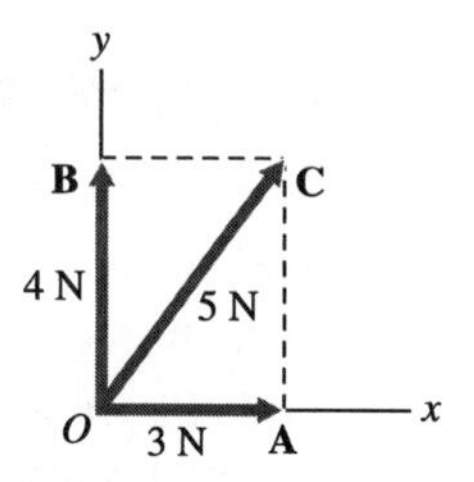

Figure P2.5

2.2 *Newton's Third Law*

2.6 Consider a weight W supported by a rope that hangs from a truss, as shown in Fig. P2.6a. The system consisting of the weight W, isolated from the rope, is shown in Fig. P2.6b. The actions **T** and **W** that act on the system are also shown. In a sketch, show the reactions of the actions **T** and **W** and the systems on which they act; state how they are created.

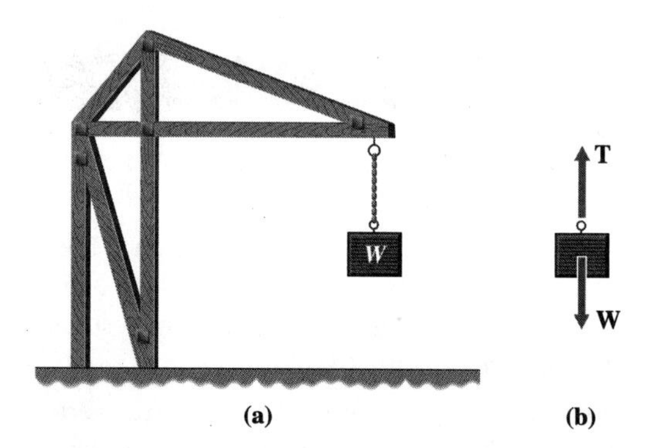

Figure P2.6

2.7 Imagine that this book rests on your desk, which in turn rests on the floor. Describe and illustrate in sketches the *reactions* for the following actions:

a. the action that gravity exerts on the book

b. the actions that the legs of your desk exert on the floor

c. the action that gravity exerts on your desk

d. the action that the book exerts on your desktop

2.8 According to Newton's third law, in a tug of war between two teams A and B, the pull of team A on team B is equal in magnitude and opposite in sense to the pull of team B on team A. If team A can never exert a greater force (pull) on team B than team B can exert on team A, how can either team win the tug of war?

2.9 An apple rests on a table that sits on a floor (see Fig. P2.9).

a. What forces act on the apple? Show them in a sketch, and explain their origins.

b. Identify the reactions of the forces that act on the apple and their origins. Show the reactions in a sketch.

c. What forces act on the table? Show them in a sketch, and explain their origins.

d. Identify the reactions of the forces that act on the table and their origins. Show the reactions in a sketch.

Figure P2.9

2.3 *Force Resultants and Force Components*

2.10 The parallelogram construction of the resultant **R** of two forces **A** and **B** is represented in Fig. P2.10.

a. By trigonometry, derive formulas for the magnitude R of **R** and the angle ϕ in terms of A, B, and θ.

b. Check your answers for the special cases $\theta = 0°$, 90°, 180°.

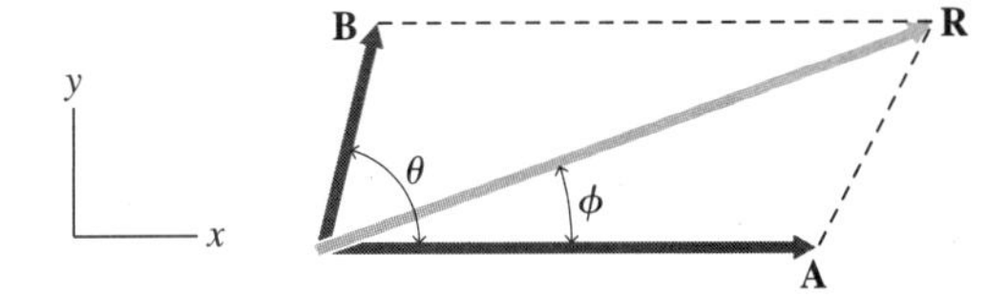

Figure P2.10

2.11 Determine the magnitude and direction of the resultant for each of the four force systems shown in Fig. P2.11 by the parallelogram construction and each of the following:

a. direct measurement of a scaled drawing

b. trigonometry

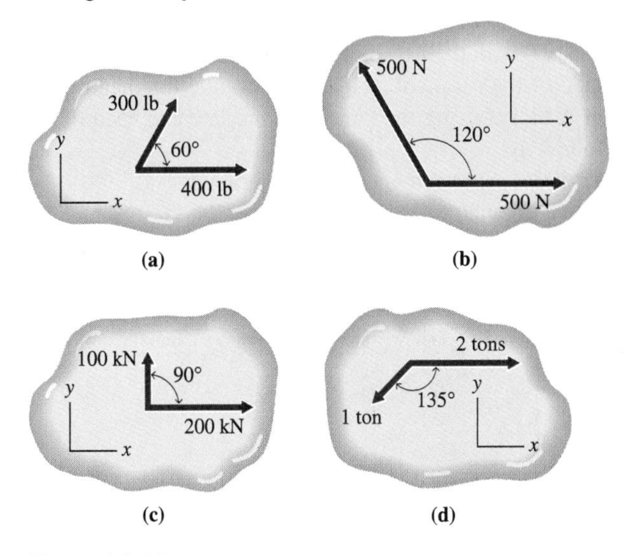

Figure P2.11

2.12 A particle moves around a circular ring. It is attracted toward the center of the ring by a force of 1 kN. A second force of 3 kN, with fixed direction toward the right, acts as shown in Fig. P2.12.

a. Derive a formula for the magnitude of the resultant force that acts on the particle, in terms of the position angle θ.

b. Derive a formula for the angle ϕ that the resultant force forms with the 3 kN force.

c. Evaluate your formulas for the cases $\theta = 0°, 90°, 180°$.

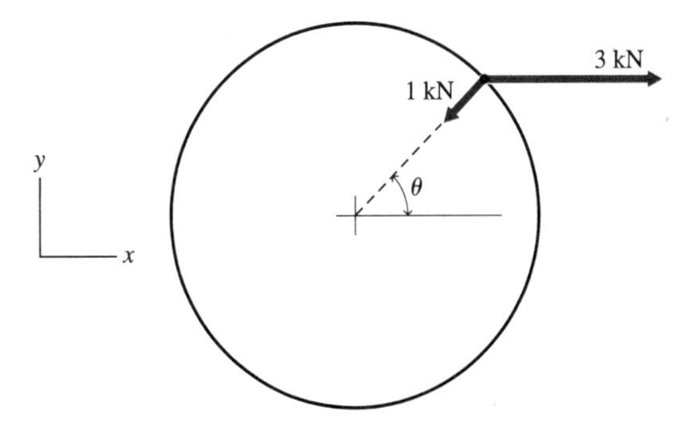

Figure P2.12

2.13 Solve Problem 2.12 for the case where the 3 kN force is directed toward the left.

2.14 Solve Problem 2.12 for the case where the 3 kN force is directed upward.

2.15 Solve Problem 2.12 for the case where the 3 kN force is directed downward.

2.16 Two forces **A** and **B** form the respective angles 45° and 60° with their resultant (see Fig. P2.16). The magnitude of the resultant is 1 kN. Determine the magnitudes of **A** and **B** using the parallelogram law and each of the following:

a. direct measurement of a scaled drawing

b. trigonometry

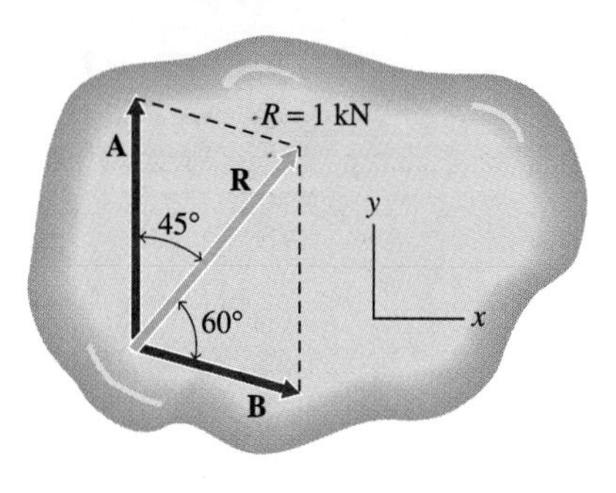

Figure P2.16

2.17 The resultant **R** of the two forces **A** and **B** is directed along the x axis, as shown in Fig. P2.17. The magnitude of **B** is $B = 10$ lb. Determine the magnitudes A and R of **A** and **R**.

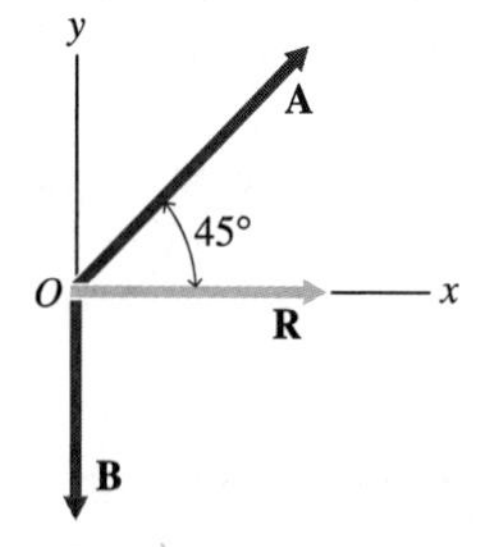

Figure P2.17

2.18 The resultant **R** of two nonzero forces **A** and **B** has a magnitude $R = 1000$ lb (see Fig. P2.18). The magnitude of **A** is also 1000 lb. The resultant forms an angle of 45° with **B**. Determine θ and B by the parallelogram construction and each of the following:

a. direct measurement of a scaled drawing

b. trigonometry

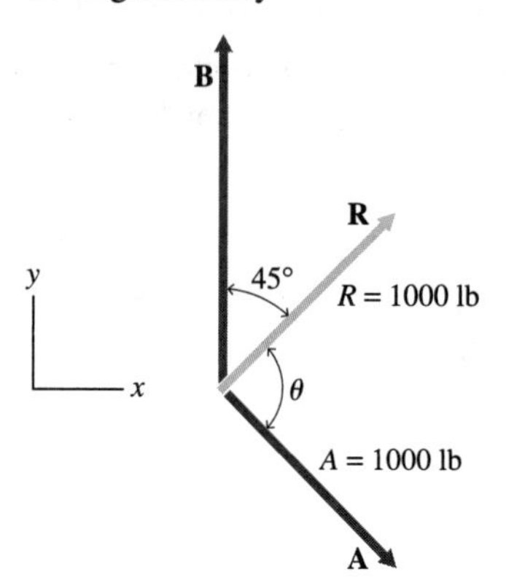

Figure P2.18

2.19 The magnitude of the resultant **R** of two concurrent forces **A** and **B**, which include an angle of 60°, is $R = 30$ kN (see Fig. P2.19). The magnitude of **B** is 10 kN. Determine the magnitude of **A** and the direction of **R** by the parallelogram construction and each of the following:

a. direct measurement of a scaled drawing

b. trigonometry

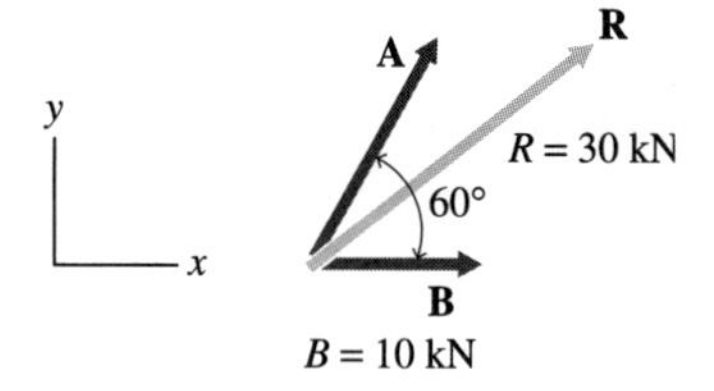

Figure P2.19

2.20 A radio tower is constrained by two cables *AC* and *BC* that connect its top to the ground and lie in a common vertical plane (see Fig. P2.20). The tensions in the cables are $T_A = 10$ kips and $T_B = 15$ kips. Determine the magnitude and the direction of the resultant force **R** that the cables exert on the tower by the parallelogram construction and each of the following:
a. direct measurement of a scaled drawing
b. trigonometry

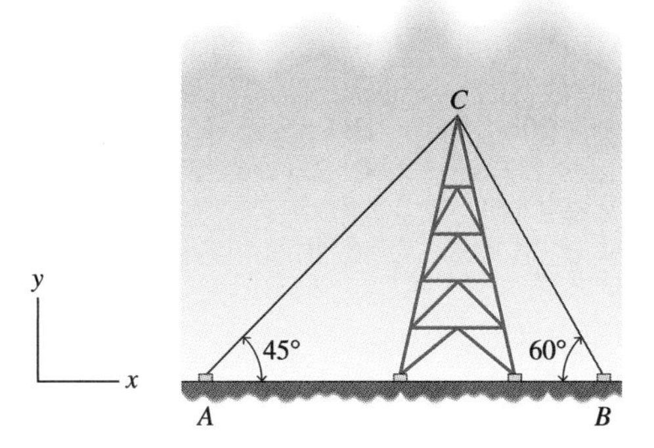

Figure P2.20

2.21 a. Using the parallelogram construction, graphically resolve the force **F** in Fig. P2.21 into two components, one in the direction of line *OA* and the other in the direction of line *OB*.
b. By trigonometry of the parallelogram, compute the magnitudes of the two components in terms of *F*.

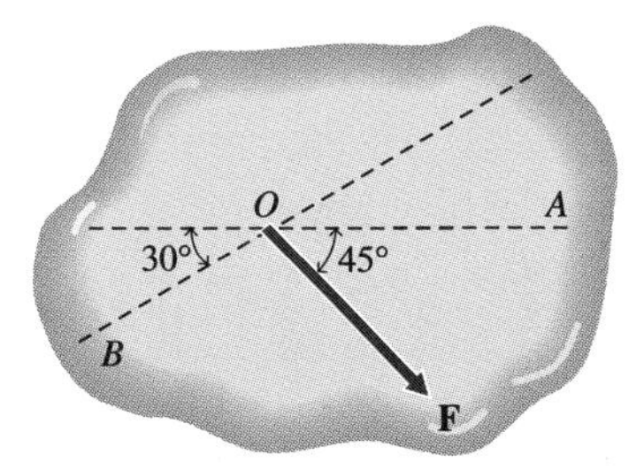

Figure P2.21

2.22 A two-bar boom holds a 1000 lb force as shown in Fig. P2.22. Resolve the 1000 lb force into its components along bars *AP* and *BP* by the parallelogram construction and each of the following:
a. direct measurement of a scaled drawing
b. trigonometry

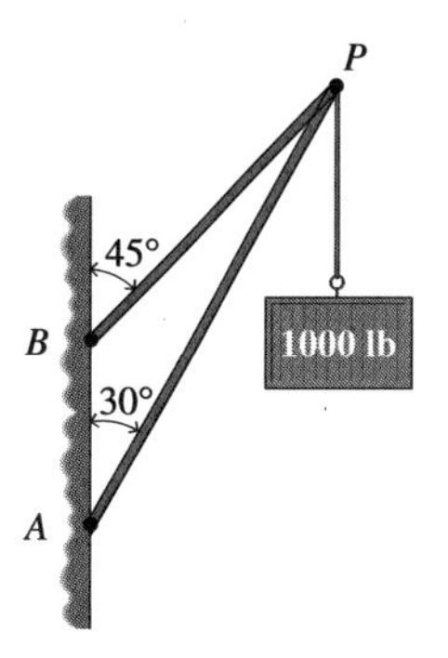

Figure P2.22

2.23 Two forces $\mathbf{F}_1$ and $\mathbf{F}_2$ act on a body at point *P*, as shown in Fig. P2.23. A force **F** is applied at *P* such that the resultant **R** of $\mathbf{F}_1$, $\mathbf{F}_2$, and **F** is zero. Determine the magnitude and the direction of the force **F** by the parallelogram construction and each of the following:
a. direct measurement of a scaled drawing
b. trigonometry

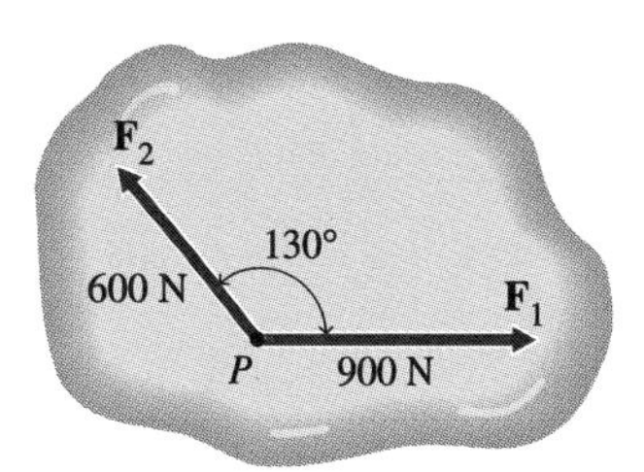

Figure P2.23

2.24 Determine the resultant **R** of the coplanar forces shown in Fig. P2.24 by two means:

a. direct measurement of two force triangles

b. trigonometry

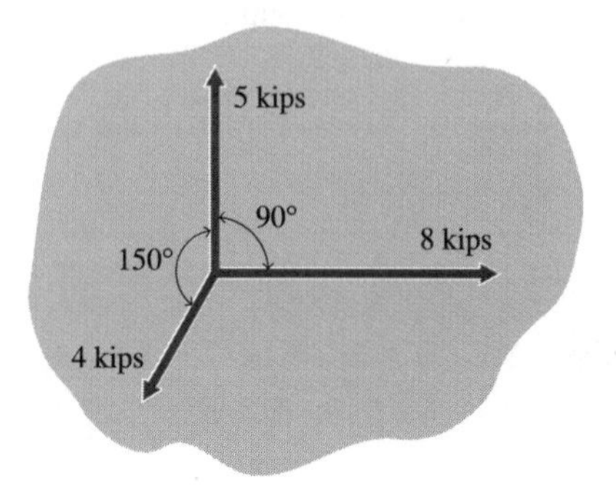

Figure P2.24

2.25 Three coplanar, concurrent forces, each of magnitude F, form angles of 120° with each other (see Fig. P2.25). Show that their resultant is zero.

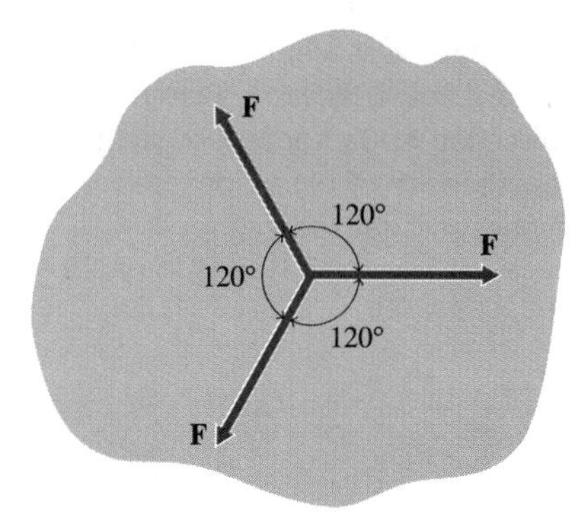

Figure P2.25

2.26 Two 40 lb weights hang by a rope, as shown in Fig. P2.26. Determine the resultant forces exerted on the pegs A and B. Neglect the weight of the rope, and assume that the tension in the segment AB of the rope is 40 lb.

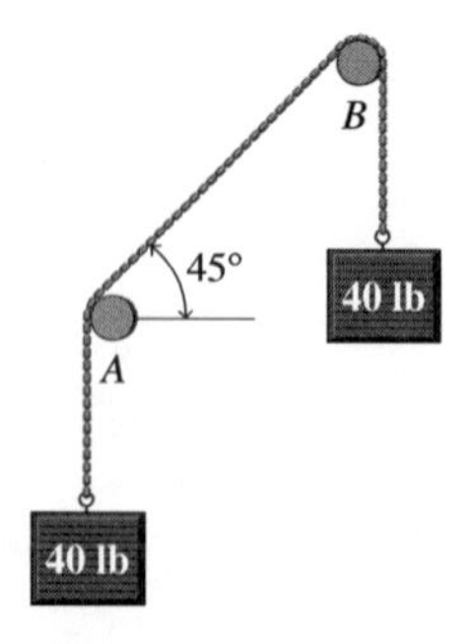

Figure P2.26

The forces listed in Problems 2.27–2.33 are coplanar and concurrent. For each force, the magnitude is given first and then the orientation, with the direction angles measured counterclockwise from the positive x axis. Choose a suitable scale and determine the resultant force for each problem graphically, by the polygon construction.

2.27 **A**: 100 N, 30° **B**: 200 N, 120° **C**: 300 N, 270°

2.28 **A**: 220 N, 45° **B**: 120 N, 60° **C**: 300 N, 210° **D**: 100 N, 330°

2.29 **A**: 4 lb, 20° **B**: 10 lb, 130° **C**: 6 lb, 60° **D**: 8 lb, 225°

2.30 **A**: 0.2 kip, 90° **B**: 1.0 kip, 45° **C**: 2.0 kips, 60° **D**: 2.0 kips, 270° **E**: 0.5 kip, 240°

2.31 **A**: 50 N, 0 rad **B**: 30 N, $\pi/6$ rad **C**: 40 N, $3\pi/4$ rad **D**: 80 N, $7\pi/6$ rad **E**: 100 N, $3\pi/2$ rad

2.32 **A**: 360 lb, $\pi/6$ rad **B**: 100 lb, $\pi/2$ rad **C**: 180 lb, π rad **D**: 200 lb, $4\pi/3$ rad **E**: 140 lb, $11\pi/6$ rad

2.33 **A**: 5 kN, $\pi/4$ rad **B**: 10 kN, $\pi/3$ rad **C**: 2 kN, $3\pi/2$ rad **D**: 8 kN, $3\pi/4$ rad **E**: 6 kN, $7\pi/4$ rad

2.34 A swing is held at 30° from the vertical by the father of a child (see Fig. P2.34). The child weighs 40 lb. The forces that act on the swing are **F** (exerted by the father), **T** (due to the swing rope), and **W** (the weight of the child). The resultant of these forces is zero. Determine the magnitudes of **F** and **T** using the polygon construction and each of the following:

a. direct measurement of a scaled drawing

b. trigonometry

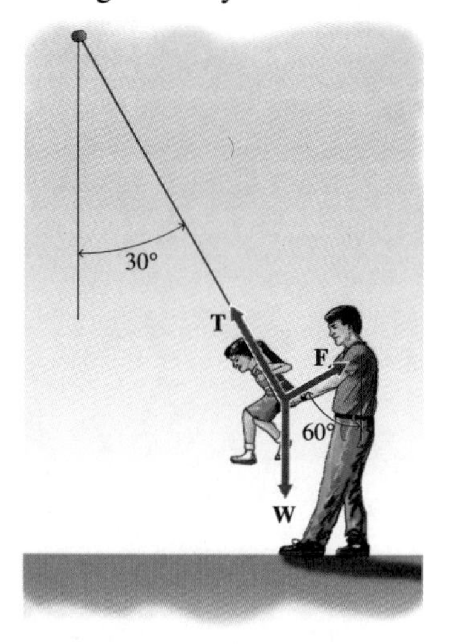

Figure P2.34

2.35 An engine block that weighs 2400 N is suspended by two ropes as shown in Fig. P2.35. The resultant of the tensions $\mathbf{T}_1$ and $\mathbf{T}_2$ and the weight **W** of the block is zero. Determine the tensions in the ropes by the polygon construction and each of the following:

a. direct measurement of a scaled drawing
b. trigonometry

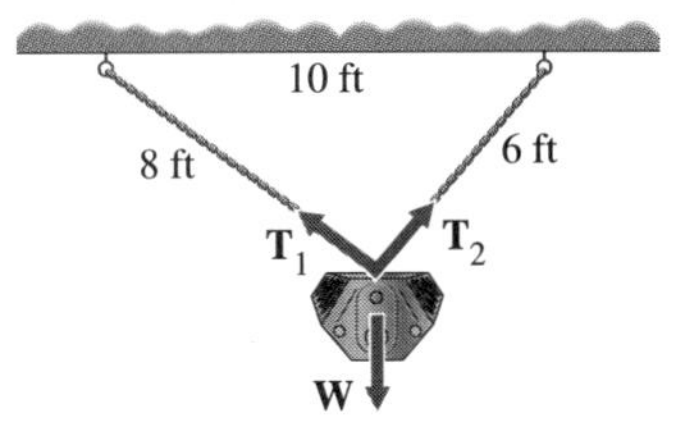

Figure P2.35

2.36 The force **R** is the resultant of four concurrent forces (see Fig. P2.36). Determine the magnitudes of **R** and **F** by the polygon construction and each of the following:

a. direct measurement of a scaled drawing
b. trigonometry

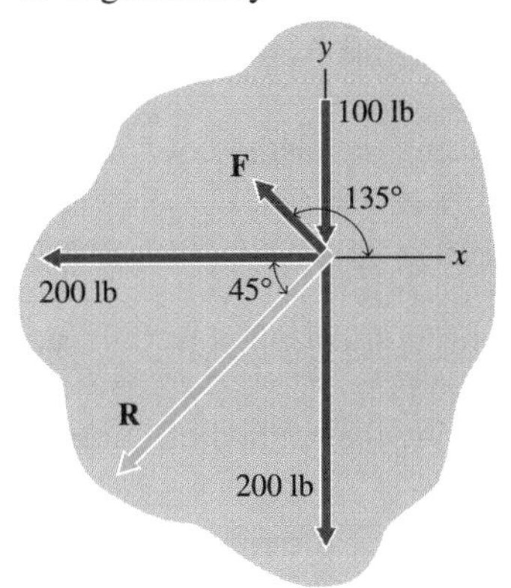

Figure P2.36

2.37 The resultant of the four forces in Fig. P2.37 is zero. By the polygon construction, determine the magnitudes of forces **P** and **Q**.

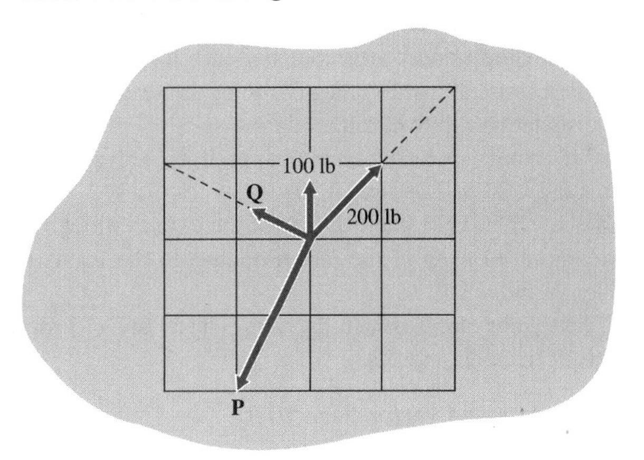

Figure P2.37

2.38 The resultant of three concurrent, coplanar forces **A**, **B**, and **C** is zero; $A = 1000$ N, $B = 1200$ N, and $C = 800$ N. Determine the angles inside the force polygon by two methods:

a. direct measurement of a scaled drawing
b. trigonometry

2.39 Use the polygon construction to determine the force that must be added to the four forces shown in Fig. P2.39 to make the resultant force zero.

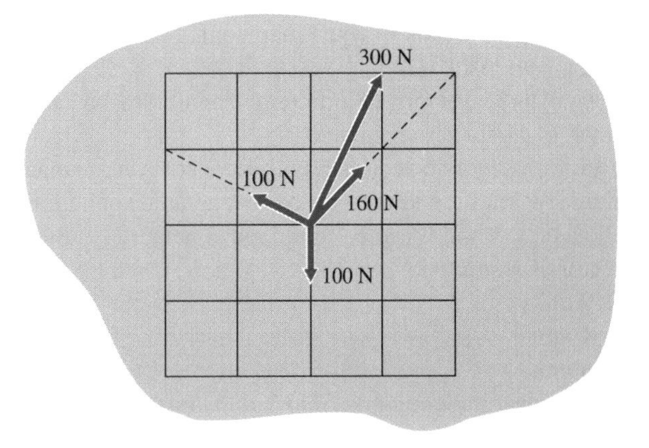

Figure P2.39

2.40 a. Develop a spreadsheet to find the resultant of up to five concurrent, coplanar forces. The given data consist of the magnitude of each force and its orientation (in degrees) relative to the x axis. The spreadsheet should determine the x and y projections of each force, the sums of those projections, the magnitude of the resultant, and its orientation relative to the x axis.
b. Test your spreadsheet by using it to solve Problems 2.27, 2.28, 2.29, and 2.30.

2.41 a. Repeat Problem 2.40a for the case in which the orientation of each of the forces relative to the x axis is given in radians.
b. Test your spreadsheet by using it to solve Problems 2.31, 2.32, and 2.33.

2.4 *Vectors and Vector Quantities*

2.42 Reduce each vector equation to its simplest form.
a. $2(3\mathbf{A} - \mathbf{B}) - \mathbf{C} + 6(\mathbf{B} - 4\mathbf{C}) = 0$
b. $\mathbf{F} = 3[4(\mathbf{A} + 2\mathbf{B} + 3\mathbf{C}) - (\mathbf{A} + 5\mathbf{C}) - 16\mathbf{B}]$
c. $13\mathbf{A} + 7\mathbf{B} - 6(4\mathbf{B} - 3\mathbf{A}) = 0$

2.43 Vector **A** has magnitude 5 units and direction 90° relative to the x axis. What are the magnitudes and directions of the following vectors?
a. $6\mathbf{A}$
b. $-5\mathbf{A}$

2.44 a. Can two vectors of equal magnitude be combined to give a zero resultant?
b. Can two vectors of different magnitudes be combined to give a zero resultant?
c. Can three vectors of equal magnitude be combined to give a zero resultant?
d. Can three vectors of different magnitudes be combined to give a zero resultant?

2.45 Consider two vectors **A** and **B**. Vector **A** has a magnitude of 3 units, and vector **B** has a magnitude of 4 units. Show how **A** and **B** can be combined to obtain a resultant vector of magnitude
a. 7 units
b. 1 unit
c. 5 units
d. any magnitude ranging from 1 to 7 units

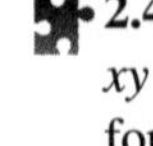
2.46 Three concurrent vectors **A**, **B**, and **C** lie in the xy plane. Each has a magnitude of 100 units, and they form angles of 45°, 210°, and 330°, respectively, with the x axis. Using the polygon construction, determine the magnitude and direction of
a. $\mathbf{A} + \mathbf{B} + \mathbf{C}$
b. $\mathbf{A} + \mathbf{B} - \mathbf{C}$
c. vector **D**, such that $\mathbf{A} + \mathbf{B} + \mathbf{C} + \mathbf{D} = 0$
d. vector **D**, such that $\mathbf{A} + \mathbf{B} - \mathbf{C} - \mathbf{D} = 0$

2.47 Given the coplanar vectors **A**, **B**, and **C** in Fig. P2.47, use the polygon construction to determine the magnitude of
a. $(\mathbf{A} + \mathbf{B}) + \mathbf{C}$
b. $(\mathbf{B} + \mathbf{C}) + \mathbf{A}$
c. $\mathbf{A} + \mathbf{B} + \mathbf{C}$
d. $\mathbf{B} + \mathbf{A} + \mathbf{C}$

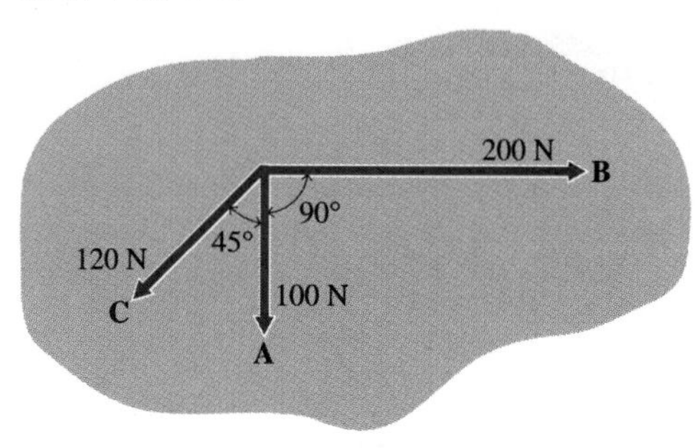

Figure P2.47

2.5 *Rectangular Components of a Vector*

2.48 Solve Problem 2.10 using the method of Cartesian projections.
2.49 Solve Problem 2.11 using the method of Cartesian projections.
2.50 Solve Problem 2.12 using the method of Cartesian projections.
2.51 Solve Problem 2.13 using the method of Cartesian projections.
2.52 Solve Problem 2.14 using the method of Cartesian projections.
2.53 Solve Problem 2.15 using the method of Cartesian projections.
2.54 Solve Problem 2.16 using the method of Cartesian projections.
2.55 Solve Problem 2.17 using the method of Cartesian projections.
2.56 Solve Problem 2.18 using the method of Cartesian projections.
2.57 Solve Problem 2.19 using the method of Cartesian projections.
2.58 Solve Problem 2.20 using the method of Cartesian projections.

Use the method of Cartesian projections in Problems 2.59–2.69.

2.59 Determine the resultant of the four forces shown in Fig. P2.59.

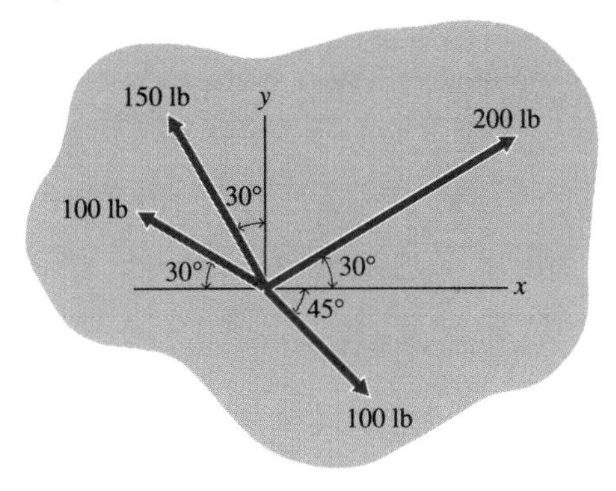

Figure P2.59

2.60 Three concurrent, coplanar forces act on a body at point O (see Fig. P2.60). Determine two additional forces, directed along lines OA and OB, respectively, such that the resultant of the five forces is zero.

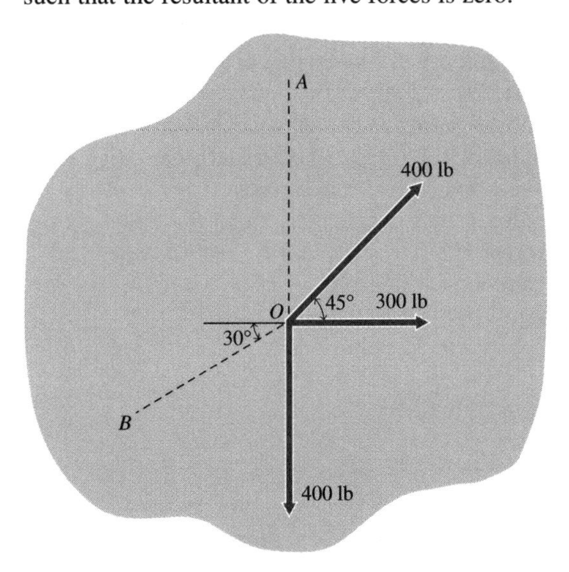

Figure P2.60

2.61 Newton's law of universal gravitation states that the force of attraction between any two particles with masses m and m' separated by a distance r has magnitude $F = kmm'/r^2$, where k is the gravitational constant. The force is directed along the line connecting the two particles. In Fig. P2.61, A and B are fixed particles, each of mass m. Particle C of mass $m/2$ travels on a circular path about mass A. Determine the x and y projections, as a function of angle θ, of the forces of attraction $\mathbf{F}_{CA}$ and $\mathbf{F}_{CB}$ that act on particle C.

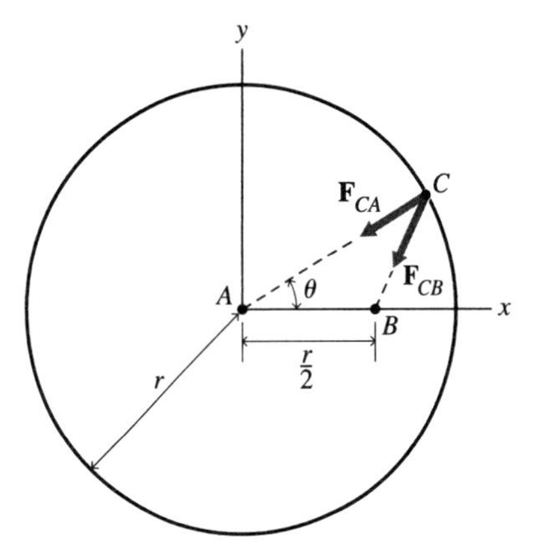

Figure P2.61

2.62 Let **A** and **B** be two vectors in the xy plane whose projections on the x and y axes satisfy the equation $A_xB_x + A_yB_y = 0$. Prove that the vectors **A** and **B** are perpendicular to each other.

2.63 A force **F** forms equal angles θ with the x, y, and z axes (see Fig. P2.63). Determine the angle θ.

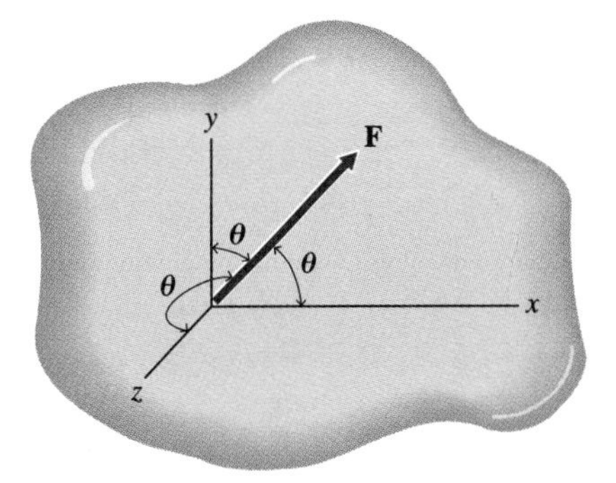

Figure P2.63

2.64 The (x, y, z) projections, in pounds, of two concurrent forces **A** and **B** are $(A_x, A_y, A_z) = (7.0, -2.0, -4.0)$ and $(B_x, B_y, B_z) = (-5.0, 6.0, 8.0)$. Determine the resultant in the form $\mathbf{R} = (R_x, R_y, R_z)$.

2.65 Two vectors, **A** and **B**, have the following projections relative to rectangular *xyz* coordinate axes: $(A_x, A_y, A_z) = (6.0, -1.0, -3.0)$ and $(B_x, B_y, B_z) = (-4.0, 5.0, 7.0)$.

a. Determine the direction cosines of the vector $\mathbf{R} = \mathbf{A} + \mathbf{B}$.

b. Determine the angles that **R** forms with the *x, y,* and *z* axes.

2.66 Each of two cords *AB* and *AC* in Fig. P2.66 is subjected to a tension of 100 N.

a. Determine the resultant force that the cords exert on the ring *A*.

b. Determine the direction angles between this resultant and the *x, y,* and *z* axes.

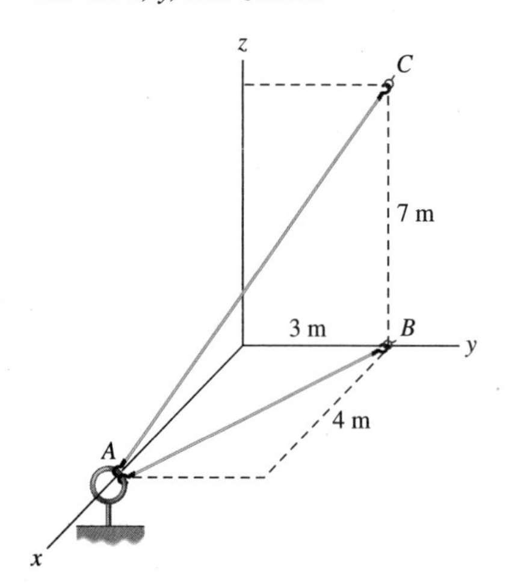

Figure P2.66

2.67 The vector $\mathbf{R} = \mathbf{A} + k\mathbf{B}$ is defined in terms of the two vectors **A** and **B** and the scalar *k*. Vector **A** has a fixed orientation along the *x* axis; vector **B** can be rotated relative to **A**. Figure P2.67 shows the case when $\theta = 90°$ and $k = 1$. Determine the locus of the tip of **R** as *k* varies; that is, let *k* take on the values 0, ±1, ±2, ±3, . . . , and plot the location of point *P* in a sketch.

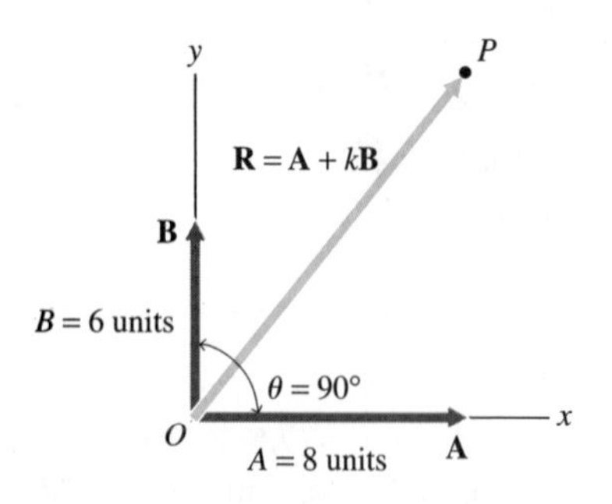

Figure P2.67

2.68 Solve Problem 2.67 for the following cases:

a. $\theta = 60°$

b. $\theta = 45°$

c. $\theta = 0°$

2.69 The three forces shown in Fig. P2.69, plus two additional forces, one in the *u* direction and one in the *v* direction, combine to produce a zero resultant. Determine the magnitudes of the forces in the *u* and *v* directions.

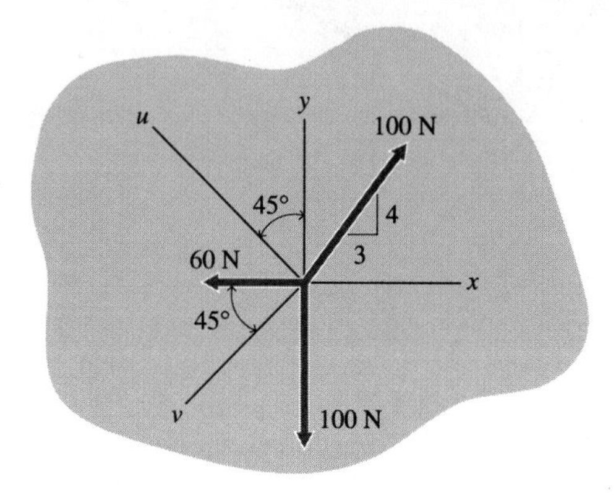

Figure P2.69

2.6 *Unit Vectors*

2.70 The force **F** forms angles of 45° with the *x* and *y* axes (see Fig. P2.70).

a. Resolve **F** into *x* and *y* components, and express **F** in the form $\mathbf{F} = F_x\mathbf{i} + F_y\mathbf{j}$, where **i** and **j** are unit vectors along the *x* and *y* axes, respectively.

b. Resolve **F** into components along the *x* and *u* axes, and express **F** in the form $\mathbf{F} = F_x\mathbf{i} + F_u\mathbf{n}$, where **n** is a unit vector along the *u* axis.

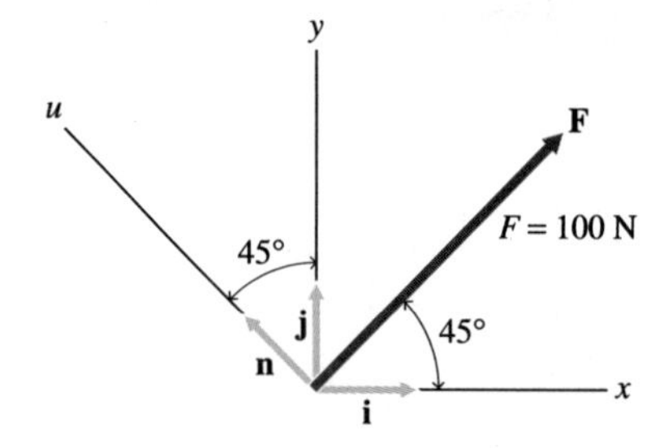

Figure P2.70

2.71 A force **F**, in pounds, is given as $\mathbf{F} = 2\mathbf{i} - 3\mathbf{j} + 6\mathbf{k}$, where **i**, **j**, and **k** are unit vectors along rectangular *xyz* coordinate axes (Fig. P2.71). Determine the following:
a. the magnitude of **F**
b. the angles that **F** forms with the *x*, *y*, and *z* axes

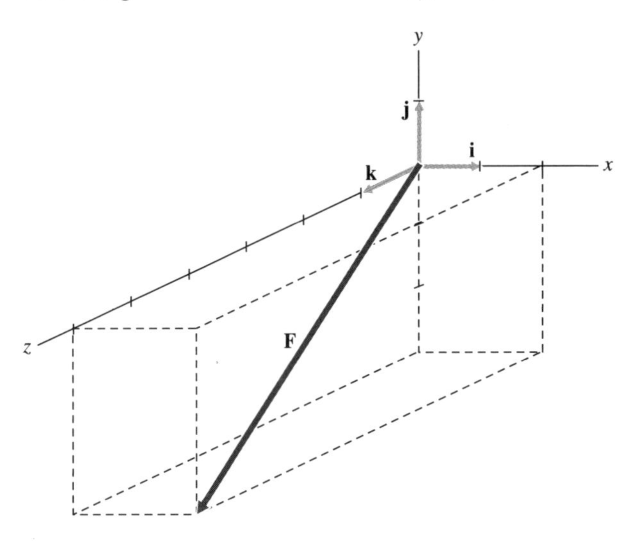

Figure P2.71

2.72 Let **A** and **B** be vectors whose (x, y) projections are $(A_x, A_y) = (10, 20)$ and $(B_x, B_y) = (8, -6)$. Express the vectors $\mathbf{A} + \mathbf{B}$, $\mathbf{A} - \mathbf{B}$, $\mathbf{A} + 3\mathbf{B}$, and $3\mathbf{A} - \mathbf{B}$ in the form $P_x\mathbf{i} + P_y\mathbf{j}$, where **i** and **j** are unit vectors along the *x* and *y* coordinate axes, respectively.

2.73 Determine the (x, y, z) projections of the following vectors:
a. $\mathbf{A} = (3\mathbf{i} - 2\mathbf{j} + \mathbf{k}) - (-4\mathbf{i} + 6\mathbf{j} - 2\mathbf{k})$
b. $\mathbf{B} = (5\mathbf{i} + 7\mathbf{j} + 2\mathbf{k}) - [(3\mathbf{i} - 2\mathbf{j} + \mathbf{k}) - (-4\mathbf{i} + 3\mathbf{j} + 6\mathbf{k})]$

The forces listed in Problems 2.74–2.79 are noncoplanar and concurrent. For each force, the magnitude is given first and then the direction angles θ_x, θ_y, and θ_z, with the direction angles measured from the respective axes. Use unit vectors to determine the resultant force for each problem.

2.74 A: 100 N; 0, 90°, 90°
B: 200 N; 90°, 90°, 180°
C: 160 N; 90°, 30°, 60°

2.75 A: 5.0 kips; 90°, 180°, 90°
B: 4.0 kips; 120°, 45°, 120°
C: 3.0 kips; 54.74°, 54.74°, 54.74°

2.76 A: 6.0 kN; 45°, 90°, 135°
B: 4.0 kN; 90°, 45°, 45°
C: 8.0 kN; 30°, 75°, 64.67°
D: 12.0 kN; 135°, 135°, 90°

2.77 A: 4 kN; $\pi/4$ rad, $\pi/2$ rad, $\pi/4$ rad
B: 6 kN; $3\pi/4$ rad, $\pi/3$ rad, $\pi/3$ rad
C: 3 kN; $\pi/2$ rad, $\pi/2$ rad, π rad

2.78 A: 16 lb; $\pi/3$ rad, $7\pi/6$ rad, $3\pi/2$ rad
B: 20 lb; $\pi/2$ rad, $3\pi/4$ rad, $3\pi/4$ rad
C: 10 lb; $2\pi/3$ rad, $3\pi/4$ rad, $3\pi/4$ rad

2.79 A: 350 N; $\pi/2$ rad, $\pi/3$ rad, $\pi/6$ rad
B: 200 N; 0, $\pi/2$ rad, $\pi/2$ rad
C: 400 N; $\pi/3$ rad, $\pi/4$ rad, $5\pi/6$ rad
D: 300 N; $3\pi/4$ rad, $\pi/3$ rad, $2\pi/3$ rad

2.80 a. Develop a spreadsheet to find the resultant of up to five concurrent, noncoplanar forces. The given data consist of the magnitude of each force and its direction angles (in degrees). The spreadsheet should determine the direction cosines of each force, the projections of each force on the *x*, *y*, and *z* axes, the sum of the projections, the magnitude of the resultant, and its direction angles.
b. Test your spreadsheet by using it to solve Problems 2.74, 2.75, and 2.76.

2.81 a. Repeat Problem 2.80a for the case in which the direction angles of each force are given in radians.
b. Test your spreadsheet by using it to solve Problems 2.77, 2.78, and 2.79.

REVIEW QUESTIONS

2.82 What are the characteristics of a force?
2.83 What is a projection of a vector? What are rectangular Cartesian projections of a vector?
2.84 What is the direction line of a force? The line of action of a force? The sense of a force?
2.85 What is meant by *direction of a force*?
2.86 What is a concentrated force, a point force, a distributed force, and a body force?
2.87 What are concurrent forces, coplanar forces, and collinear forces?
2.88 Are concurrent forces always coplanar? Are collinear forces always coplanar?
2.89 What is a reaction of a force? What is Newton's law of reaction? What does it mean?
2.90 State Newton's third law. What does it mean?
2.91 What force is the reaction to the weight of a falling body in a vacuum?
2.92 What is an external force? An internal force? Give examples of each.

2.93 What is a resultant force? What is meant by *composition of forces*, *component of a force*, and *resolution of a force*? Can a force be resolved into nonperpendicular components? If so, how?

2.94 What is the difference between a force projection and a force component?

2.95 What is the parallelogram law? What is the polygon law? How are the parallelogram and polygon constructions related?

2.96 Explain how the parallelogram law can be used to obtain the resultant of more than two concurrent forces.

2.97 What is the triangle construction of concurrent forces? How does it relate to the polygon construction?

2.98 What is a vector? A vector quantity? How are vectors added? Subtracted?

2.99 What are the characteristics of a vector? What is a fixed vector, a unit vector, and a free vector?

2.100 What is meant by the term *scalar*? What are the characteristics of a scalar?

2.101 How are two vectors added? How does vector addition differ from scalar addition?

2.102 What is meant by *a nonvector quantity*? List two physical quantities that can be represented by arrows but are not vector quantities.

2.103 What is meant by *the product of a vector and a scalar*? How does this product differ from the original vector? Explain all possibilities.

2.104 What is an axis? What is meant by *positive sense of an axis* and *negative sense of an axis*?

2.105 What are axes? What are rectangular Cartesian coordinate axes? What is a rectangular Cartesian coordinate system?

2.106 Define and distinguish between right-handed and left-handed Cartesian coordinate systems in space and in a plane.

2.107 How is the angle between two nonintersecting axes in space defined?

2.108 What is meant by *direction of a vector*, *direction angle*, and *direction cosines*?

2.109 What is the projection of a vector on an axis? Write a formula for it. How is a positive projection defined? A negative projection?

2.110 What are the Cartesian components of a vector? How do they differ from the projections of the vector on Cartesian coordinate axes? Explain the distinction with the aid of unit vectors **i**, **j**, and **k**.

2.111 What is the commutative law of vector addition? Explain its significance with reference to the polygon construction of a vector sum.

2.112 What is meant by the associative law of vector addition?

2.113 How are vectors added using rectangular Cartesian projections? Using unit vectors?

Chapter 3 EQUILIBRIUM OF A PARTICLE

A Look Forward

IN THIS CHAPTER, the conditions under which a particle is in equilibrium are developed. The concept of equilibrium is interpreted from Newton's first law of motion and is used to determine the forces acting on bodies that are at rest. The critical need for proper free-body diagrams is discussed, and examples of their use are illustrated. Particle equilibrium is used to solve many practical problems in mechanics, including those involving booms, systems of cables, and rigid bodies subjected to concurrent forces. With some extension, it can even be used to form the analytical basis for analysis of trusses (Chapter 6).

After you have studied this chapter, you should have an understanding of the concept of particle equilibrium and how to interpret equilibrium in different situations. You should know when and how to apply the concepts of particle equilibrium to a particular situation, and you should be able to solve for the forces in statically determinate systems to which particle equilibrium applies. You should also be able to apply the concept of particle equilibrium to a rigid body subjected to concurrent forces.

To find the forces in the sections of rope, you could model the climber's fall protection anchors as particles, lacking significant mass. In some cases, the climber could also be modeled as a particle. Obviously, the mass of the climber would be significant.

Survey Questions

- What conditions are necessary for a particle to be in equilibrium?
- Do the two interpretations of particle equilibrium represent the same concept?
- What is a free-body diagram?
- Can a rigid body be regarded as a particle?
- What is a two-force member?

3.1 CONCEPT OF PARTICLE EQUILIBRIUM

After studying this section, you should be able to:

- Explain the two interpretations of particle equilibrium.
- Write the equations of equilibrium for a particle in terms of either force projections or force components.

Key Concept

From Newton's first law of motion, a particle is in equilibrium if no net force acts on it.

The concept of equilibrium of a particle follows from Newton's first law (introduced in Sec. 1.2):

> In the absence of applied forces, a particle originally at rest or moving with constant speed in a straight line will remain at rest or continue to move with constant speed in a straight line.

Thus, Newton's first law makes no distinction between a particle that *is at rest* and one that *moves with constant velocity.*[1] A particle that is at rest or that moves with constant velocity is said to be in *equilibrium.* Newton's first law implies that either there is no force acting on the particle or the forces that are acting on it have a zero resultant.

equilibrium

particle

In a limited sense, you may think of a *particle* as a point of concentrated mass—that is, as a body of infinitesimal size. In many situations, however, a body of finite size may be treated as a particle; see Sec. 3.7 and the comment following Theorem 3.5. Therefore, the forces that act on a particle are concurrent (see Sec. 2.1); they act at the point of concentrated mass.[2] The concept of equilibrium of a particle can be interpreted in two ways. But it is important to remember that these interpretations are simply different representations of Newton's first law.

> *Interpretation 1: Forces acting on a particle have a zero resultant.* If the resultant of several forces that act on a particle is zero, the force polygon closes (see Sec. 2.3). Consequently, the forces produce no change in the motion of the particle, and the particle is said to be in equilibrium under the action of these forces.
>
> *Interpretation 2: Forces acting on a particle produce no change in the motion of the particle.* Conversely, if several forces, including the weight of the particle, act on a particle but produce no change in motion of the particle, their resultant is zero. Again, the particle is said to be in equilibrium under the action of these forces.

1. The question of whether a body is at rest or moving with constant velocity depends on the reference frame in which the body's motion is observed. In other words, Newton's first law is only valid relative to a certain reference frame. Engineers generally refer to a reference frame in which Newton's laws are valid as a *Newtonian frame;* physicists usually refer to such a frame as an *inertial frame.* We shall examine the question of reference frames when we study kinematics of a particle (Chapter 13). The present discussion assumes that the reference frame is one in which Newton's first law holds.

2. In some situations, you may represent a particle to be a point without mass and, therefore, without weight. Also, if a particle is in outer space, far removed from other bodies, its weight is zero, even though it has mass.

The study of statics treats bodies that are motionless. Therefore, for statics problems, the above interpretations are summarized by the following theorem.

Theorem 3.1

A particle that is initially motionless is in equilibrium if, and only if, the vector sum of the forces that act on it is zero.

In Chapter 14, we will see that a particle moving with a constant velocity continues to move with a constant velocity if, and only if, the vector sum of the forces that act on it is zero. This is a more general statement of particle equilibrium; a motionless particle is simply a special case for which the velocity is zero.

In general, with the word "particle" replaced by "body," the above statements hold true for a body subjected to concurrent forces, regardless of its size, shape, and physical properties. In particular, these statements hold true for a rigid body (Sec. 3.4) subjected to concurrent forces, including the weight of the body (Sec. 3.7); that is, a rigid body subjected to concurrent forces may be treated as a particle with regard to its equilibrium.

Usually, the objective in the study of statics is to determine certain unknown forces that act on or within a system that is in equilibrium. In this chapter, we study systems composed of one or more particles to which either of the two interpretations of particle equilibrium can be applied. We will *assume* that the system is in equilibrium and then use the equations of equilibrium to determine the unknown quantities (forces or geometric quantities) that ensure equilibrium.

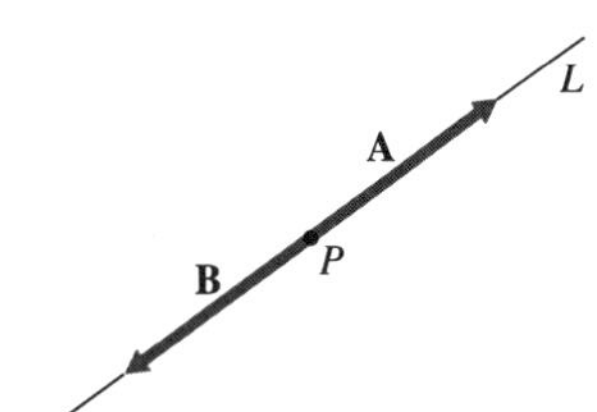

Figure 3.1 Particle subjected to collinear forces.

As an example, consider a particle P that is in equilibrium under the action of two forces **A** and **B**. Let the forces have the same magnitude ($A = B$) and the same line of action L but opposite senses (Fig. 3.1). The force polygon degenerates into two equal and oppositely directed forces (Fig. 3.2). Their resultant is therefore zero.

If a particle is in equilibrium under the action of more than two forces, the force polygon must again close; that is, the *resultant of the forces is zero.* This is true for noncoplanar as well as coplanar force systems. Since the resultant force is zero, the sum of the projections of the forces on any line is zero. Thus, the sums of the projections on the rectangular Cartesian xyz axes are zero. For example, for coplanar forces (discussed in Secs. 2.1 and 2.5) that are referred to rectangular coordinate xy axes (Fig. 3.3), the necessary and sufficient conditions for equilibrium of a particle acted on by those forces are expressed by the equations

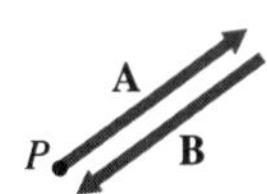

Figure 3.2 Force parallelogram for collinear forces.

$$\sum F_x = 0 \qquad \sum F_y = 0 \tag{3.1}$$

where F_x and F_y are the x and y projections, respectively, of the forces that act on the particle. The Greek letter Σ (capital sigma) denotes the summation, or sum, of the projections.

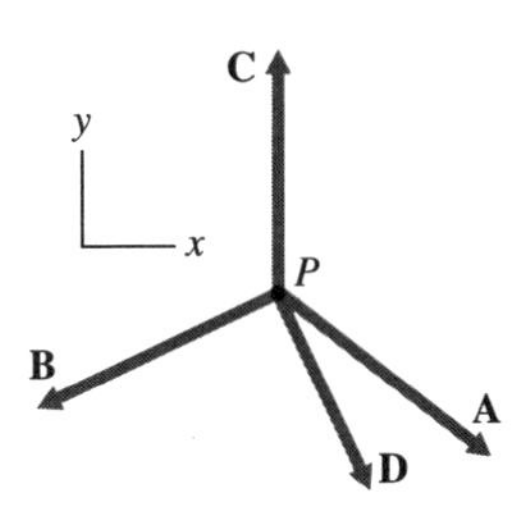

Figure 3.3 Particle subjected to coplanar forces.

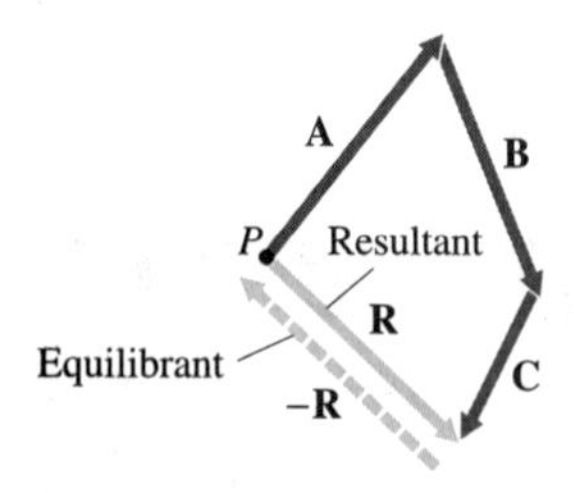

Figure 3.4 The equilibrant $-\mathbf{R}$ of the unbalanced forces **A**, **B**, and **C** is collinear with their resultant **R** and equal in magnitude but opposite in sense to it.

For noncoplanar forces (Sec. 2.5), the necessary and sufficient conditions for equilibrium of the particle are

$$\sum F_x = 0 \qquad \sum F_y = 0 \qquad \sum F_z = 0 \tag{3.2}$$

where F_x, F_y, and F_z are the *x, y,* and *z* projections, respectively, of the forces that act on the particle. Equations (3.1) and (3.2) may be written in vector form as $\Sigma \mathbf{F} = 0$. For coplanar forces, we can write

$$\sum \mathbf{F} = \left(\sum F_x\right)\mathbf{i} + \left(\sum F_y\right)\mathbf{j} = 0 \tag{3.3}$$

since $\Sigma F_x = 0$ and $\Sigma F_y = 0$. Likewise, for noncoplanar forces,

$$\sum \mathbf{F} = \left(\sum F_x\right)\mathbf{i} + \left(\sum F_y\right)\mathbf{j} + \left(\sum F_z\right)\mathbf{k} = 0 \tag{3.4}$$

where **i**, **j**, and **k** denote unit vectors in the positive *x, y,* and *z* directions.

equilibrant

If several unbalanced forces act on a particle, the force that is required to establish equilibrium of the particle is called the *equilibrant* of the unbalanced forces. Thus, the equilibrant is a force that is equal in magnitude but *opposite in direction* to the resultant of the other forces. In other words, the equilibrant is the force $-\mathbf{R}$ that is required to close the force polygon of the unbalanced forces (see Fig. 3.4).

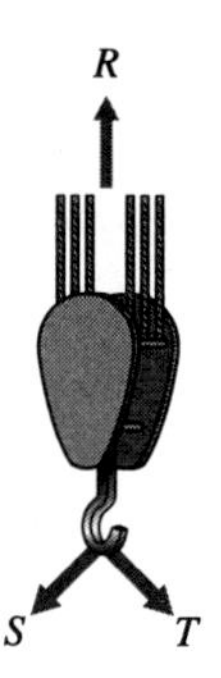

A timber beam is held aloft by a crane. Since the system is in equilibrium, all portions of it are also in equilibrium. So, the vector sum of forces on the pulley is zero.

3.2 FREE-BODY DIAGRAMS

After studying this section, you should be able to:

- Draw an accurate free-body diagram of a particle.
- Solve the equations of equilibrium for a particle to determine the unknown quantities, with the aid of a free-body diagram.
- Classify a system as statically determinate or statically indeterminate.

In Chapter 2, we discussed how to obtain the resultant of a system of concurrent forces that act at a point (or on a particle). In some cases, a part of a physical system may be modeled as a particle that has mass or as a particle that has no mass. Then, Newton's laws apply. Newton's first law applies to the equilibrium of a particle. Therefore, it applies to each individual particle of a system that is in equilibrium. In a particular problem, you must decide which part (or particle) of the system you wish to examine. In some cases, a complete solution to the problem will require examination of two or more particles. The choice of a particle is the essential first step in the process of solving for the equilibrium conditions of the system. Once you have chosen a particle in the system, you must identify all the forces that act on it. This identification is not a trivial matter, since it is easy to get confused between the forces that act on the particle and the forces that the particle exerts on some other part of the system—that is, between actions and reactions.

Key Concept An accurate free-body diagram is an essential component of the problem-solving process in mechanics.

free-body diagram

Unless you follow a systematic approach, you might easily overlook some forces that act on a particle. To help to identify the forces that act on a particle, you should construct a *free-body diagram.* A free-body diagram is a sketch of the particle that shows the particle (by itself, *free* of the other parts of the system) and all the forces applied to it—that is, all the forces exerted on it by other parts of the system. You must be careful to show *all* the forces acting on the particle. You must also be careful *not* to show any of the forces that the particle exerts on other parts of the system. Specifically, two forces in an action-reaction relationship must *never* be shown in the same free-body diagram, since they act on different bodies.

Construction of a free-body diagram is considered by many engineers to be one of the most important steps, if not the most important step, in the problem-solving process in mechanics.

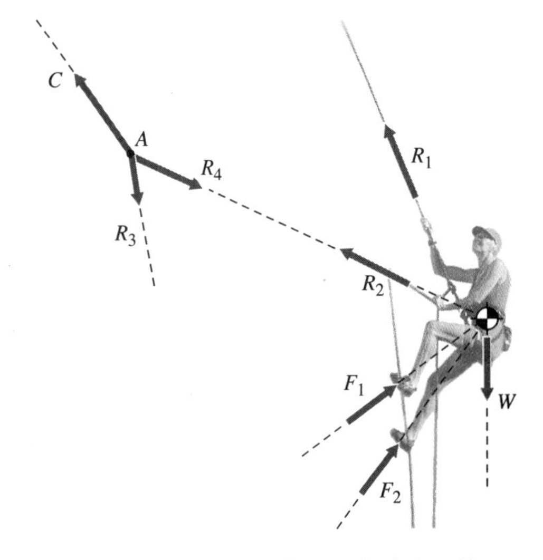

A free-body diagram is essential when solving equilibrium problems. The diagram includes all forces that act on the particle in question. Here the climber is modeled as a particle with mass concentrated at her center of gravity. The fall protection anchor *A* is a separate particle, with its own free-body diagram.

There are important reasons for using clear, accurate free-body diagrams in the problem-solving process:

- They help you reduce real systems to accurate, idealized models that may be analyzed mathematically.
- They help you see and understand separate parts of a problem.
- They help you identify known and unknown quantities in a problem.
- They help you efficiently outline your approach to the solution of a problem.
- They help you make accurate, simplifying approximations.
- They help you progress step-by-step through a solution.
- They make it easier for you to present and explain your solution to others, particularly your teacher or employer.

As was noted in Sec. 3.1, for equilibrium of a particle, the vector sum of all the forces that act on the particle must be zero—that is, $\sum \mathbf{F} = 0$. Therefore, it is essential that you identify all the forces that act on the particle (don't forget its weight). This identification is greatly facilitated by the use of a free-body diagram of the particle. For example, consider a weightless particle P in equilibrium, subjected to three concurrent forces **A**, **B**, and **C** of equal magnitude F (Fig. 3.5). The particle is represented by a large dot, and the concurrent forces are shown with their proper magnitudes and directions. The condition of equilibrium of the particle is determined graphically by the fact that the force polygon is closed (Fig. 3.6).

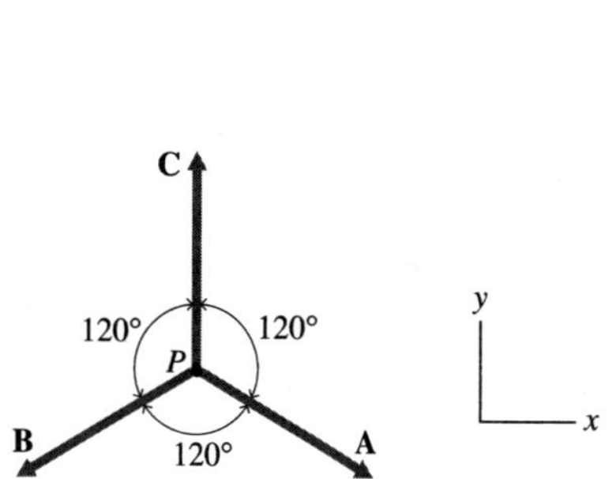

Figure 3.5 Three equal-magnitude, concurrent forces in equilibrium.

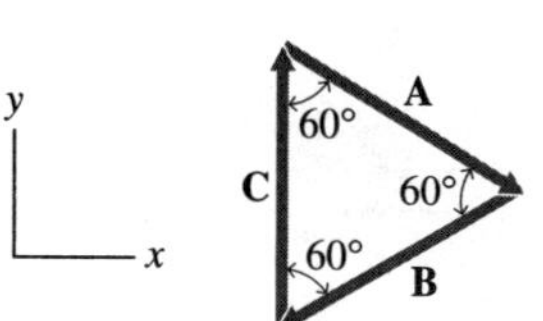

Figure 3.6 Polygon of the three forces in Fig. 3.5.

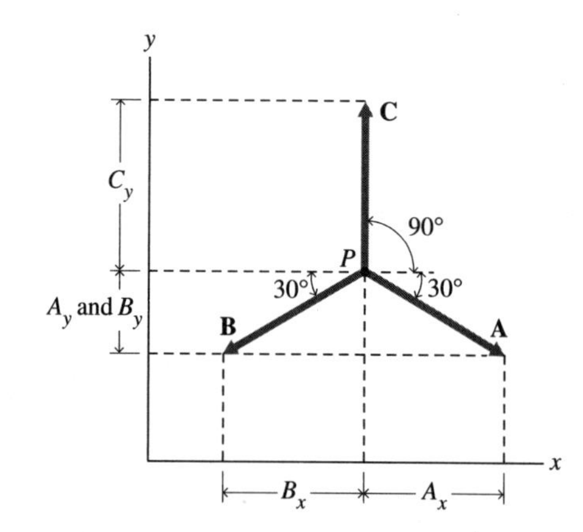

Figure 3.7 Projections of the three forces in Fig. 3.5 on xy axes.

Alternatively, the equilibrium of the particle can be shown by summing the projections of the forces on the x and y axes, as shown in Fig. 3.7. Thus, for equilibrium of the particle P, where F denotes the magnitude of each of the forces **A**, **B**, and **C**, we have

$$\begin{aligned}\sum F_x &= A_x + B_x + C_x \\ &= F\cos 30^\circ - F\cos 30^\circ + F\cos 90^\circ \\ &= 0.866F - 0.866F + 0 = 0\end{aligned}$$

$$\begin{aligned}\sum F_y &= A_y + B_y + C_y \\ &= -F\sin 30^\circ - F\sin 30^\circ + F\sin 90^\circ \\ &= -0.5F - 0.5F + F = 0\end{aligned}$$

Proper construction of the free-body diagram for a particle depends on Newton's third law, since the forces (actions) that act on the particle are produced by other bodies in the system. The reactions are the forces that the particle exerts on these other bodies.

It is essential that you correctly decide what particle, or particles, of a system you must consider when solving a particular problem. For some problems, the selection of a particle may be obvious; for other problems, it may not be so obvious. In any case, once you have selected a body, you must identify all the forces *acting on it.* For equilibrium, the vector sum of these forces is zero.

Key Concept

A statically determinate system is one that can be analyzed using only the equations of equilibrium.

In some cases, the magnitudes and the directions of all the forces that act on a particle are not known. In those cases, the equilibrium equation $\Sigma \mathbf{F} = 0$ may be used to help determine the unknowns. Depending on how many unknowns there are in a problem, *free-body diagrams for more than one particle of the system may be required to obtain a sufficient number of equations to solve for the unknowns.* When the number of *independent* equations (each of which contains one or more of the unknowns and that together include all the unknowns) obtained from the equilibrium conditions equals the number of unknowns, the system is said to be *statically determinate.* Those equations can be solved for the unknowns; that is, the equations of equilibrium are sufficient to analyze a statically determinate system. A system that is *statically indeterminate* has more unknowns than independent equilibrium equations. The equations cannot be solved for all the unknowns. Analysis of such systems requires application of additional principles from mechanics. These facts and concepts will become more apparent in the examples that follow and in subsequent sections.

statically determinate

statically indeterminate

PROBLEM-SOLVING TECHNIQUE

Equilibrium of a Particle—Statically Determinate Problems

To solve a problem involving equilibrium of a particle:

1. Draw an accurate sketch of the mechanical system or structure described in the problem, showing all important dimensions, including angles.
2. Select the particle of interest. If several particles must be considered, the conditions of equilibrium are applied to each particle individually. Recall that, for each particle, the equations of equilibrium provide for determination of only two unknowns in two-dimensional (coplanar) problems [Eqs. (3.1)] and only three un-

(continued)

PROBLEM-SOLVING TECHNIQUE *(CONT.)*

knowns in three-dimensional (noncoplanar) problems [Eqs. (3.2)]. If there are several particles, it may be convenient to write equations for the particles in a sequence that allows you to evaluate one or more of the unknowns immediately, rather than collecting all the equations for all the particles and solving them simultaneously. For relatively simple problems, this method is often convenient. For more complex problems, the equations may have to be solved simultaneously.

3. Select appropriate reference axes, and draw a free-body diagram of the particle. That is, make a sketch of the particle and identify the known and unknown forces. Label the known forces with their magnitudes and senses. Label the unknown forces with a vector symbol such as $\mathbf{F}$, $\vec{F}$, or $\underline{F}$. If a force's magnitude is unknown but its direction is known, use a letter such as F to denote its magnitude. If both the magnitude and direction of a force are unknown, show the force as an arrow of magnitude F with an assumed line of action and sense relative to the reference axes. Recall that the magnitude F of a force is assumed to be a positive number. Then, if F is negative in the solution, the force really acts in the sense opposite to that assumed.
4. Use the equation $\Sigma\mathbf{F} = 0$ with the polygon construction method to define the equilibrium conditions of the particle. Alternatively, with appropriate axes, use $\Sigma F_x = 0$ and $\Sigma F_y = 0$ with the force projection method to define equilibrium for coplanar forces. For noncoplanar forces, use the equations $\Sigma F_x = 0$, $\Sigma F_y = 0$, and $\Sigma F_z = 0$. For each particle, these equations can be used to solve for two unknowns in a coplanar problem and three unknowns in a noncoplanar problem.
5. With the force projection method, the proper selection of a coordinate system, showing the origin and positive directions of the axes, often simplifies the solution. For example, if a problem involves a body resting on an inclined plane, it may be simplest to make two of the coordinate axes parallel and perpendicular to the plane. Be consistent with the signs of projections. Once you have defined positive directions for the axes, the projection of a force on an axis is positive if the force has a component that points in the positive sense of the axis; otherwise, the projection is negative (see Sec. 2.5 and Fig. 2.22). If a force is perpendicular to an axis, its projection on that axis is zero (refer to Fig. 3.7, in which force $\mathbf{C}$ has projection $C_x = 0$). When a force has been replaced by its projections, cross it out so that you don't count it twice.
6. If there are more unknowns than equations (from Step 4), you will need to consider other bodies in the system and repeat Steps 2 through 5. If the bodies interact with one another, use Newton's third law to relate the forces they exert on one another. To solve for the unknowns in a statically determinate problem, you need to write as many *independent* equations (each of which contains one or more unknowns and that together include all the unknowns) as there are unknowns. Then, solve these equations for the unknowns.
7. When you have evaluated the unknowns, examine the results to see whether they make sense. For example, do the results satisfy the equilibrium conditions? Are the units consistent? Are the orders of magnitude reasonable; that is, have you avoided misplacing a decimal point?

Example 3.1

Load Supported by a Construction Derrick

Problem Statement A window frame is suspended by a cable from a construction derrick (see Fig. E3.1a). Draw a free-body diagram of the frame.

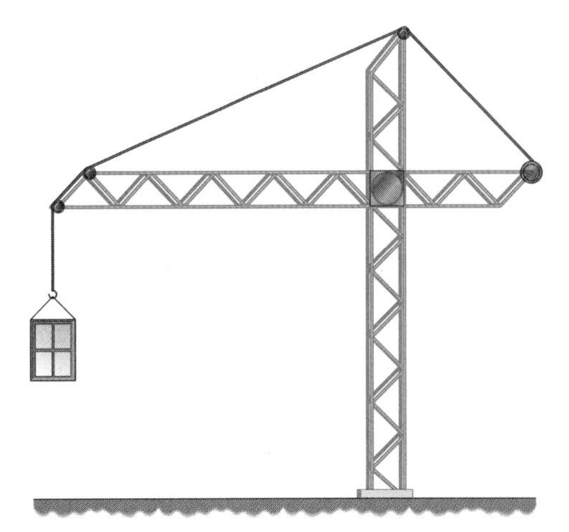

Figure E3.1a

Solution We assume that the frame may be treated as a particle. To create a free-body diagram of the frame, we first isolate it from all other parts of the system that produce significant forces on it—in this case, the cable and the earth. Ordinarily, the earth is not shown in a sketch of such a system; however, *you must not overlook the force of gravity due to the earth.* The isolated part of the system is then represented diagrammatically (Fig. E3.1b). Included in this diagram are the significant forces exerted by other bodies on the isolated part. We assume that the forces due to the mutual gravitational attractions between the frame, the cable, and the derrick are small compared to the contact force **T** due to the cable and the body force **W** due to the gravitational attraction of the earth (contact and body forces are discussed in Sec. 3.3). In Fig. E3.1b, the forces (vectors) **T** and **W** are represented as arrows. The direction and senses of **T** and **W** are given by the arrows. Consequently, the free-body diagram shows the frame *free of* (isolated from) the other parts of the system and acted on by the forces **T** and **W**.

Figure E3.1b

Since the cable is taut, the force **T** due to the cable is an upward pull on the frame of magnitude T. The force **W** due to gravity is a pull of magnitude W directed downward. If the mass of a body (particle) is given, we can use $W = mg$ to find the weight. As noted earlier, the direction of the weight is directed toward the center of the earth (usually downward in the free-body diagram).

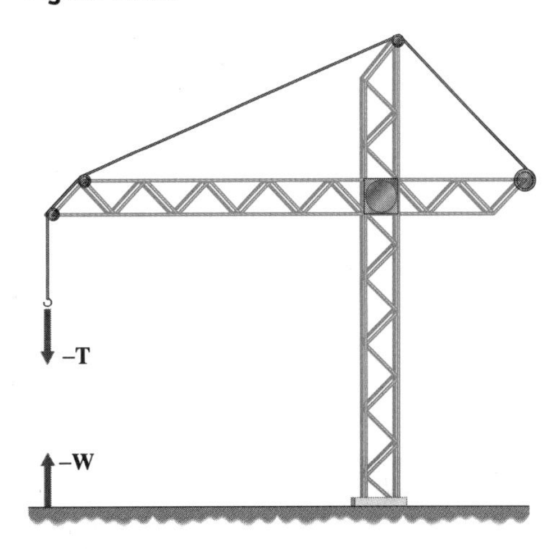

Figure E3.1c

The reactions of the forces **T** and **W** are the forces $-$**T** and $-$**W**, shown in Fig. E3.1c. These reactions are the forces exerted by the frame on the cable and on the earth. The reactions are not shown in the free-body diagram of the frame. Note, however, that if Figs. E3.1b and E3.1c are superimposed, the forces **T** and **W** in Fig. E3.1b cancel the reactions $-$**T** and $-$**W** in Fig. E3.1c, and Fig. E3.1a is obtained. This recombining of the free-body diagrams of various parts of the system to obtain the original sketch of the system is perfectly general. It serves as a check of the correctness of the free-body diagrams.

The forces **T** and **W** in Fig. E3.1b are *external forces,* since they are produced by bodies external to the frame. Considering the system as a whole, including the earth, the forces **T** and **W** are *internal forces* (Fig. E3.1a). (Internal forces are not ordinarily shown in the overall sketch of a system.) Thus, depending on the system or the part of the system for which a free-body diagram is drawn, a given force may be external or internal.

Example 3.2 Equilibrium of a Crate

Problem Statement A crate that weighs 4.5 kN hangs from a pair of hoists on a loading dock. As the crate is moved, the hoist cables are tensioned so that the crate assumes the position shown in Fig. E3.2a.

a. Draw a free-body diagram of the crate, treating it as a particle.

b. Determine the tensions in the cables.

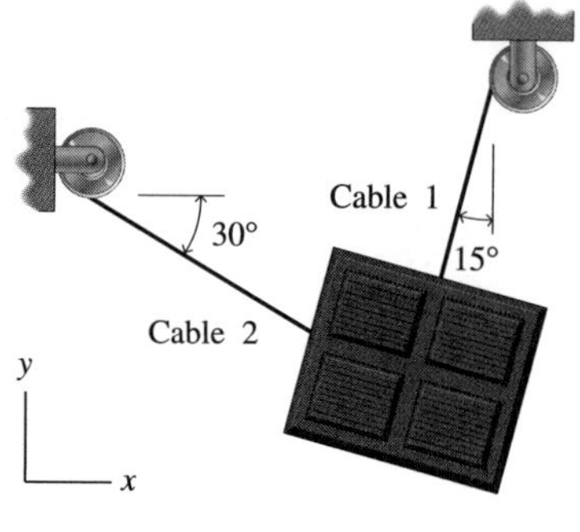

Figure E3.2a

Solution **a.** The free-body diagram of the crate is shown in Fig. E3.2b. Note that only three forces act on the crate. They are the tensions T_1 and T_2 in the cables and the weight of the crate.

b. From Fig. E3.2b, the equilibrium equations for the crate, written in terms of force projections, are

$$\sum F_x = T_1 \sin 15° - T_2 \cos 30° = 0 \quad \text{(a)}$$

$$\sum F_y = T_1 \cos 15° - T_2 \sin 30° - 4.5 = 0 \quad \text{(b)}$$

Equations (a) and (b) are solved simultaneously to obtain

$$T_1 = 4.03 \text{ kN} \qquad T_2 = 1.21 \text{ kN}$$

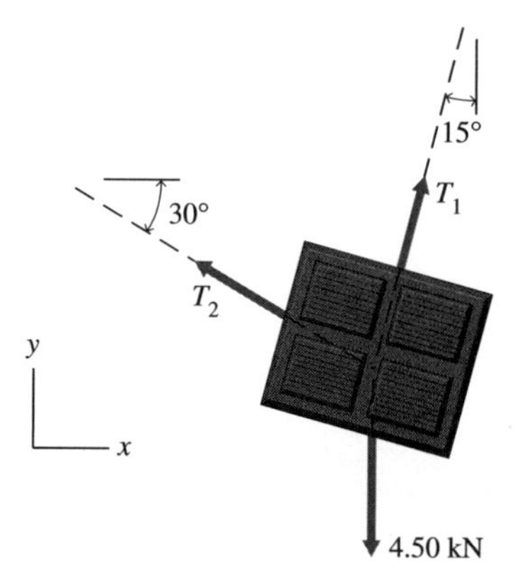

Figure E3.2b

Example 3.3 Weight Supported by Wires

Problem Statement A weight of 1000 N is held in equilibrium by three wires *PA, PB,* and *PC,* as shown in Fig. E3.3a. The wires *PA* and *PB* form angles of 30° and 45°, respectively, with the horizontal direction. Determine the force in each of the wires.

Solution The junction *P* of the wires may be regarded as a particle. Since the junction is small relative to the remainder of the system, we assume that it has no weight. We isolate the particle *P* and draw a free-body diagram of it (Fig. E3.3b). The magnitudes of the forces that the wires exert on *P* are designated as T_1, T_2, and T_3. The magnitude T_3 and the sense of $\mathbf{T}_3$ may be obtained from the free-body diagram of the 1000 lb weight (Fig. E3.3c) by ignoring the weight of cable *PC* and summing the projections on the *y* axis:

$$\sum F_y = T_3 - 1000 = 0$$

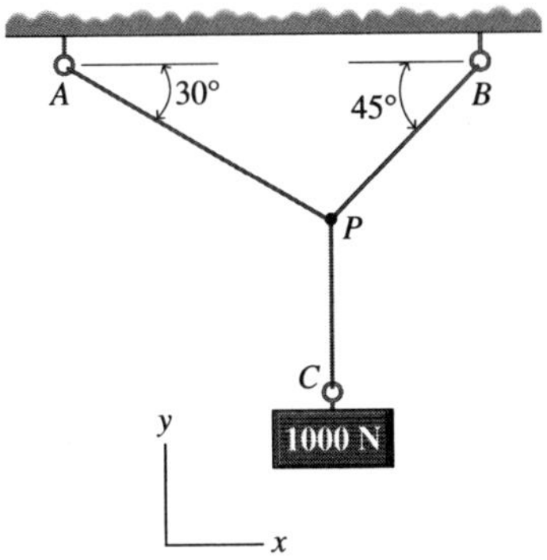

Figure E3.3a

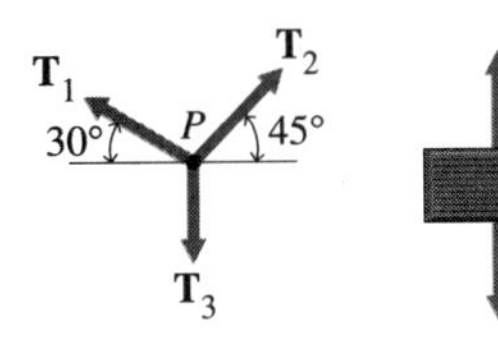

Figure E3.3b

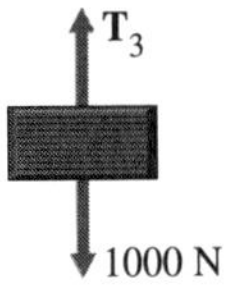

Figure E3.3c

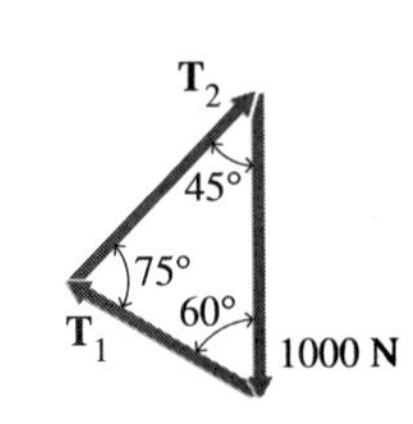

Figure E3.3d

Therefore, $T_3 = 1000$ N, and the sense of $\mathbf{T}_3$ is correct as shown. Also, since the particle P is in equilibrium, the sums of the x and y projections of the forces acting on P *must be zero*. Hence, by Eq. (3.1) and Fig. E3.3b,

$$\sum F_x = -T_1 \cos 30° + T_2 \cos 45° = 0 \quad \text{(a)}$$

$$\sum F_y = T_1 \sin 30° + T_2 \sin 45° - 1000 = 0 \quad \text{(b)}$$

Solution of Eqs. (a) and (b) gives $T_1 = 732.0$ N and $T_2 = 896.6$ N. Therefore, the forces that the wires exert on P are $T_1 = 732.0$ N, $T_2 = 896.6$ N, and $T_3 = 1000$ N.

Alternatively, once T_3 is known, we can solve for the forces T_1 and T_2 by graphical means. The polygon of forces that act on particle P closes, since P is in equilibrium. The force polygon is shown in Fig. E3.3d. By the law of sines,

$$\frac{T_1}{\sin 45°} = \frac{T_2}{\sin 60°} = \frac{1000}{\sin 75°}$$

Again, $T_1 = 732.0$ N and $T_2 = 896.6$ N.

CHECKPOINT

1. A particle is in equilibrium if
   a. the polygon of forces that act on the particle closes.
   b. forces that act on the particle produce no change in the motion of the particle.
   c. both a and b are true.
2. ***True or False:*** Equations of equilibrium for a particle can be written in terms of force vectors or force projections.
3. ***True or False:*** The equilibrant of a force system is identical to the resultant of the forces.
4. ***True or False:*** Construction of a free-body diagram is a fundamental part of the problem-solving process in mechanics.
5. ***True or False:*** A free-body diagram of a particle includes the action of the other bodies on the particle but not the reaction of the particle on the other bodies.

Answers: 1. c; 2. True; 3. False; 4. True; 5. True

3.3 TYPES OF FORCES

After studying this section, you should be able to:

- Correctly represent contact forces, body forces, and support reactions in a free-body diagram.

> ***Key Concept***
> The equilibrium of a body is affected by all the external forces, in the form of body forces and contact forces, that act on the body.

contact force

body force

As noted in Example 3.1, forces may be classified as internal or external. Forces may also be classified in other ways. For example, in Fig. E3.1b, the force **T** results from the contact between the suspended frame and the cable. Consequently, it is called a *contact force.* On the other hand, the gravitational force **W** in Fig. E3.1b is transmitted through space. No physical contact between bodies is required for this force to exist. Other kinds of forces are exerted without physical contact—notably, the forces produced by electromagnetic effects. Forces that are exerted with no physical contact are called *body forces,* since they are distributed throughout the body on which they act.

normal force

frictional force

In general, if one body presses against another, the contact forces that exist between the bodies are not solely *normal forces* (forces directed perpendicular to the surface of contact). They generally have components that are both normal to and tangent to the surface. The tangential component of the contact force is called the *frictional force.* However, at a point of contact between two frictionless bodies, the force one body exerts on the other is normal to the surface of contact. Frictional forces are considered in Chapter 10.

center of gravity

The force of gravity that acts on a system is the resultant of the weights of all the particles of the system. This resultant is represented by a vector that passes through a point in the system called the *center of gravity.* In diagrams in this book, the center of gravity is denoted by this symbol:

The center of gravity of a homogeneous body with three mutually perpendicular planes of symmetry coincides with the point of intersection of these planes. Thus, the center of gravity of a homogeneous sphere is located at its center. Likewise, the center of gravity of a homogeneous circular disk is located at its center. The centers of gravity of several simple geometric shapes are shown in Fig. 3.8. Refer to this figure as you solve problems involving the weights of objects with these shapes. Center of gravity is discussed in greater detail in Chapter 8.

Frequently, a body is attached to supports that restrain its motion. The action of a support on the body is a contact force. This book uses special symbols to represent the method by which a body is attached to its supports:

In these symbols, the brown base represents a rigid support surface. The first two symbols are different representations of a roller support; the third symbol represents a frictionless pin or a hinge.

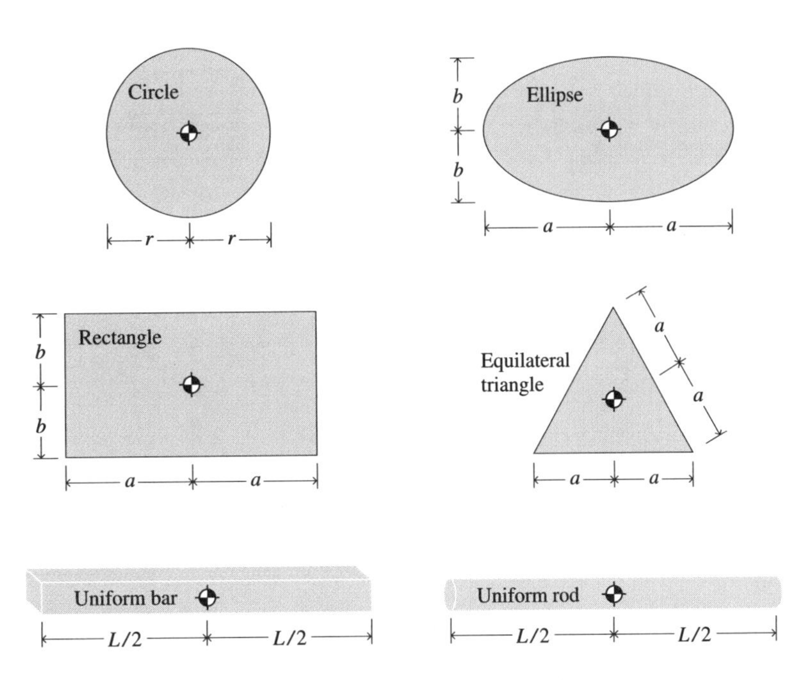

Figure 3.8 Center of gravity for some simple geometric shapes.

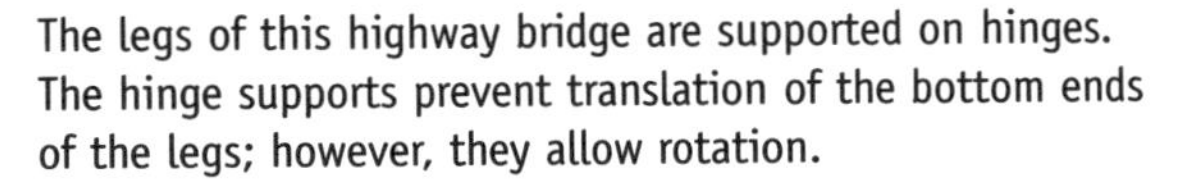

The legs of this highway bridge are supported on hinges. The hinge supports prevent translation of the bottom ends of the legs; however, they allow rotation.

support reaction

These symbols and the corresponding *support reactions* are illustrated in Fig. 3.9.[3] This book follows the convention (common in structural analysis) that a roller support cannot resist a force that is directed parallel to the rigid support surface but can resist a force directed perpendicular to the rigid support surface (Figs. 3.9a and 3.9b). This perpendicular force may have a sense either toward or away from the surface. Hence, a roller symbol represents a support that can resist either a push against the rigid support surface or a pull away from it. A pin or hinge symbol represents a support that can resist a force in any

3. A support reaction is a force that a support exerts on a body. It is the reaction to the force (action) that the body exerts on the support.

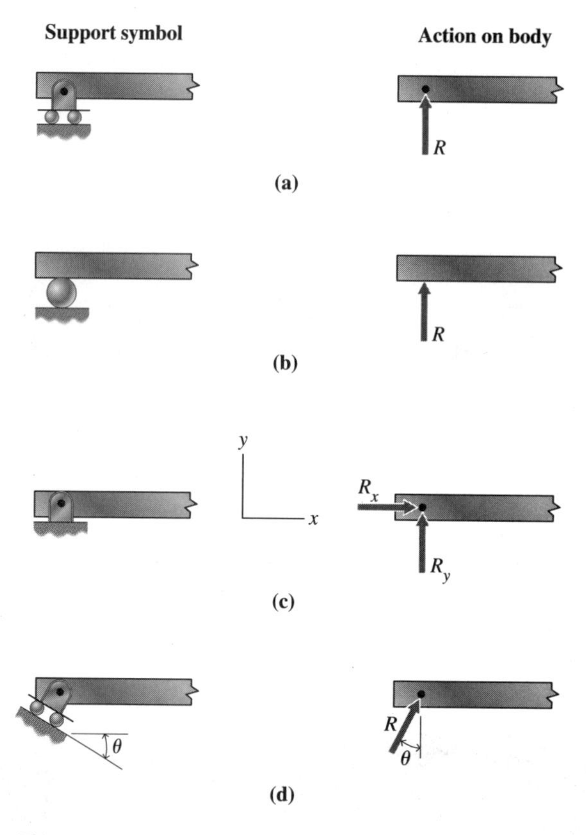

Figure 3.9 Support symbols: **(a)** roller support; **(b)** roller support; **(c)** pinned (hinged) support; **(d)** inclined roller support.

direction (Fig. 3.9c). The rigid support surface, represented by the brown base, need not be horizontal. Occasionally, it is inclined at some angle relative to the horizontal. In such a case, the symbol for the support is inclined at the same angle, as shown in Fig. 3.9d.

For a roller support, the support reaction (the force the support exerts on the body) has known direction but unknown magnitude. So, the reaction at a roller support can be expressed as a single unknown quantity, such as R. For a pin or hinge support, both the magnitude R and direction θ of the support reaction are unknown. Alternatively, these two unknown quantities can be expressed as two rectangular force components R_x and R_y. It is often said that a roller support exerts a single reaction on the structure and a pin or hinge support can exert two reactions.

3.4 CONCEPT OF A RIGID BODY

After studying this section, you should be able to:

- Understand the concept of a rigid body.

Key Concept

A rigid body does not change size or shape when loads are applied to it. A rigid body is an idealization of a real body.

rigid body A body is said to be a *rigid body* if it does not deform under the action of forces. In other words, the distances between the particles of a rigid body remain constant, irrespective of the forces that act on the body. In reality, a rigid body, like a particle, is an idealization. All bodies deform under the action of loads. However, the deformation of a body is often so small that it does not significantly affect the body's equilibrium. In such cases, the body can be treated as if it were rigid. For example, a steel shaft, a wooden beam, a cable, or a pane of glass can sometimes be considered to be rigid, depending on the application. However, you must be careful not to apply the idea of a rigid body carelessly. Some of the early studies of the internal forces in beams were in error because the scientists overlooked the effects of elastic deformations. The suitability of the rigid-body approximation depends on the situation.

3.5 EQUILIBRIUM OF A TWO-FORCE MEMBER

After studying this section, you should be able to:

- Recognize a two-force member in a system.
- Use the concept of two-force members to help solve equilibrium problems.

Key Concept

When two collinear forces having equal magnitudes and opposite directions act on a rigid body, the body is in equilibrium.

self-equilibrating forces

If several forces act on a rigid body and do not change the motion or the equilibrium of the body, the forces are *self-equilibrating.* For example, several concurrent forces whose vector sum is zero are self-equilibrating (see Fig. 3.10). If a motionless rigid body is subjected to only two collinear forces having equal magnitudes and opposite directions, then the body is in equilibrium (see Fig. 3.11).[4] This principle can be proved using Newton's laws of motion but is simply adopted here as a hypothesis. It is confirmed by everyday experience.

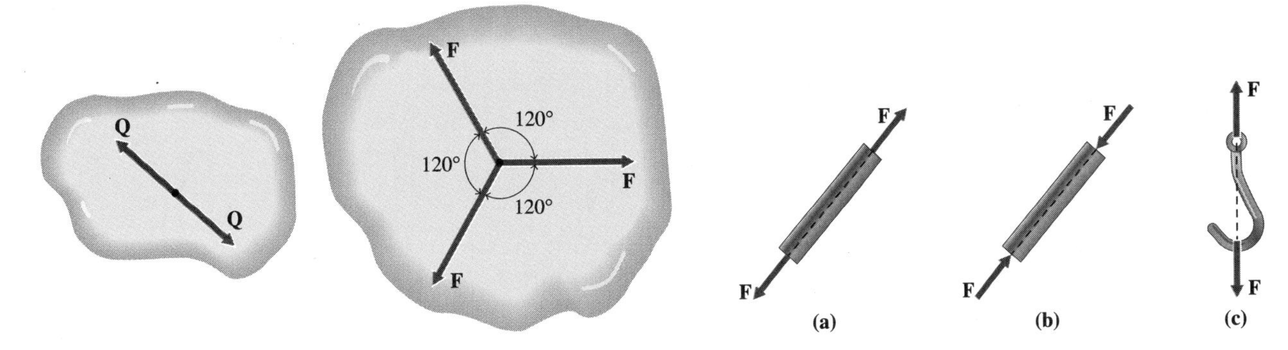

Figure 3.10 Examples of self-equilibrating forces.

Figure 3.11 Self-equilibrating pairs of forces.

Conversely, if a body is in equilibrium under the action of only two external forces, the magnitudes of the forces are equal, and the forces are acting along the same line of action but with opposite senses. This fact is true for straight bars (Figs. 3.11a and 3.11b) and for curved bars (Fig. 3.11c). This principle is also taken as a hypothesis. In summary, we have the following equilibrium principle.

Theorem 3.2

Equilibrium under Collinear Forces

A body subjected to two collinear forces is in equilibrium if, and only if, the forces are equal in magnitude and opposite in sense.

The cases illustrated by Figs. 3.11a and 3.11b differ with regard to the internal effects produced in the bar. In the first case, the bar is subjected to a *pull;* in the second case, the

4. As previously noted, the straight line along which a force vector lies is called the *direction line* or the *line of action* of the force. If two or more forces have a common line of action, they are *collinear.*

tension **compression** **strut** **column**

bar is subjected to a *push*. The first type of loading is called a *tension*, or a *tensile force;* the second type is called a *compression*, or a *compressive force*. For example, a pull that is transmitted along a taut cable or wire is a tension. A wire or a cable cannot transmit appreciable compressive force since its flexibility makes it unstable (it buckles) under small compressive loads. However, a compressive force may be transmitted by a stiff member, such as a steel bar. A member that carries mainly compressive loads is called a *strut* or a *column* (Fig. 3.11b). The load on the hook shown in Fig. 3.11c is not a case of simple tension or compression, since the forces cause bending (see Sec. 11.2) in the curved part of the hook.

Consider a straight bar AB that is loaded at its ends by tensile forces that are equal in magnitude and opposite in direction (Fig. 3.12a). The bar is in equilibrium under the action of these forces. Since the entire bar is in equilibrium, every part of the bar is in equilibrium. If we pass an imaginary plane *a–a* through a cross section of the bar at C, the parts of the bar above and below plane *a–a* are in equilibrium. We may imagine that the bar is separated into two bars, AC and BC, both of which are in equilibrium (Fig. 3.12b). Therefore, the tensile force at C has magnitude F. Since we could pass the imaginary plane *a–a* through any cross section of the bar AB, it is clear that the internal tensile force at any cross section of the bar has magnitude F. This internal force is called the *tension* in the member. If the directions of the forces at the ends of the bar are reversed (Fig. 3.12c), the bar is subjected to compression. Such a compressive force is also referred to as the *thrust* in the member. Alternatively, a tension or a thrust in the bar can be referred to as an *internal force*, or *internal action*.

thrust

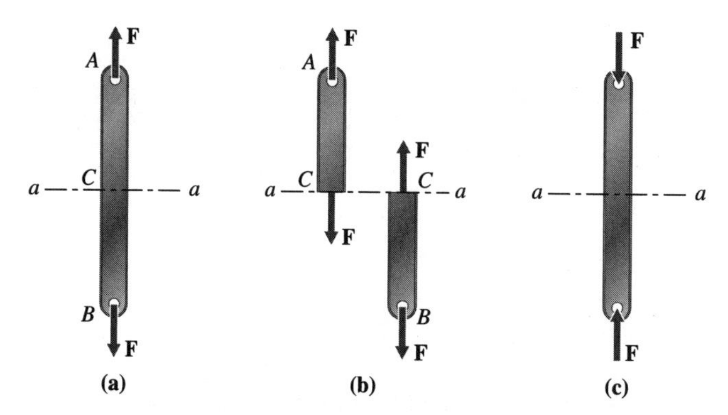

Figure 3.12 Bar subjected to a pair of collinear forces: **(a)** tension loading; **(b)** bar separated into two parts; **(c)** compression loading.

The internal force in the bar is actually *distributed* over its cross-sectional area. The ratio of the magnitude F of the force to the cross-sectional area A of the bar is called the *stress* σ in the bar:

stress

$$\sigma = \frac{F}{A} \tag{3.5}$$

Like the internal force in the bar, the stress may be either tensile or compressive.

two-force member

A structural member that is subjected to the actions of only two forces acting at separate points is called a *two-force member*. If a two-force member is in equilibrium, the resultant of the two forces must be zero. Hence, the forces have equal magnitudes, the same line of action, and opposite senses. Many structural systems, particularly trusses (Chapter 6), contain two-force members. Note that the definition of a two-force member is a convenience that allows the principles of equilibrium of a particle to be applied to structural members. If you overlook a two-force member and treat it as a general rigid body, the principles of equilibrium for rigid bodies (Chapters 4 and 5) can be applied to solve the problem. Note also that a two-force member need not be straight. For instance, the hook in Fig. 3.11c is a two-force member in equilibrium when loaded as shown. The following examples consider such systems.

PROBLEM-SOLVING TECHNIQUE

Two-Force Members

To solve a problem that involves two-force members in equilibrium:

1. Identify all the two-force members. That is, identify each member that has only two forces acting on it at separate points. A two-force member is often a straight rod, strut, or cable, though it need not be (see Fig. 3.11c).
2. Recall that the lines of action of the two forces acting on a two-force member in equilibrium are collinear and pass through the two points at which the forces act. For a straight member, the line of action of the forces lies along the member.
3. Represent the contact force that exists between the member and another part of the system (such as a support or another member) as a single force, rather than as two or three rectangular components. Otherwise, you might not identify the member as a two-force member.
4. Recognize that you can treat the two-force member using the principles of equilibrium of a particle subjected to collinear forces. Also, the points of application of the loads on a two-force member can be regarded as particles. Thus, the problem-solving technique for equilibrium of a particle (Sec. 3.2) can be followed.

Example 3.4

Equilibrium of a Boom

Problem Statement A body of weight W hangs motionless from the end of a boom that is inclined at an angle θ to the horizontal (Fig. E3.4a). The boom, whose weight is negligible compared to W, is supported at its upper end by the cable BC, which forms an angle ϕ with the horizontal.

a. Determine the tension T in the cable and the thrust P in the boom in terms of the weight W and the angles θ and ϕ.

b. Calculate T and P for the case where $\theta = 30°$, $\phi = 45°$, and $W = 100$ kN.

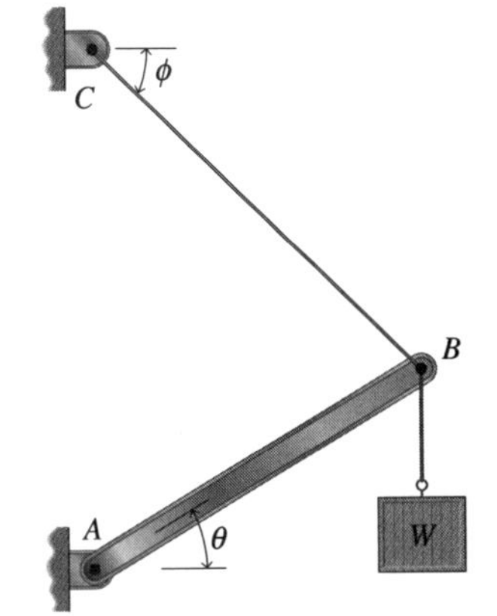

Figure E3.4a

Solution **a.** Consider the pin B (of negligible weight) at the tip of the boom to be a particle. The free-body diagram of B is shown in Fig. E3.4b. Since the weight of the boom is negligible, *the boom is a two-force member in equilibrium.* Hence, the pin connections at the ends of the boom exert forces that are directed along the axis of the boom. The boom therefore exerts a force P on pin B that is directed along the boom. By Fig. E3.4b and Eqs. (3.1), the equilibrium conditions for pin B are

$$\sum F_x = P\cos\theta - T\cos\phi = 0 \tag{a}$$

and

$$\sum F_y = P\sin\theta + T\sin\phi - W = 0 \tag{b}$$

We multiply Eq. (a) by $\sin\phi$ and Eq. (b) by $\cos\phi$, and add them:

$$P(\sin\phi\cos\theta + \cos\phi\sin\theta) = W\cos\phi \qquad \text{(c)}$$

By trigonometry, $\sin\phi\cos\theta + \cos\phi\sin\theta = \sin(\phi + \theta)$. Therefore, we can rewrite Eq. (c):

$$P\sin(\phi + \theta) = W\cos\phi$$

From this, we obtain

$$P = \frac{W\cos\phi}{\sin(\phi + \theta)} \qquad \text{(d)}$$

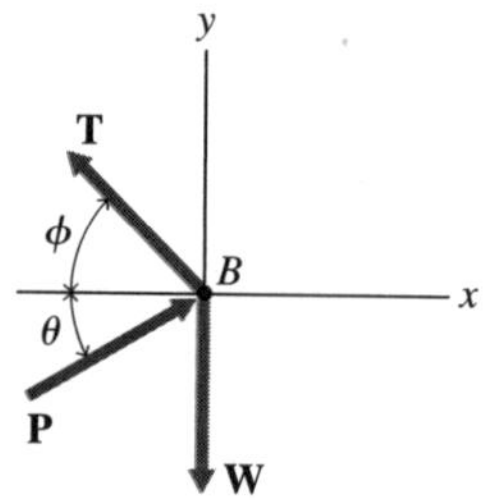

Figure E3.4b

Substitution of Eq. (d) into Eq. (a) yields

$$T = \frac{W\cos\theta}{\sin(\phi + \theta)} \qquad \text{(e)}$$

b. For $\theta = 30°$, $\phi = 45°$, and $W = 100$ kN, Eqs. (d) and (e) yield $P = 73.20$ kN and $T = 89.66$ kN.

Example 3.5 *Equilibrium of a Tripod*

Problem Statement An application involving the noncoplanar equilibrium of a particle is a tripod, a structural system that consists of three legs connected at the top, point P (see Fig. E3.5a). The other ends of the legs are connected to a horizontal surface at points A, B, and C. All connections are assumed to be pinned, so they exert forces directed along the axes of the legs. In other words, *each leg of the tripod is a two-force member.* The tripod is subjected to an external force at point P. The projections of the force relative to xyz coordinate axes are $F_x = 4$ kN, $F_y = 5$ kN, $F_z = -12$ kN. Determine the forces in the legs of the tripod by the force projection method.

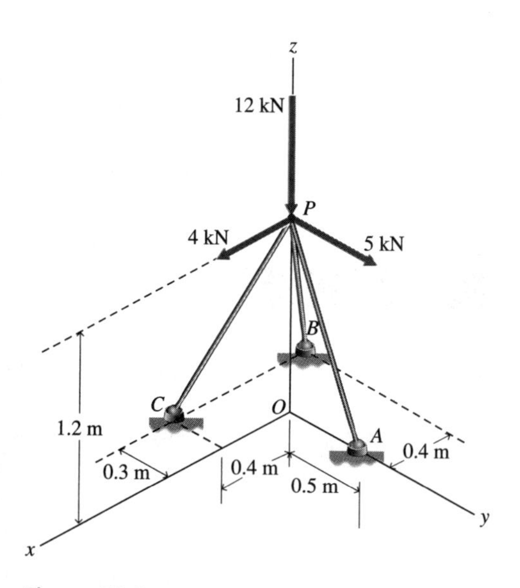

Figure E3.5a

Solution From Fig. E3.5a, we see that the coordinates of the connection points (with units of meters) are

P: (0.0, 0.0, 1.2)

A: (0.0, 0.5, 0.0)

B: (−0.4, −0.3, 0.0)

C: (0.4, −0.3, 0.0)

By Eq. (2.12), the lengths of the legs are

$$PA = \sqrt{0.5^2 + 1.2^2} = 1.3 \text{ m}$$

$$PB = \sqrt{0.4^2 + 0.3^2 + 1.2^2} = 1.3 \text{ m}$$

$$PC = \sqrt{0.4^2 + 0.3^2 + 1.2^2} = 1.3 \text{ m}$$

The forces that the legs of the tripod exert on point P are **A**, **B**, and **C**, with magnitudes A, B, and C. These forces are directed along the legs PA, PB, and PC, respectively, since the legs are two-force members. We arbitrarily assume that each of the legs is in tension, which means that each pulls on point P. If the solution yields positive values for A, B, and C, this assumption is correct. However, if A, B, or C turns out to be negative, it means that the corresponding leg is in compression. Since the forces **A**, **B**, and **C** are directed along the members PA, PB and PC, respectively, they have the same direction cosines as the members. Consequently, by Eqs. (2.13), the direction cosines of vectors **A**, **B**, and **C** are

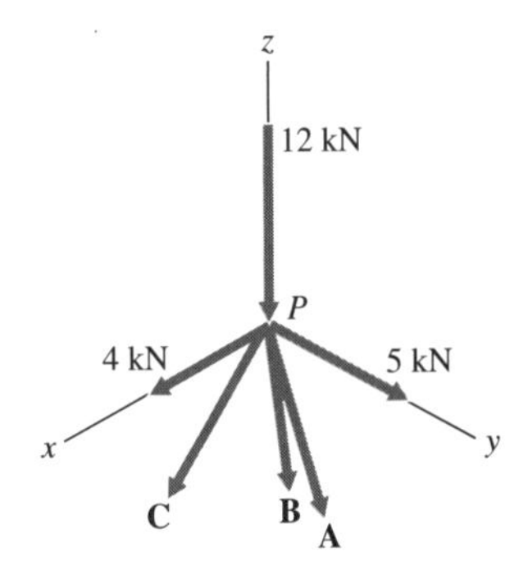

Figure E3.5b

$$\begin{aligned} \mathbf{A}&: \left(0, \tfrac{5}{13}, -\tfrac{12}{13}\right) \\ \mathbf{B}&: \left(-\tfrac{4}{13}, -\tfrac{3}{13}, -\tfrac{12}{13}\right) \\ \mathbf{C}&: \left(\tfrac{4}{13}, -\tfrac{3}{13}, -\tfrac{12}{13}\right) \end{aligned} \tag{a}$$

By Eqs. (2.1) and Eqs. (a), the equilibrium equations for point P (Fig. E3.5b) are

$$\begin{aligned} \sum F_x &= -\tfrac{4}{13}B + \tfrac{4}{13}C + 4 = 0 \\ \sum F_y &= \tfrac{5}{13}A - \tfrac{3}{13}B - \tfrac{3}{13}C + 5 = 0 \\ \sum F_z &= -\tfrac{12}{13}A - \tfrac{12}{13}B - \tfrac{12}{13}C - 12 = 0 \end{aligned} \tag{b}$$

The solution to Eq. (b) is $A = -13$ kN, $B = 6.5$ kN, and $C = -6.5$ kN. Since A and C are negative, the assumption that legs PA and PC are in tension is wrong. Legs PA and PC are in compression (see Sec. 3.5). Since B is positive, leg PB is in tension. Thus, **A** and **C** are compressive forces with magnitudes 13 kN and 6.5 kN, respectively, and **B** is a tensile force with magnitude 6.5 kN.

CHECKPOINT

1. ***True or False:*** A free-body diagram of a particle includes contact forces but not body forces.
2. ***True or False:*** A roller support represented by the symbol 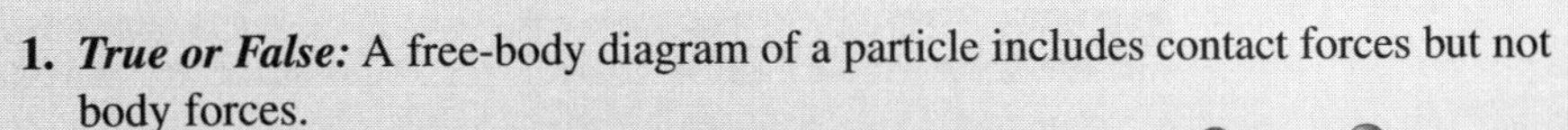 or can resist a force directed toward the rigid surface.
3. ***True or False:*** A roller support represented by the symbol or can resist a force directed away from the rigid surface.
4. ***True or False:*** Rigid bodies are convenient idealizations to make the study of mechanics easier.
5. ***True or False:*** Several forces acting on a rigid body must be coplanar in order to be self-equilibrating.
6. ***True or False:*** If a rigid body is in equilibrium, every part of the rigid body is in equilibrium.
7. ***True or False:*** A straight two-force member that is in equilibrium must be subjected to a compressive force at one end and a tensile force at the other end.

Answers: 1. False; 2. True; 3. True; 4. True; 5. False; 6. True; 7. False

3.6 TRANSMISSIBILITY OF FORCES THAT ACT ON A RIGID BODY

After studying this section, you should be able to:

- Understand the applications and the limitations of the principle of transmissibility.

Key Concept
Dynamically equivalent forces produce identical effects on the motion or the equilibrium of a body.

dynamically equivalent forces

statically equivalent forces

Two different sets of forces that act on a rigid body are *dynamically equivalent* if they produce identical effects on the motion or the equilibrium of the body. For example, several concurrent forces are dynamically equivalent to their resultant. The phrase "dynamically equivalent" is preferable to the more conventional expression *statically equivalent,* since the replacement of a set of forces by its resultant is valid for both the statics and the dynamics of rigid bodies.

Key Concept
If the point of application of a force is displaced along its line of action, the equilibrium of a rigid body on which the force acts is not affected.

Figure 3.13a represents a rigid body that is loaded by several external forces. Consider the force **F** that acts at point *P.* Nothing is altered if we introduce two self-equilibrating forces at another point *Q* that lies on the line of action of **F**. Let the forces at point *Q* also have magnitude *F,* and let them be collinear with the force at point *P* (see Fig. 3.13b). Then the force at point *P* and the right-hand force at *Q* are self-equilibrating. Since these forces cancel each other, we discard them. We are then left with the loading shown in Fig. 3.13c. Accordingly, the force systems shown in Figs. 3.13a and 3.13c are dynamically equivalent. Hence, the following result is obtained.

Theorem 3.3

Principle of Transmissibility
The equilibrium (or the motion) of a rigid body is not altered if the point of application of any force that acts on the body is displaced along the line of action of the force.

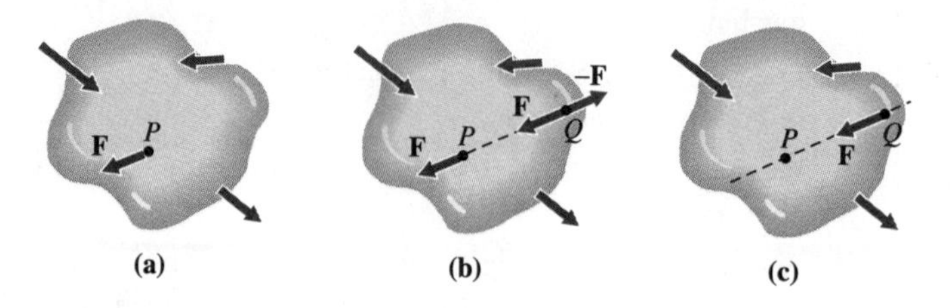

Figure 3.13 Demonstration of transmissibility.

principle of transmissibility

This result is known as the *principle of transmissibility* of force. It signifies that a force that acts on a rigid body can be represented by a slide vector (see Sec. 2.4).

The principle of transmissibility allows you to transfer the point of application of a force to a location outside the body on which the force acts. You may wish to do this, for instance, when several forces act on a body and you want to show that they are concurrent. For example, Fig. 3.14a illustrates two self-equilibrating forces, **F** and −**F**, acting on a rigid block. If you wanted to transfer the left-hand force farther to the left, you could provide a rod to transmit the force, as shown in Fig. 3.14b. However, you may transmit the force *without* providing the rod or any other equivalent structure. Supplementary structures of this kind are unnecessary, since you are not really transferring the forces. You are merely imagining their transfer in order to simplify the analysis of their effects.

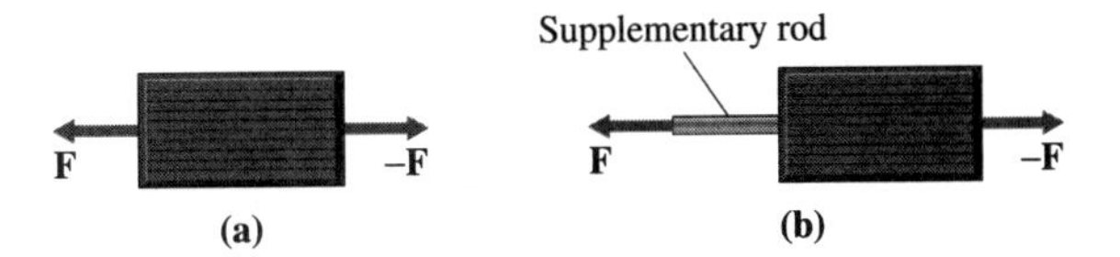

Figure 3.14 Transmission of a force to a point outside a body.

The principle of transmissibility does not apply to a body for which internal forces or deformations are to be determined.

You should always remember that the principle of transmissibility merely asserts that moving the point of application of a force along its line of action does not alter the motion or the equilibrium of a *rigid body.* In general, moving the point of application of a force alters the distribution of *internal* forces and the *stresses* in a body. For example, a bridge truss may sometimes be regarded as a rigid body. If an external force that acts on the truss is shifted along its line of action from one joint to another, the overall equilibrium of the truss is not affected. However, the internal forces in the members of the truss do change.

For internal effects, we can consider the rigid bars shown in Figs. 3.11a and 3.11b. Figure 3.11b may be obtained from Fig. 3.11a by sliding the end forces along the bar to opposite ends. In both figures, the bars are in equilibrium, and Theorem 3.3 still holds. However, the internal force is a tension in Fig. 3.11a, whereas it is a compression in Fig. 3.11b. For deformations, consider a toy balloon that is in equilibrium under the action of two diametrically opposed forces (Fig. 3.15a). These forces produce pronounced cusps of deformation at their points of application. If the force that acts on the left side of the balloon is transferred to the point of application of the force on the right, the deformation is eliminated (Fig. 3.15b). Hence, the deformation of the balloon caused by the true force system is unrelated to the deformation caused by the transferred force system. In other words, internal effects and deformations must be evaluated using the true force system.

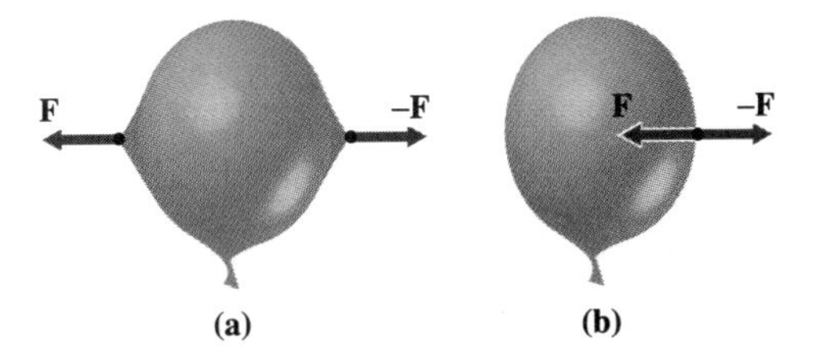

Figure 3.15 Transmission of forces altering the strains in a body.

3.7 EQUILIBRIUM OF A RIGID BODY SUBJECTED TO CONCURRENT FORCES

After studying this section, you should be able to:

- Draw an accurate free-body diagram of a rigid body.
- Recognize when a rigid body is subject to a system of concurrent forces.
- Apply the equations of equilibrium to solve problems involving a rigid body subjected to a system of concurrent forces.

> ***Key Concept***
>
> A rigid body subjected to forces having concurrent lines of action can be treated as a particle with respect to equilibrium.

resultant axis

Consider a rigid body that is subjected to several forces whose lines of action intersect at point P (see Fig. 3.16a). By the principle of transmissibility, the points of application of these forces may be transferred to point P without altering the equilibrium or the motion of the body (Fig. 3.16b). Since the forces are concurrent after they are transferred to point P, they may be combined into *a single resultant force* **R** by vector addition (see Sec. 2.4). The line of action of the resultant force **R** through point P is called the *resultant axis* (Fig. 3.16c). Thus, the principle of transmissibility along with simple planar geometry leads to the following observation.

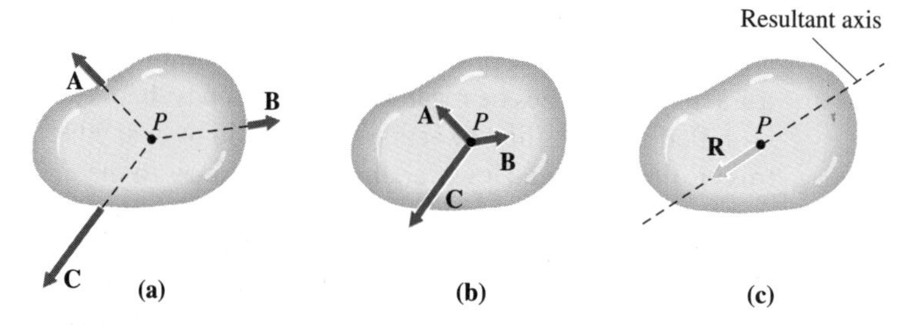

Figure 3.16 Forces with concurrent lines of action.

> ***Theorem 3.4***
>
> Any two nonparallel coplanar forces that act on a rigid body may be combined into a single resultant force, since the lines of action of the two forces must intersect.

In the more general case (for more than two forces), the principle of transmissibility can be combined with the concept of equilibrium.

> ***Theorem 3.5***
>
> If the lines of action of all the forces that act on a motionless rigid body intersect at the same point, equilibrium exists if, and only if, the vector sum of the forces is zero.

Thus, the requirements for equilibrium of a rigid body that is subject to forces with concurrent lines of action are the same as the requirements for equilibrium of a particle (Sec. 3.1). This result is a special case of the general three-dimensional theory of equilibrium (Chapter 5).

EQUILIBRIUM OF A THREE-FORCE MEMBER

Based on the foregoing discussion, the following theorem can be stated.

Theorem 3.6

If a rigid body is in equilibrium under the action of three nonparallel forces, the forces are coplanar and their lines of action are concurrent; that is, their lines of action intersect at a common point.

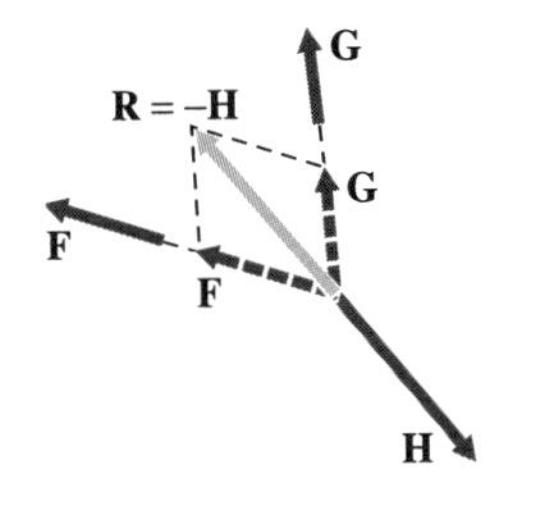

Figure 3.17 Three concurrent forces: **F**, **G**, and **H**.

To prove this theorem, let the three forces be **F**, **G**, and **H**. The equilibrium condition $\mathbf{F} + \mathbf{G} + \mathbf{H} = 0$ shows that the forces are coplanar, since this vector equation signifies that the three force vectors form a triangle.[5] If any two of the forces (say, **F** and **G**) are not parallel, they may be displaced to the point of intersection of their lines of action, and there they may be combined by the parallelogram construction into a single resultant **R** (Fig. 3.17). Since the force **R** is balanced by the single force **H**, it is equivalent to $-\mathbf{H}$. Accordingly, the lines of action of the forces **F**, **G**, and **H** are concurrent.

A rigid body under the action of three nonparallel forces is commonly referred to as a *three-force member.* The word "member" might not be entirely accurate for describing

three-force member

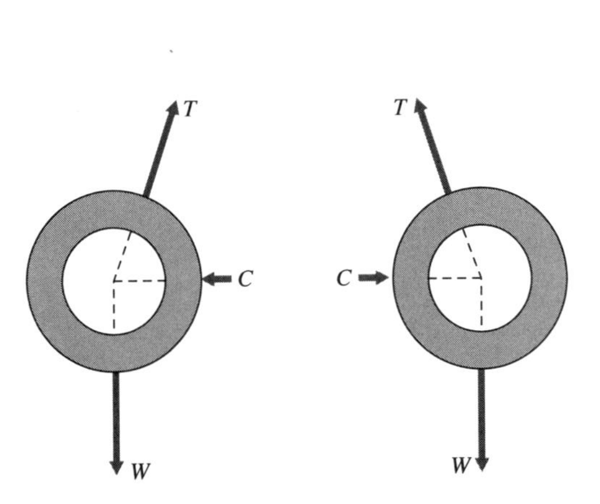

The spools of sheet steel being lifted by this crane can be treated as three-force members. Each spool is acted on by three forces: its weight *W*, a contact force *C* with the other spool, and the tension *T* in the rigging chain.

5. Although the requirement that the vector sum of the forces be zero has been shown to be a necessary condition for equilibrium only in the case of coplanar forces, it remains valid for three-dimensional force systems (see Chapter 5).

such a rigid body, since it tends to suggest a part of a structure, such as a truss or a frame. In certain cases, a mechanical component of a system (such as a wheel) may be treated as a three-force member. Nevertheless, it can be helpful in solving equilibrium problems to identify three-force members, since the principles of particle equilibrium can be applied to them.

Like that of a two-force member, the definition of a three-force member is established for convenience in applying the principles of particle equilibrium to a rigid body. If you prefer to treat a three-force member as a general rigid body, the conditions of equilibrium for rigid bodies (Chapters 4 and 5) can be applied to solve the problem.

PROBLEM-SOLVING TECHNIQUE

Three-Force Members

To solve a problem that involves three-force members in equilibrium:

1. Identify all the three-force members. Don't forget to include the weight of a member if it is significant. Also recall that any contact force should be represented as a single force (see the Problem-Solving Technique in Sec. 3.5).
2. For each three-force member, apply the principle of transmissibility to locate the intersection point of the three lines of action of the forces.
3. Locate a particle at the intersection of the lines of action of the three forces that act on the member. Then apply the principles of equilibrium to the particle (see the Problem-Solving Technique in Sec. 3.2).

Example 3.6

Equilibrium of a Rigid Ring

Problem Statement A rigid, weightless ring is supported by two radially directed wires *AB* and *CD* (Fig. E3.6a). A radially directed pull of 100 N is applied to the ring at point *E*. Under the actions of the wires and the pull, the ring is in equilibrium. Determine the tensions in the wires.

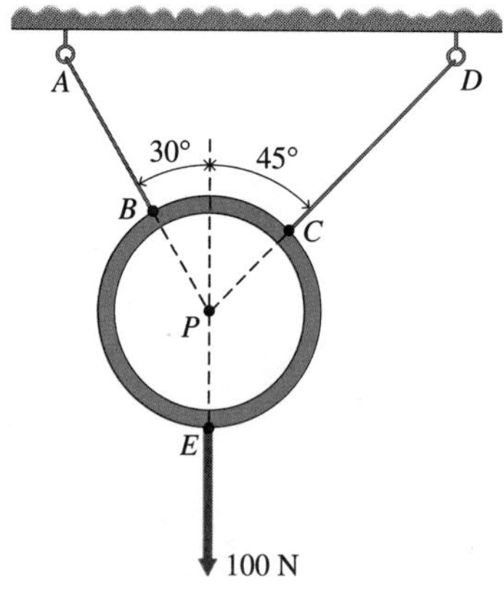

Figure E3.6a

Solution Since all three external loads on the ring act radially, the lines of action of the forces are concurrent at the center of the ring. Accordingly, the forces may be transmitted to the center of the ring, the point of concurrence. The equilibrium of the ring may then be treated as the equilibrium of a particle *P* that exists at the center of the ring (see Fig. E3.6b). Hence, the tensions T_1 and T_2 are determined by the conditions of equilibrium of a particle. By Eqs. (3.1) and Fig. E3.6b, we have

$$\sum F_x = T_2 \sin 45° - T_1 \sin 30° = 0$$

$$\sum F_y = T_2 \cos 45° + T_1 \cos 30° - 100 = 0$$

The solution of these equations is $T_1 = 73.21$ N and $T_2 = 51.77$ N.

Information Item Although the analysis in this example appears similar to that in Example 3.3 (see Figs. E3.3b and E3.6b), the problems are different conceptually. Example 3.3 treats the equilibrium of a particle. This example treats the equilibrium of a rigid body subjected to three forces. Using the principle of transmissibility of forces, we reduced the rigid-body problem to an equivalent particle problem.

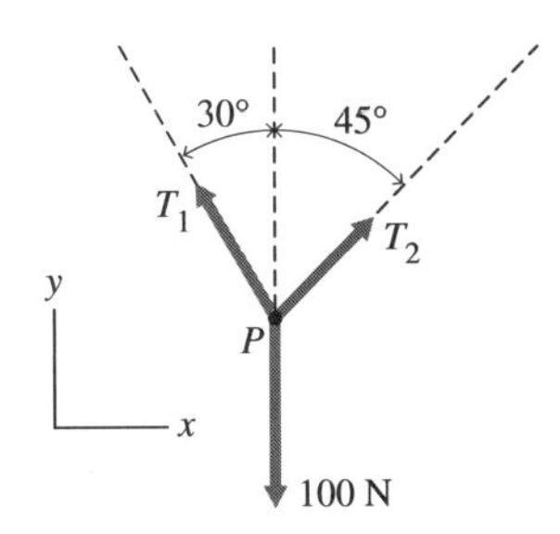

Figure E3.6b

Example 3.7

Maximum Allowable Load of Wires

Problem Statement Each of the wires AB and CD in Fig. E3.6a can sustain a maximum tension of 1 kN. Determine the maximum load W that can be applied to the ring and the corresponding tensions in each of the wires.

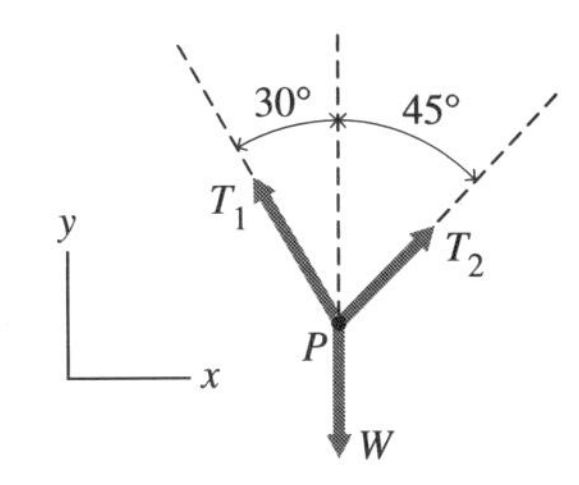

Figure E3.7

Solution As in Example 3.6, the ring may be treated as a particle P that exists at the center of the ring. Here, however, the load W is unknown (see Fig. E3.7). By Fig. E3.7 and Eqs. (3.1), we have

$$\sum F_x = T_2 \sin 45° - T_1 \sin 30° = 0 \tag{a}$$

and

$$\sum F_y = T_2 \cos 45° + T_1 \cos 30° - W = 0 \tag{b}$$

From Eq. (a), we find

$$T_1 = T_2 \frac{\sin 45°}{\sin 30°} = 1.414 T_2 \tag{c}$$

Therefore, $T_1 > T_2$, and the maximum load W is reached when $T_1 = 1$ kN. Thus, from Eqs. (a), (b), and (c), with $T_1 = 1$ kN, we obtain

$$T_1 = 1 \text{ kN} \qquad T_2 = 0.707 \text{ kN} \qquad W = 1.366 \text{ kN}$$

In other words, when the maximum load $W = 1.366$ kN is reached, the tensions in the wires AB and CD are 1 kN and 0.707 kN, respectively.

Example 3.8

System of Three Wheels

Problem Statement A system of three wheels, A, B, and C, arranged as shown in Fig. E3.8a, is in equilibrium. Wheels B and C rest on a frictionless horizontal surface S and are connected to the link DE by pins at D and E. Wheel A rests on wheels B and C. The surfaces of the wheels are smooth. (Note that the term "smooth" is often used to denote a frictionless surface.) The weights of wheels A, B, and C are 500 lb, 300 lb, and 600 lb, respectively, and the three wheels are the same thickness. The weight of the link DE is negligible. Determine the contact forces that exist among the wheels, the surface S, and the link.

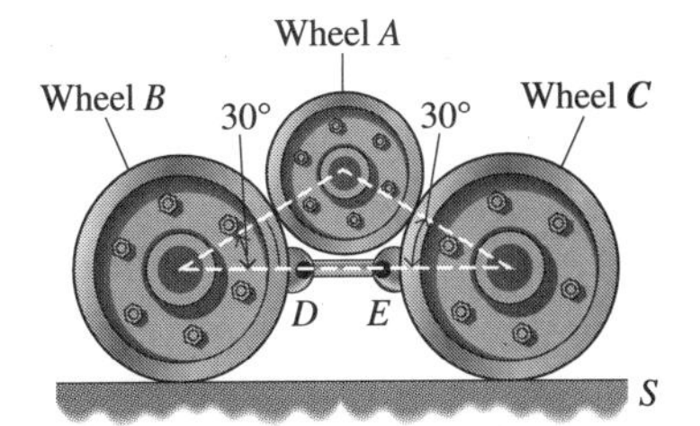

Figure E3.8a

Solution Since the weight of DE is negligible, the link is a two-force member. The pins therefore exert forces on the link that have lines of action directed along DE. Likewise, the contact forces that the smooth surface S exerts on the wheels are normal to the wheel surfaces; that is, they are directed through the centers of the wheels (see Sec. 3.3 for a discussion of types of forces). Also, since the wheels are homogeneous, the weight vectors are directed through the centers of the wheels. Hence, the forces acting on each wheel have lines of action that are concurrent at the center of the wheel, and each wheel may be treated as a particle.

Having made these observations, we are ready to solve the problem. We know that equilibrium of a particle in two dimensions (Sec. 3.1) is defined by two equations: $\sum F_x = 0$ and $\sum F_y = 0$. So, for each particle in the system, we have two equations available to solve for two unknowns. If possible, you should analyze a system particle by particle in a sequence such that no more than two unknown forces act on each particle. For example, as we examine the system in Fig. E3.8a, we note that wheel A is subjected to only two unknown forces. These forces occur at the points of contact between wheel A and the other two wheels. Accordingly, we should analyze wheel A first. The free-body diagram of wheel A is shown in Fig. E3.8b. The contact forces that wheels B and C exert on wheel A are denoted by $\mathbf{P}_B$ and $\mathbf{P}_C$, respectively. The lines of action of the contact forces and the weight (500 lb) are concurrent at the center of wheel A.

Summation of the x projections of the forces on wheel A (Fig. E3.8b) yields

$$\sum F_x = P_B \cos 30° - P_C \cos 30° = 0$$

Therefore, $PB = PC$. Similarly, summation of y projections yields

$$\sum F_y = P_B \sin 30° + P_C \sin 30° - 500 = 0 \quad \text{(a)}$$

Since $P_B = P_C$, Eq. (a) yields $P_B = P_C = 500$ lb.

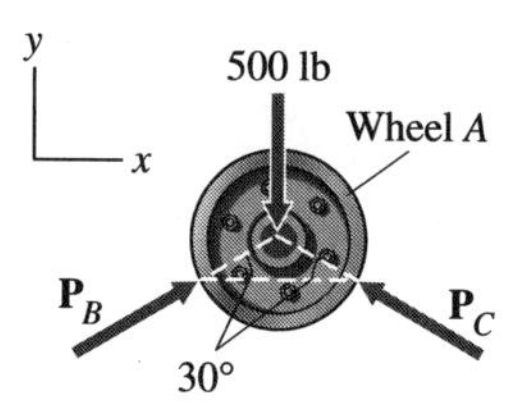

Figure E3.8b

Next, we can consider either wheel B or wheel C, because each is now subjected to only two unknown contact forces. For instance, consider the free-body diagram of wheel B shown in Fig. E3.8c. The 500 lb oblique force is the contact force that wheel A exerts on wheel B. The force $\mathbf{T}$ is the contact force that the link DE exerts on wheel B (the link is assumed to be in tension). Force $\mathbf{R}_B$ is the contact force that surface S exerts on B. Now, summation of the y projections of the forces that act on wheel B yields

$$\sum F_y = R_B - 300 - 500 \sin 30° = 0 \quad \text{(b)}$$

Equation (b) yields the result $R_B = 550$ lb. Similarly,

$$\sum F_x = T - 500 \cos 30° = 0$$

Hence, $T = 433$ lb.

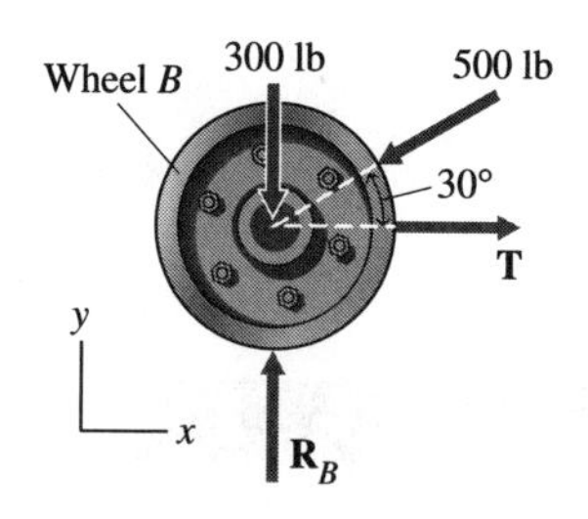

Figure E3.8c

Finally, we consider the free-body diagram of wheel C (Fig. E3.8d). Since only one unknown force acts on the wheel, we need apply only one of the equations of equilibrium. Summation of the y projections yields

$$\sum F_y = R_C - 600 - 500 \sin 30° = 0$$

from which we obtain $R_C = 850$ lb.

As we look back over the solution, we observe that the equation $\sum F_x = 0$ for wheel C was not needed to solve

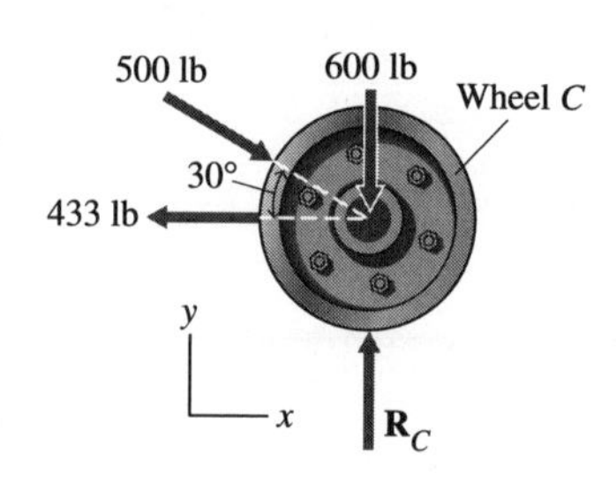

Figure E3.8d

the problem. However, since this equation must be satisfied, we can use it as a check of the consistency of the solution:

$$\sum F_x = 500\cos 30° - 433 = 0 \quad ✓$$

Consistency is satisfied, to the degree of accuracy of our computations.

Alternatively, we could have solved this problem by expressing the x and y equilibrium equations for wheels A and B and the y equilibrium equation for wheel C (five independent equations) in terms of the five unknowns P_B, P_C, T, R_B, and R_C. Then, by linear algebra (method of determinants), we could have solved for the five unknowns simultaneously.

Information Item The techniques employed in this example are typical of those used in the analysis of equilibrium of rigid bodies subjected to forces with concurrent lines of action. You should therefore study this example carefully and understand the solution techniques. In later examples, it will be assumed that you understand these techniques and can use them without extensive explanation.

Example 3.9

Equilibrium of a Wall Bracket

Problem Statement A rigid wall bracket is used to support a 4 kN load, as shown in Fig. E3.9a. The wall bracket is attached at A by a frictionless pin and at B by a roller. Determine the reactions of the supports at A and B.

Solution We start by drawing a free-body diagram of the bracket as a whole (Fig. E3.9b). Because three nonparallel external forces act on the bracket, Theorem 3.6 applies, and we conclude that the forces P, A, and B are concurrent at point C. Therefore, by the principle of transmissibility, we may transfer the forces to point C (Fig. E3.9c). Then, by the conditions of equilibrium for concurrent forces, we have

$$\sum F_x = -B + A\sin\theta = 0 \tag{a}$$

and

$$\sum F_y = A\cos\theta - P = 0 \tag{b}$$

From Fig. E3.9b, we see that $AC = \sqrt{3^2 + 4^2} = 5$ m; thus, $\sin\theta = 3/5$ and $\cos\theta = 4/5$. Consequently, by Eq. (b), $A = 5P/4 = 5$ kN, and by Eq. (a), $B = 3A/5 = 3$ kN. Furthermore, since A and B are positive, the forces act as shown in Fig. E3.9b.

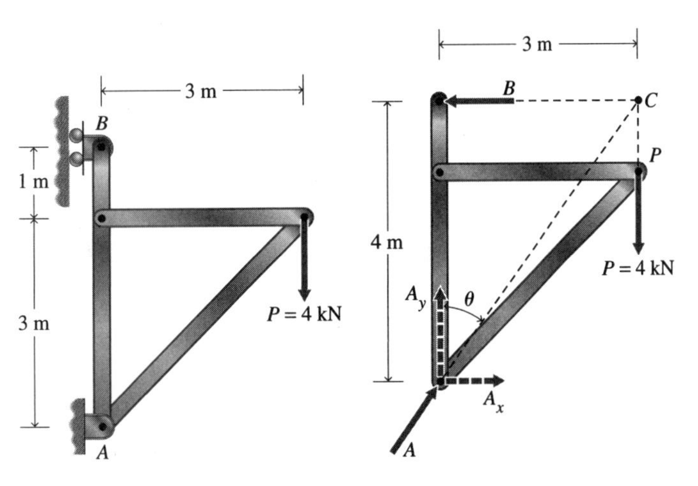

Figure E3.9a **Figure E3.9b** **Figure E3.9c**

Information Item At first glance, you might not notice that only three external forces act on the bracket. Usually, a reaction at a pin support, such as A in this case, is represented by its x and y projections (A_x, A_y). In fact, however, the vector sum of the components of the force at A gives a single force directed as shown in Fig. E3.9b. In effect, recognition that Theorem 3.6 applies eliminates θ as one of the unknowns.

CHECKPOINT 

1. The principle of transmissibility permits a force acting on a rigid body to be displaced
   a. along its line of action.
   b. normal to its line of action.
   c. in any direction.
2. Application of the principle of transmissibility does not alter the deformation or the motion of
   a. a rigid body.
   b. a deformable body.
   c. either a rigid body or a deformable body.
3. ***True or False:*** A set of forces that act on a rigid body with concurrent lines of action but different points of application can be considered to be a system of concurrent forces.

Answers: 1. a; 2. a; 3. True

Chapter Highlights

- A particle is in equilibrium if the resultant of the forces acting on it is zero or it remains motionless or it maintains constant velocity under the action of external forces.
- A free-body diagram is a sketch of a particle (or rigid body) that shows the particle (or body) free from other parts of the system and shows all forces acting on it. Construction of a free-body diagram is an essential step in the problem-solving process in mechanics.
- A statically determinate system is one that can be analyzed using the laws of equilibrium alone.
- The principle of transmissibility allows the point of application of a force to be displaced along its line of action without altering the state of equilibrium of a rigid body.
- The equilibrium of a rigid body subjected to a set of concurrent forces can be analyzed using the same equilibrium principles that apply to a particle.
- If a rigid body is in equilibrium under the action of three nonparallel forces, the forces are coplanar and their lines of action are concurrent.

Problems

On Design Problems

In much of this book, you will focus your problem-solving efforts on *analysis* of mechanical systems. That is, you will determine how a well-defined system behaves under the action of certain forces. For instance, you will solve problems to find support reactions and member forces of a system, given the geometry of the system and the loads applied to it. The principles of equilibrium will be your primary tools for doing this analysis. As you progress through your engineering coursework, you will also be expected to solve problems involving *design.* Design differs from analysis in that the system is not well defined. That is, some aspect of the geometry or material properties of the system must be determined by the design process. This determination is made based on the loads and other performance criteria that are specified for the system.

Design can be a complex, iterative task, without step-by-step procedures to follow. Engineering design requires judgment, along with application of basic principles, to achieve a result that might not always be unique. Generally, the design process starts with a *preliminary design*. That is, engineers begin the iterative design process by making a preliminary (initial) design selection. Preliminary design selections are usually based on reasonable assumptions and perhaps an approximate analysis. Then, once the preliminary design is selected, engineers proceed with comprehensive and accurate analysis to validate the assumptions and verify their design decisions.

In order to help engineering students become better designers, increased emphasis is being placed on the design process in introductory engineering courses, including statics. In this book, you will be guided through the first steps in the design process. Design problems may prescribe one or two simple conditions that a structure must meet and ask you to make corresponding design selections, or they may describe a general engineering system and ask you to make reasonable assumptions and develop a preliminary design. As you progress through your engineering coursework, you will be called upon to design more complex systems that require more judgment and allow greater freedom in decision making.

3.2 *Free-Body Diagrams*

3.1 Draw a free-body diagram of each of the following particles. Be sure to include the known or unknown magnitudes and directions of all forces that act on the particle.

a. point B in Fig. P3.2
b. point B in Fig. P3.3
c. point B in Fig. P3.4
d. point B in Fig. P3.5
e. pendulum bobs B and C in Fig. P3.10; include the weight of the bobs
f. points B and C in Fig. P3.12
g. particles A and B in Fig. P3.15; include the weight of the particles
h. point A in Fig. P3.17; include the weight of the cylinder
i. point D in Fig. P3.29
j. point A in Fig. P3.30

3.2 a. Show that the tensions in the cables AB and BC are equal (see Fig. P3.2) and determine the tension T in the cables for $\theta = 0, 30°, 60°, 85°$, and $89°$.
b. Plot a curve of the tension T versus the angle θ. How does the tension T behave as θ approaches $90°$?

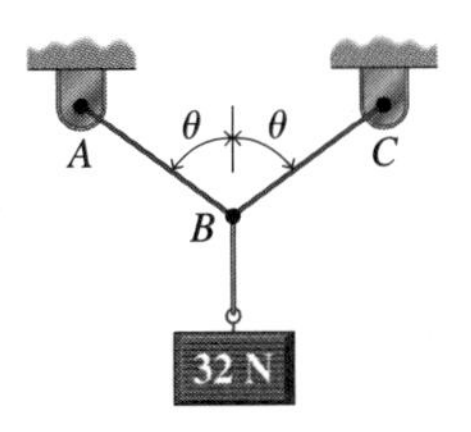

Figure P3.2

3.3 A crate weighing 360 lb is attached to two cables AB and BC (see Fig. P3.3). A third cable is attached at B and pulls horizontally with force P to keep the crate from swinging back against the wall. Determine the range of values of P for which cables AB and BC remain taut. (*Hint*: A cable is no longer taut when its tension drops to zero.)

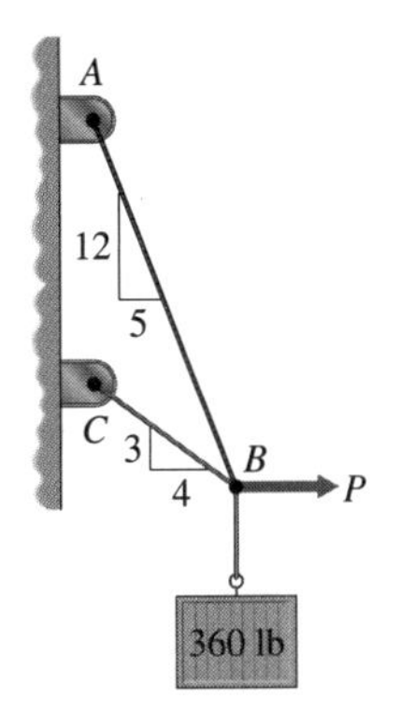

Figure P3.3

3.4 For the system shown in Fig. P3.4, assume that the magnitude of the oblique force F is allowed to vary. Determine the range of values of F such that cables AB and BC remain taut. (*Hint*: A cable is no longer taut when its tension drops to zero.)

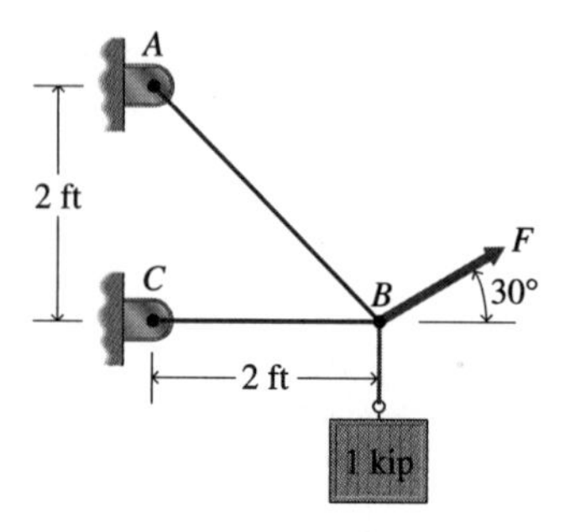

Figure P3.4

3.5 For the system shown in Fig. P3.5, assume that the horizontal force P is allowed to vary in magnitude. Determine the range of values of P such that cables AB and BC remain taut. (*Hint*: A cable is no longer taut when its tension drops to zero.)

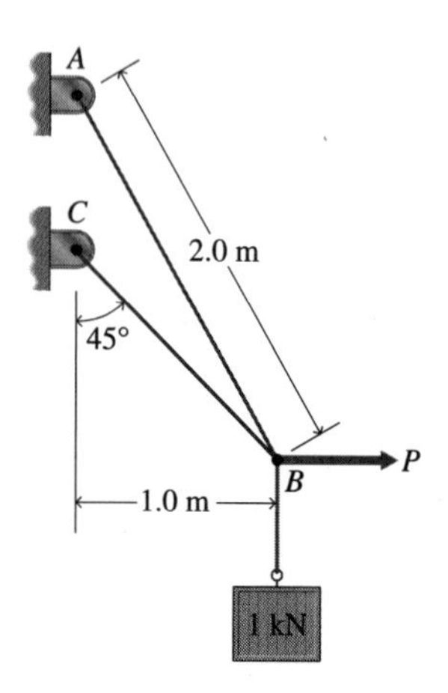

Figure P3.5

3.6 A batter hits a pitched baseball with a bat.

a. Draw a free-body diagram of the ball when the bat and the ball are in contact.

b. Draw a free-body diagram of the ball after it leaves the bat.

c. Is the ball in equilibrium in part a? Explain.

d. Is the ball in equilibrium in part b? Explain.

3.7 Two small electrostatic spheres, each weighing 100 dynes (10^5 dynes $= 1$ N), have equal electric charges e (in electrostatic units, or esu) (see Fig. P3.7). They are suspended from the same point P by silk fibers 600 mm long. The repulsion between the spheres keeps them 80 mm apart. The force of repulsion between two electric charges e and e' (in esu) that are separated by a distance r (in cm) is ee'/r^2 (in dynes). Recall that 1 cm $=$ 10 mm. Calculate the charges on the spheres in electrostatic units.

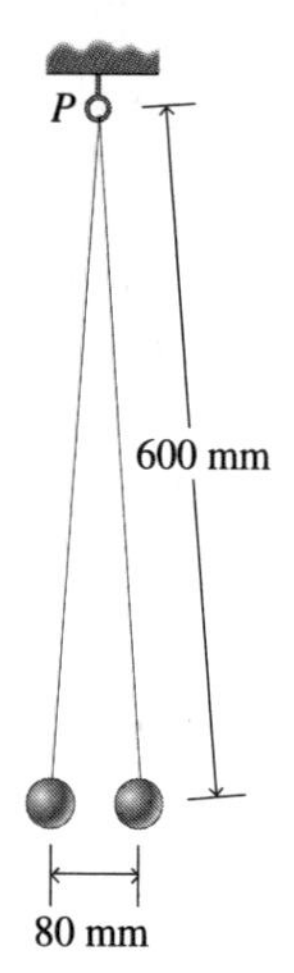

Figure P3.7

3.8 An automobile is stuck in a ditch (see Fig. P3.8). The driver attaches one end of a rope to the automobile and the other end to a stump and then pulls on the midpoint of the rope with a force of 100 lb. The 100 lb force produces a tensile force T in the rope.

a. Derive a formula for the force T that the rope exerts on the automobile and on the stump, in terms of the angle θ.

b. Calculate values of T for $\theta = 10°$, 20°, and 45°. What do you conclude from these results?

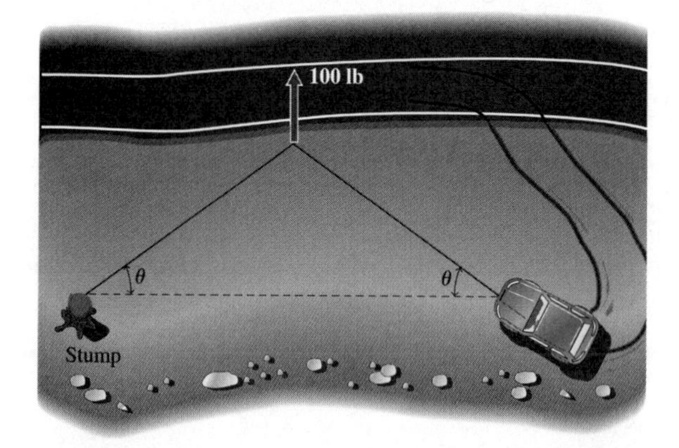

Figure P3.8

3.9 To pull a rowboat out of a creek, a boy ties a rope from the boat to a tree. He grips the rope tightly and pulls laterally with force of 450 N, as shown in Fig. P3.9. This pull produces a force T in the section of rope between the boy and the boat.

a. For a given angle ϕ, determine the angle θ such that the initial pull T on the boat is a maximum.

b. Calculate the maximum value of T for $\phi = 170°$, 160°, and 150°.

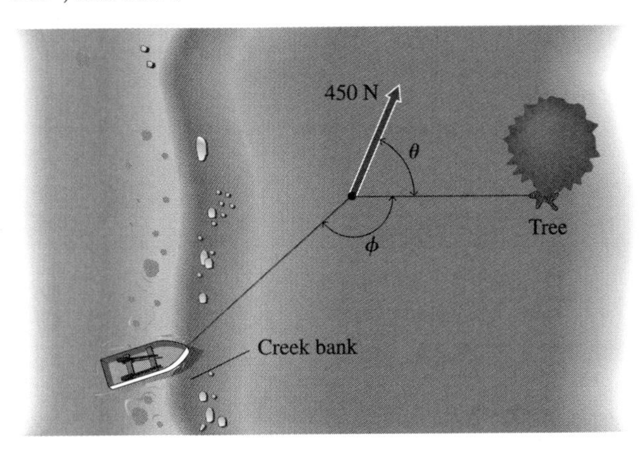

Figure P3.9

3.10 A double pendulum is held in equilibrium in the position shown in Fig. P3.10 by the horizontal force F. The bobs B and C each weigh 15 N. Determine the force F, the tensions in the strings AB and BC, and the angle θ.

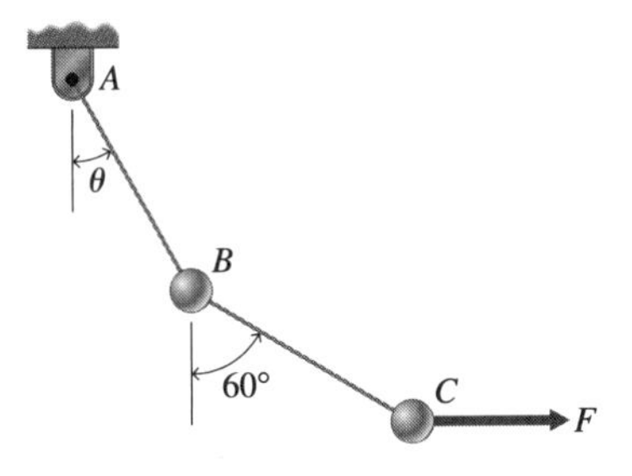

Figure P3.10

3.11 A weight W is suspended from a cable, as shown in Fig. P3.11. Show that

$$T_1 = \frac{W \sin \beta}{\sin(\alpha + \beta)} \quad \text{and} \quad T_2 = \frac{W \sin \alpha}{\sin(\alpha + \beta)}$$

where T_1 is the force in cable AB and T_2 is the force in cable AC.

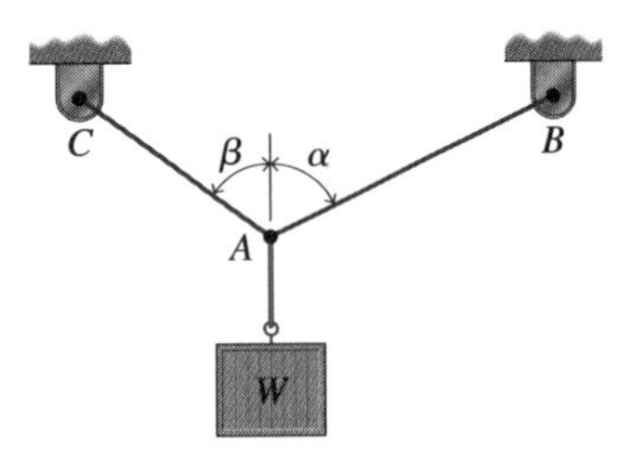

Figure P3.11

3.12 A weightless rope that is fastened at two points A: (9, 0) and D: (0, 0) carries a 500 N weight and an unknown weight W, as shown in Fig. P3.12. The points of attachment of the weights to the rope lie at B: (7, 3) and C: (4, 4). Determine the tensions T_1, T_2, and T_3 in the three segments of the rope and the weight W.

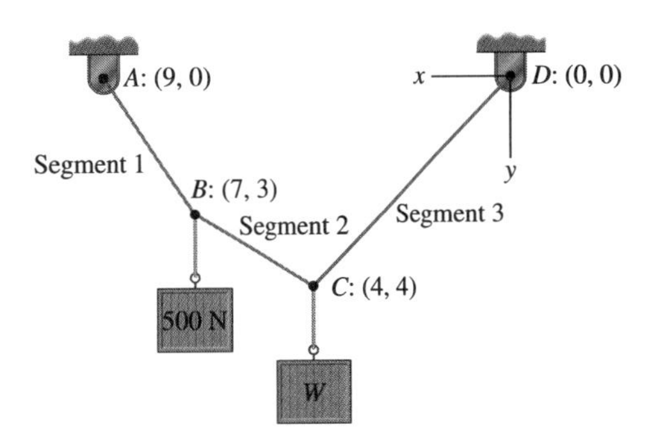

Figure P3.12

3.13 The length of the trolley cable in Fig. P3.13 is 50 m. Neglecting the weight of the cable and the weight and friction of the small roller, determine the distance x necessary for equilibrium. (*Hint:* Assume that the tension in the cable has the same magnitude on both sides of the small roller.) Use a computer-based equation solver to solve the simultaneous equations.

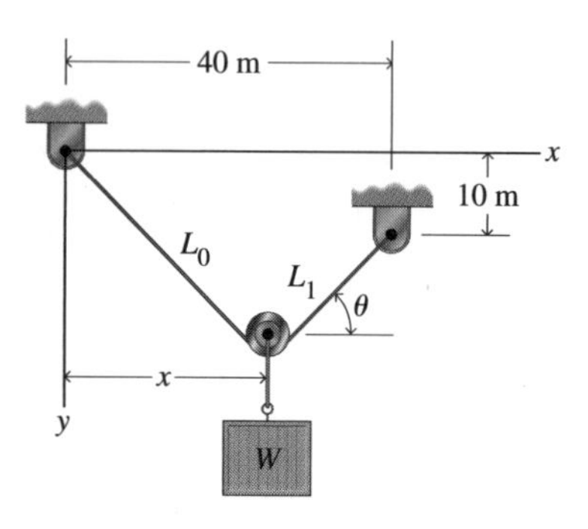

Figure P3.13

3.14 The cargo box shown in Fig. P3.14 weighs 900 N. The force $P = 200$ N is just sufficient to lift the box off the floor. Determine the angle θ.

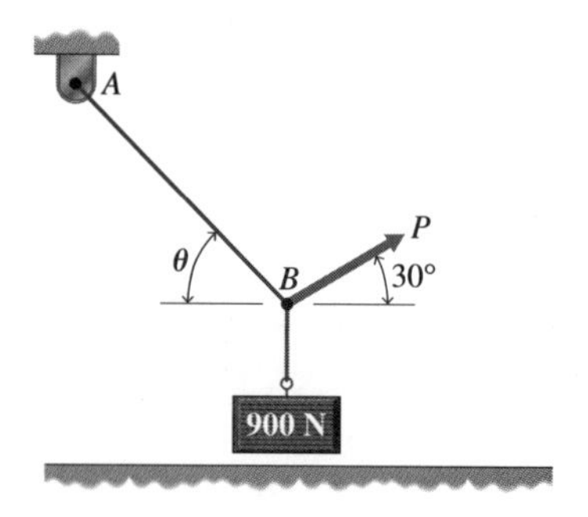

Figure P3.14

3.15 Particles A and B each weigh 1 N (see Fig. P3.15). They are held in equilibrium in the positions shown by the cords. Determine the tension in each of the cords. Neglect the weight of the cords.

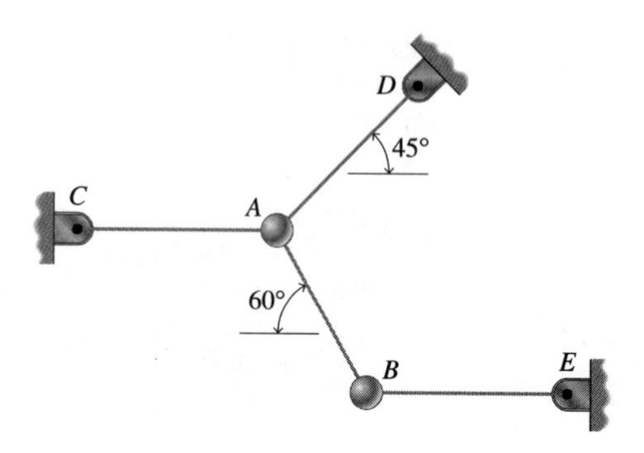

Figure P3.15

3.16 Each of the two ropes AB and AC is subject to a tensile force of 100 N (see Fig. P3.16).

a. Determine the equilibrant of the force that the ropes exert on point A.

b. Determine the angle between the equilibrant and the positive x axis.

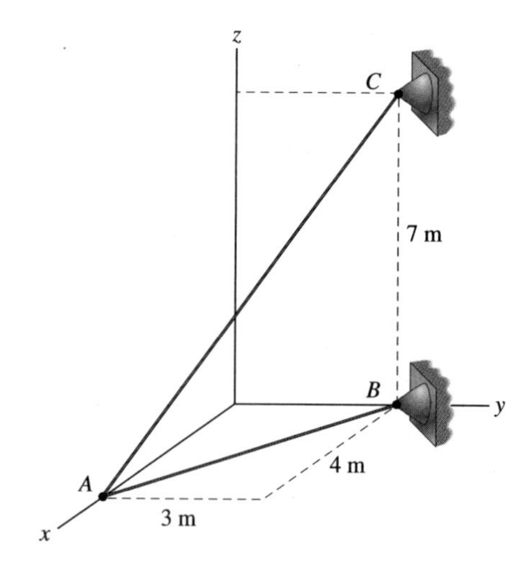

Figure P3.16

3.17 A cylindrical oil drum is hung from three cables, as shown in Fig. P3.17. The coordinates of points *A*, *B*, *C*, and *D* are given in meters. The drum weighs 1.6 kN. Determine the tension in each of the cables.

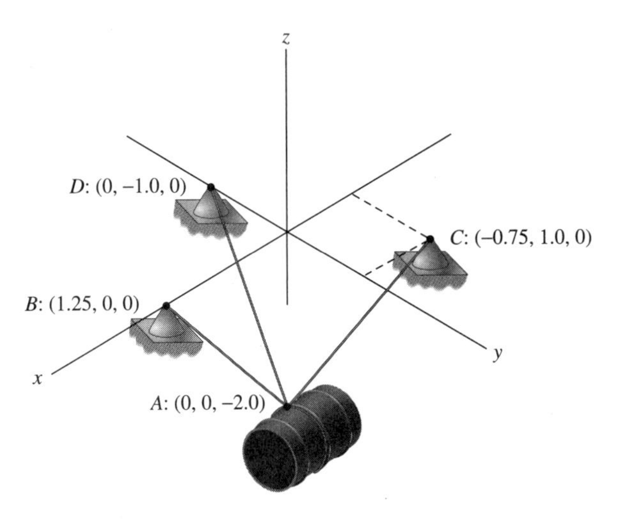

Figure P3.17

3.18 Before the 445 N load is applied, the wire *ABC* in Fig. P3.18 is straight and horizontal and the tension in it is negligible. After the load is applied, the increase in length δ (in mm) of the wire caused by the tension T (in N) in the wire is given by $\delta = 0.0114T$.

a. Calculate the vertical deflection x caused by the 445 N load.

b. Calculate the corresponding tension T in the wire. (*Hint:* Express the equilibrium condition and, hence, the overall length $L' = L + \delta$ of the wire in terms of θ. The condition $\cos\theta = L/L'$, where $L = 1500$ mm, then yields a trigonometric equation in θ that may be solved for θ by a computer-based equation solver. Alternatively, solve the trigonometric equation with a spreadsheet.)

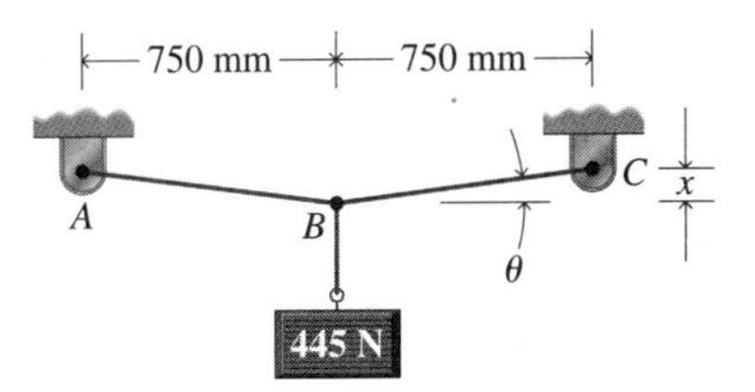

Figure P3.18

3.19 A weight $W = 100$ lb hangs from a roller that rolls along a cable *ABC* (see Fig. P3.19). The horizontal force F (in lb) moves the roller slowly along the cable. The roller is frictionless. Plot the force F as a function of the distance x for the range $0 \le x \le S = 20$ ft, in increments of 0.5 ft. The length of the cable is $L_1 + L_2 = 24$ ft. Assume that the tension T in the cable is the same on both sides of the small roller.

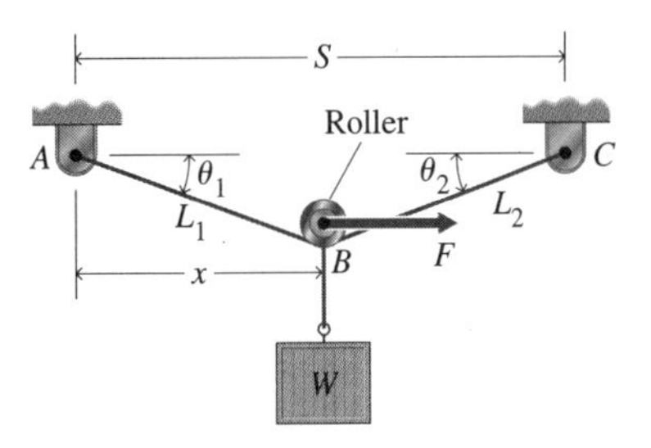

Figure P3.19

3.20 A basketball scoreboard that weighs 2000 lb is to be hung from the roof framing of an arena using a cable *ABC* (Fig. P3.20). The ends of the cable are attached to hinges at points *A* and *C*. Before the scoreboard is attached at *B*, the midpoint, cable *ABC* is slightly slack (it is longer than 10 ft). The weight of the scoreboard is to be applied slowly at *B* until equilibrium is reached. At equilibrium, point *B* on the cable is a distance *d* below line *AC* as the cable becomes taut (its length does not change, however). The cable can safely carry a maximum tension of 4500 lb. Design limitations require that point *B* drop not more than 12 in below line *AC* as the scoreboard is attached. As a consulting engineer, you are asked to design the system. Analyze the given data to see whether you can satisfy the requirements. If you can, specify the precise length of the cable and the value of θ to use. If you cannot, what might you recommend?

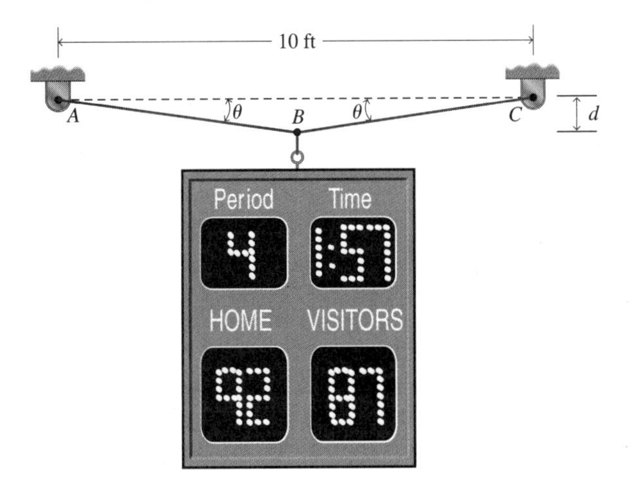

Figure P3.20

3.5 *Equilibrium of a Two-Force Member*

3.21 For the structure shown in Fig. P3.21, determine the forces exerted on joint B by members AB and BC.

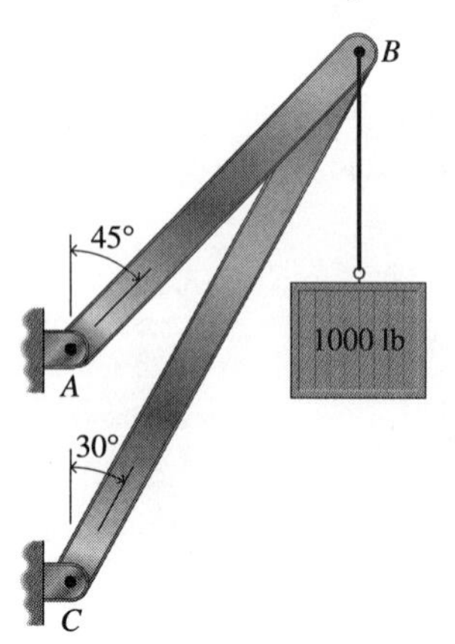

Figure P3.21

3.22 Determine the forces that the members exert on joint B of the structure shown in Fig. P3.22.

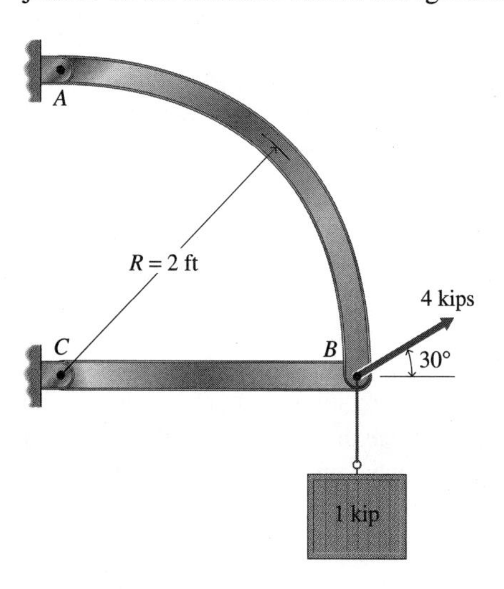

Figure P3.22

3.23 Determine the forces that the members exert on joint B of the structure shown in Fig. P3.23.

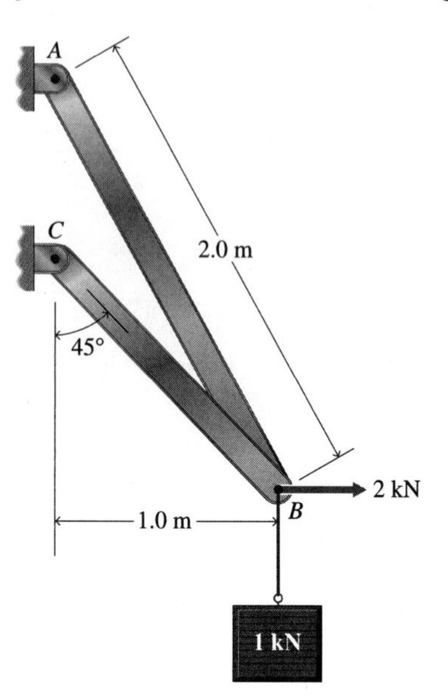

Figure P3.23

3.24 Determine the forces that the members exert on joint B of the structure shown in Fig. P3.24, where distances are in meters.

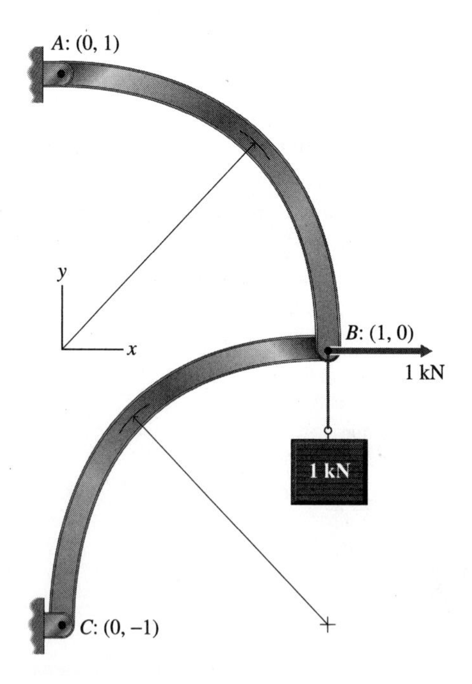

Figure P3.24

3.25 Determine the forces that the members exert on joint B of the structure shown in Fig. P3.25, where distances are in meters.

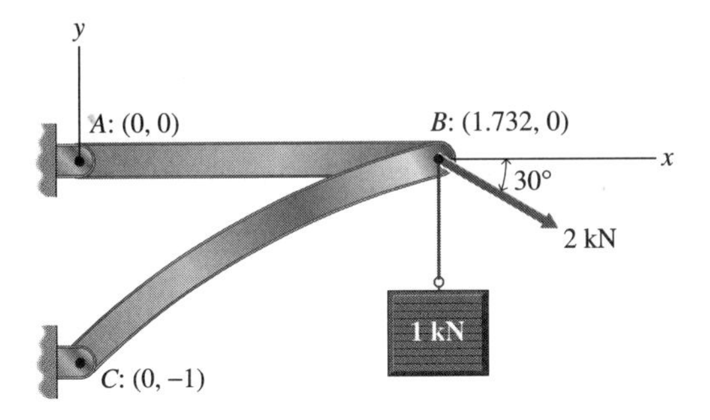

Figure P3.25

3.26 Determine the forces that the members exert on joint B of the structure shown in Fig. P3.26.

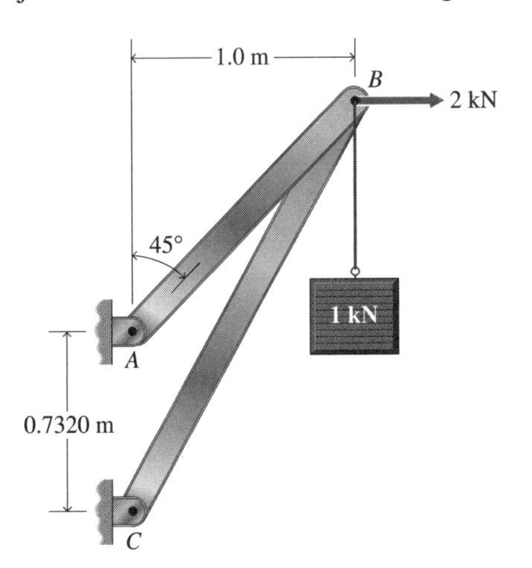

Figure P3.26

3.27 Determine the forces that the members exert on joint B of the structure shown in Fig. P3.27.

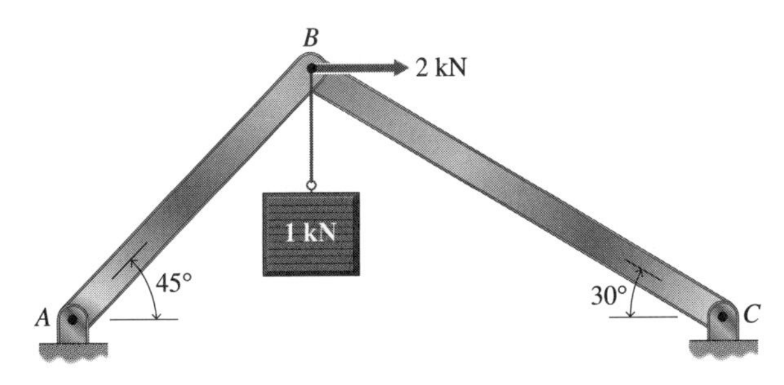

Figure P3.27

3.28 The frame shown in Fig. P3.28 is a regular octahedron (all edge lengths are equal). Determine the tension in each of the members meeting at point O, using the fact that symmetry makes these tensions equal. All joints are pinned, so that each member is a two-force member.

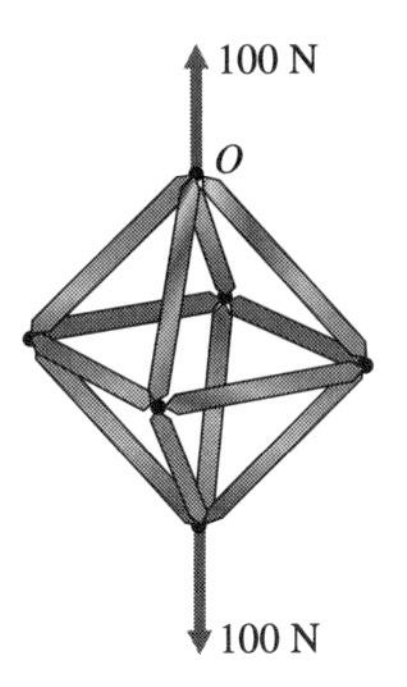

Figure P3.28

3.29 Determine the forces in the three members of the wall bracket $ABCD$ in Fig. P3.29. In the figure, point A lies on the z axis, points B and C on the y axis, and point D on the x axis.

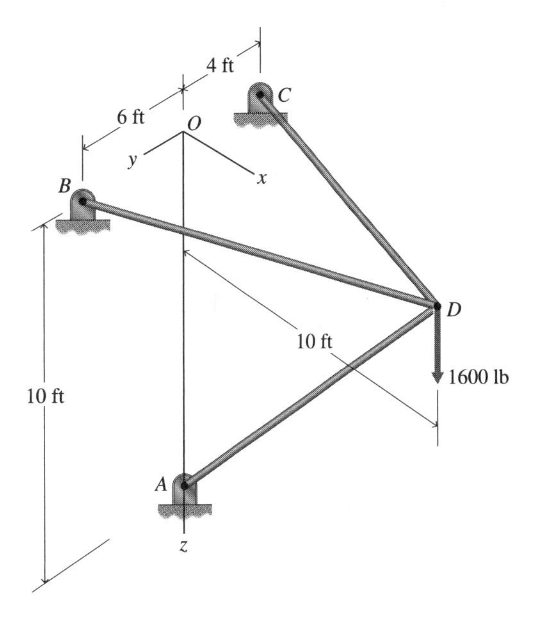

Figure P3.29

3.30 Two bars *AB* and *AC*, each 1 m long, and a cable *AD*, also 1 m long, are assembled into the boom shown in Fig. P3.30. A weight *W* is hung from the boom at point *A*. Determine the force in each bar and the force in the cable in terms of *W*.

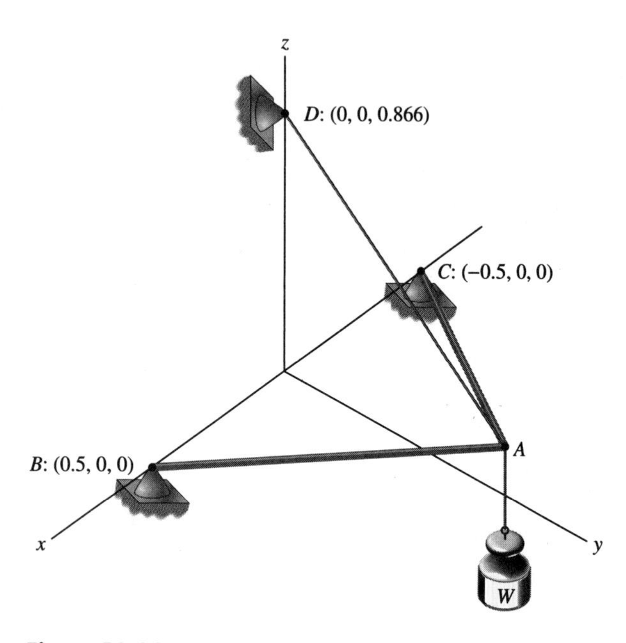

Figure P3.30

3.31 For the boom described in Problem 3.30, assume that cable *AD* is attached to a hoist mechanism at *D*. The length *L* of the cable can be varied from 0 (when the boom is pulled up vertically) to 1.732 m (when the boom hangs down vertically). Let $W = 18$ kN. Calculate the forces in the bars and the cable and plot them over the interval $0 < L < 1.732$ m, breaking this interval into twenty increments. (*Note:* The problem is statically indeterminate at $L = 0$ and at $L = 1.732$ m.)

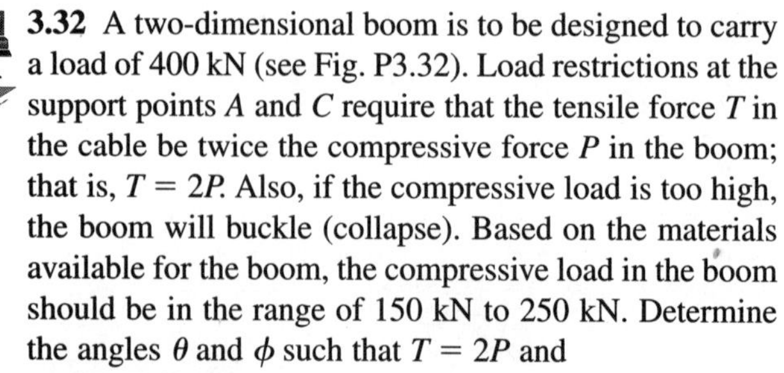

3.32 A two-dimensional boom is to be designed to carry a load of 400 kN (see Fig. P3.32). Load restrictions at the support points *A* and *C* require that the tensile force *T* in the cable be twice the compressive force *P* in the boom; that is, $T = 2P$. Also, if the compressive load is too high, the boom will buckle (collapse). Based on the materials available for the boom, the compressive load in the boom should be in the range of 150 kN to 250 kN. Determine the angles θ and ϕ such that $T = 2P$ and

a. $P = 150$ kN
b. $P = 200$ kN
c. $P = 250$ kN

(*Hint:* Write the equations of equilibrium in terms of the unknown angles θ and ϕ. Then use a computer-based equation solver to find the angles for a given value of *P*.)

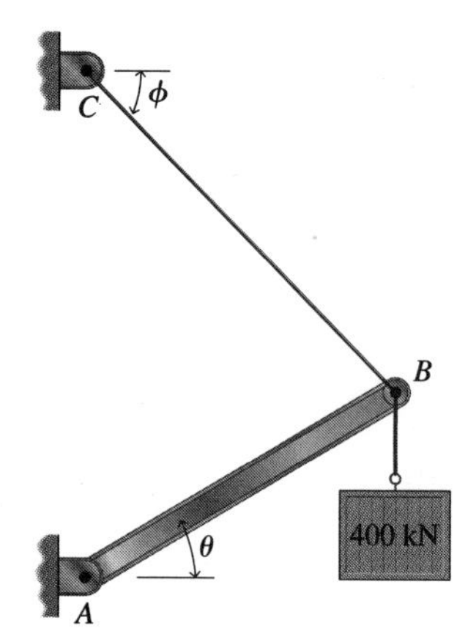

Figure P3.32

3.7 *Equilibrium of a Rigid Body Subjected to Concurrent Forces*

It was shown in Sec. 3.7 that if a rigid body is in equilibrium under the action of three nonparallel forces, the lines of action of the forces are coplanar and concurrent. This fact may help you solve some of the following problems.

3.33 Determine the resultant of the three forces that act on the triangular plate in Fig. P3.33.

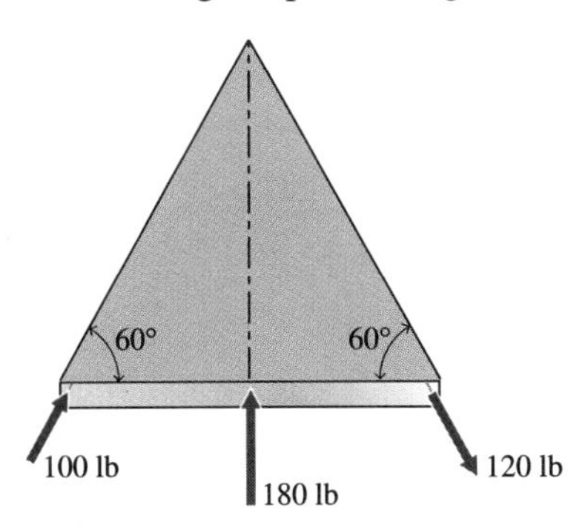

Figure P3.33

3.34 Express in terms of R_1, R_2, W, and P the projections (F_x, F_y) of the resultant force that acts on the lawn roller in Fig. P3.34.

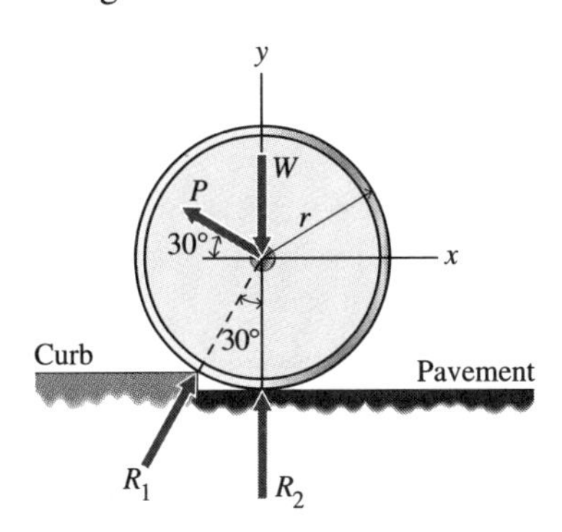

Figure P3.34

3.35 What minimum force P, written in terms of W, is required to make the lawn roller of Problem 3.34 start to roll over the curb?

3.36 The bar AB in Fig. P3.36 is held in equilibrium by the three concurrent forces. Determine θ, F, and P.

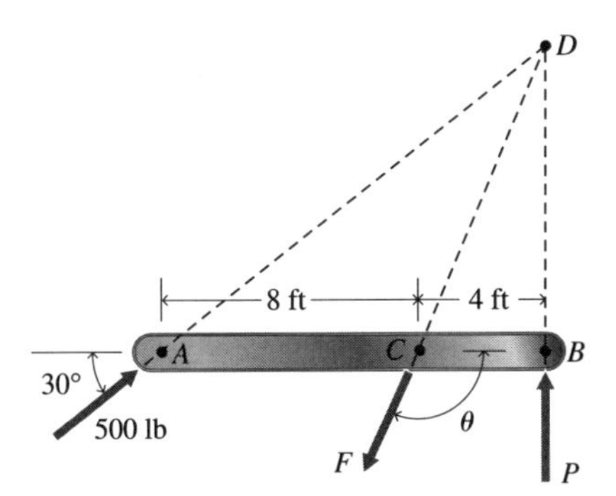

Figure P3.36

3.37 A plate is held in equilibrium by the forces shown in Fig. P3.37. Calculate the forces T_1 and T_2 in two ways:

a. projecting forces onto the x and y axes

b. projecting forces in the directions of forces T_1 and T_2

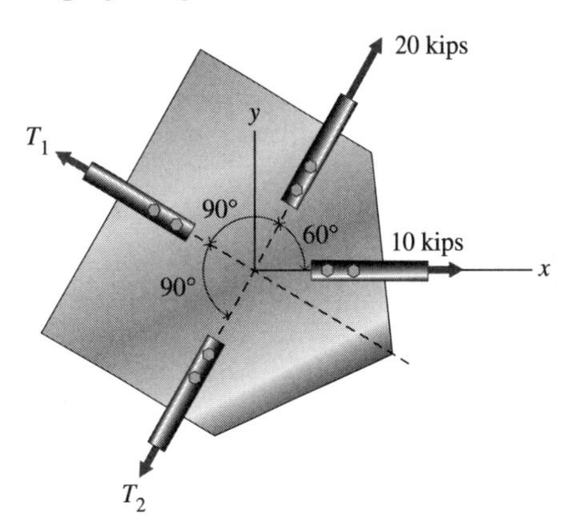

Figure P3.37

3.38 The large cylinder in Fig. P3.38 weighs 50 kN. Calculate the reactions of the rollers on the cylinder. Assume that these forces act normal to the cylinder.

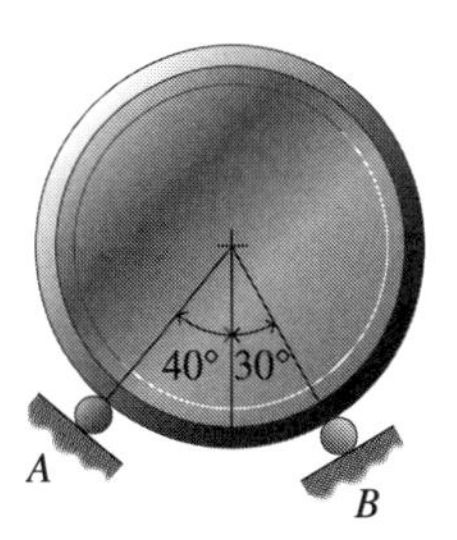

Figure P3.38

3.39 Each of the two balls in Fig. P3.39 weighs 1 kN. Calculate the tension in each of the three cords.

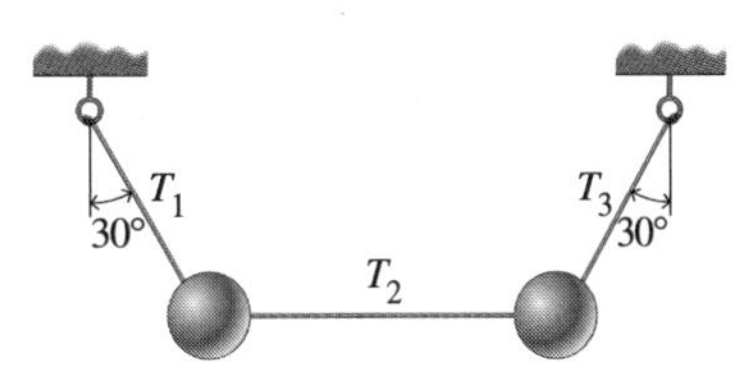

Figure P3.39

3.40 A cylinder of mass m_1 is held in equilibrium on an inclined plane by a mass m_2 attached to the cylinder by a cord that passes over a wheel (see Fig. P3.40). All surfaces are smooth, and the wheel revolves on a frictionless pin. Therefore, the tension in the cord is constant throughout its length, and the reaction of the inclined plane on the cylinder is directed normal to the plane.

a. Draw a free-body diagram of m_1.
b. Draw a free-body diagram of m_2.
c. Determine the relationship between m_1 and m_2.
d. Compute the tension in the cord in terms of m_1.
e. Compute the reaction of the inclined plane on the cylinder in terms of m_1.

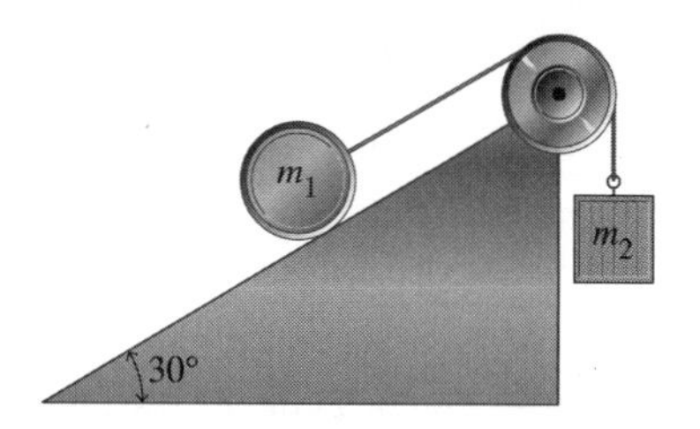

Figure P3.40

3.41 Two balls, whose masses are m_1 and m_2, are connected by a bar of negligible mass and set on a smooth cylindrical drum of radius R (see Fig. P3.41). The reactions of the drum on the balls are directed normal to the surface of the drum. For angle θ, the balls are in equilibrium.

a. Draw a free-body diagram of each ball.
b. In terms of m_1, m_2, and appropriate angles θ, α, β, and γ, determine the reactions of the drum on the balls and the force in the bar. Express the reactions in simplified form.

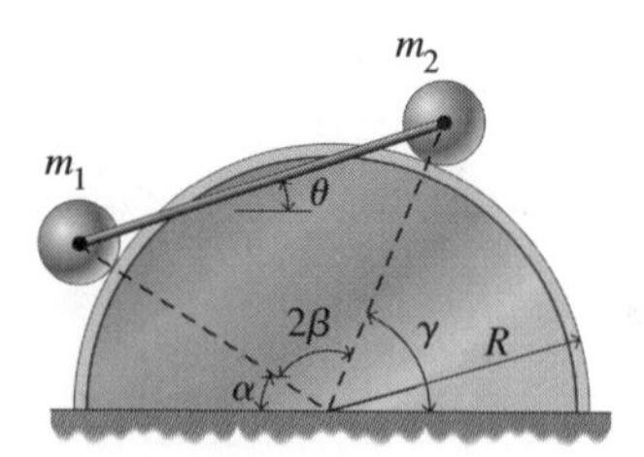

Figure P3.41

3.42 A cable is wrapped onto a spool. A force T is applied to the cable, as shown in Fig. P3.42. The spool does not skid.

a. Draw the free-body diagram of the spool.
b. What is the angle θ (expressed in terms of r_1 and r_2) such that the spool does not roll?

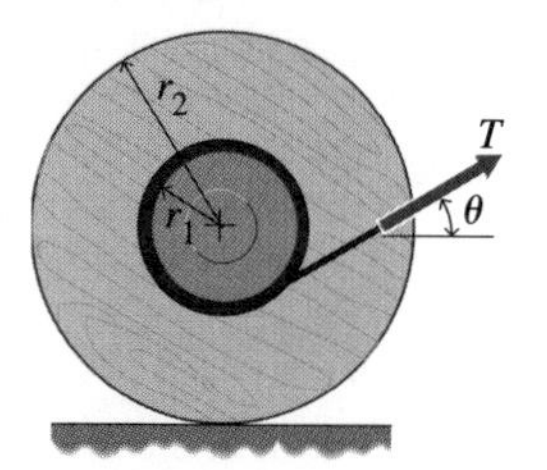

Figure P3.42

3.43 The two identical cylinders shown in Fig. P3.43 weigh 200 lb each, and the cubical block weighs 300 lb. Neglecting friction, construct the necessary free-body diagrams and calculate all the contact forces.

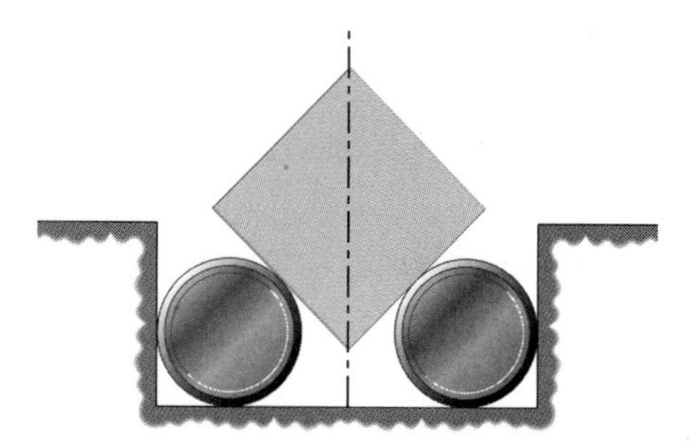

Figure P3.43

3.44 The cylindrical rollers in Fig. P3.44 weigh 50 N each. Neglecting friction, construct free-body diagrams of the individual rollers and determine all the contact forces.

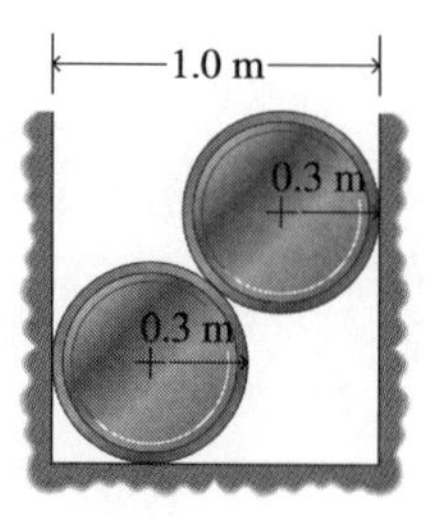

Figure P3.44

3.45 A square picture, 1 m on a side, is hung by a wire that is attached to the upper corners, as shown in Fig. P3.45. If the picture hangs crookedly at 30° to the horizontal, then the points *A, B,* and *C* lie on a straight line. Determine the length of the wire *BCD*. Neglect any friction from the peg at point *C*.

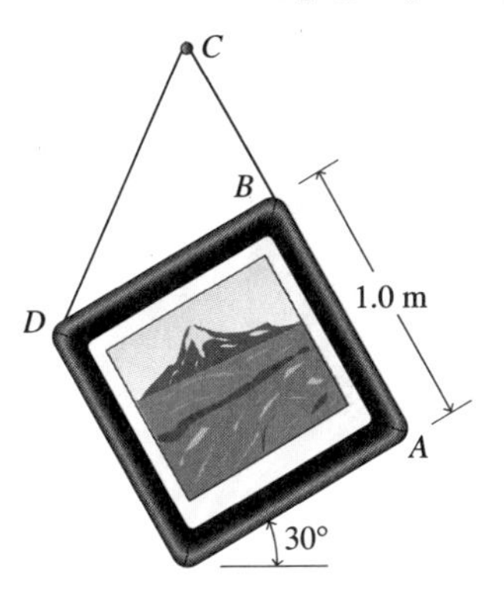

Figure P3.45

3.46 The bed of a loaded dump truck weighs 50 kN (see Fig. P3.46). It is held in equilibrium in a tilted position by a hydraulic cylinder whose axis lies 15° from the vertical. Using the given dimensions, compute the horizontal and vertical reactions A_x and A_y of the frictionless pin *A* and the thrust *P* of the hydraulic cylinder. (*Hint:* See Theorem 3.6.)

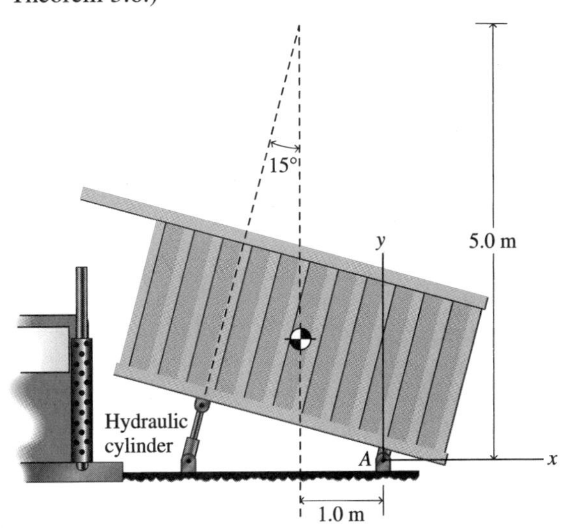

Figure P3.46

3.47 The roller in Fig. P3.47 weighs 100 lb. The force *P* parallel to the inclined plane holds the roller in equilibrium on the plane. Determine the force *P* and the contact force between the roller and the plane by

a. projecting forces onto the *x* and *y* axes

b. projecting forces onto lines parallel and perpendicular to the inclined plane

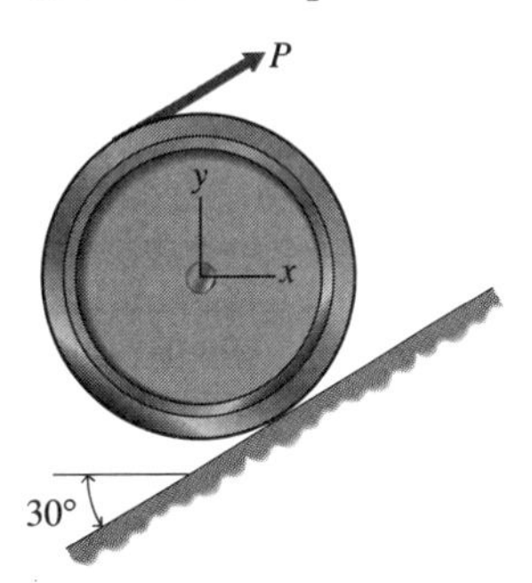

Figure P3.47

3.48 The ring in Fig. P3.48 weighs 400 N and is supported by two cords that pass around two frictionless pulleys. Since the pulleys are frictionless, the tension in each cord does not change as the cord passes over the pulley. Determine the angles α and β. Solve the simultaneous equations with the aid of a computer-based equation solver.

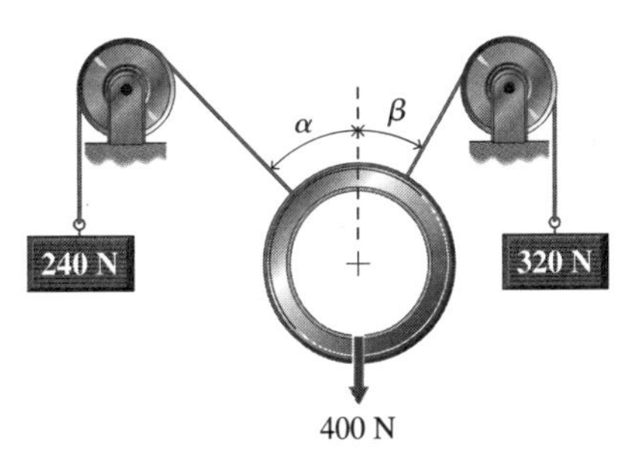

Figure P3.48

3.49 The cylinders B and C in Fig. P3.49 weigh 1 kN each. Calculate the tension in each of the three cords AB, AC, and BD. (*Hint:* See Theorems 3.5 and 3.6.)

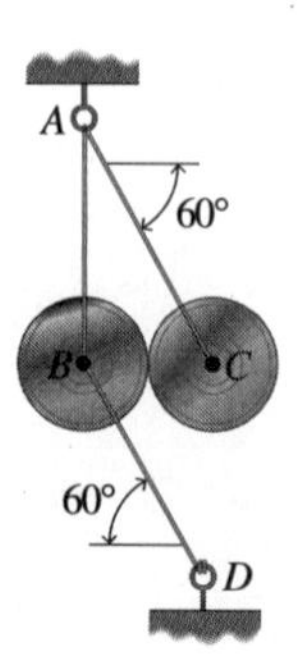

Figure P3.49

3.50 The roller in Fig. P3.50 weighs 100 lb. The force P holds it in equilibrium on the inclined plane. Determine the force P and the contact force between the roller and the smooth plane by the following methods (see Theorem 3.6):

a. projecting the forces onto the x and y axes

b. projecting the forces onto lines parallel and perpendicular to the inclined plane

c. the polygon construction

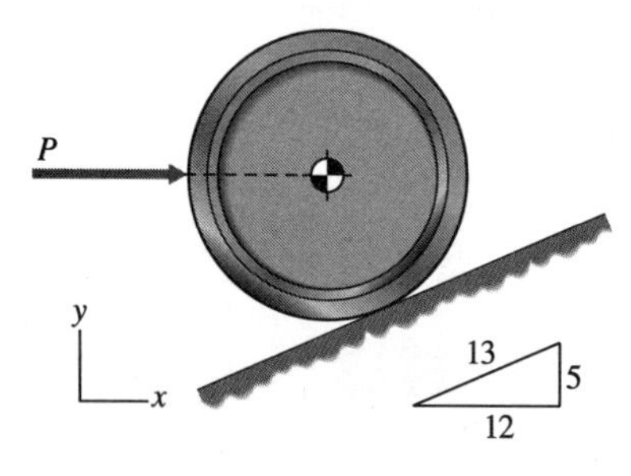

Figure P3.50

3.51 The homogeneous rectangular slab shown in Fig. P3.51 weighs 35 kN. It is suspended by Chains 1, 2, and 3. Calculate the tensions T_1, T_2, and T_3 in the chains.

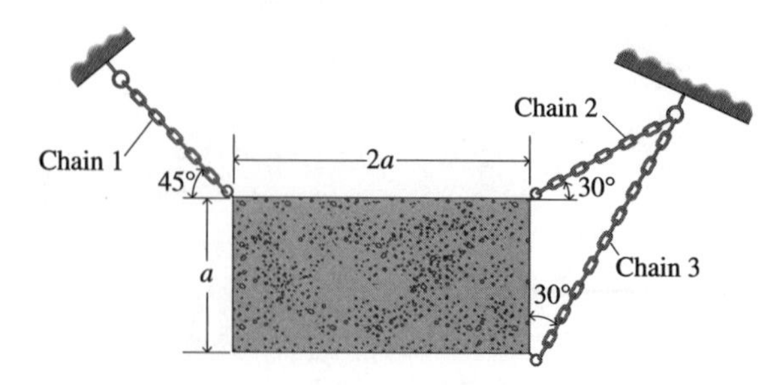

Figure P3.51

3.52 The crate in Fig. P3.52 weighs 800 lb. It is held by Chains 1, 2, and 3. Calculate the tension in each chain.

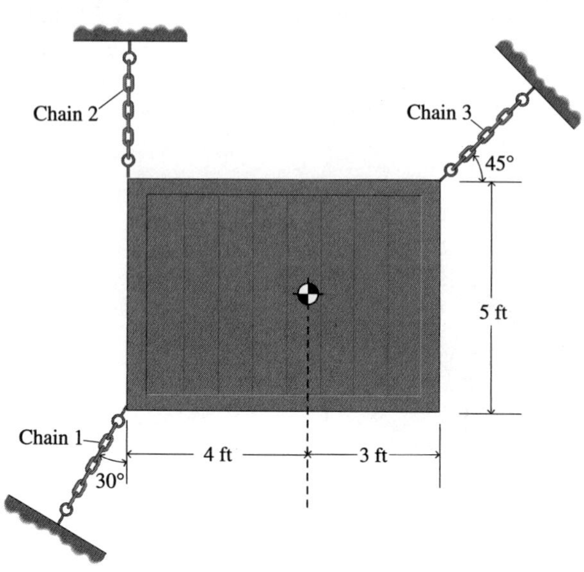

Figure P3.52

3.53 Two cylindrical rollers each have weight W (see Fig. P3.53). Neglecting friction, construct a free-body diagram of each cylinder and determine the contact forces at A, B, C, and D in terms of W. (*Hint:* See Theorems 3.5 and 3.6.)

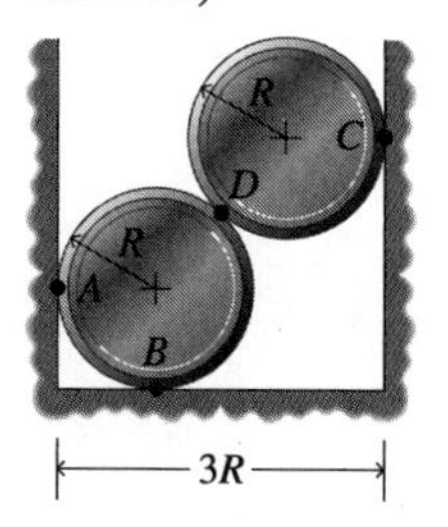

Figure P3.53

3.54 The structure (tripod) shown in Fig. P3.54 is used to tie down a small dirigible. The force in the tie rope has magnitude $F = 45$ kN. While it is tied down, the dirigible moves, causing the direction of the force (which remains essentially constant in magnitude) to vary in the yz plane such that the angle θ between the force and the y axis varies over the range $-20° \leq \theta \leq 20°$. This means that the tip of the force vector moves along the arc of a circle in the yz plane (see Fig. P3.54). Plot the forces in the members PA, PB, and PC as functions of θ over the range $-20° \leq \theta \leq 20°$, in increments of 1°.

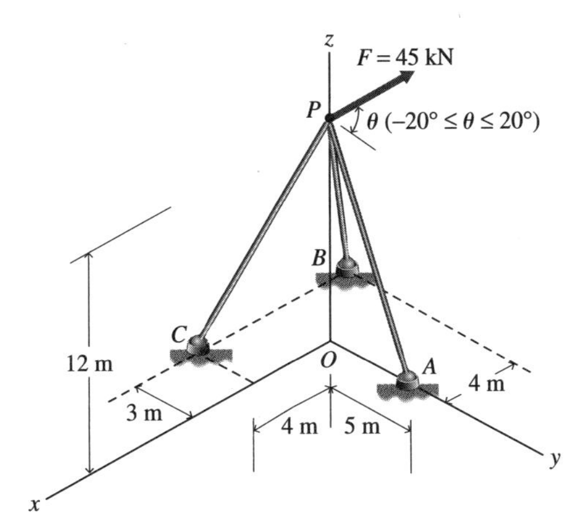

Figure P3.54

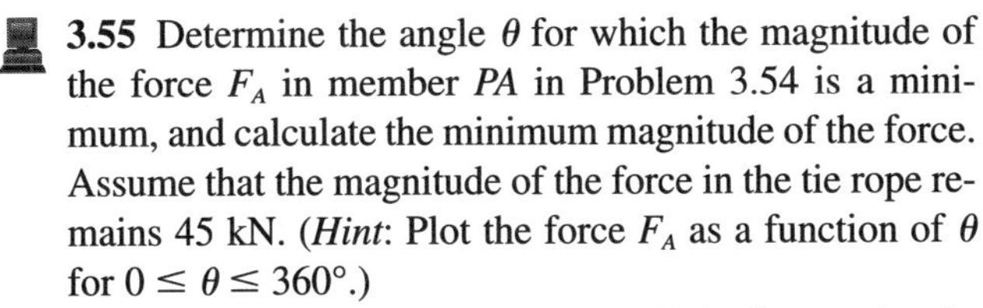

3.55 Determine the angle θ for which the magnitude of the force F_A in member PA in Problem 3.54 is a minimum, and calculate the minimum magnitude of the force. Assume that the magnitude of the force in the tie rope remains 45 kN. (*Hint*: Plot the force F_A as a function of θ for $0 \leq \theta \leq 360°$.)

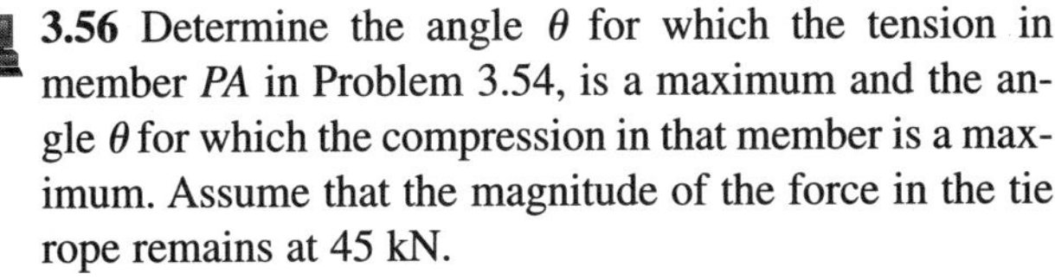

3.56 Determine the angle θ for which the tension in member PA in Problem 3.54, is a maximum and the angle θ for which the compression in that member is a maximum. Assume that the magnitude of the force in the tie rope remains at 45 kN.

REVIEW QUESTIONS

3.57 If a motionless particle remains motionless, what can you say about the forces that act on it?

3.58 State the two general algebraic conditions of equilibrium for concurrent, coplanar forces in terms of (x, y) projections.

3.59 State the three general algebraic conditions of equilibrium for concurrent, noncoplanar forces in terms of (x, y, z) projections.

3.60 State the equilibrium criterion for a rigid body that is subjected to forces with concurrent lines of action.

3.61 Define *contact forces* and *body forces*.

3.62 What is a free-body diagram? What does it show?

3.63 *True or False*: The principle of transmissibility allows you to move a force to a parallel line of action.

3.64 *True or False*: The resultant of two concurrent forces may replace the two forces without altering the motion of a rigid body on which they act.

3.65 *True or False*: Two collinear forces have a zero resultant provided they have the same magnitude.

3.66 How can the proper selection of xy coordinate axes simplify the solution of an equilibrium problem in two dimensions?

3.67 Four mechanical systems are shown on the left in Fig. P3.67. Each body of each system depicted weighs 50 N, and each weight acts at the center of gravity of the body. Assume that all bodies are in equilibrium and that all connections and surfaces are frictionless. Supply the necessary forces to complete the free-body diagrams of the bodies in the right-hand column. Indicate the directions and senses of the forces in the free-body diagrams.

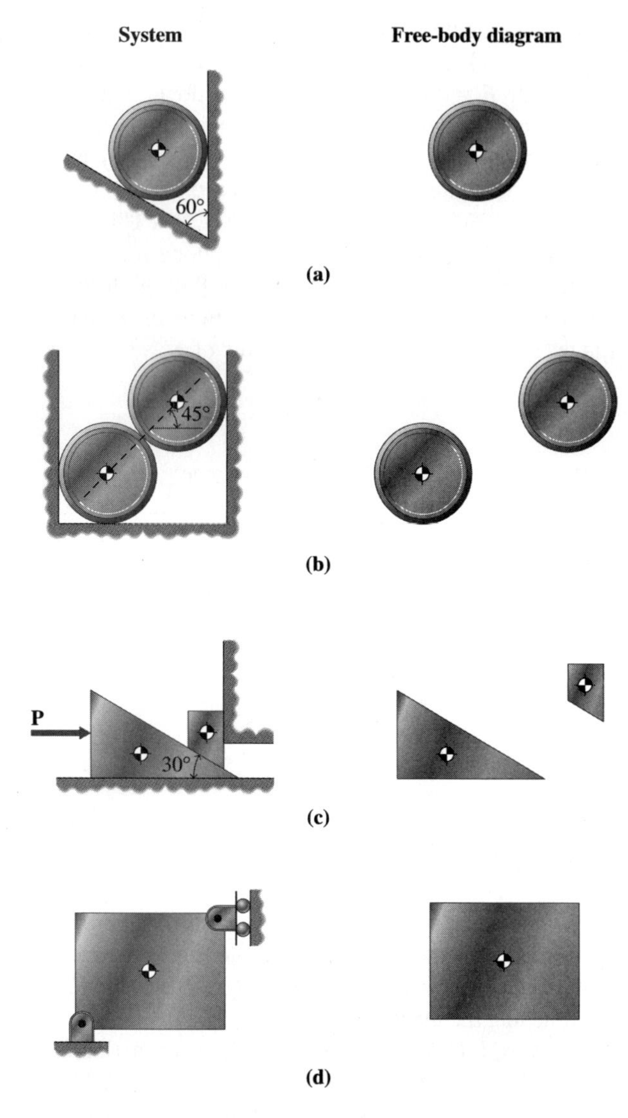

Figure P3.67

Chapter 4 TWO-DIMENSIONAL FORCES, COUPLES, AND RIGID-BODY EQUILIBRIUM

The woman at the helm of this sailboat applies a tangential force to the wheel. That force creates a moment about the axis of the wheel, causing it to turn.

A Look Forward

IN CHAPTER 3, we considered the principles of equilibrium of a particle. That discussion included a study of the equilibrium of rigid bodies subjected to concurrent forces. In this chapter, we will extend the concept of equilibrium to include rigid bodies under the action of coplanar forces that are not concurrent.

Intuitively, you know that a force can cause a body to translate (to move in the direction of the force). However, depending on how a force is applied to a body, it can also cause the body to rotate. The action that causes rotation is called a *moment.* In the first part of this chapter, the concept of a moment of a force is defined. You will learn how to compute the moment of a force about an axis and how to determine the resultant moment of several coplanar forces. Two parallel noncollinear forces with equal magnitudes but opposite senses form a couple. For a couple, the resultant force is zero, but the moment is nonzero.

After we discuss the effects of coplanar forces on rigid bodies, we will establish the conditions for which a rigid body is in equilibrium under the action of coplanar forces. Then, we will consider the special case of three-dimensional equilibrium of a rigid body subjected to parallel noncoplanar forces, as a prelude to the general theory of three-dimensional equilibrium of a rigid body treated in Chapter 5.

After you have studied this chapter, you should be able to determine the equilibrium conditions for rigid bodies subjected to coplanar forces and to parallel noncoplanar forces. When you perform such an equilibrium analysis for a rigid body, you presume that the body is in equilibrium. Then, by the requirements for equilibrium, you can obtain a set of equations, whose solution determines the unknown quantities in the problem.

Survey Questions

- How does rigid-body equilibrium differ from particle equilibrium?
- What is a moment of a force with respect to an axis?
- What is meant by a couple?
- What is the difference between the moment of a force and the moment of a couple?
- What are the conditions of equilibrium for a rigid body subjected to coplanar forces?

4.1 MOMENT OF COPLANAR FORCES WITH RESPECT TO AN AXIS

After studying this section, you should be able to:

- Define *moment of a force* and *moment arm.*
- Determine the moment of a force about an axis normal to a plane containing the force.
- Determine the moment of several coplanar forces about any point in their plane.

Key Concept A force **F** that lies in a plane Q causes a moment **M** with respect to an axis that is perpendicular to plane Q. The magnitude M of the moment **M** is Fr, where F is the magnitude of **F** and r is the perpendicular distance from the axis to the line of action of **F**.

Consider a force **F** that lies in a given plane Q (say, the xy plane of the body in Fig. 4.1a). The force **F** acts at point P. Let the z axis be perpendicular to plane Q and intersect Q at point O. Then, the x, y, and z axes form a right-handed rectangular coordinate system with origin O. The vectors **i**, **j**, and **k** are unit vectors along the x, y, and z axes, respectively.

Intuitively, you can see that force **F** would tend to cause the body in Fig. 4.1a to rotate about the z axis; this rotation would be counterclockwise when viewed from a position along the positive z axis (see Fig. 4.1b). The action of a force that tends to rotate a body about an axis is called a *moment.* The force **F** causes a moment **M** with respect to the z axis, defined by the equation[1]

moment

$$\mathbf{M} = \pm M\mathbf{k} = \pm Fr\mathbf{k} \tag{4.1}$$

1. In Chapter 5, it is shown that the moment of a force is a vector. Hence, the moment is denoted by the boldface symbol **M**. The vector **M** has magnitude $M = Fr$ and direction perpendicular to plane Q. The sense of **M** is either the same as that of the unit vector **k** (the plus sign) or opposite to that of **k** (the minus sign). In Fig. 4.1b, the plus sign applies; in Fig. 4.1c, the minus sign applies.

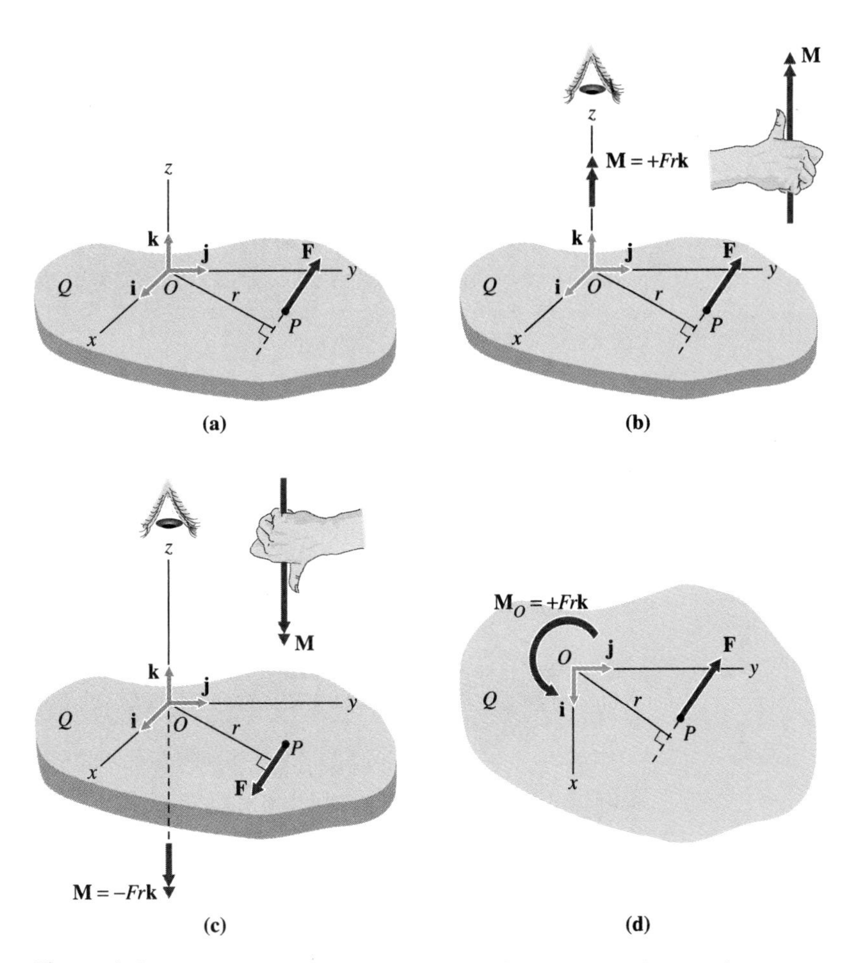

Figure 4.1 **(a)** Moment arm of a force. **(b)** Positive moment about z axis. **(c)** Negative moment about z axis. **(d)** Curved arrow representation of moment.

where $M = Fr$ is the magnitude of the moment **M**, F is the magnitude of the force **F**, and r is the perpendicular distance from the line of action of **F** and the z axis (point O). The distance r is called the *moment arm* of force **F** with respect to the z axis.

moment arm

To establish the sign in Eq. (4.1), it is arbitrarily specified that the moment is positive if it has a counterclockwise sense of rotation about point O (Fig. 4.1b). Then, $\mathbf{M} = +Fr\mathbf{k}$, and the moment vector **M** is directed in the positive z direction; it has the same sense as the unit vector **k**. The moment **M** is represented by a double-headed arrow to distinguish it from a force vector (see Fig. 4.1b). Thus, for a positive moment, the double-headed arrow has the same sense as **k** (the positive direction of the z axis). If the moment **M** is negative, it has a clockwise sense of rotation about the point O (Fig. 4.1c). Then, by Eq. (4.1), $\mathbf{M} = -Fr\mathbf{k}$, and the double-headed arrow is directed in the $-\mathbf{k}$ direction.

Generally speaking, if a rigid body is subjected to a force **F** that lies in the xy plane and produces a positive moment about point O (or the z axis), the force **F** tends to cause counterclockwise rotation of the body about the z axis, when the body is viewed from above—that is, from the positive part of the z axis (Fig. 4.1b). If **F** produces a negative moment about point O, it tends to produce a clockwise rotation of the body, when viewed from the positive part of the z axis (Fig. 4.1c).

For coplanar problems, the z axis is represented adequately by point O, regarded as an end view of the positive z axis (Fig. 4.1d). Consequently, instead of referring to the z axis and unit vector **k**, we can simply say that Eq. (4.1) represents the moment of force **F** about point O.[2] Also for simplicity, in the coplanar case, Eq. (4.1) can be written as

$$M = \pm Fr \tag{4.2}$$

with the understanding that M represents a moment, with magnitude Fr and with either positive sense (plus sign) or negative sense (minus sign). Also, the vector **M** may be represented by a curved arrow M_O in the plane Q (Fig. 4.1d). The subscript O signifies that the moment is with respect to point O.

Key Concept

When several concurrent coplanar forces act, the sum of the moments of the forces about a point O equals the moment of the resultant force about O.

When several coplanar forces act, each can produce a moment about a point O. The following definition applies to such a case.

DEFINITION: *The moment of several coplanar forces about a point O that lies in their plane is defined as the algebraic sum of the moments of the individual forces about O.*

Note that the moment of any one of the coplanar forces about a given axis (point O) is not changed if the force is displaced along its line of action. In other words, the moments of the forces that act on a rigid body are not affected when the principle of transmissibility is applied.

Varignon's theorem

The concept of moment of a force derives its usefulness from the following theorem, known as *Varignon's theorem.*

Theorem 4.1

Varignon's Theorem

The moment of several concurrent coplanar forces about any point O in their plane equals the moment of their resultant about point O.

To prove this theorem, we let point O be the origin of rectangular xy coordinates that lie in plane Q (see Fig. 4.2). The force **F** acts at the point P: (x, y) and lies in plane Q. Let the line segment ON be perpendicular to the line of action of **F**. Line PS is perpendicular to the x axis, line RS is perpendicular to the extension of line ON, and line PT is perpendicular to the extension of line RS. Hence, the moment arm r from the line of action of **F** to point O is

$$r = ON = OR - PT$$

where, from Fig. 4.2,

$$OR = x \cos\theta$$

$$PT = y \sin\theta$$

2. The general definitions of the moment of a force about an axis and about a point are given in Chapter 5.

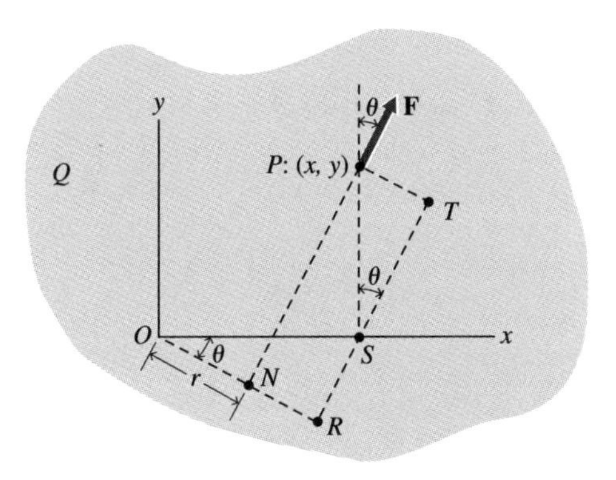

Figure 4.2 Demonstration of Varignon's theorem.

and θ is the angle that ON (which is normal to **F**) forms with the x axis. Since r is considered to be a positive number, we have for all values of x, y, and θ,

$$r = |x \cos \theta - y \sin \theta| \tag{a}$$

Multiplying Eq. (a) by the magnitude F of force **F** and noting Eq. (4.2), we obtain

$$M = \pm Fr = \pm F|x \cos \theta - y \sin \theta| \tag{b}$$

The sign of the moment M in Eq. (b), is positive for the force **F** illustrated in Fig. 4.2. More generally, for all values of x, y, and θ,

$$M = F(x \cos \theta - y \sin \theta) \tag{c}$$

Then, the sign of M, corresponding to the sign of the term $(x \cos \theta - y \sin \theta)$, is correct for all values of x, y, and θ.

The projections of the force **F** on the x and y axes (Fig. 4.2) are

$$F_x = F \sin \theta \qquad F_y = F \cos \theta \tag{d}$$

Consequently, by Eqs. (c) and (d),

$$M = xF_y - yF_x \tag{4.3}$$

Equation (4.3) is valid irrespective of the sense of the force **F** and of the quadrant of the xy coordinate system in which the point P lies; that is, the sign of the moment M is determined automatically by the signs of x, y, F_x, and F_y. The magnitude of the moment is $|M| = |xF_y - yF_x|$.

To interpret Eq. (4.3), we resolve the force **F** into components $F_x\mathbf{i}$ and $F_y\mathbf{j}$, where **i** and **j** are unit vectors along the coordinate axes (see Fig. 4.3 and Sec. 2.6). The moment arm of force $F_x\mathbf{i}$ with respect to the origin is y, if point P lies in the first quadrant. Then, the moment of force $F_x\mathbf{i}$ about the origin is negative. Thus, by Eq. (4.2), the moment of force $F_x\mathbf{i}$ about the origin is $-yF_x$. Similarly, the moment of force $F_y\mathbf{j}$ about the origin is $+xF_y$. Therefore, Eq. (4.3) means that the moment of force **F** about the origin equals the moment of its components $F_x\mathbf{i}$ and $F_y\mathbf{j}$ about the origin. This conclusion is an important special case of Varignon's theorem.

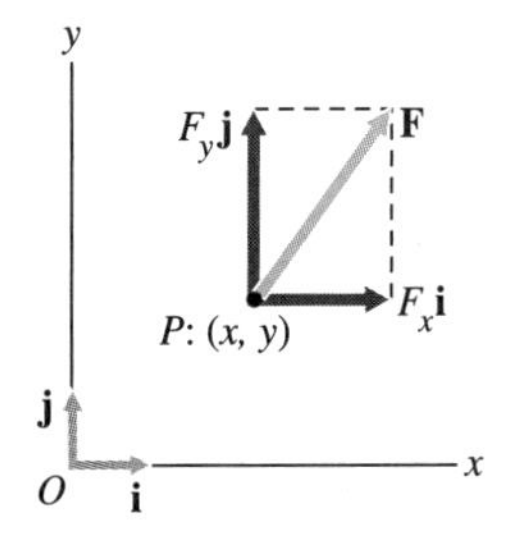

Figure 4.3 Components of a force **F**.

To generalize the result, we let several coplanar forces, $\mathbf{F}_1$, $\mathbf{F}_2$, . . . , act at point P (Fig. 4.4). The resultant of these forces is **F**. We denote the x and y projections of these forces by (F_{1x}, F_{1y}), (F_{2x}, F_{2y}), ... Then, by Eq. (4.3), the following formula for the moment of all the forces about the origin is obtained:

$$M = xF_{1y} - yF_{1x} + xF_{2y} - yF_{2x} + \cdots$$

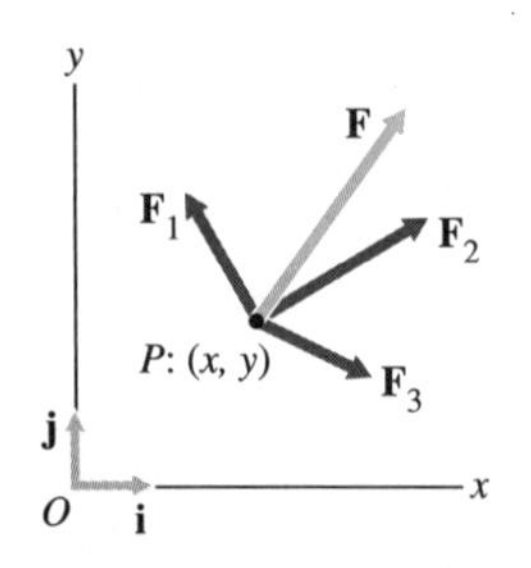

Figure 4.4 Resultant of coplanar concurrent forces.

This equation may be written as

$$M = x(F_{1y} + F_{2y} + \cdots) - y(F_{1x} + F_{2x} + \cdots) \tag{e}$$

Since the projections (F_x, F_y) of the resultant force are determined by the equations

$$F_x = F_{1x} + F_{2x} + \cdots$$

$$F_y = F_{1y} + F_{2y} + \cdots$$

Eq. (e) yields

$$M = xF_y - yF_x \tag{f}$$

This conclusion verifies Varignon's theorem, since, by Eq. (4.3), the right side of Eq. (f) represents the moment of the resultant force **F** about the origin (Fig. 4.4). This moment is equal to M, the moment of the components of force **F**, according to Eq. (e).

The fundamental unit of moment in SI is the newton-meter (N·m). In the U.S. Customary System, the fundamental unit of moment is the pound-foot (lb·ft).

PROBLEM-SOLVING TECHNIQUE

Moment of a Force about an Axis

To determine the moment of a force with respect to an axis through point O:

1. Review the definition of moment of a force given by Eq. (4.1).
2. Determine the moment arm r of the force **F** relative to point O. Use Eq. (4.2) to calculate the moment M_O about point O.
3. Alternatively, select xy axes with origin at O and determine the projections (F_x, F_y) of the force **F** and the (x, y) coordinates of the point of application of **F**. Then, use Eq. (4.3) to compute the moment M_O about point O.

Example 4.1 Moment Exerted on a Nut

Problem Statement A wrench is used to tighten a nut on a machine part (see Fig. E4.1a). Determine the moment exerted about the center O of the nut by the force **F**.

Solution Using the definition of moment given in Eq. (4.1), we need to determine the moment arm r of the force **F** relative to point O. From Fig. E4.1b, the moment arm is

$$r = a \sin \theta \tag{a}$$

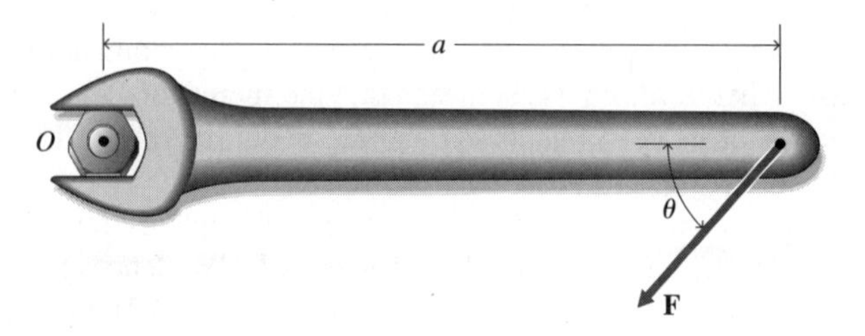

Figure E4.1a

Since the sense of the moment of **F** is clockwise, by Eqs. (4.1) and (a), the moment about point O is

$$\mathbf{M}_O = -Fr\mathbf{k}$$
$$= -(Fa \sin \theta)\mathbf{k}$$

where F is the magnitude of **F**, $M_O = Fa|\sin \theta|$ is the magnitude of $\mathbf{M}_O$, and **k** is a unit vector directed perpendicular to the plane of the wrench at O.

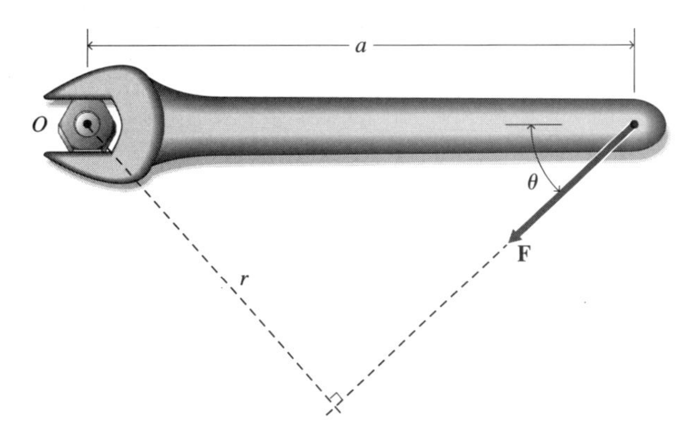

Figure E4.1b

Alternatively, we can calculate the moment by considering the projections (F_x, F_y) of the force **F** (see Fig. E4.1c). Then, since the problem is coplanar, by Eq. (4.3) and Fig. E4.1c, we obtain the moment M_O as

$$M_O = xF_y - yF_x$$
$$= (a)(-F \sin \theta) - (0)(-F \cos \theta)$$
$$= -Fa \sin \theta$$

The magnitude of the moment is $M_O = Fa|\sin \theta|$, as before.

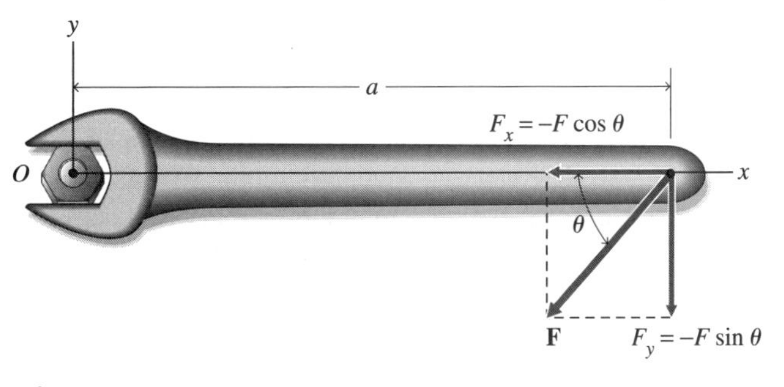

Figure E4.1c

Example 4.2

Moment of Two Forces about a Point

Problem Statement Figure E4.2a represents two forces **P** and **Q** that act at point (600, 900) in the xy plane (distances expressed in millimeters). Determine the moment of the forces about the origin O.

Solution The result is obtained most easily with the aid of Varignon's theorem. The x and y projections of the forces **P** and **Q** are

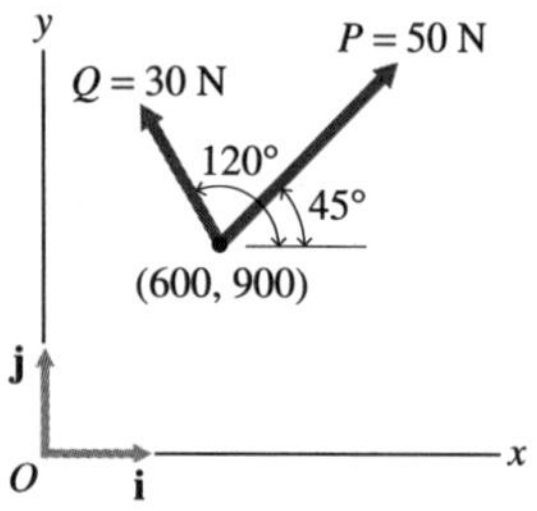

$$P_x = 50\cos 45^\circ = 35.36 \text{ N}$$

$$P_y = 50\sin 45^\circ = 35.36 \text{ N}$$

$$Q_x = 30\cos 120^\circ = -15.00 \text{ N}$$

$$Q_y = 30\sin 120^\circ = 25.98 \text{ N}$$

Figure E4.2a

Hence, the x and y projections of the resultant force **F** are

$$F_x = P_x + Q_x = 20.36 \text{ N}$$

$$F_y = P_y + Q_y = 61.34 \text{ N}$$

The moment of the force $F_x\mathbf{i}$ about the origin is $-0.900F_x = -18.32$ N·m. The moment of the force $F_y\mathbf{j}$ about the origin is $0.600F_y = 36.80$ N·m. Hence, by Eq. (4.3), the moment of forces **P** and **Q** about the origin is

$$M_O = 36.80 - 18.32 = 18.48 \text{ N·m}$$

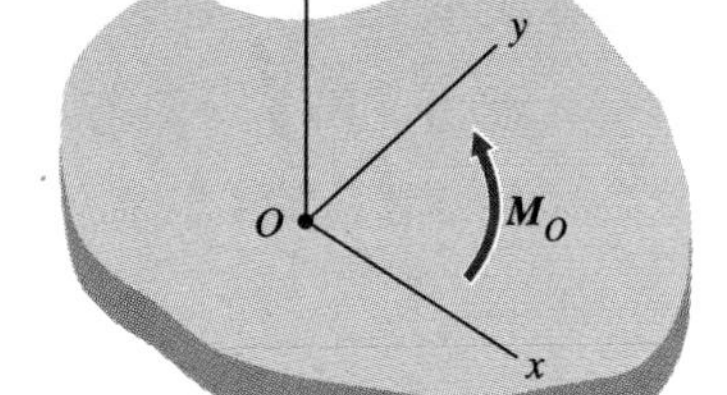

Figure E4.2b

Since the moment is positive, its sense may be represented by a curved arrow indicating a counterclockwise rotation about the z axis, as shown in Fig. E4.2b. ■

4.2 RESULTANT OF COPLANAR FORCES THAT ACT ON A RIGID BODY

After studying this section, you should be able to:

- Determine the resultant force and the resultant axis for a system of coplanar forces.

Sometimes, several distinct forces act on a rigid body. To determine the response of the body to the forces, you may find it convenient to first determine the resultant of the force system. In particular, under some conditions, any number of coplanar forces that act on a rigid body may be combined by the parallelogram law into a single resultant force with a specific line of action. Consider a case with five coplanar forces. The most obvious procedure is first to slide any two nonparallel force vectors along their lines of action to the point of intersection of these lines by the principle of transmissibility (Sec. 3.6) and then combine the two forces by parallelogram construction. This process may be repeated with any two of the remaining four forces to yield a system of three forces, and so on. If the system does not reduce to a set of parallel forces, it will eventually yield a single resultant force. (Since the lines of action of parallel forces do not intersect, parallel forces require special consideration.)

An alternative method for finding the resultant force vector is to sum the projections of the force vectors on rectangular coordinate xy axes (Sec. 2.5), since this method provides the same x and y projections as the parallelogram construction. Likewise, polygon construction (Sec. 2.3) may be used to determine the magnitude, direction, and sense of the resultant force. However, the latter two methods leave the line of action of the resultant force undetermined.

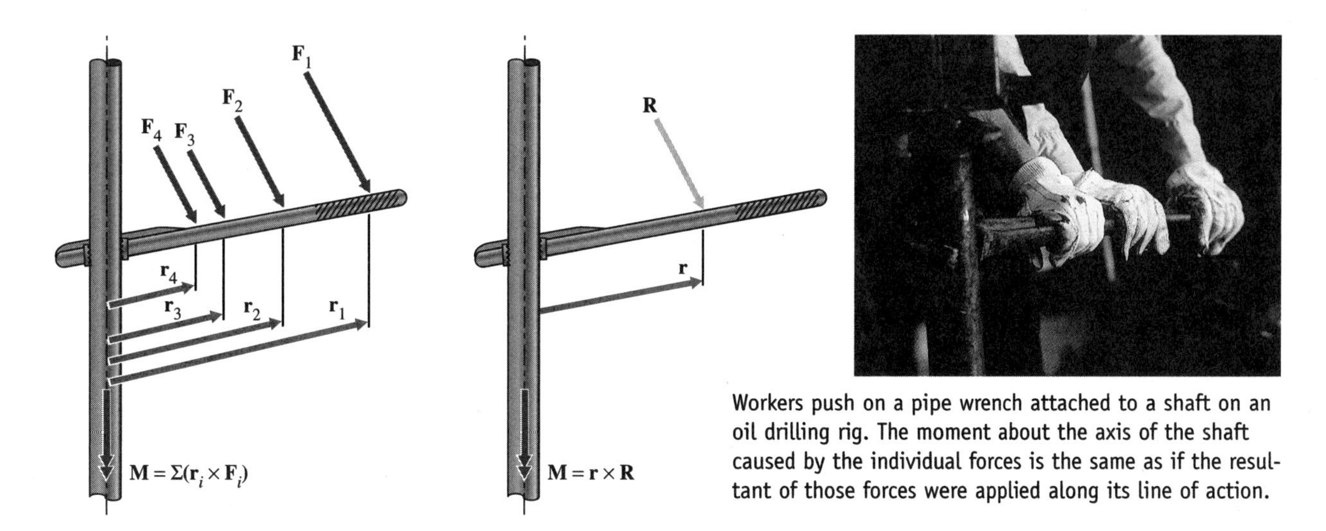

Workers push on a pipe wrench attached to a shaft on an oil drilling rig. The moment about the axis of the shaft caused by the individual forces is the same as if the resultant of those forces were applied along its line of action.

Instead of determining the line of action of the resultant force by a cumbersome method such as parallelogram construction, you can utilize the theory of moments that was developed in Sec. 4.1. As was noted there, the moment of a force about a point O is not altered by a displacement of the force along its line of action. Also, Varignon's theorem asserts that the resultant of two or more concurrent coplanar forces exerts the same moment about point O as the separate forces do. Therefore, the parallelogram constructions described above provide a resultant force that exerts the same moment about any point O as the original forces. Varignon's theorem serves to locate the line of action of the resultant force—that is, the *resultant axis.* Consequently, we have the following theorem.

resultant axis

Theorem 4.2

Resultant-Axis Theorem

If any number of coplanar forces, $\mathbf{F}_1, \mathbf{F}_2, \ldots,$ act on a rigid body, their resultant $\mathbf{F}$ is the vector sum of all the forces. If the force $\mathbf{F}$ *is not zero,* its line of action is chosen so that $\mathbf{F}$ produces the same moment about any point in the plane of the forces as the original forces $\mathbf{F}_1, \mathbf{F}_2, \ldots$. Then the resultant force $\mathbf{F}$ is dynamically equivalent to its components $\mathbf{F}_1, \mathbf{F}_2, \ldots$. If the resultant force is zero, a resultant axis does not exist.

In Sec. 4.3, we will see that if parallelogram constructions reduce several forces to two parallel forces that are noncollinear, equal in magnitude, and opposite in sense, the resultant force $\mathbf{F}$ is zero, but the moment of the forces is nonzero. In such a case, parallelogram construction fails to yield a resultant axis. Hence, this case requires special treatment.

PROBLEM-SOLVING TECHNIQUE

Resultant of Coplanar Forces

To determine the resultant $\mathbf{F}$ of a system of coplanar forces, $\mathbf{F}_1, \mathbf{F}_2 \ldots$:

1. Select appropriate *xy* axes and determine the *x* and *y* projections of the forces $\mathbf{F}_1, \mathbf{F}_2, \ldots$.

(continued)

PROBLEM-SOLVING TECHNIQUE *(CONT.)*

2. Sum the x projections of the forces to obtain F_x; likewise, sum the y projections of the forces to obtain F_y.
3. Calculate the magnitude of the resultant force **F** as $F = \sqrt{F_x^2 + F_y^2}$. If $F = 0$, the resultant force **F** is zero, and no resultant axis exists.
4. Determine the (x, y) coordinates of the points of application of the forces $\mathbf{F}_1, \mathbf{F}_2, \ldots$.
5. With the x and y projections of the forces $\mathbf{F}_1, \mathbf{F}_2, \ldots$ and the (x, y) coordinates of their points of application, use Eq. (4.3) to determine the moments of $\mathbf{F}_1, \mathbf{F}_2, \ldots$ about the origin O of the xy axes.
6. Sum the moments of $\mathbf{F}_1, \mathbf{F}_2, \ldots$, and, by Varignon's theorem, equate this sum to the moment of the force **F** about the origin O to locate the line of action of **F** (see Example 4.3). Then, with magnitude F and line of action known, the resultant **F** is determined.

Example 4.3 Resultant of Coplanar Forces That Act on a Body

Problem Statement Figure E4.3 represents a rigid body under the action of four forces $\mathbf{F}_1$, $\mathbf{F}_2$, $\mathbf{F}_3$, and $\mathbf{F}_4$. Determine the x and y projections of the resultant of these forces, and locate the resultant axis.

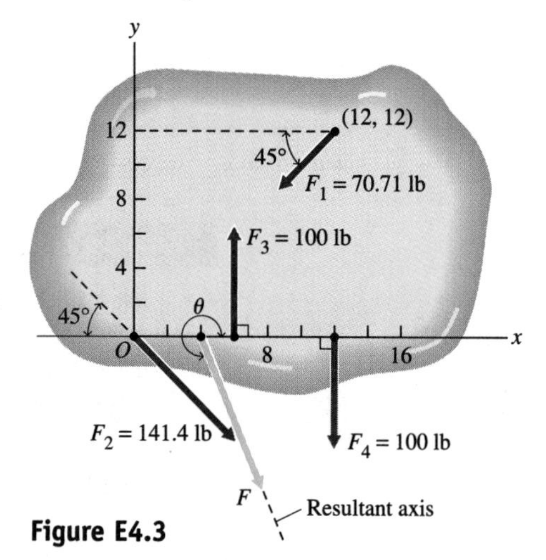

Figure E4.3

Solution The x and y projections of the resultant force may be determined by summation of the projections of forces $\mathbf{F}_1$, $\mathbf{F}_2$, $\mathbf{F}_3$, and $\mathbf{F}_4$ on xy axes:

$$\begin{aligned} F_x &= \sum F_{ix} \\ &= -70.71\cos 45^\circ + 141.4\cos 45^\circ \\ &= 50 \text{ lb} \end{aligned}$$

$$\begin{aligned} F_y &= \sum F_{iy} \\ &= -70.71\sin 45^\circ - 141.4\sin 45^\circ \\ &\quad + 100 - 100 \\ &= -150 \text{ lb} \end{aligned}$$

where the subscript i is the index indicating the force ($i = 1, 2, 3$, or 4). Therefore, the magnitude of the resultant force is

$$F = \sqrt{50^2 + 150^2} = 158.1 \text{ lb}$$

The direction angle θ of the resultant force is determined from

$$\tan\theta = \frac{F_y}{F_x} = \frac{-150}{50} = -3.00$$

Consequently, $\theta = 288.43^\circ$.

By Eq. (4.3), the moment about the origin of the force $\mathbf{F}_1$ is $M_1 = x_1F_{1y} - y_1F_{1x}$, where (x_1, y_1) are the coordinates of the point of application of $\mathbf{F}_1$, and (F_{1x}, F_{1y}) are the

x and y projections of $\mathbf{F}_1$. The moments about the origin of forces $\mathbf{F}_2$, $\mathbf{F}_3$, and $\mathbf{F}_4$ may be written similarly:

$$M_1 = (12)(-70.71 \sin 45°) - (12)(-70.71 \sin 45°) = 0$$

$$M_2 = (0)(-141.4 \sin 45°) - (0)(141.4 \sin 45°) = 0$$

$$M_3 = (6)(100) - (0)(0) = 600 \text{ lb·ft}$$

$$M_4 = (12)(-100) - (0)(0) = -1200 \text{ lb·ft}$$

Adding these moments yields the moment about the origin of all the forces: $M_O = -600$ lb·ft. By Varignon's theorem, this moment is also equal to the moment about the origin of the resultant force $\mathbf{F}$. To locate the axis of the resultant force $\mathbf{F}$, we shift the force along its line of action so that it acts at a point on the x axis. Then, by Eq. (4.3), the moment of the resultant force about the origin is $M_O = xF_y$, where x is the x-axis intercept of the resultant axis. Consequently, $-600 = -150x$, or $x = 4.0$ ft. The resultant force and the resultant axis are shown in Fig. E4.3.

Information Item It should be clear that the moment about the origin of a force or one of its components is zero when the moment arm of the force or component is zero or the force component is zero. In this example, the moment arms for $\mathbf{F}_2$ are both zero, since the point of application of the force is the origin O. These zero values for moment arms are shown in the equation for M_2. As a result, $M_2 = 0$. In addition, both $\mathbf{F}_3$ and $\mathbf{F}_4$ are parallel to the y axis, and thus y_3, y_4, F_{3x}, and F_{4x} are zero. For completeness, these zero values are also shown in the equations for M_3 and M_4. Finally, since the line of action of $\mathbf{F}_1$ passes through the origin, that force causes no moment about O ($M_1 = 0$). In general, when it is clear that the line of action of a force, or of a force component, passes through a point about which moments are taken, the examples in this book will utilize the fact that the force produces no moment and will not include the zero value of moment in the computations.

In analyzing equilibrium of rigid bodies, you may choose any point about which to take moments. Often, you may find it possible to select the point such that it lies on the line of action of one or more forces under consideration. Then, the moments of these forces about that point are zero, and your computations are simplified. ■

Example 4.4

Forces on the Bottom of a Flying Boat

Problem Statement Determine the resultant force that acts on the hull of a flying boat at a particular instant during landing (see Fig. E4.4a). For computation of the resultant force, the forward part of the hull has been partitioned into five segments by specified cross-

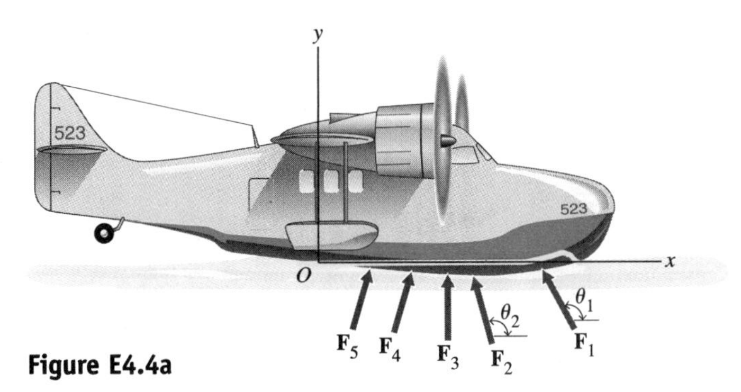

Figure E4.4a

sectional planes, and the forces $\mathbf{F}_1, \ldots, \mathbf{F}_5$ that the water exerts on the separate segments of the bottom of the flying boat have been estimated by hydrodynamic theory. Table E4.4 gives the force magnitudes F_i, their direction angles θ_i, and the coordinates (x_i, y_i) of their points of application with respect to rectangular coordinate axes with origin at the step of the hull. The x and y projections of each force $\mathbf{F}_i$ are $F_{ix} = F_i \cos\theta_i$ and $F_{iy} = F_i \sin\theta_i$. These projections have been computed and are also recorded in Table E4.4.

Table E4.4

SEGMENT OF HULL	F_i (lb)	θ_i (deg)	x_i (ft)	y_i (ft)	F_{ix} (lb)	F_{iy} (lb)	M_i (lb·ft)
1	1200	120	40	−1	−600	1039	40,960
2	1800	100	30	−2	−312	1773	52,570
3	2400	90	20	−2	0	2400	48,000
4	3200	80	12	−1	556	3151	38,370
5	3500	80	6	0	608	3447	20,680
				Totals:	252	11,810	200,580

Solution The sums of the terms in the F_{ix} and F_{iy} columns of the table are the (F_x, F_y) projections of the resultant force. Hence, $F_x = 252$ lb and $F_y = 11{,}810$ lb, and the magnitude of the resultant force is

$$F = \sqrt{252^2 + 11{,}810^2} = 11{,}813 \text{ lb}$$

The direction angle θ of the resultant force is determined by

$$\tan\theta = \frac{F_y}{F_x} = \frac{11{,}810}{252} = 46.86$$

Consequently, $\theta = 88.78°$.

By Eq. (4.3), the moment of the force $\mathbf{F}_i$ about the origin is $M_i = x_i F_{iy} - y_i F_{ix}$. This formula has been used to calculate the moments of the forces $\mathbf{F}_1, \ldots, \mathbf{F}_5$ about the origin, tabulated in the last column of the table. The sum of the terms in this column is the moment of all the forces about the origin. That is, $M_O = 200{,}580$ lb·ft. By Varignon's theorem, the moment of the resultant force $\mathbf{F}$ about the origin is also equal to M_O. We can slide the resultant force along its line of action so that it acts at a point on the x axis. Then, by Eq. (4.3), the moment of the resultant force about the origin is $M = xF_y$, where x is the x-axis intercept of the resultant axis. Therefore, $200{,}580 = 11{,}810x$. Consequently, $x = 16.98$ ft. The resultant force and the resultant axis are shown in Fig. E4.4b.

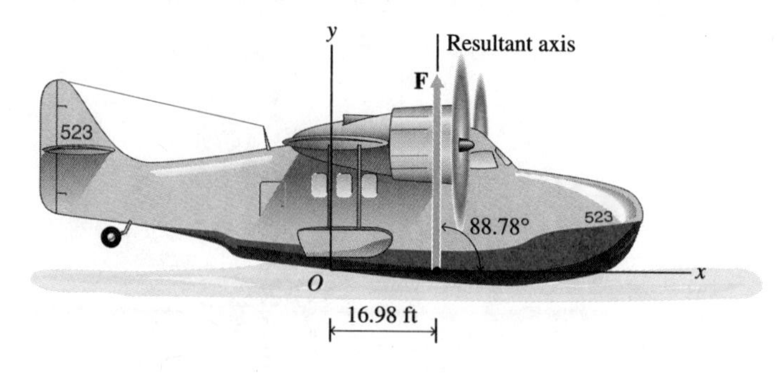

Figure E4.4b

CHECKPOINT 

1. ***True or False:*** The moment of a force with respect to an axis has dimension $[F/L]$.
2. ***True or False:*** The moment arm of a force is always a positive number that is independent of the axis about which the moment acts.
3. ***True or False:*** Varignon's theorem depends on the selected point about which moments are taken.
4. ***True or False:*** If a force acting on a rigid body lies in a plane Q and has positive moment with respect to a point O in plane Q, the force will tend to rotate the body about point O in the clockwise sense, as viewed from above the body.

Answers: 1. False; 2. False; 3. False; 4. False

4.3 PARALLEL COPLANAR FORCES AND COUPLES

After studying this section, you should be able to:

- Determine the resultant force and the resultant axis for a system of parallel coplanar forces.
- Describe the characteristics of a couple.

PARALLEL COPLANAR FORCES

parallel forces

The parallelogram constructions discussed in Sec. 4.2 cannot be performed directly if the force system reduces to two *parallel forces.* So, a resultant of parallel forces cannot be obtained through parallelogram construction. However, except for one case (to be discussed in the next subsection), parallel forces may be converted into nonparallel forces, for which a resultant can be determined. The following graphical technique illustrates how this is accomplished.

Let parallel forces $\mathbf{F}_1$ and $\mathbf{F}_2$ act at points P_1 and P_2 of a rigid body, where the line segment P_1P_2 is perpendicular to the force vectors (see Fig. 4.5).[3] To combine $\mathbf{F}_1$ and $\mathbf{F}_2$ into a single resultant force, we imagine that self-equilibrating forces $\mathbf{f}$ and $-\mathbf{f}$ act at points P_1 and P_2 such that the line of action of these forces is the line P_1P_2. The magnitude f of the imaginary forces $\mathbf{f}$ and $-\mathbf{f}$ is arbitrary. Addition of these forces is permissible, since self-equilibrating forces produce no dynamical effects on a rigid body. The resultants $\mathbf{R}_1$ and $\mathbf{R}_2$ of the concurrent pairs of forces are formed by parallelogram construction (Fig. 4.5). Since the forces $\mathbf{f}$ and $-\mathbf{f}$ cancel each other, the forces $\mathbf{R}_1$ and $\mathbf{R}_2$ are *dynamically equivalent* to the forces $\mathbf{F}_1$ and $\mathbf{F}_2$. Thus, the parallel forces $\mathbf{F}_1$ and $\mathbf{F}_2$ are replaced by equivalent nonparallel forces $\mathbf{R}_1$ and $\mathbf{R}_2$.

The resultant force is $\mathbf{R} = \mathbf{R}_1 + \mathbf{R}_2$. But $\mathbf{R}_1 = \mathbf{F}_1 + \mathbf{f}$ and $\mathbf{R}_2 = \mathbf{F}_2 - \mathbf{f}$. Consequently, $\mathbf{R} = (\mathbf{F}_1 + \mathbf{f}) + (\mathbf{F}_2 - \mathbf{f})$, or $\mathbf{R} = \mathbf{F}_1 + \mathbf{F}_2$. The line of action of the resultant $\mathbf{R}$ is parallel to the original force vectors $\mathbf{F}_1$ and $\mathbf{F}_2$ and passes through point P_3, the intersection of the lines of action of $\mathbf{R}_1$ and $\mathbf{R}_2$ (Fig. 4.5). Example 4.5 demonstrates this procedure and also

3. If the force vectors are not applied at points P_1 and P_2, the principle of transmissibility allows you to slide the force vectors along their lines of action so that they do act at P_1 and P_2.

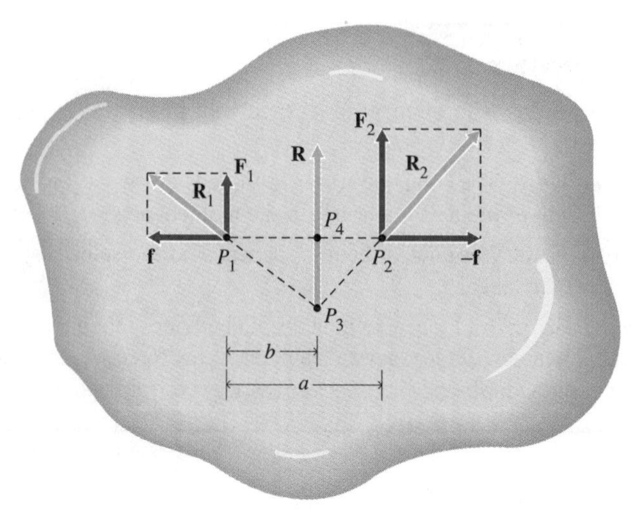

Figure 4.5 Graphical construction of nonparallel forces that are dynamically equivalent to given parallel forces.

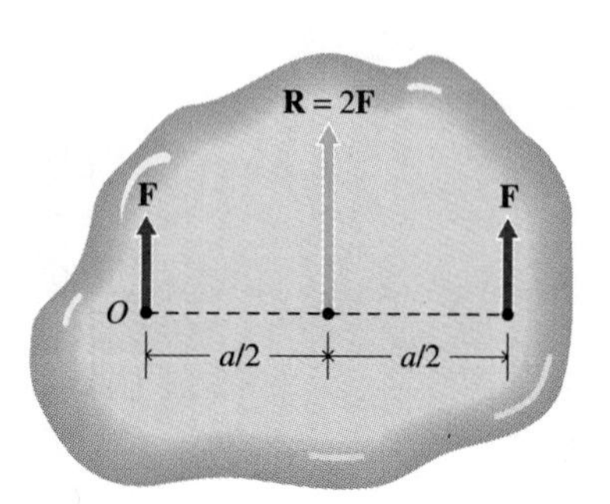

Figure 4.6 Resultant of equal parallel forces.

shows that the moment about a fixed point of the resultant of two parallel forces is the same as the sum of the moments about the fixed point of the two individual forces.

If the forces $\mathbf{F}_1$ and $\mathbf{F}_2$ are equal (say, $\mathbf{F}_1 = \mathbf{F}$ and $\mathbf{F}_2 = \mathbf{F}$), the resultant is $\mathbf{R} = \mathbf{F} + \mathbf{F} = 2\mathbf{F}$. The resultant axis passes midway between the forces $\mathbf{F}_1$ and $\mathbf{F}_2$, since this condition ensures that the moment of the resultant $\mathbf{R}$ about any point in the plane of the forces (say, point O in Fig. 4.6) is the same as the sum of the moments of the two forces $\mathbf{F}_1$ and $\mathbf{F}_2$ about that point.

The technique just described is graphical. Alternatively, we can determine $\mathbf{R}$ analytically in the following way. First, we sum the given parallel forces $\mathbf{F}_1$ and $\mathbf{F}_2$. Then we require that the moment of the resultant about any point in the plane (a point on the line of action of one of the parallel forces is often convenient) be dynamically equivalent to the sum of the moments of $\mathbf{F}_1$ and $\mathbf{F}_2$ about the point. Thus, from Fig. 4.5, summing parallel forces $\mathbf{F}_1$ and $\mathbf{F}_2$ and taking moments about point P_1, we have

$$\sum \mathbf{F} = \mathbf{F}_1 + \mathbf{F}_2 = \mathbf{R} \quad \text{(a)}$$

$$\sum M_{P_1} = aF_2 = bR \quad \text{(b)}$$

where b is the perpendicular distance from point P_1 to the line of action of $\mathbf{R}$. By Eq. (b), we find

$$b = \frac{aF_2}{R}$$

Consequently, the resultant force $\mathbf{R}$ is parallel to $\mathbf{F}_1$ and $\mathbf{F}_2$, and its line of action passes through the points P_3 and P_4.

Example 4.5

Parallel Forces on a Beam

Problem Statement A horizontal beam is supported by a hinge at its left end A and by an inclined cable at its right end D. Two crates rest on the beam, as shown in Fig. E4.5a. The weight of crate B is 0.5 kN and that of crate C is 2.0 kN.

a. Determine the resultant force on the beam due to the weights of the crates.

b. Show that the resultant force produces the same moment about point A as the forces due to the weights of the crates.

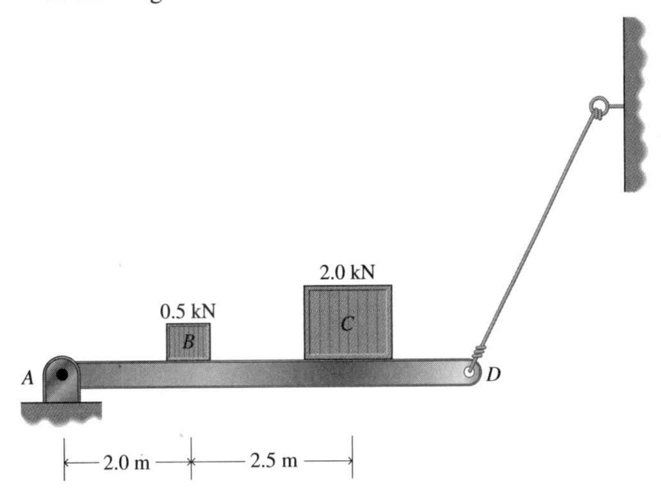

Figure E4.5a

Solution

a. The vertical forces due to the weights of the crates are shown in Fig. E4.5b on the isolated beam; the effects of the hinge support and the cable are not shown. Imaginary loads of 0.5 kN directed to the left and to the right are applied at points B and C, respectively. Parallelogram construction is used to find the lines of action of the resultant forces at B and C. The lines of action of these resultants intersect at point E (Fig. E4.5b). Since the horizontal components cancel, the resultant force $\mathbf{R}$ is directed vertically through point E. Its magnitude is $R = 2.0 + 0.5 = 2.5$ kN. By graphical construction, we find that the line of action of $\mathbf{R}$ is 0.5 m to the left of point C.

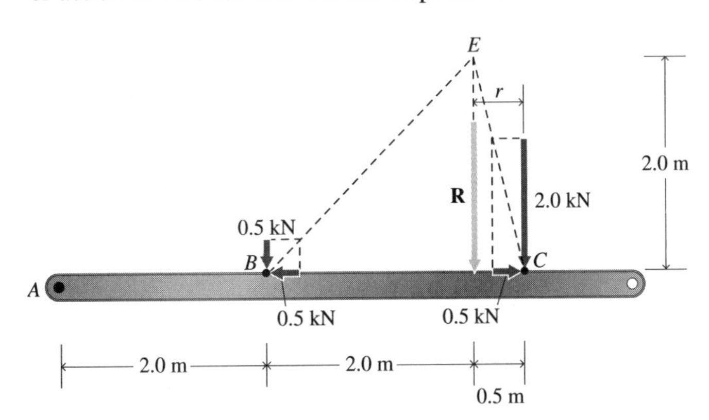

Figure E4.5b

b. The clockwise moment of the two weight forces about point A is

$$M_A = (2.0)(0.5) + (4.5)(2.0) = 10.0 \text{ kN·m}$$

Likewise, the clockwise moment of the resultant $\mathbf{R}$ about A is given by

$$M_A = (4.0)(2.5) = 10.0 \text{ kN·m}$$

Thus, the equivalence of the moment of the force pair to that of its resultant is demonstrated.

Alternatively, we can determine **R** analytically, by summing forces and moments. Thus, from Fig. E4.5b, we have

$$R = 0.5 + 2.0 = 2.5 \text{ kN}$$

$$rR = \sum M_C = (0.5)(2.5) = 1.25 \text{ kN·m} \tag{a}$$

where r is the perpendicular distance from point C to the line of action of **R**. Then, by Eqs. (a), we obtain

$$R = 2.5 \text{ kN}$$

$$r = 0.5 \text{ m}$$

Thus, again, the line of action passes through point E. ■

COUPLES

Key Concept

A couple consists of two noncollinear parallel forces of equal magnitudes and opposite senses.

There is one case for which use of imaginary forces **f** and −**f** (Fig. 4.5) is not effective for finding the resultant of two parallel forces. This occurs when the noncollinear forces $\mathbf{F}_1$ and $\mathbf{F}_2$ have equal magnitudes and opposite directions (that is, when $\mathbf{F}_1 = \mathbf{F}$ and $\mathbf{F}_2 = -\mathbf{F}$). Then, the forces $\mathbf{R}_1$ and $\mathbf{R}_2$ are also parallel (see Fig. 4.7). In this case, the forces $\mathbf{F}_1$ and $\mathbf{F}_2$ are said to constitute a *force couple,* or simply a *couple.* For a couple, the resultant force has zero magnitude, since $\mathbf{R} = \mathbf{F}_1 + \mathbf{F}_2 = \mathbf{F} - \mathbf{F} = 0$. However, a couple produces a constant nonzero moment about any fixed point in the plane.

couple

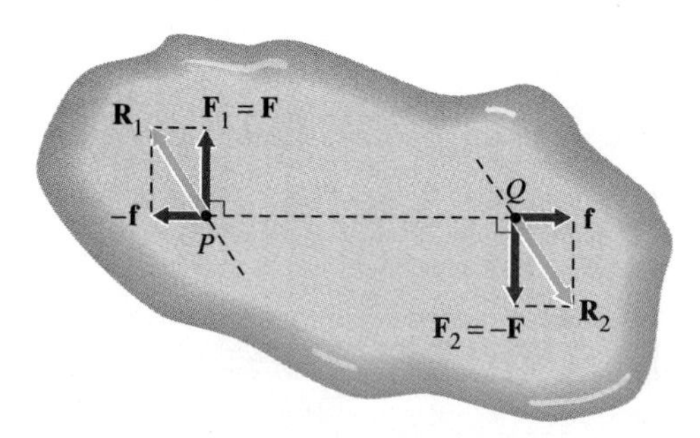

Figure 4.7 Demonstration of the impossibility of replacing a couple by dynamically equivalent nonparallel forces.

If the forces that form a couple do not act at points P and Q in Fig. 4.7, by the theorem of transmissibility of force (Sec. 3.6), we can displace the forces along their lines of action so that they do act at P and Q and the line segment PQ is perpendicular to the force vectors. Then the line segment PQ is called the *arm of the couple.*

arm of a couple

It should be apparent that a couple tends to rotate a rigid body on which it acts. But, contrary to what you might expect, the center of rotation is not ordinarily the midpoint of the arm of the couple. Rather, the center of rotation is a special point in the body, called the *center of mass.* The location of this point depends only on the distribution of matter in the body; it is independent of the couple itself. We will not discuss this concept further until additional dynamical principles have been introduced. Just be sure that you do **not** draw erroneous conclusions concerning the rotation produced by a couple.

center of mass

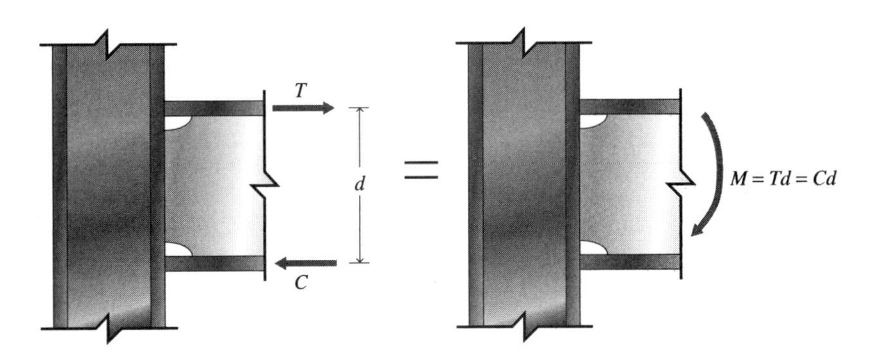

The flanges of a steel beam are welded to the flange of a column. Equal and opposite forces T and C in the beam flanges form a couple with moment M that is transferred into the column.

Louis Poinsot (1777–1859) was the originator of the theory of couples. The following treatment of couples draws on the presentation in his book *Elements de Statique,* published in 1803.

Displacement of a Couple in the Direction of One of Its Forces The theorem of transmissibility of force allows a couple to be displaced in the direction of either of its forces. For example, in Fig. 4.8, the forces **F** and $-$**F** at points P and Q are displaced to points R and S, respectively. The couple with arm RS is dynamically equivalent to the couple with arm PQ.

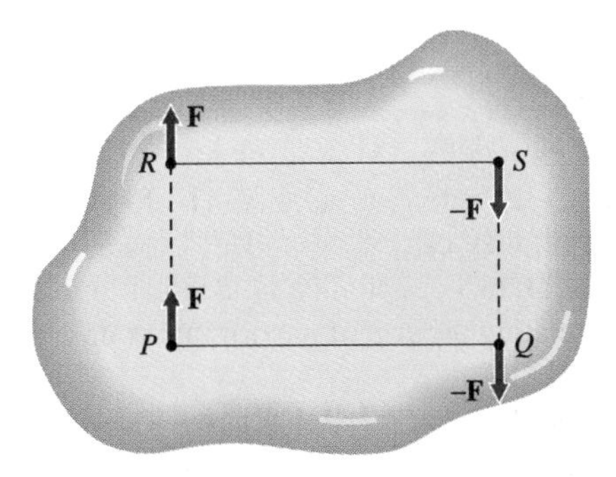

Figure 4.8 Displacement of a couple in the direction of one of its forces.

Displacement of a Couple in the Direction of Its Arm The dynamical effect of a couple on a rigid body is not changed if the arm of the couple is displaced along the line on which it lies. To see this, we consider a couple consisting of forces **F** and $-$**F** and having arm *PQ*. Let *RS* be a line segment that is collinear with *PQ* and that has the same length as *PQ* (see Fig. 4.9). Introduce self-equilibrating pairs of forces **F** and −**F** at points *R* and *S*. The forces **F** at points *P* and *S* may be replaced by their resultant 2**F** at point *O*, midway between points *P* and *S* (see Fig. 4.6). Likewise, the forces −**F** at points *Q* and *R* may be replaced by their resultant −2**F** at point *O*. The forces 2**F** and −2**F** at point *O* cancel each other. Thus, all the forces are eliminated, except the force **F** at point *R* and the force −**F** at point *S*. These forces constitute a couple that is identical in effect to the original couple. Effectively, then, the arm of the original couple is displaced from *PQ* to *RS*.

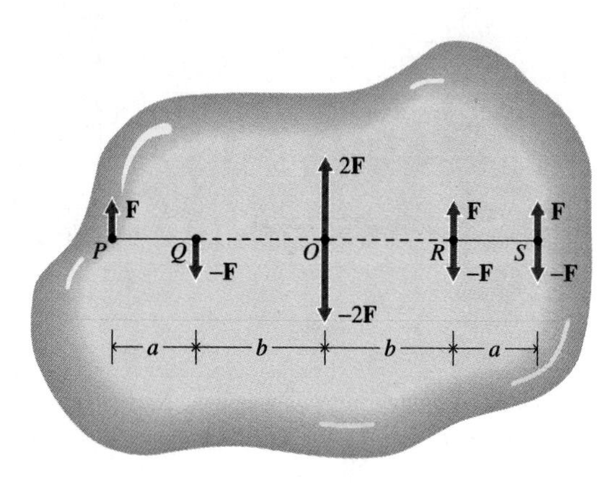

Figure 4.9 Displacement of a couple in the direction of its arm.

Rotation of the Arm of a Couple The dynamical effect of a couple on a rigid body is not changed if the arm of the couple is rotated about its midpoint *O* in the plane of the couple. In Fig. 4.10, forces **F** and −**F** of magnitude *F* acting at points *P* and *Q* constitute a couple. Line segment *RS* lies in the plane of the couple and has the same length as line segment *PQ*. The line segments *PQ* and *RS* intersect at their midpoints *O*. Self-equilibrating pairs of forces $\mathbf{F}_1$ and $-\mathbf{F}_1$ of magnitude *F* are introduced at the points *R* and *S*. These forces are perpendicular to the line segment *RS* and lie in the plane of the original couple. By the theorem of transmissibility, the forces **F** at *P* and $-\mathbf{F}_1$ at *R* may be displaced to point *M* on the bisector of the angle θ. There, they may be combined by the parallelogram construction to form their resultant $\mathbf{F}_2$. Likewise, the forces −**F** at *Q* and $\mathbf{F}_1$ at *S* have the resultant $-\mathbf{F}_2$ acting at point *N*. The forces $\mathbf{F}_2$ at *M* and $-\mathbf{F}_2$ at *N* are self-equilibrating. Therefore, they can be discarded, leaving only the forces $\mathbf{F}_1$ at *R* and $-\mathbf{F}_1$ at *S*. These forces constitute a couple that is identical in effect to the original couple and that was obtained by rotating the arm of the original couple through the angle θ (Fig. 4.10).

Displacement of a Couple to a Parallel Plane The dynamical effect of a couple on a rigid body is not changed if the couple is displaced to another plane that is parallel to its original plane. Let forces **F** and −**F** at points *P* and *Q* constitute a couple in plane *T* (see Fig. 4.11). Lines *PR* and *QS* are perpendicular to plane *T* and to plane *U*, and *PR* = *QS*. Then, line *RS* lies in plane *U*, which is parallel to plane *T*, and line *RS* is parallel to line *PQ*. Self-equilibrating pairs of forces **F** and −**F** are introduced in plane *U* at points *R* and *S*. The parallel forces **F** at points *P* and *S* may be replaced by their resultant 2**F** at the midpoint *O* of the diagonal line *PS*. The parallel forces −**F** at points *R* and *Q* may be replaced by their resultant −2**F** at point *O*. The forces 2**F** and −2**F** at point *O* cancel

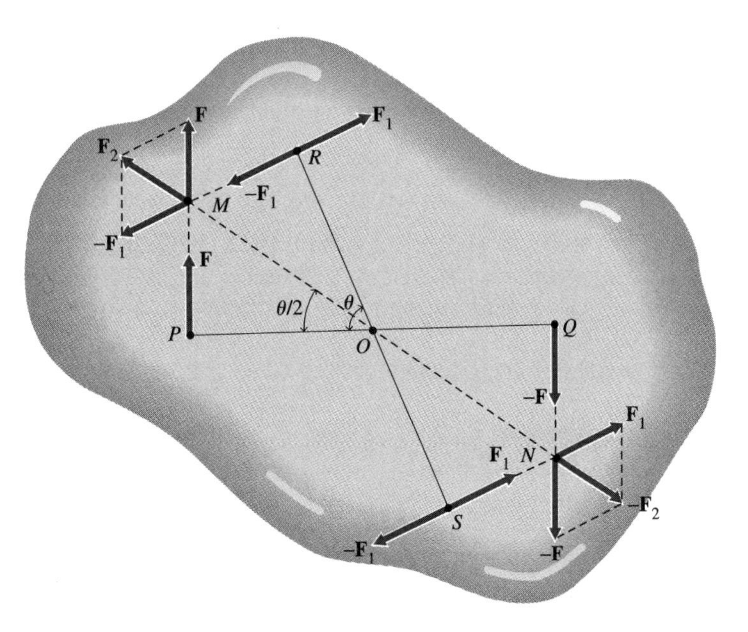

Figure 4.10 Rotation of the arm of a couple in its plane.

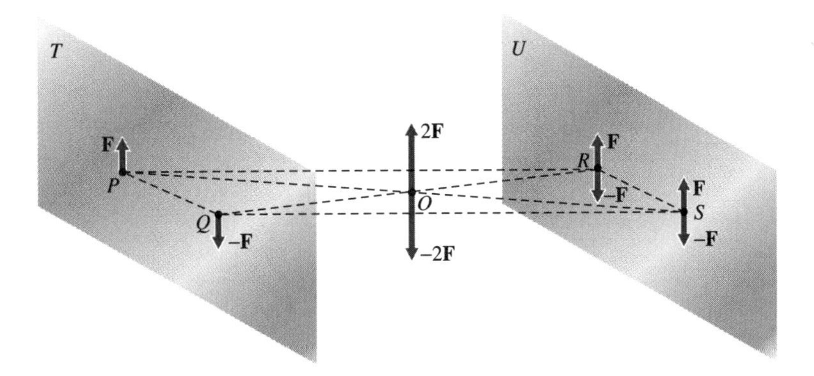

Figure 4.11 Displacement of a couple to a parallel plane.

each other, leaving only the force **F** at point R and the force $-\mathbf{F}$ at point S in plane U. These forces constitute a couple that was obtained by displacing the arm of the original couple from PQ to RS—that is, by displacing the original couple from plane T to the parallel plane U.

The preceding conclusions are summarized by the following theorem.

Theorem 4.3

The dynamical effect of a couple on a rigid body is not changed if the couple is displaced or rotated in its plane or if the couple is displaced from its plane to a parallel plane.

Therefore, the location of the plane of a couple is irrelevant. Only the orientation of the plane has significance.

4.4 MOMENT OF A COUPLE

After studying this section, you should be able to:

- Determine the moment and the sense of a couple.
- Determine the resultant of a system of coplanar couples.

As was mentioned earlier, the effect of a couple on a rigid body is to cause rotation of the body. Since the resultant of the forces of a couple has zero magnitude, the magnitudes of the individual forces are usually of little interest. By itself, the length of the arm of the couple is also of little consequence. Since the effect of a couple is to cause rotation, the quantity of interest is the moment caused by the couple.

Key Concept

The magnitude of the moment of a couple is the product Fa, where F is the magnitude of either force of the couple and a is the length of the arm of the couple.

The effect (the moment) of a couple on a rigid body is not altered if the magnitude of the forces of the couple is changed and the length of the arm of the couple is changed accordingly. Consider the couple consisting of forces $\mathbf{F}$ and $-\mathbf{F}$ acting at points P and Q (see Fig. 4.12). Collinear self-equilibrating forces $-\mathbf{f}$ and $\mathbf{f}$ can be introduced at points P and Q. Let $\mathbf{R}$ and $-\mathbf{R}$ be the resultants of the concurrent pairs of forces at points P and Q, respectively. Displace the vectors $\mathbf{R}$ and $-\mathbf{R}$ along their lines of action to points M and N, where the line segment MN is perpendicular to the vectors $\mathbf{R}$ and $-\mathbf{R}$. Then, the forces $\mathbf{R}$ and $-\mathbf{R}$ at points M and N constitute a couple that is dynamically equivalent to the original couple. The lengths of the arms of the two couples are a and b. Figure 4.12 shows that $b = a\cos\theta$ and $F = R\cos\theta$. Consequently, $Rb = Fa$.

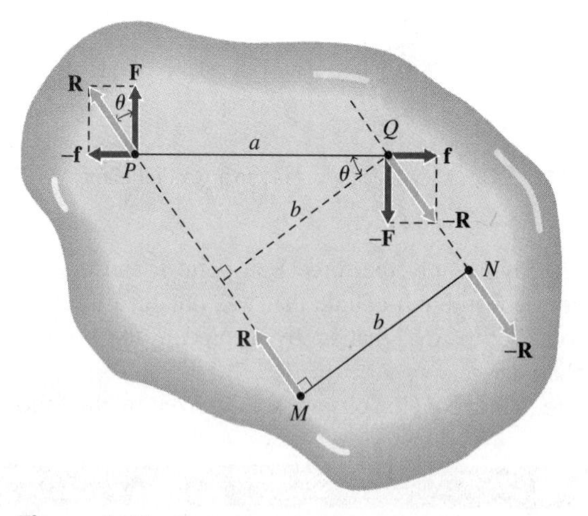

Figure 4.12 Transformation of a couple into a dynamically equivalent couple with a different arm length.

moment of a couple

The product Fa, with appropriate sign, is called the *moment of the couple.* The preceding construction allows you to transform any couple into a dynamically equivalent couple with a prescribed arm length. Since $Rb = Fa$, the transformation leaves the moment of the couple unchanged.

SENSE OF A COUPLE

sense of a couple A couple is not determined solely by its plane and the magnitude of its moment. The *sense of the couple* is also important. Figure 4.13 shows two couples that are identical except for their sense. Obviously, these couples would cancel each other if they acted on the same rigid body. In problems involving coplanar force systems, you can designate the sense of a couple by the following sign convention: *The moment of a couple is positive if it tends to produce a counterclockwise rotation about any axis perpendicular to its plane* (say, the $+z$ axis in Fig. 4.14a). This convention is ordinarily associated with a right-handed coordinate system (refer to Fig. 2.20a).

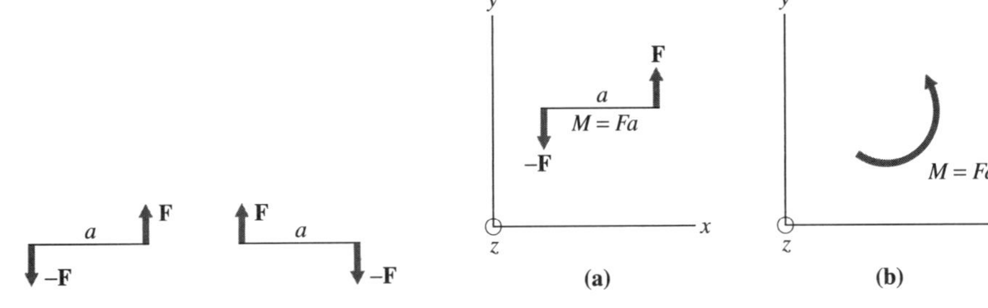

Figure 4.13 Couples with opposite senses.

Figure 4.14 Two ways of representing a couple: **(a)** couple with positive sense; **(b)** curved arrow representation of a couple.

REPRESENTATION OF A COUPLE

Instead of being represented explicitly by its forces and arm (Fig. 4.14a), a couple can be represented by a curved arrow, as shown in Fig. 4.14b. The arrowhead indicates the sense of the couple. The curved arrow is considered to lie in the plane of the couple. The location of the curved arrow in the plane is immaterial, since a couple may be displaced freely in its plane. The letter M in Fig. 4.14b represents the moment of the couple. As with the moment of a force in two-dimensional problems (Sec. 4.1), $M = Fa$ represents the magnitude of the moment of the couple [see Eqs. (4.1) and (4.2)].

MOMENT OF THE FORCES OF A COUPLE WITH RESPECT TO A POINT

Let's calculate the moment of the forces of a couple with respect to any point O in the plane of the couple. The theorem of transmissibility permits us to relocate the couple so that point O lies on the extended line of the arm of the couple (see Fig. 4.15). Then the sum of the moments of the two forces (the moment of the couple) about point O is

$$M = F(a + b) - Fb$$

or

$$M = Fa$$

This result leads to the following theorem.

Theorem 4.4

The forces of a couple exert the same moment (the moment of the couple) about all points in their plane.

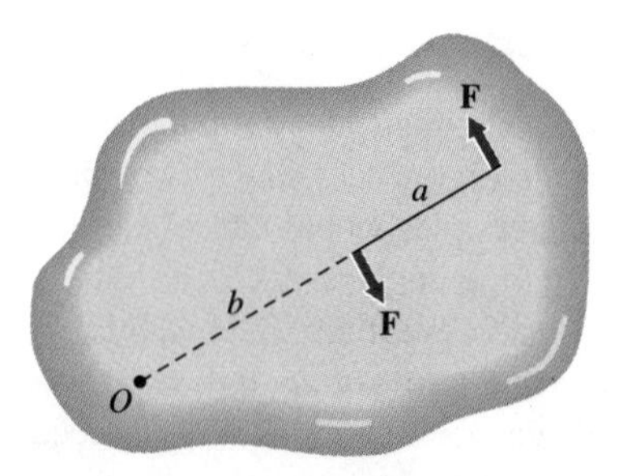

Figure 4.15 Moment of the forces of a couple.

RESULTANT OF SEVERAL COUPLES

Several couples that lie in the *xy* plane or in parallel planes may be combined into a single resultant couple. The process is quite similar to composition of concurrent forces. Since the couples lie in the *xy* plane or in parallel planes, they all cause rotation about the same axis, the *z* axis. Also, since any couple may be moved freely in its plane and may be transferred from its plane to any parallel plane (Sec. 4.3), the couples may be placed in coincidence with one another. The moments M_1, M_2, . . . of the superimposed couples may be added algebraically, since the forces of couples combine by algebraic addition.[4] Hence, the moment of the resultant couple is $M = M_1 + M_2 + \cdots$.

PROBLEM-SOLVING TECHNIQUE

Resultant Force and Couple for Coplanar Force Systems

To find the resultant force and resultant couple for a coplanar system of forces and couples:

1. Identify a point in the plane of the forces and couples that will be used as the origin of the *xy* coordinate system. Establish a convenient set of *xy* axes and corresponding unit vectors **i** and **j**.
2. Replace each force by its *x* and *y* components at the point of application of the force.
3. For the *x* and *y* components of each force (from Step 2), calculate the moment about the origin *O* of the *xy* axes, using Eq. (4.3).
4. With due regard for signs, sum the moments of the forces (from Step 3) with the moments of all the couples to obtain the total moment **M** about the origin *O*.
5. Calculate the sum of the force projections for each axis, with due regard for signs. Label the sums R_x and R_y, respectively. The moment **M** determined in Step 4 and the resultant $\mathbf{R} = R_x\mathbf{i} + R_y\mathbf{j}$ applied at the origin *O* comprise the resultant couple **M** and the resultant force **R** relative to point *O*. (*Note:* The use of a table like that in Example 4.4 may facilitate the summations of Steps 4 and 5, if a large number of forces is involved.)

4. "Algebraically" means that a sign must be attached to M_1, M_2, . . . , either + indicating a positive moment or − indicating a negative moment.

Example 4.6

Resultant of a Force and a Couple

Problem Statement A cubical rigid body is loaded by a force and a couple, as shown in Fig. E4.6a. Determine the resultant of this system.

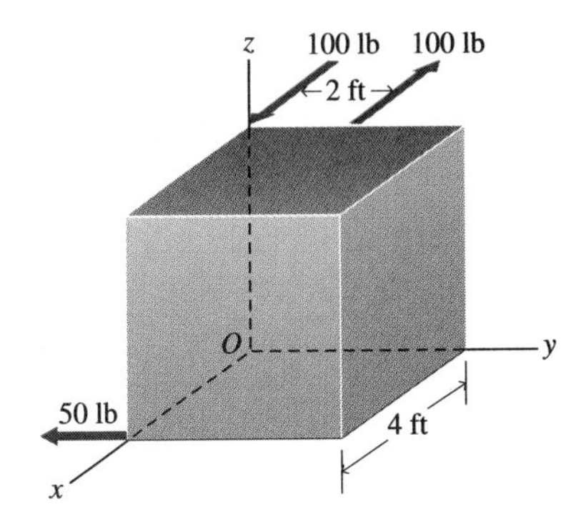

Figure E4.6a

Solution The force lies in the xy plane, and the couple lies in a plane parallel to that plane. The moment of the couple is (100)(2) = 200 lb·ft, and the couple acts in the counterclockwise sense about the z axis. The pair of forces is replaced with a curved arrow in Fig. E4.6b. The moment of the 50 lb force about point O is (50)(4.0) = 200 lb·ft. This moment acts in a clockwise sense about the z axis. Hence, as shown in Fig. E4.6c, we can translate the 50 lb force to point O and add a couple of 200 lb·ft acting clockwise. Since the couples have equal magnitudes and opposite senses and lie in parallel planes, their effects cancel each other. Therefore, the force system can be reduced to a single 50 lb force at O (see Fig. E4.6d).

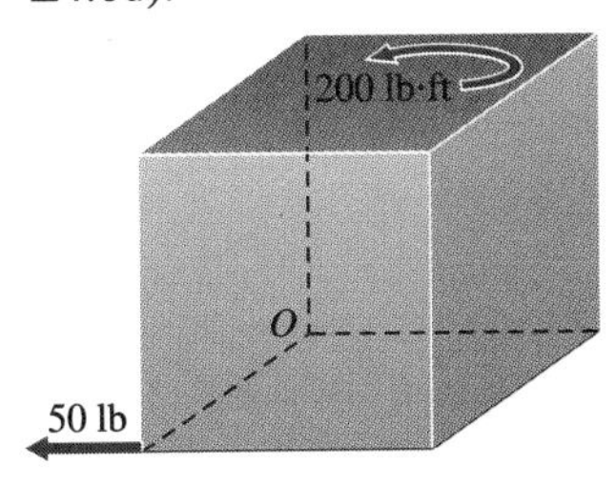

Figure E4.6b

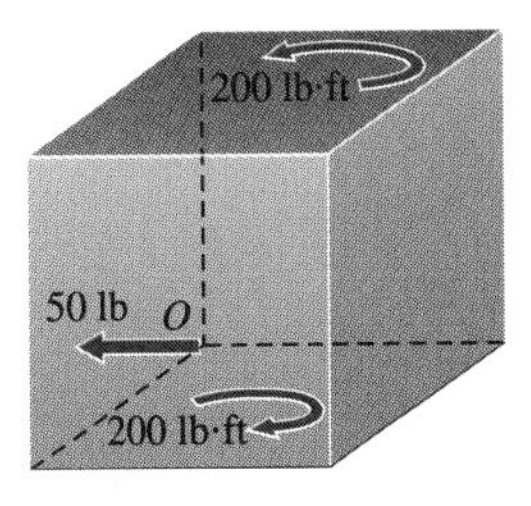

Figure E4.6c

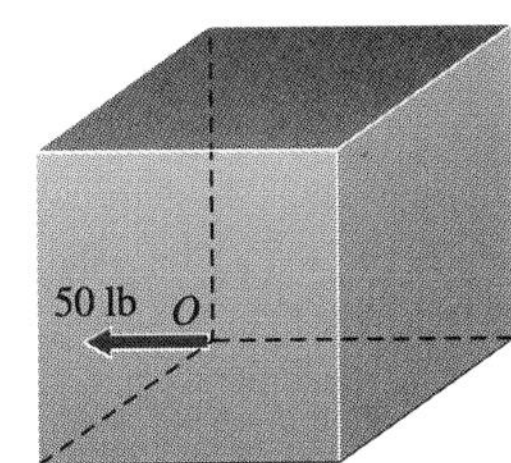

Figure E4.6d

Information Item In general, any force and couple that lie in the same plane or in parallel planes can be reduced to a *single* force that is dynamically equivalent to the original force-couple pair, as long as the line of action of the single force is properly located. This principle is discussed in Sec. 4.5.

Example 4.7

Resultant of Several Couples That Lie in Parallel Planes

Problem Statement Three couples act on a rigid body, as shown in Fig. E4.7a. Determine the resultant couple of this system.

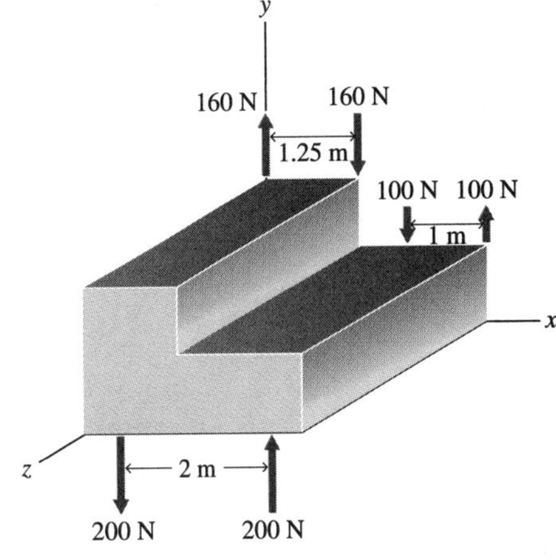

Figure E4.7a

Solution Since the couples lie in parallel planes, their moments may be added algebraically. The respective moments of the couples are

$$M_1 = (100)(1.0) = 100 \text{ N·m}$$

$$M_2 = (-160)(1.25) = -200 \text{ N·m}$$

$$M_3 = (200)(2.0) = 400 \text{ N·m}$$

The signs associated with the moments represent the senses of the couples. The positive moments M_1 and M_3 represent couples

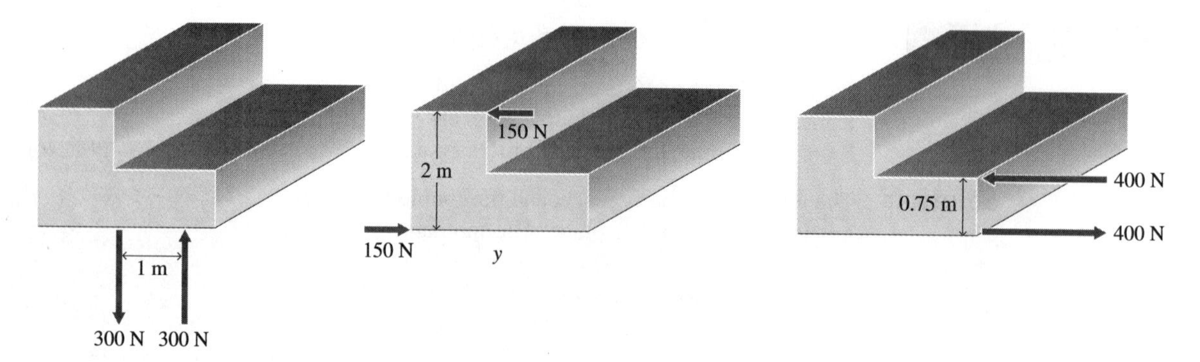

Figure E4.7b

that cause counterclockwise rotation, as viewed from the positive z axis; the negative moment M_2 represents clockwise rotation. Accordingly, the moment of the resultant couple is

$$M = M_1 + M_2 + M_3 = 100 - 200 + 400 = 300 \text{ N·m}$$

The resultant couple may be expressed in various ways by pairs of forces. Three dynamically equivalent resultant couples are shown in Fig. E4.7b. ■

CHECKPOINT 

1. ***True or False:*** A couple is composed of two noncollinear parallel forces that have equal magnitudes and opposite senses.
2. ***True or False:*** The arm of a couple is the same as the moment arm of a force.
3. ***True or False:*** A couple always tends to rotate a rigid body on which it acts about the midpoint of the arm of the couple.
4. ***True or False:*** A couple that acts on a rigid body produces the same effect if it is displaced anywhere in its plane.
5. ***True or False:*** A couple that acts on a rigid body produces the same effect on the body if it is moved to a plane normal to its original plane.
6. ***True or False:*** Several couples that act on a rigid body and lie in the same plane or in parallel planes may be combined into a single resultant couple without changing the effect on the body.

Answers: 1. True; 2. False; 3. False; 4. True; 5. False; 6. True

4.5 LATERAL DISPLACEMENT OF FORCES

After studying this section, you should be able to:

- Replace a force acting at one point with an equivalent force acting at a different point and a compensating couple.

Key Concept A force can be shifted laterally from its line of action if a compensating couple is added.

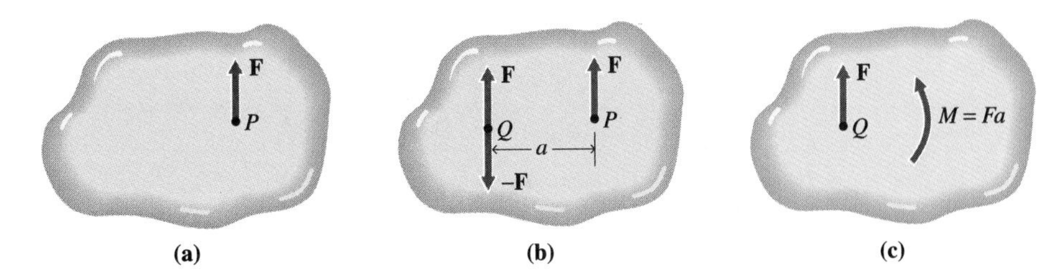

Figure 4.16 Lateral displacement of a force.

In Example 4.6, a force and a couple lying in the same plane were combined to obtain a single force with a line of action different from that of the original force. In effect, the original force was shifted laterally (moved sideways) so that its new line of action remained parallel to its original line of action to counteract the effect of (that is, to eliminate) the couple. In general, a force that acts at a given point of a rigid body may be moved sideways to any other point, provided that a *compensating couple* is introduced. This transformation is demonstrated below.

compensating couple

Consider a rigid body that is subjected to a force **F** at point P (Fig. 4.16a). Introduce self-equilibrating forces **F** and −**F** at any other point Q (Fig. 4.16b). The force **F** at point P and the force −**F** at point Q constitute a couple with moment Fa, where a is the distance between the lines of action of the forces. Consequently, the force systems illustrated by Figs. 4.16a, 4.16b, and 4.16c are dynamically equivalent. The original force **F** has been displaced to point Q, and a compensating couple with moment $M = Fa$ has been introduced. The moment $M = Fa$ is identical to the moment of the original force **F** about point Q. This conclusion leads immediately to the following theorem.

Theorem 4.5

Several coplanar forces $\mathbf{F}_1, \mathbf{F}_2, \ldots$ that act on a rigid body are dynamically equivalent to their resultant force **F**, acting at an arbitrary point Q in the plane of the forces, and a single couple, where the resultant force **F** is the vector sum of the forces $\mathbf{F}_1, \mathbf{F}_2, \ldots$ and the moment M of the associated couple is the sum of the moments of the forces $\mathbf{F}_1, \mathbf{F}_2, \ldots$ about point Q.

To prove this fundamental theorem, we transfer the forces $\mathbf{F}_1, \mathbf{F}_2, \ldots$ to point Q and introduce compensating couples with moments $M_1 = \pm F_1 a_1$, $M_2 = \pm F_2 a_2, \ldots,$ as explained above. Since the forces at point Q are concurrent, they may be added vectorially, that is, $\mathbf{F} = \mathbf{F}_1 + \mathbf{F}_2 + \cdots$. In other words, the resultant force **F** is obtained by the polygon construction or, alternatively, by algebraic addition of the projections of the individual forces on Cartesian coordinate axes.

Also, the moments $M_1, M_2, \ldots$ may be combined into a single moment by addition, with proper regard for the sense (+ or −) of each moment (see Sec. 4.4). That is, the moment M of the couple that is associated with force **F** at point Q is $M = M_1 + M_2 + \cdots$. Accordingly, M is the sum of the moments of the original forces about point Q. The resultant-axis theorem (Theorem 4.2) is a special case of Theorem 4.5 that results if point Q is located so that $M = 0$.

We have seen that the resultant **F** of several coplanar forces $\mathbf{F}_1, \mathbf{F}_2, \ldots$ may be assigned an arbitrary point of action Q in the plane of the forces but that the moment M of

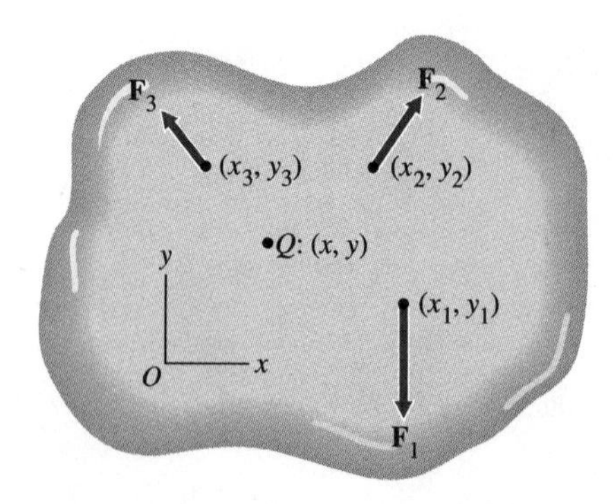

Figure 4.17

the associated couple generally depends on the location of point Q. Now, we will see that *if* $\mathbf{F} = 0$, *the associated couple does not depend on the location of point Q.*

Let the coordinates of point Q be (x, y) and let the forces $\mathbf{F}_1, \mathbf{F}_2, \ldots$ act at the respective points $(x_1, y_1), (x_2, y_2), \ldots$ relative to xy axes (see Fig. 4.17). By Varignon's theorem, the moment M of all the forces about point Q is

$$M = (x_1 - x)F_{1y} - (y_1 - y)F_{1x} + (x_2 - x)F_{2y} - (y_2 - y)F_{2x} + \cdots$$

or

$$M = y(F_{1x} + F_{2x} + \cdots) - x(F_{1y} + F_{2y} + \cdots) + (x_1F_{1y} - y_1F_{1x}) + (x_2F_{2y} - y_2F_{2x}) + \cdots$$

Since, by hypothesis, $\mathbf{F}_1 + \mathbf{F}_2 + \cdots = 0$, this reduces to

$$M = (x_1F_{1y} - y_1F_{1x}) + (x_2F_{2y} - y_2F_{2x}) + \cdots$$

However, by Varignon's theorem, the right side of this equation represents the moment of all the forces about the origin O, irrespective of the point Q. This result provides the following theorem (which will be generalized for three-dimensional force systems in Chapter 5).

Theorem 4.6

If the vector sum of a set of coplanar forces is zero, the forces exert the same moment about all points in their plane.

Example 4.8

Resultant of General Coplanar Force System

Problem Statement A rigid T-bar is subjected to several coplanar forces (see Fig. E4.8a). Determine the resultant force on the system.

Solution Our objective is to reduce a system of six forces to a single resultant force with a specific line of action—that is, to a single force that is dynamically equivalent to the six forces. Examination of Fig. E4.8a reveals that the horizontal forces at E and F form a couple and the vertical forces at B and C form another couple. The moments associated with these two couples are 1600 lb·ft and −600 lb·ft, respectively. This leaves a 100 lb vertical force at D and a 200 lb horizontal force A. We displace the 100 lb vertical force laterally from D to A. The moment of the compensating couple is $(-7)(100) = -700$ lb·ft. Hence, after the 100 lb force has been transferred to point A, the moment about point A of all the forces is

$M_A = -700 - 600 + 1600 = 300$ lb·ft. This is the moment of the resultant couple of the force system, relative to point A.

Thus, the original force system is dynamically equivalent to that shown in Fig. E4.8b. The magnitude of the resultant force is $F = \sqrt{100^2 + 200^2} = 223.6$ lb. The direction of the resultant force is determined by $\tan \theta = 100/200 = 1/2$, which yields $\theta = 26.56°$. We can simplify the force system further by displacing the resultant force laterally so that it produces the same moment about point A as all the original forces did—that is, so that it produces a moment of 300 lb·ft about point A. The lateral displacement h (see Fig. E4.8c) is determined by the equation $223.6h = 300$, giving $h = 1.342$ ft. Thus the original force system has been reduced to the single force shown in Fig. E4.8c. The line L is the resultant axis.

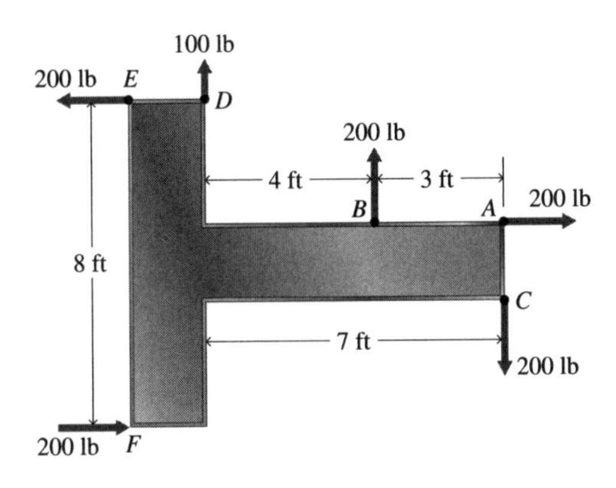

Figure E4.8a

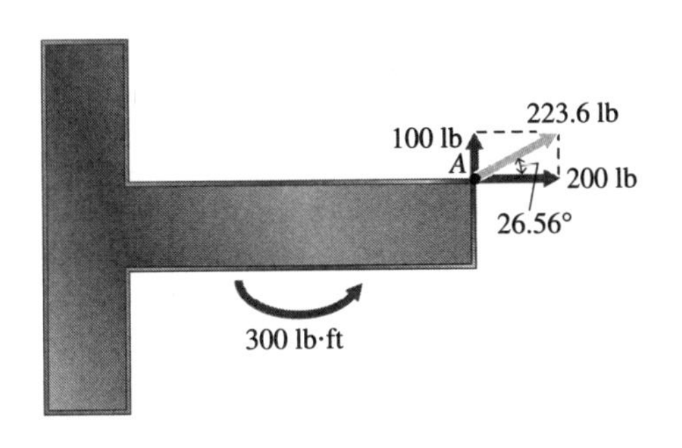

Figure E4.8b

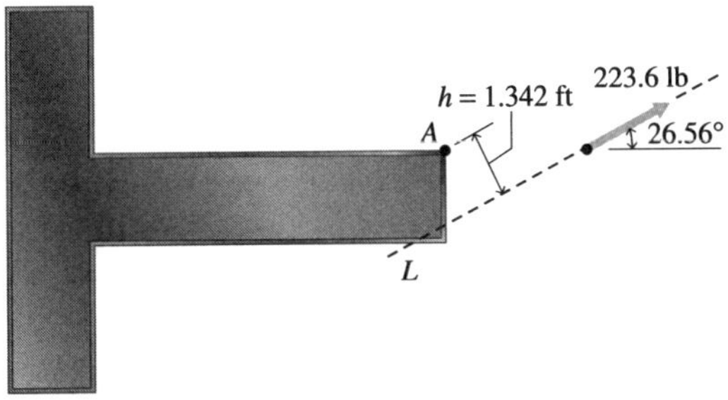

Figure E4.8c

Example 4.9

Simplification of the Loading of a Beam

Problem Statement The right end of a beam is subjected to an axial force of 8 kips and to a couple with moment 20 kip·in, as shown in Fig. E4.9a. Relocate the axial force to eliminate the couple.

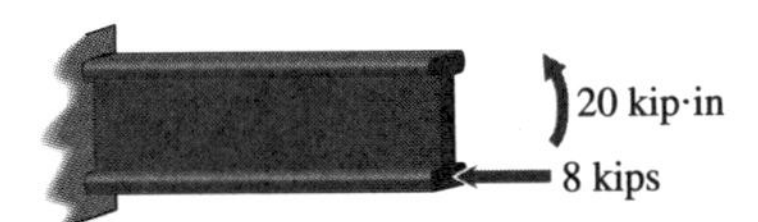

Figure E4.9a

Solution Elimination of the couple can be accomplished simply. We introduce self-equilibrating forces of magnitude 8 kips at a point P located a distance h above the bottom edge of the beam (see Fig. E4.9b). The three forces then are equivalent to an axial compression force of 8 kips at point P and a couple with moment $-8h$. This couple opposes the original couple. Hence, the net moment of the couples is $20 - 8h$. Since we wish to eliminate the couple, we write $20 - 8h = 0$, from which we compute $h = 2.5$ in. Therefore, the loading shown in Fig. E4.9c is dynamically equivalent to that shown in Fig. E4.9a.

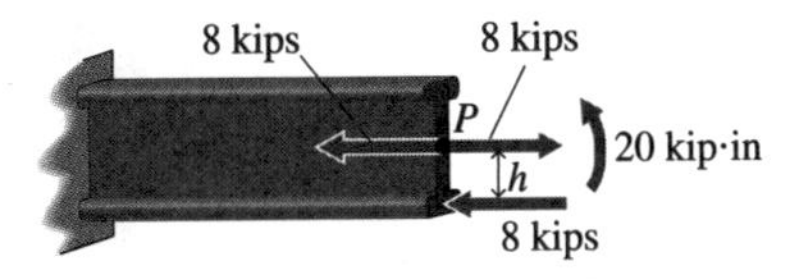

Figure E4.9b

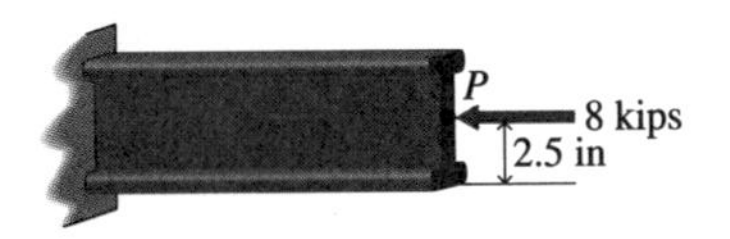

Figure E4.9c

CHECKPOINT 

1. ***True or False:*** A force **F** that acts on a rigid body may be displaced laterally, provided that a compensating couple is introduced.
2. ***True or False:*** A set of coplanar forces that act on a rigid body exerts the same moment about all points in the rigid body.
3. ***True or False:*** If the resultant of a system of coplanar forces is zero, the moment of the resultant couple is independent of the reference point *O*.

Answers: 1. True; 2. False; 3. True

4.6 EQUILIBRIUM OF A RIGID BODY SUBJECTED TO COPLANAR FORCES

After studying this section, you should be able to:

- Write and solve equations of equilibrium for a rigid body subjected to coplanar forces.

Key Concept

A rigid body subjected to coplanar forces is in equilibrium if the net force on the body is zero and the net moment about any point in the plane of the forces is zero.

A single force that acts through the center of mass of a body tends to make the body translate. In such cases, the conditions of equilibrium for a particle that were presented in Chapter 3 may be applied to the body. That is, the body is in equilibrium if, and only if, the force is zero. However, when a rigid body is subjected to a force whose line of action does not pass through the center of mass, the body tends to translate and rotate. Likewise,

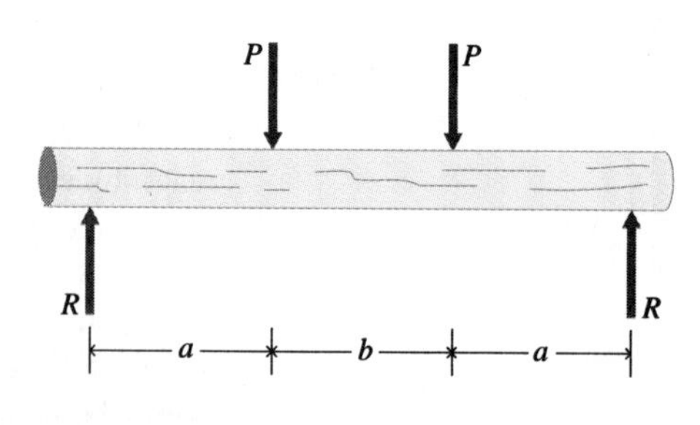

The bending strength of a wood peg is determined by using the load P required to cause material failure in a free-body diagram to find the bending moment in the peg and thus the bending stress at failure.

when a couple is applied, a rigid body tends to rotate about its center of mass. Consequently, a rigid body should be in equilibrium if, and only if, the resultant force and the resultant couple that act on the body are zero. This conjecture may be verified using momentum principles (see Chapter 16).

Recall that the resultant couple of a coplanar system of forces is independent of the reference point Q if the resultant force is zero (Sec. 4.5). Consequently, the following theorem summarizes the entire theory of equilibrium of rigid bodies subjected to coplanar force systems.

Theorem 4.7

A rigid body that is subjected to coplanar forces is in equilibrium if, and only if, the vector sum of all the external forces is zero and the moment of the external forces about any point in their plane is zero.

EQUILIBRIUM CONDITIONS REFERRED TO RECTANGULAR COORDINATES

Key Concept

The equilibrium conditions for a rigid body subjected to coplanar forces can be written in terms of two independent force equations and one moment equation, each of which involves rectangular projections of the coplanar forces.

The equilibrium principle of Theorem 4.7 is expressed by the equations

$$\begin{aligned} \sum \mathbf{F} &= 0 \\ \sum M &= 0 \end{aligned} \tag{4.4}$$

where $\Sigma \mathbf{F}$ is the *vector sum* of all the external forces that act on the rigid body and ΣM is the total moment of those forces about any point in their plane. To apply this principle, we can refer the body to rectangular xy coordinate axes. Then, Eqs. (4.4) are equivalent to the equations

$$\begin{aligned} \sum F_x &= 0 \\ \sum F_y &= 0 \\ \sum M &= 0 \end{aligned} \tag{4.5}$$

where ΣF_x and ΣF_y are the x and y projections of the vector sum $\Sigma \mathbf{F}$. In turn, ΣF_x and ΣF_y are determined by the equations

$$\sum F_x = F_{1x} + F_{2x} + F_{3x} + \cdots$$
$$\sum F_y = F_{1y} + F_{2y} + F_{3y} + \cdots$$

where F_{1x}, F_{1y}, F_{2x}, F_{2y}, ... are the x and y projections of the various external forces, $\mathbf{F}_1$, $\mathbf{F}_2$, ..., that act on the body.

Since ΣF_x and ΣF_y vanish for a body in equilibrium, the external forces produce the same moment about all points in their plane (Sec. 4.5). Consequently, in applying the condition $\Sigma M = 0$, *you may take moments about any convenient point.* The expression for the moment of the external forces about the origin is given by Eq. (4.3). Equations (4.5) may be written as follows:

$$\sum F_x = \sum F_{ix} = 0 \tag{4.6a}$$

$$\sum F_y = \sum F_{iy} = 0 \tag{4.6b}$$

$$\sum M = \sum (x_i F_{iy} - y_i F_{ix}) = 0 \tag{4.6c}$$

where $i = 1, 2, \ldots, N$ and N is the number of forces acting. Also, (x_i, y_i) is the point of application of the ith force with projections (F_{ix}, F_{iy}). Equations (4.6) indicate that the sum of the N force projections and the sum of the moments of the N forces acting on the body equal zero.

When using the equation $\Sigma M = \Sigma (x_i F_{iy} - y_i F_{ix}) = 0$, you must determine the correct sign for each of the force projections (F_{ix}, F_{iy}) and for the coordinates (x_i, y_i) of the points of application of the forces. Using tables to organize the computations, as in Example 4.4, may prove helpful. However, regardless of your approach, you must be careful to avoid sign errors.

The moment equation [Eq. (4.6c)] uses the sign convention established in Sec. 4.1: A positive moment tends to cause counterclockwise rotation about point O (the z axis) for forces and couples that lie in the xy plane. But, you do not have to use this particular sign convention. You may wish to take the positive moment as causing clockwise rotation. For equilibrium to exist, the moments must sum to zero, regardless of sign convention. That is, the choice of sign convention (positive direction for rotation) is arbitrary. However, once you choose a sign convention for moments in a particular problem, you should use it consistently throughout the problem.

Information Item As you develop your problem-solving skills, you might prefer to dispense with the moment equilibrium equation [Eq. (4.6c)] in favor of an approach that is based directly on the definition of moment of a force. For example, as an alternative to using that equation for moment equilibrium in coplanar problems, consider the following approach. First, choose a point in the plane about which moments are to be computed. Then, for each force (or force component), determine its moment arm r. Find the magnitude of its moment using Eq. (4.2). Establish a moment sign convention. Then, examine each force (or force component) to determine whether it causes a positive or a negative moment. Assign the correct sign to the moment of each force, and algebraically sum the moments for equilibrium—that is, form the sum $\Sigma M_i = 0$. You may find it helpful to draw a symbol such as $\circlearrowright\!+$ or $\circlearrowleft\!+$ next to the equilibrium equation for moment to indicate your sign convention. This direct approach employs the definition of a moment [a force times a perpendicular distance; see Eq. (4.2)] and an understanding of the sense in which the force rotates the body.

PROBLEM-SOLVING TECHNIQUE

Equilibrium of a Rigid Body Subjected to Coplanar Forces

To solve a problem involving equilibrium of a rigid body subjected to coplanar forces:

1. First, review the discussion of free-body diagrams in Sec. 3.2.
2. Select a rigid body (a part of the system) to which the equilibrium conditions are to be applied. You may find it necessary to select several parts (rigid bodies) from the system in order to solve for all the unknowns.

3. Sketch a free-body diagram of the body. Be sure to include all the forces acting *on* the body, and be equally careful to exclude forces exerted *by* the body on some other part of the system.
4. Choose appropriate coordinate axes, and draw them next to the free-body diagram. (If additional free-body diagrams of other parts of the system are needed, you may find it helpful to use a separate set of axes for each diagram.)
5. Use algebraic symbols to label the magnitudes of unknown forces and dimensions.
6. If one of the unknown forces is the weight of a body, label it with the symbol W.
7. Sum the force projections relative to the xy axes chosen in Step 4, and set the sums equal to zero [see Eqs. (4.6a) and (4.6b)]. In the equilibrium equations, the signs of the projections *must* be consistent with the chosen axes.
8. Sum the moments of the forces about an appropriate reference point in the plane, using either Eq. (4.6c) or the alternative approach outlined in the preceding Information Item.
9. If free-body diagrams of other parts of the system are needed to solve for the unknowns, repeat Steps 3 through 8 for each body.
10. Collect the independent equilibrium equations derived in Steps 7 and 8 for each body. These equations will generally involve relations among forces that act on several parts of the system if more than one part is involved in the problem.
11. Solve the equations of Step 10 to determine the unknowns.
12. Where possible, check particular cases or extreme values of the unknowns. Consider these results, and ask yourself, "Do the results make sense?" If you cannot answer "yes," review your work to be sure you are correctly applying fundamental concepts and correctly using the equations of equilibrium.

Example 4.10

Reactions on a Truss

Problem Statement A truss (a structure built of straight two-force members; see Chapter 6) is subjected to coplanar forces, as shown in Fig. E4.10a. Each of the members in the truss is 10 ft long. Determine the forces that the supports at A and G exert on the truss.

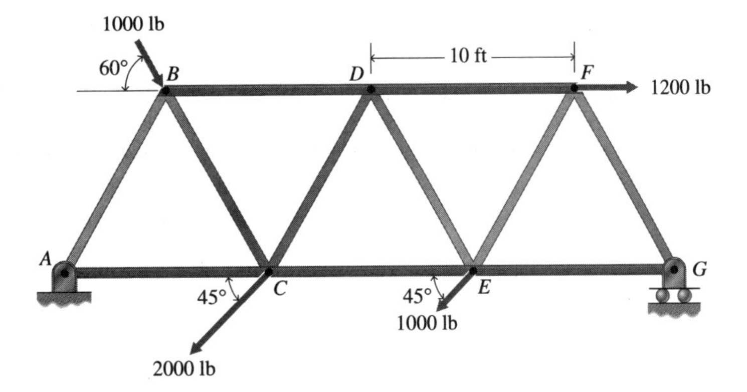

Figure E4.10a

Solution

Since our objective is to find support reactions on the truss, we take the entire truss as a free body. The free-body diagram of the truss is shown in Fig. E4.10b. Point A is selected as the origin of the xy coordinate axes, with the x axis directed along line AG. The support reaction at point A is represented in the free-body diagram by its Cartesian components of magnitudes A_x and A_y. The support reaction at point G is directed vertically upward. The magnitude of this reaction is denoted by G. (The use of the same letter to represent a point and an external force that is applied at the point is convenient and should cause no confusion.)

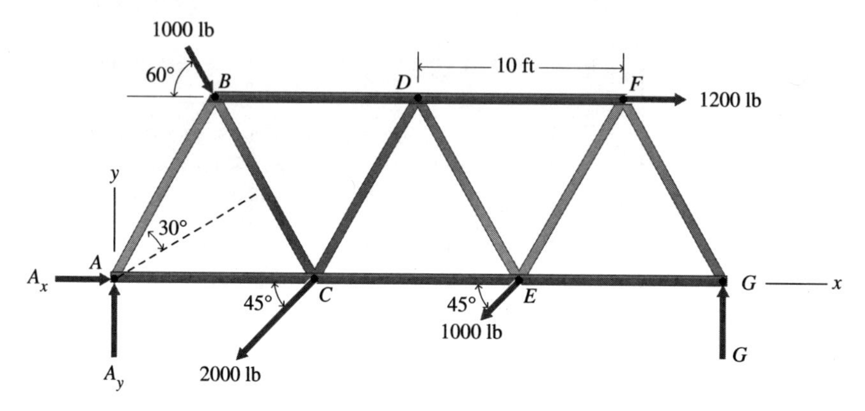

Figure E4.10b

The equilibrium equations for the entire truss are given by Eqs. (4.6). First, we consider the force equilibrium equations. From Fig. E4.10b,

$$\sum F_x = A_x + 1000\cos 60° - 2000\cos 45° - 1000\cos 45° + 1200 = 0 \qquad \text{(a)}$$

$$\sum F_y = A_y + G - 1000\sin 60° - 2000\sin 45° - 1000\sin 45° = 0 \qquad \text{(b)}$$

Equation (a) yields $A_x = 421.3$ lb. However, the vertical reactions at A and G cannot be found without consideration of the moment equilibrium equation. To satisfy the moment equilibrium equation [Eq. (4.6c)], we organize the forces and the coordinates of their points of application in table form; see Table E4.10. If we take moments about point A, we obtain

$$\circlearrowleft^{+} \sum M_A = 30G - 47{,}330 = 0$$

which gives $G = 1578$ lb. Now, this value for G may be substituted into Eq. (b), the equilibrium equation for force projections in the y direction, to obtain $A_y = 1409$ lb. Since the

Table E4.10

JOINT	F_{ix} (lb)	F_{iy} (lb)	x_i (ft)	y_i (ft)	$x_iF_{iy} - y_iF_{ix}$ (lb·ft)
A	A_x	A_y	0.0	0.0	0
B	500	−866	5.0	8.66	−8660
C	−1414	−1414	10.0	0.0	−14,140
E	−707	−707	20.0	0.0	−14,140
F	1200	0	25.0	8.66	−10,390
G	0	G	30.0	0.0	$30G$
				Total:	$\Sigma M_A = 30G - 47{,}330$

values for A_x, A_y, and G are positive, the senses of the support reactions are correct as shown in Fig. E4.10b.

As a check, we use the alternative approach, described earlier in the Information Item, by writing moments about joint A in terms of each force magnitude times its moment arm. In this case, this approach is somewhat simpler than using Eqs. (4.6), and the moment equilibrium equation is

$$\circlearrowleft\!+ \sum M_A = 30G - (1000)(10\cos 30°) - (2000)(10\sin 45°) - (1000)(20\sin 45°) - (1200)(10\sin 60°) = 0$$

From this equation, we get $G = 1578$ lb. Substitution of this value into Eq. (b) gives $A_y = 1409$ lb. Thus, our results check. ■

Example 4.11

Hyatt Regency Hotel Walkways

Preliminary Information During construction or manufacture of a product, changes are often made to the original design. As an engineer in charge of a project, you will be responsible for ensuring that any changes made in the design during the construction or manufacturing process do not affect the integrity of the final product. The elevated walkways (also known as *skywalks*) of the Hyatt Regency Hotel in Kansas City, Missouri, provide an example of a case in which a change in design seriously affected a project.

The failed fourth-floor and second-floor walkways landed one on top of another in the lobby of the Hyatt Regency. Failure was initiated when the nut and washer of a hanger rod pulled through the box beam supporting the fourth-floor walkway.

Figure E4.11a shows a simplified cross-section of the walkway as designed. (*Note:* The figures are not to scale, and the concrete slab has been omitted for clarity.) The walkways were intended to be held by continuous rods *ABC* and *DEF*, with washer-nut supports at *B, C, E,* and *F.* The two elevated walkways spanned the lobby of the hotel. During construction, a change was made in the design, and the walkways were built as shown in Fig. E4.11b. In particular, each of the continuous rods was replaced by two rods: *GH* and *IJ* replaced *ABC* and *KL* and *MN* replaced *DEF*, with washer-nut supports at *H, I, J, L, M,* and *N.* On July 18, 1981, a dance was held in the lobby. The walkways were crowded with spectators watching the event. Under the weight of the spectators, the walkways fell into the lobby—111 people were killed and 188 others were injured. Let's examine why this happened.

Problem Statement Assume that the weight of the spectators plus the weight of the walkway forms a uniformly distributed load with a total magnitude *W* on each walkway

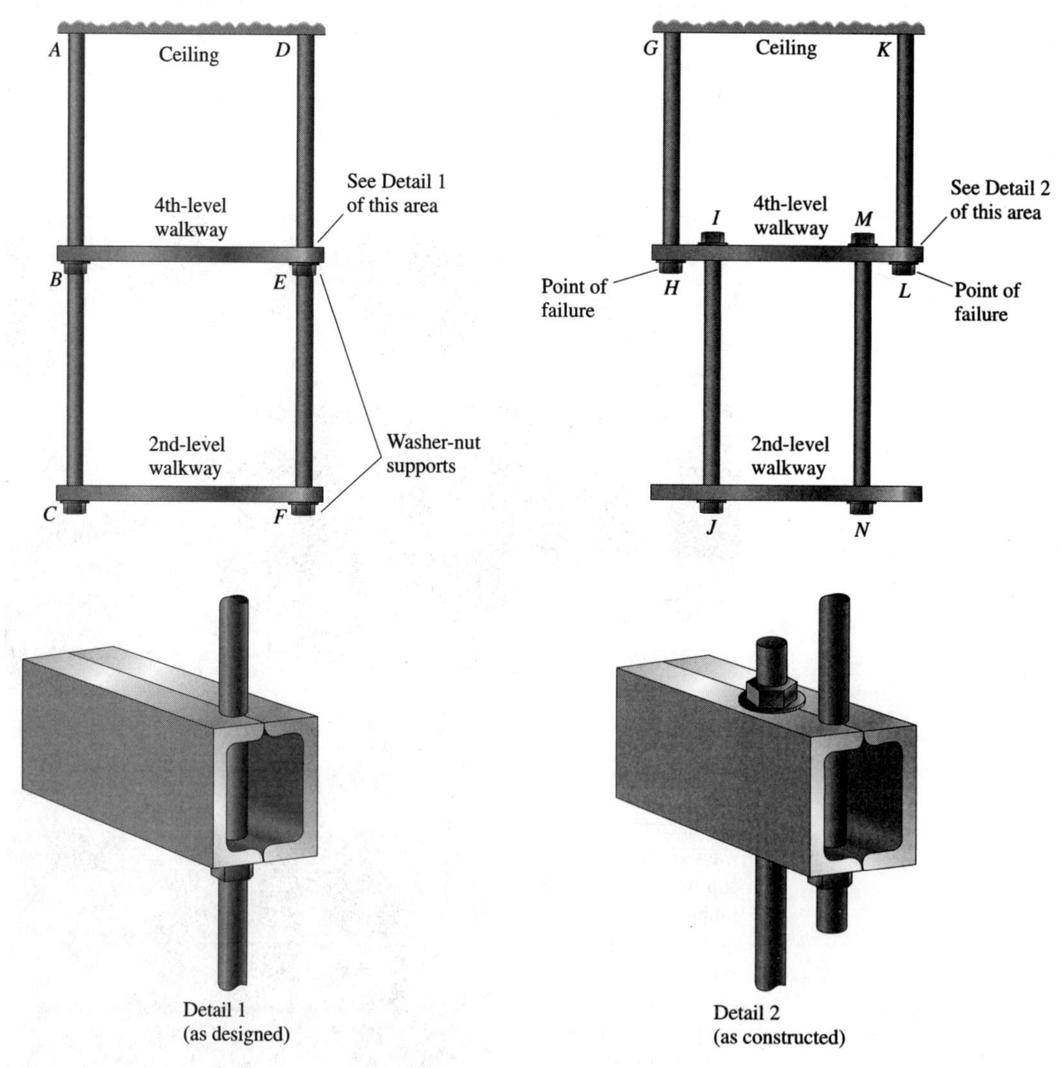

Figure E4.11a

Figure E4.11b

surface. Compute the loads carried by the rods and by the washer-nut supports on each of the walkways for the following conditions:

a. for the walkways as designed

b. for the walkways as constructed

Solution **a.** To determine the loads supported by the rods and the washer-nut supports, we imagine that the walkways are disengaged from the rods and washer-nut supports and draw free-body diagrams of the individual parts. Since the loads are assumed to be uniform and the structure is symmetric, each rod supports a load W at A and D—that is, half of the total load $2W$. These loads are balanced by the loads $W/2$ carried by the washer-nut supports at B, C, E, and F (Fig. E4.11c). For each walkway to be in equilibrium the two washer-nut reactions $W/2$ must oppose the weight W.

b. For the walkways as constructed (Fig. E4.11b), we again disassemble the walkway system and draw free-body diagrams of the individual parts (Fig. E4.11d). By inspection of these free-body diagrams, we see that the washer-nut supports at H and L must each carry a load of W. This load is twice the magnitude of the load carried by each of the corresponding washer-nut supports (at B and E) in the originally designed walkways (Fig. E4.11c). Thus, the as-built system of walkways failed when the washers and nuts were pulled through the rod holes in the box beam at H and L (Fig. E4.11b). In subsequent legal actions, the project engineer's license was revoked, and, in addition, he was assessed a large monetary fine.

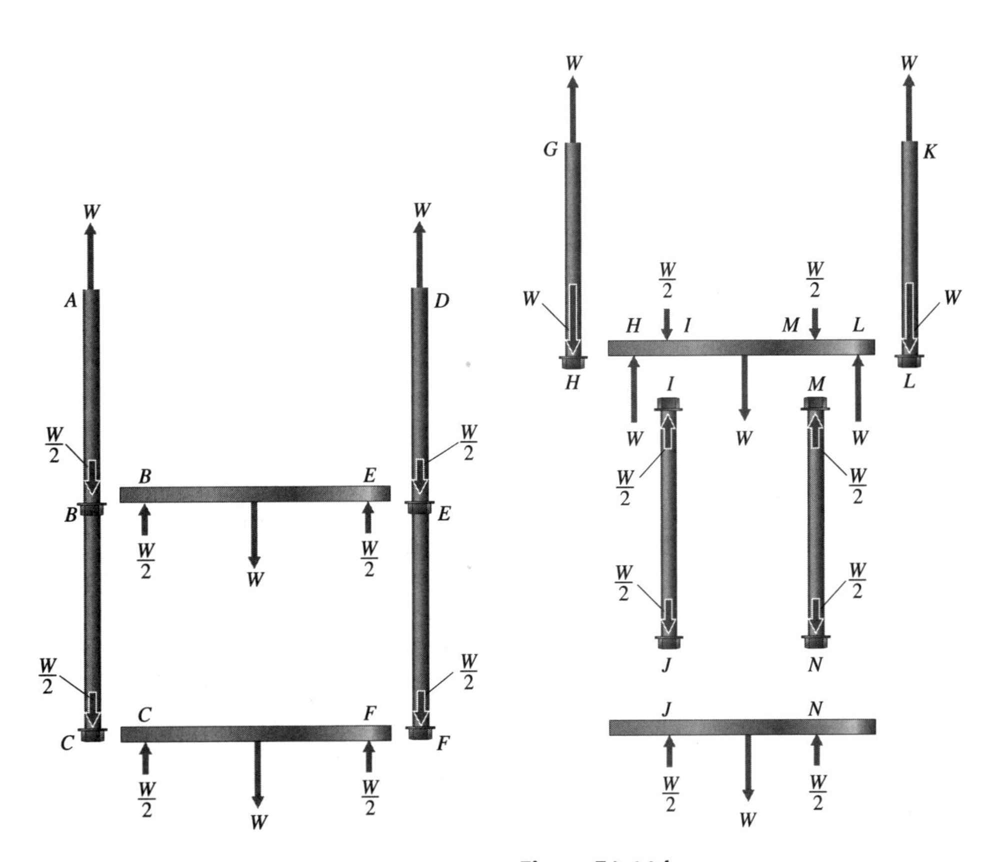

Figure E4.11c **Figure E4.11d**

Information Item As an engineer in charge of a project, you have the responsibility of approving any changes to a design. Therefore, you should scrutinize all suggested changes and make sure you analyze the effects of the change. Did anyone draw a free-body diagram of the rods of the walkway after the design change, as in Fig. E4.11d? An accurate free-body diagram and application of the condition for equilibrium of forces ($\Sigma \mathbf{F} = 0$) would have revealed the flaw in the design change. Indeed, it was not necessary to even consider equilibrium of moments. So, remember, when you approve changes, *you* become responsible for any consequences.

Discussion Item Why do you suppose the connection detail was changed from that shown in Fig. E4.11a to that shown in Fig. E4.11b? Can you conceive a better connection than that used in Fig. E4.11b?

Example 4.12

Equilibrium of a Flying Boat

Problem Statement Figure E4.12 represents a flying boat that is planing at constant speed in calm water. The boat is in equilibrium under the actions of the propeller thrust T, the lift force L on the wing, the weight W, the air drag D, the water drag R, the buoyant force B of the water, and the tail force F. Although these forces are distributed over the surfaces of the boat, only their resultants are shown. When the speed of the boat is 60 ft/s, $T = 2400$ lb, $D = 700$ lb, and $B = 10{,}500$ lb. The weight of the boat is $W = 15{,}000$ lb. Determine the forces L, R, and F.

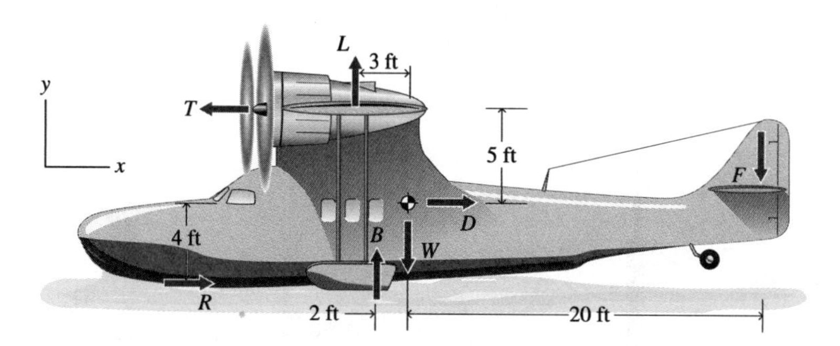

Figure E4.12

Solution The forces L, R, and F are determined from the conditions of equilibrium for a coplanar force system. Since each force is parallel to either the x axis or the y axis, the moment equation is relatively simple. The three equilibrium equations are

$$\sum F_x = -T + R + D = 0$$

$$\sum F_y = L + B - W - F = 0$$

$$\circlearrowleft\!+ \sum M_{cg} = -3L - 20F - 2B + 4R + 5T = 0$$

Here, ΣM_{cg} denotes moments about the center of gravity, the point at which the resultant weight acts. The forces L, F, and B produce negative moments about the center of gravity; R and T cause positive moments. The forces W and D act through the center of gravity, so they cause no moments about that point. Introducing the numerical values for T, D, W, and

B and solving the resulting equations for R, L, and F, we obtain $R = 1700$ lb, $L = 3817$ lb, and $F = -683$ lb.

Since F is negative, the correct sense of the force acting on the tail is opposite to that shown in Fig. E4.12. That is, the tail force is directed upward. ■

OTHER SPECIAL CASES

If *any* number of coplanar forces that act on a rigid body have concurrent lines of action, as in Fig. 4.18, the equation $\Sigma M = 0$ is satisfied automatically if moments are taken about the point of concurrency P. Also, if the sum of the concurrent forces is zero, there is no moment about *any* point. In this case, the moment equation is superfluous. So, again, the principles of particle equilibrium apply.

Another simplification results if a rigid body is subjected to parallel coplanar forces. If all of the forces are parallel to the y axis, then the condition $\Sigma F_{ix} = 0$ is satisfied automatically. Hence, Eqs. (4.6) reduce to $\Sigma F_{iy} = 0$ and $\Sigma M = 0$.

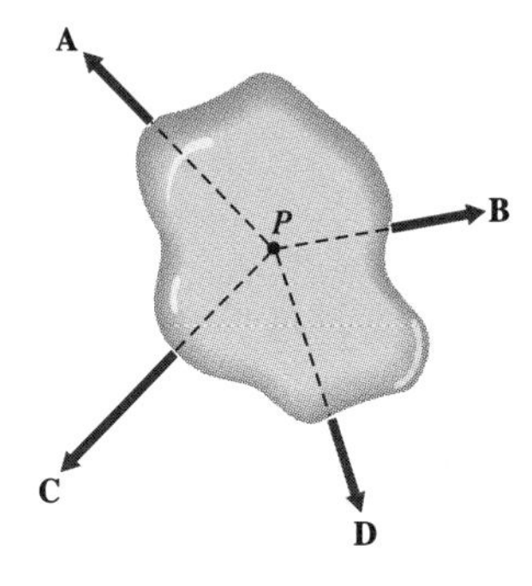

Figure 4.18 Forces with concurrent lines of action.

Example 4.13 Teeter Totter

Problem Statement A uniform teeter totter board weighs 45 N (Fig. E4.13a). Two children weighing 400 N and 300 N are balanced on the board. The center of gravity of the board is directly over the support O (called the *fulcrum*). The 400 N child is 1.5 m from the fulcrum.

a. Determine the distance x of the 300 N child from the fulcrum.

b. Determine the reaction force of the fulcrum.

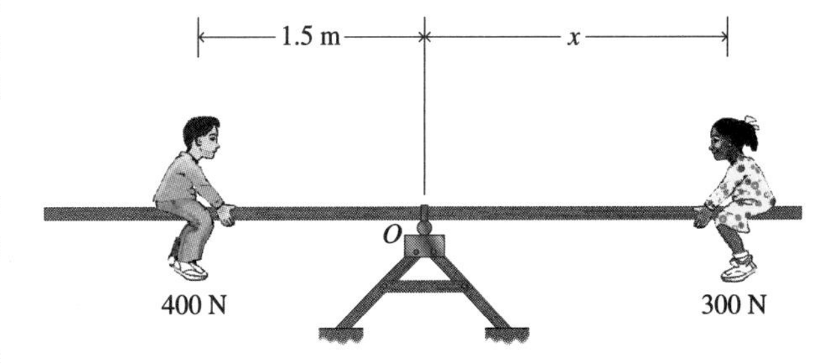

Figure E4.13a

Solution The free-body diagram of the balanced teeter-totter board is shown in Fig. E4.13b.

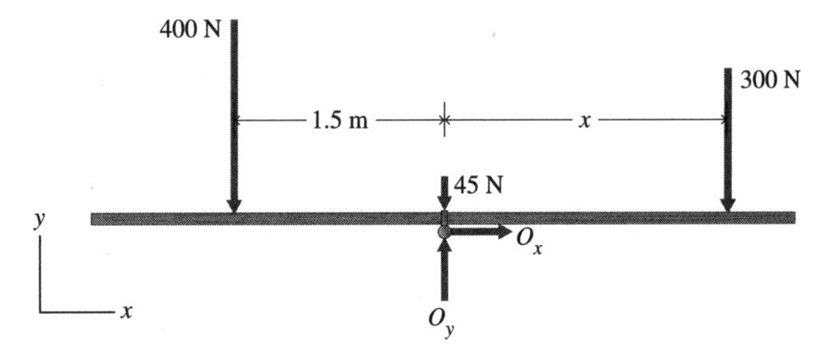

Figure E4.13b

a. To determine the distance x, we use Eq. (4.6c):

$$\circlearrowleft\!+ \sum M_O = 400(1.5) - 300x = 0$$

Thus, $x = 2$ m.

b. To determine the reaction force O_y, we use Eq (4.6b):

$$\sum F_y = O_y - 400 - 45 - 300 = 0$$

Thus, $O_y = 745$ N. Note that the condition $\Sigma F_x = 0$ yields $O_x = 0$, since there are no other forces acting horizontally (in the x direction). ■

4.7 ALTERNATIVE FORMULATIONS OF EQUILIBRIUM FOR COPLANAR FORCES

After studying this section, you should be able to:

- Write and solve equations of equilibrium for a rigid body subjected to coplanar forces, using two independent moment equations and one force equation or using three independent moment equations.

Key Concept

The equilibrium conditions for a rigid body subjected to coplanar forces can be written in terms of one force equation and two independent moment equations.

The fundamental statement of the equilibrium criteria for a rigid body is that the vector sum of the external forces is zero and the moment of those forces about any point in their plane is zero [see Theorem 4.7 and Eq. (4.4) or (4.5)]. Occasionally, it is more convenient to express the equations of equilibrium in some form other than as two force equations and one moment equation. In this section, we consider two alternative statements of the equilibrium principle that are useful for certain problems. The first of these alternatives is expressed by the following theorem.

Theorem 4.8

A motionless rigid body that is subjected to coplanar forces is in equilibrium if, and only if, the following conditions are satisfied:

1. The algebraic sum of the projections of the forces on an axis L in the plane of the forces is zero.
2. The forces produce no moments about two separate points A and B that lie in the plane of the forces on a line that is not perpendicular to the axis L.

In equation form, the conditions of Theorem 4.8 may be written as

$$\sum F_L = 0$$
$$\sum M_A = 0 \qquad (4.7)$$
$$\sum M_B = 0$$

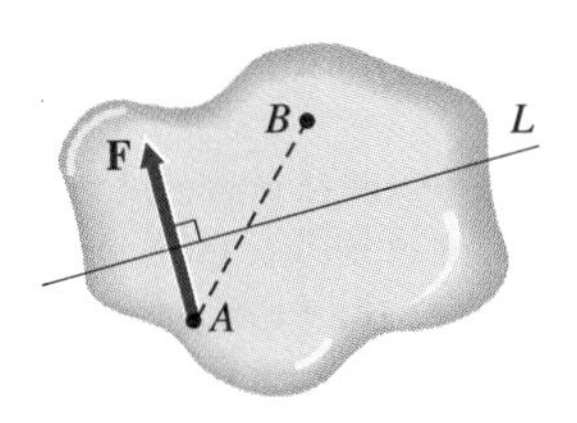

Figure 4.19

where L denotes the axis on which all forces are projected and A and B identify two separate points that lie in the plane of the forces on a line AB that is not perpendicular to L (Fig. 4.19).

To prove Theorem 4.8, we make the following argument. In Sec. 4.5, we proved that a system of coplanar forces acting on a rigid body can be reduced to a dynamically equivalent force-couple system. The resultant force **F** may be applied at any point in the plane, and the moment of the couple is equal to the moment of all the external forces about that point. Likewise, it is possible to move the force **F** to a point A in the plane, about which the moment of all external forces is zero. Now, suppose that a particular rigid body (Fig. 4.19) is subject to a system of coplanar forces. As noted above, the force system can be reduced to a single resultant force **F** acting at point A on line AB such that the resultant couple vanishes; that is, the condition described in Theorem 4.8 for point A exists. Since, by the first condition in Theorem 4.8, the sum of the projections of the coplanar forces on the line L is zero, either the force vector **F** is perpendicular to line L or $\mathbf{F} = 0$. Now, since the line AB is not perpendicular to the line L, the point B does not lie on the line of action of force **F**. So, in order for the moment of the force **F** about point B to be zero, we must have $\mathbf{F} = 0$. Therefore, the conditions $\mathbf{F} = 0$ and $M = 0$ are satisfied. According to the conditions defined by Eqs. (4.4), equilibrium must exist. Thus, Theorem 4.8 is proved, and Eqs. (4.7) express the equilibrium conditions in an alternative form.

Key Concept

The equilibrium conditions for a rigid body subjected to coplanar forces can be written in terms of three independent moment equations.

Another formulation of the equilibrium principle is the following theorem.

Theorem 4.9

A motionless rigid body that is subjected to coplanar forces is in equilibrium if, and only if, the moments of the forces about any three noncollinear points in the plane of the forces are zero.

Expressed in equation form, the conditions of Theorem 4.9 become

$$\sum M_O = 0$$
$$\sum M_P = 0 \qquad (4.8)$$
$$\sum M_Q = 0$$

where O, P, and Q are noncollinear points in the plane about which moments are taken (see Fig. 4.20).

To show that these conditions are equivalent to those in Eqs. (4.4), we let the moments of the forces about three noncollinear points O, P, and Q that lie in the plane of the forces be zero. Since the moment about point O vanishes, the force system is equivalent to a single force that acts at point O (Fig. 4.20). Since the moment about point P vanishes, point P lies on the line of action of force **F**, or else $\mathbf{F} = 0$. The same remark applies to point Q. Since the points O, P, and Q are not collinear, the points P and Q cannot both lie on the line of action of force **F**. Therefore, the only possibility is $\mathbf{F} = 0$. Consequently, by Eq. (4.4), equilibrium exists. Theorem 4.9 is therefore proved, and Eqs. (4.8) express the equilibrium conditions in another form. Thus, Theorems 4.7, 4.8, and 4.9, and the corre-

Figure 4.20

sponding Eqs. (4.5), (4.7), and (4.8), constitute three equivalent representations of equilibrium of a rigid body subjected to coplanar forces.

Each of the preceding formulations of the equilibrium principle leads to three algebraic equations. It is not possible to obtain more than three independent equilibrium equations for a single rigid body that is subjected to coplanar forces. If more than three equations are derived, one or more of them are redundant.

Example 4.14 Reactions on a Truss

Problem Statement Determine the support reactions on the truss in Example 4.10 in two ways:

a. by Theorem 4.8

b. by Theorem 4.9

Solution

The truss loading and support conditions are shown in Fig. E4.14a. Note that the free-body diagram for this example (Fig. E4.14b) is identical to that used in Example 4.10. This demonstrates that the concept of a free-body diagram is independent of the mechanics principles that are applied to it.

a. By Theorem 4.8, we write one force equation and two moment equations. Since the truss has only one horizontal reaction, that reaction can be determined directly by project-

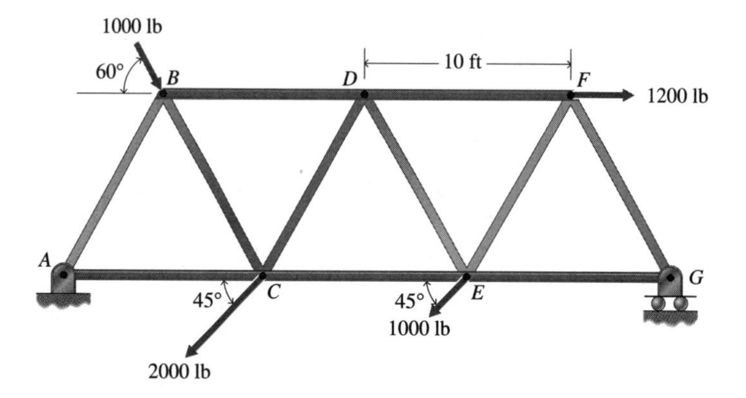

Figure E4.14a

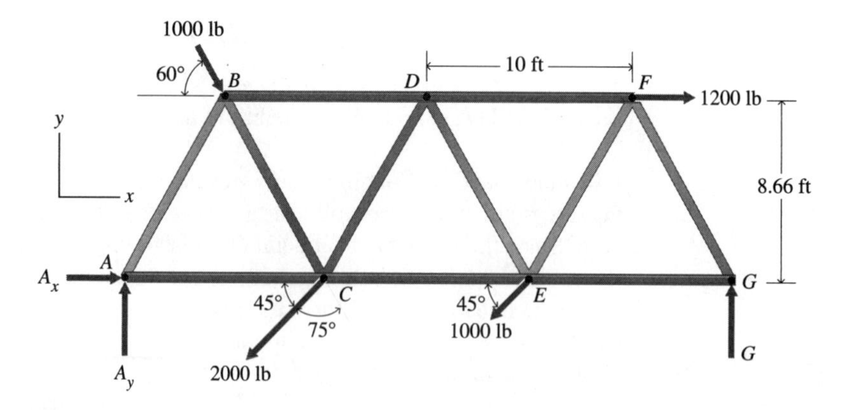

Figure E4.14b

ing all forces onto line AG, which is parallel to the x axis. The equilibrium equation is identical to Eq. (a) in Example 4.10:

$$\sum F_{AG} = A_x + 1000\cos 60^\circ - 2000\cos 45^\circ - 1000\cos 45^\circ + 1200 = 0$$

Thus, $A_x = 421.3$ lb. Again, as in Example 4.10, the reaction at G can be obtained by taking moments about point A. The equation

$$\circlearrowleft^{+} \sum M_A = 30G - (1000)(10\cos 30^\circ) - (2000)(10\sin 45^\circ) - (1000)(20\sin 45^\circ) - (1200)(10\sin 60^\circ) = 0$$

gives $G = 1578$ lb. Finally, we can take moments about another point to determine the vertical reaction at A. If we choose point C, we see that the force at C, the force at B, and the horizontal reaction at A cause no moment about that point. Hence, the equilibrium equation contains relatively few terms. Using counterclockwise rotation as an indication of positive moment, we have

$$\circlearrowleft^{+} \sum M_C = 20G - (10)(1000\sin 45^\circ) - 10A_y - (8.66)(1200) = 0$$

With $G = 1578$ lb, we obtain $A_y = 1409$ lb. Note that with this formulation of the equilibrium principle, we took moments about points C and G, which form a line that is not perpendicular to line AG, along which force projections were taken (see Theorem 4.8).

b. By Theorem 4.9, we take moments about three noncollinear points. Points C and G can be used to obtain $A_y = 1409$ lb and $G = 1578$ lb. To find A_x, we must take moments about a third point that cannot lie along line AG. An examination of the free-body diagram (Fig. E4.14b) reveals that if we were to take moments about a point along line AG, then the reaction A_x would not even appear in the equation. We decide that point B is a convenient choice, since the forces at B and F cause no moment about that point. Choosing clockwise rotation about B as positive, we have

$$\circlearrowright^{+} \sum M_B = 5A_y - 8.66A_x + (10)(2000\sin 75^\circ) + (8.66)(1000\cos 45^\circ) + (15)(1000\sin 45^\circ) - 25G = 0$$

With $A_y = 1409$ lb and $G = 1578$ lb, this equation yields $A_x = 421.3$ lb.

Information Item The equilibrium equation for moments about B is written using the sign convention that clockwise rotation about that point is considered a positive moment. This is done to demonstrate the arbitrary nature of the sign convention. In general, however, it is a good idea to select a sign convention and stick with it; avoid switching from one sign convention to another.

CHECKPOINT

1. ***True or False:*** If the resultant force of three concurrent forces that act on a rigid body is zero, the body is in equilibrium.
2. ***True or False:*** A rigid body subjected to coplanar forces is in equilibrium if the projections of the forces along an axis sum to zero and the moments about any two points are zero.
3. ***True or False:*** A rigid body subjected to coplanar forces is in equilibrium if the moments about any three points in the plane are zero.

Answers: 1. True; 2. False; 3. False

4.8 GENERAL THEORY OF PARALLEL FORCES

After studying this section, you should be able to:

- Write and solve equations of equilibrium for a rigid body subjected to noncoplanar parallel forces.

The previous sections in this chapter have treated the equilibrium of a rigid body under the action of a coplanar force system. In this section, we consider the equilibrium of a rigid body under the action of a noncoplanar parallel force system. Although this case is not two-dimensional, we are again able to express the equilibrium conditions for the body in terms of three equilibrium equations. A fully general treatment of the statics of a rigid body in space, leading to six independent equilibrium equations, is given in Chapter 5.

LATERAL DISPLACEMENT OF A FORCE

Let a rigid body be subjected to a force **F** that is parallel to the z axis. Let the positive sense of the force **F** agree with the positive sense of the z axis. By the theorem of transmissibility, the force **F** may be considered to act at a point P: (x, y) in the xy plane (see Fig. 4.21a). Let Q be a point on the x axis such that line PQ is perpendicular to the x axis. Self-equilibrating forces **F** and $-$**F** are introduced at point Q (Fig. 4.21a). The three forces at points P and Q are equivalent to a single force **F** at point Q and a couple with moment $M_x = yF$, consisting of force **F** at point P and force $-$**F** at point Q. The plane of the couple lies parallel to the yz plane. The moment of the couple is equal to the moment of force **F** at point P about the x axis.

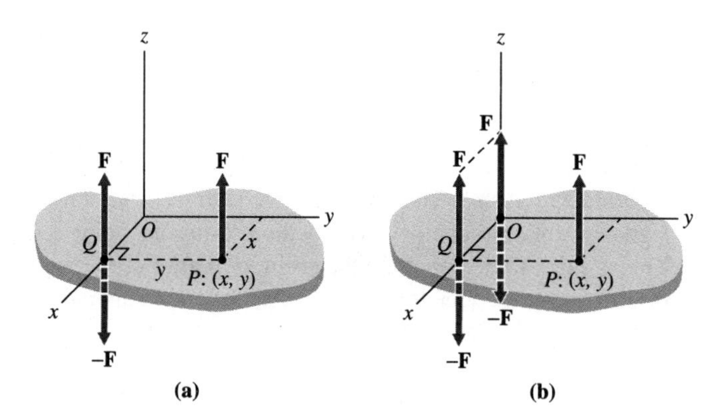

Figure 4.21

By introducing self-equilibrating forces **F** and $-$**F** at the origin O, we can transfer the force **F** from point Q to point O (Fig. 4.21b). The compensating couple for this transformation lies in the xz plane. Its moment $M_y = -xF$ is equal to the moment of the original force **F** about the y axis. Accordingly, the given force **F** at point P is dynamically equivalent to a force **F** at the origin O and two couples whose planes lie parallel to the yz and xz planes. The moments M_x and M_y of the two couples are, respectively, the moments of the original force **F** about the x and y axes. This conclusion is valid irrespective of the quadrant in which the point P lies. Also, it remains valid if the sense of force **F** is reversed.

COMPOSITION OF PARALLEL FORCES

Let a rigid body be subjected to several forces, $\mathbf{F}_1, \mathbf{F}_2, \ldots$, that are parallel to the z axis. The forces may all be transferred to the origin, provided that compensating couples are introduced. Then, the forces may be combined into a single resultant:

$$\mathbf{F} = \mathbf{F}_1 + \mathbf{F}_2 + \cdots$$

Since the forces all have the same direction (though not necessarily the same sense), their vector sum reduces to an algebraic sum.

Besides the resultant force $\mathbf{F}$ at the origin, the process of composition introduces two couples $\mathbf{M}_x$ and $\mathbf{M}_y$ that lie in the yz and xz planes.[5] The moments of these couples are denoted by M_x and M_y, respectively. They are the sums of the respective moments of all the original forces $\mathbf{F}_1, \mathbf{F}_2, \ldots$ about the x and y axes:

$$M_x = +\sum y_i F_i \qquad \text{and} \qquad M_y = -\sum x_i F_i$$

RESULTANT AXIS OF A SYSTEM OF PARALLEL FORCES

As shown above, several forces, $\mathbf{F}_1, \mathbf{F}_2, \ldots$, that are parallel to the z axis may be resolved into a dynamically equivalent system of a single force $\mathbf{F}$ at the origin and two couples with moments M_x and M_y. Suppose such an operation has been performed. By reversing the transformation illustrated in Fig. 4.21, we can displace the force $\mathbf{F}$ from the origin to a point (a, b) in the xy plane. If $\mathbf{F}$ is not zero, we may choose the point (a, b) so that the moments M_x and M_y are canceled. Then, the force $\mathbf{F}$ is called the resultant force $\mathbf{R}$. The resultant $\mathbf{R}$ exerts the same moments about the x and y axes as all the original forces $\mathbf{F}_1, \mathbf{F}_2, \ldots$. The line of action of $\mathbf{R}$ is called the *resultant axis of the force system.*

The foregoing conclusions are summarized by the following theorem.

Theorem 4.10

If several forces, $\mathbf{F}_1, \mathbf{F}_2, \ldots$, that act on a rigid body are parallel to the z axis and if their vector sum $\mathbf{R}$ is not zero, the given forces are dynamically equivalent to the single force $\mathbf{R}$, provided that this force is located so that it produces the same moments about the x and y axes as do all the original forces $\mathbf{F}_1, \mathbf{F}_2, \ldots$. Then, the force $\mathbf{R}$ is called the resultant force and is said to lie on the resultant axis of the force system.

Example 4.15

Resultant of a System of Noncoplanar Parallel Forces Acting on a Slab

Problem Statement Figure E4.15a represents a rigid slab that is loaded by noncoplanar parallel forces that are perpendicular to the face of the slab. The magnitudes and senses of the forces are listed in Table E4.15.

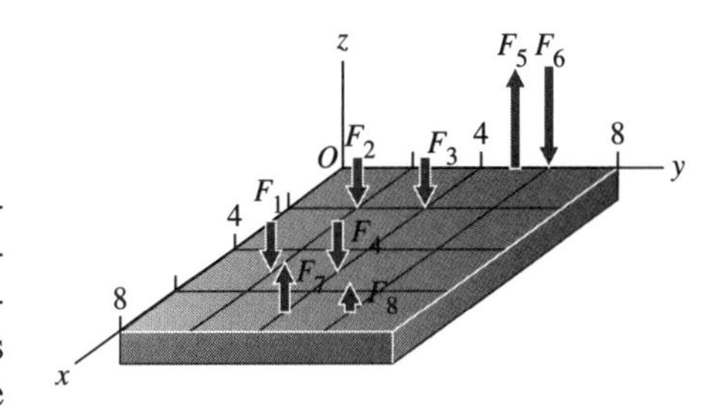

Figure E4.15a

5. In Chapter 5, we will see that couples are free vectors (Sec. 2.4). Hence, the two couples are denoted by vector notation as $\mathbf{M}_x$ and $\mathbf{M}_y$.

Table E4.15

z PROJECTION (kN)	x (m)	y (m)	M_x (kN·m)	M_y (kN·m)
$F_1 = -4$	5	2	−8	20
$F_2 = -4$	2	2	−8	8
$F_3 = -4$	2	4	−16	8
$F_4 = -4$	5	4	−16	20
$F_5 = 8$	0	5	40	0
$F_6 = -8$	0	6	−48	0
$F_7 = 4$	7	4	16	−28
$F_8 = 2$	7	6	12	−14
$\Sigma F_z = -10$			$\Sigma M_x = -28$	$\Sigma M_y = 14$

a. Show that the given forces are equivalent to a force **F** at the origin and two couples $\mathbf{M}_x$ and $\mathbf{M}_y$, with moments M_x and M_y, that lie in the yz and xz planes, respectively.

b. Determine the magnitude and the line of action of the resultant force **R** that is dynamically equivalent to the forces that act on the slab; that is, determine the resultant force and the resultant axis of the given forces.

Solution

a. The force **F** is the algebraic sum of the forces that act on the slab. The moments M_x and M_y of couples $\mathbf{M}_x$ and $\mathbf{M}_y$ are the moments of all the forces about the x axis and the y axis, respectively. Table E4.15 lists the z projections of the forces, the coordinates of the points of applications of the forces, and the moments of the forces about the x and y axes. The force **F** and the associated couples are the sums of the columns in the table. Thus, the force **F** is directed in the negative sense of the z axis, and its magnitude is 10 kN. The couples $\mathbf{M}_x$ and $\mathbf{M}_y$ have moments of magnitudes 28 kN·m and 14 kN·m. Couple $\mathbf{M}_x$ is directed along the negative x axis, whereas couple $\mathbf{M}_y$ is directed along the positive y axis, as indicated in Fig. E4.15b. In other words, $\mathbf{M}_x$ produces *clockwise* rotation about the x axis (when viewed from the positive part of the x axis), and $\mathbf{M}_y$ produces *counterclockwise* rotation about the y axis (when viewed from the positive part of the y axis). (See the discussions of the right-hand screw rule in Sec. 2.4 and of right-handed coordinate systems in Sec. 2.5.)

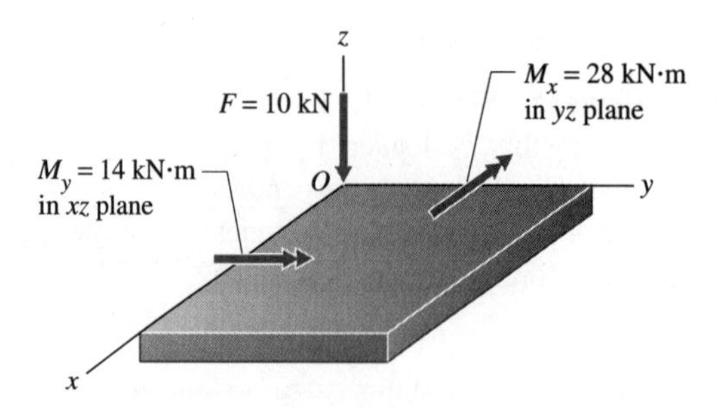

Figure E4.15b

b. The resultant force **R** has the same magnitude and sense as force **F**. However, its line of action differs from that of **F**. The axis of **R** is determined by displacing force **F** from the origin to a point (a, b) such that this single force produces the moments M_x and M_y. Hence,

$$M_x = +b\sum F_z$$
$$M_y = -a\sum F_z \qquad \text{(a)}$$

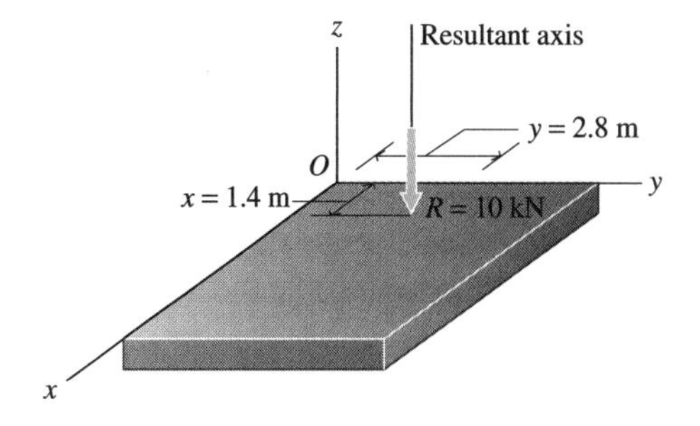

Figure E4.15c

The coordinates (a, b) of the resultant axis are determined by Eqs. (a). Setting $\sum F_z = -10$ kN, $M_x = -28$ kN·m, and $M_y = 14$ kN·m, we obtain $a = 1.4$ m, $b = 2.8$ m. Accordingly, the resultant force, $R = 10$ kN, acts downward at $x = 1.4$ m, $y = 2.8$ m. The resultant axis is perpendicular to the plane of the slab (Fig. E4.15c).

EQUILIBRIUM OF A RIGID BODY UNDER THE ACTION OF PARALLEL FORCES

Key Concept A rigid body that is subjected to parallel forces is in equilibrium if, and only if, the algebraic sum of the forces is zero and the sums of the moments of the forces about any two intersecting axes that are perpendicular to the forces are zero.

Consider a rigid body subjected to forces that are parallel to the z axis of a rectangular Cartesian coordinate system. Note that the body is not restricted to the xy plane; the shape of the body is irrelevant here. Since each force $\mathbf{F}_i$ acting on the body is parallel to the z axis, it can be expressed uniquely in terms of its sense (+ or −) and magnitude F_i. The equilibrium equations for the body are then

$$\sum F_i = 0$$
$$\sum M_x = \sum(+y_i F_i) = 0 \qquad (4.9)$$
$$\sum M_y = \sum(-x_i F_i) = 0$$

where (x_i, y_i) are the coordinates at which the line of action of $\mathbf{F}_i$ intersects the xy plane.

This principle is a special case of a general law of equilibrium for rigid bodies that is introduced in Sec. 5.8.

Example 4.16

Reactions on Airplane Landing Wheels

Problem Statement An airplane that weighs 40,000 lb rests on a tricycle landing gear (see Fig. E4.16). One wing tank contains 10,000 lb of fuel; the other tank contains 5000 lb of fuel. Determine the reactions R_1, R_2, and R_3 of the landing wheels.

Solution For the xyz axes shown in Fig. E4.16, the z projections of the forces, the (x, y) coordinates of the lines of action of the forces, and the moments of the forces about the x and y axes are given in Table E4.16. By Eqs. (4.9), the equations of equilibrium are

$$\sum F_z = R_1 + R_2 + R_3 - 55{,}000 = 0$$
$$\sum M_x = -12R_2 + 12R_3 + 125{,}000 = 0$$
$$\sum M_y = -17R_1 + 3.5R_2 + 3.5R_3 - 45{,}000 = 0$$

Solving these equations for R_1, R_2, and R_3, we obtain

$$R_1 = 7195 \text{ lb}$$
$$R_2 = 29{,}111 \text{ lb}$$
$$R_3 = 18{,}694 \text{ lb}$$

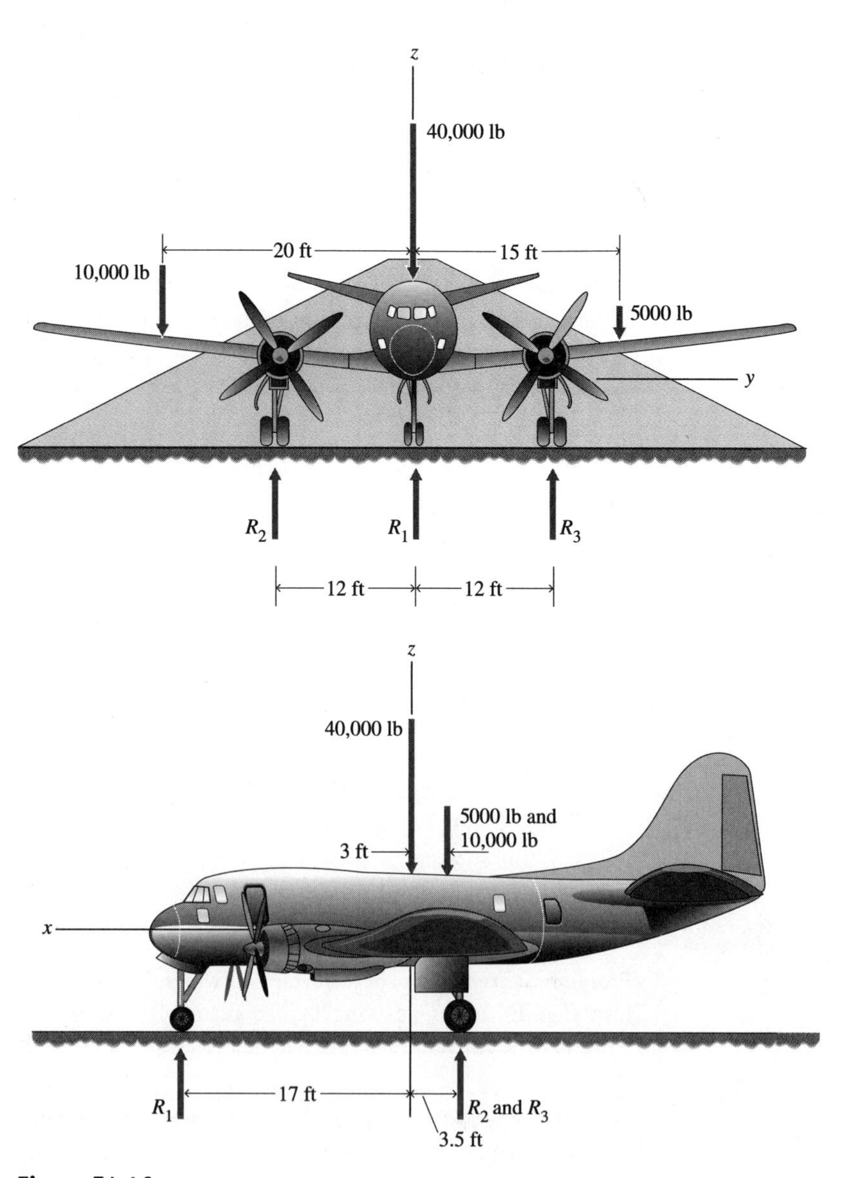

Figure E4.16

Table E4.16

z PROJECTIONS (lb)	x (ft)	y (ft)	M_x (lb·ft)	M_y (lb·ft)
−40,000	0.0	0.0	0	0
−10,000	−3.0	−20.0	200,000	−30,000
−5000	−3.0	15.0	−75,000	−15,000
R_1	17.0	0.0	0	$-17R_1$
R_2	−3.5	−12.0	$-12R_2$	$3.5R_2$
R_3	−3.5	12.0	$12R_3$	$3.5R_3$
$\sum F_x =$ $R_1 + R_2 + R_3 - 55{,}000$			$\sum M_x =$ $-12R_2 + 12R_3 + 125{,}000$	$\sum M_y =$ $-17R_1 + 3.5R_2 + 3.5R_3 - 45{,}000$

CHECKPOINT 

1. ***True or False:*** A force **F** that acts on a rigid body at point P in the xy plane and is parallel to the z axis can be displaced laterally to the origin O of the xy coordinate system, provided that a compensating couple is introduced with moment equal to the moment of the original force about the z axis.
2. Any number of forces that act on a rigid body and that are parallel to the z axis can be reduced to a single force **R** at the origin of the xy axes and two couples that lie in the yz and xz planes. What additional conditions must be applied?
3. ***True or False:*** A rigid body that is subjected to noncoplanar parallel forces is in equilibrium if, and only if, the algebraic sum of the forces is zero and the sum of the moments about any two intersecting axes is zero.

Answers: 1. False; 2. $R \neq 0$; 3. False

Chapter Highlights

- If a point O and the line of action of a force **F** lie in the same plane, the moment M of force **F** with respect to point O is defined by the equation $M = \pm Fr$, where F denotes the magnitude of **F** and r is the moment arm of the force—that is, the perpendicular distance from point O to the line of action of the force.
- Varignon's theorem allows the sum of the moments, about a point O, of any number of concurrent forces to be represented by the moment, about point O, of their resultant force.
- The line of action of the nonzero resultant force of any number of coplanar forces is the resultant axis of the force system.
- A couple is defined to be a system of two noncollinear parallel forces with equal magnitudes and opposite senses.
- The arm of a couple is the perpendicular distance between the lines of action of the two forces of the couple.
- A couple that acts on a rigid body can be displaced and rotated in its plane, and it can be translated to a parallel plane without changing its effect on the body.
- The magnitude of a couple is the product of the magnitude of one of its forces times its arm.
- For coplanar force systems, the sense of a couple may be chosen arbitrarily by using a sign convention. In this book, the sense of the couple is usually taken to be positive if the moment of the couple tends to produce a counterclockwise rotation *when viewed from above the plane of the couple.* Otherwise, the sense is negative.

- The forces of a couple exert the same moment about all points in their plane.
- A force that acts at a given point in a rigid body may be displaced laterally to any other point not on its line of action, provided that a compensating couple is introduced.
- If the vector sum of a set of coplanar forces is zero, the forces exert the same moment about all points in their planc.
- A motionless rigid body is said to be in equilibrium. If the body is subjected to coplanar forces, it will remain in equilibrium if, and only if, the vector sum of the forces is zero and the moment of the forces about any point in their plane is zero.
- With reference to xyz axes, the nonzero resultant force of a system of forces parallel to the z axis is a force parallel to the z axis. This force may be located at any point (a, b) in the xy plane, provided that its moments about the x and y axes are the same as the sums of the moments of the original forces about the x and y axes.

Problems

On Design Problems and the Design Process

Most of the problems in this book deal with bodies or systems for which the geometry (dimensions) and loads have been well defined. Usually in these problems, unknown support reactions and/or forces in members of the system are to be determined by statics. Chapter 3 posed several problems (labeled "design problems") in which you were required to make certain choices regarding the geometry or the magnitude of a force that a member could support. Some of the remaining problems in this book will expand on this type of problem and include slightly more complex requirements that will broaden the range of possible solutions. In other words, they will take you one small step further into the design process.

The *design process* requires that you find a suitable system to perform a given function. In a complete design, you must also consider the safety and cost of the system. These two considerations, however, lie outside the scope of this book, which focuses mainly on the use of statics to design a statically determinate system that will perform an intended function. In this type of problem, the solution is usually not unique; that is, more than one satisfactory solution often exists. The question of which solution is "best" depends on many factors—economics, environmental issues, social issues, and so on. Thus the question is addressed in more advanced courses in design.

Although design is a creative process that often proceeds in nonpredictable ways, it is beneficial to recognize certain general steps in the process.

1. Clearly and accurately define the problem, including all constraints.
2. Propose a feasible solution to the problem, recognizing that there may be more than one workable solution.
3. Analyze the solution (or solutions). This analysis may reveal flaws in the solution(s) that require further analysis of the system. This step may have to be repeated several times. In other words, the design process is an iterative process, with each modification requiring additional justification by analysis.
4. Select the best solution.
5. Implement the best solution; that is, build, manufacture, or install the system to perform its function.

In your more advanced engineering courses and in your engineering career, you will be required to design complete systems. Statics is only one of the tools that you will need, since many design problems involve various technical topics, such as dynamics, mechanics of materials, fluid mechanics, thermodynamics, electronics, and chemistry, as well as economical, social, and political constraints.

4.1 *Moment of Coplanar Forces with Respect to an Axis*

4.1 Referring to Problem 3.34, write the equations for the moment of all the forces about the point at which the roller touches the curb. Express all pertinent lengths in terms of the radius r of the roller.

4.2 Determine the moment of the forces acting at point P: (3, 2) about the point Q: (−3, 5) in Fig. P4.2. Lengths are expressed in meters.

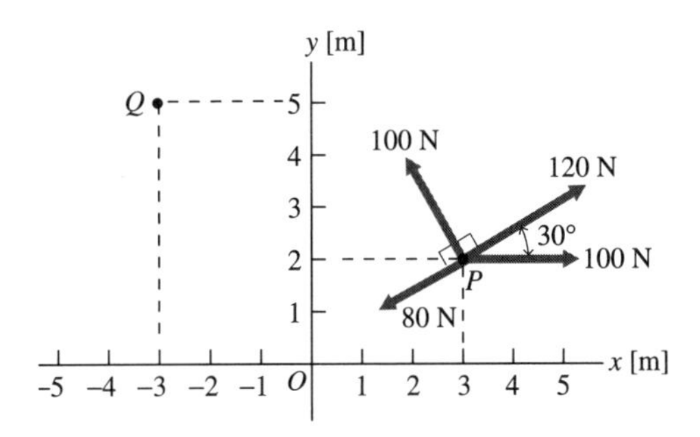

Figure P4.2

4.3 In Fig. P4.3, lengths are expressed in inches; each division represents 1 in.

a. Determine the moment of the five forces about the origin O.

b. Determine the moment of the five forces about the point (3, −1).

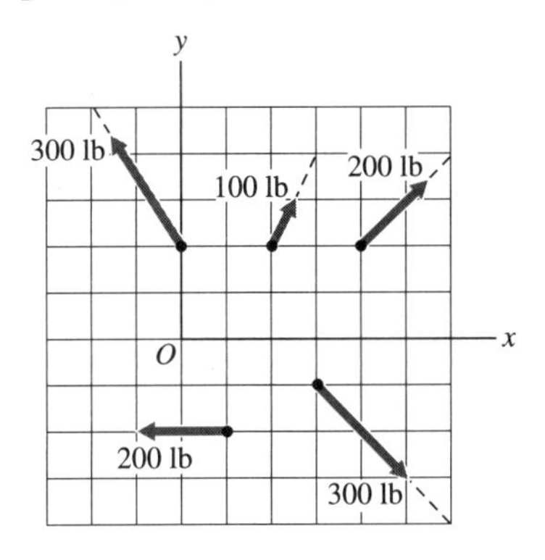

Figure P4.3

4.4 A body is subjected to five coplanar forces whose x and y projections and points of application are given in Table P4.4. Draw a figure, and determine the resultant moments of these forces about the respective points (0, 0), (3, 7), and (−2, 5).

Table P4.4

FORCE PROJECTION (lb)		POINT OF APPLICATION (ft)	
F_x	F_y	x	y
100	80	2	4
−50	90	3	1
100	60	−4	3
−120	−180	−2	−5
80	−60	3	−5

4.5 In arbitrary units, a force of magnitude 200 acts in the xy plane at the point (9, 4). The angle between the line of action of the force and the x axis is $\sin^{-1}(4/5)$. Draw a figure, and compute the moment of the force about the point (3, 2) by the following two methods:

a. Calculate the perpendicular distance from the point (3, 2) to the line of action of the force, and use the definition of a moment [Eq. (4.2)]. From analytic geometry, the perpendicular distance s from a point (x_i, y_i) to a line $y = mx + b$ is given by the formula

$$s = \frac{|mx_i - y_i + b|}{\sqrt{1 + m^2}}$$

b. Resolve the force into (x, y) components, and compute the moments of these components about point (3, 2) using Eq. (4.3).

4.6 Three coplanar forces act on a square block, as shown in Fig. P4.6. Determine the moment of the forces about the edge O (along the z axis) by the following two methods:

a. Compute the distances from the edge O to the lines of action of the forces, and apply the definition of a moment [Eq. (4.2)].

b. Resolve the forces into (x, y) components, and calculate the moments of the components about axis O.

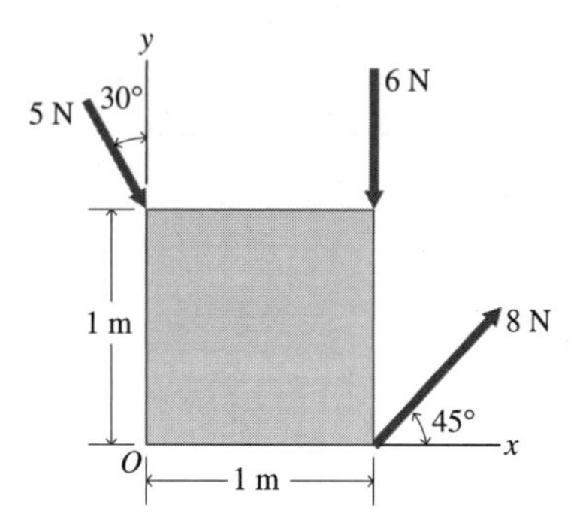

Figure P4.6

4.7 Show that the moment of force **F** in Fig. P4.7 about the point O is represented by twice the area of the triangle OAB.

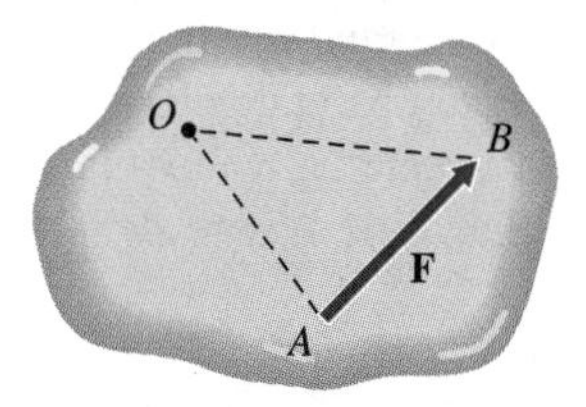

Figure P4.7

4.8 Two forces, with magnitudes 150 N and 400 N, act in the xy plane at the respective points (8, 3) and (2, −1); distances are in meters. The direction angles of the forces with respect to the x axis are 30° and −60°, respectively. Draw a figure, use the formula from analytic geometry for the distance from a point to a line (see Problem 4.5a), and calculate the moment of the two forces about the point (−2, 3) by the following three methods:

a. Calculate the perpendicular distances from the point (−2, 3) to the lines of action of the forces, and use the definition of a moment [Eq. (4.2)].

b. Determine the resultant of the two forces, calculate the perpendicular distance from the point (−2, 3) to the resultant axis, and use the definition of a moment.

c. Resolve the forces into (x, y) components, and calculate the moments of these components about the point (−2, 3) using Eq. (4.3).

4.2 *Resultant of Coplanar Forces That Act on a Rigid Body*

For Problems 4.9–4.14, the x and y projections of several forces that act on a rigid body and the coordinates of the points of application of the forces are given in a table. For each of these systems of forces:

a. Determine the magnitude F of the resultant force, the direction angle θ of the resultant axis, and the y-axis intercept b of the resultant axis.

b. Draw a figure showing the resultant force and the resultant axis.

c. Write the equation for the resultant axis.

4.9

FORCE PROJECTION (lb)		POINT OF APPLICATION (ft)	
F_x	F_y	x	y
300	500	2	1
400	−100	5	−3
600	800	−1	−2

4.10

FORCE PROJECTION (lb)		POINT OF APPLICATION (in)	
F_x	F_y	x	y
200	0	60	20
−800	−100	−20	100
−1000	500	80	−60
0	700	50	0

4.11

FORCE PROJECTION (kips)		POINT OF APPLICATION (ft)	
F_x	F_y	x	y
3	−2	5	2
2.5	0	−8	2
7	4	0	0
3	−1	4	8
−6	3	0	6

4.12

FORCE PROJECTION (N)		POINT OF APPLICATION (mm)	
F_x	F_y	x	y
40	10	80	−60
25	45	100	−60
−80	50	−200	−400
30	35	0	−150

4.13

FORCE PROJECTION (N)		POINT OF APPLICATION (m)	
F_x	F_y	x	y
500	−600	30	2
320	−100	6	6
120	100	7	9
80	300	−4	−5
150	125	3	7
−400	−600	−5	−8

4.14

FORCE PROJECTION (N)		POINT OF APPLICATION (m)	
F_x	F_y	x	y
300	0	20	6
−200	0	−4	7
−300	0	12	−8
100	0	0	0
−500	0	40	−12

4.15 a. For the forces in Fig. P4.3, determine the magnitude F of the resultant force, the direction angle θ of the resultant axis, and the y-axis intercept b of the resultant axis.
b. Draw a figure showing the resultant force **F** and the resultant axis.
c. Write the equation for the resultant axis.

4.16 a. Determine the resultant of the three forces that act on the angle bar in Fig. P4.16.
b. Determine where the resultant axis intersects the line AB. Show the resultant force and the resultant axis on a drawing.

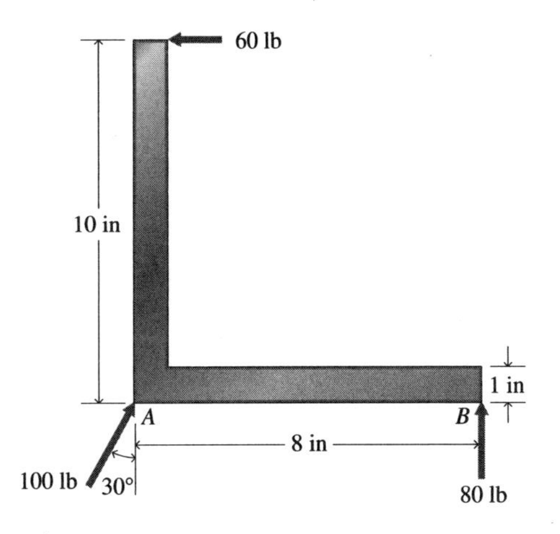

Figure P4.16

4.17 a. Determine the resultant of the coplanar forces that act on the curved bar in Fig. P4.17.
b. Calculate the distance to the resultant axis from the center O, and show the resultant force and the resultant axis on a drawing.

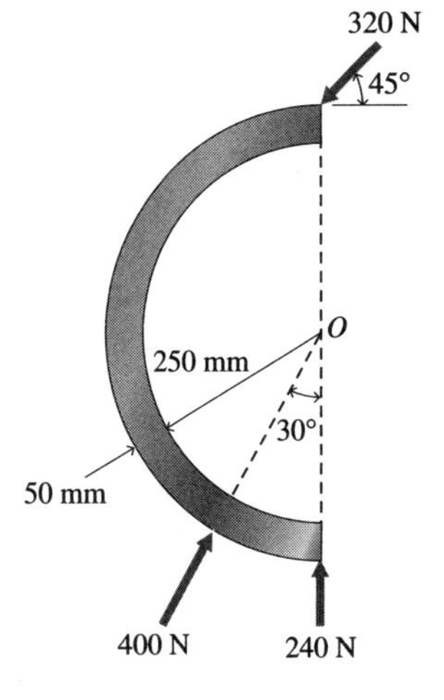

Figure P4.17

4.18 a. Determine the resultant of the four forces shown in Fig. P4.18. Locate the resultant axis, and show it on a drawing.
b. Remove the 20 N force, and determine the resultant of the remaining three forces. Locate the resultant axis, and show it on a drawing.

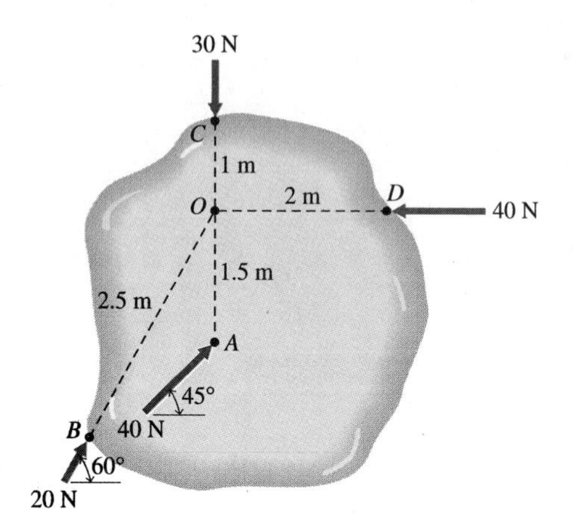

Figure P4.18

4.19 Determine the resultant of each of the following sets of forces in Fig. P4.19 and the intercept of each resultant axis with the line *AE*. In each case, show the resultant force and the resultant axis on a drawing.
a. the forces at points *A*, *B*, and *D*
b. the forces at points *B*, *C*, *D*, and *E*
c. all five forces

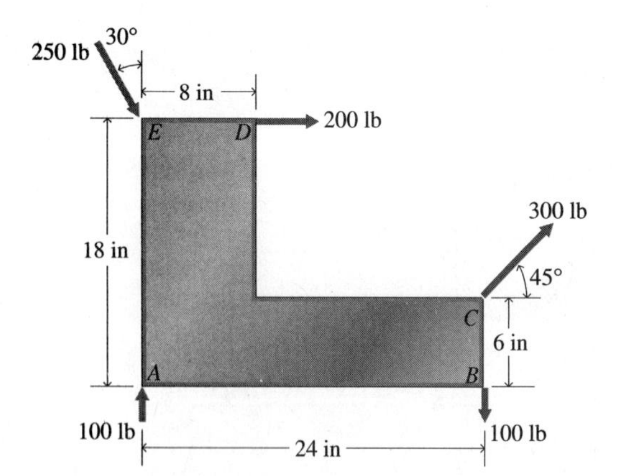

Figure P4.19

4.20 a. Determine the resultant of all the forces that act on the disk in Fig. P4.20.
b. Determine the distance to the resultant axis from the center *C* of the disk.
c. Show the resultant force and the resultant axis on a drawing.

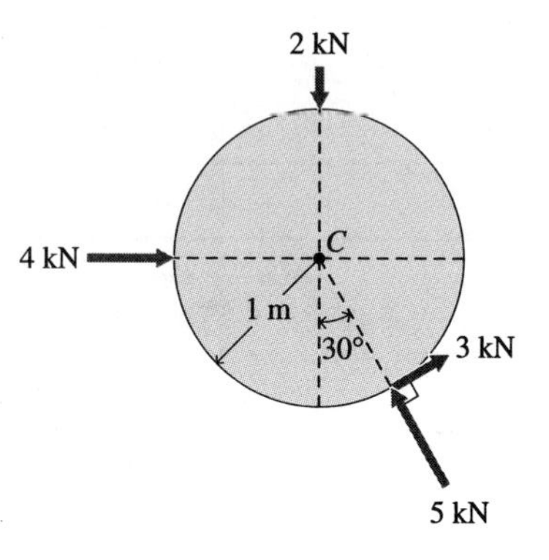

Figure P4.20

4.21 Figure P4.21 represents wind forces concentrated at the floor levels of a tall building. The values of h give the elevations of the respective floors above the street.

a. Calculate the resultant force and the elevation of the resultant axis.

b. Show the resultant force and the resultant axis on a drawing.

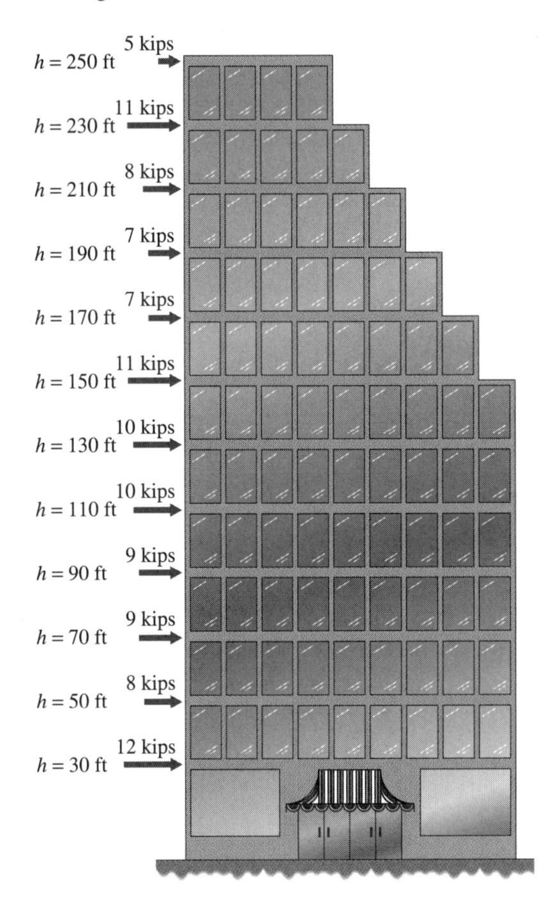

Figure P4.21

4.22 At some instant, the three forces on a boat result from the propeller thrust, the weight, and the action of the water (see Fig. P4.22; distances are in feet). Determine the resultant force and the x-axis intercept of the resultant axis.

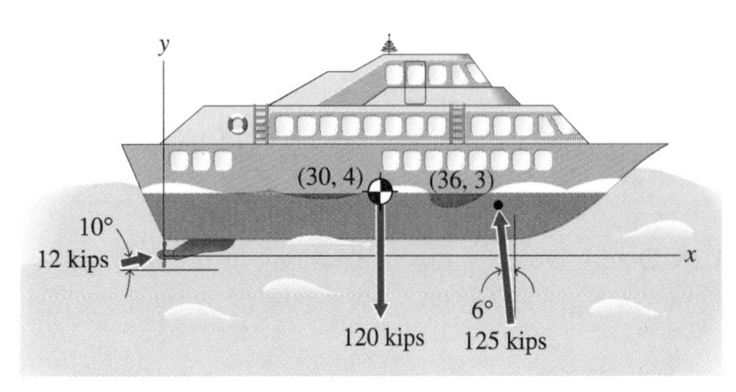

Figure P4.22

4.3 *Parallel Coplanar Forces and Couples*

4.23 Determine the simplest resultant of the two forces shown in Fig. P4.23, and show it on a drawing.

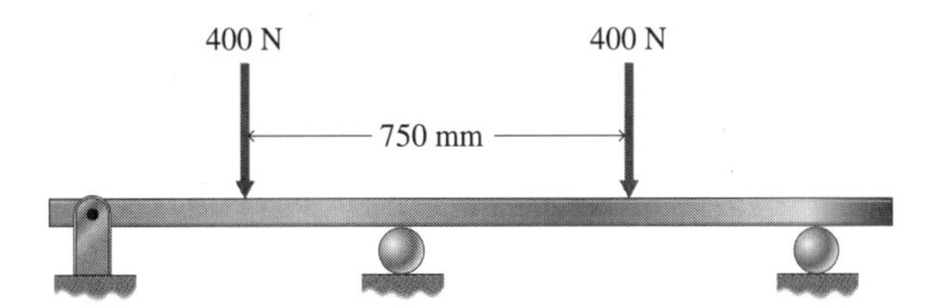

Figure P4.23

4.24 Determine the simplest resultant of the three forces shown in Fig. P4.24, and show it on a drawing.

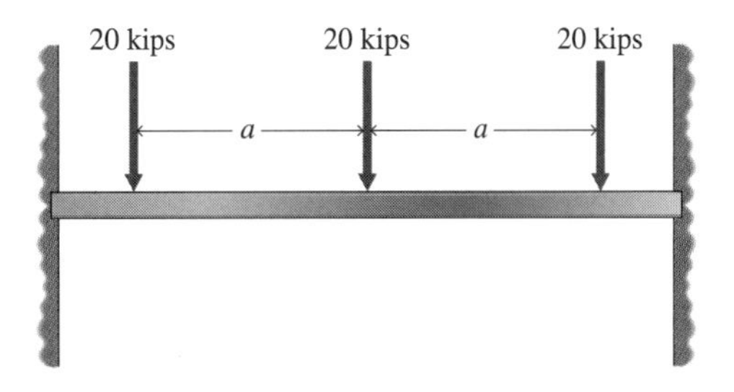

Figure P4.24

4.25 Determine the simplest resultant of the four forces shown in Fig. P4.25, and show it on a drawing.

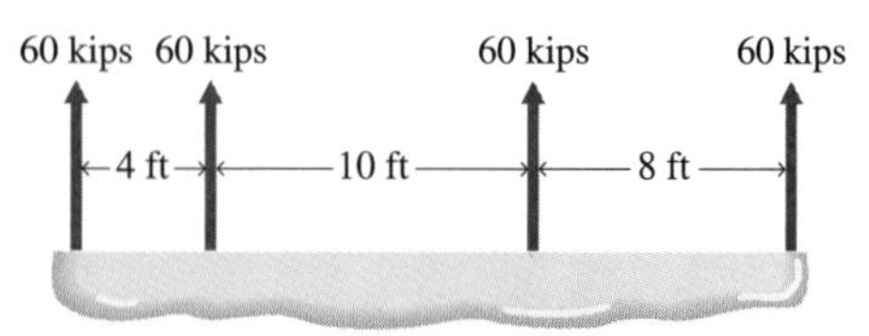

Figure P4.25

4.26 Determine the simplest resultant of the four forces shown in Fig. P4.26, and show it on a drawing.

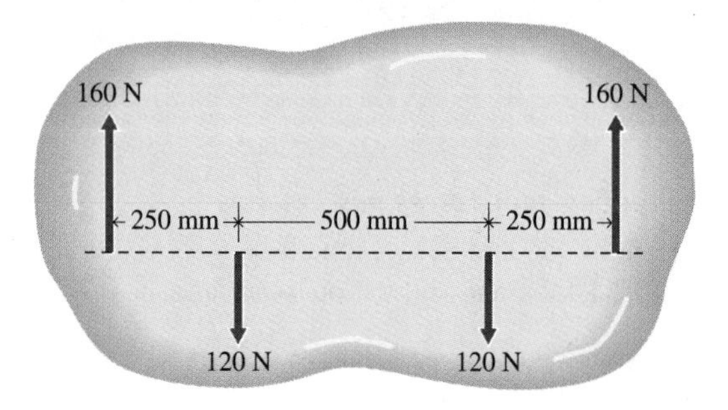

Figure P4.26

4.27 Determine the simplest resultant of the four forces shown in Fig. P4.27, and show it on a drawing.

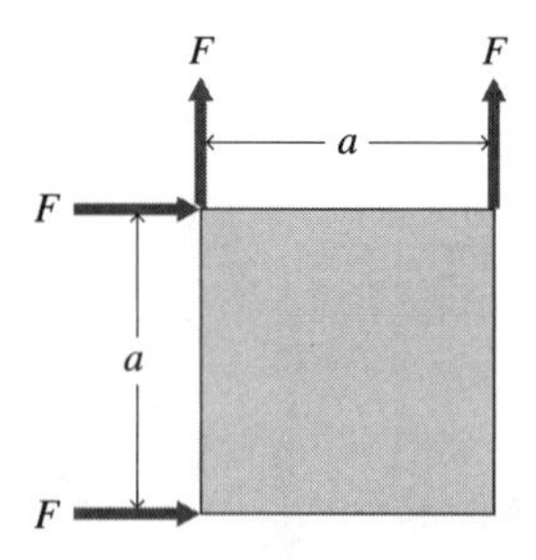

Figure P4.27

4.28 **a.** Convert the parallel forces shown in Fig. P4.28 to dynamically equivalent nonparallel forces by introducing two collinear forces of equal magnitude and opposite sense.

b. Determine the resultant of the original forces graphically by the parallelogram construction.

c. Compare the result of part b with the analytical solution obtained by the method of moments (Sec. 4.2).

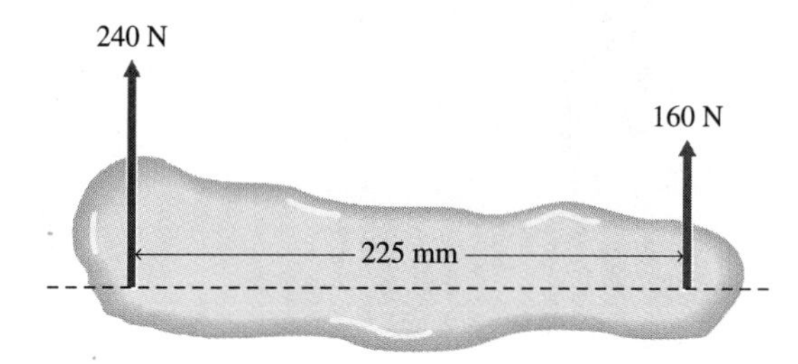

Figure P4.28

4.29 Solve Problem 4.28 using Fig. P4.29.

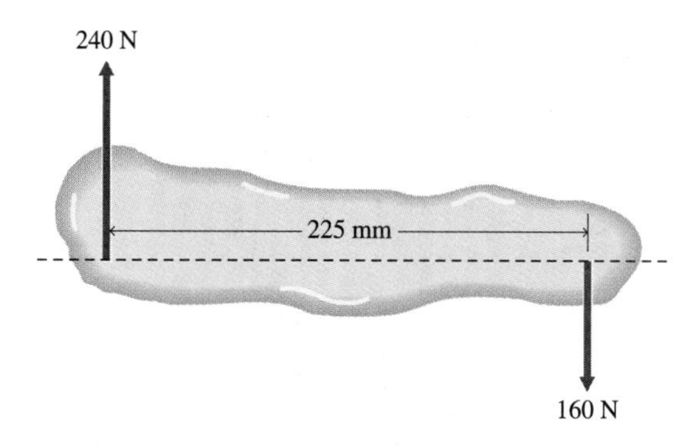

Figure P4.29

4.30 Figure P4.30 represents a socket wrench. In using the wrench to change a car tire, a mechanic applies equal parallel forces that have opposite senses and are directed perpendicular to the plane of the page at points A and B. The wrench produces a turning moment about the axle of the car.

a. Does the turning moment about the axle depend on the distance of the nut from the center of the wheel?

b. Would the wrench work better if the socket were located near the middle of the bar AB? Explain.

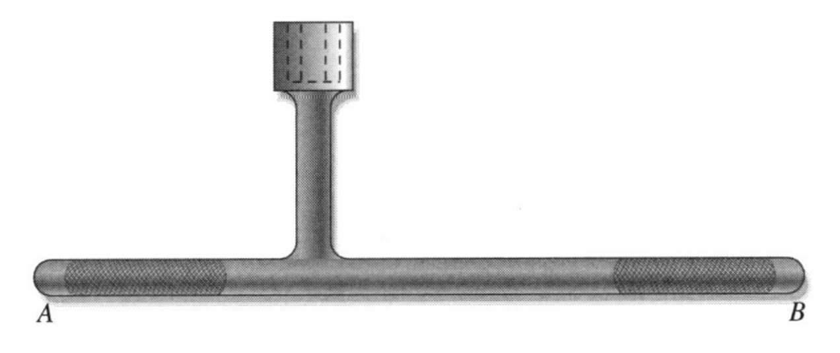

Figure P4.30

4.4 *Moment of a Couple*

4.31 Determine the simplest resultant of the three forces shown in Fig. P4.31, and show it on a drawing.

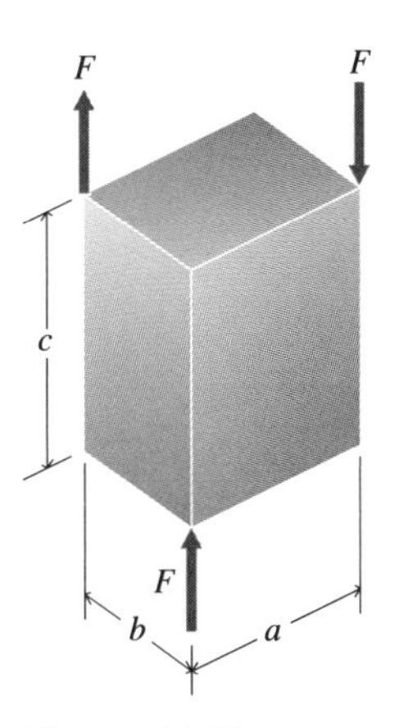

Figure P4.31

4.32 Determine the simplest resultant of the six forces shown in Fig. P4.32 and show it on a drawing.

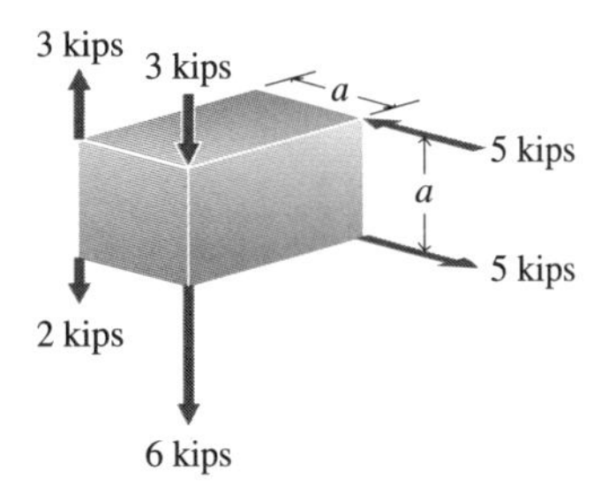

Figure P4.32

4.33 Determine the simplest resultant of the five forces shown in Fig. P4.33, and show it on a drawing.

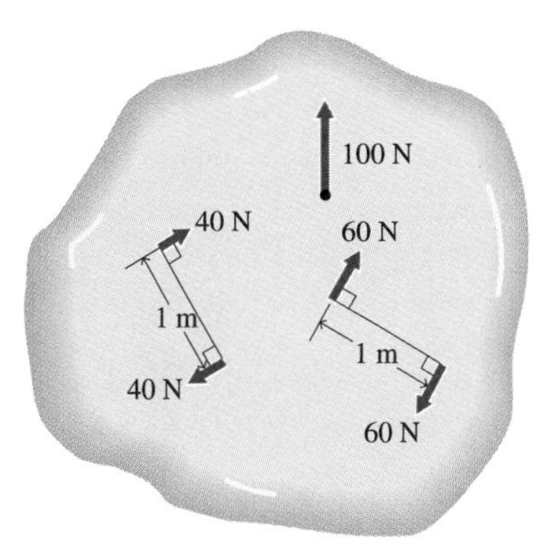

Figure P4.33

4.34 Each of the square plates in Fig. P4.34 is subjected to a force-couple system as shown. Each plate is 1 m on a side, and the force magnitudes are $P = 40$ N, $F = 30$ N, $Q = 20$ N, and $C = 20$ N·m. Determine the resultant force or resultant couple for each plate, and show it on a drawing.

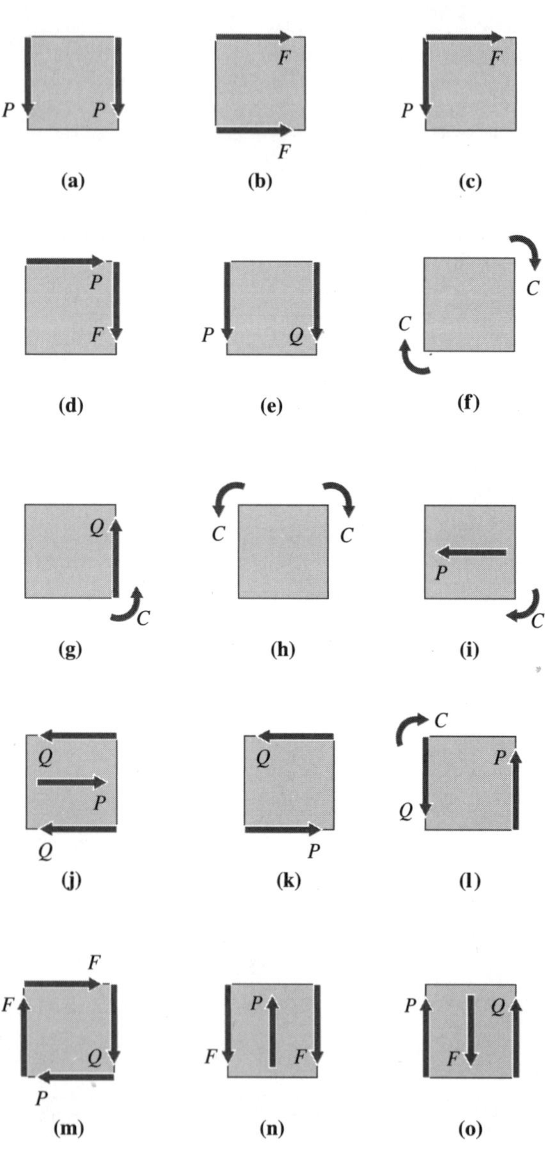

Figure P4.34

4.35 Determine the simplest resultant of the four forces shown in Fig. P4.35, and show it on a drawing.

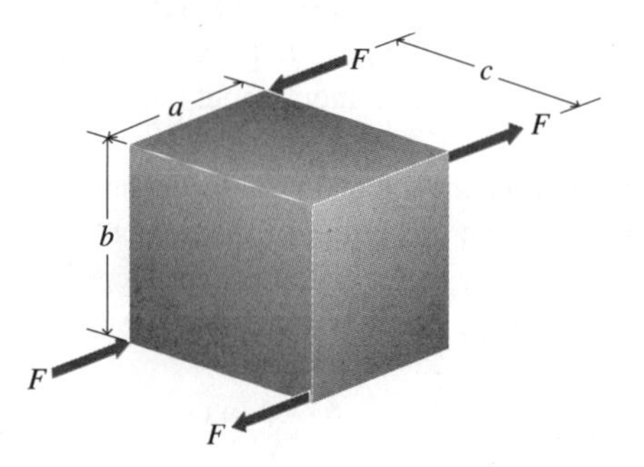

Figure P4.35

4.36 Determine the simplest resultant of the three forces shown in Fig. P4.36, and show it on a drawing.

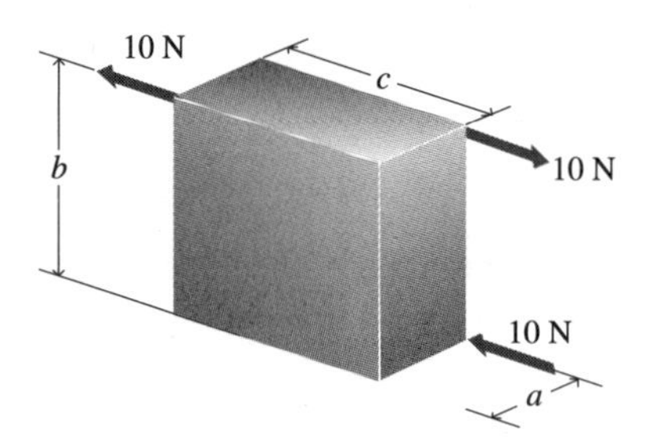

Figure P4.36

4.37 In Fig. P4.37, apply additional forces at points A and B such that the resultant force and the resultant couple are zero.

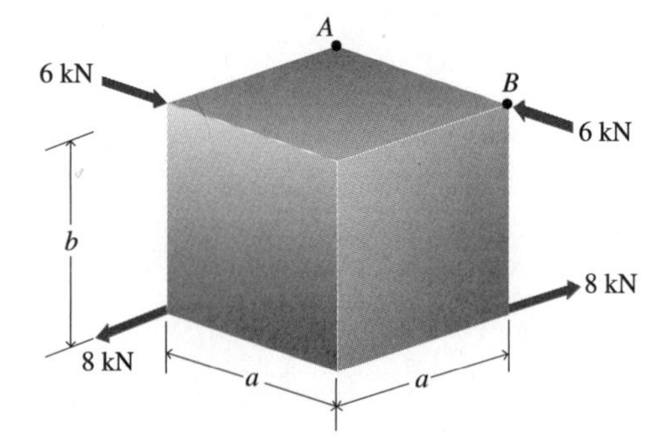

Figure P4.37

4.38 Replace the four forces on the plate in Fig. P4.38 by a dynamically equivalent system of forces acting at points A and B.

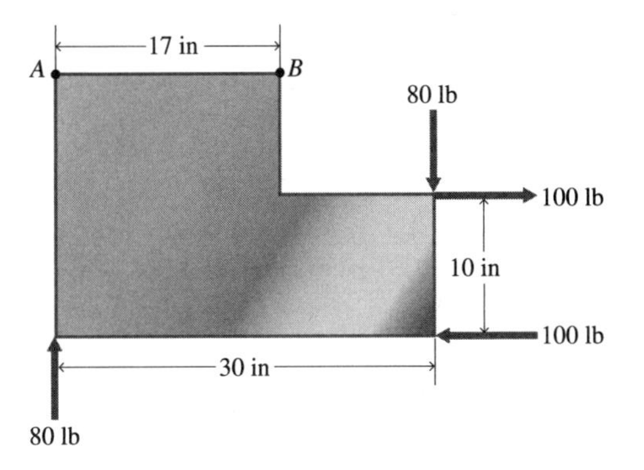

Figure P4.38

4.39 Calculate the resultant moment of the forces that act on the semicircular ring in Fig. P4.39, and represent it by a curved arrow.

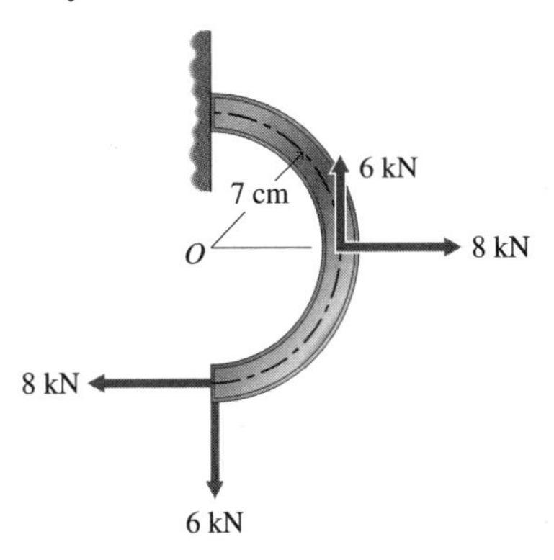

Figure P4.39

4.40 Determine the resultant couple for the force system shown in Fig. P4.40, and represent it by a curved arrow.

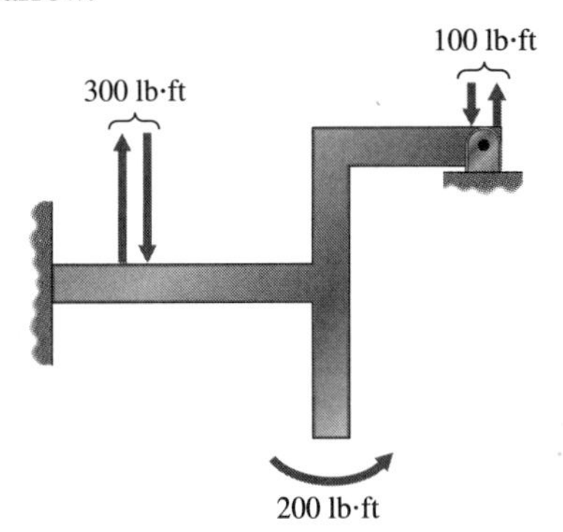

Figure P4.40

4.41 a. Determine the resultant of the force system that acts on the disk in Fig. P4.41.

b. Calculate the distance to the resultant axis from point O.

c. Show the resultant force and the resultant axis on a drawing.

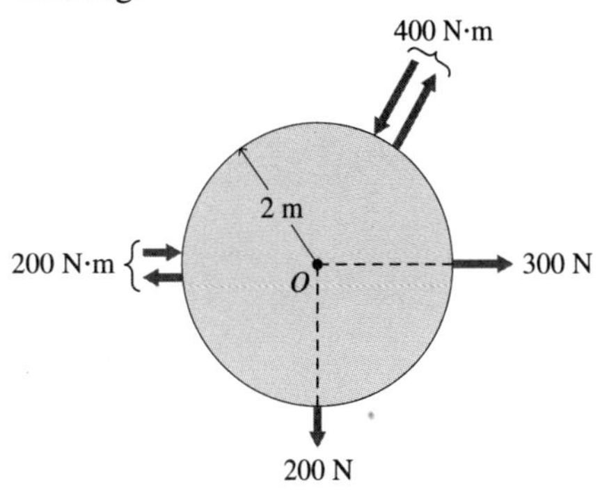

Figure P4.41

4.42 Determine the force that is dynamically equivalent to the system of forces acting on the cube in Fig. P4.42.

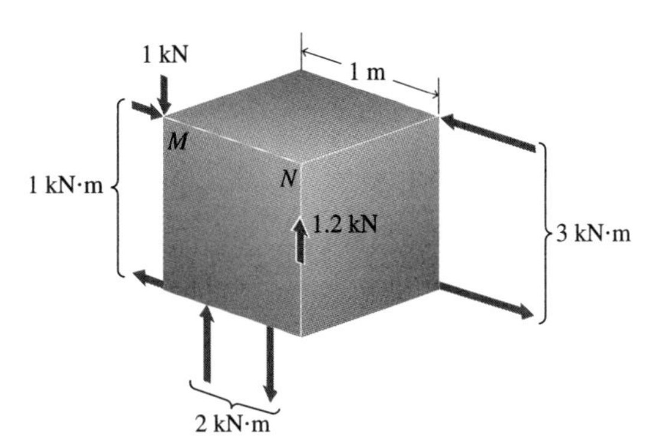

Figure P4.42

4.43 The resultant moment of the forces acting on the pulley in Fig. P4.43 is zero. Determine the force P.

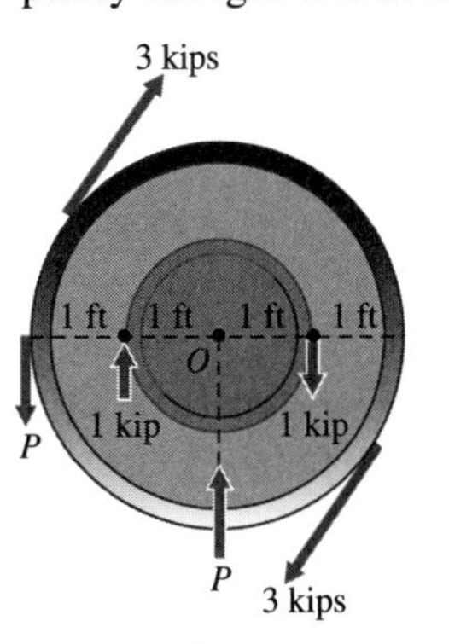

Figure P4.43

4.44 The elliptic disk in Fig. P4.44 is subjected to uniformly distributed tangential forces of intensity τ (in N/mm). Explain why the moment of these forces about any point P in the plane of the disk does not depend on the location of point P.

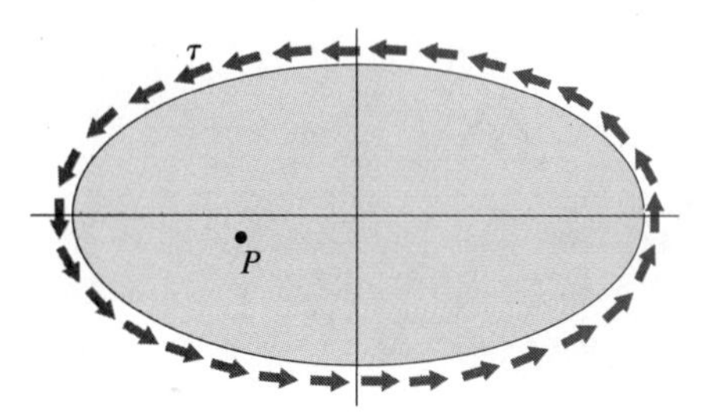

Figure P4.44

4.45 The circular disk in Fig. P4.45 is subjected to uniformly distributed tangential forces of intensity 300 N/mm. Calculate the moment of these forces about any point O in the plane of the disk.

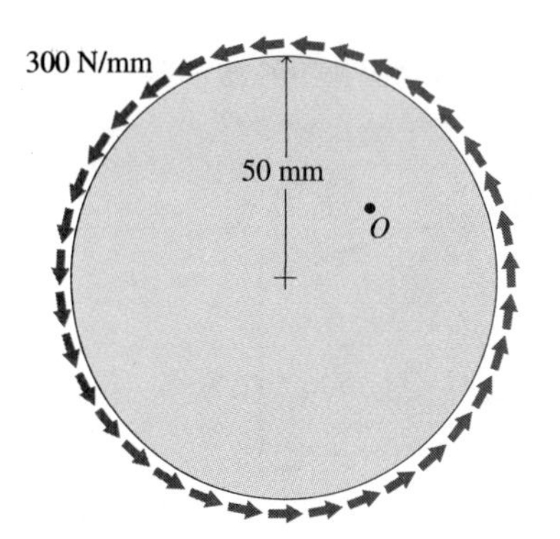

Figure P4.45

4.5 *Lateral Displacement of Forces*

4.46 A rigid body is subjected to five forces in the xy plane, summarized in Table P4.46.

Table P4.46

FORCE PROJECTION (lb)		POINT OF APPLICATION (in)	
F_x	F_y	x	y
300	500	4	8
700	−650	−7	3
−800	150	−5	4
−200	400	10	−2
0	−400	−7	0

a. Calculate the moment of the forces about the point (x_O, y_O).
b. What is the significance of the fact that this moment does not depend on x_O or y_O?

4.47 The resultant force on the airplane in Fig. P4.47 is zero. Determine the resultant couple.

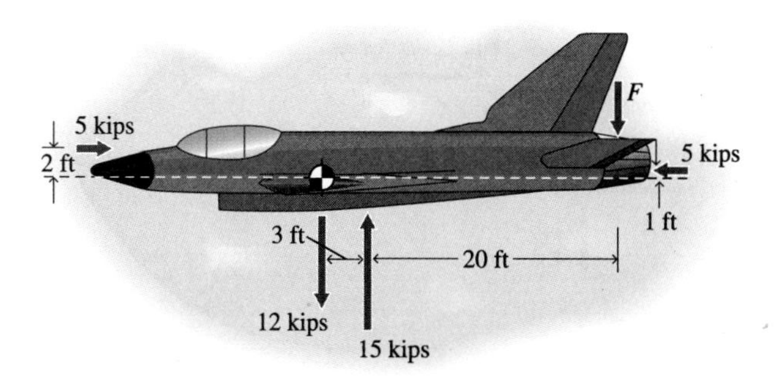

Figure P4.47

4.48 a. Replace the three forces that act on the rigid bar in Fig. P4.48 by a dynamically equivalent force at point O and a couple.
b. Locate the resultant axis of the three forces, and show it on a drawing.

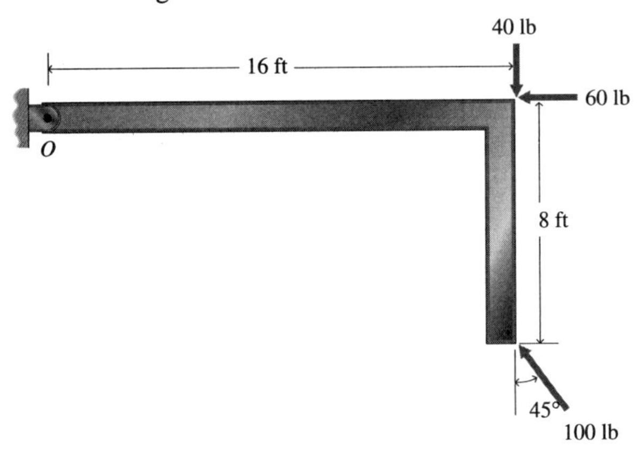

Figure P4.48

4.49 a. Replace the four forces shown in Fig. P4.49 by a single force that acts at point O and a couple.
b. Locate the resultant axis of the force system, and show it on a drawing.

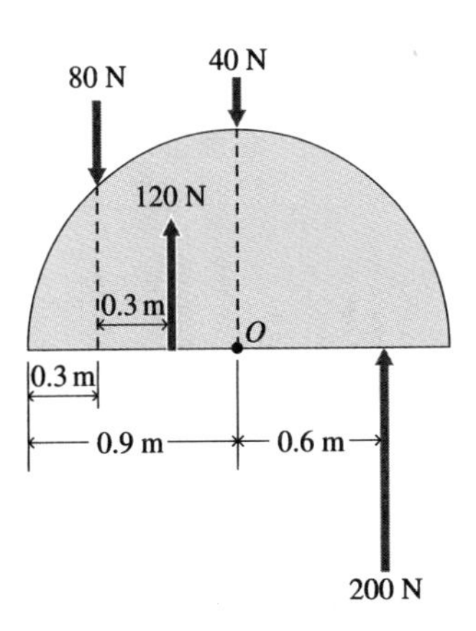

Figure P4.49

4.50 a. Replace the forces that act on the bracket shown in Fig. P4.50 by a single force that acts at point O and a couple.
b. Locate the resultant axis, and show it on a drawing.

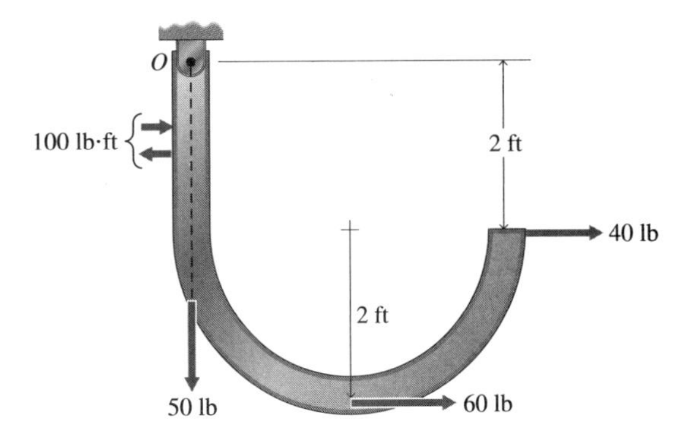

Figure P4.50

4.51 The forces that act on the end of the beam in Fig. P4.51 are dynamically equivalent to the force and the couple shown. Transfer the force to point A, and modify the couple appropriately.

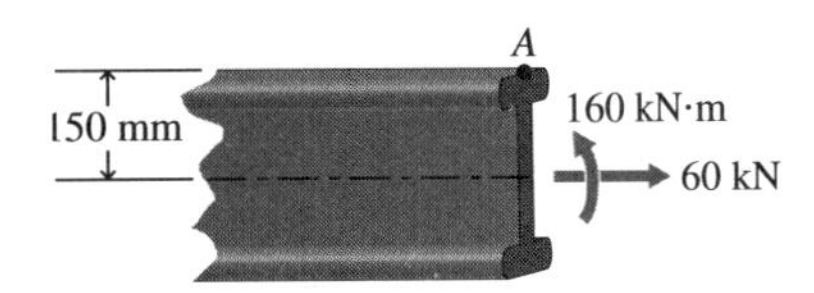

Figure P4.51

4.52 Reduce the force system in Fig. P4.52 to a single force that acts at point O and a couple.

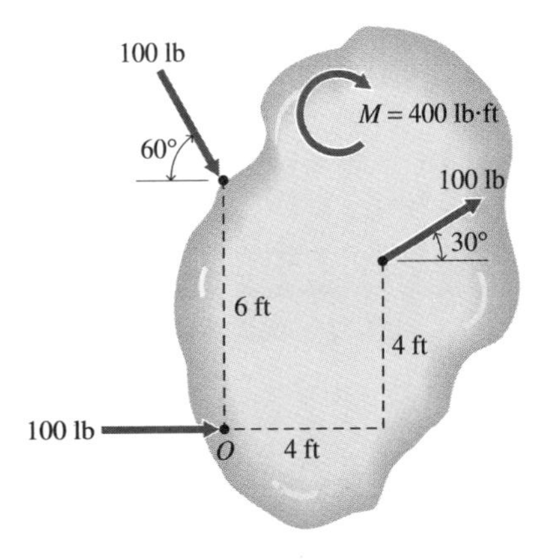

Figure P4.52

4.53 Replace the force system that acts on the body of Problem 4.18 with a force at point A and a couple.

4.6 *Equilibrium of a Rigid Body Subjected to Coplanar Forces*

4.7 *Alternative Formulations of Equilibrium for Coplanar Forces*

The following problems may be worked using Eqs. (4.6), (4.7), or (4.8). Refer also to the Problem-Solving Technique of Sec. 4.6.

4.54 What force must be added to the force system of Problem 4.49 to attain equilibrium?

4.55 To the force system that acts on the body of Problem 4.18, add a force at point O and a couple to put the body in equilibrium.

4.56 The 100 lb uniform ladder of length L shown in Fig. P4.56 rests against a smooth wall at A and a smooth horizontal floor at B. The reactions of the wall and the floor on the ladder are normal to the wall and the floor at A and B. The cord holds the ladder in equilibrium. Determine the forces exerted on the ladder at A and B.

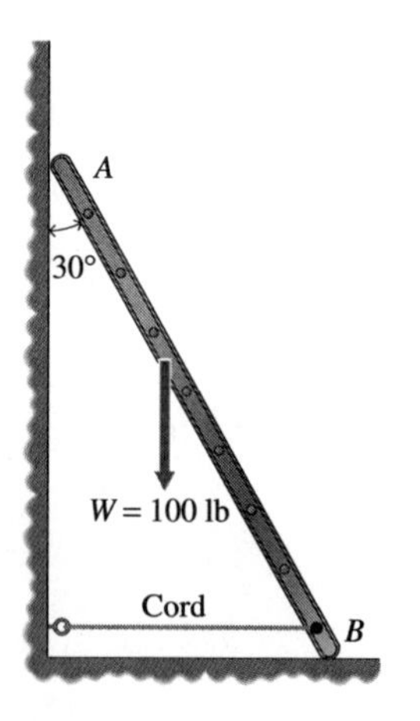

Figure P4.56

4.57 The beam shown in Fig. P4.57 weighs 50 lb/ft and rests on two support rollers. Calculate the minimum load P that will cause the beam to lift off the left support.

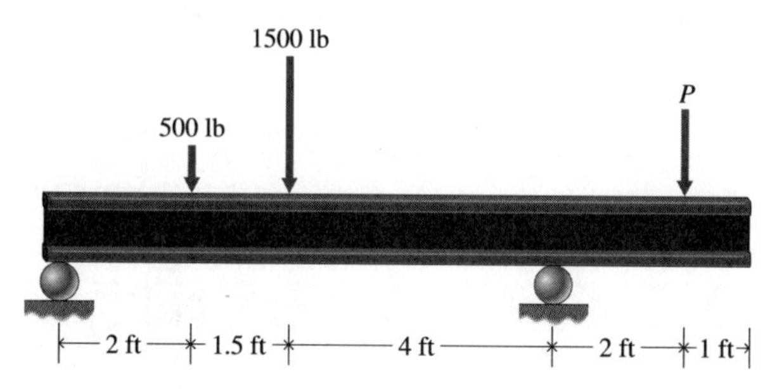

Figure P4.57

4.58 The weight of the beam shown in Fig. P4.58 is negligible. Calculate the force F and the couple M for the beam to be in equilibrium.

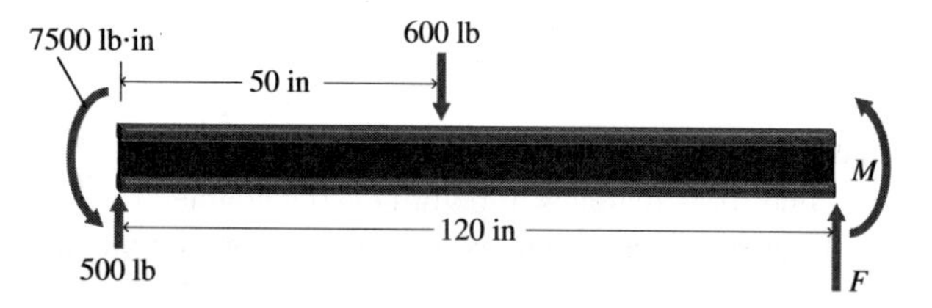

Figure P4.58

4.59 The uniform concrete beam shown in Fig. P4.59 weighs 2250 N/m. Calculate the reactions at the supports A and B.

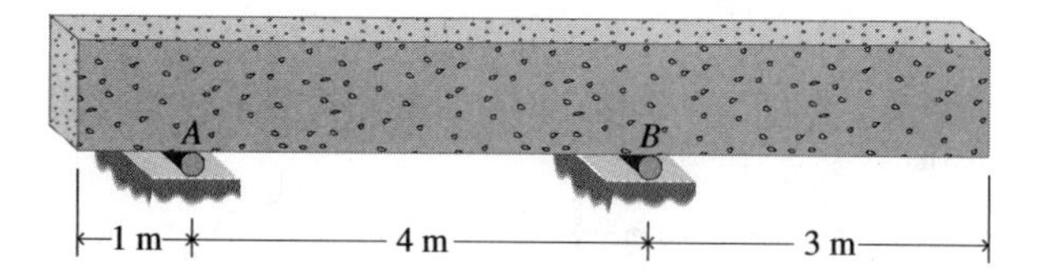

Figure P4.59

4.60 A uniform beam weighs 600 N/m and is loaded as shown in Fig. P4.60. Calculate the reactions at the supports A and B.

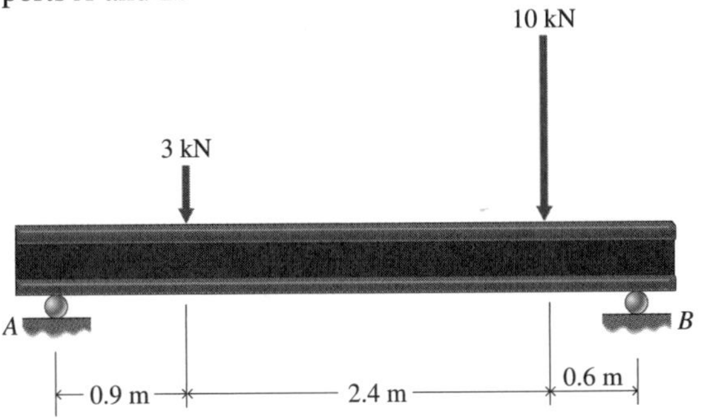

Figure P4.60

4.61 The beam shown in Fig. P4.61 weighs 30 lb/ft. Calculate the force F and the couple M for the beam to be in equilibrium.

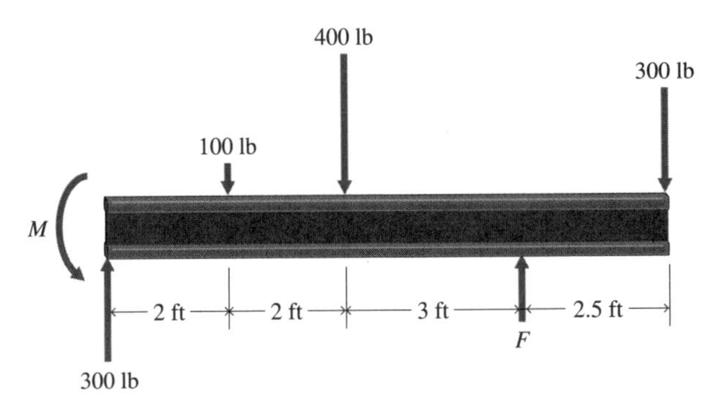

Figure P4.61

4.62 Calculate the tension T in the wire and the projections (R_x, R_y) of the reaction of the hinge in Fig. P4.62. Neglect the weights of the curved bar and the wire.

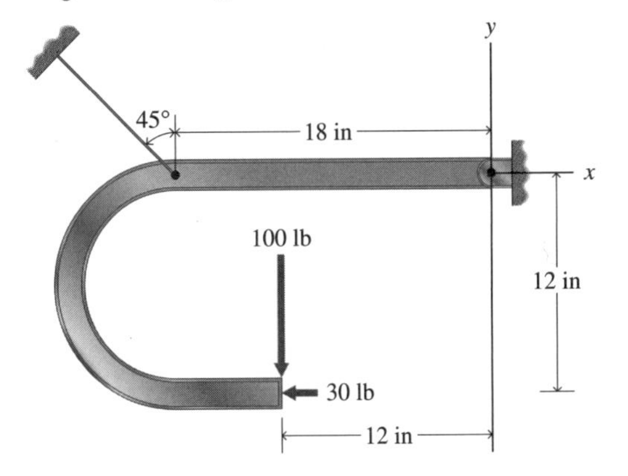

Figure P4.62

4.63 Calculate the thrust force in the pawl of the mechanism shown in Fig. P4.63. Consider the pawl to be a two-force member. Neglect the weight of the mechanism.

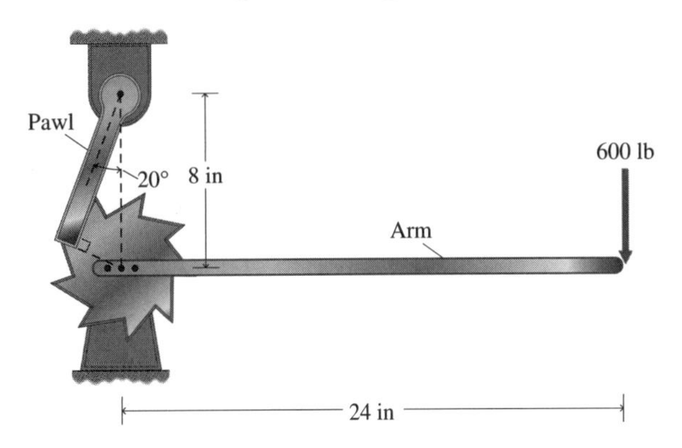

Figure P4.63

4.64 Draw a free-body diagram, and determine the reactions of the supports of the truss shown in Fig. P4.64.

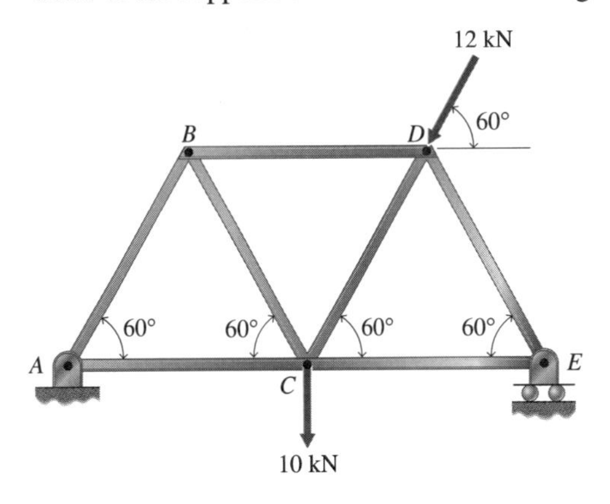

Figure P4.64

4.65 Draw a free-body diagram, and determine the reactions of the supports of the square truss shown in Fig. P4.65.

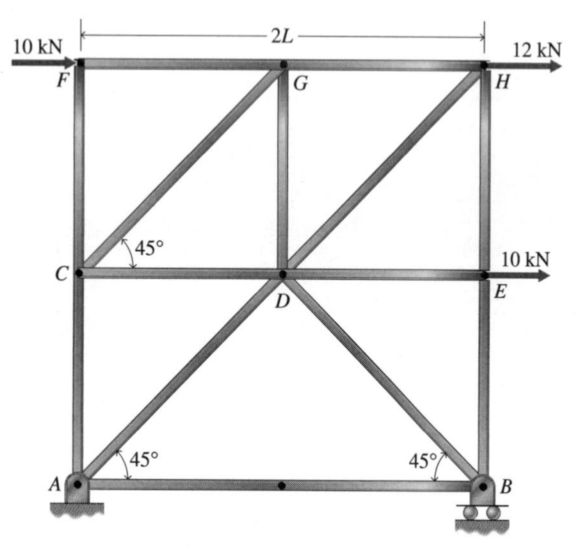

Figure P4.65

4.66 Draw a free-body diagram, and determine the reactions of the supports of the plane arch shown in Fig. P4.66.

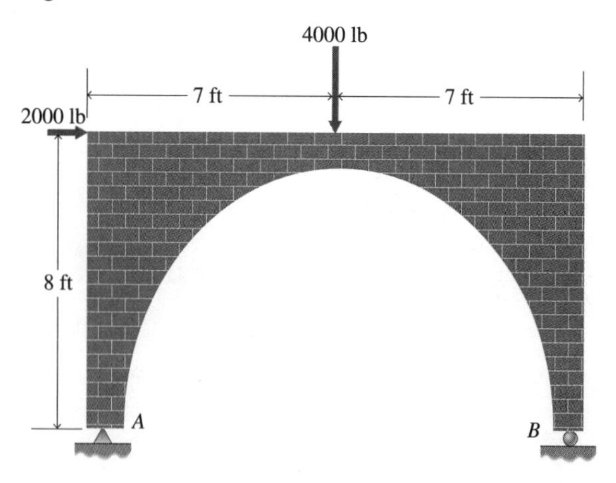

Figure P4.66

4.67 Draw a free-body diagram, and determine the reactions of the supports for the wall truss shown in Fig. P4.67.

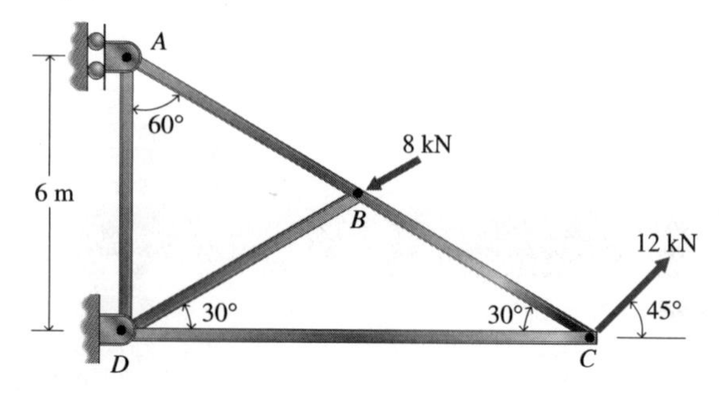

Figure P4.67

4.68 Draw a free-body diagram, and determine the reactions of the supports for the skewed truss shown in Fig. P4.68.

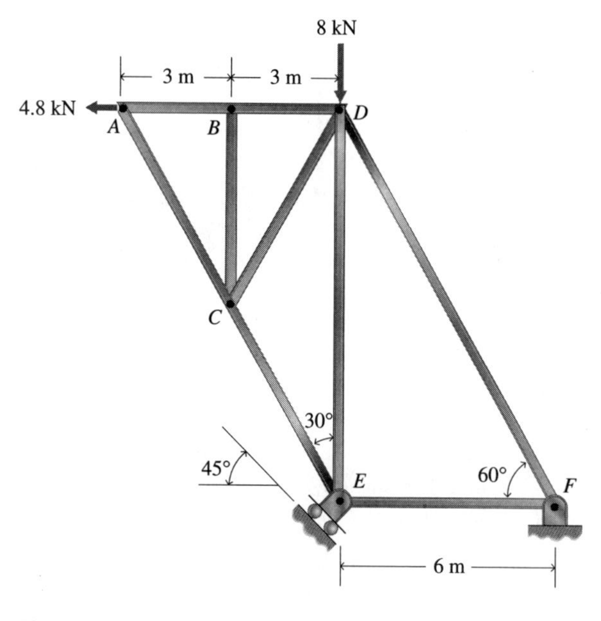

Figure P4.68

4.69 The homogeneous cubical block in Fig. P4.69 weighs 5 kN. What force **F**, applied as shown, will cause the block to tip counterclockwise? Assume that friction prevents the block from sliding down the plane.

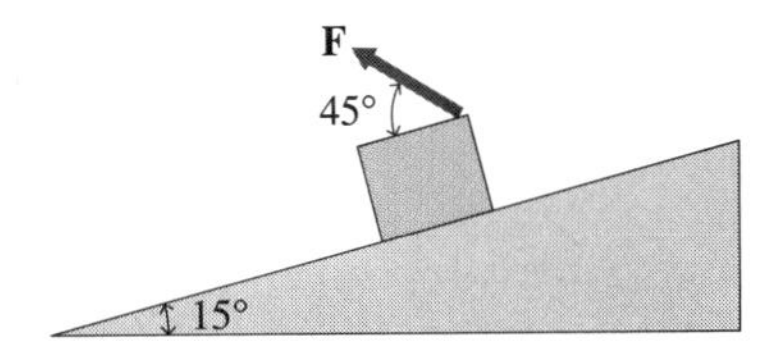

Figure P4.69

4.70 A truck is at rest on a hill (see Fig. P4.70). It is braked on the rear wheels only. Consequently, the reaction of the pavement on the front wheels is normal to the pavement. Calculate the reactions of the pavement on the wheels (R_1, R_2, and F) in terms of W and θ.

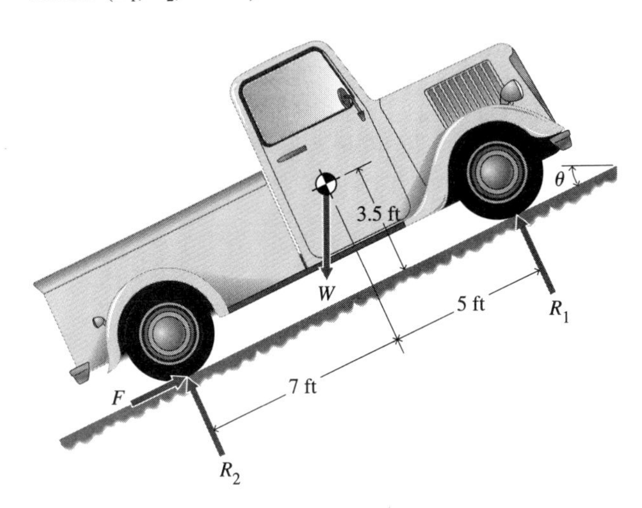

Figure P4.70

4.71 The boom of a jib crane is uniform and weighs 1.5 kN. It is hinged at the left end and supports a 6.0 kN load.

a. Draw a free-body diagram of the boom, and calculate the tension in the tie rod for the conditions shown in Fig. P4.71.

b. Calculate the thrust force P and the vertical reaction R of the hinge at the left end of the boom. Neglect the weight of the chain hoist.

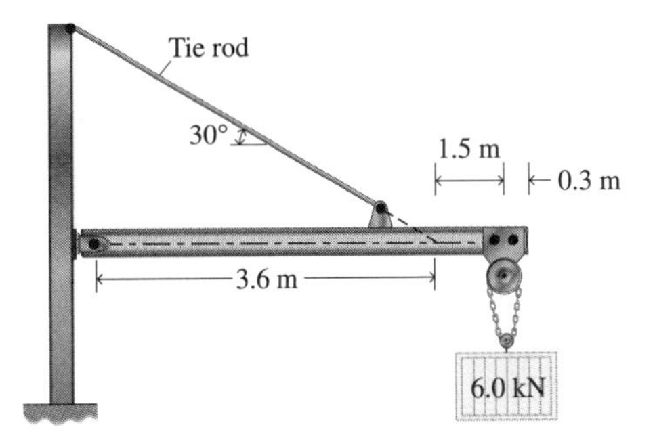

Figure P4.71

4.72 a. Create free-body diagrams of the pump handle OBA and the link BC in Fig. P4.72. Neglect weights.

b. Determine the vertical and horizontal forces that act on the rod OD at point O.

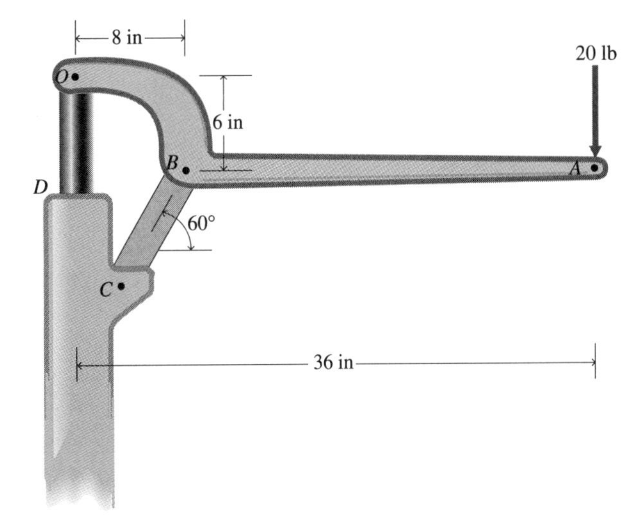

Figure P4.72

4.73 Create free-body diagrams of the curved bars shown in Fig. P4.73, and determine the forces that act on each bar.

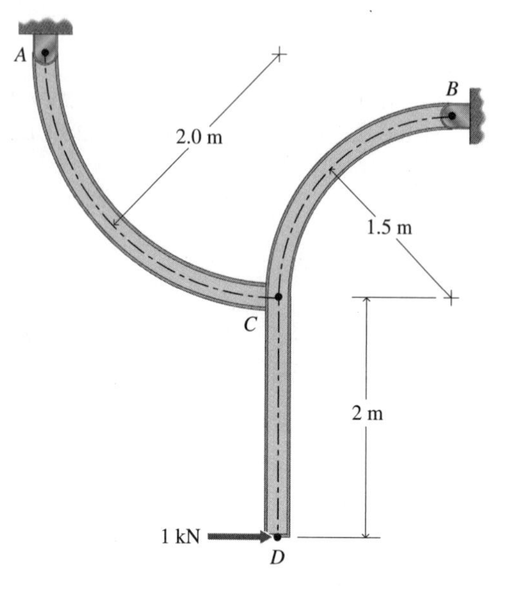

Figure P4.73

4.74 Three slabs of concrete, with specific weight 150 lb/ft^3, are set up as shown in Fig. P4.74. The slabs are 2 ft wide (the dimension perpendicular to the page). Under the action of the horizontal force **F**, the vertical slabs start to tip about points *A* and *B*. For the instant when tipping starts, calculate the forces acting on the slabs at *A* and *B* and the magnitude *F* of **F**.

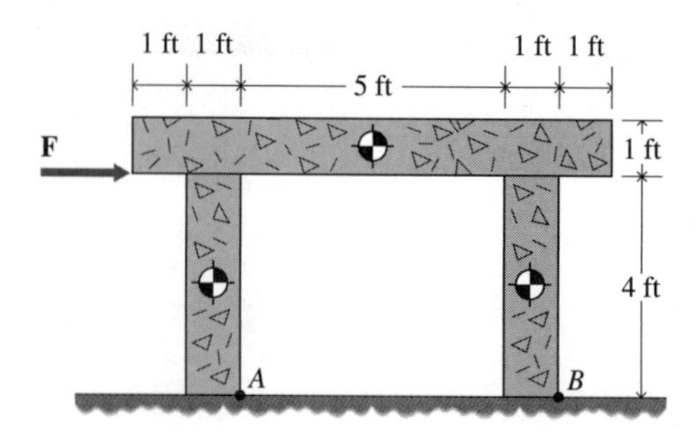

Figure P4.74

4.75 Five pennies are arranged in a skewed stack (see Fig. P4.75). The offset of the top penny is *kd*, where *d* is the diameter of a penny. Calculate the maximum value of *k* that is possible without toppling the stack. (*Hint:* Do the experiment with five pennies first.)

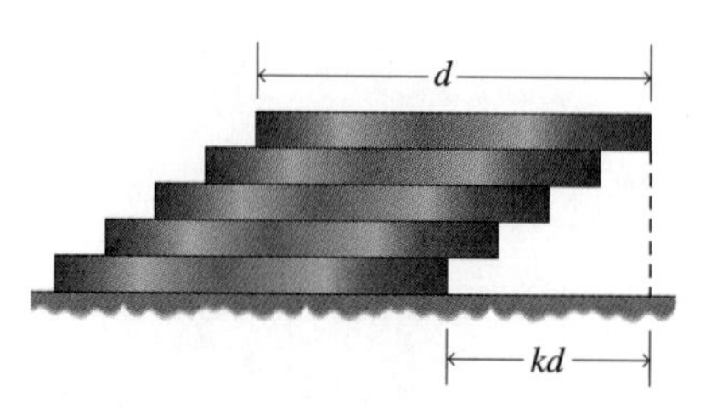

Figure P4.75

4.76 The solid rectangular blocks in Fig. P4.76 are made of a homogeneous material and have the same width (the dimension perpendicular to the page). Block *A* rests on block *B*, which rests on a horizontal floor. If the load *P* is increased gradually, block *A* topples off by rotating about point *O*, but block *B* remains motionless. Show that, for this to happen, $a/b > 1 + \sqrt{2}/2$. Assume that no sliding occurs.

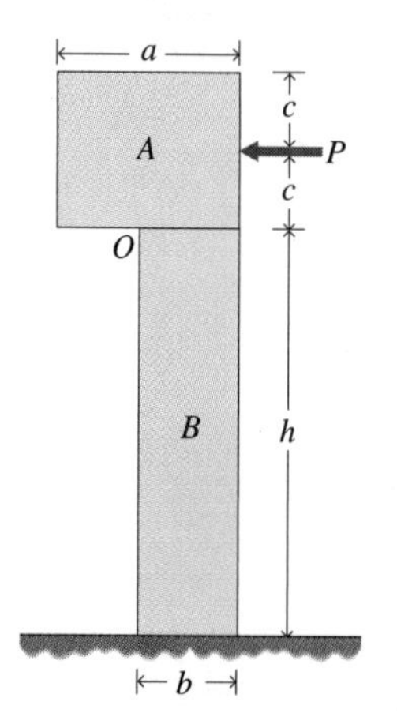

Figure P4.76

4.77 Figure P4.77 represents a buckled portal frame. The vertical members are hinged at their lower ends.

a. Express the vertical reactions at A and B in terms of the applied load P, the width b, and the deflection d.

b. Solve for the horizontal reactions at A and B. Explain any difficulty you encounter.

Figure P4.77

4.78 For the jib crane in Fig. P4.71, plot the magnitude of the force in the tie rod as a function of the distance x (in meters) of the 6.0 kN load from the left end of the boom, for the range $0 \leq x \leq 5.1$.

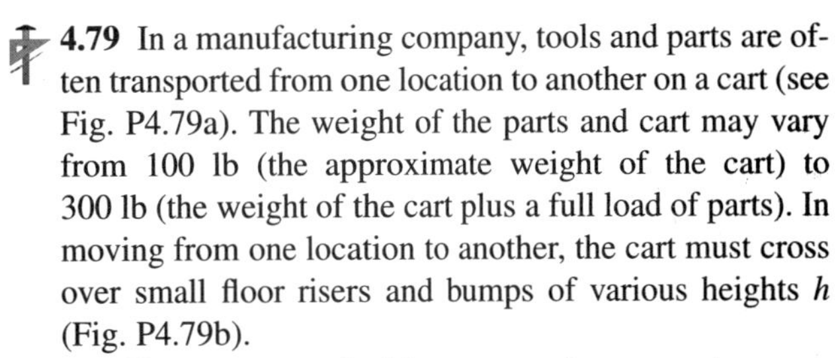

4.79 In a manufacturing company, tools and parts are often transported from one location to another on a cart (see Fig. P4.79a). The weight of the parts and cart may vary from 100 lb (the approximate weight of the cart) to 300 lb (the weight of the cart plus a full load of parts). In moving from one location to another, the cart must cross over small floor risers and bumps of various heights h (Fig. P4.79b).

Your company decides to manufacture such a cart, and your supervisor wants you to do a preliminary design. You are told that a person exerting a horizontal pull or push on the cart should be able to deliver parts from one location to another. Using statics, select a wheel size that will enable an average person to start the wheels rolling over the risers and bumps. Your company has access to wheels ranging in diameter from 3 in to 18 in. Because of cost, it is desirable to use the smallest wheel that will meet the design requirement of allowing the average person to deliver the parts.

The risers and bumps (which you may assume to be rectangular in cross section, as in Fig. P4.79b) vary in height from 0.25 in to 0.50 in. Note that there are several conditions you must establish: How much pulling or pushing force can the average person easily exert? Does it make a difference whether the cart is pushed or pulled? How is the load on the cart distributed, and does it make a difference? Are there other considerations? (Depending on the choices you make for these conditions, various solutions may be feasible.)

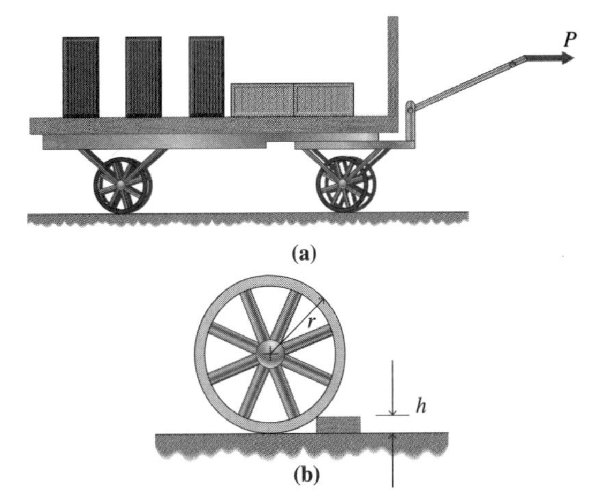

Figure P4.79

4.80 Your company is asked to design a jib crane (see Fig. P4.80) to pick up truck engines that weigh 6 kN and lower them into crates for shipping. In general, a jib crane consists of a uniform beam (usually a beam with an I-shaped cross section) pinned at one end to a vertical column and supported by a cable or tie rod (a two-force member) at another point along its length. The beam and tie rod are connected to the column in such a way that the beam can rotate about the axis of the column through ~180°. The engines are to be picked up by a chain hoist that may move along the beam to position the engines in their crates. The chain hoist is free to move along the beam to within 0.3 m of the unpinned end. Because of jars and shocks (dynamic effects) that occur in picking up and moving an engine, as a design requirement, the tie rod must be capable of withstanding a tensile force 2.5 times the static load that an engine places on the rod.

Your supervisor asks you to do a preliminary design of the system so that the force in the tie rod is minimized in terms of the lengths *a*, *b*, and *h*. She also tells you that you will need a beam that is about 6 m long and suggests that you use an I-beam that weighs about 1.8 kN.

a. With the above specifications, design the system so that the tension in the tie rod is minimized, subject to the further conditions that $a \geq 3$ m and $h \leq 4$ m.

b. Write a brief report for your supervisor, summarizing the results of your study. Include in your report any improvements in the jib crane that might have occurred to you.

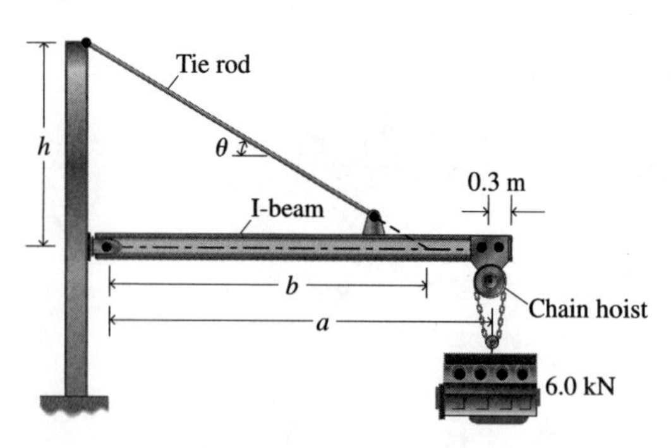

Figure P4.80

4.81 One aspect of the design process is to consider several ways of doing a task. Outline one or two different ways to perform the task in Problem 4.80 of crating truck engines. Illustrate each idea with sketches, and briefly discuss its feasibility.

4.82 Your supervisor assigns you to help in the design of a single-span, single-lane bridge that will be subjected to car and truck traffic on a country road. The loads due to contact between the tires and the bridge deck may be considered as parallel forces perpendicular to the bridge deck (Fig. P4.82). As an initial assignment, your supervisor asks you to estimate the maximum and minimum support reactions for the cases where a 4-wheel car crosses the bridge and an 18-wheel semitrailer-truck crosses the bridge.

Information obtained from this type of preliminary study is important in selecting appropriate design parameters. Outline the steps you would take to obtain the desired data. List important parameters you need to carry out your assignment. Estimate how accurately you can determine these parameters. How would you extend your results to the case where several vehicles are on the bridge at the same time?

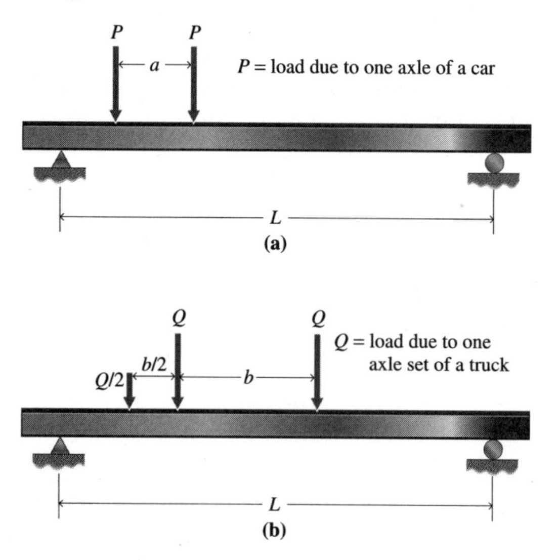

Figure P4.82

4.8 *General Theory of Parallel Forces*

For Problems 4.83–4.86, magnitudes of forces that lie parallel to the z axis and the points of application of these forces in the xy plane are tabulated.

a. Determine the resultant force and the coordinates of the point at which the resultant axis passes through the xy plane.

b. If the resultant force has zero magnitude, determine the resultant couples in the xz and yz planes.

c. Make a drawing showing the resultant force and the resultant axis (or the resultant couples for a zero resultant force) in each case.

4.83

FORCE PROJECTION (N)	x COORDINATE (mm)	y COORDINATE (mm)
−400	50	75
600	−25	125
−1200	50	75
1600	−75	−150
2400	0	0
−2000	25	100

4.84

FORCE PROJECTION (kips)	x COORDINATE (ft)	y COORDINATE (ft)
1.0	1	1
2.5	0	2
4.0	−4	−1
−5.0	−1	5
3.0	2	0
10.0	2	−3
−15.5	2	−3

4.85

FORCE PROJECTION (kN)	x COORDINATE (m)	y COORDINATE (m)
5	1	3
8	2	2
10	−4	2
−6	−2	−4
−1.85	0	4
2.6	5	−3

4.86

FORCE PROJECTION (N)	x COORDINATE (m)	y COORDINATE (m)
450	2	2
625	−2	1
−625	3	4
−850	0	0
−450	−4	−2
850	6	−3

4.87 Determine the resultant of the four forces and the couple shown in Fig. P4.87. Dimensions are in feet.

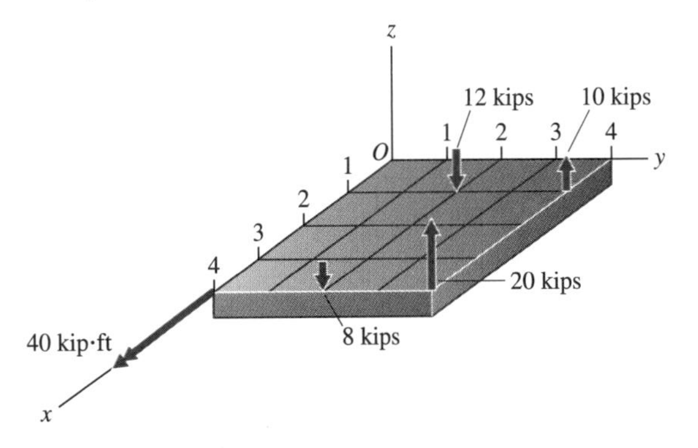

Figure P4.87

4.88 Replace the 500 N force that acts at corner A of the slab in Fig. P4.88 by a dynamically equivalent system of three parallel forces that act at corners B, C, and D.

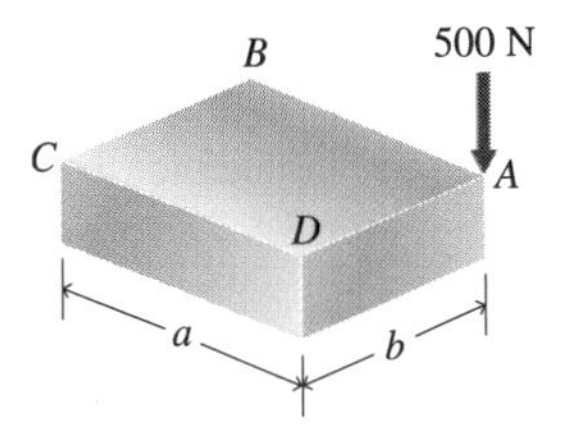

Figure P4.88

4.89 A table rests on three legs whose feet are located in the xy plane at the points (0, 0), (4, 0), and (0, 3). Determine the coordinates (x, y) of the center of gravity of the table, if the legs at the points (4, 0) and (0, 3) carry the respective loads $5W/12$ and $W/3$, where W is the total weight of the table.

4.90 A lathe rests on three feet A, B, and C, located in the xy plane at the respective points (0, 0), (8, 2), and (7, −1). The center of gravity of the lathe is at the point (4, 0.5). Determine the fraction of the total weight W carried by each foot.

4.91 A cylindrical vat (shown in plan view in Fig. P4.91) weighs 5 kN when empty. It is supported by three legs at A, B, and C. It is filled to a depth of 4 m with a liquid having a specific weight of 10.68 kN/m³. Calculate the force in each leg.

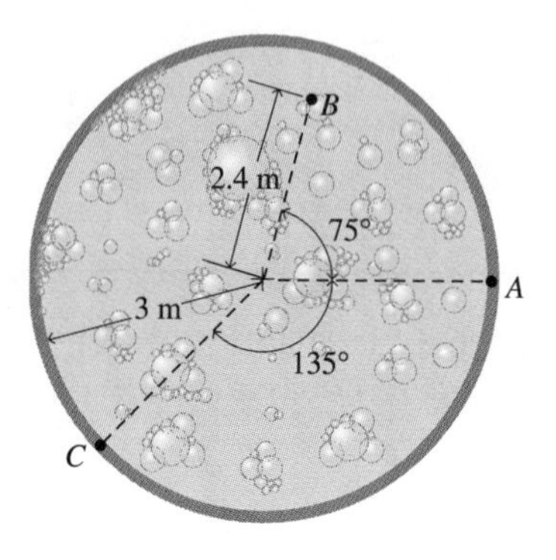

Figure P4.91

4.92 a. Develop a spreadsheet to determine the resultant and the coordinates of the point at which the resultant passes through the xy plane of a system of n parallel forces that lie parallel to the z axis and that act at points $(x_1, y_1), (x_2, y_2), \ldots, (x_n, y_n)$.
b. Use the spreadsheet you developed in part a to solve Problems 4.83–4.86.

4.93 A uniform rectangular platform weighs 1 kN and carries three concentrated forces of magnitudes 2 kN, 4 kN, and 6 kN, as shown in Fig. P4.93. The platform is supported by two fixed legs A and B and a movable leg C, which exert parallel forces acting perpendicular to the platform. Plot the magnitudes of the forces in the three legs as they vary with the (x, y) coordinates of leg C so as to keep the platform in equilibrium. Consider the following conditions:
a. $x = 1.5$ m, $0 < y < 2$ m
b. $y = 1.0$ m, $0 < x < 3$ m

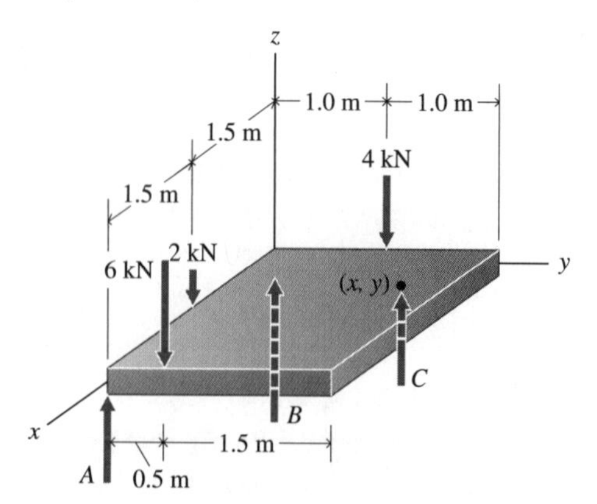

Figure P4.93

4.94 Consider the platform described in Problem 4.93. Determine the range of values of x and y that will satisfy the conditions that each leg carry a load no greater than 6 kN and that $0 < x < 3$ m and $0 < y < 2$ m. Check your result by hand for the case where the loads A, B, and C are equal.

REVIEW QUESTIONS

4.95 Define *moment of a force about an axis.*
4.96 State Varignon's theorem.
4.97 Several coplanar forces $\mathbf{F}_1, \mathbf{F}_2, \ldots,$ act on a rigid body. The vector sum of the forces is not zero. Explain how the resultant force is determined. Explain how the resultant axis is located.
4.98 Define *resultant of the forces that act on a rigid body.*
4.99 Construct a diagram to show how two noncollinear parallel forces may be replaced by dynamically equivalent nonparallel forces. What happens to the construction if the parallel forces have equal magnitudes and opposite senses?
4.100 What is the resultant of two equal noncollinear parallel forces having the same sense?
4.101 Define *couple.*
4.102 Explain the difference between a moment and a couple.
4.103 If a motionless rigid body is subjected only to a couple, about what point does it begin to rotate?
4.104 Prove that a couple may be displaced in the direction of its forces.
4.105 Prove that a couple may be displaced in the direction of its arm.
4.106 Prove that a couple may be rotated in its plane.
4.107 Prove that a couple may be displaced to a parallel plane.
4.108 Prove that two coplanar couples with the same moment and sense are dynamically equivalent.
4.109 Prove that the moment of the forces of a couple about any point in the plane of the couple is Fa, where a is the arm of the couple and F is the magnitude of either force.
4.110 Three couples with moments 40, 50, and −20 N·mm act in the same plane. Compute the moment of the resultant couple.
4.111 Denote three parallel edges of a cube by A, B, and C. A 50 N force acts along edge A, a 50 N force acts along B with sense opposite to that of the force on edge A, and a 100 N force acts along edge C in the same sense as the force on edge A. Replace the three forces by a single dynamically equivalent force.

4.112 A 100 N force acts on a rigid body. Displace the line of action of the force 75 mm and introduce the required compensating couple.

4.113 Several coplanar forces **F**′, **F**″, . . . act on a rigid body.

a. State the theorem that allows these forces to be replaced by a single force **F** at any given point O and a couple M.

b. How is the force calculated?

c. How is the moment M calculated?

4.114 The angle θ is such that the vector sum of the three forces shown in Fig. P4.114 is zero. Calculate the moment of the three forces about point P.

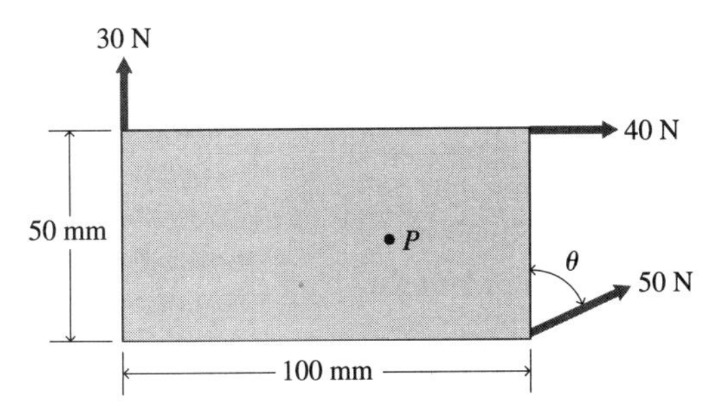

Figure P4.114

4.115 State the general equilibrium principle for a rigid body that is subjected to coplanar forces. Then, state two alternative forms of this principle. Represent the three forms as algebraic equations.

4.116 State the equilibrium principle for a rigid body in a special form that is applicable if only three nonparallel forces act on the body.

4.117 A rigid body is subjected to several parallel noncoplanar forces. State the conditions for equilibrium of the body.

4.118 *True or False*: A nonzero couple always has a nonzero moment with respect to a given point.

4.119 *True or False*: The moment of the forces of a couple with respect to any point in the plane of the couple depends on the position of the point.

4.120 *True or False*: Two coplanar couples may be combined into a single dynamically equivalent couple.

4.121 *True or False*: Coplanar forces that act on a rigid body are always reducible to a single force or a single couple.

4.122 How can the solution of an equilibrium problem be simplified by proper selection of xy axes? By proper selection of moment centers?

4.123 Each force shown in Fig. P4.123 has magnitude 100 N, and each disk is 1 m in radius. Replace each system by the simplest possible dynamically equivalent system.

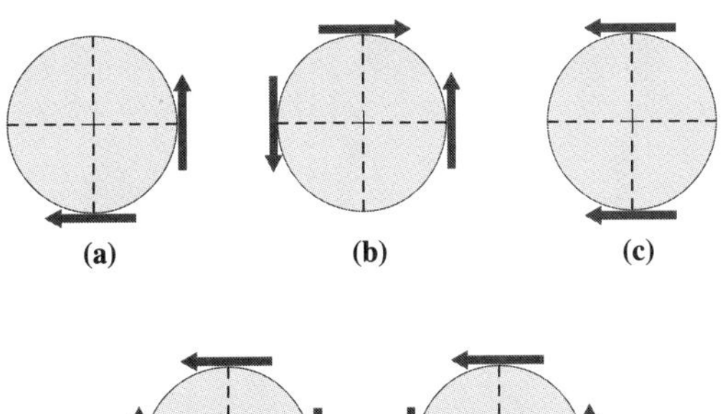

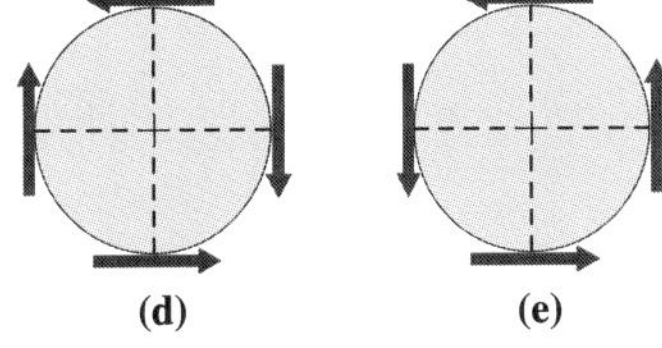

Figure P4.123

Chapter 5

THREE-DIMENSIONAL FORCES, COUPLES, AND RIGID-BODY EQUILIBRIUM

The Atomium, a three-dimensional architectural representation of an atom, was the main attraction at the Universal Exhibition in Brussels in 1958.

A Look Forward

IN CHAPTER 4, we studied the equilibrium of rigid bodies subjected to coplanar force systems. In this chapter, we will extend such force systems to three dimensions. Three-dimensional equilibrium of rigid bodies is just like coplanar equilibrium in that the concepts of moment of force, couples, composition of forces, and resultant axis play important roles. However, the geometry of three-dimensional problems is more complex. This complexity is overcome to some extent by the use of vector algebra. In Chapter 2, you learned how to represent vectors and how to add and subtract vectors. In this chapter, we will examine additional vector algebra concepts that will facilitate analysis of the three-dimensional equilibrium of a rigid body.

Upon completing this chapter, you should be able to use vector algebra to express the moment of a force, to represent a couple as a vector product, to perform the composition of a three-dimensional force system, to obtain the resultant force and the resultant couple of a three-dimensional force system, and to analyze the equilibrium conditions of a body subjected to a three-dimensional force system.

Survey Questions

- What is a scalar product of vectors?
- What is a vector product of vectors?
- What is the difference between a moment and a couple?
- What is the fundamental principle of equilibrium of a rigid body?

5.1 VECTOR ALGEBRA

After studying this section, you should be able to:

- Calculate the scalar (dot) product of two vectors.
- Use the scalar product to calculate the projection of a vector on a line.
- Calculate the vector (cross) product of two vectors.

In this section, you will learn how to compute the scalar product of two vectors and the vector product of two vectors. You will also learn some additional properties of vector algebra. You will use these concepts of vector algebra to represent forces and couples in three-dimensional space and to determine the resultant axis of a force system.

SCALAR PRODUCT OF TWO VECTORS

Key Concept

The scalar product of two vectors **A** and **B** is given by the scalar equation

$$\mathbf{A} \cdot \mathbf{B} = AB\cos\theta = A_xB_x + A_yB_y + A_zB_z$$

scalar product

The *scalar product* is a product of two vectors **A** and **B** that produces, as the name implies, a scalar quantity (a number). In vector notation, the scalar product is written $\mathbf{A} \cdot \mathbf{B}$, and it is defined to be the quantity $AB\cos\theta$. Thus, we have

$$\mathbf{A} \cdot \mathbf{B} = AB\cos\theta \tag{5.1}$$

where A and B are the magnitudes of vectors **A** and **B** and θ is the angle between these vectors (Fig. 5.1a). Since the notation $\mathbf{A} \cdot \mathbf{B}$ is read "A dot B," the scalar product is also called the *dot product*.

dot product

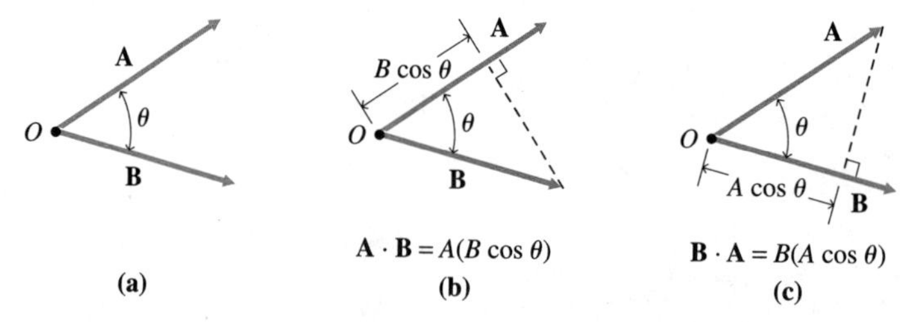

Figure 5.1 The scalar (dot) product of two vectors.

From Eq. (5.1), we see that the scalar product is zero if either A or B is zero or if the vector **A** is perpendicular to the vector **B** (then $\cos\theta = \cos 90° = 0$). So, for nonzero A and B, the scalar product may be zero, a positive number, or a negative number, depending on the value of θ.

Key Concept

The scalar product of two vectors is commutative; that is, $\mathbf{A} \cdot \mathbf{B} = \mathbf{B} \cdot \mathbf{A}$.

By manipulating Eq. (5.1), we see that the scalar product is *commutative* (see Figs. 5.1b and 5.1c). That is,

$$\mathbf{A} \cdot \mathbf{B} = AB\cos\theta = BA\cos\theta = \mathbf{B} \cdot \mathbf{A} \tag{5.2}$$

The scalar product can also be expressed in terms of the projections (A_x, A_y, A_z) and (B_x, B_y, B_z) of vectors **A** and **B** on *xyz* coordinate axes. It can be written in the form

$$\mathbf{A} \cdot \mathbf{B} = A_xB_x + A_yB_y + A_zB_z \tag{5.3}$$

We can show that Eq. (5.3) is equivalent to Eq. (5.2) as follows: We translate vectors **A** and **B** so that their tails are placed at the origin O of the *xyz* coordinate system. Then, we connect the tips of the vectors by line segment C. The resulting triangle has vertices at the points (0, 0, 0), (A_x, A_y, A_z), and (B_x, B_y, B_z) (see Fig. 5.2). By the law of cosines of plane trigonometry,

$$C^2 = A^2 + B^2 - 2AB\cos\theta$$

we find

$$AB\cos\theta = \frac{1}{2}(A^2 + B^2 - C^2) \tag{a}$$

Also, by the Pythagorean theorem,

$$\begin{aligned} A^2 &= A_x^2 + A_y^2 + A_z^2 \\ B^2 &= B_x^2 + B_y^2 + B_z^2 \\ C^2 &= C_x^2 + C_y^2 + C_z^2 \end{aligned} \tag{b}$$

$$C^2 = (B_x - A_x)^2 + (B_y - A_y)^2 + (B_z - A_z)^2$$

Now, if we substitute Eqs. (b) into Eq. (a) and simplify the result, we obtain

$$AB\cos\theta = A_xB_x + A_yB_y + A_zB_z \tag{5.4}$$

With Eq. (5.1), this verifies Eq. (5.3).

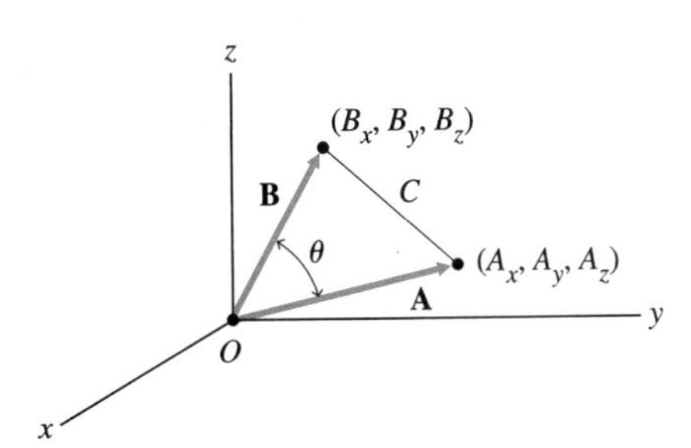

Figure 5.2

Special Cases of the Scalar Product Several special cases of the scalar product deserve attention.

Case 1 If **A** and **B** are the same vector, Eq. (5.3) yields

$$\mathbf{A} \cdot \mathbf{A} = A_x^2 + A_y^2 + A_z^2 = A^2$$

Accordingly, the Pythagorean theorem, Eq. (2.2), is a special case of the scalar-product relation. Frequently, the scalar product $\mathbf{A} \cdot \mathbf{A}$ is denoted by $\mathbf{A}^2$. Hence, $\mathbf{A}^2 = A^2$; the scalar product of a vector with itself is the square of its magnitude.

projection of a vector

Case 2 An important case arises if **B** is a unit vector—that is, if $B = 1$. Then, $A \cos \theta$ is the *projection of vector* **A** on a line in the direction of vector **B**, and Eq. (5.4) yields the following conclusion:

> The projection of a vector **A** on a line with the direction and sense of a *unit* vector **B** is $\mathbf{A} \cdot \mathbf{B} = A \cos \theta$ (see Fig. 5.1c, with $B = 1$).

Case 3 If vectors **A** and **B** are both unit vectors, their projections on *xyz* axes are given by their direction cosines (see Sec. 2.5). Hence, Eq. (5.4) yields

$$\cos \alpha_A \cos \alpha_B + \cos \beta_A \cos \beta_B + \cos \gamma_A \cos \gamma_B = \cos \theta \tag{5.5}$$

where α_A, β_A, γ_A and α_B, β_B, γ_B are the direction angles of vectors **A** and **B**. Equation (5.5) expresses the angle θ between any two directed lines in terms of the direction cosines of the lines.

direction numbers

The direction of a line is uniquely determined by its direction angles α, β, and γ relative to *xyz* axes. (Recall that $\cos^2 \alpha + \cos^2 \beta + \cos^2 \gamma = 1$.) If the direction of a line is specified by three numbers *a, b,* and *c,* the numbers are called the *direction numbers,* or *direction ratios.* In terms of the direction numbers, the direction cosines are given by the formulas

$$\cos \alpha = \frac{a}{\sqrt{a^2 + b^2 + c^2}}$$

$$\cos \beta = \frac{b}{\sqrt{a^2 + b^2 + c^2}}$$

$$\cos \gamma = \frac{c}{\sqrt{a^2 + b^2 + c^2}}$$

Case 4 If vectors **A** and **B** are perpendicular to each other, $\theta = 90°$. Then, Eq. (5.4) yields $A_xB_x + A_yB_y + A_zB_z = 0$, or $\mathbf{A} \cdot \mathbf{B} = 0$. Accordingly, we have the following theorem.

Theorem 5.1

If the angle between two vectors is 90°, the scalar product of the vectors is zero. Conversely, if the scalar product of two vectors is zero and if neither vector has zero magnitude, the angle between the vectors is 90°.

Scalar Products of Unit Vectors Since **i**, **j**, and **k** are unit vectors, they have magnitude 1. Hence, by Case 1, we have

$$\mathbf{i} \cdot \mathbf{i} = \mathbf{j} \cdot \mathbf{j} = \mathbf{k} \cdot \mathbf{k} = 1 \tag{c}$$

Also, since the vectors **i**, **j**, and **k** are mutually perpendicular (see Case 4),

$$\mathbf{i} \cdot \mathbf{j} = \mathbf{j} \cdot \mathbf{k} = \mathbf{k} \cdot \mathbf{i} = 0 \tag{d}$$

Equations (d) also follow immediately from Theorem 5.1.

As noted in Sec. 2.6, we may represent any vector in terms of the unit vectors **i**, **j**, and **k** and the (*x, y, z*) projections of the vector. Hence, as an example of the use of Eqs. (c) and (d), we may write

$$\mathbf{A} \cdot \mathbf{B} = (A_x\mathbf{i} + A_y\mathbf{j} + A_z\mathbf{k}) \cdot (B_x\mathbf{i} + B_y\mathbf{j} + B_z\mathbf{k}) = A_xB_x + A_yB_y + A_zB_z$$

Other Properties of the Scalar Product The scalar product of vectors has other properties in common with the product of numbers. For example, the scalar product of vectors obeys the *distributive* law:

$$\mathbf{A} \cdot (\mathbf{B} + \mathbf{C}) = \mathbf{A} \cdot \mathbf{B} + \mathbf{A} \cdot \mathbf{C} \tag{e}$$

To verify this identity, we use the definition of the scalar product. Noting that the (x, y, z) projections of vector $\mathbf{B} + \mathbf{C}$ are $B_x + C_x$, $B_y + C_y$, $B_z + C_z$, respectively, we obtain, by Eq. (5.3),

$$\mathbf{A} \cdot (\mathbf{B} + \mathbf{C}) = A_x(B_x + C_x) + A_y(B_y + C_y) + A_z(B_z + C_z)$$

Hence,

$$\mathbf{A} \cdot (\mathbf{B} + \mathbf{C}) = (A_xB_x + A_yB_y + A_zB_z) + (A_xC_x + A_yC_y + A_zC_z)$$

Since Eq. (5.3) shows that the sums on the right are equal to $\mathbf{A} \cdot \mathbf{B}$ and $\mathbf{A} \cdot \mathbf{C}$, Eq. (e) is verified.

Equation (e) may be generalized as follows:

$$(\mathbf{A} + \mathbf{B}) \cdot (\mathbf{C} + \mathbf{D}) = (\mathbf{A} + \mathbf{B}) \cdot \mathbf{C} + (\mathbf{A} + \mathbf{B}) \cdot \mathbf{D}$$

Hence,

$$(\mathbf{A} + \mathbf{B}) \cdot (\mathbf{C} + \mathbf{D}) = \mathbf{A} \cdot \mathbf{C} + \mathbf{B} \cdot \mathbf{C} + \mathbf{A} \cdot \mathbf{D} + \mathbf{B} \cdot \mathbf{D} \tag{f}$$

This formula may be generalized further to provide expansions of expressions such as $(\mathbf{A} + \mathbf{B} + \mathbf{C} + \cdots) \cdot (\mathbf{P} + \mathbf{Q} + \mathbf{R} + \cdots)$. The formulas are exactly like those of elementary algebra. In particular, Eq. (f) yields

$$(\mathbf{A} + \mathbf{B}) \cdot (\mathbf{A} - \mathbf{B}) = \mathbf{A}^2 - \mathbf{B}^2 = A^2 - B^2$$

The following examples illustrate some of the vector concepts discussed in this section.

Example 5.1

Projection of a Vector on a Given Directed Line in Space

Problem Statement Calculate the projection of vector $\mathbf{F} = 30\mathbf{i} + 50\mathbf{j} + 70\mathbf{k}$ on a line in the direction of vector $\mathbf{A}$, which extends from point (0, 0, 0) to point (3, 2, 1).

Solution Vector $\mathbf{A}$ is defined by $\mathbf{A} = A_x\mathbf{i} + A_y\mathbf{j} + A_z\mathbf{k} = 3\mathbf{i} + 2\mathbf{j} + \mathbf{k}$. The magnitude of vector $\mathbf{A}$ is

$$A = \sqrt{A_x^2 + A_y^2 + A_z^2} = \sqrt{3^2 + 2^2 + 1^2} = \sqrt{14} = 3.742$$

The projection of $\mathbf{F}$ on a line in the direction of $\mathbf{A}$ is $F\cos\theta$, where θ is the angle between $\mathbf{F}$ and $\mathbf{A}$. The scalar product [Eq. (5.4)] yields

$$AF\cos\theta = A_xF_x + A_yF_y + A_zF_z$$

Consequently,

$$3.742(F\cos\theta) = (3)(30) + (2)(50) + (1)(70) = 260$$

Hence, the projection of $\mathbf{F}$ in the direction of vector $\mathbf{A}$ is $F\cos\theta = 69.49$.

Alternatively, we can compute the projection of $\mathbf{F}$ by using the unit vector $\mathbf{a}$ in the direction of $\mathbf{A}$,

$$\mathbf{a} = \frac{3}{\sqrt{14}}\mathbf{i} + \frac{2}{\sqrt{14}}\mathbf{j} + \frac{1}{\sqrt{14}}\mathbf{k}$$

and forming the scalar product

$$\mathbf{a} \cdot \mathbf{F} = F \cos \theta = \frac{3}{\sqrt{14}} 30 + \frac{2}{\sqrt{14}} 50 + \frac{1}{\sqrt{14}} 70 = 69.49$$

Example 5.2

Law of Cosines

Problem Statement Let the vectors **A**, **B**, and **C** form a triangle as shown in Fig. E5.2, so that $\mathbf{A} + \mathbf{B} = \mathbf{C}$. Prove that

$$C^2 = A^2 + B^2 - 2AB \cos \theta$$

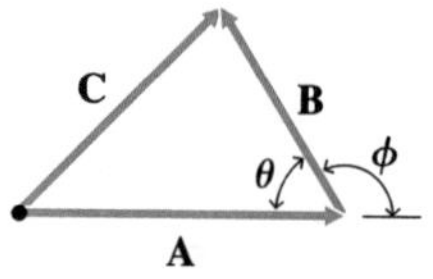

Figure E5.2

Solution Using the fact that $\mathbf{A} + \mathbf{B} = \mathbf{C}$, we square both sides of the given equation to obtain

$$(\mathbf{A} + \mathbf{B})^2 = (\mathbf{A} + \mathbf{B}) \cdot (\mathbf{A} + \mathbf{B}) = \mathbf{C}^2 = C^2 \tag{a}$$

By algebraic expansion, Eq. (a) yields

$$A^2 + B^2 + 2(\mathbf{A} \cdot \mathbf{B}) = C^2 \tag{b}$$

Now $\mathbf{A} \cdot \mathbf{B} = AB \cos \phi$, where ϕ is the angle between vectors **A** and **B** (when they are placed tail-to-tail; see Fig. E5.2). Hence, $\mathbf{A} \cdot \mathbf{B} = AB \cos \phi = -AB \cos \theta$, where $\theta = \pi - \phi$ is supplementary to angle ϕ. Therefore, Eq. (b) becomes

$$A^2 + B^2 - 2AB \cos \theta = C^2$$

Thus, the well-known law of cosines for a triangle can be derived by vector algebra.

Example 5.3

Theorem on Parallelograms

Problem Statement Using the scalar product, prove that the sum of the squares of the sides of a parallelogram equals the sum of the squares of the diagonals.

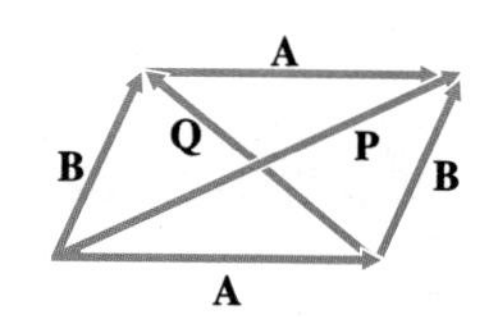

Figure E5.3

Solution Figure E5.3 shows that the diagonal vectors **P** and **Q** of a parallelogram are related to the side vectors **A** and **B** by the relations $\mathbf{A} + \mathbf{B} = \mathbf{P}$ and $\mathbf{A} + \mathbf{Q} = \mathbf{B}$, or $\mathbf{B} - \mathbf{A} = \mathbf{Q}$. Consequently, by squaring and adding the expressions for **P** and **Q**, we have

$$(\mathbf{A} + \mathbf{B})^2 + (\mathbf{B} - \mathbf{A})^2 = \mathbf{P}^2 + \mathbf{Q}^2$$

Carrying out the vector multiplication, we obtain the desired result:

$$2A^2 + 2B^2 = P^2 + Q^2$$

Example 5.4

Work Done by a Constant Force

Problem Statement A particle travels on a straight line *OA* that forms a 30° angle with respect to the *x* axis (see Fig. E5.4). It is acted on by a constant force

$$\mathbf{F} = 4\mathbf{i} + 6\mathbf{j} \quad [\text{N}] \tag{a}$$

The integral

$$U = \int \mathbf{F} \cdot d\mathbf{s} \tag{b}$$

represents the *work* that the force does on the particle. The particle travels 2 m along line *OA* starting from the origin *O*. Determine the work done by the force.

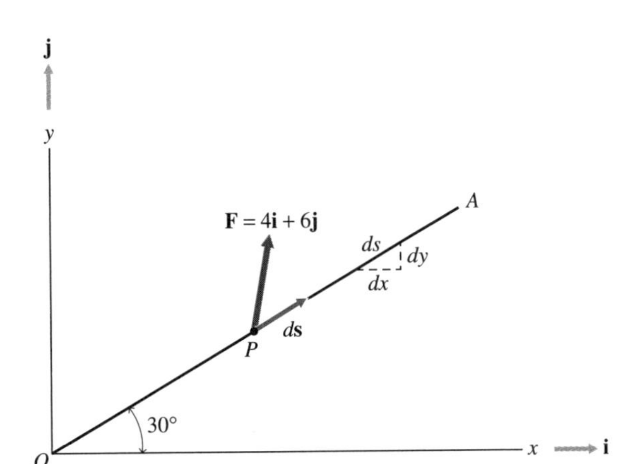

Figure E5.4

Solution When the particle travels an infinitesimal distance along the line OA, the corresponding increments dx and dy are

$$dx = \cos 30^\circ\, ds = \frac{\sqrt{3}}{2}\, ds$$

$$dy = \sin 30^\circ\, ds = \frac{1}{2}\, ds$$

Therefore, the infinitesimal vector $d\mathbf{s}$ is

$$\begin{aligned} d\mathbf{s} &= (dx)\,\mathbf{i} + (dy)\,\mathbf{j} \\ &= \tfrac{1}{2}(\sqrt{3}\,\mathbf{i} + \mathbf{j})\, ds \end{aligned} \tag{c}$$

By Eqs. (a), (b), and (c), the work done by force $\mathbf{F}$ on the particle is

$$\begin{aligned} U &= \tfrac{1}{2}\int_0^2 (4\mathbf{i} + 6\mathbf{j}) \cdot (\sqrt{3}\mathbf{i} + \mathbf{j})\, ds \\ &= \tfrac{1}{2}\int_0^2 (4\sqrt{3} + 6)\, ds \\ &= 12.93\ \text{N·m} \end{aligned}$$

■

CHECKPOINT ✓

1. What is meant by "the scalar product of two vectors"?
2. ***True or False:*** The dot product is the same as the scalar product of vectors.
3. ***True or False:*** The scalar product of any two of the unit vectors **i**, **j**, and **k** is 1.
4. ***True or False:*** The scalar product of vectors is commutative.
5. Describe how to calculate the projection of a force on a line.
6. What is the scalar product of two perpendicular vectors?

Answers: 1. Two vectors combined to obtain a scalar result; 2. True; 3. False; 4. True; 5. Take the scalar product of the force with the unit vector for the line; 6. Zero

VECTOR PRODUCT OF TWO VECTORS

Key Concept

The vector product of two vectors **A** and **B** is a vector **C** that is perpendicular to the plane of vectors **A** and **B**. The vector **C** has magnitude $C = AB\sin\theta$, where A and B are the magnitudes of vectors **A** and **B** and θ is the angle between vectors **A** and **B**.

Equation (5.4) provides an expression for $AB\cos\theta$, in which A and B are the magnitudes of two vectors **A** and **B** that include the angle θ. We now consider an important transformation of this equation; namely, the *vector product* $\mathbf{C} = \mathbf{A} \times \mathbf{B}$ of two vectors **A** and **B**. The vector product is also referred to as the *cross product,* because the expression $\mathbf{A} \times \mathbf{B}$ is usually read "A cross B."

vector product

cross product

The vector product is useful in the representation of the moment of a force (see Secs. 5.2 and 5.3) and of couples (see Secs. 5.4 and 5.5). To interpret the meaning of the vector product, we start with Eq. (5.4) and derive expressions for the (x, y, z) projections of **C** in terms of the (x, y, z) projections of **A** and **B**. Then, we show that **C** is perpendicular to the plane formed by **A** and **B**. Finally, we show that the vectors **A**, **B**, and **C** form a triad of vectors ordered in the same way as rectangular coordinate xyz axes. The final result of this process is expressed by the following theorem, which uses the *right-hand screw rule* (see Sec. 2.4 and Fig. 2.14).

right-hand screw rule

Theorem 5.2

Consider two vectors **A** and **B** separated by an angle θ. The vector product $\mathbf{A} \times \mathbf{B}$ is a vector **C**, whose magnitude is $AB\sin\theta$, whose direction is perpendicular to the vectors **A** and **B**, and whose sense is such that a right-hand screw, laid along the vector $\mathbf{C} = \mathbf{A} \times \mathbf{B}$, advances in the positive sense of vector **C** when the screw is twisted from vector **A** toward vector **B**.

To begin, we note that Eq. (5.4) may be written

$$\cos\theta = \frac{A_xB_x + A_yB_y + A_zB_z}{AB} \tag{a}$$

Consequently, the trigonometric identity, $\sin\theta = \sqrt{1 - \cos^2\theta}$, and Eq. (a) yield the result

$$AB\sin\theta = \sqrt{A^2B^2 - (A_xB_x + A_yB_y + A_zB_z)^2}$$

Hence, by the Pythagorean theorem,

$$AB\sin\theta = \sqrt{(A_x^2 + A_y^2 + A_z^2)(B_x^2 + B_y^2 + B_z^2) - (A_xB_x + A_yB_y + A_zB_z)^2}$$

Regrouping terms under the radical, we obtain

$$AB\sin\theta = \sqrt{(A_yB_z - A_zB_y)^2 + (A_zB_x - A_xB_z)^2 + (A_xB_y - A_yB_x)^2} \tag{b}$$

Let the quantities C_x, C_y, C_z be defined as follows:

$$
\begin{aligned}
C_x &= A_y B_z - A_z B_y \\
C_y &= A_z B_x - A_x B_z \\
C_z &= A_x B_y - A_y B_x
\end{aligned}
\tag{5.6}
$$

By the Pythagorean theorem, we may consider the terms C_x, C_y, and C_z to be the (x, y, z) projections of a vector **C**, called the vector product of **A** and **B**. Since $C = \sqrt{C_x^2 + C_y^2 + C_z^2}$ where C is the magnitude of **C**, Eqs. (b) and (5.6) yield

$$C = AB \sin \theta \tag{5.7}$$

Equation (5.7) determines the magnitude of the vector **C**.

We can also show that *vector* **C** *is perpendicular to both vector* **A** *and vector* **B**. This relation is easily proved by means of Eq. (5.4). If vectors **A** and **C** are perpendicular, $\mathbf{A} \cdot \mathbf{C} = 0$ (Case 4, Sec. 5.1), or

$$A_x C_x + A_y C_y + A_z C_z = 0 \tag{c}$$

Likewise, if vectors **B** and **C** are perpendicular, $\mathbf{B} \cdot \mathbf{C} = 0$, or

$$B_x C_x + B_y C_y + B_z C_z = 0 \tag{d}$$

Substitution of the expressions for C_x, C_y, and C_z from Eqs. (5.6) into Eqs. (c) and (d) satisfies both of the latter identically. Thus, it is verified that vector **C** is perpendicular to vectors **A** and **B**.

The magnitude and the direction of vector **C** have now been determined. Finally, we can show that the sense of vector **C** is such that the three vectors **A**, **B**, and **C**, in this order, form a right-handed triad; that is, **A**, **B**, and **C** are oriented in the same way as rectangular coordinate *xyz* axes (see Sec. 2.5).

To prove this relation, we consider first the case in which vectors **A** and **B** lie in the *xy* plane, with their tails at the origin *O*. Let the directions of these two vectors be specified by angles α and β, measured counterclockwise from the positive x axis (see Fig. 5.3). Then,

$$
\begin{aligned}
A_x &= A \cos \alpha \qquad & A_y &= A \sin \alpha \qquad & A_z &= 0 \\
B_x &= B \cos \beta \qquad & B_y &= B \sin \beta \qquad & B_z &= 0
\end{aligned}
$$

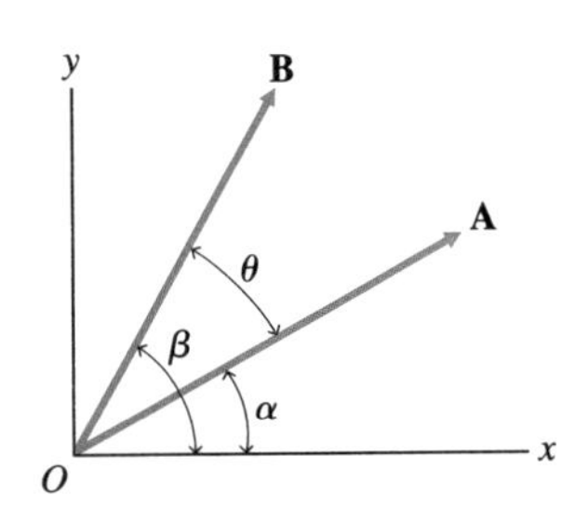

Figure 5.3

Hence, Eqs. (5.6) yield

$$
\begin{aligned}
C_x &= C_y = 0 \\
C_z &= AB \sin(\beta - \alpha) = AB \sin \theta
\end{aligned}
\tag{e}
$$

The sign of C_z, determined by Eq. (e), is such that the three vectors **A**, **B**, and **C** form a right-handed system. This result holds for the case in which vectors **A** and **B** are parallel to the *xy* plane. If we displace **A** and **B** in any way without changing the angle between them, the magnitude of **C** remains constant, according to Eq. (5.7). Since vector **C** remains perpendicular to vectors **A** and **B**, it maintains a constant sense with respect to these vectors. Hence, the result is valid in all cases.

For a right-handed coordinate system, the above results can be summarized by Theorem 5.2. If the *xyz* coordinate system is left-handed, the sense of vector **C** is determined by the left-hand screw rule (see Sec. 2.5 and Fig. 2.20).

If the vectors **A** and **B** act at the same point, they form two sides of a parallelogram (see Fig. 5.4). The vector $\mathbf{C} = \mathbf{A} \times \mathbf{B}$ is perpendicular to the plane of the parallelogram. By Eq. (5.7), its magnitude is equal to the area of the parallelogram.

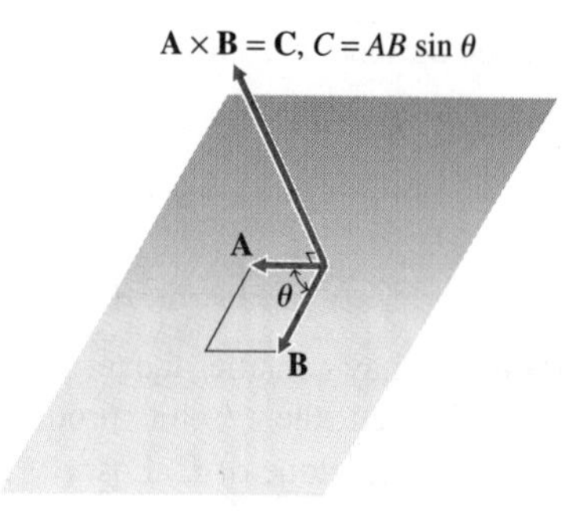

Figure 5.4 Vector product $\mathbf{C} = \mathbf{A} \times \mathbf{B}$.

In determinant notation, Eqs. (5.6) are expressed as follows (see Appendix A):

$$C_x = \begin{vmatrix} A_y & A_z \\ B_y & B_z \end{vmatrix} \qquad C_y = \begin{vmatrix} A_z & A_x \\ B_z & B_x \end{vmatrix} \qquad C_z = \begin{vmatrix} A_x & A_y \\ B_x & B_y \end{vmatrix} \tag{5.8}$$

Consequently, if **i**, **j**, and **k** are unit vectors along the *xyz* axes, the vector **C** may be expressed as follows:

$$\mathbf{C} = \mathbf{A} \times \mathbf{B} = \begin{vmatrix} \mathbf{i} & \mathbf{j} & \mathbf{k} \\ A_x & A_y & A_z \\ B_x & B_y & B_z \end{vmatrix} \tag{5.9}$$

Key Concept

The vector product is not commutative. That is, $\mathbf{A} \times \mathbf{B} \neq \mathbf{B} \times \mathbf{A}$; but rather $\mathbf{A} \times \mathbf{B} = -\mathbf{B} \times \mathbf{A}$.

The vector product is *not* commutative. In fact, from Eq. (5.9) or the right-hand screw rule, the vector **C** reverses its sense when the order of vectors **A** and **B** is reversed; that is, $\mathbf{A} \times \mathbf{B} = \mathbf{C}$, whereas $\mathbf{B} \times \mathbf{A} = -\mathbf{C}$ (see Figs. 5.4 and 5.5).

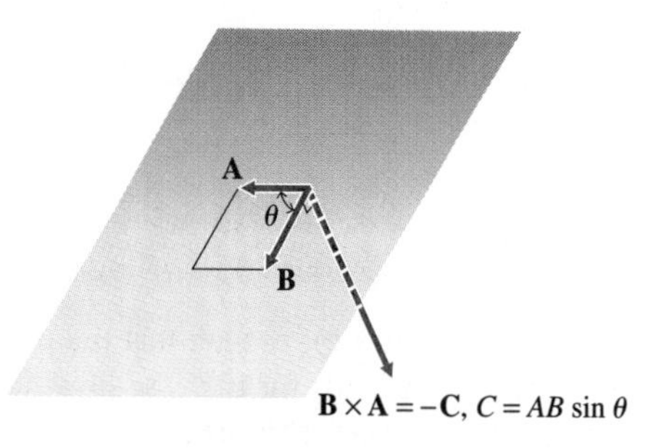

Figure 5.5 Vector product $\mathbf{B} \times \mathbf{A}$ is opposite in sense to $\mathbf{A} \times \mathbf{B}$.

The vector product of two parallel vectors is zero, for, if the vectors are parallel, $\theta = 0$ and $\sin \theta = 0$ [see Eq. (5.7)]. Conversely, if the vector product of two vectors is zero and if neither vector has zero magnitude, the vectors are parallel.

Since any vector **a** is parallel to itself, $\mathbf{a} \times \mathbf{a} = 0$. Similarly, for the unit vectors **i**, **j**, and **k** along rectangular coordinate *xyz* axes, we have the relations

$$\mathbf{i} \times \mathbf{i} = \mathbf{j} \times \mathbf{j} = \mathbf{k} \times \mathbf{k} = 0$$

Also, by Eq. (5.9) $\mathbf{i} \times \mathbf{j} = \begin{vmatrix} \mathbf{i} & \mathbf{j} & \mathbf{k} \\ 1 & 0 & 0 \\ 0 & 1 & 0 \end{vmatrix} = +\mathbf{k}$ and $\mathbf{j} \times \mathbf{i} = \begin{vmatrix} \mathbf{i} & \mathbf{j} & \mathbf{k} \\ 0 & 1 & 0 \\ 1 & 0 & 0 \end{vmatrix} = -\mathbf{k}$

Similar relations are obtained for the other products of the coordinate unit vectors. In summary,

$$\begin{array}{lll} \mathbf{i} \times \mathbf{j} = \mathbf{k} & \mathbf{j} \times \mathbf{k} = \mathbf{i} & \mathbf{k} \times \mathbf{i} = \mathbf{j} \\ \mathbf{j} \times \mathbf{i} = -\mathbf{k} & \mathbf{k} \times \mathbf{j} = -\mathbf{i} & \mathbf{i} \times \mathbf{k} = -\mathbf{j} \\ \mathbf{i} \times \mathbf{i} = \mathbf{0} & \mathbf{j} \times \mathbf{j} = \mathbf{0} & \mathbf{k} \times \mathbf{k} = \mathbf{0} \end{array} \tag{5.10}$$

These relations also follow immediately from Theorem 5.2.

As an example of the use of Eqs. (5.10), consider the vector identity

$$\mathbf{A} \times \mathbf{B} = (A_x\mathbf{i} + A_y\mathbf{j} + A_z\mathbf{k}) \times (B_x\mathbf{i} + B_y\mathbf{j} + B_z\mathbf{k})$$

Expansion of this vector identity yields, with Eqs. (5.10),

$$\mathbf{A} \times \mathbf{B} = (A_yB_z - A_zB_y)\mathbf{i} + (A_zB_x - A_xB_z)\mathbf{j} + (A_xB_y - A_yB_x)\mathbf{k}$$

This expression is identical to the determinant in Eq. (5.9).

Example 5.5

Vector Products of Sums of Vectors

Problem Statement In spite of the fact that $\mathbf{A} \times \mathbf{B} = -\mathbf{B} \times \mathbf{A}$, the vector product has many properties similar to scalar products. For example, the vector product is distributive:

$$\mathbf{R} \times (\mathbf{A} + \mathbf{B}) = \mathbf{R} \times \mathbf{A} + \mathbf{R} \times \mathbf{B} \tag{a}$$

and

$$(\mathbf{A} + \mathbf{B}) \times \mathbf{R} = \mathbf{A} \times \mathbf{R} + \mathbf{B} \times \mathbf{R} \tag{b}$$

A straightforward generalization of Eq. (a) is

$$\begin{aligned} (\mathbf{A} + \mathbf{B}) \times (\mathbf{C} + \mathbf{D}) &= (\mathbf{A} + \mathbf{B}) \times \mathbf{C} + (\mathbf{A} + \mathbf{B}) \times \mathbf{D} \\ &= \mathbf{A} \times \mathbf{C} + \mathbf{B} \times \mathbf{C} + \mathbf{A} \times \mathbf{D} + \mathbf{B} \times \mathbf{D} \end{aligned}$$

A similar generalization of Eq. (b) can be written. Prove Eq. (a).

Solution By Eq. (5.9), we can write

$$\mathbf{R} \times (\mathbf{A} + \mathbf{B}) = \begin{vmatrix} \mathbf{i} & \mathbf{j} & \mathbf{k} \\ R_x & R_y & R_z \\ (A_x + B_x) & (A_y + B_y) & (A_z + B_z) \end{vmatrix} \tag{c}$$

By a theorem of determinants, we can write Eq. (c) in the form

$$\mathbf{R} \times (\mathbf{A} + \mathbf{B}) = \begin{vmatrix} \mathbf{i} & \mathbf{j} & \mathbf{k} \\ R_x & R_y & R_z \\ A_x & A_y & A_z \end{vmatrix} + \begin{vmatrix} \mathbf{i} & \mathbf{j} & \mathbf{k} \\ R_x & R_y & R_z \\ B_x & B_y & B_z \end{vmatrix} \tag{d}$$

Then, by Eqs. (5.9) and (d), we have

$$\mathbf{R} \times (\mathbf{A} + \mathbf{B}) = \mathbf{R} \times \mathbf{A} + \mathbf{R} \times \mathbf{B}$$

Thus, Eq. (a) is proved. Equation (b) can be verified in a similar manner.

Note that the original order of the terms is retained in these expansions since, for example, $\mathbf{A} \times \mathbf{C} \neq \mathbf{C} \times \mathbf{A}$.

Example 5.6

Shortest Distance between Two Lines

Problem Statement A surveyor must determine the minimum distance between two lines L_1 and L_2 in mountainous terrain. Line L_1 passes through the points A: (50, −75, 25) and B: (25, 0, −50); line L_2 passes through the points C: (100, 50, 25) and D: (−25, −50, 25). All distances are in meters. Help the surveyor solve this problem.

Solution We note that if we take a vector $\mathbf{r}_{AB} = -25\mathbf{i} + 75\mathbf{j} - 75\mathbf{k}$ from point A to point B and a vector $\mathbf{r}_{CD} = -125\mathbf{i} - 100\mathbf{j}$ from point C to point D, the vector

$$
\begin{aligned}
\mathbf{r} &= \mathbf{r}_{AB} \times \mathbf{r}_{CD} \\
&= -7500\mathbf{i} + 9375\mathbf{j} + 11\,875\mathbf{k} \quad [\text{m}] \qquad \text{(a)}
\end{aligned}
$$

is perpendicular to both $\mathbf{r}_{AB}$ and $\mathbf{r}_{CD}$. Then, we observe that if we take any vector from line L_1 to line L_2—say, vector $\mathbf{q}$—the projection of $\mathbf{q}$ on $\mathbf{r}$ gives the minimum distance between the lines. To obtain the projection of $\mathbf{q}$ on $\mathbf{r}$, we form the scalar product $\mathbf{e} \cdot \mathbf{q}$, where $\mathbf{e}$ is a unit vector along $\mathbf{r}$. From Eq. (a), we obtain

$$\mathbf{e} = -0.4441\mathbf{i} + 0.5552\mathbf{j} + 0.7032\mathbf{k}$$

Next, let's take the vector $\mathbf{q}$ from point C on line L_2 to point B on line L_1; that is,

$$\mathbf{q} = -75\mathbf{i} - 50\mathbf{j} - 75\mathbf{k}$$

Hence, the minimum distance between lines L_1 and L_2 is

$$
\begin{aligned}
d &= |\mathbf{e} \cdot \mathbf{q}| \\
&= |(-0.4441\mathbf{i} + 0.5552\mathbf{j} + 0.7032\mathbf{k}) \cdot (-75\mathbf{i} - 50\mathbf{j} - 75\mathbf{k})| \\
&= 47.2 \text{ m}
\end{aligned}
$$

CHECKPOINT

1. What is meant by "the vector product of two vectors"?
2. ***True or False:*** The scalar product of two vectors is a special case of the vector product.
3. ***True or False:*** The cross product is the same as the vector product.
4. ***True or False:*** The vector product is a scalar.
5. What is the vector product of two parallel vectors with opposite senses?
6. ***True or False:*** The vector product of any two of the unit vectors **i**, **j**, and **k** is 1.
7. ***True or False:*** The thumb, the index finger, and the middle finger of your right hand, held as shown, form a right-handed triad.

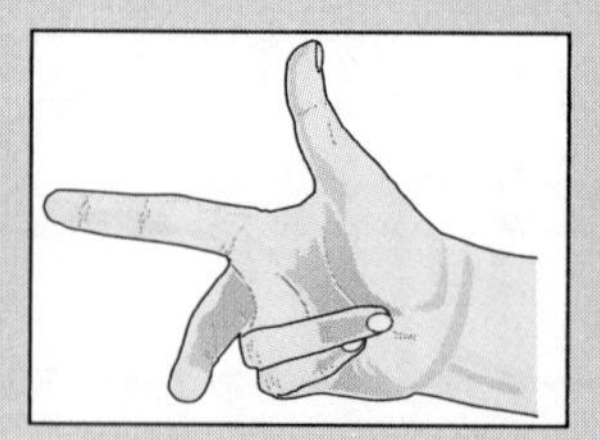

Answers: 1. Two vectors combined to obtain a vector result; 2. False; 3. True; 4. False; 5. Zero; 6. False; 7. True

5.2 MOMENT OF A FORCE ABOUT A POINT

After studying this section, you should be able to:

- Calculate the moment of a force about a point.

Key Concept The moment of a force about a point is a vector.

moment of a force about a point

Let a be the perpendicular distance from a reference point O to the line of action of force $\mathbf{F}$ (see Fig. 5.6). The *moment of force* $\mathbf{F}$ *about point* O is defined to be a vector $\mathbf{M}$ with magnitude aF. The vector $\mathbf{M}$ is defined to be perpendicular to the plane Q determined by force $\mathbf{F}$ and point O. The sense of vector $\mathbf{M}$ is defined by the right-hand screw rule.

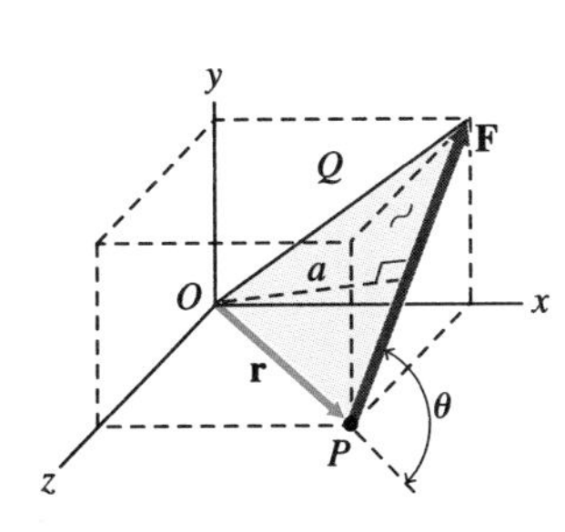

Figure 5.6

Let the position vector *from* the reference point O *to* the point P on the line of action of force $\mathbf{F}$ (Fig. 5.6) be denoted by $\mathbf{r}$. Then the conditions of the preceding definition are fulfilled by the vector equation

$$\mathbf{M} = \mathbf{r} \times \mathbf{F} \tag{5.11}$$

Sliding the force $\mathbf{F}$ along its line of action does not change its moment about point O (Fig. 5.6). Consequently, $\mathbf{r}$ may be defined as the vector from point O to any point on the line of action of $\mathbf{F}$.

In the theory of coplanar forces (Sec. 4.1), the moment of a force about a point is defined to be a scalar. This is adequate in planar problems because if the force vector $\mathbf{F}$ and the reference point O lie in the xy plane, the moment vector $\mathbf{M}$ is perpendicular to that plane. Its direction is always known, so it can be completely described as a scalar. In three dimensions, however, the moment of a force about a point is interpreted as a vector. Thus, the moment is specified by both its magnitude and its direction.

Physically, the moment of a force about a point may be regarded as the rotational effect that the force tends to produce on a rigid body having point O fixed. The magnitude of the moment is a measure of this effect, and the direction of the moment corresponds to the axis and sense of rotation it tends to produce.

MOMENT OF A FORCE ABOUT THE ORIGIN

If the reference point O in Fig. 5.6 is the origin of rectangular coordinates and the coordinates of point P are (x, y, z), the vector $\mathbf{r}$ is given by $\mathbf{r} = x\mathbf{i} + y\mathbf{j} + z\mathbf{k}$. Then, in view of Eqs. (5.11) and (5.9), the moment vector $\mathbf{M}$ about O is determined by the equation

$$\mathbf{M} = \mathbf{r} \times \mathbf{F} = \begin{vmatrix} \mathbf{i} & \mathbf{j} & \mathbf{k} \\ x & y & z \\ F_x & F_y & F_z \end{vmatrix} \tag{5.12}$$

Alternatively, we can write

$$\mathbf{M} = M_x\mathbf{i} + M_y\mathbf{j} + M_z\mathbf{k} \tag{5.13}$$

where, by expansion of the determinant in Eq. (5.12), the projections M_x, M_y, and M_z of $\mathbf{M}$ on xyz axes are given by

$$M_x = yF_z - zF_y$$
$$M_y = zF_x - xF_z$$
$$M_z = xF_y - yF_x$$

Example 5.7

Moment of a Force about a Point

Problem Statement The line of action of the force

$$\mathbf{F} = 10\mathbf{i} - 10\mathbf{j} + 20\mathbf{k} \tag{a}$$

passes through the point P: (4, 2, −2), relative to xyz axes (force in kips and distances in feet).

a. Determine the moment $\mathbf{M}_A$ of $\mathbf{F}$ about the point A: (2, 3, −1).

b. Determine the projections of $\mathbf{M}_A$ on the xyz axes.

Solution

a. First, we determine the vector $\mathbf{r}$ from point A to point P:

$$\mathbf{r} = 2\mathbf{i} - \mathbf{j} - \mathbf{k} \tag{b}$$

Then, from Eqs. (a) and (b),

$$\begin{aligned} \mathbf{M}_A &= \mathbf{r} \times \mathbf{F} \\ &= \begin{vmatrix} \mathbf{i} & \mathbf{j} & \mathbf{k} \\ 2 & -1 & -1 \\ 10 & -10 & 20 \end{vmatrix} \\ &= -30\mathbf{i} - 50\mathbf{j} - 10\mathbf{k} \end{aligned} \tag{c}$$

b. By Eq. (c), the projections of $\mathbf{M}_A$ on xyz axes are

$$M_x = -30 \text{ kip·ft}$$
$$M_y = -50 \text{ kip·ft}$$
$$M_z = -10 \text{ kip·ft}$$

Example 5.8

Moment of a Force about the Origin

Problem Statement A rancher uses the rope-pulley system shown in Fig. E5.8 to keep a corral gate from sagging and to pull the gate closed when it is left open. With the gate in the position shown, the force in the rope is

$$\mathbf{F} = 50\mathbf{i} + 100\mathbf{j} - 30\mathbf{k} \quad [\text{N}] \tag{a}$$

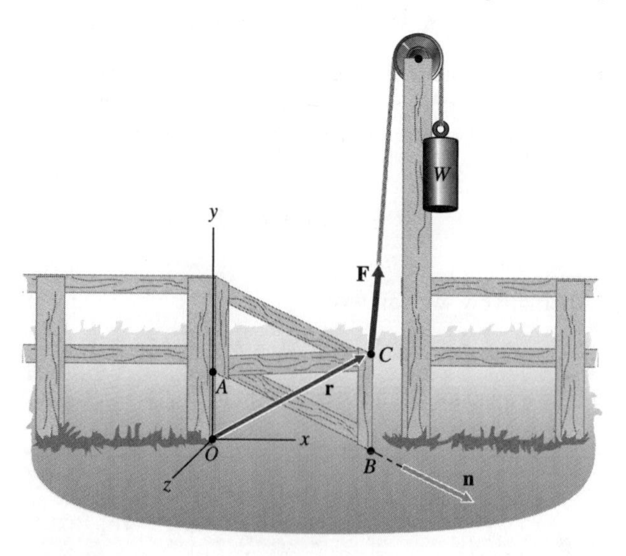

Figure E5.8

This force acts at C, the corner of the gate. The gate swings about the y axis. Points A, B, and C are located at A: (0, 0.75, 0), B: (1.6, 0.75, 1.2), and C: (1.6, 1.5, 1.2) (distances in meters).

a. Determine the moment of **F** about the origin O.

b. What is the moment tending to close the gate?

Solution **a.** The moment $\mathbf{M}_O$ of the force about O is

$$\mathbf{M}_O = \mathbf{r} \times \mathbf{F} \tag{b}$$

where, from Fig. E5.8,

$$\mathbf{r} = 1.6\mathbf{i} + 1.5\mathbf{j} + 1.2\mathbf{k} \quad [\text{m}] \tag{c}$$

Hence, by Eqs. (a), (b), and (c),

$$\begin{aligned} \mathbf{M}_O &= \mathbf{r} \times \mathbf{F} \\ &= \begin{vmatrix} \mathbf{i} & \mathbf{j} & \mathbf{k} \\ 1.6 & 1.5 & 1.2 \\ 50 & 100 & -30 \end{vmatrix} \\ &= -165\mathbf{i} + 108\mathbf{j} + 85\mathbf{k} \quad [\text{N}\cdot\text{m}] \end{aligned} \tag{d}$$

b. The magnitude of the moment tending to close the gate is the magnitude M_y of the y component of M_O. Therefore, by Eq. (d),

$$M_y = 108\ \text{N}\cdot\text{m}$$

5.3 MOMENT OF A FORCE ABOUT AN AXIS

After studying this section, you should be able to:

- Calculate the moment of a force about an axis or a directed line.

Key Concept The moment of a force about an axis is a scalar quantity.

The moment of a force about an axis perpendicular to the plane of the force was defined in Sec. 4.1. A generalization of this idea to noncoplanar forces is useful. This generalization is obtained as follows:

Let **F** be a force that acts on a rigid body. Let z be any axis or directed line in space. If the line of action of force **F** does not intersect the z axis, there exists a line segment OP that is perpendicular to both the line and the axis (see Fig. 5.7a). The length r of the line segment OP is the minimum distance between the line and the axis (see Example 5.6). This distance is called the *moment arm* of force **F** with respect to the z axis. It is a nonnegative number. By the theorem of transmissibility (Sec. 3.6), the force **F** may be displaced along its line of action to point P. There, it may be resolved into two components $\mathbf{F}_z$ and $\mathbf{F}_n$, parallel and perpendicular, respectively, to the z axis (Fig. 5.7b). The *moment M_z of force* **F** *about the z axis* is defined by the equation

moment arm

moment of a force about an axis

$$M_z = \pm r F_n \tag{5.14}$$

where F_n is the magnitude of force $\mathbf{F}_n$. Hence, the moment of a force about an axis is a scalar—that is, a signed magnitude. Note that if $r = 0$, force **F** intersects the z axis. Then, $M_z = 0$.

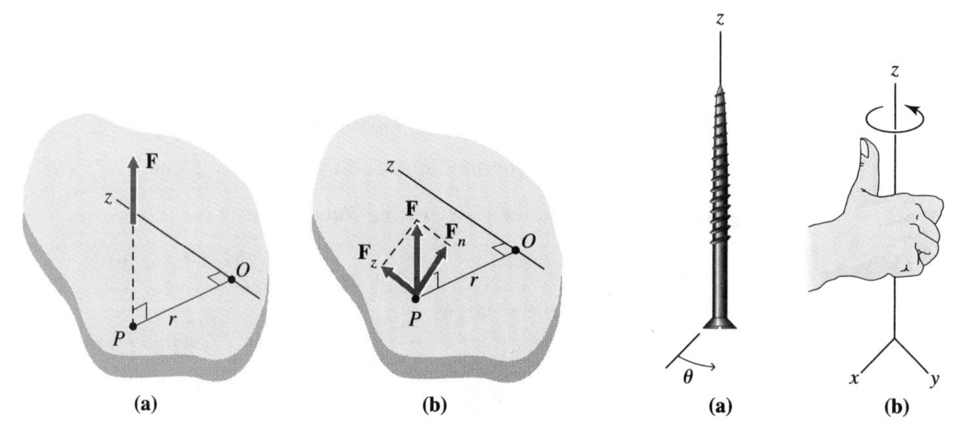

Figure 5.7

Figure 5.8 **(a)** Right-hand screw rule. **(b)** Right-hand screw rule demonstrated with right hand.

By definition [Eq. (5.14)], force $\mathbf{F}_z$, which is parallel to the z axis, exerts no moment about the z axis. To establish the sign in Eq. (5.14), we specify that M_z is positive if vector $\mathbf{F}_n$ would cause a rotation about the z axis in the sense that a right-hand screw would be twisted such that it would advance in the positive z direction. (See Theorem 5.2 and Fig. 5.8a.) Alternatively, the right-hand screw rule may be visualized as follows. Imagine that you grasp the z axis in your right hand with your thumb pointing in the positive sense of the axis, as shown in Fig. 5.8b. Your fingers then curl around the z axis in the positive sense of rotation.

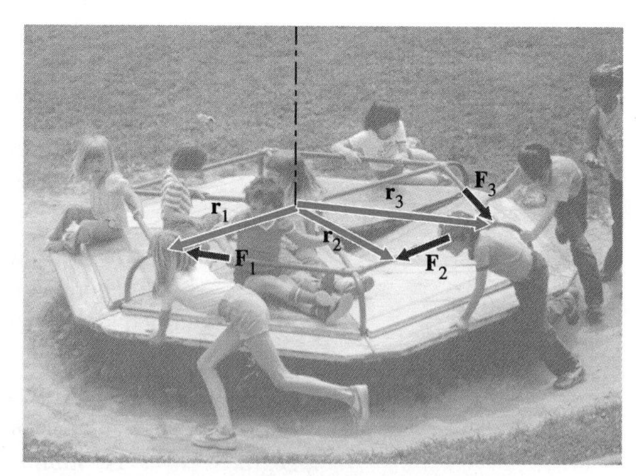

Each child pushing on this merry-go-round causes a moment about the vertical axis.

Let's now consider the moments of force $\mathbf{F}$ about xyz axes. We denote these moments by M_x, M_y, and M_z, respectively. Force $\mathbf{F}$ may be resolved into components ($F_x\mathbf{i}$, $F_y\mathbf{j}$, $F_z\mathbf{k}$) parallel to the x, y, and z axes, respectively. From Fig. 5.9, we see that the moment of force $F_z\mathbf{k}$ about the x axis is $+yF_z$, where P: (x, y, z) is the point of application of force $\mathbf{F}$. Similarly, the moment of force $F_y\mathbf{j}$ about the x axis is $-zF_y$. The moment of force $F_x\mathbf{i}$ about the x axis is zero, since the vector $F_x\mathbf{i}$ is parallel to the x axis. Hence, by Varignon's theorem, the moment of force $\mathbf{F}$ about the x axis is $M_x = yF_z - zF_y$. The moments of force $\mathbf{F}$

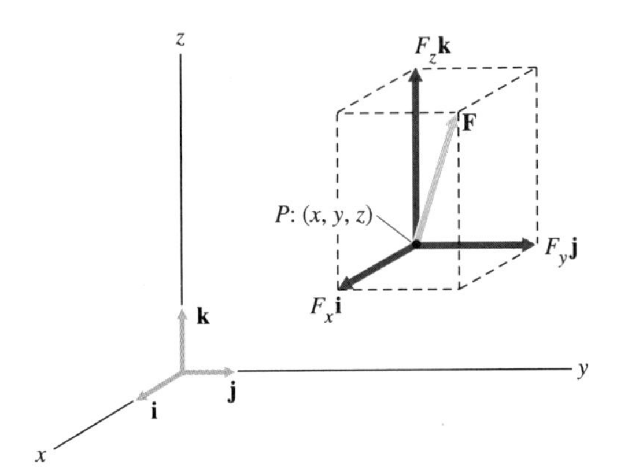

Figure 5.9

about the y and z axes are derived similarly. Therefore, the moments of M_x, M_y, and M_z of force **F** about xyz coordinate axes are given by

$$\begin{aligned} M_x &= yF_z - zF_y \\ M_y &= zF_x - xF_z \\ M_z &= xF_y - yF_x \end{aligned} \tag{5.15}$$

In applying Eqs. (5.15), you must take into account the signs of the (x, y, z) coordinates of the point of application of **F** and of the projections F_x, F_y, and F_z (Fig. 5.9). Equations (5.15) are three-dimensional analogues of Eq. (4.3).

We can write Eqs. (5.15) in determinant notation [Eq. (5.8)]:

$$M_x = \begin{vmatrix} y & z \\ F_y & F_z \end{vmatrix} \qquad M_y = \begin{vmatrix} z & x \\ F_z & F_x \end{vmatrix} \qquad M_z = \begin{vmatrix} x & y \\ F_x & F_y \end{vmatrix} \tag{5.16}$$

MOMENT OF A FORCE ABOUT A LINE WITH SPECIFIED DIRECTION COSINES

As shown above [see Eq. (5.14)], the moment of a force about an axis or a directed line is a scalar. The method used earlier to compute the moment required that the force **F** be resolved into two components $\mathbf{F}_z$ and $\mathbf{F}_n$, parallel and perpendicular to the axis, respectively. Here, we present an alternate method of computing the moment of a force about an axis or a directed line.

Suppose we wish to find the moment M_L of force **F** about an axis or a directed line L (see Fig. 5.10). For M_L not to be zero, the line of action of **F** must neither intersect L nor be parallel to L. Without loss of generality, we let L lie in the xy plane and pass through the origin O of the xyz axes. We also translate **F** along its line of action so that it intersects the xy plane at point P. Let **n** be a unit vector in the direction of L; the sense of **n** is arbitrary. Let **r** be a vector *from* the reference point O *to* point P. Force vector **F** forms an angle ϕ with the line S, which is normal to the xy plane (Fig. 5.10). Then, the projection F_S of **F** on S is

$$F_S = F\cos\phi$$

The perpendicular distance from L to point P is $r\sin\theta$, where θ is the angle between **r** and L. Thus, the moment of **F** about L is [see Eq. (5.14)]

$$M_L = \pm F_S(r\sin\theta) = \pm(F\cos\phi)(r\sin\theta) = \pm F(r\sin\theta)\cos\phi \tag{a}$$

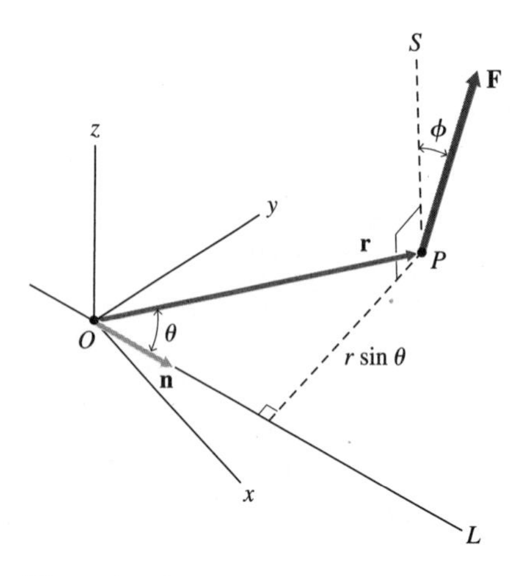

Figure 5.10

We note that the vector product $\mathbf{n} \times \mathbf{r}$ is a vector that has magnitude $r \sin\theta$ and is directed perpendicular to the xy plane. Therefore, in view of Eq. (a) and the properties of the scalar product, the moment M_L of force F about L is

$$M_L = \mathbf{F} \cdot (\mathbf{n} \times \mathbf{r}) \tag{5.17}$$

The sign of M_L, determined by Eq. (5.17), is consistent with the sign specified using the right-hand screw rule.

scalar triple product

mixed triple product

The expression $\mathbf{F} \cdot (\mathbf{n} \times \mathbf{r})$ is called the *scalar triple product* of the vectors **F**, **n**, and **r**. Sometimes the term *mixed triple product* is used to indicate that a vector product and a scalar product both occur in Eq. (5.17). With reference to a rectangular Cartesian coordinate system, the vector $\mathbf{n} \times \mathbf{r}$ has the following projections on the xyz axes:

$$\begin{aligned} x \text{ projection: } & n_y r_z - n_z r_y \\ y \text{ projection: } & n_z r_x - n_x r_z \\ z \text{ projection: } & n_x r_y - n_y r_x \end{aligned} \tag{b}$$

Consequently, with Eqs. (b) and (5.3), Eq. (5.17) yields

$$M_L = F_x(n_y r_z - n_z r_y) + F_y(n_z r_x - n_x r_z) + F_z(n_x r_y - n_y r_x) \tag{c}$$

Equation (c) is expressed more compactly by means of determinant notation, as follows:

$$M_L = \mathbf{F} \cdot (\mathbf{n} \times \mathbf{r}) = \begin{vmatrix} F_x & F_y & F_z \\ n_x & n_y & n_z \\ r_x & r_y & r_z \end{vmatrix} \tag{5.18}$$

ALTERNATIVE FORMS FOR EXPRESSING THE MOMENT OF A FORCE ABOUT A LINE

If two rows of a determinant are interchanged, only the sign of the determinant is changed. Its magnitude is not affected. Therefore, two consecutive interchanges of two rows of a determinant leave the determinant unchanged. Consequently, Eq. (5.17) may be expressed in any one of the following alternative forms:

$$M_L = \mathbf{F} \cdot (\mathbf{n} \times \mathbf{r}) \tag{5.19a}$$

$$M_L = \mathbf{r} \cdot (\mathbf{F} \times \mathbf{n}) \tag{5.19b}$$

$$M_L = \mathbf{n} \cdot (\mathbf{r} \times \mathbf{F}) \tag{5.19c}$$

The formulas for the moments of a force about the coordinate axes [Eqs. (5.15)] are special cases of Eqs. (5.18) and (5.19). For example, if line L is the x axis and if point O is the origin, then $n_x = 1$, $n_y = n_z = 0$, and $r_x = x$, $r_y = y$, $r_z = z$. Hence, by Eq. (5.18) or Eq. (5.19a)

$$M_x = \begin{vmatrix} F_x & F_y & F_z \\ 1 & 0 & 0 \\ x & y & z \end{vmatrix} = yF_z - zF_y$$

This result agrees with Eqs. (5.15). Alternatively, by Eq. (5.19b),

$$M_x = \begin{vmatrix} x & y & z \\ F_x & F_y & F_z \\ 1 & 0 & 0 \end{vmatrix} = yF_z - zF_y$$

Similarly, Eq. (5.19c) yields the same result.

In Sec. 5.2, the moment of force $\mathbf{F}$ about point O was defined by the equation $\mathbf{M} = \mathbf{r} \times \mathbf{F}$. Then, by Eq. (5.19c), the moment of force $\mathbf{F}$ about L is

$$M_L = \mathbf{n} \cdot (\mathbf{r} \times \mathbf{F}) = \mathbf{n} \cdot \mathbf{M}$$

This result yields the following theorem.

Theorem 5.3

The moment of a force $\mathbf{F}$ about a line L is equal to the projection of the moment vector $\mathbf{M} = \mathbf{r} \times \mathbf{F}$ on L where $\mathbf{M}$ is the moment of force $\mathbf{F}$ about any point on L.

Example 5.9

Moment of a Force about a Line

Problem Statement A force (in kilonewtons) is defined by $\mathbf{F} = 1.00\mathbf{i} + 0.50\mathbf{j} + 2.50\mathbf{k}$. With the meter as the unit of length, the force $\mathbf{F}$ acts at the point Q: (2.00, 2.25, 1.00). A line L, with direction cosines $(n_x, n_y, n_z) = (0.316, -0.316, 0.895)$, passes through the point P: (−1.25, 0.75, 0.50). Determine the moment M_L of $\mathbf{F}$ about L.

Solution Let $\mathbf{r}$ be the vector *from* the reference point P on L *to* point Q. Then, $\mathbf{r} = 3.25\mathbf{i} + 1.50\mathbf{j} + 0.50\mathbf{k}$. Hence, by Eq. (5.18), we obtain the moment M_L of $\mathbf{F}$ about L as

$$\begin{aligned} M_L &= \mathbf{F} \cdot (\mathbf{n} \times \mathbf{r}) \\ &= \begin{vmatrix} 1.00 & 0.50 & 2.50 \\ 0.316 & -0.316 & 0.895 \\ 3.25 & 1.50 & 0.50 \end{vmatrix} = 3.63 \text{ kN·m} \end{aligned}$$

Example 5.10

Moment of a Force about the Base of a Gate

Problem Statement Referring to Fig. E5.8, determine the moment M_{AB} of force $\mathbf{F} = 50\mathbf{i} + 100\mathbf{j} - 30\mathbf{k}$ [N] about the bottom rail of the gate (line AB).

Solution The line AB passes through the points (0, 0.75, 0) and (1.6, 0.75, 1.2) (distances in meters). Therefore, the unit vector $\mathbf{n}$ in the direction of AB is

$$\mathbf{n} = \frac{1.6\mathbf{i} + 1.2\mathbf{k}}{2.0} = 0.8\mathbf{i} + 0.6\mathbf{k} \qquad \text{(a)}$$

We choose the vector $\mathbf{r}_{BC} = 0.75\mathbf{j}$ to locate $\mathbf{F}$ relative to line AB. Then, by Eqs. (5.18) and (a), with the data of Example 5.8,

$$M_{AB} = \mathbf{F} \cdot (\mathbf{n} \times \mathbf{r})$$

$$= \begin{vmatrix} 50 & 100 & -30 \\ 0.8 & 0 & 0.6 \\ 0 & 0.75 & 0 \end{vmatrix}$$

$$= -40.5 \text{ N·m}$$

Alternatively, without using determinants, we can solve the problem by considering the components of $\mathbf{F}$ in the plane of the gate and perpendicular to the gate. For example, the force component $\mathbf{F}_n$ in the plane of the gate and in the direction of $\mathbf{n}$ has magnitude

$$F_n = \mathbf{n} \cdot \mathbf{F}$$

$$= (0.8\mathbf{i} + 0.6\mathbf{k}) \cdot (50\mathbf{i} + 100\mathbf{j} - 30\mathbf{k})$$

$$= 22 \text{ N}$$

Similarly, the force component $\mathbf{F}_y$ in the plane of the gate and in the y direction (perpendicular to $\mathbf{F}_n$) has magnitude

$$F_y = \mathbf{j} \cdot (50\mathbf{i} + 100\mathbf{j} - 30\mathbf{k})$$

$$= 100 \text{ N}$$

Hence, for the force component perpendicular to the gate, we have

$$F_p = \sqrt{F^2 - F_n^2 - F_y^2}$$

$$= \sqrt{13\,400 - 484 - 10\,000}$$

$$= 54 \text{ N}$$

and, as before, the moment about line AB is

$$M_{AB} = -0.75(54) = -40.5 \text{ N·m}$$

CHECKPOINT

1. ***True or False:*** The moment of a force about a point is a vector.
2. ***True or False:*** The moment of a force about an axis is a scalar.
3. ***True or False:*** The moment of a force about a directed line can be represented by the scalar product of vectors.
4. ***True or False:*** The mixed triple product is a vector.
5. Must the vector product always be found before the scalar product when determining a scalar triple product? Explain.

Answers: 1. True; 2. True; 3. False; 4. False; 5. Yes—if the scalar product is found first, the vector product will be impossible to form.

5.4 VECTOR REPRESENTATION OF COUPLES

After studying this section, you should be able to:

- Determine the resultant of a system of couples.

In Sec. 4.3, we defined a couple. In Sec. 4.4, we defined the properties of couples and obtained the resultant of several couples that lie in a plane or in parallel planes. In this section, we extend the study of couples to three dimensions. In particular, we again observe that a couple can be displaced arbitrarily in its plane or to a parallel plane. The moment of a couple is thus independent of any point in its plane, unlike the moment of a force, which depends on the point O about which it is taken (see Sec. 5.2). In Sec. 5.2, we showed that the moment of a force about point O is a vector. In this section, we will show that a couple is a vector as well.

direction of a couple

The *direction of a couple* is the direction perpendicular to its plane. The essential properties of a couple—its magnitude, direction, and sense—may be represented by a double-headed arrow. Figure 5.11 illustrates the vector representation of a couple. The vector **M** that represents the couple is perpendicular to the plane of the couple. The magnitude of vector **M** is $M = aF$. The sense of vector **M** must be related to the sense of the couple. To do this, we apply the right-hand screw rule: The sense of **M** is the sense in which a right-handed screw advances when it is twisted in the sense indicated by the couple. (See Sec. 5.3 and Fig. 5.8 for two interpretations of the right-hand screw rule.)

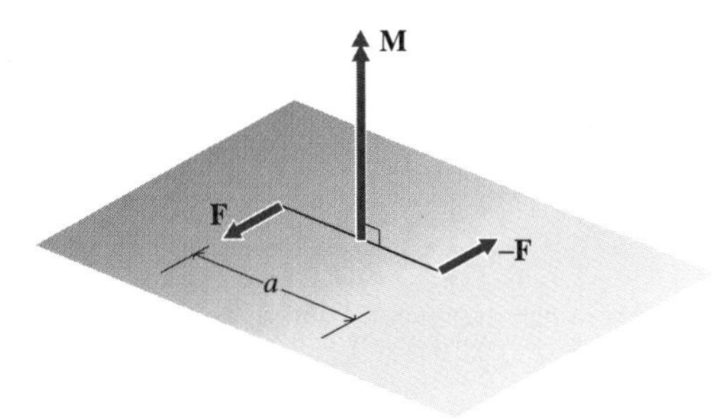

Figure 5.11 Vector representation of a couple.

Key Concept

A couple is a free vector.

Since a couple may be displaced arbitrarily in its plane, vector **M** has no definite point of application. Also, **M** may be displaced axially, since the plane of a couple may be displaced parallel to itself. Accordingly, we can shift vector **M** anywhere, provided that we do not change its direction or its sense. Therefore, a couple is a *free vector* (Sec. 2.4).

free vector

To verify that couples are vector quantities, we must prove that they combine by the parallelogram rule. First, we consider two couples $\mathbf{M}_1$ and $\mathbf{M}_2$ that lie in parallel planes A and B (see Fig. 5.12). We let the couple arms have the same length, since the lengths of the arms are adjustable (see Sec. 4.4). The couples may be moved to a common plane C, where the forces in the couples may be aligned with each other. Since their forces combine algebraically, the couples can be added to obtain a single couple. The force vectors that are

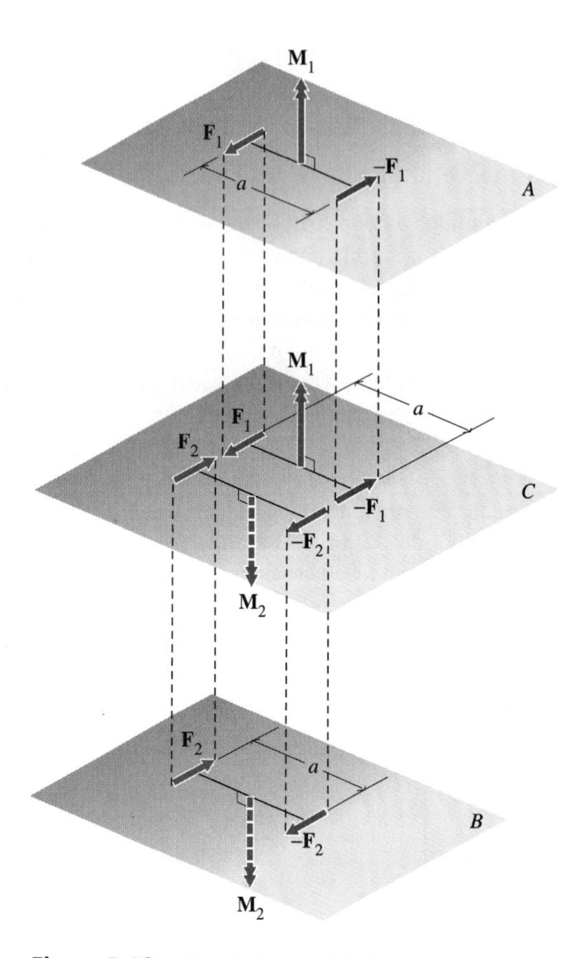

Figure 5.12 Couples in parallel planes A and B.

combined have the same lines of action, so algebraic addition is equivalent to vector addition in this case.

If the two couples $\mathbf{M}_1$ and $\mathbf{M}_2$ do not lie in parallel planes, their planes must intersect. Let's displace the couples in their respective planes so that they have a common arm PQ that lies on the line of intersection of the two planes (Fig. 5.13a). Let the length of the arm be a. Thus, couple $\mathbf{M}_1$ consists of forces $\mathbf{F}_1$ and $-\mathbf{F}_1$ with lever arm a; likewise, couple $\mathbf{M}_2$ consists of forces $\mathbf{F}_2$ and $-\mathbf{F}_2$ with lever arm a. By the parallelogram construction, forces $\mathbf{F}_1$ and $\mathbf{F}_2$ at point P combine to form a resultant $\mathbf{F}$. Likewise, forces $-\mathbf{F}_1$ and $-\mathbf{F}_2$ at point Q provide the resultant $-\mathbf{F}$. Forces $\mathbf{F}$ and $-\mathbf{F}$ at points P and Q, respectively, constitute the resultant couple $\mathbf{M}$.

resultant of noncoplanar couples

To see that the *resultant of the noncoplanar couples* is obtained by vector addition of the couples, view the force parallelogram at point P orthogonally, as indicated by the eye in Fig. 5.13a. The orthogonal view is represented by Fig. 5.13b. By definition, the couples $\mathbf{M}_1$ and $\mathbf{M}_2$ (Fig. 5.13b) are perpendicular to the force vectors $\mathbf{F}_1$ and $\mathbf{F}_2$. Consequently, the angles of the parallelogram $PABC$ are equal, respectively, to the angles of the parallelogram $PXYZ$. Also, by definition, the magnitudes of $\mathbf{M}_1$ and $\mathbf{M}_2$ are aF_1 and aF_2, respectively. Therefore, the parallelograms $PABC$ and $PXYZ$ are similar. Hence, $\mathbf{M}$ is perpendicular to vector $\mathbf{F}$, and its magnitude M is aF. By the parallelogram construction of vectors (Fig. 5.13b), $\mathbf{M} = \mathbf{M}_1 + \mathbf{M}_2$, and it is formed by the forces $\mathbf{F}$ and $-\mathbf{F}$, with couple arm a. This completes the proof that couples are vector quantities.

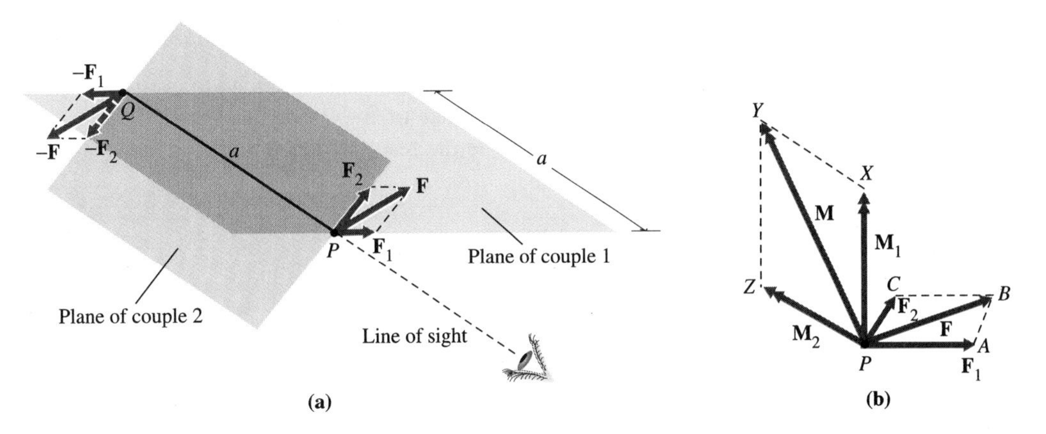

Figure 5.13

Example 5.11

Coplanar Couples and a Jet-Propelled Sled

Problem Statement An experimental jet-propelled sled that weighs 14 kN moves at constant speed along a track (see Fig. E5.11). The resistive force D due to wind and friction is balanced by the thrust T of the jet, and the weight of the sled is balanced by the vertical reaction force N of the track. If the thrust T required to propel the sled at constant speed is 1.4 kN, determine the reaction couple $\mathbf{M}_s$ that the track exerts on the sled.

Solution Since the sled moves at constant velocity along the track, $T = D = 1.4$ kN and $N = W = 14$ kN. The couples due to the forces T and D and the forces N and W are

$$\begin{aligned} \mathbf{M}_{TD} &= (0.30)(1.4)(-\mathbf{k}) = -0.42\mathbf{k} \\ \mathbf{M}_{NW} &= (1.5)(14)(\mathbf{k}) = 21.00\mathbf{k} \end{aligned} \tag{a}$$

Therefore, since the sled is in equilibrium,

$$\mathbf{M}_s + \mathbf{M}_{TD} + \mathbf{M}_{NW} = 0 \tag{b}$$

or, with Eqs. (a) and (b),

$$\mathbf{M}_s = -\mathbf{M}_{TD} - \mathbf{M}_{NW} = -20.58\mathbf{k}\ \text{kN·m}$$

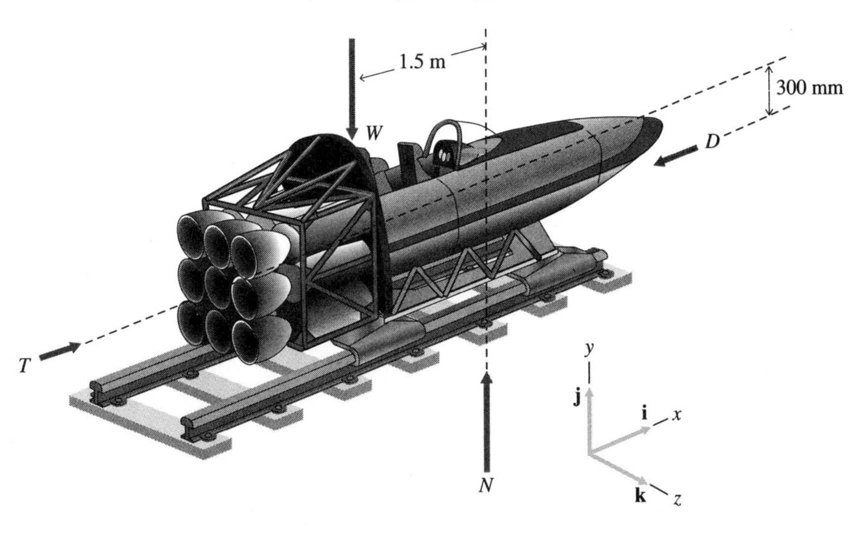

Figure E5.11

Example 5.12 *Resultant of a System of Noncoplanar Couples*

Problem Statement A rigid T-bar is designed to transmit loads to a support structure O. Several forces are applied to the T-bar, as shown in Fig. E5.12a (support structure not shown). Determine the resultant force and resultant couple transmitted to the structure at O.

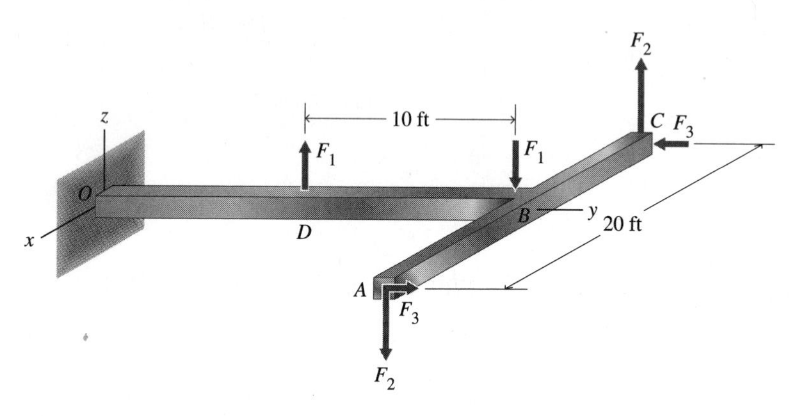

Figure E5.12a

Solution

Note that the forces applied at A, B, C, and D are equivalent to three couples. The magnitudes of the forces are $F_1 = 2$ kips, $F_2 = 3$ kips, $F_3 = 1.5$ kips. Therefore, the magnitudes of the moments of the three couples are $M_1 = (10)(2) = 20$ kip·ft, $M_2 = (20)(3) = 60$ kip·ft, and $M_3 = (20)(1.5) = 30$ kip·ft. The vectors $\mathbf{M}_1$, $\mathbf{M}_2$, and $\mathbf{M}_3$ that represent the three couples have magnitudes M_1, M_2, and M_3, and they are directed along the negative x axis, the positive y axis, and the positive z axis, respectively (Fig. E5.12b). Since the couples are free vectors, they may be moved to a common point—say, the origin O. Then, they can be added vectorially. The resultant vector is $\mathbf{M} = \mathbf{M}_1 + \mathbf{M}_2 + \mathbf{M}_3$ (Fig. E5.12c). The (x, y, z) projections of this vector are $M_x = -20$ kip·ft, $M_y = 60$ kip·ft, $M_z = 30$ kip·ft. By Case 1 of Sec 5.1, the magnitude of the resultant couple is computed as follows:

$$\mathbf{M} \cdot \mathbf{M} = M^2 = (-20\mathbf{i} + 60\mathbf{j} + 30\mathbf{k}) \cdot (-20\mathbf{i} + 60\mathbf{j} + 30\mathbf{k})$$
$$= (-20)^2 + 60^2 + 30^2 = 4900$$

or $M = 70$ kip·ft. In vector form, the resultant couple is

$$\mathbf{M} = -20\mathbf{i} + 60\mathbf{j} + 30\mathbf{k}$$

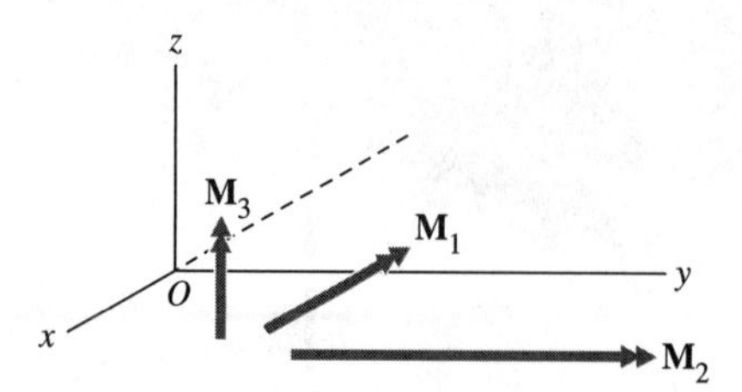

Figure E5.12b

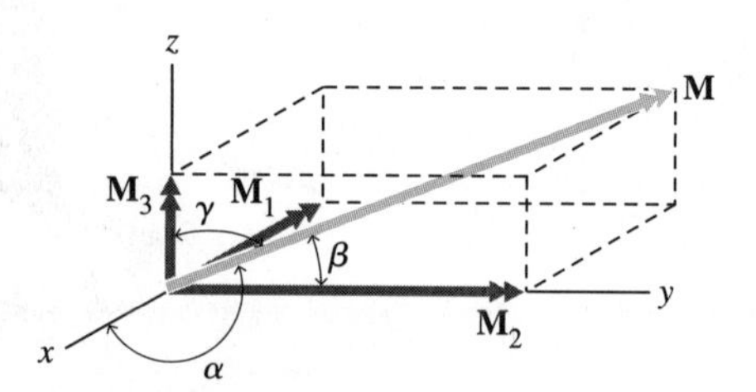

Figure E5.12c

The direction cosines of the resultant couple are, by Eq. (2.1),

$$\cos\alpha = \frac{M_x}{M} = -\frac{20}{70} = -0.28571$$

$$\cos\beta = \frac{M_y}{M} = +\frac{60}{70} = +0.85714$$

$$\cos\gamma = \frac{M_z}{M} = +\frac{30}{70} = +0.42857$$

$$\cos^2\alpha + \cos^2\beta + \cos^2\gamma = 1$$

and the direction angles are $\alpha = 106.60°$, $\beta = 31.00°$, and $\gamma = 64.62°$.

Finally, by Fig. E5.12a, the resultant force **F** is zero; that is,

$$\mathbf{F} = (0)\mathbf{i} + (F_3 - F_3)\mathbf{j} + (F_1 - F_1 + F_2 - F_2)\mathbf{k} = 0$$

CHECKPOINT

1. ***True or False:*** A couple is a slide vector.
2. Can a couple ever be represented by a scalar? Explain.
3. ***True or False:*** The resultant of a system of noncoplanar couples is a single couple.

Answers: 1. False; 2. No—the magnitude of a couple (known as the moment of a couple) is a scalar, but a couple is always a vector; 3. True

5.5 REPRESENTATION OF A COUPLE AS A VECTOR PRODUCT

After studying this section, you should be able to:

- Use the vector product to determine the couple of two forces.

Key Concept A couple **M** may be represented by the vector product $\mathbf{M} = \mathbf{r} \times \mathbf{F}$.

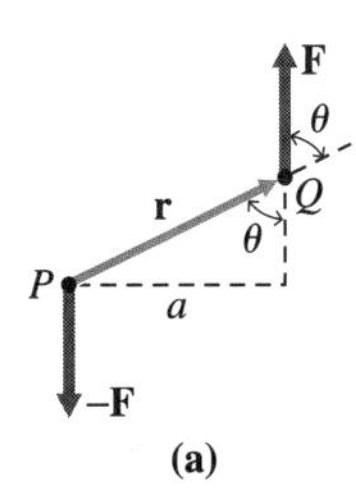

(a)

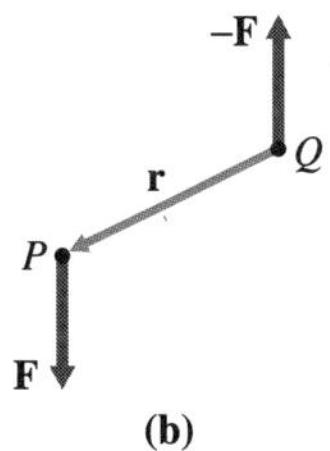

(b)

Figure 5.14

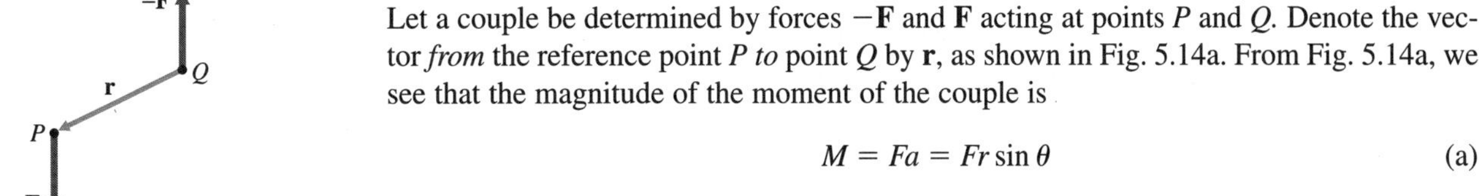

Let a couple be determined by forces $-\mathbf{F}$ and $\mathbf{F}$ acting at points P and Q. Denote the vector *from* the reference point P *to* point Q by **r**, as shown in Fig. 5.14a. From Fig. 5.14a, we see that the magnitude of the moment of the couple is

$$M = Fa = Fr\sin\theta \qquad \text{(a)}$$

It was shown in Sec. 5.4 that the couple **M** is a vector, with magnitude equal to the moment of the couple and with direction perpendicular to the plane of the forces of the couple. The sense of the couple is determined by the right-hand screw rule. Hence, in view of Eq. (a) and the properties of the vector product (Sec. 5.1), the couple **M** represented in Fig. 5.14a is determined by the vector equation

$$\mathbf{M} = \mathbf{r} \times \mathbf{F} \tag{5.20}$$

Note that the vector **r** is drawn from the point of application of force $-\mathbf{F}$ (point P) to the point of application of force **F** (point Q). This convention applies irrespective of the designations of the signs of the force vectors. If Fig. 5.14a is replaced by Fig. 5.14b, the sense of **M**, as determined by Eq. (5.20), is not altered. Note that in Fig. 5.14b the vector **r** is again drawn from the point of application of $-\mathbf{F}$ (now point Q) to the point of application of **F** (now point P). In general, the points P and Q may be located anywhere on the lines of action of the forces.

Since a couple is a vector, it may be resolved into components parallel to the coordinate axes. The (x, y, z) projections of a couple **M** are denoted by M_x, M_y, and M_z. So,

$$\mathbf{M} = M_x\mathbf{i} + M_y\mathbf{j} + M_z\mathbf{k} \tag{b}$$

It is readily shown by Eqs. (5.15) and (5.20) that M_x, M_y, and M_z are identical to the moments of the forces of the couple about the respective coordinate axes. For example, if point P in Fig. 5.14a is taken as the origin of xyz coordinate axes and point Q is located at (x, y, z), $\mathbf{r} = x\mathbf{i} + y\mathbf{j} + z\mathbf{k}$. Then, Eq. (5.20) in determinant form is

$$\mathbf{M} = \begin{vmatrix} \mathbf{i} & \mathbf{j} & \mathbf{k} \\ x & y & z \\ F_x & F_y & F_z \end{vmatrix} \tag{5.21}$$

where F_x, F_y, and F_z are the (x, y, z) projections of **F**. Expansion of Eq. (5.21) yields

$$\mathbf{M} = (yF_z - zF_y)\mathbf{i} + (zF_x - xF_z)\mathbf{j} + (xF_y - yF_x)\mathbf{k} \tag{c}$$

Comparison of Eqs. (b) and (c) yields [see also Eq. (5.15)]

$$M_x = yF_z - zF_y$$
$$M_y = zF_x - xF_z$$
$$M_z = xF_y - yF_x$$

Example 5.13

Vector Product Representation of a Couple

Problem Statement A couple consists of a force **F** that acts at point (2, 9, 7) and a force $-\mathbf{F}$ that acts at point $(-5, 1, -2)$. Force **F** is given by the equation $\mathbf{F} = 100\mathbf{i} + 300\mathbf{j} - 500\mathbf{k}$ (units are pounds and feet).

a. Define the vector **r** from reference point $(-5, 1, -2)$ to point (2, 9, 7).

b. Express the couple formed by the forces **F** and $-\mathbf{F}$ in determinant form.

c. Express the couple in terms of the unit vectors **i**, **j**, and **k**.

Solution

a. The vector **r** is determined as follows:

$$\mathbf{r} = [2 - (-5)]\mathbf{i} + (9 - 1)\mathbf{j} + [7 - (-2)]\mathbf{k} = 7\mathbf{i} + 8\mathbf{j} + 9\mathbf{k}$$

b. By Eq. (5.20), the couple is $\mathbf{M} = \mathbf{r} \times \mathbf{F}$. Hence, by Eq. (5.21),

$$\mathbf{M} = \begin{vmatrix} \mathbf{i} & \mathbf{j} & \mathbf{k} \\ 7 & 8 & 9 \\ 100 & 300 & -500 \end{vmatrix} \tag{a}$$

c. Expansion of Eq. (a) yields

$$\mathbf{M} = -6700\mathbf{i} + 4400\mathbf{j} + 1300\mathbf{k}$$

or

$$M_x = -6700 \text{ lb·ft}$$
$$M_y = +4400 \text{ lb·ft}$$
$$M_z = +1300 \text{ lb·ft}$$

Example 5.14

Composition of Space Couples

Problem Statement A rigid box is subjected to the force system shown in Fig. E5.14.

a. Determine the resultant couple of the force system in terms of F_1, F_2, and F_3.

b. For $F_1 = 10$ kips, $F_2 = 6$ kips, and $F_3 = 8$ kips, determine the magnitude M of the resultant couple.

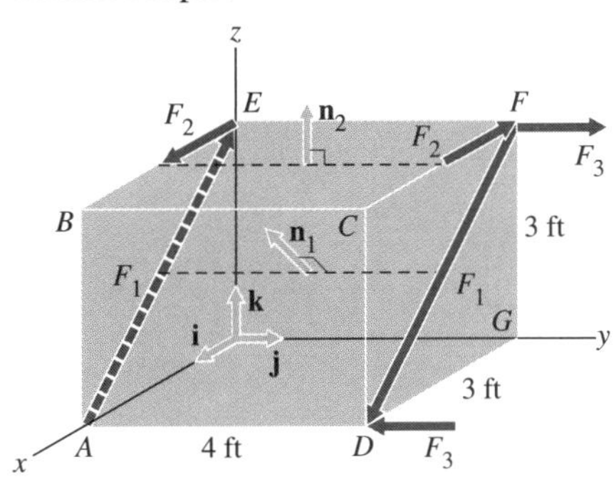

Figure E5.14

Solution

a. In Fig. E5.14, we see that the forces F_1, F_2, and F_3 in pairs form three couples. Therefore, the resultant couple of the force system is the sum of these three couples. Forces F_1 and F_3 lie in the plane $AEFD$ that forms angles of 45° with respect to the x and z axes and is parallel to the y axis. Therefore, the unit normal vector to plane $AEFD$ is

$$\mathbf{n}_1 = \frac{1}{\sqrt{2}}\mathbf{i} + \frac{1}{\sqrt{2}}\mathbf{k}$$

Also, forces F_2 lie in the plane $BEFC$, whose unit normal vector is $\mathbf{n}_2 = \mathbf{k}$. Therefore, the couples due to forces F_1, F_2, and F_3 are

$$\mathbf{M}_1 = -4F_1\mathbf{n}_1 = -\frac{4}{\sqrt{2}}F_1(\mathbf{i} + \mathbf{k})$$

$$\mathbf{M}_2 = +4F_2\mathbf{n}_2 = +4F_2\mathbf{k}$$

$$\mathbf{M}_3 = -3\sqrt{2}F_3\mathbf{n}_1 = -3F_3(\mathbf{i} + \mathbf{k})$$

Hence, the resultant couple is

$$\mathbf{M} = \mathbf{M}_1 + \mathbf{M}_2 + \mathbf{M}_3$$

$$= -\left(\frac{4}{\sqrt{2}}F_1 + 3F_3\right)\mathbf{i} + \left(-\frac{4}{\sqrt{2}}F_1 + 4F_2 - 3F_3\right)\mathbf{k} \qquad \text{(a)}$$

b. For $F_1 = 10$ kips, $F_2 = 6$ kips, and $F_3 = 8$ kips, Eq. (a) yields

$$\mathbf{M} = -52.28\mathbf{i} - 28.28\mathbf{k}$$

$$M = +59.44 \text{ kip·ft}$$

5.6 COMPOSITION OF FORCES THAT ACT ON A RIGID BODY

After studying this section, you should be able to:

- Transform a system of forces that act on a rigid body into a single force acting at an arbitrary point and a single couple.

Key Concept A three-dimensional force system can be resolved into a dynamically equivalent single force and resultant couple.

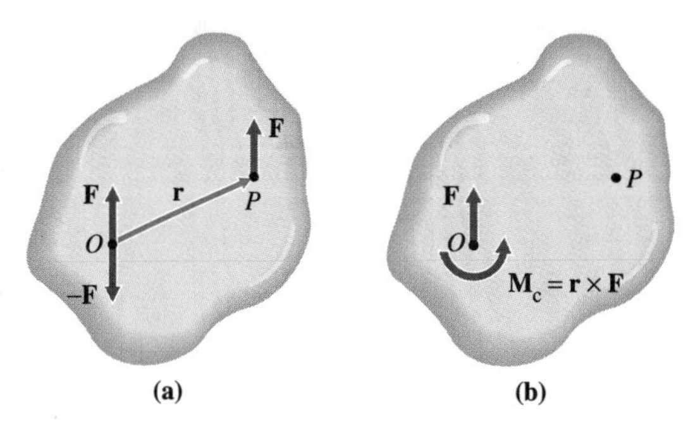

Figure 5.15

compensating couple

Let a force $\mathbf{F}$ act at a point P of a rigid body (see Fig. 5.15a). No dynamical effects are produced by adding self-equilibrating forces $\mathbf{F}$ and $-\mathbf{F}$ at another point O. Force $\mathbf{F}$ at point P and force $-\mathbf{F}$ at point O constitute a couple, $\mathbf{M}_c = \mathbf{r} \times \mathbf{F}$, where $\mathbf{r}$ is the distance vector from reference point O to point P [see Fig. 5.15a and Eq. (5.20)]. Thus, force $\mathbf{F}$ at point P is transferred to point O by this technique, and a *compensating couple* $\mathbf{M}_c$ is introduced (Fig. 5.15b). This transformation is the three-dimensional analogue of a procedure that we employed in the analysis of coplanar forces (Sec. 4.5). It is now apparent that any number of forces, $\mathbf{F}_1, \mathbf{F}_2, \ldots$, that act on a rigid body may be transferred to a common point O, provided that compensating couples $\mathbf{r}_1 \times \mathbf{F}_1, \mathbf{r}_2 \times \mathbf{F}_2, \ldots$ are introduced. The vectors $\mathbf{r}_1$, $\mathbf{r}_2, \ldots$ are vectors *from* the reference point O *to* the original points of application of the forces $\mathbf{F}_1, \mathbf{F}_2, \ldots$, respectively.[1] Forces $\mathbf{F}_1, \mathbf{F}_2, \ldots$, being concurrent at point O, may be added vectorially to provide a single resultant force $\mathbf{F}$:

$$\mathbf{F} = \mathbf{F}_1 + \mathbf{F}_2 + \cdots \tag{5.22}$$

Here, we assume that $\mathbf{F} \neq 0$. The couples may be added vectorially to provide a single compensating couple $\mathbf{M}_c$:

$$\mathbf{M}_c = \mathbf{r}_1 \times \mathbf{F}_1 + \mathbf{r}_2 \times \mathbf{F}_2 + \cdots \tag{5.23}$$

By Eq. (5.23), we see that $\mathbf{M}_c$ depends on the reference point O, since vectors $\mathbf{r}_1, \mathbf{r}_2, \ldots$ emanate from point O. Accordingly, we have the following theorem.

1. Alternatively, the position vectors $\mathbf{r}_1, \mathbf{r}_2, \ldots$ can be taken from the reference point O to any points along the lines of action of $\mathbf{F}_1, \mathbf{F}_2, \ldots$; see Sec. 5.2.

Theorem 5.4

Any number of forces, $\mathbf{F}_1, \mathbf{F}_2, \ldots$, that act on a rigid body may be transformed into an equivalent system, consisting of a single force $\mathbf{F}$ acting at an arbitrary point O and a single compensating couple $\mathbf{M}_c$. The force $\mathbf{F}$ and the compensating couple $\mathbf{M}_c$ are determined by Eqs. (5.22) and (5.23), respectively. Couple $\mathbf{M}_c$ is the moment of forces $\mathbf{F}_1, \mathbf{F}_2, \ldots$ about point O. For $\mathbf{F} \neq 0$, the moment $\mathbf{M}_c$ depends on point O.

FORCES WITH ZERO RESULTANT

We saw earlier in this section that the resultant $\mathbf{F}$ of forces $\mathbf{F}_1, \mathbf{F}_2, \ldots$ that act on a rigid body may be assigned an arbitrary point of action O but that the compensating couple $\mathbf{M}_c$ depends on the location of point O, if $\mathbf{F} \neq 0$. Let's now show that if $\mathbf{F} = 0$, the compensating couple $\mathbf{M}_c$ does not depend on the location of point O.

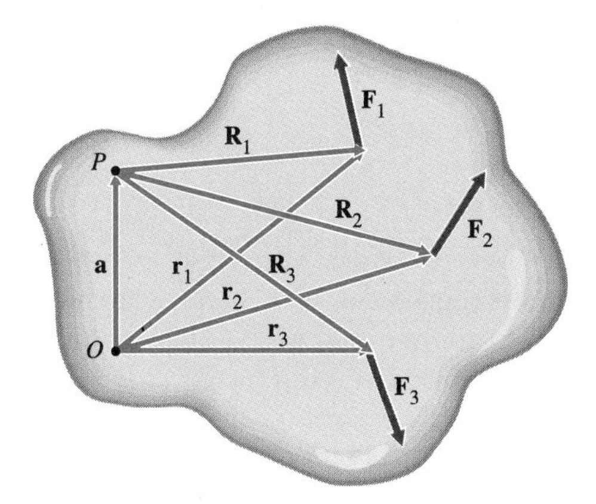

Figure 5.16

To show this, we first let $\mathbf{r}_1, \mathbf{r}_2, \ldots$ be the vectors *from* the arbitrary reference point O *to* the points of application of forces $\mathbf{F}_1, \mathbf{F}_2, \ldots$. Also, we let $\mathbf{R}_1, \mathbf{R}_2, \ldots$ be vectors *from* a second arbitrary reference point P *to* the points of application of the same forces (see Fig. 5.16). We denote the vector from reference point O to point P as $\mathbf{a}$. Then, $\mathbf{r}_1 = \mathbf{a} + \mathbf{R}_1$, $\mathbf{r}_2 = \mathbf{a} + \mathbf{R}_2, \ldots$. Also, the compensating couple of the forces relative to the reference point O is $\mathbf{M}_c = \mathbf{r}_1 \times \mathbf{F}_1 + \mathbf{r}_2 \times \mathbf{F}_2 + \cdots$. Hence,

$$\mathbf{M}_c = (\mathbf{a} + \mathbf{R}_1) \times \mathbf{F}_1 + (\mathbf{a} + \mathbf{R}_2) \times \mathbf{F}_2 + \cdots$$

Collecting terms, we may write

$$\mathbf{M}_c = \mathbf{a} \times (\mathbf{F}_1 + \mathbf{F}_2 + \cdots) + (\mathbf{R}_1 \times \mathbf{F}_1 + \mathbf{R}_2 \times \mathbf{F}_2 + \cdots) \tag{a}$$

However, our initial assumption was that $\mathbf{F} = \mathbf{F}_1 + \mathbf{F}_2 + \cdots = 0$. Also, the moment of the forces about the reference point P is $\mathbf{M}_P = \mathbf{R}_1 \times \mathbf{F}_1 + \mathbf{R}_2 \times \mathbf{F}_2 + \cdots$. Hence, by Eq. (a), $\mathbf{M}_c = \mathbf{M}_P$. Since point P is arbitrary, the compensating couple $\mathbf{M}_c$ is the same for any point. Consequently, we have the following theorem.

Theorem 5.5

If the vector sum of a set of forces is zero, the forces exert the same moment about all points.

Example 5.15 *Composition of Noncoplanar Forces*

Problem Statement A rigid body is subjected to the forces $\mathbf{F}_1 = 2\mathbf{i} + 2\mathbf{j} + 4\mathbf{k}$, $\mathbf{F}_2 = 7\mathbf{i} + 6\mathbf{j} - 4\mathbf{k}$, and $\mathbf{F}_3 = -5\mathbf{i} - 4\mathbf{j} + 7\mathbf{k}$ at the points (8, 10, 4), (−5, 1, 7), and (6, −11, −2), respectively. The unit of force is the kilonewton, and the unit of length is the meter. Determine the magnitude and direction of a single force **F** acting at the point (4, 5, 6) and a compensating couple $\mathbf{M}_c$ that are dynamically equivalent to the given force system.

Solution To solve this problem, we sum the forces $\mathbf{F}_1$, $\mathbf{F}_2$, and $\mathbf{F}_3$ to obtain the resultant force **F**:

$$\begin{aligned}\mathbf{F}_1 &= +2\mathbf{i} + 2\mathbf{j} + 4\mathbf{k}\\ \mathbf{F}_2 &= +7\mathbf{i} + 6\mathbf{j} - 4\mathbf{k}\\ \mathbf{F}_3 &= -5\mathbf{i} - 4\mathbf{j} + 7\mathbf{k}\end{aligned} \tag{a}$$

$$\mathbf{F} = 4\mathbf{i} + 4\mathbf{j} + 7\mathbf{k} \tag{b}$$

Now, by Eq. (b), the magnitude and the direction of **F** are determined as

$$F^2 = 4^2 + 4^2 + 7^2 = 81$$

$$F = 9 \text{ kN}$$

and $\cos\alpha = \frac{4}{9} = 0.4444 \qquad \cos\beta = \frac{4}{9} = 0.4444 \qquad \cos\gamma = \frac{7}{9} = 0.7778$

or, $\alpha = \beta = 63.61°$ and $\gamma = 38.94°$. As a check, note that $\cos^2\alpha + \cos^2\beta + \cos^2\gamma = 1$.

Next, we place **F** at the reference point (4, 5, 6), maintaining its magnitude and its original direction. Recall that couple $\mathbf{M}_c$ depends on the moment reference point. Therefore, to determine the moment $\mathbf{M}_c$, we form distance vectors $\mathbf{r}_1$, $\mathbf{r}_2$, and $\mathbf{r}_3$ *from* the reference point (4, 5, 6) *to* the initial points of application of forces $\mathbf{F}_1$, $\mathbf{F}_2$, and $\mathbf{F}_3$ and sum the moments of these forces to obtain the resultant moment $\mathbf{M}_c$. Thus,

$$\begin{aligned}\mathbf{r}_1 &= +4\mathbf{i} + 5\mathbf{j} - 2\mathbf{k}\\ \mathbf{r}_2 &= -9\mathbf{i} - 4\mathbf{j} + \mathbf{k}\\ \mathbf{r}_3 &= +2\mathbf{i} - 16\mathbf{j} - 8\mathbf{k}\end{aligned} \tag{c}$$

Then, by Eqs. (a), (c), and (5.23),

$$\mathbf{M}_c = (\mathbf{r}_1 \times \mathbf{F}_1) + (\mathbf{r}_2 \times \mathbf{F}_2) + (\mathbf{r}_3 \times \mathbf{F}_3) \tag{d}$$

By Eqs. (d) and (5.9), in determinant notation, we have

$$\mathbf{M}_c = \begin{vmatrix}\mathbf{i} & \mathbf{j} & \mathbf{k}\\ 4 & 5 & -2\\ 2 & 2 & 4\end{vmatrix} + \begin{vmatrix}\mathbf{i} & \mathbf{j} & \mathbf{k}\\ -9 & -4 & 1\\ 7 & 6 & -4\end{vmatrix} + \begin{vmatrix}\mathbf{i} & \mathbf{j} & \mathbf{k}\\ 2 & -16 & -8\\ -5 & -4 & 7\end{vmatrix} \tag{e}$$

Expansion of Eq. (e) yields

$$\mathbf{M}_c = -110\mathbf{i} - 23\mathbf{j} - 116\mathbf{k}$$

Therefore, $M_x = -110 \text{ kN·m} \qquad M_y = -23 \text{ kN·m} \qquad M_z = -116 \text{ kN·m}$ ■

Example 5.16 *Crank Handle*

Problem Statement A crank handle of a machine lies in the *xy* plane and is subjected to the force system shown in Fig. E5.16. Replace the force system by a single resultant **F** acting at point *O* and a single couple **M**.

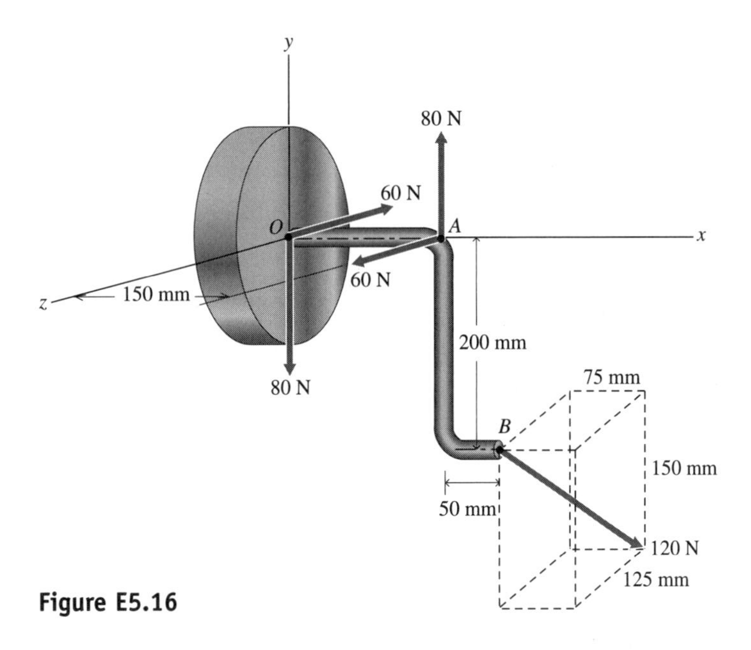

Figure E5.16

Solution The force system consists of two couples that act on the handle segment OA and a force at B. To transfer the force at B to point O, we first determine a unit vector that lies along the line of action of the force. By geometry (see Fig. E5.16), we obtain

$$\begin{aligned}\mathbf{n} &= \frac{75}{209.16}\mathbf{i} - \frac{150}{209.16}\mathbf{j} - \frac{125}{209.16}\mathbf{k} \\ &= 0.359\mathbf{i} - 0.717\mathbf{j} - 0.598\mathbf{k}\end{aligned} \tag{a}$$

Hence, the resultant force is

$$\mathbf{F} = 120\mathbf{n} = 43.03\mathbf{i} - 86.06\mathbf{j} - 71.71\mathbf{k}\ [\mathrm{N}]$$

To calculate the compensating couple associated with the transfer of $\mathbf{F}$ from point B to the reference point O, we need to determine the vector from point O to point B. From Fig. E5.16, we find

$$\mathbf{r}_B = 0.20\mathbf{i} - 0.20\mathbf{j} \quad [\mathrm{m}]$$

Then, the compensating couple for the transfer of $\mathbf{F}$ from B to O is

$$\begin{aligned}\mathbf{M}_O = \mathbf{r}_B \times \mathbf{F} &= \begin{vmatrix} \mathbf{i} & \mathbf{j} & \mathbf{k} \\ 0.20 & -0.20 & 0 \\ 43.03 & -86.06 & -71.71 \end{vmatrix} \\ &= 14.34\mathbf{i} + 14.34\mathbf{j} - 8.61\mathbf{k} \quad [\mathrm{N{\cdot}m}]\end{aligned} \tag{b}$$

To obtain a single couple for the equivalent force system, we must add to $\mathbf{M}_O$ the couples of the 60 N forces and the 80 N forces (Fig. E5.16):

$$\mathbf{M}_{60} = -9.00\mathbf{j} \quad [\mathrm{N{\cdot}m}] \tag{c}$$

$$\mathbf{M}_{80} = +12.00\mathbf{k} \quad [\mathrm{N{\cdot}m}] \tag{d}$$

Hence, adding Eqs. (b) and (c), we obtain the single couple

$$\mathbf{M} = 14.34\mathbf{i} + 5.34\mathbf{j} + 3.39\mathbf{k} \quad [\mathrm{N{\cdot}m}]$$

5.7 RESULTANT AXIS AND THE WRENCH

After studying this section, you should be able to:

- Locate the resultant axis for a system of forces that act on a rigid body.

In Examples 4.8 and 4.9, we reduced a system of coplanar forces to a single force acting at a particular point, so that the resultant couple is zero. In this section, we generalize the concept of resultant axis to three dimensions. However, *in three-dimensional problems, we cannot generally find a resultant axis such that the resultant couple is zero. The best we can do is to locate the resultant force so that the magnitude of the resultant couple is a minimum.*

Several forces that act on a rigid body can be resolved into a dynamically equivalent force **F** and a compensating couple $\mathbf{M}_c$. Once a line of action of **F** has been selected, **F** is represented by a slide vector and the compensating couple is represented by a free vector (see Sec. 5.4). If $F \neq 0$, the compensating couple depends on the arbitrarily selected point O on the line of action of **F**. If $F = 0$, the compensating couple is independent of O; that is, the forces exert the same moment about all points (see Sec. 5.6 and Theorem 5.5).

Consider a situation in which we have resolved a system of forces into a resultant force **F** that acts at a reference point O and a resultant couple **M**. The couple **M** is the sum of all the couples in the system plus the compensating couple required to transfer all forces in the system to point O (see, for instance, Example 5.16). Since **M** is a free vector, its line of action can be moved until it intersects the line of action of **F**. Then, the plane formed by these two lines contains both **F** and **M** (see Fig. 5.17). We proceed by resolving the couple into two components $\mathbf{M}_p$ and $\mathbf{M}_n$, which are parallel and normal to **F**, respectively. In the case of coplanar force systems (Chapter 4), all the forces lie in a plane and the associated couple vectors are normal to that plane and, therefore, normal to the forces. In that case, only $\mathbf{M}_n$ existed.

By the theorem of transmissibility, couple **M** is unchanged as force **F** is displaced along its line of action. Hence, only lateral displacements of force **F** affect couple **M**.

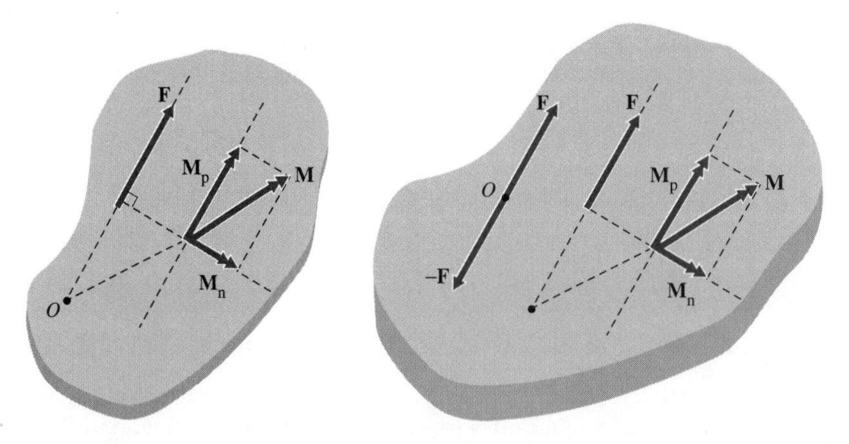

Figure 5.17 **Figure 5.18**

Suppose that vectors **F** and **M** lie in the plane of the page (Fig. 5.18). We may displace the force **F** laterally in the plane of the page by introducing self-equilibrating forces **F** and −**F** at an arbitrary reference point O in the plane (Fig. 5.18), as explained in Sec. 5.6. The

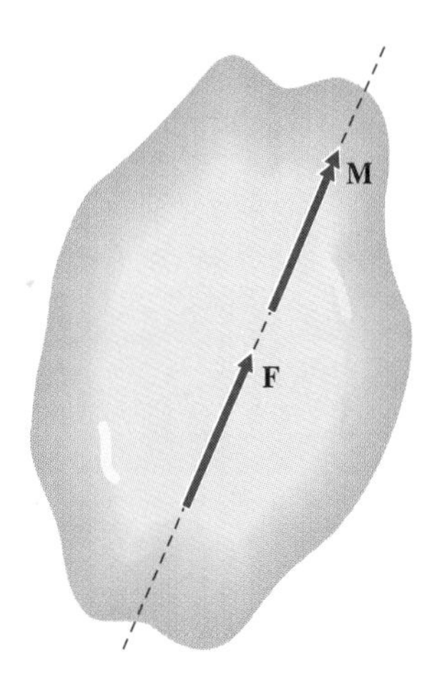

Figure 5.19
The wrench.

compensating couple that is introduced by this transformation is represented by a vector perpendicular to the plane of the page and thus perpendicular to **M**. Combining this couple vectorially with **M**, we obtain a couple of magnitude greater than **M**. Therefore, the magnitude of couple **M** cannot be reduced by a lateral displacement of force **F** in the plane of the page.

If we displace the force **F** laterally in the direction perpendicular to the plane of the page, the compensating couple is represented by a vector in the same direction as the vector $\mathbf{M}_n$. The sense of the displacement can be chosen so that the compensating couple counteracts the couple $\mathbf{M}_n$. (For the case illustrated in Fig. 5.17, force **F** should be moved away from you, into the plane of the page, to counteract $\mathbf{M}_n$.) If the magnitude of the displacement is $r = M_n/F$, couple $\mathbf{M}_n$ is canceled completely, and only couple $\mathbf{M}_p$ remains. Then, the vector $\mathbf{M} = \mathbf{M}_p$ is parallel to **F**, and its magnitude is a minimum, since a displacement of force **F** can never change couple $\mathbf{M}_p$.

Key Concept

When the line of action of the resultant **F** of a force system is located so that **F** and the resultant couple **M** are parallel, **F** is said to lie on the resultant axis of the force system.

From these considerations, we conclude that the line of action of the force **F** may be located so that force **F** and couple **M** are parallel. The magnitude of couple **M** is then a minimum. This is known as the *condition of parallelism*. Under this condition, the force **F** is said to lie on the *resultant axis* of the force system. The parallel force system (couple **M** and force **F**) is called a *wrench*. Furthermore, since **M** is a free vector, we can place it so that its line of action coincides with the line of action of **F** (see Fig. 5.19).

condition of parallelism

resultant axis

wrench

The preceding technique can be used to locate the resultant axis of a force system relative to rectangular coordinates (x, y, z). Alternatively, we can determine the resultant axis by the vector equation $\mathbf{F} \times \mathbf{M} = 0$, which signifies that the resultant axis is parallel to the vector **M**.

Example 5.17

Resultant Axis of a Force System

Problem Statement The force $\mathbf{F} = 10\mathbf{i} - 5\mathbf{j} + 8\mathbf{k}$ acts on a rigid body at the point $(3, 2, -5)$. The body is also subjected to a couple $\mathbf{M}_O = 120\mathbf{i} + 60\mathbf{j} - 80\mathbf{k}$. The units are pounds and feet.

a. Replace this force system by a dynamically equivalent system, consisting of the force **F** at the point $(a, b, 0)$ and a couple $\mathbf{M} = \mathbf{M}_O + \mathbf{M}_c$, where $\mathbf{M}_c$ is the compensating couple required to transfer **F** from point $(3, 2, -5)$ to point $(a, b, 0)$.

b. Write the determinate form of the condition of parallelism for vectors **F** and **M**—namely, $\mathbf{F} \times \mathbf{M} = 0$—to obtain a vector equation relating a and b. Then, determine the direction cosines of the resultant axis.

c. Put the vector equation obtained in part b in scalar form, and solve any two of the three resulting algebraic equations for the constants a and b.

d. Show that the third equation is then satisfied automatically.

Solution **a.** To solve this problem, we first move force **F** to point $(a, b, 0)$. Next, we determine the compensating couple required for this move. To do this, we form the vector **r**, representing the moment arm of force **F** *from* $(a, b, 0)$ *to* $(3, 2, -5)$ (see Sec. 5.6 and Fig. 5.15a). Then, we calculate the compensating couple $\mathbf{M}_c = \mathbf{r} \times \mathbf{F}$. Finally, we determine the required couple $\mathbf{M} = \mathbf{M}_O + \mathbf{M}_c$.

Following the method outlined above, we find $\mathbf{r} = (3 - a)\mathbf{i} + (2 - b)\mathbf{j} + (-5)\mathbf{k}$. Then, by Eq. (5.9),

$$\mathbf{M}_c = \mathbf{r} \times \mathbf{F} = \begin{vmatrix} \mathbf{i} & \mathbf{j} & \mathbf{k} \\ (3-a) & (2-b) & -5 \\ 10 & -5 & 8 \end{vmatrix} \tag{a}$$

$$= -(8b + 9)\mathbf{i} + (8a - 74)\mathbf{j} + (5a + 10b - 35)\mathbf{k}$$

Hence, with $\mathbf{M}_O$ and Eq. (a), we have

$$\mathbf{M} = \mathbf{M}_O + \mathbf{M}_c = -(8b - 111)\mathbf{i} + (8a - 14)\mathbf{j} + (5a + 10b - 115)\mathbf{k} \tag{b}$$

Thus, couple **M**, given by Eq. (b), and force **F** at point $(a, b, 0)$ constitute the dynamically equivalent force system.

b. From Eq. (b), with **F**, we have

$$\mathbf{F} \times \mathbf{M} = \begin{vmatrix} \mathbf{i} & \mathbf{j} & \mathbf{k} \\ 10 & -5 & 8 \\ (111 - 8b) & (8a - 14) & (5a + 10b - 115) \end{vmatrix} = 0 \tag{c}$$

Or, by expansion of Eq. (c), we can write

$$(687 - 89a - 50b)\mathbf{i} + (2038 - 50a - 164b)\mathbf{j} + (415 + 80a - 40b)\mathbf{k} = 0 \tag{d}$$

Equation (d) represents the condition of parallelism for vectors **F** and **M**. Thus, the resultant axis of the force system is parallel to vector **M** and passes through point $(a, b, 0)$. The direction numbers of the resultant axis are given by the (x, y, z) projections of **F**—that is, by $(10, -5, 8)$. Hence, the direction cosines and the direction angles of the resultant axis are

$$\cos\alpha = +\frac{10}{\sqrt{10^2 + 5^2 + 8^2}} = +0.7274; \ \alpha = 43.33°$$

$$\cos\beta = -\frac{5}{\sqrt{10^2 + 5^2 + 8^2}} = -0.3637; \ \beta = 111.33° \tag{e}$$

$$\cos\gamma = +\frac{8}{\sqrt{10^2 + 5^2 + 8^2}} = +0.5819; \ \gamma = 54.41°$$

c. In scalar form, Eq.(d) is

$$89a + 50b = 687 \tag{f}$$

$$50a + 164b = 2038 \tag{g}$$

$$80a - 40b = -415 \tag{h}$$

From Eqs. (f) and (g), we find

$$a = 0.8902 \text{ ft}$$
$$b = 12.155 \text{ ft} \tag{i}$$

Therefore, the resultant axis passes through the point $(0.8902, 12.155, 0)$ in the direction given by α, β, and γ in Eqs. (e).

From Eqs. (b) and (i), the couple **M** is

$$\mathbf{M} = 13.76\mathbf{i} - 6.88\mathbf{j} + 11.00\mathbf{k} \quad [\text{lb·ft}] \tag{j}$$

With Eq. (j), we can verify that the direction cosines of **M** are the same as those of **F** [Eqs. (e)], since **M** and **F** are parallel.

d. Substitution of Eqs. (i) into Eq. (h) verifies, to the order of accuracy of the computation, that Eq. (h) is satisfied. That is,

$$80(0.8902) - 40(12.155) = -414.98 \approx -415$$

Example 5.18

The Wrench

Problem Statement Determine the wrench of the forces that act on the crank of Example 5.16.

Solution In Example 5.16, the force system of Fig. E5.16 was transferred to the point O as

$$\mathbf{F} = 43.03\mathbf{i} - 86.06\mathbf{j} - 71.71\mathbf{k} \quad [\text{N}] \tag{a}$$

and

$$\mathbf{M}_O = 14.34\mathbf{i} + 5.34\mathbf{j} + 3.38\mathbf{k} \quad [\text{N·m}] \tag{b}$$

To determine the wrench for this force system, we follow the procedure of Example 5.17. Thus, we replace the force system given by Eqs. (a) and (b) by a dynamically equivalent system consisting of the force **F** located at the (unknown) point $(a, b, 0)$ in the xy plane and a couple $\mathbf{M} = \mathbf{M}_O + \mathbf{M}_c$, where $\mathbf{M}_c$ is the compensating couple required to transfer **F** from the origin O in Fig. E5.16 to the point $(a, b, 0)$. To determine $\mathbf{M}_c$, we follow the method outlined in Example 5.16 and calculate the vector **r** from the reference point $(a, b, 0)$ to the origin $(0, 0, 0)$, where **F** is applied:

$$\mathbf{r} = -a\mathbf{i} - b\mathbf{j} \tag{c}$$

Then, with Eqs. (a) and (c), the compensating couple is

$$\mathbf{M}_c = \mathbf{r} \times \mathbf{F} = \begin{vmatrix} \mathbf{i} & \mathbf{j} & \mathbf{k} \\ -a & -b & 0 \\ 43.03 & -86.06 & -71.71 \end{vmatrix} \tag{d}$$

$$= (71.71b)\mathbf{i} - (71.71a)\mathbf{j} + (86.06a + 43.03b)\mathbf{k} \quad [\text{N·m}]$$

Hence, by Eqs. (b) and (d) and since $\mathbf{M} = \mathbf{M}_O + \mathbf{M}_c$, we obtain

$$\mathbf{M} = (14.34 + 71.71b)\mathbf{i} + (5.34 - 71.71a)\mathbf{j} + (3.38 + 86.06a + 43.03b)\mathbf{k} \quad [\text{N·m}] \tag{e}$$

Couple **M**, given by Eq. (e), and force **F**, given by Eq. (a), constitute the dynamically equivalent force system.

To determine the constants a and b, we require that **F** and **M** be parallel; that is,

$$\mathbf{F} \times \mathbf{M} = \begin{vmatrix} \mathbf{i} & \mathbf{j} & \mathbf{k} \\ 43.03 & -86.06 & -71.71 \\ (14.34 + 71.71b) & (5.34 - 71.71a) & (3.38 + 86.06a + 43.03b) \end{vmatrix} = 0 \tag{f}$$

Expansion of Eq. (f) yields

$$\begin{aligned} &(92.05 - 12\,549a - 3703b)\mathbf{i} \\ &+ (-1174 - 3703a - 6994b)\mathbf{j} \\ &+ (1464 - 3086a + 6171b)\mathbf{k} = 0 \end{aligned} \tag{g}$$

Equations (f) and (g) represent the condition of parallelism for vectors **F** and **M**. This condition locates the resultant axis of the wrench. The resultant axis of the force system is parallel to vectors **F** and **M** and passes through point $(a, b, 0)$. Since force **F** maintains the same direction as it is transferred from point (0, 0, 0) to point $(a, b, 0)$, its direction is the same as that of the unit vector [see Eq. (a) from Example 5.16]

$$\mathbf{n} = 0.358\mathbf{i} - 0.717\mathbf{j} - 0.598\mathbf{k} \tag{h}$$

To determine the values of a and b, we write Eq. (g) in scalar form as

$$\begin{aligned} 12\,549a + 3703b &= 92.05 \\ 3703a + 6994b &= -1174 \\ 3086a - 6171b &= 1464 \end{aligned} \tag{i}$$

Using any two of Eqs. (i), we solve for a and b and obtain

$$\begin{aligned} a &= +0.0674 \text{ m} \\ b &= -0.2037 \text{ m} \end{aligned} \tag{j}$$

Consequently, the wrench consists of force **F** [Eq. (a)] in the direction $\mathbf{n} = 0.358\mathbf{i} - 0.717\mathbf{j} - 0.598\mathbf{k}$ at the point (0.0674, −0.2037, 0) and, by Eqs. (e) and (j), couple $\mathbf{M} = -0.267\mathbf{i} + 0.507\mathbf{j} - 2.966\mathbf{k}$ [N·m]. ■

CHECKPOINT 

1. What is the resultant of a three-dimensional system of forces called?
2. ***True or False:*** The resultant axis of a force system is the axis of the resultant force obtained by adding the forces of the system vectorially.
3. ***True or False:*** If the resultant force of a system of forces that acts on a rigid body is not zero, the resultant may be moved to any point in the body. There it is dynamically equivalent to the system of forces.
4. ***True or False:*** If the resultant force of a system of forces that acts on a rigid body is zero, the sum of the moments of the forces is the same about every point in the body.
5. ***True or False:*** Any force system that acts on a rigid body can be replaced by a dynamically equivalent system consisting of a single force and single moment.

Answers: 1. A wrench; 2. False; 3. False; 4. True; 5. True

5.8 THE FUNDAMENTAL PRINCIPLE OF EQUILIBRIUM OF A RIGID BODY

After studying this section, you should be able to:

- Draw an accurate free-body diagram of a rigid body subjected to a system of forces, couples, and support conditions.
- Write and solve a system of equilibrium equations for a rigid body subjected to a system of forces and couples.

Key Concept

A motionless rigid body is in equilibrium if, and only if, the resultant force and the resultant couple that act on the body are both zero.

equilibrium of a rigid body

Recall that if the resultant force is zero, the resultant couple is independent of the reference point O (Theorem 5.5). Consequently, we can state the fundamental principle of *equilibrium of a rigid body* in the form of the following theorem.

Theorem 5.6

A motionless rigid body is in equilibrium if, and only if, the vector sum of all the external forces is zero and the moment of the external forces about *any* point is zero.

The proof of this theorem must be deferred until we deal with momentum principles (see Secs. 16.2 and 16.6).

EQUILIBRIUM CONDITIONS REFERRED TO RECTANGULAR COORDINATES

Theorem 5.6 is expressed by the equations

$$\mathbf{F} = 0, \qquad \mathbf{M} = 0 \tag{5.24}$$

where $\mathbf{F}$ is the vector sum of all the external forces that act on a rigid body and $\mathbf{M}$ is the moment of those forces about any point. To apply this principle, we may work with the projections of $\mathbf{F}$ and $\mathbf{M}$ on rectangular coordinate xyz axes. Then, Eqs. (5.24) are equivalent to the equations

$$F_x = 0 \qquad F_y = 0 \qquad F_z = 0$$

$$M_x = 0 \qquad M_y = 0 \qquad M_z = 0$$

where F_x, F_y, F_z and M_x, M_y, M_z are the projections of the vectors $\mathbf{F}$ and $\mathbf{M}$ on the xyz coordinate axes. The force projections F_x, F_y, and F_z are determined by the equations

$$F_x = F_{1x} + F_{2x} + \cdots$$

$$F_y = F_{1y} + F_{2y} + \cdots$$

$$F_z = F_{1z} + F_{2z} + \cdots$$

where (F_{1x}, F_{1y}, F_{1z}), (F_{2x}, F_{2y}, F_{2z}), . . . are the (x, y, z) projections of the various external forces that act on the body.

If F_x, F_y, and F_z vanish, the external forces produce the same moment about *all* points (Sec. 5.6). Consequently, in applying the condition $\mathbf{M} = 0$, we may take moments about the origin. In other words, the moment projections M_x, M_y, and M_z are identical to the sum of the moments of the original forces about the respective axes. The moments of the original forces about the coordinate axes are determined by Eqs. (5.15). Therefore, designating the ith force by (F_{ix}, F_{iy}, F_{iz}) and its point of application by (x_i, y_i, z_i), we have the following six equilibrium equations:

$$\sum F_{ix} = 0$$
$$\sum F_{iy} = 0 \tag{5.25}$$
$$\sum F_{iz} = 0$$

$$\sum M_{ix} = \sum (y_i F_{iz} - z_i F_{iy}) = 0$$
$$\sum M_{iy} = \sum (z_i F_{ix} - x_i F_{iz}) = 0 \tag{5.26}$$
$$\sum M_{iz} = \sum (x_i F_{iy} - y_i F_{ix}) = 0$$

Equations (5.25) and (5.26) are generalizations of the equilibrium criteria that were derived for coplanar forces (Sec. 4.6). By the reasoning of Sec. 4.6, they are applicable not only to isolated rigid bodies but also to systems of rigid bodies and to deformable bodies.

THREE-DIMENSIONAL SUPPORTS

In order to construct an accurate three-dimensional free-body diagram, you must represent the support reactions accurately. In practice, three-dimensional supports may be quite complex, and they may exert distributed forces (Chapter 8) rather than just point forces on a body. However, the major effects of the supports on the equilibrium of a body can be modeled by replacing the supports by point forces and moments. Then, the model of the most general support reaction consists of a force **F** and a moment **M**, each with unknown magnitude and direction. Alternatively, the force and the moment may be represented by three force components $\mathbf{F}_x$, $\mathbf{F}_y$, and $\mathbf{F}_z$ and three moment components $\mathbf{M}_x$, $\mathbf{M}_y$, and $\mathbf{M}_z$, relative to xyz axes.

Several types of supports that we will use in this book are shown in Table 5.1, along with their idealized support reactions. Often three-dimensional problems use both two-

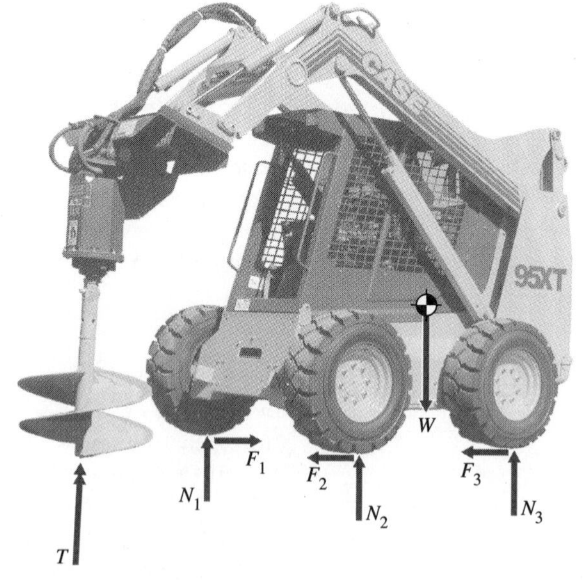

The forces that act on the skid-steer loader and its auger include the weight W of the unit, the normal forces N_i and friction forces F_i on the tires, and the torque T on the auger.

dimensional and three-dimensional support symbols. Therefore, you should review the two-dimensional supports shown in Fig. 3.9. Frequently, the problem statement of an example or homework problem describes the features of a particular support in order to avoid ambiguity in interpretation of the graphic symbol.

Note that, since a three-dimensional support can have up to three independent components of force and three independent components of moment, the support conditions represented in Table 5.1 are just a small selection of the possible combinations that exist in three dimensions.

PROBLEM-SOLVING TECHNIQUE

Equilibrium of Statically Determinate Rigid-Body Systems

As has been mentioned several times previously, drawing a free-body diagram is one of the most important steps (if not *the* most important step) in obtaining solutions in mechanics (see Sec. 3.2 for several reasons why free-body diagrams are so important). In the phrase "free-body diagram," the word "body" may refer to an entire structure or machine, a portion of a structure or machine, a single member or part, a particle, and so on. Also, the word "free" refers to the fact that the body has been freed (isolated) from its surrounding parts, with the effects of these parts on the body replaced by forces and couples. Thus, the fundamental concept of a free-body diagram is to isolate a significant part of a system (solid body, particle, cable, pulley, or any combination of these things) and show the forces and couples that act on the part. In brief,

> **In a free-body diagram, you must isolate the significant part of the system.**

To paraphrase W. F. Osgood (1937), the person who first conceived this idea deserves a *monumentum aere* (Latin for "bronze monument").

As a general problem-solving technique for the solution of equilibrium problems involving rigid-body systems, use the following steps. You should also review the Problem-Solving Technique for particles at the end of Sec. 3.2.

1. Construct free-body diagrams of the significant parts of the system. Specifically, construct free-body diagrams of parts that contain the desired quantities to be determined or that contain quantities that lead to the solution for the desired quantities. This step requires that you first plan your approach to solving the problem.
2. Assign a coordinate system to each free-body diagram so that you can sum forces and moments efficiently.
3. Apply the equilibrium conditions to each free body, relative to the coordinate systems assigned in Step 2. In general, the equilibrium conditions yield a set of simultaneous linear algebraic equations that you must solve for the unknown quantities.
4. Solve the set of simultaneous equations obtained in Step 3 for the unknowns.
5. Substitute your answers back into the equilibrium equations to verify that the equations are satisfied.
6. As a check, it is a good idea to solve the problem in two different ways. This check gives you some confidence that your solution is correct.

Table 5.1
Three-dimensional supports

DESCRIPTION	SUPPORT	SUPPORT REACTIONS
Ball-and-socket joint: reaction force has components in each of the coordinate directions; no moment is resisted.	Member, Ball, Smooth socket	y, x, z, F_x, F_y, F_z
Two smooth (frictionless) surfaces in contact: reaction force is normal to the plane of contact.	Smooth surfaces	N
Member resting on a smooth (frictionless) surface: reaction force is normal to the surface.	Member, Smooth surface	y, x, z, F_y
Member resting on a rough surface (with friction): reaction force has components in each of the coordinate directions; no moment is resisted.	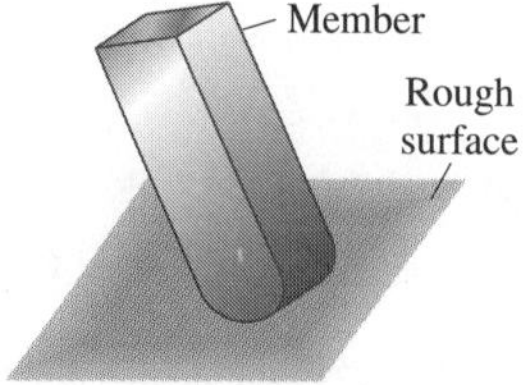	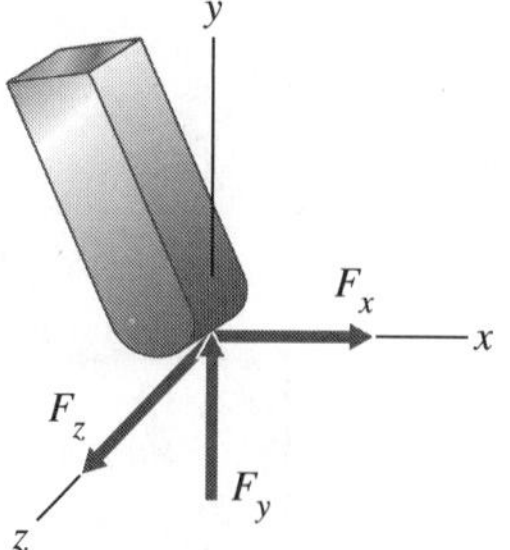

DESCRIPTION	SUPPORT	SUPPORT REACTIONS
Member built into a rigid base: reaction force has components in each coordinate direction; reaction moment also has three components.		
Member rigidly attached to a cylinder that slides on a smooth shaft: reaction force and moment both have two rectangular components; force along the shaft and moment about the axis of the shaft cannot be resisted.		
Member attached to a three-dimensional pin: reaction force has three rectangular components; no moment can be resisted.		
Smooth member attached to a thrust bearing: reaction force has three rectangular components; reaction moment has two components; moment about the axis of the member cannot be resisted.		

Example 5.19 Reactions in Three Dimensions

Problem Statement A rigid block of weight W is supported by a pin at A and by rollers at B, C, and D (see Fig. E5.19a). The weight vector lies along a vertical line G that passes through the center of gravity, located as shown. The three-dimensional pin at A resists forces in the three coordinate directions. The roller B resists forces in the z direction. Likewise, the rollers at C and D resist forces in the x and y directions, respectively. Determine the (x, y, z) projections of the reactions of the supports. Units are newtons and meters.

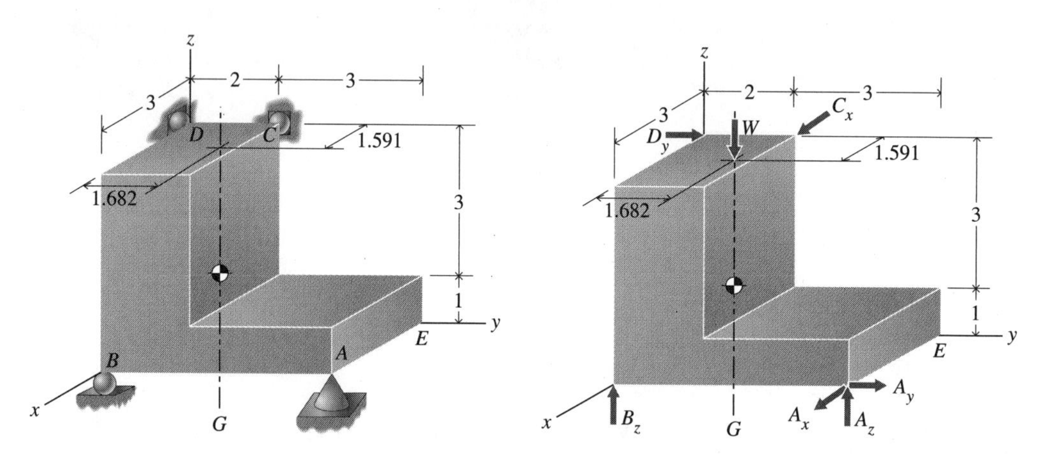

Figure E5.19a **Figure E5.19b**

Solution

The free-body diagram of the block is shown in Fig. E5.19b. Since the reaction force **B** is parallel to the z axis, $B_x = B_y = 0$. Likewise, $C_y = C_z = D_x = D_z = 0$. There are only six unknown force projections—namely, A_x, A_y, A_z, B_z, C_x, and D_y. The six equations of equilibrium for a rigid body are sufficient to determine these six unknowns. The equations of equilibrium of force for the block are [Eqs. (5.25)]

$$\sum F_{ix} = A_x + C_x = 0 \tag{a}$$

$$\sum F_{iy} = A_y + D_y = 0 \tag{b}$$

$$\sum F_{iz} = A_z + B_z - W = 0 \tag{c}$$

By Eqs. (5.26), with reference to Fig. E5.19b, the equations of equilibrium of moments about the coordinate axes are

$$\sum M_{ix} = 5A_z - 4D_y - 1.682W = 0 \tag{d}$$

$$\sum M_{iy} = 4C_x + 1.591W - 3A_z - 3B_z = 0 \tag{e}$$

$$\sum M_{iz} = 3A_y - 5A_x - 2C_x = 0 \tag{f}$$

We can solve this set of six simultaneous equations as follows: First, we multiply Eq. (c) by 3, and add the result to Eq. (e). Then, we solve for C_x to obtain $C_x = 0.3523W$. Next, we substitute this value into Eq. (a) to obtain $A_x = -0.3523W$. We continue to substitute known quantities into the remaining equations, in the order Eq. (f), (b), (d), and (c), to solve for the remaining support reactions. In this manner, we obtain the complete solution:

$$
\begin{aligned}
A_x &= -0.3523W \\
A_y &= -0.3523W \\
A_z &= 0.6182W \\
B_z &= 0.3818W \\
C_x &= 0.3523W \\
D_y &= 0.3523W
\end{aligned}
\tag{g}
$$

Substitution of Eqs. (g) into the equilibrium equations [Eqs. (a) through (f)] will verify that the equilibrium equations are satisfied.

Information Item It is wise to perform this type of check when you solve a set of simultaneous equations. Alternatively, you could solve the set of simultaneous equations using a computer-based solver or an advanced scientific calculator. As yet another approach, you could solve this problem by taking moments about line *AC* (Fig. E5.19b) and determining B_z directly. Taking moments about line *AB* would allow you to solve for C_x directly, and taking moments about line *AE* would allow you to solve for D_y. Finally, by summing forces in the (x, y, z) directions, you could solve for A_x, A_y, and A_z.

Example 5.20

Forces in the Struts of an Airplane Engine Mount

Problem Statement The cylindrical body in Fig. E5.20a represents an airplane engine. Its weight is $W = 18$ kN, and its center of gravity is located at the point shown. The engine is supported by pin-connected struts, as shown. The engine and its support struts are symmetric with respect to the *xz* plane. Determine the forces in the struts.

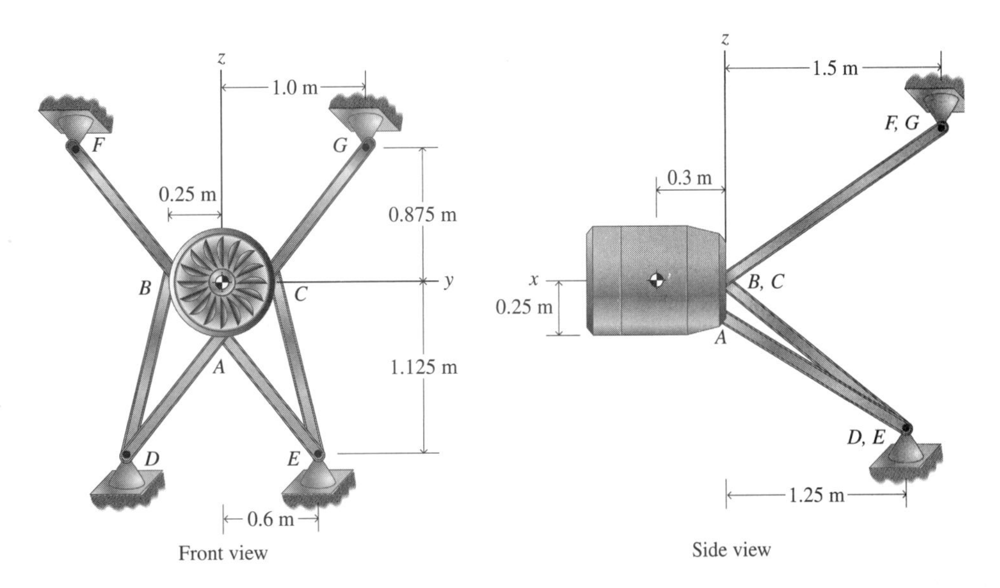

Figure E5.20a

Solution Front and side views of the free-body diagram of the engine are shown in Fig. E5.20b. We assume that all the struts are in tension.

We first determine the direction cosines of the struts. We therefore require the lengths of the struts. For example, consider strut AE. The coordinates of points E and A are $(x_E, y_E, z_E) = (-1.25, 0.60, -1.125)$ and $(x_A, y_A, z_A) = (0, 0, -0.25)$. Hence, the length of member AE [see Eq. (2.12)] is

$$L_{AE} = \sqrt{(x_E - x_A)^2 + (y_E - y_A)^2 + (z_E - z_A)^2}$$
$$= \sqrt{(-1.25)^2 + (0.60)^2 + (-0.875)^2} = 1.640 \text{ m}$$

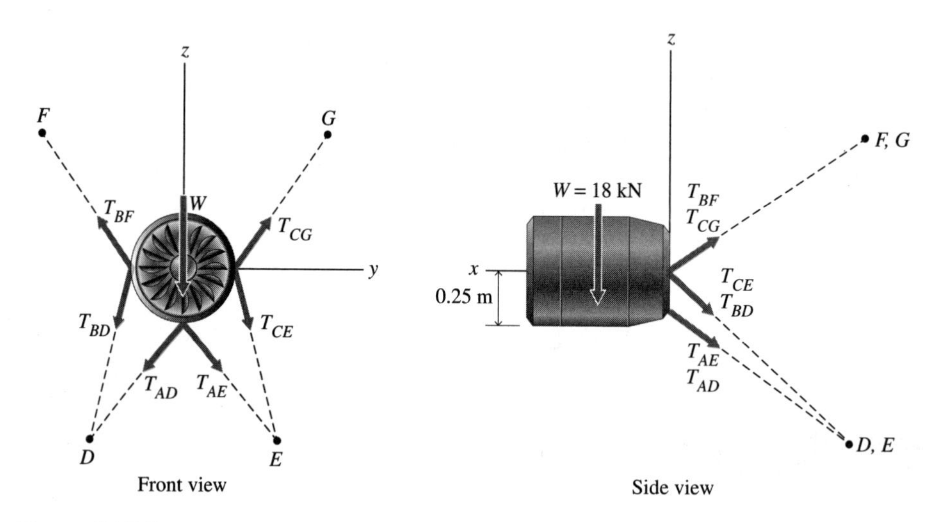

Figure E5.20b

Therefore, the direction cosines of member AE [see Eq. (2.13)] are

$$\cos\alpha_{AE} = \frac{x_E - x_A}{L_{AE}} = -0.7622$$

$$\cos\beta_{AE} = \frac{y_E - y_A}{L_{AE}} = 0.3659$$

$$\cos\gamma_{AE} = \frac{z_E - z_A}{L_{AE}} = -0.5335$$

$$\cos^2\alpha_{AE} + \cos^2\beta_{AE} + \cos^2\gamma_{AE} = 1$$

The lengths and direction cosines of the other members may be determined in a similar manner. The results are given in Table E5.20.

Table E5.20

MEMBER	LENGTH (m)	$\cos\alpha$	$\cos\beta$	$\cos\gamma$
AE	1.640	−0.7622	0.3659	−0.5335
CE	1.718	−0.7276	0.2037	−0.6548
CG	1.892	−0.7930	0.3965	0.4626

The forces in the members are denoted by T_{AE}, T_{CE}, T_{CG}, . . . (taken to be tension); see Fig. E5.20b. Note that $T_{AE} = T_{AD}$, $T_{CE} = T_{BD}$, and $T_{CG} = T_{BF}$, by symmetry. For equilibrium, the resultant force and the resultant couple that act on the engine must each be zero [see Eqs. (5.25) and (5.26)]. Consequently, for equilibrium of forces in the x direction, we have

$$\sum F_x = 2T_{AE}\cos\alpha_{AE} + 2T_{CE}\cos\alpha_{CE} + 2T_{CG}\cos\alpha_{CG} = 0 \quad \text{(a)}$$

By symmetry, the forces in the y direction cancel and $\Sigma F_y = 0$ is satisfied identically. For equilibrium of forces in the z direction, we have

$$\sum F_z = 2T_{AE}\cos\gamma_{AE} + 2T_{CE}\cos\gamma_{CE} + 2T_{CG}\cos\gamma_{CG} - W = 0 \quad \text{(b)}$$

By symmetry, the equilibrium equations for moments of the forces about the x and z axes are identically zero and yield no new information. Since $T_{AE}\cos\alpha_{AE}$ is the x projection of T_{AE}, the condition of equilibrium for moments about the y axis is

$$\sum M_y = 0.3W - 0.25(2T_{AE}\cos\alpha_{AE}) = 0 \quad \text{(c)}$$

Equation (c) yields $T_{AE} = -14.170$ kN. Hence, struts AE and AD are in compression. Equations (a) and (b) now yield

$$0.7276T_{CE} + 0.7930T_{CG} = 10.80$$

$$-0.6548T_{CE} + 0.4626T_{CG} = 1.440$$

Solution of these equations gives $T_{CE} = 4.50$ kN and $T_{CG} = 9.49$ kN. Thus, struts CE, BD, CG, and BF are in tension. ■

Example 5.21

Auto Engine Lifted by a Crane

Problem Statement An auto engine that weighs 8 kN is lifted by a crane (see Fig. E5.21a). The crane is pinned at O and tied down with cables CF and CG. The weights of the crane and the cables are small compared to that of the auto engine and, as a first approximation, can be neglected. The boom ADE lies in the xy plane.

a. Determine the tensions in cables CF, CG, and DB.

b. Determine the forces at A that act on the boom.

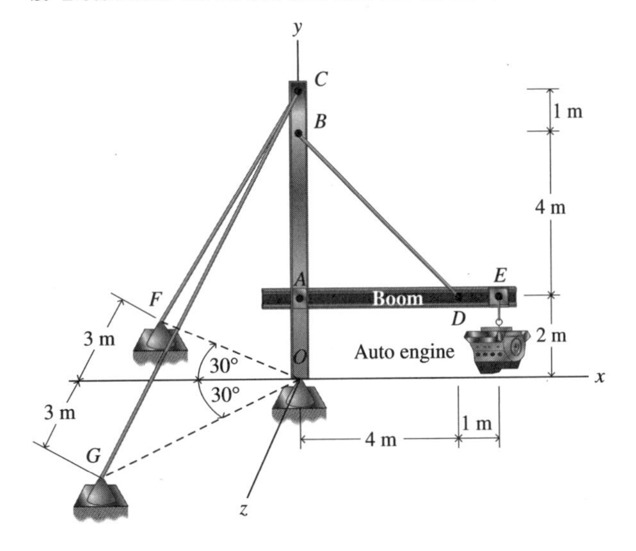

Figure E5.21a

Solution **a.** First, we determine the direction cosines of the cables. To do this, we require the lengths of the cables. Consider cable *CF*. The coordinates of points *C* and *F* are $(x_C, y_C, z_C) = (0, 7, 0)$ and $(x_F, y_F, z_F) = (-5.196, 0, -3)$. Therefore, the length L_{CF} of cable *CF* (and, by symmetry, of cable *CG*) is

$$L_{CF} = \sqrt{(x_F - x_C)^2 + (y_F - y_C)^2 + (z_F - z_C)^2}$$

$$= \sqrt{(-5.196)^2 + (-7)^2 + (-3)^2}$$

$$= 9.219 \text{ m}$$

So, the direction cosines of cable *CF* are

$$\cos \alpha_{CF} = \frac{x_C - x_F}{L_{CF}} = +0.5636$$

$$\cos \beta_{CF} = \frac{y_C - y_F}{L_{CF}} = +0.7593$$

$$\cos \gamma_{CF} = \frac{z_C - z_F}{L_{CF}} = +0.3254$$

$$\cos^2 \alpha_{CF} + \cos^2 \beta_{CF} + \cos^2 \gamma_{CF} = 1$$

Similarly, the direction cosines of cable *CG* are

$$\cos \alpha_{CG} = \frac{x_C - x_G}{L_{CG}} = +0.5636$$

$$\cos \beta_{CG} = \frac{y_C - y_G}{L_{CG}} = +0.7593$$

$$\cos \gamma_{CG} = \frac{z_C - z_G}{L_{CG}} = -0.3254$$

$$\cos^2 \alpha_{CG} + \cos^2 \beta_{CG} + \cos^2 \gamma_{CG} = 1$$

Since the angle between boom *ADE* and cable *DB* is 45° and the boom lies in the *xy* plane, the direction cosines of cable *DB* are

$$\cos \alpha_{DB} = +0.7071$$

$$\cos \beta_{DB} = -0.7071$$

$$\cos \gamma_{DB} = 0$$

$$\cos^2 \alpha_{DB} + \cos^2 \beta_{DB} + \cos^2 \gamma_{DB} = 1$$

The tensions in the cables are denoted by T_{DB}, T_{CF}, and T_{CG}. Also, by symmetry, $T_{CF} = T_{CG}$. To determine the tension in cable *DB*, we construct a free-body diagram of the boom (Fig. E5.21b). For overall equilibrium of the boom, both the sum of the forces and the sum of the moments that act on the boom must be zero. Therefore, from Fig. E5.21b, we determine T_{DB} by

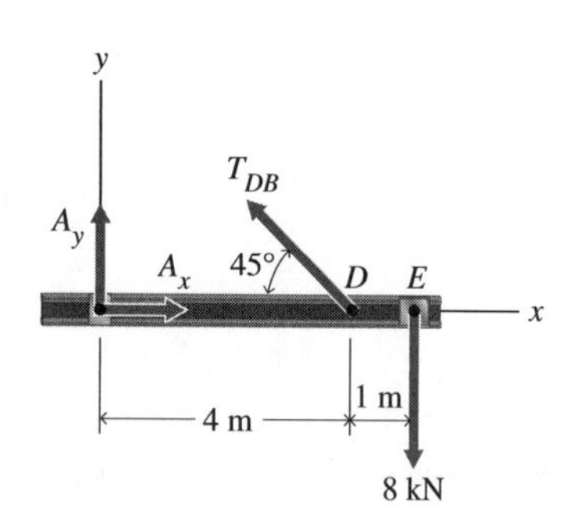

Figure E5.21b

taking moments about point A. Thus, since $T_{DB} \sin 45°$ is the x projection of T_{DB},

$$\sum M_A = (4)(T_{DB} \sin 45°) - (5)(8) = 0$$

$$T_{DB} = 14.14 \text{ kN}$$

To determine the tensions in cables CF and CG, we construct a free-body diagram of the crane (Fig. E5.21c) and take moments about point O. Then, recalling that $T_{CF} = T_{CG}$ and noting that $\cos \alpha_{CF} = \cos \alpha_{CG}$, we have (taking moments about the positive z axis as positive and noting that $T_{CF} \cos \alpha_{CF}$ is the x projection of T_{CF})

$$\sum M_O = (7)(2T_{CF} \cos \alpha_{CF}) - (5)(8) = 0 \qquad \text{(a)}$$

With the value of $\cos \alpha_{CF}$ given above, Eq. (a) yields T_{CF} $(= T_{CG}) = 5.07$ kN. In summary,

$$T_{CF} = 5.07 \text{ kN}$$

$$T_{CG} = 5.07 \text{ kN}$$

$$T_{DB} = 14.14 \text{ kN}$$

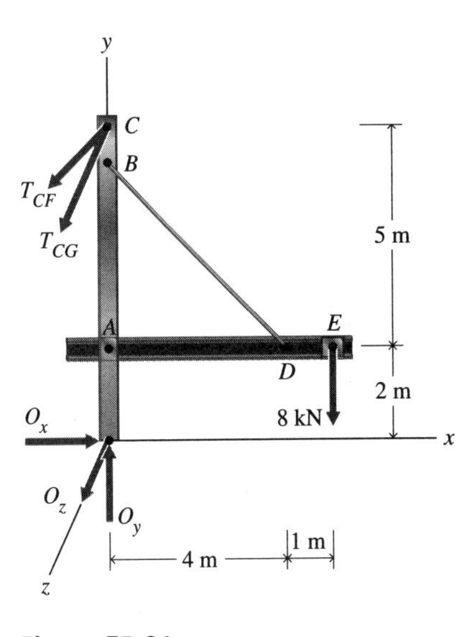

Figure E5.21c

b. To determine the forces at A, we return to the free-body diagram of the boom (Fig. E5.21b). From Fig. E5.21b, summing forces in the x and y directions, we obtain

$$\begin{aligned} \sum F_x &= A_x - T_{DB} \cos 45° = 0 \\ \sum F_y &= A_y + T_{DB} \sin 45° - 8 = 0 \end{aligned} \qquad \text{(b)}$$

With $T_{DB} = 14.14$ kN, the solution of Eqs. (b) is

$$A_x = +10.00 \text{ kN}$$

$$A_y = -2.00 \text{ kN}$$

Information Item To solve for the tensions T_{CF} and T_{CG}, we used only $\cos \alpha_{CF}$ and symmetry with respect to the xy plane. If the forces O_x, O_y, and O_z that act at point O in Fig. E5.21c were required, they could be obtained by the conditions $\Sigma F_x = 0$, $\Sigma F_y = 0$, and $\Sigma F_z = 0$. Then, we would also need $\cos \beta_{CF}$. More generally, if the cables CF and CG were tied off at unequal angles (not 30° for both), all the direction cosines would be required.

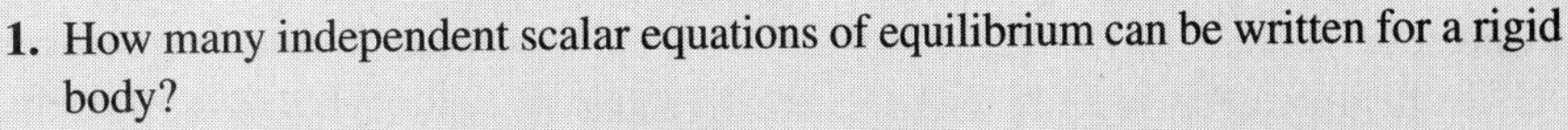

CHECKPOINT

1. How many independent scalar equations of equilibrium can be written for a rigid body?
2. ***True or False:*** If the resultant force of a system of forces that acts on a rigid body is zero, the moment of the forces about the origin of rectangular coordinates must be zero for equilibrium.

Answers: 1. Six; 2. True

Chapter Highlights

- The scalar product of two vectors **A** and **B** is defined as $\mathbf{A} \cdot \mathbf{B} = AB\cos\theta$, where A and B are the magnitudes of **A** and **B**, respectively, and θ is the angle between **A** and **B**.
- The scalar product of two vectors **A** and **B** can be written as $\mathbf{A} \cdot \mathbf{B} = A_xB_x + A_yB_y + A_zB_z$, where A_x, A_y, A_z and B_x, B_y, B_z are the (x, y, z) projections of vectors **A** and **B**, respectively.
- The vector product **C** of two vectors **A** and **B** is written as $\mathbf{C} = \mathbf{A} \times \mathbf{B}$. Vector **C** is perpendicular to the plane of vectors **A** and **B**, and the magnitude of **C** is $C = AB\sin\theta$, where A and B are the magnitudes of **A** and **B**, respectively, and θ is the angle between **A** and **B**.
- A couple is a free vector. It can be moved freely, provided that its direction and sense are not changed.
- A couple **M** that consists of forces **F** and $-\mathbf{F}$ acting at points P and Q, respectively, is represented by the vector product $\mathbf{M} = \mathbf{r} \times \mathbf{F}$, where **r** is the distance vector from point Q to point P.
- If a force **F** acts at a point P, its moment **M** about another point O is given by the vector product $\mathbf{M} = \mathbf{r} \times \mathbf{F}$, where **r** is the vector from point O to point P.
- If a force **F** acts at a point P, its scalar moment M_L about a line L is given by the scalar triple product $M_L = \mathbf{n} \cdot (\mathbf{r} \times \mathbf{F})$, where **n** is a unit vector in the direction of L and **r** is a vector from a point O on L to point P. Since $\mathbf{M} = \mathbf{r} \times \mathbf{F}$ is the moment of **F** about point O, $M_L = \mathbf{n} \cdot \mathbf{M} = M\cos\theta$, where M is the magnitude of **M** and θ is the angle between **n** and **M**. In other words, M_L is the projection of vector **M** on L.
- Any system of forces that acts on a rigid body is dynamically equivalent to a single resultant force acting at an arbitrary point O and a single resultant couple. The resultant force is the vector sum of the system of forces, and the resultant couple is the moment of the forces about point O.
- If the forces that act on a body sum to zero, the forces exert the same moment about all points in the body.
- If **F** and **M** are the single resultant force and the single resultant couple, respectively, which together are dynamically equivalent to a force system, the resultant axis of that force system is the line along which the resultant force **F** lies and for which $\mathbf{F} \times \mathbf{M} = 0$.
- A motionless rigid body is in equilibrium if, and only if, the vector sum of the external forces is zero and the moment of the external forces about any point is zero.

Problems

5.1 *Vector Algebra: Scalar Product*

5.1 Given the force $\mathbf{F} = 2\mathbf{i} - 3\mathbf{j} + 6\mathbf{k}$, determine
a. the magnitude of **F**.
b. the angle between **F** and the x axis.
c. the direction cosines for the line of action of **F**.

5.2 A force is given by $\mathbf{F} = 3\mathbf{i} + 4\mathbf{j} + 2\mathbf{k}$ [N].
a. Determine the projection of this force along the axis with direction numbers (1, 1, 1).
b. Determine the projection of this force along the axis with direction numbers (−1, 1, 1).

5.3 Given the force $\mathbf{F} = 4\mathbf{i} - 2\mathbf{j} + 4\mathbf{k}$ relative to rectangular coordinates (x, y, z), determine
a. the magnitude of **F**.
b. the angle between **F** and the y axis.
c. the projection of **F** along the line that passes through the points (1, 2, 3) and (7, 4, 0).

5.4 Compute the projection of vector $-5\mathbf{i} + 6\mathbf{j} - 2\mathbf{k}$ on the line that is parallel to vector $2\mathbf{i} - 3\mathbf{j} + \mathbf{k}$.

5.5 Three forces are given by

$$\mathbf{A} = 8\mathbf{i} + 10\mathbf{j} - 6\mathbf{k}$$
$$\mathbf{B} = 4\mathbf{i} + 3\mathbf{j} + 10\mathbf{k}$$
$$\mathbf{C} = -6\mathbf{i} - 4\mathbf{j} + 6\mathbf{k}$$

a. Determine the resultant **F** of these forces.
b. Determine the sum of the projections of these forces along an axis with direction numbers (4, 5, 2).

5.6 Two axes have direction numbers (1, 2, 3) and (6, −1, 4), respectively.
a. Determine the angle between the axes.
b. If a force of 10 N acts along the first axis, what is its projection on the second axis?
c. If the 10 N force acts along the second axis, what is its projection on the first axis?

5.7 a. For what value of n are the vectors from the origin to the points $(1, -2, 3)$ and $(2, 1, n)$ perpendicular?
b. For what value of n is the cosine of the angle between the vectors equal to $\frac{4}{9}$?
5.8 Four vectors are given as follows:

$$\mathbf{A} = \mathbf{i} + 3\mathbf{j} - \mathbf{k}$$

$$\mathbf{B} = 4\mathbf{i} + 2\mathbf{j}$$

$$\mathbf{C} = 8\mathbf{i} - 4\mathbf{k}$$

$$\mathbf{D} = -3\mathbf{i} + 2\mathbf{j} + 2\mathbf{k}$$

a. Determine $\mathbf{A} \cdot \mathbf{B}$, $\mathbf{A} \cdot \mathbf{C}$, $(\mathbf{B} - \mathbf{C}) \cdot \mathbf{D}$, and $(\mathbf{A} - \mathbf{B}) \cdot \mathbf{C}$.
b. Determine the angles between **A** and **B**, between **B** and **C**, and between $(\mathbf{A} - \mathbf{B})$ and $(\mathbf{C} - \mathbf{D})$.
c. Determine the projection of the vector $(\mathbf{B} - \mathbf{D})$ along the axis of the vector **A**.
5.9 The (x, y, z) projections of vectors **A** and **B** are $(1, 2, 3)$ and $(-1, 5\ -2)$, respectively.
a. Determine the angle between **A** and **B**.
b. Determine the projection of **A** on a line in the direction of **B**.
5.10 Determine the projection of a force with a magnitude of 100 N and with direction cosines $(2/\sqrt{14}, -3/\sqrt{14}, 1/\sqrt{14})$ on a line whose direction cosines are $(1/\sqrt{6}, 2/\sqrt{6}, -1/\sqrt{6})$.
5.11 a. Resolve the force **F** shown in Fig. P5.11 into its (x, y) components; use the unit vectors **i** and **j** along the x and y axes, respectively.
b. Resolve the force **F** into its components parallel to the x and U axes; **u** is a unit vector along axis U.

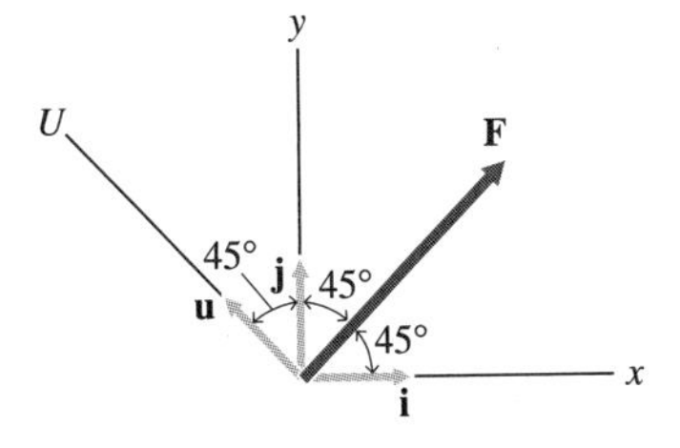

Figure P5.11

5.12 Given the force $\mathbf{F} = 2\mathbf{i} - 3\mathbf{j} + 6\mathbf{k}$, determine
a. the projection of **F** on the line $4\mathbf{i} + 3\mathbf{k}$.
b. the component of **F** parallel to the line $4\mathbf{i} + 3\mathbf{k}$.

5.13 Let **a** and **b** be unit vectors in the xy plane (see Fig. P5.13).
a. Show that $\mathbf{a} = \cos\alpha\mathbf{i} + \sin\alpha\mathbf{j}$ and $\mathbf{b} = \cos\beta\mathbf{i} + \sin\beta\mathbf{j}$.
b. Form the scalar product $\mathbf{a} \cdot \mathbf{b}$, and show that it yields the trigonometric identity $\cos(\beta - \alpha) = \cos\beta\cos\alpha + \sin\beta\sin\alpha$.

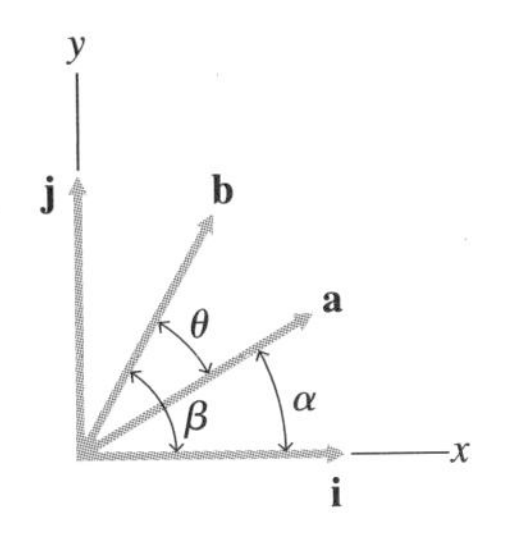

Figure P5.13

5.14 Using the concept of the scalar product of two vectors, determine the unit vector that is perpendicular to the vectors $\mathbf{r}_1 = \mathbf{i} + 4\mathbf{j} + 2\mathbf{k}$ and $\mathbf{r}_2 = 3\mathbf{i} - \mathbf{j} + \mathbf{k}$.
5.15 Let (x, y, z) and (X, Y, Z) be two systems of rectangular coordinates with the same origin. The cosine of the angle between two axes in the different systems is the number in the corresponding column and row of Table P5.15. For example, the cosine of the angle between the Y axis and the z axis is n_2.
a. Show that the sum of the squares of the numbers in any column or any row is 1. For instance, show that $l_1^2 + m_1^2 + n_1^2 = 1$. (*Hint:* Let $\mathbf{e}_1 = l_1\mathbf{i} + m_1\mathbf{j} + n_1\mathbf{k}$ be a unit vector along the X axis, where **i**, **j**, and **k** are unit vectors along xyz axes.)

Table P5.15

	x	y	z
X	l_1	m_1	n_1
Y	l_2	m_2	n_2
Z	l_3	m_3	n_3

b. Show that the sum of the products of the numbers in any row (or column) with the corresponding numbers in any other row (or column) is zero. For instance, show that $l_1 m_1 + l_2 m_2 + l_3 m_3 = 0$.
5.16 A line that is perpendicular to a given plane forms equal angles with the x, y, and z axes. A vector that lies in the plane forms a 60° angle with the z axis. The magnitude of the vector is 30 units. Determine two different vectors (a_x, a_y, a_z) and (b_x, b_y, b_z) that fulfill these conditions.

5.17 Let **F** be any given vector, and let **n** be a unit vector.

a. Let $\mathbf{F} = \mathbf{F}_1 + (\mathbf{F} \cdot \mathbf{n})\mathbf{n}$, and show that $(\mathbf{F} \cdot \mathbf{n})\mathbf{n}$ is the component of **F** in the direction of **n** and that $\mathbf{F}_1$ is the component of **F** perpendicular to **n**.

b. Let $\mathbf{F} = 5\mathbf{i} + 8\mathbf{j} + 2\mathbf{k}$. Determine the component $\mathbf{F}_1$ that is perpendicular to the line with direction cosines $(1/\sqrt{3}, 1/\sqrt{3}, 1/\sqrt{3})$.

5.18 Let xyz be oblique (nonrectangular) Cartesian coordinate axes (see Fig. P5.18). Let α, β, and γ be the angles between the pairs of axes y and z, z and x, and x and y, respectively. Let **i**, **j**, and **k** be unit vectors along the xyz axes. Let **r** be the vector from the origin to any point P with coordinates (x, y, z).

a. Show that $\mathbf{r} = x\mathbf{i} + y\mathbf{j} + z\mathbf{k}$.

b. By forming the scalar product $\mathbf{r} \cdot \mathbf{r}$, derive a formula for the distance r from the origin to point P, in terms of (x, y, z) and (α, β, γ).

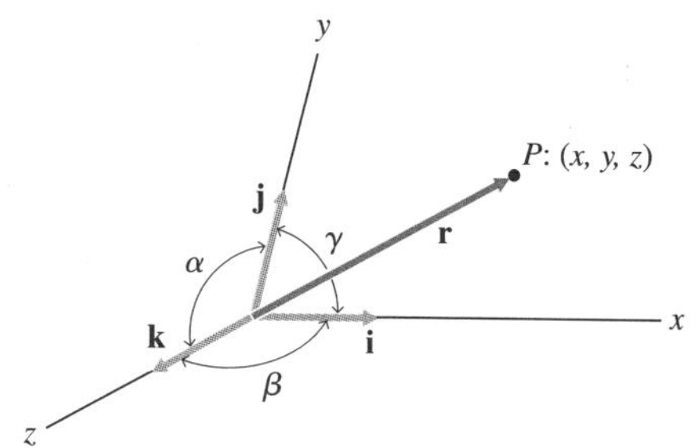

Figure P5.18

5.19 Consider a point A: (x_A, y_A, z_A) and a line BC through the points B: (x_B, y_B, z_B) and C: (x_C, y_C, z_C) in a Cartesian xyz coordinate system (see Fig. P5.19).

a. Express the vector $\mathbf{r}_{BA}$ from B to A in terms of the (x, y, z) coordinates of B and A.

b. Determine the projection BP of $\mathbf{r}_{BA}$ on the vector $\mathbf{r}_{BC}$ from B to C.

c. In terms of the magnitudes BA and BP, express the minimum distance d from point A to line BC. (*Note:* The distance d can be used in summing the moments about line BC of a force applied at A. To determine the distance d, you could also use the vector from C to A and its projection on the vector $\mathbf{r}_{BC}$.)

d. For A: $(30, 10, -10)$, B: $(20, 30, 0)$ and C: $(-10, 20, 40)$, calculate d. Lengths are in inches.

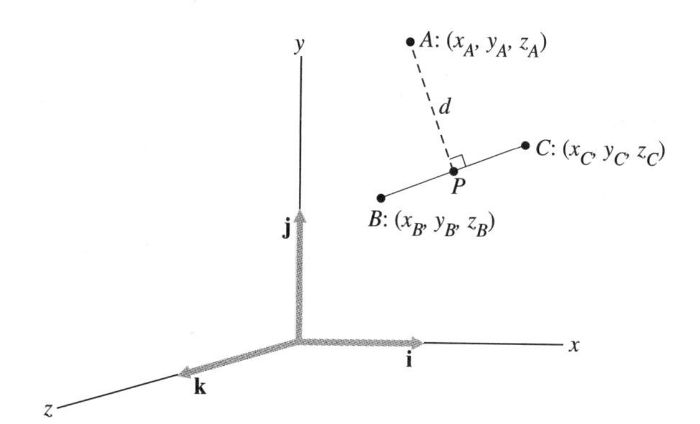

Figure P5.19

The following problems are not true design problems, but rather parts of design problems. They illustrate that a given force in a design problem related to rigid bodies may be replaced by any one of several equivalent force systems that act at the point of application of the given force. That is, an equivalent force system may not be unique.

5.20 Consider a vector $\mathbf{B} = 4\mathbf{i} + 2\mathbf{j}$. Determine two vectors in the xy plane whose resultant is equivalent to **B** and whose magnitudes are 3 units and 4 units.

Method: Consider two circles with radii of 3 and 4 centered at *each* end of **B** (that is, a total of four circles). A vector from the tail of **B** to any point of intersection between circles of radii 3 and 4 plus a vector from the point of intersection to the tip of **B** is a solution. How many solutions are possible?

5.21 For the vector **B** in Problem 5.20, determine three vectors in the xy plane whose resultant is equivalent to **B** and whose magnitudes are 2 units, 3 units, and 4 units.
Method: Consider two circles with radii equal to the magnitudes of two of the required forces (say, circles with radii 2 and 4) centered at *each* end of **B** (that is, four circles in all). Draw a vector from the tail of **B** to one of the intersection points of the circles with radii 2 and 4. Draw a circle of radius 3 with origin at the intersection point of the circles. Consider the intersection points of this circle with the other circles, and draw the appropriate force from an intersection point to the tip of **B**. How many solutions are possible?

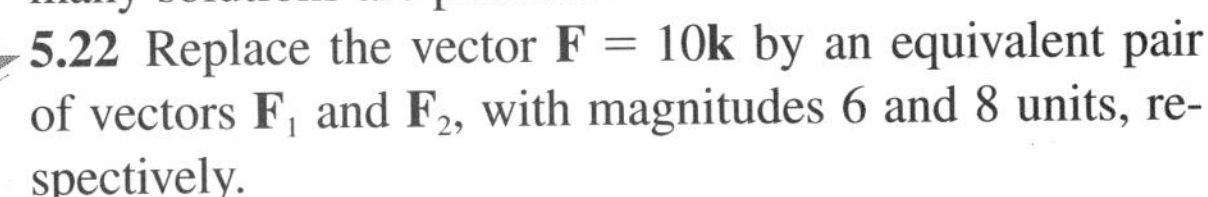

5.22 Replace the vector $\mathbf{F} = 10\mathbf{k}$ by an equivalent pair of vectors $\mathbf{F}_1$ and $\mathbf{F}_2$, with magnitudes 6 and 8 units, respectively.
Method: Attach the tail of one of the vectors—say, $\mathbf{F}_1$—to the tail of **F**, and rotate $\mathbf{F}_1$ about the z axis. The tip of $\mathbf{F}_1$ thus generates a circle around the z axis. Adjust the circle so that the distance from the circle to the tip of **F** is 8 units. Force $\mathbf{F}_2$ may be directed from any point on this circle to the tip of **F**. Then, select a pair of vectors $\mathbf{F}_1$ and $\mathbf{F}_2$ out of the infinite number of pairs that are equivalent to **F**. (*Note:* You could also start with $\mathbf{F}_2$ and rotate it about the z axis with its tail at the tail of **F**. Alternatively, note that vectors **F**, $\mathbf{F}_1$, and $\mathbf{F}_2$ form a triangle that can lie in any plane containing **F**. You can choose any convenient triangle to determine a pair of forces $\mathbf{F}_1$ and $\mathbf{F}_2$.)

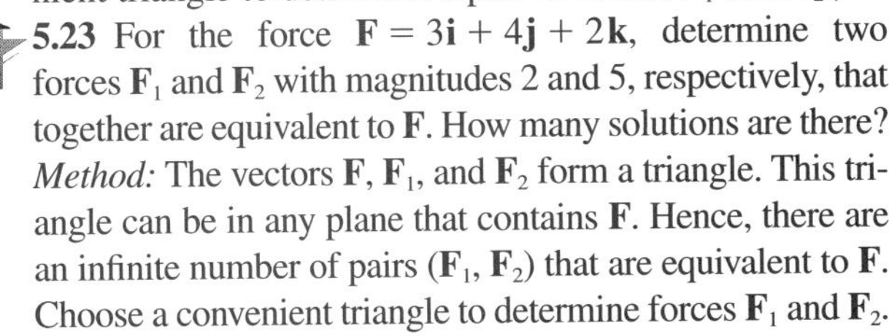

5.23 For the force $\mathbf{F} = 3\mathbf{i} + 4\mathbf{j} + 2\mathbf{k}$, determine two forces $\mathbf{F}_1$ and $\mathbf{F}_2$ with magnitudes 2 and 5, respectively, that together are equivalent to **F**. How many solutions are there?
Method: The vectors **F**, $\mathbf{F}_1$, and $\mathbf{F}_2$ form a triangle. This triangle can be in any plane that contains **F**. Hence, there are an infinite number of pairs ($\mathbf{F}_1$, $\mathbf{F}_2$) that are equivalent to **F**. Choose a convenient triangle to determine forces $\mathbf{F}_1$ and $\mathbf{F}_2$.

5.1 *Vector Algebra: Vector Product*

5.24 Using the concept of the vector product of two vectors, determine the unit vector that is perpendicular to the vectors $\mathbf{i} + 4\mathbf{j} + 2\mathbf{k}$ and $3\mathbf{i} - \mathbf{j} + \mathbf{k}$.

5.25 Show that if you take vector **q** in Example 5.6 from point A to point C, you get the same result for d.

5.26 A straight line passes through the points (3, 1, −2) and (6, 8, 7). Another straight line passes through the points (−4, 0, −2) and (20, 56, 70). Show that the two lines are parallel.

5.27 Let **a** and **b** be unit vectors in the xy plane in Fig. P5.13. Form the vector product $\mathbf{a} \times \mathbf{b}$, and show that it yields the trigonometric identity

$$\sin(\beta - \alpha) = \sin\beta\cos\alpha - \cos\beta\sin\alpha$$

5.28 Four vectors are given by the equations

$$\mathbf{A} = 4\mathbf{i} + \mathbf{j} - \mathbf{k}$$
$$\mathbf{B} = 3\mathbf{i} + 6\mathbf{j}$$
$$\mathbf{C} = 3\mathbf{j} + 2\mathbf{k}$$
$$\mathbf{D} = -2\mathbf{i} - 4\mathbf{j} - 8\mathbf{k}$$

Determine the following quantities:
a. $\mathbf{A} \times \mathbf{B}$
b. $\mathbf{C} \times \mathbf{D}$
c. $(\mathbf{D} + \mathbf{B}) \times (\mathbf{A} + \mathbf{C})$
d. $\mathbf{j} \times \mathbf{B}$

5.29 Given the two vectors $\mathbf{A} = 3\mathbf{i} + \mathbf{j}$ and $\mathbf{B} = \mathbf{j} + 2\mathbf{k}$, calculate the following quantities:
a. $\mathbf{A} + \mathbf{B}$
b. $\mathbf{A} - \mathbf{B}$
c. $\mathbf{A} \cdot \mathbf{B}$
d. $\mathbf{C} = \mathbf{A} \times \mathbf{B}$
e. a unit vector in the direction of **C**, from part d
f. $(\mathbf{A} + \mathbf{B}) \cdot (\mathbf{A} - \mathbf{B})$
g. $(\mathbf{A} + \mathbf{B}) \times (\mathbf{A} - \mathbf{B})$
h. $(\mathbf{A} - \mathbf{B}) \times (\mathbf{A} + \mathbf{B})$

5.30 Use the vector product of two vectors to determine the area of a parallelogram, two adjacent sides of which are given by the vectors $\mathbf{A} = 14\mathbf{i} - 6\mathbf{j}$ and $\mathbf{B} = 4\mathbf{i} + 10\mathbf{j}$.

5.31 A triangle is formed by the origin of the xyz coordinate system and the points (1, 2, 3) and (−2, 4, 1). Use the vector product of two vectors to determine the area of the triangle. Lengths are measured in feet.

5.32 Given three vectors,

$$\mathbf{A} = 2\mathbf{i} - \mathbf{j} + 4\mathbf{k}$$
$$\mathbf{B} = 6\mathbf{i} + 2\mathbf{j} - \mathbf{k}$$
$$\mathbf{C} = -4\mathbf{i} + \mathbf{j}$$

determine the following vector products:
a. $\mathbf{A} \times \mathbf{B}$
b. $\mathbf{B} \times \mathbf{A}$
c. $\mathbf{A} \times (\mathbf{B} \times \mathbf{C})$
d. $\mathbf{i} \times \mathbf{B}$
e. $\mathbf{C} \times \mathbf{k}$

5.33 Consider two vectors $\mathbf{a} = 3\mathbf{i} + 5\mathbf{j} - 2\mathbf{k}$ and $\mathbf{b} = -2\mathbf{i} + 6\mathbf{j} - 4\mathbf{k}$.
a. Evaluate $\mathbf{c} = \mathbf{a} \times \mathbf{b}$ by expanding the product $(3\mathbf{i} + 5\mathbf{j} - 2\mathbf{k}) \times (-2\mathbf{i} + 6\mathbf{j} - 4\mathbf{k})$.
b. Check the result by means of Eq. (5.9).
c. Show, by means of the scalar product, that **c** is perpendicular to both **a** and **b**.

5.34 a. Prove that the magnitude of the vector product $\mathbf{r} \times \mathbf{s}$ is equal to the area of the parallelogram whose two adjacent sides are formed by vectors **r** and **s**.
b. Determine the area of the parallelogram having two adjacent sides $\mathbf{r} = \mathbf{i} + 2\mathbf{j} + 3\mathbf{k}$ and $\mathbf{s} = -3\mathbf{i} - 2\mathbf{j} + \mathbf{k}$. Lengths are in meters.

5.35 Let vectors **a**, **b**, and **c** form the sides of any triangle with interior angles α, β, and γ opposite to **a**, **b**, and **c**, respectively. Arrange the vectors so that $\mathbf{a} + \mathbf{b} = \mathbf{c}$. The vector product of $(\mathbf{a} + \mathbf{b})$ with **c** gives $(\mathbf{a} + \mathbf{b}) \times \mathbf{c} = 0$, or $\mathbf{a} \times \mathbf{c} = \mathbf{c} \times \mathbf{b}$. Using Theorem 5.2, express this result in scalar form, and show that it is equivalent to the law of sines for triangles:

$$\frac{a}{\sin \alpha} = \frac{b}{\sin \beta}$$

5.36 Determine the minimum distance (units of feet) between a line L_1 that passes through the points (−50, 100, 75) and (950, −200, 0) and a line L_2 that passes through the points (25, −75, 125) and (100, 25, −175).

5.37 Determine the minimum distance (units of meters) between a line L_1 that passes through the points (50, 75, 25) and (0, −25, 50) and a line L_2 that passes through the points (25, 50, 125) and (−75, 25, 0).

5.38 Determine the minimum distance (units of inches) from a plane that passes through the points *A:* (25, −25, 50), *B:* (75, 50, 100), and *C:* (150, 50, −50) to the point *P:* (25, 25, −75).

5.39 Determine the equation of a plane that passes through the points *A:* (1, 0, 2), *B:* (2, 3, 0), and *C:* (−1, −2, 5), where *P:* (*x*, *y*, *z*) is a general point in the plane. Verify that points *A*, *B*, and *C* satisfy the equation. (Distances are in meters.)

5.40 a. Determine the equation of a plane that passes through points *A:* (2, −3, 4) and *B:* (1, 2, 3) and is perpendicular to the plane $2x + 3y - 2z + 8 = 0$. (*Hint:* Note that both the vector $\mathbf{r}_{AB}$ and $\mathbf{r} = 2\mathbf{i} + 3\mathbf{j} - 2\mathbf{k}$ are parallel to the required plane.)
b. Verify that the points *A* and *B* lie in the plane and that it is perpendicular to the plane $2x + 3y - 2z + 8 = 0$.

5.2 *Moment of a Force about a Point*

5.41 A force $\mathbf{F} = \mathbf{i} + 2\mathbf{j} + \mathbf{k}$ [kN] passes through the point *A:* (1, 1, 2) [m]. Determine the moment of the force about the point *B:* (−1, 2, −1) [m].

5.42 A force $\mathbf{F} = \mathbf{i} + 3\mathbf{k}$ [lb] passes through the point *A:* (−1, 0, 2) [ft].
a. Determine the moment $\mathbf{M}_O$ about the origin (0, 0, 0).
b. Calculate the direction cosines of $\mathbf{M}_O$.

5.43 A force of magnitude $F = 50$ N is directed along a line from point *A:* (1, 2, −3) to point *B:* (2, 0, 1) [m].
a. Determine the moment of the force about the point (0, 0, 0).
b. Determine the moment of the force about the point (0, 1, 2).

5.44 The following forces act on a rigid body at the given point:

$$\mathbf{F} = 4\mathbf{i} - 3\mathbf{j} + 6\mathbf{k} \text{ at } (2, 1, -4)$$
$$\mathbf{G} = -4\mathbf{i} + 2\mathbf{j} + 5\mathbf{k} \text{ at } (-2, 3, 1)$$
$$\mathbf{H} = 3\mathbf{i} - 4\mathbf{j} - 5\mathbf{k} \text{ at } (3, 2, -1)$$

Calculate the sum of the moments of the forces about the point (4, 5, 6).

5.45 Force **P** acts along line *AF* of the wedge in Fig. P5.45. Its magnitude is $P = 170$ N.
a. Determine the moment of the force about point *E*.
b. Determine the moment of the force about point *D*.
c. Determine the angle between **P** and the *z* axis.

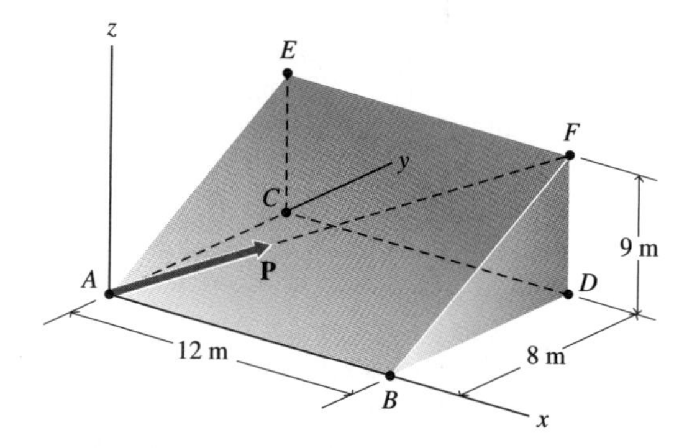

Figure P5.45

5.46 The force **F** acts along the diagonal of the box in Fig. P5.46. Its magnitude is $F = 3000$ lb.
a. Express the force in terms of the unit vectors **i**, **j**, and **k**.
b. Determine the moment of the force about point *A*.

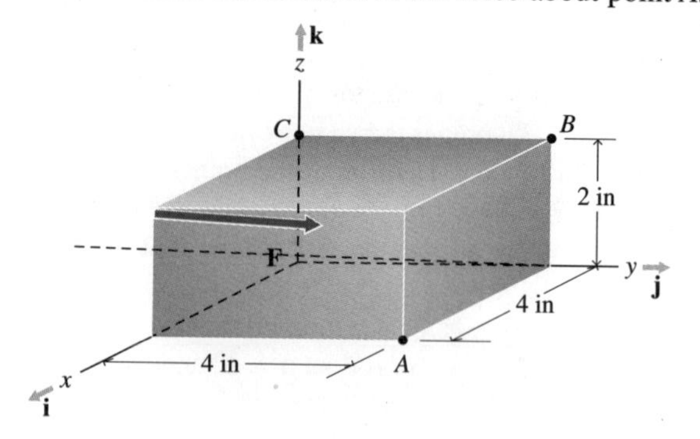

Figure P5.46

5.47 The force **F** acts along the diagonal of the box in Fig. P5.47. Its magnitude is $F = 2$ kN.
a. Determine the moment of the force about point A.
b. Determine the moment of the force about point C.

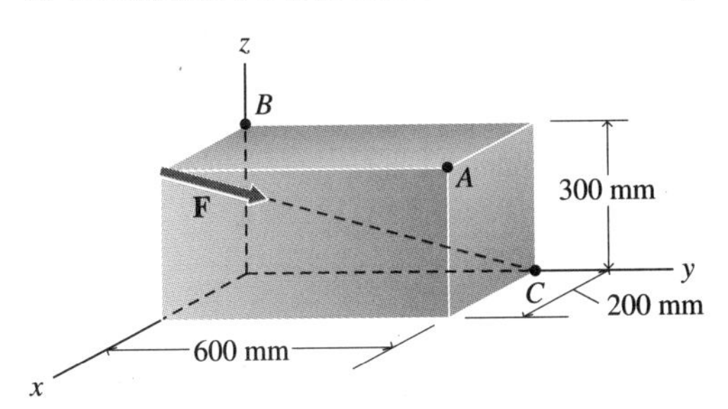

Figure P5.47

5.48 Three concurrent forces **F**, **G**, and **H** act at the point P: (x, y, z). Show by means of properties of the vector product that the sum of the moments of these forces about the point Q: (x_O, y_O, z_O) is equal to the moment of their resultant about Q.

5.49 The following forces act on a rigid body at the given points:

$$\mathbf{F}_1 = 3\mathbf{i} - 3\mathbf{j} + 5\mathbf{k} \text{ at } (1, 2, -4)$$

$$\mathbf{F}_2 = -2\mathbf{i} + 2\mathbf{j} + 6\mathbf{k} \text{ at } (-3, 2, 1)$$

$$\mathbf{F}_3 = 6\mathbf{i} - 8\mathbf{j} - 5\mathbf{k} \text{ at } (4, 2, -4)$$

a. Compute the moment of each force about the point (5, 6, 8).
b. Determine the magnitude and the direction cosines of the resultant moment **M**.

5.50 The forces $8\mathbf{i} + 11\mathbf{j} + 15\mathbf{k}$ and $-6\mathbf{i} + 9\mathbf{j} + 10\mathbf{k}$ act at the respective points $(2, 4, -1)$ and $(4, -2, 2)$. Show that it is impossible to place a force $\mathbf{F} = F_x\mathbf{i} + F_y\mathbf{j} + F_z\mathbf{k}$ at the point $(5, 10, -20)$ so that the moment of the three forces about the origin is zero.

5.3 *Moment of a Force about an Axis*

5.51 A force $\mathbf{F} = 3\mathbf{i} + 4\mathbf{j} - \mathbf{k}$ acts at the point (2, 4, 6) in a rectangular xyz coordinate system. Determine the moments of this force about the x, y, and z axes. (Units are newtons and meters.)

5.52 Given the two forces $\mathbf{F} = \mathbf{i} - 4\mathbf{j} + 8\mathbf{k}$ acting at point (0, 1, 4) and $\mathbf{G} = 6\mathbf{i} - 3\mathbf{j} + 2\mathbf{k}$ acting at point (2, 2, 2) in a rectangular xyz coordinate system, determine the sum of the moments of the two forces about the x, y, and z axes. (Units are kilonewtons and meters.)

5.53 The force $\mathbf{F} = 20\mathbf{i} + 60\mathbf{j} + 90\mathbf{k}$ [N] acts at point (0, 40, 0) [mm]. Determine the moment of the force about a line through the origin that has direction cosines $(1/\sqrt{3}, 1/\sqrt{3}, 1/\sqrt{3})$.

5.54 The force $\mathbf{F} = 90\mathbf{i} + 60\mathbf{j} + 20\mathbf{k}$ acts at the point (0, 0, 40). Compute the moment of the force about a line through the origin that has direction cosines $(1/\sqrt{3}, 1/\sqrt{3}, 1/\sqrt{3})$. (Units are in pounds and inches.)

5.55 a. Determine the moment of the force **F** in Fig. P5.46 about a line extending from point A to point B. The magnitude of **F** is 3000 lb.
b. Determine the moment of the force **F** about a line extending from point A to point C.

5.56 Show that the moment M_L in Example 5.9 is also obtained by Eqs. (5.19b) and (5.19c).

5.57 Force $\mathbf{F} = 3\mathbf{i} + 4\mathbf{j} + 5\mathbf{k}$ acts at the point (1, 2, 3) in a rectangular xyz coordinate system. Determine the moments of the force about the x, y, and z axes. (Units are pounds and feet.)

5.58 Force $\mathbf{F} = 10\mathbf{i} + 20\mathbf{j} + 5\mathbf{k}$ acts at point $(3, 0, -4)$. Calculate the moment of the force **F** about the line through point $(-4, -7, -2)$ in the direction of the vector $3\mathbf{i} + \mathbf{j} - 2\mathbf{k}$. (Units are kilonewtons and meters.)

5.59 A force $\mathbf{F} = F_x\mathbf{i} + F_y\mathbf{j} + F_z\mathbf{k}$ acts at the tip of the vector $\mathbf{r} = x\mathbf{i} + y\mathbf{j} + z\mathbf{k}$, whose tail is at the origin. Prove that the moments of the force about the x, y, and z axes are equal to the respective projections of the vector $\mathbf{r} \times \mathbf{F}$ on these axes.

5.60 Show that the moment M_{AB} obtained in Example 5.10 is also obtained by Eqs. (5.19b) and (5.19c).

5.61 The force $\mathbf{F} = 10\mathbf{i} + 20\mathbf{j} + 5\mathbf{k}$ [lb] acts at point $(3, 0, -4)$ [ft]. Calculate the moment of the force about the line through the point $(-4, -7, -2)$ [ft] in the direction of the vector $3\mathbf{i} + \mathbf{j} - 2\mathbf{k}$.

5.62 a. Determine the moment of the force **F** in Fig. P5.47 about a line extending from point A to point B.
b. Determine the moment of the force **F** about a line extending from point A to point C.

5.63 Three edges of a parallelepiped are formed by line segments drawn from the origin to the points (2, 1, 3), (6, 2, 4), and (5, 4, 0) [ft]. Determine the volume of the parallelepiped, which is given by the scalar triple product $\mathbf{a} \cdot (\mathbf{b} \times \mathbf{c})$, where vectors **a**, **b**, and **c** are three conterminous edges of the parallelepiped.

5.64 If vectors **a**, **b**, and **c** are all parallel to the same plane, then $\mathbf{a} \cdot (\mathbf{b} \times \mathbf{c}) = 0$.
a. Use this fact to prove that the vectors from the origin to the three points $(1, -3, 2)$, $(-2, 4, 3)$, and $(0, -4, 14)$ are parallel to the same plane.
b. Determine the unit normal vector **n** of that plane.

5.65 a. Show that the absolute value of the scalar triple product $\mathbf{a} \cdot (\mathbf{b} \times \mathbf{c})$ represents the volume of the parallelepiped of which the vectors **a**, **b**, and **c** are three conterminous edges.
b. Show that $\mathbf{a} \cdot (\mathbf{b} \times \mathbf{c}) = 0$ if, and only if, the nonzero vectors **a**, **b**, and **c** are all parallel to the same plane.

5.5 *Representation of a Couple as a Vector Product*

5.66 A couple consists of a force at the origin and a force $\mathbf{F} = 3\mathbf{i} + 4\mathbf{j} - \mathbf{k}$ [kN] at the point (2, 4, 6) [m] in a rectangular xyz coordinate system. Determine the moment of this couple.

5.67 A force **F** with a magnitude of 800 N and with direction cosines (0.36, −0.48, 0.80) [m] acts at the origin. Another force −**F** acts at the point (9, 8.4, −3). Calculate the (x, y, z) projections of the resultant couple **M**.

5.68 A driver applies two forces of magnitude 50 N to the steering wheel shown in Fig. P5.68. Determine the torque [the projection of the couple (**F**, −**F**) on the z axis] that is turning the steering wheel.

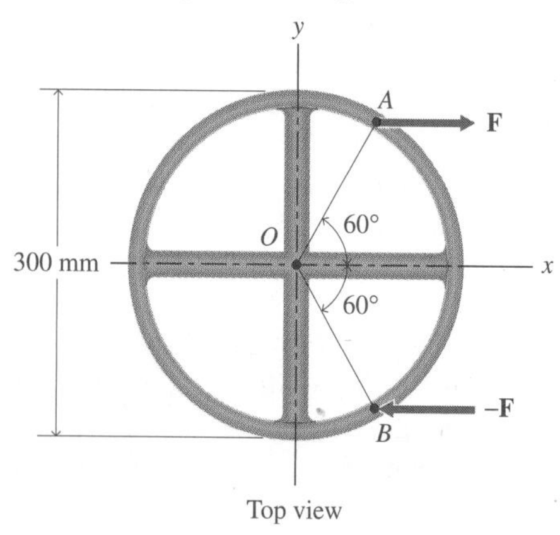

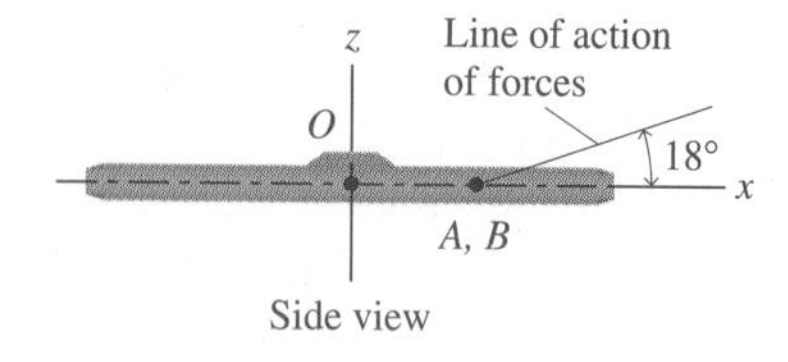

Figure P5.68

5.69 The gear box in Fig. P5.69 is subject to torques (in N·m)

$$\mathbf{M}_1 = +3\mathbf{i} - \mathbf{j} + 2\mathbf{k}$$
$$\mathbf{M}_2 = +4\mathbf{i} + 5\mathbf{j}$$
$$\mathbf{M}_3 = -2\mathbf{i} + \mathbf{j} + 4\mathbf{k}$$

a. Determine the resultant couple that acts on the gear box, in terms of the unit vectors **i**, **j**, and **k**.
b. Determine the magnitude and the direction angles of the resultant couple.

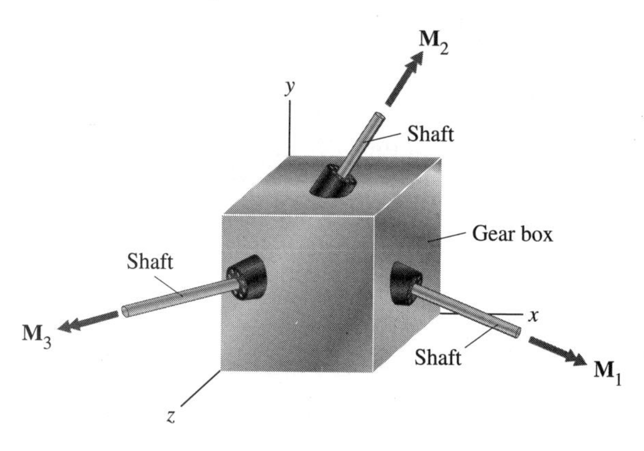

Figure P5.69

5.70 Two couples are given by $\mathbf{A} = 3\mathbf{i} + 4\mathbf{j} - \mathbf{k}$ and $\mathbf{B} = 8\mathbf{i} - 6\mathbf{j} + a\mathbf{k}$, where a is an unknown constant.
a. Determine a such that the couples are perpendicular.
b. Using the results of part a, determine the resultant of the couples.

5.71 Three couples

$$\mathbf{M}_1 = 6\mathbf{i} + 5\mathbf{j} - 8\mathbf{k}$$
$$\mathbf{M}_2 = \mathbf{i} - 3\mathbf{j} + \mathbf{k}$$
$$\mathbf{M}_3 = -4\mathbf{i} + 2\mathbf{j} - 6\mathbf{k}$$

act on a rigid body at the points (1, 2, 3), (−1, −1, −1), and (8, 7, −4), respectively, in a rectangular xyz coordinate system. Determine the resultant of these couples. (Units are kilonewtons and meters.)

5.72 Determine the resultant of the three couples shown in Fig. P5.72.

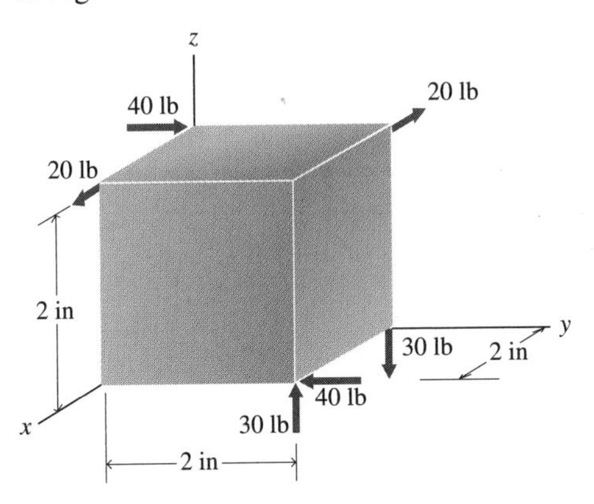

Figure P5.72

5.73 The forces shown in Fig. P5.73 are acting on a rigid flat concrete slab.

a. Determine the resultant couple $\mathbf{M} = M_x\mathbf{i} + M_y\mathbf{j} + M_z\mathbf{k}$.

b. Determine the magnitude and the direction cosines of **M**.

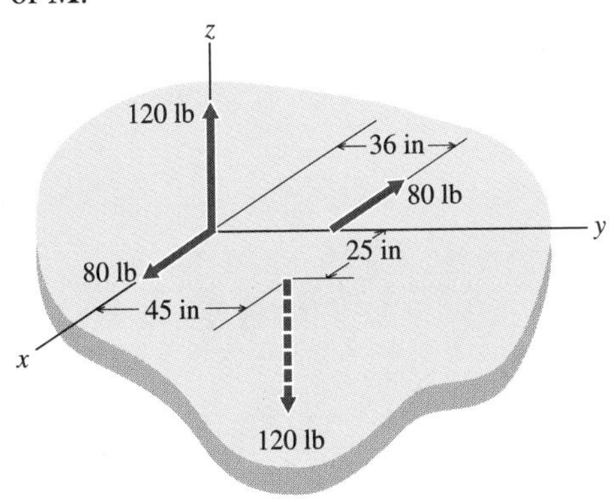

Figure P5.73

5.74 Three forces

$$\mathbf{F} = 7\mathbf{i} - 2\mathbf{j} + 4\mathbf{k}$$

$$\mathbf{G} = -8\mathbf{i} + 3\mathbf{j} + 10\mathbf{k}$$

$$\mathbf{H} = -10\mathbf{i} + 4\mathbf{j} - 6\mathbf{k}$$

act on a rigid body at the respective points (3, 1, 4), (6, 8, 2), and (1, −4, −3). The forces $-\mathbf{F}$, $-\mathbf{G}$, and $-\mathbf{H}$ act at the respective points (6, 1, −7), (8, −4, −10), and (7, 0, 1). Determine the resultant couple $\mathbf{M} = M_x\mathbf{i} + M_y\mathbf{j} + M_z\mathbf{k}$. (Lengths are in meters, and forces are in kilonewtons.)

5.75 The gear box shown in Fig. P5.75 is subjected to torques $M_1 = 10$ kN·m and $M_2 = 8$ kN·m. Determine the resultant couple **M** [kN·m] that acts on the gear box. Express the resultant couple in terms of its magnitude and direction angles relative to *xyz* axes.

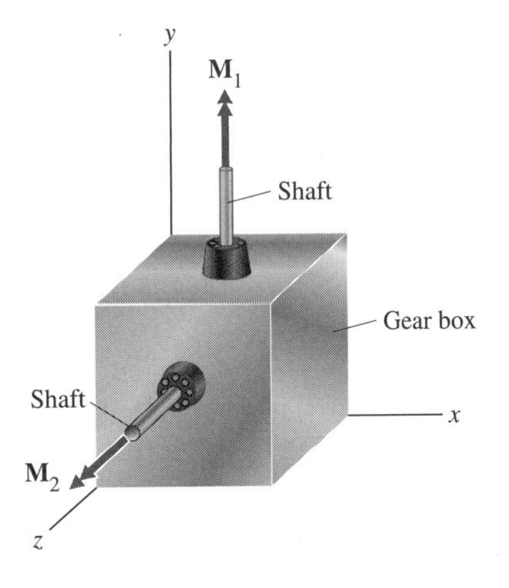

Figure P5.75

5.6 *Composition of Forces That Act on a Rigid Body*

5.76 A force $\mathbf{F} = 6\mathbf{i} + 5\mathbf{j} - 3\mathbf{k}$ [kN] acts on a rigid body at the point (1, 1, −2) [m]. Place the force at the origin, and determine the required compensating couple $\mathbf{M}_c$ so that the effect on the rigid body is not changed.

5.77 In a test of an airplane wing, engineers apply forces to the wing as shown in Fig. P5.77. The forces lie in the *xy* plane. Determine the dynamically equivalent force and couple relative to the origin *O*.

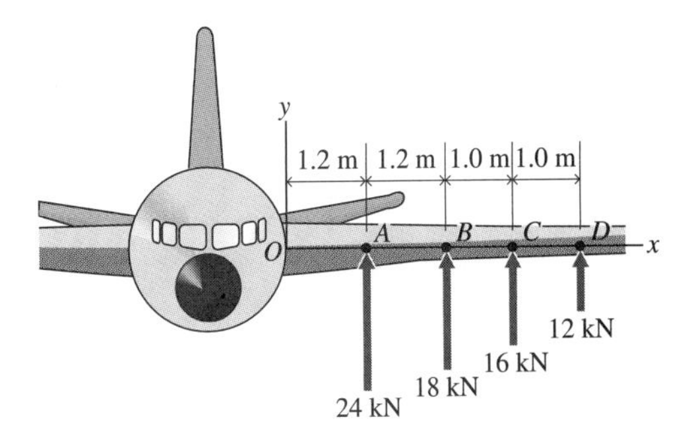

Figure P5.77

5.78 Two forces $\mathbf{F}_1 = 3\mathbf{i} + 4\mathbf{j}$ at (1, 2, 3) and $\mathbf{F}_2 = \mathbf{i} + \mathbf{j} + \mathbf{k}$ at (−1, 1, 4) and a couple $\mathbf{M} = 40\mathbf{i} + 20\mathbf{k}$ act on a rigid body. Replace the two forces and the couple by a single force acting through the origin and a couple $\mathbf{M}_O$. (Lengths are in meters, and forces are in kilonewtons.)

5.79 The following forces and couples act on a rigid body at the points given:

$$\mathbf{F}_1 = 600\mathbf{i} - 350\mathbf{j} + 90\mathbf{k} \text{ at } (3, 8, 21)$$

$$\mathbf{F}_2 = -1000\mathbf{i} + 260\mathbf{j} - 720\mathbf{k} \text{ at } (7, -9, -40)$$

$$\mathbf{F}_3 = 300\mathbf{i} - 950\mathbf{j} + 850\mathbf{k} \text{ at } (60, -3, 25)$$

$$\mathbf{M}_1 = 4000\mathbf{i} - 2800\mathbf{j} + 10{,}000\mathbf{k} \text{ at } (1, 14, 28)$$

$$\mathbf{M}_2 = -1600\mathbf{i} - 24{,}000\mathbf{j} - 8000\mathbf{k} \text{ at } (-4, -6, -80)$$

Replace this force system by a dynamically equivalent system that consists of a force **F** that acts at point (5, 9, −35) and a couple **M**. (Lengths are in inches, and forces are in pounds.)

5.80 The following forces act on a rigid body:

$$\mathbf{F}_1 = 600\mathbf{i} - 350\mathbf{j} + 90\mathbf{k} \text{ at point } (3, 8, 21)$$

$$\mathbf{F}_2 = -1000\mathbf{i} + 260\mathbf{j} - 720\mathbf{k} \text{ at point } (7, -9, -40)$$

$$\mathbf{F}_3 = 300\mathbf{i} - 950\mathbf{j} + 850\mathbf{k} \text{ at point } (60, -3, 25)$$

Replace these forces by a dynamically equivalent force system consisting of a force **F** that acts at point (5, 9, −35) and a couple **M**. (Lengths are in feet, and forces are in pounds).

5.81 Replace the two forces and the couple that act on the rigid block in Fig. P5.81 by a dynamically equivalent system consisting of a force at point *A* and a couple.

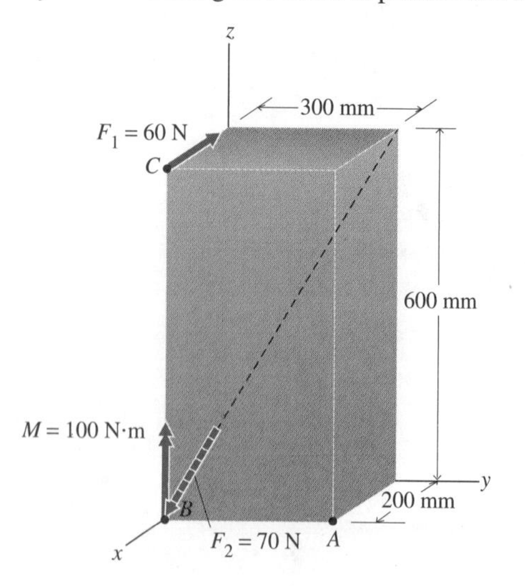

Figure P5.81

5.82 Replace the force system acting on the block in Fig. P5.82 by a dynamically equivalent system consisting of a force **F** acting at the origin *O* and a couple **M**.

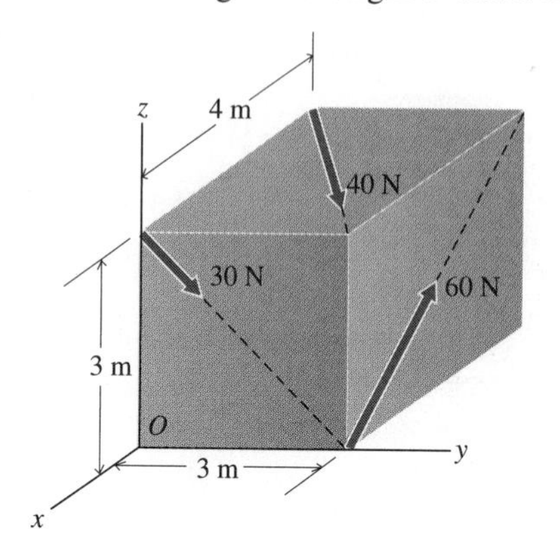

Figure P5.82

5.83 The forces **F** and −**F** act on the rigid block shown in Fig. P5.83. The magnitude of the forces is $F = 10\sqrt{45}$ N.

a. Determine the moment of the forces about point *A*.

b. Determine the moment of the forces about point *B*.

c. Determine the moment of the forces about the line that extends from point *B* to point *A*.

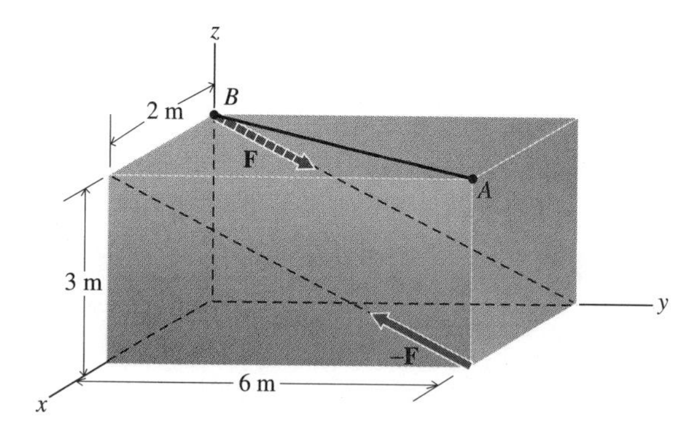

Figure P5.83

5.84 A rigid body is subjected to two forces $\mathbf{F}_1 = 4\mathbf{i} - 3\mathbf{j} + 4\mathbf{k}$ at (0, 1, 4) and $\mathbf{F}_2 = -\mathbf{i} + 4\mathbf{j} - 6\mathbf{k}$ at (−1, 2, 0) and a couple $\mathbf{M} = 8\mathbf{i} - 5\mathbf{j} + 18\mathbf{k}$. (Lengths are in feet, and forces are in kips.)

a. Replace the forces and the couple by a dynamically equivalent system consisting of a force **F** through the point (8, 0, −2) and a couple $\mathbf{M}_1$.

b. Determine the moment $\mathbf{M}_2$ of the original forces and couple about a line passing through the origin *O* and having direction numbers (1, 2, 2).

5.85 Replace the force system acting on the rigid body in Fig. P5.85 by a dynamically equivalent force system consisting of a force **P** that acts at the origin O and a couple **M**.

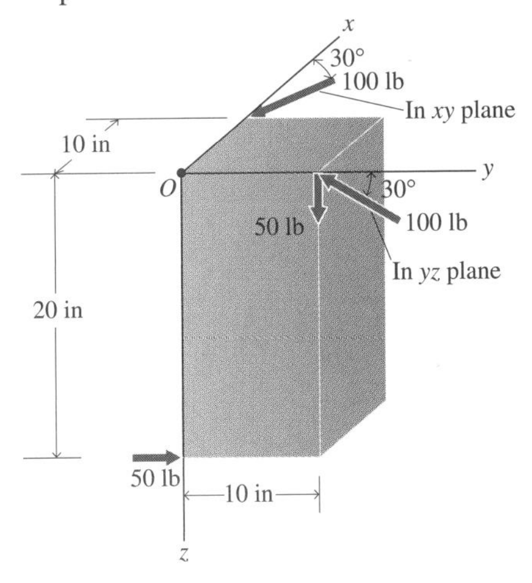

Figure P5.85

5.86 Replace the force system acting on the rigid body in Fig. P5.86 by a dynamically equivalent force system, consisting of a force **P** that acts at the origin O and a couple **M**.

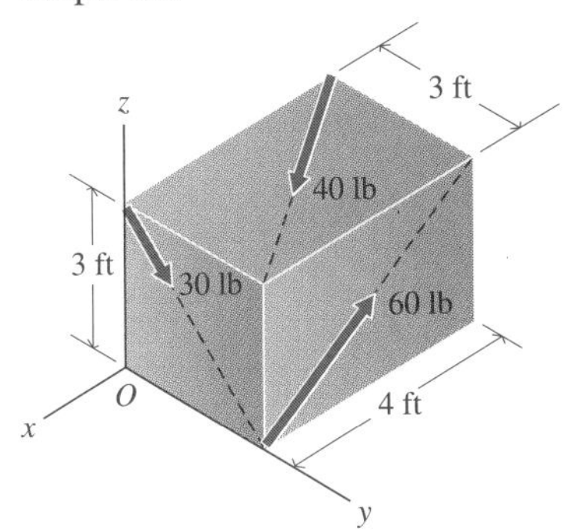

Figure P5.86

5.87 a. Replace the forces that act on the L-shaped bar in Fig. P5.87 by a dynamically equivalent system that consists of a force **F** at point O and a couple **M**.
b. Determine the magnitudes and the direction cosines of **F** and **M**.

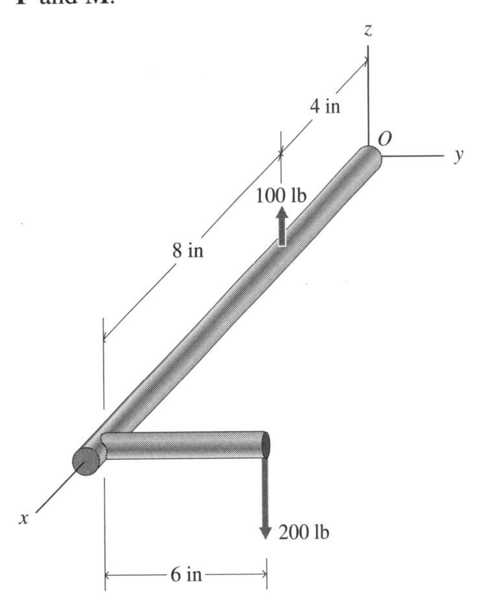

Figure P5.87

5.88 A rigid body is subjected to forces $\mathbf{F}_1$, $\mathbf{F}_2$, and $\mathbf{F}_3$ and to couples $\mathbf{M}_1$ and $\mathbf{M}_2$, defined as follows:

$$\mathbf{F}_1 = 500\mathbf{i} - 340\mathbf{j} - 650\mathbf{k} \text{ at point } (9, 7, 23)$$

$$\mathbf{F}_2 = -360\mathbf{i} + 550\mathbf{j} + 720\mathbf{k} \text{ at point } (30, 8, -22)$$

$$\mathbf{F}_3 = 860\mathbf{i} - 25\mathbf{j} - 150\mathbf{k} \text{ at point } (-60, 45, 10)$$

$$\mathbf{M}_1 = 10\,500\mathbf{i} - 6400\mathbf{j} + 13\,600\mathbf{k}$$

$$\mathbf{M}_2 = 16\,000\mathbf{i} + 9500\mathbf{j}$$

Replace this force system by a dynamically equivalent system that consists of a force **F** acting at point (0, 0, 20) and a couple **M**. (Lengths are in centimeters, and forces are in newtons.)

5.89 A pipe wrench is clamped to the piping system shown in Fig. P5.89, and a plumber applies to the wrench handle a force $P = 80$ N directed into the plane of the page. All of the pipes lie in that plane.

a. Determine the equivalent resultant couple and force that act on the hot-water heater at A.

b. Calculate the magnitude and direction angles of the resultant couple.

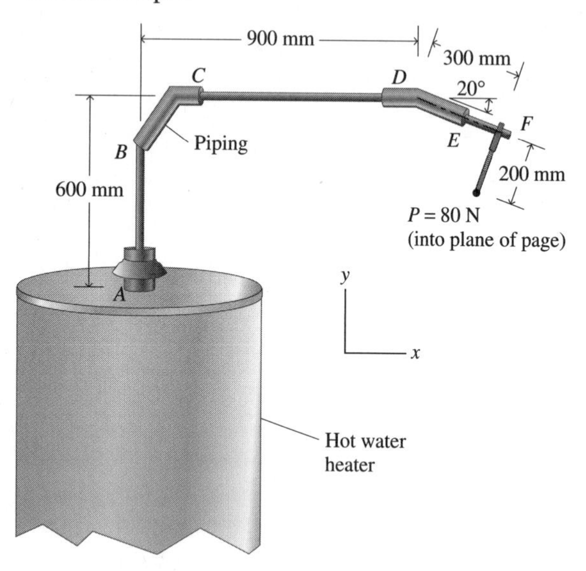

Figure P5.89

5.90 Replace the four forces acting on the rigid block in Fig. P5.90 by a force at O and a couple.

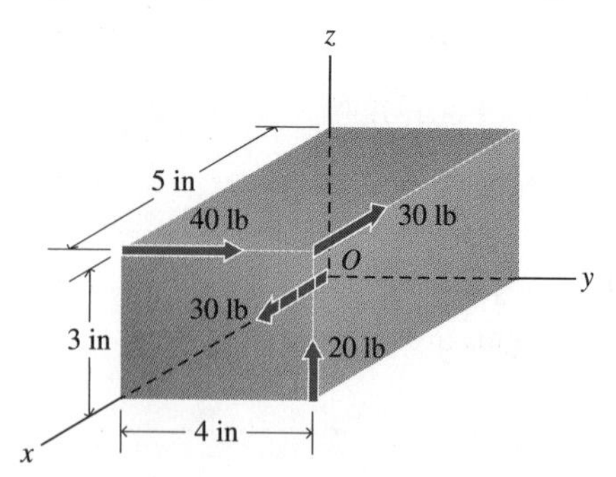

Figure P5.90

5.91 A rigid body is subjected to forces $\mathbf{F}_1 = 3\mathbf{i} + 4\mathbf{j} - \mathbf{k}$ at $(-1, 4, 0)$ and $\mathbf{F}_2 = -2\mathbf{i} + 5\mathbf{j} + 5\mathbf{k}$ at $(0, 8, 6)$ and a couple $\mathbf{M} = 4\mathbf{i} + 6\mathbf{k}$. Replace the two forces and the couple by a dynamically equivalent system consisting of a force $\mathbf{F}$ through the origin and a couple $\mathbf{M}_O$. (Lengths are in feet, and forces are in pounds.)

5.92 A bar is subjected to two forces and a couple, as shown in Fig. P5.92. Determine a dynamically equivalent system consisting of a force at point A and a couple.

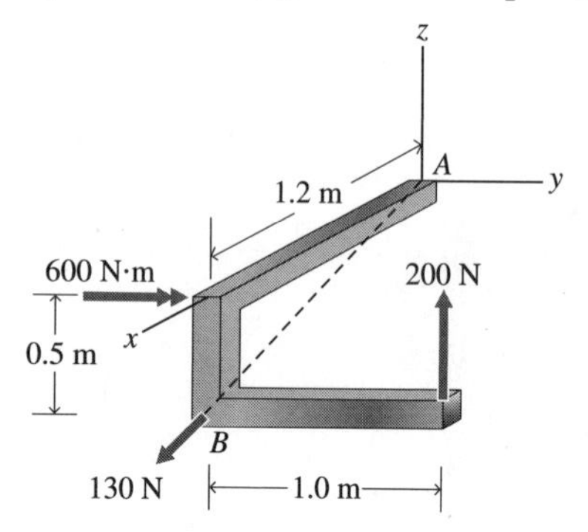

Figure P5.92

5.93 Replace the force system acting on the L-shaped bar in Fig. P5.93 by a dynamically equivalent system consisting of a force $\mathbf{F}$ acting at the origin O and a couple $\mathbf{M}$.

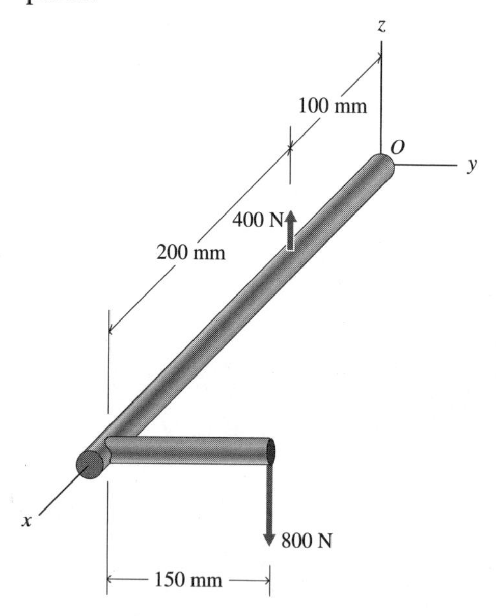

Figure P5.93

5.94 Replace the two forces that act on the rigid block in Fig. P5.94 by a force that acts at point A and a couple.

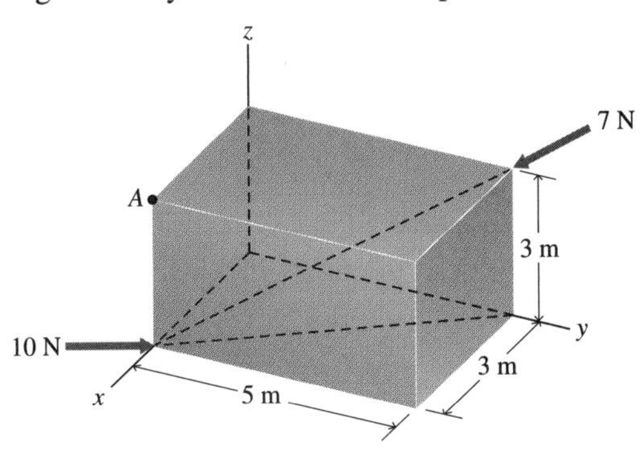

Figure P5.94

5.95 a. Replace the force system acting on the rigid block in Fig. P5.95 by a dynamically equivalent system consisting of a force at the origin O and a couple. The magnitudes of $\mathbf{F}$, $\mathbf{C}_1$, and $\mathbf{C}_2$ are $F = 100$ N, $C_1 = 500$ N·m, $C_2 = 200$ N·m.
b. Compute the moment of the equivalent force system about point A.
c. Determine the moment of the equivalent force system about the line from point A to point E.

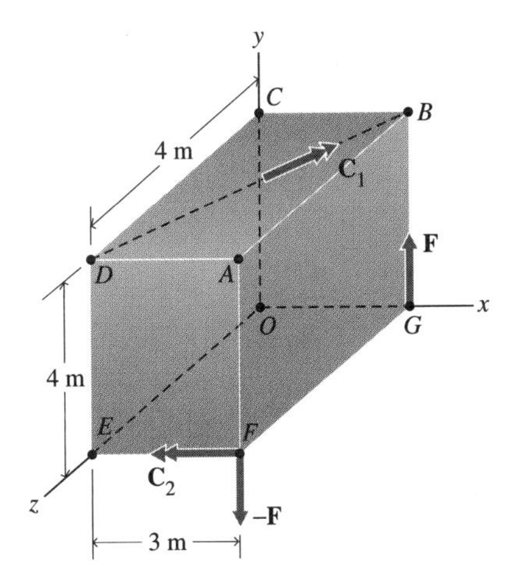

Figure P5.95

5.96 In a test of a wing of a jet airplane, engineers apply forces to the wing at points A, B, C, and D, perpendicular to the xz plane (see Fig. P5.96). The positive y axis is directed out of the plane of the page. The forces at A, B, C, and D are 6000 lb in the positive y direction, 4500 lb in the negative y direction, 4000 lb in the positive y direction, and 3000 lb in the positive y direction, respectively.
a. Determine the dynamically equivalent force and couple relative to the origin O.
b. Determine the moments of the forces about the x and z axes.

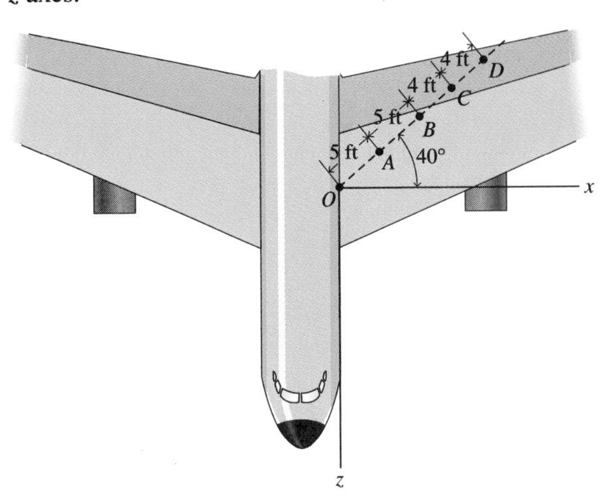

Figure P5.96

5.97 A tubular-bar frame supports a sign (see Fig. P5.97). A strong wind blows against the sign, causing it to form an angle of 20° with the vertical. The cables attached at A and B exert tensile forces of 2 kN each on the frame.

a. For these cable forces, determine the dynamically equivalent force and couple relative to point O, the base of the frame.

b. Determine the moment of the cable forces about the line CD.

c. Determine the moment of the cable forces about the line DE.

d. Treat the uniform sign as a particle, and determine the wind force F and the weight W of the sign.

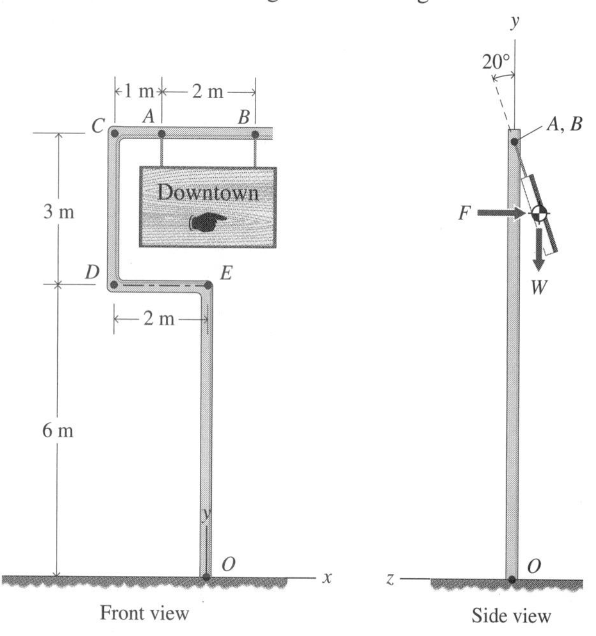

Figure P5.97

5.7 *Resultant Axis and the Wrench*

5.98 Four forces act at the origin of a rectangular xyz coordinate system. The force magnitudes are 4, 6, 3, and 10, and their direction numbers are (3, 2, 1), (1, 1, 1), (3, −2, −1), and (0, 1, 2), respectively. Determine the magnitude of the resultant and the direction cosines of the resultant axis. (Lengths are in meters and forces are in kilonewtons.)

5.99 The following three forces act on a rigid body:

$$\mathbf{F} = -50\mathbf{j} \text{ at } (1, 4, 2)$$

$$\mathbf{G} = 40\mathbf{j} \text{ at } (3, 1, -5)$$

$$\mathbf{H} = 25\mathbf{j} \text{ at } (-10, 0, 6)$$

Determine the resultant force $\mathbf{R}$, the resultant couple $\mathbf{M}$, and the point of intersection of the resultant axis with the xz plane. (Lengths are in feet, and forces are in pounds.)

5.100 A rigid body is subjected to three forces,

$$\mathbf{F} = \mathbf{i} - \mathbf{j} + 0\mathbf{k} \text{ at } (1, 1, 1)$$

$$\mathbf{G} = 0\mathbf{i} + 2\mathbf{j} + \mathbf{k} \text{ at } (-1, 0, 4)$$

$$\mathbf{H} = -3\mathbf{i} + \mathbf{j} + \mathbf{k} \text{ at } (3, 4, 0)$$

and a couple, $\mathbf{C} = 10\mathbf{i} - 5\mathbf{j} + 4\mathbf{k}$. (Lengths are in meters, and forces are in kilonewtons.)

a. Determine the simplest dynamically equivalent system that consists of a force $\mathbf{R}$ and a couple $\mathbf{M}$ parallel to $\mathbf{R}$.

b. Locate the intersection $(a, b, 0)$ of the resultant axis of the equivalent system with the xy plane, where a and b are determined by the condition that $\mathbf{M}$ is parallel to $\mathbf{R}$.

5.101 The force $\mathbf{F} = 3\mathbf{i} - 2\mathbf{j} + \mathbf{k}$ [kN] acts on a rigid body at the point (1, 4, 0) [m]. The body is also subjected to a couple $\mathbf{M}_O = 20\mathbf{i} - 10\mathbf{j} + 30\mathbf{k}$ [kN·m].

a. Replace this force system by a dynamically equivalent system consisting of the force $\mathbf{F}$ at point $(a, b, 0)$ and a couple $\mathbf{M} = M_x\mathbf{i} + M_y\mathbf{j} + M_z\mathbf{k}$.

b. Write the condition of parallelism for the vectors $\mathbf{F}$ and $\mathbf{M}$ in the form $\mathbf{F} \times \mathbf{M} = 0$.

c. Write the vector result obtained in part b in scalar form, and solve any two of the resulting equations for a and b.

d. Show that the third equation of part c is satisfied automatically.

5.102 Solve Problem 5.101 for the conditions $\mathbf{F} = -10\mathbf{i} + 5\mathbf{j} - 8\mathbf{k}$ [kip] at point (3, 2, −5) [ft] and $\mathbf{M}_O = 120\mathbf{i} + 60\mathbf{j} - 80\mathbf{k}$ [kip·ft].

5.103 The force $\mathbf{F} = 10\mathbf{i} - 5\mathbf{j} + 8\mathbf{k}$ [lb] acts on a rigid body at the point (3, 2, −5) [in]. The body is also subjected to a couple $\mathbf{M}_O = 120\mathbf{i} + 60\mathbf{j} - 80\mathbf{k}$ [lb·in].
a. Replace this force system by a dynamically equivalent system consisting of the force **F** at the point $(a, 0, c)$ [in] and a couple $\mathbf{M} = M_x\mathbf{i} + M_y\mathbf{j} + M_z\mathbf{k}$ [lb·in].
b. Write the condition of parallelism for the vectors **F** and **M** in the form $\mathbf{F} \times \mathbf{M} = 0$; that is, determine the resultant axis.
c. Put the equation from part b in scalar form, and solve any two of the resulting equations for the constants a and c.
d. Show that the third equation of part c is satisfied automatically.

5.104 Solve Example 5.17 for $\mathbf{F} = 3\mathbf{i} - 2\mathbf{j} + \mathbf{k}$ and $\mathbf{M}_O = 20\mathbf{i} - 10\mathbf{j} + 30\mathbf{k}$. (Units are pounds and feet.)

5.105 Solve Problem 5.103 using $\mathbf{F} = 3\mathbf{i} - 2\mathbf{j} + \mathbf{k}$ at the point (1, 2, −3) and $\mathbf{M}_O = 20\mathbf{i} - 10\mathbf{j} + 30\mathbf{k}$, with units in kilonewtons and meters.

5.106 Determine the wrench for the force system shown in Fig. P5.106. (*Hint:* See Example 5.18.)

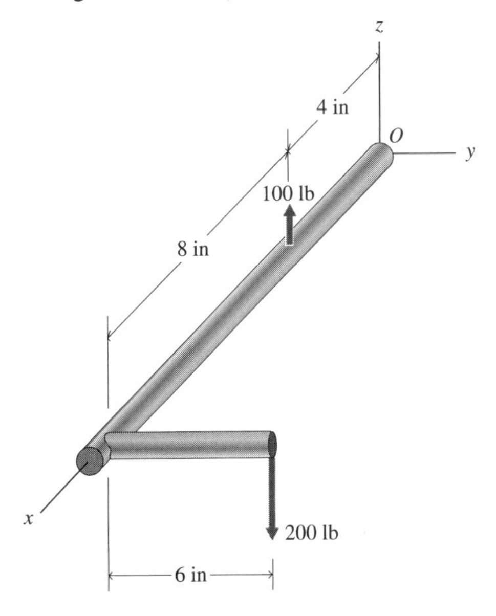

Figure P5.106

5.107 Determine the wrench at the point $(0, b, c)$ for the forces given in Problems 5.80.

5.108 Determine the wrench at the point $(a, 0, c)$ for the forces given in Problem 5.88.

5.109 Determine the wrench for the force system acting on the rigid block in Fig. P5.109.

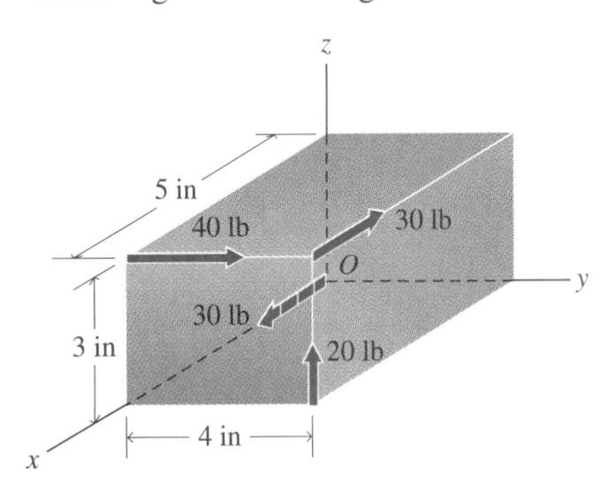

Figure P5.109

5.110 Determine the wrench for the force system acting on the rigid block in Fig. 5.110.

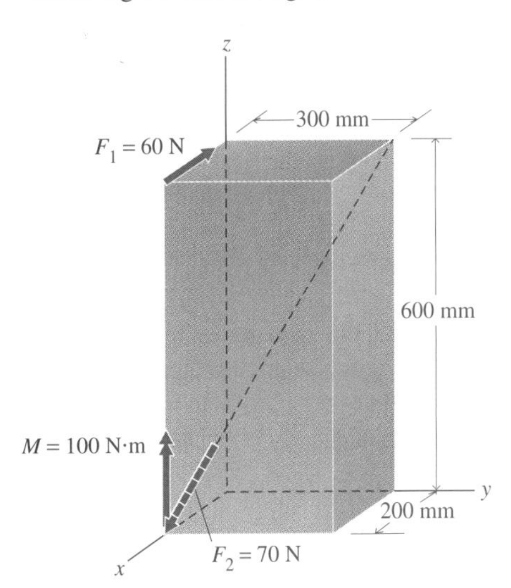

Figure P5.110

5.111 Determine the wrench for the force system acting on the rigid hanger-bar in Fig. P5.111.

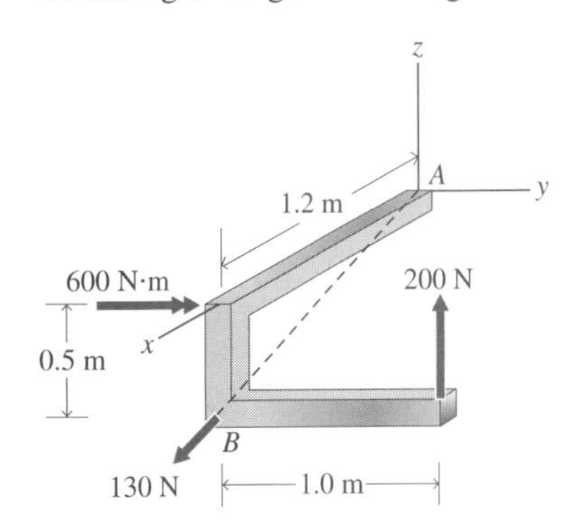

Figure P5.111

5.112 Determine the wrench for the forces acting on the rigid L-shaped bar in Fig. 5.112.

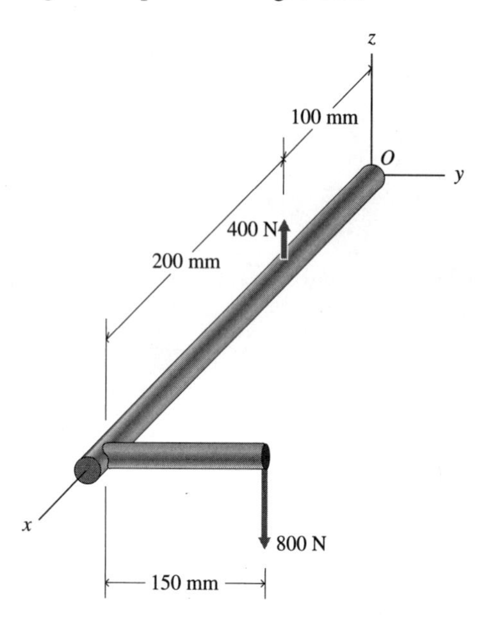

Figure P5.112

5.8 *The Fundamental Principle of Equilibrium of a Rigid Body*

In Problems 5.113–5.125, the systems are in equilibrium. Sketch the free-body diagram of each indicated part of a system. Where the weight of the body is to be considered, the location of the center of gravity is shown in the figure.

5.113 Rod *OAB* in Fig. P5.113

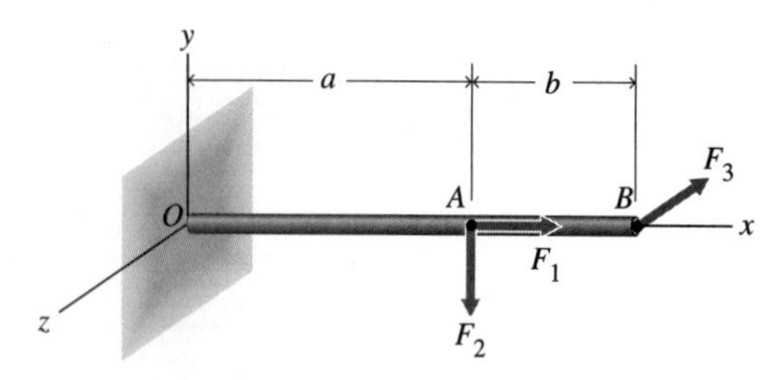

Figure P5.113

5.114 Spool *S* in Fig. P5.114

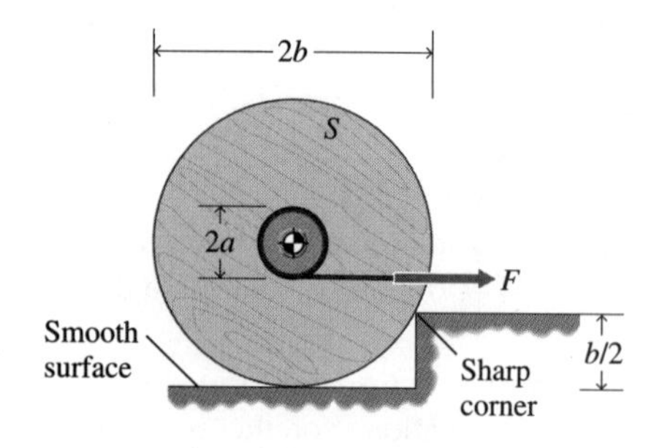

Figure P5.114

5.115 Bar *OA* (of weight W_b) and weight *W* in Fig. P5.115

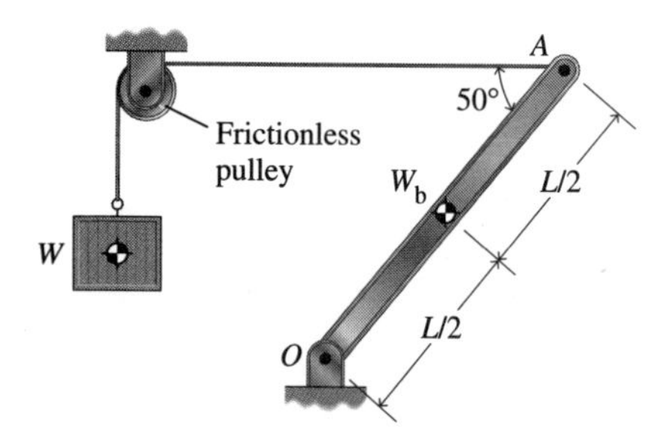

Figure P5.115

5.116 The sign (side view) in Fig. P5.97

5.117 The frame *OEDCAB* in Fig. P5.97, with the sign displaced as shown in the side view

5.118 The bar in Fig. P5.92, taking point *A* as fixed to a rigid support

5.119 The L-shaped bar in Fig. P5.93, taking point *O* as fixed to a rigid support

5.120 Bar *OAB* in Fig. P5.120, assuming that the pin at *O* and the roller at *B* are frictionless

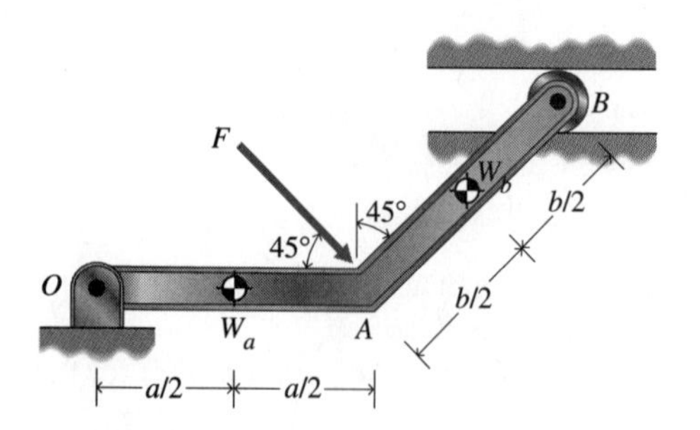

Figure P5.120

5.121 Crank bar *ABCDO* in Fig. P5.121, where segments *OD* and *BA* are parallel to the *x* axis, *CB* is parallel to the *y* axis, and *DC* is parallel to the *z* axis

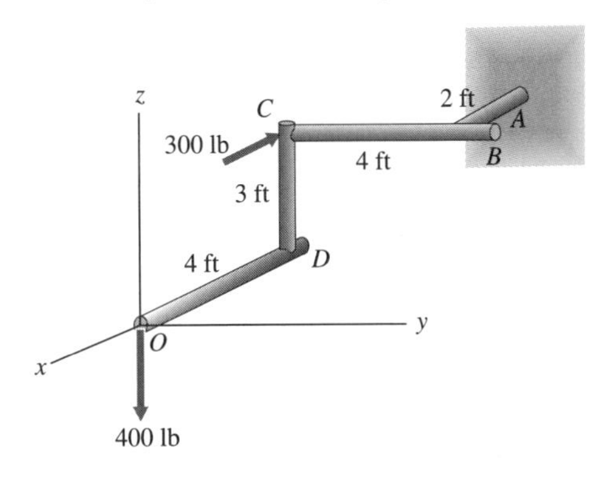

Figure P5.121

5.122 The identical uniform bars *AB* and *BC* in Fig. P5.122 as a single member, *ABC* (that is, do not separate them). Each bar is of weight *W*. They are hinged at *B*.

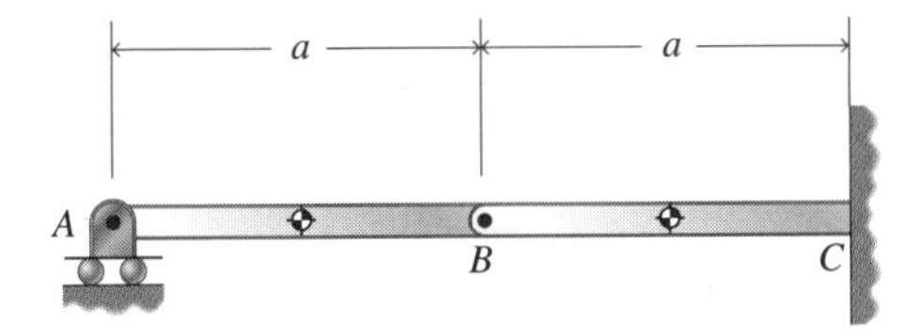

Figure P5.122

5.123 The identical uniform bars *AB* and *BC* in Fig. P5.122, taken as separate bars

5.124 The uniform car hood *OABC* in Fig. P5.124. The rod *DE* is pinned at both ends so that it is a two-force member. The supports at *O* and *C* cannot exert moments. Also, the support at *C* cannot exert a force in the *y* direction.

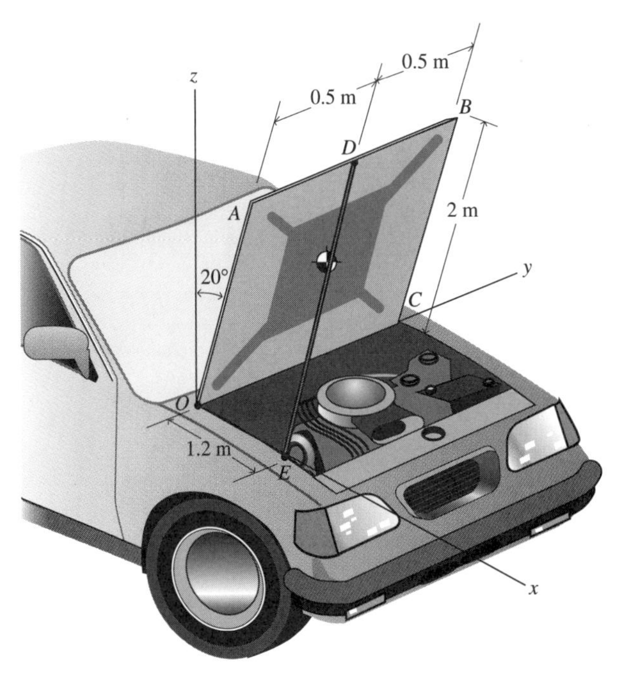

Figure P5.124

5.125 The uniform barn door of weight W in Fig. P5.125. The door is hinged at A and B. The hinges are incapable of resisting a moment about any axis.

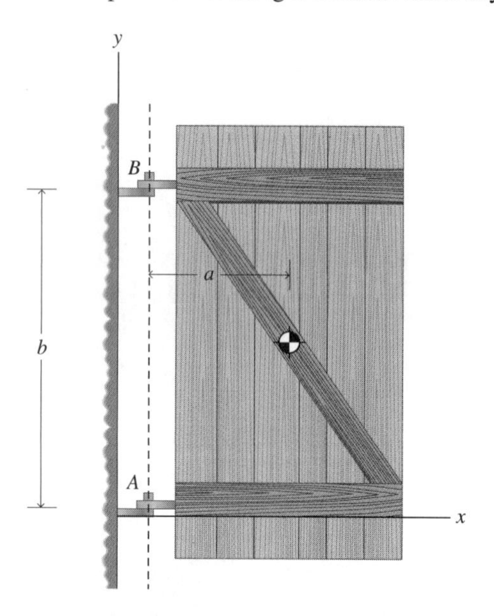

Figure P5.125

5.126 The hinge at B in Fig. P5.125 supports no vertical load. Neither of the hinges is capable of resisting moments. Determine the support reactions at A and B. Let $a = 1.0$ m, $b = 2.0$ m, and the weight of the door be $W = 900$ N.

5.127 The hinge at A in Fig. P5.125 supports no vertical load. Neither of the hinges is capable of resisting moments. Determine the support reactions at A and B. Let $a = 2.5$ ft, $b = 6$ ft, and the weight of the door be $W = 160$ lb.

5.128 The L-shaped bar ABC in Fig. P5.128 is fixed to a wall at A. The leg BC is parallel to the y axis. A force $\mathbf{F} = -40\mathbf{i} + 60\mathbf{j} + 30\mathbf{k}$ [N] is applied at C.

a. Replace this force by a force at B and a couple.

b. Determine the force and the moment exerted on the bar by the wall at A.

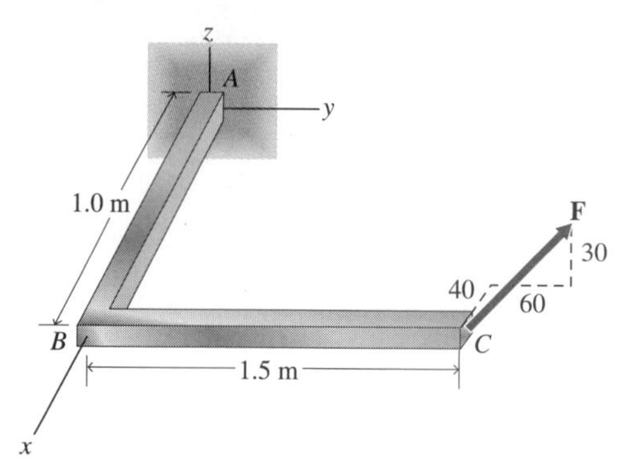

Figure P5.128

5.129 Bar ABC in Fig. P5.129 lies in the xz plane. A 50 N force acts at point C on the bar and lies in the xy plane. The bar is clamped at point A. Determine the reactions F_x, F_y, F_z and M_x, M_y, M_z at the clamp.

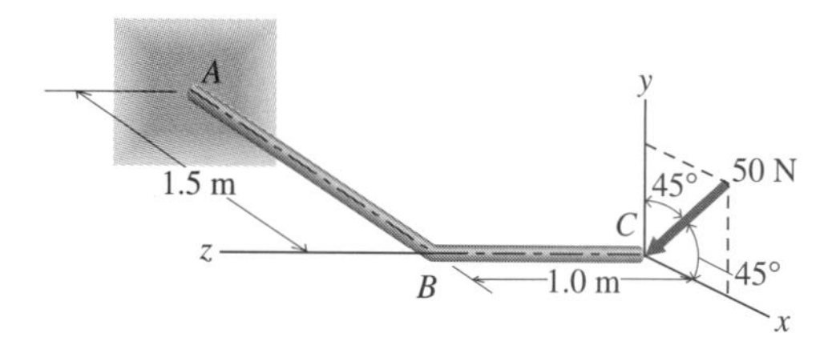

Figure P5.129

5.130 The crank bar $ABCDO$ in Fig. P5.121 is held in equilibrium by a rigid wall at A. The wall is parallel to the yz plane. The 300 lb force is parallel to the x axis, and the 400 lb force is parallel to the z axis. Determine the force $\mathbf{F}_A$ and the couple $\mathbf{M}_A$ that the wall exerts on the bar. Neglect the weight of the bar.

5.131 The segment AB of the crank bar in Fig. P5.121 is cut off, and the remaining segment $BCDO$ is supported by a rigid wall at B parallel to the xz plane. Determine the force $\mathbf{F}_B$ and the couple $\mathbf{M}_B$ that the wall exerts on the bar. Neglect the weight of the bar.

5.132 The structure $ABCDE$ shown in Fig. P5.132 has a constant cross-sectional area and weighs 140 kN. It is held in equilibrium in the xy plane by a ball-and-socket joint at A, a bearing at E, and a flexible cable CF. (The z axis is vertical.) The ball-and-socket joint exerts reactions in the x, y, and z directions, and the bearing exerts reactions in the y and z directions. Determine the tension in the cable CF. The cross-sectional dimensions of the bars are small compared to their lengths.

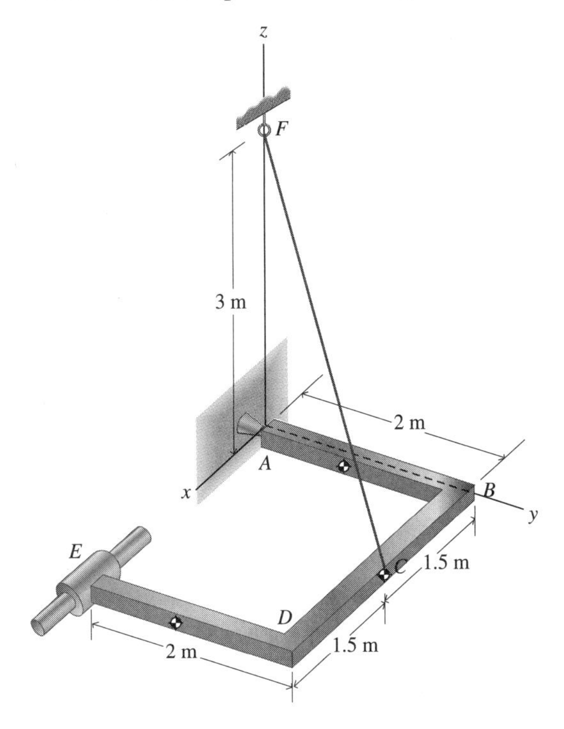

Figure P5.132

5.133 A uniform slender straight bar of weight W and length L is supported by forces **F** and **G**, applied at its ends (see Fig. P5.133). No moment is exerted at either end of the bar. Let xyz be rectangular coordinate axes, with the z axis directed vertically upward. Let the ends of the bar be located at the points (0, 0, 0) and (x, y, z).

a. Write the six equations of equilibrium in terms of the force projections F_x, F_y, F_z and G_x, G_y, G_z, assuming that **F** is applied at (0, 0, 0) and **G** is applied at (x, y, z).

b. Solve for the terms G_x, G_y, and G_z in terms of F_x, F_y, F_z, and W, and show that one of the remaining equations is redundant (see Appendix A).

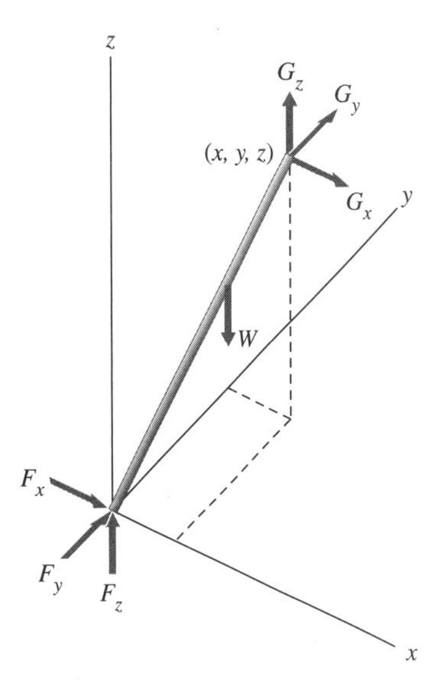

Figure P5.133

5.134 Determine the support reactions at point O on the frame in Fig. P5.97.

5.135 Determine the support reactions at O in Fig. P5.113.

5.136 The spool in Fig. P5.114 is on the verge of rolling over the sharp corner. Determine all the forces that act on the spool. The spool weighs 100 lb, $b = 1$ ft, and $a = 0.25$ ft.

5.137 The bar OA in Fig. P5.115 weighs 1.5 kN and $L = 3$ m.

a. Determine all the forces that act on bar OA, and show them in a diagram.

b. Determine the weight W required to maintain equilibrium of the system.

5.138 In Fig. P5.120, let $a = 1.0$ m, $b = 1.2$ m, and $F = 2$ kN. The uniform bar weighs 0.5 kN/m. Determine the support reactions on the bar at O and B. The cross-sectional dimensions of the bar are small compared to a and b.

5.139 In Fig. P5.122, each of the identical and homogeneous bars AB and BC has weight W. Determine the force system that acts on each bar in terms of a and W.

5.140 Each of the homogeneous bars AB and BC in Fig. P5.140 weighs 40 lb and has length $a = 4$ ft. Determine the force system that acts on each bar. The bars are hinged together at B.

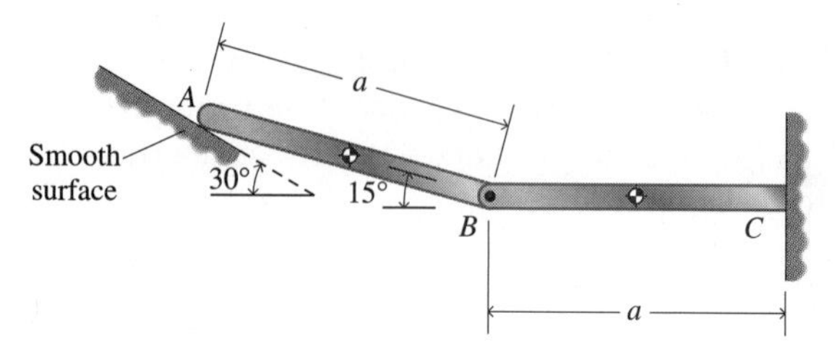

Figure P5.140

5.141 A uniform horizontal boom OD is fastened by a three-dimensional pin to a vertical wall at point O (see Fig. P5.141). The boom weighs 3.60 kN. Determine the tensions in the supporting cables AB and AC and the projections R_x, R_y, and R_z of the support reaction at point O.

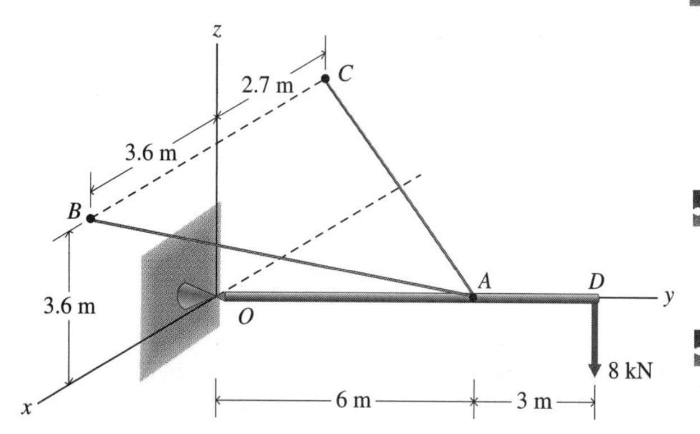

Figure P5.141

5.142 Figure P5.142 represents the nose landing gear of an airplane. The shock strut AC is a continuous rod. The links BD and BE are pin-connected. Also, joint C is pinned. For a side load of 67.50 kN and a vertical load of 42.75 kN, determine the forces in the links BD and BE. The thickness of ABC is small compared to the lengths of the members.

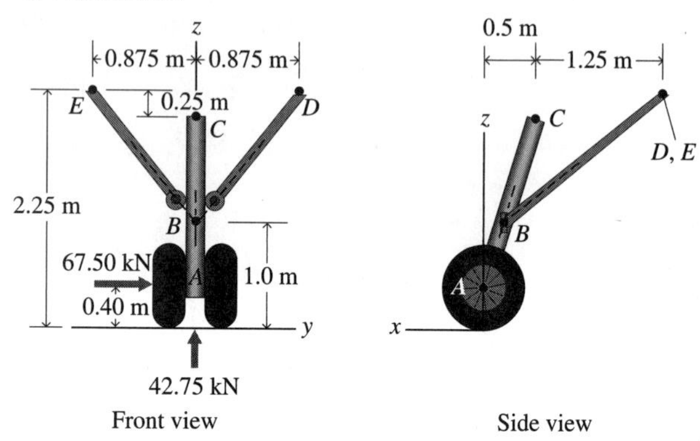

Figure P5.142

5.143 The car hood in Fig. P5.124 weighs 175 N. The rod DE is shortened, and its upper end D is pinned to the side OA of the hood, 0.3 m from point O. The center of gravity of the hood is located at the centroid of the area $OABC$. (Assume that the hood is flat.) Determine the support reactions at O and C and the force in the rod DE.

5.144 Referring to Problem 5.143 and Fig. P5.124, suppose the supports at O and C are moved to new locations at $y_O = 0.2$ m and $y_C = 0.8$ m. Determine the support reactions at O and C and the force in the rod DE.

5.145 Point F in Fig. E5.21a is moved in the xz plane so that the angle between the negative x axis and line OF is increased to 45°. The length of cable CF is not changed.

a. Determine the tensions in cables DB, CF, and CG.

b. Determine the forces that act on the boom ADE at A.

c. Determine the support reactions at O.

5.146 A horizontal flagpole (see Fig. P5.146) weighs 35 lb and is supported by a ball-and-socket joint at *A* and two cables attached at *B*. The flag exerts a force of 20 lb and a couple of 40 lb·ft at *C*. The couple is parallel to the *x* axis. The cable *BE* is attached at point *E*, a horizontal distance *d* from a vertical line through *A*.

a. Plot the force required in cable *BE* to maintain equilibrium of the flagpole for the range $0 \leq d \leq 6$ ft.

b. What restrictions must be placed on *d* if the cable *BE* can support a maximum load of 100 lb?

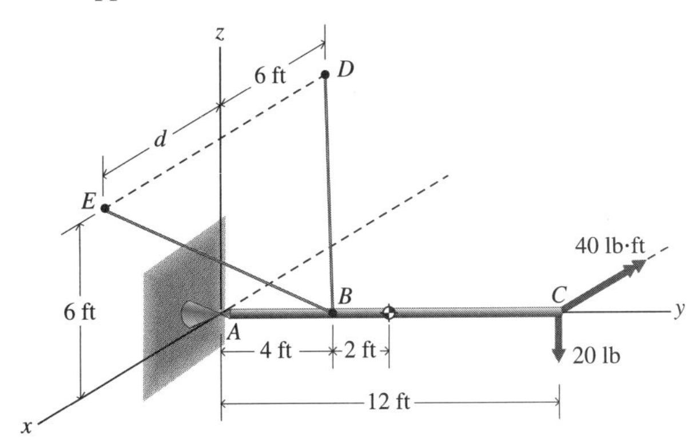

Figure P5.146

5.147 Figure P5.147 represents a mechanism that is part of a robot's arm and that supports a weight *W* in various positions. The pins at *A, B, C,* and *D* are a part of member *ABCD*.

a. Express the pin reactions at *A* and *B* in terms of the weight *W* and the angle θ, measured in the *xy* plane. Assume that the pin reactions are tangent to the surfaces of member *ABCD* and that the vertical forces that act on *ABCD* are at *B* and *C;* that is, there are no vertical forces at *A* and *D*. Also, neglect the weight of the members.

b. For $W = 200$ N, plot the pin reactions at *A* and *B* as a function of angle θ in the range $0 \leq \theta \leq 150°$, where θ is the angle that arm *CDEF* rotates about *CD*.

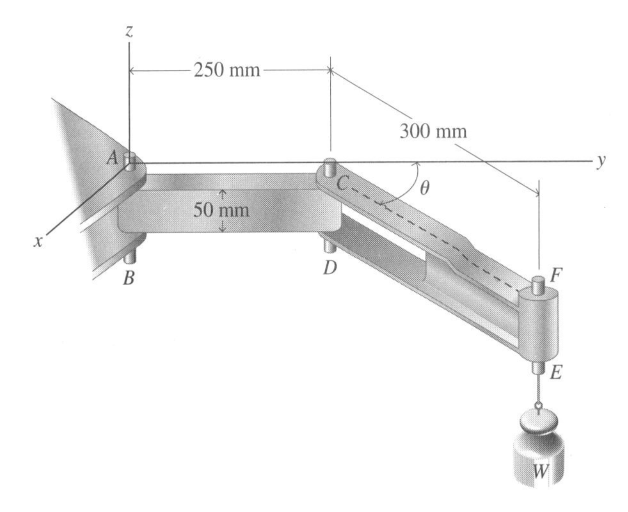

Figure P5.147

5.148 A robot mechanism similar to that of Problem 5.147 is subjected to horizontal forces **P** and −**P** at E and F, which lie in a vertical plane (see Fig. P5.148).

a. For $\theta = 30°$, determine the (x, y, z) projections of the pin reactions at A and B in terms of the magnitude P of the forces and the angle ϕ. Use the same assumptions as in part a of Problem 5.147.

b. For $P = 200$ lb, plot the (x, y, z) projections of the pin reactions at A and B as a function of the angle ϕ, where $0 \le \phi \le 90°$.

c. For $P = 200$ lb, plot the magnitude of the pin reactions at A and B as a function of the angle ϕ, where $0 \le \phi \le 90°$. Could you have anticipated this result? Explain.

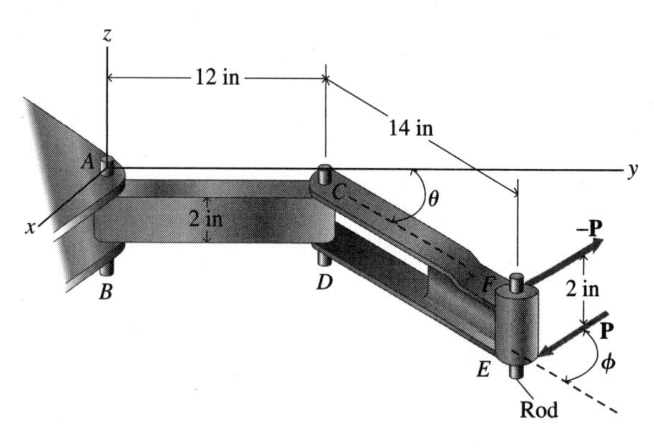

Figure P5.148

5.149 In the preliminary design of a large uniform sign that weighs 1.6 kN, it is suggested that the sign be supported by two cables and a ball-and-socket joint at A, as shown in Fig. P5.149.

a. Consider the geometry of the system as shown, and suggest two practical alternative placements of the cables that could reduce the tensions in them.

b. Write a short paragraph on the advantages and disadvantages of your suggestions as compared to the original suggestion.

Method: Analyze the system in the present configuration. Then move the cables as per one suggestion and reanalyze the system. Repeat the analysis for the second suggestion. Then do part *b.*

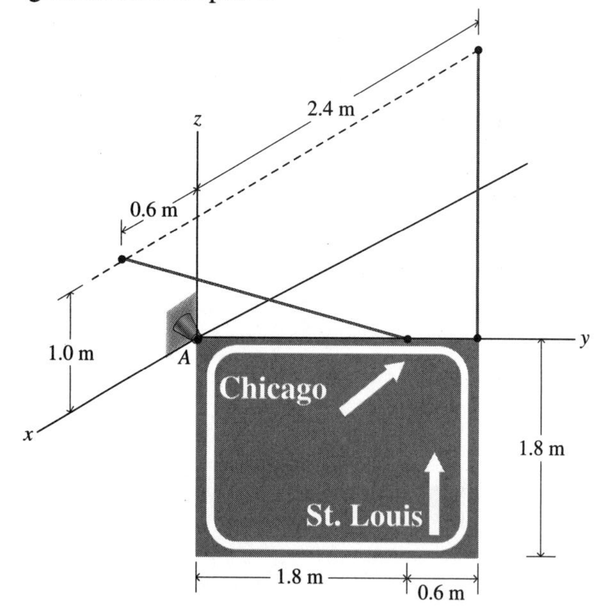

Figure P5.149

5.150 For the crane in Fig. E5.21, the tension in cable DB was found to be considerably larger than the tensions in cables CF and CG. Your supervisor asks you to propose other possible locations for the end supports of the cables—that is, the supports at B, C, D, F, and G—that would make the tensions in cables DB, CF, and CG more equal. However, you must not change the lengths of the column OC or the boom AE.

Method: Consider a possible change of location for one or two supports, and estimate the effects of the change. That is, will it result in more equal tensions in the cables? Determine the tensions to see whether it does. Ideally, to impress your supervisor, you should try to make the tensions in the cables equal, or nearly so. This solution may require several changes and reanalyses.

REVIEW QUESTIONS

5.151 Define *direction cosines.* Express the (x, y, z) projections of a vector $\mathbf{F}$ in terms of the magnitude F and the direction cosines of the vector. Show that the sum of the squares of the direction cosines is 1.

5.152 Write the formulas for the distance between two points in space in terms of the rectangular coordinates of the points.

5.153 Define *free vector* and *slide vector.* Give a physical example of each type of vector.

5.154 Explain how a couple may be represented by a vector. Is it a free vector or a slide vector?

5.155 What is the moment about an axis in space of a force $\mathbf{F}$ that acts at the point (x, y, z)? Write the equations for the moments of the force about the coordinate axes.

5.156 Define *scalar product of two vectors.* Explain how it is used to determine the angle between two vectors.

5.157 Using the scalar product, derive a formula for the angle between two lines with given direction cosines.

5.158 Explain how the scalar product is used to determine the projections of a vector on a line with given direction cosines.

5.159 Two vectors are defined by their (x, y, z) projections. How would you check to determine whether the vectors are perpendicular to each other?

5.160 How could you verify that the scalar product follows the ordinary rules of algebra?

5.161 Define *vector product of two vectors.* What is the magnitude of this product? What is its direction? What is its sense?

5.162 Calculate the angle between the vectors $\mathbf{A} = 2\mathbf{i} + 3\mathbf{j} + \mathbf{k}$ and $\mathbf{B} = 5\mathbf{i} - 2\mathbf{j} - 4\mathbf{k}$.

5.163 If $\mathbf{C} \neq 0$, what can be concluded from the equation $(\mathbf{A} - \mathbf{B}) \cdot \mathbf{C} = 0$?

5.164 A particle is displaced through the distance $\sqrt{201}$ m. The direction cosines of the line along which the particle is displaced are $(4/\sqrt{201}, -4/\sqrt{201}, 13/\sqrt{201})$. Determine the projections of this displacement on xyz axes.

5.165 The rectangular coordinates of a point P are $(4, -3, 0)$. Determine the direction cosines and the length of the straight line segment PO that joins P to the origin.

5.166 How is the right-hand screw rule used to determine the sense of a vector product?

5.167 Represent the vector product of two vectors by a determinant.

5.168 Show that the vector product is not commutative.

5.169 Show that the magnitude of the vector product $\mathbf{a} \times \mathbf{b}$ is equal to the area of the parallelogram of which vectors $\mathbf{a}$ and $\mathbf{b}$ form two conterminous sides.

5.170 Write the nine formulas for the vector products of pairs of the unit vectors $\mathbf{i}$, $\mathbf{j}$, and $\mathbf{k}$.

5.171 Explain how a couple may be represented as a vector product.

5.172 Define *moment of a force about a point.* Explain how it is represented as a vector product.

5.173 Show how the moment of a force about an axis with given direction cosines may be represented as a scalar triple product. Express the relation in the form of a determinant.

5.174 Explain how the line of action of a force may be displaced if a compensating couple is introduced.

5.175 A rigid body is subjected to the action of several forces and couples. State and prove the general theorem that permits this system to be replaced by a dynamically equivalent system consisting of a single force acting at an arbitrary point O and a couple.

5.176 Show that, if the vector sum of a set of forces is zero, the forces exert the same moment about all points.

5.177 Define the *resultant axis* of noncoplanar forces that act on a rigid body. Explain how the existence of the resultant axis is demonstrated.

5.178 State the fundamental principle of equilibrium of a rigid body that is subjected to noncoplanar forces.

5.179 *True or False:* If a system of forces that act on a rigid body is reducible to a force and a couple, it can be reduced further to a single force.

Chapter 6 TRUSSES

The Navajo Bridge spans 834 feet across the Colorado River, 467 feet above the water at the mouth of the Grand Canyon in northern Arizona. The bridge is a three-hinge truss arch in the Pratt configuration.

A Look Forward

IN PREVIOUS CHAPTERS, we covered many of the fundamental principles of statics. Much of the rest of this book will be devoted to the application of those principles in the analysis of structural and mechanical systems. In this chapter, we will study a particular kind of structure called a *truss*. We will analyze trusses using the principles of particle and rigid-body equilibrium introduced in Chapters 3 and 4.

After completing this chapter, you should understand the basic assumptions that are made in the analysis of planar (two-dimensional) trusses. You should also be able to analyze a truss by either the method of joints or the method of sections. You should be able to identify simple, compound, and complex trusses. Finally, you should have an understanding of space (three-dimensional) trusses and how they are analyzed.

Survey Questions

- Is a member of a truss a particle or a rigid body?
- What is a simple truss?
- What is the method of joints?
- What is the method of sections?
- What is a compound truss?
- What is a complex truss?

6.1 THE TRUSS—A SYSTEM OF TWO-FORCE MEMBERS

After studying this section, you should be able to:

- Recognize some truss configurations used in roofs and bridges.
- Identify the characteristics of an ideal truss.

truss

A *truss* is an assembly of slender, straight members that carries loads *primarily* through axial forces (tension or compression) in the members. The arrangement of members in a truss makes it an efficient system for carrying loads. That is, a truss can carry heavy loads compared to its own weight. Several types of trusses used in the supporting structures of roofs and in steel and timber bridges are shown schematically in Fig. 6.1. These trusses are called *planar trusses* because all members and loads lie in the same plane. Each of the trusses in Fig. 6.1 has a particular name that is associated with its geometric configuration. These and other truss configurations also appear in radio towers, in special structures like the Epcot dome at Disney World, and in innumerable other engineering and architectural structures.

planar truss

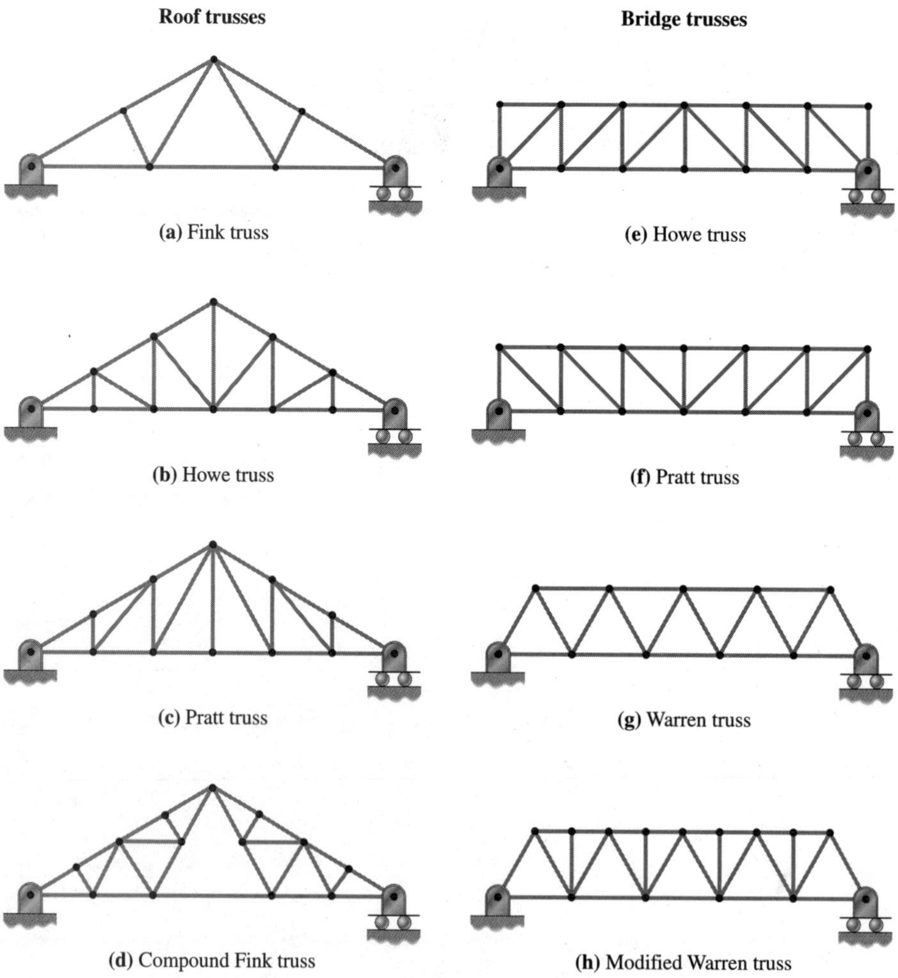

Roof trusses

(a) Fink truss

(b) Howe truss

(c) Pratt truss

(d) Compound Fink truss

Bridge trusses

(e) Howe truss

(f) Pratt truss

(g) Warren truss

(h) Modified Warren truss

Figure 6.1 Types of trusses used in roofs (left column) and bridges (right column).

Open-web joists are two-dimensional trusses commonly used for roof systems of buildings. They are lighter in weight but deeper than wide-flange beams of equivalent strength.

Trusses are constructed in a wide variety of ways. The materials used in trusses include timber, steel, and aluminum. The members in a truss can be joined by bolts, metal nail plates (truss plates), welds, hinges in the form of a single large pin, or other means. Joints in a truss are usually located at the ends of the members. However, sometimes a member will be continuous (unbroken) through a joint. Trusses are usually designed so that the loads (other than the weights of the individual members) are concentrated at the joints.

The key feature of a truss that distinguishes it from other structural forms is that it resists loads primarily through axial forces in its members. That is, members of a truss are designed to carry tension or compression. Whereas bending and shear forces (discussed in

This riveted truss bridge in Boston uses eyebars (see blow-up) for its tension members. The joint at the end of the eyebar closely represents the pin in an ideal truss.

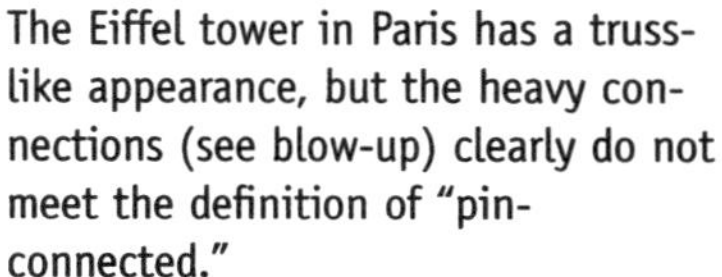

The Eiffel tower in Paris has a truss-like appearance, but the heavy connections (see blow-up) clearly do not meet the definition of "pin-connected."

Sec. 11.2) might exist, their effect on a truss member is usually of secondary importance. In contrast, beams carry loads primarily through bending and shear forces rather than axial force.

HISTORICAL REVIEW OF TRUSSES

The trusses shown in Fig. 6.1 represent some of the classic configurations used since the Industrial Revolution of the 19th century. Because of their simplicity of form and unique characteristics, they are often included in textbooks on statics and structural analysis. At the time of their development, these trusses were distinguished by both their configurations and the materials used in their construction. Today, the configurations of these trusses retain the names of the individuals who developed them, whereas the materials used in their construction are not specified. To help you recognize the significance of the various designs, we will briefly review the development of these and other trusses used in bridges.

Truss forms have been used for many centuries to span relatively long distances. In fact, the Roman engineer Apollodorus constructed a multi-span, truss-type bridge across the Danube River about 105 A.D. Each span of the bridge took a form similar to the arch shape shown in Fig. 6.2 (Turner & Goulden, 1981). Surprisingly, though, the design of trusses for use in bridges was not based on general mathematical and physical principles until the Industrial Revolution. It was during this time that the availability of wrought iron in Europe and the westward expansion of the railroads in the United States pressured engineers to develop more rational truss designs for light-weight, long-span bridges.

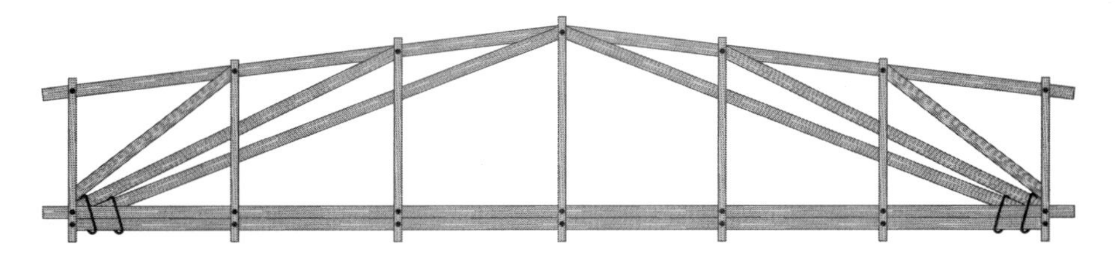

Figure 6.2 Apollodorus's bridge across the Danube (adapted from Turner & Goulden, 1981).

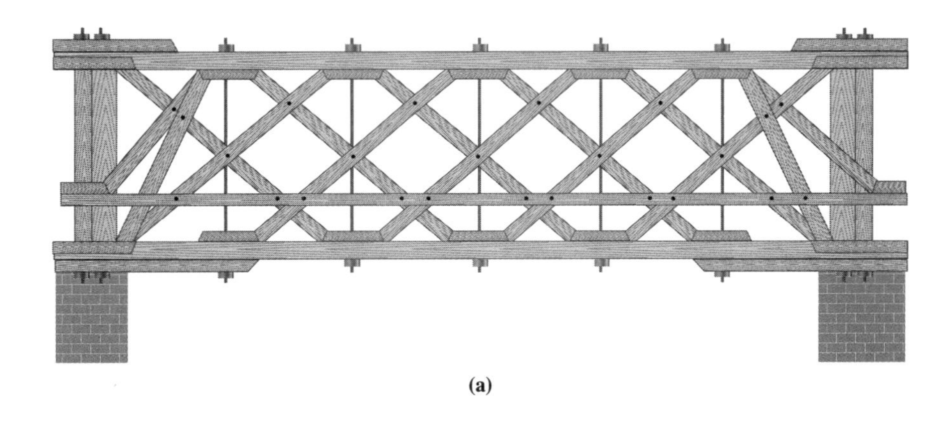

(a)

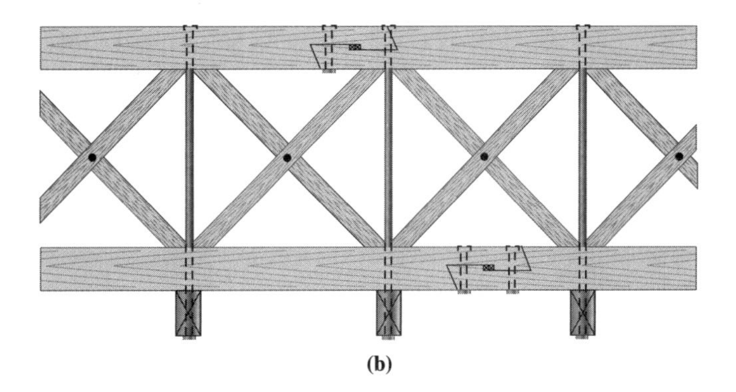

(b)

Figure 6.3 Howe trusses: **(a)** bridge truss (adapted from Jacobs & Neville, 1968); **(b)** stiffening truss for suspension bridge (adapted from Peters, 1987).

Howe truss

One of these truss forms, the *Howe truss,* was patented by William Howe in 1840 (Jacobs & Neville, 1968). Howe's design was an advance over a previous design that used a latticework of diagonal pieces. Howe simplified the design and added wrought-iron vertical members (see Fig. 6.3a). When loads are applied along the bottom chord of the Howe truss, as in a thru-truss bridge,[1] the vertical members are in tension and the diagonals are in compression. Howe pioneered the use of wrought iron in bridges because of its superior ability to carry tension (timber is more suitable under compression). The form of the Howe truss shown in Fig. 6.3b was widely used to stiffen the roadways of

1. A thru-truss bridge has its roadway at the level of the bottom chord. Hence, one travels through the truss rather than on top of it.

suspension bridges in Europe in the early 19th century (before Howe received his patent) (Peters, 1987). In the statically determinate forms of the Howe truss (Figs. 6.1b and 6.1e), the vertical members are in tension and the diagonals are in compression.

Pratt truss

In 1844, the Howe truss design was modified by Thomas Pratt and his father Caleb to create the *Pratt truss* (see Figs. 6.1c and 6.1f). Their design reversed the orientation of the diagonals, which had the effect of placing those members in tension and the vertical members in compression. This change came in response to the growing use of iron as the construction material for an entire bridge. In the Pratt truss, the compression members (the verticals) are shorter than in the Howe truss, which leads to greater inherent stability of those members. Like the Howe truss, the Pratt truss is noted today for the orientation of its members, not for the materials used to build it.

Warren truss

In 1848, a truss with horizontal top and bottom chords and diagonals inclined at 60° (see Fig. 6.1g) was patented in England by James Warren and W. T. Manzoni (Singer et al., 1958). Construction of the *Warren truss* involved use of castings to form triangular units consisting of the top chord and two diagonals of one bay. These units were then joined with wrought-iron threaded rods, forming the bottom chord. A modification of the basic form results from the addition of a vertical member in each bay (see Fig. 6.1h) or in alternating bays. These verticals serve to brace the top chord against buckling under compression forces.

Fink truss

Albert Fink, a German-born engineer and railroad businessman, made further refinements to all-iron railroad bridges and achieved a design that remains popular today. The configuration he patented in 1851 is known as the *Fink truss* (see Fig. 6.4a). This configuration is actually a compound truss made of simple elements (see Fig. 6.4b). Fink realized that such an assembly of elements made the bottom chord of the truss superfluous; the roadway could simply be placed on the top chord of the truss. The basic element of Fink's design is commonly used in roof trusses for residential (Fig. 6.1a) and industrial (Fig. 6.1d) construction.

bow-string truss

The *bow-string truss* (Fig. 6.5) was developed by Squire Whipple, a surveyor, who published the first mathematical analysis of bridge design in his 1847 book *A Work on*

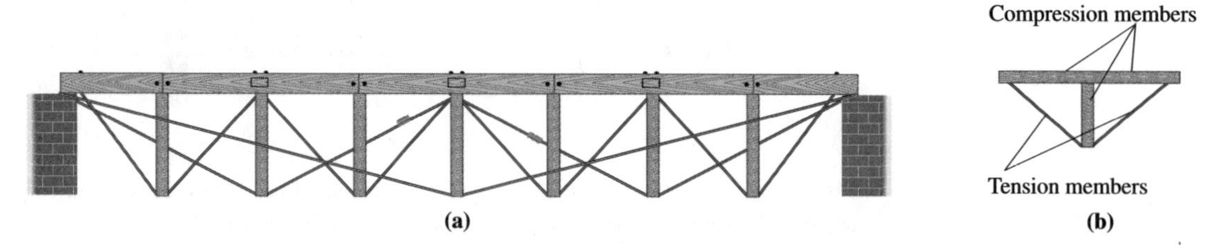

Figure 6.4 Fink truss: **(a)** compound Fink deck truss (adapted from Jacobs & Neville, 1968); **(b)** basic element of the Fink truss.

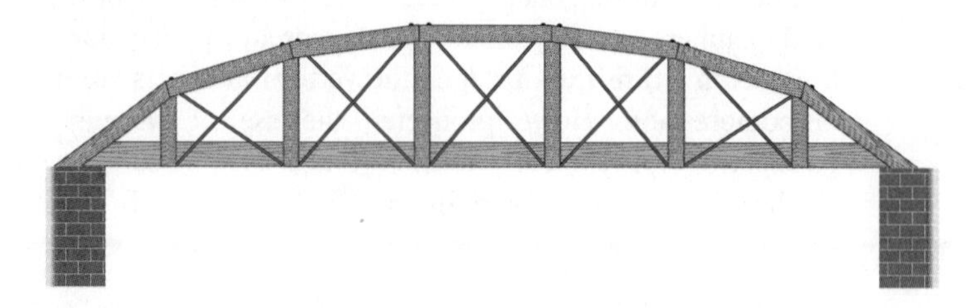

Figure 6.5 Bow-string truss (adapted from Jacobs & Neville, 1968).

Bridge Building. Whipple's bow-string truss combines the principles of the arch with those of the truss. The truss shown in Fig. 6.5 also has a feature known as *counter-diagonals* (see Sec. 6.3).

IDEALIZATION OF TRUSSES

Key Concept

A real truss can often be analyzed as an ideal truss.

Line diagrams of trusses, as illustrated in Fig. 6.1, are useful for analysis purposes. In some trusses, such as those with counterdiagonals, two members may cross each other without a connection. Therefore, the joints (connections) at the ends of members are identified by solid black dots, as shown in Fig. 6.1, to distinguish them from locations at which members cross without connections. Since the primary load-carrying mechanism in a truss is axial force in the members, it is customary to make certain assumptions about a truss to simplify its analysis. The following assumptions permit the use of relatively simple concepts from statics alone to analyze a truss. Without use of these assumptions, truss analysis requires complex procedures that are not practical without the aid of a computer.

Assumptions for an Ideal Truss

1. All members of a truss are straight and can be represented by lines (which have no width).
2. Joints occur only at the ends of the members. Joints can be represented by points (which have no size).
3. All joints are formed by frictionless pins.
4. The weight of each member is applied at the ends of the member, or the weight of each member is negligible. In this book, we will usually assume that the member weight can be ignored.[2]
5. Only concentrated loads can be applied to a truss, and they are applied at the joints.
6. For a planar (two-dimensional) truss, all members and loads lie in the same plane. For a space (three-dimensional) truss, the members are not coplanar and the directions of the loads are arbitrary. We will focus mainly on planar trusses; space trusses are discussed briefly in Sec. 6.7.

ideal truss

These assumptions establish the characteristics of an *ideal truss.* We can thus conclude that an ideal truss is a system of straight two-force members. As discussed in Sec. 3.5, a two-force member in equilibrium is one that is subjected to two forces that are collinear and equal in magnitude but opposite in sense (see Fig. 6.6). Thus, since the pins of a truss are smooth and the loads are applied at the joints, each member of a truss is a two-force member; it is either in tension (Fig. 6.6a) or in compression (Fig. 6.6c). The frictionless pin is not capable of resisting a couple, so it does not cause bending in the members of a truss.

In general, a two-force member may be curved (see Fig. 3.11c), in which case it is subjected to shear force, bending moment, and tension or compression. However, each member of a truss is a straight two-force member, and the force in the member (tension or com-

2. Frequently, the weight of a member has a component perpendicular to the member's axis. The member then bends under its own weight, causing its axis to deform slightly from the original straight line. Since this bending is very small, it is usually ignored in truss analysis.

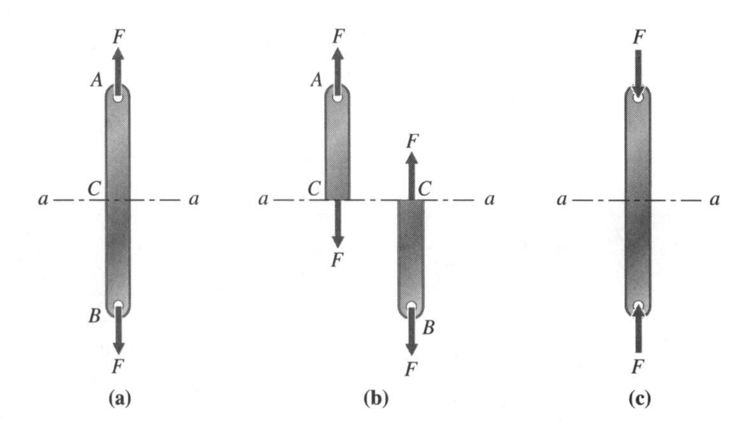

Figure 6.6 Bar subjected to a pair of collinear forces: **(a)** tension loading; **(b)** bar separated into two parts; **(c)** compression loading.

pression) is constant along the member's length. This fact is illustrated in Fig. 6.6b, where the location of section *a–a* is arbitrary (refer also to the discussion in Sec. 3.5 and Fig. 3.12).

In some cases, a *real* truss (one that has actually been built) will conform very closely to the ideal characteristics described above. For instance, a truss with long, slender members has little shear force and bending moment in its members. In other cases, significant differences exist between the real truss and the ideal case. Trusses with short, stocky members and heavy connections behave more like frames (Chapter 7) than like ideal trusses. So, if we analyze a real truss using the assumptions that apply to an ideal truss, the results we obtain will be approximate. Even so, the concept of an ideal truss is often used to perform a preliminary analysis and design; the secondary effects of shear and bending in the members are examined later in the design process. In many practical situations, the error due to this idealization has minimal effect on the overall design of the truss. In the remainder of this chapter, the word "truss" means an ideal truss, but it is understood that real trusses are represented in approximate form.

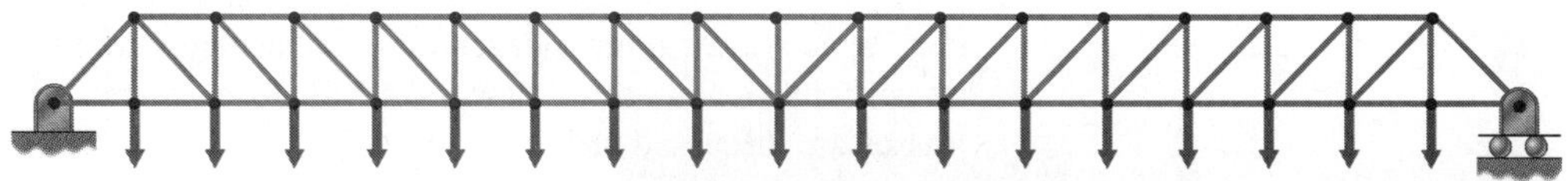

This pedestrian footbridge on a mountain trail in Wyoming has a Pratt truss configuration. The longitudinal deck planks are supported by transverse beams. The beams are attached to the truss at the panel points along the bottom chord to produce the concentrated joint loads that place the diagonals in tension. The joints in this truss are fully welded, yet such a truss can be designed as though the joints were pin-connected. An idealized model of this truss is shown in the line diagram.

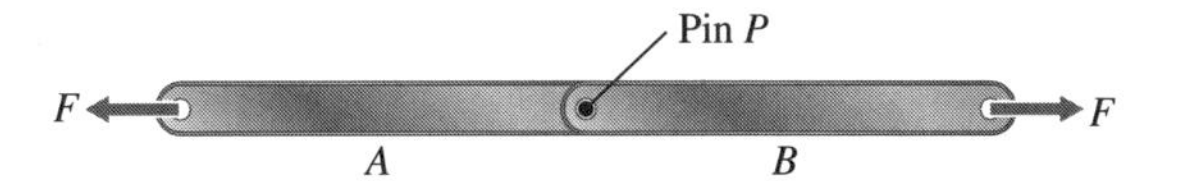

Figure 6.7 Pin-connected bars in equilibrium, under the action of two collinear forces.

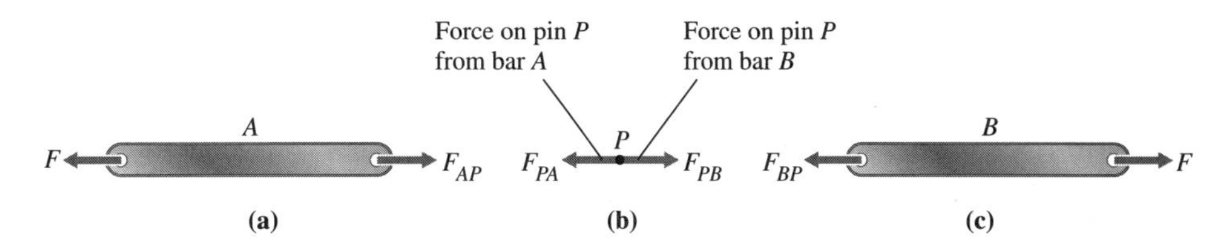

Figure 6.8 Free-body diagrams of the bars and pin of Fig. 6.7.

If a pin in the joint of a truss is smooth (frictionless), the force it exerts on a member is directed along the axis of the member. To see this, consider two members A and B that are pinned together and subjected to self-equilibrating forces of magnitude F (see Fig. 6.7). The frictionless pin P that joins members A and B transmits the force F from one member (say, member A) to the other member (member B). In doing so, the pin is subjected to self-equilibrating forces F. This fact can be demonstrated by separating the members from the pin and constructing free-body diagrams of all three, as shown in Fig. 6.8. The forces F_{AP} and F_{BP} at the right end of member A and the left end of member B, respectively, are the contact forces (actions) exerted by the pin P on the members A and B. The reactions of these forces act on the pin. These are forces F_{PA} and F_{PB} that members A and B exert on the pin (see Fig. 6.8b).

Since the members and the pin are in equilibrium, by the conditions of equilibrium and Newton's third law relating actions and reactions, we have

$$F = F_{AP} = F_{PA} = F_{PB} = F_{BP} = F$$

Accordingly, the force exerted by a smooth pin on a member in a truss is directed along the axis of the member.

6.2 STABILITY AND STATICAL DETERMINACY

After studying this section, you should be able to:

- Sketch a simple truss using basic elements.
- Determine whether a truss is stable or unstable.
- Determine whether a truss is statically determinate or statically indeterminate.

Consider an assembly of three members, AB, BC, and AC, that are pin-connected at their ends to form a truss consisting of one triangular element (see Fig. 6.9a). This element comprises a *stable truss*, in that it does not change its shape under the actions of force F_B applied at joint B and the corresponding support reactions at A and C. (The word "rigid" is also used to describe a stable truss.)

stable truss

In contrast, an assembly of four pin-connected members joined to form a rectangle is not stable (or rigid); see Fig. 6.9b. Under the action of force F_A, this system undergoes a change in shape; its members undergo large motions relative to each other. This will lead ultimately to collapse of the system. Consequently, the system is said to constitute an *unstable truss*. More generally, any system of four or more pin-connected members that

unstable truss

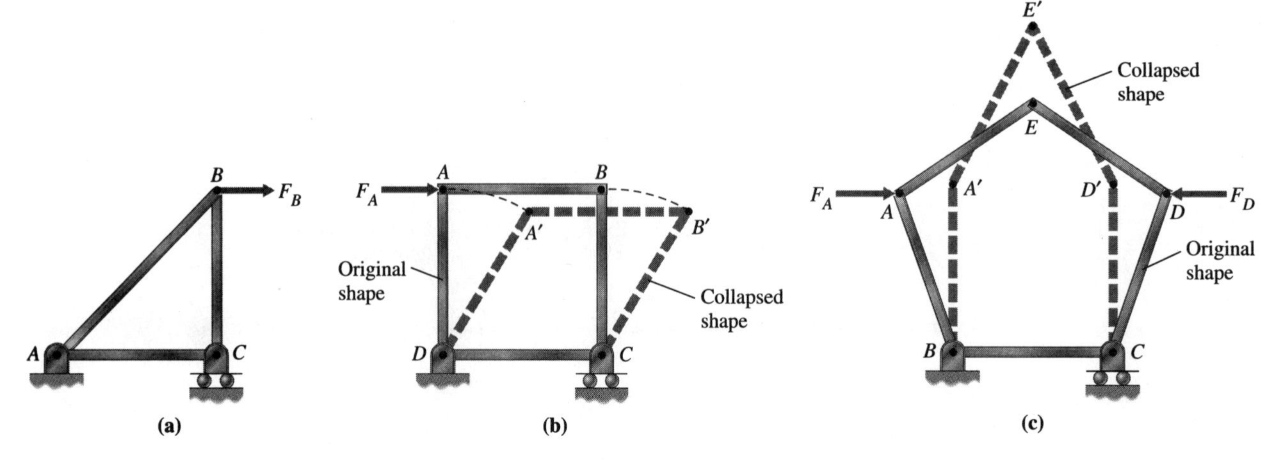

Figure 6.9 Systems of bars connected to form geometric elements: **(a)** triangular element; **(b)** rectangular element; **(c)** polygonal element.

form a polygon is not stable; it will collapse under some combination of loads (see also Fig. 6.9c).

Starting with the fundamental rigid triangular element shown in Fig. 6.9a, we can add two more noncollinear members to obtain a new joint (joint D in Fig. 6.10). The resulting truss $ABCD$ is rigid. This process can be continued to obtain a further expansion of the basic triangular truss (Fig. 6.11). A truss that is assembled in this manner is called a *simple truss*. Each of the trusses in Figs. 6.9a, 6.10, and 6.11 satisfies the equation

simple truss

$$m + r = 2j \tag{6.1}$$

where m is the number of members, r is the total number of support reactions (two at a pin support, one at a roller), and j is the total number of joints. For example, for the truss in Fig. 6.10, $m = 5$, $r = 3$, and $j = 4$. In Eq. (6.1), $m + r$ is the total number of unknown quantities (member forces plus reactions) and $2j$ is the number of independent equilibrium equations available to determine the unknowns. You can see that any truss assembled in the above manner will be rigid and will satisfy Eq. (6.1).

statically determinate

A simple truss is also *statically determinate;* that is, the support reactions and the member forces can be determined using only the equations of equilibrium (statics). If members are added to a statically determinate truss without adding joints or if support reactions are added, the truss becomes *statically indeterminate.* A statically indeterminate

statically indeterminate

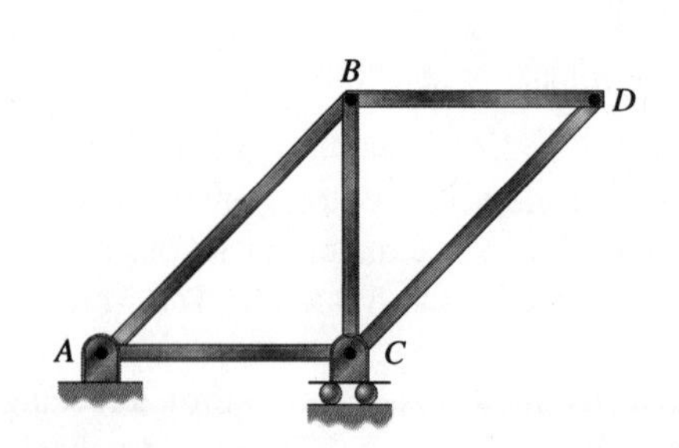

Figure 6.10 Extension of basic triangular element.

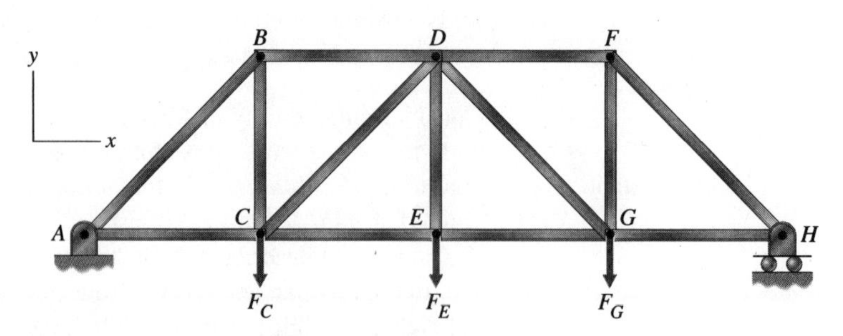

Figure 6.11 Simple truss: combination of triangular elements.

truss is one for which all the support reactions and all the member forces *cannot* be determined solely from the equations of equilibrium.[3] The deformations (change in length) of the members must be considered in the analysis of a statically indeterminate truss. For example, the trusses shown in Fig. 6.12 are both modifications of the truss in Fig. 6.10 and are statically indeterminate. In Fig. 6.12a, an additional member (member AD) has been added; in Fig. 6.12b, an additional support (at joint D) has been added. With these changes, we see that the trusses in Fig. 6.12 satisfy the inequality

$$m + r > 2j \tag{6.2}$$

Specifically, in Fig. 6.12a, $m = 6$, $r = 3$, and $j = 4$, which give nine unknowns and only eight equations of equilibrium. Although the trusses in Fig. 6.12 are statically indeterminate, they are also stable (rigid).

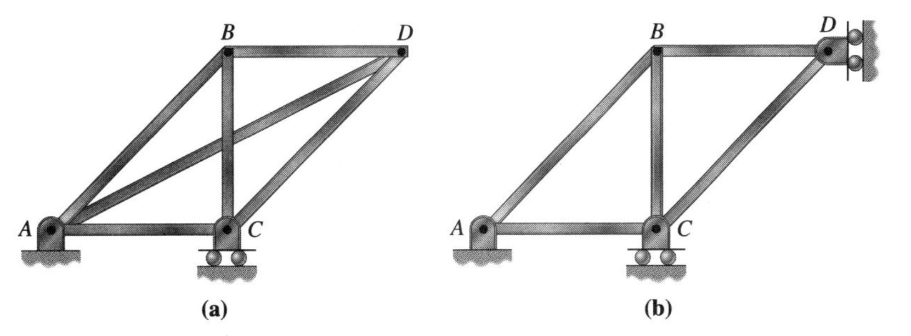

Figure 6.12 Statically indeterminate trusses.

Equation (6.1) is a necessary condition for stability of statically determinate trusses. However, it is not sufficient to guarantee stability of any truss. For example, the members must also be properly arranged to ensure rigidity of the entire truss. For instance, the truss in Fig. 6.13a satisfies Eq. (6.1), but it is not stable because the upper portion could collapse when loads are applied. In addition, the supports for the truss must be properly located. For instance, if the supports are arranged so that the reaction forces are concurrent, the equation for equilibrium of moments cannot be satisfied; the truss will rotate under arbitrary loads. Problem 6.1 at the end of the chapter addresses this point.

If a particular truss does not have enough members or support reactions to prevent motion of the joints, then the truss is unstable. This condition is expressed by the inequality

$$m + r < 2j$$

The trusses in Figs. 6.13b and 6.13c meet this condition. Simply stated, the number of unknowns $(m + r)$ is smaller than the number of equations $(2j)$. From your study of algebra, you know that such a system of equations is inconsistent. In general, such a system does not have a solution.

We have seen that the system of members in Fig. 6.9b is unstable. It contains $m = 4$ members and $r = 3$ support reactions (two at D and one at C), for a total of $m + r = 7$ unknown quantities. There are $2j = 8$ equilibrium equations (two each for the four joints A,

3. The terms "statically determinate" and "statically indeterminate" apply to all structures, not just trusses. In general, any structure is statically determinate if its support reactions and member forces can be determined using the equations of static equilibrium. If additional equations are required, the structure is statically indeterminate. The additional equations are based on concepts from mechanics of materials, taking into account the deformation of the members.

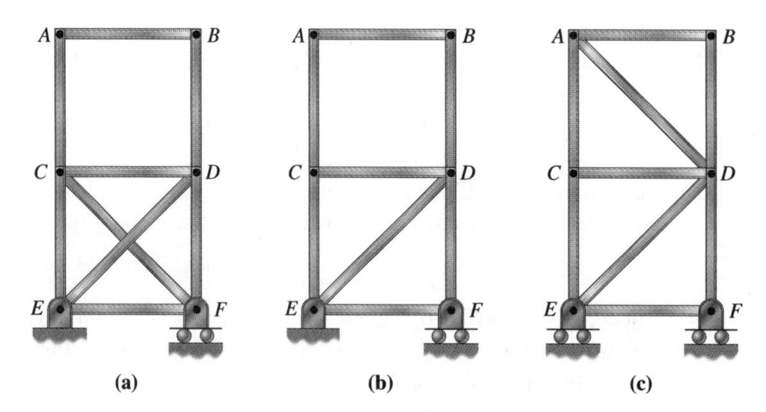

Figure 6.13 Unstable trusses: **(a)** improper configuration; **(b)** inadequate number of members; **(c)** inadequate number of support reactions

B, C, and *D*) to satisfy. Likewise, the system of members in Fig. 6.9c is also unstable. For this system, there are $2j = 10$ equilibrium equations, but $m = 5$ and $r = 3$ for a total of eight unknowns.

Analysis of a truss for statical determinacy or indeterminacy can be summarized as follows:

> If the forces in the members of a truss can be determined by the equations of equilibrium, the truss is statically determinate; otherwise, it is statically indeterminate. If a truss has r unknown support reactions and m unknown member forces, $m + r$ independent equations of equilibrium are required to solve for the unknowns. If a truss has j joints, $2j$ equations of equilibrium are available. If $m + r = 2j$, the unknowns can all be determined by solving the equations of equilibrium; the truss is *statically determinate.* If $m + r > 2j$, the unknowns cannot all be determined by solving the equations of equilibrium; the truss is *statically indeterminate.*

CHECKPOINT 

1. Which of the following are not characteristics of an ideal truss?
   - **a.** All members are straight.
   - **b.** All members are connected at their ends with smooth pins.
   - **c.** Except for the weight of each member, all loads on the truss are applied at the joints.
   - **d.** Each truss member must be either in tension or in compression.
2. Consider a truss with 6 joints, 9 members, and 3 support reactions. Draw a sketch of such a truss to help answer the following:
   - **a.** How many independent equations of equilibrium are available to use for analysis?
   - **b.** How many unknown quantities are there to determine?
   - **c.** Is the truss statically determinate or statically indeterminate?
   - **d.** Is the truss stable or unstable?
3. ***True or False:*** A two-force member is acted on by two collinear forces of equal magnitude and opposite sense.

Answers: 1. c; 2. a. 12 equations, b. 12 unknowns, c. and d. Answer depends on the arrangement of the members; 3. True

6.3 METHODS OF TRUSS ANALYSIS

After studying this section, you should be able to:

- Identify zero-force members in a truss.
- Recognize counter-diagonals in a truss.

The analysis of a truss involves determining the forces in the members, the forces acting on the pins, and the forces exerted on the truss by its supports. This analysis may be carried out in several ways. Two particular methods that are widely used are the method of joints and the method of sections. Section 6.4 describes the method of joints, which employs the concept of particle equilibrium. Section 6.5 discusses the method of sections, which requires the use of the concept of moments (Sec. 4.1) and rigid-body equilibrium.

Ordinarily, if the forces in all the members of a statically determinate truss are required, the method of joints is more efficient. On the other hand, if the forces in only a few interior members (say, members *BD* and *CD* in Fig. 6.11) are required, the method of sections may be more efficient. However, the method of joints and the method of sections are not mutually exclusive. The solution method you choose should suit the objectives of your analysis. You might find that a combination of the two methods will result in the most efficient solution procedure.

ZERO-FORCE MEMBERS

zero-force member

Under the action of a system of applied loads, axial forces are produced in the members of a truss. For a given set of loads, some members may carry no internal force; that is, they are *zero-force members*. For example, for the loads shown in Fig. 6.14a, members *BC* and *EF* are zero-force members. This can be seen by considering the free-body diagram of joint *B* shown in Fig. 6.14b. For convenience, a coordinate system is chosen so that the *x* axis is parallel to members *AB* and *BD;* member *BC* need not be aligned with the *y* axis. The equilibrium condition for joint *B* requires that the summation of force projections in the *y* direction be zero. Hence, from Fig. 6.14b, we have $\Sigma F_y = T_{BC} \sin\theta = 0$; the force T_{BC} in member *BC* is zero. In other words, member *BC* is a zero-force member. In a like manner, we can show that member *EF* is also a zero-force member. The identification of zero-force members helps to simplify the analysis of a truss.

Note that zero-force members exist because of the geometry of a truss and the nature of the load applied to it. If a different pattern of loads were applied to the truss in Fig. 6.14a—such as vertical loads at joints *B* and *F*—members *BC* and *EF* would not be zero-force members. Also, zero-force members can often be identified by inspection. A visual examination of a particular truss might be enough to allow you to locate the zero-force

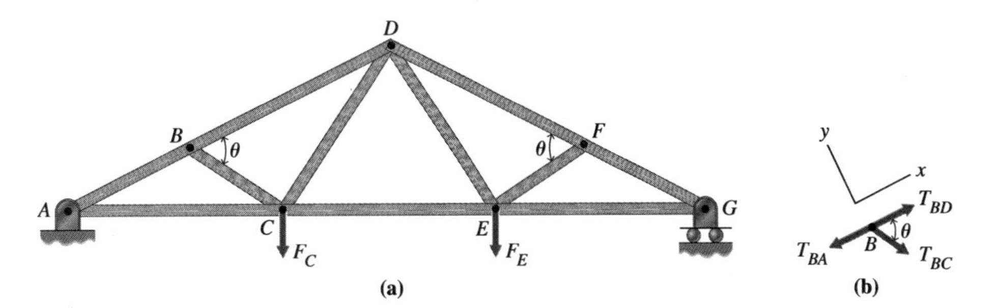

Figure 6.14 **(a)** Truss with zero-force members. **(b)** Free-body diagram of joint *B*.

members. In other cases, you will be able to identify them only through formal analysis of the truss.

COUNTER-DIAGONALS

counter-diagonals

Frequently, a truss is rendered indeterminate by the presence of so-called counter-diagonals. *Counter-diagonals* are a pair of diagonal members that bisect the same panel in a truss. For instance, the bow-string truss in Fig. 6.5 contains counter-diagonals, as does the indeterminate truss in Fig. 6.12a. Counter-diagonals are usually two relatively thin members (rods, bars, or cables) appearing in each panel. Counter-diagonals are common in trusses and frames because they permit the use of lighter-weight members.

When both diagonals in a panel are in tension, they share the load in the panel; in such a case, the truss is indeterminate and must be analyzed as such. However, when one of the diagonals is in tension and the other is in compression, the one in compression typically goes slack or buckles. Then the load in the panel is resisted mainly by the remaining diagonal in tension. For instance, consider the two-story truss shown in Fig. 6.15a, which has thin counter-diagonals. Under a horizontal load, one of the diagonals in each panel will buckle (or slacken), as in Fig. 6.15b. The horizontal load is then resisted by the diagonals in tension. If we assume that the load carried by the buckled members is negligible, we can analyze the truss as if it were statically determinate, with the buckled members removed entirely (see Fig. 6.15c).

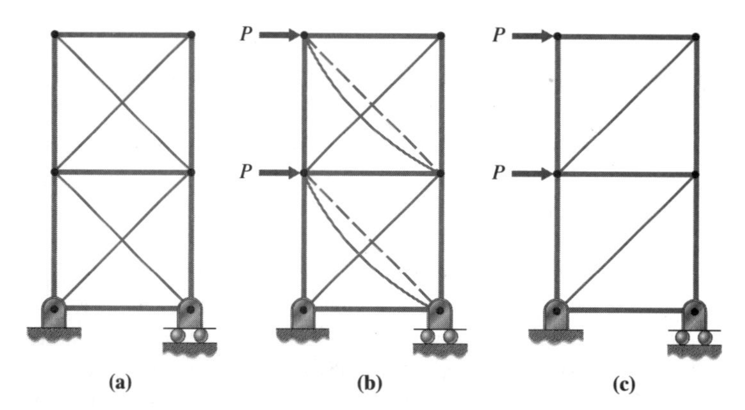

Figure 6.15 **(a)** Two-story truss with counter-diagonals. **(b)** Compression diagonal in buckled shape. **(c)** Simplified statically determinate truss.

If the horizontal load can change direction, as in the case of a wind load on a building, the compression (buckled) diagonal may go into tension, and the tension diagonal may buckle. Hence, each counter-diagonal must be able to carry the applied tension load.

6.4 THE METHOD OF JOINTS

After studying this section, you should be able to:

- Use the method of joints to analyze a truss efficiently.

Key Concept The method of joints for truss analysis is based on particle equilibrium of the joints of the truss.

The assumptions used to model a truss are discussed in Sec. 6.1. In this section, we discuss application of an important method of analysis to a simple, statically determinate truss. The *method of joints* is a method of analyzing a statically determinate truss by writing and solving equations of equilibrium for the joints of the truss. As we noted previously, there are two independent equations of equilibrium for each joint of a simple truss. If these equations are sufficient to determine all the member forces and all the support reactions, the truss is statically determinate. If not, the truss is statically indeterminate (see Sec. 6.2).

method of joints

In the method of joints, free-body diagrams of the joints of the truss are drawn, and the equations of equilibrium are written for each joint.[4] In a statically determinate truss, the solution of these equations yields all the forces acting on the pins and, hence, in the truss members. Recall that in Sec. 6.1, it was shown that the force (either tension or compression) in a truss member is constant throughout the length of the member (refer to Fig. 6.6b). Also, by Newton's third law, the force a member exerts on a pin is equal and opposite to the force the pin exerts on the member.

In the free-body diagram of a pin, the force that a tension member exerts on the pin is directed away from the pin. If a member is in compression, the force exerted on a pin is directed toward the pin. Thus, if a member is known to be in tension (or in compression), we can show in the free-body diagram the force that it exerts on the pin, with the correct sense. However, in analyzing a truss with more than a few members, you may find it difficult to guess whether a member is in tension or compression. Therefore, you should assume any unknown force to be tensile when drawing the free-body diagram of a pin and label the force with a symbol, such as F, denoting its magnitude. Then, write the equations of equilibrium for the pin using this assumption. If F is a positive number in the solution of the equations of equilibrium, the force is tensile, as assumed. If F is a negative number, the force is compressive.

PROBLEM-SOLVING TECHNIQUE

Method of Joints for Truss Analysis

Truss analysis by the method of joints involves the same techniques as those used to solve problems involving particle equilibrium. Each pin in the truss is considered to be a particle. Since the force in each member is aligned with the axis of the member, the force exerted by a member on a pin is directed along the axis of the member. Forces acting on a pin can be the result of the actions of members, the action of a support, and the effects of concentrated loads.

To analyze a truss using the method of joints:

1. Draw an accurate sketch of the truss, showing all important dimensions and quantities, including angles. If possible, identify the zero-force members by inspection of the truss geometry and the loads. Refer to the Problem-Solving Technique in Sec. 3.2 to review the solution of problems involving particle equilibrium.

(continued)

4. In fact, the equations of equilibrium are written for the pins within the joints of the truss. The pins are treated as particles, and the forces acting on the pins are concurrent with the pin. Nevertheless, this analysis method is known as the method of joints. Thus, we will use the words "joint" and "pin" interchangeably.

PROBLEM-SOLVING TECHNIQUE *(CONT.)*

2. Based on the members of interest, select a joint to begin the analysis. If possible, choose a joint at which only two unknown member forces act. Recall that, for each joint, the equations of equilibrium provide for solution of only two unknowns in two-dimensional problems. It is therefore convenient to select the joints in a sequence that allows you to evaluate one or more of the unknowns immediately, rather than collecting equations for all the joints and solving them simultaneously.
3. If you cannot identify a suitable joint at which to start the analysis, it might be helpful to find the support reactions for the truss. Once the support reactions are known, one of the supported joints might be a suitable starting point.
4. Draw a free-body diagram of the selected joint. Use it to write and solve the equilibrium equations for that joint.
5. Repeat Step 4 for each of the succeeding joints, until you have determined the member forces of interest.

Note: You will often be required to find only a few of the member forces in a truss. In such cases, spend a few minutes at the outset to develop a strategy that follows the most efficient path to the solution.

The following examples illustrate the method of joints.

Example 6.1 *Analysis of a Simple Truss by the Method of Joints*

Problem Statement A vertical load of 30 kN is applied to joint A of the statically determinate truss shown in Fig. E6.1a. Determine the member forces and the support reactions.

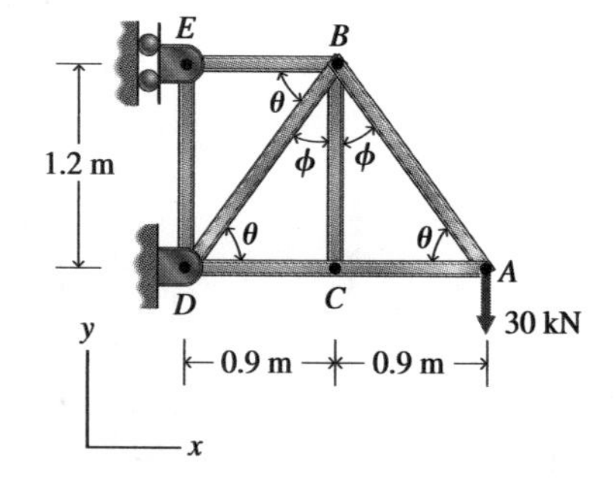

Figure E6.1a

Solution

From the given dimensions in Fig. E6.1a, we find that $\sin\theta = \cos\phi = 0.8$, and $\cos\theta = \sin\phi = 0.6$. The forces in members AB, BC, . . . are denoted by T_{AB}, T_{BC}, ... Since the force in a member is constant along its length, $T_{AB} = T_{BA}$, and likewise for member BC and the rest.

We start the analysis by looking for zero-force members. We see that joint C connects the two horizontal members AC and CD and one vertical member, BC. By inspection, we see that BC must be a zero-force member in order to satisfy the equilibrium equation $\Sigma F_y = 0$ for joint C. A less obvious zero-force member is DE. Since the roller at joint E can resist only a horizontal load, there can be no vertical force on the joint from member DE and, hence, no force in DE. For completeness, we will show that members BC and DE are zero-force members by solving the equilibrium equations for all the joints. For any isolated joint, the equilibrium equations are $\Sigma F_x = 0$ and $\Sigma F_y = 0$. In the free-body diagrams of the joints, we assume that the forces in the members are tension forces, denoted by T_{AB}, T_{BC}, and so on.

Joint A: Joint A is considered first since only two members are joined there. From the free-body diagram in Fig. E6.1b, the equilibrium equations are

$$\sum F_x = -T_{AB}\cos\theta - T_{AC} = 0$$
$$\sum F_y = T_{AB}\sin\theta - 30 = 0 \qquad \text{(a)}$$

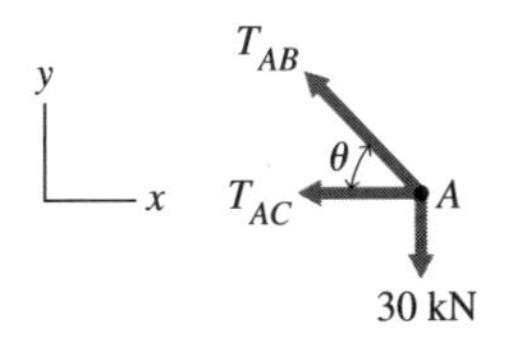

Figure E6.1b

The solutions of Eqs. (a) are $T_{AB} = 37.5$ kN and $T_{AC} = -22.5$ kN.

Joint C: From Fig. E6.1c, the equilibrium equations are

$$\sum F_x = T_{CA} - T_{CD} = 0$$
$$\sum F_y = T_{CB} = 0 \qquad \text{(b)}$$

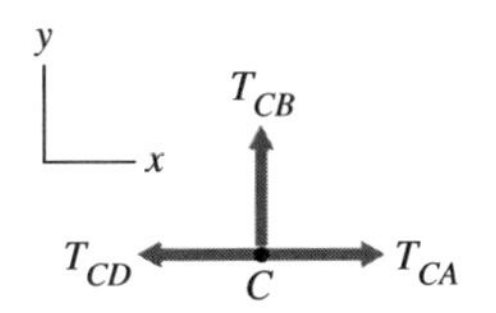

Figure E6.1c

Solution of Eqs. (b) gives $T_{CB} = 0$ (as expected) and $T_{CA} = T_{CD} = -22.5$ kN.

Joint B: From Fig. E6.1d, the equilibrium equations are

$$\sum F_x = -T_{BE} - T_{BD}\cos\theta + T_{BA}\cos\theta = 0$$
$$\sum F_y = -T_{BA}\cos\phi - T_{BD}\cos\phi = 0 \qquad \text{(c)}$$

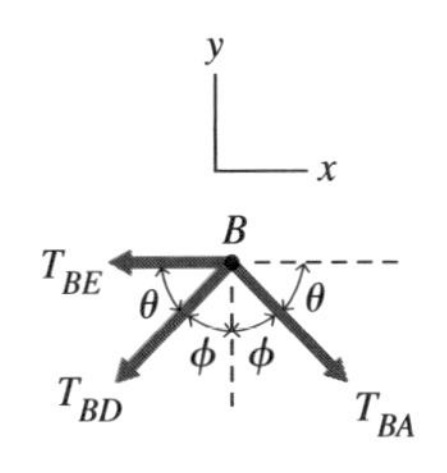

Figure E6.1d

With the previous results, we solve Eqs. (c) to obtain $T_{BD} = -37.5$ kN and $T_{BE} = 45$ kN.

Since all the forces were assumed to be tensile, those for which the computed magnitudes are negative are actually compressive. The support reactions and the force in member DE are determined by considering the free-body diagrams of joints E and D.

Joint E: From Fig. E6.1e, the equilibrium equations are

$$\sum F_x = E_x + T_{EB} = 0$$
$$\sum F_y = -T_{ED} = 0 \qquad \text{(d)}$$

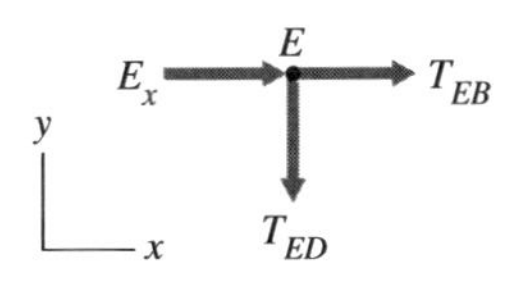

Figure E6.1e

Thus, Eqs. (d) give the support reaction at E as $E_x = -45$ kN and the force in member DE as $T_{ED} = 0$; that is, the support at joint E exerts a horizontal force directed to the left, and member DE is a zero-force member.

Joint D: From Fig. E6.1f, the equilibrium equations are

$$\sum F_x = D_x + T_{DC} + T_{DB}\cos\theta = 0$$
$$\sum F_y = D_y + T_{DE} + T_{DB}\sin\theta = 0 \qquad \text{(e)}$$

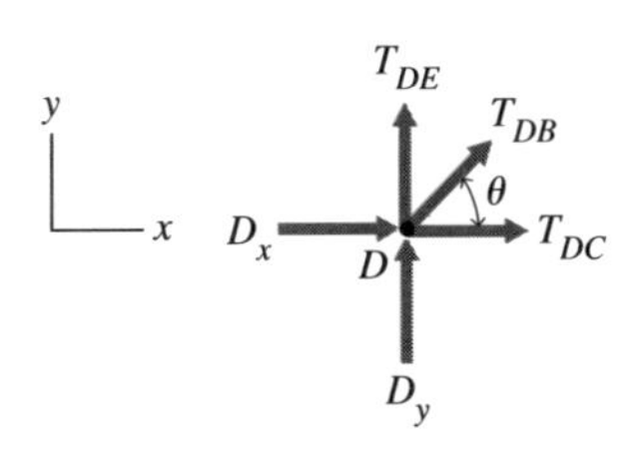

Figure E6.1f

Solution of Eqs. (e) gives $D_x = 45$ kN and $D_y = 30$ kN. Therefore, the support at joint D exerts a horizontal force to the right and a vertical force upward on the truss.

Information Item Note that we were able to find the support reactions for this truss using the method of joints. As an alternative, we could have started the analysis by viewing the truss as a rigid body and writing three equations of equilibrium to find the three support reactions, D_x, D_y, and E_x. These equations can be used as a check on our results. That check is left as an exercise.

Example 6.2 Storage Structure Analysis

Problem Statement An agricultural storage structure, with the cross-section shown in Fig. E6.2a, is proposed by a building contractor. The contractor wants to hire an engineer to evaluate her design concept. One aspect of the evaluation is an analysis of the structure to find its support reactions and member forces. The loads on the structure are shown in Fig. E6.2a. Determine the support reactions and member forces.

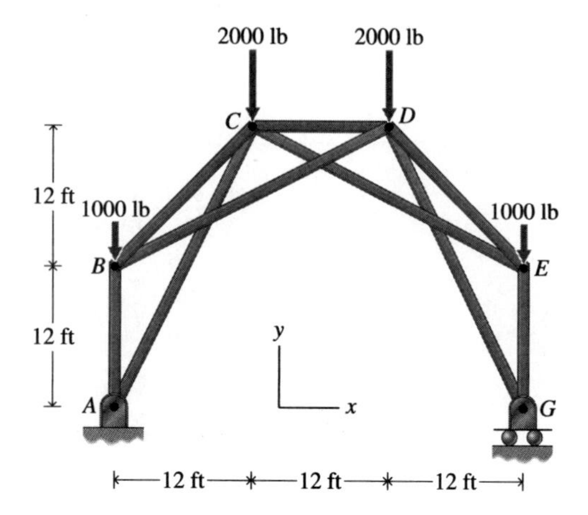

Figure E6.2a

Solution

Examination of the structure allows us to make the following sequence of observations and calculations:

1. The structure is a statically determinate, simple truss; $m = 9$, $r = 3$, and $j = 6$, so $m + r = 2j$.
2. The truss and its loads are symmetric. Hence, the support reactions and members forces are also symmetric.
3. There are no joints at which only two unknown forces act. So, we must find the support reactions first in order to have a starting point for the method of joints.
4. From inspection of the free-body diagram of the truss in Fig. E6.2b, we can see that the reactions are $A_y = G_y = 3000$ lb and $A_x = 0$.

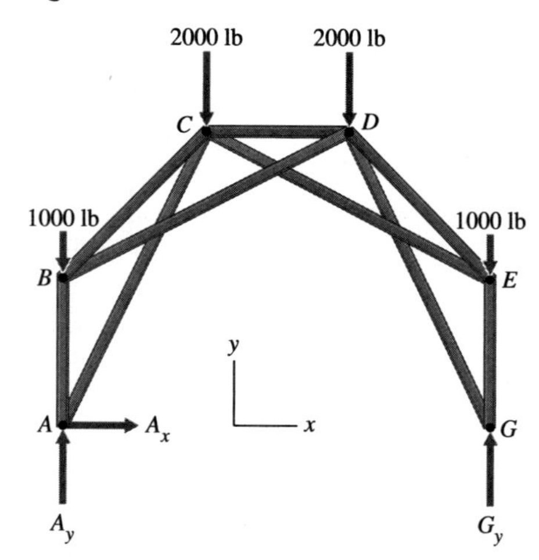

Figure E6.2b

5. At joint G, member DG is the only member capable of carrying a force component in the x direction. However, there are no horizontal loads or reactions at that joint. Hence, $T_{DG} = 0$; that is, member DG is a zero-force member.
6. From the free-body diagram of joint G (Fig. E6.2c) and equilibrium of forces in the y direction

$$\sum F_y = 3000 + T_{EG} = 0$$

which gives

$$T_{EG} = -3000 \text{ lb (compression)}$$

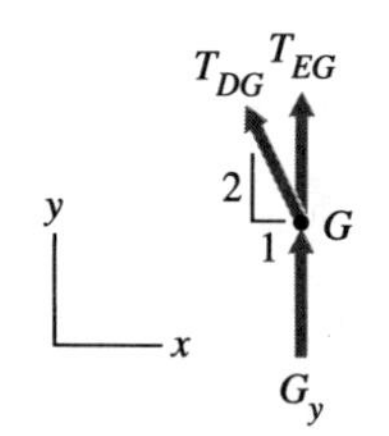

Figure E6.2c

7. Next, we consider joint E, whose free-body diagram is shown in Fig. E6.2d. The equilibrium equations for this joint are

$$\sum F_x = -\frac{1}{\sqrt{2}}T_{DE} - \frac{2}{\sqrt{5}}T_{CE} = 0$$

$$\sum F_y = \frac{1}{\sqrt{2}}T_{DE} + \frac{1}{\sqrt{5}}T_{CE} - 1000 - T_{EG} = 0$$

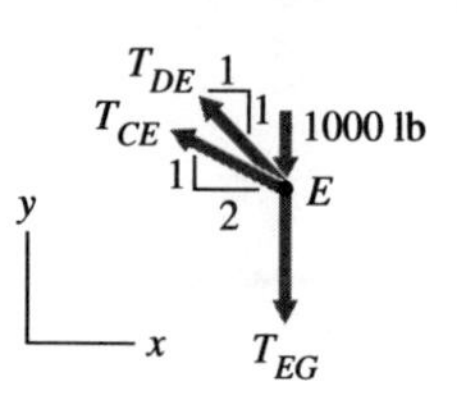

Figure E6.2d

The solution to these equations is

$$T_{CE} = 4472 \text{ lb (tension)}$$

$$T_{DE} = -5657 \text{ lb (compression)}$$

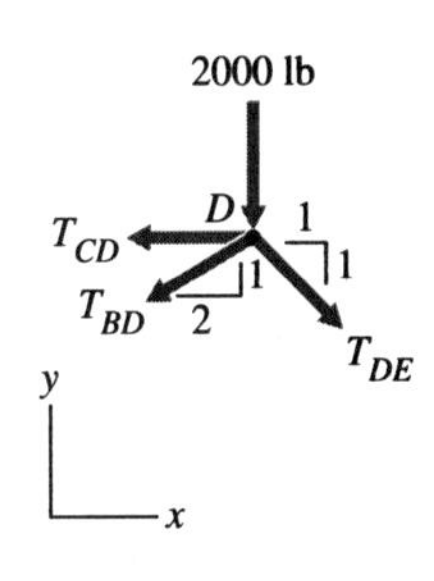

Figure E6.2e

8. Because symmetry exists, we need to examine only one more joint; we select joint D. Its free-body diagram is shown in Fig. E6.2e. Since $T_{DG} = 0$, $T_{DE} = -5657$ lb, and $T_{BD} = T_{CE} = 4472$ lb, the only unknown at joint D is T_{CD}. It is found using the following equilibrium equation:

$$\sum F_x = \frac{1}{\sqrt{2}} T_{DE} - T_{CD} - \frac{2}{\sqrt{5}} T_{BD} = 0$$

So, finally,

$$T_{CD} = -8000 \text{ lb (compression)}$$

9. We apply symmetry to find the remaining member forces. In summary, the reactions and member forces for the truss are

$$A_y = G_y = 3000 \text{ lb}$$

$$A_x = 0$$

$$T_{AC} = T_{DF} = 0$$

$$T_{AB} = T_{EG} = -3000 \text{ lb (compression)}$$

$$T_{BC} = T_{DE} = -5657 \text{ lb (compression)}$$

$$T_{BD} = T_{CE} = 4472 \text{ lb (tension)}$$

$$T_{CD} = -8000 \text{ lb (compression)}$$

CHECKPOINT

1. ***True or False:*** In the method of joints, a member of a truss is treated as a particle.
2. ***True or False:*** In the method of joints, a joint (pin) of a truss is treated as a particle.
3. ***True or False:*** The method of joints allows you to solve for up to three unknown forces acting at each joint.

Answers: 1. False; 2. True; 3. False

6.5 THE METHOD OF SECTIONS

After studying this section, you should be able to:

- Use the method of sections to analyze a truss efficiently.

Key Concept The method of sections for truss analysis is based on rigid-body equilibrium of a portion of the truss.

In Sec. 6.4, the method of joints was used to calculate the forces in the members of a statically determinate truss; the equilibrium of each joint was examined. If only the reactions of the supports or the forces in some interior members are required, the method of joints will often entail unnecessary calculations of other quantities. In Example 4.10, we determined the reactions at the supports of a truss by applying the conditions of equilibrium of a rigid body to the entire truss. The same conditions can be used to help determine the internal forces in a statically determinate truss. The analytic procedure based on this idea is called the *method of sections.*

method of sections

If an entire truss is in equilibrium under the action of a set of coplanar forces, any part of the truss is also in equilibrium. Hence, we can cut the truss into two or more parts, each of which is a rigid body in equilibrium. Then, we can draw free-body diagrams of the parts, showing the forces in the members at the cut. Since the truss members are two-force members, the member forces are directed along their axes. As in the method of joints (Sec. 6.4), we may assume unknown member forces to be tensile.

PROBLEM-SOLVING TECHNIQUE

Method of Sections for Truss Analysis

To analyze a truss using the method of sections:

1. Draw a free-body diagram of the entire truss, and write the equilibrium equations. Solve these equations to determine the support reactions. Occasionally, this step may be omitted from the procedure, but it is usually necessary to determine support reactions as part of the solution of the problem.
2. Locate the members in the truss for which forces are desired. Mark each with two short lines drawn across the member like this:

   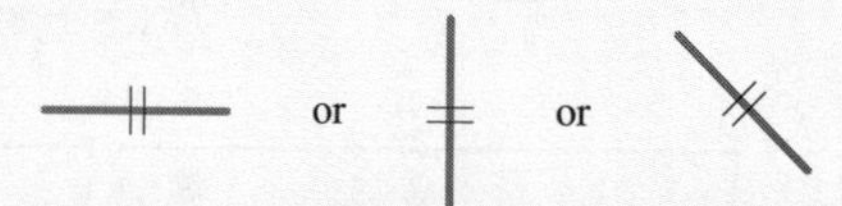

   Examine the truss and its loads to identify zero-force members.
3. Pass a section line (cut) through the truss to separate it into two parts. If possible, the section line should pass through no more than three members with unknown member forces. The section line need not be straight but must separate the truss into two appropriate parts. In some cases, it may be necessary to cut more than three members to separate the truss into two parts. Also, note that each part of the truss must contain at least one complete (uncut) member. Otherwise, only a single joint would be isolated, as in the method of joints.
4. Select one of the truss parts sectioned in Step 3, and draw a free-body diagram of it. Unless you know otherwise, assume that the unknown member forces are tensile.
5. Write the equilibrium equations for the part selected in Step 4. If it was necessary to cut more than three members with unknown member forces in Step 3, you may have to consider additional parts of the truss or perhaps individual joints in order to determine the unknowns. If so, draw appropriate free-body diagrams of the additional parts or joints, and write their equilibrium equations to obtain the same number of independent equilibrium equations as the number of unknowns.
6. Solve the set of equations derived in Step 5 to determine the unknown forces.
7. Repeat Steps 3 through 6 as required to complete the analysis.

Example 6.3

Analysis of a Truss by the Method of Sections

Problem Statement Consider the truss of Example 6.1, a free-body diagram of which is shown in Fig. E6.3a. Determine the forces in members *AB* and *BC.*

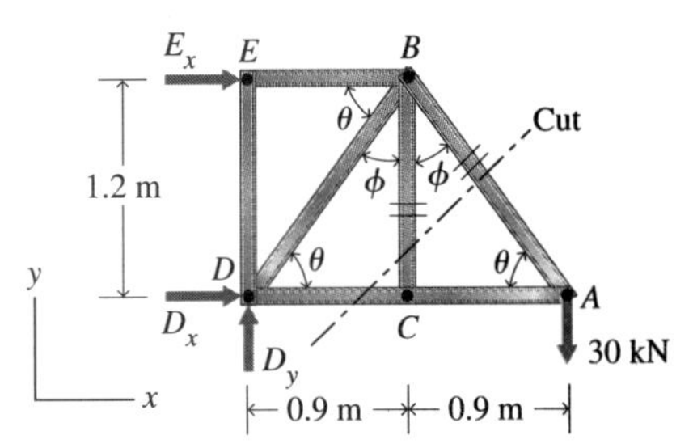

Figure E6.3a

Solution The members of interest, members *AB* and *BC,* are marked with two short lines on the free-body diagram. Normally, we would begin solving this problem by determining the reactions at *D* and *E.* In this case, however, we see that if we pass a cut (shown as a dashed line on the free-body diagram of Fig. E6.3a) through three members—specifically, members *AB, BC,* and *CD*—we can isolate the right part of the truss as a rigid body and the unknown support reactions do not enter into our computations.

By inspection, we see that member *BC* is a zero-force member, since joint *C* connects it to collinear members *AC* and *CD,* and no loads are applied at *C* (Fig. E6.3a). Member *DE* is also a zero-force member, since the reaction at *E* is perpendicular to *DC.*

The free-body diagram of the isolated part of the truss is shown in Fig. E6.3b. The forces in the cut members are directed along their longitudinal axes. Since it is not obvious whether the forces at the cut are tensile or compressive, we assume that all members are in tension.

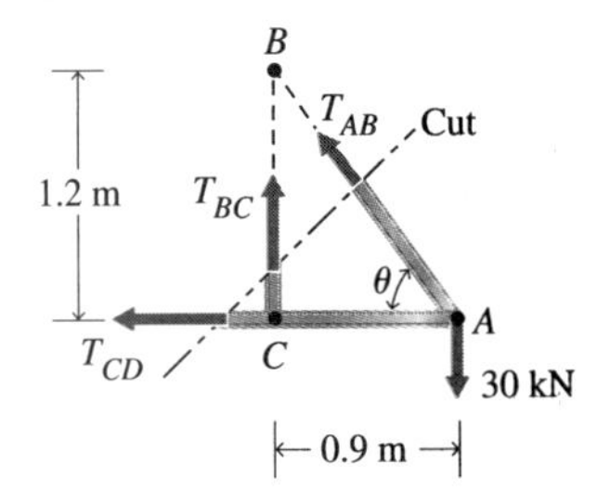

Figure E6.3b

The isolated part of the truss is in equilibrium under the action of a system of nonparallel, nonconcurrent, coplanar forces. Hence, rigid-body equilibrium applies, and three independent equations of equilibrium are available. Consequently, only three unknown quantities (T_{AB}, T_{BC}, and T_{CD}) can be determined from the equilibrium equations for the isolated part. In this case, we cut only three members. So it will not be necessary to consider other parts or joints to solve for the unknowns.

First, to determine the force T_{BC}, we consider moments about point *A,* the intersection of the lines of action of forces T_{AB} and T_{CD}. The equation of equilibrium of moments about this point is

$$\circlearrowleft\!+ \sum M_A = 0.9T_{BC} = 0$$

This equation yields $T_{BC} = 0$, which confirms our initial observation from inspection of the truss.

Since we know that member *BC* is a zero-force member, we can find T_{AB} by either of two equations. We could consider equilibrium of moments about joint *C or* equilibrium of forces in the *y* direction. If we choose to consider force equilibrium, we have

$$\sum F_y = T_{AB}\sin\theta + T_{BC} - 30 = 0$$

Substituting $T_{BC} = 0$ into this equation and noting that $\sin\theta = 0.8$, we have $T_{AB} = 37.5$ kN. Since the result is positive, our assumption that T_{BC} is a tensile force is correct. As expected, these results agree with the values that were obtained by the method of joints in Example 6.1.

Information Item Having identified member *BC* as a zero-force member at the outset, we could have used the method of joints on joint *A* to find the force in member *AB,* which would have completed the solution. That approach would have been faster than the method of sections in this case. Once you gain experience in analyzing trusses, you will be able to identify the method that leads most efficiently to the problem solution.

Example 6.4 *Analysis of a Modified Warren Truss*

Problem Statement The chief engineer of your firm needs to know the forces in members *BD, BE, BC, CE,* and *DE* of the modified Warren truss shown in Fig. E6.4a. She asks you, a junior engineer who works for her, to analyze the truss. Carry out the analysis to determine the required member forces.

Solution You decide to proceed as follows. You identify the members for which the forces are required by drawing two short lines across each such member in Fig. E6.4a. By inspection, you observe that *BC* is a zero-force member (as is *FG*).

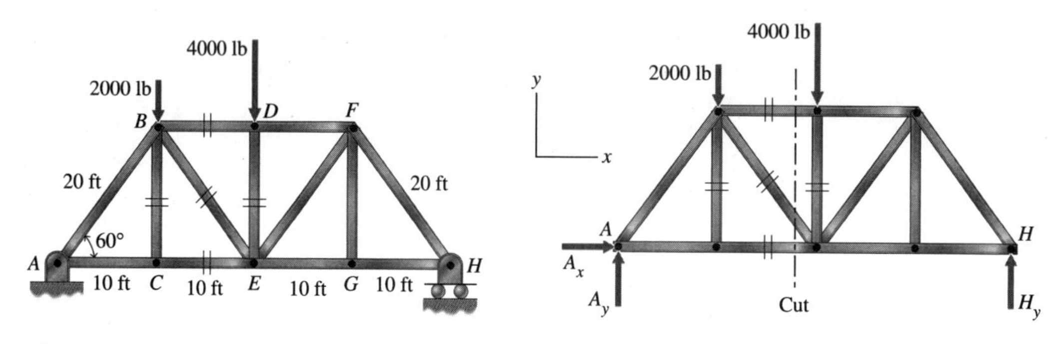

Figure E6.4a **Figure E6.4b**

From the free-body diagram of the entire truss (Fig. E6.4b), you determine the reactions at joints *A* and *H* using rigid-body equilibrium:

$$\sum F_x = A_x = 0$$

$$\sum F_y = A_y + H_y - 2000 - 4000 = 0 \tag{a}$$

$$\circlearrowleft^{+} \sum M_A = 40H_y - 10(2000) - 20(4000) = 0 \tag{b}$$

You solve Eqs. (a) and (b) to determine $H_y = 2500$ lb and $A_y = 3500$ lb.

You decide that the force in member *DE* is determined readily by the method of joints and proceed as follows.

Joint D: From the free-body diagram in Fig. E6.4c, the equilibrium equation for vertical force projections yields $-T_{DE} - 4000 = 0$, or $T_{DE} = -4000$ lb. Hence, member *DE* is in compression.

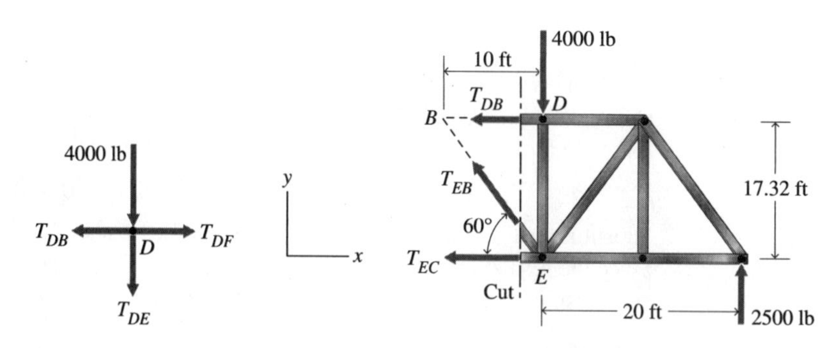

Figure E6.4c **Figure E6.4d**

The forces in members *BD, BE,* and *CE* remain to be calculated. You observe that these members may be cut by a vertical section passed between joints *B* and *D*. The isolated part of the truss shown in Fig. E6.4d is chosen as the free body. So, by the equilibrium equations for the isolated part, you find

$$\sum F_x = -T_{DB} - T_{EB}\cos 60° - T_{EC} = 0$$
$$\sum F_y = -T_{EB}\sin 60° - 4000 + 2500 = 0 \qquad \text{(c)}$$
$$\circlearrowleft + \sum M_E = 17.32T_{DB} + 20(2500) = 0$$

Equations (c) yield $T_{DB} = -2887$ lb, $T_{EB} = 1732$ lb, and $T_{EC} = 2021$ lb. Since T_{DB} is negative, member *DB* is in compression; members *EB* and *EC* are in tension.

This completes your solution. However, being a cautious engineer, you decide to check your solution by the following alternative method.

Alternative Solution You observe that the forces T_{DB}, T_{EB}, and T_{EC} may be determined using Theorem 4.8 and that the equilibrium conditions for the section shown in Fig. E6.4d may be written as

$$\circlearrowleft + \sum M_B = -10(4000) + 30(2500) - 17.32T_{EC} = 0$$
$$\circlearrowleft + \sum M_E = 17.32T_{DB} + 20(2500) = 0$$
$$\sum F_y = -T_{EB}\sin 60° - 4000 + 2500 = 0$$

where the subscripts *B* and *E* in the moment equations refer to joints *B* and *E*. You then solve these equations to obtain $T_{DB} = -2887$ lb, $T_{EB} = 1732$ lb, and $T_{EC} = 2021$ lb. These values agree with your previous solution, and you feel good about that. However, as another check, you decide to check the sum of forces in the *x* direction:

$$\sum F_x = -T_{DB} - T_{EC} - T_{EB}\sin 30° = 2887 - 2021 - 1732(0.5) = 0$$

This is indeed correct.

You note that the advantage of this alternative solution over the previous one is that T_{DB}, T_{EC}, and T_{EB} are determined independently from single equations. Thus, the solution of a set of simultaneous equations is avoided. Also, since T_{DB}, T_{EC}, and T_{EB} are determined independently, the equation $\sum F_x = 0$ may be used as a check on your answers.

Information Item When you applied the method of sections to find the forces in members *BD, BE,* and *CE,* you separated the truss into two parts with a vertical cut. In general, the process of separating a truss into two parts does not require use of a straight cut. Any bent or curved section line may be passed through a truss to separate it into two parts. Keep in mind, though, that no more than three members with unknown forces should be cut by any one section line. Also, note that either part of a truss can be selected as the free body for analysis by the method of sections. In this case, the portion of the truss to the right of the cut was used. However, the left portion could have been used to obtain equivalent results. ■

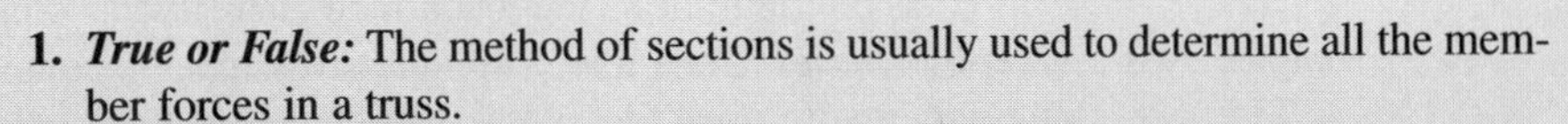

CHECKPOINT

1. ***True or False:*** The method of sections is usually used to determine all the member forces in a truss.
2. ***True or False:*** The method of sections recommends cutting no more than three members having unknown forces because there are only three equations of equilibrium to help determine those unknowns.
3. ***True or False:*** The method of joints and the method of sections can be used together for efficient analysis of a truss.

Answers: 1. False; 2. True; 3. True

6.6 COMPOUND AND COMPLEX TRUSSES

After studying this section, you should be able to:

- Classify a truss as simple, compound, or complex.

As discussed above, a *simple truss* is one that can be assembled by successively adding two noncollinear members to an initial triangular element. As each pair of members is added, another joint is added. Many truss configurations, though, cannot be assembled in this manner. These configurations are often composed of two or more simple trusses that are joined together. Such structures are known as *compound trusses*. Specifically, a compound truss is composed of two or more simple trusses joined together by one or more common joints or by additional members.

compound truss

Examples of compound trusses are shown in Fig. 6.16, with the simple truss portions shaded. In Fig. 6.16a, two simple trusses are joined at a common joint, A. To make the resulting compound truss stable, two hinge supports (four support reactions) are required. In Fig. 6.16b, two simple trusses are tied together with three additional members. Stability is assured with the usual three support reactions.

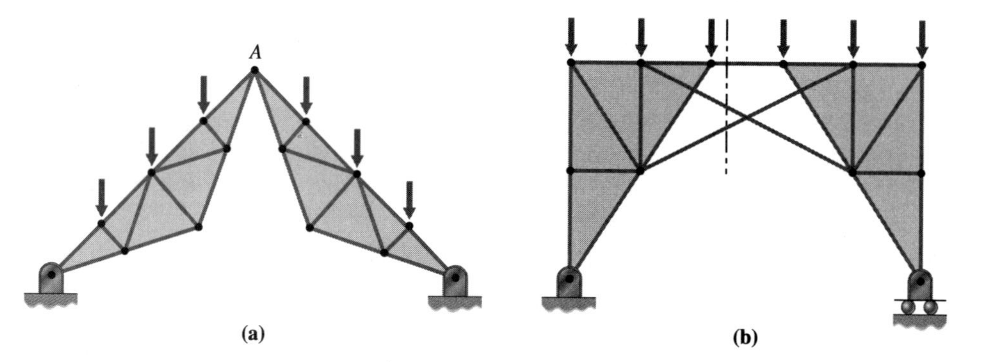

Figure 6.16 Compound trusses.

Analysis of a compound truss can be performed using either the method of joints or the method of sections. If the method of sections is used, we begin by breaking the truss into its simple truss components. For instance, the pin at joint A of the compound truss in Fig. 6.16a can be removed to separate that truss into two simple trusses. Likewise, the compound truss in Fig. 6.16b can be separated into two parts by cutting along the dashed line. Then the simple trusses can be analyzed using either the method of joints or the method of sections. Regardless of which method you use, you are likely to be solving simultaneous equations when analyzing a compound truss.

complex truss

Truss configurations that cannot be classified as simple or compound are known as *complex trusses*. Generally, a complex truss can be composed of any combination of triangular, quadrilateral, and polygonal elements. A complex truss often has members that cross over one another without being joined. Examples of complex trusses are shown in Fig. 6.17 (in both examples, members cross without joints). In the analysis of a complex truss, the method of joints or the method of sections may be used to generate equilibrium equations, which are then solved simultaneously for the member forces. Note that each of the trusses shown in Figs. 6.16 and 6.17 is statically determinate and stable [see Eq. (6.2)].

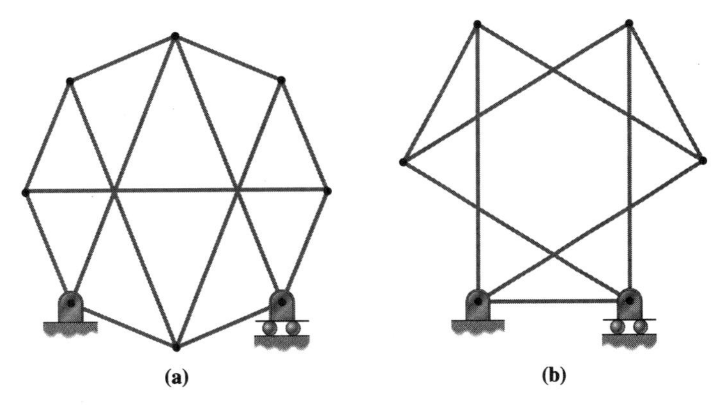

Figure 6.17 Complex trusses, with members that are joined only at dots.

Example 6.5

Compound Whipple Truss

Problem Statement Figure E6.5a represents a compound truss.[5] Joints *A, B, C, D, E,* and *F* are pin-connected. The other three crossings of members are not joints. The truss supports are a roller at joint *A* and a hinge at joint *D*. Determine the member forces and support reactions.

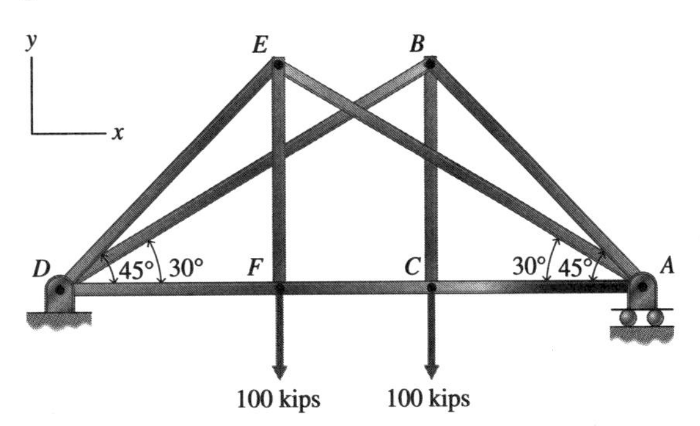

Figure E6.5a

Solution The forces in members *AB, AC,* . . . are denoted by the symbols T_{AB}, T_{AC}, ... We assume that these forces are tensile. Since the structure and the loads are symmetric, the vertical reactions A_y and D_y are equal. Symmetry conditions can also be used to relate certain of the member forces:

$$T_{AC} = T_{DF} \qquad T_{AB} = T_{DE} \qquad T_{AE} = T_{BD}$$

Since we must find all member forces, we will use the method of joints. We would like to begin the analysis by selecting a joint on which no more than two unknown forces act. However, you should recall that in some truss analyses this may not be possible. Indeed, it

5. This truss configuration was introduced by Squire Whipple. In his 1847 book, *A Work on Bridge Building,* Whipple presented the first rational treatment of the analysis of trusses. Note that this truss is composed of two simple three-member trusses *ABCA* and *DEFD* joined by three additional members *AE, BD,* and *CF.*

is not possible in this example. Thus, we begin by considering the equilibrium conditions for joint C, on which three unknown forces act.

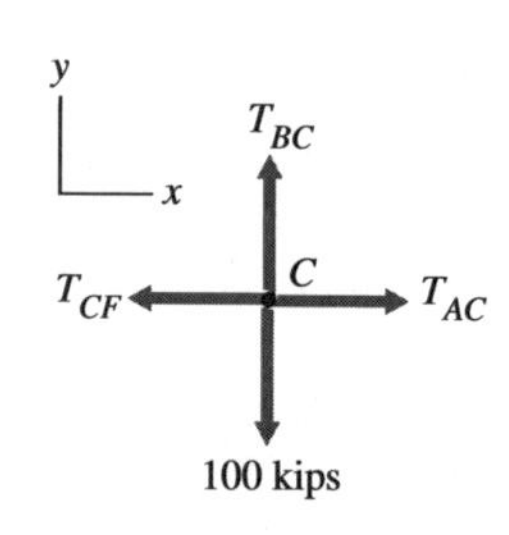

Figure E6.5b

Joint C: From the free-body diagram in Fig. E6.5b, the equilibrium equations for joint C are

$$\sum F_x = T_{AC} - T_{CF} = 0$$
$$\sum F_y = T_{BC} - 100 = 0 \quad \text{(a)}$$

Equations (a) yield the solution $T_{AC} = T_{CF}$, and $T_{BC} = 100$ kips. Also, by symmetry we see that $T_{AC} = T_{DF}$, and $T_{EF} = T_{BC} = 100$ kips. Thus, $T_{AC} = T_{CF} = T_{DF}$. That is, the tension is constant throughout the bottom chord of the truss.

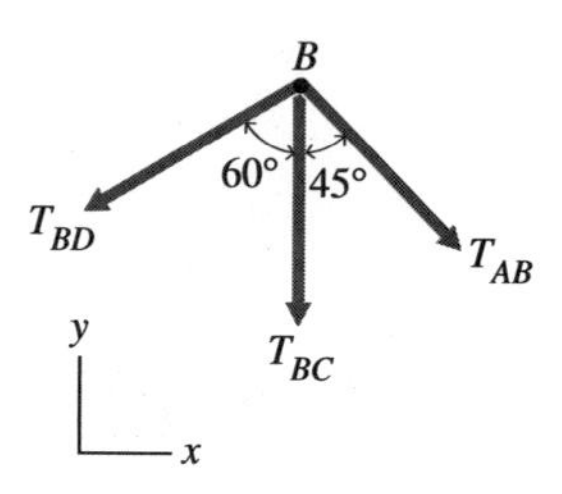

Figure E6.5c

Joint B: From Fig. E6.5c, the equilibrium equations are

$$\sum F_x = T_{AB} \sin 45° - T_{BD} \sin 60° = 0 \quad \text{(b)}$$

$$\sum F_y = -T_{BC} - T_{BD} \cos 60° - T_{BA} \cos 45° = 0 \quad \text{(c)}$$

Since $T_{BC} = 100$ kips, Eq. (c) reduces to

$$T_{AB} \cos 45° + T_{BD} \cos 60° = -100 \quad \text{(d)}$$

Subtracting Eq. (b) from Eq. (d), we obtain $T_{BD} = -73.21$ kips. Since T_{BD} is negative, member BD is in compression. With T_{BD} known, from Eq. (b), we find $T_{AB} = -89.66$ kips. Hence, member AB is also in compression. Again applying symmetry, we obtain $T_{AE} = -73.21$ kips (member AE is in compression), and $T_{DE} = -89.66$ kips (member DE is in compression).

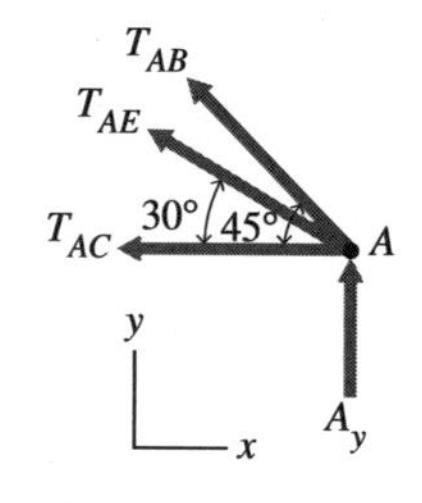

Figure E6.5d

Joint A: From Fig. E6.5d, the equations of equilibrium are

$$\sum F_x = -T_{AB} \cos 45° - T_{AE} \cos 30° - T_{AC} = 0 \quad \text{(e)}$$

$$\sum F_y = A_y + T_{AB} \sin 45° + T_{AE} \sin 30° = 0 \quad \text{(f)}$$

Since T_{AB} and T_{AE} are known, by Eq. (e), $T_{AC} = 126.8$ kips. Accordingly, the horizontal members AC, CF, and DF are all subject to a tensile force of 126.8 kips. Then, from Eq. (f), we find $A_y = 100$ kips.

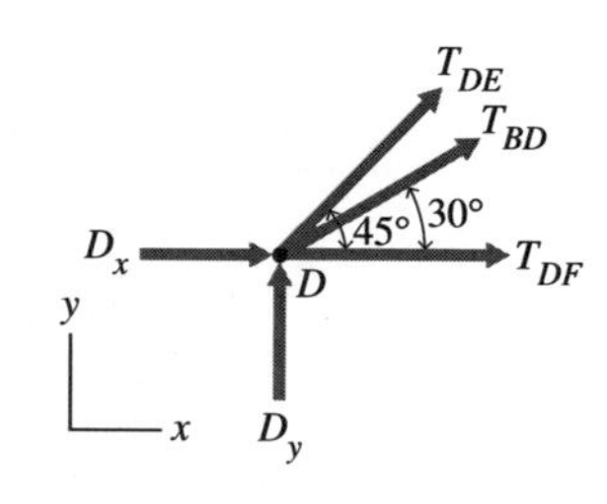

Figure E6.5e

Joint D: From Fig. E6.5e, since the vertical reaction D_y is equal to A_y, the only unknown at joint D is the horizontal reaction D_x. Therefore, summation of forces in the horizontal direction yields

$$\sum F_x = D_x + T_{DE} \cos 45° + T_{BD} \cos 30° + T_{DF} = 0 \quad \text{(g)}$$

Substitution of the known values of T_{DE}, T_{BD}, and T_{DF} into Eq. (g) yields $D_x = 0$. ■

Example 6.6

Compound Fink Truss

Problem Statement Figure E6.6a shows a compound truss formed by joining two basic elements of the Fink truss (see Fig. 6.4b). All joints are pin-connected. Each triangular panel forms a 30°–60°–90° triangle. The truss is supported by hinges at joints A and E.

Members AF and GE are collinear. Determine the support reactions and tensile forces in members AF, CF, CG, and EG.

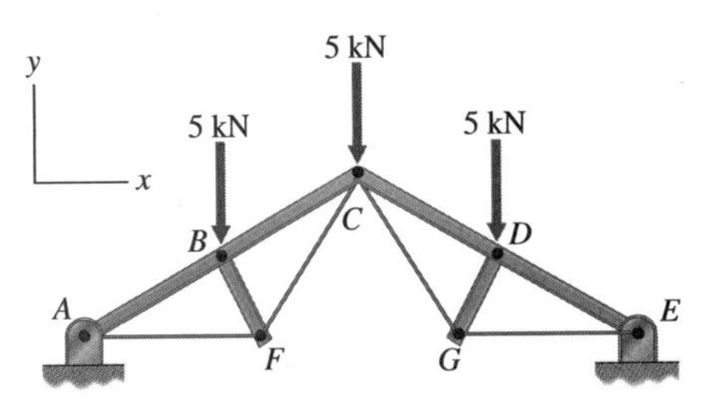

Figure E6.6a

Solution To analyze this compound truss, we must first remove the pin at C to obtain two simple trusses (Fig. E6.6b). In the free-body diagrams of these elements, we see that each has four unknown forces—A_x, A_y, C_x, and C_y on the left element and C_x, C_y, E_x, and E_y on the right. However, we observe that the truss elements *are not* statically indeterminate, since the pin forces C_x and C_y are equal and opposite actions of one element on the other. So, in fact, there are only six independent forces—A_x, A_y, C_x, C_y, E_x, and E_y—that can be evaluated using the six equations of equilibrium (three for each truss element). Note that in separating the compound truss into two elements, we placed the load at joint C on the left element. The load could have been placed on the right element, or even divided in half and placed half on each element, without changing the member forces.

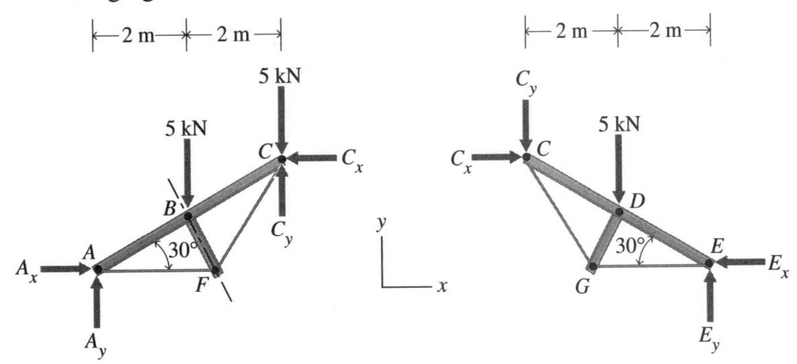

Figure E6.6b

To find A_x, A_y, E_x, and E_y, we write the equilibrium equations for the left and right trusses. First, for the left:

$$\sum F_x = 0: A_x - C_x = 0 \tag{a}$$

$$\sum F_y = 0: A_y + C_y - 10 = 0 \tag{b}$$

$$\circlearrowleft^{+} \sum M_C = 0: 4A_y - 2(5) - (4\tan 30°)(A_x) = 0 \tag{c}$$

And for the right, we have

$$\sum F_x = 0: C_x - E_x = 0 \tag{d}$$

$$\sum F_y = 0: E_y - C_y - 5 = 0 \tag{e}$$

$$\circlearrowright^{+} \sum M_C = 0: 4E_y - 2(5) - (4\tan 30°)(E_x) = 0 \tag{f}$$

Addition of Eqs. (a) and (d) gives

$$A_x = E_x \tag{g}$$

which we might have anticipated, since the structure is symmetric. Likewise, addition of Eqs. (b) and (e) gives

$$A_y + E_y - 15 = 0 \tag{h}$$

which is the equilibrium equation of vertical (y direction) forces for the whole truss. Substituting for A_x and A_y from Eqs. (g) and (h) into Eq. (c) and simplifying, we obtain

$$50 - 4E_y - (4\tan 30°)(E_x) = 0 \tag{i}$$

Finally, we add Eqs. (f) and (i) to obtain $E_x = 8.660$ kN. With successive substitutions into Eqs. (g), (c), and (h), we find the remaining reactions to be $A_x = 8.660$ kN, $A_y = 7.50$ kN, and $E_y = 7.50$ kN.

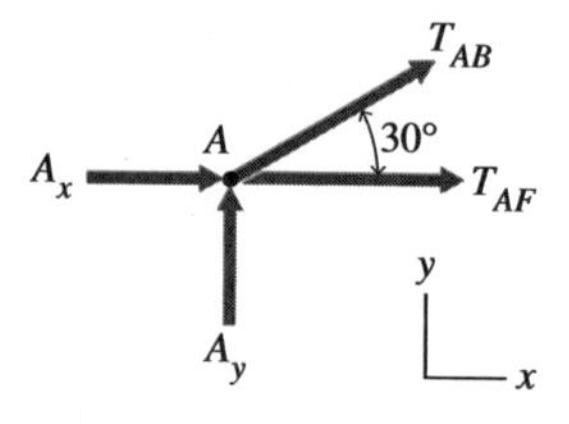

Figure E6.6c

To find the tension in members *AF, CF, CG,* and *EG,* we use the method of joints and examine joint *A*. From the free-body diagram of joint *A* (Fig. E6.6c), the first equilibrium equation for the joint is

$$\sum F_y = 0{:}\ T_{AB}\sin 30° + A_y = 0 \tag{j}$$

From this, we obtain $T_{AB} = -15$ kN (the member is in compression). The second equation of equilibrium for joint *A* is

$$\sum F_x = 0{:}\ A_x + T_{AB}\cos 30° + T_{AF} = 0 \tag{k}$$

This gives $T_{AF} = 4.33$ kN (tension). Symmetry of the members joining at *F* about a line along member *BF* (refer to Fig. E6.6b) yields $T_{CF} = T_{AF}$. Likewise, symmetry of the whole truss about a vertical line through joint *C* gives $T_{CF} = T_{CG}$ and $T_{AF} = T_{EG}$. So each of these four members is subjected to a tension $T = 4.33$ kN.

To summarize the solution, the support reactions and the tensions in the members are shown in Fig. E6.6d.

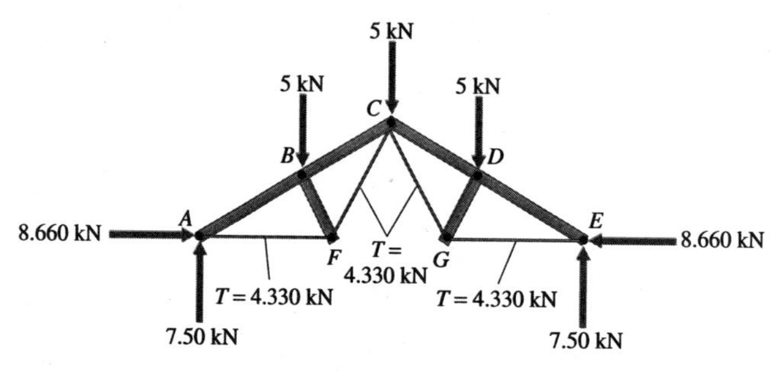

Figure E6.6d

CHECKPOINT

1. ***True or False:*** A compound truss is a combination of simple trusses that are joined together.
2. ***True or False:*** A complex truss is a truss that is unusually difficult to analyze.

Answers: 1. True; 2. False

6.7 SPACE TRUSSES

After studying this section, you should be able to:

- Analyze a space truss using the method of joints, the method of sections, or a combination of the two.

Key Concept A space truss can be analyzed by the same methods used to analyze a planar truss.

space truss

So far in this chapter, our discussion has been limited to planar (two-dimensional) trusses. In this section, we briefly consider space (three-dimensional) trusses. A *space truss* is an assembly of straight, two-force members in which the arrangement is not limited to planar configurations. Both the method of joints and the method of sections, with only minor extensions, apply to the analysis of space (three-dimensional) trusses.

The fundamental rigid element of a space truss is the tetrahedron, shown in Fig. 6.18a. This element contains six members and four joints. The basic truss is extended by successively adding three noncoplanar members and a common joint (see Figs. 6.18b and 6.18c). If a space truss is assembled in this manner and constrained by six support reactions (the minimum number for equilibrium of a three-dimensional body), then it is statically determinate and stable. As with planar trusses, a space truss is statically determinate when the total number of unknowns (consisting of m unknown member forces plus r support reactions) equals the number of independent equilibrium equations for the joints $3j$—that is,

$$m + r = 3j \tag{6.3}$$

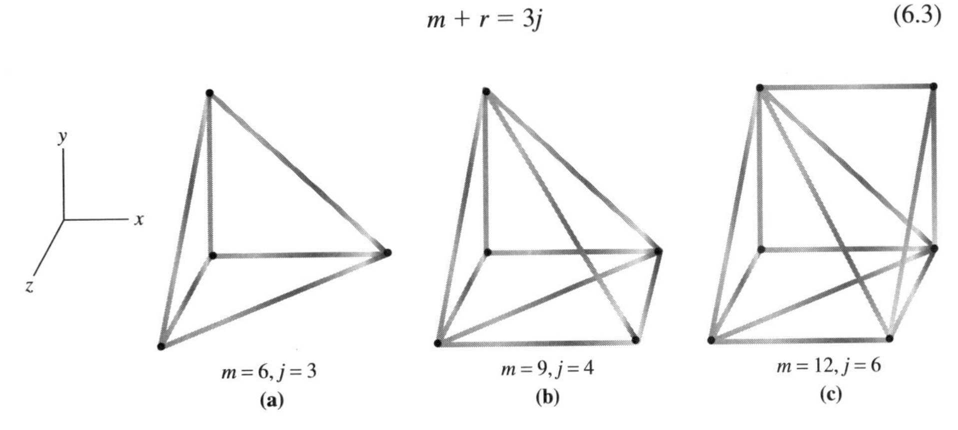

Figure 6.18 Simple space truss: **(a)** fundamental element, a tetrahedron; **(b)** first extension of the fundamental element; **(c)** second extension of the fundamental element.

If more members are added to a space truss without addition of more joints, the truss becomes statically indeterminate. Then this inequality applies:

$$m + r > 3j$$

If one or more members are removed from a statically determinate space truss, Eq. (6.3) is not satisfied, and the truss becomes unstable.

The method of joints can be used to analyze a space truss, with the provision that three equations of equilibrium must be satisfied at each joint (taken as a particle). Likewise, the method of sections (Sec. 6.5) can be used, with the understanding that no more than six members with unknown forces should be cut to isolate a part since six equations of equilibrium apply to a rigid body in three dimensions.

A tubular-steel space truss is used as a children's climbing structure in a public park in Frankfurt, Germany. An irrigation sprinkler system also uses space trusses to span the large distances between wheel units.

Example 6.7

Loading Dock Boom

Problem Statement A rotating boom truss is used to lift cargo from ships and place it on a dock (see Fig. E6.7a). The truss can pivot about a vertical axis (parallel to the y axis) through joint A, which is hinged to the dock. Joints B and C are supported by rollers on the dock. For a load $P = 50$ kips, determine the forces in the compression struts BE and CE and in the tension tie AD.

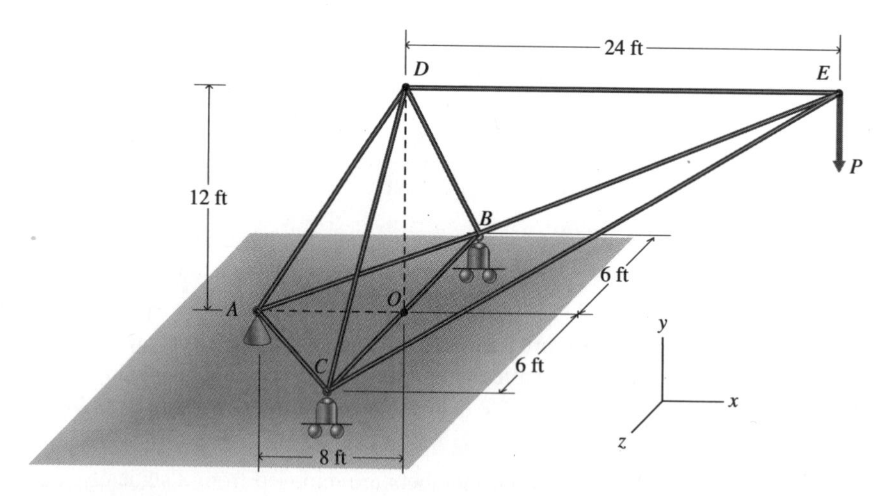

Figure E6.7a

Solution We must first decide on a solution strategy. We can use the method of joints, the method of sections, or a combination of the two. If we choose the method of joints, we could start at joint E and find the compressive forces T_{BE} and T_{CE} immediately, along with T_{DE}. Then we could proceed to joint D to find T_{AD}. By the method of sections, we can determine all three member forces of interest, T_{BE}, T_{CE}, and T_{AD}, with a single section that cuts members

AD, BD, CD, BE, and CE to isolate member DE. It appears that the computational effort is about the same with either approach. So, we choose the method of sections. The solution by the method of joints is left to you as an exercise.

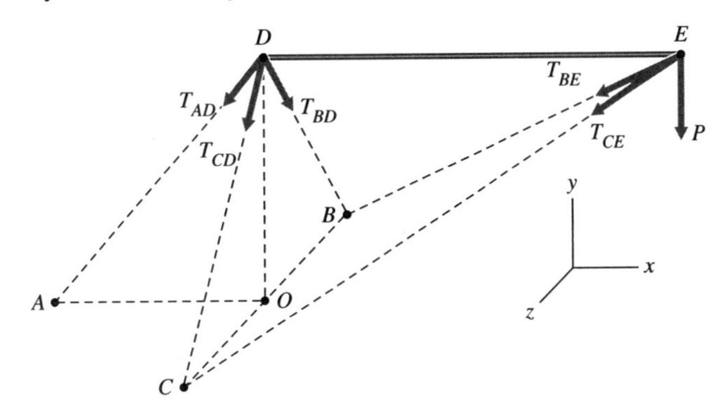

Figure E6.7b

By cutting through members AD, BD, CD, BE, and CE, we isolate member DE, as shown in the free-body diagram of Fig. E6.7b. Using point O as the origin of the xyz coordinate system, we can establish distance vectors from point to point and, hence, direction cosines for the unknown forces. The coordinates of the truss joints, the distance vectors, and the direction cosines for the forces are summarized in Table E6.7.

Table E6.7

POINT	COORD. (ft) x	y	z	LINE	DISTANCE VECTOR $\mathbf{r}$ (ft)	LENGTH r (ft)	DIRECTION COSINES θ_x	θ_y	θ_z
A	−8	0	0	DA	$-8\mathbf{i} - 12\mathbf{j} + 0\mathbf{k}$	14.42	−0.5547	−0.8321	0
B	0	0	−6	DB	$0\mathbf{i} - 12\mathbf{j} - 6\mathbf{k}$	13.42	0	−0.8944	−0.4472
C	0	0	6	DC	$0\mathbf{i} - 12\mathbf{j} + 6\mathbf{k}$	13.42	0	−0.8944	0.4472
D	0	12	0	EB	$-24\mathbf{i} - 12\mathbf{j} - 6\mathbf{k}$	27.50	−0.8729	−0.4364	−0.2182
E	24	12	0	EC	$-24\mathbf{i} - 12\mathbf{j} + 6\mathbf{k}$	27.50	−0.8729	−0.4364	0.2182

Before writing equations of equilibrium for member DE, we note that the symmetry of the structure and its loading will produce symmetric member forces. Therefore, $T_{BD} = T_{CD}$, and $T_{BE} = T_{CE}$. We also express the load (in kips) in vector form as $\mathbf{P} = -50\mathbf{j}$. Now, with careful selection of the equations of equilibrium, we can avoid solving simultaneous equations. We begin with an equation for moments about point D:

$$\circlearrowleft^{+} \sum \mathbf{M}_D = 0: \mathbf{r}_{ED} \times \mathbf{P} + \mathbf{r}_{ED} \times \mathbf{T}_{BE} + \mathbf{r}_{ED} \times \mathbf{T}_{CE} = 0 \quad \text{(a)}$$

where $\mathbf{r}_{ED}$ is the distance vector from joint D to joint E; $\mathbf{r}_{ED} = 24\mathbf{i}$. Evaluating Eq. (a) and noting that $T_{BE} = T_{CE}$, we obtain

$$24(-50)\mathbf{k} + 24[2T_{BE}(-0.4364)]\mathbf{k} + 24(0.2182T_{BE})\mathbf{j} + 24(-0.2182T_{BE})\mathbf{j} = 0$$

which gives $T_{BE} = T_{CE} = -57.29$ kips. Thus, members BE and CE are in compression.

Now, to find T_{AD}, we observe that the scalar equilibrium equation $\sum F_x = 0$ for member DE will have only one unknown, that being T_{AD}. That equation takes the form:

$$\sum F_x = 0: 2T_{BE}(-0.8729) + T_{AD}(-0.5547) = 0 \quad \text{(b)}$$

Solution of Eq. (b) yields $T_{AD} = 180.31$ kips (member AD is in tension). ■

Chapter Highlights

- A truss is a system of coplanar two-force members connected by frictionless pins and loaded at its joints.
- The method of joints is a procedure for truss analysis that applies the conditions for particle equilibrium to each joint of a truss.
- The method of sections is a procedure for truss analysis that is based on rigid-body equilibrium.

Problems

6.2 *Stability and Statical Determinacy*

6.1 Classify each of the trusses shown in Fig. P6.1 as stable or unstable. For those that are stable, further classify them as statically determinate or statically indeterminate.

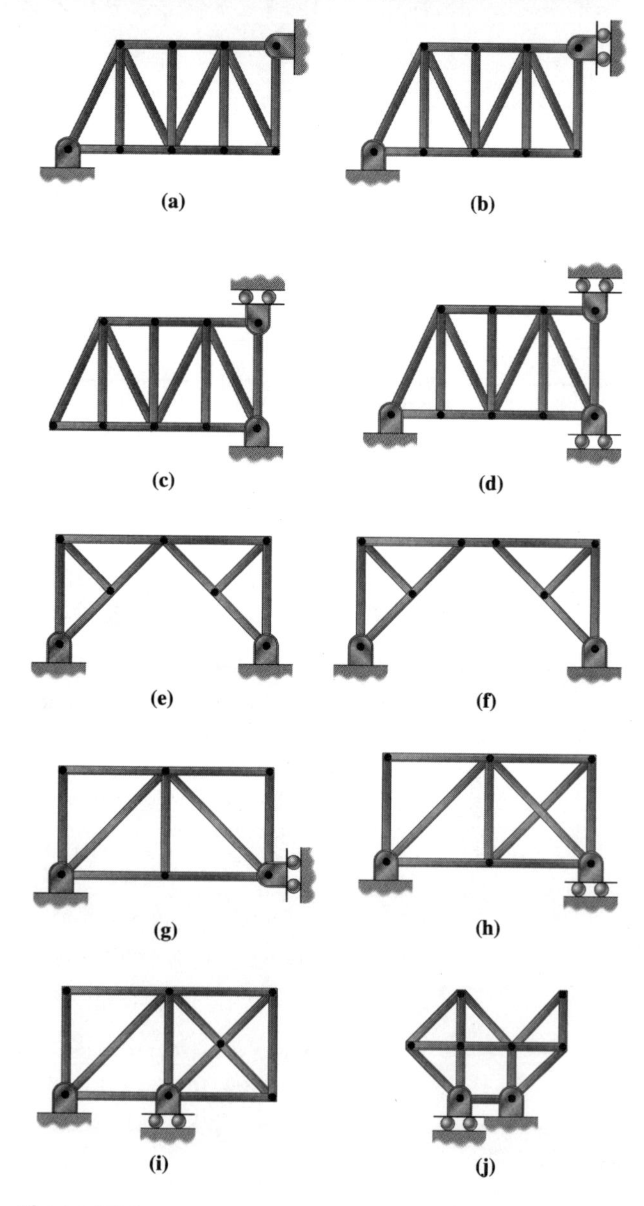

Figure P6.1

6.4 *The Method of Joints*

In each of the following problems, use the method of joints to determine the member forces.

6.2 In Fig. P6.2, the member with the turnbuckle has an internal force of 3 kN tension. Calculate the forces in the other members.

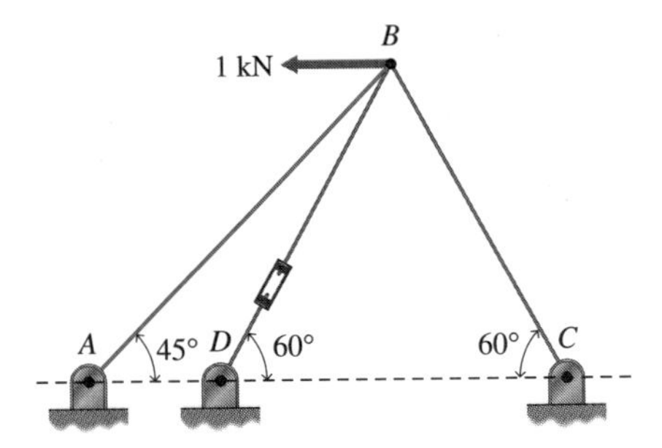

Figure P6.2

6.3 Express the forces in the members of the platform truss in Fig. P6.3 in terms of a, b, and F.

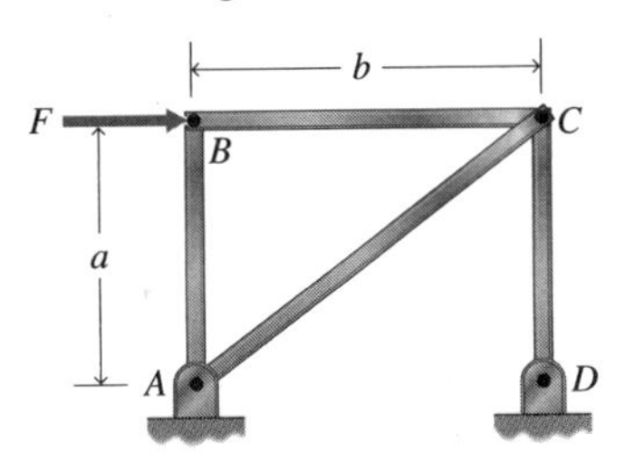

Figure P6.3

6.4 Express the forces in the members of the kite truss in Fig. P6.4 in terms of α, β, and F.

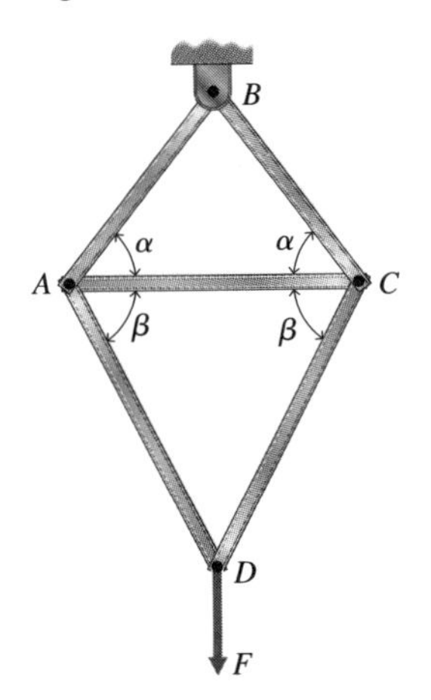

Figure P6.4

6.5 Determine the support reactions and the force in member CE for the truss shown in Fig. P6.5.

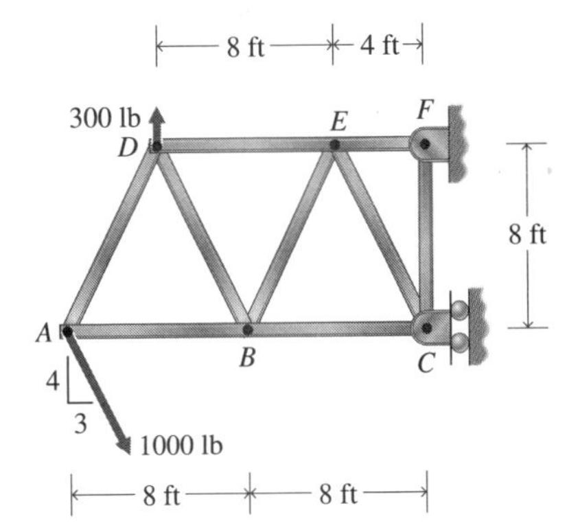

Figure P6.5

6.6 Determine the forces in the members of the roof truss in Fig. P6.6.

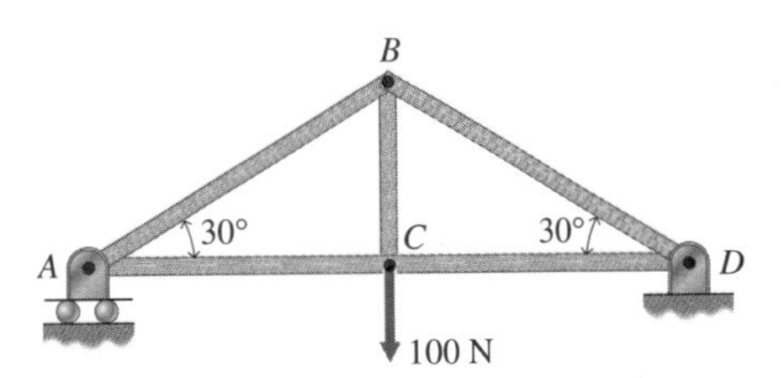

Figure P6.6

6.7 Determine the forces in the members of the bridge truss in Fig. P6.7. Make use of symmetry conditions.

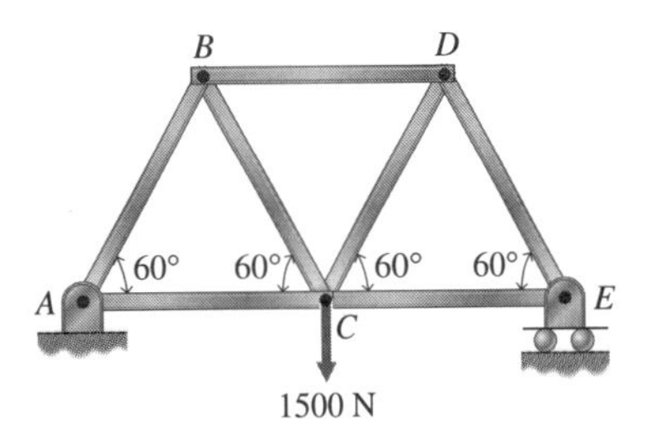

Figure P6.7

6.8 All the members of the Warren truss shown in Fig. P6.8 have the same length. Calculate the forces in the members. Make use of symmetry conditions.

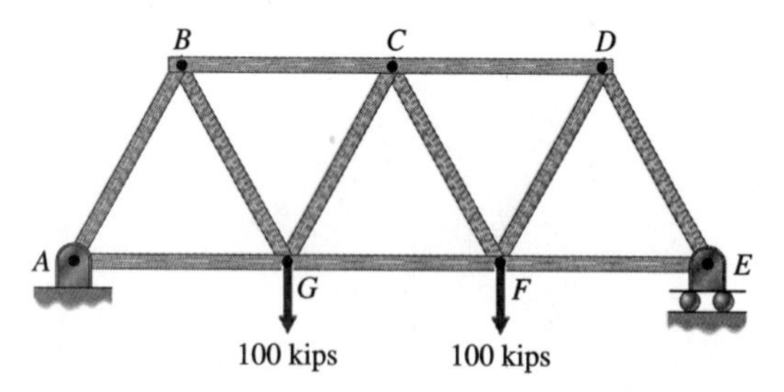

Figure P6.8

6.9 The mechanism of a rock crusher is shown schematically in Fig. P6.9. If a 10 kN force is required to crush a rock, what is the required magnitude of the horizontal driving force F applied at D? Neglect friction.

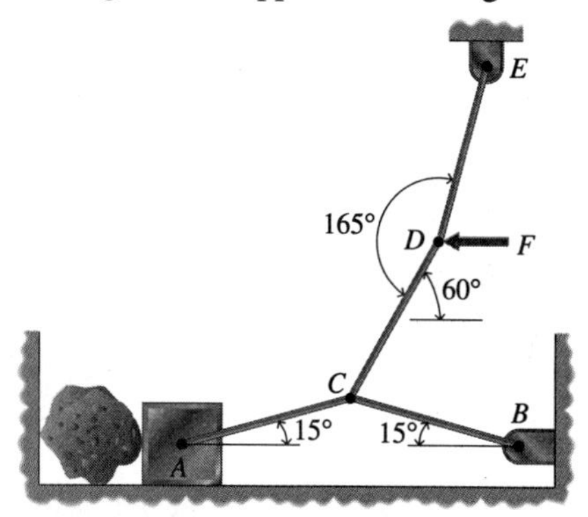

Figure P6.9

6.10 Analyze the truss shown in Fig. P6.10. Express the member forces and the support reactions in terms of P and L.

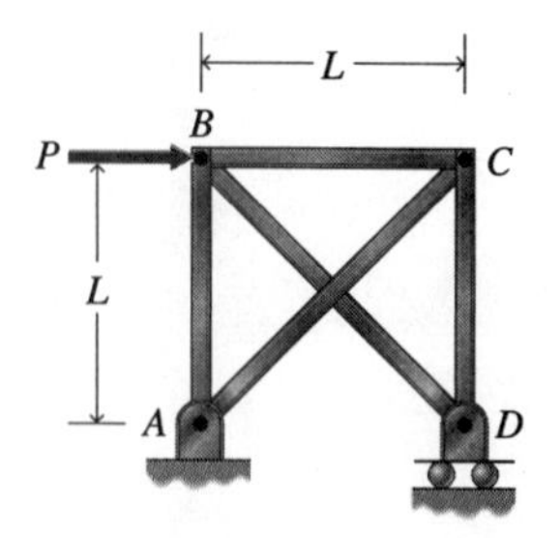

Figure P6.10

6.11 Determine the forces in the members of the wall crane in Fig. P6.11.

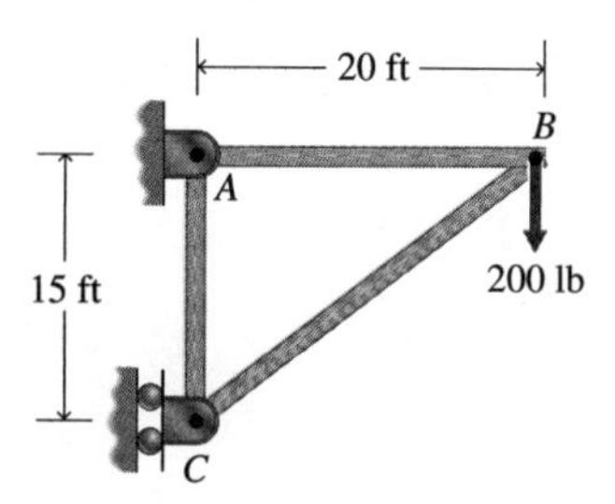

Figure P6.11

6.12 For the truss shown in Fig. P6.12, determine the forces in members AC, AD, BD, and CD. Begin by finding the support reactions on the truss.

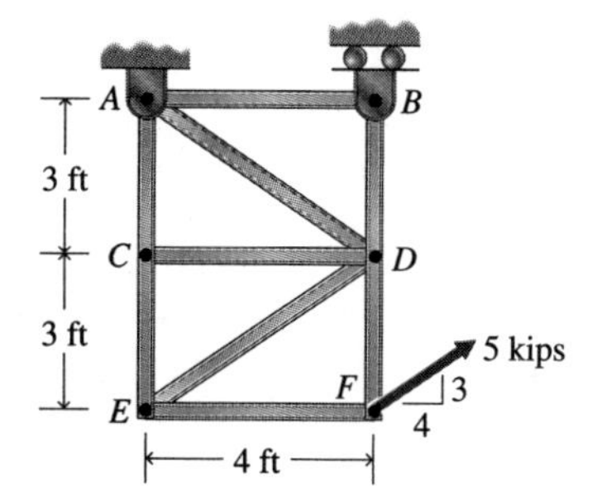

Figure P6.12

6.13 Calculate the forces in the members of the wall truss in Fig. P6.13.

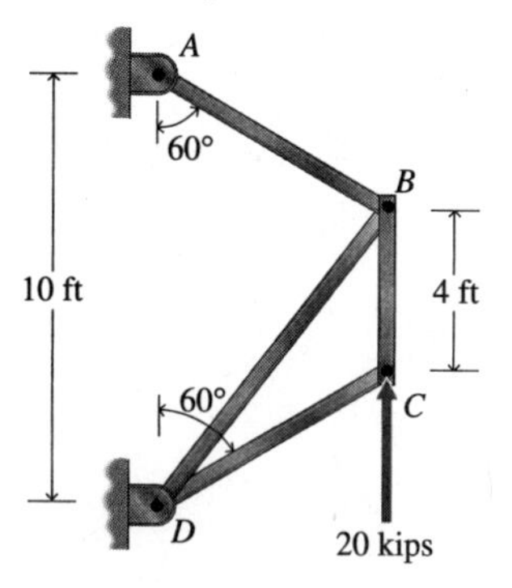

Figure P6.13

6.14 Calculate the forces in the members of the crane in Fig. P6.14.

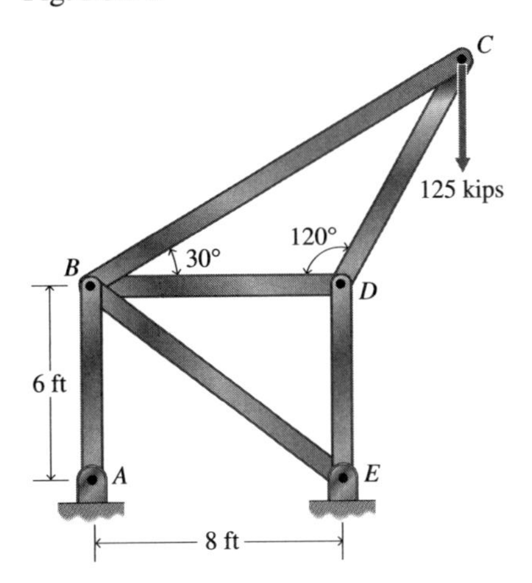

Figure P6.14

6.15 For the truss in Fig. P6.15, determine
a. the support reactions
b. the forces in all members

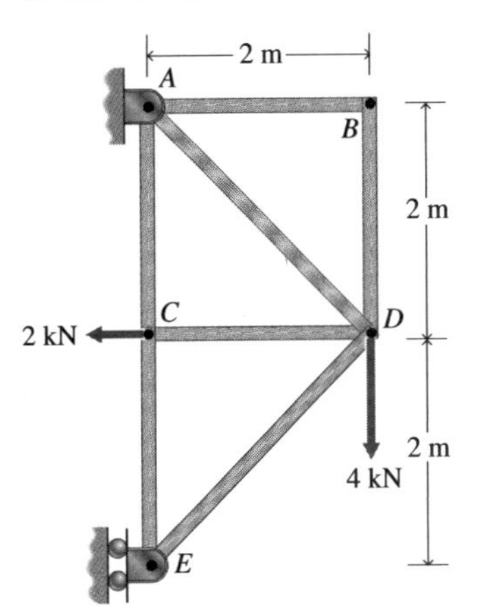

Figure P6.15

6.16 Determine the forces in the members *BD, CD,* and *CE* of the wall crane in Fig. P6.16.

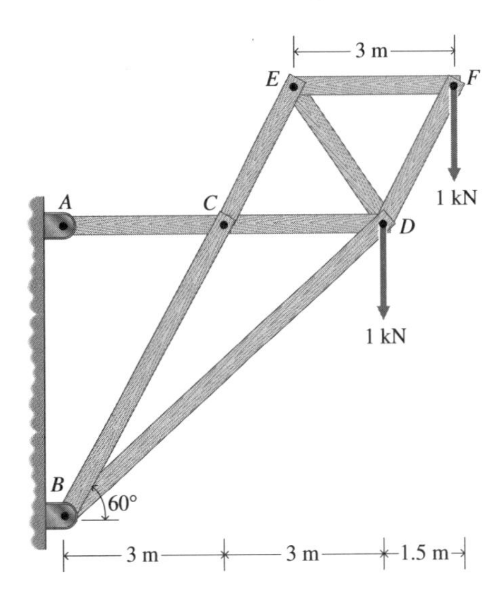

Figure P6.16

6.17 Determine the force in member *BD* of the cantilever truss shown in Fig. P6.17.

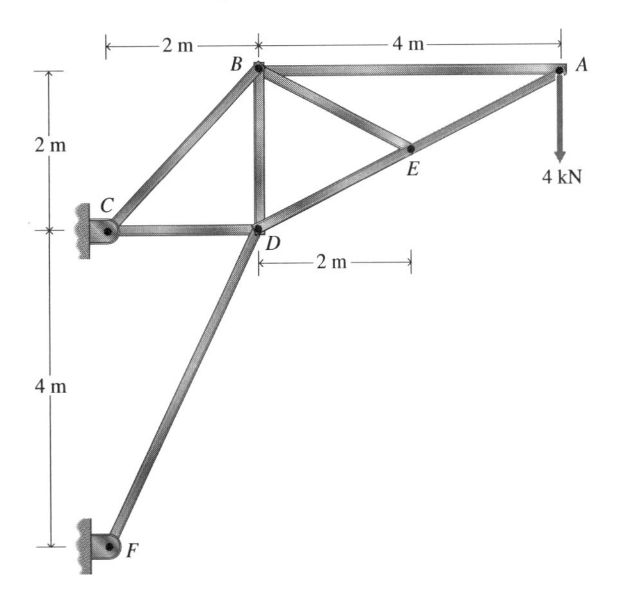

Figure P6.17

6.18 Determine the forces in the members of the square airplane truss in Fig. P6.18.

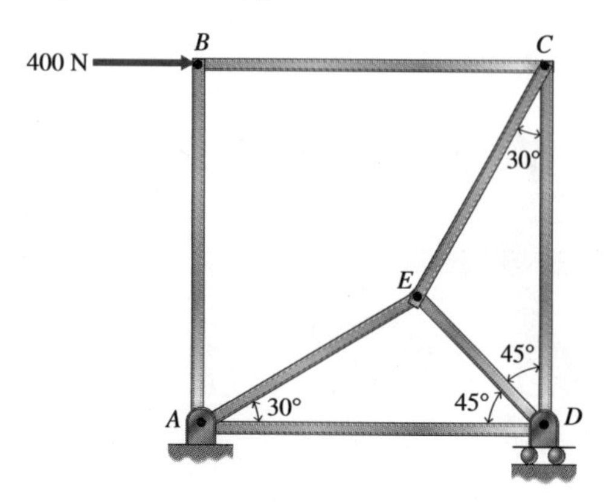

Figure P6.18

6.19 The Fink truss in Fig. P6.19 is used to support a roof for a residential garage. The design loads due to the weight of the roofing and snow on the roof result in joint loads of 520 lb at each joint along the top chord of the truss. Determine the support reactions and the forces in all members.

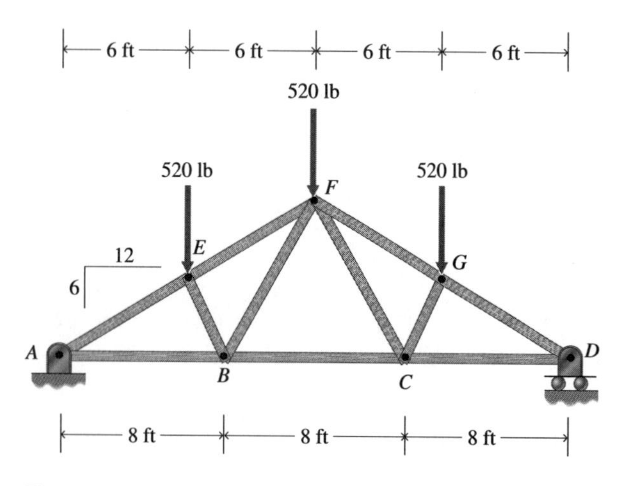

Figure P6.19

6.20 Determine the angle θ and the forces in cords AB, BC, and CD in Fig. P6.20.

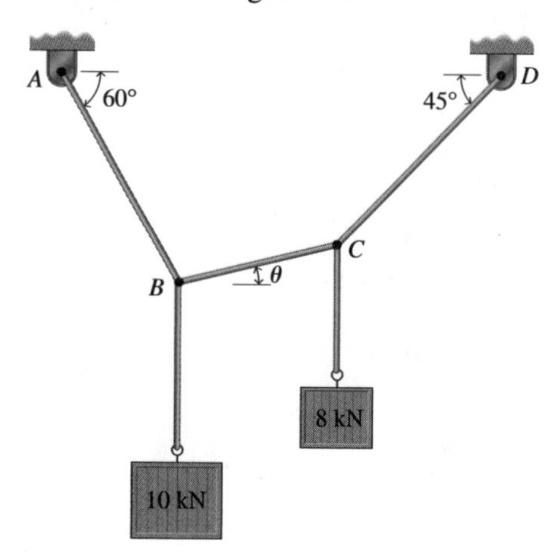

Figure P6.20

6.5 *The Method of Sections*

In each of the following problems, use the method of sections to determine the member forces. If it is necessary to cut through more than three members to separate a truss appropriately, you may wish to use a combination of the method of sections and the method of joints.

6.21 Determine the forces in members CD, BD, and BE of the truss shown in Fig. P6.21.

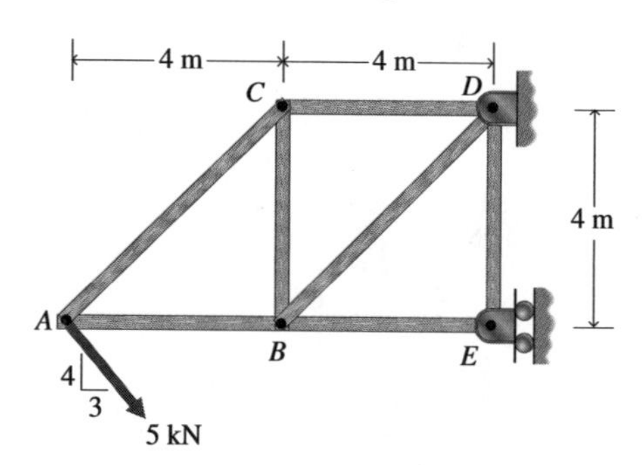

Figure P6.21

6.22 For the truss shown in Fig. P6.22, determine the forces in members *AB*, *BD*, and *DE*.

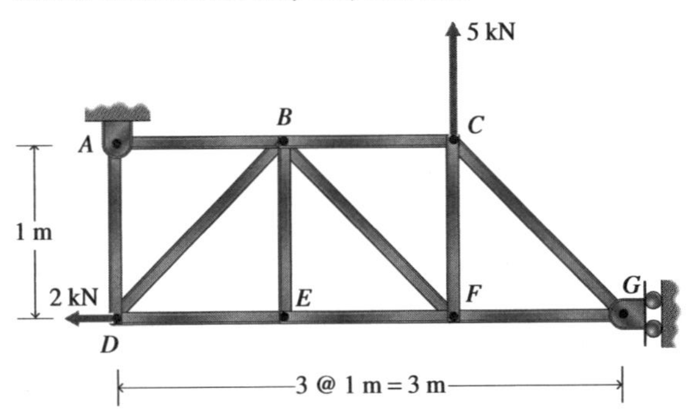

Figure P6.22

6.23 Determine the forces in members *BC*, *BD*, and *BE* of the bridge truss in Fig. P6.23.

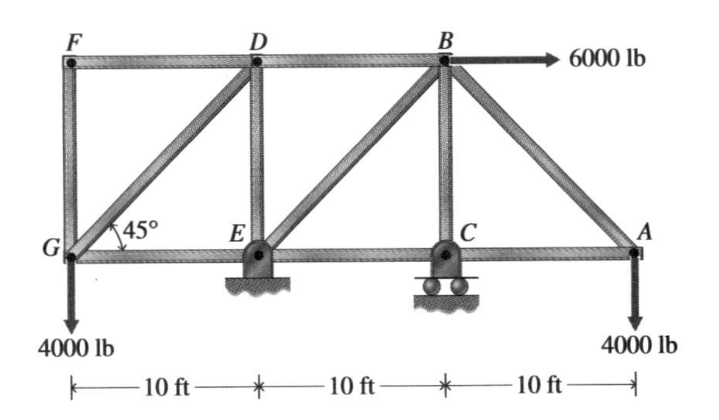

Figure P6.23

6.24 Determine the tensions in strings *BD*, *CD*, and *CE* by applying equilibrium conditions to the entire part of the system below and to the right of the dashed line in Fig. P6.24.

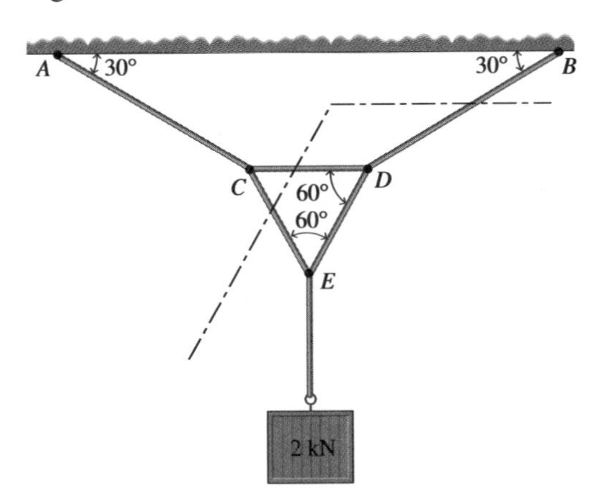

Figure P6.24

6.25 Determine the forces in members *AB*, *AE*, and *ED* of the wall bracket in Fig. P6.25.

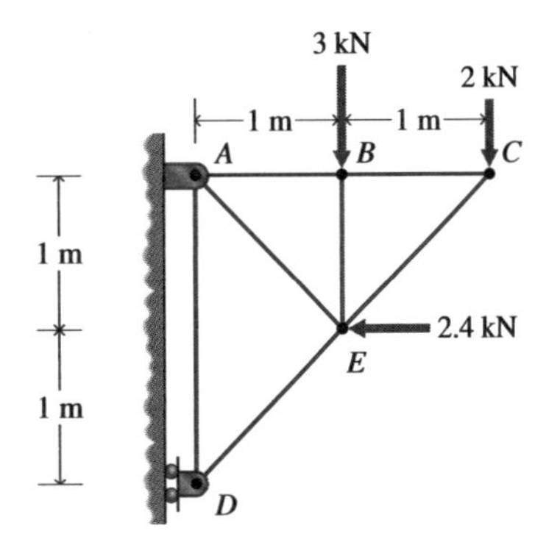

Figure P6.25

6.26 Determine the forces in members *CE*, *DF*, and *EF* of the truss shown in Fig. P6.26.

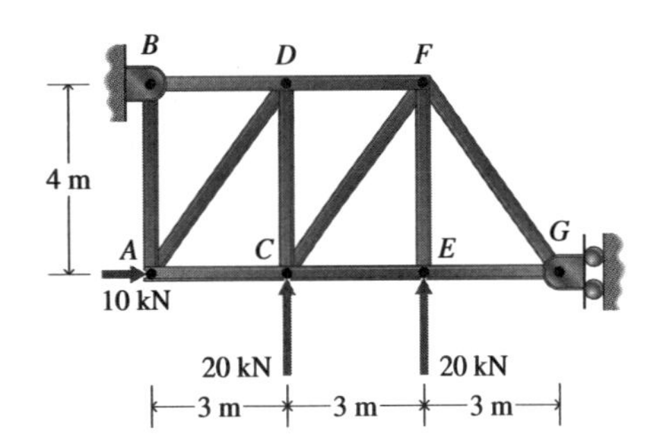

Figure P6.26

6.27 Determine the forces in members *BC*, *BE*, and *EF* of the roof truss in Fig. P6.27.

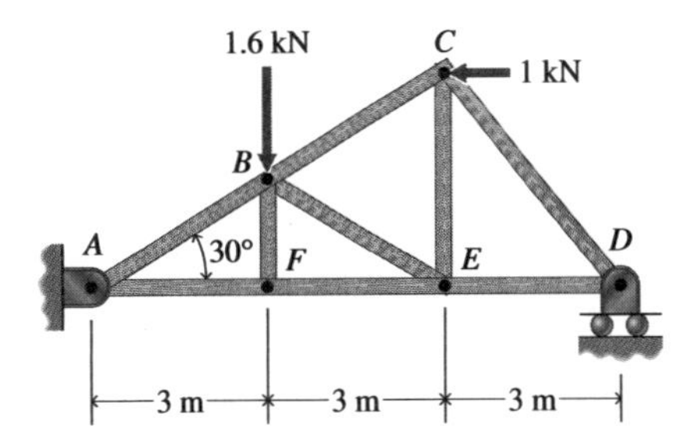

Figure P6.27

6.28 Determine the forces in members *EF, BC,* and *BE* of the hanging truss in Fig. P6.28.

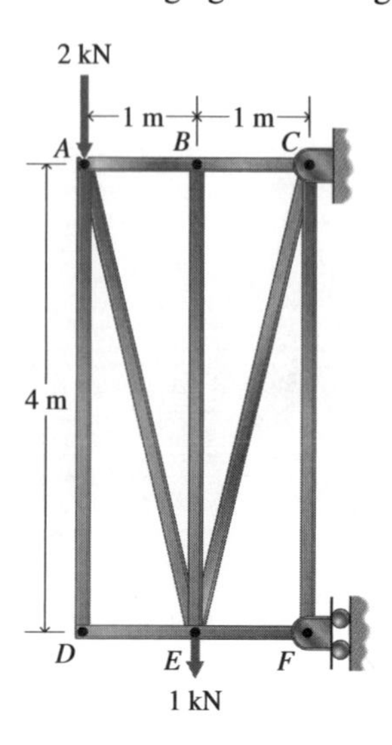

Figure P6.28

6.29 Determine the forces in members *CE, CD,* and *BD* of the truss of Problem 6.16.

6.30 Determine the forces in members *BC, BD,* and *BE* of the bridge truss in Fig. P6.30.

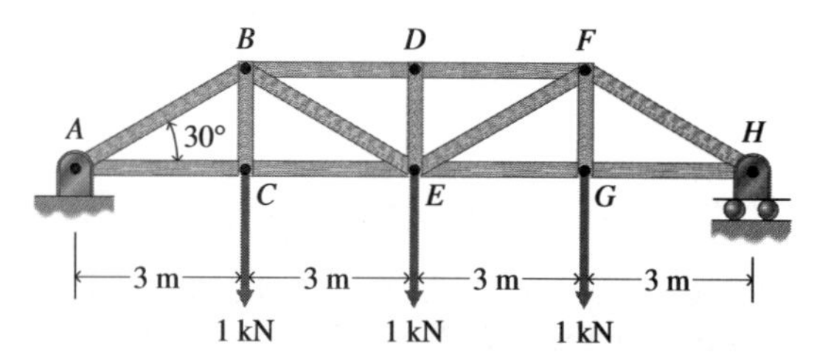

Figure P6.30

6.31 Compute the reactions and member forces for the truss shown in Fig. P6.31.

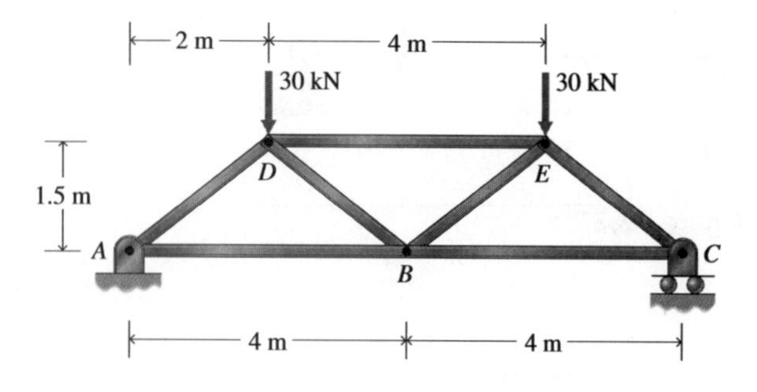

Figure P6.31

6.32 For the bridge truss shown in Fig. P6.32, determine the forces in the members cut by section *A–A*.

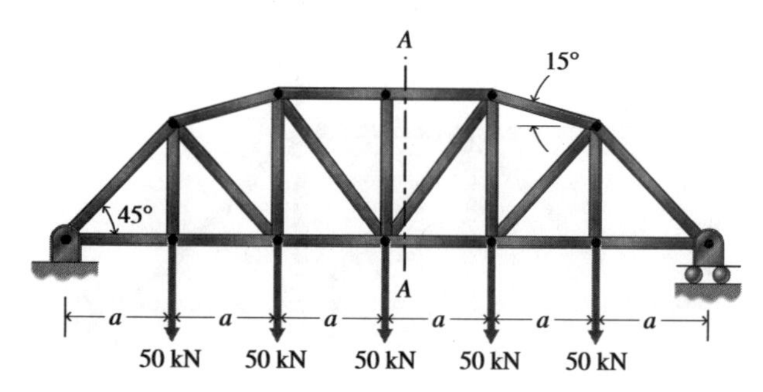

Figure P6.32

6.33 Determine the force in member *CD* of the truss in Fig. P6.33.

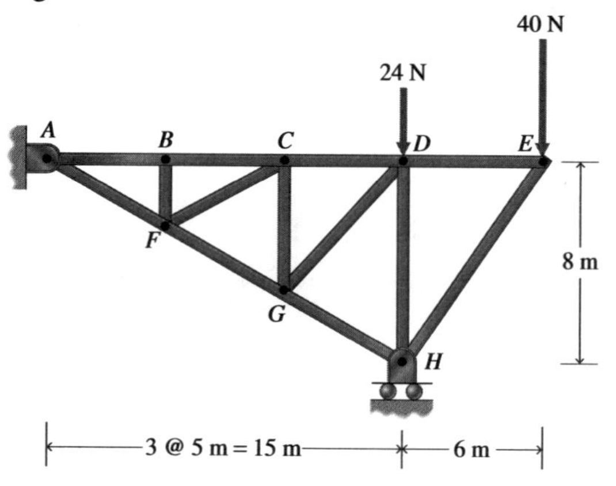

Figure P6.33

6.34 Determine the forces in members *BC, CD, DF,* and *CE* of the roof truss in Fig. P6.34.

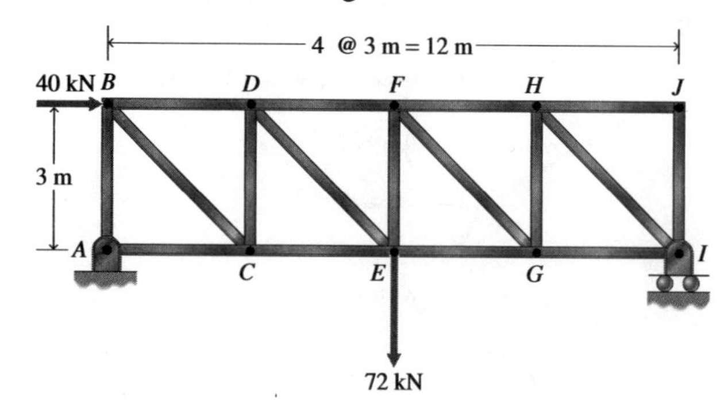

Figure P6.34

6.35 Determine the forces in members *BC, CG, FG, CF,* and *EF* of the wall truss in Fig. P6.35.

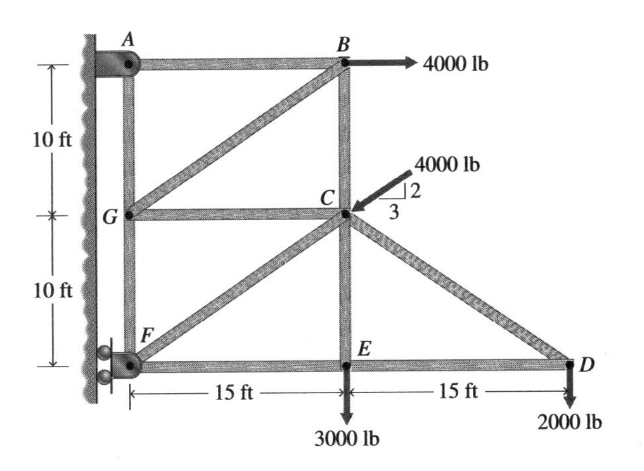

Figure P6.35

6.36 Determine the forces in members *BD, CD, CE,* and *BC* of the truss in Fig. P6.36.

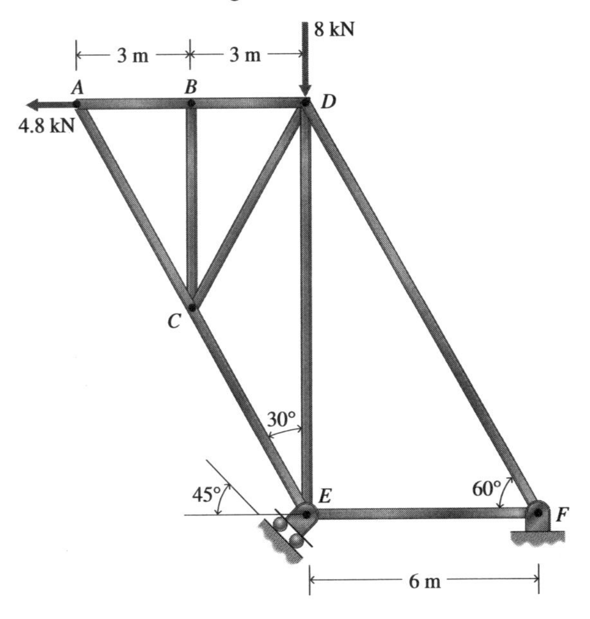

Figure P6.36

6.37 The truss shown in Fig. P6.37 contains four zero-force members. Locate them.

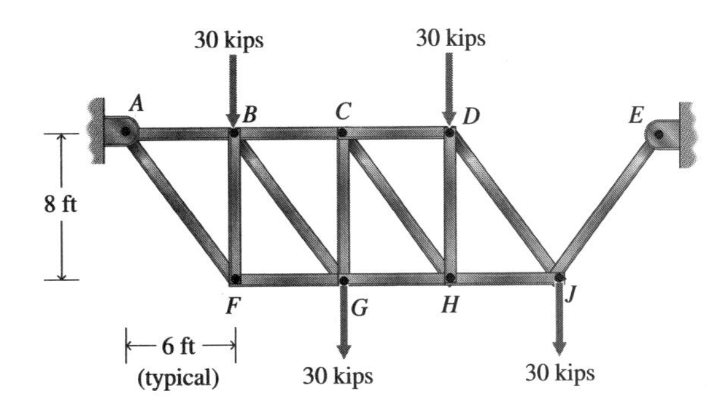

Figure P6.37

6.38 Determine the forces in members *BC, CD, DE,* and *CE* of the Warren truss shown in Fig. P6.38.

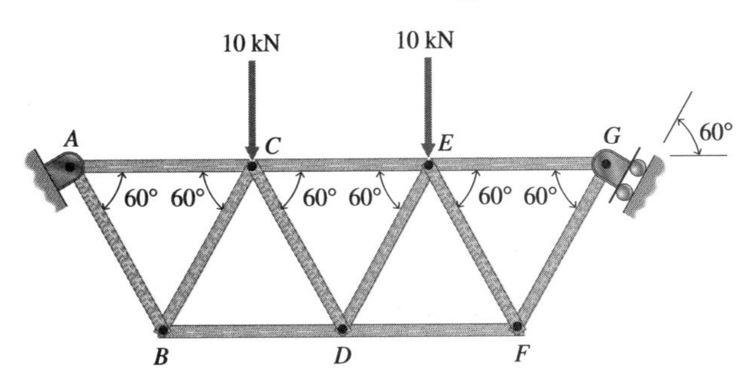

Figure P6.38

6.39 For the truss in Fig. P6.39, find the forces in members *GH, GL, DF, AD,* and *CD.*

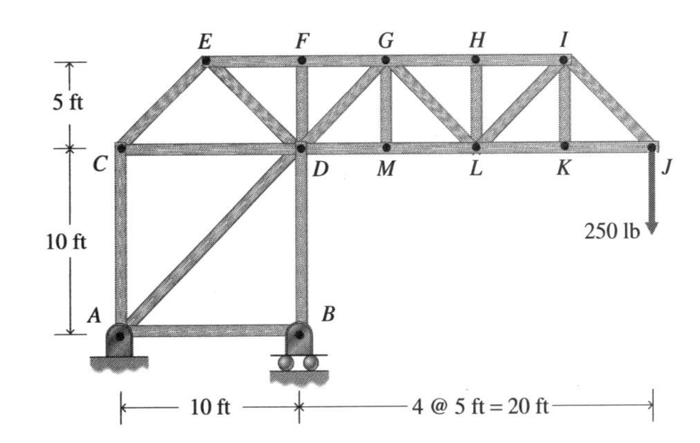

Figure P6.39

6.40 For the truss shown in Fig. P6.31, move the roller support from joint *C* to joint *B*. Compute the reactions and member forces under the same loading.

6.41 A truck-mounted hoist system (see Fig. P6.41) has been proposed to lift and move aircraft parts (engines, wing sections, propellers, and so on). The hoist is to support loads as large as $W = 2000$ lb. Because of height and length restrictions, distance AE is not to exceed 10 ft and member DE must not be longer than 8 ft. To ensure that stability of the truss is properly maintained, T_{BC}, the tension in member BC, must be held within the range 500 lb $< T_{BC} <$ 900 lb. Determine the geometry of the truss $ABCD$ (the range of lengths for members AB, AC, BD, and BC) that will ensure that this requirement is met and the corresponding range of forces in the members.

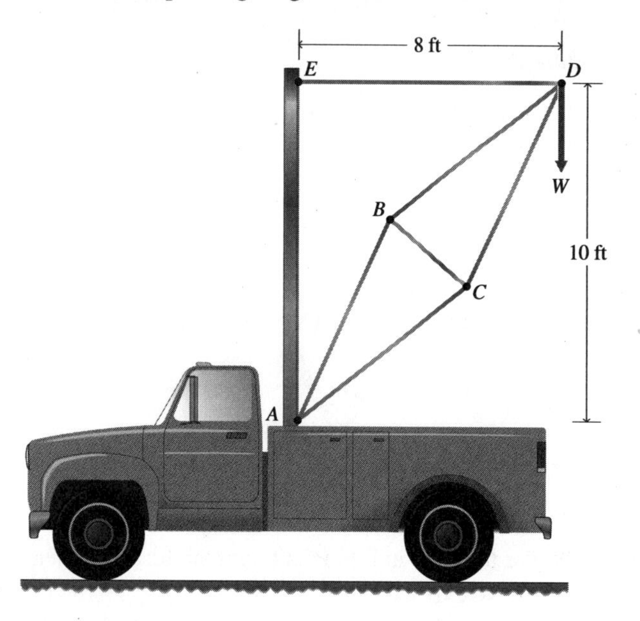

Figure P6.41

6.42 A truss with outside dimensions shown in Fig. P6.42 is required for a roof system for a small warehouse. Loads are applied at the points shown along the top chord. The warehouse owner hires you to design a low-cost truss configuration for this application. You can buy tension members for \$5 per meter. Zero-force members also cost \$5 per meter. Long compression members cost more per foot than short compression members because they are more likely to buckle. A compression member costs (\$5)L^2, where L is the length of the member in meters. For instance, a 2-meter-long compression member costs \$20. Your task is to design a truss with the perimeter dimensions shown. Specify the number of members, the joint locations for the top and bottom chords of the truss, and the arrangement of the interior members. Determine the cost of your truss. Submit at least two alternative designs for consideration by the owner. Assume that the support roller and pin for the roof truss are located at the tops of the warehouse walls.

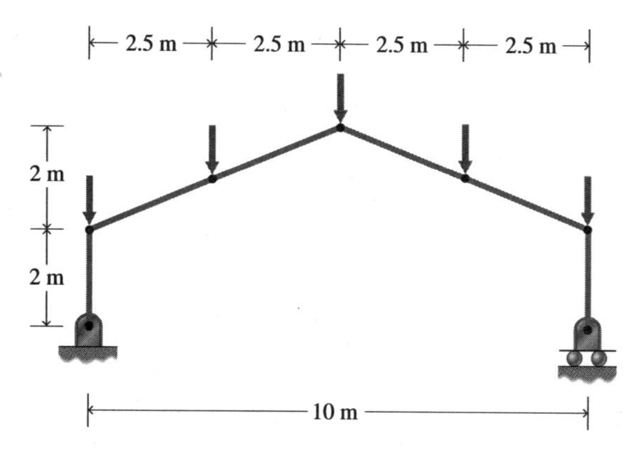

Figure P6.42

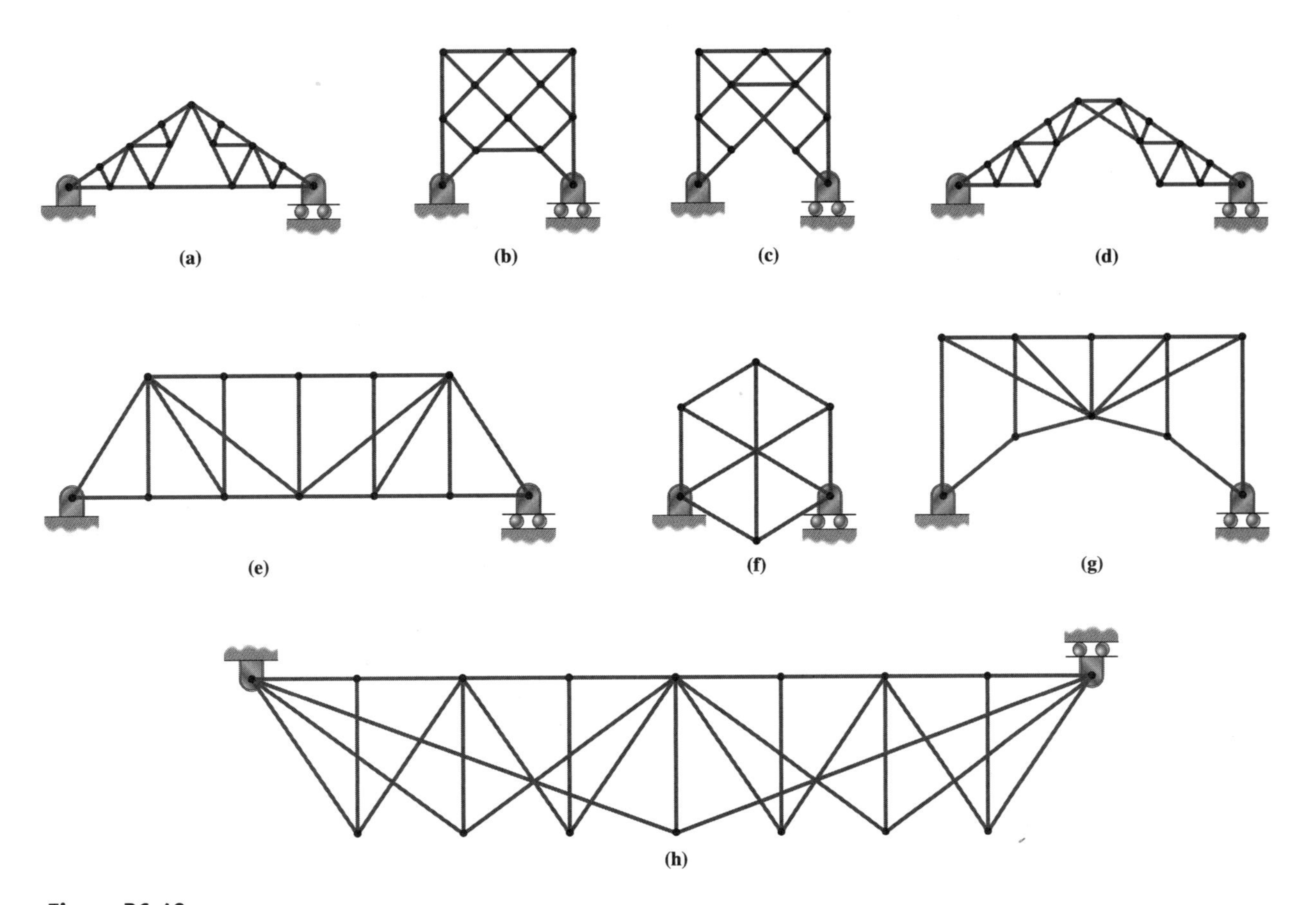

Figure P6.43

6.6 *Compound and Complex Trusses*

6.43 Identify each of the trusses shown in Fig. P6.43 as simple, compound, or complex.

6.44 Reposition a single member of the complex truss in Fig. P6.44 so that it becomes a simple truss. How many different solutions are there?

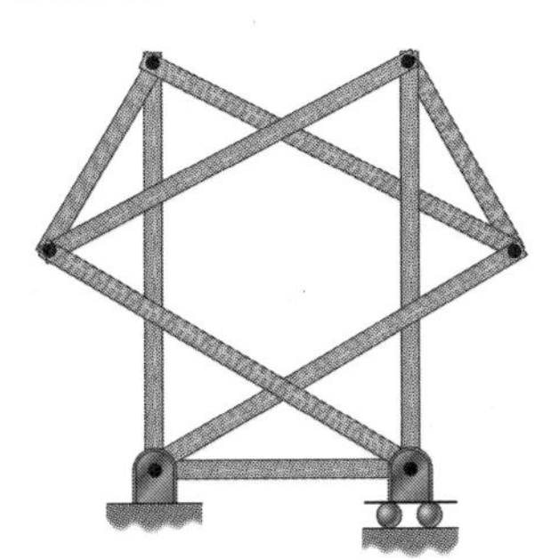

Figure P6.44

6.45 Determine the reactions of the supports on the truss in Fig. P6.45.

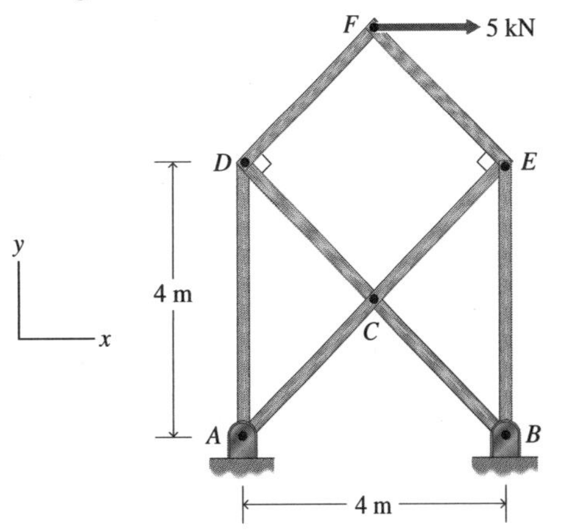

Figure P6.45

6.46 Determine the forces in the members of the X-frame in Fig. P6.46 for each of the following two cases:
a. A 10 N force is applied vertically upward at joint *B*.
b. A 10 N force is applied vertically upward at joint *D*.
Is the fact that the member forces are different in these two cases consistent with the theorem of transmissibility? Explain.

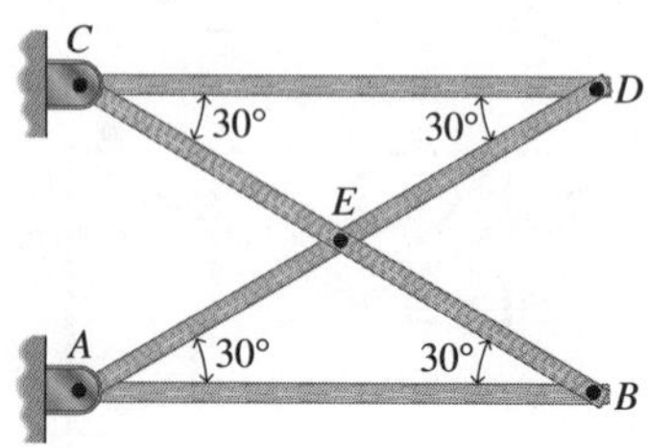

Figure P6.46

6.7 *Space Trusses*

6.47 A wall-mounted truss is used to lift cargo from trucks in a loading bay (see Fig. P6.47). The truss is attached to a wall at *B* with a ball-and-socket joint, at *C* with a roller that prevents movement in the *x* direction, and at *D* with a roller that prevents movement in the *x* direction. A strut *DE* prevents movement in the *z* direction. The cargo is lifted with a rope that passes vertically over a pulley and then extends out horizontally in the *z* direction. Assume that the pulley is small enough that the vertical and horizontal loads *P* can be taken to act at joint *A*. Determine the support reactions and member forces for the truss.

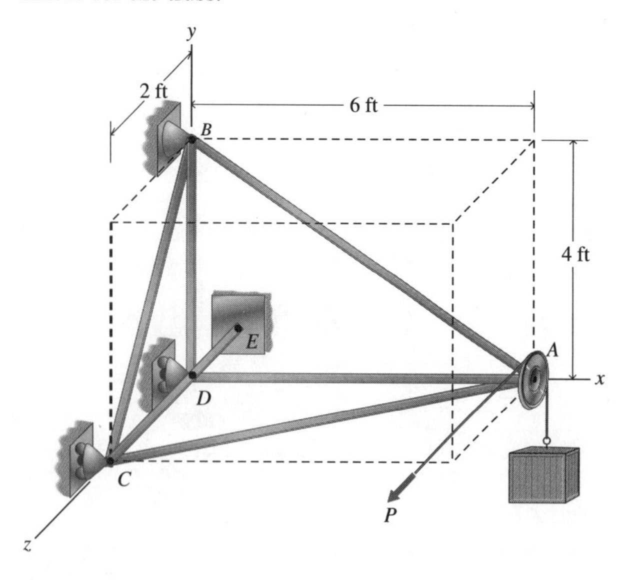

Figure P6.47

6.48 The vertical segment of rope in Problem 6.47 is rotated through an angle of 30° in the *xy* plane because the truck driver begins to pull away from the wall before the equipment is removed. The horizontal force remains in the *z* direction. Determine the support reactions and member forces for the truss.

6.49 The truss in Fig. P6.49 is a regular octahedron (all edge lengths are equal), with member *CE* added to maintain stability. Determine the forces in members *AB, BC,* and *CE.*

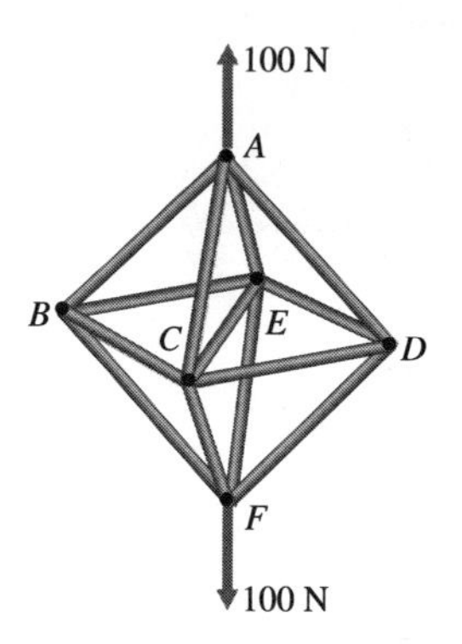

Figure P6.49

6.50 Determine the force in the 3-meter-long top chord of the truss in Fig. P6.50.

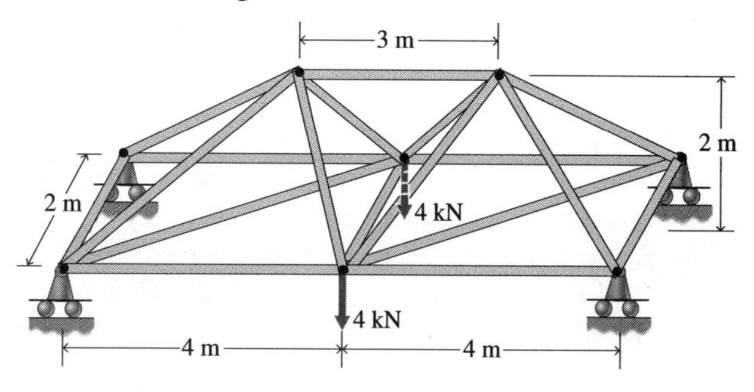

Figure P6.50

6.51 A space truss is used as a derrick over a large-diameter shaft (see Fig. P6.51). Joints A, B, C, D, E, and F lie in the xy plane, with origin at O. Members DE, EF, and DF are each 2 m long. Members AD, AF, BD, BE, CE, and CF are each 3 m long. Joint G lies along the z axis at a height of 3 m. Lines OA, OB, and OC form 120° angles with each other. Joints A, B, and C are constrained so that they ride in horizontal guide slots parallel to lines OA, OB, and OC, respectively. A load $P = 10$ kN is applied to joint G. Determine the forces in members AD, AG, and DF. Note that the truss is symmetric with respect to vertical planes through lines OA, OB, and OC.

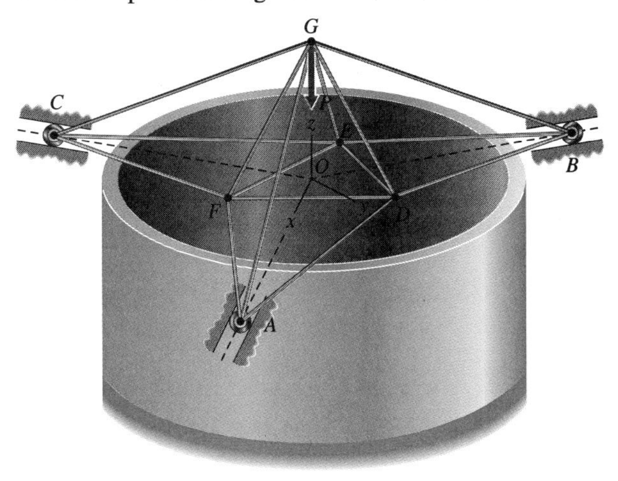

Figure P6.51

REVIEW QUESTIONS

6.52 Does the transmission of a force along its line of action from one joint of a truss to another affect the forces in the truss members? Explain.

6.53 What are the assumptions used to define an ideal truss?

6.54 Describe the characteristics of a statically determinate truss and a statically indeterminate truss.

6.55 Explain the method of joints for analyzing trusses.

6.56 In the analysis of trusses, what are the essential differences between the method of joints and the method of sections?

6.57 The three-member truss in Fig. P6.57 is loaded with a couple at joint A. Try to use the method of joints to analyze the structure, and then use equilibrium of the entire truss to check the reactions. Why can't you analyze the structure? What fundamental characteristic of a truss is violated? Describe what would happen physically to such a truss loaded as shown.

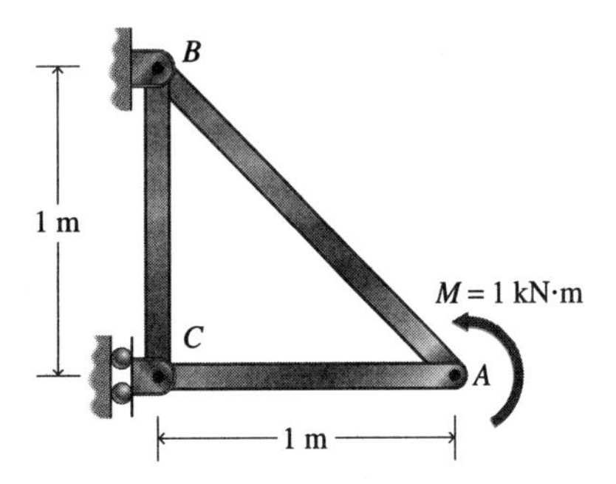

Figure P6.57

Use the method of joints, the method of sections, or both to solve each of the following problems.

6.58 Determine the forces in the interior diagonals BD and BE of the truss in Fig. P6.58.

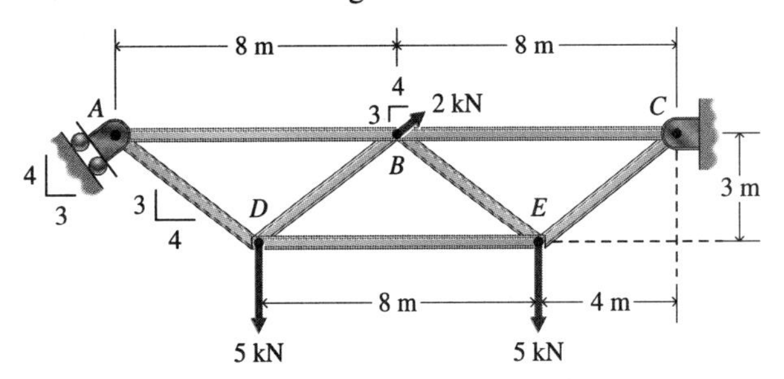

Figure P6.58

6.59 Determine the forces in members *BC, CD,* and *DF* of the truss in Fig. P6.59.

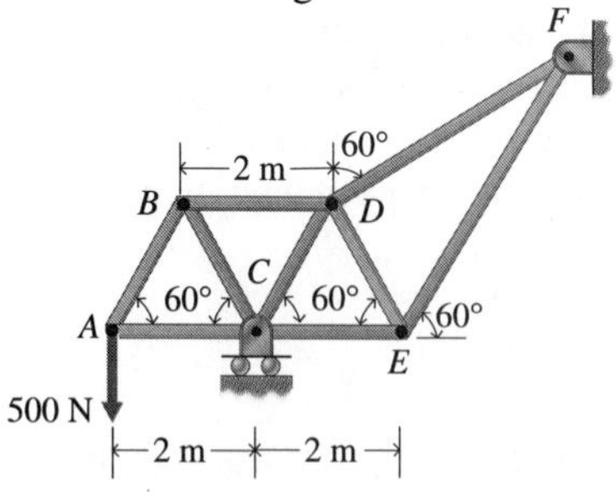

Figure P6.59

6.60 Assume that the Howe truss shown in Fig. 6.1e has an overall length of L and a height of $L/6$. Apply a downward vertical load of magnitude P to each joint along the bottom chord, and show that all of the diagonal members are in compression. Utilize symmetry.

6.61 Assume that the Pratt truss shown in Fig. 6.1c has an overall length of L and a roof pitch of 7/12. Apply a downward vertical load of magnitude P to each joint along the top chord, and show that all of the diagonal members are in tension. Exploit symmetry.

Chapter 7 SIMPLE STRUCTURES AND MACHINES

The mechanical systems of machines as complex as this automobile are combinations of simple machines, including levers, wheels and axles, inclined planes, pulleys, and screws.

A Look Forward

IN PRECEDING CHAPTERS, we considered general theories of statics for rigid bodies. In this chapter, we apply the general theories to some simple structures and machines. For purposes of this book, a *machine* can be regarded as a mechanical device consisting of one or more movable parts that serve to transmit and/or modify force. This definition of a machine can be expanded to include chemical and electrical systems in addition to mechanical devices. Complex machines, such as a derrick and an automobile, are combinations of basic mechanical, electrical, and chemical parts.

Basic mechanical devices used in machines are generally classified as one of five types: the lever, the inclined plane and the wedge, the pulley, the screw (a circular form of inclined plane), and the wheel and axle. These devices are called *simple machines.* They serve to expand the abilities of humans by supplementing the functions of the hands and arms and making human effort more efficient. Various tools are based on these five simple machines—for example, a chisel is a sharp wedge, a block and tackle is a system of pulleys, a nutcracker is a combination of two levers, and a jackscrew is a combination of a lever and a screw.

In this chapter, you will learn about applications of levers, pulleys, systems of pulleys, and other *mechanisms* (other combinations of simple machines). In addition, you will be introduced to some basic structural elements—namely, simple frames, cables used to support suspension bridge roadways, and electric transmission cables. We will study applications of the inclined plane, the wedge, and the screw. More advanced applications involving combinations of these simple machines are discussed in Chapter 10, which deals with friction.

Survey Questions

- What is a lever?
- Does a pulley modify a force or merely transmit it?
- What is a block and tackle?
- What is the difference between a frame and a truss?
- What distinguishes a frame from a mechanism?
- How does a suspension bridge cable differ from a catenary?

7.1 THE LEVER

After studying this section, you should be able to:

- Determine the mechanical advantage of a lever or a system of levers.

Key Concept A lever is a device that changes the effect of a force.

lever

fulcrum

A *lever* is one of the simplest of all mechanical devices. It consists of a rigid bar, straight or curved, that is hinged at one point. The hinged point is called the *fulcrum*. Levers are used in many engineering applications. Figure 7.1 illustrates several types of levers used in machines; in these levers, the fulcrum is located between the forces that act on the lever: E (the effort, or input) and L (the load, or output). Fulcrums are denoted by these symbols:

•  

lever arm

The perpendicular distances a and b from the fulcrum to the lines of action of the forces are called the *lever arms*.

mechanical advantage

The purpose of a lever is to transmit effort E into output L. Ordinarily, in applications of the lever, $L > E$. The ratio L/E is called the *mechanical advantage* of the lever. If the lever is in equilibrium, the ratio L/E is determined by the equilibrium of moments about the fulcrum. Hence, if a and b are the lever arms of forces E and L, respectively, then

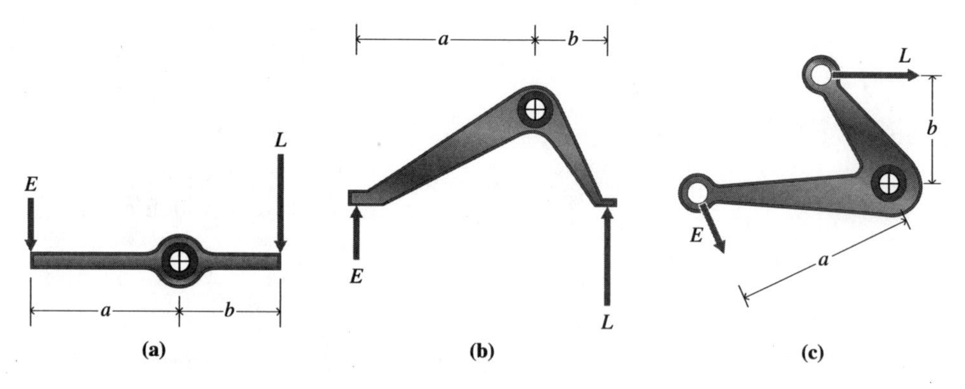

Figure 7.1 (a) Straight lever; (b) rocker arm; (c) bell crank.

The vertical arm of this backhoe is a first-class lever. The upper arm provides the effort, and the bucket carries the load.

The L-shaped lifting arm (the one with the CASE logo) of this skid-steer loader is a third-class lever. The hydraulic ram provides the effort to lift the load of concrete blocks on the pallet. The fulcrum is at the left end of the arm.

$aE = bL$ (a relation discovered by the ancient Greeks). The mechanical advantage m of the lever may then be expressed as

$$m = L/E = a/b \tag{7.1}$$

first-class lever

second-class lever

third-class lever

A lever whose fulcrum lies between the applied forces is called a *first-class lever* (see Fig. 7.2a). A *second-class lever* is one in which the output (load) lies between the fulcrum and the input (effort) (Fig. 7.2b). A *third-class lever* is one in which the input lies between the fulcrum and the output (Fig. 7.2c). Tools or simple machines that function as first-class levers include a teeter board, a crowbar, a weight balance, a pair of scissors, and a pair of pliers. The nutcracker and the wheelbarrow are second-class levers. A baseball batter's arm exerts effort (the input) through muscles between the shoulder (the fulcrum) and the hands (the output) and therefore acts as a third-class lever. The boom of an excavator is another example of a third-class lever.

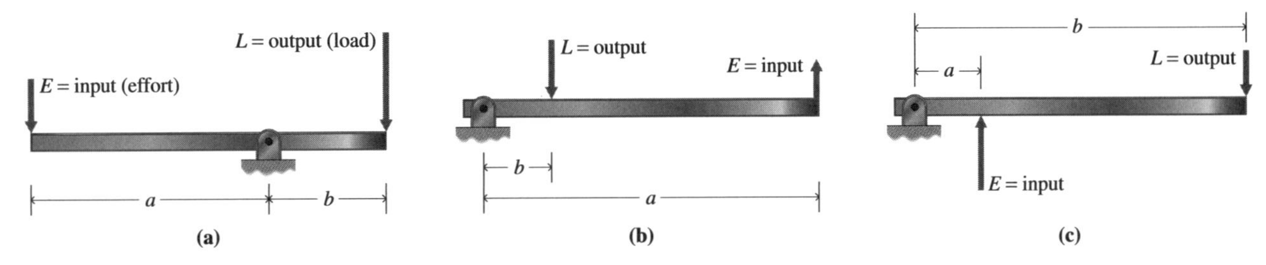

Figure 7.2 Classes of levers: **(a)** A first-class lever has its fulcrum between the effort and the load. **(b)** A second-class lever has its load between the fulcrum and the effort. **(c)** A third-class lever has its effort between the fulcrum and the load.

Example 7.1

System of Bars That Function as a Lever

Problem Statement The structural system shown in Fig. E7.1a is used to support loads at *A*. Note that the system of bars *ABCE* and *CD* functions as a lever with fulcrum at *E*. Determine the support reactions at *G* and *H* and the mechanical advantage of the lever.

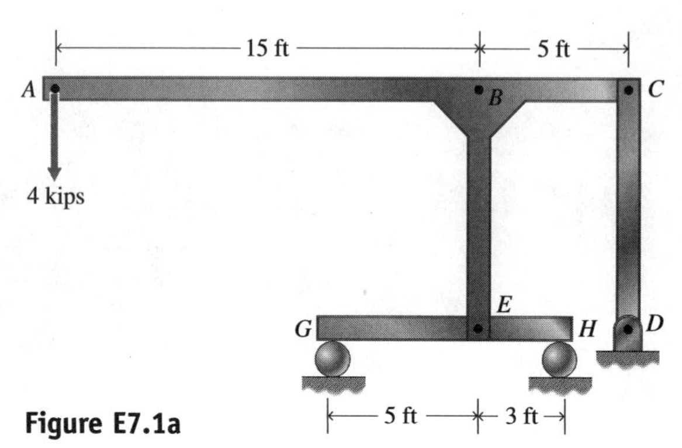

Figure E7.1a

Solution To begin the solution, we consider the free-body diagram of the rigid T-bar *ABCE* shown in Fig. E7.1b. The conditions of equilibrium for the T-bar may be written as

$$\sum F_y = 0 \tag{a}$$

$$\sum M_C = 0 \tag{b}$$

Equation (b) yields

$$\circlearrowleft\!+ \sum M_C = (20)(4) - 5R = 0$$

Hence, $R = 16$ kips. By Eq. (a) and Fig. E7.1b,

$$\sum F_y = R - C - 4 = 0$$

which gives $C = 12$ kips.

Figure E7.1c is the free-body diagram of the beam *GEH*. Note that we have used the symbol *R* in this diagram with sense *opposite* to that in Fig. E7.1b—that is, *we have invoked the principle of action and reaction* (Newton's third law). The conditions of equilibrium for this beam are

$$\begin{aligned} &\sum F_y = G + H - R = 0 \\ \circlearrowleft\!+ &\sum M_H = -8G + 3R = 0 \end{aligned} \tag{c}$$

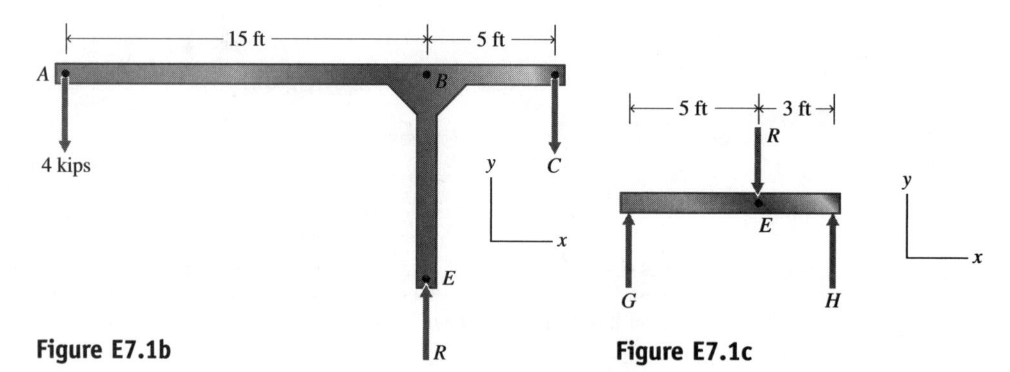

Figure E7.1b **Figure E7.1c**

Solving Eqs. (c) for G and H, we obtain $G = 6$ kips and $H = 10$ kips.

By Fig. E7.1b, the mechanical advantage m of this lever is $C/4 = 15/5$, or

$$m = \frac{12 \text{ kips}}{4 \text{ kips}} = \frac{15 \text{ ft}}{5 \text{ ft}} = 3.0$$

Example 7.2

Differential Lever

Problem Statement The differential lever shown in Fig. E7.2a is a device for measuring large loads. It consists of a lever AB, with a fulcrum at C, and a cross beam DE connected to the lever by two vertical hanger rods. The pin connections are frictionless. A large crate is attached to the cross beam and balanced by a counterweight of 15 N. Determine the weight W of the crate. Ignore the weight of the members.

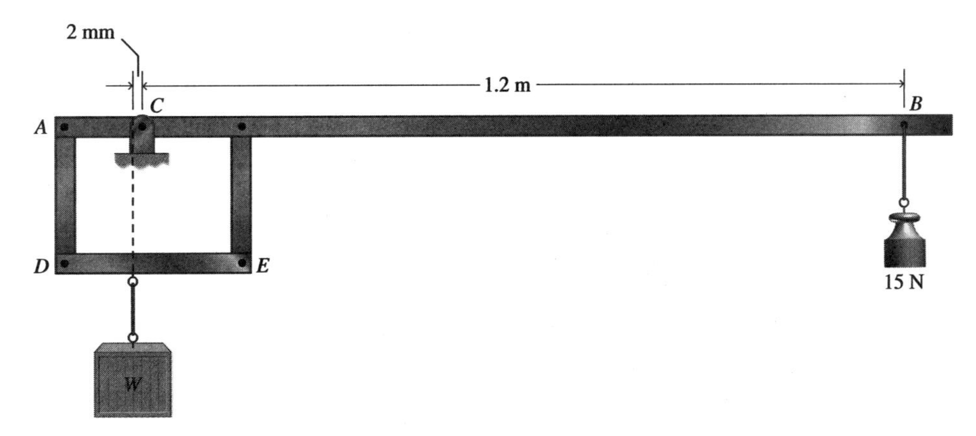

Figure E7.2a

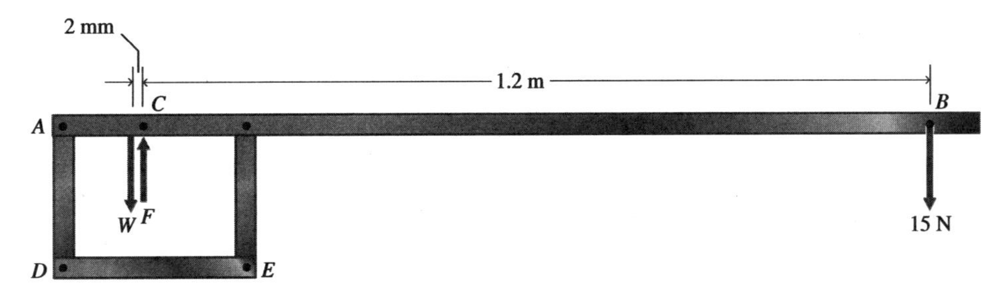

Figure E7.2b

Solution

Using the principle of transmissibility, we construct the free-body diagram of the lever (Fig. E7.2b). Then, taking moments about the fulcrum C, we obtain

$$\sum M_C = (1.2)(15) - (0.002)W = 0$$

$$W = 9 \text{ kN}$$

Information Item To solve this problem, we could have used the equilibrium of the cross beam DE and considered the forces exerted on the lever AB by the two vertical hangers. Equilibrium of that lever would yield the weight W. You can solve the problem this way as a check. By using the principle of transmissibility, we bypassed the cross beam.

7.2 THE PULLEY

After studying this section, you should be able to:

- Determine the forces in a mechanical system that contains one or more pulleys and levers.

Key Concept

A pulley is a device for changing the direction of a force or magnifying the effect of a force.

pulley The *pulley* is a wheel that is supported by an axle or shaft. The outer surface of the wheel's rim may be flat or grooved, depending on its use. If the rim is grooved, the wheel is called **sheave** a *sheave.* A rope (belt, flexible cable, or chain) is turned around the wheel (see Fig. 7.3a). One end of the rope is fastened to a load W. A force (pull) T is applied to the other end of the rope. The direction θ of the pull may vary. Effectively, the pulley behaves like a first-class lever. The axle O of the pulley is the fulcrum. The distances OA and OB are the lever arms. As for a lever, the pull T is the input and the load W is the output. When used as **simple pulley** shown in Fig. 7.3a, a pulley is referred to as a *simple pulley.*

Figure 7.3b is the free-body diagram of the pulley shown in Fig. 7.3a. It is assumed that the wheel of the pulley is free to rotate about a frictionless axle O. Hence, the reaction of the axle on the wheel is represented by the components R_x and R_y, which are assumed to act in the positive x and y directions, respectively. Since the weights of the pulley and rope are ordinarily much less than the loads, they are not included in Fig. 7.3b.

From Fig. 7.3b, since the tension in the rope at A is W, the equations of equilibrium of the pulley may be written as

$$\begin{aligned} \sum F_x &= R_x + T\cos\theta = 0 \\ \sum F_y &= R_y - W - T\sin\theta = 0 \\ \circlearrowleft\!+ \sum M_O &= aW - aT = 0 \end{aligned} \qquad \text{(a)}$$

Solving Eqs. (a), we obtain $T = W$, $R_x = -W\cos\theta$, and $R_y = W(1 + \sin\theta)$. Thus, when the pulley is in equilibrium, the tension T in the rope is transmitted around the pul-

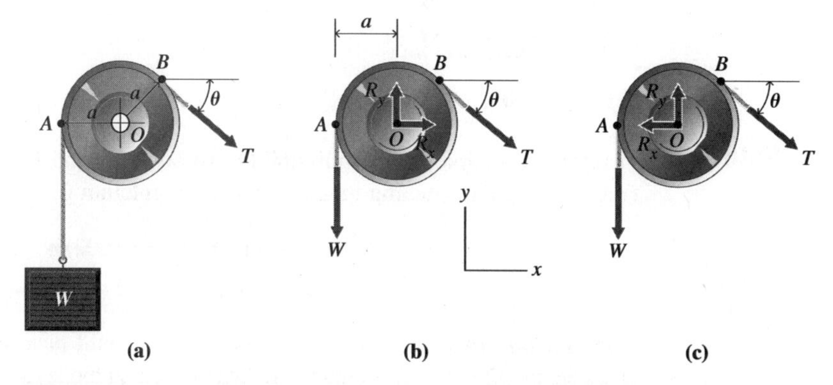

Figure 7.3 **(a)** Simple pulley; **(b)** free-body diagram of the pulley with assumed senses of axle reactions; **(c)** diagram showing correct senses of axle reactions.

ley; that is, the tension at point B is the same as the tension at point A (refer to Fig. 7.3b). Note that since the sign of R_x is negative, R_x acts opposite to the sense assumed in Fig. 7.3b; that is, it acts in the negative x direction (see Fig. 7.3c). Since the sign of R_y is positive, the sense of R_y is correct as assumed. Also, since $T = W$, the mechanical advantage of the simple pulley is $m = W/T = 1.0$. For instance, to keep a weight of 200 N in equilibrium, you would need to exert a pull of 200 N. A simple pulley allows the user to change the direction of the pull, but not to magnify its effect.

Key Concept

A block and tackle is a combination of two or more pulleys with a mechanical advantage greater than 1.

block and tackle

The mechanical advantage of the simple pulley can be increased by altering the pulley's arrangement (see Fig. 7.4) or by combining the effects of several pulleys (see Fig. 7.5). When several pulleys are combined, as in Fig. 7.5, the assembly is called a *block and tackle.* The pulley arrangement shown in Fig. 7.4 has the mechanical advantage $m = 2.0$, as can be seen by considering the equilibrium of the system below the section line a–a. The block and tackle shown in Fig. 7.5, however, has the mechanical advantage $m = 3.0$. Again, this can be seen by considering the equilibrium of the system below section line b–b (ignoring the slight incline of the rope segment AB).

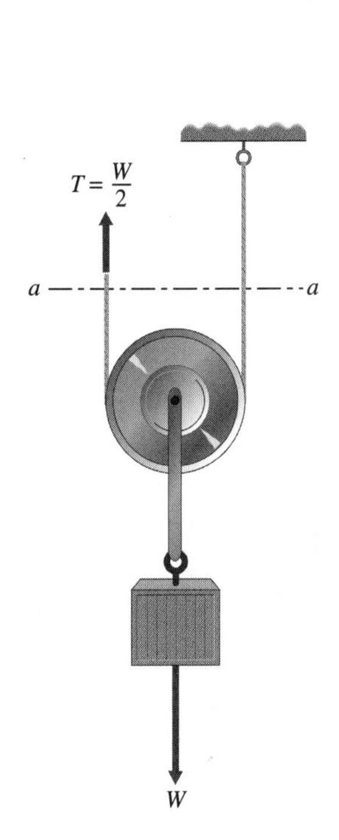

Figure 7.4 Pulley with mechanical advantage $W/T = 2.0$.

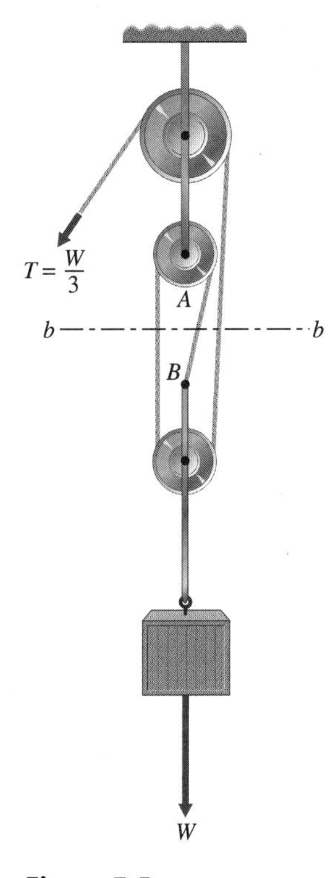

Figure 7.5 Block and tackle with mechanical advantage $W/T = 3.0$.

The six-line pulley on this construction crane gives the hoist a mechanical advantage of 6. The weight of the load is 6 times more than the tension in the line.

Example 7.3

Hydraulic Elevator System

Problem Statement Figure E7.3a is a schematic diagram of a hydraulic elevator system. (It does not show the elevator system as actually constructed but merely portrays a mechanically equivalent system, or model.) The cage is raised and lowered by means of the system of pulleys attached to the stationary hydraulic cylinder A and movable piston B. The cage is raised when the piston B is forced downward by the hydraulic pressure in the cylinder A. The maximum force that can be exerted on the piston is 48 kips. Neglecting friction, determine the maximum weight W of the cage that can be lifted by the system.

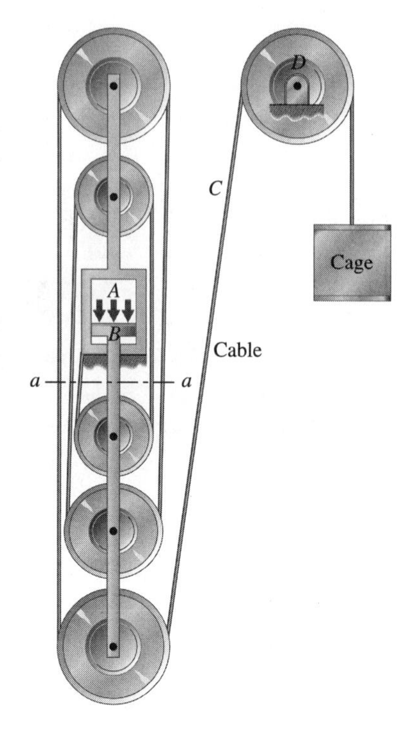

Figure E7.3a

Solution To solve this problem, we first consider the free-body diagram of the cage shown in Fig. E7.3b. When the cage is in equilibrium, the tension T in the cable is equal to the weight W of the cage. The tension T is transmitted around the pulley D to the cable at C.

Next, by passing a cut through the rod and cables at section a–a, we consider the free-body diagram of the piston rod B and the three lower pulleys taken as a unit (see Fig. E7.3c). Neglecting the slight incline of the cables from the vertical, we use equilibrium of forces in the vertical direction to obtain

$$\sum F_y = 6T - 48 = 0$$

$$T = 8 \text{ kips}$$

Hence, $W = T = 8$ kips. Therefore, if the cage is to be raised, the total weight of the cage must be less than 8 kips. Note that friction, which has been neglected in this example, may considerably reduce the allowable load.

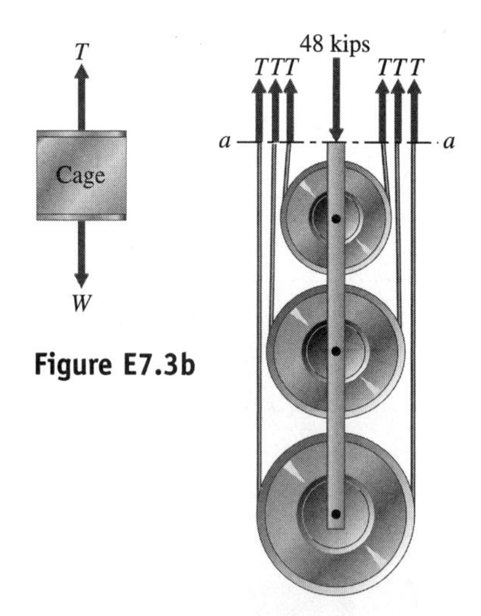

Figure E7.3b

Figure E7.3c

Example 7.4

Bell Crank Lever

Problem Statement The arms of the bell crank ABC in Fig. E7.4a form an angle of 90°. Cords AD and CE are attached to the crank at A and C, pass over frictionless pulleys, and have weights W_1 and W_2 attached to their ends. Arms AB and BC have lengths of 1 m and 0.40 m, respectively. The crank can rotate freely in the plane of the page about its frictionless bearing B. Arm AB and cord AD subtend an angle of 75°.

a. Determine the angle θ subtended by arm BC and cord CE, in terms of the weights W_1 and W_2 and the weight W of the crank.

b. For $W_1 = 80$ N, $W_2 = 240$ N, and $W = 64$ N, calculate the value of θ.

c. If arm AB makes an angle $\phi = 45°$ with respect to the horizontal, determine the bearing reaction.

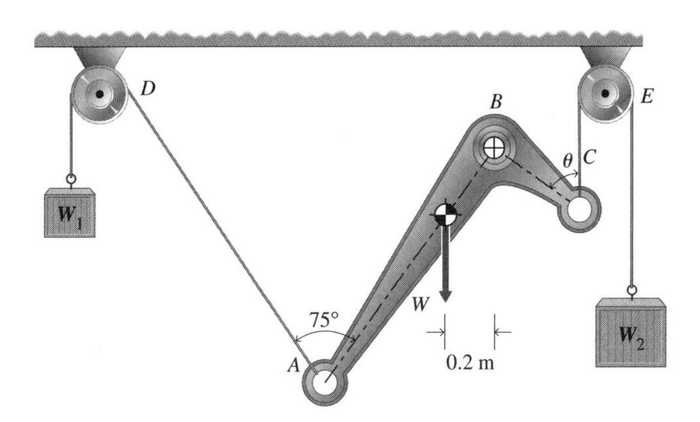

Figure E7.4a

Solution

a. The free-body diagram of the bell crank is shown in Fig. E7.4b. Since weights W_1 and W_2 are transmitted around the pulleys, the tensions in cords AD and CE are equal to W_1 and W_2, respectively. Therefore, the cords exert forces W_1 and W_2 on the bell crank. Noting that we wish to determine only the angle θ, we see that the equilibrium of moments with respect to B is given by

$$\sum M_B = (0.2)W + (0.4)(W_2 \sin\theta) - (1.0)(W_1 \sin 75°) = 0$$

$$\sin\theta = \frac{(1.0)(W_1 \sin 75°) - (0.2)W}{(0.4)W_2} \qquad \text{(a)}$$

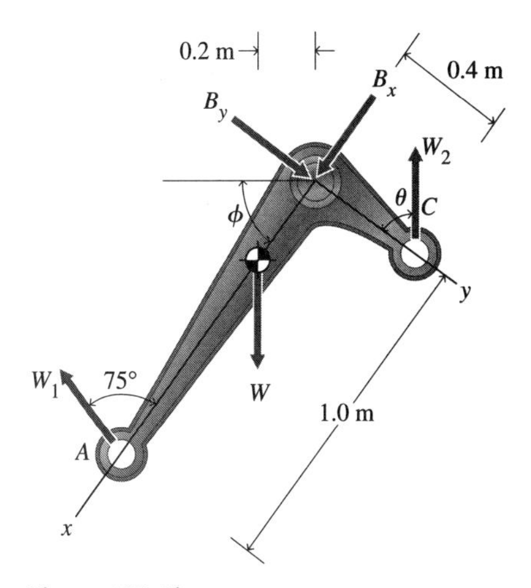

Figure E7.4b

b. For $W_1 = 80$ N, $W_2 = 240$ N, and $W = 64$ N, Eq. (a) yields

$$\sin\theta = 0.6716$$

$$\theta = 42.19° \text{ or } 137.81°$$

c. To calculate the bearing reaction, we find it convenient to select xy axes as shown in Fig. E7.4b and determine the (x, y) components of the bearing reaction. Hence, by the equilibrium conditions in the (x, y) directions, we have

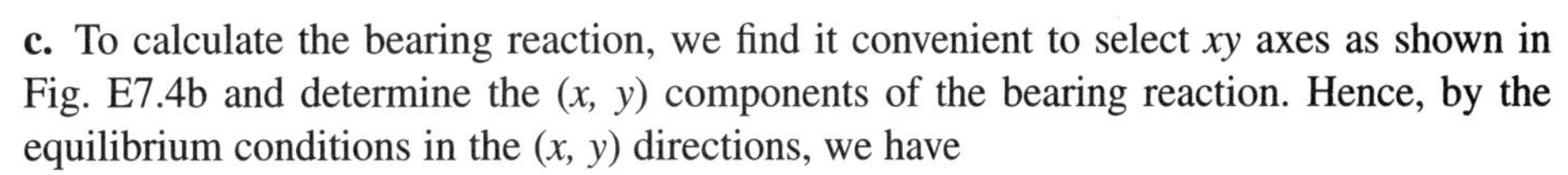

$$\sum F_x = B_x + W\cos 45° - W_1 \cos 75° - W_2 \sin 42.19° = 0$$
$$\sum F_y = B_y + W\sin 45° - W_1 \sin 75° - W_2 \cos 42.19° = 0 \qquad \text{(b)}$$

With the given values of W, W_1, and W_2, Eqs. (b) yield

$$B_x = 136.63 \text{ N}$$

$$B_y = 209.84 \text{ N}$$

or $$B = \sqrt{136.63^2 + 209.84^2} = 250.40 \text{ N}$$

at an angle relative to the x axis of

$$\beta = \tan^{-1}\left(\frac{209.84}{136.63}\right) = 56.93°$$

See Fig. E7.4c.

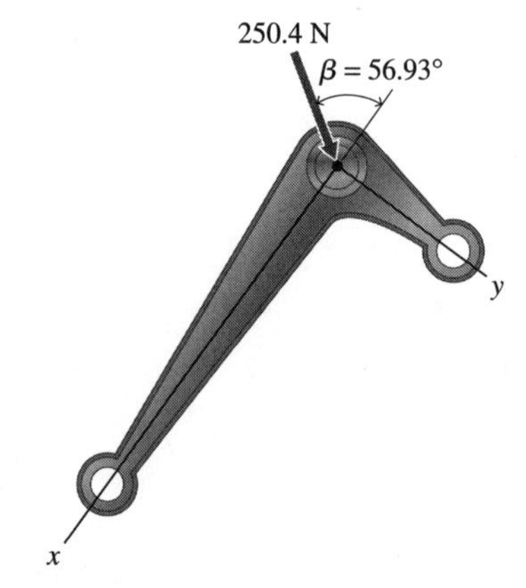

Figure E7.4c

CHECKPOINT

1. ***True or False:*** A machine is defined as a device that transmits or modifies the effect of a force.
2. ***True or False:*** A screw and a baseball bat are examples of simple machines.
3. How many classes of levers are there?
4. ***True or False:*** The input to a lever is always smaller than the output.
5. What is the mechanical advantage of a lever?
6. ***True or False:*** A simple pulley is used to magnify the effect of an input.

Answers: 1. True; 2. True; 3. Three; 4. False; 5. Ratio of output to input; 6. False

7.3 SIMPLE PLANE FRAMES

After studying this section, you should be able to:

- Determine the support reactions and member forces for a simple plane frame.

Key Concept In general, a structure that is built of bars is called a frame.

frame A *frame* is defined as a structure that is built of bars. According to this definition, a pin-connected truss is a frame. However, it is not the most general type of frame. The joints of a frame are not necessarily pin-connected, and the members of a frame need not be straight two-force members. The members of a frame may be curved and may be pinned to other members anywhere along their lengths (see Fig. 7.6). Furthermore, the loads that act on a frame need not be applied at the joints. The loads may act at locations anywhere along a member of a frame, and they may be inclined relative to the axis of the member. A frame member may also be subjected to a couple at any point along its length (see Fig. 7.6).

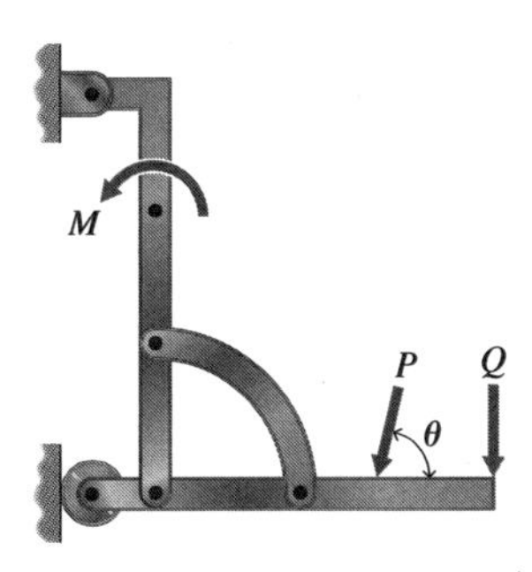

Figure 7.6 Wall-mounted equipment-support frame containing straight and curved members.

Consequently, a frame is distinguished from a pin-connected truss in that one or more of its members may be subjected to transverse forces and couples, and its members may be pinned, riveted, bolted, or welded together, not necessarily at their ends. However, we will consider only pin-connected frames. (The analysis of riveted, bolted, or welded frames—so-called rigid frames—is beyond the scope of this book. This topic is treated in structural analysis courses.) Thus, a frame member is generally a multiforce member (it is acted on by one or more nonaxial forces and, possibly, one or more couples).

A frame may contain one or more straight or curved two-force members for which the directions of the forces are known. (See, for example, the curved brace in Fig. 7.6.) Recall that for a two-force member in equilibrium, the lines of action of the forces that act on the member pass through the two points at which the forces are applied (see Fig. 7.7). In the analysis of a frame that contains two-force members, it may be necessary to use this fact in order to solve for member forces.

In this section, we consider plane frames (two-dimensional frames), whose members, supports, and forces all lie in the same plane. We will not consider space frames (three-dimensional frames). Many engineering applications involve the use of plane frames. Also, plane frames can be connected to form three-dimensional frames. A *plane frame* is a rigid structure, in contrast to a mechanism or a simple machine (see Sec. 7.4). That is, a plane frame maintains its shape when separated from its supports. For example, the A-frame in Fig. 7.8—so called because it resembles the letter *A*—is not free to change its shape if it is removed from its supports (see Fig. 7.9). In contrast, a *structural mechanism* is free to change shape if it is removed from its supports. If the supports are removed from the structural mechanism in Fig. 7.10, it will collapse, as shown in Fig. 7.11, because bars *AC* and *BCD* are free to rotate relative to one another about pin *C*. A structural mechanism is therefore nonrigid, or collapsible.

plane frame

structural mechanism

If a frame is in equilibrium, each member of the frame must satisfy the conditions of equilibrium for a rigid body (see Sec. 4.6). If all the forces that act on the members of a

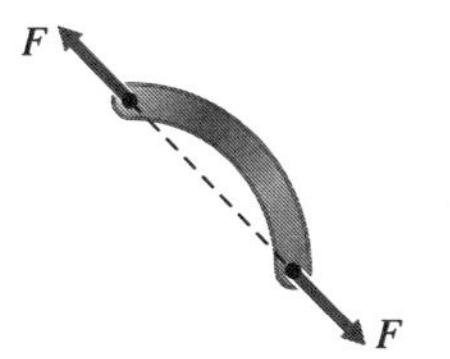

Figure 7.7 Curved two-force member.

The Ronald McDonald House in Chattanooga, Tennessee is built from a series of plane frames. The frame members carry axial and shear forces as well as bending moments.

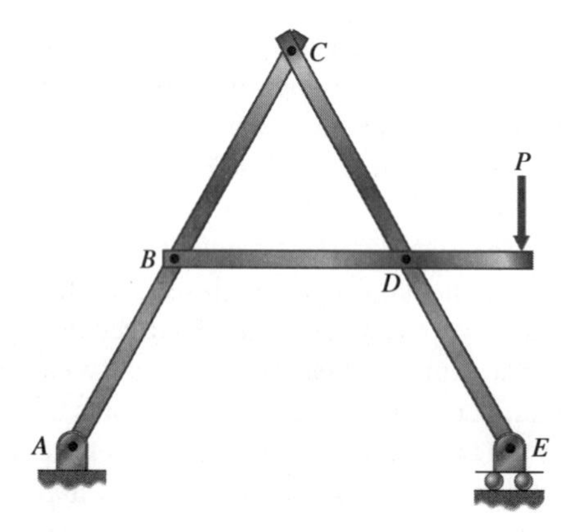

Figure 7.8 A-frame.

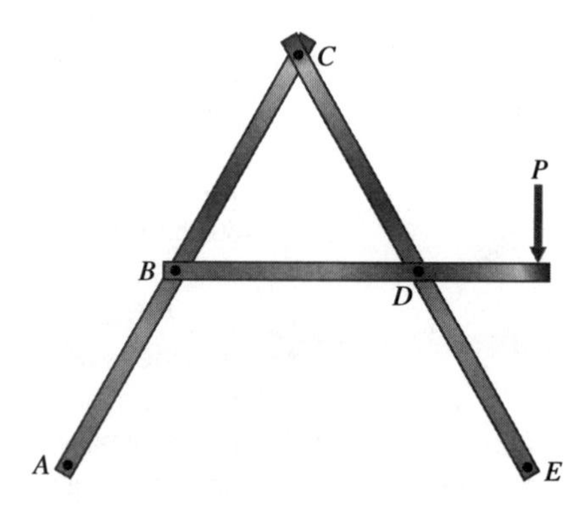

Figure 7.9 A-frame without supports.

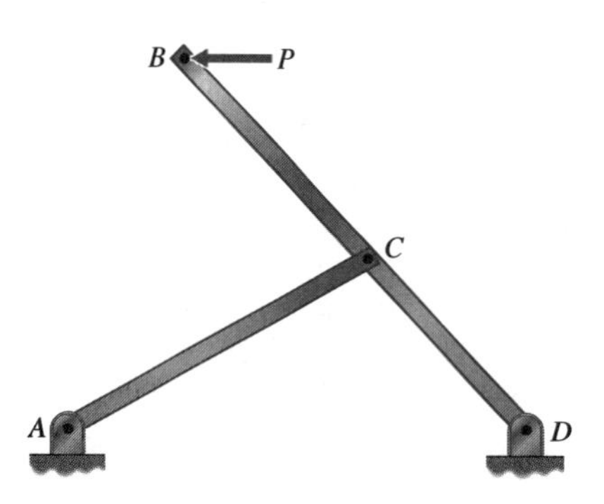

Figure 7.10 Mechanism with supports.

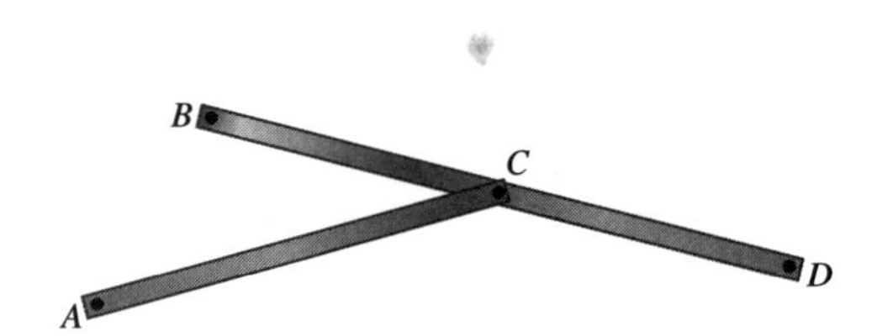

Figure 7.11 Collapsed mechanism.

statically determinate

simple frame

frame and all the reactions of the supports can be determined by the equations of static equilibrium, the frame is *statically determinate.* Such a frame is known as a *simple frame.*

Like the support reactions of a truss, the support reactions of a simple plane frame may be determined by drawing a free-body diagram of the frame and applying the conditions of equilibrium. Once the support reactions have been determined, the members of the frame may be isolated and their free-body diagrams drawn.

As each member or part of a frame is isolated, its free-body diagram is drawn. Since at least one member of a frame is a multiforce member, the directions of the forces exerted on that member by a support or a connection to another member may not be known. If the direction of a force is not known, the force should be represented by its components relative to the chosen reference axes or by a force at an unknown angle. In construction of the free-body diagrams, the *principle of action and reaction (Newton's third law) must be imposed.* For example, in Fig. 7.12, the force **P** exerted on member B by member A is opposite to the force **Q** exerted on member A by member B. If we call the force **P** the action exerted on B by A, the force **Q** that acts on A is said to be the reaction of force **P**. The collinear forces **P** and **Q** have equal magnitudes, $P = Q$, but opposite senses. Recall that the free-body diagram of a given member must show *only* the actions of other members on that member. Hence, in the free-body diagrams of members A and B, we may use the same symbol, **P**, to denote the contact forces between the members A and B (see Fig. 7.13).

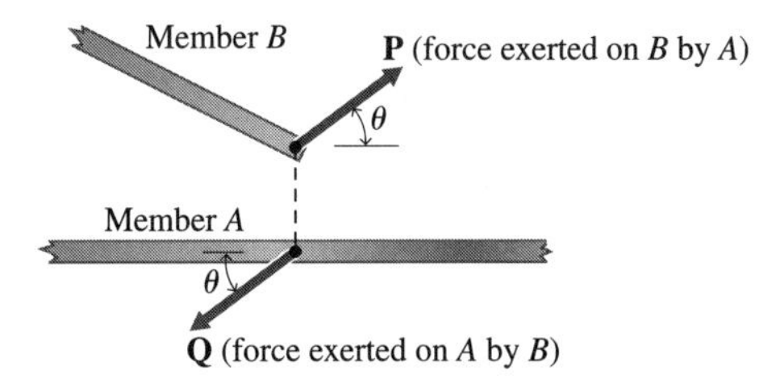

Figure 7.12 Action and reaction between contiguous members.

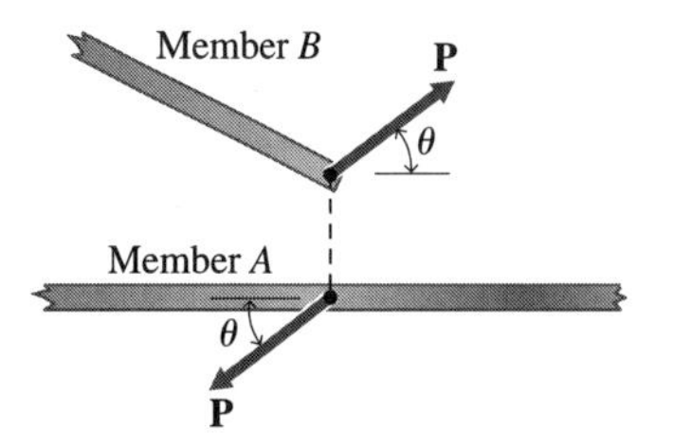

Figure 7.13 Equal and opposite contact forces on contiguous members.

In Chapter 4 we studied the equilibrium of a single rigid body subjected to a system of coplanar forces, but here we consider the equilibrium of interconnected rigid bodies, paying particular attention to the principle of action and reaction.

When we isolate a member of a frame and draw its free-body diagram, the sense of a force that an adjacent member exerts on the isolated member may not be apparent; that is, it may not be obvious whether the adjacent member exerts a pull or a push on the isolated member. In such cases, we assume a sense for the force and use a letter such as P to label the magnitude of the force. In the solution of the equations of equilibrium for the system of forces, *if P is found to be a positive number, the assumed sense is correct; if P is found to be a negative number, the correct sense is opposite to that assumed.*

Recall that in trusses the loads are applied at the joints, and in the method of joints (see Sec. 6.4), free-body diagrams of the truss pins are used to determine the forces in the members of the truss. A frame member may have forces applied anywhere along its length (refer to Fig. 7.6). As a result, the direction of the force that the member exerts on a pin is not always known. Therefore, it is not always practical to consider free-body diagrams of pins in a frame, since the directions of the forces on the pins may not all be known. In such cases, we may assume that a pin is part of one of the members that it joins. In the free-body diagram of this member, we take the pin to be an integral part of the member. Ordinarily, it does not matter to which member the pin is attached. However, there are certain cases in which it is convenient to attach the pin to a particular member:

1. *A support is joined to two or more members by a pin.* In this case, consider the pin to be a part of one of the members, and apply the support reactions to the pin. Construct the free-body diagrams for the members accordingly.
2. *A force is applied directly to a pin that joins two or more members.* In this case, consider the pin to be a part of one of the members, with the load applied to the pin. The reactions of the other members on the pin should also be shown in terms of components relative to the chosen reference axes.
3. *A pin joins three or more members.* This case is similar to case 1, with the support being replaced by a third member. Again, consider the pin to be a part of one of the members, and construct the free-body diagrams of the members accordingly.

In all cases, if a frame has multiforce members, consider the pin to be a part of one of the multiforce members. In certain cases, if the directions of the forces acting on the pin are known (for example, for two-force members), it may be best to draw a free-body diagram of the pin and consider the equilibrium conditions for the pin. The following problem-solving technique summarizes these concepts, and then two example problems illustrate the technique.

PROBLEM-SOLVING TECHNIQUE

Simple Plane Frames

The equilibrium analysis of simple frames involves the determination of unknown forces in interconnected members. One or more members of a frame are generally subject to laterally directed (nonaxial) forces that may be applied anywhere along the lengths of the members. By definition, simple plane frames are statically determinate and consist of interconnected coplanar members subjected to forces in the plane.

To analyze a simple plane frame for support reactions and member forces:

1. Draw a free-body diagram of the frame, and select an appropriate set of xy reference axes for summing forces. If the direction of a support reaction is not known, represent the reaction by its (x, y) components or by a force at an unknown angle.
2. Determine the support reactions, using the conditions of equilibrium.
3. Isolate each member of the frame in turn, and construct appropriate free-body diagrams, paying particular attention to the application of the principle of action and reaction (Newton's third law) between interconnected members and among pins, members, and supports.
4. Draw xy reference axes to aid in summing forces. In some cases, it is advisable to choose separate reference axes for each free-body diagram. To determine any unknown forces, apply the conditions of equilibrium to the appropriate free-body diagram.
5. Solve the equations of equilibrium for the unknowns. First, examine these equations to see whether it is possible to solve for an unknown directly from one equation; it may also be possible to solve for additional unknowns directly. (For example, by a proper choice of moment center, an unknown may be determined directly.) Such inspection may reduce the number of equations to be solved. If there are three or more simultaneous equations involving the unknowns, the use of a computer may be appropriate, particularly if the solution is required for different sets of loads.

Example 7.5

Analysis of an A-Frame

Problem Statement For the A-frame shown in Fig. E7.5a, determine the forces exerted on member *BDF* by the pins at points *B* and *D*.

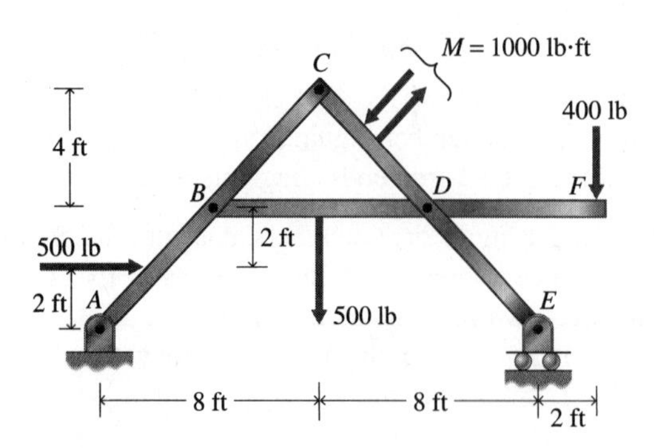

Figure E7.5a

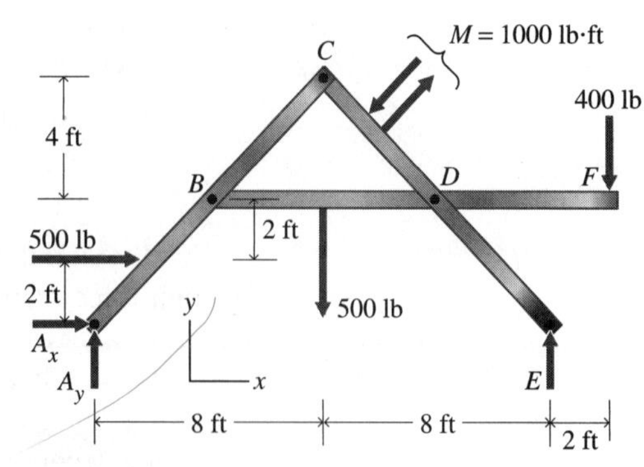

Figure E7.5b

Solution Figure E7.5b is a free-body diagram of the A-frame. The equation of equilibrium of moments about point A is

$$\circlearrowleft^{+} \sum M_A = 16E - 2(500) - 18(400) - 8(500) + 1000 = 0$$

Hence, $E = 700$ lb.

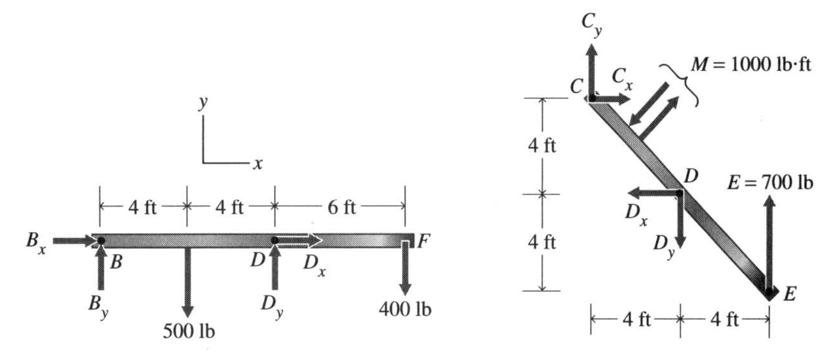

Figure E7.5c **Figure E7.5d**

Figures E7.5c and E7.5d are free-body diagrams of bars *BDF* and *CDE*, respectively. The equilibrium equations for bar *BDF* may be written as

$$\sum F_x = B_x + D_x = 0$$

$$\sum F_y = B_y + D_y - 500 - 400 = 0 \qquad \text{(a)}$$

$$\circlearrowleft^{+} \sum M_B = 8D_y - 14(400) - 4(500) = 0$$

Equations (a) contain four unknowns, B_x, B_y, D_x, and D_y. These equations yield

$$D_y = 950 \text{ lb}$$

$$B_y = -50 \text{ lb}$$

$$B_x = -D_x \qquad \text{(b)}$$

To determine B_x and D_x, we consider Fig. E7.5d. For bar *CDE*, the equation of equilibrium of moments about point C is

$$\circlearrowleft^{+} \sum M_C = -4D_x - 4D_y + 8E + 1000 = 0 \qquad \text{(c)}$$

Since $E = 700$ lb and $D_y = 950$ lb, Eq. (c) yields $D_x = 700$ lb. Equation (b) now yields $B_x = -700$ lb.

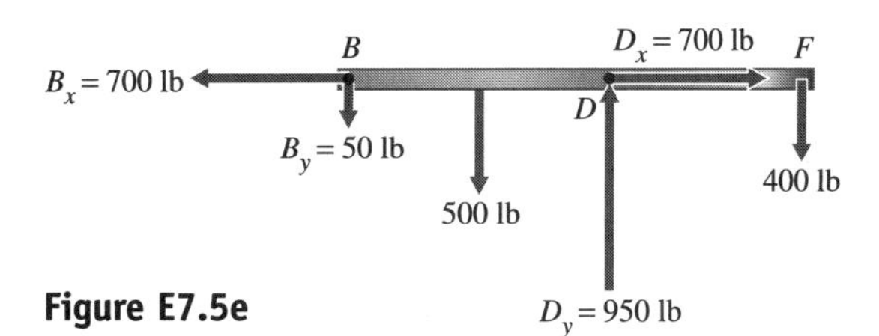

Figure E7.5e

Figure E7.5e is a diagram showing the forces on bar *BDF*, with their correct senses. Since only the forces that act on member *BDF* at points B and D were required, the analysis was set up so that the forces at points A and C did not enter into the calculations.

Example 7.6

A Hoist Frame

Problem Statement The hoist frame shown in Fig. E7.6a is used to lift the weight W. As shown, the pull T on the pulley balances the weight W. For $W = 4$ kN, determine all the forces acting on the members at joints A, B, C, D, and E. Indicate the correct directions of the forces on sketches of the members. Assume that the weights of the members are small compared to the load and that the wall is smooth.

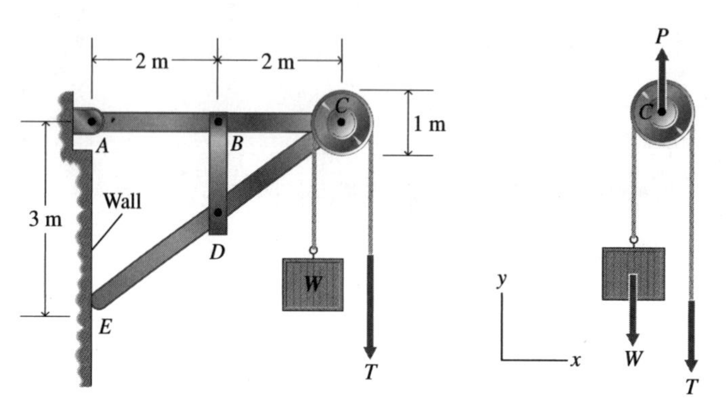

Figure E7.6a **Figure E7.6b**

Solution

Initially, we draw the free-body diagram of the pulley and weight W (Fig. E7.6b), in which P is the force exerted on the pulley at C by the pin (considered to be a part of member ABC). By Fig. E7.6b, the equilibrium equations may be written as

$$\sum F_y = P - T - W = 0$$
$$\circlearrowleft\!+ \sum M_C = 0.5T - 0.5W = 0 \qquad \text{(a)}$$

Solution of Eqs. (a) yields

$$T = W = 4 \text{ kN}$$
$$P = 8 \text{ kN}$$

As recommended in the problem-solving technique for analysis of simple plane frames, we proceed as follows: We draw a free-body diagram of the frame removed from its supports. The force $Q = 8$ kN is the reaction of the force P. It acts on the pin at C, which is part of member ABC (see Fig. E7.6c). The forces A_x, A_y, and E are support reactions. The equilibrium conditions for the frame may be written as

$$\sum F_x = A_x + E = 0$$
$$\sum F_y = A_y - 8 = 0 \qquad \text{(b)}$$
$$\circlearrowright\!+ \sum M_A = 3E - 4(8) = 0$$

The solution of Eqs. (b) is

$$A_x = -10.67 \text{ kN}$$
$$A_y = 8 \text{ kN}$$
$$E = 10.67 \text{ kN}$$

Since A_x is negative, its assumed sense in Fig. E7.6c is wrong. The correct sense of A_x is in the negative x direction.

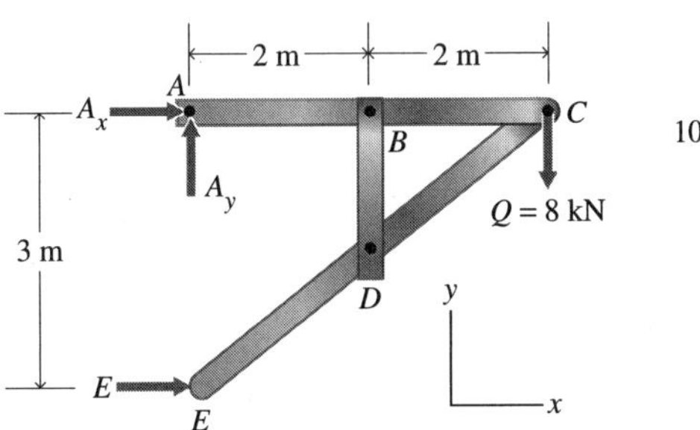

Figure E7.6c

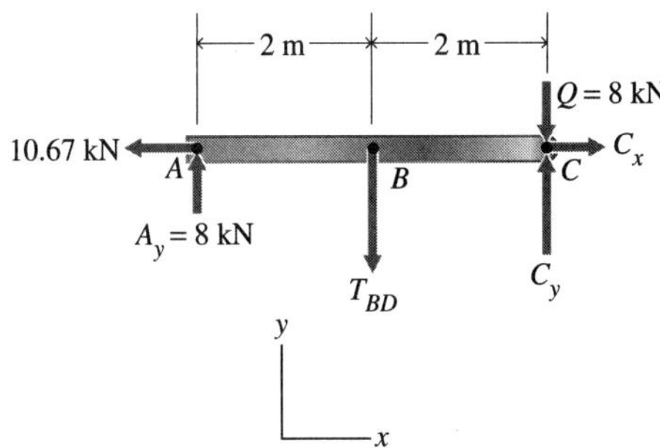

Figure E7.6d

We next isolate member *ABC* and draw its free-body diagram (Fig. E7.6d), with the 10.67 kN force shown in the correct sense. We take the pin at *A* to be a part of member *ABC*. Therefore, the vertical support reaction ($A_y = 8$ kN) acts on member *ABC* at *A*. The forces C_x and C_y are the actions of the member *CDE* on the pin at *C*, considered to be a part of member *ABC*. The force T_{BD}, assumed to be directed downward, is the action of the two-force member *BD* on member *ABC*. The force T_{BD} is directed axially along member *BD*. Finally, we include the force $Q = 8$ kN, which is the action of the pulley on the pin at *C*. The equilibrium equations for member *ABC* may be written as

$$\sum F_x = C_x - 10.67 = 0$$

$$\sum F_y = 8 - T_{BD} + C_y - 8 = 0 \qquad \text{(c)}$$

$$\circlearrowleft\!+ \sum M_C = 2T_{BD} - 4(8) = 0$$

The solution of Eqs. (c) is

$$T_{BD} = 16 \text{ kN}$$

$$C_x = 10.67 \text{ kN}$$

$$C_y = 16 \text{ kN}$$

Note again that C_x and C_y are the horizontal and vertical forces, respectively, exerted on the pin at *C* by member *EDC*.

Next, we draw the free-body diagram of member *CDE* (Fig. E7.6e). The equilibrium equations for member *CDE* are

$$\sum F_x = 10.67 - 10.67 = 0$$

$$\sum F_y = 16 - 16 = 0 \qquad \text{(d)}$$

$$\circlearrowleft\!+ \sum M_E = (10.67)(3) + 16(2) - 16(4) = 0$$

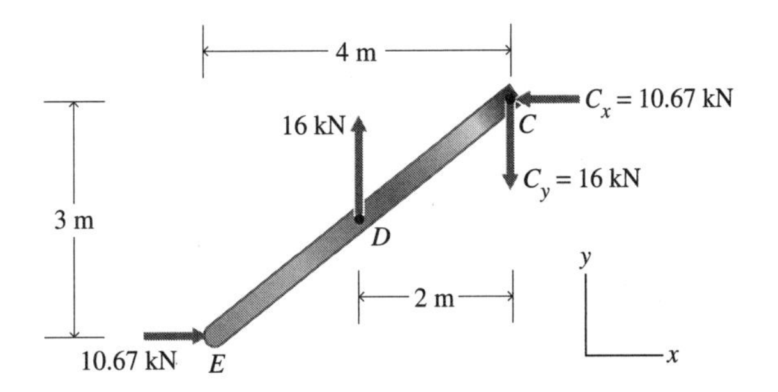

Figure E7.6e

We see from Eqs. (d) that the equilibrium equations for member *CDE* are satisfied identically. This serves as a check on the accuracy of our solution. The member forces are shown in Fig. E7.6f, with their correct senses, and the resultant forces at *A* and *C* are shown dashed. Note that each of the members *ABC* and *CDE* is in equilibrium under the action of three forces. These forces are therefore concurrent (see Sec. 4.6). Note also that the net vertical force exerted on member *ABC* at *C* is 8 kN upward. This force is the sum of the upward force of 16 kN exerted by member *EDC* on *ABC* and the downward force of 8 kN exerted on *ABC* by the pulley (refer to Fig. E7.6d).

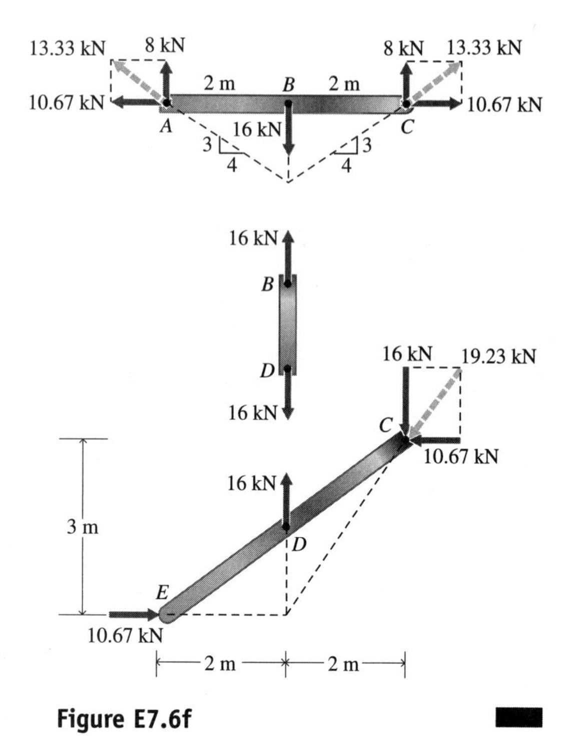

Figure E7.6f

CHECKPOINT

1. ***True or False:*** A truss is a frame.
2. ***True or False:*** The members of a simple frame are joined by pins.
3. ***True or False:*** A frame can be treated as a rigid body.
4. ***True or False:*** A frame will collapse if its supports are removed.
5. ***True or False:*** In the free-body diagram of a pin-connected frame, the pins do not require special attention.
6. ***True or False:*** In the analysis of a simple plane frame, it is often necessary to determine support reactions before considering the equilibrium of any member of the frame.

Answers: 1. True; 2. True; 3. True; 4. False; 5. False; 6. True

7.4 MECHANISMS

After studying this section, you should be able to:

- Determine support reactions and member forces for a structural mechanism.
- Determine the mechanical advantage and member forces for a tool mechanism.

Key Concept

A mechanism is a nonrigid mechanical system whose components are capable of relative motion when not attached to supports.

mechanism A *mechanism* is a nonrigid mechanical system whose components are capable of relative motion when not attached to supports. Besides structural mechanisms (see Sec. 7.3), **tool mechanism** other simple mechanisms include *tool mechanisms,* such as pliers, clamps, bolt cutters, scissors, and dividers, which are not generally attached to supports. Yet, the parts of such tools are free to move relative to each other. When these tools are used, however, they are often held in a stationary position and are therefore in equilibrium. In such cases, the equations of equilibrium may be applied to the tool as a whole and to each of its individual parts.

Tool mechanisms are usually assembled from two or more simple machines. This book uses the terms "mechanism" and "tool" interchangeably.

The usefulness of a mechanism is often based on the fact that its various parts can move relative to one another. This movement is permitted by joints that are loosely pinned, slotted, or geared or by some other feature that permits relative motion. If a mechanism is in equilibrium, we may analyze it as a rigid body. Then, the forces on the mechanism must conform to the laws of equilibrium for a rigid body.

The design of a mechanism often requires determining the magnitudes of the forces within the mechanism and their relation to the forces applied to the mechanism. For instance, many hand tools provide a mechanical advantage: The clamping force of the jaws of a pliers is greater than the gripping force you apply with your hand. To analyze such a system, we must disassemble the mechanism in such a manner that the clamping force appears in the free-body diagram and in the equilibrium equations. If the mechanism as a whole is in equilibrium, each part of the mechanism is also in equilibrium. The mechanism may then be separated into its basic parts, each of which must satisfy the equations of equilibrium. This process is demonstrated in the following examples.

Example 7.7

Mechanical Advantage of Pliers

Problem Statement Pliers are used to hold a bolt head (see Fig. E7.7a). The pliers are formed by two curved handles (levers) joined by a pin (a fulcrum). Gripping forces of magnitude 45 lb are applied to the handles of the pliers. Determine the clamping force exerted by the jaws on the bolt head and the mechanical advantage of the pliers. Neglect friction.

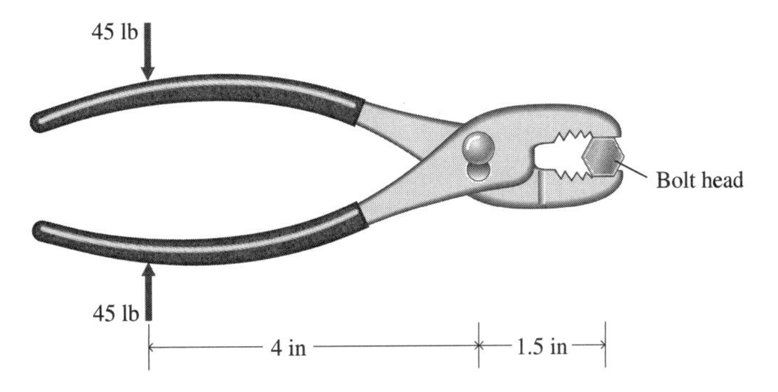

Figure E7.7a

Solution Since the pliers and the bolt are in equilibrium as a unit, we see that the gripping forces constitute a pair of collinear forces of equal magnitude but opposite senses. Hence, they are self-equilibrating. Consequently, from the free-body diagram of Fig. E7.7a, we cannot directly determine the clamping force exerted by the jaws on the bolt.

To determine the force on the bolt, we remove the pin that holds the two handles of the pliers together and draw a free-body diagram of one of the handles, shown in Fig. E7.7b. The force exerted by the bolt on one jaw is denoted by B. Since friction is neglected, there is no resisting couple at the pin. Hence, the (x, y) components of the pin force, denoted by P_x and P_y, are assumed to act at the center of the pin hole. Note that the handle is a lever, with the fulcrum at the center of the pin hole.

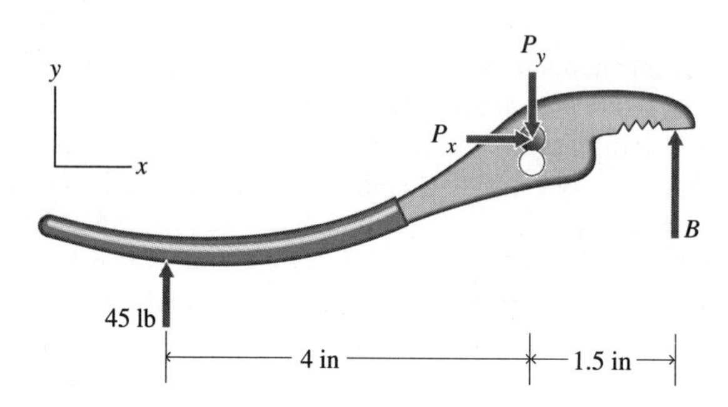

Figure E7.7b

The equilibrium conditions for the handle are

$$\sum F_x = P_x = 0$$
$$\sum F_y = 45 + B - P_y = 0$$
$$\circlearrowleft + \sum M_P = 4(45) - 1.5B = 0$$

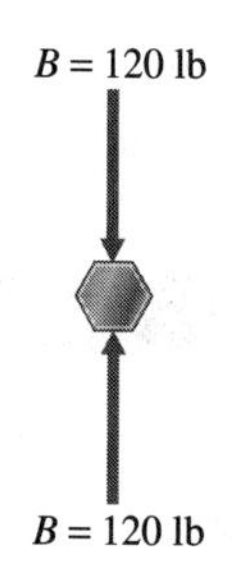

Figure E7.7c

Solution of these equations gives $P_x = 0$, $P_y = 165$ lb, and $B = 120$ lb. Thus, the bolt is clamped by a pair of collinear forces of equal magnitude (120 lb) but opposite senses (see Fig. E7.7c). In turn, the bolt exerts a force of $B = 120$ lb on the jaw (see Fig. E7.7b). The mechanical advantage of the pliers is thus 120 lb/45 lb = 4 in/1.5 in = 2.67 [see Sec. 7.1 and Eq. (7.1)].

Information Item The pin forces P_x and P_y were assumed to act at the center of the pin hole in Fig. E7.7b. In reality, however, the forces are distributed around the contact surface between the pin and the handle. Do you think this fact is significant? Explain.

Example 7.8

Ore-Loading Platform

Problem Statement Figure E7.8a is a diagram of a platform used for loading iron ore into railroad cars. The maximum capacity of the ore loader is 20 kips. Determine the forces acting on the platform, the forces in bars *BD*, *CD*, and *CE*, and the support reactions for the case in which the ore loader is loaded to capacity and is at rest at the left end of the platform.

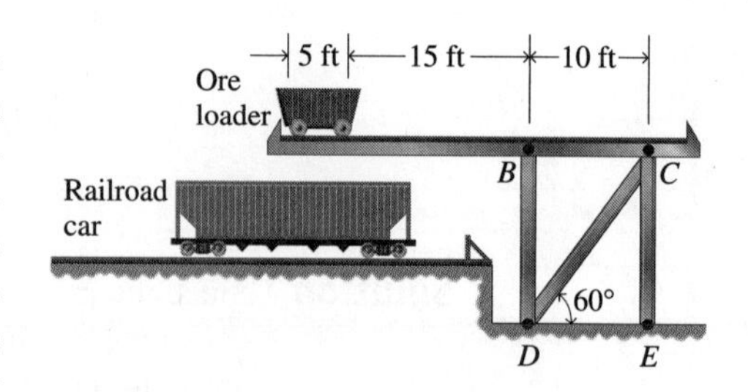

Figure E7.8a

Solution We note that if the platform is removed from its supports, member *CE* is free to rotate, and the structure is then free to change shape. Therefore, according to the definition in Sec. 7.3, the platform is a structural mechanism. Figure E7.8b is the free-body diagram of the platform. In this case, members *CE, CD,* and *BD* are two-force members. We assume that the ore in the ore loader is distributed so that each wheel carries the same load. Then, at maximum capacity, the ore loader produces equal loads of 10 kips on the platform at each pair of wheels, as shown in Fig. E7.8b.

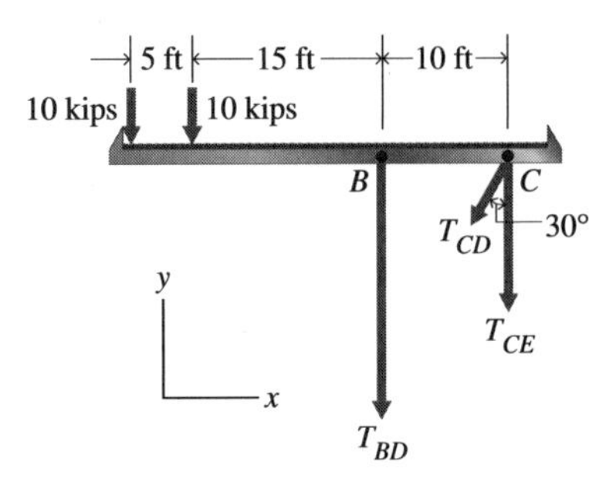

Figure E7.8b

The equations of equilibrium for the platform are

$$\sum F_x = -T_C \sin 30° = 0$$

$$\sum F_y = -T_{BD} - T_{CE} - T_{CD} \cos 30° - 20 = 0$$

$$\circlearrowleft^{+} \sum M_C = 30(10) + 25(10) + 10T_{BD} = 0$$

These equations yield $T_{CD} = 0$, $T_{BD} = -55$ kips, and $T_{CE} = 35$ kips.

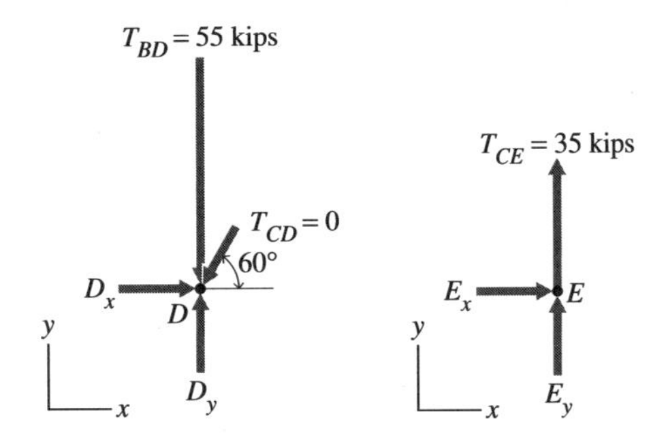

Figure E7.8c

Figure E7.8d

The reactions of supports *D* and *E* are determined by considering the free-body diagrams of the support pins *D* and *E,* shown in Figs. E7.8c and E7.8d. From these figures, we obtain $D_x = 0$, $D_y = 55$ kips, $E_x = 0$, and $E_y = -35$ kips.

Example 7.9

A Sign Structure

Problem Statement A heavy sign that weighs $W = 4.5$ kN is hung from a structural mechanism, as shown in Fig. E7.9a. All the pins are frictionless. Determine the force in bar *AB* and the (x, y) force components that act on the pins at *B* and *C.* Neglect the weights of the bars.

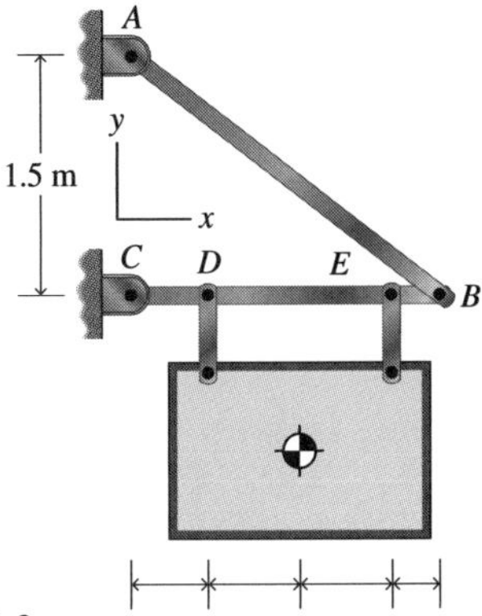

Figure E7.9a

Solution We first apply the principle of transmissibility and transfer the force W to bar BC. In so doing, we effectively replace the hanger forces at D and E by their resultant. We also observe that bar AB is a two-force member, and so the direction of the support reaction at A is known. The free-body diagram of mechanism ABC is shown in Fig. E7.9b.

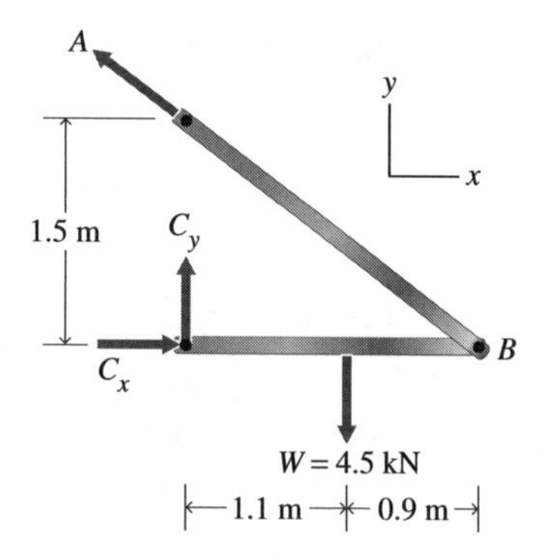

Figure E7.9b

From Fig. E7.9b, the length of member AB is 2.5 m. Therefore, the equilibrium conditions may be expressed as

$$\sum F_x = C_x - \left(\frac{2}{2.5}\right)A = 0$$

$$\sum F_y = C_y + \left(\frac{1.5}{2.5}\right)A - 4.5 = 0 \qquad \text{(a)}$$

$$\circlearrowleft\!+ \sum M_A = (1.5)C_x - (1.1)(4.5) = 0$$

The solution of Eqs. (a) is

$$A = 4.125 \text{ kN}$$

$$C_x = 3.30 \text{ kN} \qquad \text{(b)}$$

$$C_y = 2.025 \text{ kN}$$

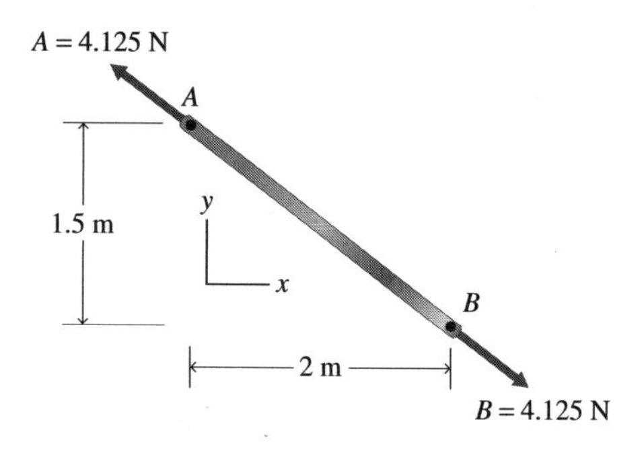

Figure E7.9c

Since bar AB is a two-force member, the force that acts on the pin at B is equal to the force that acts at A, but has opposite sense (see Fig. E7.9c). Hence, the (x, y) components of the force that acts on the pin at B are

$$B_x = -4.125\left(\frac{2}{2.5}\right) = -3.30 \text{ kN}$$

$$B_y = +4.125\left(\frac{1.5}{2.5}\right) = +2.475 \text{ kN}$$

The force in bar AB is thus 4.125 kN.

Note that with C_x and C_y known [Eqs. (b)], we could have determined B_x and B_y by considering a free-body diagram of bar BC. ■

CHECKPOINT

1. ***True or False:*** A structural mechanism is a structure that can maintain its shape when removed from its supports.
2. ***True or False:*** A mechanism may be analyzed by the same methods used to analyze a simple plane frame.
3. What distinguishes a tool mechanism from a structural mechanism?

Answers: 1. False; 2. True; 3. A tool mechanism has no supports, a structural mechanism does.

7.5 PARABOLIC CABLES AND THE SUSPENSION BRIDGE

After studying this section, you should be able to:

- Determine the tension force in, and the length of, a parabolic cable.

Key Concepts

Cables are long, slender, flexible structural members that are designed to carry axial loads in tension. They will buckle under compression.

A suspension-bridge cable, loaded by a uniform horizontal load, takes the shape of a parabola.

suspension bridge

hangers

A *suspension bridge* is a bridge whose horizontal roadway is supported by many vertical *hangers.* The hangers are suspended from heavy, flexible steel cables that are anchored at each end of the bridge (see Fig. 7.14). If the horizontal roadway is subjected to a uniformly distributed load due to gravity (its weight), the load on the cables is also assumed to be uniformly distributed horizontally. This is a reasonable assumption because the spacing between the hangers is usually small compared to the distance between the supports. Also, the weight of the roadway is many times larger than the weight of the cables and the hangers that transmit the load to the cables. So, in an initial analysis for such applications, the weights of cables and hangers are neglected.

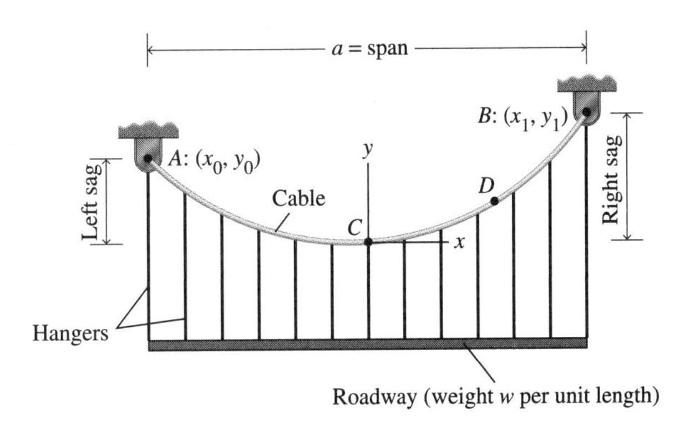

Figure 7.14 Schematic diagram of a suspension bridge with supports at different levels.

In this section, we consider a suspension-bridge cable and see that if the weight and the elongation of the cable are neglected, the cable hangs in a parabolic curve. In contrast, a cable that serves as an electrical power transmission line is subject to a weight that is uniformly distributed along the arc length of the cable; it takes on the shape of a catenary (see Sec. 7.6).

We will restrict our analysis of cables to those that are suspended between two supports (also called *anchorages*). The horizontal distance between the supports is called the *span,* and the vertical distance from a support to the lowest point of the cable, where it becomes tangent to the horizontal, is called the *sag* (refer to Fig. 7.14). In general, the supports (at A and B in Fig. 7.14) may be at different elevations. Thus, the sag in a cable may be measured relative to either the left or right support. When an actual cable is loaded and stressed, it will stretch. This elongation and its effect on the sag and, hence, on the tension

span

sag

The 2300 ft main span of the George Washington Bridge across the Hudson River between New York and New Jersey was twice as long as that of any other suspension bridge when it was completed in 1931.

in the cable are usually small. In this discussion, we will neglect elongation; we will assume that the cable is inextensible.[1]

Supports at Different Levels Consider a cable with supports at A: (x_0, y_0) and B: (x_1, y_1) and with horizontal span a (refer to Fig. 7.14). The cable supports a uniformly distributed roadway load of w (force per unit length of roadway). The elongation and the weight of the cable are neglected. To determine expressions for the tension T at any point D of the cable, consider the equilibrium of part CD of the cable, where C is the lowest point of the cable (see the free-body diagram in Fig. 7.15). This segment is subject to the horizontal force H at C, the tension T at D, and a distributed load whose resultant wx bisects the horizontal distance x from C to D. (A discussion of the resultant of distributed loads is presented in Chapter 8.) Since segment CD is in equilibrium, the force polygon of the three forces H, T, wx must close, and their lines of action must be concurrent. Hence, from Fig. 7.15,

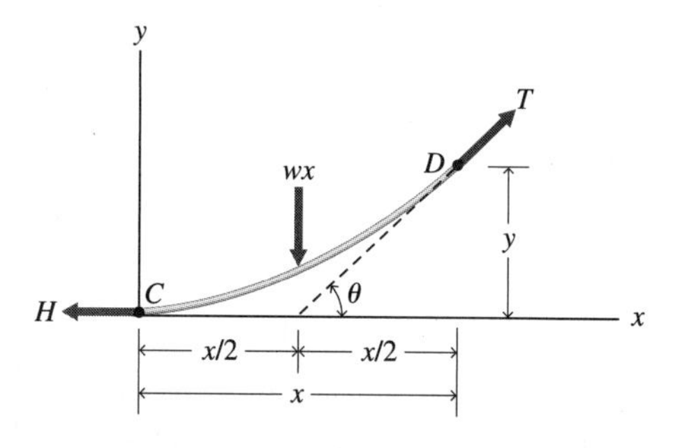

Figure 7.15 Free-body diagram of segment CD of cable.

$$\sum F_x = T\cos\theta - H = 0$$
$$\sum F_y = T\sin\theta - wx = 0$$

or

$$T\cos\theta = H$$
$$T\sin\theta = wx \tag{a}$$

Thus,

$$\tan\theta = \frac{wx}{H} \tag{7.2}$$

1. Determination of the effect of cable extension requires a knowledge of the stress-strain relations of the cable, a topic considered in courses on mechanics of materials.

Also, since the forces are concurrent, by geometry (Fig. 7.15),

$$\tan\theta = \frac{y}{(x/2)} = \frac{2y}{x} \tag{b}$$

From Eqs. (7.2) and (b), we obtain

$$y = \frac{wx^2}{2H} \tag{7.3}$$

Equation (7.3) represents a parabola whose vertex is at C and whose axis is vertical.

Squaring and adding Eqs. (a) and taking the square root of the result, we obtain

$$T = \sqrt{w^2x^2 + H^2} \tag{7.4}$$

We see from Eq. (7.4) that the minimum tension in the cable occurs at $x = 0$ and is equal to H.

Since $y = y_0$ at $x = x_0$ and $y = y_1$ at $x = x_1$, Eq. (7.3) yields

$$H = \frac{wx_0^2}{2y_0} = \frac{wx_1^2}{2y_1} \tag{7.5}$$

Equations (7.4) and (7.5) yield

$$T = w\sqrt{x^2 + \frac{x_0^4}{4y_0^2}} = w\sqrt{x^2 + \frac{x_1^4}{4y_1^2}} \tag{7.6}$$

Equation (7.6) expresses the tension T at any point D of the cable.

If $|x_1| > |x_0|$, by Eq. (7.6), the maximum tension occurs at support B, and its magnitude is given by the formula

$$T_{\max} = wx_1\sqrt{1 + \frac{x_1^2}{4y_1^2}} \tag{7.7}$$

Supports on the Same Level If the supports are on the same level, $y_1 = y_0 = d$, where d is the sag of the midpoint C of the cable. Also, $x_1 = -x_0 = a/2$, where a is the span (see Fig. 7.16). Then, by Eq. (7.5), the minimum tension in the cable is

$$H = \frac{wa^2}{8d} \tag{7.8}$$

By Eqs. (7.4) and (7.8), the tension T in the cable at x is

$$T = w\sqrt{x^2 + \frac{a^4}{64d^2}} \tag{7.9}$$

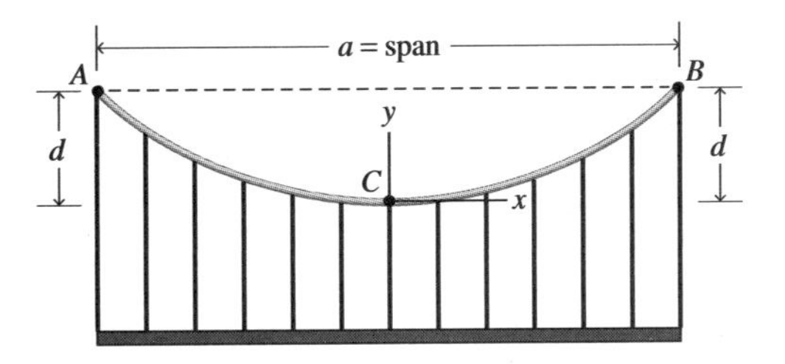

Figure 7.16 Schematic diagram of suspension bridge with supports on the same level.

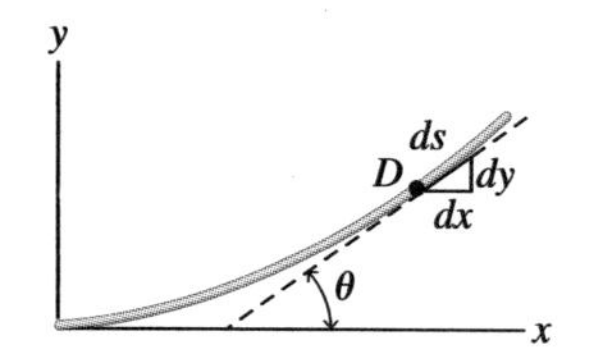

Figure 7.17 Element ds of cable.

Hence, for supports on the same level, Eq. (7.9), with $x = a/2$, yields the maximum tension in the cable:

$$T_{\max} = \frac{wa}{2}\sqrt{1 + \frac{1}{16}\left(\frac{a}{d}\right)^2} \tag{7.10}$$

Length of a Cable To determine the length L of a cable, consider an infinitesimal element ds of the cable at any point D (refer to Figs. 7.15 and 7.17). By geometry, $ds/dx = \sec\theta$. Since $\sec\theta = \sqrt{1 + \tan^2\theta}$, Eqs. (7.2) and (7.8) yield

$$ds = \sqrt{1 + \frac{64d^2x^2}{a^4}}\,dx$$

Hence, the total length L is given by

$$L = \int ds = 2\int_0^{a/2}\left(\sqrt{1 + \frac{64d^2x^2}{a^4}}\right)dx \tag{c}$$

Integration of Eq. (c) yields[2]

$$L = \frac{a}{2}\left(\sqrt{1 + 16\left(\frac{d}{a}\right)^2} + \frac{1}{4}\left(\frac{a}{d}\right)\ln\left[4\left(\frac{d}{a}\right) + \sqrt{1 + 16\left(\frac{d}{a}\right)^2}\right]\right) \tag{7.11}$$

For small ratios of sag d to span a, a convenient approximation of the length L can be obtained by expanding the integrand of Eq. (c) by the binomial series and retaining only the first few terms:

$$\sqrt{1 + \frac{64d^2x^2}{a^4}} = 1 + \frac{32d^2}{a^4}x^2 - \frac{512d^4}{a^8}x^4 + \cdots \tag{d}$$

With Eq. (d), Eq. (c) yields

$$L = a\left[1 + \frac{8}{3}\left(\frac{d}{a}\right)^2 - \frac{32}{5}\left(\frac{d}{a}\right)^4 + \cdots\right] \tag{7.12}$$

Equation (7.12) is valid only for cables whose supports are at the same level. For small ratios of sag d to span a, the series of Eq. (7.12) converges so rapidly that the first two or three terms give a sufficiently close approximation of the length L. In fact, for a ratio $d/a = 0.1$, the length L obtained from the first three terms in Eq. (7.12) agrees to three significant figures with that obtained from Eq. (7.11). If the sag to span ratio is large, however, Eq. (7.11) must be used to calculate L or more terms of Eq. (7.12) must be retained.

If a cable is stretched rather tightly, so that it is approximately straight, the weight of a segment over an interval Δx is approximately $k\Delta x$, where k is a constant. Consequently, the theory of the parabolic cable that has been developed in this section may be applied as an *approximation* to a cable that is loaded by its own weight, *provided that the sag is small compared with the span length of the cable* (see Sec. 7.6). If the ratio of sag to span is 0.15 or smaller, the error in this approximation is less than 5%.

2. Refer to any good table of integrals for the integration.

Example 7.10

A Suspension-Bridge Cable

Problem Statement A suspension-bridge roadway is supported by a cable anchored at (−15 m, 3 m) and (30 m, 12 m), relative to xy axes with origin at C, the lowest point on the cable (see Fig. E7.10a). The roadway weighs 3 kN per meter of horizontal length. Determine the following quantities:

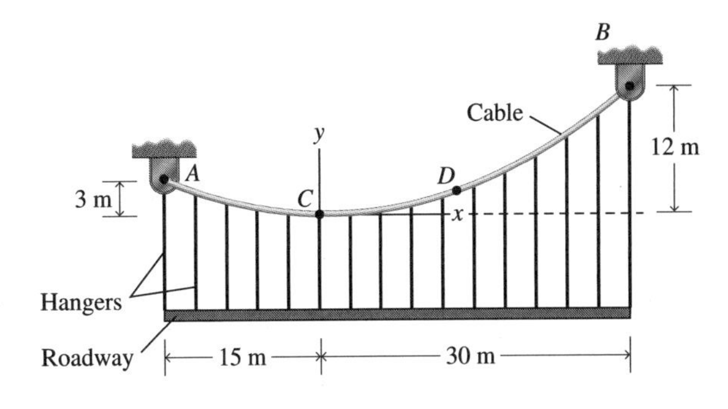

Figure E7.10a

a. the minimum tension in the cable

b. the cable tensions at the supports

c. the maximum tension in the cable

d. the length of the cable

Solution

a. The tension T in the cable at any point D (see Fig. E7.10b) is given by Eq. (7.6):

$$T = w\sqrt{x^2 + \frac{x_1^4}{4y_1^2}} \tag{a}$$

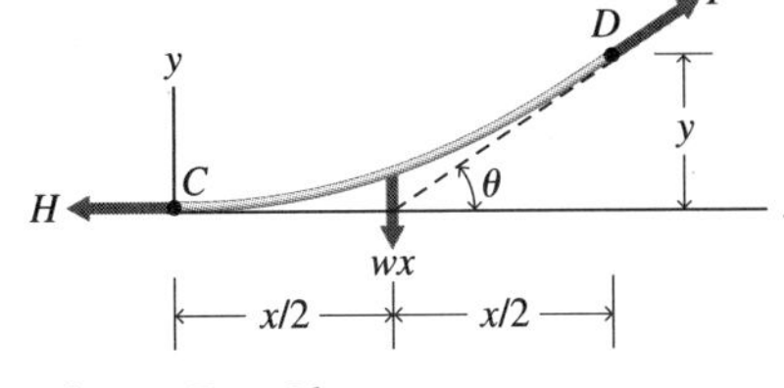

Figure E7.10b

where $(x_1, y_1) = (30 \text{ m}, 12 \text{ m})$ are the coordinates of support B. Thus, for any point on the cable, Eq. (a) yields

$$T = w\sqrt{x^2 + 1406.25} \tag{b}$$

From Eq. (b), with $x = 0$ and $w = 3$ kN/m, we find the minimum tension $H = T|_{x=0} =$ 112.50 kN. [Note that the value of H could also have been obtained directly from Eq. (7.5).]

b. Similarly, for $x = -15$ m and $x = 30$ m, Eq. (b) yields the tensions in the cable at supports A and B, respectively. Thus, from Eq. (b), for $x = -15$ m, $T_A = 121.17$ kN, and for $x = 30$ m, $T_B = 144.07$ kN.

c. The maximum tension in the cable occurs for the largest value of x [examine Eq. (b)]. Therefore, the maximum tension occurs at $x = 30$ m (at support B) and is $T_{\max} =$ 144.07 kN.

d. To determine the length L of the cable, we begin by noting that the supports are at different levels. We must therefore consider the lengths to the left and to the right of the lowest point C. Following the procedure outlined earlier for supports on the same level, we find by Eqs. (b) and (7.5) that

$$\tan^2\theta = \left(\frac{4y_1^2}{x_1^4}\right)x^2 = \left(\frac{4y_0^2}{x_0^4}\right)x^2 = \frac{x^2}{1406.25}$$

and, therefore, $ds = \sec\theta\,dx = \left(\sqrt{1 + \tan^2\theta}\right)dx = (1/37.5)\left(\sqrt{x^2 + 1406.25}\right)dx$ (refer to Fig. 7.17). Consequently, the length L of the cable is given by

$$L = \frac{1}{37.5}\int_{-15}^{30}\left(\sqrt{x^2 + 1406.25}\right)dx$$

From a table of integrals, we find

$$L = \frac{1}{75}\left[x\sqrt{x^2 + 1406.25} + 1406.25\ln\left(x + \sqrt{x^2 + 1406.25}\right)\right]_{-15}^{30}$$

Evaluation of this expression yields $L = 48.34$ m. Thus, the sag in the cable requires that the cable be 3.34 m longer than the span $d = 45$ m.

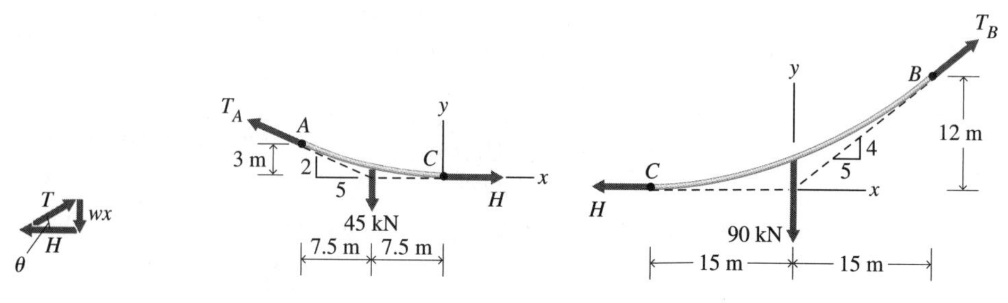

Figure E7.10c **Figure E7.10d** **Figure E7.10e**

Alternative Solution for Tensions T_A, T_B, and H Since any segment *CD* of the cable is in equilibrium under the action of the three forces *T, H, wx,* the forces are concurrent (see Fig. E7.10b). Thus, the action line of *T* is determined by geometry (see Fig. E7.10c). Using this fact, we draw free-body diagrams of *AC* and *CB,* the left and right parts of the cable (Figs. E7.10d and E7.10e).

For the left part of the cable, the equilibrium equations are

$$\sum F_x = H - \frac{5}{\sqrt{29}}T_A = 0$$

$$\sum F_y = \frac{2}{\sqrt{29}}T_A - 45 = 0$$

Solution of these equations gives $T_A = 121.17$ kN and $H = 112.50$ kN.

For the right part of the cable, the equilibrium equations are

$$\sum F_x = \frac{5}{\sqrt{41}}T_B - H = 0$$

$$\sum F_y = \frac{4}{\sqrt{41}}T_B - 90 = 0$$

Solution of these equations yields $T_B = 144.07$ kN and $H = 112.50$ kN. These values of T_A, T_B, and H agree with the results obtained in parts a and b. ■

Example 7.11

Suspension Bridge with Cable Supports at the Same Level

Problem Statement A uniformly distributed roadway is carried by the cables of a suspension bridge. Each cable supports $w = 1200$ lb per horizontal foot, with span $a = 880$ ft and sag $d = 48$ ft. Each cable is anchored at its ends at the same elevation (refer to Fig. 7.16).

a. Determine the tension in a cable at a support and the minimum tension in the cable.

b. Calculate the length of the cable by Eq. (7.11).

c. Calculate the length of the cable by Eq. (7.12).

d. What can be concluded from a comparison of the results of parts b and c?

Solution **a.** The tension at the supports is the maximum tension in the cable. Therefore, from Eq. (7.10), we find

$$T_{max} = \frac{wa}{2}\sqrt{1 + \frac{1}{16}\left(\frac{a}{d}\right)^2}$$

$$= \frac{(1200)(880)}{2}\sqrt{1 + \frac{1}{16}\left(\frac{880}{48}\right)^2}$$

$$= 2{,}477{,}000 \text{ lb}$$

Also, from Eq. (7.8), the minimum tension in the cable is

$$H = \frac{wa^2}{8d}$$

$$= \frac{(1200)(880)^2}{(8)(48)}$$

$$= 2{,}420{,}000 \text{ lb}$$

b. From Eq. (7.11), the length of the cable is

$$L = \frac{a}{2}\left(\sqrt{1 + 16\left(\frac{d}{a}\right)^2} + \frac{1}{4}\left(\frac{a}{d}\right)\ln\left[4\left(\frac{d}{a}\right) + \sqrt{1 + 16\left(\frac{d}{a}\right)^2}\right]\right)$$

$$= \frac{880}{2}\left(\sqrt{1 + 16\left(\frac{48}{880}\right)^2} + \frac{1}{4}\left(\frac{880}{48}\right)\ln\left[4\left(\frac{48}{880}\right) + \sqrt{1 + 16\left(\frac{48}{880}\right)^2}\right]\right)$$

$$= 886.93 \text{ ft}$$

c. From Eq. (7.12), the length of the cable is

$$L = a\left[1 + \frac{8}{3}\left(\frac{d}{a}\right)^2 - \frac{32}{5}\left(\frac{d}{a}\right)^4 + \cdots\right]$$

$$= 880\left[1 + \frac{8}{3}\left(\frac{48}{880}\right)^2 - \frac{32}{5}\left(\frac{48}{880}\right)^4\right]$$

$$= 886.93 \text{ ft}$$

d. Since the lengths calculated by Eqs. (7.11) and (7.12) are equal, to five significant figures, the ratio of the sag d to the span a may be assumed to be small. In fact, in this example, $d/a = 0.0545$.

Example 7.12

Force on a Cable Tower

Problem Statement A suspension-bridge roadway is supported by two cables AB and BC (see Fig. E7.12a). One end of each cable is pinned to a tower at B. The other ends of the cables are anchored at the same elevation at A and C, respectively, where the cables have horizontal slopes. The roadway load is 4 kips/ft.

a. Determine the resultant of the forces exerted on the tower by the cables and the forces exerted on the supports at A and C.

b. Determine the lengths of cables AB and BC.

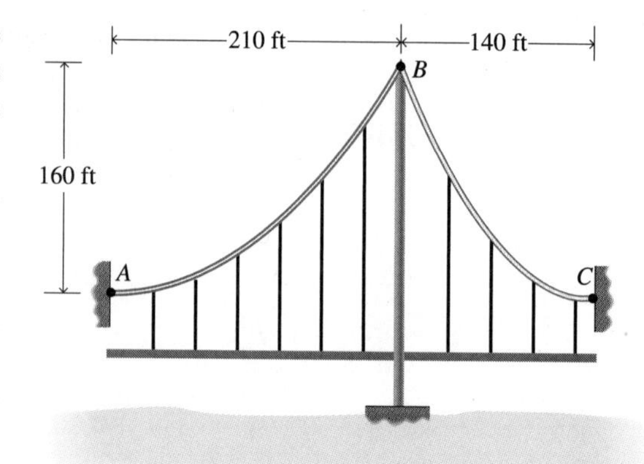

Figure E7.12a

Solution

a. To solve this problem, we consider cables AB and BC separately. For cable AB, we have the following data:

$$\begin{aligned} a &= 420 \text{ ft} \\ d &= 160 \text{ ft} \\ w &= 4 \text{ kips/ft} \end{aligned} \tag{a}$$

The maximum tension in cable AB occurs at B and, from Eq. (7.10), is

$$\begin{aligned} T_{\max(AB)} &= \frac{wa}{2}\sqrt{1 + \frac{1}{16}\left(\frac{a}{d}\right)^2} \\ &= \frac{(4)(420)}{2}\sqrt{1 + \frac{1}{16}\left(\frac{420}{160}\right)^2} \\ &= 1004.7 \text{ kips} \end{aligned}$$

From Eq. (7.8), the force exerted on the support at A is

$$\begin{aligned} H_A &= \frac{wa^2}{8d} \\ &= \frac{(4)(420)^2}{(8)(160)} \\ &= 551.25 \text{ kips} \end{aligned} \tag{b}$$

Similarly, for cable BC, we have

$$\begin{aligned} a &= 280 \text{ ft} \\ d &= 160 \text{ ft} \\ w &= 4 \text{ kips/ft} \end{aligned}$$

The maximum tension in cable BC also occurs at B and, from Eq. (7.10), is

$$\begin{aligned} T_{\max(BC)} &= \frac{wa}{2}\sqrt{1 + \frac{1}{16}\left(\frac{a}{d}\right)^2} \\ &= \frac{(4)(280)}{2}\sqrt{1 + \frac{1}{16}\left(\frac{280}{160}\right)^2} \\ &= 611.3 \text{ kips} \end{aligned}$$

From Eq. (7.8), the force exerted on the support at C is

$$H_C = \frac{wa^2}{8d}$$

$$= \frac{(4)(280)^2}{(8)(160)}$$

$$= 245.0 \text{ kips}$$

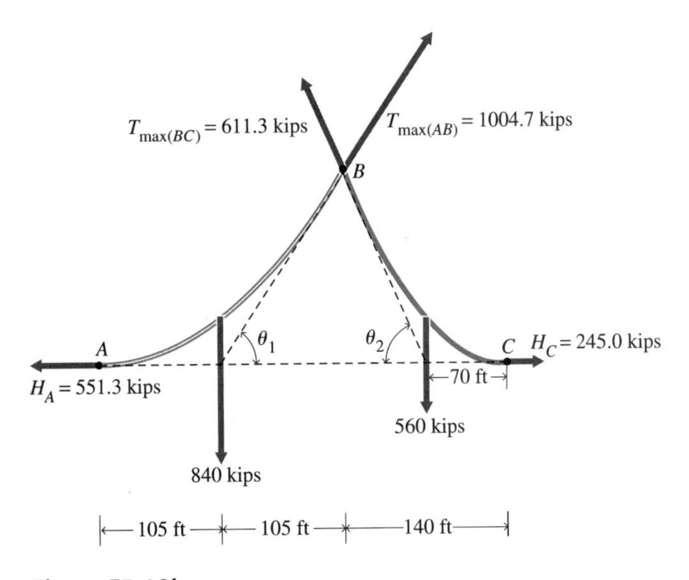

Figure E7.12b

To determine the resultant force on the tower at B, we need to calculate the slope of the cables at B (refer to Fig. E7.12b). For cable AB, with Eq. (7.2), Eqs. (a) and (b), and Fig. E7.12b, we find

$$\tan\theta_1 = \frac{wx}{H_A}$$

$$= \frac{(4)(210)}{551.25}$$

$$= 1.5238$$

$$\theta_1 = 56.72°$$

Similarly, for cable BC, we have

$$\tan\theta_2 = \frac{wx}{H_C}$$

$$= \frac{(4)(140)}{245.0}$$

$$= 2.2857$$

$$\theta_2 = 66.37°$$

Hence, the (x, y) components of the forces exerted on the tower by the cables are (see Fig. E7.12c)

$$\sum F_x = +(611.25)(\cos 66.37°) - (1004.73)(\cos 56.72°) = -306.26 \text{ kips}$$
$$\sum F_y = -(611.25)(\sin 66.37°) - (1004.73)(\sin 56.72°) = -1399.95 \text{ kips}$$

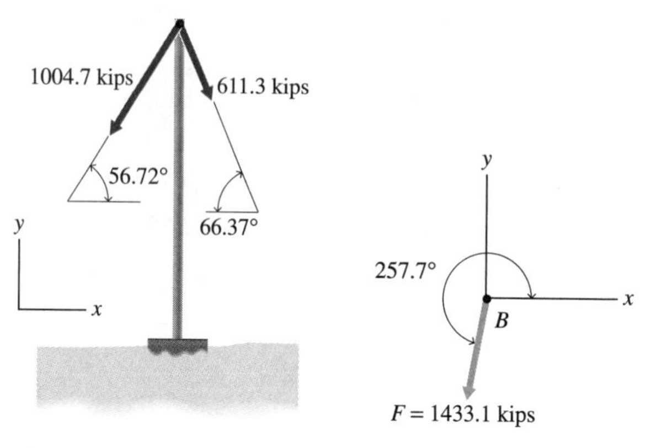

Figure E7.12c **Figure E7.12d**

Then, the magnitude of the resultant force that acts on the tower and its direction angle are (see Fig. E7.12d)

$$F = \sqrt{306.26^2 + 1399.95^2} = 1433.1 \text{ kips}$$

$$\theta = \tan^{-1}\left(\frac{-1399.95}{-306.26}\right) = 257.7°$$

b. Since the sag-to-span ratio for each of the cables is large, we use Eq. (7.11) to calculate L:

$$L_{AB} = \frac{a}{2}\left(\sqrt{1 + 16\left(\frac{d}{a}\right)^2} + \frac{1}{4}\left(\frac{a}{d}\right)\ln\left[4\left(\frac{d}{a}\right) + \sqrt{1 + 16\left(\frac{d}{a}\right)^2}\right]\right)$$

$$= \frac{420}{2}\left(\sqrt{1 + 16\left(\frac{160}{420}\right)^2} + \frac{1}{4}\left(\frac{420}{160}\right)\ln\left[4\left(\frac{160}{420}\right) + \sqrt{1 + 16\left(\frac{160}{420}\right)^2}\right]\right)$$

$$= 549.22 \text{ ft}$$

Similarly, we obtain $L_{BC} = 445.11$ ft. If we had used Eq. (7.12) with three terms in the series, we would have obtained $L_{AB} = 525.93$ ft and $L_{BC} = 332.74$ ft. Note that for larger d/a ratios, the error increases rapidly.

7.6 THE CATENARY—A FREELY HANGING CABLE

After studying this section, you should be able to:

- Determine the tension in, and the length of, a catenary cable.

Key Concept

A cable that hangs under its own weight takes the shape of a catenary.

catenary

The curve formed by an inextensible sagging rope or flexible cable that is loaded only by its own weight is called a *catenary.* To derive the equation of a catenary, we consider a uniform flexible cable, as shown in Fig. 7.18a. The cable supports its own weight, which is distributed uniformly along its arc length. Consider the free-body diagram of a segment of the cable, with its left end at the minimum point (see Fig. 7.18b). Let s be the arc length of the curved segment, measured positively to the right from the lowest point of the curve. The forces at the ends of the cable segment are H and T, respectively.

Since they hang under their own weight, electrical power transmission lines drape in the shape of a catenary.

The weight of the segment is ws, where w is the weight per unit length of the cable (a constant) and s is the length of the segment. Consequently, the equations of equilibrium for horizontal and vertical components of the forces on the free body are

$$\sum F_x = T\cos\theta - H = 0 \qquad \sum F_y = T\sin\theta - ws = 0 \tag{a}$$

Solving Eqs. (a) for T and θ in terms of H, w, and s, we obtain

$$T = \sqrt{H^2 + (ws)^2} \tag{7.13}$$

$$\theta = \tan^{-1}\left(\frac{ws}{H}\right) \tag{7.14}$$

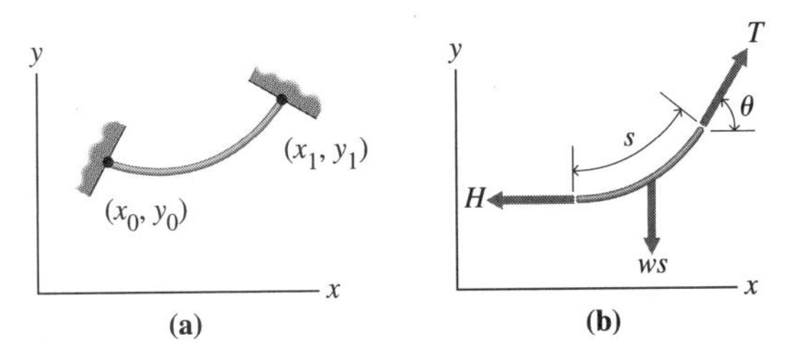

Figure 7.18 **(a)** A uniform flexible cable. **(b)** Free-body diagram of segment of cable.

Note the similarity between Eq. (7.4) and Eq. (7.13). In Eq. (7.13), the distance s is measured along the cable from its lowest point, whereas in Eq. (7.4), the distance x is measured horizontally from the lowest point of the cable. Also, in Eq. (7.13), w denotes the weight per unit length of the cable, and in Eq. (7.4), w is the weight per unit horizontal distance. Since w is constant, the ratio w/H is a constant for a given cable. Consequently, differentiation of Eq. (7.14) yields

$$\sec^2\theta\frac{d\theta}{ds} = \frac{w}{H} \tag{b}$$

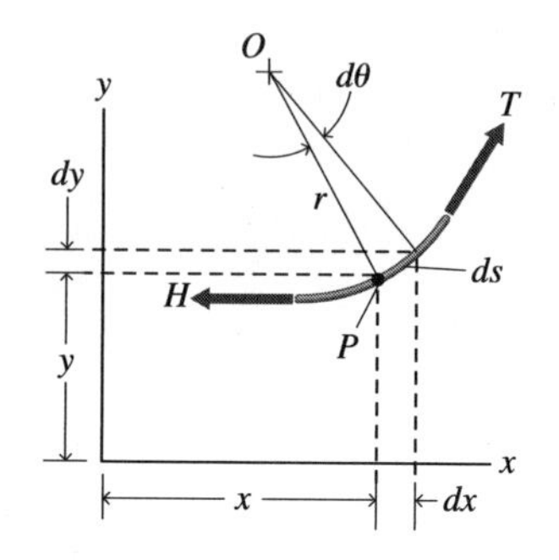

Figure 7.19 Infinitesimal cable length *ds*.

Equation (b) is the differential equation of the cable segment, in terms of θ as a function of s. To obtain a differential equation for the segment as a function $y(x)$, we must rewrite Eq. (b) in terms of x. To determine $\sec^2\theta$ as a function of $y(x)$, note that $\tan\theta = y'$, where y' denotes dy/dx. Hence,

$$\sec^2\theta = 1 + \tan^2\theta = 1 + (y')^2 \tag{c}$$

To determine $d\theta/ds$ as a function of $y(x)$, consider a point P on the curve and a differential element ds at P, subtended by the infinitesimal angle $d\theta$ that emanates from O, the center of curvature of the curve (see Fig. 7.19). The radius r is the radius of curvature of the curve. From Fig. 7.19, $ds = r\,d\theta$. Therefore,

$$\frac{d\theta}{ds} = \frac{1}{r} = \kappa$$

where $1/r = \kappa$ is the curvature of the curve at point P. The formula for the curvature κ in terms of $y(x)$ is derived in calculus books:

$$\frac{d\theta}{ds} = \kappa = \frac{y''}{[1 + (y')^2]^{3/2}} \tag{7.15}$$

Consequently, Eqs. (b), (c), and (7.15) yield

$$\frac{y''}{\sqrt{1 + (y')^2}} = \frac{w}{H} \tag{d}$$

To solve Eq. (d) for $y(x)$, we set $y' = p$. Then Eq. (d) becomes

$$\frac{dp}{\sqrt{1 + p^2}} = \frac{w}{H}dx \tag{e}$$

Integration of Eq. (e) yields

$$p = \sinh\left(\frac{wx}{H} + C\right)$$

where C is a constant of integration. Therefore,

$$\frac{dy}{dx} = y' = \sinh\left(\frac{wx}{H} + C\right) \tag{7.16}$$

Integrating Eq. (7.16), we obtain the general equation of the curve $y = y(x)$ as

$$wy = H\cosh\left(\frac{wx}{H} + C\right) + K \tag{7.17}$$

where K is a constant of integration. This curve is a catenary.[3]

If $y = y' = 0$ for $x = 0$, Eq. (7.17) becomes

$$wy = H\left(\cosh\frac{wx}{H} - 1\right) \tag{7.18}$$

Equation (7.17) is simplified further if xy coordinate axes are chosen such that for $x = 0$, $y = H/w$, and $y' = 0$. Then Eq. (7.17) reduces to the simpler form

$$wy = H\cosh\frac{wx}{H} \tag{7.19}$$

Equation (7.17) is the general equation of the curve that a cable forms when it is subjected to w, its own weight per unit length. It contains two constants, C and K, that may be adjusted to satisfy end conditions of the cable. Also, the constant w/H may be unknown. When a cable hangs freely, as in Fig. 7.20, it is in equilibrium under the internal forces (w, H, and T) [Eqs. (a)], and there are no external forces tending to change the shape of the cable. Therefore, if an arch is built so that it has the shape of a hyperbolic cosine curve—that is, in the shape of a catenary [see Eqs. (7.17) and (7.18) or (7.19)]—no forces exist to distort the arch. If the arch is turned over (see Fig. 7.21), all forces are reversed and the arch remains in equilibrium (retains its shape). This fact led architects and engineers to choose a catenary curve for the Gateway Arch in St. Louis, Missouri.

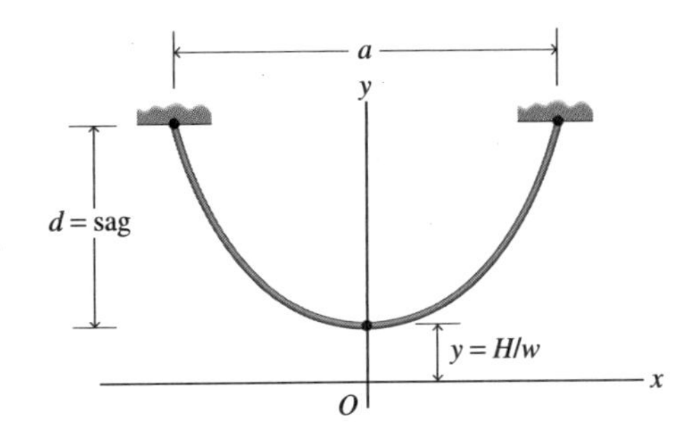

Figure 7.20 Freely hanging cable.

Figure 7.21 One of the most impressive modern arches is the Gateway Arch in St. Louis, Missouri. Designed by Saarinen, this upright catenary has both span (the distance shown by the dashed line) and height of 192 m.

3. A catenary is sometimes referred to as a "chain curve," since the name is derived from the Latin word *catena,* which means "chain."

To express the distance s along a cable as a function of x, we first observe from Fig. 7.19 that $ds^2 = dx^2 + dy^2$, or

$$ds = \sqrt{1 + (y')^2}\,dx \tag{f}$$

The length L of the cable is $L = \int ds$. If the ends of the cable are at the points (x_0, y_0) and (x_1, y_1), the length is

$$L = \int_{x_0}^{x_1} \left[\sqrt{1 + (y')^2}\right] dx \tag{7.20}$$

Substituting for y' from Eq. (7.16) into Eq. (7.20) and noting the identity for hyperbolic functions, $\sqrt{1 + \sinh^2 \phi} = \cosh \phi$, we obtain, by integration,

$$L = \frac{H}{w}\left[\sinh\left(\frac{w}{H}x_1 + C\right) - \sinh\left(\frac{w}{H}x_0 + C\right)\right] \tag{7.21}$$

By Eq. (7.14), we may express y' in terms of s:

$$y' = \tan\theta = \frac{ws}{H} \tag{g}$$

Then, Eqs. (f) and (g) yield

$$dx = \frac{H}{w}\left[\frac{ds}{\sqrt{\left(\frac{H}{w}\right)^2 + s^2}}\right] \tag{h}$$

Integration of Eq. (h) yields

$$\int_0^x dx = \frac{H}{w}\int_0^s \frac{ds}{\sqrt{\left(\frac{H}{w}\right)^2 + s^2}}$$

$$x = \frac{H}{w}\sinh^{-1}\left(\frac{ws}{H}\right)$$

or

$$ws = H\sinh\left(\frac{wx}{H}\right) \tag{7.22}$$

Then, by Eq. (7.13) and Eq. (7.22), we can express the tension T as a function of x:

$$T = H\sqrt{1 + \sinh^2\left(\frac{wx}{H}\right)} = H\cosh\left(\frac{wx}{H}\right) \tag{7.23}$$

Equations (7.18) [or (7.19)], (7.22), and (7.23) are sufficient for solving problems involving cables subject to uniformly distributed weight along their lengths. In particular, for a cable supported at its ends by anchors at the same elevation, Fig. 7.20 and Eq. (7.19) yield the ratio of sag d to span a as

$$\frac{d}{a} = \frac{H}{wa}\left[\cosh\left(\frac{wa}{2H}\right) - 1\right] \tag{7.24}$$

Then, for a given cable with known values of w, d, and a, Eq. (7.24) gives the minimum tension H in the cable and, with Eq. (7.23), the tension T at any point of the cable.

Example 7.13

Sagging Cable

Problem Statement As shown in Fig. E7.13a, an inextensible cable, 50 ft long, is fastened at the points (20 ft, 10 ft) and (−20 ft, 10 ft) and is loaded by its own weight ($w = 0.2$ lb/ft of cable length).

a. Determine the sag of the cable.

b. Determine the tension H in the cable at its minimum elevation.

c. Determine the maximum tension in the cable.

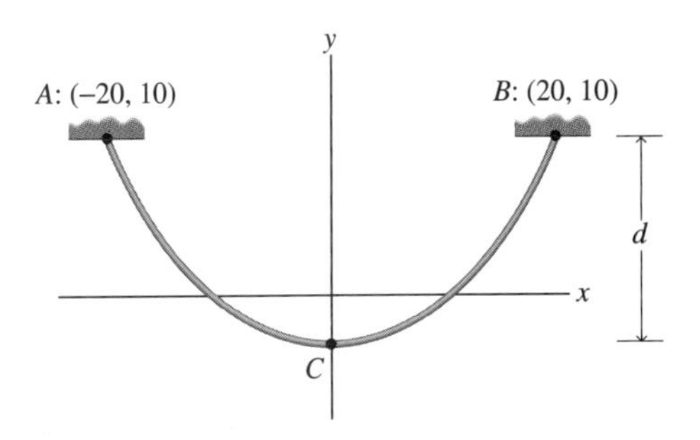

Figure E7.13a

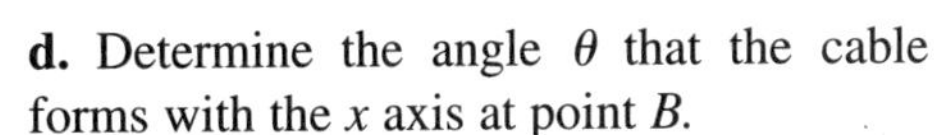

d. Determine the angle θ that the cable forms with the x axis at point B.

Solution **a.** By symmetry, the minimum elevation lies at $x = 0$. Therefore, $y' = 0$ at $x = 0$, and Eq. (7.16) shows that $C = 0$. Since $x_0 = -20$ ft and $x_1 = 20$ ft, Eq. (7.21) yields, with $L = 50$ ft,

$$50 = \frac{H}{w}\left[\sinh\left(\frac{20w}{H}\right) - \sinh\left(-\frac{20w}{H}\right)\right]$$

$$= \frac{2H}{w}\sinh\left(\frac{20w}{H}\right)$$

or

$$\frac{25w}{H} = \sinh\left(\frac{20w}{H}\right) \qquad \text{(a)}$$

A numerical method must be used to solve Eq. (b). Using values from a table of hyperbolic sines, we find that the solution is $w/H = 0.05914$.[4]

Setting $x = 20$ and $y = 10$, we now obtain, from Eq. (7.17), $K = -1.19376$. Hence, the equation of the catenary is

$$y = 16.91[\cosh(0.05914x)] - 20.18$$

Setting $x = 0$, we obtain the ordinate of the minimum point of the curve:

$$y_{\min} = 16.91 - 20.18 = -3.27 \text{ ft}$$

Therefore, the sag is $d = 10.00 + 3.27 = 13.27$ ft.

b. Since $w/H = 0.05914$, $H = 0.2/0.05914 = 3.382$ lb.

c. From the free-body diagram of the right half of the cable (Fig. E7.13b), we see that cable segment BC is in equilibrium under the action of the three forces T, $H = 3.382$ lb, and $ws = (0.2)(25) = 5$ lb. Hence, these three forces are concurrent at point I. Note that the line of action of the resultant

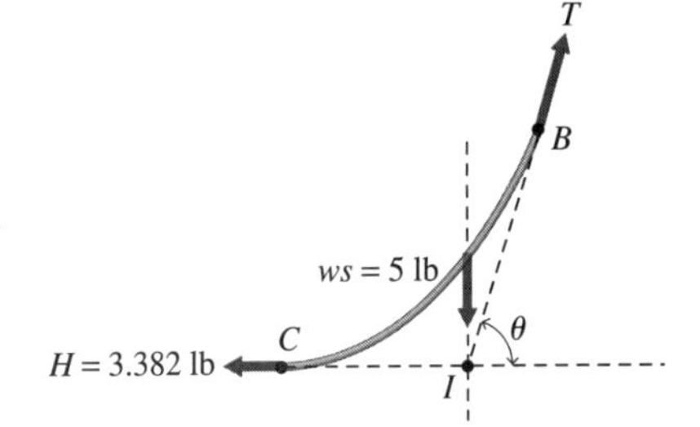

Figure E7.13b

4. Refer to any good table of numerical values of trigonometric functions. This equation can also be solved by using a computer program, by using an advanced scientific calculator, or by trial and error.

of the distributed force w is closer to point B than to C. From Fig. E7.13b, the equations of equilibrium for cable segment BC are

$$\sum F_x = T\cos\theta - H = 0$$
$$\sum F_y = T\sin\theta - ws = 0$$

Therefore, $T\cos\theta = H$ and $T\sin\theta = ws$, or

$$T^2 = H^2 + (ws)^2 \tag{b}$$

Substitution of the values of H and ws into Eq. (b) yields $T^2 = 36.437$, or $T = T_{max} = 6.036$ lb.

d. With the value of T known, the equation $T\cos\theta = H$ yields $\cos\theta = H/T = 3.382/6.036 = 0.560$, or $\theta = 55.9°$. As a check, we note that the equation $T\sin\theta = ws$ yields the result $\sin\theta = 5/6.036 = 0.828$, or $\theta = 55.9°$.

Example 7.14

A Hanging Cable

Problem Statement As noted in this section, the general equation of the catenary is simplified if xy coordinate axes are chosen such that for $x = 0$, $y = H/w$, and $dy/dx = 0$. Then, Eq. (7.17) reduces to Eq. (7.19), repeated here:

$$wy = H\cosh\frac{wx}{H} \tag{a}$$

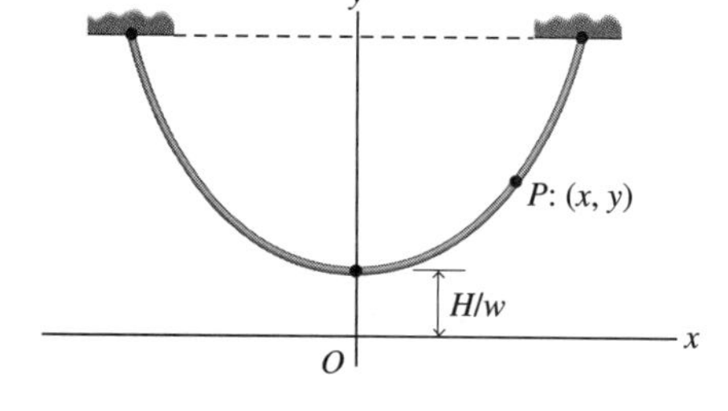

Figure E7.14a

a. Show that in this xy coordinate system (see Fig. E7.14a) the tension T at any point P: (x, y) in a hanging chain is given by the equation $T = wy$.

b. Interpret this result.

Solution

a. To show that $T = wy$, we first consider a free-body diagram of the cable from $x = 0$ to point P (see Fig. E7.14b), in order to express T in terms of the minimum tension H. Thus, from Fig. E7.14b, we have $\sum F_x = T\cos\phi - H = 0$, or

$$T = H\sec\phi = H\sqrt{1 + \tan^2\phi} \tag{b}$$

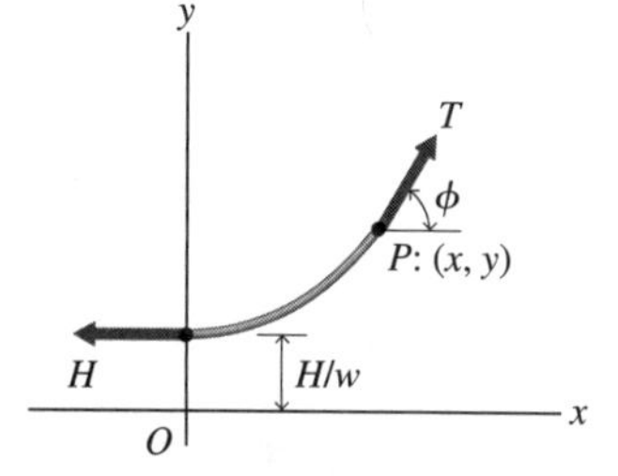

Figure E7.14b

The term $\tan\phi$ is the slope of the catenary. It can therefore be obtained by differentiation of Eq. (a):

$$\tan\phi = y' = \sinh\frac{wx}{H} \tag{c}$$

Substitution of Eq. (c) into Eq. (b) yields [see also Eq. (7.23)]

$$T = H\sqrt{1 + \sinh^2\frac{wx}{H}} = H\sqrt{\cosh^2\frac{wx}{H}} = H\cosh\frac{wx}{H} \tag{d}$$

Then, by Eqs. (a) and (d), we have

$$T = wy \tag{e}$$

b. Equation (e) means that T is equal to wy at every point along the hanging cable for the coordinate system of Fig. E7.14a (see also Fig. 7.20). This, in turn, means that if the cable is hung over a smooth peg at the point P: (x, y) (see Fig. E7.14c) and is cut off at the point Q, *where the cable crosses the x axis,* the weight of the cable from P to Q is just sufficient to keep the cable from slipping at P. A similar result holds for the segment RS. Therefore, the cable can be hung over two smooth pegs at points P and R without slipping, if the ends of the cable just reach the x axis.

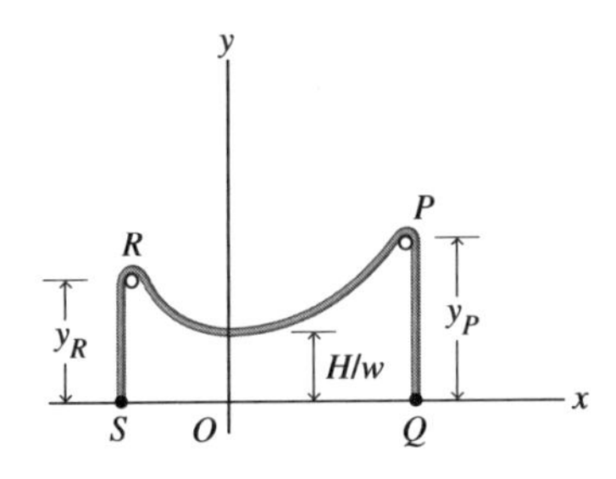

Figure E7.14c

Example 7.15

Icing of an Electrical Power Line

Problem Statement During an ice storm, ice forms uniformly along the length of an electrical power line that is strung between poles 200 m apart at the same elevation. The ice-free weight of the line is 50 N/m. How much additional weight per meter due to the ice can the cable support if the sag is not to exceed 5 m and the maximum allowable tension is 90 kN?

Solution From Fig. 7.20 and Eq. (7.24), the sag is

$$d = \frac{H}{w}\cosh\left(\frac{wx}{H}\right) - \frac{H}{w} \tag{a}$$

and, from Eq. (7.23), the tension in the cable is

$$T = H\cosh\left(\frac{wx}{H}\right) \tag{b}$$

Substitution of the given data into Eqs. (a) and (b) yields, for $x = 100$ m,

$$5 = \frac{H}{w}\cosh\left(\frac{100w}{H}\right) - \frac{H}{w} \tag{c}$$

$$90 = H\cosh\left(\frac{100w}{H}\right) \tag{d}$$

Equations (c) and (d) are two nonlinear algebraic equations in the variables H and w/H. They may be solved by trial and error or by an equation solver. The solution is

$$\frac{w}{H} = 0.000\,999 \text{ m}^{-1}$$

$$H = 89.553 \text{ kN}$$

Hence, the maximum weight that the cable can support is $w = (0.000\,999)(89\,553) = 89.463$ N/m, and the weight due to the ice is $w - 50 = 39.46$ N/m. For example, if a power line has a diameter of 50 mm and if ice with a specific weight of 9 kN/m³ forms a uniform cylinder around the power line, the outer diameter d of the cylinder of ice is given by the relationship

$$9000\left[\frac{\pi(d^2 - 0.050^2)}{4}\right] = 39.46 \text{ N/m}$$

or

$$d = 89.90 \text{ mm}$$

Example 7.16

Comparison of Parabolic and Catenary Solutions

Problem Statement An electrical power line that weighs 1 lb/ft is anchored to towers at the same elevation with a span a of 400 ft (see Fig. E7.16). The maximum design sag d is restricted to 80 ft. Determine the length of the cable and the maximum tension in the cable by these two methods and compare the solutions:

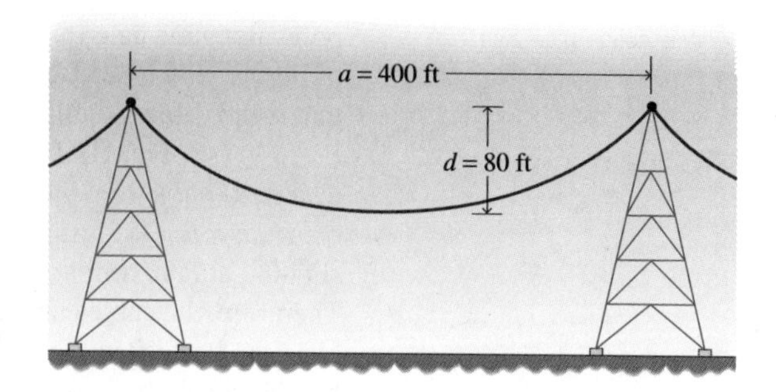

Figure E7.16

a. by the theory for a catenary cable

b. by the theory for a parabolic cable

Solution **a.** By the theory for a catenary, with $w = 1$ lb/ft and $x = 200$ ft, the sag d is related to the minimum tension H by [see Eq. (7.24) and Fig. 7.20]

$$d = 80 = H\cosh\left(\frac{200}{H}\right) - H \tag{a}$$

The solution of Eq. (a) is $H = 262.34$ lb. Then, by Eq. (7.21), the length L of the cable is

$$L = 2s\big|_{x=200} = 2H\sinh\left(\frac{200}{H}\right) = 439.9 \text{ ft}$$

Next, from Eq. (7.23) (with $w = 1$ lb/ft, $H = 262.3$ lb, and $x = 200$ ft), the maximum tension in the cable is

$$T_{\max} = H\cosh\left(\frac{wx}{H}\right) = 342.3 \text{ lb}$$

Alternatively, we could have computed $T_{\max}$ as follows [see Eq. (e) of Example 7.14]:

$$\begin{aligned} T_{\max} &= wy\big|_{x=200} \\ &= w\left(d + \frac{H}{w}\right) \qquad \text{(b)} \\ &= 80 + 262.3 \\ &= 342.3 \text{ lb} \end{aligned}$$

b. By the theory for a parabolic cable, the maximum tension is given by [see Eq. (7.11)]

$$T_{\max} = \frac{wa}{2}\sqrt{1 + \frac{1}{16}\left(\frac{a}{d}\right)^2} \tag{c}$$

With the given data, Eq. (c) yields $T_{\max} = 320.2$ lb, or 22.1 lb less than the result using the catenary theory [Eq. (b)]. Similarly, substitution of the given data into Eq. (7.12) gives the length $L = 439.3$ ft, or only 0.6 ft less than the result using the catenary theory. Note that in this example $d/a = 80/400 = 0.2$, a fairly small ratio, and a reasonably accurate result is obtained using the parabolic approximation. In general, for $a/d \geq 10$ ($d/a \leq 0.1$), the parabolic approximation is quite accurate.

CHECKPOINT 

1. ***True or False:*** A cable that is an electric power transmission line is subjected to a uniform gravity load distributed horizontally.
2. ***True or False:*** A suspension-bridge cable hangs in the form of a parabola.
3. ***True or False:*** An electrical power transmission line hangs in the form of a catenary.
4. ***True or False:*** The maximum tension in a hanging cable (parabolic or catenary) occurs at the point where the cable is tangent to the horizontal.

Answers: 1. False; 2. True; 3. True; 4. False

Chapter Highlights

- A lever is a rigid bar hinged at one point. It transfers the effect of an input force E to an output force L.
- The mechanical advantage of a lever is expressed by the equation $m = L/E = a/b$, where a and b are the lever arms of forces E and L, respectively.
- A pulley is a wheel supported by an axle or shaft. The simple pulley acts to change the direction of a force and has a mechanical advantage $m = 1.0$.
- A block and tackle is a system of pulleys that acts to change the direction of a force and to magnify its effect. It has a mechanical advantage $m > 1.0$.
- A frame is a system of structural members in which one (or more) of the members is subject to transverse forces and couples, the members being joined together, but not necessarily at their ends.
- A simple plane frame is a stable structure that maintains its shape if removed from its supports.
- A simple plane frame is statically determinate; that is, its member forces and support reactions can be determined by the equations of static equilibrium.
- A structural mechanism is a structure that is unstable (nonrigid) when it is removed from its supports; that is, it is then free to change its shape.
- A mechanism may be analyzed by the same methods as those used for simple plane frames.
- Cables used to support suspension bridges take on a parabolic shape under load, whereas cables used as electrical power transmission lines take on the shape of a catenary.

Problems

7.1 *The Lever*

7.1 A tackle on the Chicago Bears football team weighs 300 lb. In coming off his stance, he places his entire weight on the ball of his foot. In effect, his foot acts as a lever (see the simplified model of the foot and lower leg, Fig. P7.1). The weight of the tackle is transmitted to his foot through the tibia. In turn, the Achilles tendon pulls on the heel. Considering static effects only, determine the force that the tendon must withstand. (*Note:* If dynamic effects were included, the force in the tendon would be much higher.)

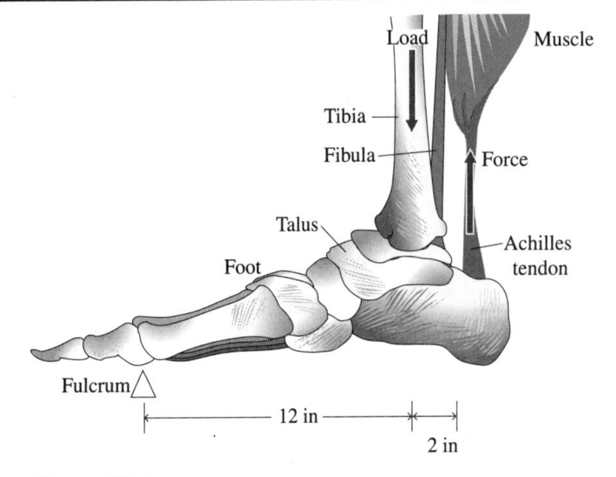

Figure P7.1

7.2 To determine the center of gravity of a nonuniform bar that weighs 24 N, an engineer suspends the bar from a wire at point A and supports it at point B on a knife edge that rests on a pan of a balance (see Fig. P7.2). A weight of 16 N is required to balance the bar in a horizontal position. Determine the horizontal distance d from point A to the center of gravity of the bar.

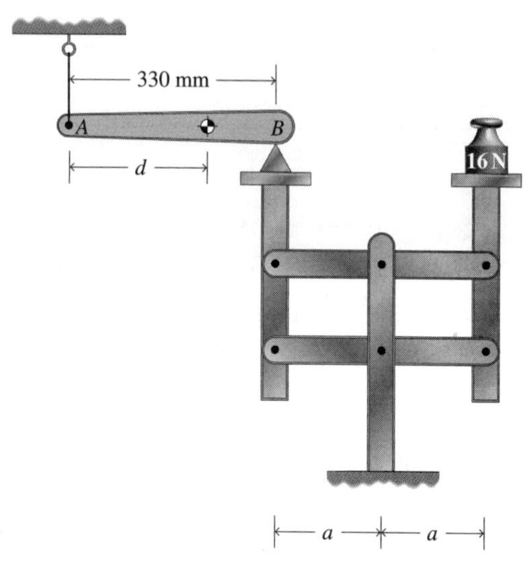

Figure P7.2

7.3 Figure P7.3 represents a safety valve of a steam boiler. The weight of the rod is negligible compared to the weight W. The cross section of the valve perpendicular to the plane of the figure is circular. Determine the weight W such that the valve will just start to open at 300 lb/in² pressure.

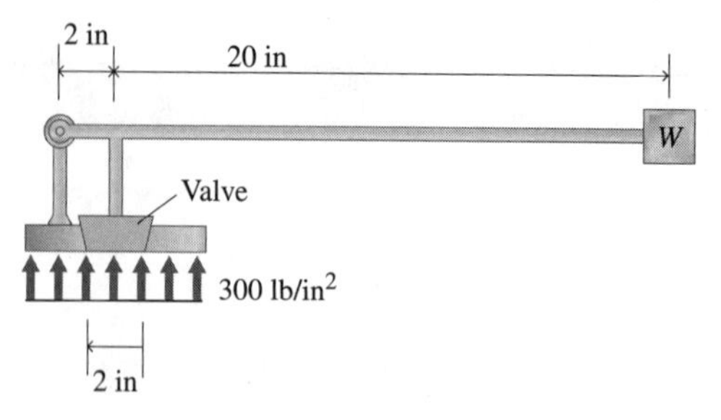

Figure P7.3

7.4 The uniform beam AB in Fig. P7.4 weighs 600 N/m. The tension in wire C with the turnbuckle is 900 N. Calculate the tensions in wires A and D.

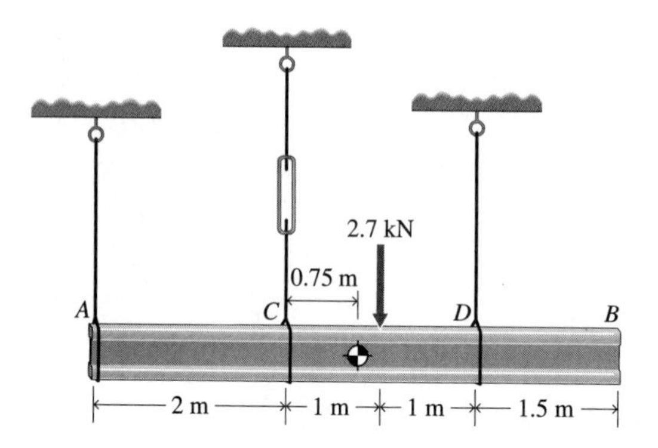

Figure P7.4

7.5 A force of 1200 lb is required at the cutter edge to cut a bolt (see Fig. P7.5). Determine the magnitude P of the forces that must be applied to the handles of the bolt cutter.

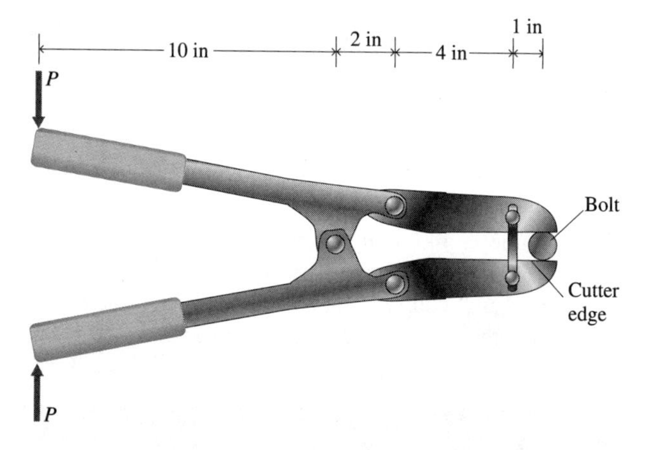

Figure P7.5

7.6 Archimedes said, "Give me a lever long enough and a place to stand and I can move the earth." If the earth weighs approximately 6.0×10^{26} N and Archimedes weighed 675 N, how long a lever (in kilometers) would he have needed to displace the earth 25 mm by turning the lever 30° about its fulcrum? Disregard the weight of the lever.

7.7 To apply different loads simultaneously to various bulkheads of an airplane wing, engineers use a "whiffletree" test jig, shown in Fig. P7.7. The load $P = 196$ kips, applied to the main cable, is to be distributed as loads (in kips) to bulkheads $A, B, \ldots, H,$ as follows: A (9), B (13), C (18), D (22), E (27), F (31), G (36), and H (40). Determine the required lever arms $a_1, a_2, a_3, a_4, a_5, a_6, a_7, b_5, b_6,$ and b_7.

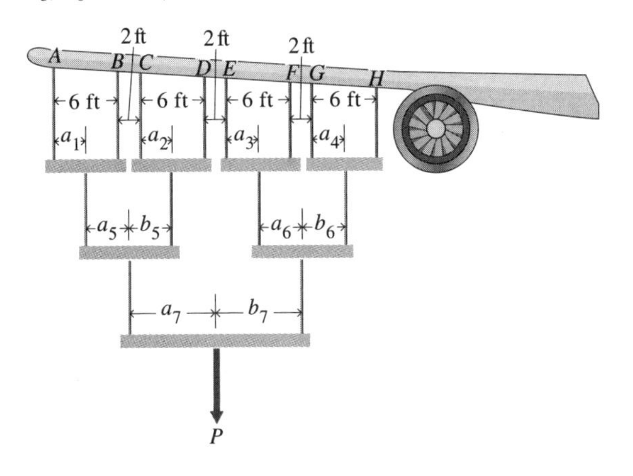

Figure P7.7

7.8 Figure P7.8 is a schematic diagram of a hydraulic lift mechanism. The cross section of the cylinder is circular. The small smooth roller B maintains contact with the rod AC and the surface DE. Determine the weight W that can be held in the designated position, if the pressure p in the cylinder is 3 N/mm². Neglect friction.

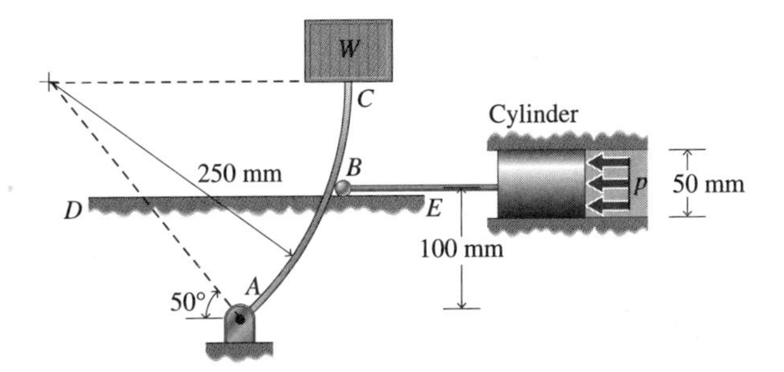

Figure P7.8

7.9 A load of 100 lb is applied to the handle of the clamping mechanism in Fig. P7.9. Calculate the force F of the clamp at A.

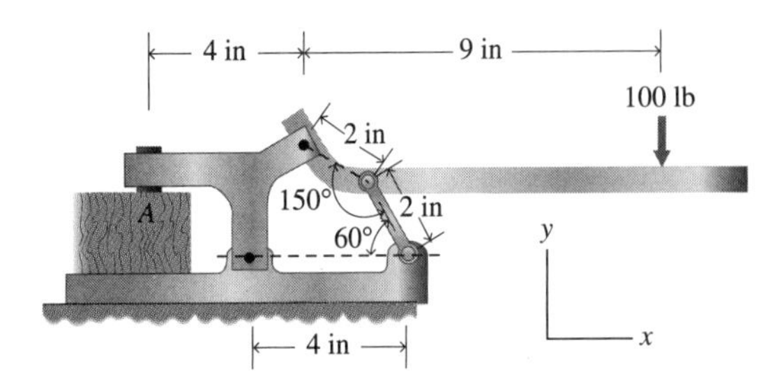

Figure P7.9

7.10 The mechanical advantage of the bolt cutter of Problem 7.5 is $m = 20$. Modify its design to ensure that the mechanical advantage is at least 22.

Method: From the solution to Problem 7.5, identify the parameters (dimensions) that affect the mechanical advantage. Then, make appropriate adjustments to ensure that $m \geq 22$.

7.11 The mechanical advantage of the bolt cutter of Problem 7.5 is $m = 20$.

a. Suggest two different design modifications for the cutter, each of which ensures a mechanical advantage of $m = 25$.

b. In a short paragraph, compare the advantages and disadvantages of the two new designs.

Method: From the solution to Problem 7.5, identify the parameters (dimensions) that affect the mechanical advantage. Then, make appropriate adjustments in each of the new designs to ensure that $m \geq 25$.

7.12 The mechanical advantage of the clamping device of Problem 7.9 is $m = 7.65$. Suggest an alternative design that increases the mechanical advantage to $m = 10$, subject to the requirement that the 9 in dimension (see Fig. P7.9) is not to be changed by more than 1 in.

Method: From the solution to Problem 7.9, identify the parameters (lengths and angles) that affect the mechanical advantage. Adjust the parameters to increase m to 10.

7.2 *The Pulley*

7.13 Neglect friction and the weights of the pulleys in the six pulley systems shown in Fig. P7.13. Determine the magnitude P of the force required to maintain equilibrium for each system.

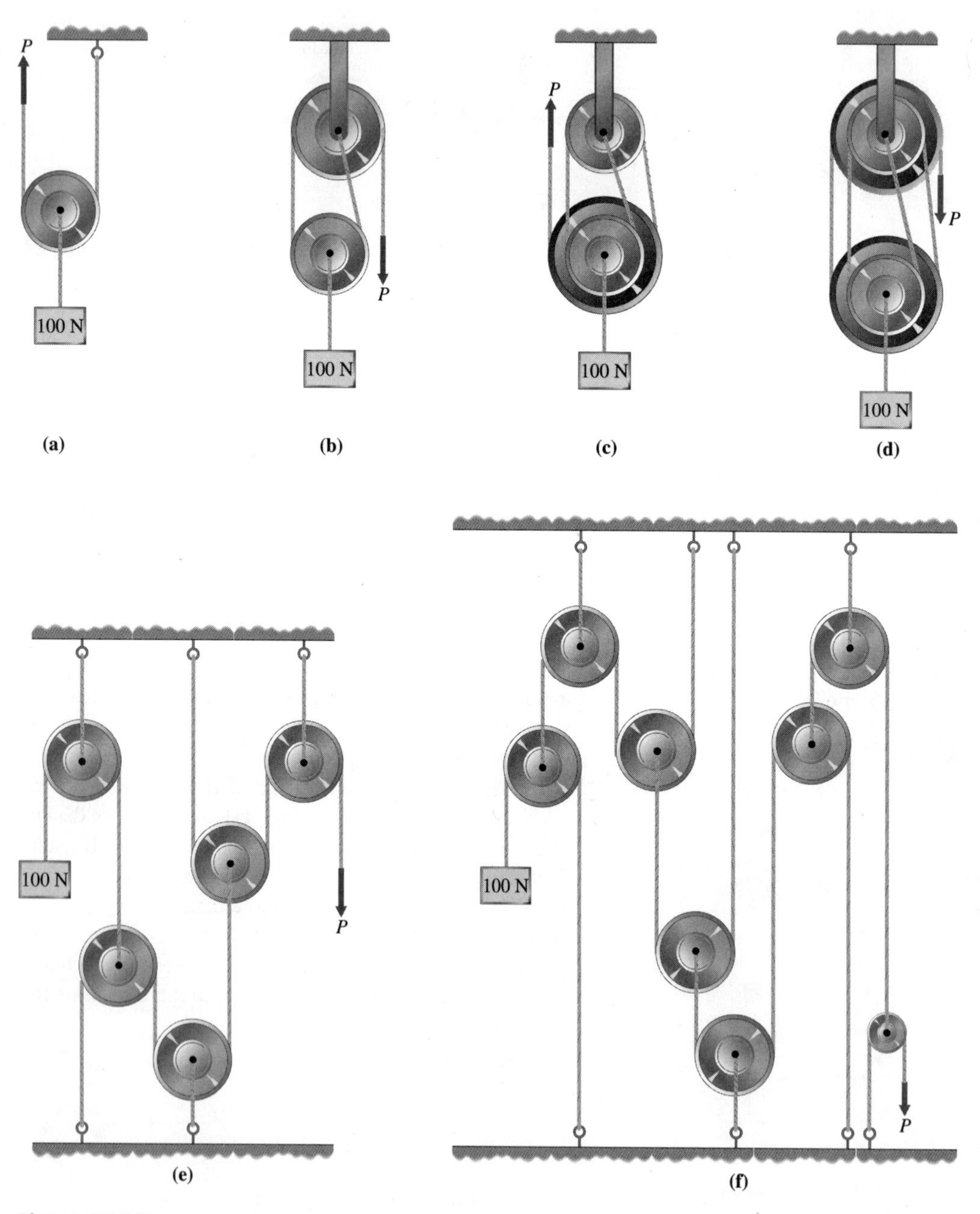

Figure P7.13

7.14 The device shown in Fig. P7.14 is called *Archimedes' system of pulleys.* Calculate the mechanical advantage of the system; that is, determine the ratio W/P. Neglect friction.

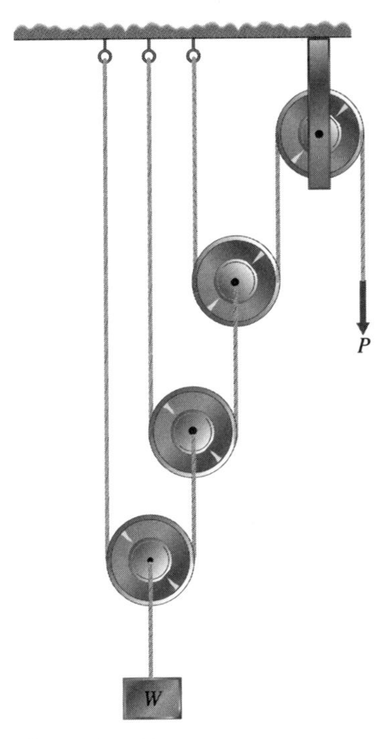

Figure P7.14

7.15 The device shown in Fig. P7.15 is a differential pulley. The chain is suspended from two sprockets that are connected and turn as a unit. The chain hangs slack from the right side of the smaller sprocket.

a. Draw separate free-body diagrams of the sprocket unit and the lower pulley.

b. Show that the mechanical advantage of the device is $W/P = 2R/(R - r)$.

Figure P7.15

7.16 The pulley-truss system shown in Fig. P7.16 supports a load of 2 kN. Determine the forces in members AB, BE, and DE.

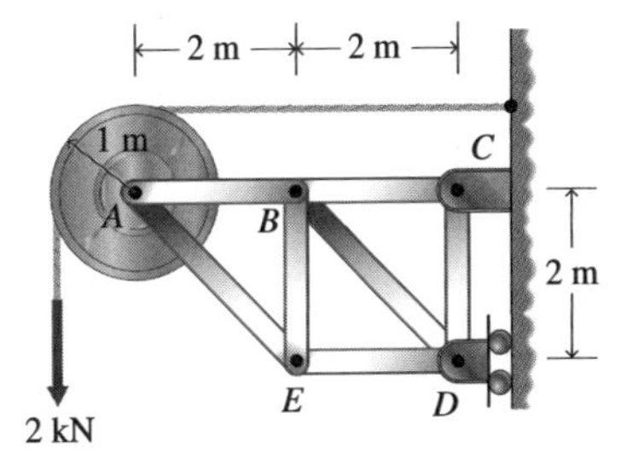

Figure P7.16

7.17 Assume that each of the pulleys of the pulley systems in Fig. P7.13 weighs 10 N. Determine the magnitude P of the force required to maintain equilibrium for each system.

7.18 For a modified version of the hydraulic elevator of Example 7.3, determine the weight of the cage if the piston B is fixed and the cylinder A is free to move, as shown in Fig. P7.18.

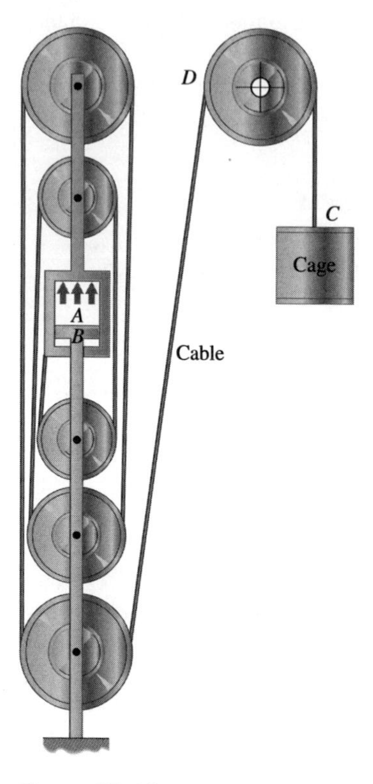

Figure P7.18

7.19 A cargo box that weighs 300 N is supported by a pulley-rope system, as shown in Fig. P7.19. The pulleys may be assumed to be frictionless, and the weights of the ropes are negligible. The box remains horizontal while it is lifted. Determine the pull P required to support the box under each of the following conditions:

a. The weights of the pulleys are negligible.

b. Each pulley weighs 15 N.

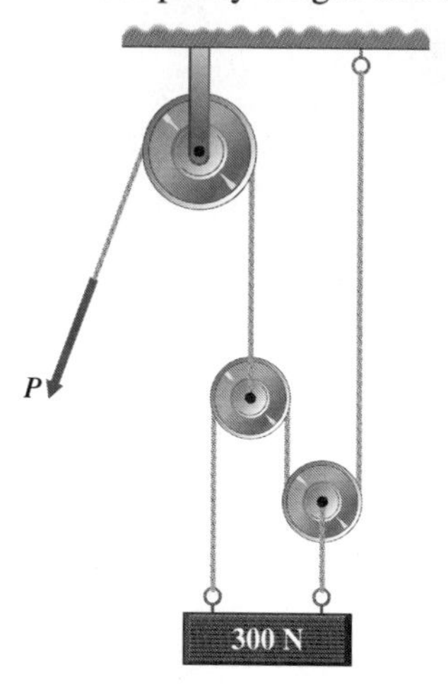

Figure P7.19

7.20 Compute the pull P required to raise the 4000 lb weight with each of the two block and tackles shown in Fig. P7.20.

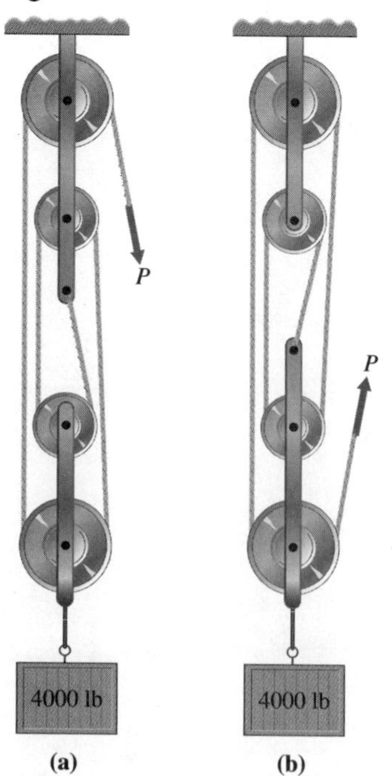

Figure P7.20

7.21 The pulley-cord system shown in Fig. P7.21 supports a horizontal slab that weighs 7 kN. The weights of the pulleys and cords are negligible, and the pulleys are frictionless. Determine the tension in each of the five cords.

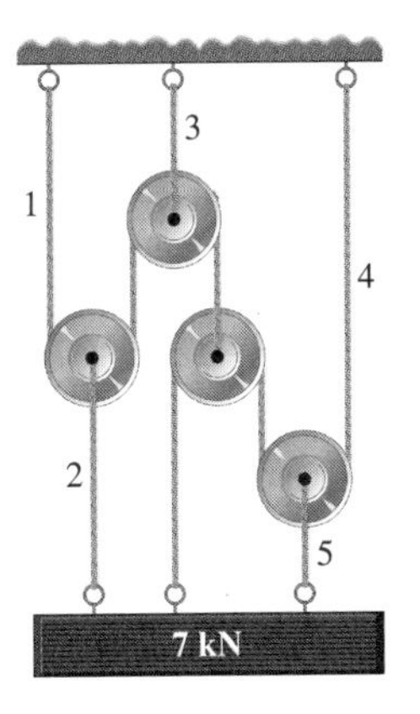

Figure P7.21

7.22 A construction worker needs to position a work platform. The worker weighs 200 lb, and the platform weighs 100 lb. Determine the pull the worker must exert to maintain equilibrium for each arrangement shown in Fig. P7.22.

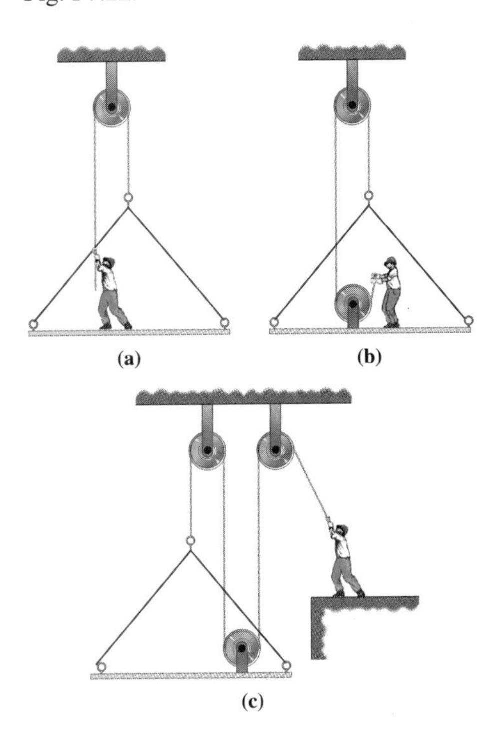

Figure P7.22

7.23 In Fig. P7.23, the horizontal rope *AB* keeps the segment *BC* vertical. Calculate the system's mechanical advantage, $m = W/P$. Neglect friction and the weight of the free pulley *D*.

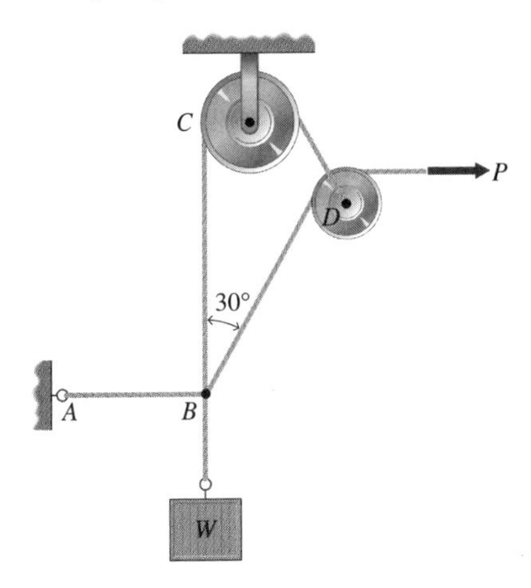

Figure P7.23

7.24 Design a system of light, frictionless, single-sheave pulleys and ropes with which a person, with minimum effort, can slowly raise a 1600 N crate. Assume that the system can be hung from an overhead support structure.

a. Design a system using two pulleys.

b. Design a system using three pulleys.

c. Explain your solutions, and estimate whether the average person could raise the crate using each design.

7.25 The pulley-truss system shown in Fig. P7.25 supports a 400 N load. Determine the forces in members *BF* and *EF* of the truss. Neglect the weight of the pulley.

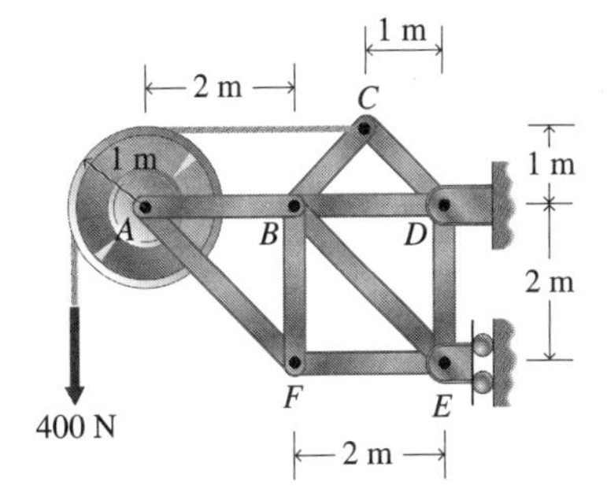

Figure P7.25

7.26 The pulley-truss system shown in Fig. P7.26 supports a load of 2 kips. Determine the forces in members *AB*, *ED*, and *EC*.

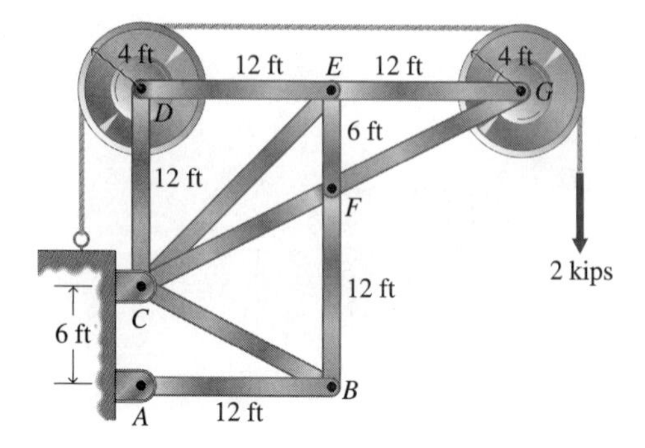

Figure P7.26

7.27 a. Determine the pull P required to support the 7500 lb weight in Fig. P7.27, assuming that the bar AB remains horizontal.

b. Let the diameter of each pulley be 2 ft. Determine the distance x such that bar AB remains horizontal. Neglect friction and the weight of the bar.

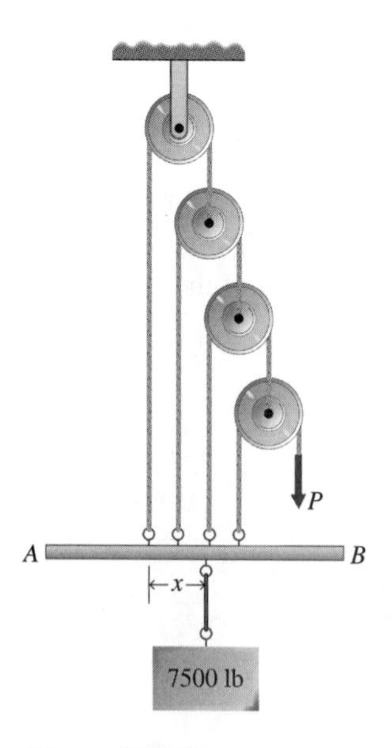

Figure P7.27

7.28 A friend of yours (who happens to be an engineer) and his brother want to pull a 250 lb engine block out of a 1957 Chevy. Your friend correctly determines that the roof truss in his garage can safely support a 400 lb load. He hangs a simple pulley from the truss. Then he and his brother tie a rope to the engine block and thread it through the pulley. They are just ready to start pulling on the rope when you show up. Your friend gives you the details of his analysis and the weight of the engine. How do you respond?

a. You help your friend and his brother pull the engine block.

b. You explain to your friend the potential risks of his plan.

c. You get out of the garage before it is too late.

Explain your choice.

7.29 For the situation in Problem 7.28, design a block and tackle that will allow your friend and his brother to pull the engine from the car safely.

7.3 *Simple Plane Frames*
7.4 *Mechanisms*

7.30 The cylinder in Fig. P7.30 weighs 1 kip, and the weight of the pin-connected frame is negligible in comparison. Determine the (x, y) components of the forces on the frame *AEFG* at points *A*, *B*, and *D*. Assume that the system is frictionless, so that the forces on the cylinder at points *B*, *C*, and *D* are perpendicular to the cylinder's surface. Express your answers using unit vectors **i** and **j**.

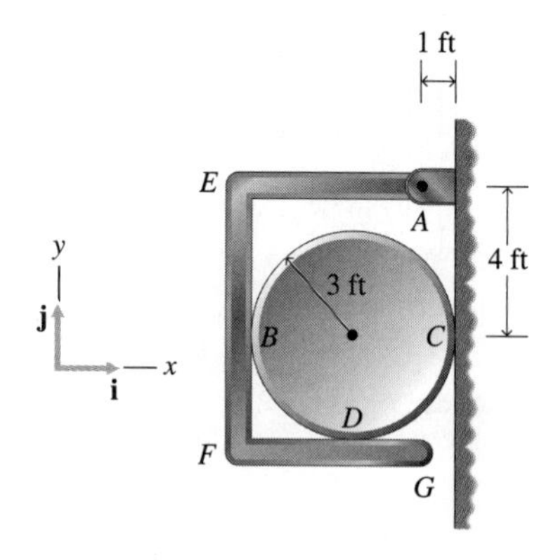

Figure P7.30

7.31 Assume that the uniform platform in Example 7.8 weighs 4 kips; that is, the weight acts at the midpoint of the platform (15 ft from points A and C).

a. Solve Example 7.8, including the weight of the platform.

b. Comparing the results of part a with those obtained in Example 7.8, discuss the significance of the weight of the platform.

7.32 The uniform bar $ABCDE$ in Fig. P7.32 weighs 1.6 kN.

a. Draw a free-body diagram of the bar.

b. Determine the support reactions acting on the bar.

c. Show the reactions with their correct senses on a free-body diagram of the bar.

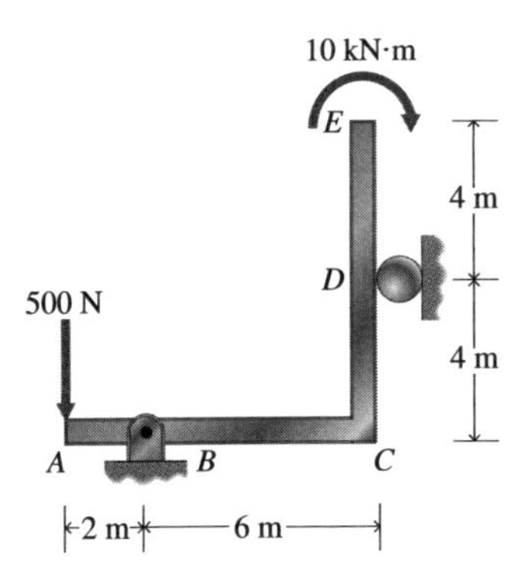

Figure P7.32

7.33 The uniform bar $ABDC$ in Fig. P7.33 weighs 10 lb/ft, and block E weighs 100 lb. Determine the tension in the cord AE and the forces acting on the bar at points A, D, and C. The pulley is frictionless and has negligible weight.

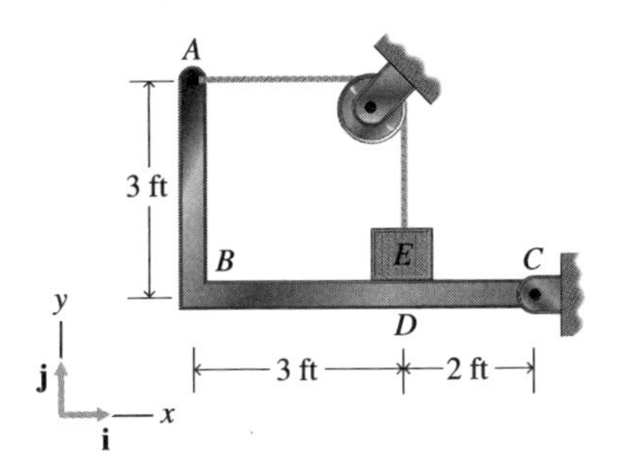

Figure P7.33

7.34 The uniform bar ABC in Fig. P7.34 weighs 400 lb and is subjected to a 300 lb force and a 1000 lb·ft couple.

a. Draw the free-body diagram of the bar.

b. Determine the reactions at A and C.

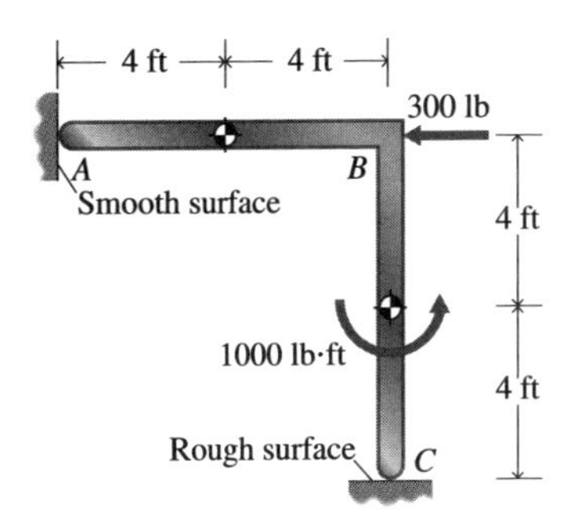

Figure P7.34

7.35 Determine the (x, y) projections of the forces exerted at points A, B, and C on member ABC of the structural mechanism shown in Fig. P7.35. Neglect friction and the weights of the pulley and bars. Express the forces using unit vectors **i** and **j**.

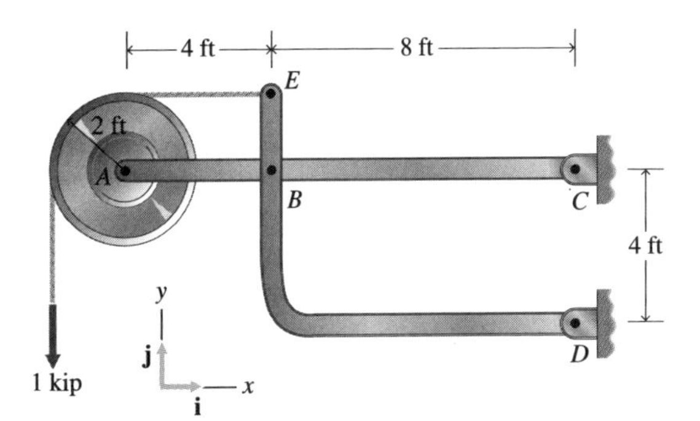

Figure P7.35

7.36 a. Draw free-body diagrams of bars *ABC* and *DEBF* and of pulley *C* of the structural mechanism shown in Fig. P7.36.
b. Determine the forces that act on member *ABC* at points *A, B,* and *C.* Express the forces using unit vectors **i** and **j**.

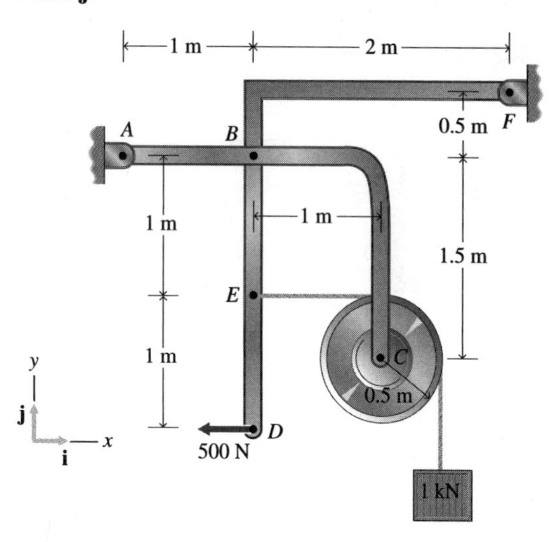

Figure P7.36

7.37 a. Solve Example 7.8 for the case in which $\frac{3}{4}$ of the load in the ore loader is carried by the wheels nearest *A* and $\frac{1}{4}$ is carried by the other wheels.
b. Solve Example 7.8 for the case where $\frac{1}{4}$ of the load is carried by the wheels nearest *A* and $\frac{3}{4}$ is carried by the other wheels.
c. Which arrangement do you think is preferable, based on the magnitudes of the forces in the members? Explain.

7.38 Assume that the uniform platform in Example 7.8 weighs 4 kips; that is, the weight acts at the midpoint of the platform (15 ft from points *A* and *C*). Solve Example 7.8, including the weight of the platform and using the following weight distributions for the ore loader:
a. Assume that $\frac{3}{4}$ of the load in the ore loader is carried by the wheels nearest *A* and $\frac{1}{4}$ is carried by the other wheels.
b. Assume that $\frac{1}{4}$ of the load is carried by the wheels nearest *A* and $\frac{3}{4}$ is carried by the other wheels.
c. Which weight distribution is preferable for the ore loader, based on the magnitudes of the forces in the members? Explain. Discuss the significance of the weight of the platform.

7.39 Determine the forces that act on member *FD* of the pulley-frame system shown in Fig. P7.39. Neglect friction and the weights of the pulley and bars.

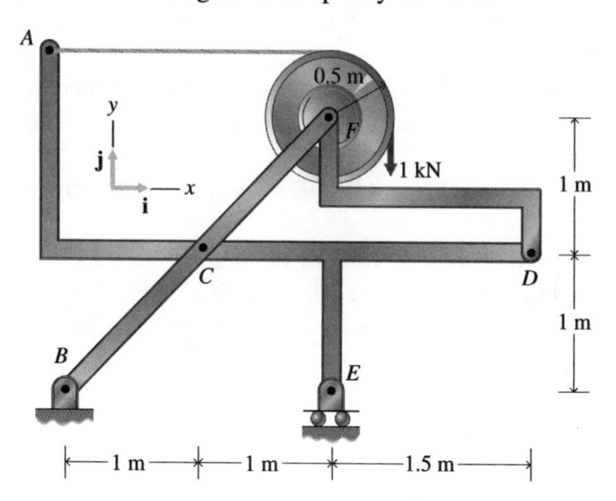

Figure P7.39

7.40 Assume that the uniform members *AB* and *BC* in Example 7.9 weigh 100 N/m each. Solve Example 7.9, including the weight of the members.

7.41 The uniform bars *AC* and *BC* in Fig. P7.41 weigh 225 N and 450 N, respectively. Determine the force in cable *EF.* Neglect friction.

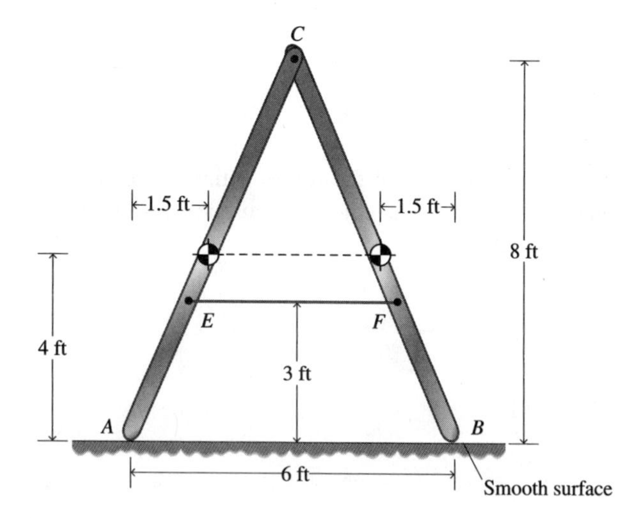

Figure P7.41

7.42 Determine the (*x*, *y*) components of the forces exerted on all the members at all the joints of the frame shown in Fig. P7.42. Member *ABC* is continuous from *A* to *C*. Express your answers using unit vectors **i** and **j**.

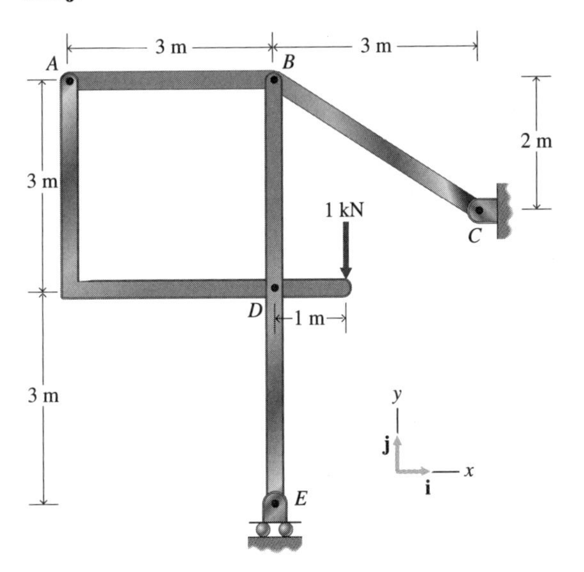

Figure P7.42

7.43 Determine the (*x*, *y*) projections of the forces that act on member *EFC* of the frame shown in Fig. P7.43. Members *ABCD* and *EFC* are continuous members.

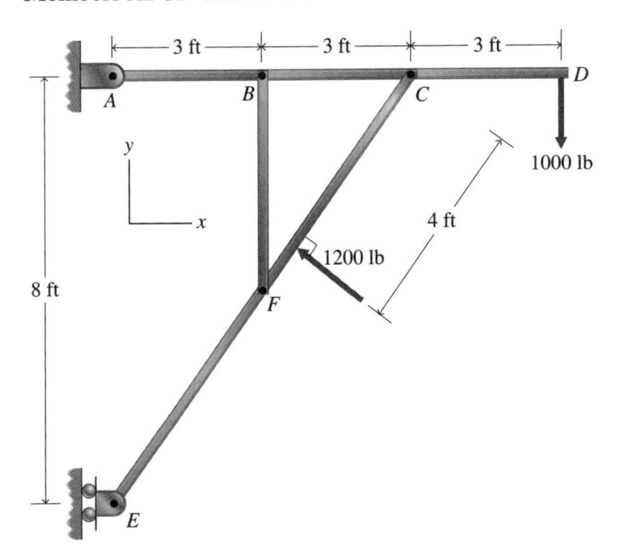

Figure P7.43

7.44 Determine the forces that act on each of the members of the plane frame shown in Fig. P7.44. Express the forces using unit vectors **i** and **j**.

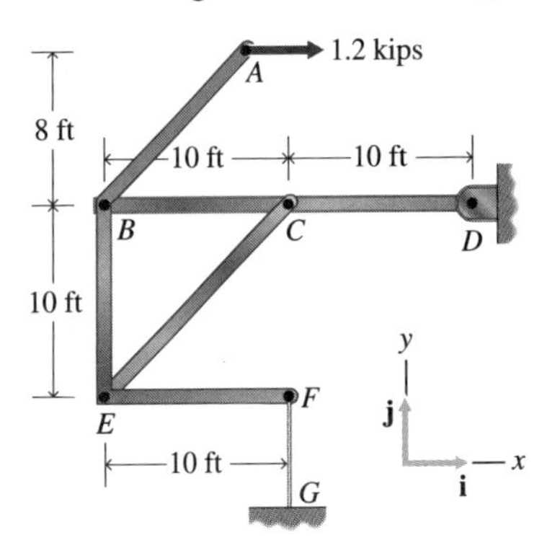

Figure P7.44

7.45 a. Calculate the tension *T* in the rod *BD* and the forces that act on each member of the pin-connected structure in Fig. P7.45.
b. According to the definition in the text, is the structure a frame? Explain.

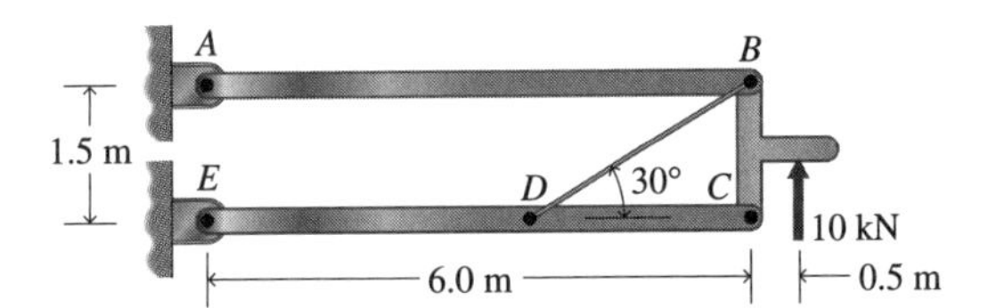

Figure P7.45

7.46 Solve Problem 7.45 for the case in which the 10 kN vertical load is applied directly to the pin of joint *C*.

7.47 Determine the force that acts at point *C* on member *ABCDE* of the structural mechanism shown in Fig. P7.47. Express your answer using unit vectors **i** and **j**.

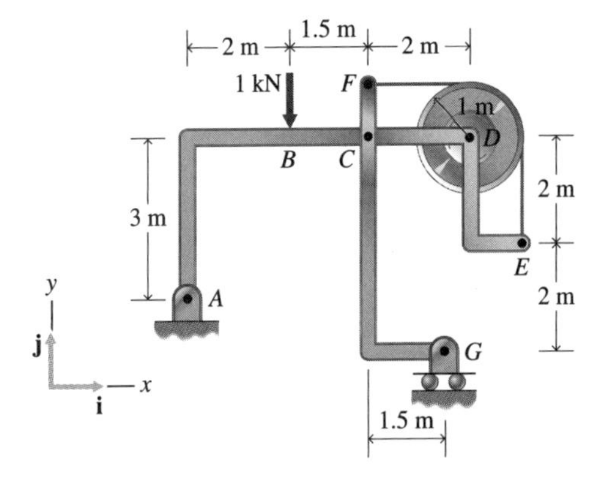

Figure P7.47

7.48 a. Derive formulas, in terms of F, a, and θ, for the reactions A_x and A_y of the support at A and for the force P in member BC of the mechanism in Fig. P7.48.
b. For $F = 1$ kN and $a = 1$ m, plot A_x, A_y, and P as functions of θ, for $0° \leq \theta \leq 360°$.

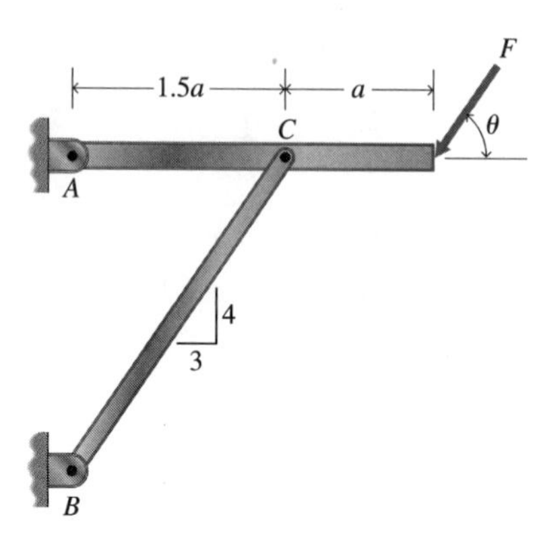

Figure P7.48

7.49 a. Compute the forces that act on members AB and BC of the structure shown in Fig. P7.49.
b. According to the definition in the text, is this structure a frame? Explain.

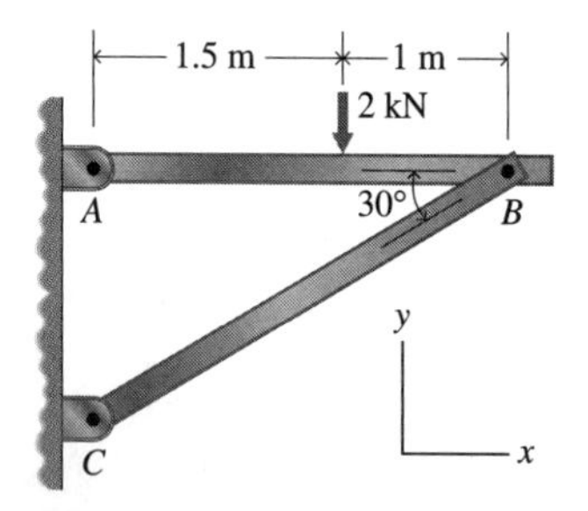

Figure P7.49

7.50 The 2 kN load in Fig. P7.49 is lowered vertically and applied to member BC.
a. Compute the forces that act on members AB and BC, and compare them to the forces obtained in Problem 7.49.
b. Do these results violate the principle of transmissibility? Explain.

7.51 a. Determine the forces that act on the two parts of the arch bridge shown in Fig. P7.51. Assume that points A and C are hinged.
b. According to the definitions in the text, is this structure a frame or a structural mechanism? Explain.

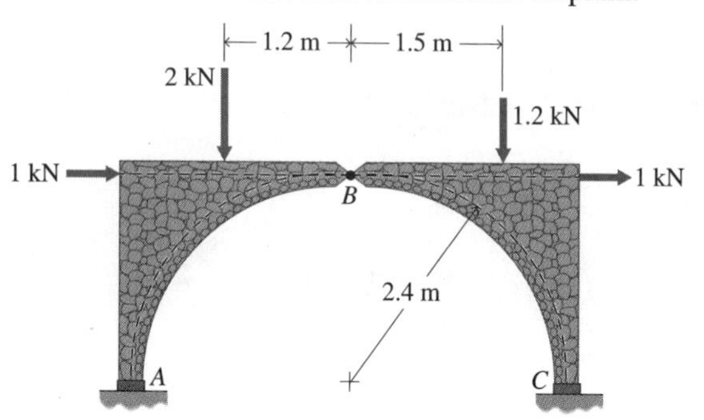

Figure P7.51

7.52 Determine the (x, y) components of the forces exerted on all the members at all the joints of the pulley-frame system shown in Fig. P7.52. Express your answers using unit vectors **i** and **j**.

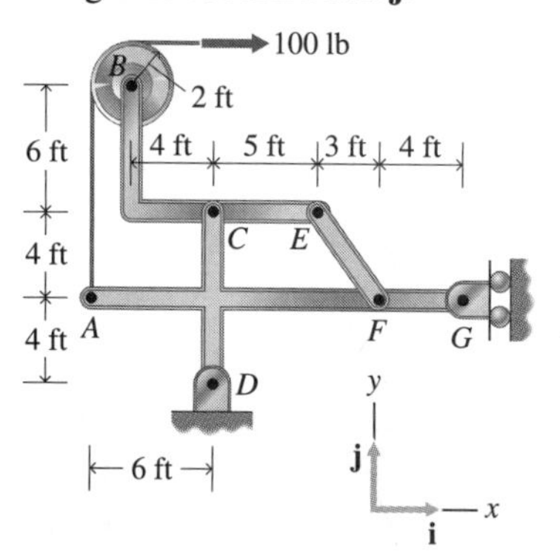

Figure P7.52

7.53 a. The semicircular arch in Fig. P7.53 is in equilibrium. Determine the force T in tie rod AB and the (x, y) projections of the forces that act on members AC and BC, which are hinged at C.
b. According to the definitions in the text, is this structure a frame or a structural mechanism? Explain.

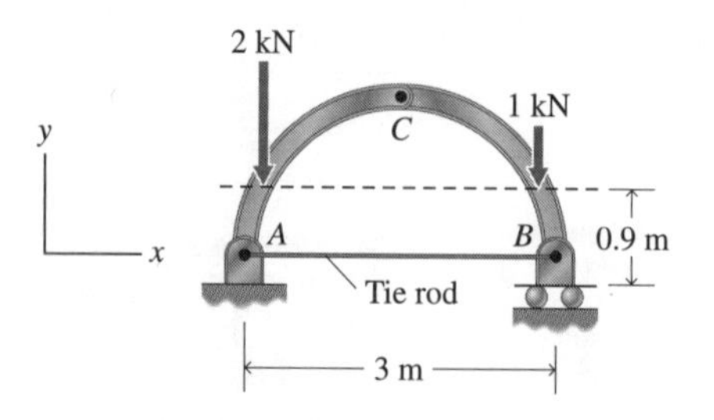

Figure P7.53

7.54 a. Calculate the forces that act on member BCF and the forces in members AD and DE of the engine mount in Fig. P7.54.

b. According to the definition in the text, is this structure a frame? Explain.

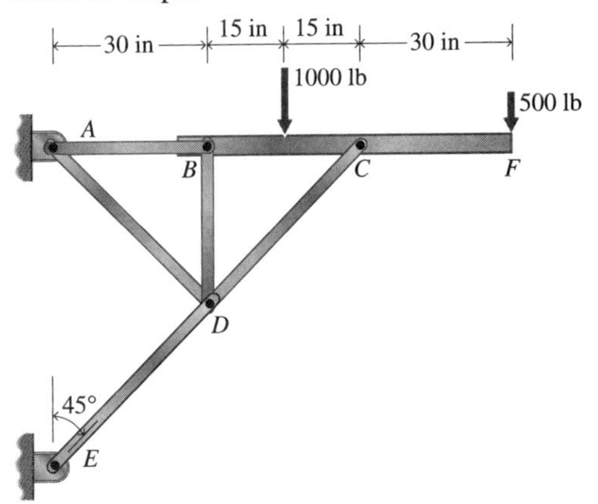

Figure P7.54

7.55 In terms of F and θ, derive expressions for the reactions A_x and A_y of support A and for the thrust force P in member BC of the structural mechanism in Fig. P7.55.

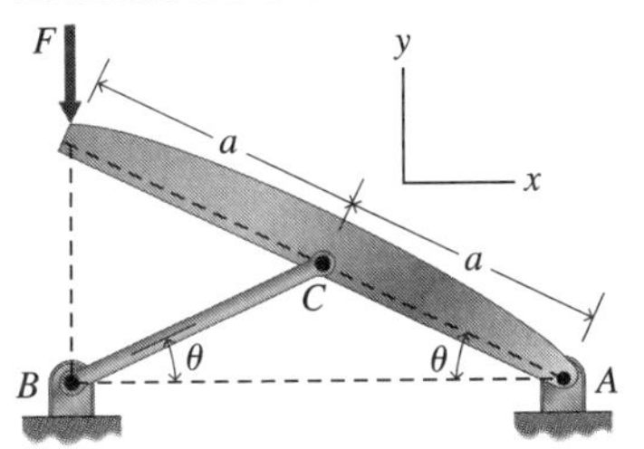

Figure P7.55

7.56 The tension in the diagonal wire of the structure in Fig. P7.56 is 3 kN. Members AB and CD are parallel and have equal lengths.

a. Determine the magnitude of the vertical load F.

b. According to the definition in the text, is the structure a structural mechanism? Explain.

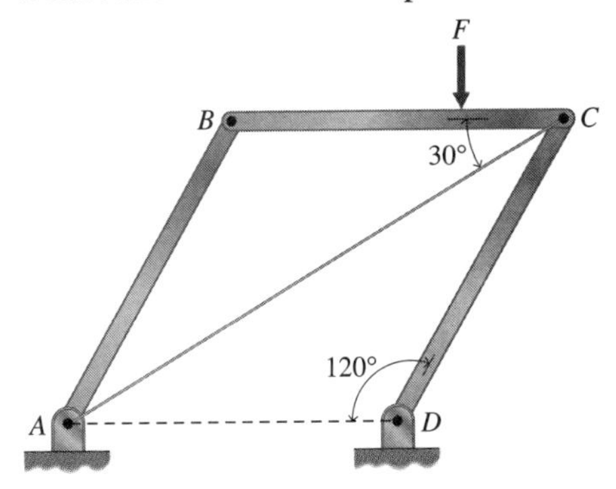

Figure P7.56

7.57 Determine the force in member CB of the pin-connected structural mechanism shown in Fig. P7.57.

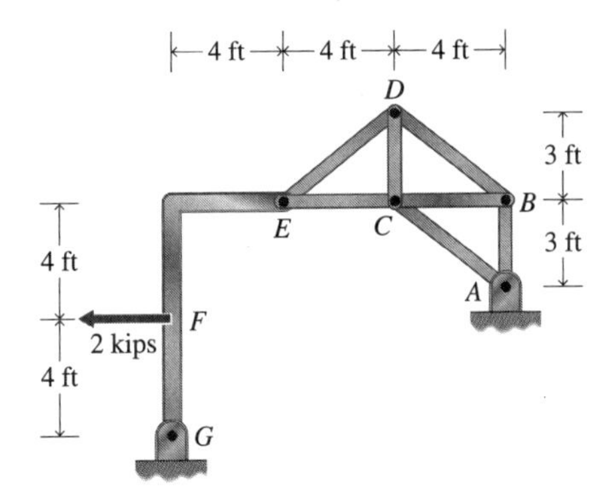

Figure P7.57

7.58 Determine the (x, y) projections of all the forces that act on the members of the wall frame in Fig. P7.58.

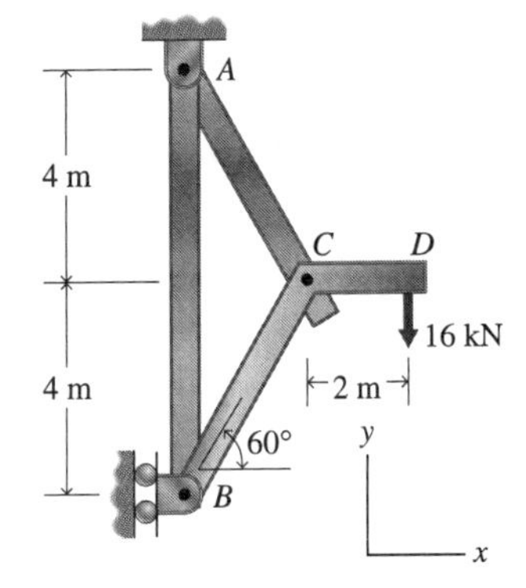

Figure P7.58

7.59 Determine the forces exerted on each of the members of the structure shown in Fig. P7.59 at points A, B, C, D, and E. Express the forces using unit vectors **i** and **j**.

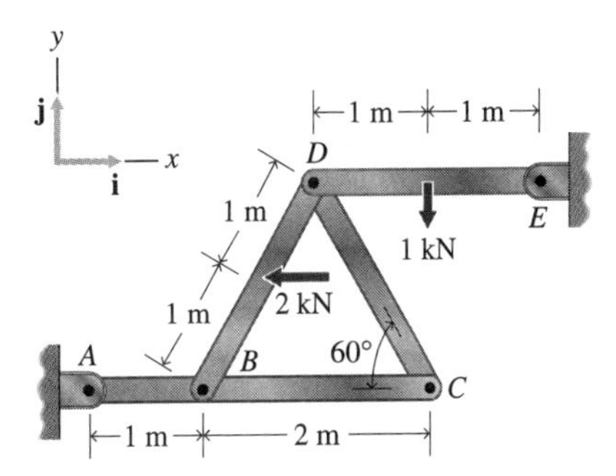

Figure P7.59

7.60 The uniform bars *ADE, BC,* and *CD,* in Fig. P7.60, each weigh 400 N/m. The hinges are frictionless. The system is loaded by its own weight.

a. Calculate the tension in the wire *CE.* Neglect the weight of the wire.

b. According to the definition in the text, is the structure a frame? Explain.

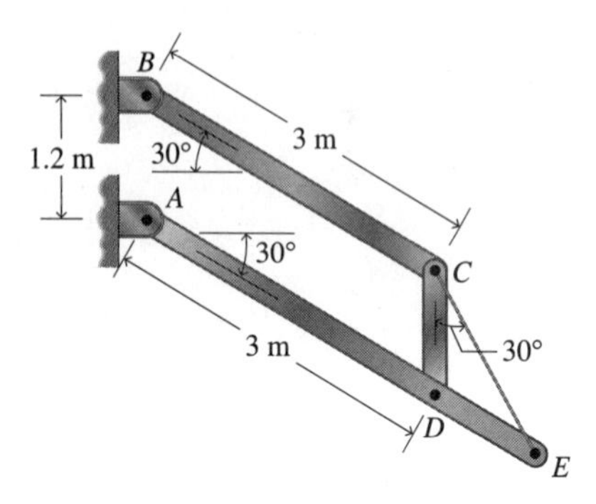

Figure P7.60

7.61 a. Design a three-member wall frame that can carry a 16 kN load but has a different structural arrangement from that in Fig. P7.58. Use the same support locations, *A* and *B,* and place the vertical load at the same horizontal distance from the supports.

b. Determine the (x, y) projections of all the forces that act on the members of your wall frame.

c. In a brief paragraph, discuss the advantages and disadvantages of your design relative to that shown in Fig. P7.58.

(*Hint:* Use the solution to Problem 7.58. Make sure that your design represents a frame and not a mechanism.)

7.62 Place a concentrated force **P** at a point other than at joint *E* of the frame in Fig. P7.43 so that the support reaction at *E* is reduced by 50%.

Method: Place a concentrated load **P** at some point on the frame and determine its effect on *E.* Repeat this process, placing **P** at another point or adjusting its magnitude or orientation, until you succeed in reducing the support reaction at *E* by 50%.

7.63 The pin *D* of the structure in Fig. P7.45 may be placed anywhere along member *CE.* Your supervisor assigns you the task of determining the "best" location for *D* and giving the reasons for your choice.

Method: Determine the forces in the member *AB* and the forces that act on the pins *C, D,* and *E.* Tabulate or plot these forces as a function of the distance x from pin *E* to pin *D.* Based on this data, choose a location for *D.*

7.64 For the structure in Fig. P7.59, assume that the 1 kN and 2 kN forces are removed and a vertical downward force $P = 10$ kN is applied at *C.* Members *BD* and *CD* will fail (buckle or collapse) if they are subjected to a compressive force greater than 13 kN. It is your task to determine whether the structure is safe under the new conditions. If it is not, redesign it so that it is safe. The design requirements are that the length of *AB* of member *ABC* be kept at 1 m, that the lengths of *BD* and *CD* be kept equal, and that *DE* remain horizontal, although its length may be changed. The distance between points *B* and *C* of member *ABC* may be changed.

Method: Calculate the forces in members *BD* and *CD* of the original design to determine whether they will collapse under the new loading. If either member will collapse, redesign the structure so that it will not collapse.

7.65 Assume that an average soda can can be crushed under a load of 240 lb. Design a can crusher that the average adult can operate. To minimize shipping costs, your can crusher must be as compact as possible. Thus, a single first- or second-class lever is not likely to be a good choice. (*Hint:* See Figs. P7.9 and P7.55 for examples of mechanisms that might be good starting points.)

7.66 For the mechanism of Fig. P7.48, assume that the force F varies in direction over the range $-135° \leq \theta \leq 45°$.

a. Plot the ratio of the force in member *BC* to the force F as a function of θ, in increments of 10°.

b. The member *BC* can withstand a maximum tension of 50,000 lb and a maximum compression of 35,000 lb. What is the maximum force F that the structure can support?

7.67 The snips in Fig. P7.67 exert a force F on a rod located $\frac{3}{2}b$ (in mm) from pin D. The two handles are joined by a pin at C. Link AB is pinned to the lower handle at A. It is not attached to the upper handle.

a. Plot the mechanical advantage $m = F/P$ (output/input) as a function of b for the range $12 \text{ mm} \leq b \leq 35$ mm, where P is the gripping force and b is the horizontal distance from the link AB to pin C.

b. From the results of part a, choose a value of b that gives the largest mechanical advantage, and explain your choice.

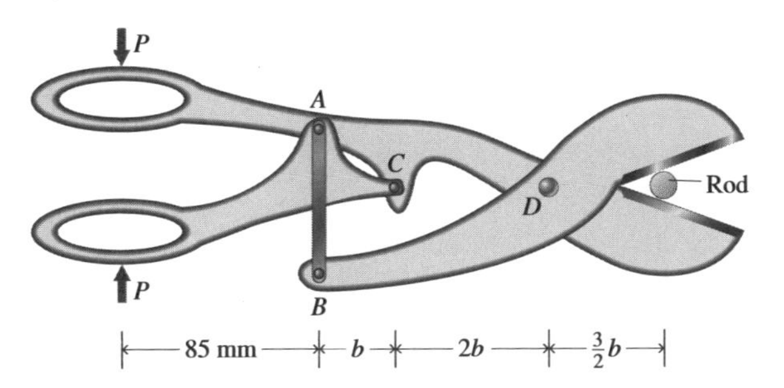

Figure P7.67

7.5 *Parabolic Cables and the Suspension Bridge*

7.68 A water conduit that crosses a canyon is suspended from a cable, as shown in Fig. P7.68. The conduit, when full of flowing water, weighs 27 kN per meter of length. Calculate the maximum tension in the cable.

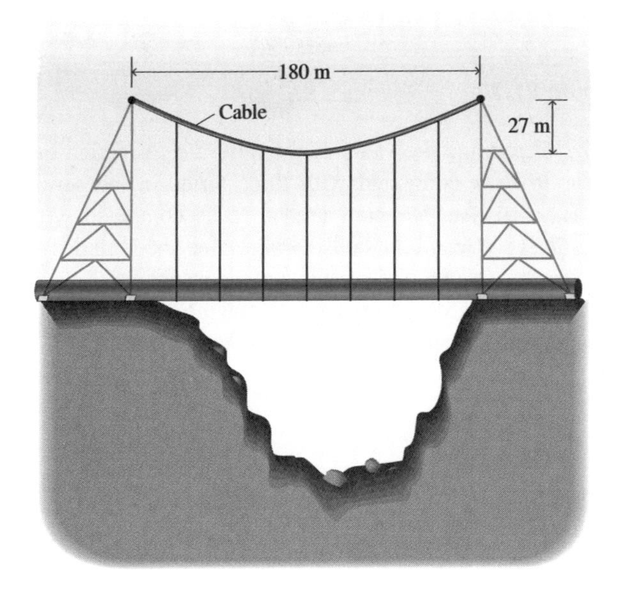

Figure P7.68

7.69 A wire weighs 1.50 N/m. It is suspended from its ends, which are 9 m apart and at the same level. The sag is 75 mm. Assume that the drape of the wire is parabolic.

a. Draw a sketch of this cable.

b. Calculate the tension at either end.

7.70 As shown in Fig. P7.70, a suspension cable supports a horizontal pipeline that weighs 3 kN/m. The cable is supported at the points (−12 m, 3 m) and (24 m, 12 m).

a. Determine the maximum tension in the cable.

b. Determine the length of the cable.

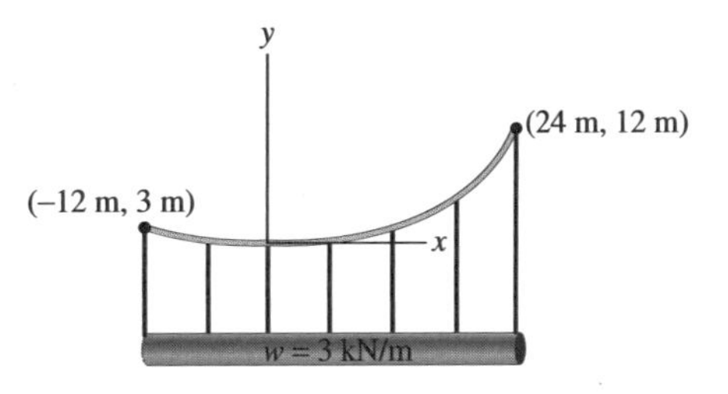

Figure P7.70

7.71 The span of a suspension bridge is 1200 ft. The roadway is suspended from two cables that are supported at the tops of towers at the ends of the span. The sag in each cable is 225 ft. The weight of the roadway is 6 kips/ft. The maximum allowable tension in each cable is 15 ksi (kips per square inch) of cross section.

a. Draw an accurate sketch of the bridge.

b. Calculate the required diameter of the cables.

7.72 A cable has a maximum allowable tension of 160 kN. The ends of the cable are anchored at the same elevation. The cable must support a uniform horizontal load over a span of 100 m, with an allowable sag of 20 m.

a. Draw a sketch of the cable.

b. What is the maximum load (in kN/m) the cable can support?

c. What length of cable is required?

7.73 A cable supports a uniform load of 6 kips/ft and is anchored at A and B (see Fig. P7.73). At A, the cable is horizontal.

a. Determine the minimum tension in the cable.
b. Determine the maximum tension in the cable.
c. Determine the length of the cable.
d. Determine the slope at B.
e. Determine the maximum distance s_{max} from the line AB to the cable.

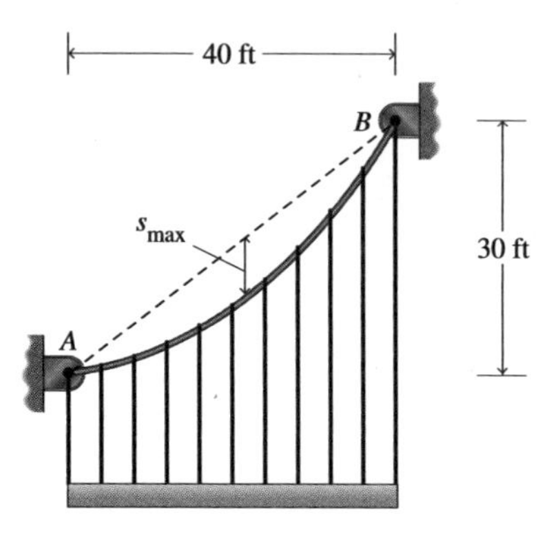

Figure P7.73

7.74 After the cable of Problem 7.72 was constructed, it was discovered that one end of the cable had been placed 6 m above the other, but the sag measured from the lower support had been maintained at 20 m. Solve Problem 7.72 for the as-constructed conditions.

7.75 Let the slope of the cable in Fig. P7.73 be 20° at support A.

a. Determine the maximum tension in the cable.
b. Determine the length of the cable.
c. Determine the maximum distance s_{max} from line AB to the cable.

7.76 A cable that weighs 360 N hangs in an arc that is approximately parabolic. With reference to (x, y) coordinates that lie in the plane of the cable (with the y axis directed upward), the end points of the cable are (0, 0) and (60 m, 30 m). Also, the point (30 m, 13.5 m) lies on the parabolic curve.

a. Draw a free-body diagram of the entire cable.
b. Determine the tensions T_0 and T_1 at the ends of the cable by applying equilibrium of forces.

(*Hint:* The general equation of a parabola that opens upward is $y = ax^2 + bx + c$, and the slope is the derivative of y.)

7.77 A pedestrian bridge in South America crosses a deep canyon (see Fig. P7.77). The bridge deck weighs 150 N/m and is suspended from two parabolic cables supported at their ends by towers whose elevations are 1350 m and 1365 m, respectively. Pedestrian traffic adds an additional 180 N/m to the deck load. The maximum allowable tension in a cable is 50 kN.

a. Determine the minimum length of each cable, assuming that there is no restriction on sag.
b. Determine the vertical force in each pair of towers and the tension in each of the tie-back cables.

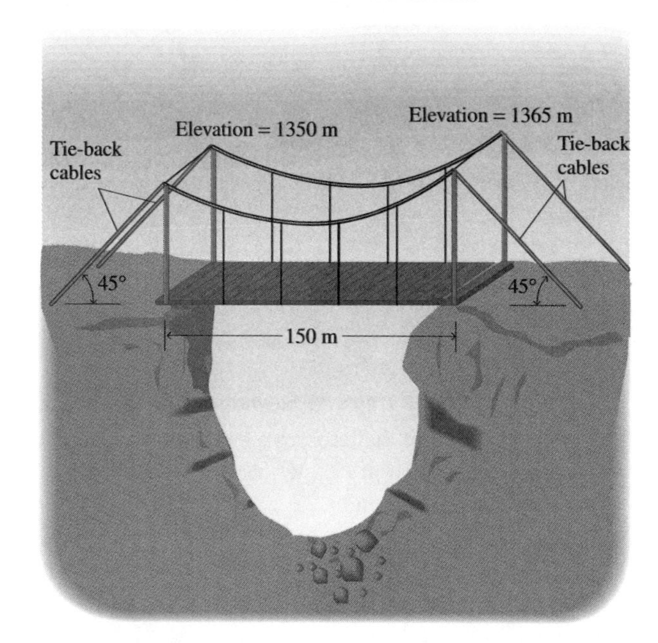

Figure P7.77

7.78 Determine the value of the ratio d/a for which the error in L as computed with three terms in the series of Eq. (7.12) is 5% compared to L as computed with Eq. (7.11). (Generally, considering other uncertainties—such as the stretch of the cable and the actual distribution of the load—a 5% difference in L may be acceptable in a design using a cable.)

7.79 A bridge cable is to be designed to carry a horizontally distributed load of $w = 30$ kN/m. It is to span a horizontal distance of 300 m, with its ends supported by towers at the same elevation (see Fig. P7.79). Determine the sag and the corresponding length of the cable for maximum cable tension T_{max} in the range 10 000 kN $\leq T_{max} \leq$ 20 000 kN.

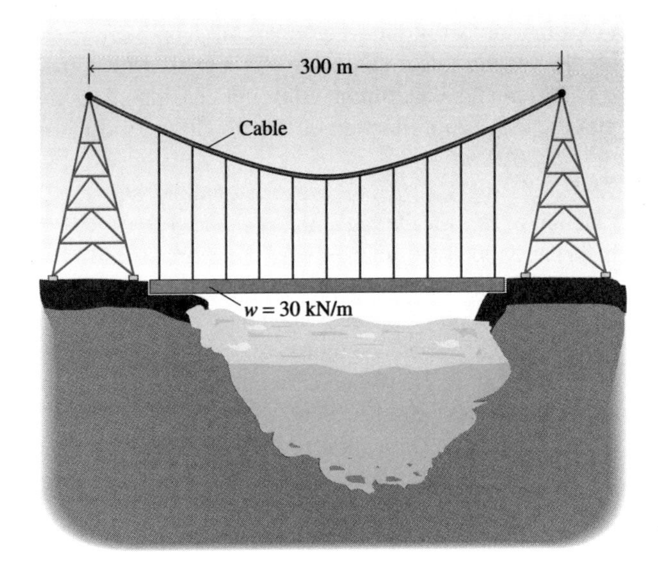

Figure P7.79

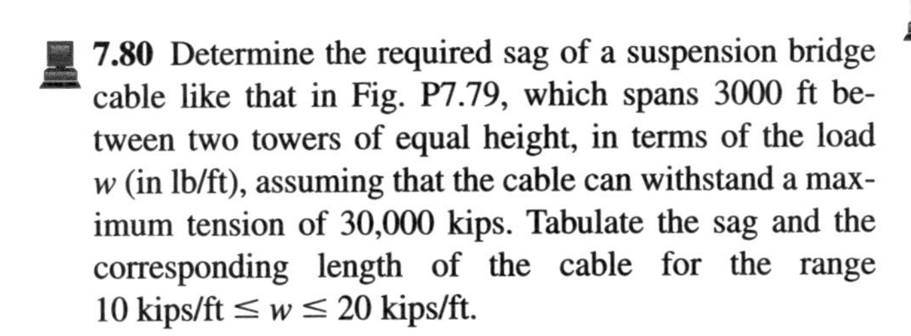

7.80 Determine the required sag of a suspension bridge cable like that in Fig. P7.79, which spans 3000 ft between two towers of equal height, in terms of the load w (in lb/ft), assuming that the cable can withstand a maximum tension of 30,000 kips. Tabulate the sag and the corresponding length of the cable for the range 10 kips/ft $\leq w \leq$ 20 kips/ft.

7.6 *The Catenary—A Freely Hanging Cable*

7.81 Show that if the lowest point of a catenary lies at the origin of the xy axes, then $C = 0$ and $K = -H$. Refer to Eq. (7.17).

7.82 A construction crew hangs an electrical power line consisting of 4-gauge copper wire between two equal-height poles that are 100 m apart. Winching the line over a pulley produces a tension of $T = 3$ kN in the line at the poles. The workers observe that the line makes an angle of approximately 2° with the horizontal at the poles.

a. Determine the minimum tension H in the line. (For such a small angle, you should expect very little difference between T and H.)

b. Specifications call for the minimum tension in the line to be no less than 50% of the tension at the poles. What is the maximum allowable angle that the cable can form with the horizontal at the poles?

7.83 A steel wire that weighs 500 N is 750 m long. It is hung between two points that are 720 m apart and at the same elevation. Determine the maximum and minimum tensions in the wire.

7.84 An electrical power line that is 20 meters long is to be placed between two towers at the same elevation. The line weighs 100 N/m. Its maximum sag is restricted to 6 meters. Determine the minimum allowable distance between the towers.

7.85 The towers that support an electrical power transmission line are at the same elevation, and they are 750 m apart. The wire weighs 3 N/m. The maximum allowable tension in the wire is 4500 N. Calculate the length of the wire and the sag that will produce this tension.

7.86 A cable 90 m long is suspended between points (0, 0) and (30 m, 45 m), relative to xy axes.

a. Make a sketch of the cable.

b. Write an expression for K/H from the equation of the curve of the cable [Eq. (7.17)], and write two additional equations that include w/H and C.

c. Using a computerized equation solver or an advanced scientific calculator, solve these equations for K/H, w/H, C.

7.87 An electrical power transmission cable is suspended by a series of towers (see Fig. P7.87). The weight per unit length of the cable is $w = 100$ N/m. Determine the resultant force the cable exerts on tower B. The cable is anchored at each tower so that the force exerted on the tower by each segment of cable differs.

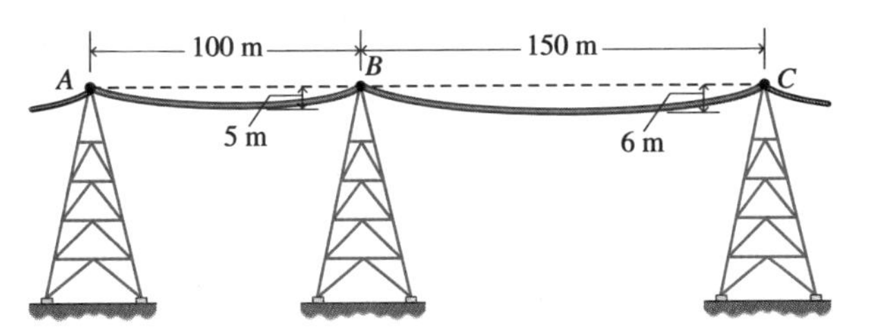

Figure P7.87

7.88 A transmission cable weighs 0.3 lb/ft and has a maximum allowable tension of 1500 lb. It is to be strung between two towers that are 2450 ft apart, horizontally, and at elevations of 7200 ft and 6500 ft, respectively (see Fig. P7.88). Determine the minimum length of cable needed.

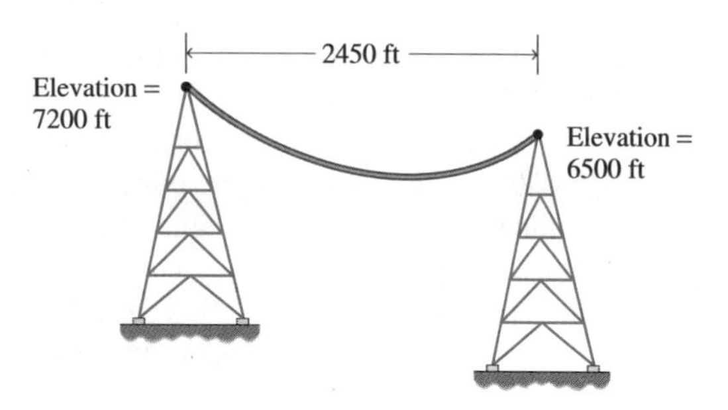

Figure P7.88

7.89 A chain weighs 2 lb/ft. It is pinned to a rigid wall at one end and passes over a frictionless pin A (see Fig. P7.89). For the chain to remain in equilibrium, what must be the length L of the overhanging segment?

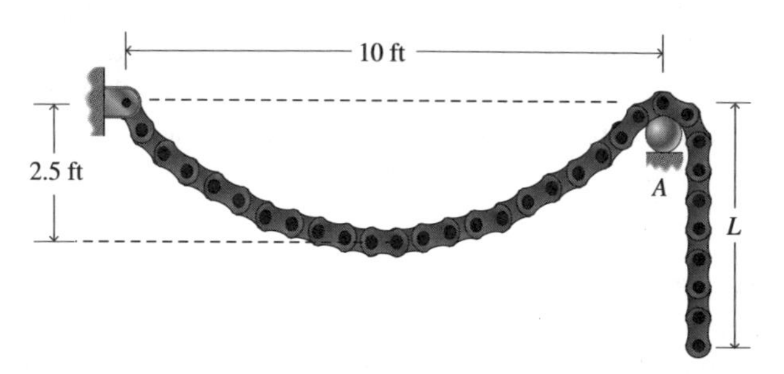

Figure P7.89

7.90 The equation of a hanging chain is $y = 8\cosh(x/8)$. With lengths in feet, the ends of a hanging chain are located at $x_0 = -10$ ft and $x_1 = 30$ ft. The weight of the chain is 2 lb/ft.

a. Calculate the coordinates y_0 and y_1 of the endpoints of the chain.

b. Calculate the length L and the weight W of the chain.

c. Calculate the tensions T_0, T_1, and H at the endpoints and at the lowest point.

7.91 A catenary is symmetrical with respect to the y axis. The arc length, measured from the lowest point, is s. Show that the curve has the following parametric equations, where the parameter $a = H/w$:

$$x = a\sinh^{-1}\frac{s}{a}$$

$$y = \sqrt{a^2 + s^2} + \text{constant}$$

7.92 A cable is to be hung between two towers that have the same elevation and are 250 m apart. The weight of the cable is 75 N/m. Plot the design sag as a function of the maximum line tension T_{max} for the range 15 kN $\leq T_{max} \leq$ 23 kN.

7.93 Regulations for electrical transmission lines in a certain state require that the sag of a freely hanging cable suspended from equal elevations at both ends be given in terms of w, the weight per unit length of the cable. Following this requirement, for a cable with a span a of 250 m and a maximum allowable tension of 16 kN, plot the sag d as a function of the weight w for 36 N/m $\leq w \leq$ 86 N/m.

7.94 a. Using Eq. (7.24), plot the quantity H/wa as a function of d/a for the range $0.1 \leq d/a \leq 1.0$.

b. What happens to your plot as d/a goes to zero? Explain this result.

(*Note:* The graph you have plotted may be used to determine the approximate minimum tension H in the cable, given d, a, and w. Then, an approximate value for the tension T in the cable can be calculated using Eq. (7.23). Alternatively, if greater accuracy is required, the graph can be used to obtain a starting value for H in a trial-and-error solution for T_{max} [Eq. (7.23) for $x = a/2$].)

REVIEW QUESTIONS

7.95 The L-shaped lever shown in Fig. P7.95 is in equilibrium under the action of a 5 kN force and a 20 kN·m couple. If the 20 kN·m couple were moved to point B on the lever, what would the magnitude of the force have to be to maintain equilibrium?

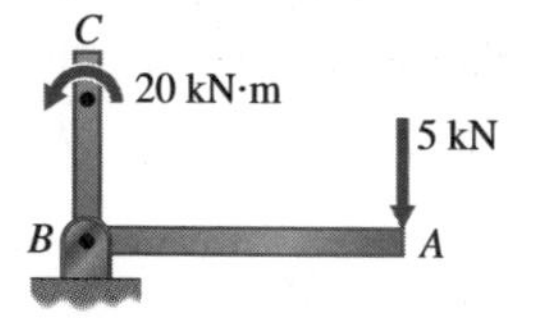

Figure P7.95

7.96 Identify the class of lever ABC in each of the parts of Fig. P7.96, where the force F is the input and the weight W is the output.

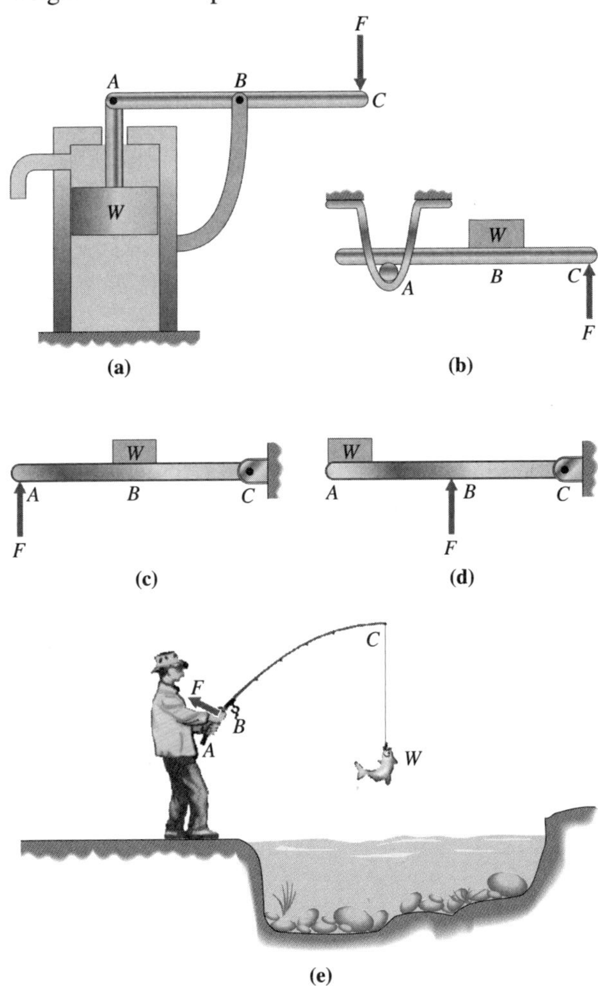

Figure P7.96

7.97 Determine the contact force between the 20 lb weight W and the 10 lb lever ABC in Fig. P7.97.

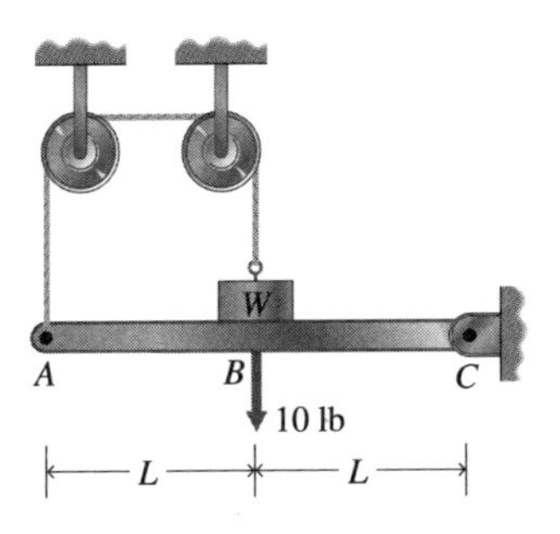

Figure P7.97

7.98 Two pulleys are welded together (see Fig. P7.98). Together the pulleys weigh 10 N. One belt is wrapped around each pulley, so that the pulleys are in equilibrium. The tension in the belt wrapped around the smaller pulley is 5 N. Determine the (x, y) components of the support reaction at point O. Assume that the belts do not slip.

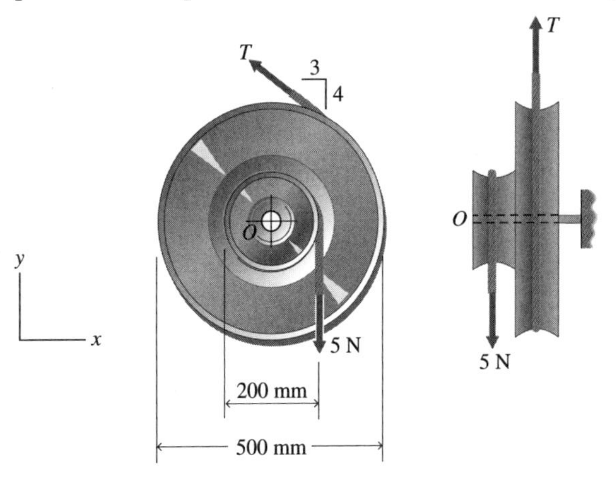

Figure P7.98

7.99 Two blocks are connected by a cord (see Fig. P7.99). The weights of the blocks are $W_A = 50$ N and $W_B = 20$ N. Assume that W_A does not slide and that friction in the pulleys is negligible.

a. Draw a free-body diagram of each of the blocks.

b. Determine the tension in the cord and the forces that act on block A.

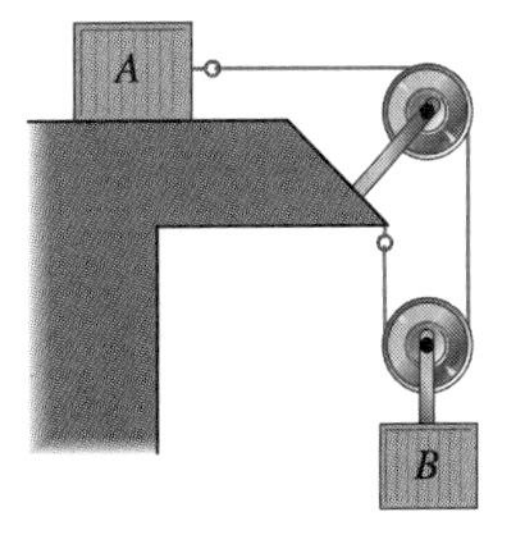

Figure P7.99

7.100 Pulleys A and B are welded together and are held in equilibrium by cord CD and the two other cords wrapped around the pulleys (see Fig. P7.100). The cords on the pulleys do not slip. Determine the force in cord CD.

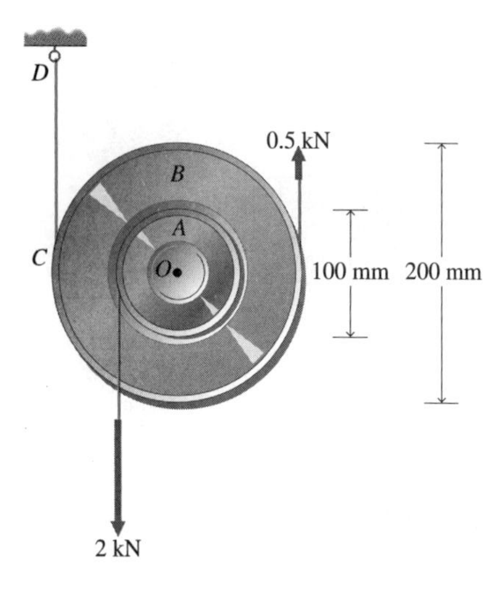

Figure P7.100

7.101 A belt-driven pulley turns freely on a shaft. Show that if the pulley is in equilibrium, the tensions in the belt on both sides of the pulley are equal.

7.102 Identify each type of structure shown in Fig. P7.102 as a truss, frame, or mechanism.

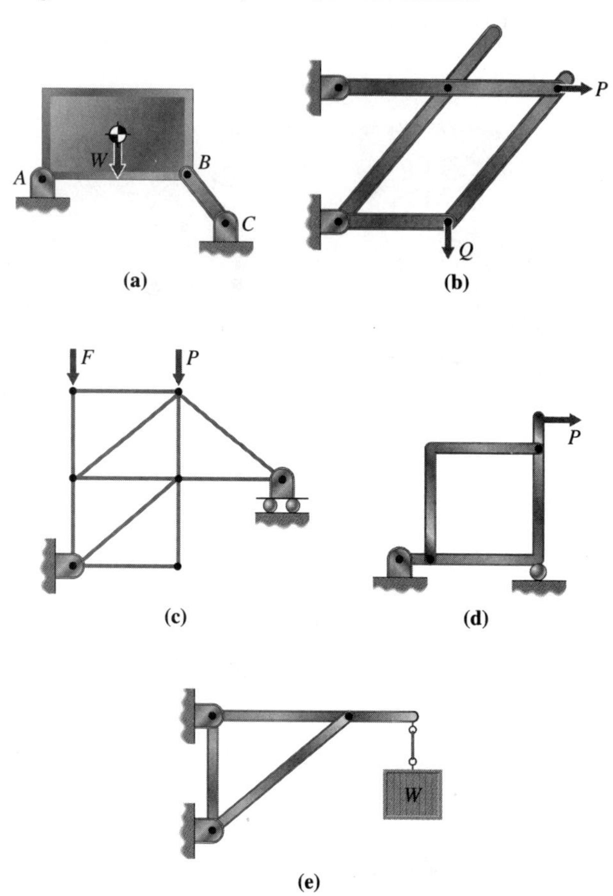

Figure P7.102

7.103 A needle-nose pliers is used to pull on a wire with a force of 50 N (see Fig. P7.103). The person holding the pliers exerts a force F on each handle, as shown.

a. Determine the force that the pin at O exerts on the member AOB.

b. Determine the clamping force exerted on the wire, assuming that the force acts at 60 mm from the pin O.

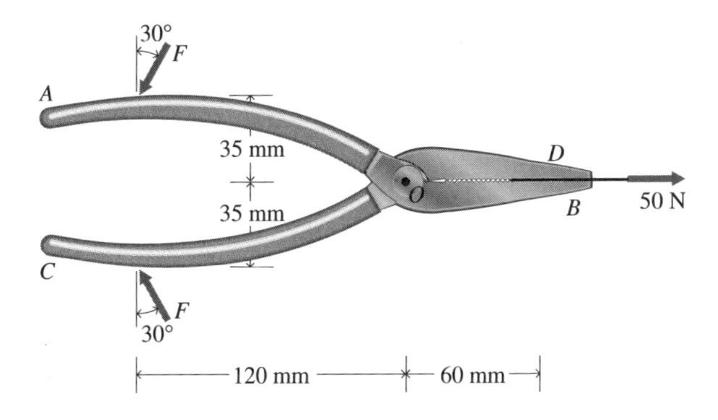

Figure P7.103

7.104 A suspended cable hangs in a parabolic form. How is the cable loaded?

7.105 Explain why a cable that is loaded only by its own weight has a curve that is approximately parabolic if the sag is small compared to the length.

7.106 A cable weighs 0.20 lb/ft. Its ends are attached to supports at the same level 100 ft apart. The sag in the cable is 3 ft. Using the parabolic approximation, determine the maximum tension in the cable.

7.107 Define *catenary,* and write the general form of its equation.

7.108 What is the form taken by a cable of a suspension bridge?

Chapter 8 CENTER OF GRAVITY, CENTROIDS, AND DISTRIBUTED FORCES

For the fiddler to remain in equilibrium, he must keep his center of gravity directly above the rope. The moment of the weight of his body about the rope is zero when his gravity axis passes through the rope.

A Look Forward

So FAR, our study of the equilibrium of rigid bodies has considered mainly forces that arise from contact between bodies. We have assumed that these forces are concentrated at a point; that is, that they are *point forces*. As noted in Sec. 2.1, however, there are no point forces in the real world, since actual contact forces between bodies are spread out (distributed) over the area of contact. The weight of a body (the gravitational force exerted by the earth on the body) is also distributed throughout the body. However, our treatment of rigid-body equilibrium has taken the weight to be concentrated at a point in the body called the *center of gravity* (see Sec. 3.3).

In this chapter, you will learn how to determine the resultant of distributed forces. You will determine the resultant of the distributed gravitational forces that act on a body; that is, you will determine the magnitude (weight W) and the line of action (the gravity axis) of the weight vector $\mathbf{W}$ of the body. You will also learn how to determine the location of the center of gravity of the body. The concept of center of gravity leads to that of the centroid, a geometric property for a homogeneous body. Finally, you will learn how to determine the effects of distributed forces on the equilibrium of a body.

Survey Questions

- What is a gravity axis?
- How is the center of gravity of a body determined?
- What is the difference between a centroid and a center of gravity?
- What is a composite body?

8.1 GRAVITY AXIS OF A BODY

After studying this section, you should be able to:

- Locate the gravity axis for a body in a particular orientation.

Key Concept

The gravity axis of a body is the line of action of the gravitational force that acts on the body.

In this section, we assume that a body in the gravitational field of the earth is small compared to the earth. Hence, we assume that g, the acceleration of gravity, is constant throughout the body and that the gravitational attraction of the earth for individual particles of the body forms a system of forces that are approximately parallel.[1]

Recall that a rigid body can be viewed as a collection of particles. The weight of each particle is the force exerted on it by the gravitational attraction of the earth. Therefore, for a rigid body, the weights of all its particles constitute a system of parallel forces (see Sec. 4.8). The resultant force is called the *weight* of the body. The resultant axis of the weight of a body is called the *gravity axis* of the body. Since the gravitational force is directed vertically downward (toward the earth), the gravity axis is vertical. The location of the gravity axis with respect to the body depends on the orientation of the body with respect to the earth. In other words, *a body has a gravity axis for each orientation relative to the earth.* Therefore, when we speak of a gravity axis of a body, a given orientation of the body is implied.

weight

gravity axis

Although the atomic theory of matter assumes that matter is composed of discrete, widely spaced particles, mechanics frequently treats bodies as though they consist of continuous space-filling material. For example, consider a cubical block that has two horizontal faces (Fig. 8.1). The material of the block is homogeneous (uniform), in the sense that parts with equal volumes have equal weights. Now suppose that we partition the block into a large number of identical prismatic elements with vertical axes. The upper ends of the prisms are represented by the small squares in Fig. 8.1. Corresponding to each of these prisms is a weight vector that coincides with the axis of the prism. Any two prisms that are located symmetrically with respect to the block's vertical central axis (axis *A–A* in Fig. 8.1)—for example, the two prisms designated by the darker squares in Fig. 8.1—provide a resultant weight vector that coincides with that central axis (see Sec. 4.3 and Fig. 4.6). Since all the prisms may be paired off in this way, the gravity axis of the cubical block coincides with its vertical central axis.

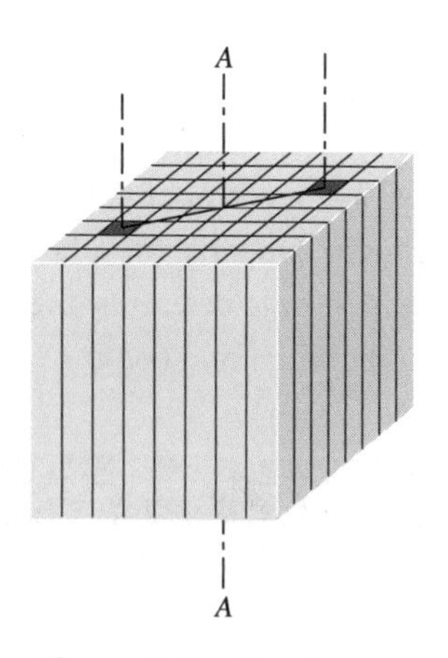

Figure 8.1 Gravity axis of a homogeneous cube.

This reasoning is not restricted to cubes; it applies equally well to cylinders, cones, spheres, ellipsoids, and other symmetric bodies. In general, we have the following theorem.

1. For many engineering problems, it is sufficiently accurate to assume that gravitational forces are parallel and constant throughout a body. Then the center of gravity coincides with the center of mass of the body (see Sec. 16.3). However, for some engineering problems, this assumption does not lead to sufficiently accurate results. For example, the stability and attitude control of a satellite that orbits the earth depend on a very accurate representation of the variation of the gravitational attraction of the earth on the satellite as a function of the orbital path (see Vigneron & Boresi, 1970; Vigneron, 1970; and Tschann & Modi, 1970).

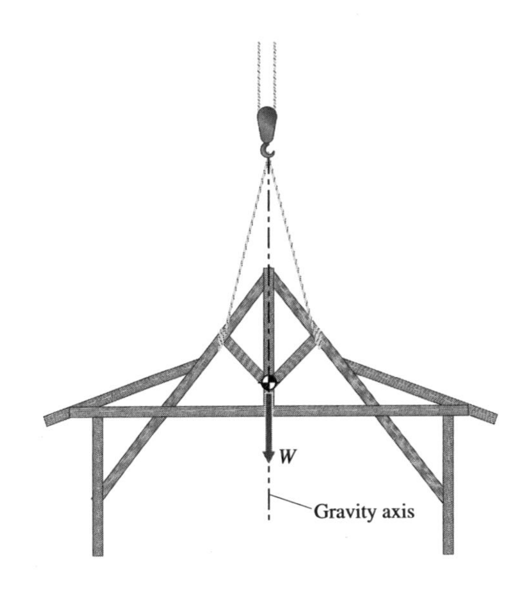

A timber frame hangs from the hoist of a construction crane. The line of action of the cable tension is a gravity axis for the frame, passing through the frame's center of gravity.

Theorem 8.1

If a homogeneous body possesses two vertical planes of symmetry, the gravity axis coincides with the line of intersection of those planes.

If a body possesses one vertical plane of symmetry, we could pair prismatic elements that are located symmetrically with respect to that plane. Accordingly, we have the following theorem.

Theorem 8.2

If a homogeneous body possesses a vertical plane of symmetry, the gravity axis lies in that plane.[2]

For example, the gravity axis of a ship or an airplane lies in the vertical plane of symmetry, if the weight is distributed symmetrically.

If a system can be divided into a finite number of parts whose gravity axes and weights are known, the gravity axis of the entire system may be determined by the theory of parallel point forces (Sec. 4.8).

2. Note that Theorem 8.1 is a corollary of Theorem 8.2.

Example 8.1

Gravity Axis of a Frame

Problem Statement Determine the gravity axis of a frame that consists of a vertical column, a horizontal beam, and a diagonal member, as shown in Fig. E8.1. Details of the joints are not significant. The members have identical cross sections.

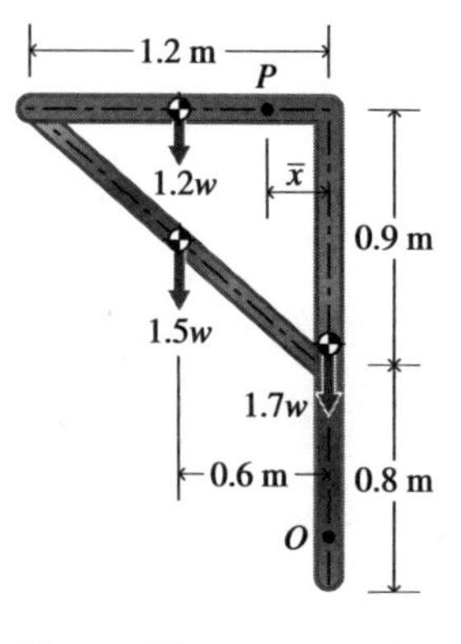

Figure E8.1

Solution Note that the gravity axes of the three members pass through their geometric centers. To verify that the gravity axis of the diagonal member passes through the center of that member, observe that infinitesimal transverse slices of the member may be paired so that the two slices in a pair lie at equal distances from the vertical axis through the center of the member.

The lengths of the members are 1.2 m, 1.5 m, and 1.7 m. Consequently, the weights of the members are $1.2w$, $1,5w$, and $1.7w$, where w is the weight in newtons per meter of any member. The weight vectors of the three members are shown in Fig. E8.1. The weight of the entire structure is $4.4w$.

We can take moments of the weights about any point in Fig. E8.1. If we choose point O on the axis of the vertical column, we obtain

$$M_O = (1.2w)(0.6) + (1.5w)(0.6) = 1.62w$$

If the vector representing the total weight of the structure is situated on the gravity axis, it also produces the moment M about point O. Therefore, let's assume that the gravity axis coincides with the vertical line that passes through the point P, located at a distance $\bar{x}$ from the center line of the vertical column (see Fig. E8.1). The coordinate $\bar{x}$ of the gravity axis is determined by the equation

$$(4.4w)\bar{x} = 1.62w$$

Hence, the gravity axis passes through point P, a distance $\bar{x} = 0.368$ m from the center line of the vertical column. The weights of the members are dynamically equivalent to a single force of magnitude $4.4w$ acting along the vertical line 0.368 m to the left of the axis of the vertical column. If the frame were lifted by a crane with its hook attached at point P, the column would remain vertical, since the system would be in equilibrium under the action of its weight and the opposite force applied by the hook. ■

Example 8.2

Gravity Axis of a Barge

Problem Statement A barge that weighs 50 tons carries a transformer that weighs 15 tons. The transformer is located unsymmetrically on the deck, as shown in the overhead view of the barge (Fig. E8.2). The gravity axes of the transformer and the barge are perpendicular to the xy plane. The coordinates of the point at which the gravity axis of the transformer intersects the xy plane are (20 ft, 8 ft). The coordinates of the point at which the gravity axis of the barge intersects the xy plane are (50 ft, 12 ft). Locate the gravity axis for the entire system.

Solution This problem may be approached in the same manner as was used in Example 8.1. Here, however, we must consider moments about both the x and y axes. Taking moments of the weights about the x and y axes, we obtain

$$\sum M_x = -50(12) - 15(8) = -720 \text{ ton·ft}$$

$$\sum M_y = 50(50) + 15(20) = 2800 \text{ ton·ft}$$

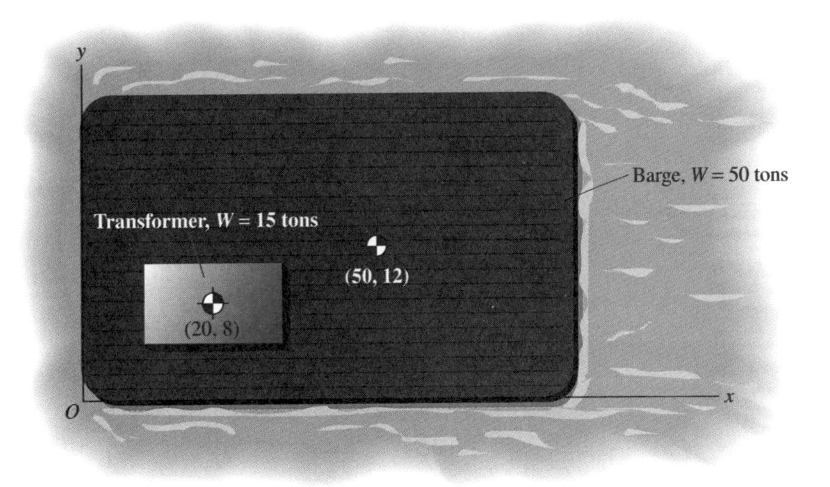

Figure E8.2

Since the moments of the total weight (65 tons) about the x and y axis are dynamically equivalent to the sum of the moments of the individual weights, the $(\bar{x}, \bar{y})$ coordinates of the point at which the gravity axis of the system intersects the xy plane are determined by the equations

$$\sum M_y = 65\bar{x} = 2800$$
$$\sum M_x = -65\bar{y} = -720$$

Hence, $\bar{x} = 43.08$ ft and $\bar{y} = 11.08$ ft.

8.2 CENTER OF GRAVITY IN CARTESIAN COORDINATES

After studying this section, you should be able to:

- Locate the center of gravity of a body using moments of weight.

Suppose that a body in a uniform, parallel gravity field is suspended from a cord at a point A (see Fig. 8.2a). The line segment formed by the cord and its extension is a gravity axis of the body (A–A). Next, let the body be suspended from the same cord but attached at point B (see Fig. 8.2b). The line formed by the cord and its extension is another gravity axis of the body (B–B). This process may be repeated any number of times. We thus obtain numerous gravity axes of the body.

Key Concept The center of gravity of a body is the point of intersection of all the gravity axes of the body.

center of gravity

All the straight lines pass through a common point, called the *center of gravity* (see Fig. 8.2c). Consequently, to determine all the gravity axes of a body, we require only the coordinates of the center of the gravity. For any orientation of the body, the gravity axis is

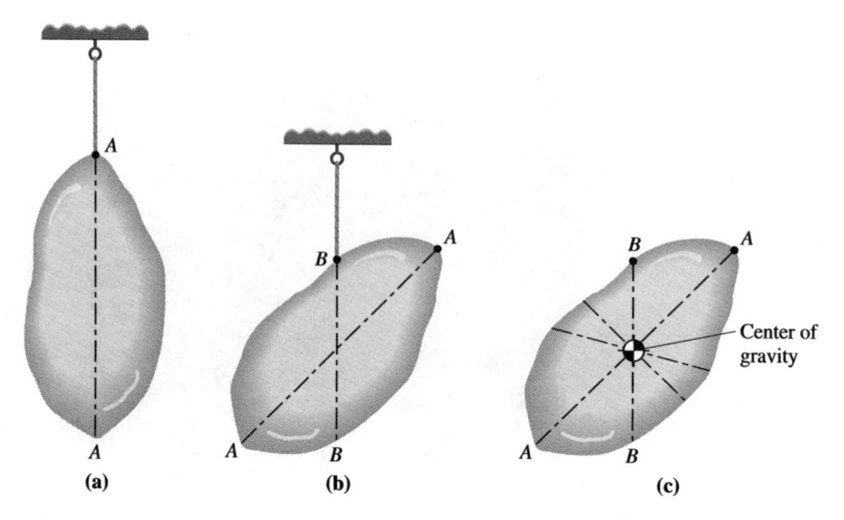

Figure 8.2 Gravity axes of a body.

the vertical line passing through the center of gravity. Thus, if a body could be suspended from its center of gravity, it would be in equilibrium in any orientation. If any gravity axis of a body is determined, the center of gravity lies on that axis. So, if we determine any two gravity axes corresponding to two different orientations of the body, the center of gravity is the point of intersection of these axes. This fact provides a method for locating the center of gravity of a body.

Instead of considering how the gravity axis of a body changes when we rotate the body, let's imagine that the body is fixed and that the direction of the force of gravity can be changed. This is an equivalent point of view, since the direction of the force of gravity relative to the body is all that matters.

Let a body be referred to Cartesian xyz coordinate axes (see Fig. 8.3a). To locate the center of gravity, we regard the body as an aggregate of particles (volume elements) with

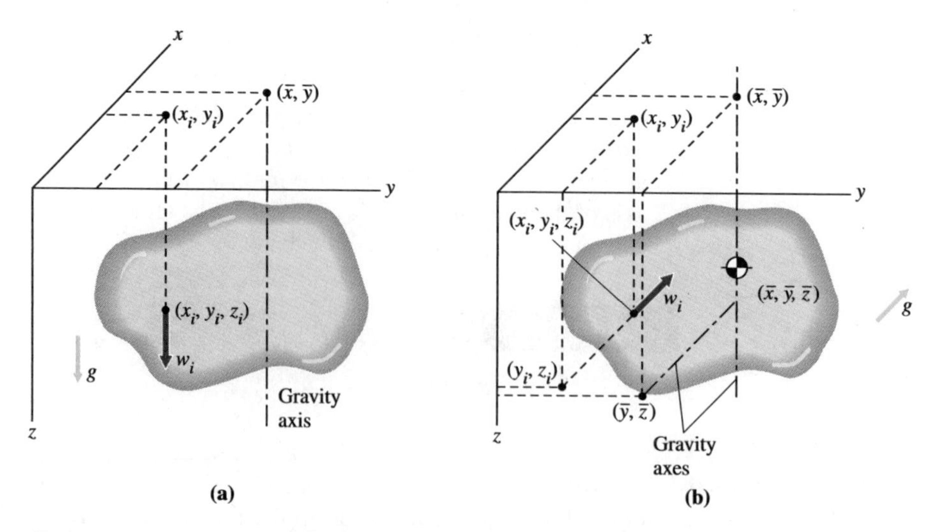

Figure 8.3 Gravity axes and center of gravity of a body.

respective weights $w_1, w_2, w_3, \ldots$. Suppose, for example, that the force of gravity acts downward in the positive z direction, as shown in the figure. Then, the moments of the weight w_i of the ith element with respect to the x and y axes are $M_{xi} = w_i y_i$ and $M_{yi} = -w_i x_i$. The moments of the weight of the entire body about the x and y axes are

$$M_x = \sum M_{xi} = \sum w_i y_i$$
$$M_y = \sum M_{yi} = -\sum w_i x_i \tag{a}$$

where the sums extend over all the elements. By the theory of parallel forces (Sec. 4.8), the resultant axis (the gravity axis) is determined by the condition that the resultant weight vector exerts the same moments about the x and y axes as does the sum of the moments of the weights of the individual elements. Accordingly, by Eqs. (a),

$$M_x = \sum w_i y_i = W\bar{y} \tag{b}$$

$$M_y = -\sum w_i x_i = -W\bar{x} \tag{c}$$

where W is the weight of the entire body and $(\bar{x}, \bar{y})$ are the coordinates of a point on the gravity axis (refer to Fig. 8.3a).

We can solve Eqs. (b) and (c) for $\bar{x}$ and $\bar{y}$, two of the three coordinates of the center of gravity of the body:

$$\bar{x} = \frac{1}{W}\sum w_i x_i \tag{d}$$

$$\bar{y} = \frac{1}{W}\sum w_i y_i \tag{e}$$

To locate the third coordinate $\bar{z}$ of the center of gravity, we assume that the force of gravity acts in the positive x direction (see Fig. 8.3b). Then, the location of the gravity axis is determined by the equations

$$M_z = -\sum w_i y_i = -W\bar{y} \tag{f}$$

$$M_y = \sum w_i z_i = W\bar{z} \tag{g}$$

Equation (f) is the same as Eq. (b). By Eq. (g), we have

$$\bar{z} = \frac{1}{W}\sum w_i z_i \tag{h}$$

In view of Eqs. (d), (e), and (h), the center of gravity of a body is determined by

$$\bar{x} = \frac{1}{W}\sum w_i x_i$$
$$\bar{y} = \frac{1}{W}\sum w_i y_i \tag{8.1}$$
$$\bar{z} = \frac{1}{W}\sum w_i z_i$$

moments of weight

The sums in Eqs. (8.1) are called the *moments of weight* with respect to the coordinate planes. For example, x_i is the distance of the ith element from the plane $x = 0$ (also called the yz plane). Therefore, $M_{yz} = \Sigma w_i x_i$ is called the moment of weight of the body with respect to the yz plane. Similarly, $M_{zx} = \Sigma w_i y_i$ and $M_{xy} = \Sigma w_i z_i$, where M_{zx}, M_{xy} are the moments of weight with respect to zx and xy coordinate planes, respectively. Therefore, in summary, we have

$$\begin{aligned} M_{yz} &= \sum w_i x_i \\ M_{zx} &= \sum w_i y_i \\ M_{xy} &= \sum w_i z_i \end{aligned} \tag{8.2}$$

where M_{yz}, M_{zx}, and M_{xy} are the moments of weight with respect to yz, zx, and xy coordinate planes, respectively.

Equations (8.1) and (8.2) may be interpreted to yield the following theorem, which serves to locate the center of gravity of a body.

Theorem 8.3

The moments of weight of a body with respect to the coordinate planes are not changed if all the weight of the body is concentrated at its center of gravity.

CHECKPOINT 

1. ***True or False:*** If a homogeneous body possesses a vertical plane of symmetry, the gravity axis lies in that plane.
2. ***True or False:*** If a homogeneous body possesses two vertical planes of symmetry, the gravity axis coincides with the line of intersection of those planes.
3. ***True or False:*** The center of gravity of a body always lies on a gravity axis of the body.
4. ***True or False:*** The center of gravity is a point in a body that is located at the intersection of any two gravity axes of the body.

Answers: 1. True; 2. True; 3. True; 4. True

8.3 CENTER OF GRAVITY BY INTEGRATION

After studying this section, you should be able to:

- Locate the center of gravity of a body by integration of the moments of weight.

Let the moments of weight of a body with respect to Cartesian coordinate planes be denoted by (M_{yz}, M_{zx}, M_{xy}), where the subscripts designate the planes. By Eqs. (8.1) and (8.2), the coordinates of the center of gravity of the body are

$$\bar{x} = \frac{M_{yz}}{W} \tag{8.3a}$$

$$\bar{y} = \frac{M_{zx}}{W} \tag{8.3b}$$

$$\bar{z} = \frac{M_{xy}}{W} \tag{8.3c}$$

where W denotes the weight of the body and M_{yz}, M_{zx}, and M_{xy} are defined by Eq. (8.2).

Key Concept The specific weight of a body is the weight per unit volume of the body. In general, the specific weight is not a constant. It may vary from point to point in a body.

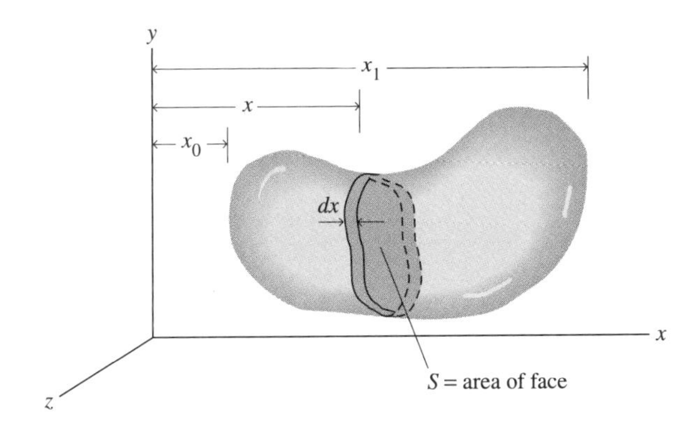

Figure 8.4 Differential slice of a body.

To determine M_{yz}, we imagine that the body is divided into thin slices parallel to the *yz* plane (see Fig. 8.4). The infinitesimal thickness of a slice is denoted by *dx*, and the area of a face of a slice is denoted by *S*. Hence, the volume of a slice is $S\,dx$. If the slice is homogeneous, its weight is $dW = \gamma S\,dx$, where γ is the *specific weight*—the weight per unit volume. Consequently, the moment of the weight of the slice with respect to the *yz* plane is $dM_{yz} = \gamma Sx\,dx$. The moment of the weight of the entire body with respect to the *yz* plane is expressed by the integral

specific weight

$$M_{yz} = \int_{x_0}^{x_1} \gamma Sx\,dx \tag{8.4}$$

where x_0 and x_1 are the extreme values of x (see Fig. 8.4). Similarly, M_{zx} and M_{xy} can be derived by taking slices parallel to the *zx* and *xy* coordinate planes, respectively. To evaluate the integral in Eq. (8.4), we must express *S* as a function of *x*.

So, in general, we can express M_{yz}, M_{zx}, and M_{xy} as integrals. For example, in Sec. 8.2, the moment M_{yz} of the weight of a system with respect to the *yz* plane was given by

$$M_{yz} = \sum w_i x_i \tag{a}$$

The relationship expressed by Eq. (a) is rigorously valid for a system of particles. It is also valid for a continuous body that is divided into a finite number of discrete parts with weights $w_1, w_2, w_3, \ldots$, provided that x_i is exactly the coordinate of the center of gravity of the *i*th part. However, if x_i is merely the coordinate of an arbitrary point in the *i*th part, Eq. (a) is an approximation. The approximation becomes better and better as the number of parts is increased, and it approaches the limit M_{yz} as all the linear dimensions of the parts approach zero. In integral calculus, the limit of this sum is denoted by $M_{yz} = \int x\,dW$, where dW represents an infinitesimal element of the total weight W. The formulas for M_{zx} and M_{xy} are similar. The complete set of formulas is

$$\begin{aligned} M_{yz} &= \int x\,dW \\ M_{zx} &= \int y\,dW \\ M_{xy} &= \int z\,dW \end{aligned} \tag{8.5}$$

where x, y, and z are the distances from the centroid of dW to the yz, zx, and xy planes, respectively (see Fig. 8.5).

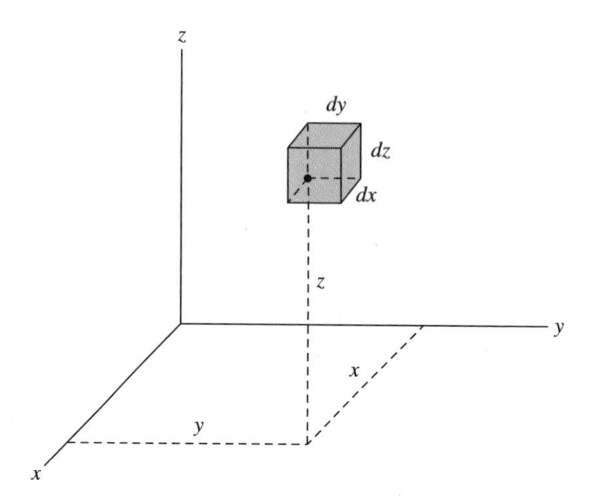

Figure 8.5 Differential element dW.

Since we can represent the weight dW in terms of the volume $dV = dx\,dy\,dz$ it occupies and the specific weight γ as $dW = \gamma dV$, we may write Eqs. (8.5) in the form

$$M_{yz} = \int \gamma x\,dV$$
$$M_{zx} = \int \gamma y\,dV \tag{8.6}$$
$$M_{xy} = \int \gamma z\,dV$$

where, in general, γ is a function of (x, y, z). If γ is a constant, the body is homogeneous.

CENTROID OF A VOLUME IN CARTESIAN COORDINATES

Key Concept

The center of gravity of a homogeneous body is called the centroid of the body. The centroid is a geometric property of a volume.

For a homogeneous body, the specific weight γ is a constant and may be placed outside the integral sign in Eqs. (8.6). Then, since the total weight W is proportional to the specific weight γ, Eqs. (8.3) and (8.6) show that the coordinates $(\bar{x}, \bar{y}, \bar{z})$ of the center of gravity of the body are independent of γ. Consequently, the location of the center of gravity of a homogeneous body is a geometric property of the volume occupied by the body. In other words, for a homogeneous body, there is a point $(\bar{x}, \bar{y}, \bar{z})$ that coincides with the center of gravity of the body. This point $(\bar{x}, \bar{y}, \bar{z})$ is called the *centroid* of the body. The centroid of a body is also referred to as the *centroid of the volume* occupied by the body. Hence, the terms "centroid of a body" and "centroid of a volume" are used interchangeably.

centroid

centroid of a volume

The equations for the centroid of a volume are obtained from Eqs. (8.3) and (8.6) by first writing

$$M_{yz} = \gamma \int x\, dV$$

$$M_{zx} = \gamma \int y\, dV$$

$$M_{xy} = \gamma \int z\, dV$$

and

$$W = \int dW = \gamma \int dV = \gamma V$$

where γ is the constant specific weight of the body that occupies volume V, and dV is an element of the volume. With these equations, Eqs. (8.3) reduce to the following form (since γ cancels):

$$\bar{x} = \frac{Q_{yz}}{V} \tag{8.7a}$$

$$\bar{y} = \frac{Q_{zx}}{V} \tag{8.7b}$$

$$\bar{z} = \frac{Q_{xy}}{V} \tag{8.7c}$$

where

$$Q_{yz} = \int x\, dV \tag{8.8a}$$

$$Q_{zx} = \int y\, dV \tag{8.8b}$$

$$Q_{xy} = \int z\, dV \tag{8.8c}$$

first moments of volume

second moments of volume

are the *first moments of volume* with respect to the yz, zx, and xy planes, respectively. They are integrals of the product of x, y, or z to the *first* degree and the volume element dV. Integrals that are the products of x, y, or z to the second degree—that is, of x^2, y^2, or z^2—and the volume are called *second moments of volume.*

In all cases, whether the body contained in the volume is homogeneous or otherwise, Eqs. (8.7) and (8.8) define the *centroid of the volume V occupied by the body.* The centroid is a geometric property of the volume occupied by the body and is independent of the specific weight of the body. Equations (8.8) are analogous to Eqs. (8.5). If the specific weight of a body is constant, then the center of gravity of the body [Eqs. (8.3)] coincides with the centroid of the volume occupied by the body [Eqs. (8.7)].

It is important to distinguish between the center of gravity of a body and the centroid of the volume occupied by the body. The center of gravity is affected by the way the weight of the body is distributed in the volume; the centroid is affected only by the geometry (shape) of the volume.

To evaluate the integrals in Eqs. (8.5) and (8.8), we must describe the position of points in the volume, as well as the weight element dW and the volume element dV, in terms of spatial coordinates. Equations (8.5) and (8.8) are expressed in Cartesian (rectangular) coordinates. Cartesian coordinates are often the coordinates of choice for numerical integration methods. However, for some shapes, other coordinate systems, such as cylindrical and spherical, are convenient to use. The use of cylindrical and spherical coordinates is discussed next.

VOLUME INTEGRALS IN CYLINDRICAL AND SPHERICAL COORDINATES

A point in space may be located in terms of many different coordinate systems. Cartesian coordinates were used above. In the following, we consider the use of two other coordinate systems—cylindrical coordinates and spherical coordinates.

cylindrical coordinates

polar coordinates

Cylindrical Coordinates Cartesian coordinates locate a point in space by three distances x, y, and z (refer to Fig. 8.5). *Cylindrical coordinates* locate a point in space by two distances r, z and an angle θ (see Fig. 8.6). Cylindrical coordinates consist of *polar coordinates* (r, θ) in the xy plane and a third coordinate (z) measured perpendicular to the xy plane. Thus, in cylindrical coordinates, a point P in space is located by the coordinates (r, θ, z), whereas, in rectangular coordinates, it is located by the coordinates (x, y, z), as shown in Fig. 8.6.

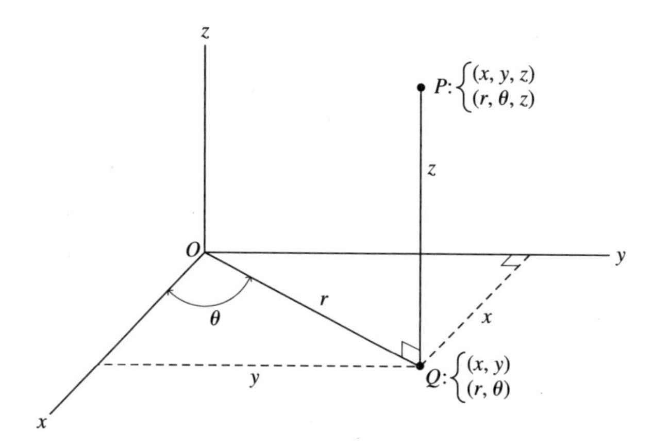

Figure 8.6 Cylindrical coordinates (r, θ, z).

The values of the two Cartesian coordinates x and y and the two polar coordinates r and θ are related by the equations

$$x = r\cos\theta$$

$$y = r\sin\theta$$

$$\theta = \tan^{-1}\left(\frac{y}{x}\right)$$

$$r^2 = x^2 + y^2$$

In cylindrical coordinates, cylinders whose axes lie along the z axis and planes that either contain the z axis or are perpendicular to the z axis have especially simple equations (see Fig. 8.7). When a body or volume involves these shapes, cylindrical coordinates may be the best coordinates to use to evaluate Eqs. (8.5) and (8.8). Note that in cylindrical coordinates the equation $r = a$ describes not just a circle in the xy plane but an entire cylinder perpendicular to the xy plane (Fig. 8.7a). Note also that in cylindrical coordinates the z axis is given by the equation $r = 0$. The equation $\theta = \theta_0$ describes the plane that contains the z axis and that forms an angle θ_0 with the positive x axis (Fig. 8.7b). The equation $z = z_0$ describes the plane that is perpendicular to the z axis and that passes through the point $(0, 0, z_0)$ (Fig. 8.7c).

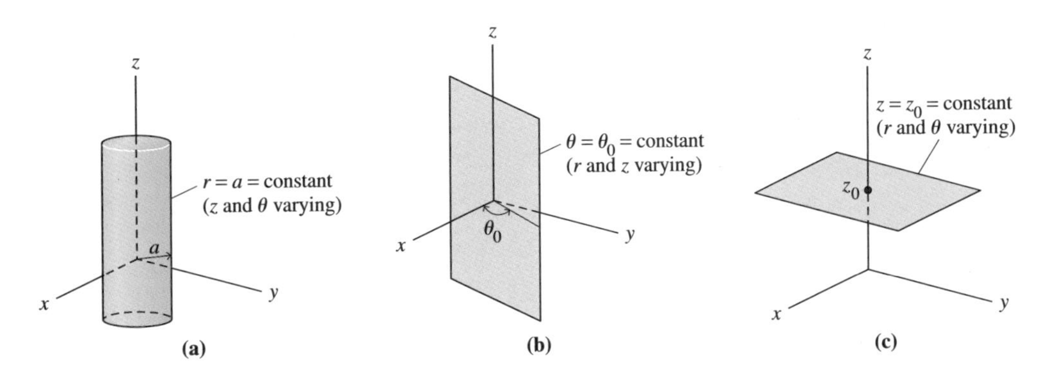

Figure 8.7 Special surfaces: **(a)** cylinder, $r = a$; **(b)** plane containing z axis, $\theta = \theta_0$; **(c)** plane perpendicular to z axis, $z = z_0$.

spherical coordinates

Spherical Coordinates To locate a point P in space, we can also use *spherical coordinates* consisting of two angles θ and ϕ and a distance ρ (see Fig. 8.8). The coordinate ρ is the distance from the origin O to point P. The coordinate ϕ is the angle that the line OP forms with the z axis; it lies in the range $0 \leq \phi \leq \pi$. The coordinate θ is the same as in cylindrical coordinates (refer to Fig. 8.6).[3]

In spherical coordinates, the equation $\rho = a$ describes a sphere of radius a with center at the origin O (Fig. 8.9a). The equation $\phi = \phi_0$ describes a cone whose vertex lies at the origin O and whose axis lies along the z axis. For $\phi_0 < \pi/2$, the cone opens upward (extends from the origin along the positive z axis). For $\pi/2 < \phi_0 < \pi$, the cone opens downward (extends from the origin along the negative z axis). The special cases $\phi_0 = 0$, $\phi_0 = \pi/2$, and $\phi_0 = \pi$ represent the positive z axis, the xy plane, and the negative z axis, respectively. As in cylindrical coordinates, the equation $\theta = \theta_0$ describes the plane that contains the z axis and that forms an angle θ_0 with the positive x axis (Fig. 8.9c). When a problem involves these shapes, spherical coordinates may be the best coordinates to use to evaluate Eqs. (8.5) and (8.8).

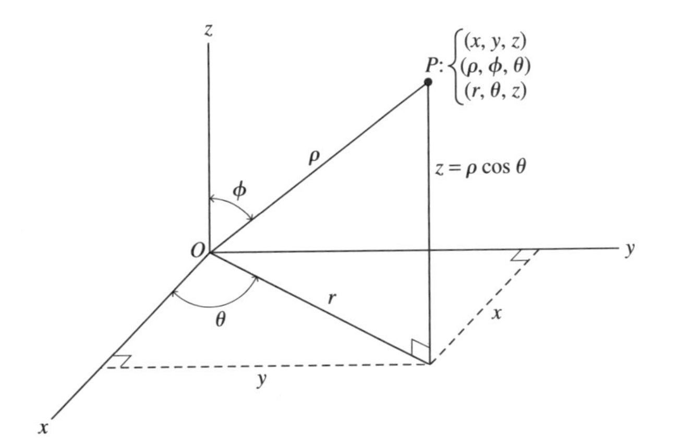

Figure 8.8 Spherical coordinates (ρ, θ, ϕ).

3. In some books, the angles θ and ϕ are reversed. Be alert to this fact when using spherical coordinates.

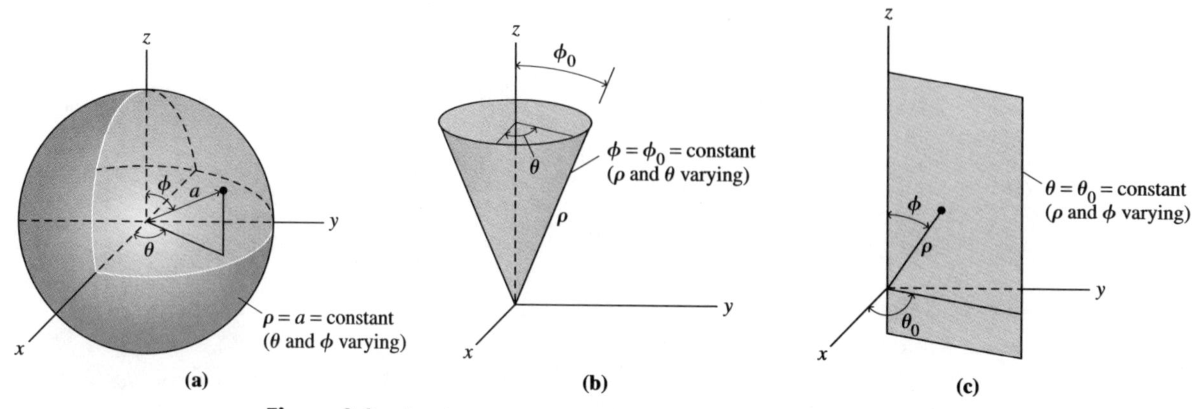

Figure 8.9 Special surfaces: **(a)** sphere with center at origin, $\rho = a$; **(b)** cone with vertex at origin, $\phi = \phi_0$; **(c)** plane that contains the z axis, $\theta = \theta_0$.

Equations relating Cartesian, cylindrical, and spherical coordinates are (refer to Figs. 8.6 and 8.8)

$$r = \rho \sin \phi$$

$$x = r \cos \theta = \rho \sin \phi \cos \theta$$

$$y = r \sin \theta = \rho \sin \phi \sin \theta$$

$$z = \rho \cos \phi$$

$$\rho = \sqrt{x^2 + y^2 + z^2} = \sqrt{r^2 + z^2}$$

For the xy plane, $z = 0$, $\phi = \pi/2$, and $r = \rho$.

Volume Elements and Integrals In Cartesian coordinates, the volume element dV in terms of dx, dy, and dz is (see Fig. 8.10)

$$dV = dx\,dy\,dz \tag{b}$$

and the volume is

$$V = \iiint dx\,dy\,dz \tag{c}$$

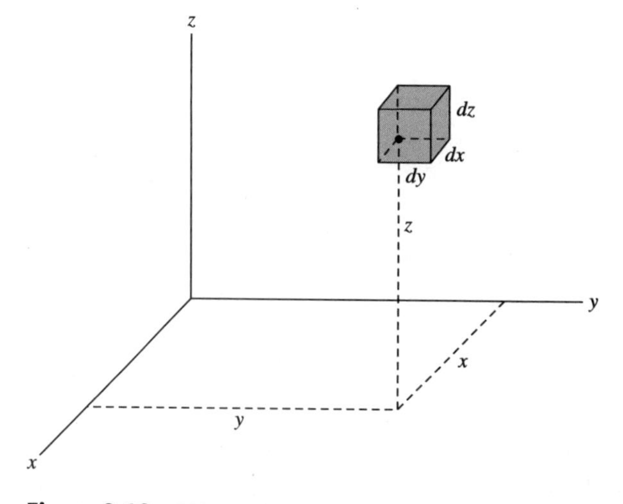

Figure 8.10 Volume element in Cartesian coordinates.

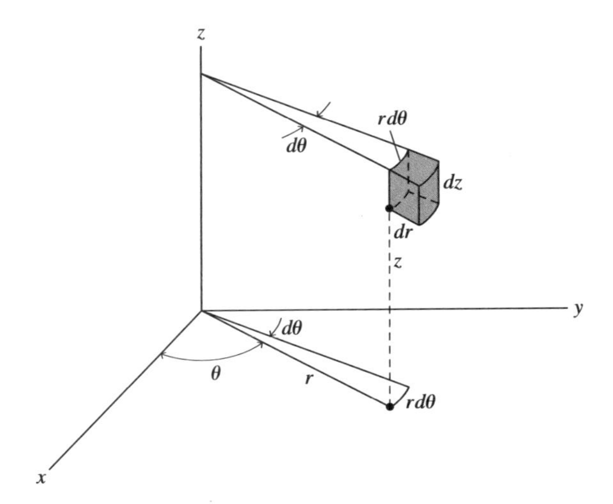

Figure 8.11 Volume element in cylindrical coordinates.

In cylindrical coordinates, in terms of dr, $d\theta$, and dz, the volume element dV is (see Fig. 8.11)

$$dV = (dr)(r\,d\theta)(dz) = r\,dr\,d\theta\,dz \tag{d}$$

and the volume is

$$V = \iiint r\,dr\,d\theta\,dz \tag{e}$$

In spherical coordinates, in terms of $d\rho$, $d\theta$, $d\phi$, the volume element dV is (see Fig. 8.12)

$$dV = (d\rho)(\rho \sin\phi\, d\theta)(\rho\, d\phi) = \rho^2 \sin\phi\, d\rho\, d\theta\, d\phi \tag{8.9}$$

and the volume is

$$V = \iiint \rho^2 \sin\phi\, d\rho\, d\theta\, d\phi$$

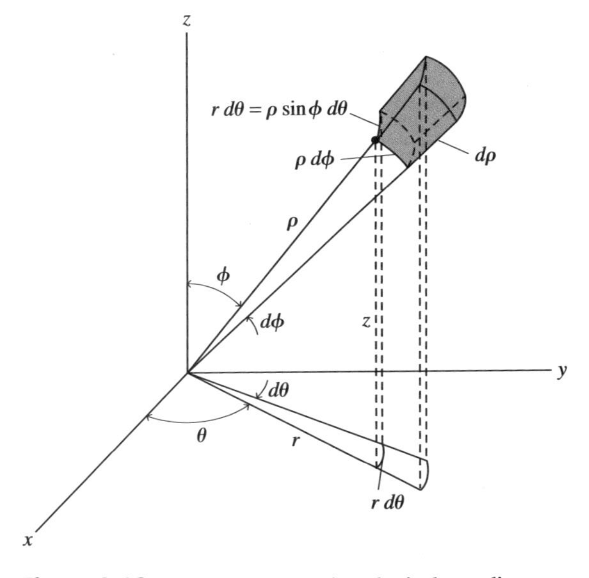

Figure 8.12 Volume element in spherical coordinates.

The integration principles discussed in this section may be used to determine the coordinates of the center of gravity or the centroid of a body. The application of these principles is summarized in the following Problem-Solving Technique and examples.

PROBLEM-SOLVING TECHNIQUE

Center of Gravity and Centroid of a Body

To determine the center of gravity or the centroid of a body by integration:

1. Sketch the body approximately to scale.
2. Choose an appropriate coordinate system for the analysis, depending on the shape of the body, and show it in the sketch (from Step 1). If the body has flat plane boundaries, Cartesian coordinates are efficient. If the body has curved boundaries, cylindrical or spherical coordinates may be better.
3. Utilize symmetry properties of the body to help determine the coordinates of the center of gravity or centroid. If the body has a line or a plane of symmetry, the center of gravity (or centroid) lies on that line or plane. If the body has two lines or planes of symmetry, the center of gravity (or centroid) lies at the intersection of the lines or on the intersection line of the planes. A coordinate axis or plane should be chosen to coincide with the line or plane of symmetry (see Step 2).
4. Choose a differential element of weight dW of the body (or a differential element of volume dV), and express the coordinates of dW (or of dV) in terms of the chosen coordinate system.
5. Determine the weight W (or the volume V) of the body by integration of dW (or dV). For the weight integration, use the appropriate specific weight γ (constant or variable).
6. Determine the first moments of weight (or of volume) by integration of Eqs. (8.6) [or of Eqs. (8.8)].
7. Divide the first moments of weight (or volume) (Step 6) by the weight W (or volume V) to determine the coordinates of the center of gravity (or of the centroid).
8. Locate the center of gravity (or centroid) in the sketch from Step 1. Check to make certain that the location makes sense. In addition, it is a good idea to check your results by solving the problem a different way, if you have time. For example, use a different differential element or a different coordinate system.

Example 8.3

Centroid of a Cone

Problem Statement Consider a solid cone of arbitrary cross section, with its vertex at the origin and its base perpendicular to the x axis (see Fig. E8.3). Determine the distance $\bar{x}$ of the cone's centroid from the yz plane.

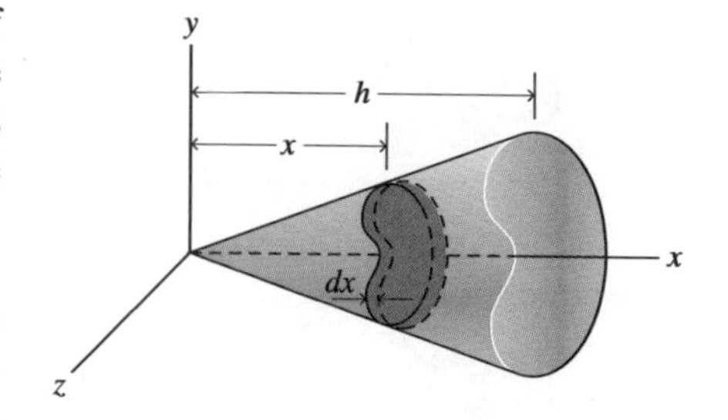

Figure E8.3

Solution Since all the cross sections of the cone have the same shape as the base and since the linear dimensions (width, diameter, etc.) of any cross section are proportional to x, the area S of a cross section is proportional to x^2; that is, $S = kx^2$, where k is a constant. Hence, an element of thickness dx has volume $dV = kx^2\,dx$. Therefore, the total volume is

$$V = \int_0^h kx^2\,dx = \tfrac{1}{3}kh^3$$

where h is the height of the cone (measured perpendicular to the base). By Eqs. (8.7) and (8.8), the centroid coordinate $\bar{x}$ is

$$\bar{x} = \frac{\int_0^h x(kx^2)\,dx}{\tfrac{1}{3}kh^3} = \frac{\tfrac{1}{4}kh^4}{\tfrac{1}{3}kh^3} = \frac{3}{4}h$$

The centroid of a solid cone is located in the cross section that lies at three-fourths of the altitude from the vertex, or one-fourth of the altitude from the base, regardless of the shape of the cross section. The remaining centroid coordinates $\bar{y}$ and $\bar{z}$ depend on the form of the cross section. Often these coordinates can be determined by symmetry conditions.

Example 8.4

Centroid of a Homogeneous Triangular Prism

Problem Statement The cross section of a homogeneous prism is a right triangle with sides a and b (see Fig. E8.4). The xy plane is a plane of symmetry for the prism. Locate the centroid $(\bar{x}, \bar{y}, \bar{z})$ of the prism.

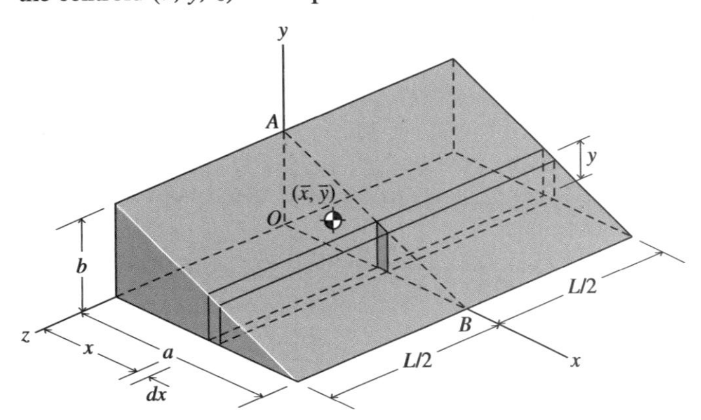

Figure E8.4

Solution Since the xy plane is the plane of symmetry of the prism, the gravity axis lies in that plane; so, $\bar{z} = 0$. We start by defining an element of the prism with volume $dV = yL\,dx$. To write y in terms of x, we need the equation of the diagonal line AB in Fig. E8.4:

$$\frac{x}{a} + \frac{y}{b} = 1 \tag{a}$$

The ordinate y of that line is

$$y = b\left(1 - \frac{x}{a}\right) \tag{b}$$

Consequently, the volume of the element shown in Fig. E8.4 is

$$dV = yL\,dx = bL\left(1 - \frac{x}{a}\right)dx \tag{c}$$

The total volume is, with Eq. (c),

$$V = \int dV = \tfrac{1}{2}abL \tag{d}$$

From Eqs. (8.7) and (8.8), we find $\bar{x}$ as

$$\bar{x} = \frac{\int_0^a x\,dV}{\frac{1}{2}abL} = \frac{bL\int_0^a x\left(1 - \frac{x}{a}\right)dx}{\frac{1}{2}abL} \tag{e}$$

Consequently, by Eq. (e), $\bar{x} = a/3$. Without performing similar calculations for the centroid coordinate $\bar{y}$, we see by the symmetry of the relations that $\bar{y} = b/3$. Accordingly, the coordinates of the center of gravity of the prism are

$$(\bar{x}, \bar{y}, \bar{z}) = \left(\frac{a}{3}, \frac{b}{3}, 0\right) \tag{f}$$

Since L does not appear in Eq. (f), we could have let L equal 1 at the outset. Then, only the geometric properties of the plane triangular area would have entered into the computations.

The point $(\bar{x}, \bar{y}, \bar{z}) = (a/3, b/3, 0)$ is the centroid of the volume occupied by the prism. Also, the point $(\bar{x}, \bar{y}) = (a/3, b/3)$ is the centroid of the triangular area OAB (refer to Fig. E8.4).

Example 8.5 Center of Gravity of a Right-Circular Cone

Problem Statement **a.** Determine the weight and the location of the center of gravity of the solid right-circular cone in Fig. E8.5a, relative to the xyz axes shown. The specific weight of the cone is γ_0 at its apex and $2\gamma_0$ at its base, and it varies linearly with coordinate z.

b. Determine the centroid of the volume of the cone and compare its location to the center of gravity determined in part a.

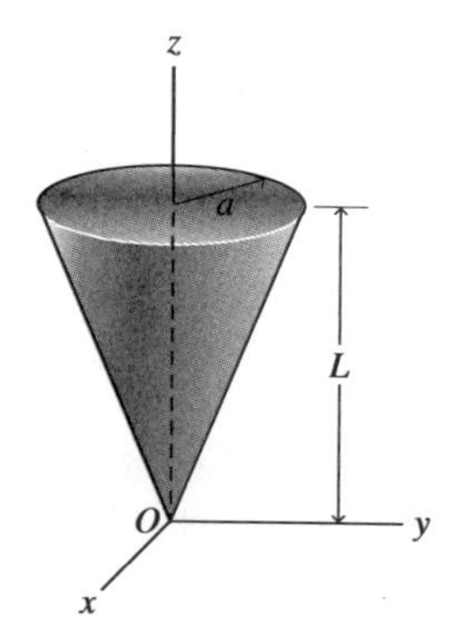

Figure E8.5a

Solution Since the cone is symmetric relative to the z axis and since the specific weight varies only with z, the center of gravity of the cone and the centroid of the volume of the cone lie on the z axis. Hence, $\bar{x} = \bar{y} = 0$ for both the center of gravity and the centroid. For this problem, using cylindrical coordinates will be most effective.

a. To calculate the center of gravity of the cone, we must determine the specific weight γ of the cone as a function of z. Since γ varies linearly with z from γ_0 at $z = 0$ to $2\gamma_0$ at $z = L$,

$$\gamma = \gamma_0\left(1 + \frac{z}{L}\right) \tag{a}$$

To determine the center of gravity, we can follow the method used in Example 8.3 and consider a cross section of the cone perpendicular to the z axis. Since all the cross sections of a right-circular cone are circular, the radius of a cross section at coordinate z is (see Fig. E8.5b)

$$r = \frac{a}{L}z$$

Hence, a differential circular cross-sectional element of thickness dz has volume

$$dV = \pi r^2\,dz = \left(\frac{\pi a^2}{L^2}\right) z^2\,dz \tag{b}$$

and weight

$$dW = \gamma\,dV = \left[\gamma_0\left(1 + \frac{z}{L}\right)\right]\left[\left(\frac{\pi a^2}{L^2}\right) z^2\,dz\right] \tag{c}$$

Integration of Eq. (c) yields

$$W = \int dW = \left(\frac{7\pi}{12}\right) a^2 L \gamma_0 \tag{d}$$

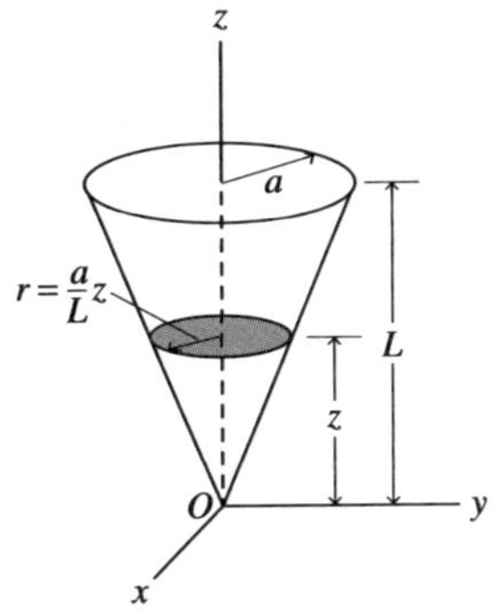

Figure E8.5b

With Eqs. (a) and (b), the first moment of weight of the cone relative to the xy plane is [see Eqs. (8.6)]

$$\begin{aligned} M_{xy} &= \int \gamma z\,dV \\ &= \int_0^L \left[\gamma_0\left(1 + \frac{z}{L}\right)\right](z)\left[\left(\frac{\pi a^2}{L^2}\right) z^2\,dz\right] \\ &= \left(\frac{9\pi}{20}\right) a^2 L^2 \gamma_0 \end{aligned} \tag{e}$$

Hence, by Eq. (8.3c), with Eqs. (d) and (e), we obtain

$$\bar{z} = \frac{M_{xy}}{W} = \tfrac{27}{35} L \approx 0.771L$$

Thus, the center of gravity of the cone is located at the point $(0, 0, 0.771L)$.

b. The centroid of the cone is obtained by application of Eq. (8.7c) and (8.8c). Thus, with Eq. (b), we obtain the volume of the cone as

$$V = \int dV = \int_0^L \left(\frac{\pi a^2}{L^2}\right) z^2\,dz = \frac{\pi}{3} a^2 L \tag{f}$$

Next, the moment of the volume relative to the xy plane is, with Eq. (b),

$$\int z\,dV = \int_0^L (z)\left(\frac{\pi a^2}{L^2} z^2\,dz\right) = \frac{\pi}{4} a^2 L^2 \tag{g}$$

By Eqs. (8.7c) and (8.8c) with Eqs. (f) and (g), we find

$$\bar{z} = \frac{Q_{xy}}{W} = \tfrac{3}{4} L = 0.750L$$

Therefore, the centroid of the volume of the cone is located at the point $(0, 0, 0.750L)$. Note that this result agrees with that of Example 8.3. Note also in Fig. E8.5c that the center of gravity and the centroid do not coincide. The center of gravity is closer to the base of the cone as a result of the higher specific weight near the base. ■

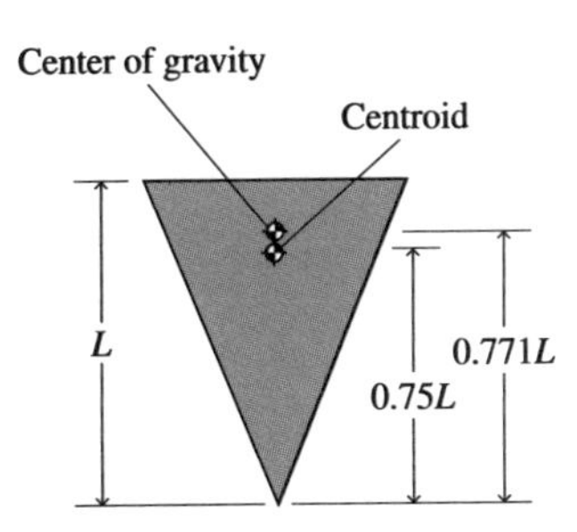

Figure E8.5c

Example 8.6 Centroid of a Hemispherical Volume

Problem Statement Determine the centroid of the hemispherical volume shown in Fig. E8.6.

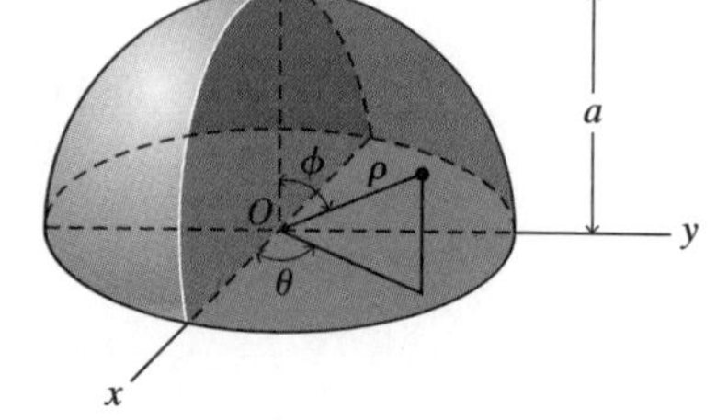

Figure E8.6

Solution

To solve this problem, we can use spherical coordinates. Note, however, that $\bar{x} = \bar{y} = 0$ for the axes chosen, since the hemisphere is symmetric relative to the z axis. Hence, we need calculate only the z coordinate of the centroid.

Equation (8.9) is

$$dV = \rho^2 \sin\phi \, d\rho \, d\theta \, d\phi \qquad \text{(a)}$$

For the hemisphere, the limits from 0 to $\pi/2$ are for ϕ, from 0 to 2π for θ, and from 0 to a for ρ (Figs. 8.8 and E8.6). Therefore, with Eq. (a), we find the volume as the triple integral

$$V = \int_0^{\pi/2}\int_0^{2\pi}\int_0^a \rho^2 \sin\phi \, d\rho \, d\theta \, d\phi \qquad \text{(b)}$$

Integration of Eq. (b) yields the well-known result[4]

$$V = \left(\frac{2\pi}{3}\right)a^3 \qquad \text{(c)}$$

The first moment Q_{xy} of volume relative to the xy plane is [with $z = \rho\cos\phi$; see Eq. (8.8c) and footnote 4]

$$\begin{aligned} Q_{xy} &= \int z\,dV \\ &= \int_0^{\pi/2}\int_0^{2\pi}\int_0^a (\rho\cos\phi)(\rho^2 \sin\phi \, d\rho \, d\theta \, d\phi) \qquad \text{(d)} \\ &= \left(\frac{\pi}{4}\right)a^4 \end{aligned}$$

Hence, by Eqs. (c) and (d),

$$\bar{z} = \frac{Q_{xy}}{V} = 0.375a$$

Therefore, the centroid is located at the point (0, 0, 0.375a).

This problem could also have been solved by taking the volume element to be a circular slice perpendicular to the z axis, with thickness dz. You can check the above result by solving for $\bar{z}$ by that method. If you do this correctly, you will need to evaluate only a single integral.

4. You may not yet have studied multiple integrals. However, in this case, the integration of Eq. (b) is the result of the three single integrations multiplied together. First integrate $\rho^2 d\rho$ from 0 to a to obtain $a^3/3$. Then, integrate $d\theta$ from 0 to 2π to obtain 2π, and $\sin\phi\, d\phi$ from 0 to $\pi/2$ to obtain 1. The volume is $V = (a^3/3)(2\pi)(1) = 2\pi a^3/3$.

CHECKPOINT 

1. ***True or False:*** The center of gravity of a body can be determined only by integration.
2. ***True or False:*** The center of gravity of any body is independent of the specific weight of the body.
3. ***True or False:*** For a homogeneous body, the center of gravity of the body and the centroid of the body coincide.

Answers: 1. False; 2. False; 3. True

8.4 CENTROIDS OF PLANE AREAS AND LINES

After studying this section, you should be able to:

- Locate the centroid of a plane area by integration of moments of area.
- Locate the centroid of a line by integration of moments of a line.

You have seen in Sec. 8.3 and Examples 8.3 and 8.4 that the center of gravity $(\bar{x}, \bar{y}, \bar{z})$ of a homogeneous body may be regarded as a geometric characteristic of the body. The point $(\bar{x}, \bar{y}, \bar{z})$ is called the centroid of the volume. A given plane area A may be regarded as the central cross section of a *homogeneous* cylindrical body with horizontal generators (that is, generators parallel to the z axis; see Fig. 8.13a). Then, as shown by Example 8.4, the center of gravity of the body lies in the given area and is called the *centroid of the plane area*. This section develops formulas for centroids of plane areas and lines.

centroid of a plane area

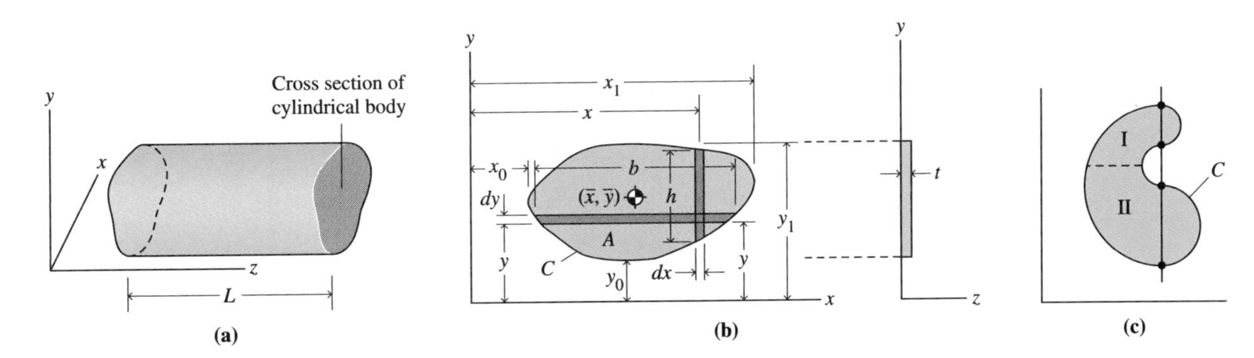

Figure 8.13 (a) Cylindrical body. (b) Plane area. (c) Segmented area.

CENTROID OF A PLANE AREA IN CARTESIAN COORDINATES

For many problems in engineering, the properties of thin slabs, disks, or plane areas are required. For example, in the study of beam bending, the centroid of the cross-sectional area of a beam plays an important role in the determination of stresses in the beam.

The centroid of a plane area was determined in Example 8.4 for the central cross section of a homogeneous triangular prism. More generally, let's consider a homogeneous cylindrical body with horizontal generators and arbitrary cross section (see Fig. 8.13a). Let the length L approach t, where t is very small compared to the cross-sectional dimensions (see Fig. 8.13b). Also, let the cross-sectional area A be bounded by a curve C. Suppose that the curve C is cut at only two points by any straight line parallel to the y axis. If the straight line parallel to the y axis cuts C at more than two points (see Fig. 8.13c), the body should be considered as two separate parts, I and II, so that the line intersects curve C at only two points in each of the two parts.

The infinitesimal area of a strip of width dx is $dA = h\,dx$. Therefore, the corresponding infinitesimal volume element of the body is $dV = th\,dx = t\,dA$. Also, the volume V of the body is $V = At$. Therefore, by Eqs. (8.7a) and (8.8a), we obtain, after canceling t,

$$\bar{x} = \frac{1}{A}\int_{x_0}^{x_1} x\,dA \tag{a}$$

where

$$dA = h\,dx \tag{b}$$

Similarly, $\bar{y}$ may be determined by taking an infinitesimal strip dy parallel to the x axis (see Fig. 8.13b). We find

$$\bar{y} = \frac{1}{A}\int_{y_0}^{y_1} y\,dA \tag{c}$$

where

$$dA = b\,dy \tag{d}$$

Summarizing Eqs. (a) through (d), we have

$$\bar{x} = \frac{1}{A}\int_{x_0}^{x_1} x\,dA = \frac{1}{A}\int_{x_0}^{x_1} xh\,dx \tag{8.10a}$$

$$\bar{y} = \frac{1}{A}\int_{y_0}^{y_1} y\,dA = \frac{1}{A}\int_{y_0}^{y_1} yb\,dy \tag{8.10b}$$

The coordinates $(\bar{x}, \bar{y})$ locate a point in the plane area A bounded by the curve C (refer to Fig. 8.13b). This point is the centroid of the area. The integrals in Eqs. (8.10) are called the *first moments of area* with respect to the y and x axes, respectively. They are denoted by Q_y and Q_x:

first moments of area

$$Q_y = \int_{x_0}^{x_1} x\,dA = \int_{x_0}^{x_1} xh\,dx \tag{8.11a}$$

$$Q_x = \int_{y_0}^{y_1} y\,dA = \int_{y_0}^{y_1} yb\,dy \tag{8.11b}$$

By Eqs. (8.10) and (8.11), the coordinates of the centroid of a plane area A are determined by the equations

$$A\bar{x} = Q_y \tag{8.12a}$$

$$A\bar{y} = Q_x \tag{8.12b}$$

or

$$\bar{x} = \frac{Q_y}{A} \tag{8.13a}$$

$$\bar{y} = \frac{Q_x}{A} \tag{8.13b}$$

where Q_y and Q_x are defined by Eqs. (8.11). The integration procedure outlined above is called the *method of strips*.

method of strips

CENTROID OF A LINE IN CARTESIAN COORDINATES

As noted at the beginning of this section, the centroid of the cross section of a beam plays an important role in the stress analysis of beams. In practice, the cross section of a beam may be a C-channel, an angle, an I-section, a semicircular channel, or another thin-wall shape (see Fig. 8.14a). The wall thickness t of such beams is usually small compared to the other dimensions. Consequently, as t approaches 0—that is, for small t—the cross sections of the beams may be approximated by lines. Then, the centroids of such sections may be approximated by the centroids of lines (or of composite lines; see Sec. 8.5).[5] You should also note that if a cross section has a plane of symmetry (see the C-channel in Fig. 8.14a), the centroid lies in the plane of symmetry, and if a cross section has two planes of symmetry (see the I-section in Fig. 8.14a), the centroid lies at the intersection of the planes.

centroid of a plane line

We may express the position of the *centroid of a plane line* by the equations

$$L\bar{x} = Q_y \tag{8.14a}$$

$$L\bar{y} = Q_x \tag{8.14b}$$

where the length L of the line replaces the area A of the cross section [see Eqs. (8.12) and (8.13)], and

$$Q_y = \int_0^L x\,ds \tag{8.15a}$$

$$Q_x = \int_0^L y\,ds \tag{8.15b}$$

first moments of a line

where s is the arc length of the curved line, ds is the length of an infinitesimal element of the line, and (x, y) are the coordinates of the element ds (see Fig. 8.14b). The integrals Q_y and Q_x, respectively, are the *first moments of the line* relative to axes y and x.

Thus, from Eqs. (8.14) and (8.15),

$$\bar{x} = \frac{1}{L}\int_0^L x\,ds \tag{8.16a}$$

$$\bar{y} = \frac{1}{L}\int_0^L y\,ds \tag{8.16b}$$

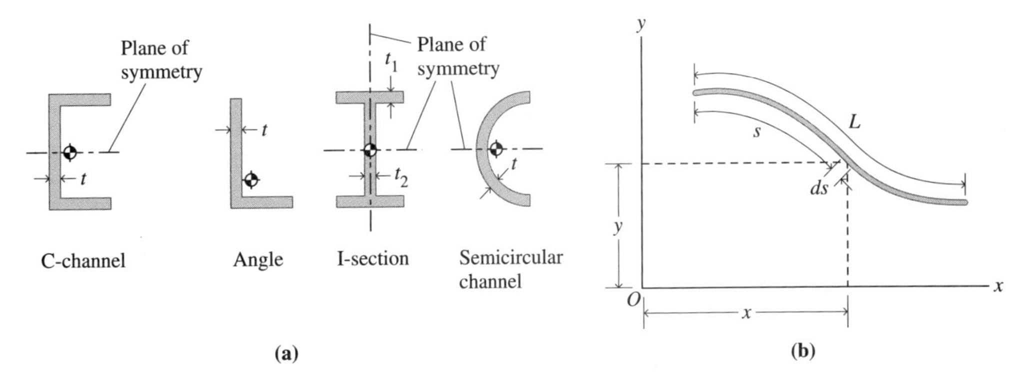

Figure 8.14 (a) Thin-wall cross sections. (b) Curved line segment.

5. Ordinarily, we will consider a line (straight or curved) to lie in a plane, in which case, two coordinates $(\bar{x}, \bar{y})$ are sufficient to locate its centroid. In general, a line could be a space curve, in which case three coordinates $(\bar{x}, \bar{y}, \bar{z})$ are required to locate its centroid.

The horizontal axle passes through the center of gravity to allow the wings of this kinetic structure to pivot freely in the wind. The wings have no preferred orientation.

Centroids of common plane areas and curves are tabulated in Appendix D.

CENTROIDS OF PLANE AREAS AND LINES IN POLAR COORDINATES

In the case of plane areas or lines that lie in the xy plane ($z = 0$), cylindrical coordinates and spherical coordinates both degenerate into polar coordinates (r, θ).

Plane Areas A plane area element in terms of dr and $d\theta$ (see Fig. 8.15) is

$$dA = (dr)(r\,d\theta) = r\,dr\,d\theta \tag{a}$$

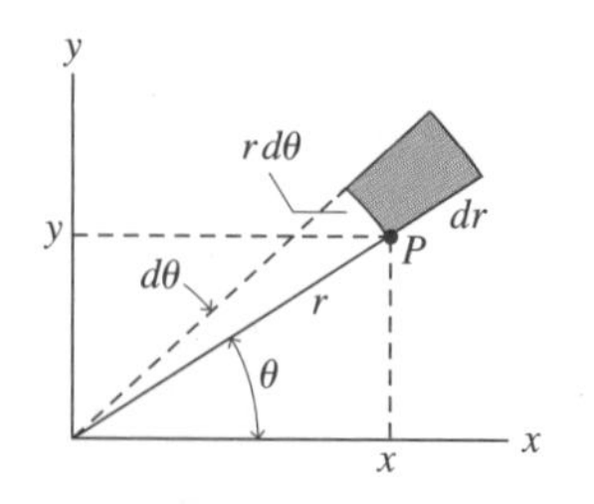

Figure 8.15 Plane area element in polar coordinates.

Also, from Fig. 8.15,

$$x = r\cos\theta \qquad y = r\sin\theta$$

Let's determine the centroid of a thin semicircular disk, a plane area for which polar coordinates are convenient to use. We select x and y axes as shown in Fig. 8.16. Since the disk is symmetrical about the y axis, the centroid lies on the y axis; that is, $\bar{x} = 0$. To determine the y coordinate of the centroid, with Eq. (a), we evaluate the area integral in polar coordinates:

$$\begin{aligned} A &= \int_0^\pi \int_0^a r\,dr\,d\theta \\ &= \frac{a^2}{2}\int_0^\pi d\theta \\ &= \left(\frac{\pi}{2}\right)a^2 \end{aligned} \tag{b}$$

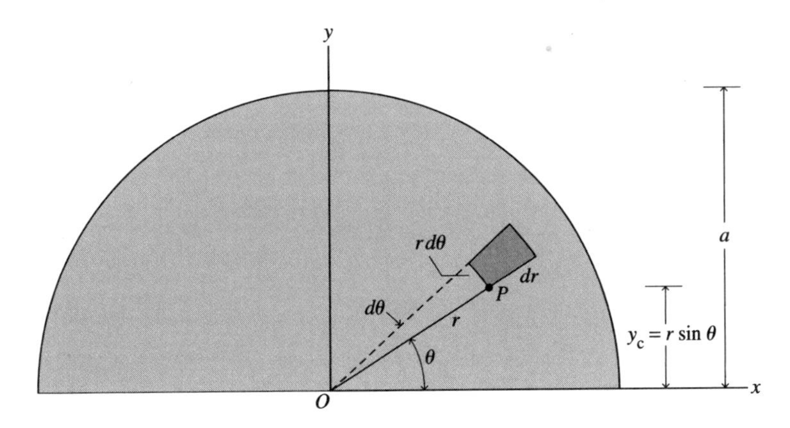

Figure 8.16 Semicircular plane area.

Then, with Fig. 8.16, we evaluate the first moment of area with respect to the x axis as

$$\begin{aligned} Q_x &= \int y_c\,dA \\ &= \int_0^{\pi}\int_0^{a} (r\sin\theta)(r\,dr\,d\theta) \\ &= \tfrac{2}{3}a^3 \end{aligned} \tag{c}$$

By Eq. (8.12b), with Eqs. (b) and (c), we obtain

$$\bar{y} = \frac{Q_x}{A} = \frac{4a}{3\pi} \tag{d}$$

So, the centroid of the semicircular area is located at the point $(\bar{x}, \bar{y}) = (0, 4a/3\pi)$.

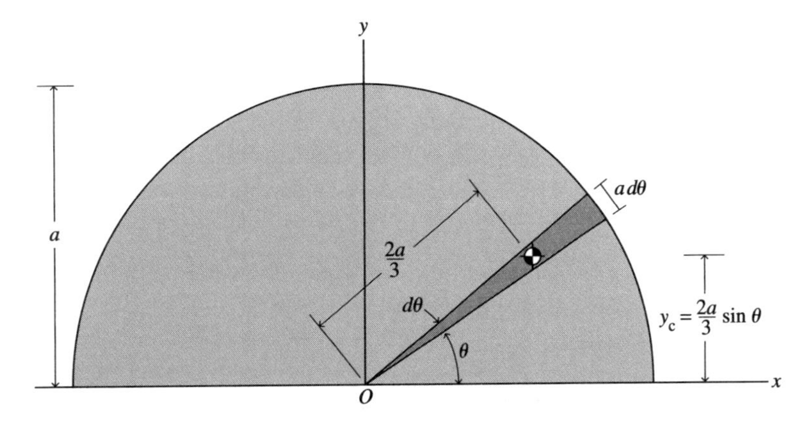

Figure 8.17 Differential area element in polar coordinates.

In the above procedure, we were required to evaluate double integrals. In this case, the integrals were easy to evaluate. In other cases, however, the evaluation of multiple integrals may be laborious. Alternatively, we can use single integrals by employing the strip method (see Fig. 8.13b). As shown in Fig. 8.17, we take a radial triangular strip with differential area dA

$$dA = \tfrac{1}{2}(a\,d\theta)(a) = \tfrac{1}{2}a^2\,d\theta \tag{e}$$

where the infinitesimal length $a\,d\theta$ is treated as a straight line. Then, the area is

$$A = \int dA = \frac{1}{2}\int_0^{\pi} a^2\,d\theta = \left(\frac{\pi}{2}\right)a^2 \qquad \text{(f)}$$

Since the centroid of the triangular strip is located at a distance $2a/3$ from the origin O, the y coordinate of the centroid of the strip is $y_c = (2a/3)(\sin\theta)$. Therefore, with Eq. (e),

$$\begin{aligned} Q_x &= \int y_c\,dA \\ &= \int_0^{\pi}\left(\frac{2}{3}a\sin\theta\right)\left(\frac{1}{2}a^2\,d\theta\right) \qquad \text{(g)} \\ &= \frac{2}{3}a^3 \end{aligned}$$

and the y coordinate of the centroid is, with Eqs. (f) and (g),

$$\bar{y} = \frac{Q_x}{A} = \frac{4a}{3\pi}$$

as before [see Eq. (d)]. The centroid is located as shown in Fig. 8.18.

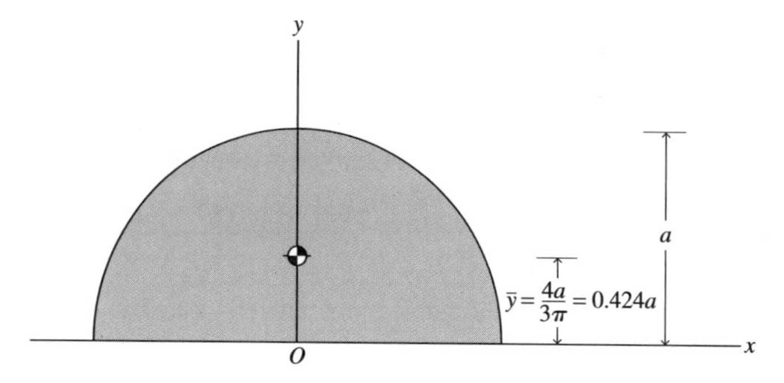

Figure 8.18 Centroid of semicircular area.

Plane Lines Similarly, we can determine the centroid of a semicircular line ABC that lies in a plane—say, the xy plane (see Fig. 8.19). Since the line ABC is symmetrically located relative to the y axis, the centroid lies on the y axis; that is, $\bar{x} = 0$.

The differential length ds of the line is $ds = a\,d\theta$. Hence, the length of the line is

$$L = \int ds = \int_0^{\pi} a\,d\theta = \pi a$$

a well-known result. Since the y coordinate of ds is $y_c = a\sin\theta$, with Eqs. (8.14b) and (8.15b), we have

$$\begin{aligned} \bar{y} &= \frac{1}{L}\int y_c\,ds \\ &= \frac{1}{\pi a}\int_0^{\pi}(a\sin\theta)(a\,d\theta) \\ &= \left(\frac{2}{\pi}\right)a \approx 0.637a \end{aligned}$$

Hence, the centroid of the line is located at the point $(\bar{x}, \bar{y}) = (0, 0.637a)$ (refer to Fig. 8.19). We could determine the centroid by integration in Cartesian coordinates as a check, but that is left as an exercise for you.

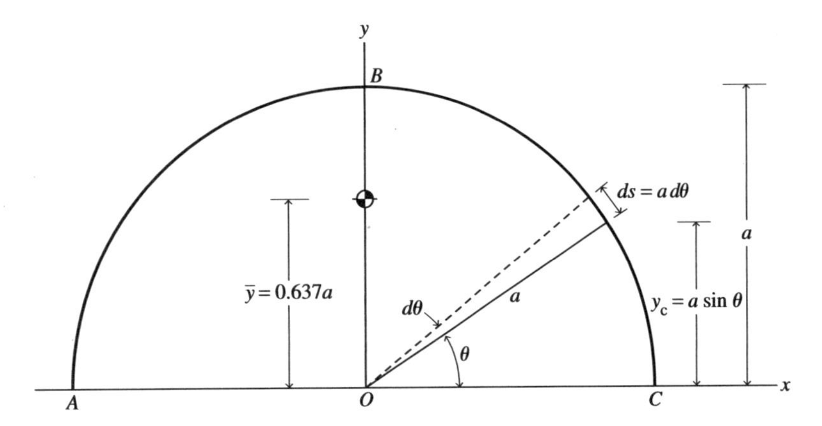

Figure 8.19 Semicircular plane line.

PROBLEM-SOLVING TECHNIQUE

Centroids of Plane Areas and Lines

The determination of the centroid of a plane area or line is similar to that for the centroid of a volume. (See the Problem-Solving Technique at the end of Sec. 8.3.) We merely replace the word "body" with "plane area" or "line," and refer to the appropriate equations. However, for your convenience, the modified Problem-Solving Technique follows.

To determine the centroid of a plane area or line:

1. Sketch the plane area or line approximately to scale.
2. Choose an appropriate coordinate system for the analysis, depending on the shape of the area or line.
3. Utilize symmetry properties of the area or line to help determine the centroid coordinates. If the area or line is symmetric with respect to an axis, the centroid lies on the axis. A coordinate axis should be chosen to coincide with the axis of symmetry (see Step 2).
4. Choose a differential element (or strip) of area dA or a differential length dL, and express the centroid coordinates of dA or dL in terms of the chosen coordinate system.
5. Determine the total area A or total length L.
6. Determine the first moments of area or line by integration [see Eqs. (8.11) for an area or Eqs. (8.15) for a line].
7. Divide the first moments of area or line, respectively, by the area A or the length L to determine the coordinates of the centroid [see Eqs. (8.13) for an area or Eqs. (8.16) for a line].
8. Locate the centroid in the sketch (Step 1) of the area or line. Make certain that the location makes sense. In addition, it is a good idea to check your results by solving the problem a different way, if you have time. For example, use a different coordinate system or differential element.

Example 8.7

Centroid of a Quadrant of an Ellipse

Problem Statement As shown in Fig. E8.7, an area A lies in the first quadrant of a Cartesian coordinate system and is bounded by the coordinate axes and the ellipse with the equation

$$\frac{x^2}{a^2} + \frac{y^2}{b^2} = 1 \tag{a}$$

Locate the centroid of the area.

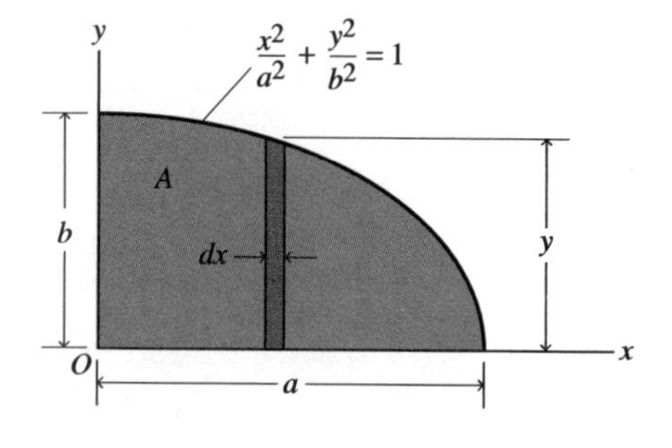

Figure E8.7

Solution The height of a vertical strip is given by $y = b\sqrt{1 - (x^2/a^2)}$ (see Fig. E8.7). The area of the strip is $y\,dx$. Hence, by Eq. (a) and Eq. (8.11a), with $h = y$, the moment of the area A about the y axis is

$$Q_y = b\int_0^a x\sqrt{\left(1 - \frac{x^2}{a^2}\right)}\,dx \tag{b}$$

The integral in Eq. (b) is evaluated readily by means of the substitution

$$u = 1 - \frac{x^2}{a^2} \tag{c}$$

Differentiation of Eq. (c) yields $du = (-2x\,dx)/a^2$; so,

$$x\,dx = -\tfrac{1}{2}a^2\,du$$

The new limits of integration are determined from Eq. (c) by the conditions $u = 1$ if $x = 0$ and $u = 0$ if $x = a$. Accordingly,

$$Q_y = -\tfrac{1}{2}ba^2\int_1^0 \sqrt{u}\,du = \tfrac{1}{3}ba^2 \tag{d}$$

Also, the area of the quadrant of the ellipse is

$$A = \frac{\pi ab}{4} \tag{e}$$

Equation (e) may be verified by integration of the relation $A = \int_0^a y\,dx$.

Equations (d) and (e) and Eq. (8.12a) now yield

$$x = \frac{4a}{3\pi}$$

By a similar computation for $\bar{y}$, we obtain

$$\bar{y} = \frac{4b}{3\pi}$$

Hence, the coordinates of the centroid of the quadrant of the elliptical area are

$$(\bar{x}, \bar{y}) = \left(\frac{4a}{3\pi}, \frac{4b}{3\pi}\right) \tag{f}$$

Note that the ellipse becomes a circle if $a = b$. Then Eq. (f) reduces to

$$\bar{x} = \bar{y} = \frac{4a}{3\pi}$$

where a is the radius of the circle.

Example 8.8

Centroid of an Area with a Parabolic Boundary

Problem Statement Determine the centroid of a region in the first quadrant that is bounded by the y axis, by the line $y = b$, and by the parabola $y = bx^2/a^2$, where a and b are constants (see Fig. E8.8a).

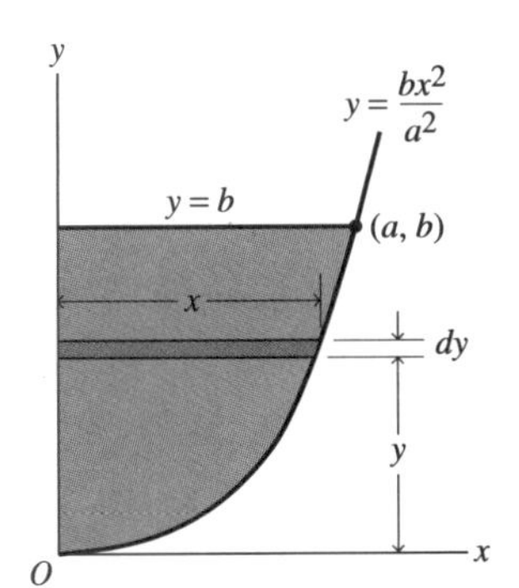

Figure E8.8a

Solution The parabola passes through the origin (0, 0) and the point (a, b). The area of a strip that is parallel to the x axis (see Fig. E8.8a) is $dA = x\,dy = a\sqrt{y/b}\,dy$. Hence, the area of the given region is

$$A = \frac{a}{\sqrt{b}}\int_0^b y^{1/2}\,dy = \tfrac{2}{3}ab \qquad \text{(a)}$$

Likewise, the moment of the area about the x axis is

$$Q_x = \frac{a}{\sqrt{b}}\int_0^b y^{3/2}\,dy = \tfrac{2}{5}ab^2 \qquad \text{(b)}$$

By Eqs. (a) and (b) and Eq. (8.13b), the ordinate of the centroid of the area is

$$\bar{y} = \frac{Q_x}{A} = \tfrac{3}{5}b$$

To determine $\bar{x}$, we consider a vertical strip (see Fig. E8.8b). The length of this strip is $b - y$. Hence,

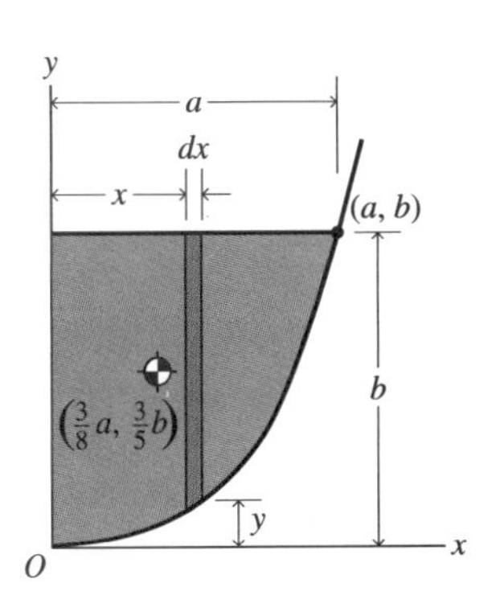

Figure E8.8b

$$dA = (b - y)\,dx = b\left(1 - \frac{x^2}{a^2}\right)dx$$

Therefore, $dQ_y = b(1 - x^2/a^2)\,x\,dx$ and

$$Q_y = b\int_0^a \left(1 - \frac{x^2}{a^2}\right)x\,dx = \tfrac{1}{4}a^2b \qquad \text{(c)}$$

Equations (a) and (c) and Eq. (8.13a) yields $\bar{x} = 3a/8$. Thus, the centroid is located at

$$(\bar{x}, \bar{y}) = \left(\tfrac{3}{8}a, \tfrac{3}{5}b\right)$$

The location of the centroid is shown in Fig. E8.8b.

Example 8.9

Centroid of a Parabolic Arc

Problem Statement Determine the centroid coordinates $(\bar{x}, \bar{y})$ of an arc of the parabola $x = ay^2$ that extends from the point (0, 0) to the point (2, 1).

Solution We note that since the parabola passes through the point (2, 1), $a = 2$ (see Fig. E8.9). Therefore,

$$x = 2y^2 \qquad \text{(a)}$$

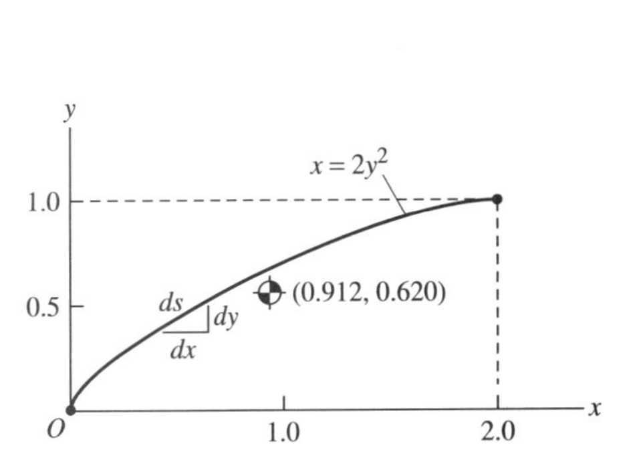

Figure E8.9

By Eqs. (8.16), the centroid coordinates $(\bar{x}, \bar{y})$ are given as

$$\bar{x} = \frac{1}{L}\int_0^L x\,ds \tag{b}$$

$$\bar{y} = \frac{1}{L}\int_0^L y\,ds \tag{c}$$

where

$$L = \int_0^L ds \tag{d}$$

To integrate Eqs. (b), (c), and (d), we first express ds in terms of dx, as follows (see Fig. E8.9):

$$ds^2 = dx^2 + dy^2 = \left[1 + \left(\frac{dy}{dx}\right)^2\right] dx^2$$

or

$$ds = \sqrt{1 + \left(\frac{dy}{dx}\right)^2}\, dx \tag{e}$$

From Eqs. (a) and (e), we find

$$ds = \frac{\sqrt{1 + 8x}}{\sqrt{8x}}\, dx \tag{f}$$

Substitution of Eq. (f) into Eq. (d) yields

$$L = \int_0^2 \frac{\sqrt{1 + 8x}}{\sqrt{8x}}\, dx$$

Integration then gives

$$L = \left[\tfrac{1}{8}\sqrt{8x + 64x^2} + \tfrac{1}{8}\ln\left(8\sqrt{8x} + 8\sqrt{1 + 8x}\right)\right]_0^2 = 2.323 \tag{g}$$

With L determined, we may compute $(\bar{x}, \bar{y})$ using Eqs. (b) and (c). By Eq. (b), with Eqs. (f) and (g), we find

$$\bar{x} = \frac{1}{(2.323)\sqrt{8}}\int_0^2 \sqrt{x + 8x^2}\, dx$$

$$= \frac{1}{(2.323)\sqrt{8}}\left[\frac{1 + 16x}{32}\sqrt{x + 8x^2} - \frac{\sqrt{2}}{128}\ln\left(\sqrt{8x} + \sqrt{1 + 8x}\right)\right]_0^2 \tag{h}$$

Evaluating Eq. (h), we obtain $\bar{x} = 0.912$.

Likewise, by Eq. (c), with Eqs. (f) and (g), we have

$$\begin{aligned}\bar{y} &= \frac{1}{2.323\sqrt{16}}\int_0^2 \sqrt{1 + 8x}\, dx \\ &= \frac{1}{2.323\sqrt{16}}\left[\frac{1}{12}\sqrt{(1 + 8x)^3}\right]_0^2\end{aligned} \tag{i}$$

Evaluating Eq. (i), we obtain $\bar{y} = 0.620$.

The centroid $(\bar{x}, \bar{y})$ is shown in Fig. E8.9. Its coordinates are (0.912, 0.620). Note that, in general, the centroid of a curved line does not lie on the line. ■

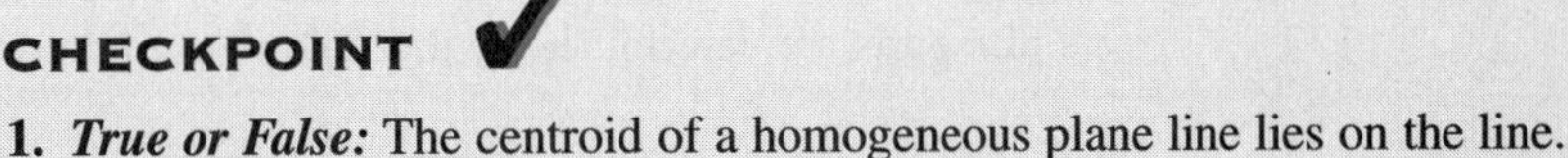

CHECKPOINT

1. ***True or False:*** The centroid of a homogeneous plane line lies on the line.
2. ***True or False:*** The centroid of a homogeneous body is always located within the body.

Answers: 1. False; 2. False

8.5 COMPOSITE BODIES

After studying this section, you should be able to:

- Locate the center of gravity or the centroid of a composite body.

Key Concept Bodies that are made up of several parts of different shapes are called composite bodies. The center of gravity of a composite body may be determined from the centers of gravity of its individual parts.

BODIES WITHOUT HOLES OR CAVITIES

composite body

A *composite body* consists of several parts whose weights and centers of gravity are known. The center of gravity of the composite body may be determined by the theory of parallel forces (Sec. 4.8), since the weight vector of each part is known. This idea has already been used in Examples 8.1 and 8.2. For example, consider a body that is divided into n parts, whose weights and centers of gravity are known. For such a body, Eqs. (8.1) may be written in the form

$$\bar{x} = \frac{w_1 x_1 + w_2 x_2 + \cdots + w_n x_n}{w_1 + w_2 + \cdots + w_n}$$

$$\bar{y} = \frac{w_1 y_1 + w_2 y_2 + \cdots + w_n y_n}{w_1 + w_2 + \cdots + w_n} \tag{8.17}$$

$$\bar{z} = \frac{w_1 z_1 + w_2 z_2 + \cdots + w_n z_n}{w_1 + w_2 + \cdots + w_n}$$

where $w_1, w_2, \ldots, w_n$ denote the weights of the individual parts and (x_1, y_1, z_1), $(x_2, y_2, z_2), \ldots, (x_n, y_n, z_n)$ denote the corresponding centers of gravity.

If each part of the body has the same specific weight γ, the weights of the individual parts may be written as γV_i, where V_i is the volume of the ith part. Then, the specific weight cancels out of Eqs. (8.17); that is, the coordinates $(\bar{x}, \bar{y}, \bar{z})$ are independent of the specific weight. Equations (8.17) define the centroid of the volume of the composite body. Equations analogous to Eqs. (8.17) hold for the centroids of composite plane areas and lines, if the quantities w_i are replaced by the areas and arc lengths, respectively, of the various parts. Thus, the centroid coordinates for a composite volume, area, and line that consists of n parts are determined by the following equations.

$$\text{For a volume} \qquad \bar{x} = \frac{\sum V_i x_i}{\sum V_i}, \quad \bar{y} = \frac{\sum V_i y_i}{\sum V_i}, \quad \bar{z} = \frac{\sum V_i z_i}{\sum V_i} \tag{8.18}$$

$$\text{For a plane area} \qquad \bar{x} = \frac{\sum A_i x_i}{\sum A_i}, \quad \bar{y} = \frac{\sum A_i y_i}{\sum A_i} \tag{8.19}$$

$$\text{For a plane line} \qquad \bar{x} = \frac{\sum L_i x_i}{\sum L_i}, \quad \bar{y} = \frac{\sum L_i y_i}{\sum L_i} \tag{8.20}$$

where the summation is taken over the n individual parts.

BODIES WITH HOLES OR CAVITIES

Many engineering systems have mechanical parts that contain holes or cavities. To determine the center of gravity of a body (or the centroid of a volume, area, or line) that contains a hole or gap, we proceed as follows: First, we consider a corresponding body without a hole and determine the moments of its weight with respect to coordinate planes [see Eqs. (8.2)]. Next, we consider the volume (or area) formed by the hole and determine the moments of its weight with respect to the coordinate planes. The moment of weight of the body with the hole is the difference of the moment of weight of the body without the hole and the moment of weight of the material that fills the volume of the hole. This procedure may also be employed to determine the centroids of volumes or areas that contain cavities or holes or the centroids of discontinuous lines. To demonstrate the procedure, we consider the shaded area A shown in Fig. 8.20a. Let part 1 be the rectangle without the hole, and let part 2 be the circular disk that is removed to form the hole (Fig. 8.20b). Let the areas of parts 1 and 2 be A_1 and A_2, respectively. Then, the centroid coordinate $\bar{x}$ of the shaded area A in Fig. 8.20a is determined by the equation

$$A\bar{x} = A_1 x_1 - A_2 x_2$$

or

$$\bar{x} = \frac{A_1 x_1 - A_2 x_2}{A} \tag{8.21}$$

where $A = A_1 - A_2$. Since the disk is removed, its area and moment are subtracted. In general, we regard the areas or weights of parts that are removed as negative quantities.

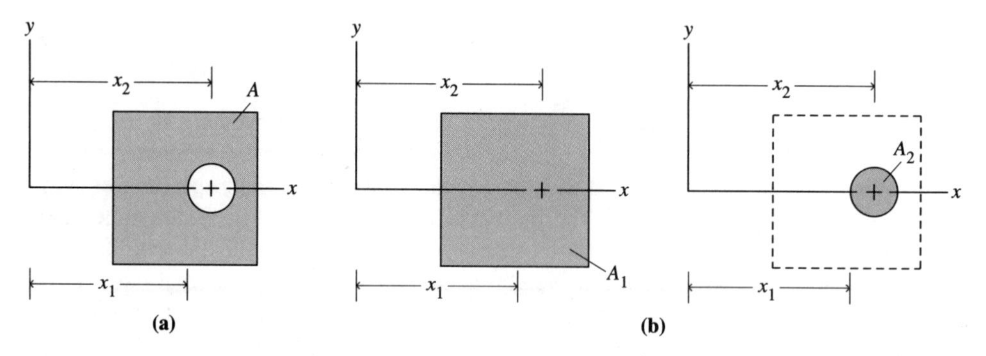

Figure 8.20 Composite body.

In summary, for composite bodies, areas, and lines, we can use the following Problem-Solving Technique. For brevity, it only describes how to determine the center of gravity of a composite body. Modifications for determining the center of gravity (or centroid) of a composite area or line are left to you.

PROBLEM-SOLVING TECHNIQUE

Center of Gravity of a Composite Body

To determine the center of gravity of a composite body:

1. Draw an accurate sketch of the body, approximately to scale.
2. If possible, divide the body into n parts, whose weights and centers of gravity are known. If the weight and center of gravity of one or more parts are not known, determine them, using the methods outlined in Secs. 8.3 and 8.4.
3. If the parts of the body have different specific weights, γ_i, Eqs. (8.17) can be used to determine the center of gravity coordinates, with $w_i = \gamma_i V_i$.
4. If the parts of the body have the same specific weight γ, the weights of the individual parts of the body may be written as γV_i, where V_i is the volume of the ith part.
5. The coordinates of the center of gravity of the composite body are obtained by dividing the sum of the first moments by the sum of the weights (see Example 8.13).
6. If a body contains a hole, first consider a corresponding body without the hole or cavity. Determine the weight and the first moments of weight of that body with respect to coordinate planes. Next, subtract from this weight and the first moments the weight and first moments of the body that fills the hole or cavity. Using these differences, repeat Step 5 to calculate the coordinates of the center of gravity.
7. Locate the center of gravity in the sketch of Step 1 to see whether it makes sense.

Example 8.10

Centroid of a Composite Area

Problem Statement Determine the coordinates $(\bar{x}, \bar{y})$ of the centroid of the C-shaped area in Fig. E8.10a.

Solution The composite area shown in Fig. E8.10a may be subdivided into three rectangles in several ways (see Figs. E8.10b, E8.10c, and E8.10d). Let's consider the composite area shown in Fig. E8.10b. The centroid of each part is located at its geometric center, so the centroid coordinates (x_i, y_i) for each part relative to the x and y axes are easily determined. It is convenient to tabulate the results of the calculations as in Table E8.10.

From Eqs. (8.19) and Table E8.10, we obtain the centroid coordinates relative to the x and y reference axes as

$$\bar{x} = \frac{25\,500}{1\,500} = 17.0 \text{ mm}$$

$$\bar{y} = \frac{40\,500}{1\,500} = 27.0 \text{ mm}$$

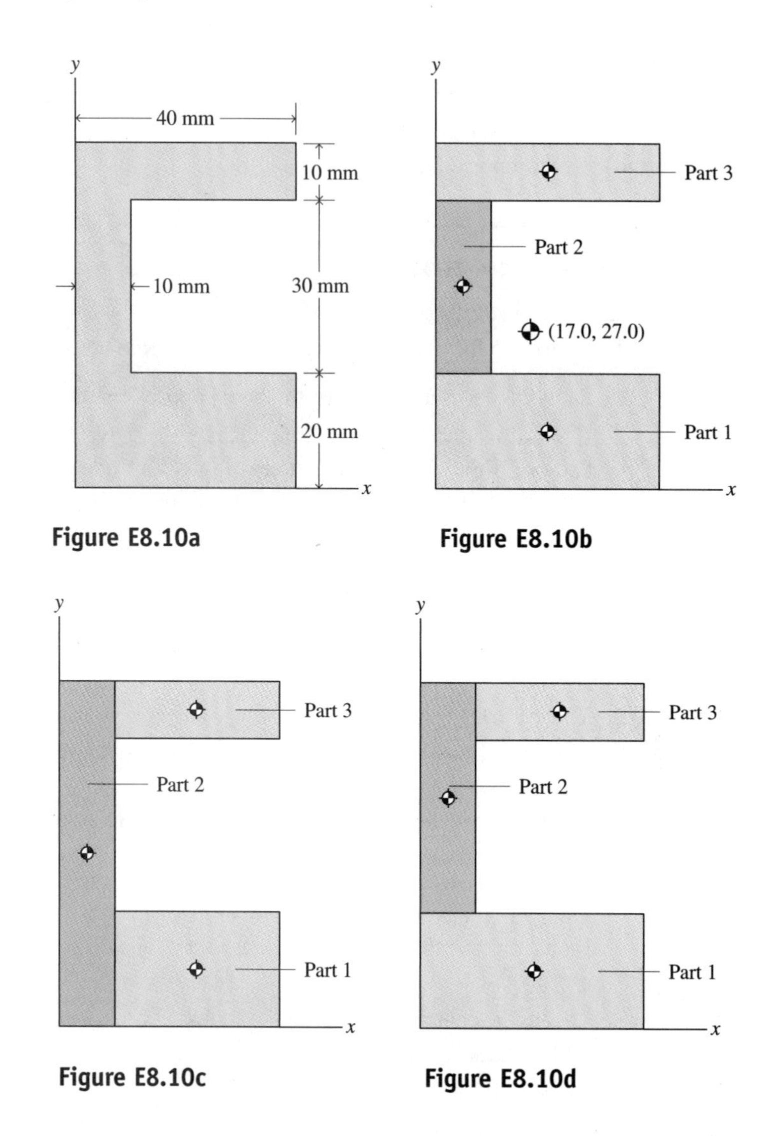

Figure E8.10a

Figure E8.10b

Figure E8.10c

Figure E8.10d

The location of the centroid is shown in Fig. E8.10b. As an exercise, determine the centroid using the composite area shown in Fig. E8.10c or E8.10d.

Table E8.10
Centroid coordinates of a composite area

PART	AREA [mm²]	CENTROID DISTANCE [mm]		MOMENT OF AREA [mm³]	
i	A_i	x_i	y_i	$A_i x_i$	$A_i y_i$
1	800	20	10	16 000	8 000
2	300	5	35	1 500	10 500
3	400	20	55	8 000	22 000
Sums	1 500	—	—	25 500	40 500

Example 8.11

Centroid of a Composite Area with a Hole

Problem Statement Let the C-shaped area of Fig. E8.10a be represented by two parts (1 and 2) as shown in Fig. E8.11. Part 1 is a rectangular area without the hole, and part 2 is a square area corresponding to the hole. Determine the centroid of the C-shaped area by considering parts 1 and 2.

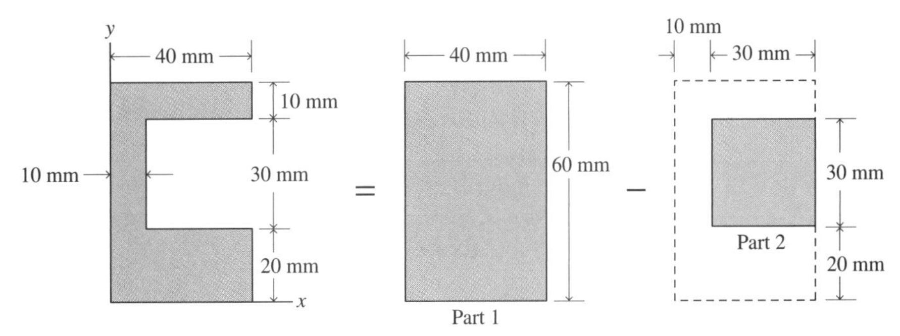

Figure E8.11

Solution From Fig. E8.11, we can tabulate the sums of the moments of the areas of parts 1 and 2 with respect to the x and y axes, as shown in Table E8.11.

Note that the sums in Table E8.11 are the same as the sums in Table E8.10. Thus, the centroid coordinates are, as in Example 8.10,

$$\bar{x} = \frac{25\,500}{1\,500} = 17.0 \text{ mm}$$

$$\bar{y} = \frac{40\,500}{1\,500} = 27.0 \text{ mm}$$

The centroidal coordinates have the same signs as those in Example 8.10, relative to the x and y reference axes. Also, the area of the hole (part 2 in Fig. E8.11) is taken as negative, and, therefore, the moments of that area are also negative [see Eq. (8.21)].

Table E8.11
Centroid coordinates of a composite area with a hole

PART	AREA [mm²]	CENTROID DISTANCE [mm]		MOMENT OF AREA [mm³]	
i	A_i	x_i	y_i	A_ix_i	A_iy_i
1	2 400	20	30	48 000	72 000
2	−900	25	35	−22 500	−31 500
Sums	1 500	—	—	25 500	40 500

Example 8.12

Centroid of the Cross Section of a Structural Member

Problem Statement A long, flat sheet of steel, 0.5 in thick, is bent into the shape shown in cross-section in Fig. E8.12a. Determine the centroid coordinates $(\bar{x}, \bar{y})$ of the cross section.

Solution The area of the cross section consists of the rectangular area AB (part 1), the rectangular area CD (part 2), and the annular area BC. Furthermore, since the centroid of a circular quadrant is known (see Appendix D), it is convenient to regard area BC as the area that remains after the circular quadrant OBC (part 4) is removed from the circular quadrant OEF (part 3) (see Fig. E8.12b). Since the area OBC is removed, its value is negative.

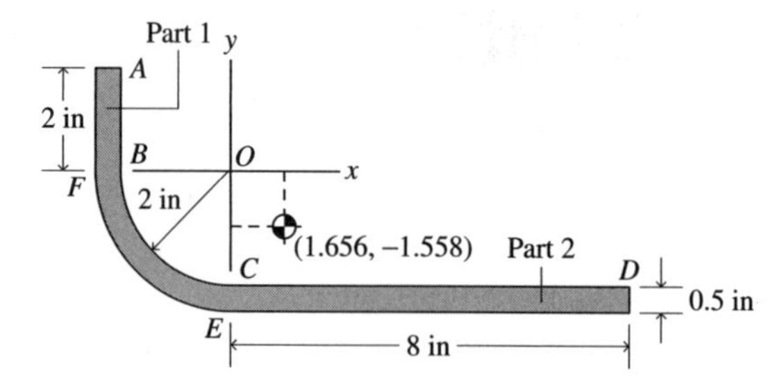

Figure E8.12a

The areas, the (x, y) coordinates of the centroids, and the moments of the areas are given in Table E8.12.

The centroids of the areas OEF and OBC may be obtained from Appendix D. The centroidal coordinates of OEF are $x_3 = y_3 = (-4)(2.5)/3\pi = -1.061$ in, and the centroidal coordinates of OBC are $x_4 = y_4 = (-4)(2)/3\pi = -0.8488$ in.

With Eq. (8.19) and the sums from Table E8.12, the centroidal coordinates of the cross section are

$$\bar{x} = \frac{11.209}{6.767} = 1.656 \text{ in}$$

$$\bar{y} = \frac{-10.541}{6.767} = -1.558 \text{ in}$$

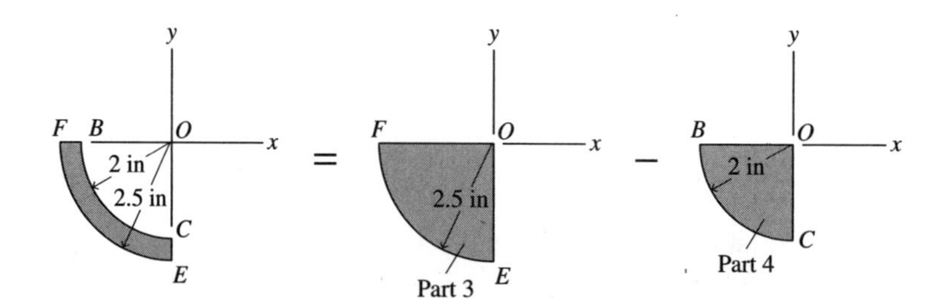

Figure E8.12b

Table E8.12
Centroid of cross section of rolled steel member

PART	AREA [in²]	CENTROID DISTANCE [in]		MOMENTS OF AREA [in³]	
i	A_i	x_i	y_i	A_ix_i	A_iy_i
1	1.000	−2.250	1.000	−2.250	1.000
2	4.000	4.000	−2.250	16.000	−9.000
3	4.909	−1.061	−1.061	−5.208	−5.208
4	−3.142	−0.8488	−0.8488	2.667	2.667
Sums	6.767	—	—	11.209	−10.541

Example 8.13

Center of Gravity of a Composite Body

Problem Statement The base of the composite machine part shown in Fig. E8.13a has specific weight $\gamma = 78$ kN/m^3. The remainder of the part has specific weight $\gamma = 26$ kN/m^3. Determine the center of gravity of the part, with respect to the xyz axes shown.

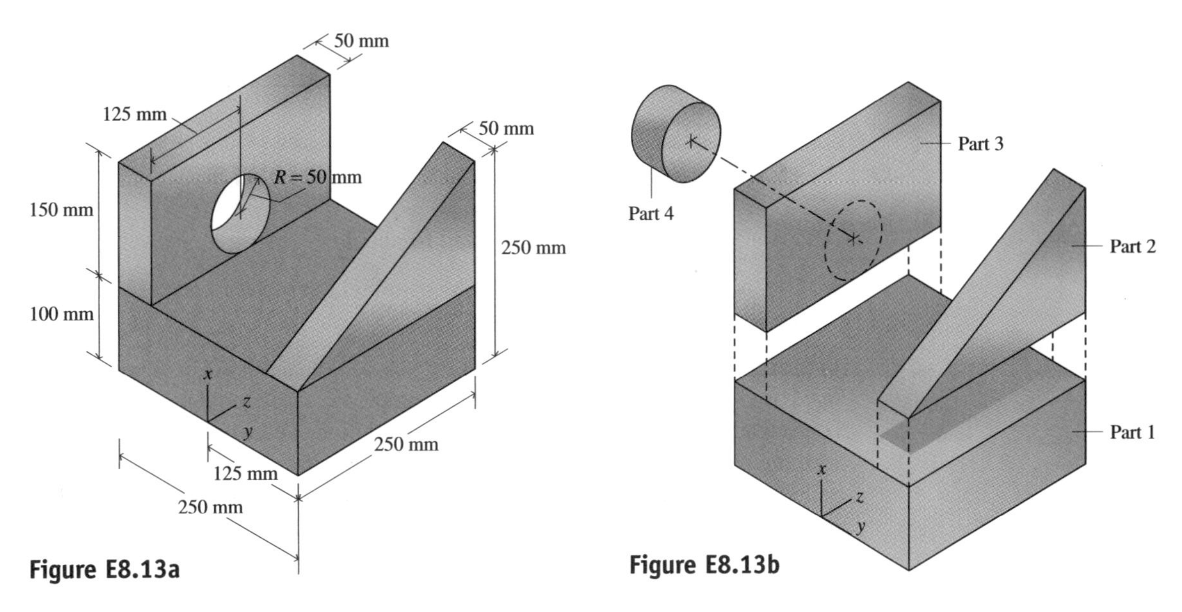

Figure E8.13a

Figure E8.13b

Solution The assembly consists of the base (part 1), the triangular prism (part 2), and the vertical rectangular slab without the hole (part 3). A circular disk forming the hole (part 4) must be removed from part 3 (see Fig. E8.13b). Since the disk is removed, its weight is listed as negative in Table E8.13, which presents the results of calculations.

Using Eq. (8.1) and the sums from Table E8.13, we obtain coordinates of the center of gravity as

$$\bar{x} = \frac{35.04}{550.42} = 63.66 \text{ mm}$$

$$\bar{y} = \frac{-1.42}{550.42} = -2.58 \text{ mm}$$

$$\bar{z} = \frac{69.81}{550.42} = 126.83 \text{ mm}$$

Table E8.13
Center of gravity of a composite body

PART	VOLUME [10^6 mm^3]	WEIGHT [N]	CENTROID DISTANCES [mm]			MOMENTS OF THE WEIGHTS [N·m]		
i	V_i	$w_i = \gamma_i V_i$	x_i	y_i	z_i	$w_i x_i$	$w_i y_i$	$w_i z_i$
1	6.2500	487.50	50	0	125.00	24.38	0.00	60.94
2	0.9375	24.38	150	100	166.67	3.66	2.44	4.06
3	1.8750	48.75	175	−100	125.00	8.53	−4.88	6.09
4	−0.3927	−10.21	150	−100	125.00	−1.53	1.02	−1.28
Sums	—	550.42	—	—	—	35.04	−1.42	69.81

CHECKPOINT 

1. ***True or False:*** The center of gravity of a composite body can be found using the theory of parallel forces.
2. ***True or False:*** In practice, the computations for determining the center of gravity of a composite body are often organized in table form for convenience.

Answers: 1. True; 2. True

8.6 THEOREMS OF PAPPUS-GULDINUS

After studying this section, you should be able to:

- Use the theorems of Pappus-Guldinus to find the surface area, volume, or centroid of a body of revolution.

In the fourth century (around 380 AD) Pappus, an Alexandrian Greek mathematician, developed two formulas that provide relatively simple means of computing the areas and volumes, respectively, of surfaces and volumes of revolution. These formulas are shortcuts for performing otherwise lengthy calculations. Generally, the Pappus-Guldinus formulas may be called (1) the Pappus-Guldinus theorem for surface areas and (2) the Pappus-Guldinus theorem for volumes.[6] The Pappus-Guldinus theorem for surface areas can be stated as follows.

Theorem 8.4

The Pappus-Guldinus Theorem for Surface Areas

If an arc C of a smooth curve that lies in a plane is rotated through an angle θ ($0 \le \theta \le 2\pi$) about an axis that also lies in the plane and that does not intersect arc C, the surface area generated by arc C as it rotates through the angle θ is equal to the length of C times the length of the path traveled by the centroid of C during the rotation θ.

Thus, if the length of the arc is L and ρ is the distance from the axis of rotation to the centroid of the arc, the surface area S generated by the arc as it rotates through the angle θ about the axis of rotation is

$$S = L\rho\theta \tag{8.22}$$

The derivation of Eq. (8.22), which proves the theorem, follows.

Consider a general continuous arc AB in the xy plane that extends from $x = a$ to $x = b$ (see Fig. 8.21a). Let the axis of rotation be the x axis. Note that the axis of rotation does not intersect the arc AB. The surface area element dS that is generated by a differential length ds of arc AB as it is rotated through θ radians about the x axis is $dS = \theta y\, ds$.

6. Since Guldinus (Paul Guldin, 1577–1643) rediscovered the theorems of Pappus, his name is often associated with those theorems. The name *Guldinus,* the Latinized version of Guldin, was introduced by later scholars.

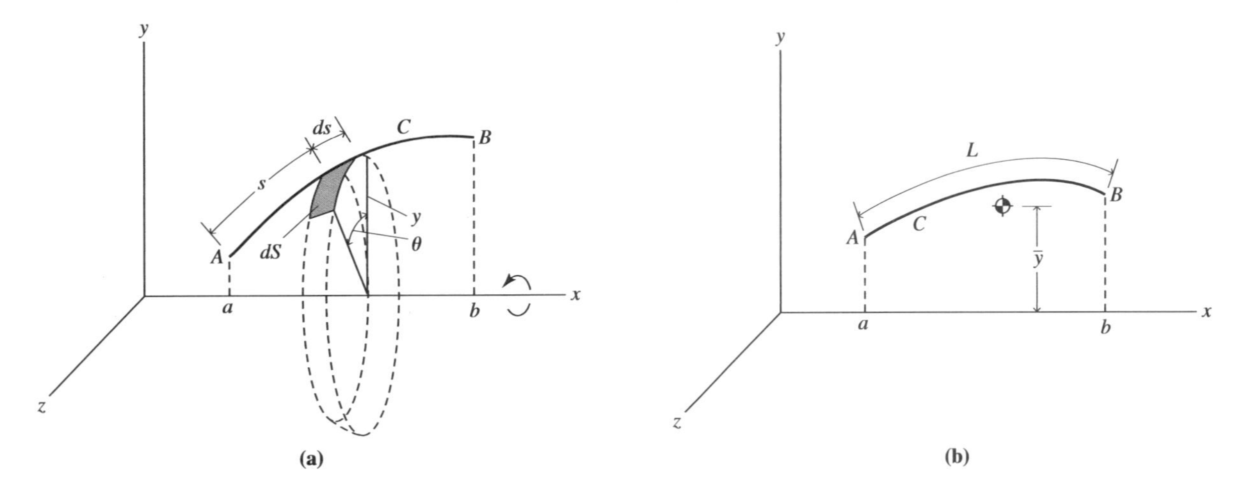

Figure 8.21

Hence, the surface area generated by the arc AB as it is rotated through the angle θ about the x axis is

$$S = \int_A^B \theta y\, ds = \theta \int_A^B y\, ds \tag{a}$$

To determine the centroid coordinate $\bar{y}$ of the arc, we have [see Eq. (8.16b) and Fig. 8.21b]

$$\int_A^B y\, ds = \bar{y}L \tag{b}$$

Substitution of Eq. (b) into Eq. (a) yields

$$S = \theta \bar{y} L \quad \text{for } 0 \leq \theta \leq 2\pi \tag{c}$$

Equation (c) is the result we set out to prove [Eq. (8.22)], with $\bar{y} = \rho$, the distance from the centroid C of the arc to the axis of rotation (the x axis). In particular, if $\theta = 2\pi$, the surface is a complete surface of revolution with area $S = 2\pi\bar{y}L$.

The Pappus-Guldinus theorem for volumes may be stated as follows.

Theorem 8.5

The Pappus-Guldinus Theorem for Volumes

If an area A that lies in a plane is rotated through an angle θ ($0 \leq \theta \leq 2\pi$) about an axis that also lies in the plane and that does not intersect area A, the volume of the solid generated by area A as it rotates through the angle θ is equal to the area of A times the length of the path traveled by the centroid of area A during the rotation θ.

If ρ is the distance from the axis of rotation to the centroid of the plane area, the volume V is given by the relation

$$V = A\rho\theta \tag{8.23}$$

The derivation of Eq. (8.23) follows.

The plane area A in Fig. 8.22a lies in the xy plane in the region from $y = a$ to $y = b$. Let the area A be rotated about the x axis through an angle θ. The element of volume dV

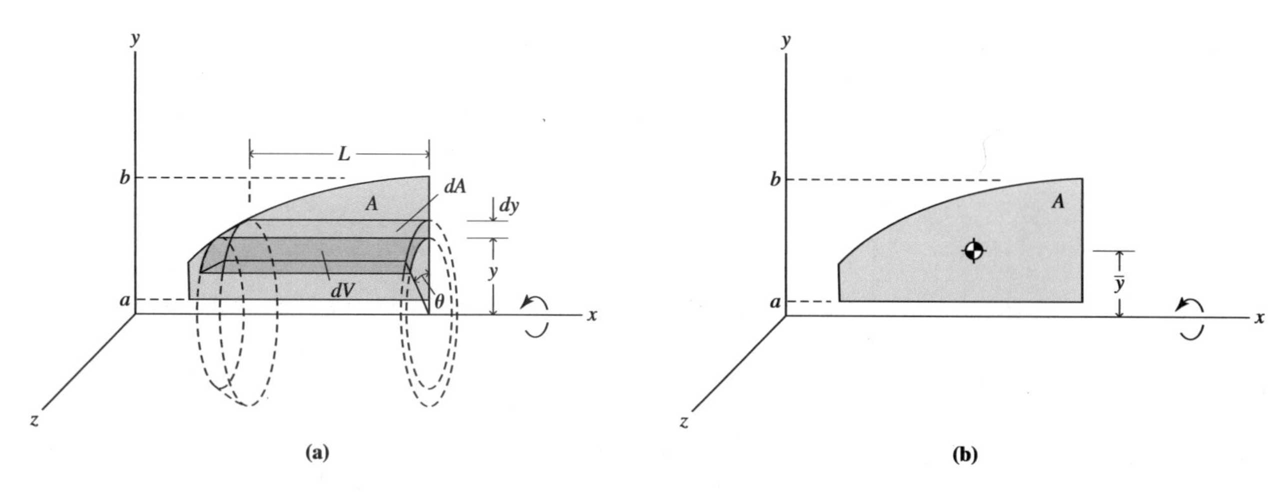

Figure 8.22

generated by the plane area $dA = L\,dy$ rotating through θ is $dV = \theta yL\,dy$. Hence, the volume V generated by the plane area A through the rotation θ $(0 \le \theta \le 2\pi)$ about the x axis is given by the formula

$$V = \int_a^b \theta yL\,dy = \theta \int_a^b yL\,dy \tag{d}$$

Substituting L for b in Eq. (8.10b), we have

$$\bar{y}A = \int_a^b yL\,dy \tag{e}$$

Substitution of Eq. (e) into Eq. (d) yields

$$V = \theta\bar{y}A \quad \text{for } 0 \le \theta \le 2\pi \tag{f}$$

Equation (f) is the result we set out to prove [Eq. (8.23)], with $\bar{y} = \rho$, the distance from the centroid of the area A to the axis of rotation (the x axis) (see Fig. 8.22b). In particular, if $\theta = 2\pi$, the volume V is a complete volume of revolution given by $V = 2\pi\bar{y}A$.

Recall that the Pappus-Guldinus theorem for surface areas restricts the arc length to a smooth function of x. Similarly, here, the region of the plane area is smooth enough that in the infinitesimal area $dA = L\,dy$, the length L varies continuously with its distance y from the axis of rotation.

The theorems of Pappus-Guldinus are still useful today in calculating the surface areas and volumes of various solids encountered in engineering problems, particularly in mechanical and civil engineering. You should note that the formulas of the theorems of Pappus-Guldinus can be used in reverse to determine the centroid of an area or a line. For example, if the volume of a given solid of revolution is known, Eq. (8.23) can be used to determine the centroid of the plane area that has generated the volume—that is, to determine the distance from the axis of rotation to the centroid of the plane area. Likewise, if the area of a given surface of revolution is known, Eq. (8.22) can be used to determine the distance from the axis of rotation to the centroid of the line that has generated the surface.

Example 8.14

Volume of a Solid Cone

Problem Statement **a.** Derive a formula for the volume V of the solid right-circular cone in Fig. E8.14a, which is generated by rotating the triangular cross section in Fig. E8.14b about the x axis.

b. For $a = 2$ m, $b = 1$ m, determine the volume V, in cubic meters.

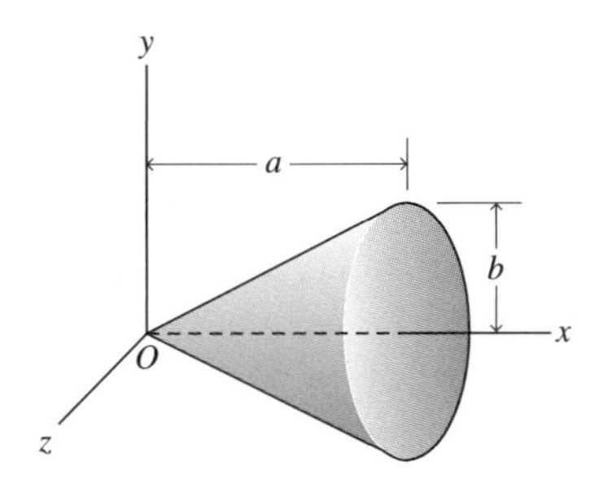

Figure E8.14a

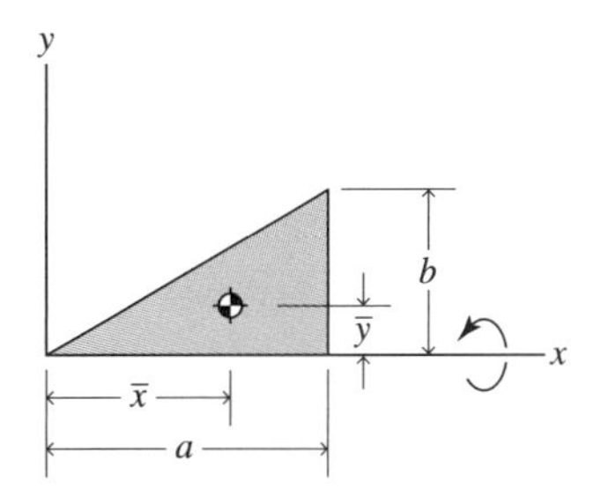

Figure E8.14b

Solution **a.** This problem may be solved using the Pappus-Guldinus theorem for volumes. From the results of Example 8.4, for the *xy* axes selected here, the *y* coordinate of the centroid of the homogeneous triangular area of Fig. E8.14b is

$$\bar{y} = \frac{b}{3} \tag{a}$$

Also, the area A of the triangle is

$$A = \frac{1}{2}ab \tag{b}$$

With Eqs. (a) and (b) and the Pappus-Guldinus theorem [Eq. (8.23)], noting that $\rho = \bar{y}$, we have

$$V = 2\pi\bar{y}A = 2\pi\left(\frac{b}{3}\right)\left(\frac{ab}{2}\right) = \frac{\pi ab^2}{3} \tag{c}$$

b. For $a = 2$ m, $b = 1$ m, Eq. (c) yields

$$V = \frac{\pi(2)(1)^2}{3} = 2.094 \text{ m}^3$$

Example 8.15 *Surface Area of a Right-Circular Cone*

Problem Statement **a.** Determine a formula for the lateral surface area S of the solid right-circular cone of Example 8.14 (Fig. E8.15a).

b. Calculate the surface area S in square meters when $a = 2$ m, $b = 1$ m.

Solution **a.** By the Pappus-Guldinus theorem for surface areas, the lateral surface S of the cone is generated by rotating the straight line OB through 2π radians about the x axis (Fig. E8.15b). Thus, since the y coordinate of the centroid of the line OB is given by

$$\bar{y} = \frac{b}{2}$$

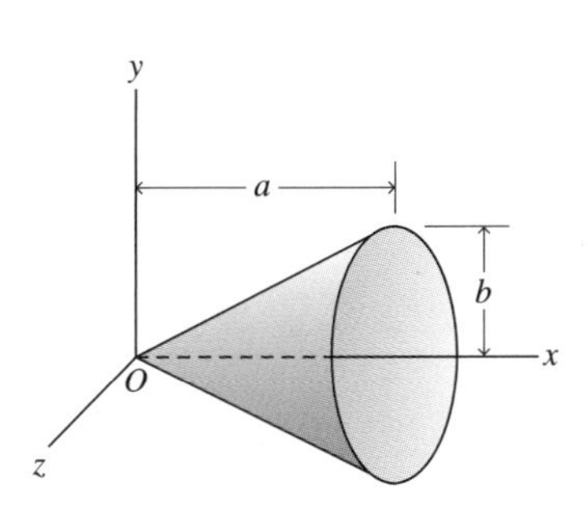

Figure E8.15a

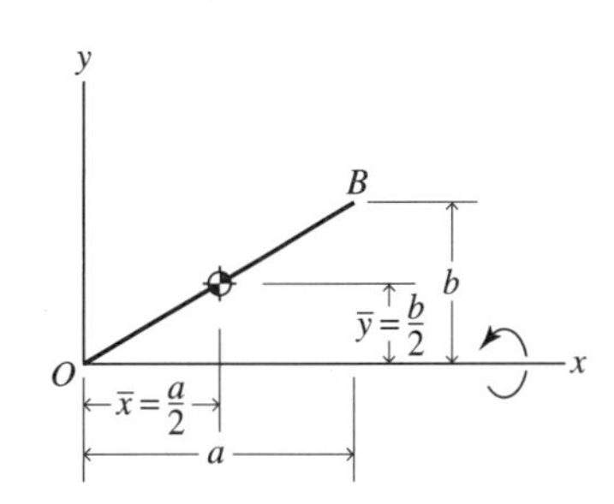

Figure E8.15b

and since the length of OB is $L = \sqrt{a^2 + b^2}$, the Pappus-Guldinus theorem [Eq. (8.22)] gives

$$S = \pi b\sqrt{a^2 + b^2} \tag{a}$$

b. For $a = 2$ m, $b = 1$ m, Eq. (a) yields $S = 7.025$ m^2.

Example 8.16

Surface Area of Sheet Metal on the Nose of an Airplane

Problem Statement An aircraft manufacturer has an order to supply 100 small airplanes to an airline. In designing the airplane, the design engineer needs to estimate the amount of thin aluminum sheet required to cover the fuselage nose of the airplane (see Fig. E8.16). The nose extends 2 m horizontally from the windshield to the tip O. The depth of the nose is also 2 m, and the cross section of the nose, at any distance x from the tip, is approximately circular.

a. How should the engineer estimate the amount of aluminum sheet needed?

b. How can he check his estimate?

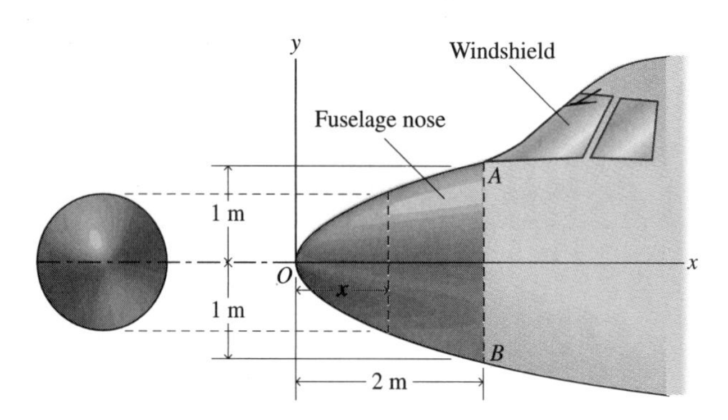

Figure E8.16

Solution **a.** After considering the shape of the nose, the engineer decides that the fuselage arc OA might be approximated by a quarter of an ellipse or possibly by a parabola, and that the surface area can be computed by the Pappus-Guldinus theorem for surface areas. However, noting that the slope at A is not horizontal (as it would be for a quarter of an ellipse from O to A), he elects to use a parabola with origin at O, as defined by the equation

$$x = ay^2$$

To determine the constant a, he notes that the parabola must pass through the points ($x = 2$ m, $y = 1$ m) and ($x = 2$ m, $y = -1$ m). Therefore, $a = 2$ m^{-1}.

To use the Pappus-Guldinus theorem, the engineer decides to rotate the parabolic arc OA about the x axis to generate the nose surface. He observes that the centroid coordinate $\bar{y}$ and the length of the arc OA are needed, which he can readily calculate (see Example 8.9):

$$\bar{y} = 0.6196 \text{ m} \tag{a}$$

$$L = 2.323 \text{ m} \tag{b}$$

With Eqs. (8.22a and 8.22b), he computes the surface area:

$$S = 2\pi\bar{y}L = 2\pi(0.6196)(2.323) = 9.044 \text{ m}^2$$

He recognizes that $S = 9.044$ m^2 is an estimate and that the flat sheet metal must be cut, fitted, and riveted to form the contour of the airplane nose. Therefore, he adds 5% and specifies that 9.50 m^2 is needed for each plane, or 950 m^2 for 100 planes.

b. Before ordering the sheet metal, the engineer decides to check his computations. Therefore, he has his technician build a full-size wood model of the nose and fit it with sheet aluminum. This requires 9.3 m^2 of sheet metal. Therefore, he feels confident that 950 m^2 of aluminum sheeting is sufficient for the 100 planes and places the order.

CHECKPOINT 

1. ***True or False:*** The theorems of Pappus-Guldinus are valid only for rotations θ in the range $0 < \theta < \pi$.
2. ***True or False:*** The theorems of Pappus-Guldinus provide a method for determining the location of the centroid of an area or a line.
3. ***True or False:*** The Pappus-Guldinus theorem for volumes may be used to determine the volume of a pyramid.

Answers: 1. False; 2. True; 3. False

8.7 RESULTANTS OF DISTRIBUTED PARALLEL FORCES

After studying this section, you should be able to:

- Determine the magnitude and line of action of the resultant of a distributed force system on a line or a plane area.

Key Concepts

The resultant and line of action of a distributed force on a line can be found by analogy to the centroid of a plane area.

The resultant and line of action of a distributed force over a plane area can be found by analogy to the centroid of a volume.

In Chapters 2 and 3, you learned that forces can be classified as point forces and distributed forces. We noted that in some cases a distributed force may be represented by its resultant—that is, by a point force. For example, we represented the weight of a rigid body by the resultant weight vector acting at the center of gravity of the body.

In some cases, the effects of a distributed force can be determined by replacing it by its resultant. For example, the resultant force has the same effect on the equilibrium (or the motion) of a rigid body as does the distributed force; in this case, the two force systems are *dynamically equivalent.* However, if a distributed force acts on a flexible body, it pro-

A large number of closely spaced concentrated loads, such as those exerted by the wheels of this old steam locomotive and its cars, may be approximated as a distributed load, $w(x)$.

duces displacements (or motion) that are quite different from those that would be produced by the resultant force.

This section examines situations in which the effects of distributed parallel forces can be accurately determined when those forces are replaced by their resultant. Distributed nonparallel forces, such as those due to water pressure acting on the curved face of a dam, are treated in Chapter 9.

DISTRIBUTED FORCES ON A STRAIGHT LINE SEGMENT

The resultant of distributed parallel forces may be determined by a slight extension of the theory of parallel point forces (Sec. 4.8). For example, let a laterally distributed load $w(x)$ (force per unit length) act on a straight rod AB (Fig. 8.23a). Consider an infinitesimal element dx of the load at a distance x from A. We may regard the corresponding infinitesimal

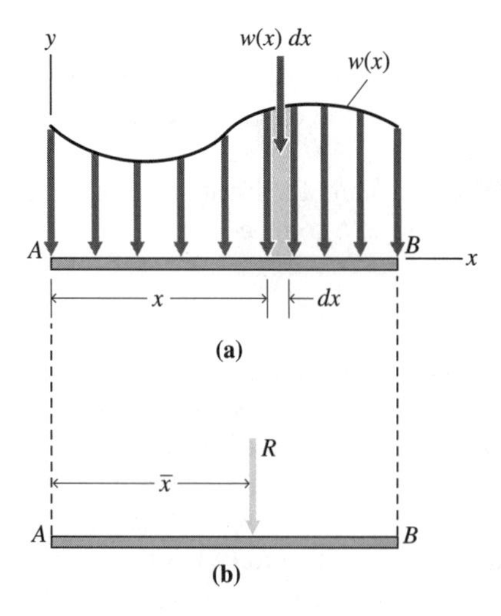

Figure 8.23

force $w(x)\,dx$ to be a point force acting at x. Hence, the resultant of the distributed force $w(x)$ is given by

$$R = \int_0^L w(x)\,dx \tag{8.24}$$

To locate the line of action of R, we note that the moment of the load $w(x)\,dx$ about point A is $xw(x)\,dx$. Hence, the net moment about point A of the distributed load is obtained by integration:

$$M = \int_0^L xw(x)\,dx \tag{8.25}$$

This moment is dynamically equivalent to the moment of the resultant R about point A. Therefore, the line of action of R, located at $\bar{x}$, is determined by the equation

$$\bar{x} = \frac{M}{R} \tag{8.26}$$

where R and M are determined by Eqs. (8.24) and (8.25), respectively. For example, if w is constant, Eqs. (8.24), (8.25), and (8.26) yield $\bar{x} = L/2$.

DISTRIBUTED LOAD ON A PLANE AREA

Consider a distributed load $q(x, y)$ that acts on a plane area A (see Fig. 8.24a). Select an infinitesimal area $dA = dx\,dy$ in area A. The infinitesimal force $[q(x, y)\,dx\,dy]$ that acts on the infinitesimal area may be considered as a point force. Hence, the resultant R of the distributed load $q(x, y)$ is the sum of all such point forces acting on area A. Thus, R is given by integration over the area A:

$$R = \iint_A q(x, y)\,dx\,dy \tag{8.27}$$

Likewise, the moments M_x and M_y of the distributed load $q(x, y)$ about the x and y axes, respectively, are given by integration as

$$M_x = \iint_A yq(x, y)\,dx\,dy \tag{8.28a}$$

$$M_y = \iint_A xq(x, y)\,dx\,dy \tag{8.28b}$$

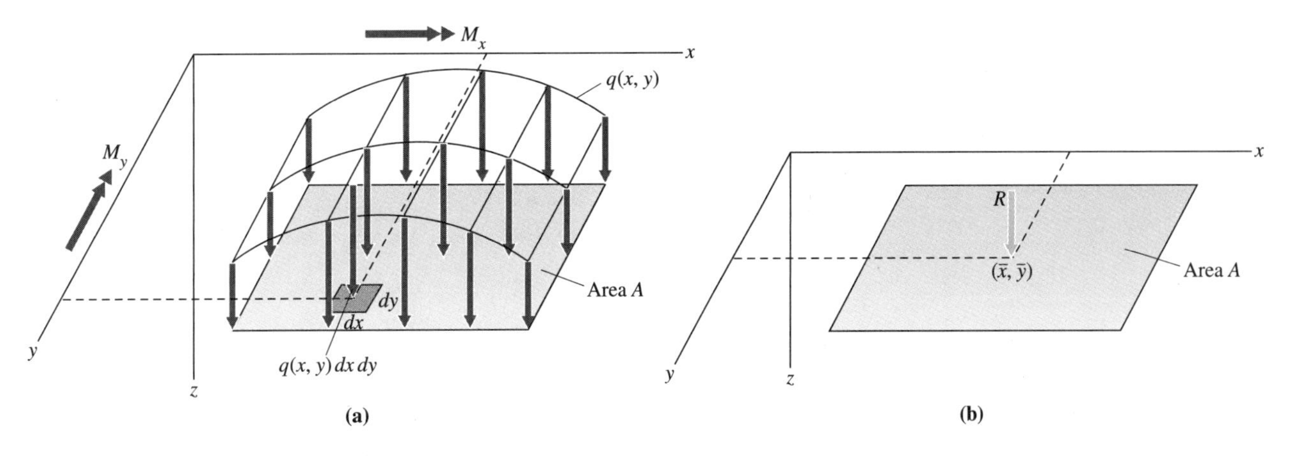

Figure 8.24

The line of action of the resultant force R is determined by the fact that moments of R about the x and y axes are dynamically equivalent to the moments M_x and M_y of the distributed load $q(x, y)$. Thus, the intersection point $(\bar{x}, \bar{y})$ of the line of action of R with the area A (see Fig. 8.24b) is given by the formulas

$$\bar{x} = \frac{M_y}{R} \qquad \bar{y} = \frac{M_x}{R} \tag{8.29}$$

where R and M_x and M_y are given by Eqs. (8.27) and (8.28), respectively.

This book uses Eqs. (8.27), (8.28), and (8.29) for rectangular areas, for which the limits on x and y are constants (see Example 8.18). For nonrectangular areas, the integration limits may be functions of x and y. The integration of these types of multiple integrals lies outside the scope of this book (see a good calculus text, such as Thomas & Finney, 1992; see also footnote 4 of this chapter).

Example 8.17

Beam Subjected to Distributed Load

Problem Statement A simple beam is subjected to a distributed lateral load $w(x)$ (force per unit length of the beam), defined by the equation $w(x) = kx^n$, where x is the distance from one end of the beam and k and n are constants (see Fig. E8.17a).

a. Determine the magnitude and the line of action of the resultant of the distributed load $w(x)$.

b. For $n = 1$, determine the support reactions of the beam.

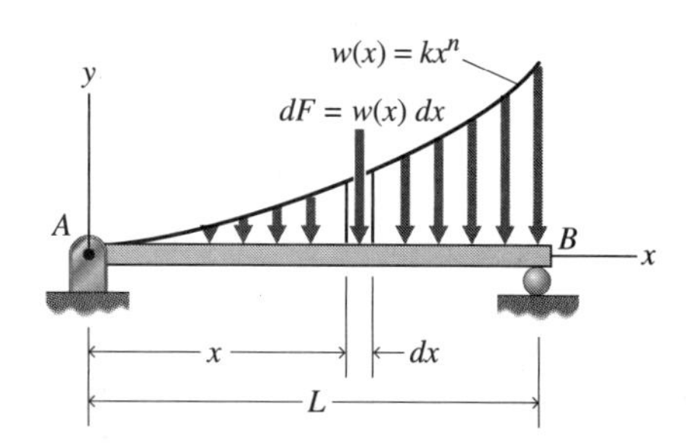

Figure E8.17a

Solution

a. The force on an element of the beam of length dx is $dF = w(x)\,dx = kx^n\,dx$. Hence, the total force F on the plate is obtained by integration:

$$F = k\int_0^L x^n\,dx = \frac{kL^{(n+1)}}{n+1} \tag{a}$$

We note that the moment of the infinitesimal force dF about the beam edge $x = 0$ is $xw(x)\,dx = x(kx^n)\,dx$. Hence, the net moment M of the distributed force about the beam edge $x = 0$ is

$$M = k\int_0^L x(x^n)\,dx = k\int_0^L x^{n+1}\,dx = \frac{kL^{(n+2)}}{n+2}$$

We assume that the line of action of F is located at $\bar{x}$ (see Fig. E8.17b). Since the moment of F is dynamically equivalent to the moment M of the distributed load $w(x)$, the line of action of F is determined by

$$\bar{x} = \frac{M}{F} = \frac{(n+1)\,L}{n+2} \tag{b}$$

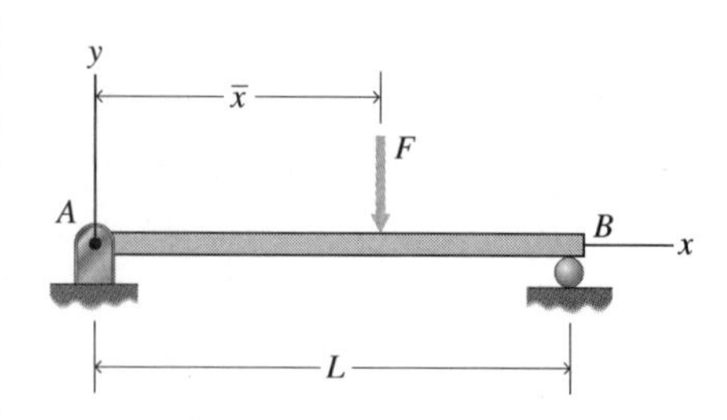

Figure E8.17b

b. If $n = 1$, the load distribution is a linear function of x (see Fig. E8.17c). Then, from

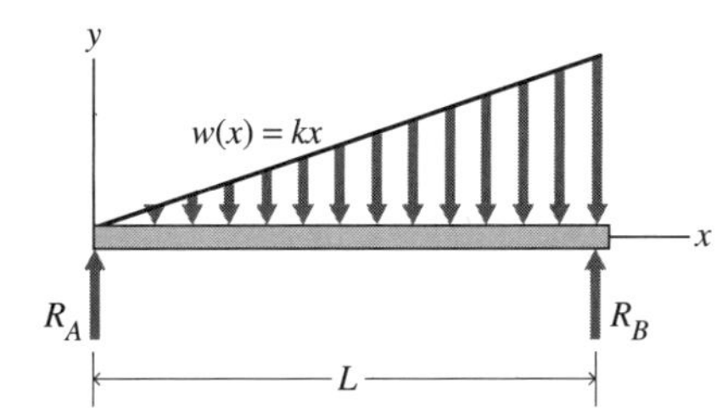

Figure E8.17c

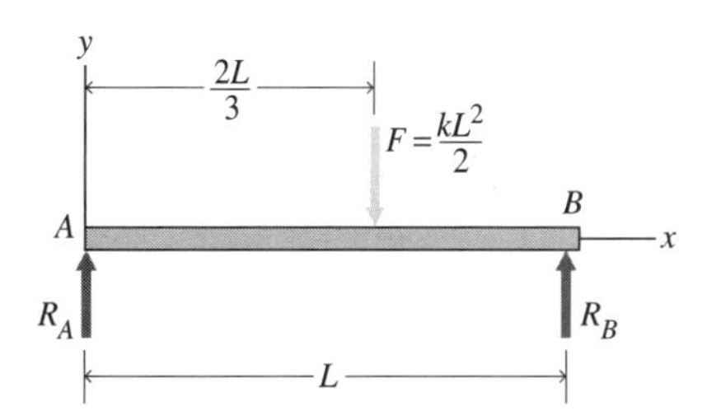

Figure E8.17d

Eqs. (a) and (b), $F = kL^2/2$ and $\bar{x} = 2L/3$. To find the support reactions, we draw the free-body diagram of the beam, replacing the distributed load by its resultant, $F = kL^2/2$ (see Fig. E8.17d). Therefore, the equations of equilibrium for the beam are

$$\sum F_y = R_A + R_B - \frac{kL^2}{2} = 0$$
$$\sum M_A = R_B L - \frac{kL^3}{3} = 0 \tag{c}$$

The solutions of Eqs. (c) are

$$R_A = \frac{kL^2}{6}$$
$$R_B = \frac{kL^2}{3} \tag{d}$$

Information Item To find the support reactions in part b, it is not necessary to determine the resultant of the distributed load first. We see from the free-body diagram of the beam in Fig. E8.17c that the equilibrium equations are

$$\sum F_y = R_A + R_B - \int_0^L w(x)\,dx = 0$$
$$\sum M_A = R_B L - \int_0^L xw(x)\,dx = 0 \tag{e}$$

Since $w(x) = kx$, integration of Eqs. (e) yields

$$\sum F_y = R_A + R_B - \frac{kL^2}{2} = 0$$
$$\sum M_A = R_B L - \frac{kL^3}{3} = 0 \tag{f}$$

Equations (f) are identical to Eqs. (c), and, therefore, the reactions are again given by Eqs. (d).

Example 8.18

Rectangular Area Subjected to Distributed Load

Problem Statement A rectangular area A is subjected to a uniformly distributed load q (see Fig. E8.18a). Determine the magnitude R and the line of action of the resultant of the distributed load.

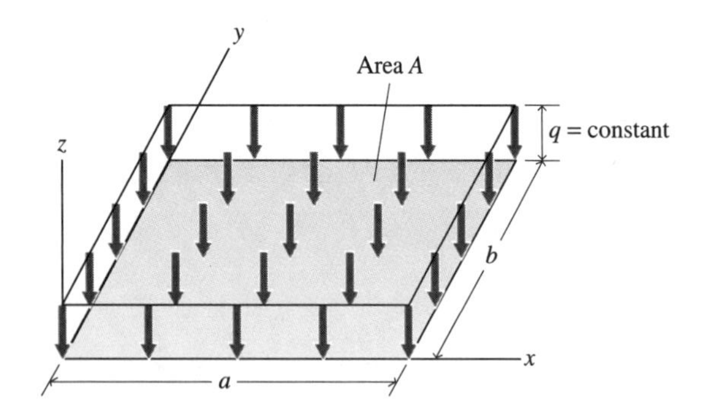

Figure E8.18a

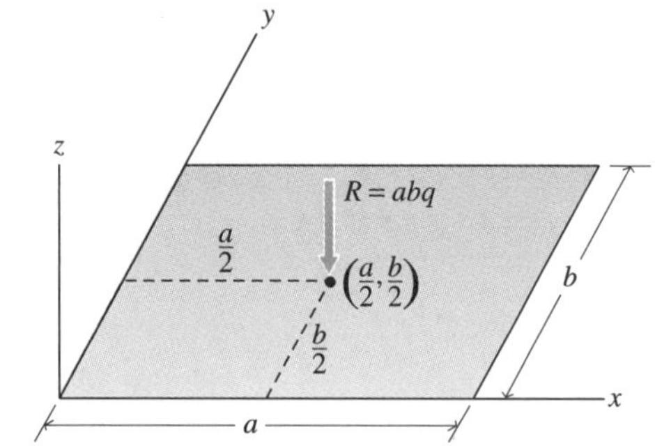

Figure E8.18b

Solution Since q is a constant, Eq. (8.27) yields

$$R = q\int_0^b\int_0^a dx\,dy = abq \tag{a}$$

And Eqs. (8.28) yield

$$M_x = q\int_0^b\int_0^a y\,dx\,dy = \frac{ab^2q}{2}$$
$$M_y = q\int_0^b\int_0^a x\,dx\,dy = \frac{a^2bq}{2} \tag{b}$$

Equations (a), (b), and (8.29) yield the coordinates of the point in the area A at which the resultant R acts: $(\bar{x}, \bar{y}) = (a/2, b/2)$. See Fig. E8.18b.

CHECKPOINT 

1. ***True or False:*** To determine support reactions of a simple beam, a system of distributed parallel forces that acts on the beam (force per unit length of the beam) may be replaced by a single, dynamically equivalent force.
2. ***True or False:*** The resultant of a system of distributed parallel forces that acts on a plane area (force per unit area) is a single force.
3. ***True or False:*** The effects of a distributed force on a body are the same as the effects of the resultant of the distributed force.

Answers: 1. True; 2. True; 3. False

Chapter Highlights

- The gravity axis of a body is the line of action of the resultant of the parallel gravitational forces exerted on the body by the earth.
- All gravity axes of a body intersect at a common point, which is the center of gravity.
- The weight of a body is the resultant force exerted on it by the earth's gravitational attraction.
- The specific weight of a body is the weight per unit volume of the body.
- If the specific weight is constant throughout a body, the body is said to be homogeneous.
- For a homogeneous body, the center of gravity is independent of the specific weight. The center of gravity of such a body is the centroid of the body. The centroid is a geometric property of the body.

- A composite body is made up of several parts whose weights and centers of gravity are known or easily determined. The center of gravity of a composite body may be determined by the theory of parallel forces, since the weight vectors of the several parts are known.
- The resultant force of a system of distributed forces that act on a rigid body is dynamically equivalent to the system of distributed forces.

Problems

8.1 *Gravity Axis of a Body*

8.1 A thin, folded homogeneous sheet of metal is oriented as shown in Fig. P8.1. Determine the gravity axis of the sheet for the orientation shown.

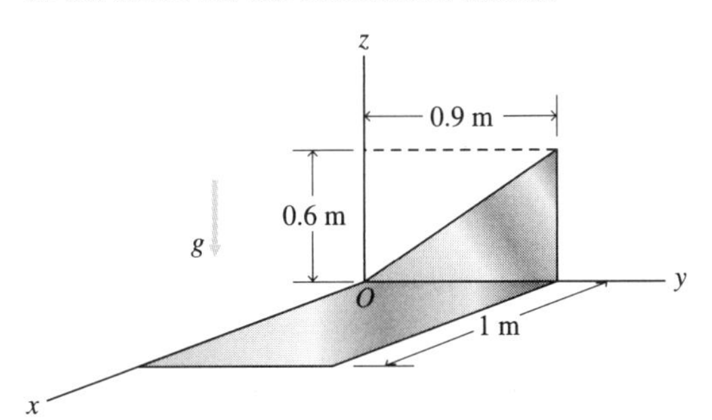

Figure P8.1

8.2 Assume that the gravity field g in Fig. P8.1 is directed along the x axis pointing in the negative direction. Determine the gravity axis of the metal sheet for this orientation.

8.3 Determine the gravity axis of the system consisting of the homogeneous cylinder C and the homogeneous block B, shown in Fig. P8.3. The specific weight of the cylinder is $\gamma_C = 0.3$ lb/in^3, and that of the block is $\gamma_B = 0.1$ lb/in^3.

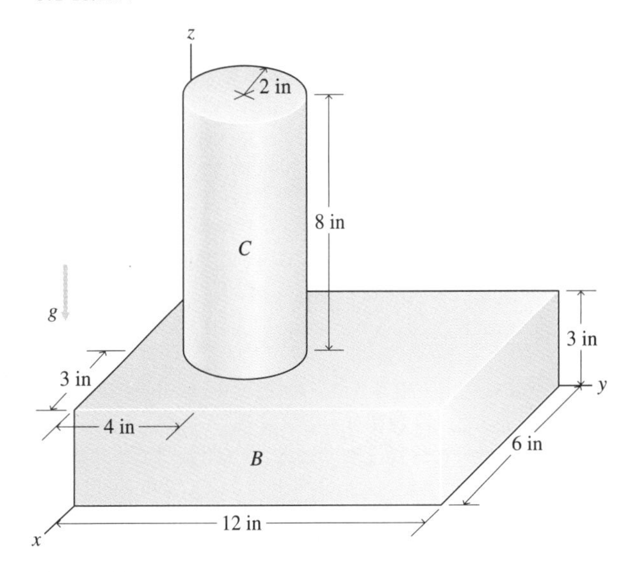

Figure P8.3

8.4 Assume that the gravity field g in Fig. P8.3 is directed along the x axis pointing in the negative direction. Determine the gravity axis of the system for this orientation.

8.5 For the truss in Fig. P8.5, the cross-sectional area of a vertical member is 75 cm^2, the cross-sectional area of a horizontal member is 100 cm^2, and the cross-sectional area of a diagonal member is 37.50 cm^2 (1 cm^2 = 100 mm^2). If the specific weights of all members are the same, where does the gravity axis of the truss lie?

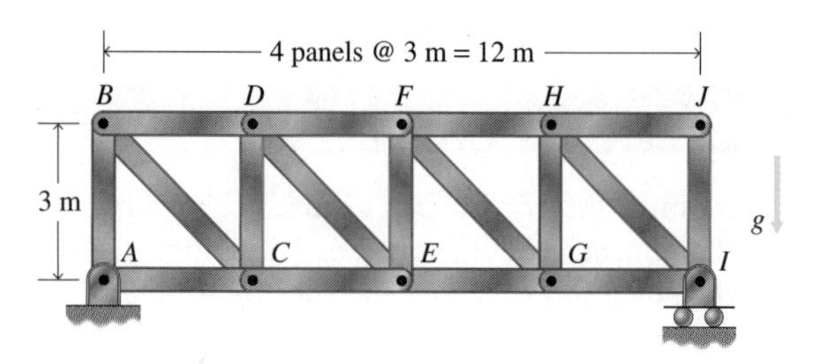

Figure P8.5

8.6 The gravity axes of three objects cut the horizontal xy plane at the points (3, 7), (−4, 6), and (0, −9). The relative weights of the objects are, respectively, 2, 5, and 4. Locate the gravity axis of the system.

8.7 An airplane weighs 225 kN. For a flight test, sandbags weighing 6750 N are placed in it. The sandbags must be placed so that the gravity axis of the loaded plane moves 225 mm back toward the tail of the plane. How far from the gravity axis of the empty plane must the gravity axis of the sandbags be located?

8.8 A cruise ship, 30 ft wide at the deck, weighs 500 tons without passengers (1 ton = 2000 lb). One hundred passengers, averaging 150 lb per person, travel on it. If the passengers, who are originally distributed symmetrically about the center of gravity of the ship, all rush to one side rail to watch a school of dolphins, how far do they displace the ship's gravity axis toward that side?

8.9 A casting of homogeneous material is in the form of a hollow cylinder with a radial fin projecting from it (see Fig. P8.9). An eye-bolt is to be screwed into the top of the casting to use for lifting it. Where should the bolt be located so that the top plane remains horizontal when the casting is lifted?

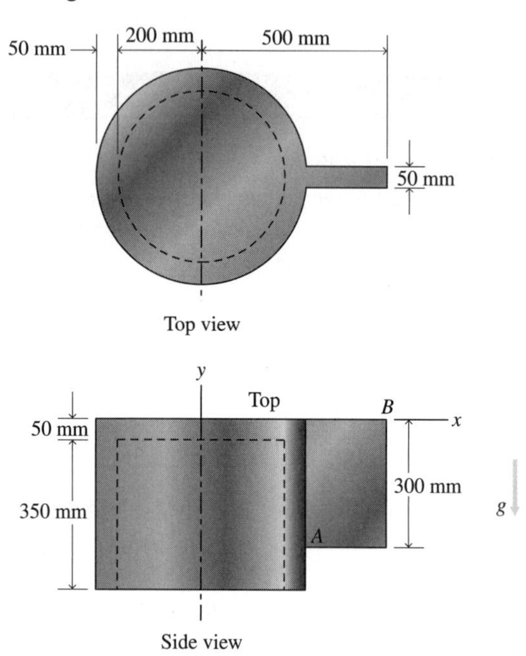

Figure P8.9

8.10 Rectangular xz axes are located in an airplane such that the z axis is the gravity axis of the plane, including its four engines. The positive x axis is directed toward the nose of the plane. The plane and engines weigh 360 kN. Each engine weighs 8.1 kN. The gravity axis of the inboard engines (the two closest to the fuselage) lies at x = 750 mm. The gravity axis of the outboard engines lies at $x = -1.0$ m. The engineering department of the plane's manufacturer is considering installing four replacement engines that weigh only 7.2 kN apiece. The gravity axis of each new engine will lie 250 mm closer to the nose than does the gravity axis of the engine that it replaces. Determine where the gravity axis of the plane will lie after the new engines are installed.

8.11 The members of the frame shown in Fig. P8.11 all have the same cross-sectional area. How far to the right of the line AB does the gravity axis of the frame lie?

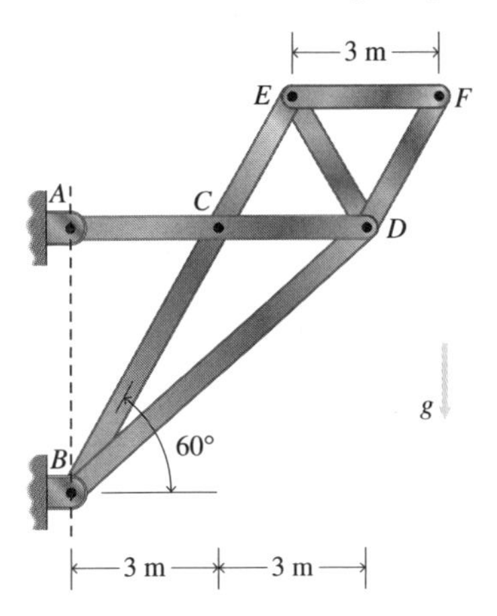

Figure P8.11

8.12 An airplane has a tricycle landing gear. With reference to rectangular xy axes in the plane of the pavement, the wheels touch the pavement at the points (0, −6 m), (0, 6 m), and (9 m, 0). By placing the wheels on scales, an engineer determines that the loads carried by the respective wheels are 135, 157.5, and 22.5 kN. Workmen plan to place equipment weighing 9 kN in the airplane 24 m behind the nose wheel (toward the tail of the airplane). Will this cause the airplane to fall back on its tail?

8.13 A solid homogeneous cube with edges of length a rests on a horizontal table top. A solid hemispherical dome of the same material as the cube is set on top of the cube, with its base circle tangent to two of the top edges of the cube (see Fig. P8.13). The diameter of the hemisphere is $3a/2$. Determine the distance between the gravity axis of the cube and the gravity axis of the combination.

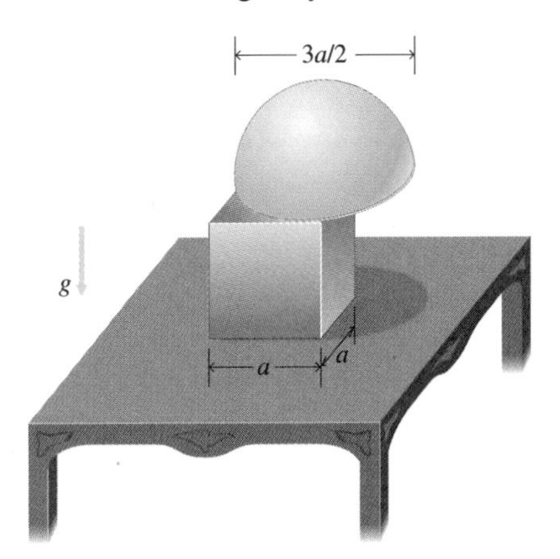

Figure P8.13

8.14 Suppose that the casting of Fig. P8.9 is to be lifted by an eye-bolt that is screwed into the outside wall of the cylinder exactly opposite the fin. How far from the top plane should the bolt be located so that the top plane of the cylinder is vertical when the casting is lifted?

8.15 A second transformer, which weighs 12 tons, must be loaded on the barge of Example 8.2. Place the transformer so that the resultant gravity axis of the barge and the two transformers intersects the deck at $\overline{x} = 50$ ft.

The following two problems illustrate an important feature of design—that there is often more than one reasonable solution to a problem.

8.16 Add two members that weigh w (in N/m) to the frame of Fig. E8.1 so that the gravity axis of the system coincides with the line $\overline{x} = 0.50$ m to the left of the axis of the vertical member.

Method: Review Example 8.1 to make certain that you understand it. Then, add the two additional members by attaching them to the original frame and determining their effect on the location of the gravity axis. If your placement of the members does not result in $\overline{x} = 0.50$ m, relocate them and/or change their lengths. Repeat the process of calculating $\overline{x}$ until you succeed in locating the gravity axis on the line $\overline{x} = 0.50$ m.

8.17 Solve Problem 8.16 so that the gravity axis is located at $\overline{x} = 0.20$ m to the right of the vertical member in Fig. E8.1.

8.2 *Center of Gravity in Cartesian Coordinates*

Solve Problems 8.18–8.23 by locating the point of intersection of two gravity axes. Use Eqs. (8.1) to check your answers.

8.18 Locate the center of gravity of the cylinder-block system of Fig. P8.3.

8.19 The cross section of a stepped prismatic casting is shown in Fig. P8.19. Determine the coordinates of the center of gravity.

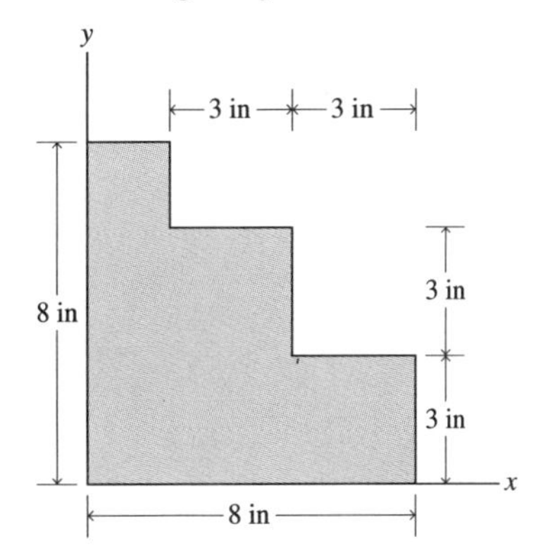

Figure P8.19

8.20 Locate the coordinates of the center of gravity of the metal sheet of Fig. P8.1.

8.21 The bench shown in Fig. P8.21 is made of homogeneous wood. The top is a rectangular solid, 2 in thick and 26 in × 50 in. Each leg measures 4 in × 4 in × 28 in. The side boards are all 1 in thick. Determine the elevation of the center of gravity of the bench.

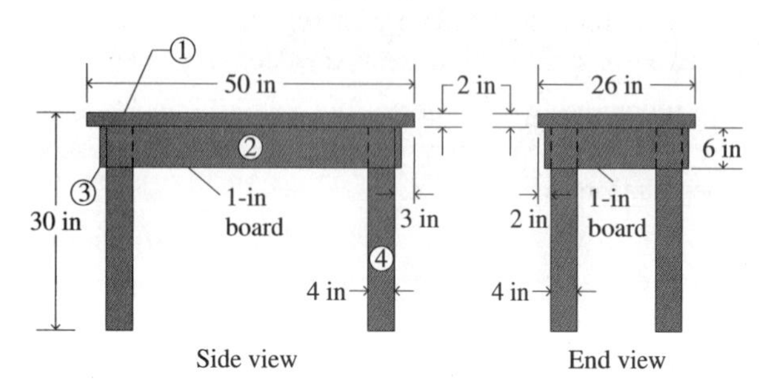

Figure P8.21

8.22 a. Locate the center of gravity of the casting of Fig. P8.9.

b. If the casting is hung from a cable so that the diagonal *AB* of the fin is vertical, how far from the line *AB* does the gravity axis lie?

8.23 A cylindrical tank of height H and diameter D is made of thin sheet metal of constant thickness t ($t << H$; $t << D$) (see Fig. P8.23). The tank is open at the top. Show that the elevation of the center of gravity relative to the base is approximately $\bar{y} = 2H^2/(4H + D)$.

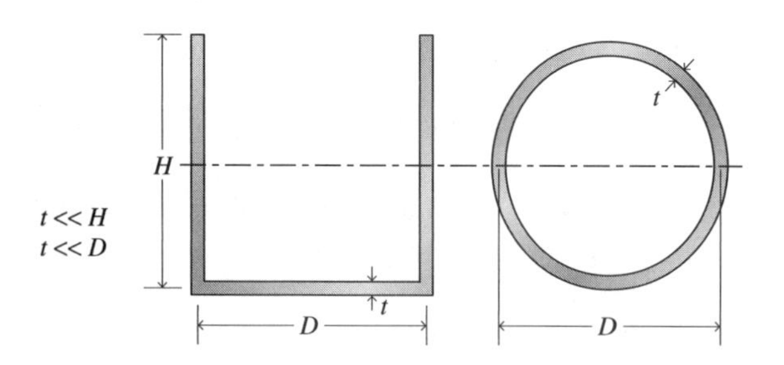

Figure P8.23

In Problems 8.24–8.29, locate the center of gravity of each structure, relative to the x and y axes shown. For each structure, assume that all members have the same cross-sectional area and the same specific weight.

8.24

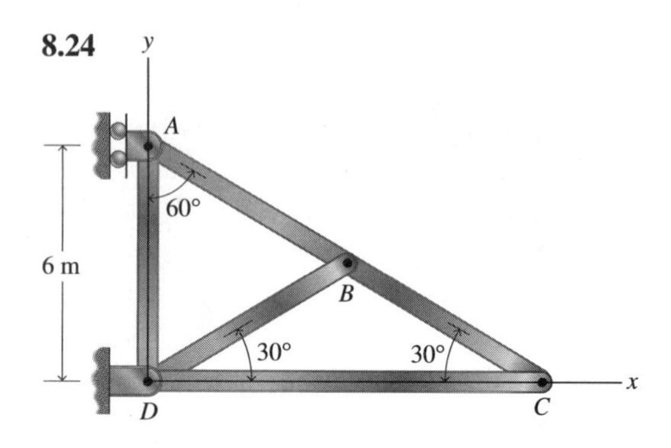

8.25

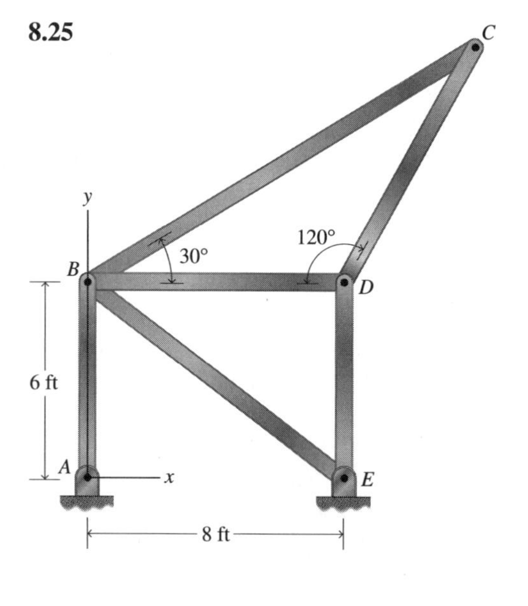

8.26

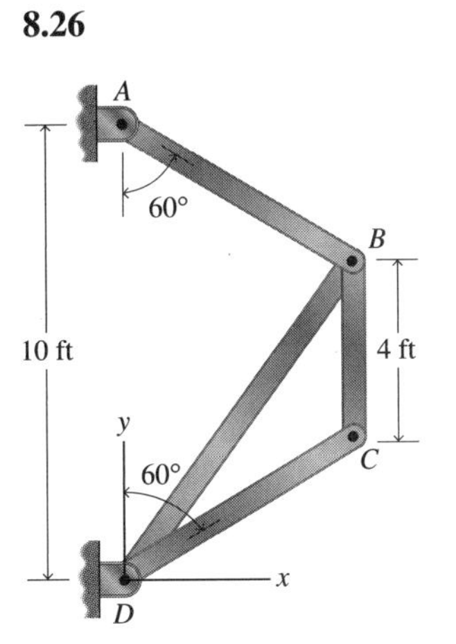

8.27

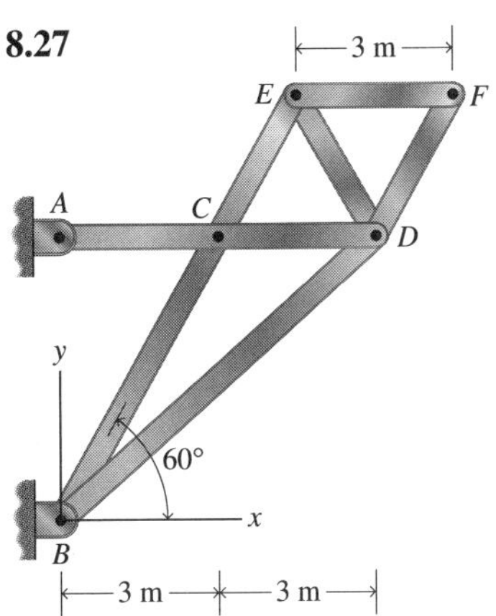

8.28

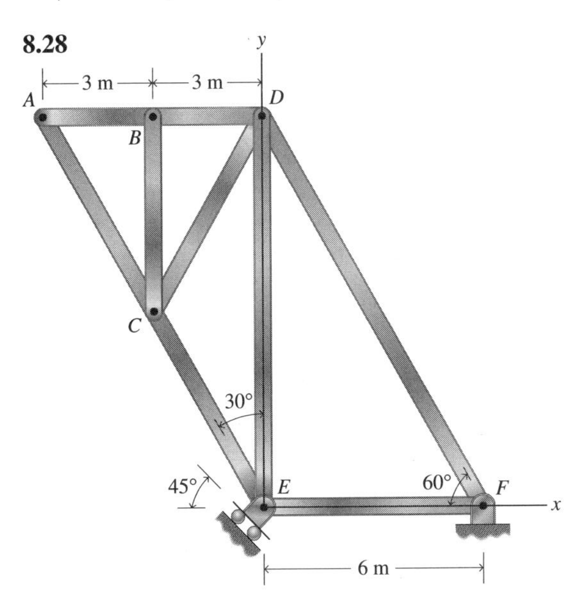

8.29

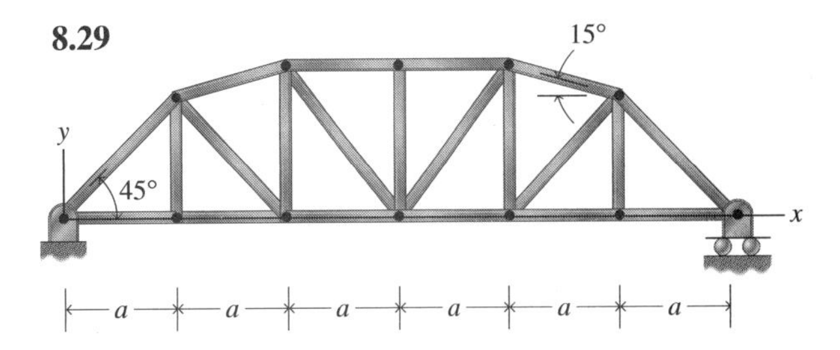

8.3 *Center of Gravity by Integration*

8.30 Determine the coordinate $\bar{x}$ of the center of gravity of a solid homogeneous hemisphere of radius a, with its base in the yz plane (see Fig. P8.30).

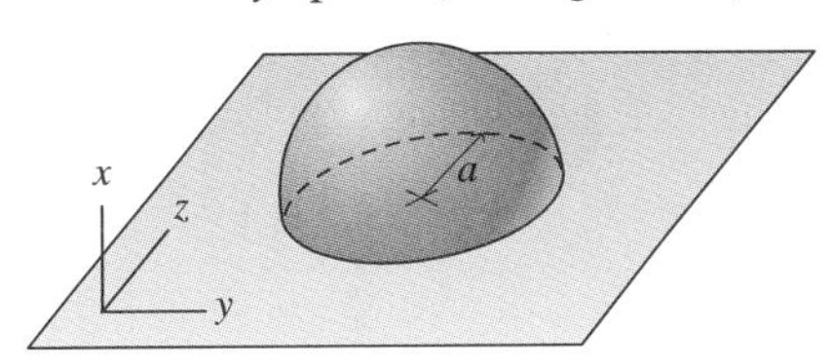

Figure P8.30

8.31 Determine the coordinate $\bar{x}$ of the center of gravity of a hollow homogeneous hemispherical dome of radius a that is thin-walled and open at the base. The base lies in the yz plane (see Fig. P8.30).

8.32 Determine the coordinate $\bar{y}$ of the center of gravity of the homogeneous paraboloidal solid of revolution in Fig. P8.32, which is bounded by the plane $y = b$ and whose surface is generated by rotating the curve $y = bx^2/a^2$ about the y axis.

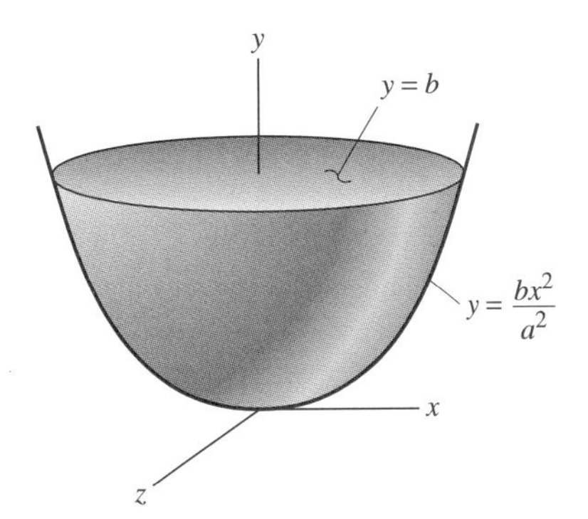

Figure P8.32

8.33 A right circular conical shell is formed from thin homogeneous sheet metal. The height of the cone is h, and the radius of its base is a. The cone is open at the base. Determine the height of the center of gravity of the shell.

8.34 Figure P8.34 represents a solid tapered shaft. Set $k = r_2/r_1$, and show that

$$\frac{\bar{x}}{L} = \frac{3k^2 + 2k + 1}{4(k^2 + k + 1)}$$

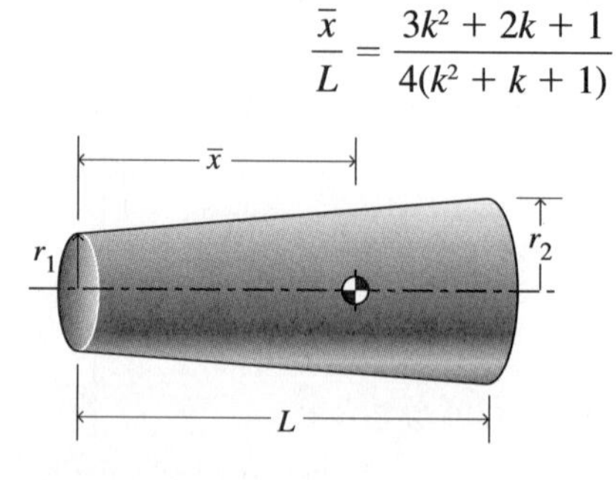

Figure P8.34

8.4 *Centroids of Plane Areas and Lines*

8.35 Determine the coordinates of the centroid of the area in the first quadrant that is bounded by the curve $y = e^{-x}$ and the coordinate axes (see Fig. P8.35).

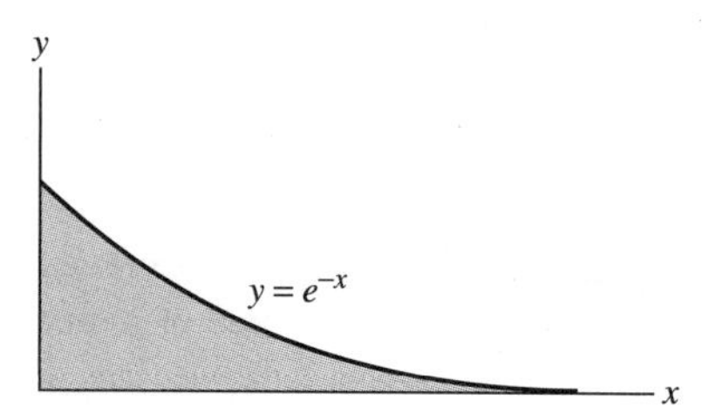

Figure P8.35

8.36 Determine the coordinates of the centroid of the area that is bounded by the curve $9y = x^3$, the line $x = 3$, and the x axis (see Fig. P8.36).

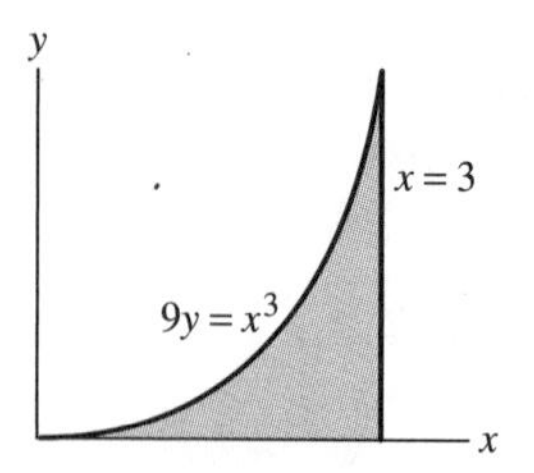

Figure P8.36

8.37 Determine the coordinates of the centroid of the area that is bounded by the straight line $y = x$, the circle $x^2 + y^2 = 1$, and the y axis (see Fig. P8.37). (*Hint:* Use polar coordinates.)

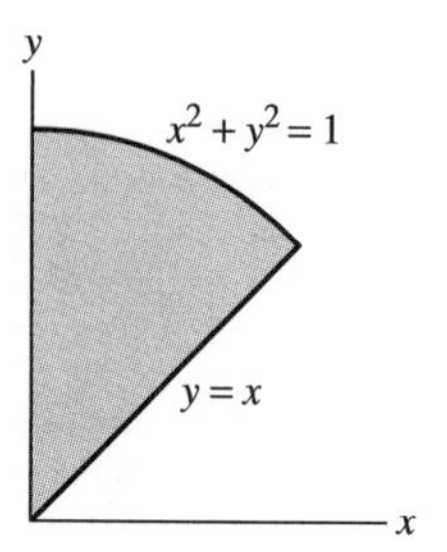

Figure P8.37

8.38 Determine the coordinates of the centroid of the area that is bounded by the circle $x^2 + y^2 = r^2$ and the line $x = h$ $(0 < h < r)$ (see Fig. P8.38).

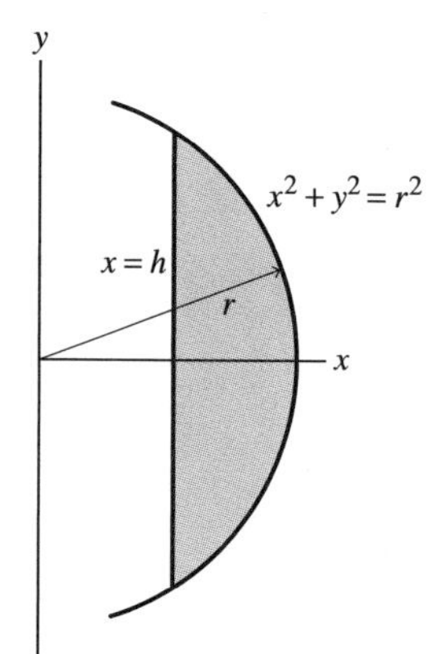

Figure P8.38

8.39 Determine the coordinates of the centroid of the area bounded by the curves $y^2 = 8x$ and $x^2 = 8y$ (see Fig. P8.39).

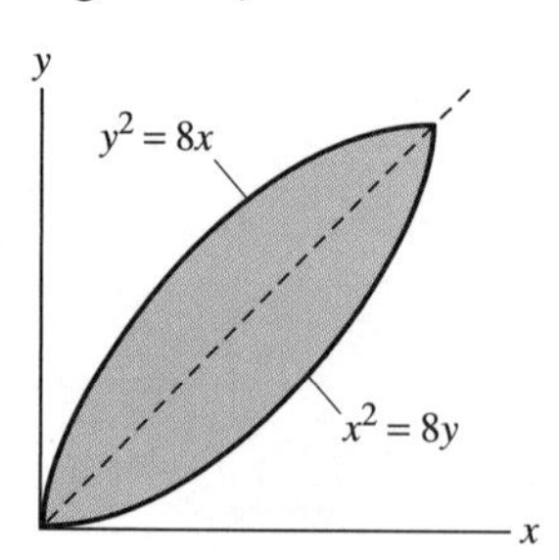

Figure P8.39

8.40 Determine the coordinates of the centroid of the area in the first quadrant that is bounded by the curve $y = e^{-a^2x^2}$, where a is a constant, and the coordinate axes (see Fig. P8.40). Use the formula

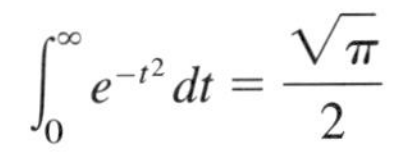

$$\int_0^\infty e^{-t^2}\,dt = \frac{\sqrt{\pi}}{2}$$

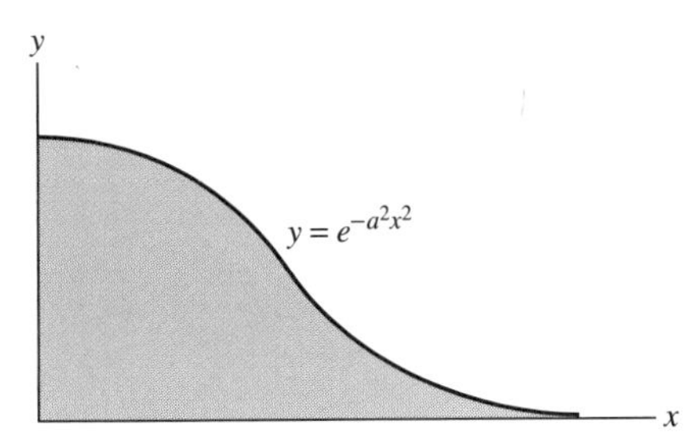

Figure P8.40

8.41 a. Derive a formula for the coordinate $\bar{x}$ of the centroid of the circular arc shown in Fig. P8.41, as a function of a and θ.
b. Specialize the formula of part a for a quarter-circular arc ($\theta = \pi/4$).
c. Specialize the formula of part a for a semicircular arc ($\theta = \pi/2$).
d. Specialize the formula for a circle ($\theta = \pi$).

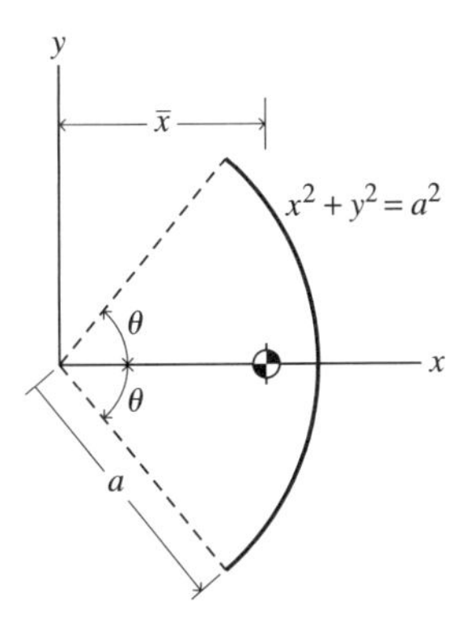

Figure P8.41

8.42 Derive a formula for the coordinate $\bar{x}$ of the centroid of the circular segment shaded in Fig. P8.42, as a function of a and θ. [The area of the segment is $a^2(2\theta - \sin 2\theta)/2$.]

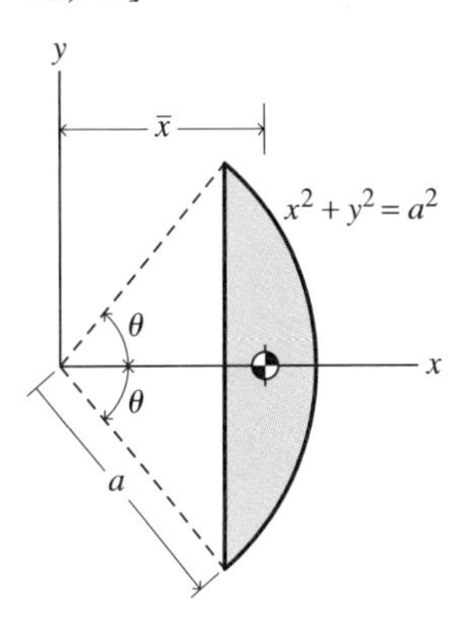

Figure P8.42

8.5 *Composite Bodies*

8.43 Determine the centroid of the thin-wall W-section in Fig. P8.43, with respect to the x and y axes shown.

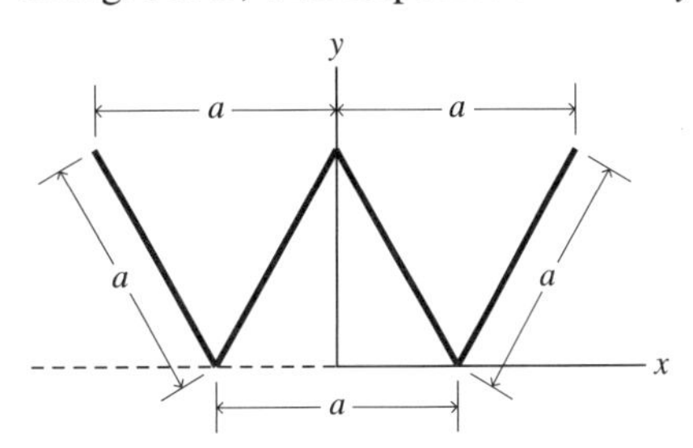

Figure P8.43

8.44 Determine the centroid of the Z-section in Fig. P8.44, with respect to the x and y axes shown.

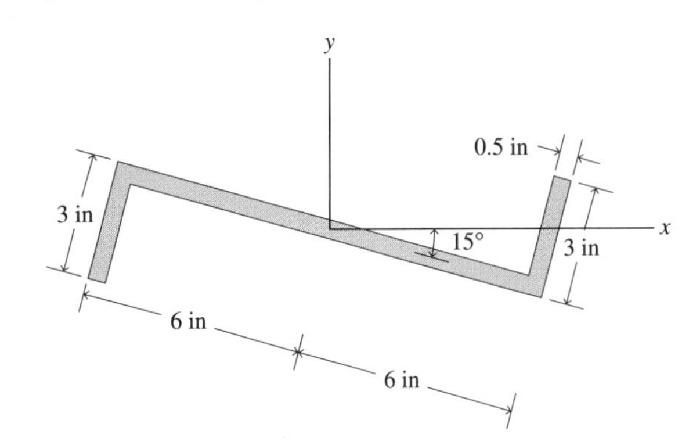

Figure P8.44

8.45 Determine the centroid of the T-section in Fig. P8.45, with respect to the x and y axes shown.

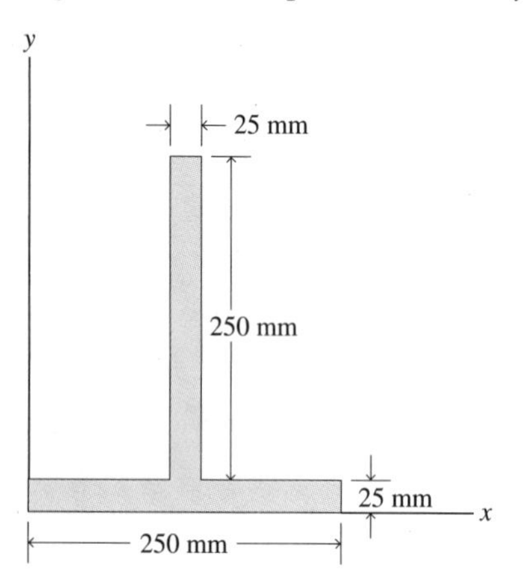

Figure P8.45

8.46 Determine the coordinate $\bar{y}$ of the centroid of the thin, uniform sheet metal shape in Fig. P8.46.

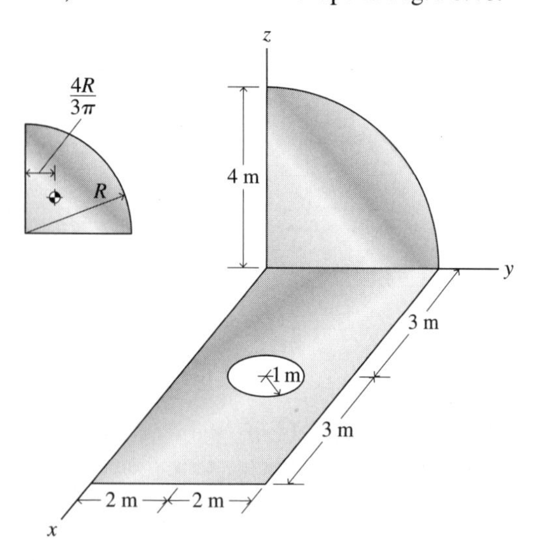

Figure P8.46

8.47 Determine the centroid of the composite line $ABCDE$ in Fig. P8.47, relative to the x and y axes shown.

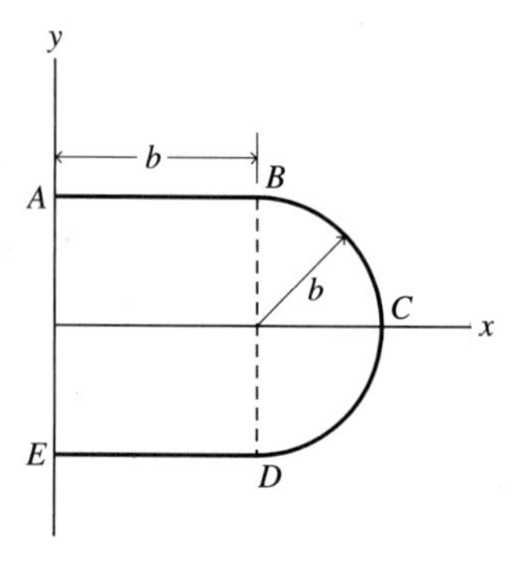

Figure P8.47

8.48 Determine the coordinates of the centroid of the shaded area in Fig. P8.48. (Compare these coordinates to those for the center of gravity from Problem 8.20.)

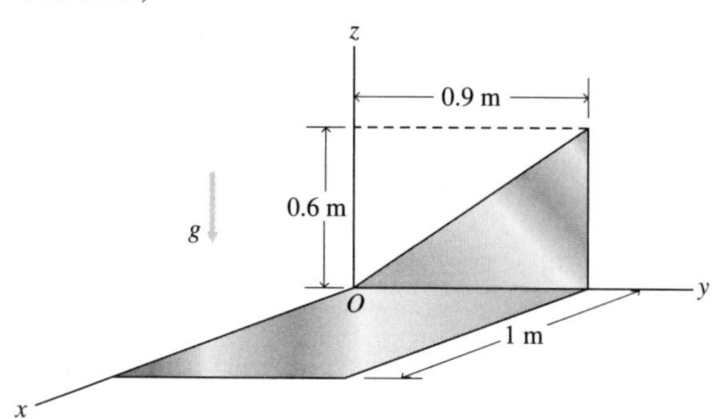

Figure P8.48

8.49 Determine the coordinates of the centroid of the shaded area in Fig. P8.49.

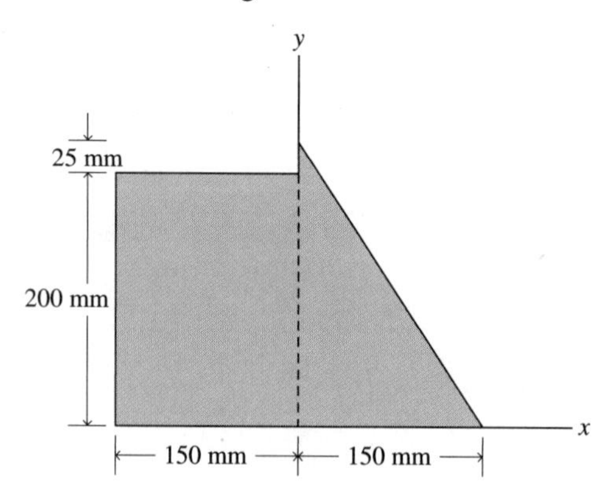

Figure P8.49

8.50 Determine the coordinates of the centroid of the shaded area in Fig. P8.50.

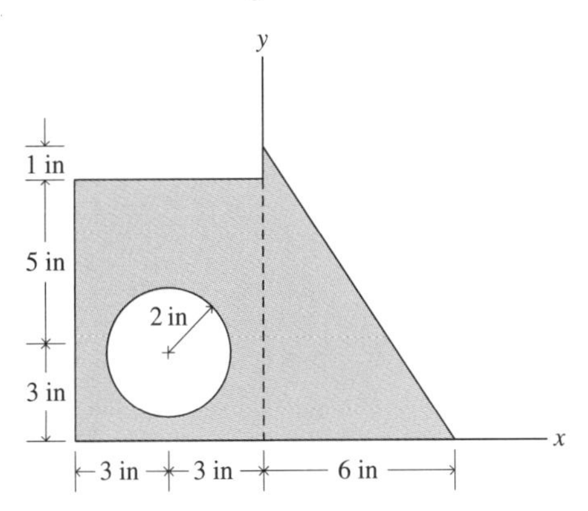

Figure P8.50

8.51 Determine the center of gravity of the composite body of Fig. P8.3.

8.52 **a.** Determine the center of gravity of the stepped shaft shown in Fig. P8.52.

b. If a 2 in diameter hole is drilled along the full length of the axis of the shaft, how far is the center of gravity of the drilled shaft displaced from that of the original shaft?

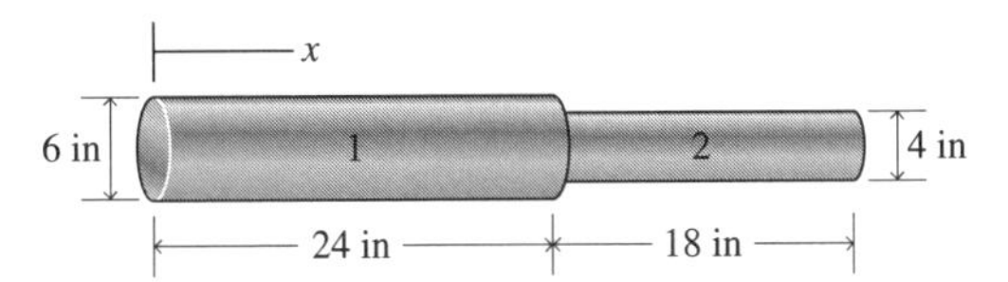

Figure P8.52

8.53 The specific weights of flywheel A and rod B in Fig. P8.53 are both 490 lb/ft^3. The specific weight of flywheel C is 150 lb/ft^3. The rod extends through flywheel C.

a. Determine the center of gravity of the system.

b. Does the center of gravity coincide with the centroid of the volume of the system? Explain.

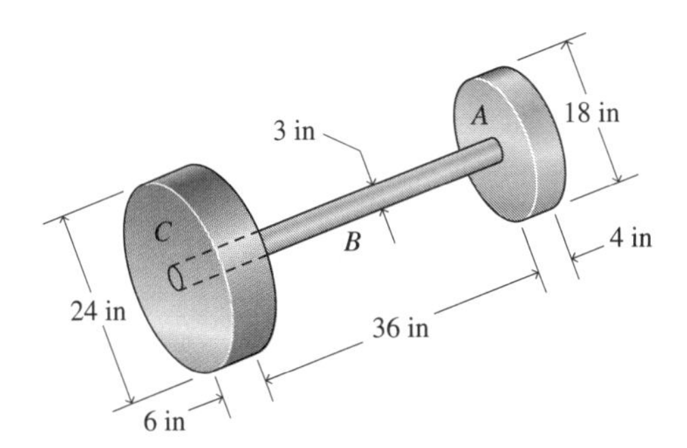

Figure P8.53

8.54 Determine the centroid of the cross section of the aluminum disk in Fig. P8.54.

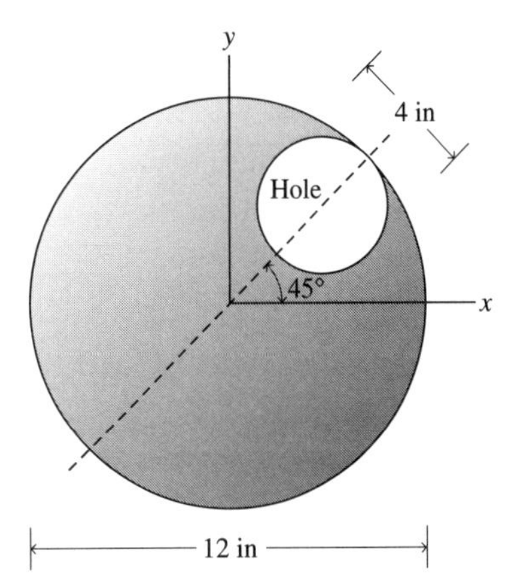

Figure P8.54

8.55 Determine the centroid of the cross section in Fig. P8.55 relative to the x and y axes shown.

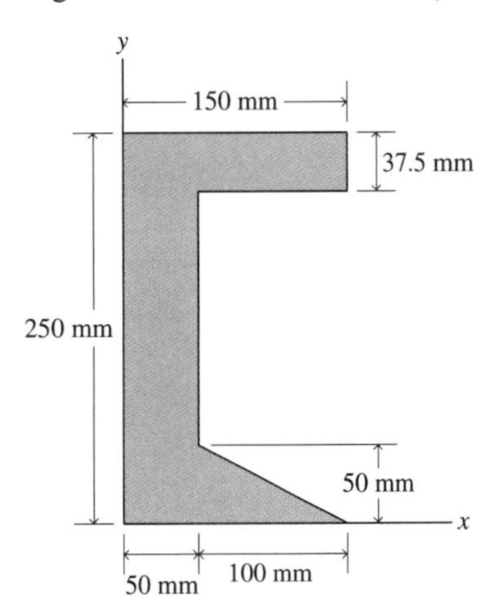

Figure P8.55

8.56 Determine the centroid of the cross section in Fig. P8.56, relative to the x and y axes shown.

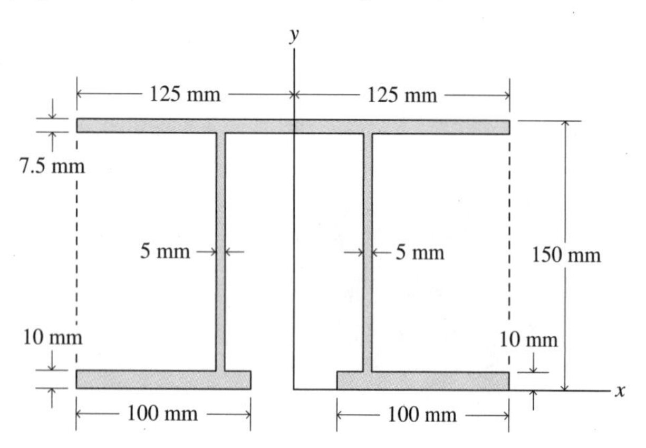

Figure P8.56

8.57 Determine the centroid of the part in Fig. P8.57, with respect to the x and y axes shown. (*Hint:* First, locate the centroid with respect to the axes X and Y.)

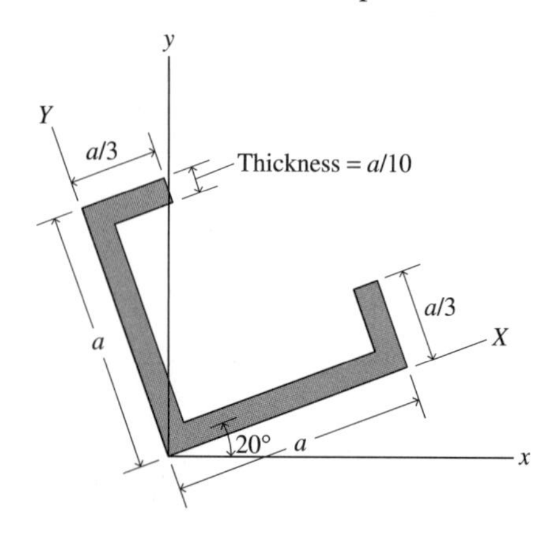

Figure P8.57

8.58 Determine the x coordinate of the center of gravity of the large homogeneous crusher shaft shown in Fig. P8.58. (*Hint:* See Example 8.3.)

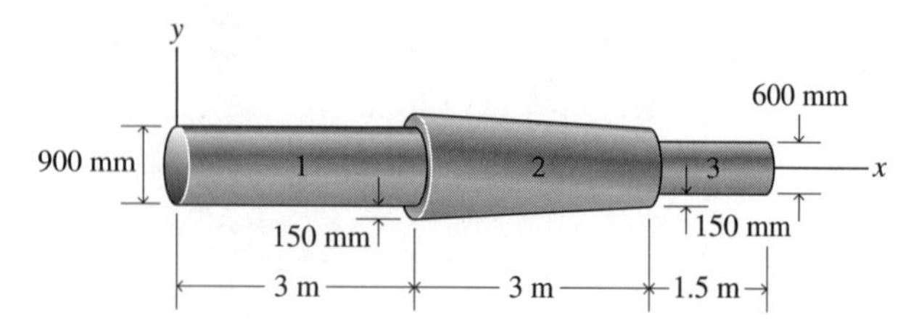

Figure P8.58

8.59 Figure P8.59 shows the cross section of a homogeneous steel flywheel. The flywheel is 0.9 m long. Locate its center of gravity. (*Hint:* See Appendix D for the centroid of the circular segment.)

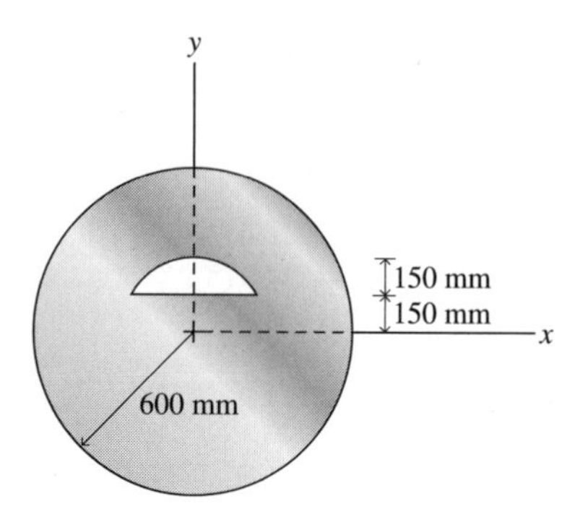

Figure P8.59

8.60 Determine the centroid of the cross section in Fig. P8.60, relative to the x and y axes shown. (*Hint:* Approximate the shape of the 6 in portions of the cross section as rectangles; that is, square off the ends of the two sections where they intersect.)

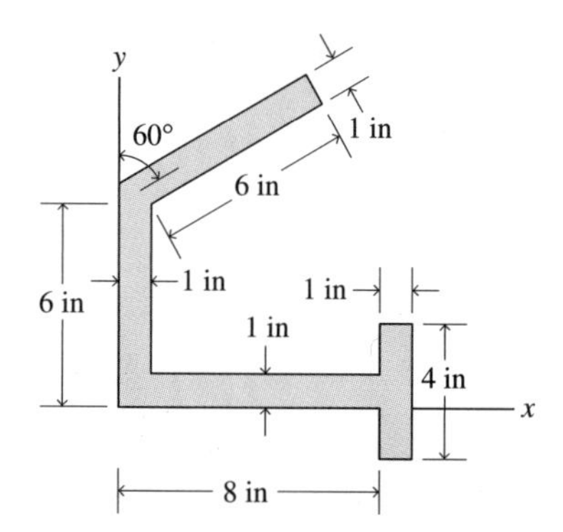

Figure P8.60

8.61 Determine the centroid of the part in Fig. P8.61 with respect to the x and y axes shown. (*Hint:* Approximate the shape of the 200 mm inclined portion of the cross section as a rectangle; that is, square off the bottom end of the section where it intersects the 250 mm horizontal section.)

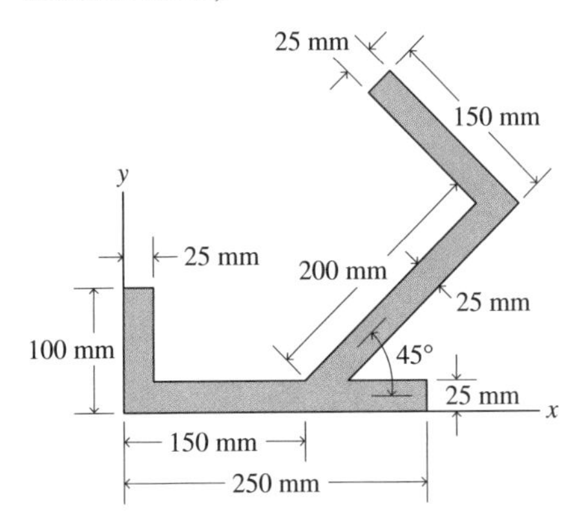

Figure P8.61

8.62 Derive a formula for the centroid of the U-section in Fig. P8.62, in terms of r and h and relative to the x and y axes.

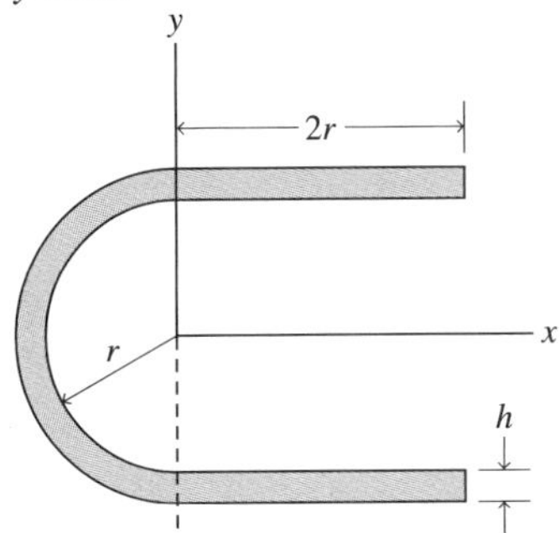

Figure P8.62

8.63 Study Examples 8.10 and 8.11 carefully so that you fully understand them. Then, add or subtract simple geometric areas to the shape in Fig. E8.10a so that the centroid of the new shape is located at $\bar{x} = 20$ mm, $\bar{y} = 20$ mm.

8.64 Solve Problem 8.63, with the new centroid at $\bar{x} = 10$ mm, $\bar{y} = 15$ mm.

8.65 Study Example 8.13 carefully so that you fully understand it. Then, add or subtract geometric shape(s) in the base and/or in the remainder of the part in Fig. E8.13a so that the new center of gravity of the system is located at $\bar{x} = 70$ mm, $\bar{y} = 10$ mm, $\bar{z} = 126.83$ mm.

Method: Choose a simple geometrical shape, place it, and determine its effect on the center of gravity. You will probably have to repeat this process—that is, select a new shape and/or adjust its location several times—before you succeed in matching the new location of the center of gravity.

8.66 Solve Problem 8.65, with the new center of gravity at $\bar{x} = 100$ mm, $\bar{y} = 0$ mm, $\bar{z} = 150$ mm.

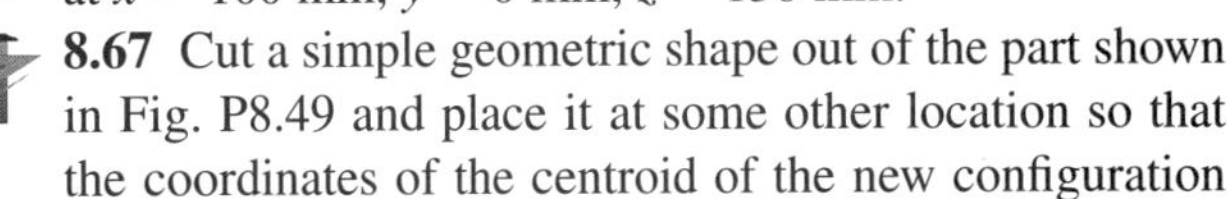

8.67 Cut a simple geometric shape out of the part shown in Fig. P8.49 and place it at some other location so that the coordinates of the centroid of the new configuration are $\bar{x} = -35.33$ mm, $\bar{y} = 107$ mm.

Method: Determine the centroid of the original part. On the original part, outline the shape you propose to cut with a dotted line. Place the cut shape at a new location on the remaining part. Then, determine the centroid of the new configuration, recognizing that the new configuration has a hole corresponding to the cut shape as well as the added shape. Repeat this process until you have succeeded in locating the centroid properly.

8.6 *Theorems of Pappus-Guldinus*

8.68 Using the Pappus-Guldinus theorem for surface areas, determine the area of the surface generated by revolving the arc AB of a circle of radius a about the x axis through 2π radians (see Fig. P8.68).

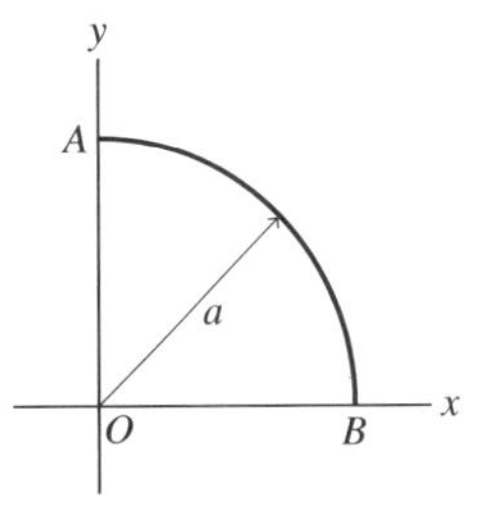

Figure P8.68

8.69 Using the Pappus-Guldinus theorem for surface areas, determine the area of the surface generated by revolving the arc AB of a circle of radius a about the y axis through 2π radians (see Fig. P8.68).

8.70 Given that the surface area of a sphere is $A = 4\pi R^2$, where R is the radius of the sphere, use the Pappus-Guldinus theorem for surface areas to determine the coordinate $\bar{y}$ of the centroid of the semicircular arc *ABC* shown in Fig. P8.70. (*Hint:* This problem uses the Pappus-Guldinus formula in reverse, to compute the centroid of a line, given the surface area generated by rotating the line about an axis.)

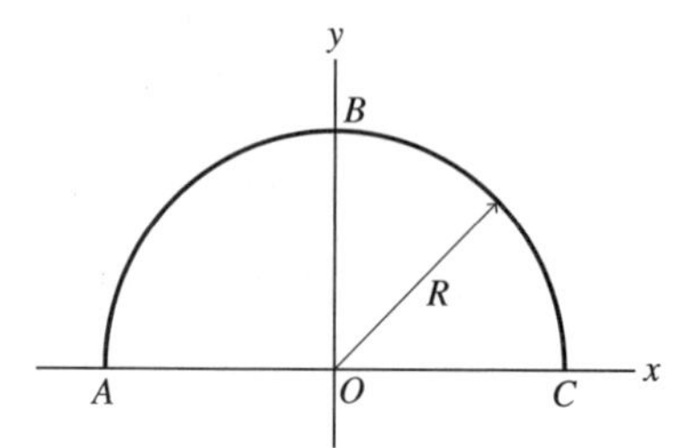

Figure P8.70

8.71 Using the Pappus-Guldinus theorem for surface areas, determine the area of the surface generated by revolving the line *ABCDE* of Fig. P8.47 about the y axis through 2π radians.

8.72 Given that the volume of an ellipsoid of revolution is $V = (4\pi/3)ab^2$, use the Pappus-Guldinus theorem for volumes to determine the centroid of the semi-elliptical area *ABCOA* bounded by the ellipse $(x^2/a^2) + (y^2/b^2) = 1$ and the x axis (see Fig. P8.72). (*Hint:* This problem uses the Pappus-Guldinus formula in reverse, to compute the centroid of an area, given the volume generated by rotating the area about an axis.)

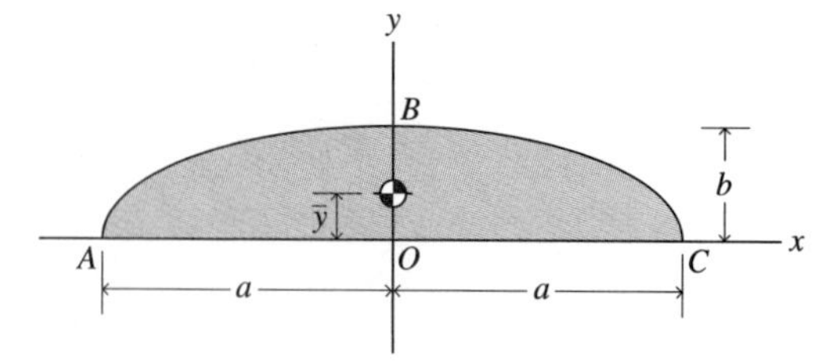

Figure P8.72

8.73 Using the Pappus-Guldinus theorem for surface areas, determine the area of the surface generated by revolving the line *ABCDE* of Fig. P8.47 about the x axis through π radians.

8.74 Given that the volume of a sphere is $V = 4\pi R^3/3$, where R is the radius of the sphere, use the Pappus-Guldinus theorem for volumes to determine:

a. the centroid coordinate $\bar{y}$ of the semicircular area *ABCOA,* shown in Fig. P8.74

b. the centroid coordinate $\bar{y}$ of the quarter circle *BCOB* (or *ABOA*)

(*Hint:* This problem uses the Pappus-Guldinus formula in reverse.)

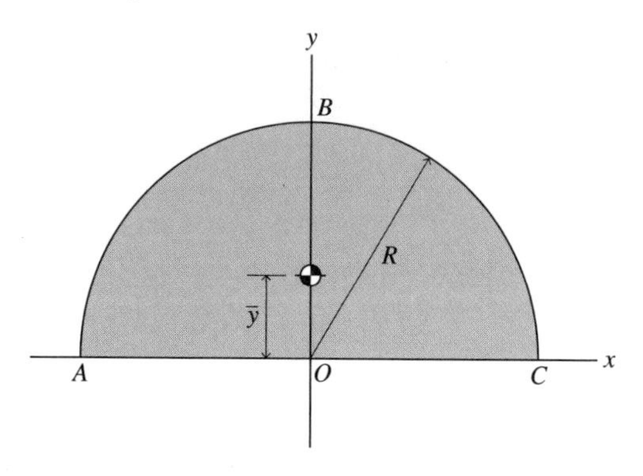

Figure P8.74

8.75 A V-shaped groove is cut out of a shaft (see Fig. P8.75). Using the Pappus-Guldinus theorems, determine:

a. the volume of the material removed

b. the surface area of the V-shaped groove

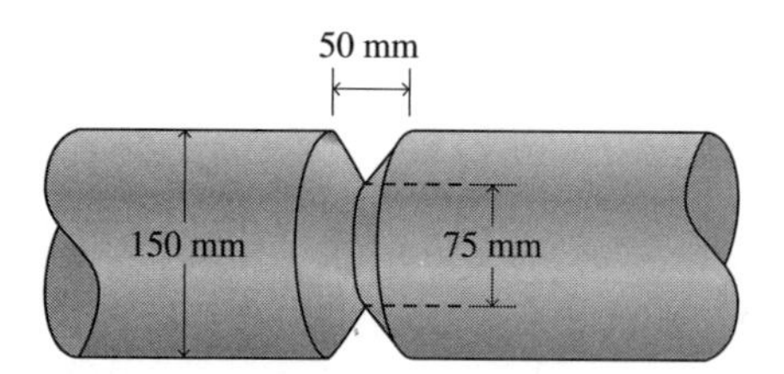

Figure P8.75

8.76 The propellant in a solid-fuel booster rocket used to launch a spacecraft is contained in a cylinder with hemispherical end caps (see Fig. P8.76). An open channel with a circular cross section running through the center of the propellant serves as the combustion chamber. Using the Pappus-Guldinus theorem for volumes, determine the volume of solid-fuel propellant contained in the booster.

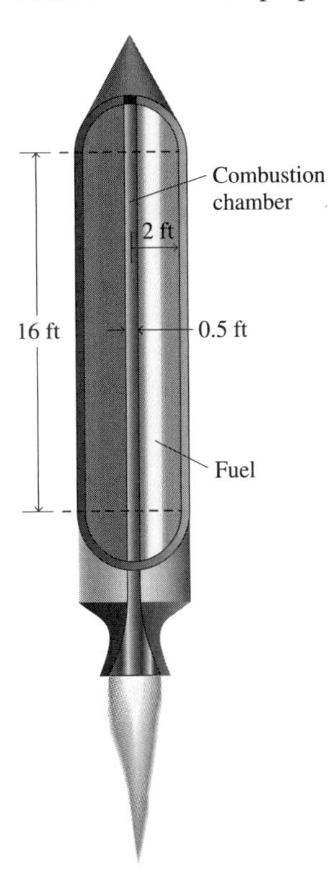

Figure P8.76

8.77 The cross section of a hot-air balloon may be approximated by a hemisphere *ABC* and a truncated cone *ACDE* (see Fig. P8.77). Using the Pappus-Guldinus theorem for volumes, determine the volume of air in the balloon when it is inflated to the dimensions shown.

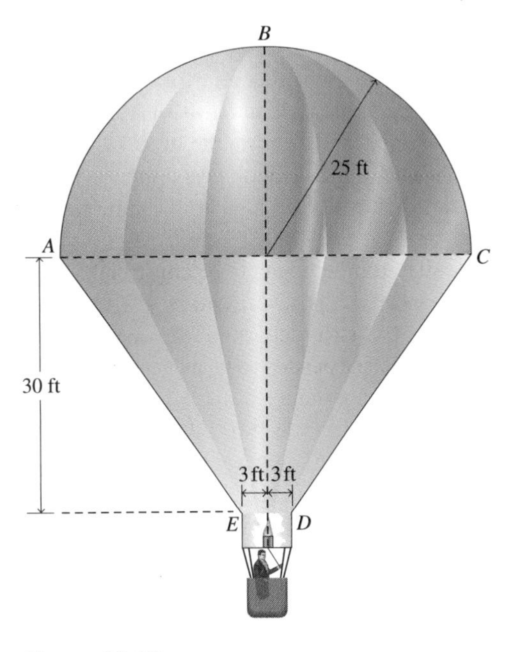

Figure P8.77

8.78 In the decade from 1980 to 1990, engineers and physicists joined to design the Super-conducting Super Collider (SSC), in which matter was to be accelerated to high speeds. The construction of the SSC required that an underground tunnel be formed in the shape of a torus with central radius R and circular cross-sectional diameter D (see Fig. P8.78).

a. From the Pappus-Guldinus theorem for volumes, derive a formula for the volume V, in terms of R and D, of soil that had to be excavated to form the tunnel.

b. From the Pappus-Guldinus theorem for surface areas, derive a formula for the surface area A of the tunnel, in terms of R and D.

c. The design called for the tunnel to be constructed with approximate dimensions $R = 43$ km and $D = 5$ m. Using the results of parts a and b, determine the approximate volume (in m^3) of soil that needed to be excavated and the surface area (in m^2) of the tunnel wall.

d. If the cost of excavation was $\$10/m^3$ and the cost of lining 30% of the tunnel wall with concrete was $\$3/m^2$, compute the approximate total cost of excavating and lining. (In addition to this cost was the cost of the experimental equipment, which was estimated to be as large, if not larger. The SSC project was canceled as the tunnel was nearing completion, because the U.S. Congress withdrew financial support.)

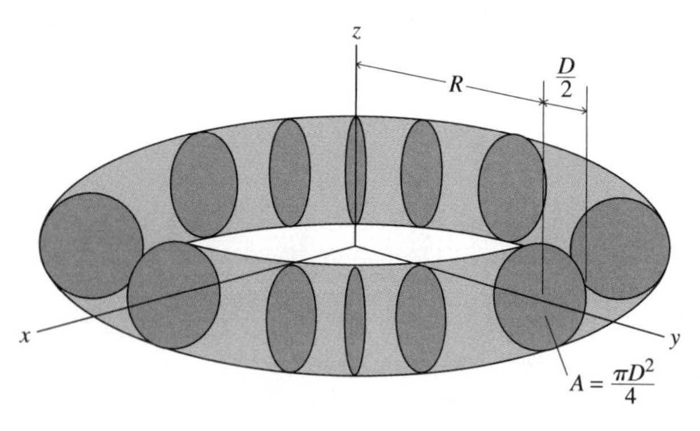

Figure P8.78

8.79 A cylindrical tube containing a liquid is rotated about its vertical axis at a constant angular speed ω. Under this rotation, the free surface of the liquid forms a paraboloidal surface (see Fig. P8.79 and Problem 8.80), with a height c at the center of the tube and a height h at the tube wall. Determine the height of the liquid in the tube before rotation began.

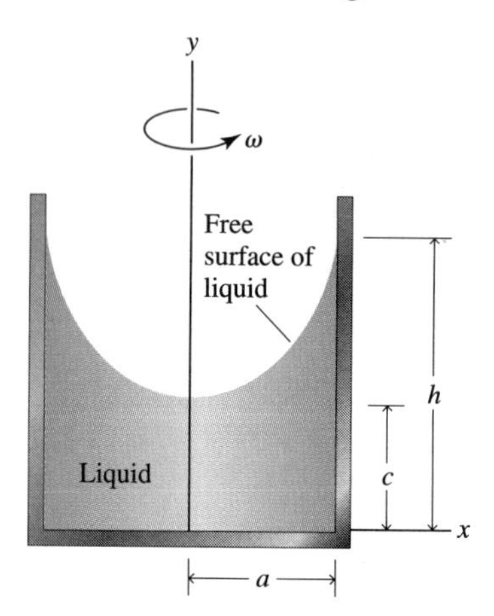

Figure P8.79

8.80 It is shown in fluid mechanics that if a cylindrical tube containing a liquid is rotated at a constant angular speed ω about the vertical axis of the tube, the free surface of the liquid forms a paraboloidal surface. For a given angular speed, the liquid rises to a height h at the wall of the tube and falls to zero height at the center of the tube (see Fig. P8.80). Determine the height of the liquid before the rotation occurred.

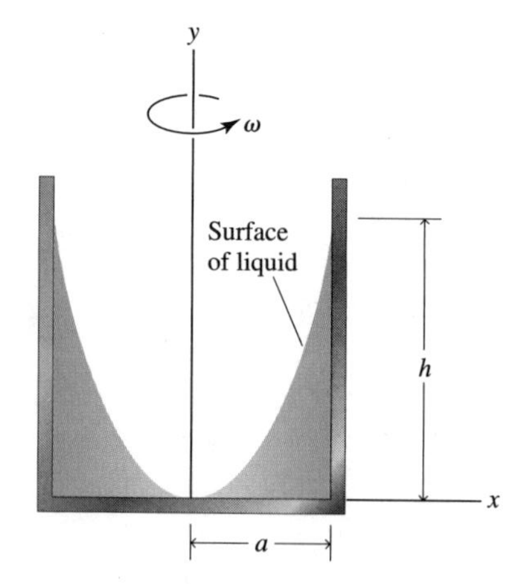

Figure P8.80

8.81 The cross section of the circular part (with inner radius c and angle $\theta = \pi$) of a crane hook is shaded in Fig. P8.81a. A separate view of the cross section is given in Fig. P8.81b.

a. Using the Pappus-Guldinus theorem for volumes, determine the volume of material required in the circular part of the hook ($0 \leq \theta \leq \pi$), in terms of the dimensions a, b, c, and d—that is, the volume generated by rotating the constant cross section about the center line O–O of the hook (see Fig. P8.81b) through the angle $\theta = \pi$.

b. Compute the volume in mm³ for the case where $a = 15$ mm, $b = 25$ mm, $c = 25$ mm, and $d = 25$ mm.

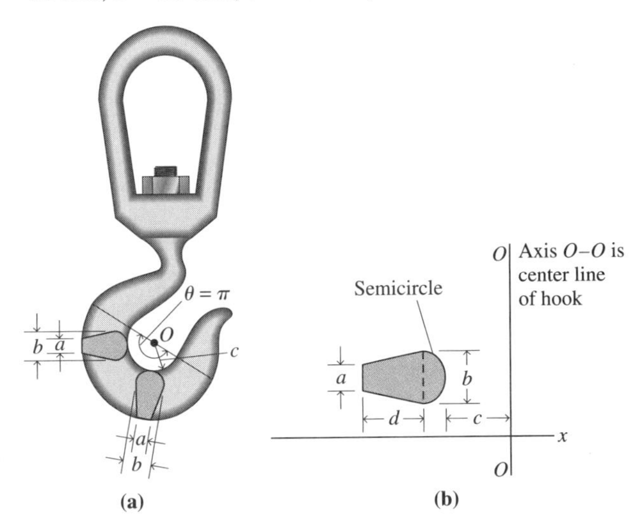

Figure P8.81

8.7 *Resultants of Distributed Parallel Forces*

8.82 a. Replace the distributed forces that act on the rigid L-beam in Fig. P8.82 by dynamically equivalent forces.

b. Draw the free-body diagram of the beam, and determine the support reactions.

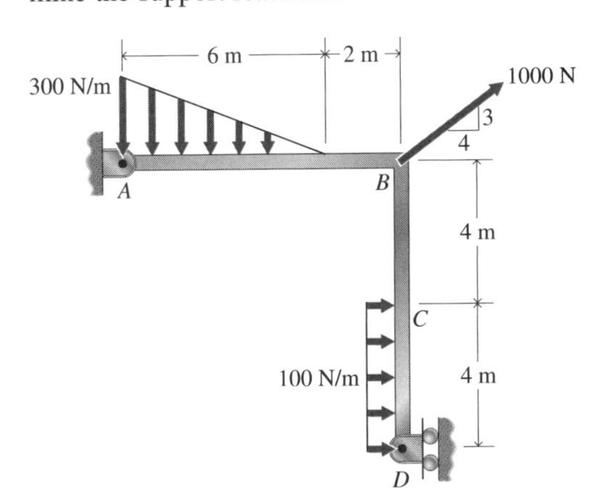

Figure P8.82

8.83 Rod AB is part of a powerboat's propeller shaft. It is subject to a sinusoidally distributed load $w(x)$ and a concentrated load P, as shown in Fig. P8.83. Determine the magnitude and the line of action of the resultant of these loads.

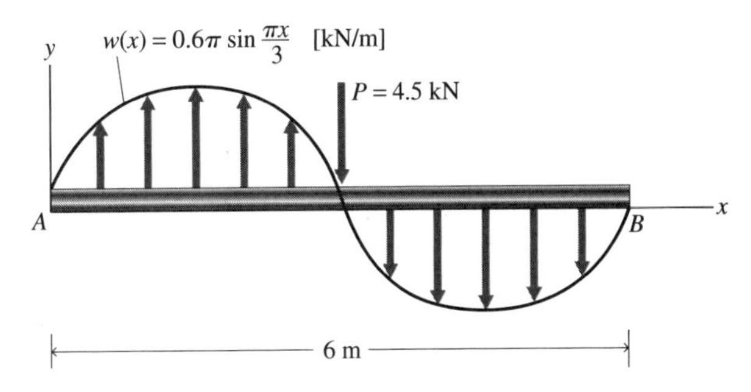

Figure P8.83

8.84 The specific weight of the aircraft panel shown in Fig. P8.84 is constant. The panel weighs 500 N.

a. Replace the distributed load by a force at C and a couple.

b. Draw a free-body diagram of the panel, and determine the reactions of the supports at A and B.

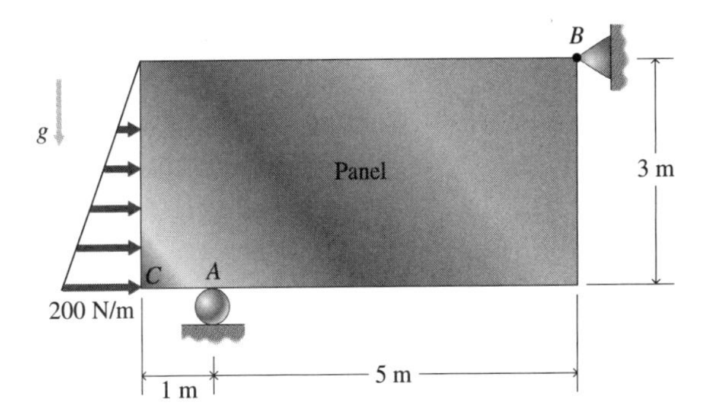

Figure P8.84

8.85 The homogeneous triangular plate in Fig. P8.85 weighs 100 lb. It is suspended in a horizontal plane by three ropes *A*, *B*, and *C*.

a. Determine the location of the center of gravity of the plate.

b. Place the weight vector of the plate at the center of gravity, and determine the forces in the three ropes.

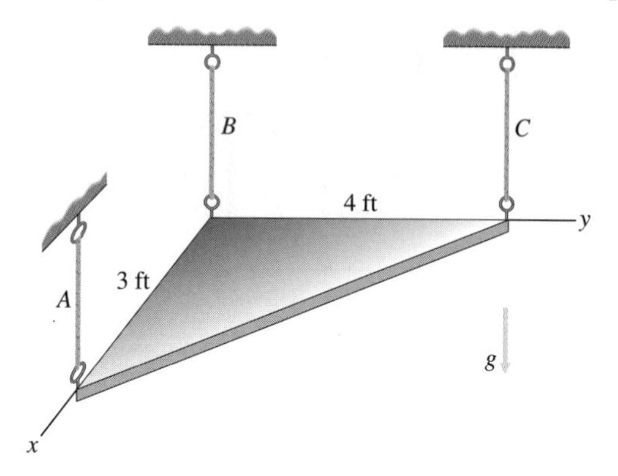

Figure P8.85

8.86 The homogeneous panel shown in Fig. P8.86 weighs 4.5 kN.

a. Replace each of the distributed loads by its resultant, and draw a free-body diagram of the panel.

b. Determine the support reactions at *A* and *B*.

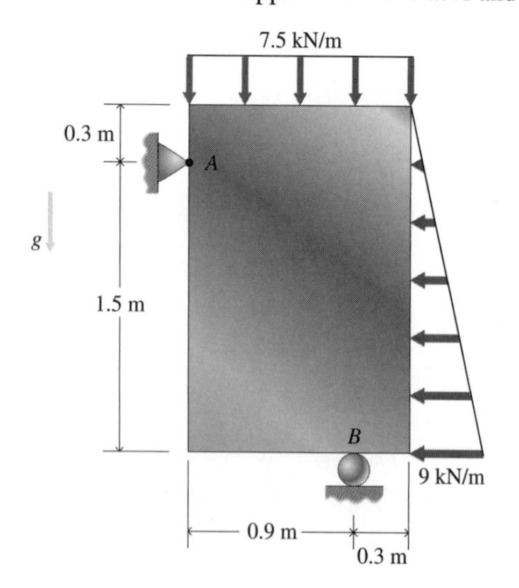

Figure P8.86

8.87 A beam used to support grain in a grain elevator is subject to a distributed load, as shown in Fig. P8.87.

a. Determine the resultant of the distributed load.

b. Replace the resultant of the distributed load by an equivalent concentrated force at *A* and a couple.

c. Determine the support reactions.

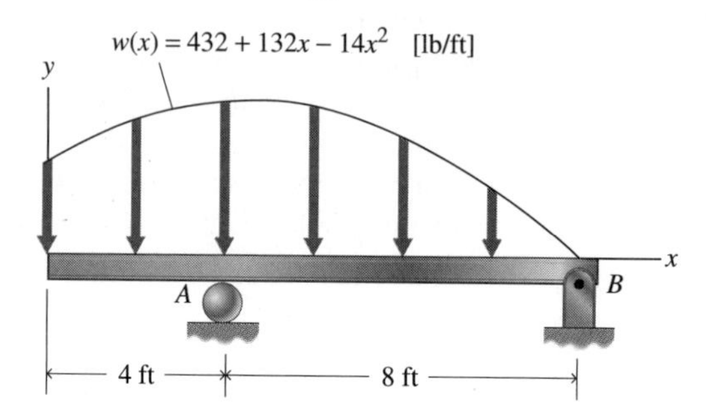

Figure P8.87

8.88 A beam is subjected to a linearly varying distributed load, as shown in Fig. P8.88.

a. Determine the resultant of the distributed load.

b. Determine the moment of the distributed load about the left end of the beam.

c. Determine the support reactions of the beam.

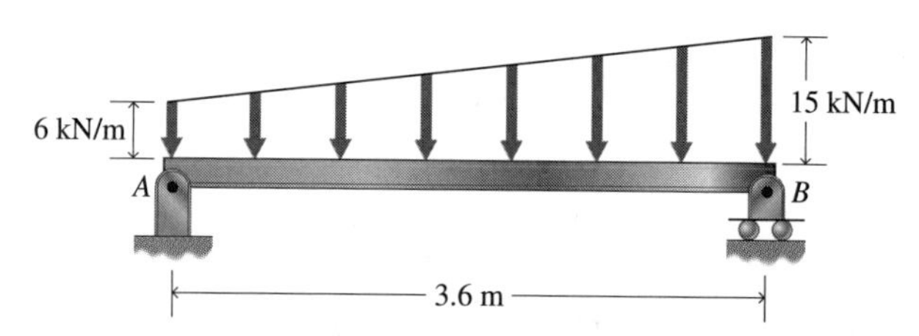

Figure P8.88

8.89 A thin airplane fuselage ring subtends an arc of 90° between radii OA and OB (see Fig. P8.89). It is subjected to a constant pressure p distributed along its arc length.
a. Determine the resultant force of the distributed pressure in terms of p and a. Ignore the thickness of the ring. In other words, assume that the pressure and support reactions act at radius a. [*Hint:* Since the pressure is symmetric with respect to the y axis, the sum of the pressure forces in the x direction cancel, and the projection of a typical differential pressure force in the y direction is $dF_y = -p(a\,d\theta)\cos(\pi/4 - \theta)$. These pressure forces form a system of distributed forces parallel to the y axis.]
b. Determine the reactions of the supports in terms of p and a.

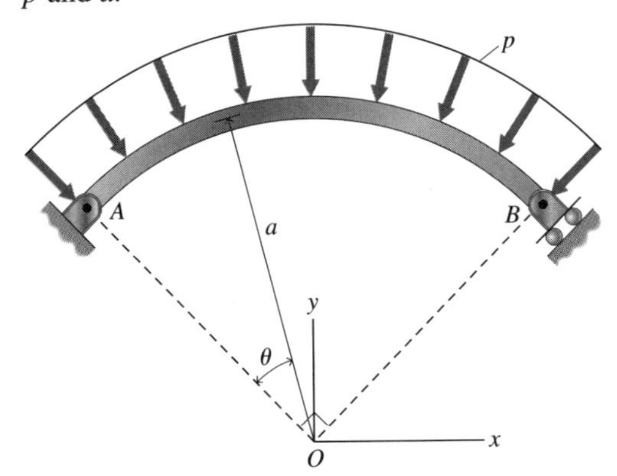

Figure P8.89

8.90 After the panel of Fig. P8.86 was designed, the design engineer discovered that the support at B had to be restricted to a maximum load of 14 kN. Examine the original design, and determine the force at B. If the force is greater than 14 kN, add an additional load to the panel to reduce the force at B to 14 kN.
Method: Consider only simple additional loads. Locate the load to reduce the force at B to 14 kN.

8.91 Add additional loading to the beam of Fig. P8.88 to reduce the support reaction at B by 50%.
Method: Consider simple additional loads.

8.92 The design engineer wishes to increase the arc AB of the fuselage ring of Fig. P8.89 from 90° to 180°. However, he is cautious about making the change from 90° to 180° in one step, since the design requires that $a = 10$ ft and $p = 20$ lb/in. Also, the supports at A and B can withstand a maximum vertical load of only 2000 lb. Therefore, he asks you to examine the arc change from $\phi = 90°$ to $\phi = 180°$, where ϕ is the arc angle between radii OA and OB, and the arc is symmetric about the y axis. Note that for an arc of ϕ radians,

$$dF_y = -pa\,d\theta\cos\left(\frac{\phi}{2} - \theta\right)$$

See Problem 8.89, where $\phi = \pi/2$ radians.
a. Plot the support reactions at A and B as a function of ϕ over the range $90° \leq \phi \leq 180°$.
b. Based on the results of part a, what would you recommend to the design engineer?

8.93 While designing the beam in Problem 8.87, you, as design engineer, decide to examine the location of the support at A. You consider the location of A in the range $0 \leq x \leq 6$ ft.
a. Plot the support reactions at A and B, as functions of x.
b. Give your supervisor your recommendation, based on the results of part a, for the best location of support A.
c. How is the corresponding support reaction at B altered from its value when support A is at 4 ft from the left end?

8.94 The weight of the aircraft panel in Problem 8.84 must be reduced to 480 N, but the length and width dimensions of the panel must not be changed. Also, the support reaction at *A* must be reduced by 20 N, and the panel must have an opening to allow electric power cables to pass through. The opening may be located anywhere in the panel, but the opening must be at least 0.5 m from any edge of the panel. The chief engineer asks you to reduce the weight by making the size of the opening such that the material removed weighs 20 N and the location of the cut results in the support reaction at *A* being reduced by 20 N.

Method: Solve the original problem for the support reaction at *A*. Then, reduce it by 20 N. Select a simple opening shape and appropriate size. Select a location for the opening. Then, draw a free-body diagram of the panel, and examine the equilibrium of the panel. If the panel is in equilibrium, and the opening is at least 0.5 m from any edge of the panel, your solution meets the requirements. If not, try again. That is, select another shape and/or location and repeat the above process. You may have to repeat the process several times before you succeed.

REVIEW QUESTIONS

8.95 Define *gravity axis of a body*.

8.96 A homogeneous rectangular table 5 ft long weighs 60 lb. A pile of books that weighs 20 lb rests on the table, with its gravity axis 2 ft from the gravity axis of the table. The gravity axis of the books lies in the longitudinal plane of symmetry of the table. How far is the gravity axis of the system from the gravity axis of the table?

8.97 Define *center of gravity of a body*. Describe an experimental procedure for locating the center of gravity of a body.

8.98 Define *moments of the weight of a body* with respect to the Cartesian coordinate planes. Explain how these moments are used to determine the location of the center of gravity of the body.

8.99 Define *centroid of a volume*.

8.100 Define *moments of a plane area* with respect to coordinate axes in the plane of the area. Explain how these moments are used to determine the location of the centroid of the area.

8.101 Where is the centroid of a right triangle located? Where is the centroid of a solid cone located? Is the formula used to determine the centroid of a solid cone restricted to cones of circular cross section?

8.102 How is the centroid of an arc of a curve determined?

8.103 How is the center of gravity of a composite body determined, if the centers of gravity of the individual parts are known?

8.104 How is the resultant force of distributed parallel loads determined?

Chapter 9 FLUID STATICS

Buoyancy forces on both the sailboat and the blimp balance their weights and allow them to float.

A Look Forward

READING THE TITLE OF THIS CHAPTER, you might wonder, "What is fluid statics, and how does it relate to rigid-body statics?" Or, more generally, "What is a fluid?" This chapter answers these questions and others.

Fluid statics, also called *hydrostatics,* is the study of fluids at rest. Thus, fluid statics is based on Newton's first and third laws and depends on the concepts of force and moment equilibrium. Using these concepts and the concept of pressure in a fluid, we can determine the forces that act on objects submerged in or floating on a fluid. We can also determine such effects as the buoyant force on a submerged body and the resultant force of a distributed pressure that acts on the face of a plane or curved surface.

You will learn what is meant by standard atmospheric conditions, absolute pressure, local atmospheric pressure, gage pressure, and buoyancy.

You will find that the concepts from this chapter on fluids at rest are important in the study of fluids in motion, covered in dynamics and later courses in fluid mechanics.

Survey Questions

- How does fluid statics differ from rigid-body statics?
- Is a gas a fluid? Is a liquid a fluid?
- Does pressure have a direction? Is it a vector quantity?
- Why are standard atmospheric conditions important in fluid statics?
- How does pressure vary with depth?
- How is fluid pressure related to the forces that act on a solid body in a fluid?
- What is the center of pressure?
- Why do some bodies float in water? What keeps them upright so they don't tip over?

9.1 DEFINITIONS OF STRESS, FLUID, AND PRESSURE

After studying this section, you should be able to:

- Describe how a fluid responds to stress.

Based upon your everyday experiences, you have a general understanding of what is meant by the word "fluids." You frequently refer to water as a liquid or a fluid. Scientists also refer to air as a fluid. Many other liquids and gases are referred to as fluids (e.g., gasoline, blood, wine, liquid nitrogen, and liquid chemicals). Fluids play a significant role in many aspects of life: We drink fluids, we breathe them, we wash in them. Fluids flow through our bodies. Fluids, when burned, provide propulsion for missiles and cars. Fluids, in the forms of winds and clouds, control the weather. Airplanes fly in fluids, and ships float on them.

Fluids form two-thirds of the substances known as matter. Ordinarily, matter is classified as being in one of three states—namely, *solid, liquid,* or *gas.*[1] Based on everyday experience, you might think of a solid as a body that has a definite volume and shape. For example, a brick has a definite size and shape. You might also think of a liquid, such as milk, as a substance that has a definite volume but no definite shape. For example, if you fill a glass with milk, the milk takes on the shape of the glass. If you then pour the glass of milk into a bowl, its volume remains unchanged, but it flows to take on the shape of the bowl. And, finally, a given amount of helium gas has neither a definite volume nor a definite shape. When a gas is placed in a container of a given volume and shape, its volume changes as it expands to take on the shape of the container.

These descriptions help in visualizing the states of matter, but they are somewhat superficial. For example, we normally consider materials such as asphalt, plastic, and glass as solids, but when they are subjected to high temperatures, they tend to flow like liquids. In fact, many substances change state depending on temperature and pressure. Also, the time it takes to cause a change in shape of a material when it is subjected to external force or pressure is often taken as a measure of whether the material is a solid, liquid, or gas.

Key Concept

A fluid at rest cannot resist shear stress. A fluid will change shape when subjected to shear stress.

How the shape of a body changes is affected by forces that are distributed inside the body. To examine the forces that act inside a body, let's consider a body subject to external loads (see Fig. 9.1a). We pass a plane Q through the body and consider the cross-sectional area A of the body cut by the plane. We designate one side of the plane Q (or of the cross-sectional area A) as positive, and the other side as negative. The material on the positive side of Q exerts a force on the material on the negative side. This force is transmitted through the plane Q by direct contact of material on the two sides of the plane. The part of this force that is transmitted through an increment of area ΔA of the cross-sectional area A of the body is denoted by $\Delta \mathbf{F}$ (see Fig. 9.1b). Since $\Delta \mathbf{F}$ lies inside the body, it is called the *internal force* in the body.

internal force

1. Physicists often include *plasma* as a fourth state of matter. Plasmas can occur when a substance is heated to very high temperatures.

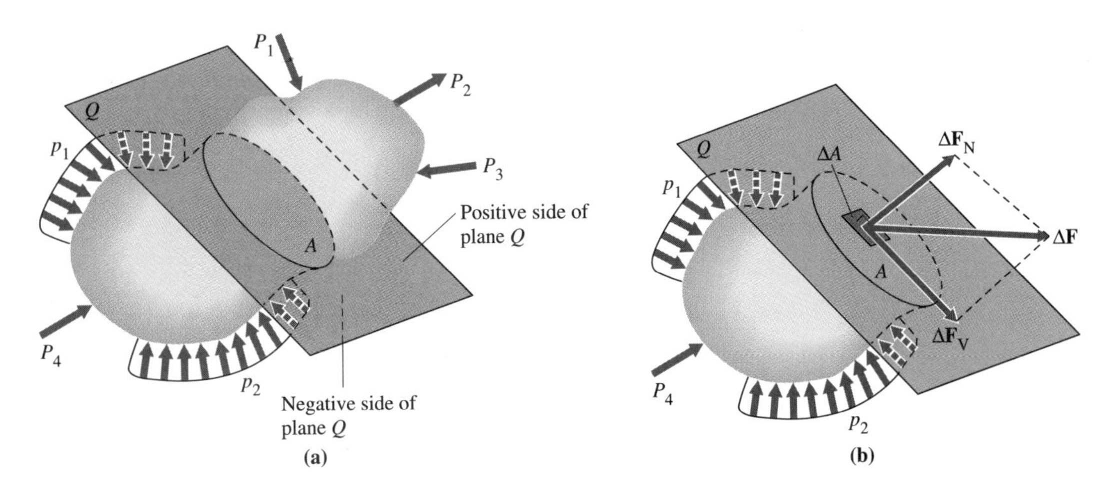

Figure 9.1

The force $\Delta\mathbf{F}$ may be resolved into components $\Delta\mathbf{F}_N$ and $\Delta\mathbf{F}_V$, such that the component $\Delta\mathbf{F}_N$ is perpendicular to plane Q and the component $\Delta\mathbf{F}_V$ is tangent to plane Q (Fig. 9.1b). The component $\Delta\mathbf{F}_N$ is called the *normal force* on the area ΔA, and the component $\Delta\mathbf{F}_V$ is called the *shear force* on the area ΔA. The foregoing concepts are equally applicable to solids, liquids, and gases.

normal force

shear force

The forces $\Delta\mathbf{F}_N$ and $\Delta\mathbf{F}_V$ depend on the size of the area ΔA. The magnitudes of the *average forces per unit area* on ΔA are $\Delta F_N/\Delta A$ and $\Delta F_V/\Delta A$. These ratios are called the *average normal stress* and the *average shear stress,* respectively, on the area ΔA. By letting the area ΔA become an infinitesimal area dA, we arrive at the concept of *stress at a point.* The forces $\Delta\mathbf{F}_N$ and $\Delta\mathbf{F}_V$ approach zero as ΔA becomes infinitesimal, but the ratios $\Delta F_N/\Delta A$ and $\Delta F_V/\Delta A$ usually approach limiting values different from zero. The limiting values of the ratios $\Delta F_N/\Delta A$ and $\Delta F_V/\Delta A$ are called the *normal stress* and the *shear stress,* respectively, on plane Q, at the point where the infinitesimal area dA is located. The Greek letters σ (sigma) and τ (tau) are often used to represent normal stress and shear stress, respectively. Hence, in equation form, the normal stress and the shear stress are written as

stress at a point

normal stress

shear stress

$$\sigma = \lim_{\Delta A \to 0} \frac{\Delta F_N}{\Delta A} \qquad \tau = \lim_{\Delta A \to 0} \frac{\Delta F_V}{\Delta A}$$

In general, these stresses depend not only on the point (coordinates) at which the infinitesimal area dA is located but also on the plane in which dA lies.

In engineering practice, liquids and gases are often referred to as *fluids.* An important difference between them is that a liquid is nearly incompressible, while a gas is easily compressed. Roughly speaking, a *fluid* is any substance that can flow. However, for our purposes, a more accurate definition of a fluid is as follows.

fluid

DEFINITION: *A fluid is a material that deforms or flows continuously in the presence of shear stress, no matter how small that shear stress.*

In contrast, a solid ordinarily undergoes a fixed deformation when it is subjected to a given shear stress. There are significant differences between the internal stresses in solids and those in fluids. For example, solids usually can sustain large internal normal stresses that are either tensile or compressive, whereas fluids generally cannot sustain large tensile

hydrostatic pressure stresses. The compressive stress in a stationary fluid is called *hydrostatic pressure.* Consequently, the normal force that is transmitted by a fluid to a body or surface is mainly a push, rather than a pull. Also, since a fluid flows continuously whenever shear stress exists in it, *shear stress cannot exist in a fluid that is at rest.* By the above definition, both liquids and gases may be classified as fluids. However, because of the compressibility of gases, they are ordinarily studied separately from liquids.

9.2 THE CONCEPT OF PRESSURE AT A POINT

After studying this section, you should be able to:

- Describe the general characteristics of hydrostatic pressure.

Let's begin our discussion of fluid statics by considering a small segment, or cell, at a point Q in a fluid at rest. We can apply the concept of the free-body diagram (Sec. 3.2) to the cell. The cell may have any desired shape—for example, a prism or a tetrahedron. The free-body diagram of the cell consists of a drawing that shows all the external forces acting on the cell. Since the entire fluid is at rest, the cell is in equilibrium. Hence, the application of the equations of equilibrium to the cell leads to relationships among the forces that act on different plane elements of the cell. Also, because the fluid is at rest, the shear stresses are zero; that is, only pressures (compressive stresses) act on the cell faces.

Key Concept The pressure at a point in a stationary fluid is the same in all directions.

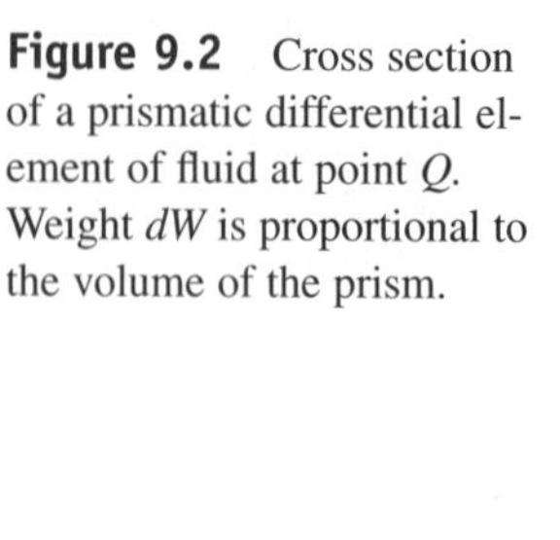

$p_2\,dA$ 2θ $p_1\,dA$ θ θ Q y x z $p_3\,dA'$

Figure 9.2 Cross section of a prismatic differential element of fluid at point Q. Weight dW is proportional to the volume of the prism.

Consider a cell in the form of an infinitesimal isosceles prism at a point Q in a fluid at rest. The cross section of the prism is shown in Fig. 9.2. Since the shear stresses are zero and since the differential weight of the cell is a higher-order quantity proportional to the differential volume (whereas the forces $p_1\,dA$, $p_2\,dA$, and $p_3\,dA'$ are proportional to differential areas), the cell is in equilibrium under the action of pressures p_1, p_2, and p_3, which act on the lateral faces of the prism. (The self-equilibrating forces on the triangular faces, acting in the z direction, are not shown. Hence, Fig. 9.2 is not a complete free-body diagram of the element.) These pressures are due to *contact* with the fluid that surrounds the cell. The resultant normal forces on the lateral faces are $p_1\,dA$, $p_2\,dA$, and $p_3\,dA'$, where dA and dA' are the differential areas of the faces perpendicular to the cross section shown in Fig. 9.2. The equation of equilibrium for forces in the x direction is

$$\sum F_x = p_2\,dA\cos\theta - p_1\,dA\cos\theta = 0$$

Hence, $p_1 = p_2 = p$. So, the pressures p_1 and p_2 on the lateral surfaces of the prism at point Q are equal; that is, they have the same magnitude p.

Since p_1 and p_2 are pressures (normal compressive stresses) on any two planes through point Q and since the position and the orientation of the infinitesimal cell in the fluid and the value of the angle θ are arbitrary, the result $p_1 = p_2 = p$ establishes the following theorem.

Theorem 9.1

If the shear stresses are zero at a point in a fluid, the normal stresses at the point have the same magnitude for all planes that pass through the point.[2]

The value p of the normal compressive stress (force per unit area) is defined to be the *pressure at point Q*.

pressure at a point

The fact that the normal stress, in the absence of shear stress, does not depend on the orientation of the plane element on which the stress acts is known as *Pascal's principle,* in honor of French mathematician Blaise Pascal (1623–1662), who performed hydrostatic experiments to verify it. Since shear stress does not exist in a fluid at rest, Pascal's principle is rigorously applicable to problems of hydrostatics.

Pascal's principle

STANDARD ATMOSPHERIC CONDITIONS

The performance of airplanes and many other engineering machines depends appreciably on atmospheric conditions. Therefore, test data for such equipment are frequently modified in order to compare equipment performance under the same atmospheric conditions. *Standard atmospheric conditions* are defined for air at sea level in terms of pressure p, temperature T, specific weight γ (or specific gravity, sg),[3] mass density ρ, and coefficient of viscosity μ_v. The standard atmospheric conditions are

standard atmospheric conditions

$$p = 14.696 \text{ psi} = 2116.2 \text{ lb/ft}^2 = 101.3 \text{ kPa} = 29.92 \text{ inHg} = 760 \text{ mmHg}$$

$$T = 59°\text{F} = 15°\text{C} = 288 \text{ K}$$

$$\gamma = 0.076510 \text{ lb/ft}^3 = 12.018 \text{ N/m}^3$$

$$\rho = 0.002378 \text{ slug/ft}^3 = 1.2256 \text{ kg/m}^3$$

$$\mu_v = 3.719 \times 10^{-7} \text{ lb·s/ft}^2 = 1.781 \times 10^{-8} \text{ kN·s/m}^2 \text{ (kPa·s)}$$

$$g = 32.1740 \text{ ft/s}^2 = 9.806\,64 \text{ m/s}^2$$

$$\gamma = \rho g$$

The *local atmospheric pressure* is the air pressure at a given location on the earth. In problems in this book, it is assumed that the local atmospheric pressure is equal to the standard atmospheric pressure, unless stated otherwise.

local atmospheric pressure

ABSOLUTE, LOCAL, AND GAGE PRESSURE

Common pressure gages measure the pressure above or below local atmospheric pressure. This pressure is called *gage pressure.* Pressure measured relative to the absolute zero of pressure is called *absolute pressure* (see Fig. 9.3). Absolute pressure is determined by adding the local atmospheric pressure to the gage pressure; that is, $p_{abs} = p_{atm} + p_g$, or

gage pressure

absolute pressure

$$p_g = p_{abs} - p_{atm} \tag{9.1}$$

where p_{abs}, p_{atm}, and p_g are the absolute, atmospheric, and gage pressures, respectively.

2. Since the equations of equilibrium are independent of the material properties of the cell, Theorem 9.1 is also valid for a solid.
3. Specific gravity (sg) of a material is the ratio of the specific weight of the material to the specific weight of water.

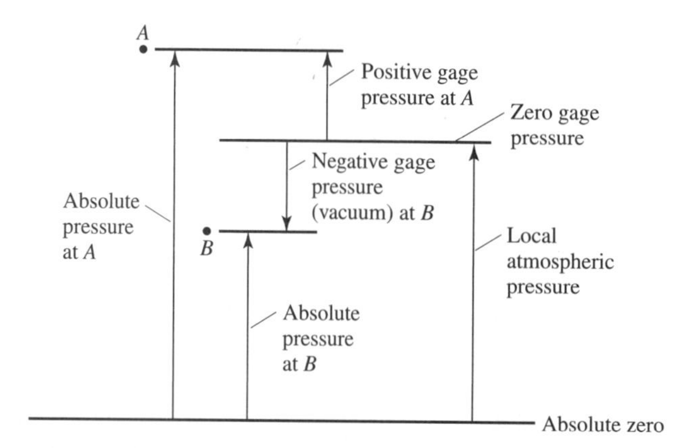

Figure 9.3 Absolute pressure = local atmospheric pressure + gage pressure ($p_{abs} = p_{atm} + p_g$). At sea level, the local atmospheric pressure is equal to the standard atmospheric pressure.

For example, if the absolute pressure is higher than the local pressure, as at point *A* in Fig. 9.3, the gage pressure at that point is positive. If the absolute pressure is lower than the local pressure, as at point *B,* the gage pressure at that point is negative (a vacuum is said to exist there).

In this book, the word "pressure," without qualification, signifies gage pressure. In cases where doubt could arise as to whether gage pressure or absolute pressure is meant, "gage pressure" or "absolute pressure" are used.

vacuum

As noted above, a negative gage pressure is called a *vacuum.* For example, a pressure that is 2 psi less than atmospheric pressure may be designated as 2 psi vacuum or as −2 psi.

Local atmospheric conditions are measured with weather instruments such as these.

PRESSURE ON THE SURFACE OF A SOLID BODY

Key Concept A submerged body in a stationary fluid is subject to hydrostatic pressure that acts normal to the surface of the body.

Figure 9.2 shows an infinitesimal cell of a fluid at rest. This cell is in equilibrium under the action of fluid pressure p. Let's now consider a finite solid body that is immersed or floating in a fluid at rest. Since the fluid is at rest, no shear stress exists in the fluid. Consequently, at the interface between the fluid and the body, no shear stress exists; the fluid exerts only a normal compressive stress (pressure) on the body's surface. The magnitude of the pressure that acts at any point on the body at the interface between the fluid and the body is equal to the pressure p in the fluid at that point. Thus, when a body is submerged in a fluid, the pressure p that exists at the fluid-body interface is transmitted to the body as a normal compressive stress.

9.3 PRESSURE VARIATION WITH DEPTH IN A FLUID

After studying this section, you should be able to:

- Determine the pressure at any depth in a fluid or a system of fluids.

Key Concept In a homogeneous fluid, hydrostatic pressure varies linearly with depth.

The pressure in a stationary fluid is constant at all points at a given depth in the fluid. This fact may be verified by considering a horizontal differential column of fluid with infinitesimal cross-sectional area dA (see Fig. 9.4, where the inverted triangle, ∇, denotes the free surface of the fluid). Let the pressures at the ends of the column be p_1 and p_2. Since there is no shear stress in a stationary fluid, the only horizontal forces that act on the column are the forces $p_1\,dA$ and $p_2\,dA$. Consequently, the condition of equilibrium of the column of fluid in the horizontal direction is $p_1\,dA - p_2\,dA = 0$, or $p_1 = p_2$. Accordingly, the pressure has the same value at any two points in the same horizontal plane, provided that the two points can be connected by a straight-line segment that lies wholly within the fluid. This conclusion remains valid if the two points can be joined by any horizontal chain of

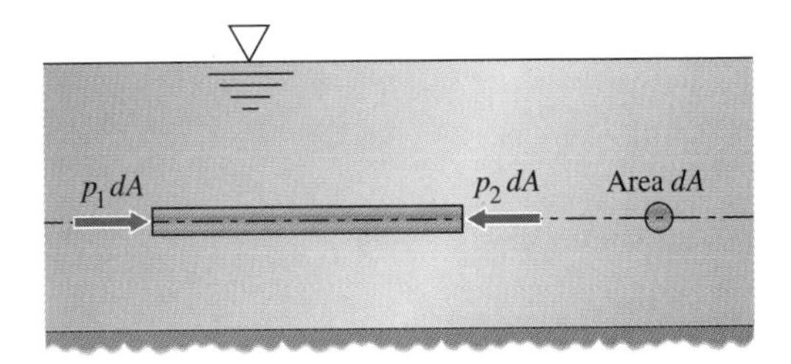

Figure 9.4 Horizontal forces that act on a horizontal differential column of fluid.

straight-line segments that lies wholly in the fluid, since the preceding proof shows that the pressure is constant on any segment of the chain and, hence, on the entire chain.

Hydrostatic pressure acts normal to the glass surface of this shark tank (or is it a person tank?). The pressure varies linearly with depth.

Consider a vertical prismatic column of a fluid at rest with cross-sectional area dA (see Fig. 9.5). Let the pressures on the upper and lower ends of the column be p_1 and p_2, respectively. There are three vertical forces that act on the column: the forces $p_1\,dA$ and $p_2\,dA$ on the ends of the column and the weight dW of the column. The weight dW is proportional to the length $h = h_2 - h_1$ and the area dA of the cross section of the column. The condition of vertical equilibrium of the column (Fig. 9.5) is

$$p_2\,dA - p_1\,dA = dW \tag{9.2}$$

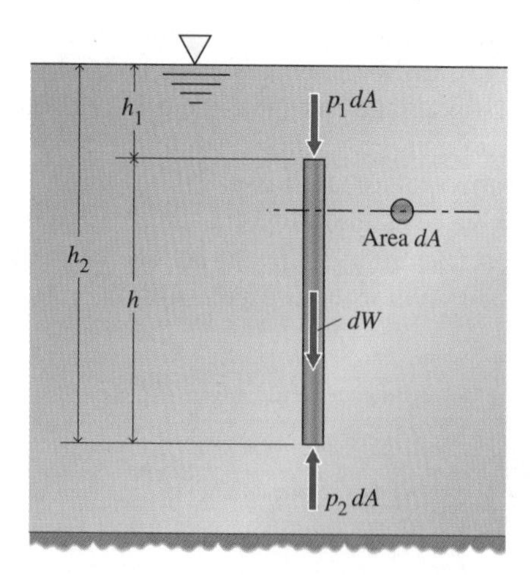

Figure 9.5 Vertical forces that act on a vertical differential column of fluid.

Equation (9.2) determines the variation of pressure with depth. This equation remains valid even if the vertical column passes through several layers of different fluids, such as oil, water, and mercury.[4] Then, the weight dW of the column is the sum of the weights of each of the layers of different fluids in the column.

If the column in Fig. 9.5 lies in a liquid of *constant specific weight* γ (weight per unit volume), the weight of the column is $dW = \gamma h\, dA$, where $h = h_2 - h_1$ is the difference in depth between the ends of the column (that is, h is the length of the column). Then Eq. (9.2) yields

$$p_2 - p_1 = \gamma h$$

or

$$p_2 = p_1 + \gamma h \tag{9.3}$$

In particular, if $h_1 = 0$, then $h = h_2$, $p_1 = p_{atm}$, and $p_2 = p_{abs}$ at a point at depth h in the fluid. Hence, the absolute pressure in the fluid is

$$p_{abs} = p_{atm} + \gamma h \tag{9.4}$$

Therefore, pressure increases linearly with depth h in a homogeneous motionless fluid. Also, comparison of Eqs. (9.1) and (9.4) yields the gage pressure at depth h as

$$p_g = \gamma h \tag{9.5}$$

SYSTEMS OF TUBES CONTAINING STATIONARY FLUIDS

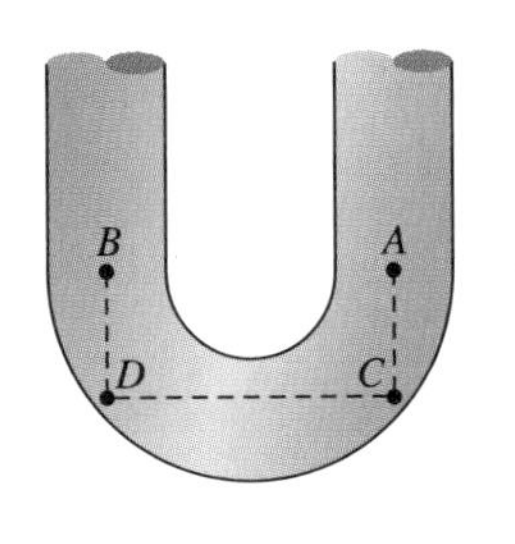

Figure 9.6 U-tube.

Consider a U-shaped tube that is filled with a homogeneous fluid (i.e., a fluid of constant specific weight γ) (see Fig. 9.6). Two points A and B that lie at the same elevation h in the fluid may be joined by a path such as $ACDB$ that lies wholly within the fluid. The pressure increases on the vertical stretch from A to C, remains constant on the horizontal stretch from C to D, and decreases on the vertical stretch from D to B. From Eq. (9.3), the pressure drop from D to B exactly cancels the pressure rise from A to C; that is, the pressures at points A and B are equal. Thus, the following general theorem is established.

Theorem 9.2

> If two points A and B in a stationary homogeneous fluid are at the same elevation and if the two points can be joined by a path that lies wholly within the fluid, the pressures at points A and B are equal.

This principle, in conjunction with Eq. (9.3), enables us to trace the pressure changes through a system of tubes containing several stationary fluids.

4. Equation (9.2) is correct for gases as well as liquids. In the case of a gas, however, the variation of density with altitude (due to the compressibility of the gas) introduces a complication into the calculation of the weight dW of the column. This difficulty is avoided if the column is short, since the weight dW of a short column of gas is ordinarily negligible compared with the forces $p_1\, dA$ and $p_2\, dA$. Consequently, the pressure in a stationary gas is practically constant for small changes of altitude. For example, atmospheric pressure does not change appreciably from one floor of a building to the next.

PROBLEM-SOLVING TECHNIQUE

Static Pressures in a Fluid

To find the pressure at any level in a fluid:

1. As for any problem, make certain at the start that you understand the problem statement—namely, what is given, what is to be determined, what principles apply, what methods apply, in what form the solution is to be presented, and so on.
2. Sketch a diagram of a vertical column of the fluid between a point in the fluid at which the pressure—say, p_1—is known to a point at which the pressure—say, p_2—is required (see Example 9.2). Often, an overall sketch of a body of fluid, with distinct regions, may be sufficient to determined the required pressures (see Examples 9.1, 9.3, and 9.4).
3. Apply Eq. (9.2) or (9.3) to determine the unknown pressure p_2.
4. Repeat Steps 2 and 3 until all unknown pressures are determined.

Example 9.1

Pressure in Stationary Liquids

Problem Statement A tank contains 3 ft of acetylene tetrabromide ($\gamma = 184.8$ lb/ft^3), 6 ft of water ($\gamma = 62.42$ lb/ft^3), and 4 ft of oil ($\gamma = 49.94$ lb/ft^3), in three separate fluid layers (see Fig. E9.1a). The top surface of the oil is exposed to the atmosphere. Plot graphs that show the gage pressure and the absolute pressure as functions of depth.

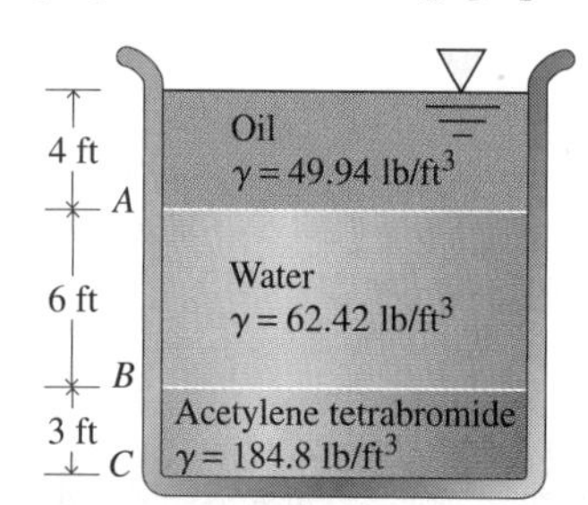

Figure E9.1a

Solution The pressure at level A results from 4 ft of oil. Hence, the gage pressure is

$$p_A = 0 + (49.94)(4) = 199.8 \text{ lb/ft}^2$$

The pressure at level B exceeds the pressure at level A by the pressure due to 6 ft of water. Hence, the gage pressure at level B is

$$p_B = 199.8 + (62.42)(6) = 574.3 \text{ lb/ft}^2$$

The pressure at level C exceeds the pressure at level B by the pressure due to 3 ft of acetylene tetrabromide. Therefore, the gage pressure at level C is

$$p_C = 574.3 + (184.8)(3) = 1{,}128.7 \text{ lb/ft}^2$$

Plotting the computed pressures p_A, p_B, and p_C at their proper levels and joining consecutive points by straight-line segments, we obtain the graph of gage pressure in Fig. E9.1b. To obtain the graph of absolute pressure (Fig. E9.1c), we superpose the atmospheric pressure $p_{\text{atm}} = 2116.2$ lb/ft^2 at all points of the graph.

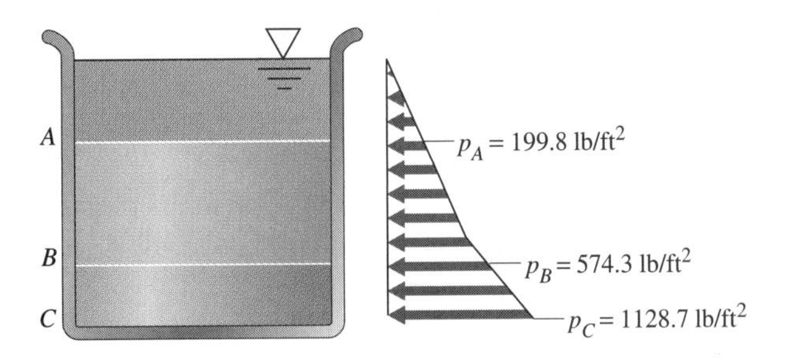

Figure E9.1b

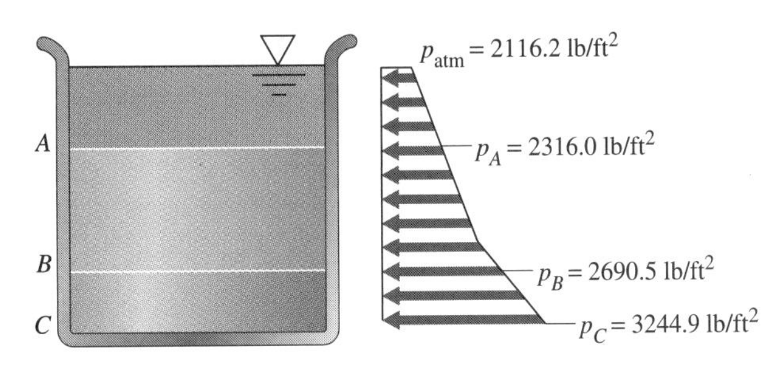

Figure E9.1c

Example 9.2

Gage Pressure in a Tank

Problem Statement The tank shown in Fig. E9.2a is filled with water and air. The gage gives the gage pressure of the air in the tank. Water fills the tube between elevation 3 and elevation 6. Mercury fills the tube between elevation 1 and elevation 3. Atmospheric pressure (p_{atm}) acts on the mercury at 1, since the tube is open at the top. Determine the gage pressure of the air in the tank.

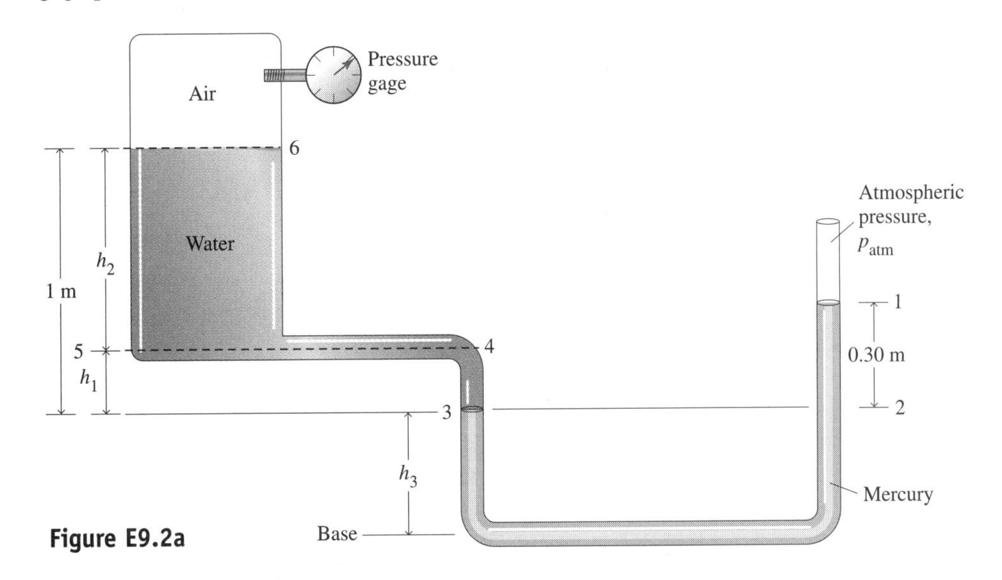

Figure E9.2a

Solution We can solve this problem by applying Eq. (9.2) or (9.3) to various sections of the tube and tank. The pressure of air in the tank may be assumed to be constant, since the specific weight of air is very small ($\gamma_{air} = 0.01202$ kN/m³) at atmospheric conditions. (Depending on the specific weights of various liquids or gases, pressure changes in air may be neglected for variations of elevation less than about 150 m.)

Consider a diagram of the mercury column between elevations 1 and 2 (Fig. E9.2b). This diagram shows only the vertical forces that act on the column of mercury, where the specific weight is 133 kN/m³. Summation of forces in the vertical direction yields

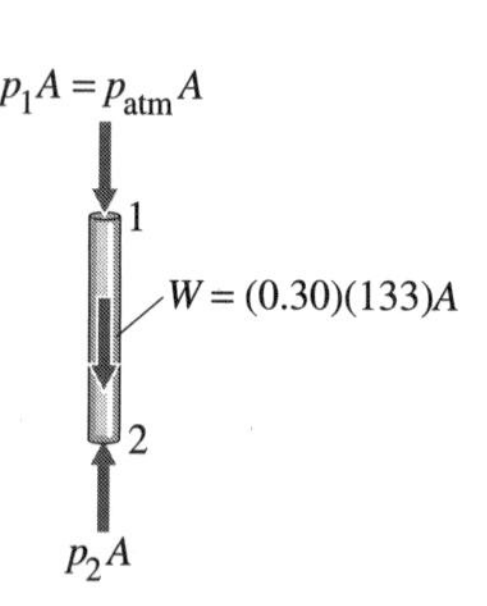

Figure E9.2b

$$\sum F_v = p_2 A - p_{atm} A - (0.30)(133)A = 0$$

where A is the cross-sectional area of the tube. Hence,

$$p_2 = p_{atm} + 39.9 \text{ kPa} \tag{a}$$

where 1 kN/m² = 1 kPa. The pressure at the base can be expressed in terms of the pressures at elevations 2 and 3:

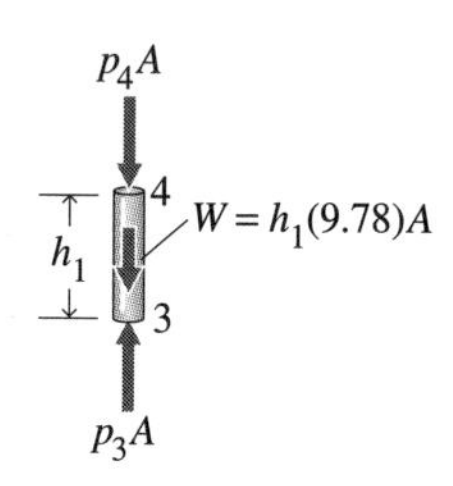

Figure E9.2c

$$p_{base} = p_2 + 133h_3 = p_3 + 133h_3$$

Consequently, $p_3 = p_2$, and, therefore, with Eq. (a),

$$p_3 = p_2 = p_{atm} + 39.9 \text{ kPa} \tag{b}$$

Next, we draw a diagram of the water column between elevations 3 and 4 (Fig. E9.2c), showing the vertical forces that act. Since the specific weight of water is $\gamma = 9.78$ kN/m³, summation of vertical forces yields

$$\sum F_v = p_3 A - p_4 A - h_1(9.78)A = 0$$

or

$$p_3 - p_4 = 9.78h_1 \tag{c}$$

Since 4 and 5 are the same elevation in the water (see Theorem 9.2),

$$p_4 = p_5 \tag{d}$$

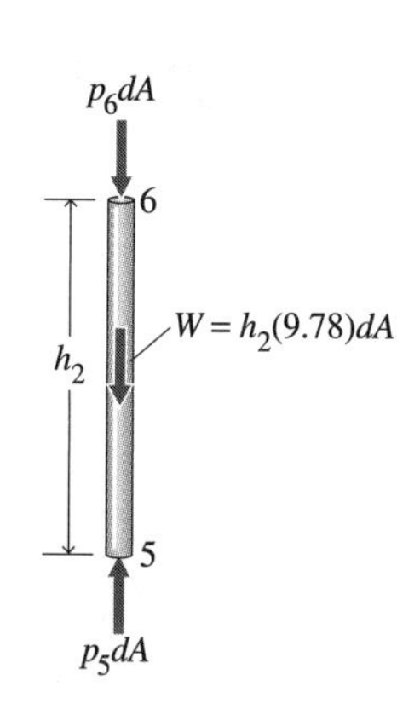

Figure E9.2d

Next, summing vertical forces on a vertical differential column of water of cross-sectional area dA between 5 and 6 (Fig. E9.2d), we have

$$\sum F_v = p_5\, dA - p_6\, dA - h_2(9.78)\, dA = 0$$

or, with Eq. (d),

$$p_5 - p_6 = p_4 - p_6 = 9.78h_2 \tag{e}$$

Adding Eqs. (c) and (e) and noting that $h_1 + h_2 = 1$ m (see Fig. E9.2a), we find

$$p_3 - p_6 = 9.78 \text{ kPa} \tag{f}$$

By Eqs. (b) and (f), we obtain the pressure at elevation 6 as $p_6 = p_{atm} + 30.12$ kPa. Then, since the gage reads the difference between p_6 and p_{atm}, the gage reading (air gage pressure in the tank) is

$$p_g = p_6 - p_{atm} = 30.12 \text{ kPa}$$

This result is independent of the values of h_1, h_2, and h_3.

Information Item The pressure difference between 3 and 6 is the same as it would be if 3 were directly below the water surface (below 6) rather than shifted laterally from 4 to 5 through the horizontal tube. Consequently, we could have calculated the difference $p_3 - p_6$ as if 6 were directly above 3; that is, $p_3 - p_6 = 9.78$ kPa. Then, with Eq. (b),

$$p_3 - p_6 = 9.78 \text{ kPa} = p_{\text{atm}} + 39.9 \text{ kPa} - p_6$$

and

$$p_g = p_6 - p_{\text{atm}} = 30.12 \text{ kPa}$$

Example 9.3

Differential Pressure Measured by a U-Tube

Problem Statement The two ends of a U-tube that is partially filled with mercury ($\gamma_{\text{Hg}} = 133$ kN/m^3) are connected to pipes A and B (see Fig. E9.3). Determine the differential pressure $p_A - p_B$ between the pipes:

a. if the pipes carry a gas with a specific weight of 0.005 kN/m^3

b. if the pipes carry water ($\gamma_{\text{H}_2\text{O}} = 9.78$ kN/m^3)

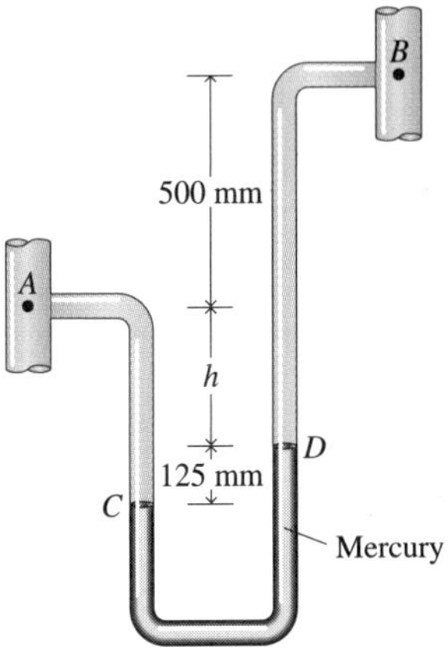

Figure E9.3

Solution First, we develop general equations that pertain to both cases, a and b. From Fig. E9.3 and Eq. (9.3), we write

$$\begin{aligned} p_D - p_B &= (h + 0.500)\gamma \\ p_C - p_D &= (0.125)\gamma_{\text{Hg}} \qquad \text{(a)} \\ p_A - p_C &= -(h + 0.125)\gamma \end{aligned}$$

where γ is the specific weight of the fluid in sections AC and BD.

Adding the pressure differences in Eqs. (a), we obtain

$$p_A - p_B = (0.375)\gamma + (0.125)\gamma_{\text{Hg}} \qquad \text{(b)}$$

Thus, the height h is arbitrary, provided that the fluids in sections AC and BD are the same.

a. For the case where a gas of specific weight 0.005 kN/m^3 is carried in the pipes, Eq. (b) yields

$$p_A - p_B = (0.375)(0.005) + (0.125)(133) = 0.001\,875 + 16.625 = 16.63 \text{ kPa}$$

b. For the case where water is carried in the pipes, $\gamma = \gamma_{\text{H}_2\text{O}} = 9.78$ kN/m^3. Then, Eq. (b) yields

$$p_A - p_B = (0.375)(9.78) + (0.125)(133) = 3.667 + 16.625 = 20.29 \text{ kPa}$$

Example 9.4

System of Tubes

Problem Statement Two closed tanks are connected by a bent tube, as shown in Fig. E9.4. The left tank contains fluid of specific weight 55 lb/ft^3, the right tank contains fluid of specific weight 80 lb/ft^3, and a part of the connecting tube contains fluid of specific weight 45 lb/ft^3. The three fluids are designated as fluids *A, B,* and *C*. The heights of the columns of fluids are indicated. The air pressure above the fluid in the left tank is 1250 lb/ft^2. Determine the pressure above the fluid in the right tank.

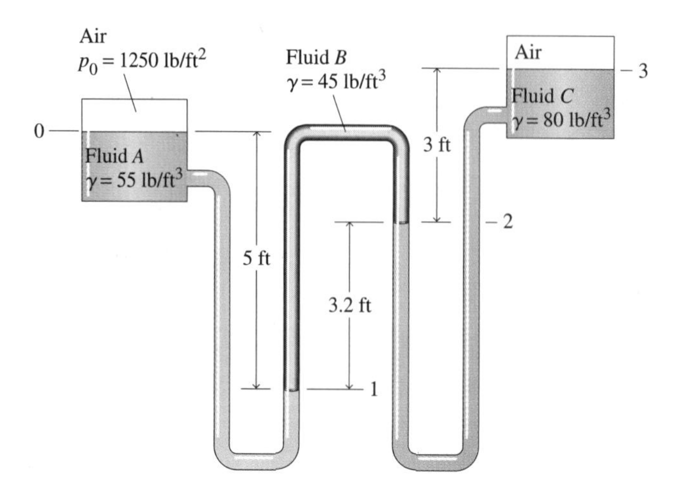

Figure E9.4

Solution In the stretch from level 0 to level 1, the increase of pressure corresponds to 5 ft of fluid *A*. Consequently, the pressure at 1 is $p_1 = 1250 + (55)(5) = 1525$ lb/ft^2. In the stretch from 1 to 2, the decrease of pressure corresponds to 3.2 ft of fluid *B*. Consequently, the pressure at 2 is $p_2 = 1525 - (45)(3.2) = 1381$ lb/ft^2. In the stretch from 2 to 3, the decrease of pressure corresponds to 3 ft of fluid *C*. Consequently, the pressure in the second tank is $p_3 = 1381 - (80)(3) = 1141$ lb/ft^2.

The preceding computations may be assembled into one equation, as follows:

$$p_3 = 1250 + (55)(5) - (45)(3.2) - (80)(3) = 1141 \text{ lb/ft}^2$$

CHECKPOINT 

1. ***True or False:*** Hydrostatics is the study of stationary fluids.
2. ***True or False:*** A vacuum is a negative absolute pressure.
3. ***True or False:*** Pressure in a stationary gas is approximately constant for small changes in altitude.
4. ***True or False:*** Pressure varies linearly with depth in a homogeneous motionless fluid.
5. ***True or False:*** If two distinct points *A* and *B* in a stationary homogeneous fluid are at different elevations, the difference in the pressures at *A* and *B* is proportional to the difference in elevation.

Answers: 1. True; 2. False; 3. True; 4. True; 5. True

9.4 NORMAL FORCE ON A FLAT SURFACE

After studying this section, you should be able to:

- Determine the magnitude and direction of the resultant force of a distributed pressure that acts on a flat surface.

skin friction

pressure force

The force that a flowing fluid exerts on a flat surface has a component normal to the surface and a component tangent to the surface. The tangential component is called *skin friction.* The normal component of the force arises from the pressure that the fluid exerts on the surface; the skin friction (tangential component) arises from the shear stresses that the flowing fluid exerts on the surface. The normal force that a fluid exerts on a flat surface is sometimes called the *pressure force.* As noted in Theorem 9.1, fluids at rest cannot sustain shear stresses. Therefore, fluids at rest exert only normal forces (pressure forces) on surfaces. In this chapter, we are concerned only with fluid statics (fluids at rest), and so this discussion is limited to pressure forces.

Key Concept The resultant force of a distributed pressure on a flat surface acts normal to the surface.

Suppose that the pressure p is known at each point on a flat surface of a submerged rigid body. We may imagine that the surface is partitioned into infinitesimal elements. The differential normal force that acts on an infinitesimal area dA is $dF = p\,dA$. Since the surface is flat, all the differential normal forces that act on the infinitesimal elements of the surface are parallel to one another. Therefore, the flat surface is subject to a system of parallel forces. Accordingly, the resultant force is perpendicular to the surface, and its magnitude F is the algebraic sum (integral) of the differential forces $p\,dA$ that act on the infinitesimal elements of area (see Sec. 4.8):

$$F = \int_A p\,dA$$

where the integral is over the entire area A.

If the pressure p is constant on a flat surface, the total normal force on the surface is pA, where A is the area of the surface. Sometimes the pressure varies in one direction only, as in the case of water pressure on the face of a dam. Then, the surface may be partitioned into small areas taking the form of narrow horizontal constant-pressure strips that run across the entire surface. Since the strips are oriented so that the pressure is constant along the length of a strip, the net force on a strip of area dA is $p\,dA$, where p is the constant pressure on the strip. For example, when calculating the total normal force on the face of a dam, we can conveniently partition the face of the dam into horizontal strips. This method is illustrated in Example 9.5.

Example 9.5

Force on the Face of a Dam

Problem Statement Figure E9.5 is a view, from downstream, of a dam of parabolic form. Assume that the face of the dam is flat. The height of the top of the dam above the vertex of the parabola is 30 m, and the equation of the parabola with reference to

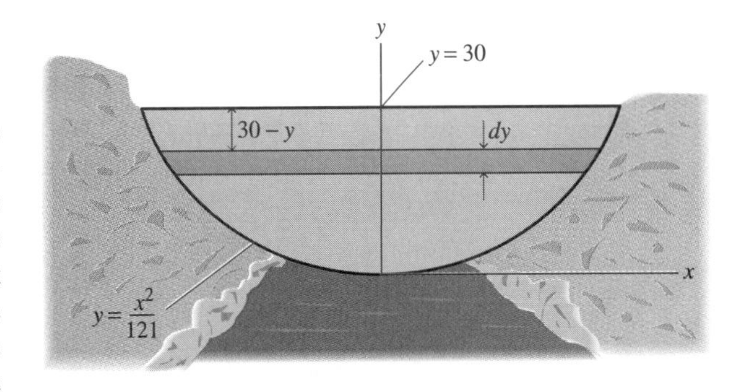

Figure E9.5

the coordinate axes shown is $y = x^2/121$. Water ($\gamma = 9.78$ kN/m^3) rises to the top of the dam. Determine the net force exerted by the water on the dam.

Solution The net force due to the water may be determined most readily by integration. Therefore, we trace a horizontal strip of infinitesimal width dy on the face of the dam (see Fig. E9.5). From the equation of the parabola, the length of the strip is $22\sqrt{y}$. Therefore, the differential area of the strip is $dA = 22\sqrt{y}\,dy$. Since the depth of the strip below the surface of the water is $30 - y$, the gage pressure on the strip is $(9.78)(30 - y)$. Hence, the differential pressure force on the strip is

$$dF = (9.78)(30 - y)22\sqrt{y}\,dy$$

The net force F due to the water pressure is the integral of this differential force:

$$F = (9.78)(22)\int_0^{30} (30 - y)\sqrt{y}\,dy$$

Integration yields $F = 282.8$ MN.

9.5 CENTER OF PRESSURE

After studying this section, you should be able to:

- Determine the line of action of the resultant force of a distributed pressure that acts on a flat surface.

Key Concept

The center of pressure, the point at which the line of action of a pressure force intersects a flat surface, is located using the theory of moments.

If a pressure acts on a flat face of a rigid body, the resulting forces on infinitesimal elements of the loaded area are parallel. The resultant axis of this system of parallel forces is perpendicular to the flat face on which the pressure acts. The point at which the resultant axis intersects the face is called the *center of pressure.* Consequently, a pressure distribution on a flat surface of a rigid body is dynamically equivalent to a single force that acts perpendicular to the surface at the center of pressure.

center of pressure

To determine the center of pressure of an area, we refer to Fig. 9.7 and consider differential strips of areas $dA_y = a\,dy$ and $dA_x = b\,dx$. The forces due to the pressure p that acts on these areas are $p\,dA_y = pa\,dy$ and $p\,dA_x = pb\,dx$. Assuming that p acts in the negative z direction, the moments of these forces about the x and y axes are $dM_x = -yp\,dA_y = -y(pa\,dy)$ and $dM_y = xp\,dA_x = x(pb\,dx)$, respectively. Hence, the total moments about the x and y axes, due to the pressure on the area A, are

$$M_x = -\int_A yp\,dA_y$$
$$M_y = +\int_A xp\,dA_x \qquad (9.6)$$

The letter A below the integral in Eqs. (9.6) means that the integration extends over the area A.

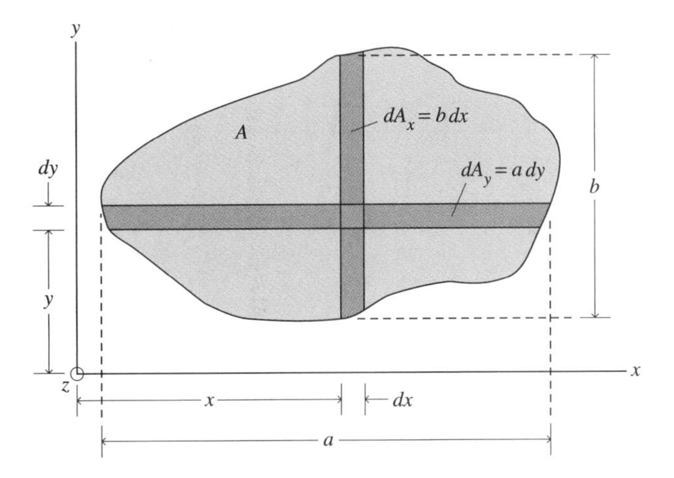

Figure 9.7

The resultant normal force that acts on the area A is

$$F = \int_A p\,dA \tag{9.7}$$

where $dA = dA_x$ or $dA = dA_y$, depending on which strip in Fig. 9.7 is used.

By the theory of moments of parallel forces (Sec. 4.8),

$$y_{cp}F = -M_x \tag{9.8a}$$

$$x_{cp}F = M_y \tag{9.8b}$$

where (x_{cp}, y_{cp}) are the coordinates of the center of pressure and F is the total normal force on the area A, as determined by Eq. (9.7). From Eqs. (9.6), (9.7), and (9.8), we obtain the center of pressure coordinates (x_{cp}, y_{cp}):

$$x_{cp} = \frac{\int_A xp\,dA_x}{\int_A p\,dA} \qquad y_{cp} = \frac{\int_A yp\,dA_y}{\int_A p\,dA} \tag{9.9}$$

The pressure force that acts on a flat surface of a rigid body is equivalent to the resultant normal force acting at the center of pressure. However, it is important to remember that replacing a distributed load by a point force alters the stresses in the body on which the load acts. For example, in the analysis of the motion of an airplane, the air pressure load on a wing may be concentrated at the center of pressure of the wing. But, this simplification of the loading is unreasonable in an analysis of the stresses in the wing, since the stresses in the wing due to the concentrated load would differ dramatically from those due to the distributed load (see Chapter 11).

The above results are summarized by the following theorem.

Theorem 9.3

A pressure p that acts on a flat face A of a rigid body is dynamically equivalent to a concentrated normal force that acts at the center of pressure of face A. The force F is determined using Eq. (9.7), and the center of pressure is determined from Eqs. (9.9).

Example 9.6 Center of Pressure on the End of a Trough

Problem Statement The end of a trough that is filled with a homogeneous liquid is a right triangular plate with horizontal edge a and vertical edge b (see Fig. E9.6). Determine the center of pressure on the end of the trough.

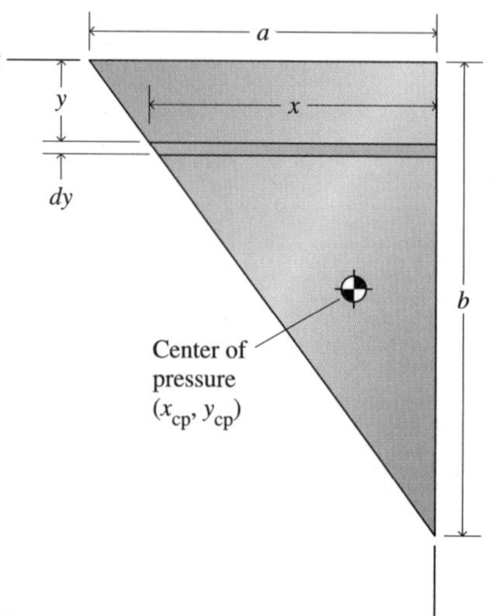

Figure E9.6

Solution The gage pressure at depth y is

$$p = \gamma y \tag{a}$$

where γ is the specific weight of the liquid.

The equation of the diagonal edge is

$$\frac{x}{a} + \frac{y}{b} = 1 \tag{b}$$

Consider an infinitesimal horizontal strip of length x and width dy (see Fig. E9.6). The area of the strip is, with Eq. (b),

$$dA = x\,dy = a\left(1 - \frac{y}{b}\right)dy \tag{c}$$

The differential force on the strip due to the gage pressure is $dF = p\,dA$. Hence, from Eqs. (a) and (c),

$$dF = a\gamma\left(1 - \frac{y}{b}\right)y\,dy \tag{d}$$

Since the pressure on the strip is constant, the resultant axis of the force that acts on the strip passes through the center of the strip. In other words, the coordinates of the center of pressure of the strip are $(x/2, y)$. In view of Eq. (b), these coordinates are

$$\left[\frac{a}{2}\left(1 - \frac{y}{b}\right), y\right] \tag{e}$$

From Eqs. (d) and (e), the moments of the force dF about the x and y axes are

$$\begin{aligned} dM_x &= -a\gamma\left(1 - \frac{y}{b}\right)y^2\,dy \\ dM_y &= \left(\frac{a^2}{2}\right)\gamma\left(1 - \frac{y}{b}\right)^2 y\,dy \end{aligned} \tag{f}$$

Integrating Eqs. (d) and (f), we obtain

$$\begin{aligned} F &= a\gamma\int_0^b\left(1 - \frac{y}{b}\right)y\,dy = \frac{\gamma ab^2}{6} \\ M_x &= -a\gamma\int_0^b\left(1 - \frac{y}{b}\right)y^2\,dy = -\frac{\gamma ab^3}{12} \\ M_y &= \frac{a^2}{2}\gamma\int_0^b\left(1 - \frac{y}{b}\right)^2 y\,dy = \frac{\gamma a^2b^2}{24} \end{aligned} \tag{g}$$

Hence, from Eqs. (9.8) and (g),

$$x_{cp} = \frac{M_y}{F} = \frac{a}{4} \qquad y_{cp} = -\frac{M_x}{F} = \frac{b}{2}$$

This point is marked on Fig. E9.6.

Example 9.7

Rectangular Area Subjected to Varying Pressure Distribution

Problem Statement A thin rectangular plate of length a and width b (the dimension perpendicular to the plane of Fig. E9.7) is subjected to a distributed pressure $p = kx^n$, where x is the distance from one end of the plate and k and n are constants. Determine the resultant force and the center of pressure.

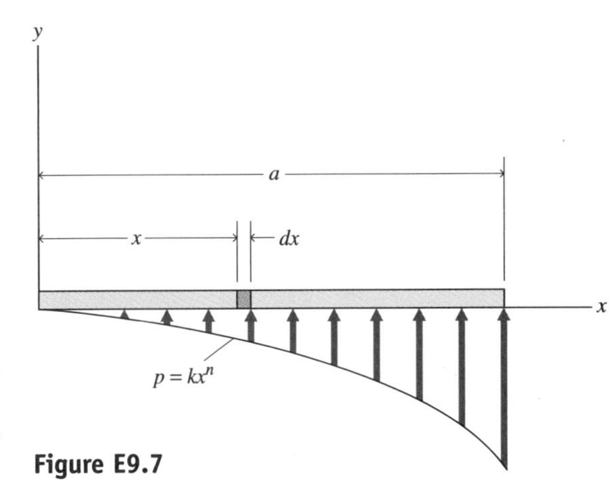

Figure E9.7

Solution The differential pressure force on a strip of width dx is $dF = pb\,dx = kbx^n\,dx$. Hence, the total force on the plate can be obtained by integration:

$$F = kb\int_0^a x^n\,dx = \frac{kba^{(n+1)}}{n+1}$$

The moment of the force dF about the edge $x = 0$ is $dM = x\,dF = kbx^{(n+1)}\,dx$. Hence, the moment of the distributed load about the edge $x = 0$ is

$$M = kb\int_0^a x^{(n+1)}\,dx = \frac{kba^{(n+2)}}{n+2}$$

Thus, the center of pressure is determined by the equation

$$x_{cp} = \frac{M}{F} = \frac{(n+1)a}{n+2}$$

For example, if the load distribution is linear, $n = 1$ and $x_{cp} = 2a/3$. If the load distribution is parabolic, $n = 2$ and $x_{cp} = 3a/4$.

Example 9.8

Water Pressure on a Gate in a Dam

Problem Statement A rectangular gate in a dam measures 10 ft by 15 ft (see Fig. E9.8a). The gate pivots about a frictionless horizontal hinge A–A passing through its center line. Water ($\gamma = 62.4$ lb/ft^3) stands 5 ft above the top of the gate on one side; the other side of the gate is subject to atmospheric pressure.

a. Determine the center of pressure of the water pressure acting on the gate.

b. What force P applied to the bottom of the gate (point C in Fig. E9.8a) is required to keep the gate from opening?

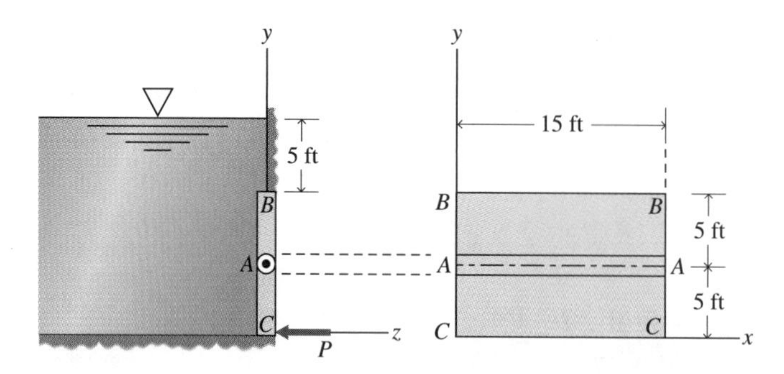

Figure E9.8a

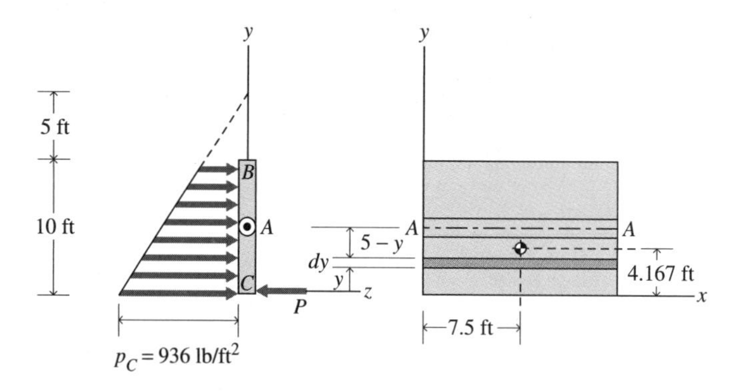

Figure E9.8b

Solution

a. The water pressure acting on the gate varies linearly with depth (see Fig. E9.8b). At the top of the gate (point B), the pressure is $p_B = 5\gamma = 5(62.4) = 312$ lb/ft². At the bottom of the gate (point C), the pressure is $p_C = 15\gamma = 15(62.4) = 936$ lb/ft². Therefore, the total pressure force acting on the gate is

$$F = \frac{936 + 312}{2}(10)(15) = 93{,}600 \text{ lb} \tag{a}$$

The coordinate y_{cp} of the center of pressure is defined by Eq. (9.8a):

$$y_{\text{cp}} = -\frac{M_x}{F} \tag{b}$$

From Fig. E9.8b, we obtain

$$M_x = -\int_0^{10} y(15p\,dy) \tag{c}$$

To evaluate M_x, we must express p as a function of y. In Fig. E9.8b, we see that

$$p = -62.4y + 936 \tag{d}$$

From Eqs. (c) and (d), we obtain

$$M_x = -15\int_0^{10} y(-62.4y + 936)\,dy = -390{,}000 \text{ lb·ft} \tag{e}$$

Hence, from Eqs. (a), (b), and (e),

$$y_{\text{cp}} = -\frac{M_x}{F} = -\frac{(-390{,}000)}{93{,}600} = 4.167 \text{ ft}$$

By symmetry, $x_{cp} = 7.5$ ft. Therefore, the center of pressure (the point at which the force F acts) is located at $x_{cp} = 7.5$ ft, $y_{cp} = 4.167$ ft (see Fig. E9.8b).

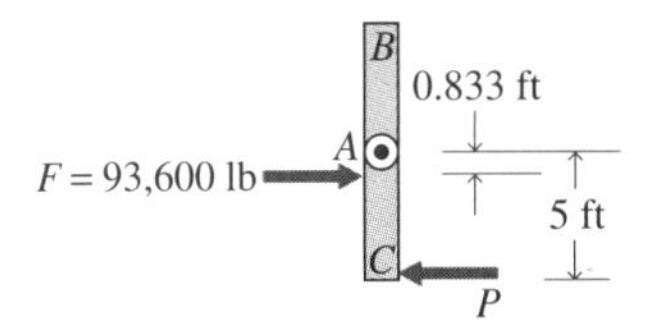

Figure E9.8c

b. To determine the force P required to keep the gate from opening, we note that the moment about hinge A–A of the forces acting on the gate must be zero. Thus, from Fig. E9.8c, we have $\Sigma M_A = 5P - (0.833)(93{,}600) = 0$, or $P = 15{,}600$ lb.

Alternative Solution of Part b The force P can also be calculated by determining the moment of the distributed pressure directly by integration, bypassing the computation of the center of pressure. Thus, from Fig. E9.8b, summation of moments about line A–A yields

$$\sum M_A = 5P - \int_0^{10} (5 - y)[(936 - 62.4y)(15)\,dy] = 0$$

Evaluating this integral and solving for P, we obtain $P = 15{,}600$ lb, as before. ■

CHECKPOINT

1. ***True or False:*** The normal force (due to a distributed pressure) that acts on a flat surface can be determined by integration of the pressure distribution.
2. ***True or False:*** The line of action of the normal force due to a distributed pressure that acts on a flat plate is perpendicular to the plate.
3. Consider the resultant axis of the concentrated force that is dynamically equivalent to a distributed pressure acting on a flat face of a rigid body. At what point does the axis pierce the face?

Answers: 1. True; 2. True; 3. At the center of pressure

9.6 DISTRIBUTED PRESSURE ON CURVED SURFACES

After studying this section, you should be able to:

- Determine the magnitude, direction, and line of action of the resultant force of a distributed pressure acting on a curved surface.

Key Concept The resultant force of a distributed pressure acting on a curved surface is found by applying the pressure to horizontal and vertical projections of the curved surface.

So far in this chapter, we have considered fluid pressures that act on flat surfaces. For a flat surface, the differential pressure force that acts on an infinitesimal area of the surface is parallel to the differential pressure force that acts on any other infinitesimal area of the surface. Hence, the surface is subjected to a system of parallel forces, and the theory of parallel forces applies (Sec. 4.8).

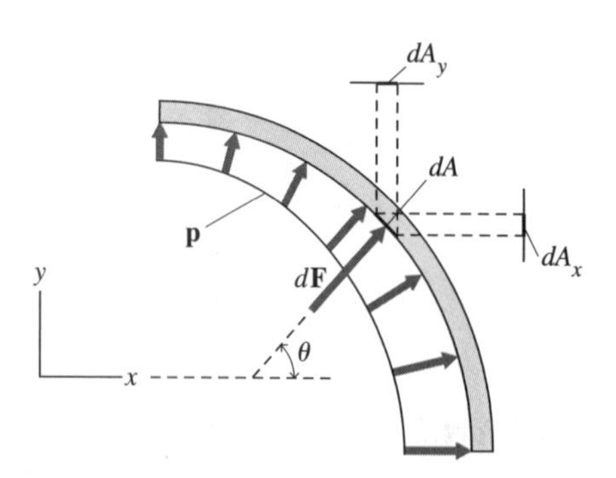

Figure 9.8 Normal force on an element of a curved surface.

On the other hand, when a fluid pressure acts on a curved surface, the pressure, which is normal to the surface, changes in both magnitude and direction from point to point on the surface. Consequently, the theory of parallel forces does not apply. To deal with this situation, we regard pressure as a vector quantity $\mathbf{p}$, since its direction varies as well as its magnitude.

To determine the effects of pressure on a curved surface, we consider the differential forces that act on horizontal and vertical projections of infinitesimal areas of the surface. Then, after integrating, we add the forces vectorially. For example, consider a curved cylindrical surface subjected to pressure on one side (see Fig. 9.8). The differential force $d\mathbf{F}$ that acts on an infinitesimal area dA of the surface is

$$d\mathbf{F} = \mathbf{p}\, dA$$

It is directed normal to the surface. The rectangular (x, y) projections of $d\mathbf{F}$ are

$$\begin{aligned} dF_x &= dF \cos\theta = p\, dA \cos\theta \\ dF_y &= dF \sin\theta = p\, dA \sin\theta \end{aligned} \tag{9.10}$$

Since the area dA is infinitesimal, it lies essentially in a plane. We can therefore develop a geometric interpretation of Eqs. (9.10) by projecting dA onto vertical and horizontal planes:

$$\begin{aligned} dA_x &= dA \cos\theta \\ dA_y &= dA \sin\theta \end{aligned} \tag{9.11}$$

where the differential areas dA_x and dA_y are perpendicular to the x and y axes, respectively (refer to Fig. 9.8). Then, Eqs. (9.10) and (9.11) yield

$$\begin{aligned} dF_x &= p\, dA_x \\ dF_y &= p\, dA_y \end{aligned} \tag{9.12}$$

The following theorem expresses the concept of the projection of an area onto a plane in more general terms.

Theorem 9.4

If an area A in a plane R is projected orthogonally onto a plane S, the projected area is equal to $A \cos\theta$, where θ is the angle subtended by planes R and S.

Clearly, this theorem is true if the area A is rectangular and one of the edges of the rectangle lies parallel to the line of intersection of the two planes. However, any plane region may be subdivided into parallel rectangular strips of infinitesimal width. Since Theorem 9.4 applies for each of these strips, it applies for their sum (or integral). Therefore, the theorem is valid irrespective of the shape of the area A. This principle is illustrated in Figure 9.9, in which a circular area of diameter d in plane R is projected onto plane S. The circle projects as an ellipse in S, with major axis $2a = d$ and minor axis $2b = d\cos\theta$. Whereas the area of the circle is $A_R = \pi d^2/4$, the area of the ellipse is $A_S = \pi ab = (\pi d^2/4)(\cos\theta) = A_R\cos\theta$.

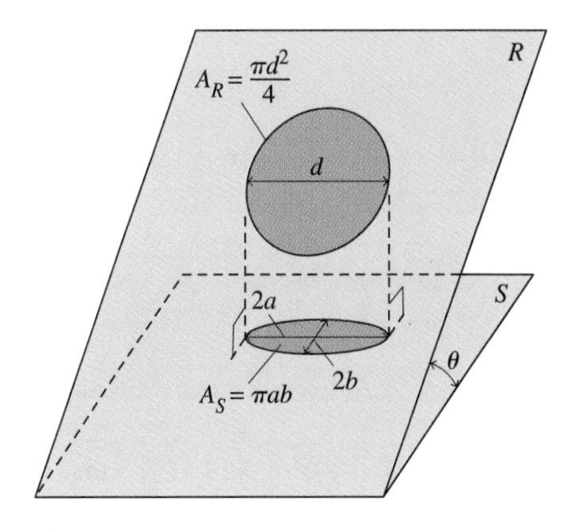

Figure 9.9

Returning to Eq. (9.12), we integrate the differential projections to obtain

$$F_x = \int_A dF_x = \int_A p\,dA_x \tag{9.13a}$$

$$F_y = \int_A dF_y = \int_A p\,dA_y \tag{9.13b}$$

where (F_x, F_y) are the (x, y) projections of the resultant force $\mathbf{F}$ that acts on the surface. Equations (9.13), in which the integrations are performed over the vertical and horizontal projections of the surfaces are analogous to Eq. (9.7), in which the integration is over the surface itself. *The pressure p in Eqs. (9.13) that acts at a point Q on a projected surface is the pressure at the point P on the curved surface that projects onto point Q.*

The approach just outlined is demonstrated below for two common cases. The approach is then generalized as a theorem.

Case 1: Constant Pressure. Let a cylindrical vessel of radius R and length L be subjected to a constant internal pressure p_0. A 90° arc of the cross section of the vessel is shown in Fig. 9.10a. To determine the net horizontal and vertical projections of the

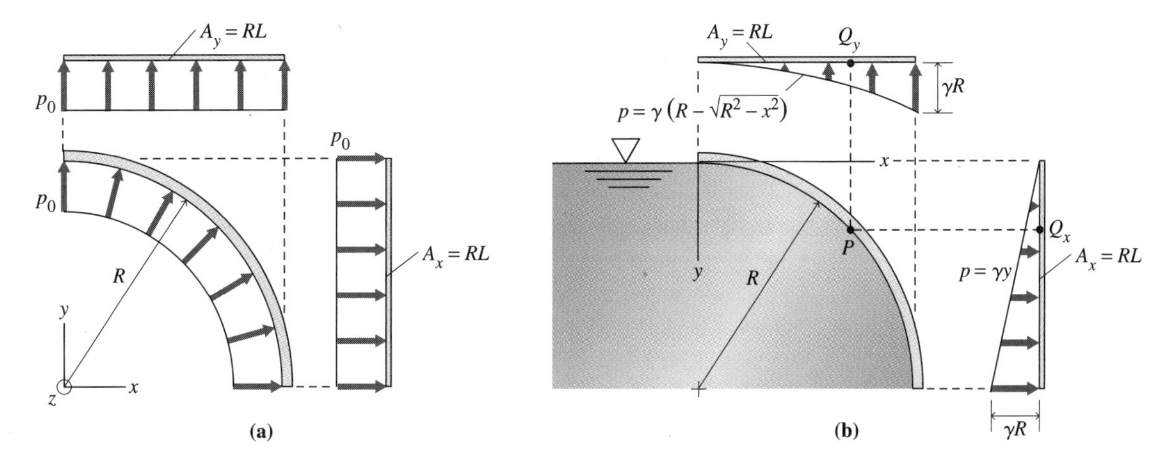

Figure 9.10 Projections of a curved surface into equivalent flat surfaces: **(a)** cylindrical arc subjected to uniform pressure; **(b)** cylindrical arc subjected to hydrostatic pressure.

pressure force that acts on the arc, we project the arc onto two planes perpendicular to the x and y axes, with corresponding areas A_x and A_y. Hence, since the length of the vessel is L (in the z direction in Fig. 9.10a) and $p = p_0 =$ constant, Eqs. (9.13) yield

$$F_x = p_0 RL \qquad F_y = p_0 RL$$

Case 2: Hydrostatic Pressure. As noted in Sec. 9.3, the hydrostatic pressure in a fluid of constant specific weight varies linearly with depth. For example, let's again consider a cylindrical surface that subtends an arc of 90° (see Fig. 9.10b) and has a length L perpendicular to the plane of the page. Let the surface be immersed in a fluid, as shown. As in Fig. 9.10a, the projections of the curved surface area on two planes perpendicular to the x and y axes, respectively, have areas $A_x = A_y = RL$. However, note that in Fig. 9.10b, the positive direction of the y axis is downward. Now, consider a point P located at coordinates (x, y) on the curved surface. The pressure p that acts on A_x at point Q_x (the projection of point P on A_x) is the same as that at P and varies linearly in the y direction. Thus, the pressure $p = \gamma y$ acts on A_x, where γ is the specific weight of the fluid. The pressure on A_y, however, varies *nonlinearly* with x. To determine the distribution of pressure on A_y as a function of x, we must express the depth y at the point P in terms of x. Since the arc is circular, for a point P: (x, y) on the arc,

$$x^2 + (y - R)^2 = R^2$$

or

$$y = R - \sqrt{R^2 - x^2};\ x \leq R$$

Therefore, the pressure on A_y at point Q_y (the projection of point P on A_y) is

$$p = \gamma y = \gamma(R - \sqrt{R^2 - x^2})$$

Hence, from Eqs. (9.13), the horizontal and vertical projections of the pressure force that acts on the curved surface are

$$F_x = \int p\,dA_x = \gamma L \int_0^R y\,dy = \frac{\gamma L R^2}{2}$$

$$F_y = \int p\,dA_y = \gamma L \int_0^R (R - \sqrt{R^2 - x^2})\,dx = \left(1 - \frac{\pi}{4}\right)\gamma L R^2$$

Information Item The force F_x can also be obtained directly from Fig. 9.10b by noting that the pressure distribution on A_x is triangular. Thus, the resultant force is given by one-half the base of the triangle times the height of the pressure distribution:

$$F_x = \left(\frac{RL}{2}\right)(\gamma R) = \frac{\gamma L R^2}{2}$$

The foregoing results may be generalized as a theorem.

Theorem 9.5

The x projection F_x of the resultant force $\mathbf{F}$ due to a pressure $\mathbf{p}$ that acts on a curved surface A is equal to the pressure force produced by p acting on area A_x, the projection of A on a plane perpendicular to the x axis. Similarly, the y projection F_y of $\mathbf{F}$ is equal to the pressure force due to $\mathbf{p}$ acting on the area A_y, the projection of A on a plane perpendicular to the y axis. The pressure p that acts on a point Q in a projected area is the pressure at point P on the curved surface that projects onto Q. Furthermore, the moment of $\mathbf{F}$ about the z axis is equal to the sum of the moments of the projections F_x and F_y about the z axis.

A limitation of Theorem 9.5 is that no two points on A project onto the same point on either A_x or A_y. In general, this limitation is not a severe one, since any curved surface may be partitioned into several parts, so that each part avoids this limitation. Then, the results for the individual parts may be combined.

Theorem 9.5 as stated applies to two-dimensional problems. However, the theorem may be extended to three dimensions by including a third projection A_z on a plane perpendicular to the z axis. The force component F_z and the moments M_x and M_y due to F_z may be determined in the same manner as for F_x and F_y.

PROBLEM-SOLVING TECHNIQUE

Resultant Force on a Curved Surface

To determine the resultant force on a curved surface subject to a distributed pressure:

1. Draw a sketch illustrating the problem, including the distributed pressure.
2. Sketch the projections A_x and A_y of the surface A onto vertical and horizontal planes, respectively.
3. Express the pressure p as a function of y, and apply that distribution to A_x. Determine F_x from Eq. (9.13a). Also, find the center of pressure coordinate y_{cp} of F_x on A_x.
4. Based on the geometry of the curved surface, express y in terms of x. Then express p as a function of x, and apply that distribution to A_y. Determine F_y from Eq. (9.13b). Also, find the center of pressure coordinate x_{cp} of F_y on A_y.
5. Apply the resultant pressure force $\mathbf{F} = F_x\mathbf{i} + F_y\mathbf{j}$ on the curved surface at the center of pressure (x_{cp}, y_{cp}). Alternatively, find the magnitude F of the resultant and its orientation θ relative to the horizontal using

$$F = \sqrt{F_x^2 + F_y^2}$$

$$\theta = \tan^{-1}\left(\frac{F_y}{F_x}\right)$$

Application of this technique is demonstrated in the following examples.

Example 9.9

Pressure on the Face of a Dam

Problem Statement The face of a dam is a parabolic cylinder defined by the equation

$$x^2 = 16y \tag{a}$$

where x and y are rectangular coordinates expressed in feet (see Fig. E9.9). The dam is 100 ft long and 80 ft high. The region upstream from the dam is filled with water to the top of the dam.

a. Determine the horizontal and vertical pressure forces (F_x, F_y) that act on the face of the dam and, hence, the total pressure force F that acts on the face.

b. Determine the lines of action of F_x and F_y.

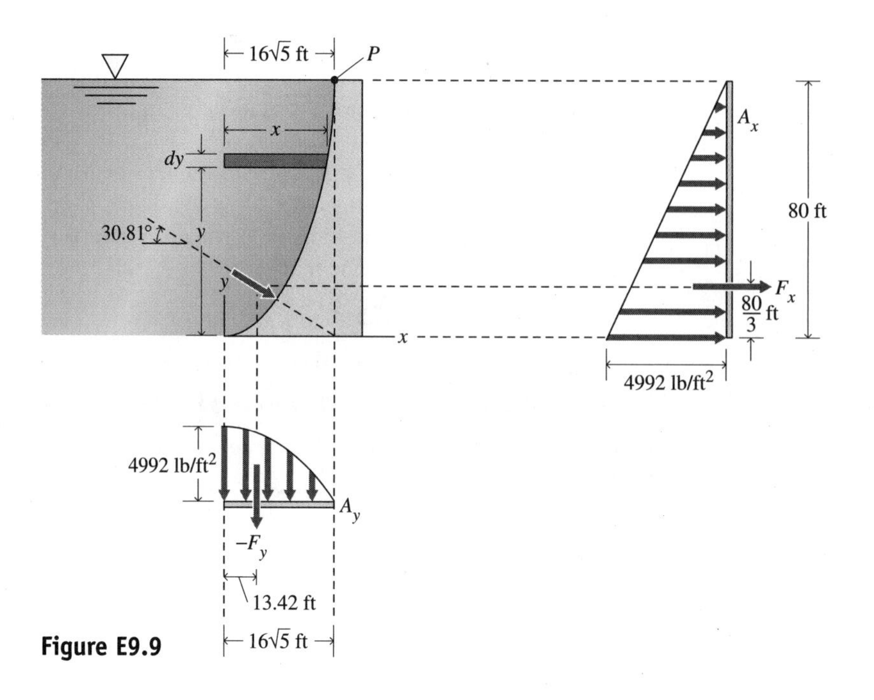

Figure E9.9

Solution **a.** Since the dam is 80 ft high, the y coordinate of point P at the top of the dam is $y =$ 80 ft, and, from Eq. (a), the x coordinate of P is $x = 16\sqrt{5}$ ft. Since the specific weight of water is 62.4 lb/ft³, the gage pressure at any point on the face of the dam is

$$p(y) = (62.4)(80 - y) \tag{b}$$

Let the face of the dam be projected onto vertical and horizontal flat surfaces A_x and A_y (see Fig. E9.9). The pressure at any point in either of these surfaces is equal to the pressure at the corresponding point on the face of the dam. Thus, Eq. (b) also gives the pressure on the vertical surface A_x. The pressure distribution on the face A_x is linear in y. Since the pressure distribution on the face A_x is linear and the length of the face is 100 ft, the total normal force on the face A_x is

$$F_x = \frac{1}{2}(4992)(80)(100) = 19{,}968{,}000 \text{ lb}$$

To compute the force on the face A_y, we must express the pressure on that face as a function of x. Equations (a) and (b) yield

$$p(x) = 62.4\left(80 - \frac{x^2}{16}\right) \tag{c}$$

In view of Eq. (c), the pressure distribution on the face A_y is parabolic (see Fig. E9.9). The vertex of the parabola is at $x = 0$. The pressure drops to zero at the top of the face—that is, $p = 0$ at $x = 16\sqrt{5} = 35.78$ ft. Since the length of the face is 100 ft, the total downward force on the face is [with Eq. (c)]

$$\begin{aligned} -F_y &= 100\int_0^{16\sqrt{5}} p\,dx \\ &= 6240\int_0^{16\sqrt{5}}\left(80 - \frac{x^2}{16}\right)dx \\ &= 11{,}907{,}000 \text{ lb} \end{aligned}$$

The forces F_x and F_y are equal to the horizontal and vertical forces on the face of the dam. Hence, the total force F on the face of the dam is

$$F = \sqrt{F_x^2 + F_y^2} = 23{,}249{,}000 \text{ lb}$$

The angle θ that the resultant force forms with the x axis is determined by the relation $\tan\theta = F_y/F_x = -0.59630$, and $\theta = -30.81°$.

b. Since the pressure distribution on the area A_x is linear, the resultant force F_x on that area acts at two-thirds of the depth of the face—that is, at the point $y = 80/3$ ft (see Fig. E9.9).

To determine the location of the resultant force F_y on the area A_y, we must find the moment about the origin of the distributed forces on A_y. In view of Eq. (c), this moment is

$$M_O = -6240\int_0^{16\sqrt{5}}\left(80 - \frac{x^2}{16}\right)x\,dx = -159{,}744{,}000 \text{ lb·ft}$$

Accordingly, the coordinate of the center of pressure of the area A_y is

$$x_{cp} = \frac{M_O}{F_y} = 13.42 \text{ ft}$$

The lines of action of the forces F_x and F_y are shown in Fig. E9.9. The distributed forces acting on the face of the dam exert the same moment about any axis as do the concentrated forces F_x and F_y acting at the centers of pressure of the respective vertical and horizontal flat surfaces.

Information Item Note that the vertical component $-F_y$ of the pressure force on the face of the dam is also equal to the weight W of the water above the face of the dam. To calculate W, we consider a horizontal strip of water across the face of the dam. This strip has an infinitesimal width and a length of 100 ft and a cross-sectional area $dA = x\,dy$, shown in Fig. E9.9. Then, we have

$$W = 100\int_0^{80}\gamma x\,dy = 100\int_0^{80}(62.4)(4\sqrt{y})\,dy = 11{,}907{,}000 \text{ lb}$$

Example 9.10

Circumferential and Axial Tensions in a Tank

Problem Statement As an application of Theorem 9.5, consider a thin-walled tank subjected to internal pressure p (see Fig. E9.10a). For a thin-walled tank, the thickness of the wall h is much less than the inside diameter D of the tank; that is, $h << D$.

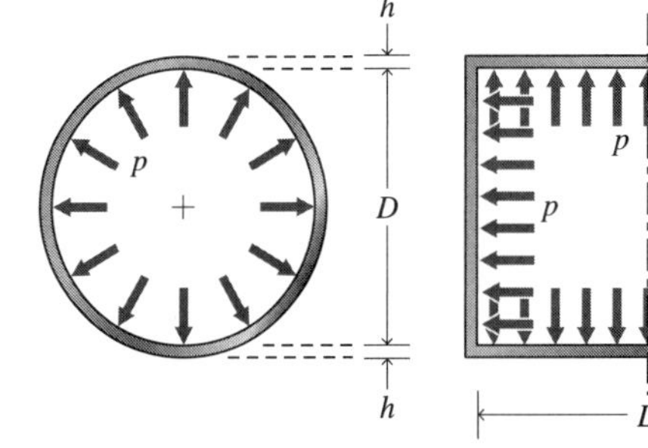

Figure E9.10a

a. For section A–A of the tank far from its ends,[5] determine the circumferential (hoop) tension T_θ and the corresponding average hoop tensile stress σ_θ in the tank in terms of p.

5. We consider such a section to get away from the ends of the tank, because the stresses near the ends are affected by end constraints, which we ignore here. End effects are studied in courses on mechanics of materials.

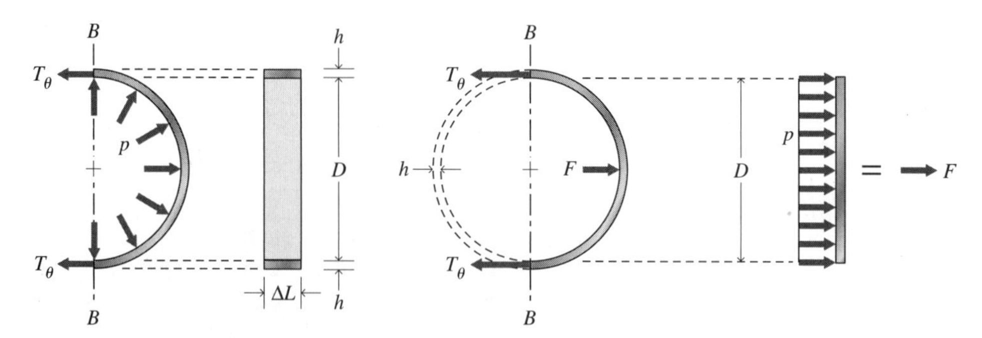

Figure E9.10b **Figure E9.10c**

b. For section A–A, determine the axial tension T_x and the corresponding average axial stress σ_x in the tank in terms of p.

Solution **a.** To determine the hoop tension and hoop stress at section A–A, consider the free-body diagram for one-half of the circumference of the tank segment of axial length ΔL (see Fig. E9.10b). Observe that the pressure p exerts a resultant pressure force F that is balanced by the hoop tension T_θ in the wall of the tank (see Fig. E9.10c). The force F is determined by projecting the interior of the half tank into a flat rectangular plate of dimensions D and ΔL, where D is the inner diameter of the tank and ΔL is the axial length of the tank segment (refer to Fig. E9.10b). Consequently, the force $F = pD(\Delta L) = 2T_\theta$. Hence, the hoop tension in the tank is

$$T_\theta = \frac{F}{2} = \frac{pD(\Delta L)}{2} \tag{a}$$

The average hoop tensile stress on section B–B (refer to Fig. E9.10b) is $\sigma_\theta = T_\theta/[(\Delta L)h]$, where h is the thickness of the wall. By substitution for T_θ from Eq. (a), the average tensile hoop stress in the tank in terms of p is

$$\sigma_\theta = \frac{pD}{2h}$$

b. To determine the axial tension in the tank, consider Fig. E9.10d, the free-body diagram of the tank to the left of section A–A. Here, we consider the pressure p as acting on the fluid at section A–A; in contrast, in Fig. E9.10b, the pressure p is applied directly to the inner circumference of the tank. Summing forces in the x direction, we obtain (since $h << D$)

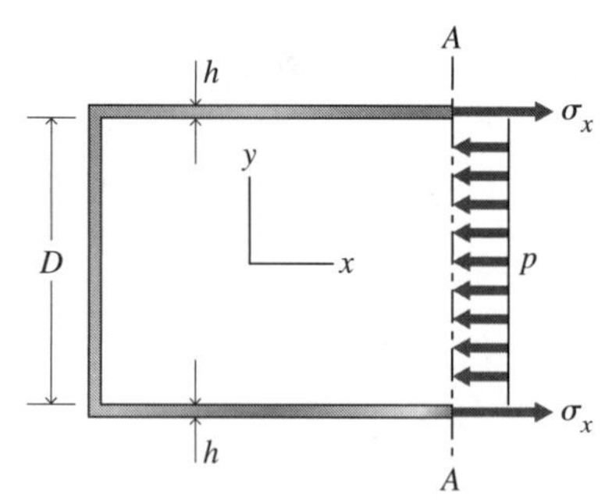

$$\sum F_x = \sigma_x(\pi Dh) - \frac{p\pi D^2}{4} = 0 \tag{b}$$

Figure E9.10d

where σ_x is the average axial stress in the tank wall. Therefore, from Eq. (b), we obtain

$$\sigma_x = \frac{pD}{4h} \tag{c}$$

The net axial tension at section A–A is equal to the average axial stress times the circumferential area of the tank. Therefore, we use Eq. (c) to obtain the axial tension:

$$T_x = \sigma_x(\pi Dh) = \frac{\pi pD^2}{4}$$

Information Item Instead of the free-body diagram of Fig. E9.10b, we could have used the free-body diagram shown in Fig. E9.10e, where the pressure p is applied to the fluid at section B–B. Then, Eq. (a) could be obtained directly from Fig. E9.10e, without any need for Fig. E9.10c.

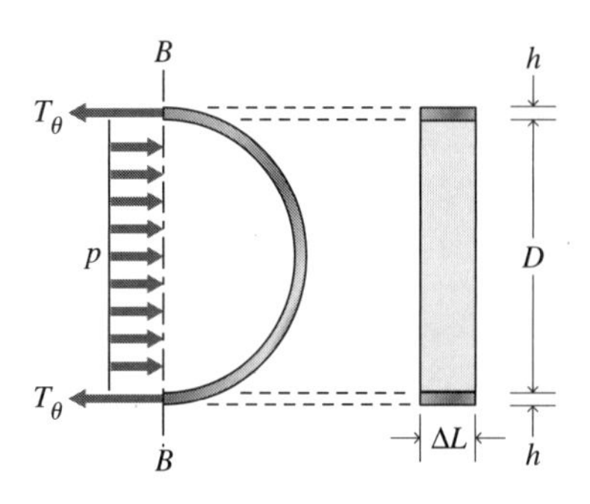

Figure E9.10e

CHECKPOINT ✓

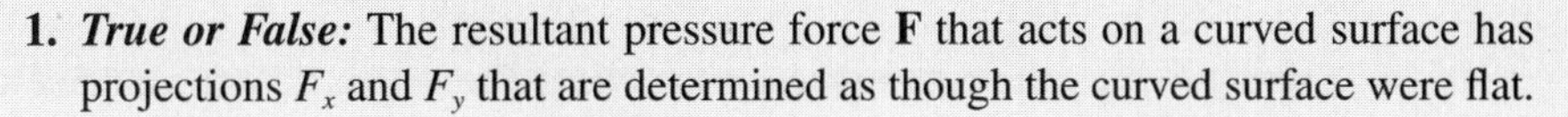

1. ***True or False:*** The resultant pressure force **F** that acts on a curved surface has projections F_x and F_y that are determined as though the curved surface were flat.

Answer: 1. True

9.7 BUOYANCY

After studying this section, you should be able to:

- Determine the buoyant force on an object that is immersed in a fluid.
- Determine whether a floating object is in stable, unstable, or neutral equilibrium.

Key Concept A body that is immersed in a fluid is subject to a buoyant force by the fluid. The buoyant force is equal to the weight of the fluid displaced by the body and acts vertically through the center of buoyancy. The center of buoyancy coincides with the center of gravity of the volume of the displaced fluid.

Consider a solid body that is immersed in a stationary fluid. The body can be any shape, but, for simplicity, let's use the tetrahedron shown in Fig. 9.11. If a horizontal prism of infinitesimal cross-sectional area dA is passed through the body, the prism cuts elements of areas dA_1 and dA_2 from two of the inclined surfaces of the body. The net horizontal force due to the pressures on the ends of the prism is

$$dF_x = p_1\, dA_1 \cos\alpha_1 - p_2\, dA_2 \cos\alpha_2$$

In Sec. 9.6 we proved that

$$dA_1 \cos\alpha_1 = dA_2 \cos\alpha_2 = dA$$

Also, since points 1 and 2 are at the same level, $p_1 = p_2$. Consequently, the net horizontal force on the prism is zero. Since this is true for any horizontal infinitesimal prism that is cut from the body, it is true for the body as a whole. Therefore, if a body is immersed in one or more stationary fluids, the net horizontal pressure force on its surface is zero.

Now, consider a vertical prism of infinitesimal cross-sectional area dA that passes through the body. This prism cuts infinitesimal areas dA_3 on an inclined face and dA_4 on

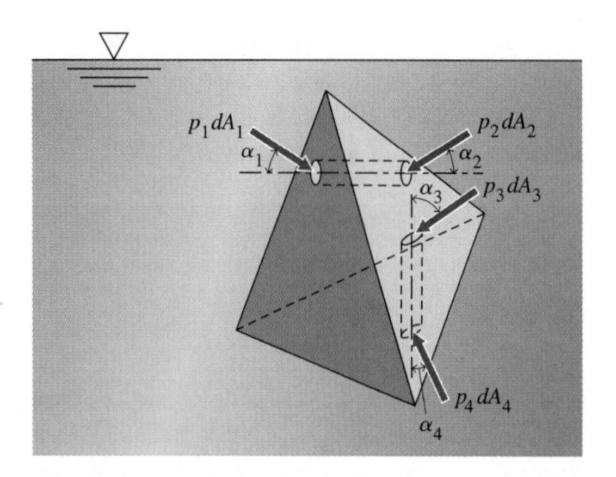

Figure 9.11 Immersed tetrahedron.

the bottom face of the body. The net vertical force due to the pressures on the two ends of the prism is (refer to Fig. 9.11)

$$dF_y = p_4 \, dA_4 \cos \alpha_4 - p_3 \, dA_3 \cos \alpha_3$$

Since $dA_4 \cos \alpha_4 = dA_3 \cos \alpha_3 = dA$, this reduces to $dF_y = (p_4 - p_3) \, dA$. By Eq. (9.2), dF_y is equal to the weight of a column of fluid that is of the same shape and size as the infinitesimal prism in the body. Since this is true for every vertical infinitesimal prism, the net vertical force on the body is equal to the weight of the displaced fluid and is directed upward. This fact is known as *Archimedes' principle*.

Archimedes' principle

Since each of the infinitesimal vertical forces dF_y is equal to the weight of the displaced infinitesimal column of fluid, the resultant vertical force F_y passes through the center of gravity of the displaced fluid, which coincides with the center of gravity of the displaced volume of the fluid. The center of gravity of the displaced fluid is called the *center of buoyancy,* and the force F_y is called the *buoyant force.*

center of buoyancy

buoyant force

The above conclusions are summarized in the following theorem, which we shall refer to as the *principle of buoyancy.*

principle of buoyancy

Theorem 9.6

The pressure on a solid body that is immersed in a stationary fluid exerts a resultant force directed vertically upward through the center of gravity of the displaced fluid (that is, through the center of buoyancy). The magnitude of the resultant force (buoyant force) is equal to the weight of the displaced fluid, and the resultant line of action of the buoyant force passes through the center of gravity of the volume of the displaced fluid.

The foregoing proof of this theorem does not require that the fluid be homogeneous. The principle of buoyancy is equally valid for liquids and gases. It may also be used to determine the buoyant force on a body that floats at the interface of two stationary fluids. However, in this case, the line of action of the buoyant force does not, in general, pass through the center of gravity of the *total* volume of displaced fluids, since the centers of gravity of the volumes of the individual fluids might not be vertically aligned. Thus, the principle of buoyancy is valid for a floating body—a body that lies partly in a liquid and partly in air. However, it *is not generally valid for a body that is not completely surrounded by one or more fluids.* To illustrate this fact, Pascal placed a well-polished face of a rectangular wooden block against the flat bottom of a vessel and then covered the block with mercury. The mercury was unable to penetrate under the block; that is, it was unable to completely surround the block. Hence, the mercury did not exert a buoyant force on the block. On the contrary, the block was pressed firmly against the bottom of the vessel by its own weight, by the weight of the mercury above, and by atmospheric pressure.

The principle of buoyancy can be easily verified on intuitive grounds, without the preceding analytical argument. An isolated portion of a stationary fluid is held in equilibrium by the pressure of the surrounding fluid on its surface. Consequently, the resultant force that the surrounding fluid exerts on the isolated portion passes vertically through its center of gravity and is equal in magnitude to the weight of the isolated portion. Now, if the isolated portion were solidified, the surrounding fluid would not alter its behavior, because the surrounding fluid would exert precisely the same force on the solid portion that it had exerted on the fluid portion. The principle of buoyancy is thus made plausible by intuitive reasoning. Arguments based on the concept of solidified fluid were used frequently by early investigators of hydrostatics.

The principle of buoyancy is the basis for the theory of stability of floating bodies, which is important in the design of ships.

STABILITY OF FLOATING AND SUBMERGED BODIES

Key Concept

If a floating body is rotated from its initial position and if a couple formed by the weight of the body and the buoyant force tends to return the body to its initial position, the body is in stable equilibrium. If no such couple is formed, the body floats in a state of neutral equilibrium. If the couple tends to overturn the body, the body is in unstable equilibrium.

A solid body floating in a fluid at rest is *stable* under a vertical displacement; that is, it will return to its original position if given a small vertical displacement and released. If the displacement is upward, the volume of fluid displaced by the body is reduced, resulting in a reduced buoyancy force, or an unbalanced downward force that tends to move the body back down to its original position. Similarly, if the displacement is downward, the buoyancy force is increased, resulting in an unbalanced upward force that tends to move the body back up to its original position.

stable equilibrium

unstable equilibrium

neutral equilibrium

A solid floating body may also be in rotationally stable, unstable, or neutral equilibrium. If a floating body in *stable equilibrium* is given any small rotation and released, it will tend to return to its original position. On the other hand, if a floating body in *unstable equilibrium* is given a small rotation and released, it will tend to continue to rotate and move further away from its original position. If a floating body in *neutral equilibrium* is given a small rotation and released, it will remain in the displaced position without any tendency to rotate further or to return to its initial position. These three states of equilibrium are illustrated in Fig. 9.12. In Fig. 9.12a, a rod (consisting of a light wooden stick with a small piece of lead glued to one end) is floating with its wooden end up. In Fig. 9.12b, the rod is floating with the lead end up. In Fig. 9.12a, the rod is in a stable equilibrium position. That is, if the rod is given any small rotation, the couple produced by the buoyant force and the weight of the rod tends to return the rod to its initial position. In Fig. 9.12b,

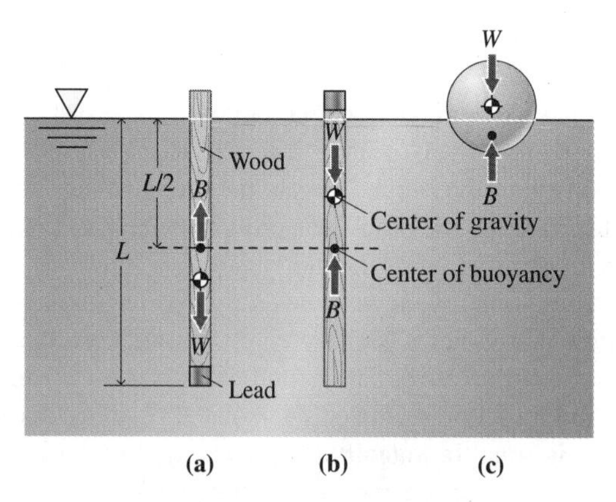

Figure 9.12 States of equilibrium: **(a)** stable equilibrium; **(b)** unstable equilibrium; **(c)** neutral equilibrium.

the rod is in an unstable state. That is, if the rod is given a small rotation, the couple produced by the buoyant force and the weight will tend to rotate the rod away from its original position and toward the stable position shown in Fig. 9.12a. The floating ball shown in Fig. 9.12c is in a neutral equilibrium state. If the ball is given a small rotation, the buoyant force and the weight remain collinear, and no couple is created to cause the ball to rotate further.

This tour boat on the Seine River in Paris, France is uniformly loaded with passengers on its decks. Since the boat is stable, it would not tip over if most of the passengers moved to one side.

As seen above, any floating body is in rotationally stable equilibrium if its center of buoyancy (the centroid of the displaced fluid volume) is above its center of gravity. By similar reasoning, it may be shown that a body that is completely submerged is stable only if the center of buoyancy is above the center of gravity of the body. However, there are floating objects that are in stable equilibrium when the center of gravity is above the center of buoyancy. Let's consider one such case.

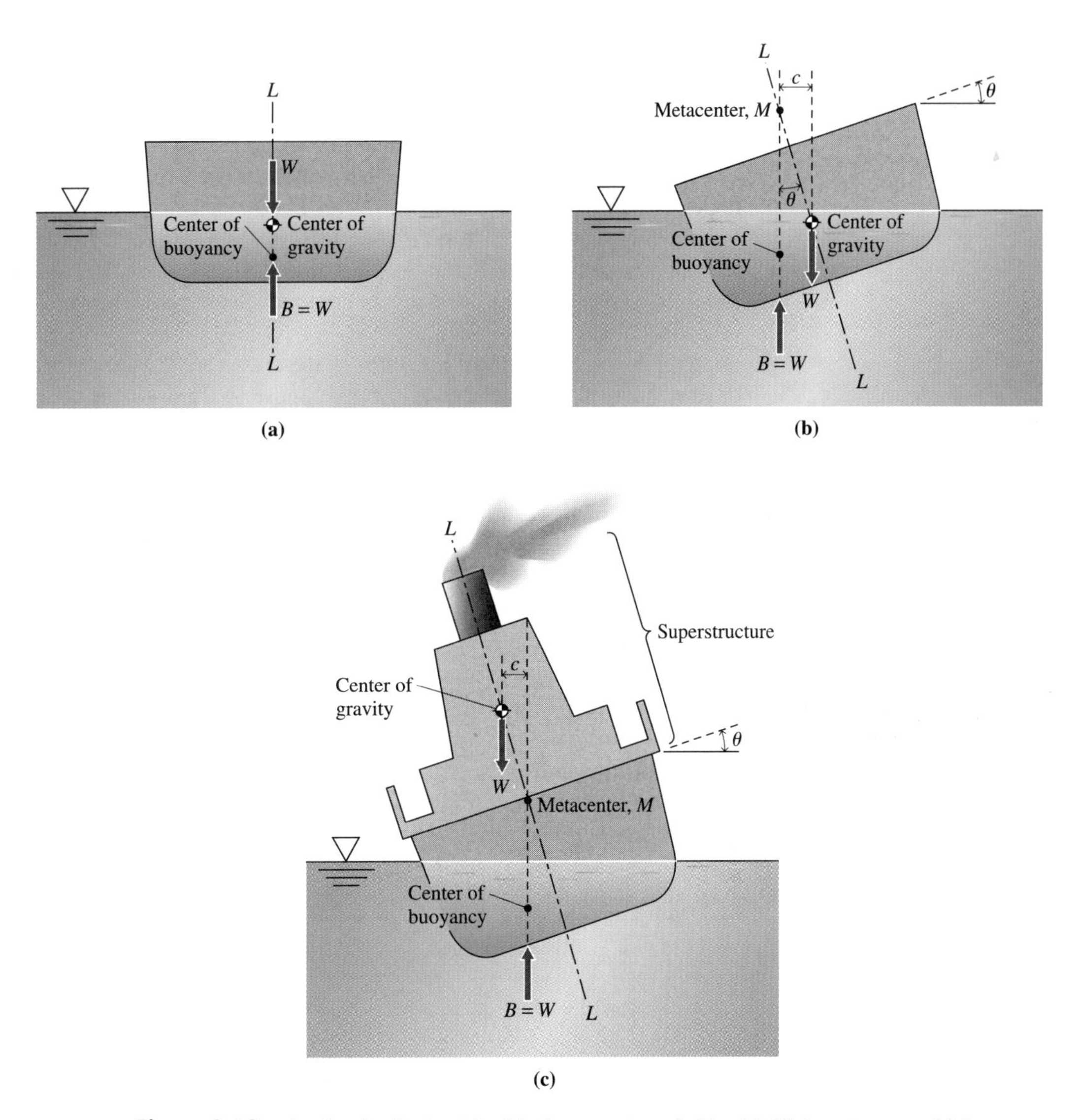

Figure 9.13 Angle of roll of a ship. **(a)** Cross section of ship. **(b)** High metacenter. **(c)** Low metacenter.

Figure 9.13a represents the cross section of a ship that is afloat in water. Assume that the cross section is constant along the ship. (For our purposes, the effect of change in cross section of the ship with length is negligible.) In the position shown, the center of gravity of the ship and cargo is directly above the center of buoyancy. Let the ship roll through an angle θ (see Fig. 9.13b). The resultant weight W of the ship is concentrated at the center of gravity of the ship, and the buoyant force B ($= W$) passes through the new center of buoyancy to the left of the center of gravity. The new center of buoyancy is located at the center of gravity of the volume of the displaced water. In the rolled position, the weight of the ship and the buoyant force form a couple that tends to right the ship; the ship is stable. The magnitude of the moment of this couple is Wc, where c is the distance between the line of action of the buoyant force and the resultant weight vector. In order to roll the ship over, wind or waves would have to produce a rolling moment on the ship, with sense opposite to that of Wc and magnitude greater than Wc.

Depending on the shape of the cross section, it is possible for the center of buoyancy to lie to the right of the center of gravity (see Fig. 9.13c), in which case the moment Wc would tend to capsize the ship. Then the equilibrium of the ship is *unstable.* Special numerical methods for calculating the arm c of the restoring couple are described in books on naval architecture. The need for these methods arises because of the difficulty in locating the center of buoyancy of a complicated body, such as a ship with a small angle of roll.

metacenter

The concept of metacenter is used in naval architecture in the design of a ship to determine whether the ship is stable or unstable. The *metacenter* is defined as follows: When a ship or other object rolls through an angle θ, a fixed vertical line L–L through the center of gravity also rotates through angle θ (Figs. 9.13a and 9.13b). In the rolled position, the line L–L intersects a vertical line through the center of buoyancy. The point of intersection of the lines is the metacenter of the body or ship (point M in Fig. 9.13b). For small angles of roll, the metacenter is approximately fixed with respect to the ship. A high metacenter implies a large moment arm c. In that case, the moment of the buoyant force B $(= W)$ acts to bring the ship back to its upright position. Consequently, a ship with a high metacenter is very *stable.* However, if a ship has a lot of superstructure, the center of gravity may be high on the ship (Fig. 9.13c) and the metacenter may be below the center of gravity. Here, the moment of the buoyant force B $(= W)$ acts to roll the ship further. In other words, the ship is *unstable,* and when it begins to roll, it may continue to roll and capsize.

Example 9.11

Rolling of a Box-Shaped Barge

Problem Statement Consider a box-shaped barge that is afloat in fresh water with specific weight $\gamma_{H_2O} = 62.4$ lb/ft^3. The barge has length l, width $2b$, and submerged depth (draft) a (see Fig. E9.11a).

a. Determine the angle of roll θ required to capsize the barge.

b. Determine the rolling moment necessary to capsize the barge.

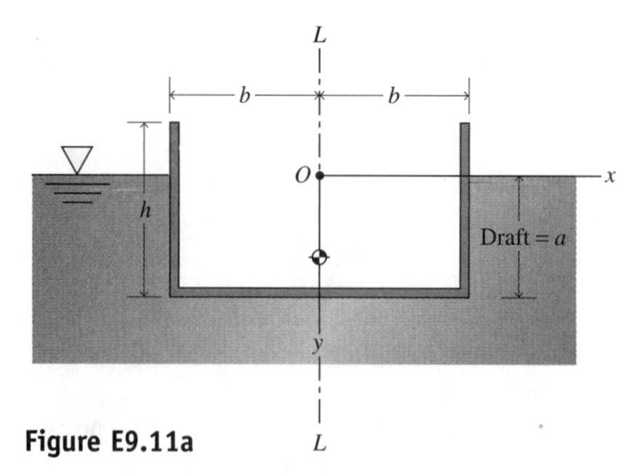

Figure E9.11a

Solution To set up the problem, we select rectangular xy coordinate axes on a cross section of the barge, with the x axis at the water level, the y axis directed vertically downward, and the origin O on the axis of symmetry of the cross section of the barge (see Fig. E9.11a). The volume of the displaced water is

$$V = a(2b)l \tag{a}$$

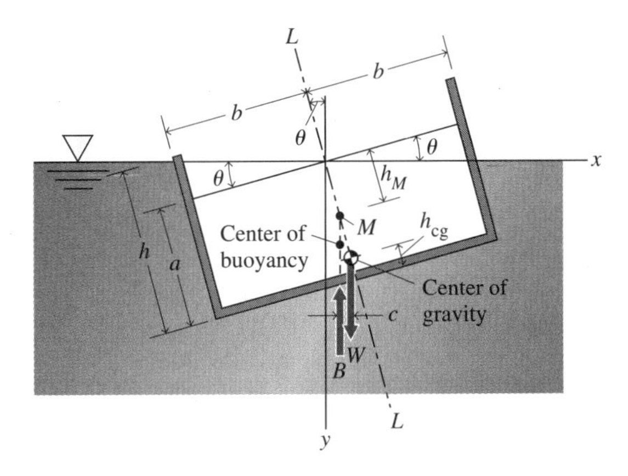

Figure E9.11b

a. Let the barge list at the angle θ, at which the water is on the verge of flowing over the side of the barge—that is, at which the barge is on the verge of capsizing (Fig. E9.11b). The dimension a is equal to the draft when the bottom is level (Fig. E9.11a), since the volume of the displaced water is the same in either case (the two triangular areas are equal, and the water from the right area is transferred to the left area). The center of buoyancy cb = (x_{cb}, y_{cb}) is the center of gravity of the volume of displaced water. In other words, cb is at the centroid of the trapezoid formed by the bottom line, the two side lines, and the x axis (Fig. E9.11b). The following formulas (from Appendix D) give the coordinates of the centroid of a trapezoid:

$$x_{cb} = \frac{a \sin\theta}{2} - \frac{b^2(2 + \tan^2\theta)\sin\theta}{6a}$$
$$y_{cb} = \frac{a\cos\theta}{2} + \frac{b^2 \sin\theta \tan\theta}{6a} \tag{b}$$

If θ is small, Eqs. (b) may be approximated by neglecting the terms $\tan^2\theta \sin\theta$ and $\sin\theta \tan\theta$, which are small compared to $\sin\theta$. Hence, we obtain

$$x_{cb} = \left(\frac{a}{2} - \frac{b^2}{3a}\right)\sin\theta$$
$$y_{cb} = \frac{a}{2} \tag{c}$$

From Fig. E9.11b, we see that the distance h_M from the origin to the metacenter M is related to x_{cb} by

$$h_M \sin\theta = x_{cb} \tag{d}$$

Hence, by Eqs. (c) and (d),

$$h_M = \frac{a}{2} - \frac{b^2}{3a}$$

Note that the line of action of the buoyant force B passes through the center of buoyancy cb and the metacenter M (Fig. E9.11b).

If the barge has no deck and the height of a side is h, the angle θ, at which the barge is on the verge of capsizing, is determined by the equation

$$\tan \theta = \frac{h - a}{b} \tag{e}$$

For a numerical example, let $h = 7$ ft, $a = 4$ ft, $b = 10$ ft, $l = 50$ ft, and $h_{cg} = 3$ ft. Then, the angle at which the barge will capsize, as determined by Eq. (e), is $\theta = 16.70°$.

b. To determine the rolling moment required to capsize the barge, we must compute the arm c of the restoring couple due to the buoyant force B and the weight W. (Note that $B = W$.) From Eqs. (b), with the numerical values given in part a, the line of action of B is located at $x_{cb} = -1.928$ ft (that is, the vertical buoyant force lies to the left of the y axis). Since $h_{cg} = 3$ ft, the abscissa of the center of gravity (from Fig. E9.11b and the values given above) is $x = (4 - 3)(\sin \theta) = 0.287$ ft. Consequently, the arm c of the restoring couple (Fig. E9.11b) is $c = 0.287 + 1.928 = 2.215$ ft. With Eq. (a), the weight W of the barge and cargo is $W = \gamma_{H_2O}V = \gamma_{H_2O}a(2b)l = (62.4)(4)(20)(50) = 249{,}600$ lb. The restoring moment is therefore $Wc = (249{,}600)(2.215) = 552{,}900$ lb·ft. The rolling moment necessary to capsize the barge must be equal to or greater than the restoring moment. That is, the rolling moment must be at least 552,900 lb·ft to capsize the barge.

Example 9.12 *Submerging of a Submarine*

Problem Statement A nuclear submarine weighs 8 MN. When the submarine floats on the surface of the ocean, the water displaced by the submarine is 90% of its volume. Determine the volume of sea water ($\gamma = 10$ kN/m^3) that must be taken into the ballast tanks to fully submerge the submarine.

Solution

Let W_1 be the weight of the submarine, and let W_2 be the weight of the submarine plus the weight of the water added to its tanks. Then,

$$W_1 = 0.90V\gamma \tag{a}$$

where V is the volume of the submarine. After the water is added to the tanks and the submarine is fully submerged, we have

$$W_2 = V\gamma \tag{b}$$

and, from Eqs. (a) and (b),

$$W_2 = \frac{W_1}{0.90} \tag{c}$$

Hence, the weight of the water added to the ballast tanks is, with Eq. (c),

$$W_2 - W_1 = \frac{W_1}{9} = \gamma V_w \tag{d}$$

where V_w is the volume of water added to the tanks. So, from Eq. (d), the volume of water added to the tanks is

$$V_w = \frac{W_1}{9\gamma} = \frac{8 \times 10^6}{9(10 \times 10^3)} = 88.9 \text{ m}^3$$

Note that this volume of sea water weighs 889 kN. This is 11.1% of the weight of the submarine.

Example 9.13

A Body Immersed in Two Fluids

Problem Statement A steel ball of radius $r = 50$ mm and of specific weight $\gamma_s = 76.8\ \text{kN/m}^3$ is suspended by a thin wire and lowered into a tank containing oil of specific weight $\gamma_{\text{oil}} = 7.8\ \text{kN/m}^3$ and mercury of specific weight $\gamma_{\text{Hg}} = 133\ \text{kN/m}^3$, until it is half submerged in oil and half in mercury (see Fig. E9.13a). Determine the tension T in the wire.

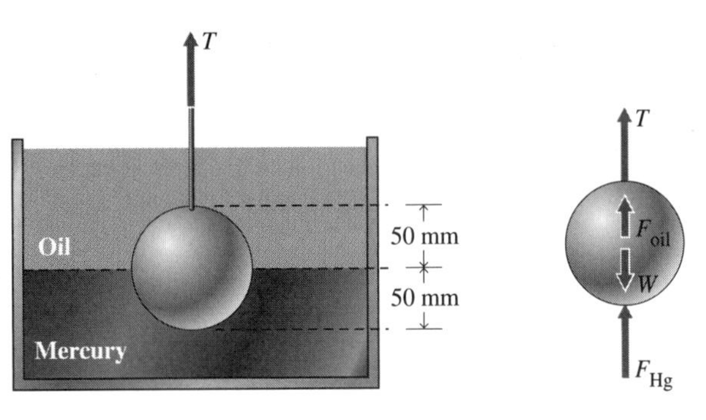

Figure E9.13a **Figure E9.13b**

Solution The volume of the ball is $V = 4\pi r^3/3 = 0.524 \times 10^{-3}\ \text{m}^3$. Therefore, the weight of the ball is

$$W = (0.524 \times 10^{-3})(76.8 \times 10^3) = 40.24\ \text{N} \tag{a}$$

The buoyant force due to the oil is

$$F_{\text{oil}} = \frac{(0.524 \times 10^{-3})(7.8 \times 10^3)}{2} = 2.04\ \text{N} \tag{b}$$

and the buoyant force due to the mercury is

$$F_{\text{Hg}} = \frac{(0.524 \times 10^{-3})(133 \times 10^3)}{2} = 34.85\ \text{N} \tag{c}$$

Summation of the vertical forces acting on the ball (Fig. E9.13b), with Eqs. (a), (b), and (c), yields

$$\sum F = T + F_{\text{oil}} + F_{\text{Hg}} - W = 0$$

or

$$T = 40.24 - 2.04 - 34.85 = 3.35\ \text{N}$$

CHECKPOINT ✓

1. What is the center of buoyancy?
2. What is the magnitude of the buoyant force?
3. What is the direction of the line of action of the buoyant force?
4. ***True or False:*** The metacenter lies above the center of gravity of a stable ship.

Answers: 1. The center of gravity of the displaced fluid; 2. The weight of the displaced fluid; 3. Vertical; 4. True

Chapter Highlights

- A fluid is a substance that deforms continuously in the presence of shear stresses, no matter how small those stresses.
- Compressive stresses in fluids are called pressures.
- Shear stresses do not exist in stationary fluids (fluids at rest).
- If the shear stresses are zero at a point in a material, the normal stresses at the point have the same magnitude for all planes that pass through the point.
- Standard atmospheric conditions are typically the pressure, temperature, specific weight, mass density, and viscosity of air at sea level.
- Local atmospheric conditions are the atmospheric conditions at a specific point on the earth's surface.
- Gage pressure is the amount by which the measured pressure differs from local atmospheric pressure. A positive gage pressure indicates that the pressure is greater than the local atmospheric pressure. A negative gage pressure (or vacuum) indicates that the pressure is less than the local atmospheric pressure.
- The gage pressure at a point in a stationary homogeneous fluid depends only on the depth h of the point below the surface of the fluid and varies linearly with h.
- If two points A and B in a stationary fluid are at the same elevation and if the two points can be connected by a path that lies wholly within the fluid, the pressures at A and B are equal.
- The normal force that a fluid exerts on a flat surface is called a pressure force.
- A pressure that acts on a flat face of a rigid body is dynamically equivalent to its resultant force—namely, a concentrated normal force (pressure force). The resultant axis of this pressure force intersects the face of the flat surface at a point called the center of pressure.
- The pressure on a solid body that is immersed in a stationary fluid exerts a resultant force (buoyant force) that is directed vertically upward through the center of gravity (center of buoyancy) of the fluid displaced by the body. The magnitude of the buoyant force is equal to the weight of the displaced fluid.

Problems

9.2 *The Concept of Pressure at a Point*
9.3 *Pressure Variation with Depth in a Fluid*

Unless otherwise stated, use the following constants: specific weight of mercury, $\gamma_{Hg} = 846$ lb/ft³ (133 kN/m³); specific weight of water, $\gamma_{H_2O} = 62.4$ lb/ft³ (9.78 kN/m³); atmospheric pressure, $p_{atm} = 14.696$ psi (101.3 kPa).

9.1 The pressure at a point in the ocean is 120 kPa. What is the depth to that point? Use $\gamma = 10.0$ kN/m³ for sea water.

9.2 The specific weight of sea water is 64 lb/ft³. Calculate the gage pressure (in lb/in²) at a depth of 3 miles below the ocean's surface.

9.3 An open tank contains crude oil ($\gamma = 54$ lb/ft³). Determine the pressure in the oil at a point 10 ft below the surface.

9.4 An open vessel (see Fig. P9.4) contains 3 ft of water ($\gamma = 62.4$ lb/ft³) above 5 ft of carbon tetrachloride ($\gamma = 99$ lb/ft³). Determine the pressure on the bottom of the vessel.

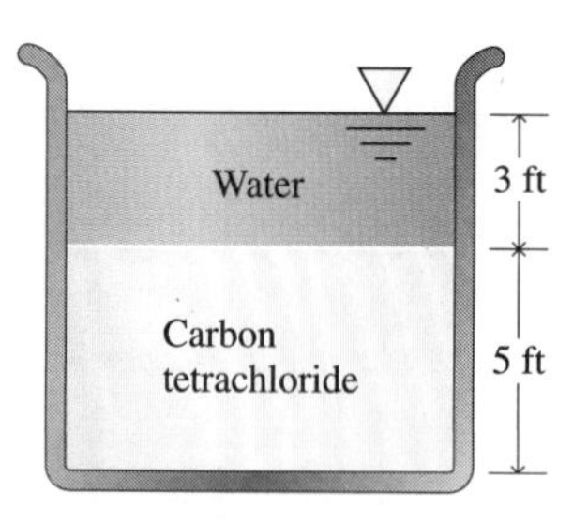

Figure P9.4

9.5 What is an air pressure of 14.696 psi equivalent to in N/m², in inches of mercury, in meters of water, and in feet of water?

9.6 The atmospheric pressure at sea level is 14.696 lb/in^2; the specific weight of air is 0.07651 lb/ft^3. Calculate the atmospheric pressure at an elevation of 700 ft, neglecting variation in the density of air.

9.7 The vertical tank of a rocket is partially filled to a depth of 9 m with liquid oxygen at a temperature of $-200°C$. At this temperature, the specific weight of liquid oxygen is $\gamma = 12$ kN/m^3. The pressure in the vapor above the liquid oxygen is maintained at atmospheric pressure (101.3 kPa). Determine the pressure at an inlet valve 2 m above the bottom of the tank.

9.8 How many inches of mercury are equivalent to a pressure of 2880 lb/ft^2? How many feet of water?

9.9 How many millimeters of mercury are equivalent to a pressure of 120 kPa? How many meters of water?

9.10 Determine the length h (in mm) of the column of mercury in the open-ended glass tube in Fig. P9.10.

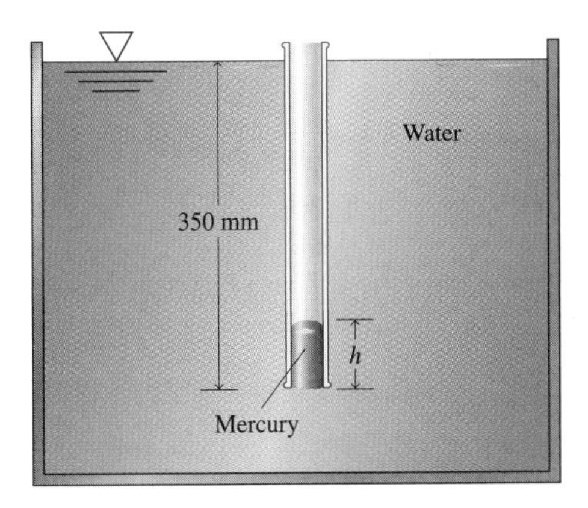

Figure P9.10

9.11 Relative to the datum, calculate the elevation x of the liquid in the right leg of the system of tubes in Fig. P9.11.

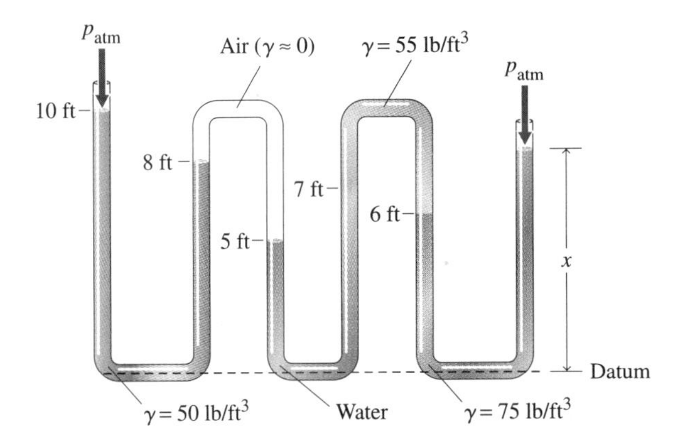

Figure P9.11

9.12 Figure P9.12 represents an inverted tank partially immersed in water. The tank contains oil, as shown in the figure. The gage reading is 20 kPa. Determine the distance h.

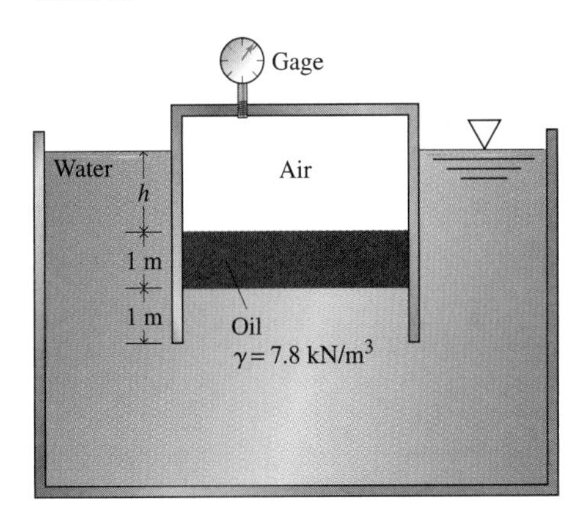

Figure P9.12

9.13 Determine the maximum pressure in each of the liquids in Fig. P9.11.

9.14 Figure P9.14 shows a U-tube for measuring gage pressure. Calculate the gage pressure (in lb/ft^2) in the pipe adjacent to the connection to the U-tube.

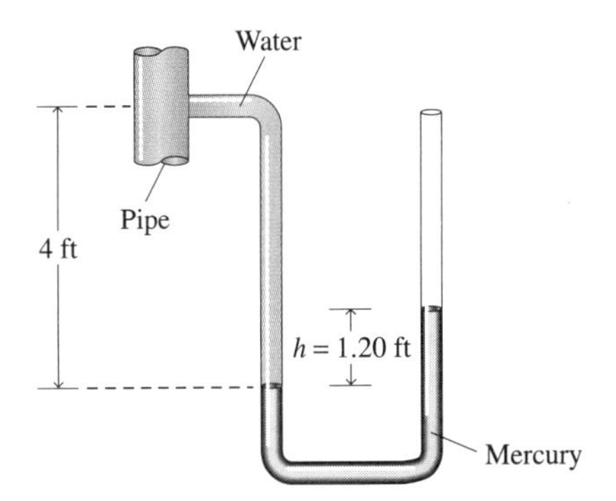

Figure P9.14

9.15 In Fig. P9.12, let $h = 3$ m and the cross-sectional area of the inverted tank be 16 m^2.

a. Determine the gage pressure [kN/m^2].

b. Determine the gage pressure required to move the oil down 1 m—that is, to begin to drive the oil from the tank.

9.16 The Goodyear blimp and a detail of its instrument panel are shown in Fig. P9.16. The cabin pressure is measured by a mercury barometer, and the air pressure outside the cabin is measured by a mercury manometer. The blimp is hovering 300 m above Coors Field in Denver, Colorado. The air pressure on the baseball field is $p_{\text{field}} =$ 625 mmHg, and the pressure inside the blimp cabin is maintained at $p_{\text{cabin}} = 675$ mmHg. Assuming that the average specific weight of air between the field and the blimp is $\gamma = 10.4$ N/m³, determine the value h of the manometer reading in millimeters.

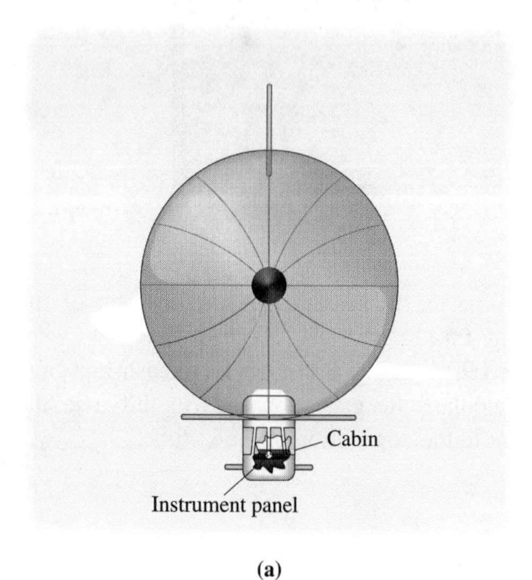

(a)

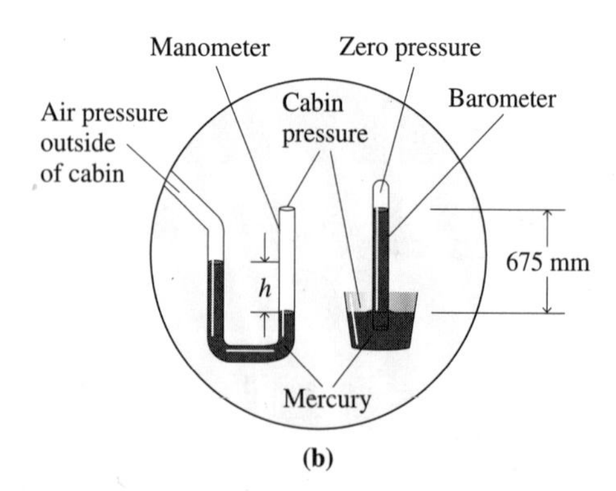

(b)

Figure P9.16

9.17 The local atmospheric pressure at the top of a hill is equivalent to 29 in of mercury. Assuming that the specific weight of air remains constant at 0.0767 lb/ft³, determine the elevation (in ft) of the hill relative to sea level (atmospheric pressure = 29.92 in Hg).

9.18 Tanks A, B, and C in Fig. P9.18 contain liquids of specific weight 50 lb/ft³, 150 lb/ft³, and 100 lb/ft³, respectively. Compute the pressure differences $p_B - p_C$, $p_C - p_A$, $p_A - p_B$ (in lb/ft²) at the level of the dashed line.

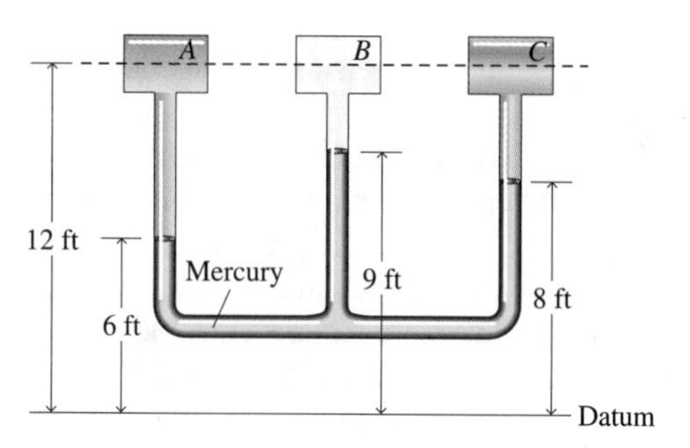

Figure P9.18

9.19 Refer to Fig. P9.18. Determine the maximum pressure in the mercury, given that the pressure in tank A is 4000 lb/in² and the distance from the datum to the horizontal tube is 3 ft.

9.20 In Fig. P9.20, the pressures p_1 and p_2 are equal to atmospheric pressure. The diameter of the vessel is $D =$ 50 mm, and the diameter of the tube is $d = 5$ mm. The liquid is mercury. An additional 2.2 N of mercury is poured into the vessel, and the scale reading increases by 200 mm. What is the angle θ of inclination of the tube with respect to the horizontal?

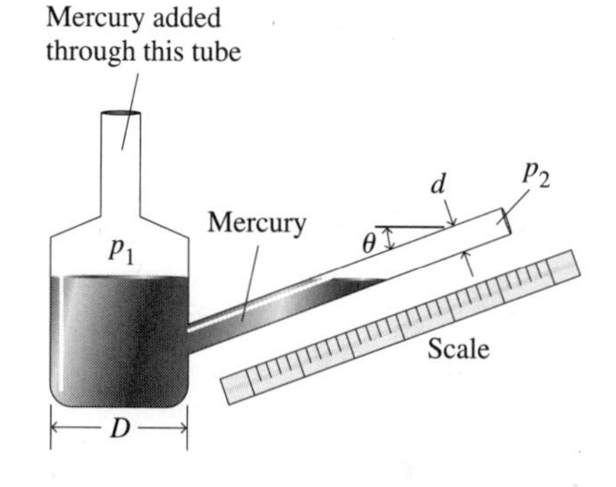

Figure P9.20

9.21 In Problem 9.20, let $\theta = 10°$, the diameter D of the bowl be 1.5 times the diameter d of the tube, and the specific weight of the liquid be 50 lb/ft³. By how much does the scale reading increase in inches if p_1 increases by 1 lb/ft² and p_2 remains constant (equal to atmospheric pressure)?

9.22 The submarine in Fig. P9.22a is submerged to a depth y in sea water ($\gamma = 10$ kN/m^3). A mercury barometer reads the internal pressure in the submarine, and a mercury manometer reads the pressure at the depth y (the barometer and manometer are not shown to scale in Fig. P9.22b, a detail of the instruments in the submarine). Determine the depth y of the submarine.

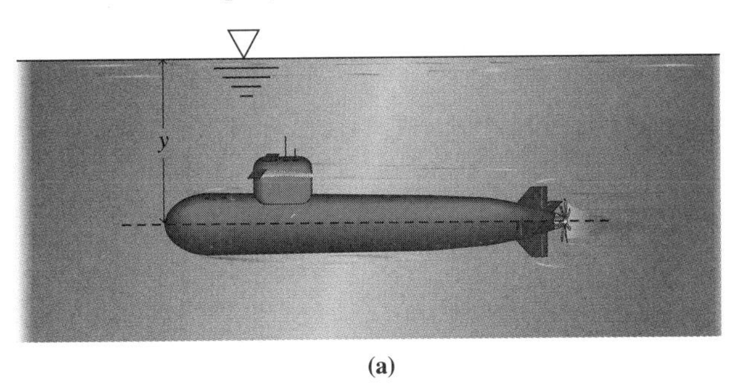

(a)

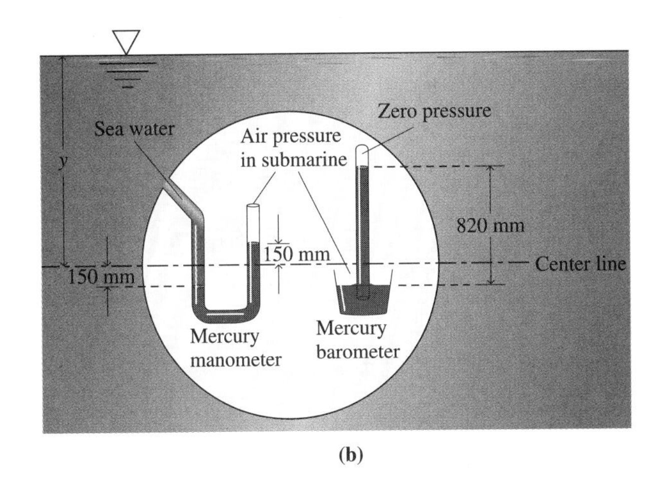

(b)

Figure P9.22

9.23 A suction pump lifts water at 82°C from a well. At this temperature, the pressure at which water boils is 51.75 kPa, and the specific weight of the water is 9.49 kN/m^3. Standard atmospheric pressure exists outside the well. Calculate the maximum height that the water can be lifted in the suction pipe.

9.24 Two tanks contain an unknown fluid of specific weight γ_3 at different levels (see Fig. P9.24). Two U-tubes containing known fluids of specific weights γ_1 and γ_2 are connected to the tanks, as shown. Determine the specific weight γ_3 of the unknown fluid in terms of the specific weights (γ_1 and γ_2) of the known fluids and the readings h_1 and h_2.

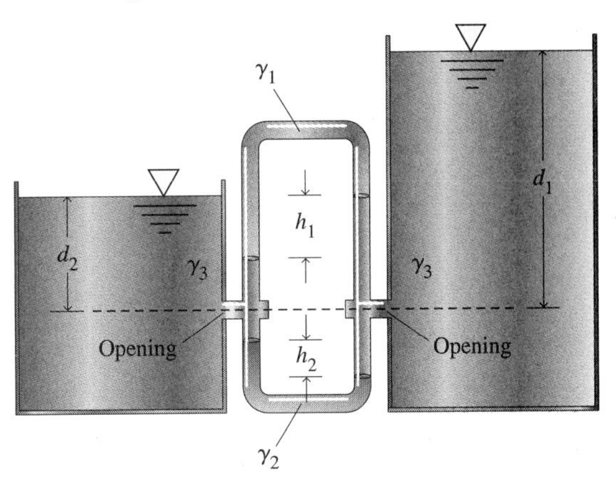

Figure P9.24

9.25 From Fig. P9.24, derive formulas for the readings h_1 and h_2 in terms of d_1, d_2, γ_1, γ_2, and γ_3.

9.26 Initially, the air pressure in the two tanks shown in Fig. P9.26 is atmospheric. The tanks contain a liquid of specific weight γ to the level *D–D*. The U-tube that is connected to both tanks contains a liquid of specific weight 1.5γ. The horizontal cross-sectional area (perpendicular to the plane of the page) of each tank is *A*, and the cross-sectional area of the tube is *a*. The pressure in tank *B* is increased by an amount Δp by the air pump.

a. Determine the distance *h* the liquid in tank *B* drops (and that in tank *C* rises), in terms of *H, A,* and *a.*

b. Determine the pressure change Δp, in terms of *H, A, a,* and γ.

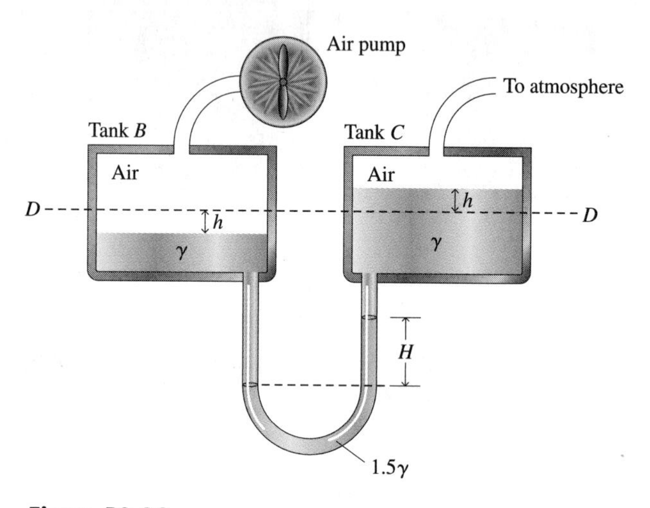

Figure P9.26

9.27 An open tank contains a liquid of unknown specific gravity. It is connected by a U-tube containing a liquid of specific gravity 1.25 to a second open tank that contains water and a liquid of specific gravity 0.8 (see Fig. P9.27). Determine the unknown specific gravity.

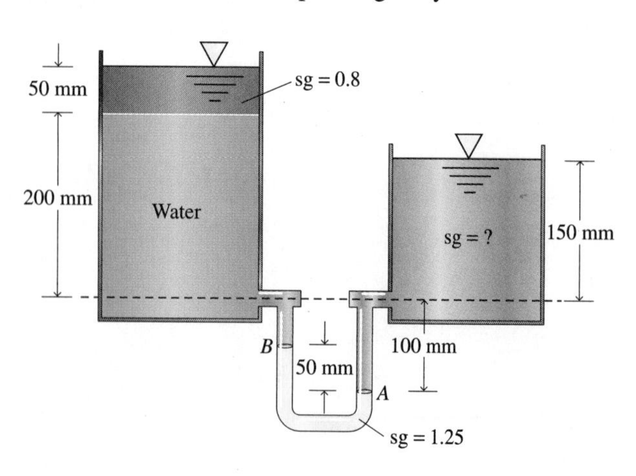

Figure P9.27

9.28 A U-tube of diameter *d* with a bowl of diameter *D* is used to measure pressure in a pipe. When the pipe is empty, as shown in Fig. P9.28a, the zero level is aligned with the mercury in the bowl and the mercury in the tube. When the pipe is filled with water, as shown in Fig. P9.28b, the mercury level drops in the bowl and rises a height *h* [ft] in the tube. Use the equation $p = kh$ [lb/ft^2], where *p* is the gage pressure in the bowl at the zero level, to derive a formula for *k* in terms of *d* and *D*.

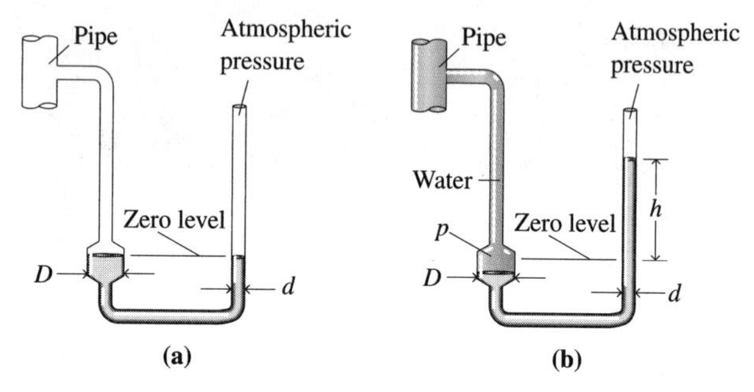

Figure P9.28

9.29 A small brass rod ($\gamma = 0.31$ lb/in^3) weighing 0.12 lb is accidently dropped into a U-tube (see Fig. P9.29). The internal diameter of the tube is 0.25 in. Calculate the pressure difference $p_B - p_A$ [lb/in^2] for the conditions shown in the figure. What is $p_B - p_A$ if the rod is removed?

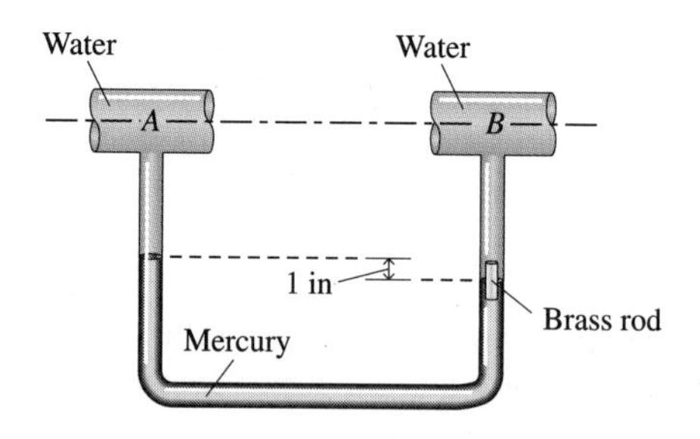

Figure P9.29

9.30 A system of tubes and conduits is shown schematically in Fig. P9.30. Calculate the pressure differences $p_B - p_C$, $p_C - p_A$, $p_A - p_B$ [N/m²].

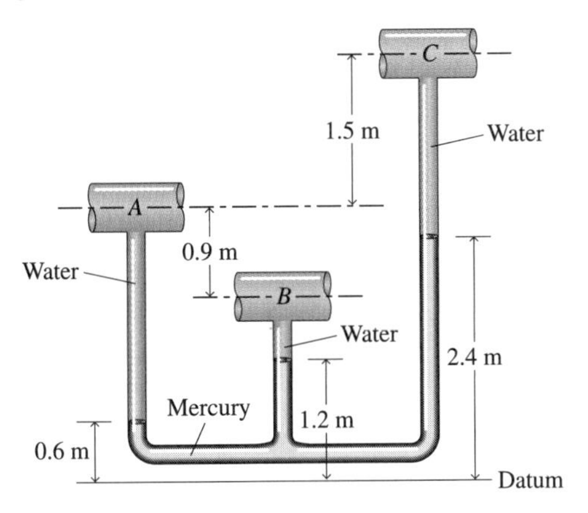

Figure P9.30

9.31 The specific weight γ of air is given by the equation $\gamma = kp$, where p is the pressure and the factor k depends on the temperature. Supposing that the temperature is constant, show that the pressure p at altitude z is determined by the formula $p = p_0 e^{-kz}$, where p_0 is the pressure at altitude $z = 0$. (*Hint:* Construct a free-body diagram of a vertical column of air of height dz. Let the pressures on the lower and upper ends of the column be p and $p + dp$, respectively.)

9.4 *Normal Force on a Flat Surface*

9.5 *Center of Pressure*

9.32 A circular plate in an airlock on a space station is subject to a pressure of 90 kN/m² on one side and to a complete vacuum on the other side. The plate has a radius $r = 400$ mm. Determine the pressure force that acts on the plate.

9.33 Solve Example 9.8 for the case where the gate is hinged at *B–B*, instead of at *A–A*.

9.34 A flat semicircle with a radius of 1 unit carries a pressure distribution defined by $p = 3r$, where r is the distance from the center. Determine the resultant force on the semicircular area and the center of pressure.

9.35 A rectangular plate is hung vertically in an open tank (see Fig. P9.35). The upper half of the plate is submerged in water ($\gamma = 9.78$ kN/m³), and the lower half in another fluid ($\gamma = 16$ kN/m³). Determine the normal pressure force due to the gage pressure that acts on one side of the plate.

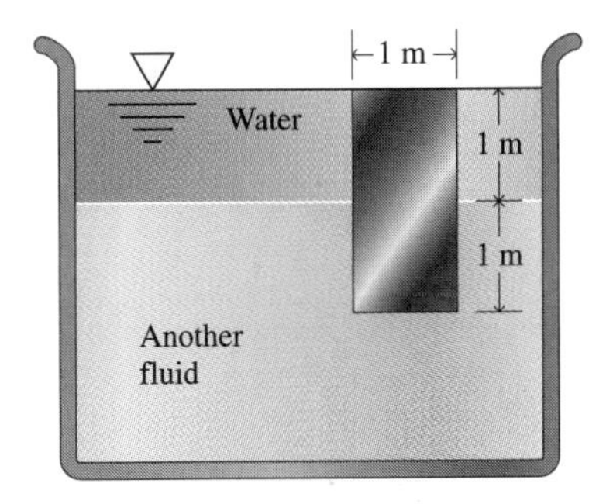

Figure P9.35

9.36 The flow rate of oil pumped downward through the tube in Fig. P9.36 is Q [volume/time]. The oil flows out between the circular disk and the fixed flat plate. The theoretical gage pressure is expressed as a function of the distance r by the equation $p = (6\mu_v Q/\pi b^3)\ln(a/r)$, in which μ_v (called the *coefficient of viscosity*) is a constant for the oil. Together, the disk and tube have weight W. Derive a formula for the gap b between the disk and the flat plate, in terms of μ_v, Q, a, and W. (Consider the inner diameter of the tube to be very small compared to a.)

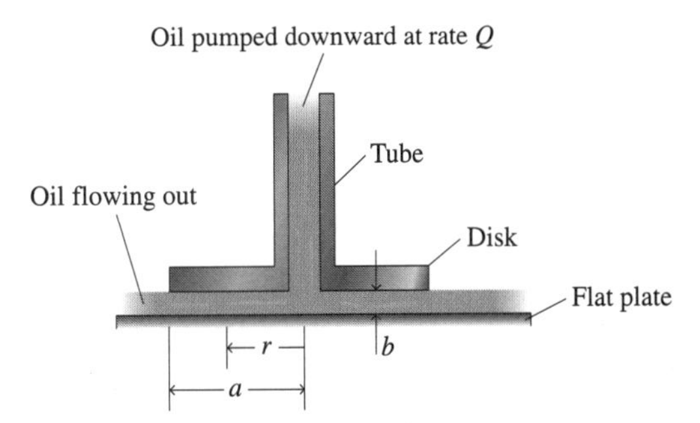

Figure P9.36

9.37 If the pressure on a plane area is constant, show that the center of pressure (x_{cp}, y_{cp}) coincides with the centroid $(\bar{x}, \bar{y})$ of the area.

9.38 A rectangular vertical plate in the hull of an aircraft carrier is submerged with its top edge 8 m below the water surface and its bottom edge 12 m below the surface (see Fig. P9.38 where the plate is marked). The plate is 3 m wide (in the direction perpendicular to the plane of the page).

a. Determine the magnitude of the normal pressure force that the sea water ($\gamma = 10$ kN/m^3) exerts on the plate.

b. Locate the center of pressure on the plate.

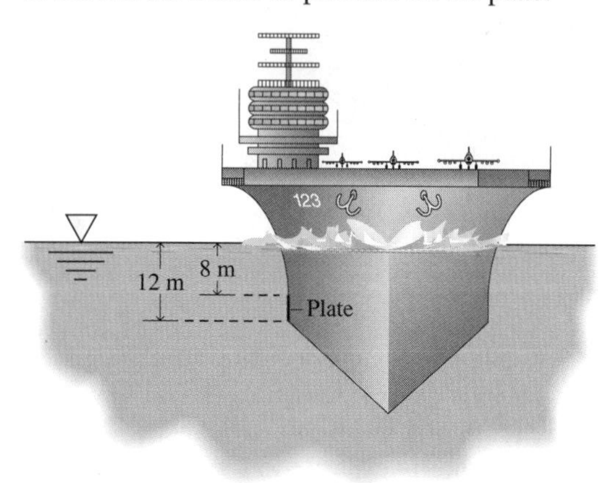

Figure P9.38

9.39 The rectangular plate *ABC* in Fig. P9.39 has a width of 4 ft (perpendicular to the plane of the page). Calculate the compressive force in the rod *BD*.

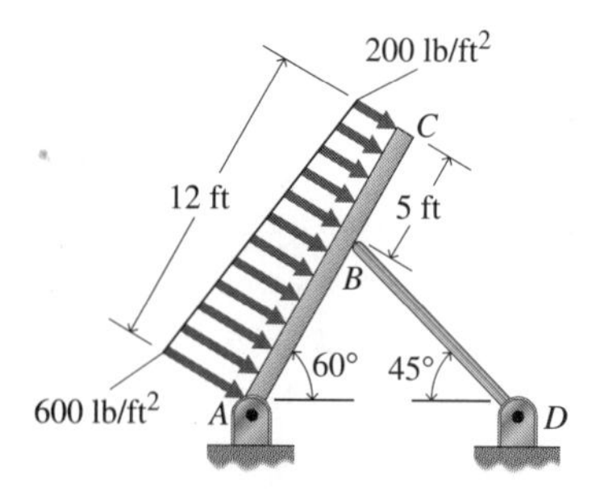

Figure P9.39

9.40 A square maintenance panel, 2 m × 2 m, is located in the hull of a ship (see Fig. P9.40, which shows the panel's location). It is inclined to the vertical by an angle of 15°. When the ship is afloat in sea water ($\gamma =$ 10 kN/m^3), the top of the panel is 2 m below the water surface.

a. Determine the magnitude of the resultant pressure force on the panel by integration.

b. Locate the center of pressure on the panel.

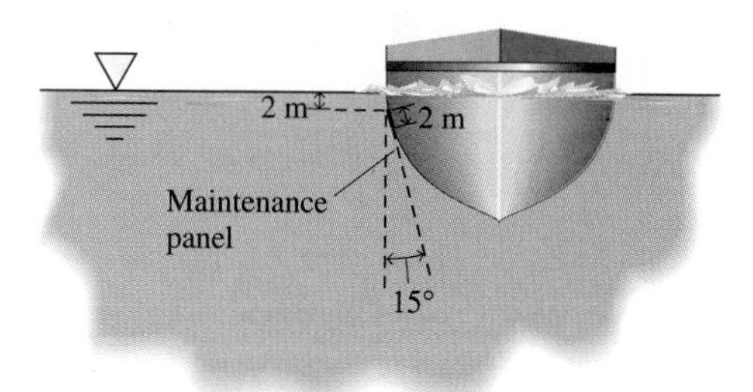

Figure P9.40

9.41 If plate *ABC* in Fig. P9.39 is in the form of a symmetrical trapezoid, 6 ft wide at *C* and 2 ft wide at *A*, what is the force in the supporting strut *BD?*

9.42 A parabolic gate is in the side of a water tank that is open at the top (see Fig. P9.42). The gate may be rotated inward about the hinge. What moment must be applied about the hinge to open the gate?

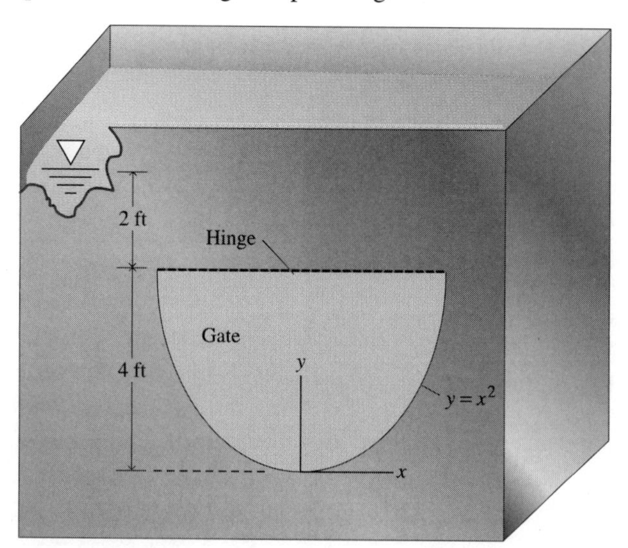

Figure P9.42

9.43 A closed cylindrical tank 600 mm in diameter lies horizontally on its side. It is half filled with a liquid of specific weight 12 kN/m³. The remaining space contains air at a gage pressure of 9.60 kN/m². Determine the total force on an end of the tank and the distance of the center of pressure below the center line of the tank.

9.44 A closed cubical tank with a horizontal bottom is 1.2 m high. The tank contains water (9.78 kN/m³) to a depth of 0.9 m. The space above the water is filled with gas at a pressure of 20 kN/m² vacuum. Calculate the resultant force and the elevation of the center of pressure on a side of the tank.

9.45 The water gate whose cross section is shown in Fig. P9.45 has a uniform specific weight of 150 lb/ft³. The gate is 20 ft wide. Determine the reaction of the hinge *A* and the force that acts on the gate at *C*. Neglect friction.

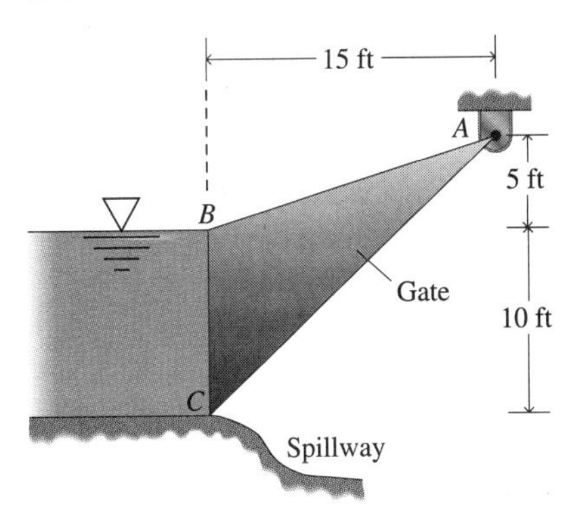

Figure P9.45

9.46 The upstream face of a dam is a trapezoid 30 m high, 60 m wide at the top, and 15 m wide at the bottom.

a. If the water level is at the top of the dam on the upstream side and if there is no water on the downstream side, what is the total force due to the pressure on the dam?

b. What is the elevation of the center of pressure above the bottom of the dam?

9.47 The rectangular homogeneous gate *AB* in Fig. P9.47 separates water ($\gamma = 9.78$ kN/m³) in two reservoirs. The gate has weight *W* and dimensions 2 m × 4 m. It is hinged at *A*. Assume that the water level in the right reservoir is below the stop at *B*.

a. Determine the force exerted by the gate on the stop at *B*, in terms of γ, *W*, d_1, d_2, and θ. Neglect friction in the hinge.

b. For $\theta = 30°$, $d_2 = 6$ m, and $d_1 = 0$ (no fluid to the right of the gate), determine the minimum weight *W* of the gate that will cause it to open.

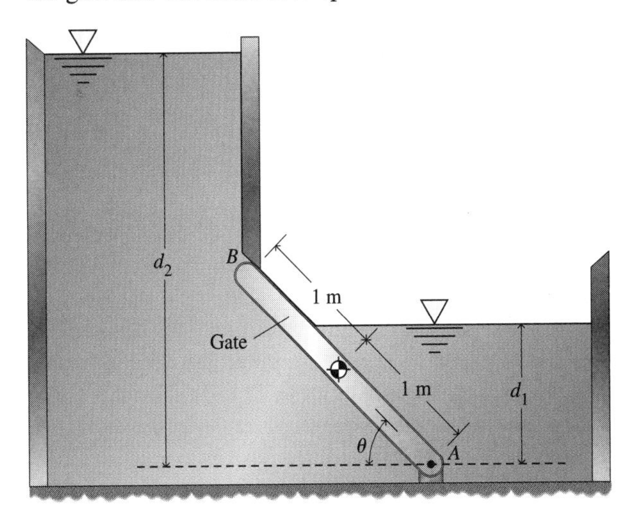

Figure P9.47

9.48 In Fig. P9.47, let $W = 9$ kN, $d_1 = 1.5$ m, $d_2 = 6$ m, and $\theta = 60°$. Determine the force exerted by the gate on the stop at *B*.

9.49 A rectangular plate with two edges horizontal and the other two edges inclined at an angle θ to the horizontal is submerged in a liquid. The center of the plate lies at the depth *h*, and the length of an inclined edge is *a*. Derive a formula for the distance between the center of the plate and the center of pressure for the pressure force acting on one face of the plate.

9.50 In Fig. P9.36, assume that the inner diameter of the tube is *ka*, where *k* is a constant.

a. Derive a formula for the gap *b* in terms of μ_v, *k*, *Q*, *a*, and *W*.

b. For $\mu_v = 0.007$ N·s/m², $Q = 0.1$ m³/s, $a = 300$ mm, and $W = 20$ N, plot the value of *b* as a function of *k*, for $0.1 < k < 0.5$.

9.51 **a.** In Example 9.8, plot the force P required to keep the gate closed as a function of h, the level of the water above point C, for the range $0 \leq h \leq 15$ ft.
b. Determine the maximum value of P and the corresponding value(s) of h.
c. Check your computer results analytically for the cases $h = 5$ ft, $h = 10$ ft, and $h = 15$ ft.

9.52 Refer to Problem 9.45. The reservoir behind the gate is filled slowly with water, and the water level rises gradually from point C to point B on the face of the gate.
a. Plot the support reactions at A and C as functions of the height h of the water above C, for the range $0 \leq h \leq 10$ ft, in increments of 0.5 ft.
b. What does your solution reveal concerning requirements for the supports at A and C?

9.6 *Distributed Pressure on Curved Surfaces*

9.53 A thin-walled spherical metal shell, of internal radius $r = 10$ in and wall thickness $h = 0.2$ in, is subjected to an internal pressure p. The shell will fracture when the stress in the metal is 60,000 psi. Determine the pressure p that will cause the shell to fracture.

9.54 An airplane fuselage of circular cross section has a radius of 4.5 ft. It contains air at 12 lb/in^2 absolute pressure. The atmospheric pressure outside the plane is 9 lb/in^2 absolute pressure. Calculate the hoop tension at this cross section caused by the difference between the internal and external pressures.

9.55 A thin-walled spherical shell, of internal radius r and wall thickness $h << r$, is subjected to an internal gas pressure p. Determine the average tensile stress in the wall in terms of p, h, and r, neglecting the weight of the gas. (*Hint:* Consider a free-body diagram of one-half of the sphere.)

9.56 A pipe with a 6 in inside diameter is cast in the shape of a U and has closed ends. The pipe is pressurized to a gage pressure of 150 psi. Determine the tensile force in either leg of the U.

9.57 The hollow conical shell shown in Fig. P9.57 weighs 450 N in air. It rests on the bottom of a tank that contains 2 m of water ($\gamma = 9.78$ kN/m^3). The air pressure inside the shell is 87.3 kPa. Determine the vertical force F necessary to pull the shell free from the bottom. Atmospheric pressure is $p_{atm} = 101.3$ kPa.

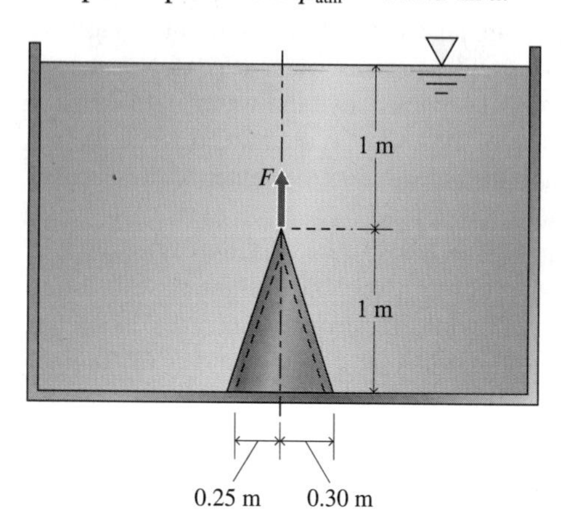

Figure P9.57

9.58 A high-pressure cylindrical tank with hemispherical end caps (see Fig. P9.58) has an internal gage pressure of 400 kPa.
a. Determine the resultant pressure force that acts on an end cap.
b. Determine the average hoop and axial stresses in the tank at section A–A.

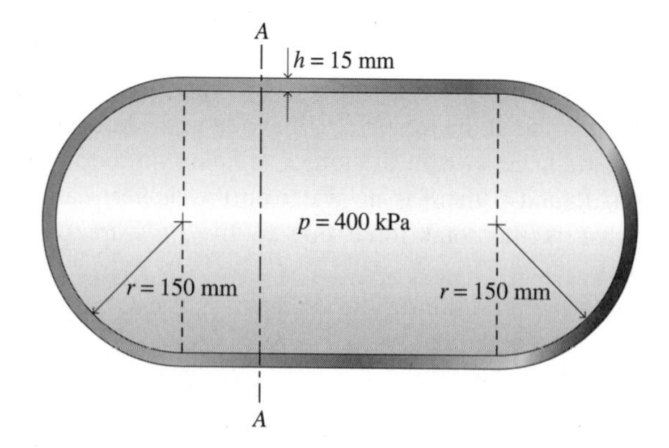

Figure P9.58

9.59 An oil tanker carries oil of specific gravity 0.84 in one of its bays (see Fig. P9.59). The upper surface of the oil is subject to atmospheric pressure. Determine the magnitudes and orientations of the pressure forces that act on segments *AB, BC,* and *CD,* per unit length (perpendicular to the plane of the page) of the bay.

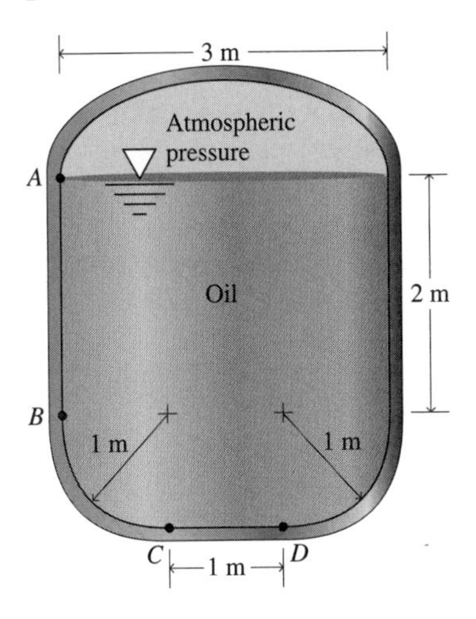

Figure P9.59

9.60 The water gate *AB* shown in Fig. P9.60 is a quarter circle and 6 m long (dimension perpendicular to the plane of the page). It is pivoted at *A*. Determine the smallest force *Q* required to keep the gate closed.

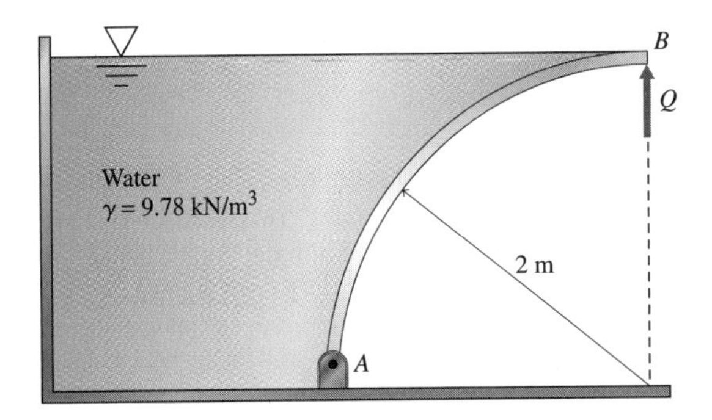

Figure P9.60

9.61 A hollow hemispherical shell is cemented to a vertical wall (see Fig. P9.61). The shell contains gas at 10 psi gage pressure. Water exerts pressure on the exterior of the shell. Calculate the net tensile force on the cemented joint.

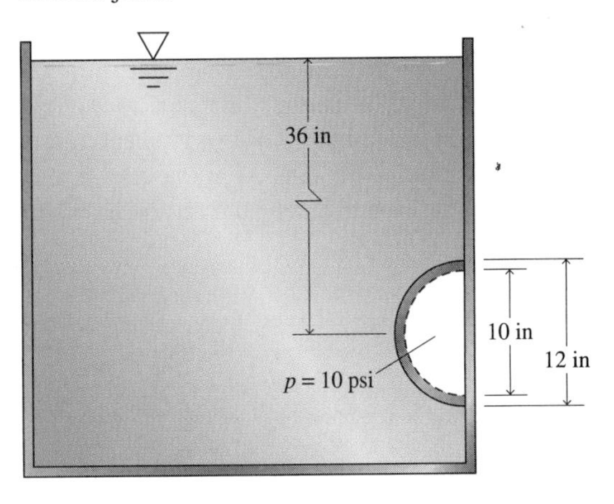

Figure P9.61

9.62 A spillway gate of a dam is mounted as shown in Fig. P9.62 (not to scale). The water level is at the top of the gate. To release water, the gate is rotated clockwise about the pivot *A*. The gate is 3 m long (dimension perpendicular to the plane of the page), weighs 256 kN, and has a cross section that is a sector of a circle.

a. Determine the magnitude and the line of action of the resultant pressure force that acts on the gate.

b. Assuming that the weight of the gate is uniformly distributed, determine the moment required to raise the gate. Neglect friction of the pivot.

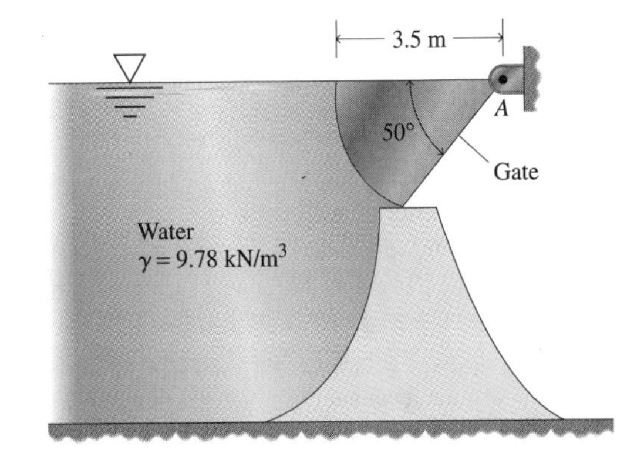

Figure P9.62

9.7 Buoyancy

9.63 In Example 9.13, remove the oil in the tank and suspend the ball, half in air and half in mercury. Determine the tension in the cord. (How significant is the buoyant force of the air?)

9.64 In World War II, submerged bombs were placed in shipping lanes to prevent ships from passing through. The bombs, approximately spherical in shape, were held in place by cables (see Fig. P9.64). A typical bomb weighed 4 kN. Determine the tension in the cable when the center of a bomb was submerged 2 m in sea water ($\gamma = 10$ kN/m^3).

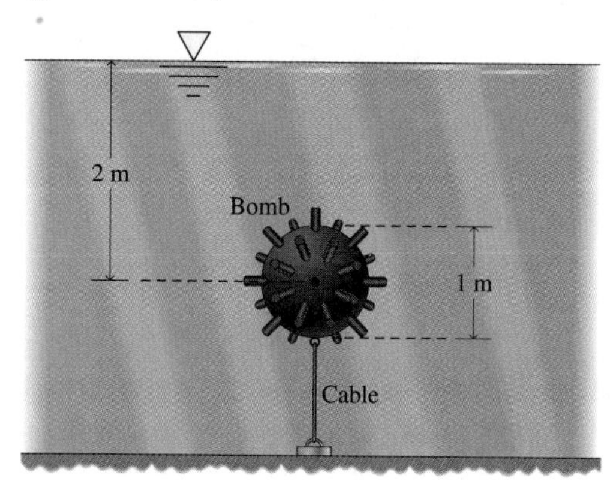

Figure P9.64

9.65 An oak block measures 75 mm × 75 mm × 150 mm and has a specific gravity of 0.60. A well-polished face of the block that measures 75 mm × 150 mm is pressed tightly against the bottom of a tank. The tank is then filled with mercury (specific gravity = 13.56) to a level of 300 mm. Determine the force that the block exerts on the bottom of the tank.

9.66 The oak block of Problem 9.65 is floated with one of its 75 mm × 150 mm faces on the surface of the mercury in the tank. Determine the depth to which the block sinks into the mercury. Ignore capillary action.

9.67 A spherical balloon weighs 0.10 lb per square foot of surface area (without gas). The balloon is to be filled with hydrogen, and the internal pressure is to be the same as the external atmospheric pressure. The specific weight of hydrogen is 6.9% of the specific weight of air at the same temperature and pressure. What is the necessary diameter of the balloon to lift 500 lb of payload to 40,000 ft altitude, if the specific weight of air at that altitude is 0.020 lb/ft^3?

9.68 A ship submerged in sea water ($\gamma = 64$ lb/ft^3) is to be raised by attaching submerged steel tanks to it and displacing the water from the tanks by means of compressed air. The average specific weight of the materials in the ship is estimated to be 360 lb/ft^3. The weight of the ship in dry dock is 30,000 kips. A tank weighs 20 kips, and its volume is 5000 ft^3.

a. What lift force does a tank exert on the ship?

b. How many tanks are needed to raise the ship high enough to bring the deck to the surface, if the superstructure weighs 5000 kips?

9.69 Two spherical floats, rigidly connected to a beam, form a dumbbell-shaped structure that weighs 8900 N (see Fig. P9.69). A vertical load F acts on the beam, 1 m from the center of the right float. What is the minimum force F that will cause the right float to be fully submerged in the water? (Neglect the buoyant force on the beam.)

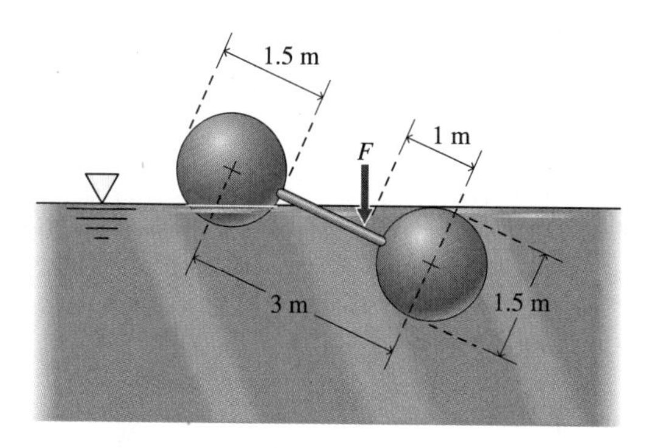

Figure P9.69

9.70 A cubical block of ice with edge length h floats in water (see Fig. P9.70). The specific gravity of ice is 0.90. What couple M is required to hold the block of ice in the position shown? (*Hint:* See Eq. (b) of Example 9.11.)

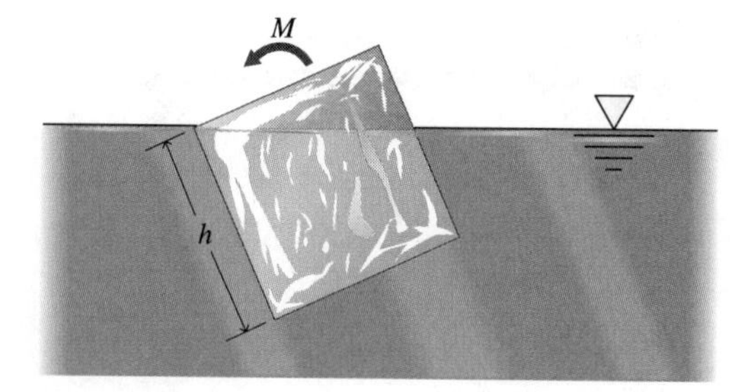

Figure P9.70

9.71 A solid body in the form of a frustum of a circular cone has its smaller end cemented to the bottom of a tank filled with 10 ft of water (see Fig. P9.71). Determine the net pressure force from the water on the body. Standard atmospheric pressure exists above the water level.

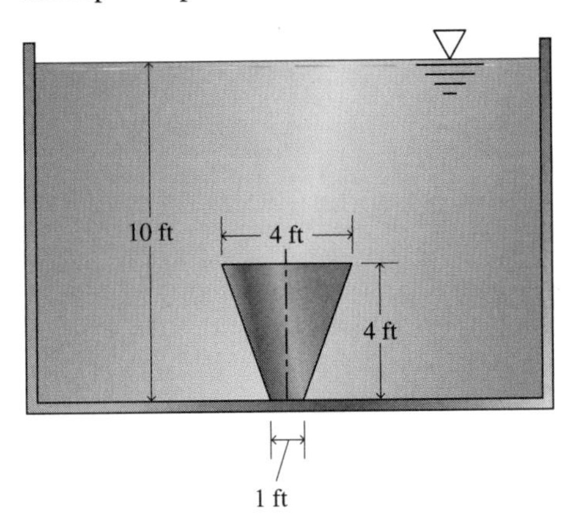

Figure P9.71

9.72 One end of the slender uniform pole (see Fig. P9.72) is submerged in water, and the other end is suspended by a string. The specific weight of the pole is 24 lb/ft^3.

a. Determine the angle θ.

b. What proportion of the weight of the pole is carried by the string?

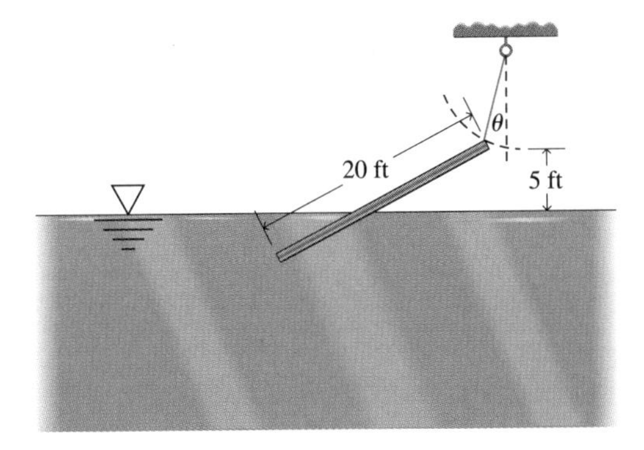

Figure P9.72

9.73 A barge floating in a fresh water lake has a hull in the shape of a half cylinder, with a diameter of 6 m and a length of 15 m. The center of gravity is on the axis of symmetry of the center cross section, 1.5 m below the central axis of the cylinder. The weight of the barge is 450 kN.

a. What is the draft (that is, the submerged depth)?

b. What is the angle of roll if one edge of the deck is at the water level?

c. What moment is required to produce this roll?

REVIEW QUESTIONS

9.74 What are compressive stresses in fluids called?

9.75 Prove that, when shear stresses are absent at a point in a material, the normal stresses on all plane elements at that point are equal.

9.76 Define *absolute pressure, gage pressure,* and *vacuum.*

9.77 Show by means of a free-body diagram how pressure varies with depth in a stationary fluid. Show that the pressure is constant on any horizontal plane.

9.78 If the pressure distribution on a flat surface is known, how is the total force on the surface determined?

9.79 Define *center of pressure of a flat area.* Explain how it is located, and write the corresponding general equations in integral form.

9.80 Explain how the force due to a distributed pressure on a curved surface can be determined by projecting the curved surface onto flat surfaces. How is the moment of the force on the curved surface with respect to a given axis determined from the forces on the projected surfaces?

9.81 Define *center of buoyancy.* Where does the resultant buoyant force act on a body? Does it ever have a horizontal component?

9.82 Draw a diagram to illustrate the righting couple on a ship that has an angle of roll. Explain where the forces of the couple act.

9.83 A small dish containing lead pellets floats in a beaker of water. What happens to the water level in the beaker as the pellets are removed from the dish and dropped into the bottom of the beaker?

Chapter 10 FRICTION

Stock car racers rely on friction to prevent their tires from slipping on the track.

A Look Forward

IN THIS CHAPTER, we will study friction. Friction may be defined broadly as the resistance that occurs between two bodies in contact when they tend to slide or roll relative to one another. We will discuss three approximate rules of friction that were established by Charles A. Coulomb (1736–1806), but are still useful today.

Frictional resistance is due in large part to the roughness of various materials, such as metals, concrete, wood, plastic, and composites. We will see that surface roughness determines, in part, the amount of friction that exists between bodies. We will analyze the performance of several mechanical devices in which friction plays a central role. These devices include the wedge, screw, and jackscrew, as well as belt-pulley assemblies, clutches, and brakes.

Friction can be simultaneously beneficial and harmful. For example, friction enables you to walk, but it also causes the soles of your shoes to wear out. Friction can be used to produce desirable wear. For example, it may be used to grind and polish metals and to "break in" a new car engine.

In your career as an engineer, you will have to deal with the two faces of friction. In some situations, friction is a help; in other situations, it is a hindrance. You will have to adapt and/or improvise to utilize its unique characteristics.

Survey Questions

- Why is friction important in statics?
- How does static friction differ from kinetic friction?
- What parameters influence the amount of friction in a system?
- How is kinetic friction, or sliding friction, affected by the speed of sliding?

10.1 FRICTIONAL FORCE

After studying this section, you should be able to:

- Describe the basic characteristics of the frictional force acting between bodies in contact.

> **Key Concept** A frictional force is a shear force that acts tangent to the surface of contact between two bodies. This force opposes sliding motion between the bodies.

If electromagnetic and gravitational effects are neglected, the force that exists between two bodies in contact is transmitted through the area of contact between the bodies. If the surface of contact between bodies A and B is flat, as shown in Fig. 10.1a, the force $\mathbf{F}$ that body A exerts on body B may be resolved into a normal component $\mathbf{F}_N$ that is perpendicular to the plane of contact and a tangential component $\mathbf{F}_V$ that lies in the plane of contact (see Fig. 10.1b). This type of resolution of forces was used in the analysis of internal forces in a body. In Sec. 9.1, the forces $\Delta\mathbf{F}_N$ and $\Delta\mathbf{F}_V$ were designated as the normal (perpendicular) component and the shear (tangential) component, respectively, of the force $\Delta\mathbf{F}$ transmitted through a cross-sectional area ΔA in a plane passed through a body (Fig. 9.1b). If the plane under consideration is a plane of contact between two bodies, rather than a cross-sectional plane in a body, the total tangential component $\mathbf{F}_V$ is called the *frictional force.* In other words, a frictional force is a shear force that acts on an *external surface* of a body. The word "friction" has the same meaning, more or less, as the phrase "frictional force," although, more generally, "friction" may signify all the phenomena associated with the occurrence of frictional forces.

frictional force

A frictional force can be viewed as a passive force, because it exists only when it is needed to oppose the forces that tend to cause one body to slide relative to another body. For example, the horizontal component of the force P in Fig. 10.1b is tending to push block B to the right, relative to block A. Therefore, the frictional force opposes the horizontal component of P; that is, $\mathbf{F}_V$ acts to the left. If there are no forces that tend to cause sliding between two bodies in contact, a frictional force will not develop.

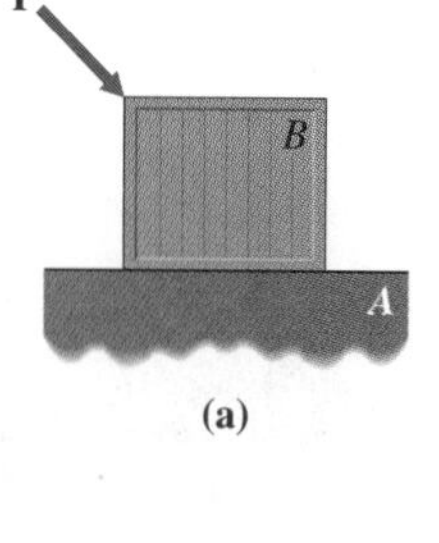

(a)

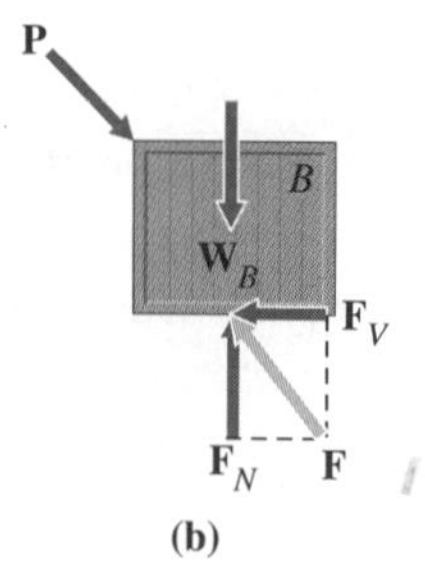

(b)

Figure 10.1 Contact forces between two bodies.

The frictional force that one body exerts on another body cannot exceed a maximum value that depends on the normal force transmitted across the surface of contact. If the force that tends to cause sliding exceeds this maximum value, by even a slight amount, the bodies will slide relative to each other. For example, in a tug-of-war, because of friction between the hands of the team members and the rope, the rope does not slip from their grasp. They can pull harder on the rope without letting it slip if they grip it more tightly. This observation indicates that the value of the frictional force at which sliding begins increases as the normal force (their grip) increases.

Some of the benefits of friction are quite evident. For example, without friction, we could not walk, run, or crawl. Many of our machines, such as brakes, belt drives, and clutches, would not work. But friction also has negative effects. For example, nearly all the energy required to operate automobiles, trains, ships, and innumerable other machines is expended to overcome friction.

FRICTION FACTS AND COEFFICIENTS OF FRICTION

The threads of a stainless steel bolt, magnified 60 times, show the surface roughness that contributes to friction.

What causes friction between two bodies that tend to move relative to one another while in contact? Metallurgists have found that at least three things do. One cause of friction is the intermeshing of microscopic, teethlike irregularities that are present on any surface. This microscopic roughness exists even on polished surfaces. A second cause of friction is adhesion between surfaces, probably due to molecular attraction. Interestingly, this factor increases in importance for highly polished surfaces. For example, this effect occurs with flat-ground gauge blocks used in machine shops. These blocks cling together if pressed and twisted against one another. Two flat pieces of glass, if wrung together to exclude air, will also stick together. A third cause of friction is an indentation on one body caused by a second, harder body. To move relative to the first body, the harder body must climb out of the hole it has created. This effect is a factor in rolling resistance, as well as in sliding friction. Other factors, such as electrostatic attraction and elastic stretching between bodies may also cause friction.

Lubricants, such as oil, wax, and graphite, reduce friction by providing a film that separates the bodies in contact. For example, oil tends to fill in gaps or irregularities on surfaces so that the bodies slide on a thin layer of oil instead of over interlocking protuberances.

Key Concept The force of static friction is maximum when the two bodies in contact are just ready to slip relative to each other.

impending sliding

The maximum value of the frictional force between two contacting bodies in static equilibrium is reached when sliding between the two bodies is about to occur. At this instant, sliding is *impending.* This maximum frictional force is determined experimentally,

A thin film of water reduces the friction between the ice and the skate blades of champion speed skaters such as Bonnie Blair.

since it depends on the nature of the contacting surfaces. For example, the maximum value of the frictional force is reduced considerably if the surfaces are lubricated. Osborne Reynolds, a pioneer in the study of fluid mechanics, attributed the slipperiness of ice to a lubricating film of water. He concluded that very cold ice would not be slippery, a conclusion that was verified by Arctic explorers. The study of lubrication of surfaces (Bowden & Tabor, 1986; Hutchings, 1992) requires an understanding of principles of fluid mechanics. Hence, it is usually taught in senior-level university courses in mechanical engineering, material sciences, or metallurgy. We restrict our study mainly to *dry friction,* between unlubricated surfaces.

dry friction

> ***Key Concept***
> The maximum force of friction increases as the normal force between the bodies increases.

On the atomic scale, all surfaces are rough. The interlocking of microscopic or submicroscopic protuberances, or asperities (bumps), on contacting bodies naturally opposes sliding motion. When sliding occurs, the tops of some of these bumps are sheared off from the softer body by the harder body, but the bodies may "ride over" most of the bumps, particularly at high speeds of sliding. This phenomenon is illustrated in Fig. 10.2. The horizontal components of the contact forces, $\mathbf{F}_1$, $\mathbf{F}_2$, $\mathbf{F}_3$, . . . , exerted by the fixed body on the moving body resist the motion of the moving body. From this interpretation of friction (which is oversimplified), you can see why the maximum value of the frictional force increases as the normal force **N** increases; that is, the larger the normal force, the more difficult it is for the bodies to "ride over the bumps."

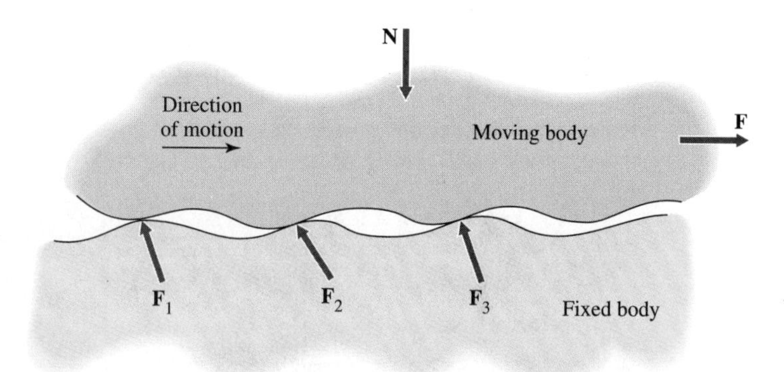

Figure 10.2 Contacting surfaces, greatly magnified (forces $\mathbf{F}_1$, $\mathbf{F}_2$, and $\mathbf{F}_3$ are exerted on the moving body by the fixed body).

> ***Key Concept***
> The frictional force that exists when two contacting bodies are about to slide relative to one another is generally larger than the frictional force when they are sliding.

The maximum value of the frictional force that exists between bodies at impending sliding is usually larger than the frictional force that exists during sliding. The frictional

static friction

kinetic friction

force between two bodies in contact that do not slide relative to one another is called *static friction,* or the *static frictional force.* The frictional force between two bodies in contact that slide relative to one another is *sliding friction,* or *kinetic friction.* The maximum value of the static frictional force is sometimes called the *friction of impending sliding. The friction of impending sliding is ordinarily greater than the friction of sliding.* A locomotive engineer utilizes this fact when he starts his train. He knows that he can gain speed more rapidly if he does not slide the wheels on the track. You may have experienced the effect of friction if you have tried to move a bookcase full of books by sliding it across the floor. You may have noticed that you had to push with a certain force before the bookcase moved, but once the bookcase started to move, you could keep it moving with less force. Of course, being a good engineering student, you know that you could have used a smaller force to move the bookcase if you first took some of the books out of it, thus reducing its normal force on the floor.

Such observations and experiences suggest a set of rules for this type of behavior. More exact experiments on friction have led to the following approximate rules for unlubricated bodies:

1. The maximum static and kinetic frictional forces are proportional to the normal force between the contacting bodies, except for very high or very low normal forces.
2. The maximum value of the static frictional force is independent of the area of contact. Furthermore, it is not affected by nonuniformity of pressure on the contact area.
3. The frictional force between sliding surfaces is independent of the speed of sliding. The maximum value of the static frictional force is somewhat larger than the frictional force between sliding surfaces (a known exception is rubber on concrete).

These rules are attributed mainly to Charles A. Coulomb. Since they are merely approximations that are applicable for limited ranges of the variables, they should be used with caution when a high degree of accuracy is required. In recognition of Coulomb's work, the theory of dry friction is also known as the theory of *Coulomb friction.*

Coulomb friction

It is customary to denote the magnitudes of the normal force $\mathbf{F}_N$ and the shear force $\mathbf{F}_V$ that act between bodies using the symbols N and F. By the rules of friction cited above, the proportionality between the maximum static frictional force F_s and the normal force N is expressed by the equation

$$F_s = \mu_s N \tag{10.1}$$

where μ_s is a parameter that depends on the nature of the contacting surfaces. For two unlubricated bodies in contact, μ_s is approximately constant. This constant is called the *coefficient of static friction.* We can write Eq. (10.1) in the form

coefficient of static friction

$$\mu_s = \frac{F_s}{N} \tag{10.2}$$

In this form, you can see that μ_s is a dimensionless ratio of two forces. Also, since the frictional force and the normal force are always perpendicular to each other, Eq. (10.2) is not a vector equation; it is merely a relation between the magnitudes of the two forces.

Since the maximum force of static friction is not usually equal to the frictional force of sliding, in some problems we might need to know two coefficients of friction—a coef-

When a bowling ball is first released, it slides on the lane and is subject to kinetic friction. Once it slows down, it rolls without sliding and is then subject to static friction. The "hook" of the ball begins at the point when it first begins to roll without sliding.

coefficient of kinetic friction

ficient of static friction μ_s, which determines the maximum value of the static frictional force, and a *coefficient of kinetic friction* μ_k, which determines the frictional force that exists while sliding is in progress.

Analogous to Eq. (10.1), with the kinetic frictional force denoted by F_k, we have

$$F_k = \mu_k N \tag{10.3}$$

or

$$\mu_k = \frac{F_k}{N} \tag{10.4}$$

As noted in the third rule for friction, the frictional force F_k for sliding and, hence, the coefficient of kinetic friction μ_k are approximately independent of the speed of sliding. Also, tests show that for hard materials and low sliding speeds, μ_k is less than μ_s.

In our study of friction, we assume that μ_k is constant, regardless of the speed of sliding. However, for many materials, it does tend to decrease slightly as sliding speed increases. An exception occurs with some combinations of soft and hard materials, for which μ_k increases at very high sliding speeds and can be larger than μ_s. For example, at a sliding speed of 100 mm/s, the value of μ_k is as high as 4.0 for rubber sliding on steel. The value of μ_k for rubber sliding on other solids ranges from 1.0 to 4.0. Tests also show that for hard materials, μ_k decreases as temperature increases (Bowden & Tabor, 1986).

Key Concept If two bodies in contact are not sliding relative to each other, the frictional force between them need not always be equal to the maximum static frictional force.

You should remember that the magnitude F of the static frictional force that acts between two bodies that do not slide relative to one another is not always equal to $F_s = \mu_s N$. The magnitude of the static frictional force need only satisfy the inequality $0 \le F \le F_s$. In other words, the static frictional force F acts to maintain the contacting bodies in a state of equilibrium under the action of applied forces. If the equilibrium conditions require that $F > F_s$, the bodies will slide relative to one another. Then, the kinetic frictional force F_k acts [see Eq. (10.3)].

Figure 10.3a, which shows a block of weight W resting on a flat surface, illustrates this point. A small horizontal force f is applied to the block and is increased slowly. The free-body diagram of the block in Fig. 10.3b shows that equilibrium exists only if the frictional force F equals the applied force f. So, F also increases slowly, until sliding is impending. When f is just large enough for sliding to be impending, $f = F = F_s$. If f is increased slightly, the block slides. Then F decreases to its kinetic value, F_k. Hence, equilibrium can no longer be maintained unless f is reduced accordingly. Depending on the material properties of the bodies, the magnitude of the frictional force F may drop below F_k at high speeds of sliding. These ranges of behavior are shown in Fig. 10.3c.

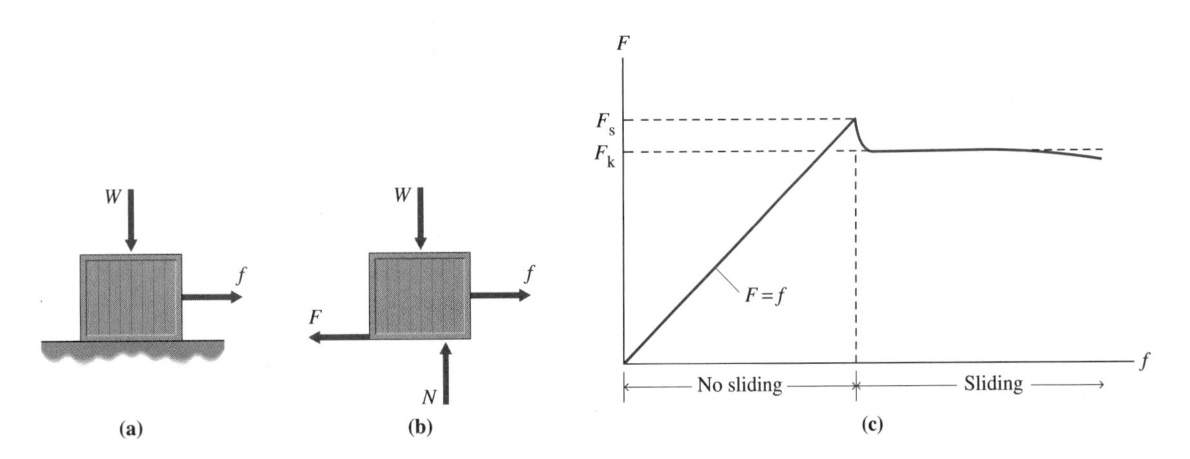

Figure 10.3

The main ideas discussed in this section can be summarized as follows:

1. The direction of the static frictional force always opposes the direction in which two bodies in contact tend to slide relative to one another. If the bodies do not slide relative to one another, the magnitude of the static frictional force F is such that $0 \le F \le F_s = \mu_s N$, where μ_s is the coefficient of static friction and N is the normal force between the bodies. That is, until sliding impends, $0 \le F < F_s$, and at impending sliding, $F = F_s$.
2. When sliding occurs, the kinetic frictional force is $F = F_k = \mu_k N$, where μ_k is the coefficient of kinetic friction. Ordinarily, $F_k < F_s$ and $\mu_k < \mu_s$.

3. Although, for many materials, kinetic friction is approximately independent of the speed of sliding, it tends to decrease slightly as sliding speed increases and as temperature increases.
4. The coefficients of static and kinetic friction, μ_s and μ_k, are approximately independent of the area of contact between bodies.

10.2 BLOCK ON AN INCLINED PLANE: THE ANGLE OF REPOSE

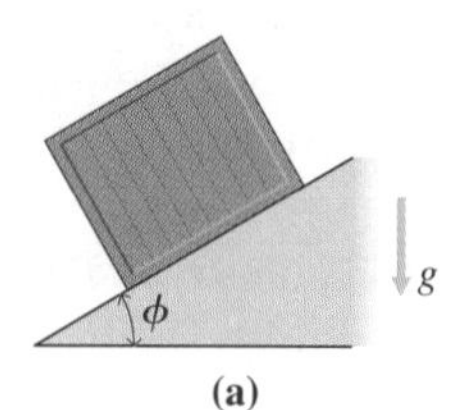

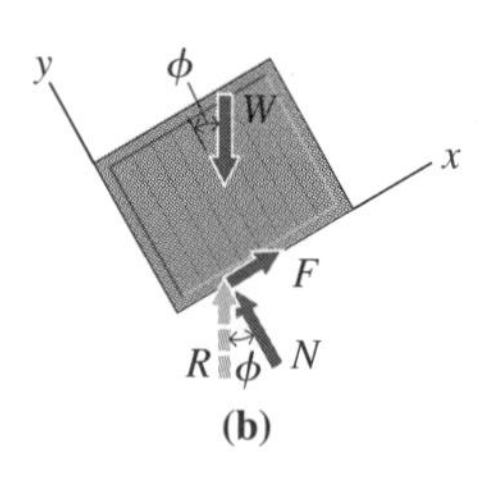

Figure 10.4 (a) Block on inclined plane. (b) Free-body diagram of block.

After studying this section, you should be able to:

- Perform an experiment to determine the coefficient of static friction for two surfaces in contact.

The coefficient of static friction μ_s may be determined experimentally in the following way. Consider a block at rest (in equilibrium) on an inclined plane (see Fig. 10.4a). To express the equations of equilibrium, it is convenient to select *xy* axes parallel and perpendicular to the inclined plane (see Fig. 10.4b). Since the block is in equilibrium, the force **R** (the resultant of the distributed forces exerted on the block by the plane) must be equal in magnitude to, and collinear with, the weight **W** of the block. The force **R** has components of magnitude $F = R\sin\phi$ and $N = R\cos\phi$, tangent and normal to the plane, respectively. With respect to the x and y axes, the equilibrium conditions for the block are

$$\sum F_x = F - W\sin\phi = 0$$

$$\sum F_y = N - W\cos\phi = 0$$

or $F = W\sin\phi$ and $N = W\cos\phi$. Hence, $F/N = \tan\phi$.

Now, let the inclined plane be tipped up slowly until the block is on the verge of sliding. At this point, at the angle $\phi = \phi_s$, the static frictional force F is equal to its maximum value F_s. That is, when $\phi = \phi_s$, $F = F_s$ and $F/N = F_s/N = \tan\phi_s$. Substituting this relation into Eq. (10.2), we obtain

$$\tan\phi_s = \mu_s \tag{10.5}$$

angle of repose

The angle ϕ_s is called the *angle of repose,* since it is the maximum value of ϕ for which the block remains at rest (in repose) on the inclined plane. So, the coefficient of static friction μ_s may be determined experimentally by measuring the angle ϕ_s at which sliding begins.

Analogous to Eq. (10.5), for sliding friction, we can write [see Eq. (10.4)]

$$\tan\phi_k = \mu_k \tag{10.6}$$

A range of experimental values of the coefficients of static friction μ_s and kinetic friction μ_k are given in Table 10.1. For a given combination of materials, the coefficients vary considerably with surface finish and environmental conditions (e.g., ultra-high vacuum conditions such as those that occur in outer space). Slight changes in the experimental conditions may lead to large changes in the coefficients of friction. Other factors also have an effect on the values. For example, the coefficients of friction for leather on wood vary with the type of wood, the direction of the wood grain, and the type of leather. For very soft metals, very large coefficients of friction are possible. For example, the coefficient of static friction for steel on indium may be an order of magnitude higher than that of steel on steel (Bowden & Tabor, 1986, Ch. XV). Consequently, the values listed in Table 10.1 are for illustrative purposes only and must be used with good judgment. At best, they allow relative comparisons between various materials in contact. In particular, to quote a single value for a coefficient of friction without a description of the experimental conditions may be misleading, since the value depends heavily on the experimental conditions.

Table 10.1
Coefficients of dry static and kinetic friction[a]

SURFACES IN CONTACT	μ_s	μ_k	SURFACES IN CONTACT	μ_s	μ_k
Steel on ice	0.04	0.01	Graphite on steel	0.1	
Rubber on ice	0.1	0.05	Graphite on graphite	0.1–0.8	
Waxed hickory on ice		0.03 (at −3°C)	Glass on glass	0.9–1.0	
Cast iron on cast iron	0.4	0.3	Wood on wood	0.4–0.7	0.2–0.5
Brass on steel	0.4–0.5	0.3	Leather on oak	0.5–0.6	0.3–0.5
Copper on copper	0.6–1.2		Leather on iron	0.2–0.4	
Copper on steel	0.6		Rubber on dry concrete	0.6–1.0	0.4–3.0
Steel on cast iron	0.4				
Brake material on cast iron	0.2–0.4		Rubber on other hard materials	0.6–1.0	1.0–4.0[c]
Mild steel on mild steel	0.2–0.5	0.03–0.09[b]	Rope on wood	0.5–0.7	0.5
			Aluminum on aluminum	0.8–1.4	
Hard steel on mild steel	0.42		Indium on indium	2.0–10.0	

a. For an extensive list of coefficients of friction for various environmental conditions, see the appendix of Bowden and Tabor (1986).
b. $\mu_k = 0.03$ at 30 m/s; $\mu_k = 0.09$ at 3 m/s.
c. $\mu_k = 4.0$ for rubber on steel at 100 mm/s (Bowden & Tabor, 1986, p. 169).

CHECKPOINT 

1. ***True or False:*** Friction is a force that exists between two bodies in contact. It is tangent to the contact surface between the bodies and opposes sliding motion between the bodies.
2. ***True or False:*** A frictional force is proportional to the normal force between contacting bodies and to the area of contact between the bodies.
3. ***True or False:*** The maximum static friction is the maximum frictional force that can exist between bodies in contact. It occurs when sliding motion between the bodies is impending.
4. ***True or False:*** The coefficient of static friction is the ratio of the maximum static frictional force to the normal force for bodies in contact.
5. ***True or False:*** The coefficient of kinetic friction is the ratio of the frictional force to the normal force for bodies in contact and sliding relative to one another.
6. ***True or False:*** The kinetic frictional force is always larger than the maximum static frictional force.

Answers: 1. True; 2. False; 3. True; 4. True; 5. True; 6. False

10.3 PROBLEMS INVOLVING FRICTION

After studying this section, you should be able to:

- Solve statically determinate problems that include friction between surfaces in contact.

Problems that involve friction can be classified into three groups:

1. There is no sliding between contacting bodies, and sliding is not impending. Then, the static frictional force F lies in the range $0 \leq F < F_s$ and is just sufficient to keep the bodies in equilibrium.
2. Sliding is impending. Then, the static frictional force is $F = F_s$ [Eq. (10.1)].
3. Sliding occurs. Then, the kinetic frictional force is $F = F_k$ [Eq. (10.3)].

Problems from each of these three groups are demonstrated in the following examples. For these examples and the problems at the end of the chapter, you should read the problem statement carefully in order to decide in which of the three groups the problem falls.

PROBLEM-SOLVING TECHNIQUE

Equilibrium with Friction

The objective of this Problem-Solving Technique is to determine the magnitude and sense of the frictional force that acts between bodies in contact. Here, we consider only two bodies in contact and focus our attention on the free-body diagram of one of them. In more complex situations, more than two bodies may be in contact and several free-body diagrams may be needed, or the problem may be statically indeterminate. In such cases, this strategy may not be sufficient to establish the magnitude and sense of all frictional forces that act on the bodies. Nevertheless, as you gain experience solving problems with friction, you should find this strategy helpful, even in more complex situations.

Problems that involve friction vary widely in form and objective. So, your task usually requires more than a determination of the force of friction between two bodies. You will frequently need the force of friction to complete the solution of a more involved problem.

This book treats statically determinate friction problems. Such problems can be separated into three general categories.

Case I: Sliding is impending. In this case, you know (or are told) that sliding between the two bodies is impending. So you may construct the free-body diagram of one of the bodies. The magnitude $F = F_s$ of the frictional force is given by Eq. (10.1), and the sense of the frictional force is opposite to the sense of impending motion of the body. Hence, you know both the magnitude of the frictional force (in terms of the normal force and the coefficient of static friction) and the sense of the frictional force. Then, you can write the equations of equilibrium for the body and solve for the unknowns, as you would for any other problem involving particle or rigid-body equilibrium.

Case II: Sliding is occurring. In this case, you know from the conditions of the problem that two bodies are sliding relative to one another. Again, you can draw the free-body diagram of one of the bodies. The magnitude $F = F_k$ of the frictional force is given by Eq. (10.3), in terms of the normal force and the coefficient of kinetic friction. The sense of the frictional force is opposite to the sense of sliding of the body. If the body slides with constant velocity, the equations of static equilibrium apply, and you can solve them for the unknown quantities. If the velocity of sliding is not constant, static equilibrium does not exist. Then, Newton's second law applies, and the problem is one of dynamics. Whether or not the body is in equilibrium, the frictional force is given by Eq. (10.3).

Case III: Both the magnitude and the sense of the frictional force are unknown. In this case, you know neither the magnitude nor the sense of the frictional force. Again, you should draw a free-body diagram of one of the bodies, and assume a magnitude F and a sense for the frictional force. Show the frictional force in the free-body diagram of the body as an arrow labeled F, with its head pointed in the assumed sense. Next, assume that static equilibrium exists, and write the equations of equilibrium for the body. Solve the equations of equilibrium for the magnitude F of the frictional force and the other unknowns. Check your solution for the following conditions:

a. If the sign of F is positive (+), your assumption regarding the sense of the frictional force is correct.

b. If the sign of F is negative (−), the assumed sense of the frictional force is incorrect, since the magnitude F of a force must be a positive number. The sense of the frictional force is thus opposite to that shown on the free-body diagram.

c. If $F < F_s$ [given by Eq. (10.1)], the body is in static equilibrium and sliding is not impending. If $F = F_s$, the body is in static equilibrium, but sliding is impending. If the solution shows that the frictional force F required for equilibrium is greater than F_s [given by Eq. (10.1)], the body is sliding. The magnitude of the actual frictional force when the body is sliding is $F = F_k$ [given by Eq. (10.3)].

Example 10.1

Equilibrium with Friction

Problem Statement Two blocks are connected by a light, flexible cable that passes over a frictionless pulley (see Fig. E10.1a). Block A weighs 500 N, and block B weighs 200 N. The coefficient of static friction between block A and the ramp is $\mu_s = 0.30$.

a. Determine whether the two blocks are in equilibrium.

b. Determine whether block A will slide down the ramp after the cable is cut.

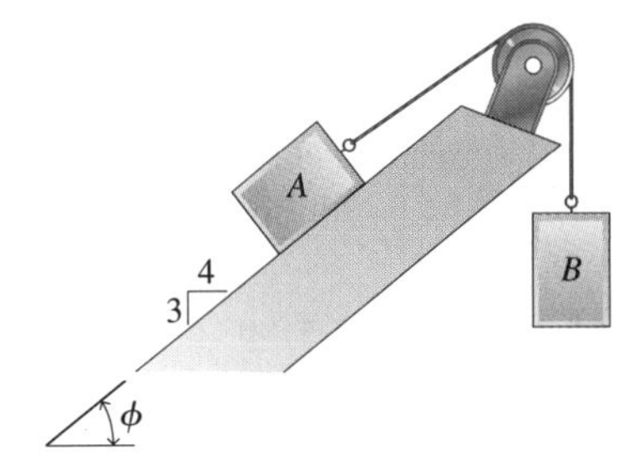

Figure E10.1a

Solution **a.** This problem falls under Case III, since neither the magnitude nor the sense of the frictional force is known. Therefore, we assume that the blocks are in equilibrium, and draw the free-body diagram of block B (see Fig. E10.1b). Summing forces in the y direction, we have $T - 200 = 0$, or $T = 200$ N. Since the sense of the frictional force that acts on block A is not immediately obvious, we draw the free-body diagram of block A, and assume that the frictional force F acts down the ramp (see Fig. E10.1c). Since the pulley is frictionless, the force that the cable exerts on block A is $T = 200$ N. From Fig. E10.1c, the equations of equilibrium for block A are

$$\sum F_x = T - F - 500 \sin \phi = 0 \tag{a}$$

$$\sum F_y = N - 500 \cos \phi = 0 \tag{b}$$

The solutions of Eqs. (a) and (b) are $N = 400$ N and $F = -100$ N. Since the sign of F is negative, the assumed sense of F is wrong. The correct sense of F is up the ramp (see Fig. E10.1d). The blocks are in equilibrium, since $F = 100$ N does not exceed the maximum static frictional force, $F_s = \mu_s N = (0.3)(400) = 120$ N.

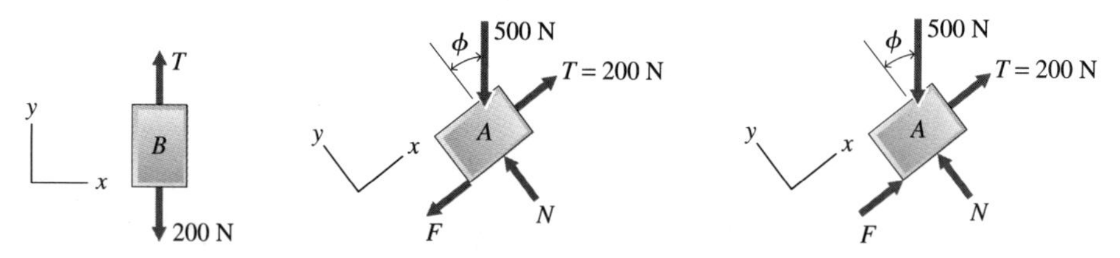

Figure E10.1b **Figure E10.1c** **Figure E10.1d**

b. To determine whether or not block A slides down the ramp after the cable is cut, we solve Eq. (a) for F, with $-F$ replaced by $+F$ and $T = 0$. Then, $F = 300$ N. Since the frictional force ($F = 300$ N) required to keep block A from sliding is greater than the maximum static frictional force ($F_s = 120$ N), block A will slide down the ramp.

Information Item The result of part b could have been obtained by comparing the ramp's angle of inclination ϕ to the angle of repose ϕ_s. From Fig. E10.1a, $\phi = \tan^{-1}(3/4) = 36.87°$. From Eq. (10.5), with $\mu_s = 0.3$, we find $\phi_s = \tan^{-1}(0.3) = 16.70°$. Since $\phi > \phi_s$, the ramp is inclined at an angle greater than the angle of repose. Therefore, block A will slide down the ramp after the cable is cut. ■

Example 10.2

The Angle of Static Friction (Sliding Impending)

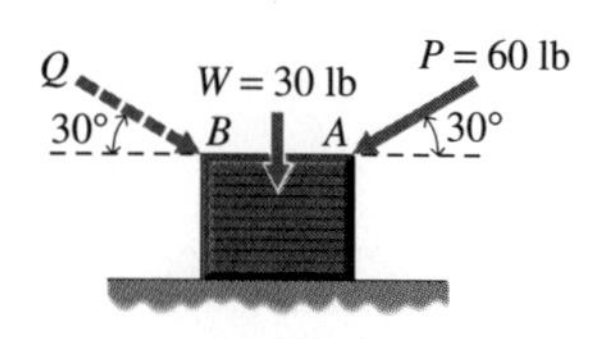

Figure E10.2a

Problem Statement A 30 lb block rests on a horizontal floor and is acted on by a force $P = 60$ lb at point A (see Fig. E10.2a).

a. Assume that sliding is impending and that the block will slide before it will tip. Determine the normal force exerted on the block by the floor and the coefficient of static friction μ_s.

b. Replace the force P with the force $Q = 60$ lb acting at point B (shown dashed in Fig. E10.2a). Repeat part a.

Solution This problem falls under Case I of the Problem-Solving Technique: Motion is impending, and it is obvious that the frictional force must act to the right for part a and to the left for part b in order to oppose the horizontal components of forces P and Q, respectively.

a. Draw the free-body diagram of the block (see Fig. E10.2b). Sliding to the left is impending. Hence, the frictional force opposes the impending motion and its magnitude is $F = F_s = \mu_s N$, where μ_s is the coefficient of static friction and N is the normal force that the floor exerts on the block. The force **R** is the resultant of forces **F** and **N**, whose magnitudes are F_s and N, respectively. The angle ϕ_s is the angle formed by the lines of action of **R** and **N**. Since sliding of the block is impending, the angle ϕ_s is called the *angle of static friction.* From Fig. E10.2b, it is evident that

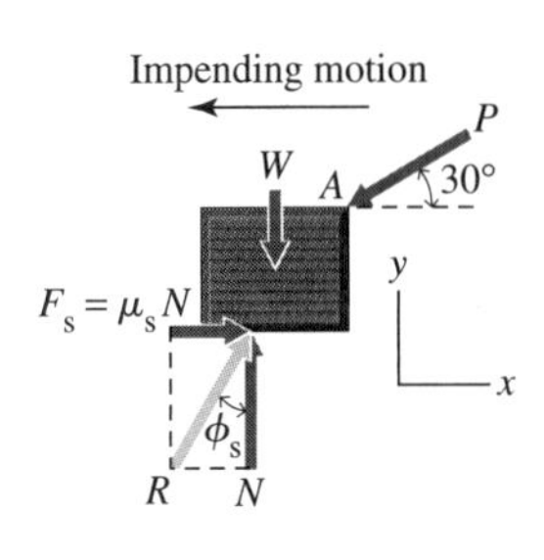

Figure E10.2b

$$F_s = R \sin \phi_s$$
$$N = R \cos \phi_s$$

Hence, $F_s/N = \tan \phi_s = \mu_s$. Comparing this result to Eq. (10.5), we see that the angle of static friction is equal to the corresponding angle of repose (Sec. 10.2).

Equilibrium conditions for the block in the x and y directions are

$$\sum F_x = \mu_s N - P \cos 30° = 0$$
$$\sum F_y = N - W - P \sin 30° = 0$$

Substituting the numerical values for W and P and solving for N and μ_s, we obtain

$$N = 60 \text{ lb}$$
$$\mu_s = 0.866$$

Since $\tan \phi = \mu_s = 0.866$, the angle of static friction for this system (or the angle of repose) is $\phi_s = 40.9°$.

b. For Q acting at point B (refer to Fig. E10.2a), the free-body diagram of the block is shown in Fig. E10.2c. Note that the sense of F_s is now directed along the negative x axis and opposes the impending sliding to the right. As in part a, from Fig. E10.2c, we have

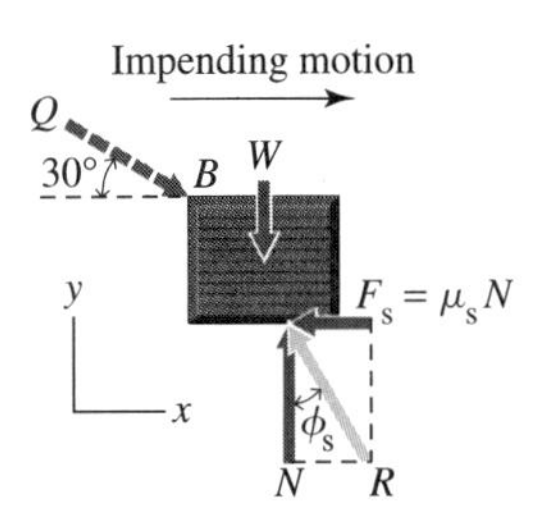

Figure E10.2c

$$F_s = R \sin \phi_s$$
$$N = R \cos \phi_s \tag{a}$$

By Eqs. (a), the coefficient of static friction μ_s is

$$\mu_s = \tan \phi_s = \frac{F_s}{N} \tag{b}$$

Also, by the equilibrium conditions for the block,

$$\sum F_x = Q \cos 30° - \mu_s N = 0$$
$$\sum F_y = N - W - Q \sin 30° = 0 \tag{c}$$

The solutions of Eqs. (c) are, as in part a,

$$N = 60 \text{ lb}$$
$$\mu_s = 0.866$$

Therefore, from Eq. (b), $\mu_s = \tan \phi_s = 0.866$, or $\phi_s = 40.9°$. Thus, the line of action of the force **R** swings through an arc subtending an angle of $2\phi_s = 81.8°$, from the case where motion is impending to the left to the case where motion is impending to the right. In two-dimensional problems, the resultant **R** of the frictional force **F** and the normal force **N** is always located within a plane circular sector of angle $2\phi_s$ (see Fig. E10.2d). In three-dimensional problems, the force **R** is always located within a cone of angle $2\phi_s$, called the *cone of friction* (see Fig. E10.2e).

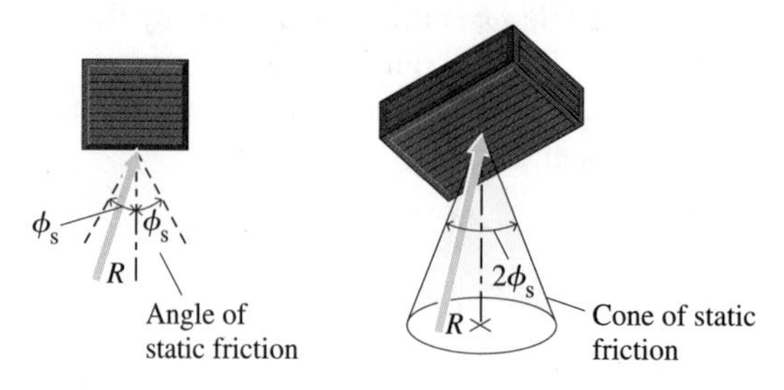

Figure E10.2d **Figure E10.2e**

Example 10.3

Equilibrium of a Shipping Crate

Problem Statement A large rectangular shipping crate of height h and width b is at rest on a floor. It is acted on by a horizontal force P, as illustrated in Fig. E10.3a. Assume that the material in the crate is uniformly distributed so that the weight acts at the centroid of the crate.

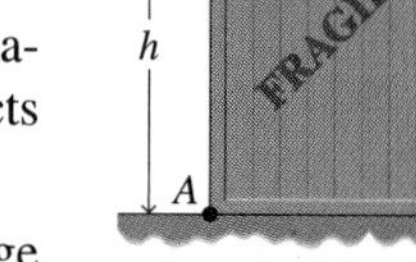

Figure E10.3a

a. Determine the conditions for which the crate is on the verge of sliding.

b. Determine the conditions under which the crate will tip about point A.

Solution Before we consider whether the crate will tip or slide, we consider the normal reaction of the floor on the crate, and draw a general free-body diagram of the crate (see Fig. E10.3b). We assume that the normal reaction is distributed over the bottom of the crate, as indicated by the dashed arrows. The resultant of these distributed parallel normal forces (see Sec. 8.7) is a normal force N that acts at some distance x from point A. For both parts a and b, we know that the force of friction F acts to the right, because it must oppose P. Under these general conditions, the equilibrium equations for the crate are

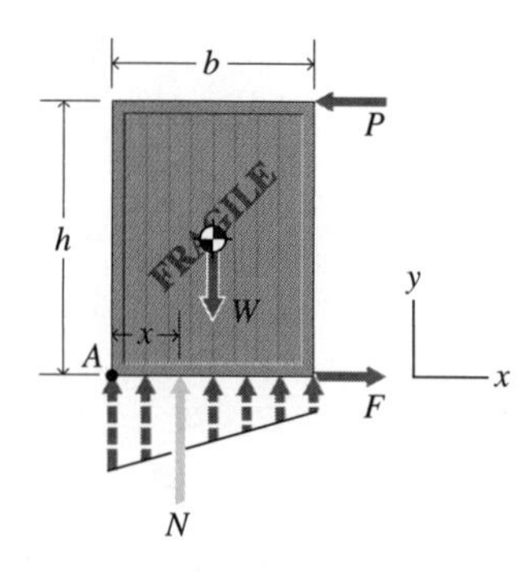

Figure E10.3b

$$\sum F_x = F - P = 0 \tag{a}$$

$$\sum F_y = N - W = 0 \tag{b}$$

$$\sum M_A = Nx - (0.5)(Wb) + Ph = 0 \tag{c}$$

Consequently, if the crate is in equilibrium, Eqs. (a), (b), and (c) yield

$$F = P \tag{d}$$

$$N = W \tag{e}$$

$$x = \frac{b}{2} - \frac{Ph}{W} \tag{f}$$

a. If the crate is on the verge of sliding, $F = F_s = \mu_s N$, where μ_s is the coefficient of static friction. By Eqs. (d) and (e), the value of P corresponding to this condition is

$$P_{\text{sliding}} = F_s = \mu_s W \tag{g}$$

b. If the crate is on the verge of tipping, it is on the verge of rotating about point A; that is, the crate and the floor are in contact only at point A. Consequently, the normal force is concentrated at A, and $x = 0$ (see Fig. E10.3b). For $x = 0$, Eq. (f) yields

$$P_{\text{tipping}} = \frac{Wb}{2h} \tag{h}$$

Note that tipping will occur before sliding, provided that $P_{\text{sliding}} > P_{\text{tipping}}$. Therefore, by Eqs. (g) and (h), if P increases until motion occurs, tipping will occur before sliding, provided that

$$\mu_s > \frac{b}{2h}$$

On the other hand, sliding will occur without tipping if $\mu_s < b/2h$. For $\mu_s = b/2h$, tipping and sliding occur simultaneously.

Example 10.4

Block on an Inclined Plane

Problem Statement The friction of a block that moves on an inclined plane is utilized in various machines. To develop the general theory of this phenomenon, let's consider a block of weight W that is pulled up an inclined plane by a force P that forms an angle θ with the plane. The plane is inclined at an angle ϕ to the horizontal (see Fig. E10.4a).

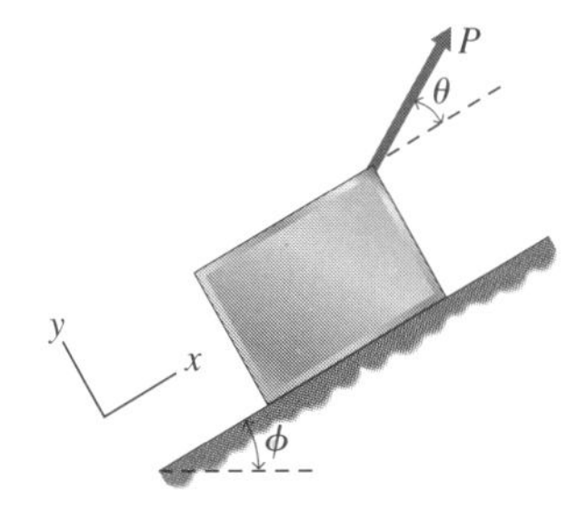

Figure E10.4a

a. Determine the force P, in terms of W and angles θ and ϕ, that will cause sliding of the block up the plane.

b. Determine the value of ϕ for which the block will be on the verge of sliding down the plane under the action of its own weight.

Solution In both parts a and b, the sense of the frictional force is known by inspection (Case I of the Problem-Solving Technique).

a. The free-body diagram of the block is shown in Fig. E10.4b. With respect to the xy axes in Fig. E10.4b, the equilibrium equations for the block are

$$\begin{aligned} \sum F_x &= P\cos\theta - W\sin\phi - F = 0 \\ \sum F_y &= P\sin\theta + N - W\cos\phi = 0 \end{aligned} \tag{a}$$

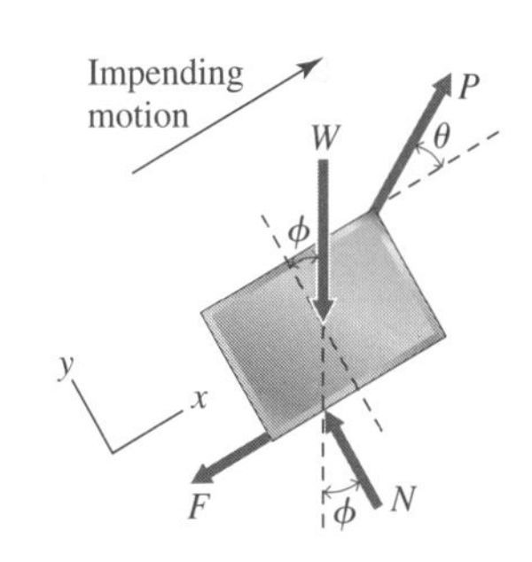

Figure E10.4b

When sliding is impending, the frictional force is $F = F_s = \mu_s N$. Substituting this relation into Eqs. (a) and solving for N and P, we obtain

$$P = W\left[\frac{\sin\phi + \mu_s\cos\phi}{\cos\theta + \mu_s\sin\theta}\right] \tag{b}$$

$$N = W\left[\frac{\cos(\theta + \phi)}{\cos\theta + \mu_s\sin\theta}\right] \tag{c}$$

Introducing the angle of static friction ϕ_s, defined by $\tan\phi_s = \mu_s$ [see Example 10.2, Eq. (b)], we can express Eqs. (b) and (c) in alternative forms, as follows:

$$P = W\left[\frac{\sin(\phi + \phi_s)}{\cos(\phi_s - \theta)}\right] \tag{d}$$

$$N = W\left[\frac{\cos\phi_s\cos(\theta + \phi)}{\cos(\phi_s - \theta)}\right] \tag{e}$$

Equation (b) or (d) defines the force P that will cause impending sliding of the block up the plane at an angle θ. Note that if $\phi = 0$, the plane on which the block slides is horizontal, and impending sliding of the block is to the right.

b. To determine the value of ϕ at which the block will slide under the action of its own weight, we set $P = 0$ in Eq. (b). Then, $\tan\phi = -\mu_s$. Note that the minus sign is due to the fact that, in part a, motion impends up the plane and $F = F_s$ is directed down the plane (refer to Fig. E10.4b), but here motion impends down the plane and $F = F_s$ is directed up the plane. In other words, in either case, $F = F_s$ opposes the impending motion. This conclusion confirms the result of Sec. 10.2. ■

Example 10.5

Wedge and Block

Problem Statement A wedge A and a block B are subject to a known load Q and a force P, as shown in Fig. E10.5a. If the force P is sufficiently large, the block is raised; if it is small, the block may be lowered. The individual weights of the wedge and the block are negligible compared to Q. The coefficients of static friction for surfaces 1, 2, and 3 are 0.1, 0.2, 0.3, respectively. The angle of the wedge is 30°. Determine the range of the force P, in terms of Q, for which there is no motion.

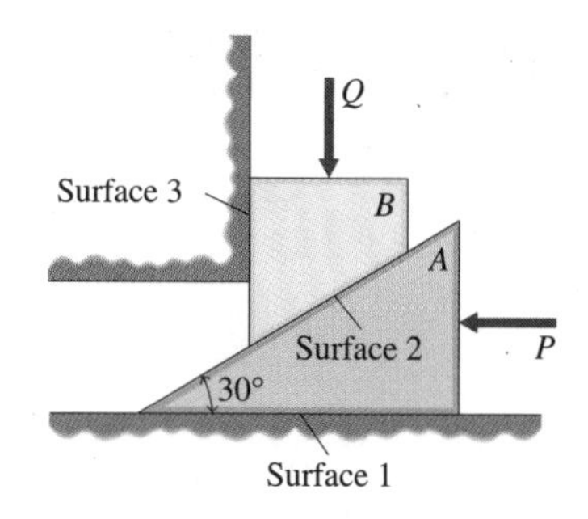

Figure E10.5a

Solution Since the sense of the impending motion of the wedge, either to the left or to the right, is known, the frictional force has the opposite sense; that is, the sense of the frictional force is known (Case I of the Problem-Solving Technique).

First, we consider the case in which the block tends to move upward; that is, motion of the block is impending upward, and motion of the wedge is impending to the left. Figure E10.5b is the free-body diagram of the block and wedge, considered as a unit. As noted in the problem statement, the weights of the bodies are negligible compared to the given

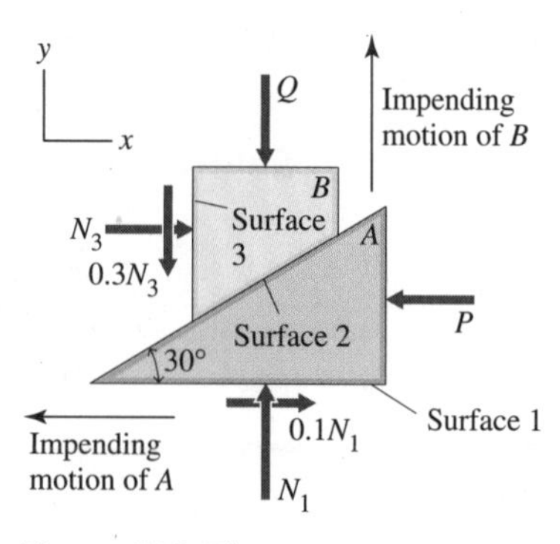

Figure E10.5b

load Q. Hence, they are ignored in Fig. E10.5b. The frictional forces at surfaces 1 and 3 are directed so as to resist the impending movement. The magnitudes of these forces are the maximum static frictional values. The equations of equilibrium of forces, from Fig. E10.5b, are

$$\sum F_x = N_3 + 0.1N_1 - P = 0 \tag{a}$$

$$\sum F_y = N_1 - 0.3N_3 - Q = 0 \tag{b}$$

Since these two equations contain three unknowns, N_1, N_3, and P, additional equations are needed. We can obtain them from the free-body diagram of the wedge (see Fig. E10.5c). (Alternatively, we could use the free-body diagram of the block.) Note that the frictional force $0.2N_2$ that acts on the inclined face of the wedge opposes the impending sliding. From Fig. E10.5c,

$$\sum F_x = 0.1N_1 + 0.2N_2 \cos 30° + N_2 \sin 30° - P = 0 \tag{c}$$

$$\sum F_y = N_1 + 0.2N_2 \sin 30° - N_2 \cos 30° = 0 \tag{d}$$

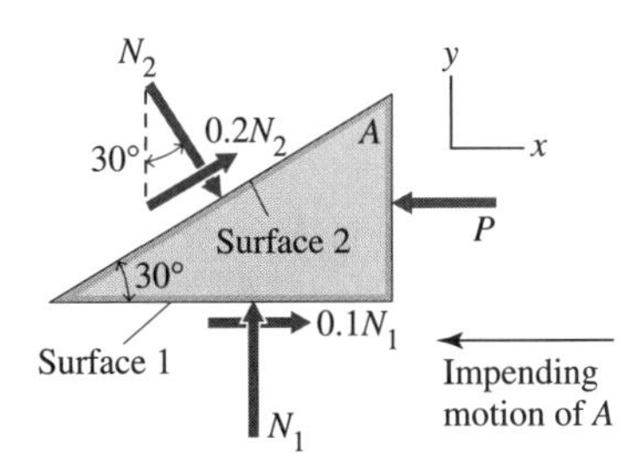

Figure E10.5c

The four equations, Eqs. (a), (b), (c), and (d), contain the four unknowns: N_1, N_2, N_3, and P. Solving for P in terms of Q, we obtain

$$P = 1.329Q \tag{e}$$

If the block B is on the verge of moving downward,

$$P = 0.2163Q \tag{f}$$

(Development of this relation is left to you as Problem 10.41. You should verify this result, noting that, in this case, the senses of the frictional forces shown in Figs. E10.5b and E10.5c are reversed.)

Hence, by Eqs. (e) and (f), the range of values that the force P may take without resulting motion is

$$0.2163Q \leq P \leq 1.329Q$$

If the coefficients of friction were somewhat larger, the lower bound for P would be zero or negative. Then the system would be *self-locking*. In order for block B in Fig. E10.5a to move downward, P would have to be directed to the right (to pull the wedge to the right). Self-locking wedges are used in various kinds of self-adjusting devices—for example, devices that automatically eliminate play caused by wear.

In this example, it was necessary to consider two free-body diagrams (Figs. E10.5b and E10.5c) because the two equations of equilibrium for either body contain three unknowns [see Eqs. (a) and (b) or Eqs. (c) and (d)]. ■

Example 10.6 Sliding Boxes

Problem Statement The boxes shown in Fig. E10.6a are connected by a light flexible cord that passes over a frictionless pulley. A pull T is exerted on box B. Box A weighs 600 N and box B weighs 900 N. The coefficients of static and kinetic friction between the bodies and the planes are $\mu_s = 0.5$ and $\mu_k = 0.2$, respectively.

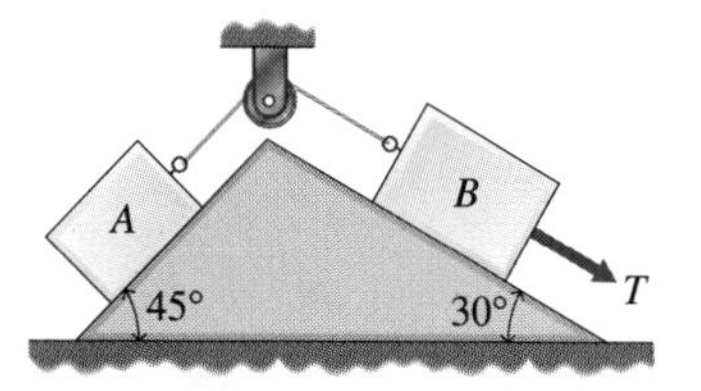

Figure E10.6a

a. Determine the force $T = T_s$ required to produce impending motion of box B down the plane.

b. Assume that sufficient force is applied to box B to start the boxes sliding. Determine the force $T = T_k$ required to keep the boxes moving at a constant speed.

Solution **a.** Since the sense of impending motion of each of the boxes is known, the sense of each frictional force that acts on the boxes is known; it is opposite to the sense of impending motion (Case I).

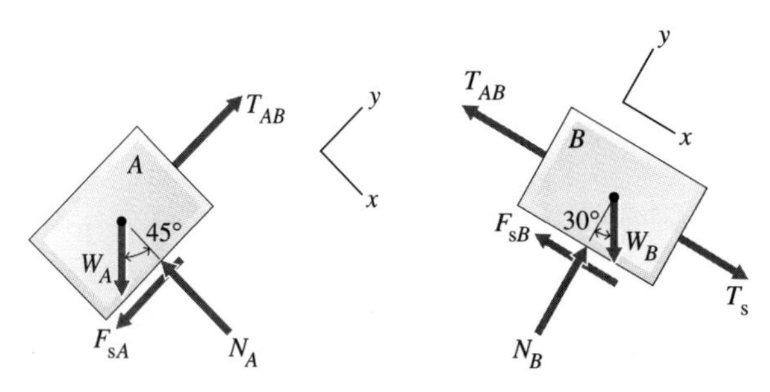

Figure E10.6b **Figure E10.6c**

For impending motion, the free-body diagrams of boxes A and B are shown in Figs. E10.6b and E10.6c, respectively. From Fig. E10.6b, summation of forces yields

$$\sum F_x = W_A \cos 45^\circ - N_A = 0$$

$$N_A = \frac{W_A}{\sqrt{2}}$$

$$\sum F_y = T_{AB} - F_{sA} - W_A \sin 45^\circ = 0$$

$$T_{AB} = F_{sA} + \frac{W_A}{\sqrt{2}}$$

Therefore,

$$F_{sA} = \mu_s N_A = \mu_s \frac{W_A}{\sqrt{2}}$$

$$T_{AB} = (1 + \mu_s)\frac{W_A}{\sqrt{2}} \tag{a}$$

Similarly, from Fig. E10.6c, we have

$$\sum F_x = T_s + W_B \sin 30^\circ - F_{sB} - T_{AB} = 0 \tag{b}$$

$$\sum F_y = N_B - W_B \cos 30^\circ = 0$$

$$N_B = W_B \frac{\sqrt{3}}{2}$$

Therefore,

$$F_{sB} = \mu_s N_B = \mu_s W_B \frac{\sqrt{3}}{2} \tag{c}$$

Then, by Eqs. (a), (b), and (c), we find

$$T_s = \frac{W_A}{\sqrt{2}}(1 + \mu_s) - \frac{W_B}{2}(1 - \sqrt{3}\mu_s) \tag{d}$$

Inserting numerical values into Eq. (d), we obtain the required pull for impending motion:

$$T_s = 576 \text{ N}$$

b. As the boxes slide with constant speed (Case II), equilibrium conditions hold. Hence, the analysis of part a is valid for sliding, except that μ_s in Eq. (d) must be replaced by μ_k. Substitution of the value $\mu_k = 0.2$, with the other values, into Eq. (d) yields the required pull for sliding with constant speed:

$$T_k = 215 \text{ N}$$

Some common features of the above examples are

- For each body, a free-body diagram is drawn. If the sense of the frictional force is known (Case I or II), the frictional force is shown with its correct sense in the diagram.
- If the sense of a frictional force is unknown (Case III), the sense of the frictional force is assumed.
- From the free-body diagram, the equations of equilibrium for each body are written and solved for the unknowns.
- The unknowns may include one or more of the following quantities: the frictional forces, the force that causes impending motion or sliding motion of a body, the minimum coefficient of static friction between two bodies for which sliding is impending, the angle of static friction of a ramp (inclined plane), and the tension in a cable or cord connecting two bodies.

In addition to noting the common features cited above, you should be careful to verify that

- The sense of impending motion or sliding is correct.
- The sense of the frictional force that acts on a body is opposite to the sense of impending motion or sliding.
- If a body is at rest, the magnitude F of the frictional force that acts on it is less than or equal to the maximum static frictional force F_s [see Eq. (10.1)].
- If sliding of a body is impending, the magnitude F of the frictional force that acts on the body is equal to the maximum static frictional force F_s [see Eq. (10.1)].
- If a body slides on another body, the magnitude F of the frictional force is equal to the kinetic frictional force F_k [see Eq. (10.3)].

10.4 SIMPLE MACHINES AND FRICTION

After studying this section, you should be able to:

- Analyze the behavior of simple machines that utilize the effects of friction.

THE WEDGE

Key Concept A wedge is used to raise or lower a load or to separate two parts of a system.

wedge A *wedge* is essentially a solid inclined plane. Ordinarily, a wedge is moved to raise or lower a load or a follower or to separate two parts of a system. A classical application of the wedge is for splitting a log or a block of wood. The principle of the wedge is also applied in common tools such as the axe, the plow, the chisel, and the air hammer.

mechanical advantage The theoretical *mechanical advantage* m of the wedge is defined as the ratio of its length b to its breadth a (see Fig. 10.5); that is, $m = b/a$. Hence, theoretically, a 75 lb force delivered to the head of a wedge for which $b = 6$ in and $a = 2$ in should produce a lateral splitting force of 225 lb in a wood block. However, because of friction, the actual mechanical advantage is much less.

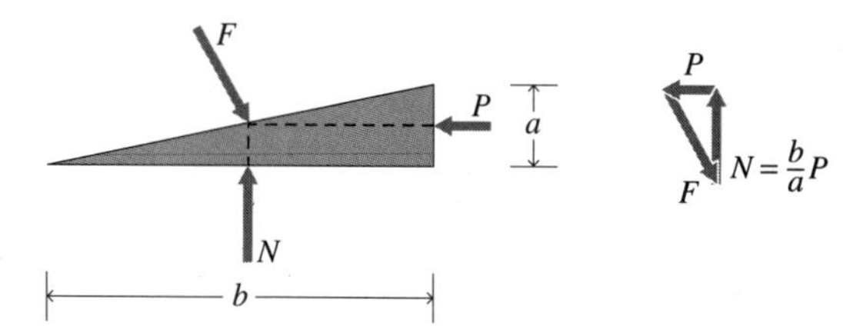

Figure 10.5 Mechanical advantage of a wedge: $m = b/a$.

THE SCREW OR SCREW THREAD

Key Concept The screw thread behaves as an inclined plane wrapped into the form of a cylinder.

If you cut out the profile of an inclined plane and wrap it into the form of a cylinder, you have the basis of the machine screw (see Fig. 10.6). The inclined edge of the plane forms a helix around the cylinder (see Fig. 10.7). The helix angle is a measure of the slope of the inclined plane.

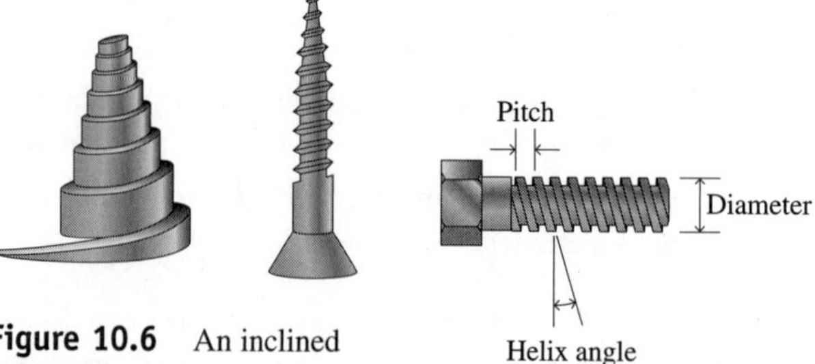

Figure 10.6 An inclined plane wrapped into the form of a cylinder is the basis of the screw.

Figure 10.7 Spiral helix of a machine screw.

screw threads *Screw threads* are used extensively in mechanical systems to convert a rotational motion into a translational motion. For example, a machine screw and nut form a device in which rotation of the screw causes an axial translation of the nut relative to the screw (see Fig. 10.8a). Many other mechanical devices use the principle of the wedge to perform their functions—for example, the rack and pinion, the jackscrew, and the C-clamp. In these examples, the screw thread is basically a wedge that has been wrapped around a cylindrical rod. In these simple machines, the wedge is called the *driver* and the object that is lifted is called the *follower* (see Fig. 10.8b).

driver

follower

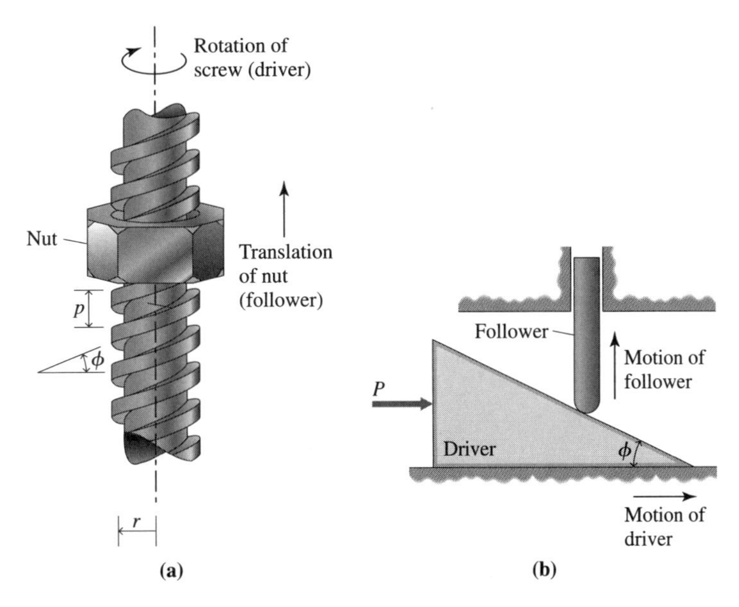

Figure 10.8

pitch angle

pitch

lead

For the screw thread and nut mechanism shown in Fig. 10.8a, the analogous wedge and follower are shown in Fig. 10.8b. The helix angle ϕ of the wedge is called the *pitch angle.* The *pitch* of the screw is defined to be the axial distance p between corresponding points on adjacent threads. The *lead l,* not shown in Fig. 10.8a, is the distance that the nut travels parallel to the screw axis when the nut is given a 360° rotation. For a single-threaded screw, as shown in Figs. 10.7 and 10.8a, the lead is the same as the pitch. A double-threaded screw has a lead twice the pitch, and a triple-threaded screw has a lead three times the pitch, and so on (Shigley & Mischke, 1989). We consider only single-threaded screws. If a single-threaded screw has n threads per unit length of the screw (see Fig. 10.7 or 10.8a), the pitch is $p = 1/n$.

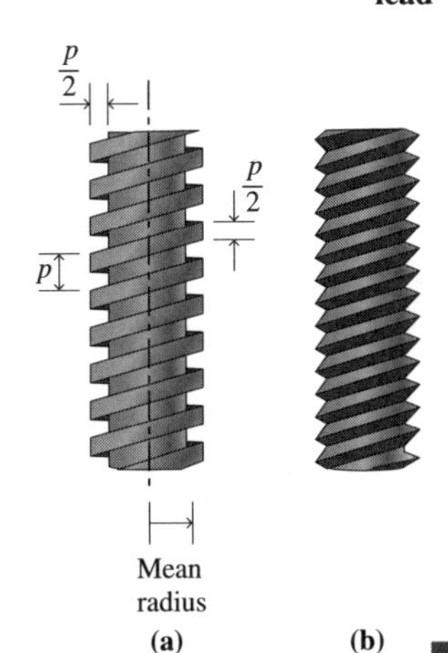

Figure 10.9 Screw profiles: **(a)** square thread; **(b)** V-thread.

Figure 10.9 illustrates two common kinds of screw threads—namely, the square thread and the V-thread. We confine our attention to the square thread. The surface of a square thread is an inclined plane on which the load is carried. Consequently, the square thread is more easily modeled as a planar wedge. A V-thread is less efficient than a square thread because it requires a greater normal force to carry the same load. However, a square thread is more difficult to manufacture (Orthwein, 1990).

THE JACKSCREW

> **Key Concept**
>
> The jackscrew applies the principles of the lever and the screw in a machine for raising and lowering large loads.

jackscrew The *jackscrew,* or power screw (see Fig. 10.10a), serves to illustrate the theory of the single-pitch screw. Jackscrews are used to raise, lower, or support large loads. The screw threads of jackscrews are usually of the square type. The load P that is supported by the screw is raised or lowered by means of a force R applied normal to a lever to rotate the screw. One full turn of the lever raises or lowers the load a distance equal to the pitch of the screw (see Fig. 10.10b). The collar separates the load from the screw head (Fig. 10.10a). We assume that the surface between the collar and the head is well lubricated so that the collar does not rotate and the head rotates freely (without friction) relative to the collar.

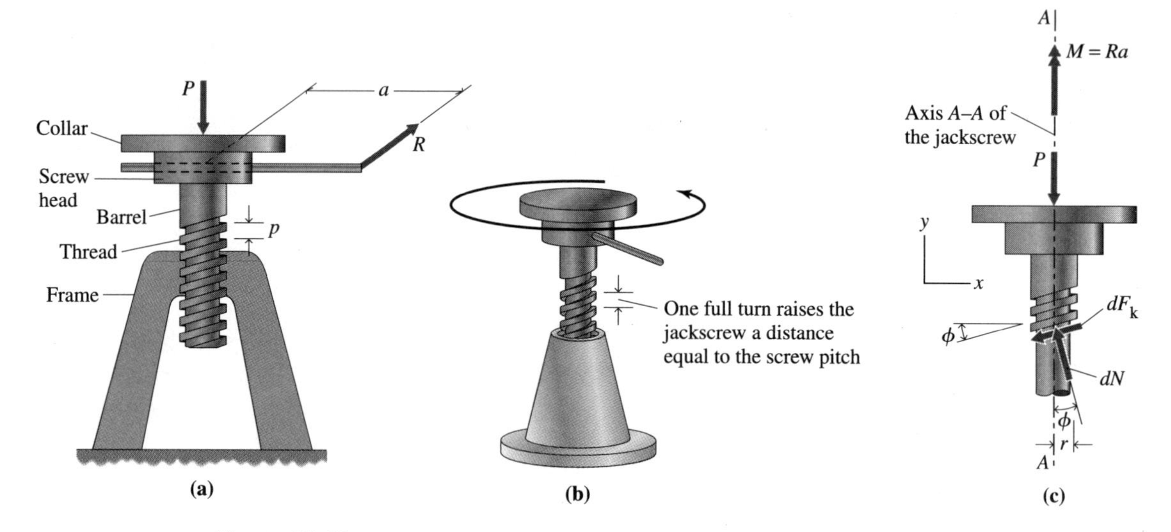

Figure 10.10 Jackscrew.

Consider the case in which the load P is raised. Figure 10.10c shows the normal force dN and the kinetic frictional force dF_k that act on an infinitesimal length of the thread. It is assumed that these forces act at the mean radius r of the screw thread. Suppose that the load is being raised at a constant speed. Then, by Newton's first law, equilibrium conditions exist, and, from Fig. 10.10c, we have

$$\begin{aligned} \sum F_y &= -\sin\phi \int dF_k + \cos\phi \int dN - P = 0 \\ \sum M_{A\text{–}A} &= -r\cos\phi \int dF_k - r\sin\phi \int dN + M = 0 \end{aligned} \tag{a}$$

where $M = Ra$ is the counterclockwise turning moment about axis A–A applied by the lever. The integrals must be taken over the length of the screw thread in contact with the frame (Fig. 10.10a). Now, from Eq. (10.3), $dF_k = \mu_k dN,$ where μ_k is the coefficient of sliding friction. Introducing this relation into Eqs. (a), eliminating $\int dN$ algebraically, and solving for $M,$ we obtain

$$M = Pr\left(\frac{\mu_k\cos\phi + \sin\phi}{\cos\phi - \mu_k\sin\phi}\right) = Pr\left(\frac{\mu_k + \tan\phi}{1 - \mu_k\tan\phi}\right) \tag{b}$$

Noting that $\mu_k = \tan\phi_k$ [see Eq. (10.6)], we may express Eq. (b) in the form

$$M = Pr\left(\frac{\tan\phi_k + \tan\phi}{1 - \tan\phi_k\tan\phi}\right) \tag{c}$$

Using the trigonometric formula for the tangent of the sum of two angles (see Appendix B), we may write Eq. (c) in the form

$$M = Pr\tan(\phi_k + \phi) \tag{10.7}$$

mechanical advantage of jackscrew

The ratio $m = P/R$ is the *mechanical advantage of the jackscrew.* Since $M = Ra$ when the load is raised, Eq. (10.7) yields

$$m = \frac{P}{R} = \frac{a}{r}\cot(\phi_k + \phi) \tag{10.8}$$

If the load P is lowered, the sense of the force R is reversed (refer to Fig. 10.10a). Then, the sense of the moment M is clockwise and the sense of the frictional force dF_k is reversed from that shown in Fig. 10.10c. In this case, we obtain

$$\begin{aligned} M &= Pr\left(\frac{\mu_k \cos\phi - \sin\phi}{\cos\phi + \mu_k \sin\phi}\right) \\ &= Pr\left(\frac{\mu_k - \tan\phi}{1 + \mu_k \tan\phi}\right) \\ &= Pr\left(\frac{\tan\phi_k - \tan\phi}{1 + \tan\phi_k \tan\phi}\right) \end{aligned}$$

Using the trigonometric formula for the tangent of the difference of two angles, we have

$$M = Pr\tan(\phi_k - \phi) \tag{10.9}$$

Since the applied turning moment is $M = Ra$, Eq. (10.9) yields the mechanical advantage for lowering the load P as

$$m = \frac{P}{R} = \frac{a}{r}\cot(\phi_k - \phi) \tag{10.10}$$

self-locking screw

Also, Eq. (10.9) shows that if $\phi_k > \phi$, the clockwise turning moment is positive. Hence, a clockwise moment must be applied to lower the load. Then the screw is *self-locking;* that is, it can sustain any load P without the need to apply a counterclockwise moment or a brake to prevent it from turning. If $\phi_k < \phi$, a counterclockwise turning moment (or a brake) is required to keep the load from moving downward under its own action.

Friction on the screw of this hand-operated car jack makes the jack self-locking. It does not lower automatically when the operator releases the crank.

The foregoing representation of a jackscrew is somewhat simplified with respect to how the lifting (or lowering) moment M is applied. Often, one or more screws are combined with a series of levers or rods to form a more efficient and more easily operated mechanism.

In designing a jackscrew or a power screw, it is useful to compare the efficiencies of various systems. For example, consider the case of raising a load by a jackscrew (refer to Fig. 10.10a). The efficiency η of a screw is calculated as the ratio of the required moment without friction to the required moment with friction. For a frictionless system $\mu_k = \tan \phi_k = 0$, which yields $\phi_k = 0$. Then, Eq. (10.7) reduces to

$$M = Pr \tan \phi \tag{10.11}$$

and the efficiency is, by Eqs. (10.7) and (10.11),

$$\eta = \frac{\tan \phi}{\tan(\phi_k + \phi)} \tag{10.12}$$

Example 10.7

The Jackscrew

Problem Statement A single-threaded jackscrew with a square thread (refer to Fig. 10.10a) has a mean diameter (pitch diameter) $d_m = 2r = 30$ mm and a pitch $p = 4$ mm. The coefficient of kinetic friction is $\mu_k = 0.08$.

a. Determine the moment required to raise a load $P = 10$ kN.

b. Determine the efficiency of the jackscrew.

Solution **a.** Since $\mu_k = \tan \phi_k = 0.08$, $\phi_k = 4.57°$. To determine the pitch angle ϕ, we imagine that the screw thread is rolled out (or developed) for exactly one turn (see Fig. E10.7). Then, $\tan \phi = p/\pi d_m = 4/(30\pi) = 0.0424$, or $\phi = 2.43°$. Now, from Eq. (10.7),

$$M = Pr \tan(\phi_k + \phi) = 10(15)[\tan(4.57° + 2.43°)] = 18.42 \text{ N·m} \tag{a}$$

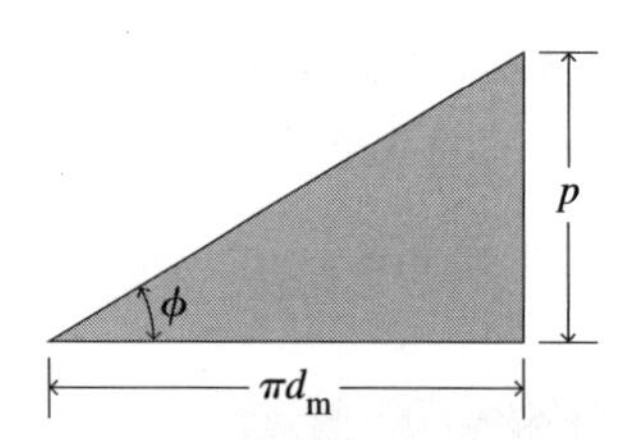

Figure E10.7

b. For $\mu_k = 0$, Eq. (10.11) yields

$$M = Pr \tan \phi = 10(15)(0.0424) = 6.36 \text{ N·m} \tag{b}$$

From Eqs. (10.12), (a), and (b), the efficiency is

$$\eta = \frac{6.36}{18.42} = 0.345 \quad \text{(or 35\%)}$$

Example 10.8

The Turnbuckle

Problem Statement A *turnbuckle* is a mechanical device that is used to increase or decrease the tension in cables or rods (see Fig. E10.8). A turnbuckle is threaded at both ends; the threads are right-handed at one end and left-handed at the other. Matching rods are screwed into the turnbuckle. Hence, when the turnbuckle is rotated, it draws the threaded ends A and B of the rods together or moves them apart.

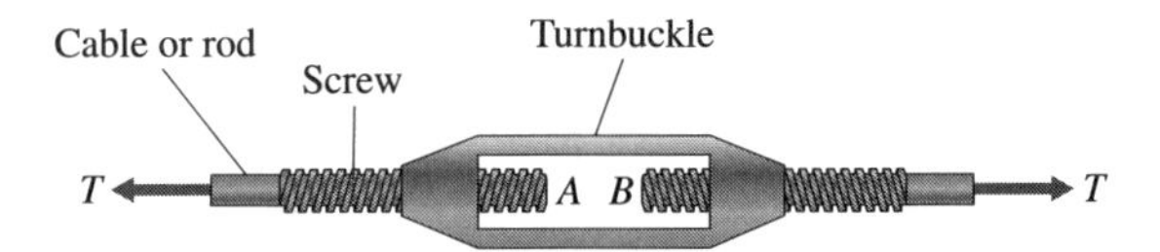

Figure E10.8

Suppose that a turnbuckle has single-threaded square screws with a pitch $p = 8$ mm and a mean radius $r = 18$ mm. Also suppose that the coefficient of static friction between the turnbuckle and the rods is $\mu_s = 0.20$. If the tension in the cable is 45 kN and the rods do not rotate when the turnbuckle is turned, determine

a. the couple that must be applied to the turnbuckle to start to increase the tension in the rods

b. the couple that must be applied to the turnbuckle to start to reduce the tension in the rods

Solution **a.** When the tension in the rods is increased, the ends A and B of the rods move toward each other, in effect raising a load in an equivalent jackscrew. To start to increase the tension, static friction must be overcome. Therefore, Eq. (10.7) applies, provided that the coefficient of kinetic friction μ_k is replaced by the coefficient of static friction μ_s. So

$$\tan \phi_s = \mu_s = 0.2$$

$$\phi_s = 0.1974 \text{ rad}$$

Also (see Fig. E10.7),

$$\tan \phi = \frac{p}{2\pi r} = \frac{8}{2\pi(18)} = 0.0707$$

$$\phi = 0.0706 \text{ rad}$$

With ϕ_s and ϕ determined, Eq. (10.7) yields

$$M = Pr \tan(\phi_s + \phi) = (45\,000)(0.018)[\tan(0.1974 + 0.0706)]$$
$$= 222 \text{ N·m.}$$

b. When the tension in the rods is decreased, Eq. (10.9) applies. Then,

$$M = Pr \tan(\phi_s - \phi) = (45\,000)(0.018)[\tan(0.1974 - 0.0706)]$$
$$= 103 \text{ N·m}$$

Example 10.9

The Compound Jackscrew

Problem Statement A weight W is lifted at a constant rate by applying a couple Rd to the compound jackscrew shown in Fig. E10.9a.

a. Derive a formula for R in terms of the angle θ, the weight W, the length d of the lever rod, the pitch p of the screw threads, the mean radius r of the screw threads, and the coefficient of kinetic friction μ_k of the screw. The friction of the vertical wall guides is negligible.

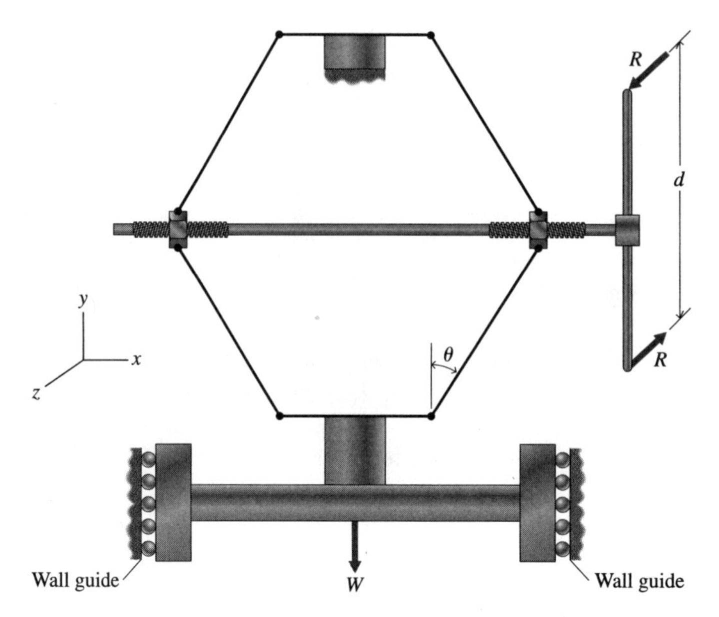

Figure E10.9a

b. For $\theta = 15°$, $W = 1000$ lb, $d = 20$ in, $r = 0.75$ in, $p = 0.333$ in, and $\mu_k = 0.15$, calculate the value of R.

Solution **a.** First, we draw a free-body diagram of the weight and the lower bars of the mechanism (see Fig. E10.9b). From Fig. E10.9b, $\Sigma F_y = 2T\cos\theta - W = 0$, or

$$W = 2T\cos\theta \tag{a}$$

Next, we draw a free-body diagram of the left nut (see Fig. E10.9c). From Fig. E10.9c, $\Sigma F_y = Q\cos\theta - T\cos\theta = 0$, or $Q = T$. Therefore, $\Sigma F_x = 2T\sin\theta - P = 0$, or

$$P = 2T\sin\theta \tag{b}$$

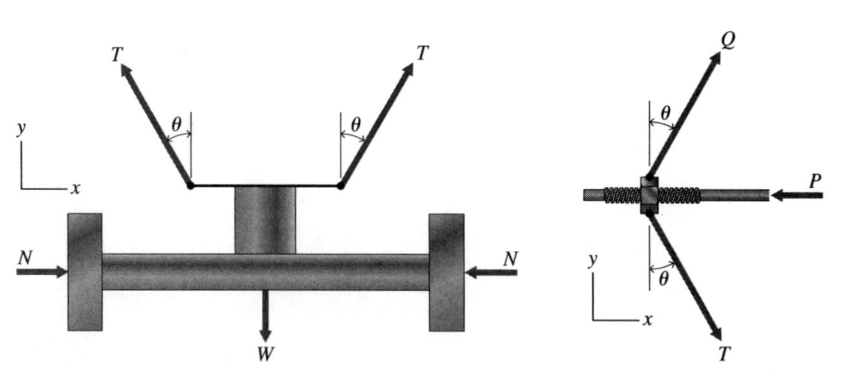

Figure E10.9b **Figure E10.9c**

Solving Eqs. (a) and (b), we obtain

$$P = W\tan\theta \tag{c}$$

Equation (c) also applies to the right nut. By symmetry, the moment applied to each nut is $M/2 = Rd/2$. Consequently, from Eqs. (c) and (10.7),

$$M = Rd = 2Pr\tan(\phi_k + \phi) = 2(W\tan\theta)r[\tan(\phi_k + \phi)] \tag{d}$$

where $\phi_k = \tan^{-1}\mu_k$ and $\phi = \tan^{-1}[p/(2\pi r)]$.

b. For $\theta = 15°$, $W = 1000$ lb, $d = 20$ in, $r = 0.75$ in, $p = 0.333$ in, and $\mu_k = 0.15$, we have

$$\tan\theta = \tan 15° = 0.2679$$

$$\phi_k = \tan^{-1}(0.15) = 0.1489 \text{ rad}$$

$$\phi = \tan^{-1}\left(\frac{p}{2\pi r}\right) = \tan^{-1}(0.07066) = 0.0705 \text{ rad}$$

$$\tan(\phi_k + \phi) = \tan(0.2196) = 0.223$$

Then, Eq. (d) yields

$$20R = 2(1000)(0.268)(0.75)(0.223)$$

$$R = 4.48 \text{ lb}$$

CHECKPOINT

1. ***True or False:*** A screw thread is effectively a wedge.
2. ***True or False:*** A V-thread is more efficient than a square thread.
3. ***True or False:*** A jackscrew consists of two parts—a driver and a follower.
4. ***True or False:*** If a jackscrew's angle of friction is less than its pitch angle, the jackscrew is self-locking.
5. ***True or False:*** For a single-threaded screw, the pitch is equal to the lead.

Answers: 1. True; 2. False; 3. True; 4. False; 5. True

10.5 BELT FRICTION

After studying this section, you should be able to:

- Determine the effect that friction between a belt and a pulley has on the behavior of a mechanical system.

Key Concept Belt friction is essential to the ability of machines to transmit torque from one part to another in a belt-pulley system.

torque

belt-pulley system

As is discussed more fully in Sec. 11.3, a *torque* is a moment that acts about the longitudinal axis of a shaft or rod, such as an automobile drive shaft or a pulley axle. In mechanical systems, torque in a drive shaft may be transmitted to a driven shaft by means of a continuous belt looped tightly around drums or pulleys attached to the shafts (see Fig. 10.11a). The operation of such a *belt-pulley system* depends on the development of frictional forces around the arc of contact between the belt and the pulleys. The cross section of the belt may be circular, rectangular, or trapezoidal (see Fig. 10.11b). Belts with a circular cross section are used to transmit relatively low torques, in record turntables and antique sewing machines, for example. Belts with a rectangular or trapezoidal cross

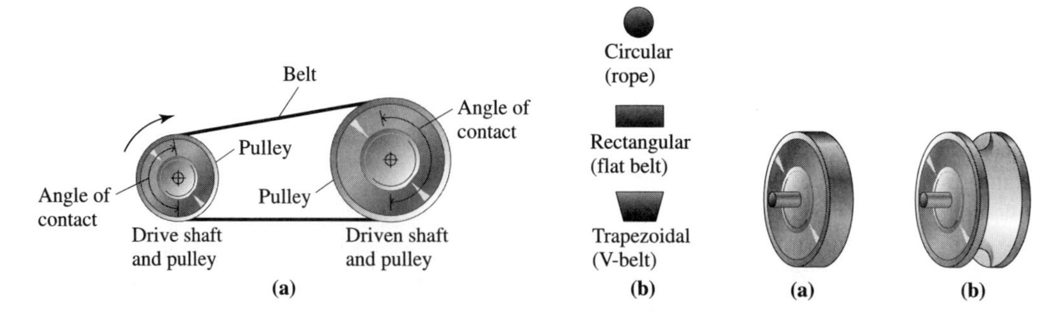

Figure 10.11 (a) Belt-pulley drive. (b) Cross sections of pulley belts.

Figure 10.12 (a) Flat-rim drum or pulley. (b) Grooved pulley.

flat belts

V-belts

section are generally used to transmit larger torques. Belts with a circular cross section are sometimes called ropes. Belts with a rectangular cross section are called *flat belts,* and belts with a trapezoidal cross section are called *V-belts.*

The rim of a pulley may be flat (or slightly crowned to produce a self-centering action) or grooved—see Fig. 10.12. Since they are the most common, we consider flat belts and V-belts in the following subsections.

FLAT BELTS

angle of wrap

There are a number of mechanical devices that utilize friction between a flat belt and a drum (or pulley) around which the belt is wrapped (see Figs. 10.13a and b). The tension T_1, acting at point *A,* where the belt initiates contact with the drum, does not necessarily equal the tension T_2, acting at point *B,* where the belt leaves the drum. The drum may be stationary or rotating. The maximum torque (twisting moment) is transmitted to the drum axis when sliding of the belt on the drum is impending. Then, the ratio T_1/T_2 depends on the coefficient of static friction μ_s and the *angle of wrap* α (sometimes called the angle of contact) between the belt and the drum. If the belt is sliding on the drum, T_1/T_2 depends on the coefficient of kinetic friction μ_k and the angle of wrap α. If the drum is rotating at high speeds, centrifugal forces on the belt may be significant (Orthwein, 1990). For low speeds, the effect of speed on the ratio T_1/T_2 is usually neglected.

To derive the relationship between T_1 and T_2, let's consider the free-body diagram of an infinitesimal segment of a flat belt in contact with the curved surface of a drum (see Fig. 10.13c). In the following derivation, we assume that the belt either is on the verge of slipping or is slipping clockwise relative to the drum. This implies $T_1 > T_2$. For cases in

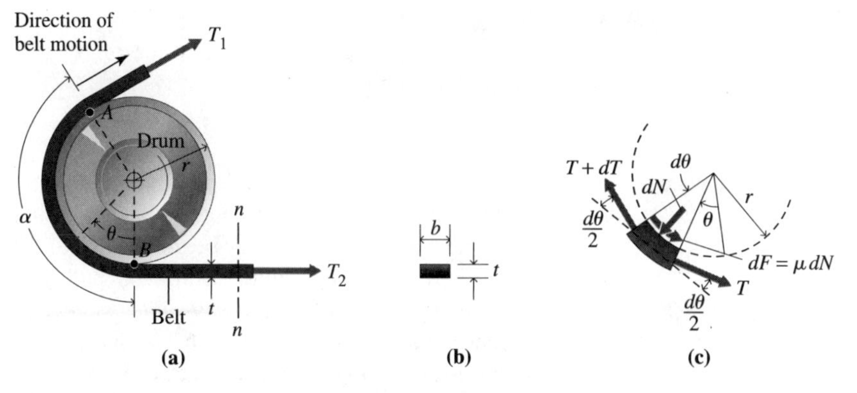

Figure 10.13 (a) Flat-belt drum system. (b) Belt cross section (section *n–n*). (c) Free-body diagram of infinitesimal element of a flat belt.

which sliding impends or occurs, $dF = \mu\, dN$, where dF and dN are the elemental frictional and normal forces, respectively, exerted by the drum on the belt. Hence, μ is either the coefficient of static friction or the coefficient of kinetic friction, depending on whether sliding is impending or occurring.

Ignoring the effects of centrifugal forces acting on the belt, we can write the equilibrium equations in the radial and tangential directions (refer to Fig. 10.13c):

$$\sum F_r = dN - (T + dT)\left[\sin\left(\frac{d\theta}{2}\right)\right] - T\sin\left(\frac{d\theta}{2}\right) = 0 \tag{a}$$

$$\sum F_\theta = (T + dT)\left[\cos\left(\frac{d\theta}{2}\right)\right] - T\cos\left(\frac{d\theta}{2}\right) - \mu\, dN = 0 \tag{b}$$

Since $d\theta$ is an infinitesimal angle, $\sin(d\theta/2) \approx d\theta/2$ and $\cos(d\theta/2) \approx 1.0$, provided that $d\theta$ is expressed in radians. Also, the term $[dT\sin(d\theta/2)]$ in Eq. (a) is small compared to the other terms and may be dropped. Hence, Eqs. (a) and (b) yield

$$dN - T\,d\theta = 0 \tag{c}$$

$$dT - \mu\, dN = 0 \tag{d}$$

Eliminating dN algebraically from Eqs. (c) and (d), we obtain

$$\frac{dT}{T} = \mu\, d\theta \tag{10.13}$$

Integration of Eq. (10.13) yields

$$\ln T = \mu\theta + C \tag{e}$$

where C is a constant of integration.

If θ is measured from point B (refer to Fig. 10.13a), then $T = T_2$ for $\theta = 0$. Hence, with $\theta = 0$ and $T = T_2$, Eq. (e) yields $C = \ln T_2$. Substituting this value of C into Eq. (e), we obtain

$$\ln\frac{T}{T_2} = \mu\theta \tag{f}$$

Equation (f) determines the tension T at any point in the angle of contact. If $\theta = \alpha$, then $T = T_1$. Introducing these values into Eq. (f), we obtain

$$\frac{T_1}{T_2} = e^{\mu\alpha} \tag{10.14}$$

Since the radius of curvature of the drum does not appear in the preceding analysis, the above equations are valid even when the drum is not circular. For example, the drum may be elliptic.

Also, from Fig. 10.13a, the turning moment (torque) transmitted to a circular drum by the belt is

$$M = (T_1 - T_2)r \tag{10.15}$$

Equation (10.14) is valid for a belt that passes around a drum at slow speeds, since centrifugal forces on such a belt are small (Orthwein, 1990).[1] Consequently, for slow speeds,

1. If centrifugal forces are significant, Eq. (10.14) becomes $T_1 - mv^2 = (T_2 - mv^2)e^{\mu\alpha}$, where m is the mass of the belt per unit length and v is the belt speed. The term mv^2 is referred to as the *centrifugal tension.* It has the effect of reducing the belt tension available for transmitting torque.

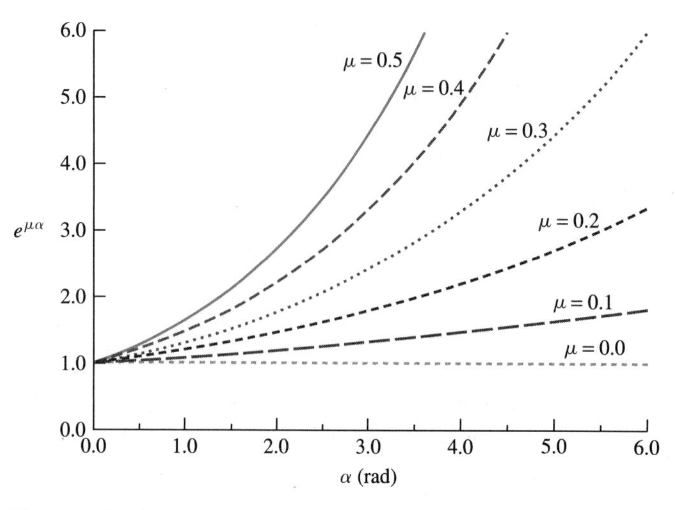

Figure 10.14 Effect of coefficient of friction and angle of wrap on belt friction.

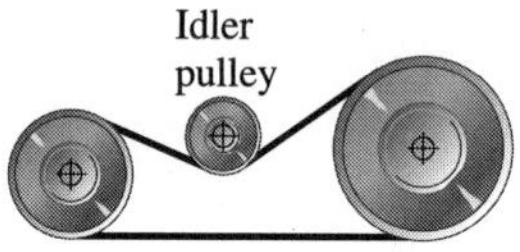

Idler pulley

Figure 10.15 Belt-pulley system with idler.

this equation is important in the design of belt drives and brakes (see Example 10.11). Since $T_1 > T_2$, T_1 is called the tension in the tight side of the belt and T_2 is called the tension in the slack side.

The value of the function $e^{\mu\alpha}$ increases rapidly as the friction coefficient μ or the wrap α increases (see Fig. 10.14). Therefore, in order to increase the maximum value of the ratio T_1/T_2, an *idler* is sometimes used in a belt-pulley system to increase the angle of wrap (see Fig. 10.15).

idler pulley

Tension-Pressure Relation The force dN (see Fig. 10.13c) may be expressed in terms of the normal pressure p (force per unit area) exerted on the belt by the drum as

$$dN = pbr\,d\theta \tag{g}$$

where b is the width of the belt (Fig. 10.13b) and $r\,d\theta$ is the arc length over which p acts (Fig. 10.13c). Equations (c) and (g) yield

$$T = pbr \tag{10.16}$$

The maximum normal pressure p_{max} occurs when $\theta = \alpha$, where $T = T_1$ is at a maximum. Thus, for $\theta = \alpha$,

$$p_{max} = \frac{T_1}{br} \tag{10.17}$$

V-BELTS

Key Concept A V-belt can transmit a higher torque than a flat belt can transmit because a V-belt grips a pulley more tightly.

In many applications, the flat belt has been replaced by the versatile V-belt (see Fig. 10.16). The sides of a V-belt remain straight between the pulleys. However, as the belt passes over the rim of a flat-faced pulley, the sides tend to bulge outward. To take advantage of this

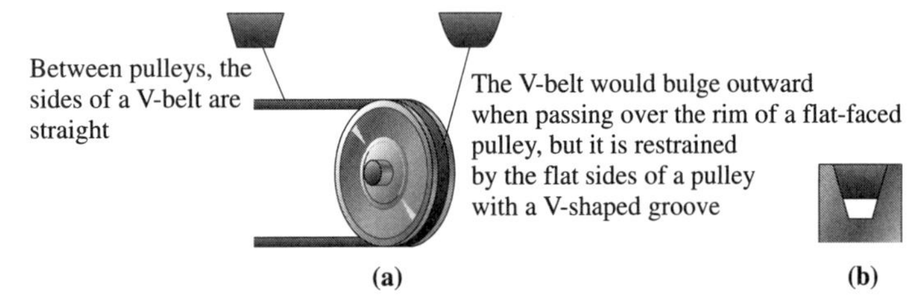

Figure 10.16 (**a**) V-belt pulley. (**b**) A V-belt grips because the flat sides of the pulley groove keep it from bulging.

outward bulging, pulleys that are used with V-belts have deep V-shaped grooves (see Fig. 10.16b). The constraint of the bulging of the belt in the V-shaped groove produces a tight gripping action between the belt and the walls of the groove.

The analysis of V-belt pulley systems is similar to that of flat-belt pulley systems, except that the forces dN' normal to the belt, and, hence, the friction forces, act on the two inclined faces of the V-belt (compare Figs. 10.13c and 10.17b).

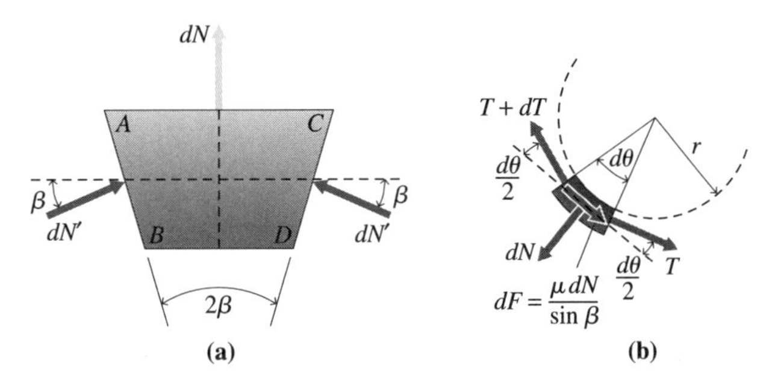

Figure 10.17 (**a**) V-belt cross section. (**b**) Infinitesimal element of a V-belt.

From Fig. 10.17a, the resultant of the forces dN' in the radial direction yields dN:

$$\sum F_r = 2\sin\beta\, dN' = dN$$

$$2\, dN' = \frac{dN}{\sin\beta} \tag{a}$$

If we assume that the belt is on the verge of slipping or is slipping, the frictional forces that act on the inclined planes AB and CD (perpendicular to the plane of the page) are both equal to $\mu\, dN'$, where μ is either the coefficient of static friction or the coefficient of kinetic friction, depending on whether slipping is impending or occurring. Thus, the total frictional force that acts on the belt is [with Eq. (a)]

$$dF = 2\mu\, dN' = \frac{\mu}{\sin\beta}\, dN \tag{b}$$

With Eqs. (a) and (b), a direct comparison can be made between Fig. 10.13c and Fig. 10.17b. Then, from Fig. 10.17b, proceeding as in the derivation of Eq. (10.13), we find

$$\frac{dT}{T} = \frac{\mu}{\sin\beta}\, d\theta \tag{10.18}$$

Integrating Eq. (10.18) and proceeding as in the derivation of Eq. (10.14), we obtain

$$\frac{T_1}{T_2} = e^{\mu\alpha/\sin\beta} \tag{10.19}$$

By comparison of Eqs. (10.14) and (10.19), we see that the relations derived for flat-belt pulley systems [Eqs. (10.13), (10.14), and (10.15)] can be used for V-belt pulley systems, provided that the coefficient of friction for the flat-belt pulley system is replaced by $\mu/\sin\beta$.

Example 10.10

Docking a Ship

Problem Statement As a ship is brought alongside a dock, its forward motion is gradually stopped by means of a rope that is tied to the dock at a point aft of the ship. The rope is wrapped (looped) in a figure-eight pattern about two stationary bitts (heavy iron posts) on the ship's deck. A sailor pulls on the rope, while allowing it to slide around the bitts, so that it is not strained severely enough to break (see Fig. E10.10). Determine the minimum number of loops required, if the end of the rope tied to the dock has a tension of 180 kN and the sailor pulls with a force of 180 N. The coefficient of kinetic friction is 0.40.

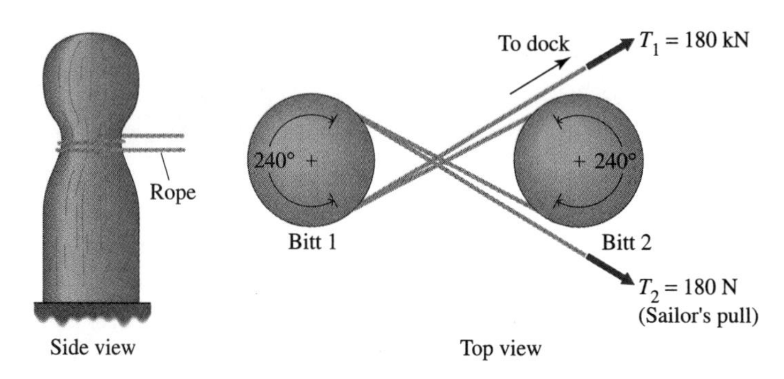

Figure E10.10

Solution The total angle of wrap is the sum of the angles of wrap around the two bitts. For each loop around a bitt, the angle of wrap is 240°, or $4\pi/3$ rad. Hence, the total angle of wrap for a whole figure-eight pattern is $8\pi/3$ rad.

Since $T_1 = 180$ kN, $T_2 = 180$ N, and the coefficient of kinetic friction $\mu_k = 0.40$, Eq. (10.14) yields

$$1000 = e^{0.4\alpha}$$

$$\alpha = 17.27 \text{ rad}$$

Since the total angle of wrap of one figure-eight is $8\pi/3 = 8.38$ rad, the number of figure-eights required is $17.27/8.38 = 2.06$, slightly more than 2. Note that if the sailor increased the angle of wrap to a value greater than 17.27 rad, a pull less than 180 N would be required. For instance, with the arrangement in Fig. E10.10, if the sailor used 2.5 figure-eights (for an angle of wrap of 20.94 rad), a pull of only 41.29 N would be required.

Example 10.11

The Band Brake

Problem Statement A manually operated band brake has a control lever ABC, shown schematically in Fig. E10.11a, where $b > c$. The pin support at B is frictionless. Determine the braking force P required to exert a restraining torque M on the drum, which rotates counterclockwise. Express the result in terms of a, b, c, μ_k, and α.

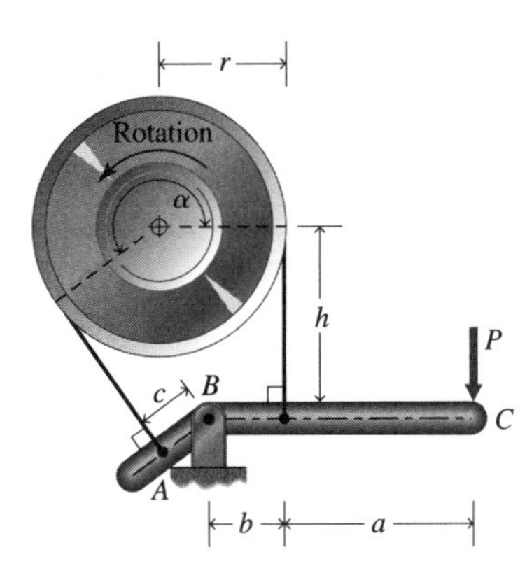

Figure E10.11a

Solution The free-body diagram of the control lever ABC is shown in Fig. E10.11b. The braking torque exerted by the band on the drum is, from Figs. E10.11a and E10.11b [see also Eq. (10.15)],

$$M = (T_1 - T_2)r \tag{a}$$

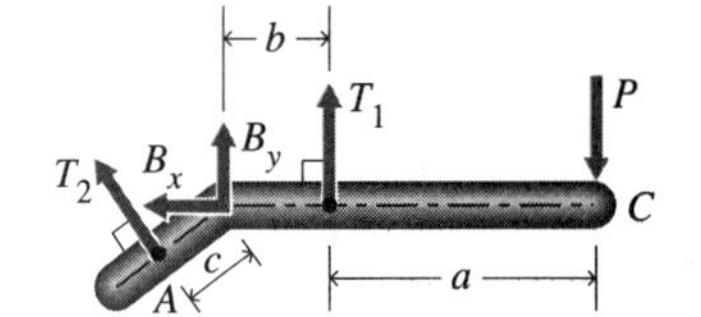

Figure E10.11b

where, by Eq. (10.14),

$$T_1 = T_2 e^{\mu_k \alpha} \tag{b}$$

The moment equilibrium equation with respect to the frictionless pin at B is

$$\sum M_B = P(a + b) - T_1 b + T_2 c = 0$$

or

$$P = \frac{\left(b - \dfrac{T_2}{T_1} c\right) T_1}{a + b} \tag{c}$$

With Eq. (b), Eq. (c) may be written as

$$P = \frac{b - ce^{-\mu_k \alpha}}{a + b} T_1 \qquad b > c \tag{d}$$

Similarly, eliminating T_2 from Eq. (a), we obtain

$$M = (1 - e^{-\mu_k \alpha}) T_1 r \tag{e}$$

Elimination of T_1 between Eqs. (d) and (e) yields the desired relation between P and M:

$$P = \frac{b - ce^{-\mu_k \alpha}}{r(1 - e^{-\mu_k \alpha})} \left(\frac{M}{a + b}\right) \tag{f}$$

From Eq. (f), if $b = ce^{-\mu_k \alpha}$, then $P = 0$; that is, no braking force is required. Also, if $b < ce^{-\mu_k \alpha}$, then Eq. (f) indicates that P is negative; that is, a force in the opposite sense of that shown in Fig. E10.11b is required to prevent the brake from acting. In other words, for $b \leq ce^{-\mu_k \alpha}$, the band brake is self-locking.

Example 10.12

Band Brake Pressure

Problem Statement The band brake shown in Fig. E10.12a has a coefficient of kinetic friction equal to 0.20. The maximum allowable normal band pressure is $p_{\max} = 0.20$ MPa, for a band width of $d = 60$ mm. The drum rotates in the clockwise sense.

a. Determine the allowable brake torque M.

b. Determine the corresponding control lever force P.

c. Determine the minimum value of c that makes the brake self-locking.

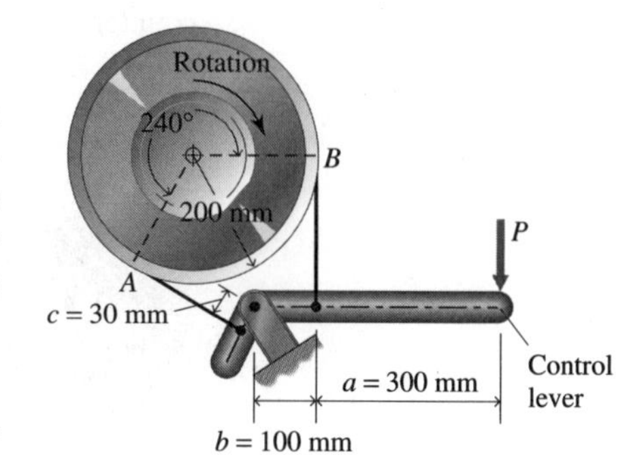

Figure E10.12a

Solution **a.** As noted earlier, the tension at point A (refer to Fig. E10.12a) is the maximum tension T_1 in the belt (the tension in the tight side of the belt). Therefore, from Eq. (10.17), we have

$$T_1 = p_{\max} dr = 0.2(60)(200) = 2400 \text{ N} \tag{a}$$

Then, from Eq. (10.14), we obtain T_2 (with $\alpha = 240° = 4\pi/3$):

$$T_2 = T_1 e^{-\mu_k \alpha} = 2400e^{-0.2(4\pi/3)} = 1038 \text{ N} \tag{b}$$

Hence, Eqs. (a), (b), and (10.15) yield the brake moment as

$$M = (T_1 - T_2)r = (2400 - 1038)(0.20) = 272 \text{ N·m}$$

b. The free-body diagram of the control lever is shown in Fig. E10.12b. Summation of moments about point R yields

$$\sum M_R = (a + b)P - bT_2 + cT_1 = 0$$

or

$$P = \frac{bT_2 - cT_1}{a + b} \tag{c}$$

Inserting known values into Eq. (c), we obtain $P = 79.5$ N.

c. For the band brake to be self-locking, $P \leq 0$. Then, from Eq. (c), the minimum value of c is

$$c_{\min} = \frac{bT_2}{T_1} = \frac{100(1038)}{2400} = 43.3 \text{ mm}$$

Therefore, the brake is self-locking for $c \geq 43.3$ mm.

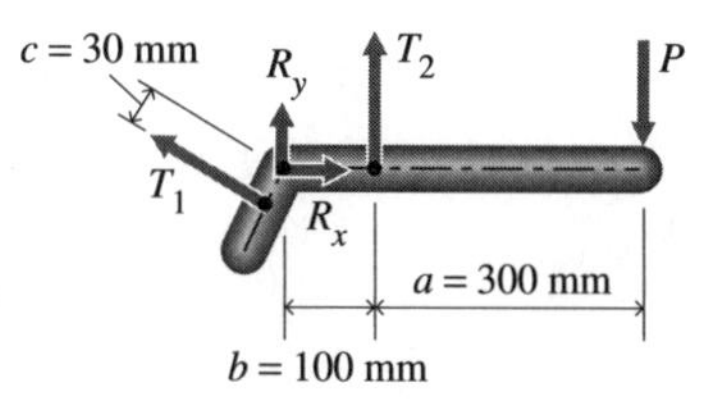

Figure E10.12b

Example 10.13

V-Belt Tension

Problem Statement A V-belt has an angle of wrap of 180° around a grooved pulley that has an effective diameter of 200 mm. The groove angle of the pulley is $2\beta = 40°$ (refer to

Fig. 10.17). For the belt and pulley, the coefficient of static friction is $\mu_s = 0.30$ and the coefficient of kinetic friction is $\mu_k = 0.15$.

a. The belt tension is adjusted so that slipping between the belt and pulley is impending and the belt transmits a torque of 6 N·m to the pulley. Determine the maximum and minimum tensions in the belt.

b. The belt tension is adjusted so that slipping between the belt and pulley occurs. The transmitted torque remains 6 N·m. Determine the maximum and the minimum tensions in the belt.

Solution **a.** The maximum and minimum tensions are T_1 and T_2 [see Eq. (10.19)]. Hence,

$$\frac{T_1}{T_2} = e^{\mu_s \alpha/\sin\beta} = e^{0.30\pi/\sin 20^\circ} = 15.731 \tag{a}$$

From Eq. (10.15), with $r = 0.10$ m, we have

$$M = (T_1 - T_2)(0.10) = 6.0 \text{ N·m} \tag{b}$$

Solving Eqs. (a) and (b), we obtain

$$T_1 = 64.07 \text{ N} \qquad T_2 = 4.07 \text{ N}$$

b. If slipping occurs, Eq. (10.19) yields

$$\frac{T_1}{T_2} = e^{\mu_s \alpha/\sin\beta} = e^{0.15\pi/\sin 20^\circ} = 3.966 \tag{c}$$

Again, from Eq. (10.15), we have

$$M = (T_1 - T_2)(0.10) = 6.0 \text{ N·m} \tag{d}$$

Solving Eqs. (c) and (d), we obtain

$$T_1 = 80.2 \text{ N} \qquad T_2 = 20.2 \text{ N}$$

We see that for the same torque, larger belt tensions are required with slipping than without. However, slipping results in increased belt wear. Hence, it is not usually economical to allow slipping to occur.

Example 10.14

Effect of Belt Tightening

Problem Statement A V-belt-pulley system transmits a torque of 7.5 lb·ft. The angle of wrap is 135° ($3\pi/4$ rad). The pulley has an effective diameter of 1.0 ft and a groove angle of $2\beta = 38^\circ$. The coefficient of static friction is $\mu_s = 0.40$.

a. The belt tension is adjusted so that slipping is impending. Determine the belt tensions T_1 and T_2.

b. An engineer decides to increase the tension in the slack side of the belt by an amount ΔT. Determine the effect on the maximum torque transmitted without slipping.

Solution **a.** From Eq. (10.19),

$$\frac{T_1}{T_2} = e^{\mu_s \alpha/\sin\beta} = e^{0.40(0.75\pi)/\sin 19^\circ} = 18.081$$

or

$$T_1 = 18.08 T_2 \tag{a}$$

From Eq. (10.15), with $r = 0.5$ ft, we find

$$M = (T_1 - T_2)(0.5) = 7.5 \text{ lb·ft}$$

or

$$T_1 - T_2 = 15 \text{ lb} \qquad \text{(b)}$$

Solving Eqs. (a) and (b), we obtain

$$T_1 = 15.88 \text{ lb} \qquad T_2 = 0.88 \text{ lb}$$

b. If the initial tension in the slack side of the belt is increased by ΔT, the tensions T_1 and T_2 are both increased. T_1 is increased by $(18.08)(\Delta T)$ and T_2 is increased by ΔT [see Eq. (a)]. Let the corresponding change in the torque be ΔM. Then, from Eq. (10.15), we obtain

$$(18.08 - 1)(\Delta T)(0.5) = \Delta M$$

or

$$\Delta M = 8.54\,\Delta T$$

Hence, tightening the belt by an amount ΔT increases the torque by $\Delta M = 8.54\,\Delta T$ [lb·ft].

CHECKPOINT 

1. ***True or False:*** The angle of wrap of a belt-pulley system is the angle of contact between the belt and the pulley.
2. ***True or False:*** For a given angle of wrap, if the coefficient of static friction is doubled, the ratio of the tension in the tight side of the belt to the tension in the slack side is doubled.
3. ***True or False:*** An idler is a pulley used to increase the total angle of wrap for a belt-pulley system.
4. ***True or False:*** The theory of belt friction is applicable only to circular drums and pulleys.

Answers: 1. True; 2. False; 3. True; 4. False

10.6 FRICTION CLUTCHES AND BRAKES

After studying this section, you should be able to:

- Analyze the performance of a friction clutch or brake.

CLUTCHES

Key Concept

A clutch is used to transmit torque from a drive shaft to a driven shaft, where the two shafts are collinear.

clutch

A *clutch* is a mechanical device used to engage and disengage a machine from its driver while the driver is running. A familiar example is the automobile clutch. For light loads and low speeds, jaw clutches are sometimes employed. These provide positive drive by means of interlocking teeth. For applications where the teeth of a jaw clutch would clash destructively during engagement, axial clutches are widely used.

disk clutch

cone clutch

A *disk clutch* is an axial clutch in which two or more annular friction plates or disks are attached to two axially aligned shafts, a drive shaft and a driven shaft (see Fig. 10.18). The disks are brought into contact (mated) by moving one of the shafts (usually the driven shaft) axially. Another type of axial clutch is the *cone clutch.* The cone clutch has been largely replaced by the disk clutch, because the latter has several advantages. In particular, the disk clutch has a large friction area that takes up little axial space. It also has large surfaces for effective heat dissipation, effective pressure distribution, and minimal centrifugal effects. The single friction-plate disk clutch in Fig. 10.18 has a pair of mated surfaces (disks). One disk, A, is fixed on the driven shaft and the other disk, B, is fixed on the drive shaft. Often, to increase the friction area, multiple pairs of friction plates are interfaced (Shigley & Mischke, 1989). For example, multiple pairs of friction plates are typically used in motorcycle clutches.

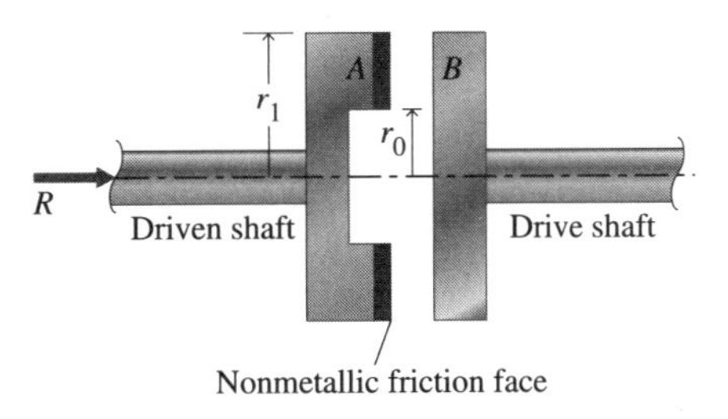

Figure 10.18 Friction clutch.

Friction on these clutch plates is necessary to transfer torque from the engine to the transmission. A Belleville spring is depressed to disengage the clutch plates.

In this book, we discuss mainly axial disk clutches. The mechanical details of such clutches vary considerably in practice, but the basic idea can be grasped by considering two shafts—the drive shaft and the driven shaft—that terminate at coaxial disks or rings. One of the disks is usually faced with a nonmetallic material that has a high coefficient of friction (the driven disk in Fig. 10.18). The driven shaft is pressed against the rotating drive shaft by an actuating force R. Typically, the actuating force R is produced by a toggle mechanism, a hydraulic device, or an electromagnet. After an initial slippage, the friction between the disks causes them to rotate together. In designing such a clutch, the designer must determine the actuating force required to produce a pressure p on the clutch disks and the maximum torque transmitted to the driven shaft. To calculate the maximum torque

$T = T_s$ that the clutch can transmit, the designer may assume that the pressure p is distributed uniformly over the contact faces.[2] The area of the annular face of the driven disk is $\pi(r_1^2 - r_0^2)$, where r_0 and r_1 are the inner and outer radii of the annular face (refer to Fig. 10.18). Hence, the pressure between the faces is

$$p = \frac{P}{\pi(r_1^2 - r_0^2)} \tag{10.20}$$

where P is the net axial force on the disk face (see Fig. 10.19). Since the driven shaft is in equilibrium in the axial direction, $P = R$.

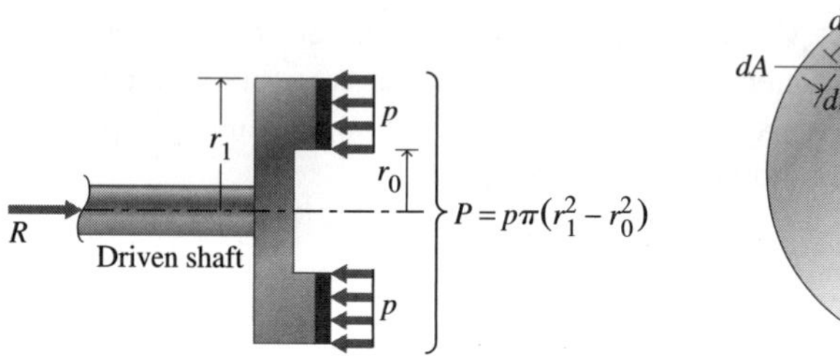

Figure 10.19 Single pair of friction plates.

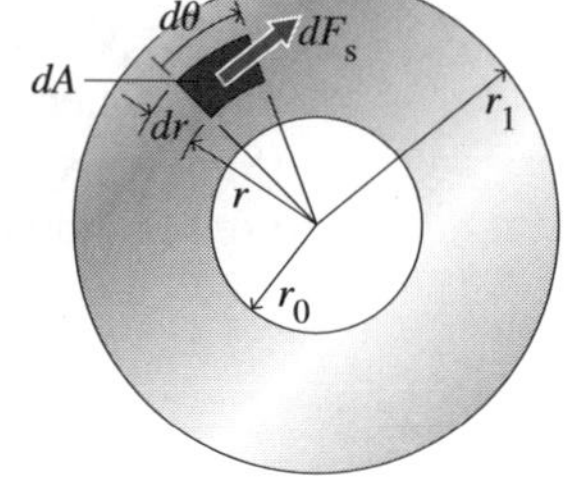

Figure 10.20

Consider a differential area dA of a disk face, at radius r (see Fig. 10.20). Using polar coordinates, we have $dA = r\,dr\,d\theta$. The maximum torque is transmitted when slipping is impending. Then, the differential frictional force dF_s on dA is $dF_s = \mu_s p\,dA$, where μ_s is the coefficient of static friction.

Hence, from Eq. (10.20), with $P = R$,

$$dF_s = \left[\frac{\mu_s R}{\pi(r_1^2 - r_0^2)}\right] r\,dr\,d\theta$$

At impending slipping, the increment of torque transmitted by the force dF_s is $dT_s = r\,dF_s$. Therefore,

$$dT_s = r\,dF_s = \left[\frac{\mu_s R}{\pi(r_1^2 - r_0^2)}\right] r^2\,dr\,d\theta$$

By integration, the maximum torque is

$$T_s = \frac{\mu_s R}{\pi(r_1^2 - r_0^2)} \int_{r_0}^{r_1} r^2\,dr \int_0^{2\pi} d\theta \tag{a}$$

Evaluation of the integrals in Eq. (a) yields

$$T_s = \frac{2\mu_s R(r_1^3 - r_0^3)}{3(r_1^2 - r_0^2)} \tag{10.21}$$

If slipping between the mated surfaces occurs, the coefficient of static friction μ_s in Eq. (10.21) must be replaced by the coefficient of kinetic friction μ_k. Also, note that the theory developed for the disk clutch may be applied in principle to other contacting flat

2. An alternative procedure is to design the clutch for uniform wear over the surface. This method assumes that the pressure is a maximum at $r = r_0$ and decreases linearly with r to $r = r_1$ (refer to Fig. 10.18) (Shigley & Mischke, 1989).

surfaces that rotate relative to one another—for example, radially symmetric collar bearings, step bearings, and flat-end pivots.

For a single pair of mated surfaces, Eq. (10.21) relates the maximum torque T_s to the actuating force R that acts on the driven shaft. In the case of a disk clutch with multiple pairs of identical mated surfaces, each pair of mated surfaces transmits a maximum torque given by Eq. (10.21). Therefore, for multiple mated surfaces, the net maximum torque is obtained by multiplying the right-hand side of Eq. (10.21) by n, the number of pairs of mated surfaces. Thus, for a multiple-disk clutch with n pairs of identical surfaces, each pair of mated surfaces is subjected to the actuating force R. So, Eq. (10.21) is modified as follows:

$$T_s = n\left[\frac{2\mu_s R(r_1^3 - r_0^3)}{3(r_1^2 - r_0^2)}\right] \tag{10.22}$$

BRAKES

Key Concept A brake is used to reduce the speed of a rotating wheel.

Like clutches, brakes come in a variety of types (Orthwein, 1986). In this book, we discuss mainly disk brakes. The *disk brake* (shown schematically in Fig. 10.21) operates in a manner similar to a disk clutch, but with a much smaller area of friction contact and, hence, with higher local pressure. Whereas a disk clutch is used to bring the driven shaft up to the same angular speed as the drive shaft, a disk brake is used to reduce the speed of a rotating shaft or wheel. Typically, the friction pad in a disk brake consists of a circular sector that subtends an angle much less than 360° (refer to Fig. 10.21). In operation, two friction pads attached to a caliper are pressed against the disk with equal normal (brake) forces, applied hydraulically or mechanically. The torque produced by the sliding of the two friction pads on the disk is proportional to twice the product of the normal force exerted by each pad and the effective radius of the pads.

disk brake

drum brake

The conventional *drum brake* consists of a drum, a brake lining, two brake shoes, and an actuating hydraulic (or mechanical) cylinder (see Fig. 10.22). The lining is bonded to the shoes, which pivot about fixed pins. When the cylinder exerts a force pushing the lining against the outer drum, a pressure is exerted on the rotating drum and the resulting friction force causes the drum to slow down.

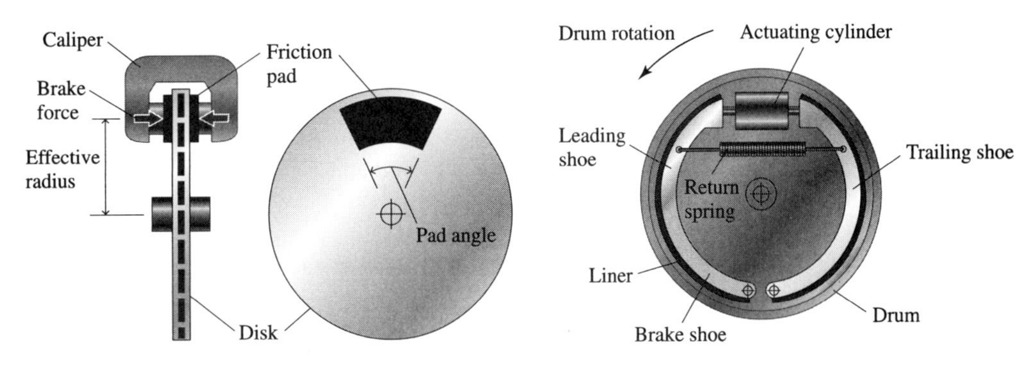

Figure 10.21 Disk brake.

Figure 10.22 Drum brake.

Disc brakes function similarly to the disk clutch. Well-designed drum brakes exert uniform outward pressure on the drum from the brake shoe.

Example 10.15

Disk Clutch with a Single Pair of Mated Surfaces

Problem Statement A disk clutch (refer to Fig. 10.18) has a single pair of mated surfaces, with radii $r_0 = 75$ mm and $r_1 = 150$ mm. The coefficient of static friction is $\mu_s = 0.30$. Determine the actuating force R and the maximum torque for a design pressure $p = 200$ kN/m^2.

Solution The net force P is exerted on the interface of the driven clutch (refer to Fig. 10.19). Therefore, the actuating force $R = P$ is related to the pressure p by Eq. (10.20) as

$$R = \pi p(r_1^2 - r_0^2) = \pi(200)(0.15^2 - 0.075^2) = 10.6 \text{ kN}$$

The maximum torque is related to the actuating force R exerted on the drive shaft by Eq. (10.21):

$$T_s = \frac{2\mu_s R(r_1^3 - r_0^3)}{3(r_1^2 - r_0^2)}$$

$$= \frac{2(0.30)(10.6 \times 10^3)(0.15^3 - 0.075^3)}{3(0.15^2 - 0.075^2)}$$

$$= 371 \text{ N·m}$$

Example 10.16

Disk Clutch with Four Pairs of Mated Surfaces

Problem Statement A disk clutch has four pairs of mated surfaces. Each disk attached to the driven shaft has an inner radius $r_0 = 40$ mm, an outer radius $r_1 = 60$ mm, and a coefficient of static friction $\mu_s = 0.20$. Determine the pressure between the faces of the disks and the maximum torque for an actuating force of $R = 13.5$ kN.

Solution The net axial force that acts on the face of each disk attached to the driven shaft is

$$R = \pi p(r_1^2 - r_0^2)$$

Then, the pressure p exerted between each interface of mated surfaces is

$$p = \frac{R}{\pi(r_1^2 - r_0^2)} = \frac{13.5 \times 10^3}{\pi(0.06^2 - 0.04^2)} = 2.15 \times 10^6 \frac{\text{N}}{\text{m}^2} = 2.15 \text{ MPa}$$

For impending sliding, the maximum torque is, by Eq. (10.22) with $n = 4$,

$$T_s = n\left[\frac{2\mu_s R(r_1^3 - r_0^3)}{3(r_1^2 - r_0^2)}\right]$$

$$= 4\left[\frac{2(0.2)(13.5 \times 10^3)(0.06^3 - 0.04^3)}{3(0.06^2 - 0.04^2)}\right]$$

$$= 547 \text{ N}\cdot\text{m}$$

CHECKPOINT 

1. ***True or False:*** A disk clutch consists of a drive shaft and a driven shaft with interlocking teeth or jaws.
2. ***True or False:*** The actuating force for a clutch with a single pair of mated surfaces is equal to the pressure that acts on the friction surface of the driven shaft.
3. ***True or False:*** The actuating force in a multiple-disk clutch is the same as the net axial force that acts on each pair of mated surfaces of the clutch.
4. ***True or False:*** The maximum torque exerted by a disk clutch occurs when the mated surfaces are slipping relative to one another.

Answers: 1. False; 2. False; 3. True; 4. False

Chapter Highlights

- A frictional force is a force that exists between two bodies in contact. It opposes the tendency of the bodies to slide relative to one another and acts tangent to the contact surface between the bodies.
- The maximum static friction is the maximum frictional force that can exist between two bodies that are in contact and on the verge of sliding relative to one another. For many practical applications, the maximum static friction is proportional to the normal force between the bodies and is independent of the area of contact.
- The coefficient of static friction for two bodies in contact is the ratio of the maximum static friction to the normal force between the bodies.
- The coefficient of kinetic (or sliding) friction for two contacting bodies sliding relative to one another is the ratio of the frictional force to the normal force between the bodies. In some cases, the coefficient of kinetic friction is independent of the speed of sliding over a wide range of speeds.
- The coefficient of static friction is usually larger than the coefficient of kinetic friction.
- Frictional forces play an important role in mechanical devices such as jackscrews, belts and pulleys, band brakes, clutches, disk brakes, and so on.

Problems

10.2 *Block on an Inclined Plane: The Angle of Repose*

10.3 *Problems Involving Friction*

10.1 The uniform bar *ABC* in Fig. P10.1 weighs 480 lb.

a. Draw a free-body diagram of the bar.

b. Write the equations of equilibrium for the bar, and determine the forces that act on the bar at *A* and *C*.

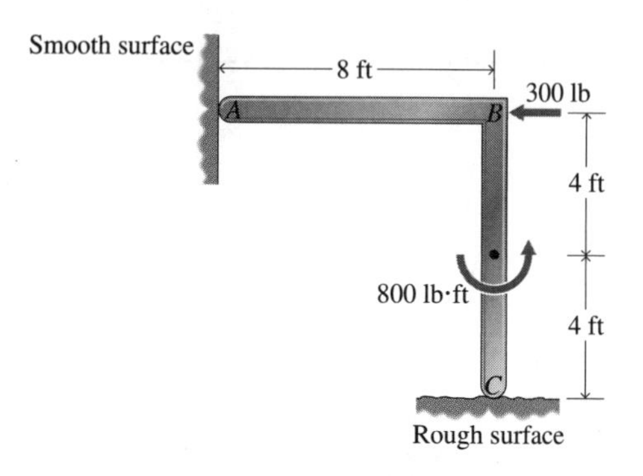

Figure P10.1

10.2 Solve Problem 10.1 for the case in which the couple shown in Fig. P10.1 is reversed; that is, it acts in the clockwise sense.

10.3 Consider the block in Fig. P10.3. Assume that the coefficient of friction is sufficiently large to prevent sliding. Determine the maximum allowable ratio of height h to width b if tipping of the block is not to occur.

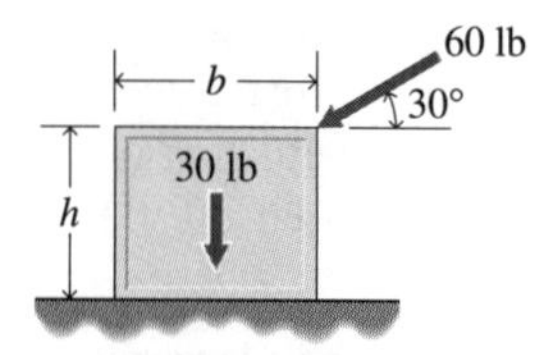

Figure P10.3

10.4 The slender uniform bar *ABC* in Fig. P10.4 weighs 300 N/m. The coefficient of static friction is $\mu_s = 0.5$ at points *A* and *C*. Determine the magnitude of the force *P* that results in impending motion of the bar.

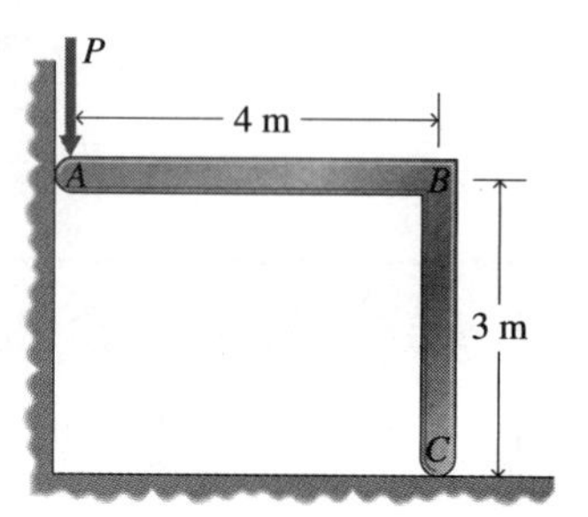

Figure P10.4

10.5 The force *P* at *A* in Fig. P10.4 is removed. A horizontal force *Q* acting to the left is applied at *C*. Determine the magnitude of the force *Q* that leads to impending motion of the bar.

10.6 Two blocks are attached by a cable that passes over a frictionless pulley, as shown in Fig. P10.6. Block *A* weighs 100 lb, and block *B* weighs 40 lb. The coefficient of static friction between block *A* and the ramp is 0.15.

a. Determine whether or not the blocks are in equilibrium.

b. Determine the minimum value of the coefficient of static friction for which the blocks are in equilibrium.

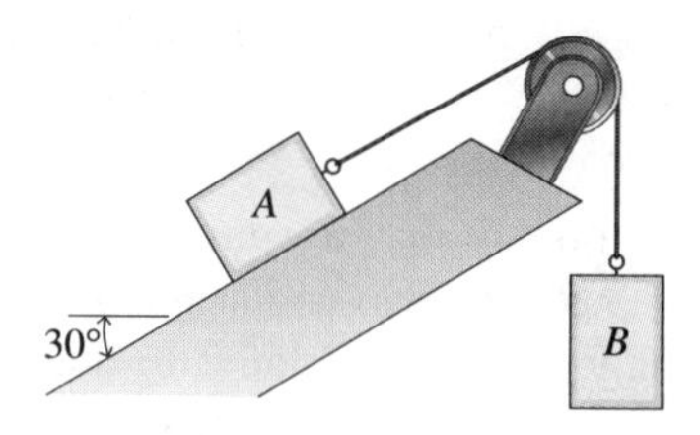

Figure P10.6

10.7 Block A weighs 250 N and is attached to hoist B by a flexible cable (see Fig. P10.7). The hoist is used to pull the block up or lower it down the ramp. The coefficient of static friction is 0.12.

a. Derive a formula, as a function of the ramp angle ϕ, for the range of values of the tension T in the cable for which the block does not move.

b. For $\phi = 30°$, determine the maximum and minimum values of T for which the block does not move.

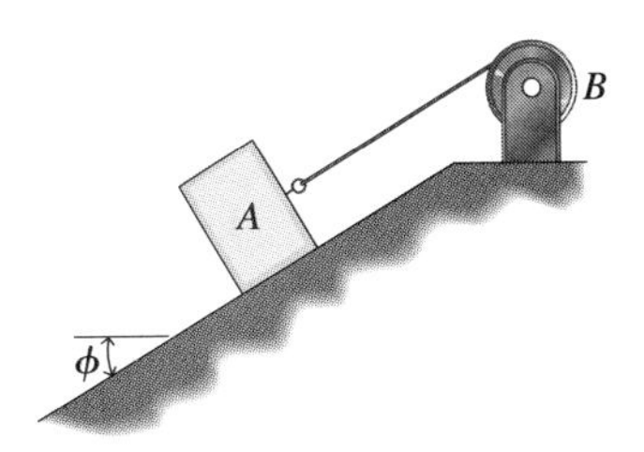

Figure P10.7

10.8 In Example 10.6, the force T is removed, and the cord between the boxes is cut.

a. Determine whether or not box A moves.

b. Determine whether or not box B moves.

10.9 In Example 10.4, let $\mu_s = 0.2$ and $\phi = 30°$. Determine the angle θ for which the ratio W/P is a maximum.

10.10 In Example 10.4, let $\mu_s = 0.2$ and $\theta = 30°$. Determine the angle ϕ for which the ratio W/P is a maximum. Explain your result.

10.11 Determine the force F required to cause impending motion of wheel B in Fig. P10.11, if the coefficient of static friction between bar AD and wheel B is 0.30. Neglect the friction of bearings C and D and the weight of bar AD.

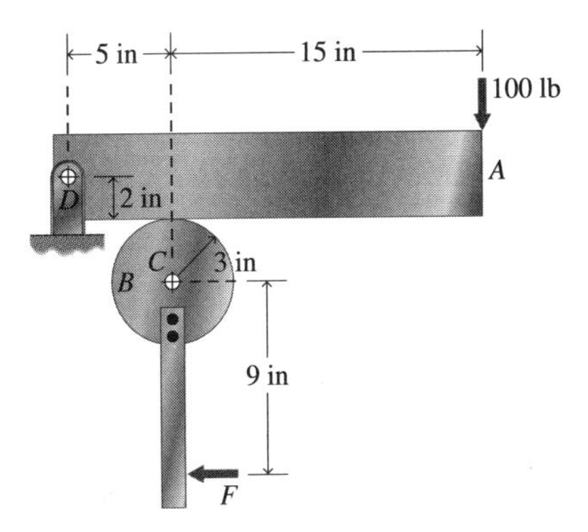

Figure P10.11

10.12 A force $P = 65$ lb is required to brake a shaft O that is subject to a torque M (see Fig. P10.12). The coefficient of static friction between the brake arm AB and the shaft is 0.40. Determine the torque M [lb·ft], neglecting friction of the bearings and the weight of the bar.

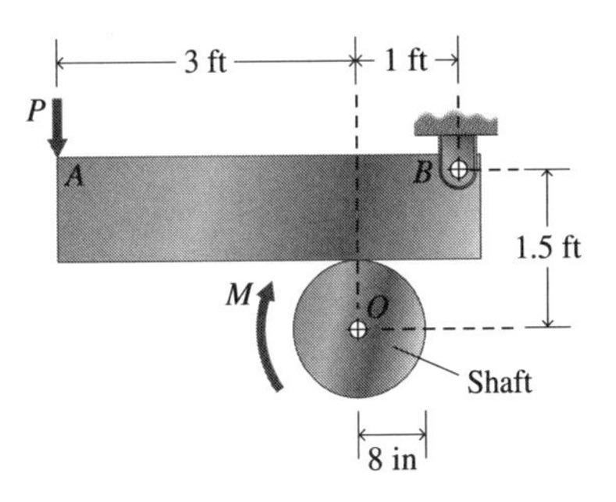

Figure P10.12

10.13 The coefficient of static friction between surface A and a scaffold hook is 0.60 (see Fig. P10.13). The hook is prevented from falling over sideways (perpendicular to the plane of the page). Determine whether the hook is safe—that is, whether it can maintain the load P in equilibrium. Neglect friction between surface B and the hook.

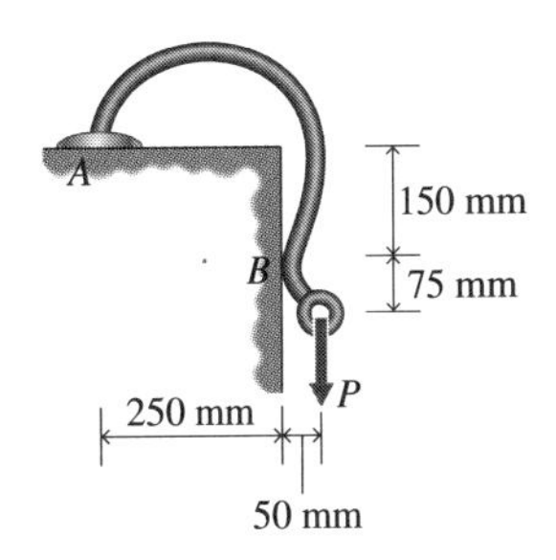

Figure P10.13

10.14 Cylinder C in Fig. P10.14 weighs 400 N. It is held in equilibrium by the horizontal force P, but it is on the verge of sliding. Determine the force P and the coefficient of static friction.

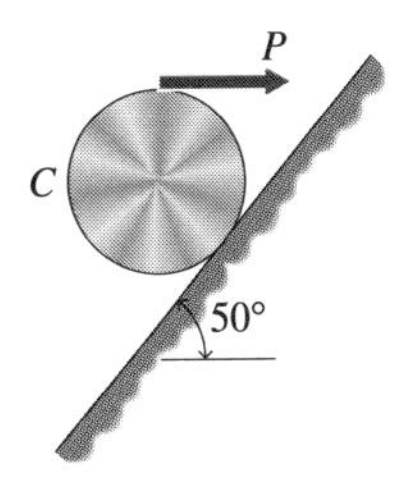

Figure P10.14

10.15 The members shown in Fig. P10.15 are made of hard steel. If the bar B is pulled down, the cam C is supposed to rotate clockwise and grip the bar so that it cannot move. Neglecting the weight of the cam and friction of the camshaft bearing, determine whether the clamp will operate properly. (From Table 10.1, $\mu_s = 0.42$.)

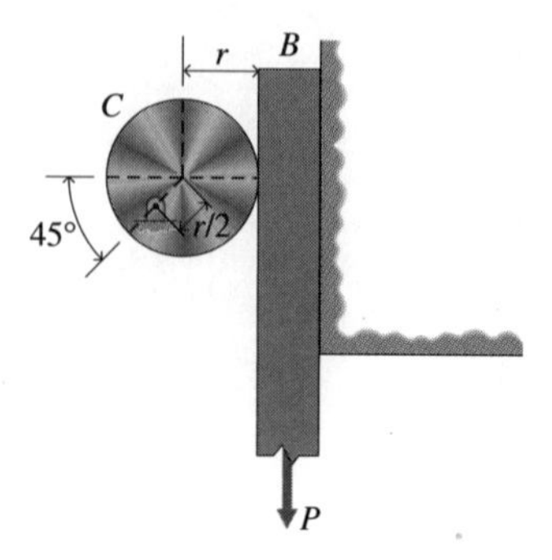

Figure P10.15

10.16 Rotation of the cylinder shown in Fig. P10.16 is impending under the action of the force P. The cylinder weighs 400 N, and the coefficient of static friction for both surfaces A and B is $\mu_s = 0.2$.

a. Draw the free-body diagram of the cylinder and label all forces.

b. Determine the magnitude of P.

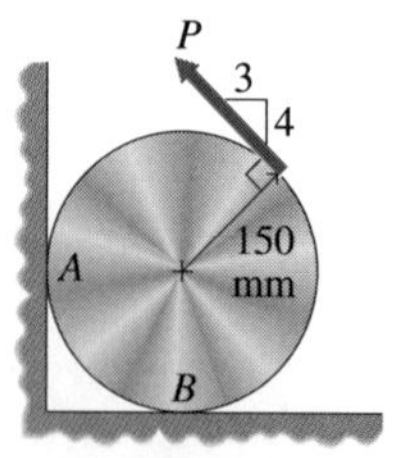

Figure P10.16

10.17 The cylinder in Fig. P10.16 weighs 400 N and $\mu_s = 0.2$ at A and B. The sense of the force P is reversed. Determine whether or not the cylinder moves.

10.18 The cylinder C in Fig. P10.18 is tied to point A with a horizontal cord AB and rests on the inclined plane. Determine, as a function of the angle ϕ, the minimum coefficient of static friction μ_s between the cylinder and the plane to keep the cylinder from slipping.

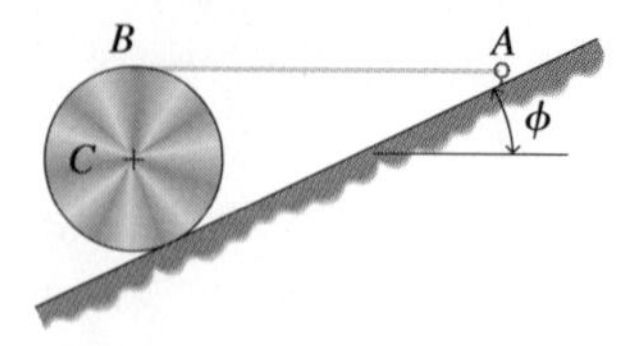

Figure P10.18

10.19 In Example 10.6, the force T is removed. Determine whether or not the boxes move.

10.20 The wheel D shown in Fig. P10.20 is supported by an axle at its center. It is on the verge of rotating counterclockwise due to the action of the couple M_D and forces exerted on it at C by the uniform brake bar ABC, which weighs 500 N and is subjected to the vertical force P.

a. Draw free-body diagrams of the wheel and the bar. The pin at A and the wheel's axle are frictionless.

b. Write the equilibrium equations needed to determine the magnitude of the force P for the impending motion.

c. The coefficient of static friction between the wheel and the bar at C is $\mu_s = 0.20$. Determine the magnitude of P.

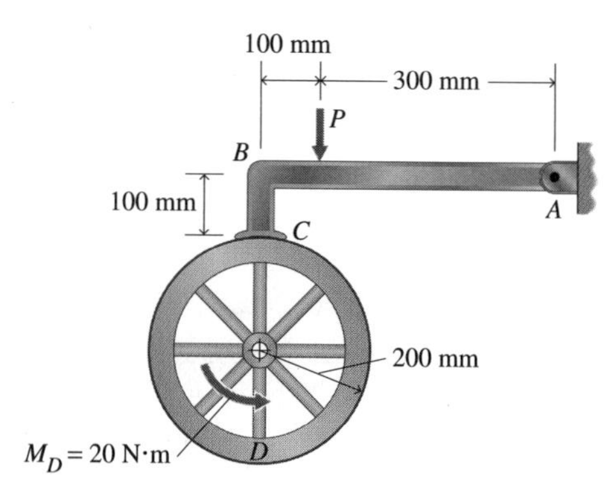

Figure P10.20

10.21 Solve Problem 10.20 for the case in which the couple M_D shown in Fig. P10.20 is directed in the clockwise sense.

10.22 The two blocks shown in Fig. P10.22 weigh 1 kN and W kN, respectively. The blocks rest on inclined planes and are connected by a light flexible rope that passes over a frictionless pulley. The coefficients of static and kinetic friction between both blocks and the planes are $\mu_s = 0.60$ and $\mu_k = 0.40$. Determine the weight W that causes sliding of the blocks to be impending.

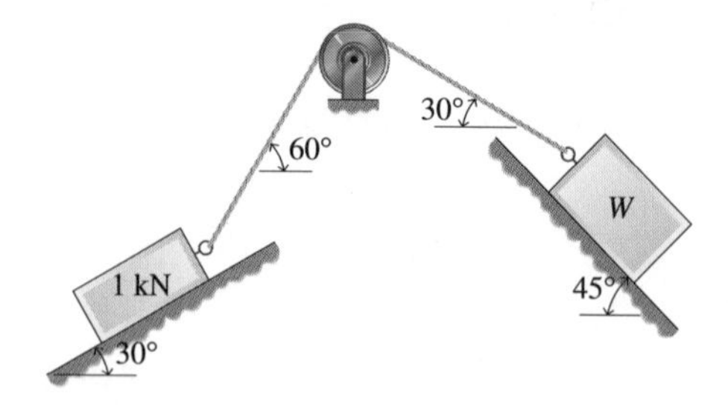

Figure P10.22

10.23 Refer to Problem 10.22, and assume that the rope is cut.

a. Determine whether or not the blocks slide.

b. Determine the frictional force that acts on each block after the rope is cut.

10.24 Refer to Example 10.4.

a. Determine the angle θ such that the force P required to start the block sliding up the plane is minimum.

b. Determine the corresponding values of P and N in terms of W, ϕ, and ϕ_s.

10.25 Rod AB is connected to blocks A and B by frictionless pins (see Fig. P10.25). The coefficient of static friction for the contacting surfaces is 0.40. Blocks A and B weigh 400 N and 600 N, respectively. Neglect the weight of the rod.

a. Determine the value of P that will cause the blocks to begin to slide up the plane.

b. Determine the minimum value of P for which the blocks can be held in the position shown.

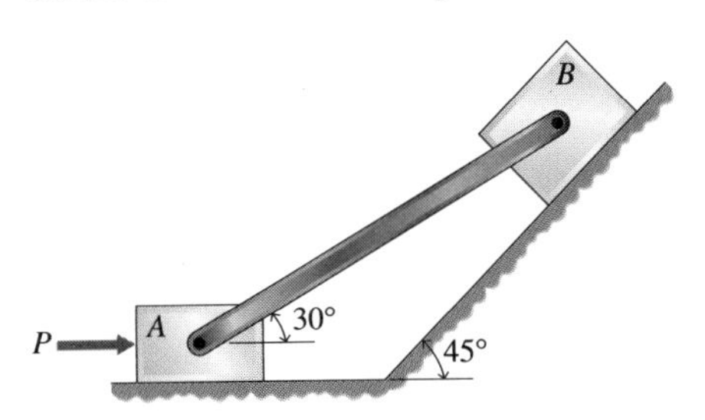

Figure P10.25

10.26 The force P causes slipping of the cylinder C to be impending (see Fig. P10.26). The cylinder's weight is W. The coefficient of static friction is μ_{sA} between the cylinder and the wall and μ_{sB} between the cylinder and the floor.

a. Draw the free-body diagram of the cylinder, and label all forces.

b. Derive a formula for the force P in terms of W, μ_{sA}, μ_{sB}, and r (the radius of the cylinder).

c. Determine P for the conditions $W = 400$ N, $\mu_{sA} = 0.1$, $\mu_{sB} = 0.2$, and $r = 300$ mm.

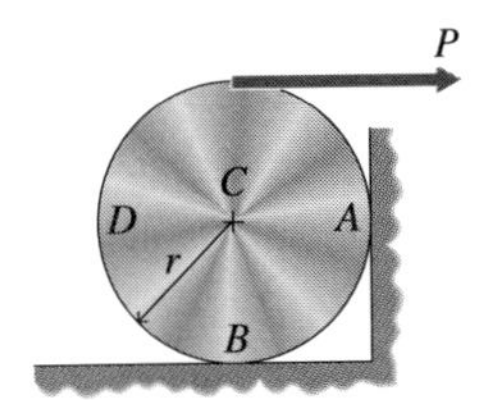

Figure P10.26

10.27 The force P in Fig. P10.26 is removed, and a force Q is applied that acts vertically upward at D. Solve Problem 10.26 for impending slipping of the cylinder C.

10.28 A thin uniform rod is bent into a square. When the rod is hung on a horizontal peg, it does not slip (see Fig. P10.28). What is the smallest coefficient of friction that can exist between the rod and the peg in terms of a and b?

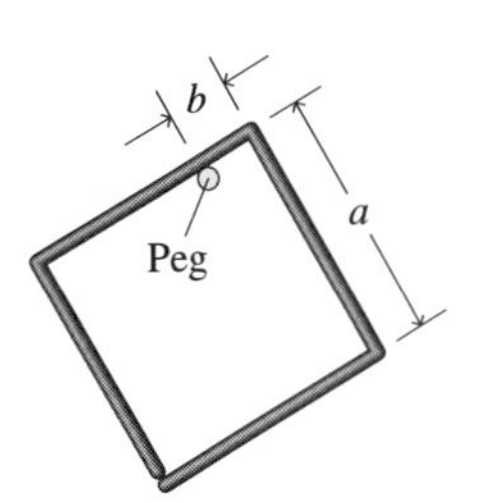

Figure P10.28

10.29 A force P causes motion of a machine part C (weight $2W$) to be impending (see Fig. P10.29). The weight of the uniform bar AB is W, and the coefficient of static friction between the bar and the part and between the part and the floor is μ_s. Assume that the pin at A is frictionless.

a. Draw free-body diagrams of the bar and the part, and label all forces.

b. Derive a formula for P in terms of W, L, θ, and μ_s.

c. For $W = 200$ lb, $L = 10$ ft, $\theta = 45°$, and $\mu_s = 0.2$, determine the magnitude of P.

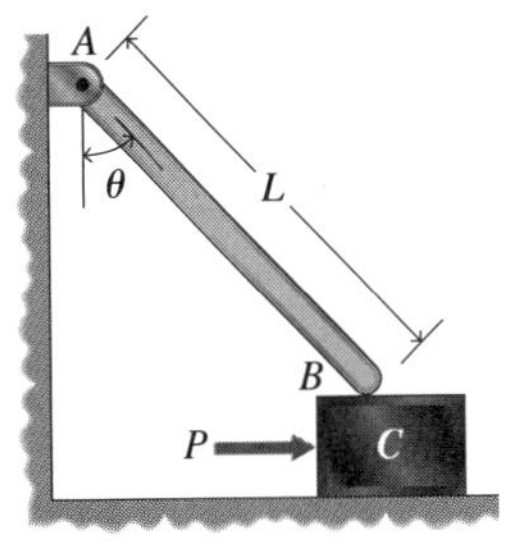

Figure P10.29

10.30 A rectangular drawer with dimensions a and b slides on horizontal frictionless guides and between parallel side walls, as shown in Fig. P10.30. If the depth a is smaller than the width b, the drawer might rotate slightly and jam between the side walls. Show that, if jamming is to be prevented, $a/b > \mu_s$, where μ_s is the coefficient of static friction for the side walls. (*Hint:* Suppose that the drawer is stuck, and examine the forces that act.)

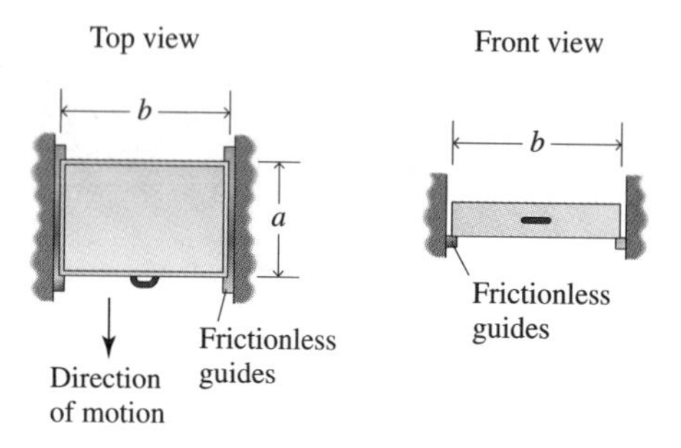

Figure P10.30

10.31 Two blocks are connected by a uniform rod that makes an angle of 30° with the horizontal (see Fig. P10.31). The weight of the rod is negligible, and block A weighs 50 N. The coefficient of static friction is $\mu_{sA} = 0.35$ between block A and the inclined plane and $\mu_{sB} = 0.20$ between block B and the vertical wall. The pins at the ends of the rod are frictionless.

a. Draw appropriate free-body diagrams, and write the equations that determine the maximum weight of block B for which the equilibrium position shown in Fig. P10.31 is maintained.

b. Determine the maximum weight of block B.

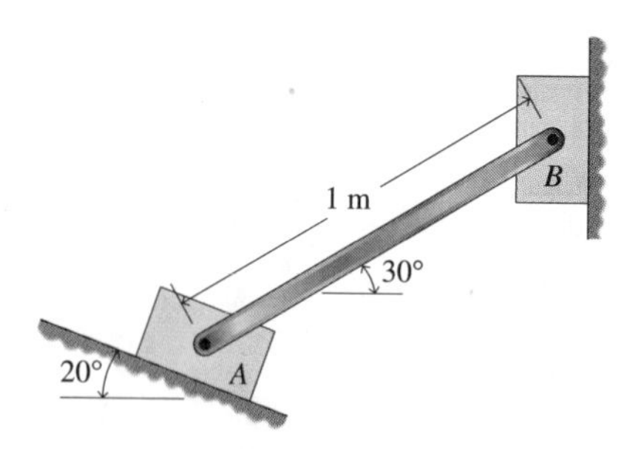

Figure P10.31

10.32 Roller B, shown in Fig. P10.32, weighs 200 N. The coefficient of static friction is the same for all contacting surfaces.

a. Neglecting the thickness and the weight of the bar, determine the least value of the coefficient of friction for which the roller does not slip.

b. At which surface is slipping more apt to occur?

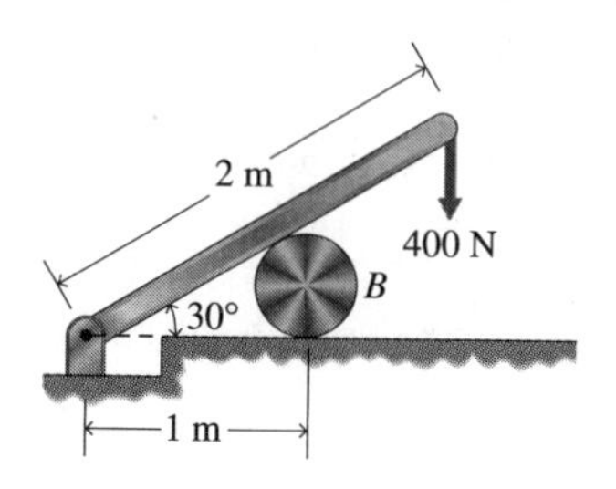

Figure P10.32

10.33 A homogeneous wedge B has been lowered into a sluice, as shown in Fig. P10.33. Surface DE is so slippery that it is practically frictionless. Workers leave the wedge in this position and go to lunch.

a. Assume the weight of the wedge is $3W$. Determine the minimum coefficient of static friction between the wedge and the surface CD to keep the wedge from slipping.

b. A boy sees the wedge and steps on it, placing his weight $2W$ at the center of the 3 ft horizontal face. If the coefficient of static friction between the wedge and surface CD is actually 0.4, determine whether or not the wedge will slip under the boy.

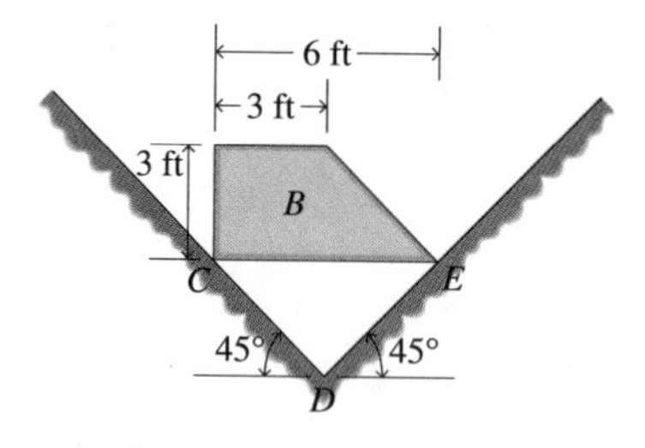

Figure P10.33

10.34 A uniform block of weight W rests on a horizontal floor (see Fig. P10.34). It is pushed by a force P, as shown.

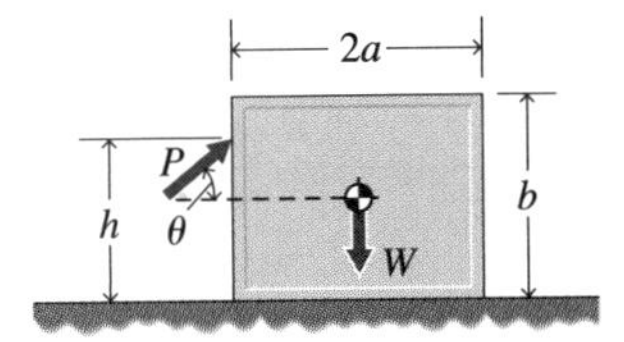

Figure P10.34

a. Show that, if the block does not slide,

$$\frac{P}{W} < \frac{\sin \phi_s}{\cos(\theta - \phi_s)}$$

where $\tan \phi_s = \mu_s$ is the coefficient of static friction between the block and the floor.
b. Show that, if the block does not tip,

$$\frac{P}{W} < \frac{a}{h \cos \theta + 2a \sin \theta}$$

10.35 In Fig. E10.5a, let the coefficients of static friction for surfaces 1, 2, and 3 be denoted by k, $2k$, and $3k$, respectively. Determine the minimum value of k to ensure that the system is self-locking.

10.36 A uniform semicircular concrete disk of constant thickness is set against a vertical wall, as shown in Fig. P10.36. The weight of the disk is W, and the coefficient of static friction is 0.30 for all contacting surfaces. Determine the maximum angle θ, in the range $0 < \theta < \pi/2$, for which equilibrium can exist. (*Hint:* Refer to Appendix D for the location of the centroid of the disk.)

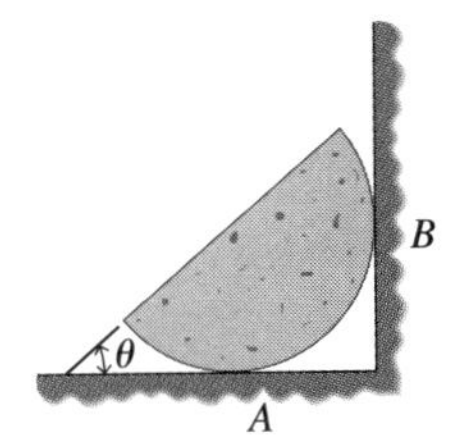

Figure P10.36

10.37 The semicircular disk in Fig. P10.36 is set against a smooth vertical wall at an angle $\theta = 30°$. The coefficient of static friction between the floor and the disk is $\mu_s = 0.30$.
a. Determine the frictional force required to keep the disk in equilibrium.
b. Determine whether or not this frictional force can exist.

10.38 The block shown in Fig. P10.38 weighs 100 N and is acted on by a force P that increases from zero. The coefficients of static and kinetic friction between the block and the surface on which it sits are $\mu_s = 0.60$ and $\mu_k = 0.20$. Plot the frictional force that acts on the block as a function of P for the range $0 \leq P \leq 200$ N.

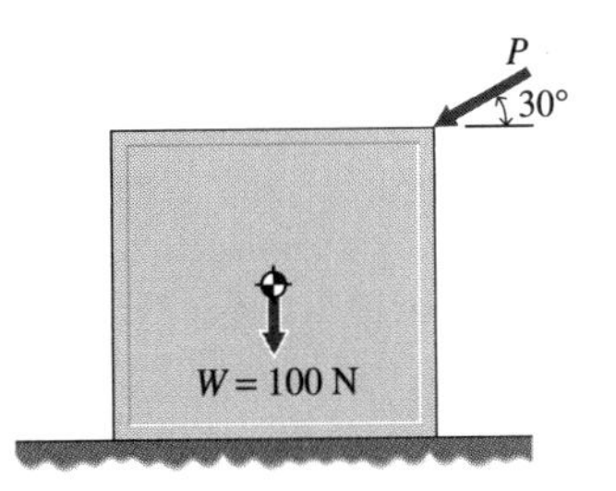

Figure P10.38

10.39 Work Problem 10.38 with the sense of P reversed—that is, with P directed up and to the right.

10.40 Refer to Example 10.4. Plot the ratios P/W and N/W as functions of the coefficient of static friction μ_s for the range $0 \leq \mu_s \leq 1.0$, with $\theta = 30°$ and $\phi = 45°$. (This plot illustrates the effect of changes in the coefficient of static friction.)

10.41 In Example 10.5, show that $P = 0.2163Q$ for the case where the block has impending motion downward.

10.42 The conveyor system shown in Fig. P10.42 is in equilibrium, and both spools A and B are on the verge of slipping. The friction of pulley P is negligible.
a. Determine the coefficients of static friction μ_{sA} and μ_{sB}, respectively, between spools A and B and the inclined surfaces.
b. Determine the ratio W_B/W_A of the weights of the spools.

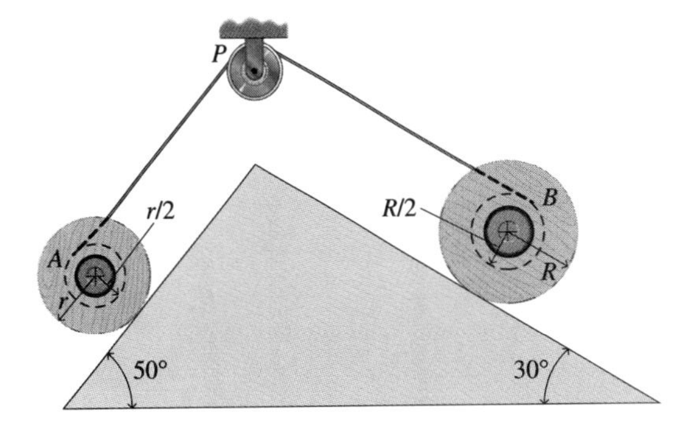

Figure P10.42

10.43 In early sailing ships, sailors used a windlass to heave up (weigh) the ship's anchor (see Fig. P10.43). Assume that a sailor is capable of exerting a force of 120 lb on the handle of a windlass. The windlass shaft has a diameter of 4 in. One end of the 1-in diameter anchor rope is fixed to the shaft. The coefficient of sliding friction of the bearing is $\mu_k = 0.30$.

a. How heavy an anchor (weight W) can the sailor raise? Assume that the bearing reactions on the shaft are equivalent to concentrated forces N (normal to the shaft) and $\mu_k N$ (tangent to the circumference of the shaft) that act on some unknown point on the shaft's circumference, and neglect the weight of the rope.

b. Calculate the ratio W/W', where W' is the weight of an anchor that the sailor can raise if the bearings are so well lubricated that friction is negligible.

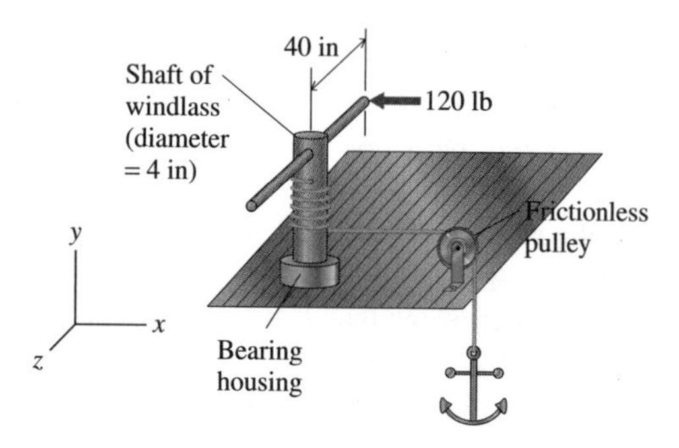

Figure P10.43

10.44 A second, identical crate is placed on top of the crate in Example 10.3 (see Fig. P10.44). The dimensions h and b are both equal to 600 mm. The horizontal force P acts at the top of the second crate and is increased gradually. Determine where the first movement occurs in this system. The coefficient of static friction is 0.20 between the lower crate and the floor and 0.35 between the two crates.

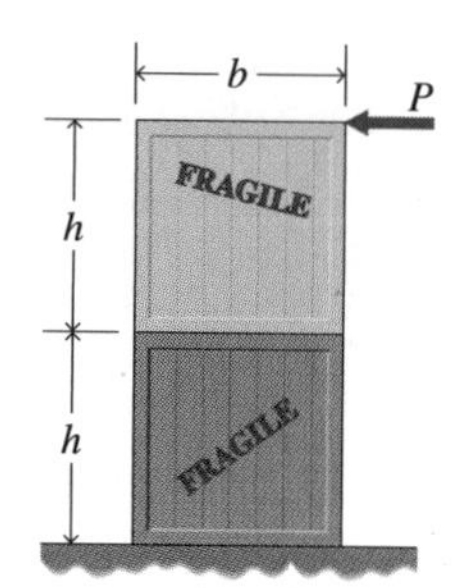

Figure P10.44

10.45 A wooden block of weight W is set on a wooden plank that is hinged at one end (see Fig. P10.45). The coefficients of static and kinetic friction are $\mu_s = 0.50$ and $\mu_k = 0.30$. Plot the ratio F/W, where F is the frictional force that acts on the block, as a function of the angle θ for $0 \leq \theta < 90°$.

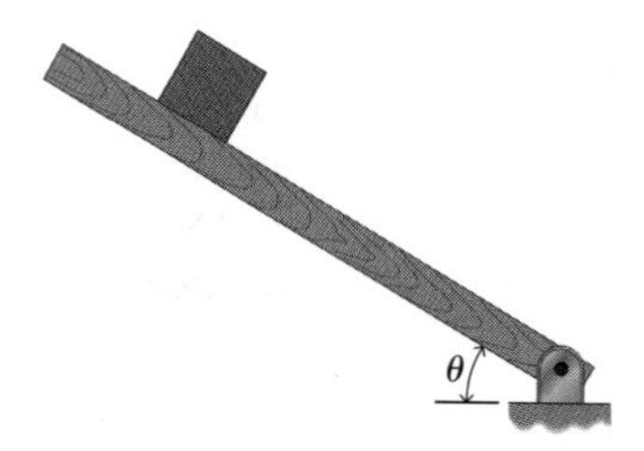

Figure P10.45

10.46 Figure P10.46a represents the jaws of a tension-testing machine that is used to determine the strength of a bar. The flat bar is clamped by wedges B and is subjected to a tension force P, as shown in Fig. P10.46a. A free-body diagram of the left-side wedge is shown in Fig. P10.46b. If the jaws are designed properly, the bar will not slip under any load P, for the greater the tension P, the more tightly the wedges grip the bar. Consequently, the wedges slide slightly on the surfaces of the retainers H as the load is applied. Denote the coefficient of static friction between the bar and each wedge B by μ_s, and denote the coefficient of sliding friction between each wedge B and retainer H by μ_k. Neglect the weight of the wedge.

a. Write the equations of equilibrium of forces for a wedge B.

b. Using the relations $P/2 < \mu_s N$ and $F = \mu_k R$, show that the angle ϕ must satisfy the following inequality if the jaws are to operate properly:

$$\tan\phi < \frac{\mu_s - \mu_k}{1 + \mu_s\mu_k}$$

c. Then, show that $\phi < \phi_s - \phi_k$, where $\mu_s = \tan\phi_s$ and $\mu_k = \tan\phi_k$.

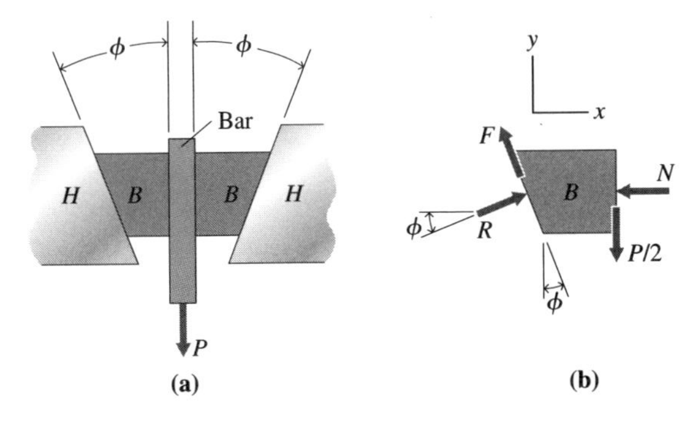

Figure P10.46

10.47 Figure P10.47 illustrates a machine for roll-finishing aluminum-alloy sheets. The two sets of rollers are driven at the same angular speed [rpm]. Consequently, since the front rollers are larger than the rear ones, the sheet is stretched continuously in the interval between the front and rear rollers. The sheet is 0.8 mm thick. The normal stress (force per unit area of cross section) in the sheet is 430 N/mm². The coefficient of static friction between a roller and the sheet is $\mu_s = 0.80$. Calculate the force N (per unit width of roller perpendicular to the plane of the page) with which the rollers must be pressed against the sheet to prevent slipping. (*Hint:* Consider a free-body diagram of a length of the sheet obtained by passing one vertical plane between the large and small rollers and a second vertical plane to the right of the larger rollers. Note that the normal stress in the sheet is 430 N/mm² between the small and large rollers and is zero to the right of the large rollers.)

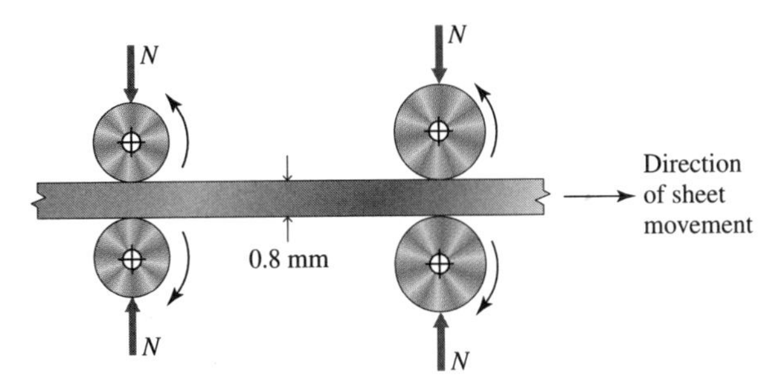

Figure P10.47

10.48 Work Example 10.5 including the weights W_A of the wedge and W_B of the block, where $W_A = W_B = Q/20$.

10.49 In Fig. P10.49, the coefficient of friction between the automobile tires and the pavement is 0.75. Assuming that slipping does not occur, calculate the angle θ of the steepest hill that the car can ascend at constant speed for the following cases:

a. The car has four-wheel drive.

b. The car has only rear-wheel drive.

c. The car has only front-wheel drive.

(Recall that the laws of statics apply for a body that moves at constant speed.)

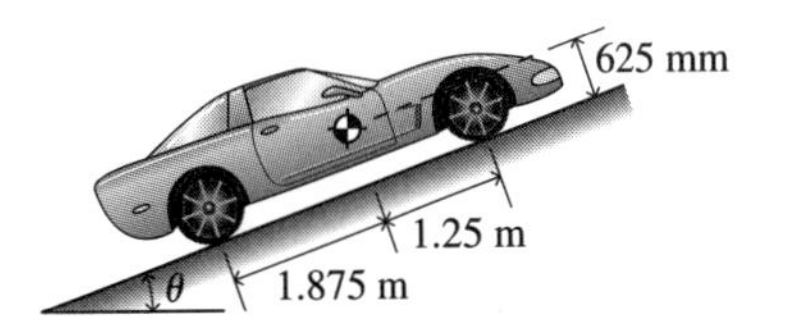

Figure P10.49

10.50 **a.** Draw a free-body diagram of the block of Example 10.4 for the case of sliding impending down the plane.
b. Derive equations for P and N analogous to Eqs. (d) and (e) of Example 10.4.

10.51 A uniform rigid, straight bar of square cross section lies on a horizontal floor (see Fig. P10.51). A horizontal force F just sufficient to cause motion of the bar to be impending is applied at one end perpendicular to the longitudinal axis of the bar. The length of the bar is L, the weight of the bar is W, and the coefficient of static friction between the bar and the floor is μ_s. Assume that the reactive pressure of the floor is uniformly distributed along the bar.
a. Show that the bar is on the verge of rotating about a vertical axis at the distance $L/\sqrt{2}$ from the end at which the force F is applied.
b. Then, show that $F = \mu_s W(\sqrt{2} - 1)$.

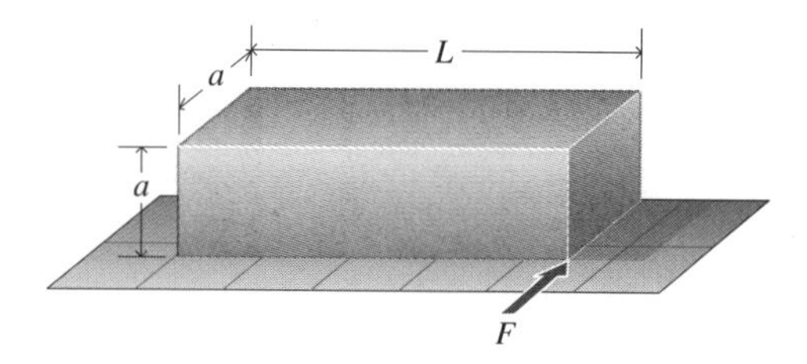

Figure P10.51

10.52 In Example 10.4, let $\mu_s = \tan \phi_s = 0.2$.
a. Plot the ratio P/W as a function of θ for the range $0 \leq \theta < 90°$, with $\phi = 30°$.
b. Plot the ratio P/W as a function of ϕ for the range $0 \leq \phi < 90°$, with $\theta = 30°$.
c. Check the results of parts a and b for the case where $\theta = \phi = 30°$.
(The results of this problem illustrate the relative significance of the inclination of the plane and the direction of the force P.)

10.53 In Example 10.5, determine the smallest coefficient of static friction μ_s between block A and the horizontal plane that will make the system self-locking.

10.54 A Roman contractor once attempted to move a stone column into Rome by mounting it as an axle between two large wheels (rigidly attached to the column) and rolling it (see Fig. P10.54). His reputation was ruined because he could not keep the column rolling straight. His difficulty occurred because the wheels had slightly different radii ($r_1 < r_2$). Suppose that the system consisting of the column and wheels weighed 30,000 lb, that the distance between the wheels was 20 ft, that the center of gravity of the system was at the distance x from the smaller wheel ($x < 10$ ft), and that the coefficient of kinetic friction was 0.60. The wheel with the lighter load skidded and the other wheel rolled, with friction forces acting between the wheels and the ground.
a. Determine which wheel skidded.
b. In terms of x, determine the couple M that must be applied to keep the column rolling straight on level ground. (*Hint:* Since the frictional force that acts on the rolling wheel is not known, sum the moments of all the forces acting on the system about the vertical axis that passes through the base of the rolling wheel. Then, note that the sum of moments must be zero to keep the column from turning about the vertical axis—that is, to keep the column rolling straight.)
c. Could the contractor have succeeded if he had shifted the column axially, so that more load was carried on one wheel? Explain.

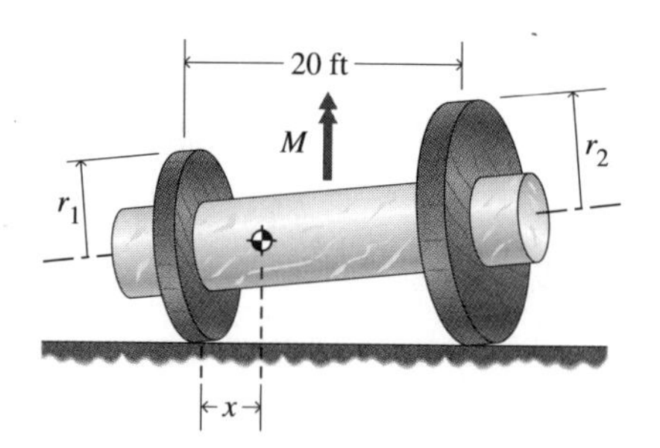

Figure P10.54

10.55 A uniform stick of length L and weight W leans at an angle θ in a vertical plane against a ledge of height h ($L > h/\sin\theta$), as shown in Fig. P10.55. The coefficient of static friction is μ_s.

a. Show that, if the stick is on the verge of slipping,

$$\cos\theta - \cos^3\theta = \frac{2\mu_s h}{(1 + \mu_s^2)L}$$

b. Then, show that, if $\mu_s h/[(1 + \mu_s^2)L] > 1/\sqrt{27}$, the stick will not slip at any angle. (*Hint:* Show that the maximum value of $\cos\theta - \cos^3\theta$ is $2\sqrt{3}/9$.)

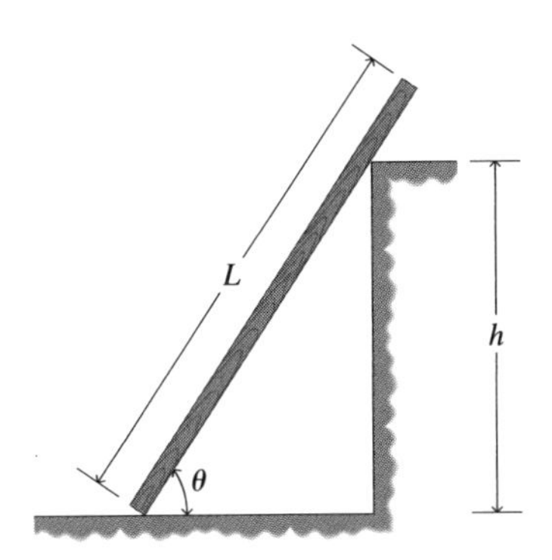

Figure P10.55

10.56 The sense of the force P in Fig. P10.29 is reversed (that is, it is directed to the left).

a. Solve Problem 10.29 for this case.

b. For $\mu_s = 0.2$, plot the ratio P/W as a function of θ, and determine a value of θ for which the machine part cannot be moved to the left.

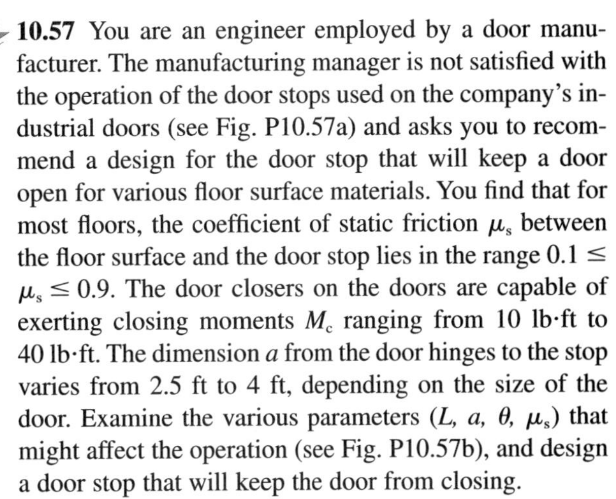

10.57 You are an engineer employed by a door manufacturer. The manufacturing manager is not satisfied with the operation of the door stops used on the company's industrial doors (see Fig. P10.57a) and asks you to recommend a design for the door stop that will keep a door open for various floor surface materials. You find that for most floors, the coefficient of static friction μ_s between the floor surface and the door stop lies in the range $0.1 \leq \mu_s \leq 0.9$. The door closers on the doors are capable of exerting closing moments M_c ranging from 10 lb·ft to 40 lb·ft. The dimension a from the door hinges to the stop varies from 2.5 ft to 4 ft, depending on the size of the door. Examine the various parameters (L, a, θ, μ_s) that might affect the operation (see Fig. P10.57b), and design a door stop that will keep the door from closing.

Method: Draw one free-body diagram of the door and door stop as a unit and another of the door stop alone. Determine the forces that the floor surface exerts on the stop. Consider various configurations of the door stop, and select one that will ensure that the door will not close for the range of parameters described in the problem statement.

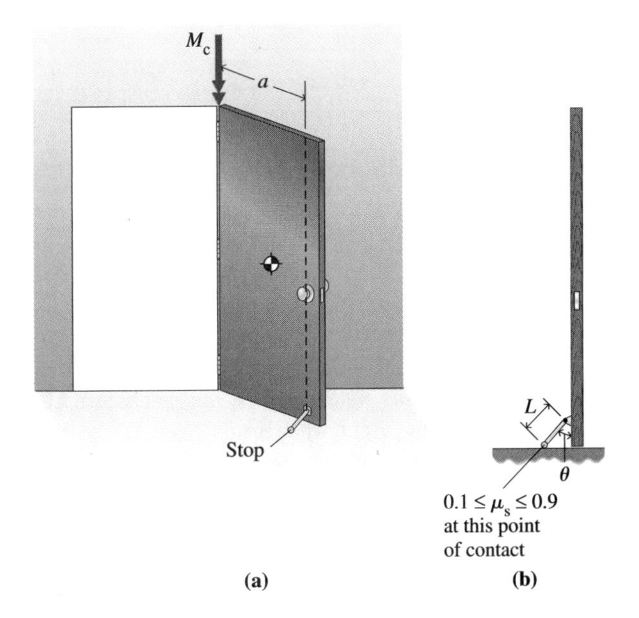

Figure P10.57

10.58 The rock crusher of Problem 6.9 is shown in Fig. P10.58. The crusher must be modified so that it can exert a crushing force of 20,000 lb on a rock. The coefficient of static friction between the crushing block A and the surface on which it rests is $\mu_s = 0.20$. The hydraulic press that exerts the force F at joint D is capable of producing a maximum force of 1 ton. The rods AC and BC must maintain angles of 15° with the horizontal, and the weight of block A is negligible. Develop a design to meet the specified requirements.
Method: Analyze the forces in the rock crusher as it is shown in Fig. P10.58. Then, make adjustments to satisfy the requirements.

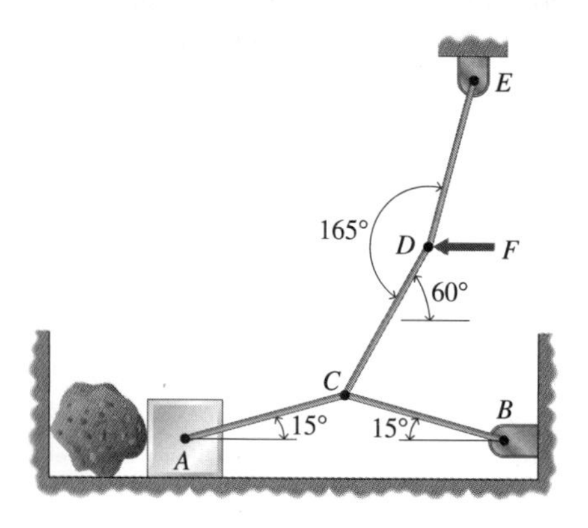

Figure P10.58

10.59 Let the weight of the bar AB in Problem 10.29 be W [kN]. The weight W_c of the part C may range from $0.5W$ to $2W$, and the coefficient of static friction μ_s between the bar and machine parts of different materials may range from 0.1 to 0.6. Assume that the surface on which part C rests is so well lubricated that its friction is negligible. Design the system so that the machine parts can be moved to the right when force P pushes to the right, but cannot be moved to the left when P pushes to the left.
Method: Assume impending motion in each case, and express P as a function of W, μ_s, and θ. Examine this function to determine conditions that satisfy the design requirements.

10.60 Let the coefficient of static friction between the block C in Problem 10.59 and the surface on which it rests also be μ_s. Your supervisor asks you to design a system that will satisfy the requirements given in Problem 10.59. Examine the problem and evaluate the feasibility of meeting the requirements.

10.61 Baggage in an airport is conveyed to the passengers by means of a ramp conveyor-belt system (see Fig. P10.61). Suitcases, packages, backpacks, and so forth are made with widely different materials. The coefficients of static friction between the various types of baggage and a proposed conveyor belt range from 0.2 to 0.9. Because of space requirements, the angle θ must be greater than 20°.
a. Design a conveyor system so that suitcase A (W_A = 75 lb, $\mu_A = 0.2$), suitcase B (W_B = 50 lb, $\mu_B = 0.6$), and suitcase C (W_C = 30 lb, $\mu_C = 0.9$) can be successfully conveyed in the order shown.
b. Discuss your design, considering the case in which the order of the suitcases is changed. Based on your considerations, what might you recommend regarding the surface material of the conveyor belt?

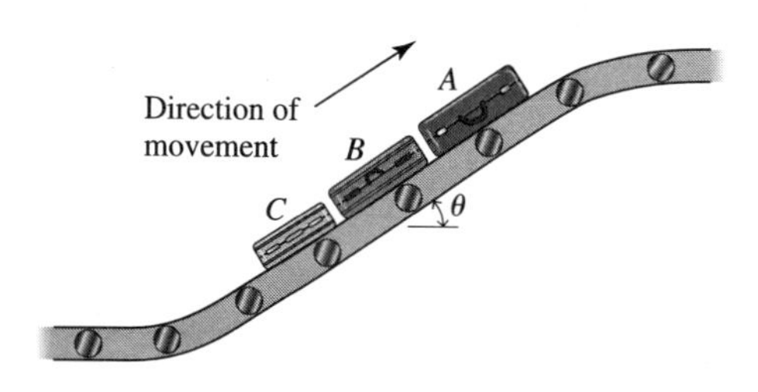

Figure P10.61

10.4 *Simple Machines and Friction*

10.62 A push screwdriver is formed by inserting the shank of a screwdriver into a handle that has a built-in screw thread of square cross section, so that the shank rotates when the handle is pushed down (see Fig. P10.62). The pitch of the thread is 1 in, and the mean diameter of the screw in the shank is $\frac{3}{8}$ in. The coefficient of kinetic friction is 0.20. If you push down on the handle with a force of 30 lb to drive a screw into a piece of wood, what torque is exerted on the screw head?

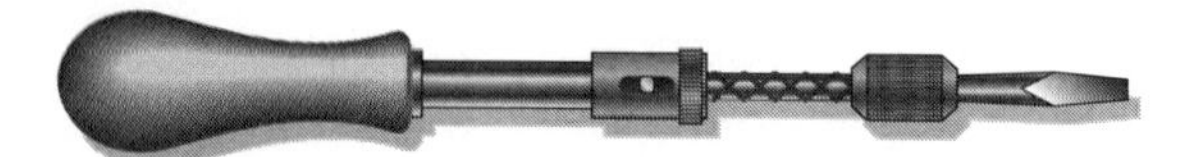

Figure P10.62

10.63 The press shown in Fig. P10.63 is used for bending or straightening steel rods. Determine the lateral force P_z, applied with a lever arm of 900 mm, required to produce a vertical force $Q_y = 100$ kN on the rod. The screw threads have a mean diameter of 50 mm and a pitch of 5 mm. The coefficient of kinetic friction between the screw and the nut is 0.20. Ignore friction between the cap and the screw shank.

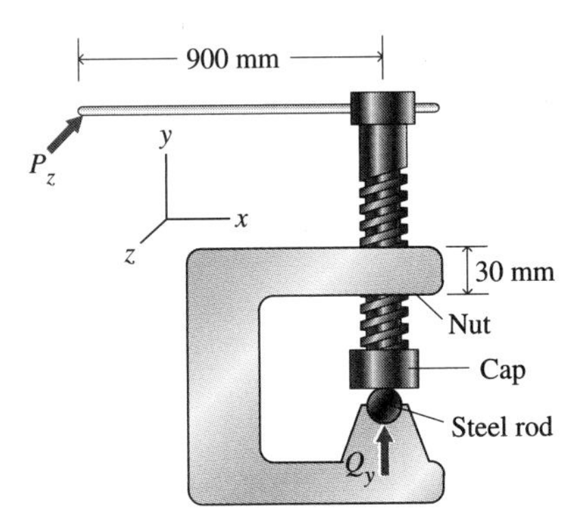

Figure P10.63

10.64 The mean radius of a single-threaded, square-thread jackscrew is 1.00 in. The pitch of the thread is 0.40 in, and the coefficient of kinetic friction for the screw and nut is 0.15.

a. What force must be applied to the jack, with a lever arm of 18 in, to raise a load of 6000 lb at a constant rate?

b. What force is required to lower the load at a constant rate?

10.65 A screw in a book press (see Fig. P10.65) has 5 threads per inch. The mean diameter of the threads is 1.2 in. The coefficient of kinetic friction is 0.12. Two forces of magnitude 20 lb are applied as shown. Determine the force that the press exerts on the book.

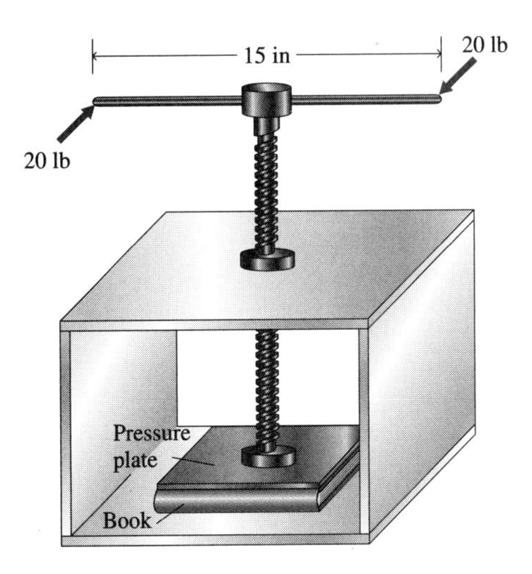

Figure P10.65

10.66 A jackscrew that is to support large loads (as much as 45 kN) has to be designed to be self-locking. When the jackscrew is used in the field, the coefficient of kinetic friction in its threads may vary from 0.10 to 0.40, depending on the amount of dirt that collects on the threads. The mean radius of the threads is restricted to $r =$ 50 mm. Determine the required pitch of the jackscrew.

10.67 A single-thread jackscrew with a square thread is used to raise the front end of a car (see Fig. P10.67). The car weighs 15 kN. The screw has a mean diameter of 38 mm and a pitch of 10 mm. The coefficient of kinetic friction of the screw is 0.1. The collar is so well lubricated that its friction is negligible. Determine the turning moment required to raise the front wheels of the car off the ground.

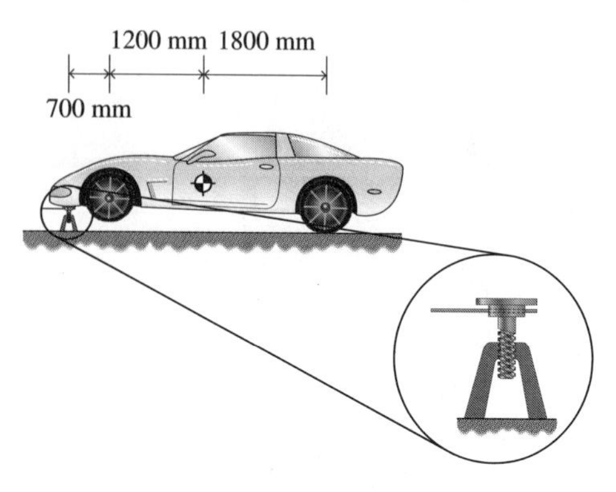

Figure P10.67

10.68 A square-thread screw, with outer and inner diameters of 1.5 in and 1.25 in, respectively, and 4 threads per inch, is used in a jack. The coefficient of kinetic friction is $\mu_k = 0.10$. A force of $R = 60$ lb is applied to the lever at a distance $a = 16$ in from the axis of the screw (see Fig. 10.10a).

a. Determine the load P that the 60 lb force can raise.

b. Determine the force R that is required to lower the load P from part a.

10.69 A 5° wedge is used to split a block of wood (see Fig. P10.69). The wedge is driven into the wood block with a force of 240 lb. The coefficient of friction between the wedge and the block is 0.25 (steel on wood).

a. Calculate the splitting force N normal to the wedge, neglecting the effect of friction. Assume that the wedge penetrates the block at constant speed.

b. Repeat part a, but include the effect of friction.

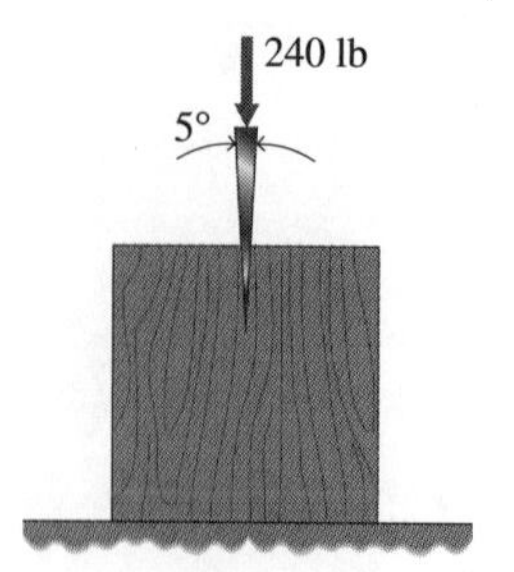

Figure P10.69

10.70 The compound jackscrew shown in Fig. P10.70 has a pitch of 1 in. The mean diameter of the screw is 1.5 in. The coefficient of kinetic friction is 0.10. Calculate the moment required to raise the 8000 lb weight.

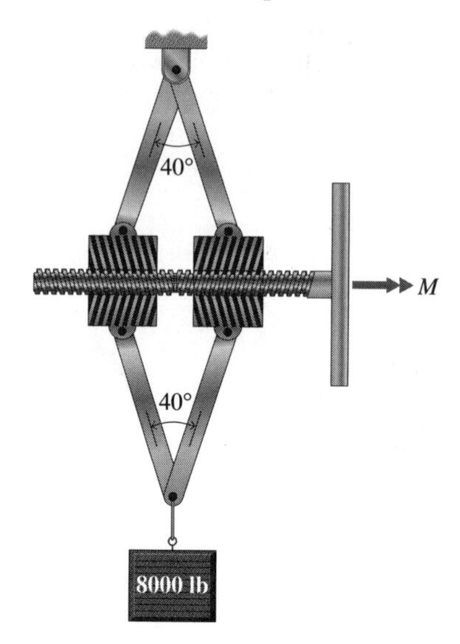

Figure P10.70

10.71 A single-thread screw with a square thread is used in a jack. The screw has 0.2 thread per millimeter. The outer and inner diameters of the thread are 18.75 mm and 14.375 mm, respectively. The coefficient of kinetic friction is $\mu_k = 0.12$.

a. Determine the efficiency η of the jack.

b. Determine whether or not the jack is self-locking.

10.72 A jackscrew has a single square thread with a mean diameter $d_m = 72$ mm and a pitch $p = 16$ mm (see Fig. P10.72a). Under the action of load P alone ($R = 0$), the screw is on the verge of moving downward (see Fig. P10.72b).

a. Determine the coefficient of static friction μ_s of the jackscrew.

b. Determine the efficiency of the jackscrew.

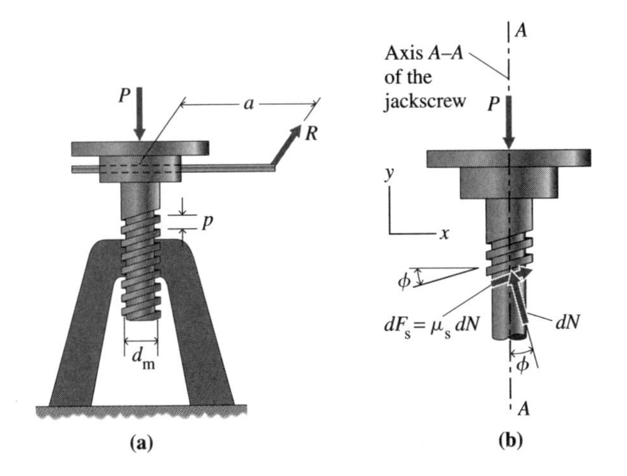

Figure P10.72

10.73 Assume that the coefficient of static friction of the jackscrew of Problem 10.72 is $\mu_s = 0.30$, and the coefficient of kinetic friction is $\mu_k = 0.10$. A load $P =$ 50 kN is applied.

a. Determine the moment $M_s = Ra$ required to cause upward motion to be impending.

b. Assume that the upward motion occurs at constant speed. Determine the moment $M_k = Ra$ required to maintain this motion.

c. What is the efficiency of the jackscrew?

d. What is the mechanical advantage of the jackscrew for lowering the load if the lever arm is $a = 400$ mm?

10.74 a. Draw a diagram showing the normal and tangential forces that act on an infinitesimal length of the thread of a jackscrew for the case in which the load P is lowered by means of a clockwise moment $M = Ra$ (see Fig. P10.72).

b. Then, derive Eq. (10.9).

10.75 A jackscrew has a pitch of 16 mm and a mean diameter of 64 mm (see Fig. P10.72). The coefficient of sliding friction is 0.40.

a. Determine the moment required to lower a load of $P = 45$ kN.

b. Determine the moment required to raise a load of $P = 45$ kN.

c. If the lever arm of the jackscrew is $a = 300$ mm, what is the mechanical advantage for raising the load?

10.76 You realize that the coefficient of kinetic friction of the jackscrew in Problem 10.75 is large. Therefore, to increase the mechanical advantage, you lubricate the screw, and you estimate that the coefficient of friction is reduced to 0.05.

a. Now what is the mechanical advantage of the jackscrew if a load is raised?

b. What moment is required to lower the load of 45 kN?

10.77 To level a floor, a carpenter needs to drive a wedge between the floor joist and a support wall (see Fig. P10.77). The wedge angle is 4°, and the top face is horizontal. The coefficient of friction is estimated to be 0.7 between all surfaces. The carpenter is able to deliver a 1 kN blow to the wedge. The vertical load on the wall is 1.5 kN.

a. Ignoring friction, determine whether or not the carpenter can drive the wedge between the floor joist and the support wall.

b. Repeat part a, but include the effect of friction.

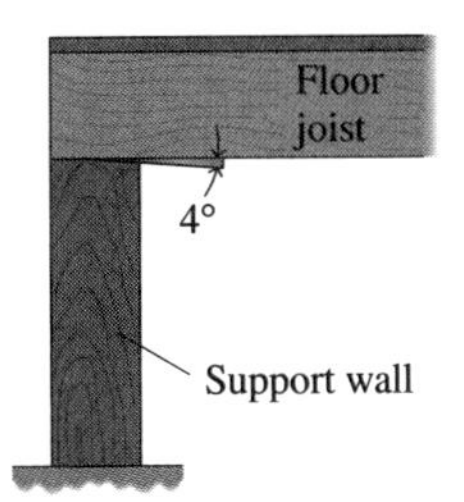

Figure P10.77

10.78 For a jackscrew, the coefficient of kinetic friction is $\mu_k = 0.10$.

a. For raising a load, plot the efficiency η_a as a function of ϕ for the range $1.1\phi_k \leq \phi \leq 1.9\phi_k$.

b. For lowering a load, plot the efficiency η_b as a function of ϕ for the range $1.1\phi_k \leq \phi \leq 1.9\phi_k$.

c. Plot the ratio η_b/η_a for the range $1.1\phi_k \leq \phi \leq 1.9\phi_k$.

d. What conclusions can you draw from these results?

10.79 For an urban renewal project, a building contractor has agreed to move 50 salvageable houses to a new site. He plans to raise the houses about 10 in off their foundations with jackscrews, so that they can be rolled onto a trailer. For this purpose, he asks you to develop specifications for a self-locking jackscrew that is capable of raising 8 tons.

a. Develop specifications for a jackscrew that meets the contractor's needs.

b. Write a one-paragraph explanation of your recommended specifications for the contractor.

Method: First, select a pitch and other parameters. Then, examine the load-carrying capacity of the jackscrew. Select reasonable parameters for a jackscrew that can travel 13 in, to allow some tolerance on the 10 in requirement.

10.5 *Belt Friction*

In the following problems, assume that the belts are flat, unless stated otherwise.

10.80 So that he can clear a blockage, a sailor who weighs 170 lb is lowered into a large ventilator shaft of a ship, as shown in Fig. P10.80. The coefficient of kinetic friction between the line and the ventilator is 0.50. *AB* is an arc of 90°. Neglecting the weight of the line and the small frictional force between the sailor and the shaft, compute the force *P* required to lower the sailor.

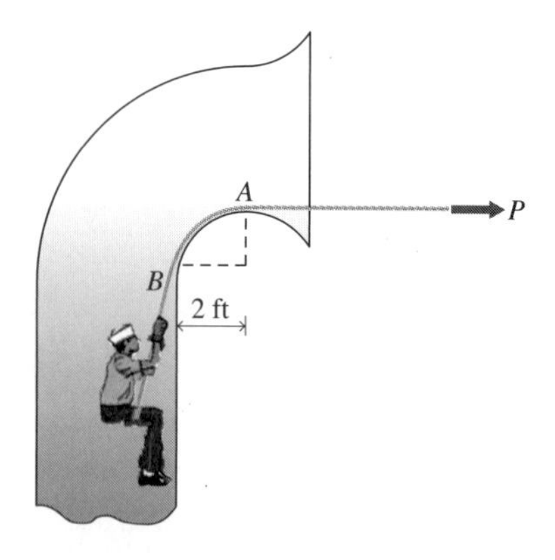

Figure P10.80

10.81 Calculate the force *P* required to hoist the sailor in Problem 10.80 up the ventilator shaft.

10.82 The cross section of a V-belt is shown in Fig. P10.82a. A single V-belt with this cross section is selected to deliver power to the wheel-drive transmission of a riding tractor. The driver sheave has an effective diameter of 6.0 in, and the driven sheave has a diameter of 12.0 in. Two idler pulleys are employed (see Fig. P10.82b) to produce a contact angle of 180° for both sheaves. The maximum torque delivered to the driven sheave is 240 lb·in when the smaller belt tension of the driven sheave is 40 lb. If slipping is impending between the belt and the driven sheave, determine the coefficient of friction between the belt and the driven sheave.

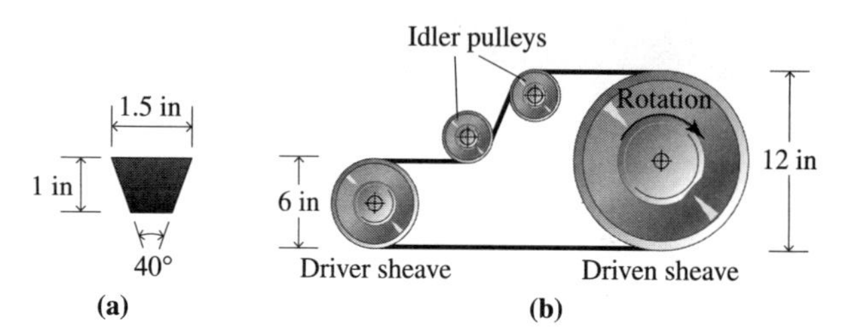

Figure P10.82

10.83 A V-belt, with an angle of 40°, has a maximum allowable tension of 400 N. It is strung between two identical pulleys of diameter 100 mm, which are 1 m apart. The coefficients of static and kinetic friction are 0.20 and 0.10, respectively. Determine the maximum torque that this belt-pulley system can transmit.

10.84 The disk of the band brake system in Fig. P10.84 rotates clockwise. The coefficient of kinetic friction between the band and the disk is 0.3. A force $P = 100$ lb is applied to the control lever. Determine the band tensions and the braking torque transmitted to the disk.

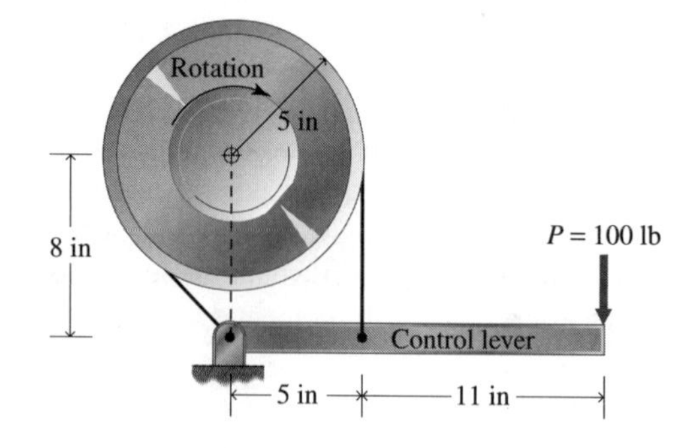

Figure P10.84

10.85 The rotation of the disk in Problem 10.84 is reversed; that is, the disk rotates counterclockwise. Repeat the problem for this case.

10.86 A torque T is applied to pulley A, which drives pulley B (see Fig. P10.86). The two pulleys have equal radii. The tension in the slack side of the belt is 2 kN. The coefficient of friction between the belt and the pulleys is 0.30.

a. What is the maximum possible torque that can be applied to pulley A?

b. What torque is transmitted to pulley B?

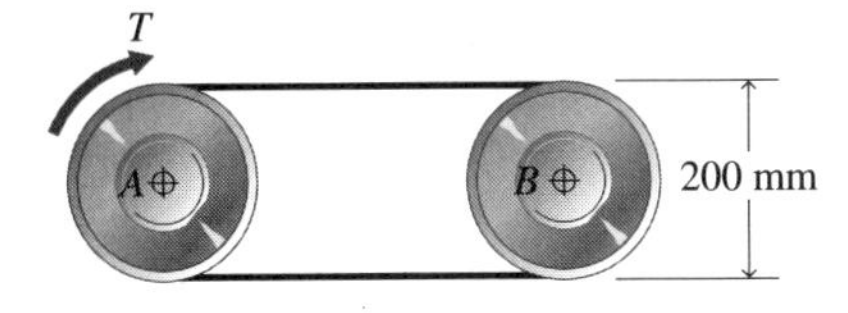

Figure P10.86

10.87 If the centers of the two pulleys in Problem 10.86 are 750 mm apart and if the diameter of driver pulley A is increased to 400 mm (see Fig. P10.87), what is the maximum possible torque that can be transmitted to pulley B?

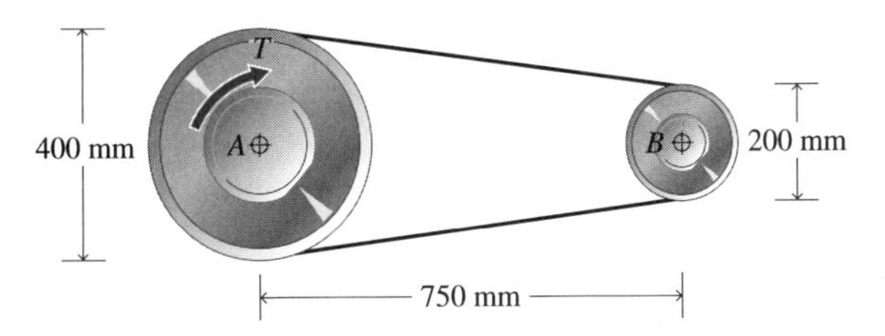

Figure P10.87

10.88 An idler pulley is installed to increase the angle of wrap of a belt-pulley system (see Fig. P10.88). A torque T is applied to the driver pulley A. The coefficient of friction is 0.30. The tension in the slack side of the belt is 2 kN.

a. Calculate the maximum torque that can be transmitted to pulley B (compare your answer with the one you obtained for Problem 10.87).

b. What conclusion can you draw concerning the benefit of using an idler pulley in this belt-pulley system? Do you see any disadvantages?

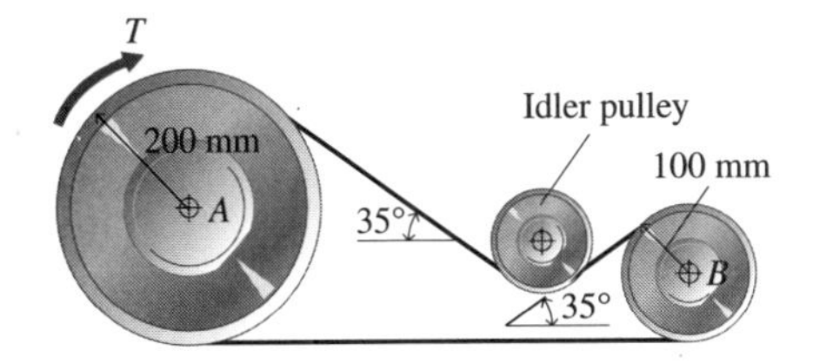

Figure P10.88

10.89 The cylindrical drums A and B in Fig. P10.89 are fixed so that they do not rotate. A worker can exert a 100 lb upward push on lever CD. The coefficients of static and kinetic friction between the belt and the drums are both 0.40.

a. Determine the maximum weight W that the worker can lift.

b. Determine the maximum weight W that the worker can hold.

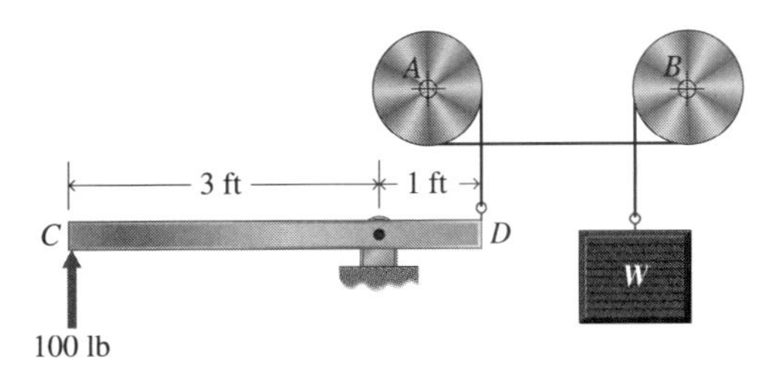

Figure P10.89

10.90 In Fig. P10.90, a torque of 1 kN·m is transmitted to pulley D by its shaft. The torque is resisted by the tensions in the belt connected at B and C, which in turn are balanced by the force F applied to the rod ABC. The coefficient of static friction between the belt and the pulley is 0.60. The shaft at O and the pin at A are frictionless.

a. Determine the magnitude of force F and the tensions in the belt at B and C.

b. Determine where the force F should be located so that the reaction at support A is zero, and determine F for this condition.

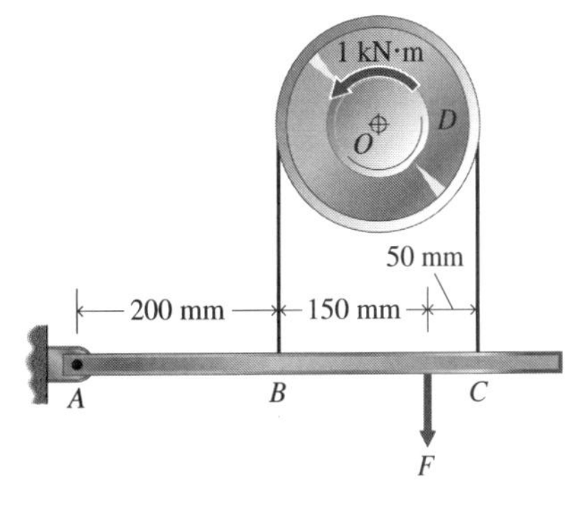

Figure P10.90

10.91 To go up or down in a tree, a tree surgeon who weighs 540 N rigs a rope to a limb, as shown in Fig. P10.91. She stands in the bight B and increases or reduces the pull on strand S to go up or down. The coefficients of kinetic friction between the rope and the limb and between the rope and her shoe are both equal to 0.40. Assume that all rope strands are vertical.

a. Calculate the pull that she must apply to S to go down slowly.

b. Calculate the pull that she must apply to S to go up slowly.

c. Comment on the results.

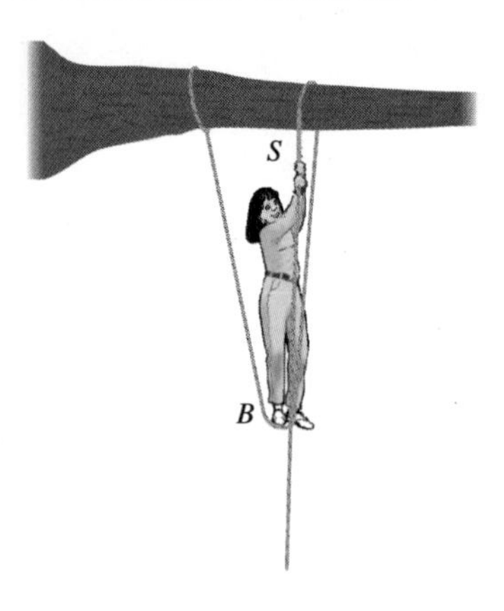

Figure P10.91

10.92 In the belt-pulley system shown in Fig. P10.92, a torque T is applied to the drive pulley A by its shaft. The maximum tension in the belt is 6 kN. The coefficient of friction between the belt and the pulleys is 0.25. Plot the maximum torque that can be transmitted to pulley B as a function of the diameter d of B, over the range 100 mm $\leq d \leq$ 600 mm.

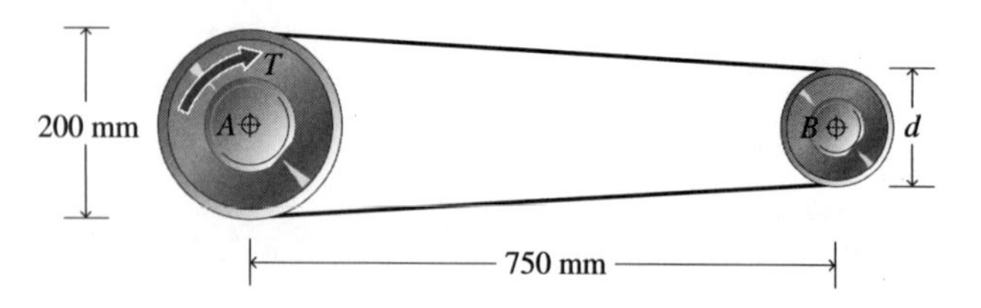

Figure P10.92

10.93 An idler pulley is installed in a belt-pulley system to increase the angle of wrap (see Fig. P10.93). To study the design of the system, an engineer applies a torque T to the driver pulley A and, by changing the length of the belt and the position of the idler pulley, computes the maximum torque that can be transmitted to pulley B.

a. If the maximum tension in the belt is 5 kN and the coefficient of friction is 0.30, plot the maximum torque that can be exerted by pulley A as a function of θ, over the range $15° \leq \theta \leq 45°$.

b. Determine the largest torque that can be transmitted to pulley B for the range $15° \leq \theta \leq 45°$.

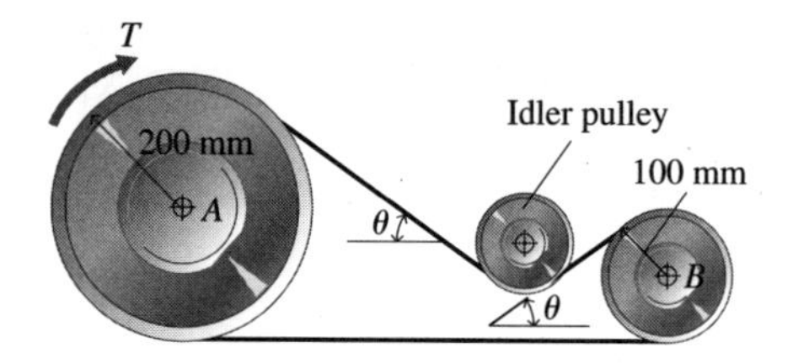

Figure P10.93

10.94 A manufacturing company wishes to use the belt-pulley system of Fig. P10.93 in its plant. However, the chief engineer wants to adjust the idler pulley so that the angle θ can be varied over the range $20° \leq \theta \leq 40°$. Also, he wants to use belts of different materials for which the coefficient of static friction μ_s lies in the range $0.1 \leq \mu_s \leq 0.5$. The chief engineer asks you to design the belt-pulley system and to determine the maximum tension in the belt, using the same pulley dimensions and a slack-side tension of $T_2 = 2$ kN.

a. Plot design values for the maximum tension T_1 in the belt as a function of the angle θ over the range $20° \leq \theta \leq 40°$ and for $\mu_s = 0.1$, 0.2, 0.3, 0.4, and 0.5 (see Fig. 10.14).

b. Write a short (one-paragraph) report for the chief engineer, explaining your design plot.

10.6 *Friction Clutches and Brakes*

10.95 An airplane model is attached to a cylindrical base 3 m in diameter, which is set on a concrete slab (see Fig. P10.95). The gravity axis of the model coincides with the geometric axis of the base. Together, the model and the base weigh 360 kN. The coefficient of static friction between the base and the slab on which it rests is 0.60, and the coefficient of kinetic friction is 0.25.

a. Assuming that the pressure between the base and the slab is uniform, determine the couple required to initiate rotation of the model about its gravity axis.

b. Determine the couple required to maintain the rotation at a constant rate.

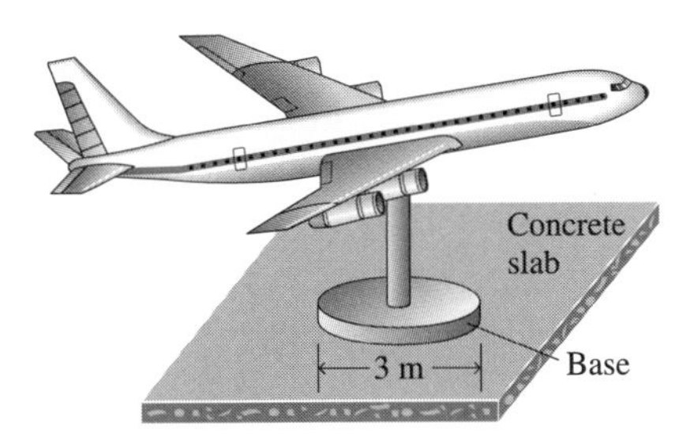

Figure P10.95

10.96 A single-collar thrust bearing is illustrated in Fig. P10.96. It consists of a shaft S, a collar C (that is rigidly attached to the shaft), and a support B. The radius of the shaft is r_S, and the radius of the collar is r_C. Assume that the pressure between the collar and the support is uniformly distributed.

a. Derive a formula for the pressure p between the collar and the support in terms of W, r_S, and r_C. (There is only a very slight clearance between the shaft and the base support.)

b. Derive a formula for the torque T required to initiate rotation of the collar and shaft relative to the support, in terms of W, r_S, r_C, and μ_s, the coefficient of static friction. [*Hint:* Note the derivation of Eqs. (10.20) and (10.21).]

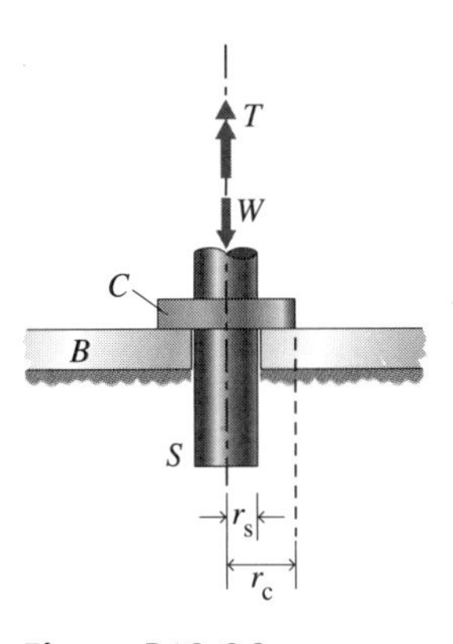

Figure P10.96

10.97 The single-collar thrust bearing shown in Fig. P10.97 is subjected to an axial load of 10,000 lb. The coefficients of static and kinetic friction between the collar C (which is rigidly attached to the shaft S) and the support B are 0.40 and 0.20, respectively.
a. Assuming that the pressure distribution on the collar is uniform, determine the torque T_k required to rotate the shaft at a constant rate.
b. What torque T_s is required to initiate rotation?

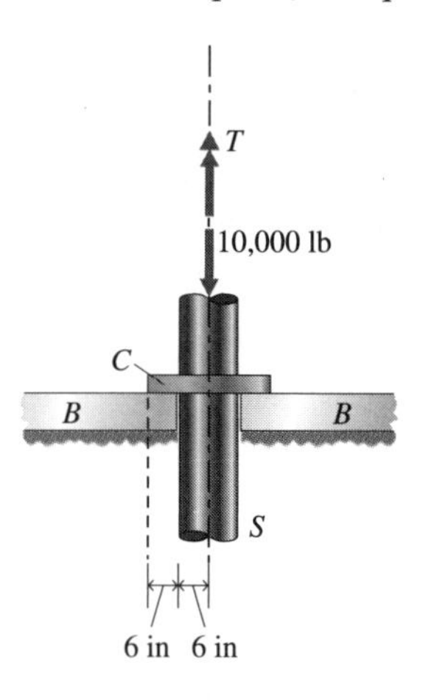

Figure P10.97

10.98 In some applications of collar bearings, multiple collars are used. Two collars are used as shown in Fig. P10.98 to support the 10,000 lb load of Problem 10.97 ($\mu_s = 0.40$ and $\mu_k = 0.20$).
a. Determine the torque T required to initiate rotation of shaft S. Assume that the load is divided equally between the collars and that the pressure on each collar is uniform.
b. Why are multiple collars used?

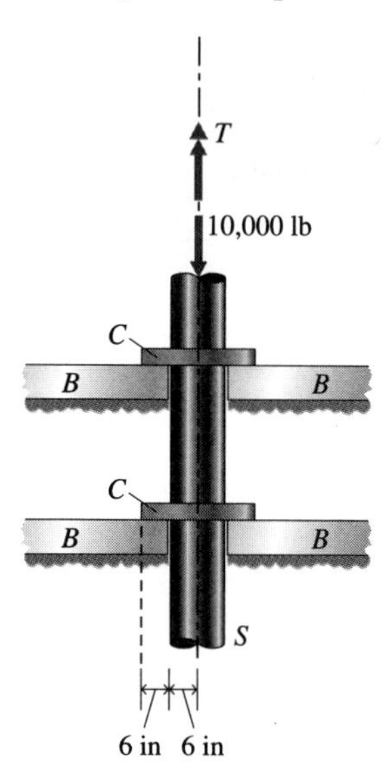

Figure P10.98

10.99 The weight that acts on a single-collar thrust bearing (refer to Fig. P10.96) is $W = 30$ kN. The shaft radius is 200 mm, the collar radius is 350 mm, and the coefficients of static and kinetic friction are 0.15 and 0.10, respectively.
a. Determine the pressure p that exists between the collar C and the support B, assuming that p is uniformly distributed.
b. Determine the torque T_s required to initiate rotation of the collar and shaft.
c. Determine the torque T_k required to maintain rotation at a constant rate.

10.100 For the disk clutch of Fig. 10.18, $r_0 = 2$ in and $r_1 = 5$ in. The pressure between the disks decreases linearly from 100 psi at $r = r_0$ to 40 psi at $r = r_1$ (see footnote 2 in Sec. 10.6). The coefficient of static friction is 0.60. Calculate the maximum torque that can be transmitted by the clutch.

10.101 Assuming that the net force due to the pressure distribution in Problem 10.100 is distributed uniformly over the disk face, $r_0 \leq r \leq r_1$, calculate the maximum torque that can be transmitted by the clutch.

10.102 The normal pressure p is assumed to be distributed uniformly over the contacting surfaces of a cone clutch. The cross section of the clutch is shown in Fig. P10.102a, and the normal pressure p is shown in Fig. P10.102b. Show that the maximum torque T_{max} transmitted by the clutch is

$$T_{max} = \frac{2\mu_s R(r_1^3 - r_0^3)}{3(\sin\alpha)(r_1^2 - r_0^2)}$$

where $2r_1 = d_1$, $2r_0 = d_0$, and the actuating force R is related to the pressure p by the formula

$$R = \pi p(r_1^2 - r_0^2)$$

Hint: Note from Fig. P10.102b that

$$dN = \frac{p(2\pi r)}{\sin\alpha}\,dr$$

$$dR = \sin\alpha\, dN$$

$$dT = r\mu_s\, dN$$

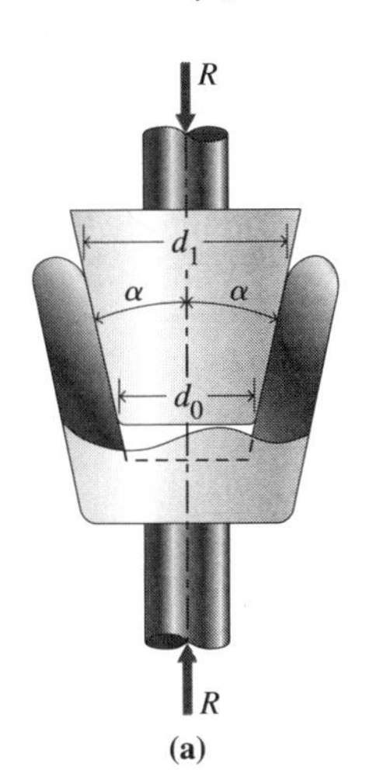

(a)

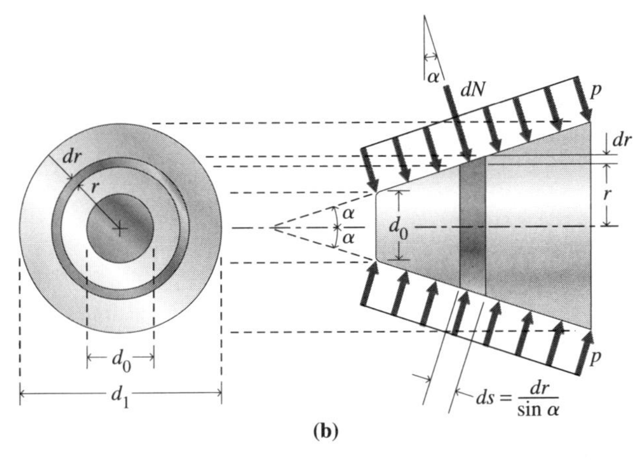

(b)

Figure P10.102

10.103 In Fig. P10.102a, $R = 18$ kN, $d_0 = 100$ mm, $d_1 = 150$ mm, $\alpha = 10°$, and $\mu_s = 0.70$. Determine the maximum torque that can be transmitted by the clutch.

10.104 The hollow flat pivot in Fig. P10.104 is subjected to a total load L, including its own weight. The coefficients of static and kinetic friction are μ_s and μ_k.

a. Show that the torque T_s required to initiate rotation about the pivot's axis is

$$T_s = \frac{2\mu_s L}{3}\left(\frac{r_2^3 - r_1^3}{r_2^2 - r_1^2}\right) \qquad \text{(a)}$$

b. What torque T_k is required to keep the pivot rotating at a constant rate?

c. Show that as $r_2 \rightarrow r_1$, where $r_2 - r_1 = t$ (the thickness of the pivot wall) and for $t << r_1$, Eq. (a) reduces to the approximate formula

$$T_s \approx \mu_s L r_1 \qquad \text{(b)}$$

[*Hint:* Eliminate r_2 from Eq. (a) by using the relation $r_2 = r_1 + t$. Note that if r_1 is eliminated from Eq. (a) by the relation $r_1 = r_2 - t$, where $t << r_2$, T_s may also be approximated by the formula $T_s \approx \mu_s L r_2$. Finally, note that for $r_1 = 0$, Eq. (a) reduces to $T_s = 2\mu_s L r_2/3$, the torque required to initiate the rotation of a solid flat pivot.]

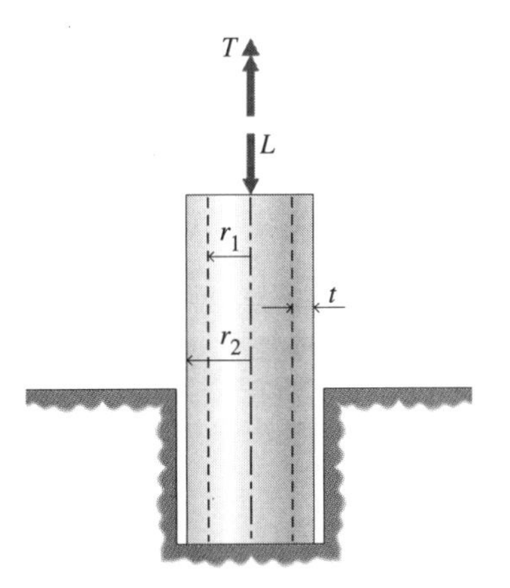

Figure P10.104

10.105 If a metal ball is pressed against a thick flat plate with force F, it flattens slightly, so that the reactive pressure is distributed over a small circle of contact with diameter $2a$ (see Fig. P10.105). By the theory of elasticity, the pressure p at the distance r from the center of the circle of contact is

$$p = \frac{3F}{2\pi a^2}\sqrt{1 - \frac{r^2}{a^2}}$$

Show that, if the ball is rotated slowly about the diametral axis perpendicular to the plate while being pressed against the plate, the required torque is $T = 3\pi\mu_k Fa/16$, where μ_k is the coefficient of kinetic friction.

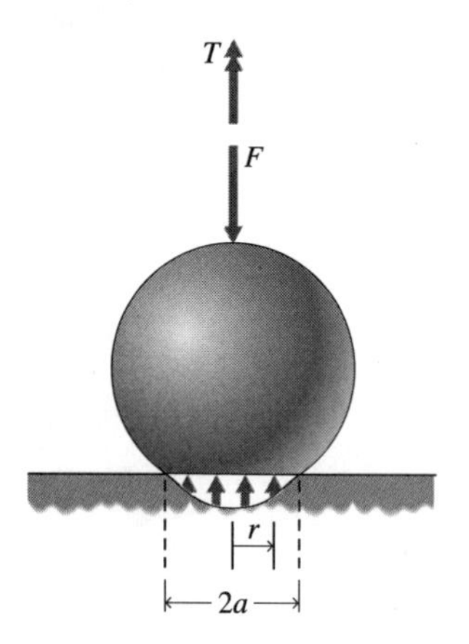

Figure P10.105

10.106 An axial force of $P = 90{,}000$ lb is required to overcome friction and press a locomotive wheel onto its axle, since the diameter of the wheel hole is slightly less than that of the axle (see Fig. P10.106a). Once pressed on, the wheel is prevented from slipping around the axle by static friction due to the pressure between the wheel and the axle. The axle has a diameter of 3.75 in. Compute the maximum torque that the axle can transmit to the wheel, assuming that the coefficients of static friction and kinetic friction are equal (see Fig. P10.106b).

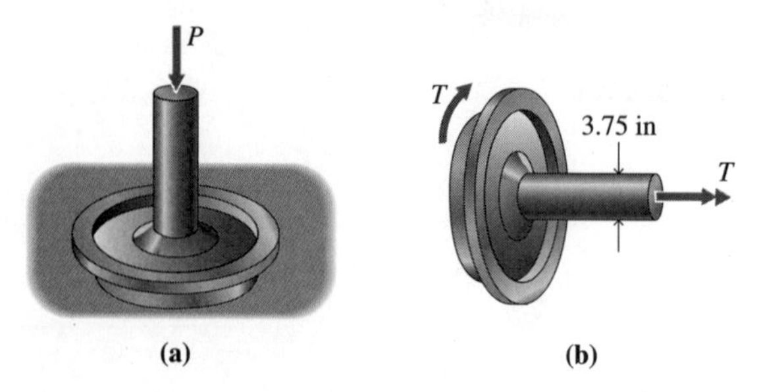

Figure P10.106

10.107 A plate shaped like a sector of a circle (see Fig. P10.107) weighs 3.6 kN. It lies on a plane horizontal surface (xy plane), and it is constrained to rotate about the vertical (z) axis through point O. The reactive pressure of the surface on which the plate lies is uniform. The coefficient of static friction is 0.30. Calculate the magnitude of the tangential force F required to move the plate.

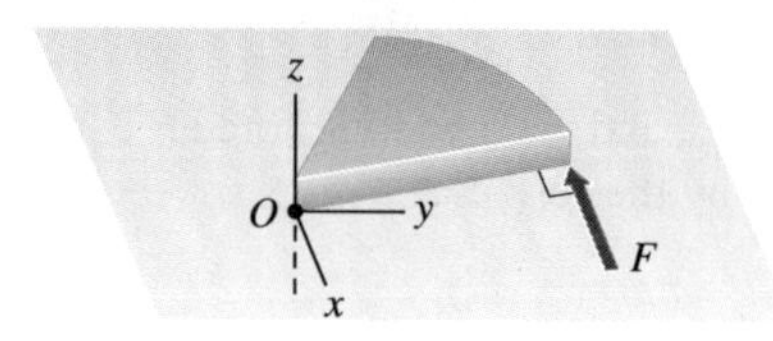

Figure P10.107

10.108 Rework Problem 10.96, assuming that the pressure p between the collar C and the support B varies linearly from $r = r_S$ to $r = r_C$. Let $p = p_S$ at $r = r_S$, and let $p = 0$ at $r = r_C$.

10.109 In a disk brake, the pads are pressed against the disk by equal and oppositely directed braking forces N (see Fig. P10.109). The disk is rigidly attached to the shaft. The disk does not slip on the pads when a torque $T = 2$ kN·m is applied to the shaft. Determine the minimum required braking force N. The coefficient of static friction is $\mu_s = 0.20$, and the effective radius is $r = 175$ mm.

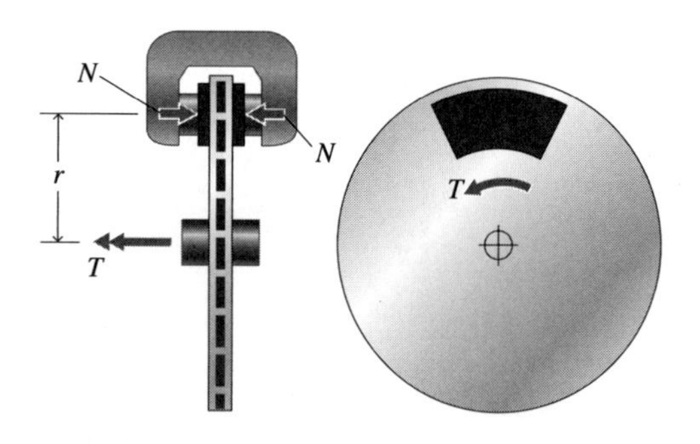

Figure P10.109

10.110 The actuating cylinder of a twin-shoe drum brake (see Fig. P10.110) pushes the shoes against the drum, exerting a pressure p on the drum. It is often assumed that the pressure exerted by each shoe varies with $\sin\theta$, where θ is measured from line OB, and that $p = 0$ at B and $p = p_A$ at A. Using this assumption, derive a formula for the total moment M_d exerted on the rotating drum by the shoes. The width (perpendicular to the page) of the lining in contact with the drum is b, and the inner radius of the drum is r. The coefficient of kinetic friction is μ_k. [Actually, the pressure distribution of the leading shoe differs from that of the trailing shoe (Orthwein, 1986).] Express M_d in terms of μ_k, p_A, b, r, and ϕ.

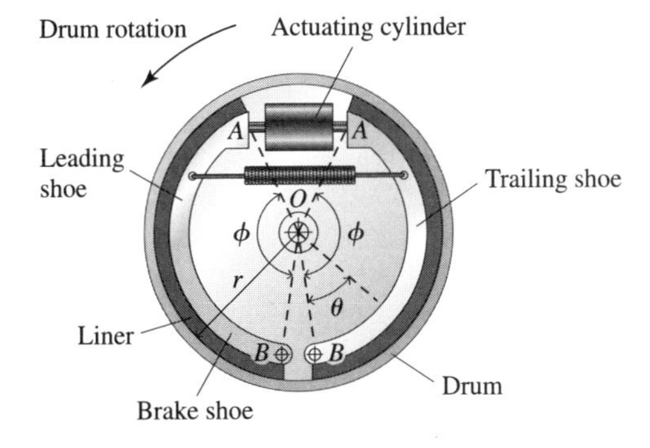

Figure P10.110

10.111 The actuating force that is to be transmitted to the cone clutch in Problem 10.102 is $R = 20$ kN.

a. For $\mu_s = 0.70$, $d_0 = 100$ mm, and $d_1 = 150$ mm, plot the maximum torque transmitted by the clutch as a function of the angle α, over the range $3° \leq \alpha \leq 15°$.

b. Discuss your results.

REVIEW QUESTIONS

10.112 Define *frictional force.* Explain the nature of friction from the molecular point of view.

10.113 State Coulomb's rules for frictional forces.

10.114 Define *coefficient of kinetic friction* and *coefficient of static friction.* Which one is usually larger?

10.115 One body rests on another body without sliding. Is the frictional force that acts on the first body equal to $\mu_s N$, where μ_s is the coefficient of static friction and N is the normal force between the bodies? Explain.

10.116 Explain how the coefficient of static friction between a block and a flat plate can be determined by setting the block on the plate and gradually tilting the plate. Define angle of repose.

10.117 Define *angle of static friction.*

10.118 Is it possible for the coefficient of friction between two bodies to be greater than 1.0?

10.119 What is a screw thread? A jackscrew? What is the purpose of a jackscrew?

10.120 What is meant by the mechanical advantage of a jackscrew? The efficiency of a jackscrew?

10.121 What is the pitch of a screw? The lead of a screw?

10.122 A rope is wrapped several times around a tree. A man holds one end of the rope, and a truck pulls on the other end. Explain, with the aid of the appropriate formula, why the man can hold the truck.

10.123 What is a belt-pulley system? What is its purpose? What is a band brake? What is its function?

10.124 What is the angle of wrap of a belt-pulley system?

10.125 What is an idler pulley? What is its function?

10.126 What distinguishes a V-belt from a flat belt?

10.127 How does the formula for the tension ratio T_1/T_2 of a V-belt pulley system differ from that of a flat-belt pulley system?

10.128 What is an axial disk clutch? An axial disk brake?

10.129 What is a single-disk clutch? A multi-disk clutch?

10.130 What is a disk brake?

10.131 What is a drum brake?

10.132 When trying to extract a cork that is stuck in a bottle, you may find it advantageous to twist the cork as you pull. Explain why.

Chapter 11 BEAMS AND SHAFTS

As a diver is poised at the end of a diving board, ready to dive, her weight causes the board—a cantilever beam—to deflect downward. The weight is counteracted by a couple and a force acting at the other (fixed) end of the board.

A Look Forward

A BEAM IS a slender bar, straight or curved, supported at one or more points along its length. Beams are widely used in mechanical and structural systems, such as buildings, bridges, and aircraft. Generally, beams are subjected to loads directed perpendicular to their longitudinal axes. These loads cause them to bend laterally. However, beams may also be subjected to axial forces or twisting moments. When the primary load on a beam is an axial force or a twisting moment, the beam is typically referred to as a shaft. Beams are supported in various ways—for example, by pins, rollers, and clamps. They are subjected to concentrated loads, to couples, and to distributed forces.

In this chapter, you will learn how to determine the forces that the supports exert on beams and how to determine internal forces in beams (axial forces, shear forces, bending moments, and twisting moments). You will learn how to draw shear and bending-moment diagrams for beams and how to determine the maximum shear force and maximum bending moment in beams. These quantities are important in the design of beams.

Survey Questions

- What are beams, and why are they important?
- What is the difference between a bending moment and a twisting moment?
- How are beams analyzed?
- Are the methods of analyzing beams different from other methods used in statics? If so, how do they differ?

11.1 SUPPORT REACTIONS: BEAMS SUBJECTED TO CONCENTRATED FORCES AND COUPLES

After studying this section, you should be able to:

- Describe the loads and support conditions for a beam.
- Determine the support reactions for a statically determinate beam.

> **Key Concept**
> A beam is a long, slender structural member that is supported at discrete locations along its length and is subjected primarily to loads that cause bending.

beam

A *beam* is a member subjected primarily to loads that cause bending. For example, a straight beam becomes curved or bent under the action of a force that is directed perpendicular to the axis of the beam (see Fig. 11.1). A beam may be initially straight or curved. In wood structures, straight timbers are often used as beams. In aircraft, beams are often curved. In modern steel structures, straight wide-flange beams are used extensively. Figure 11.2 shows the cross-sectional shape of a standard wide-flange beam, in which a large portion of the cross-sectional area is located in the flanges. This cross-section shape provides high resistance to bending.

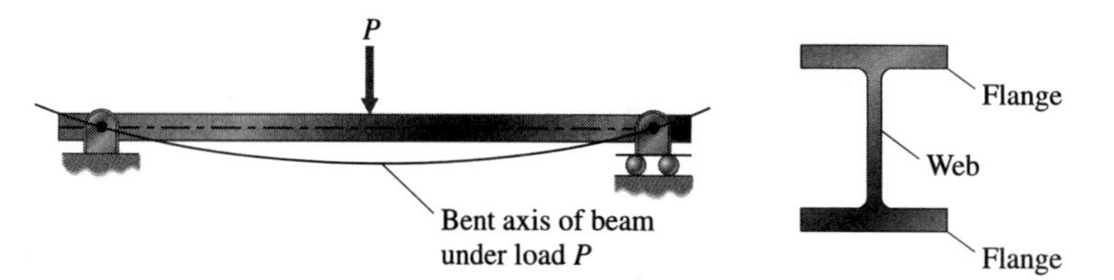

Figure 11.1 Lateral load on beam.

Figure 11.2 Cross section of a wide-flange beam.

cantilever beam

clamped end

fixed end

free end

support reactions

A beam that is supported only at one end is called a *cantilever beam.* Figure 11.3a is a diagrammatic representation of a straight cantilever beam. The left end of this beam is said to be the *clamped* end, or *fixed end,* since it is assumed that a cantilever beam does not translate or rotate at the supported end. The other end of the cantilever beam is said to be the *free end,* since it is not supported and it can translate and rotate.

The forces and the couples exerted on a beam by its supports are called the *support reactions.* They are the reactions to the forces exerted on the support by the beam. The supporting structure of a cantilever beam exerts distributed forces (see Chapter 8) on the end of the beam that are equivalent to a transverse force F and a couple M (see Fig. 11.3b). Rather than showing a cantilever beam built into its support structure, as in Figs. 11.3a and 11.3b, this book represents a fixed cantilever support and its equivalent system of forces more simply, as shown in Fig. 11.3c.

simple beam

A straight beam that is supported at one end by a roller and at the other end by a hinge is called a *simple beam* (see Fig. 11.4). As you will recall, a roller support is represented by either of these symbols: . The force exerted on a roller

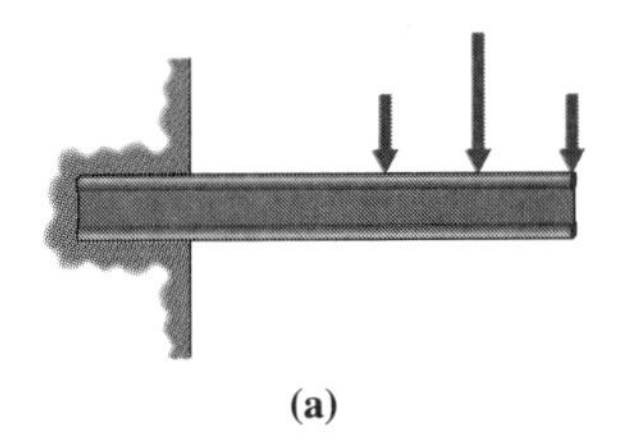

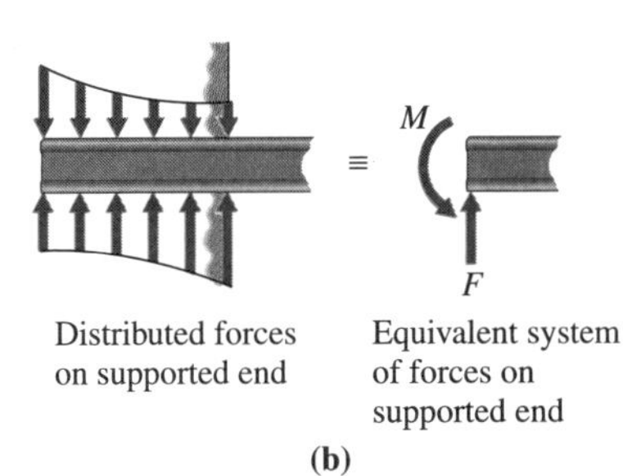

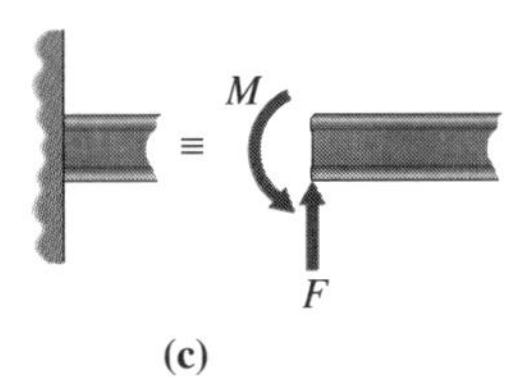

Figure 11.3 (**a**) Cantilever beam. (**b**) Forces on supported end. (**c**) Representation of fixed support.

Figure 11.4 Statically determinate simple beam.

support is perpendicular to the surface on which it acts (Figs. 3.9a and 3.9b). The symbol denotes a pin or hinge support (see Sec. 3.3 and Fig. 3.9c). The support reaction at a pin or hinge is a force acting at a particular angle or, equivalently, horizontal and vertical force components (see Fig. 3.9c).

Ordinarily, a straight beam is loaded by parallel transverse forces. The following discussion is based on this situation; in which case, only two equations of equilibrium are

A typical hinge support on a highway bridge resists horizontal and vertical forces but offers little resistance to moment. A roller support resists only vertical forces. Horizontal motion and rotation are permitted.

available for determining the support reactions (see Sec. 4.6). If the x axis is parallel to the longitudinal axis of the beam and if the y axis is transverse to the beam, these equations are $\Sigma F_y = 0$ and $\Sigma M = 0$, since in the absence of loads in the x direction, $\Sigma F_x = 0$ is satisfied identically. If there are only two unknown reactions, these equations are sufficient to determine them. In this case, the beam is *statically determinate.* If there are more than two unknown reactions, these equations are insufficient to determine them. Then, the beam is *statically indeterminate;* that is, its reactions *cannot* be determined by the equations of static equilibrium. For instance, the simple beam in Fig. 11.4 is statically determinate. However, if another roller support is added, as shown in Fig. 11.5, the beam becomes statically indeterminate.

statically determinate

statically indeterminate

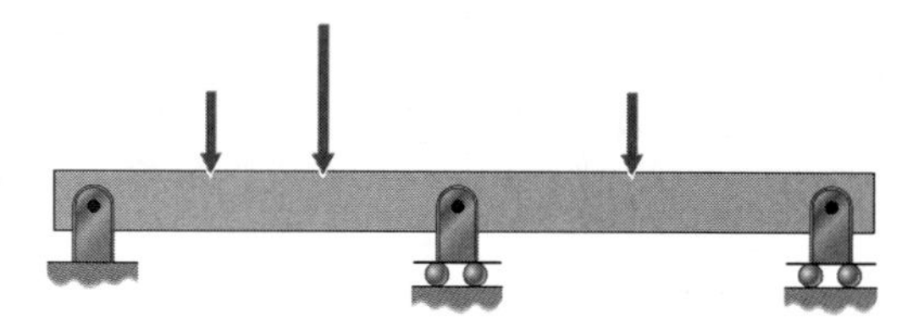

Figure 11.5 Statically indeterminate beam.

To solve problems involving statically indeterminate beams, we must investigate the deflections of the beams. However, the determination of the deflections of beams is beyond the scope of statics. It is treated in courses on strength of materials (also called mechanics of materials) and structural analysis. In this chapter, we will analyze only statically determinate beams.

PROBLEM-SOLVING TECHNIQUE

Support Reactions for Statically Determinate Beams

To determine the support reactions for a statically determinate beam:

1. Draw a free-body diagram of the beam. For a simple beam with two supports, the support reactions are forces. For a cantilever beam supported at a wall, the support reactions are a moment and a force.
2. Represent each force by a straight arrow, and label it with a letter. Represent a moment by a curved arrow, and label it with another letter. If the sense of a support reaction is not obvious, assume a sense.
3. Select xy axes, with the x axis along the axis of the beam and the y axis up. Apply the equations of equilibrium ($\Sigma F_y = 0$, $\Sigma M = 0$) to the free-body diagram.
4. Solve the equations of equilibrium for the support reactions. If a support reaction is positive in the solution, the sense chosen in Step 2 is correct. If a support reaction is negative, its correct sense is opposite to that chosen in Step 2.

Example 11.1

Simple Beam with an Overhang

Problem Statement Determine the support reactions of the wood beam shown in Fig. E11.1a.

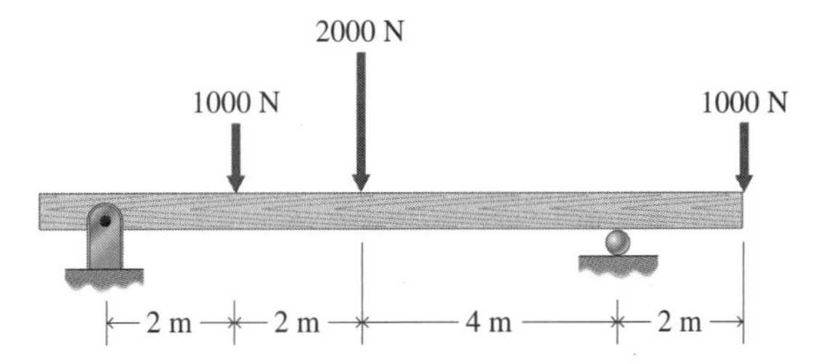

Figure E11.1a

Solution The free-body diagram of the beam and the *xy* axes are shown in Fig. E11.1b. The support reactions are represented by R_1 and R_2. Since there are no forces in the *x* direction, no horizontal reaction is shown. The equilibrium equation for the forces is

$$\sum F_y = R_1 + R_2 - 1000 - 2000 - 1000 = 0 \tag{a}$$

The equilibrium equation for moments about the left support is

$$\circlearrowleft\!+ \sum M = 8R_2 - 2(1000) - 4(2000) - 10(1000) = 0 \tag{b}$$

Equations (a) and (b) yield $R_1 = 1500$ N and $R_2 = 2500$ N.

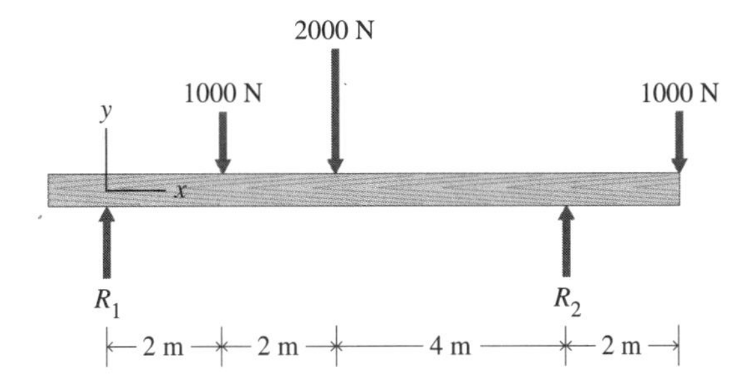

Figure E11.1b

Example 11.2

Simple Beam Subjected to Couples

Problem Statement Determine the support reactions of the simple steel beam subjected to couples, as shown in Fig. E11.2a.

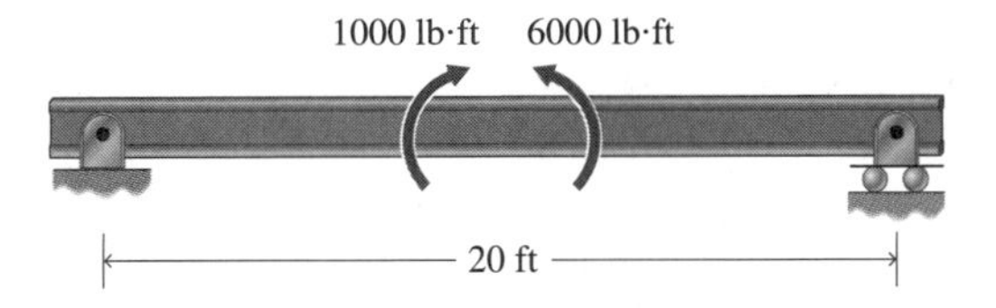

Figure E11.2a

Solution The free-body diagram of the beam and the *xy* axes are shown in Fig. E11.2b. The support reactions are represented by R_1 and R_2. Again, no horizontal reaction is shown, since no horizontal forces are applied. The equilibrium equation for forces is

$$\sum F_y = R_1 + R_2 = 0 \tag{a}$$

The equilibrium equation for moments about the left support is

$$\circlearrowleft\!+ \sum M = 20R_2 + 6000 - 1000 = 0 \qquad \text{(b)}$$

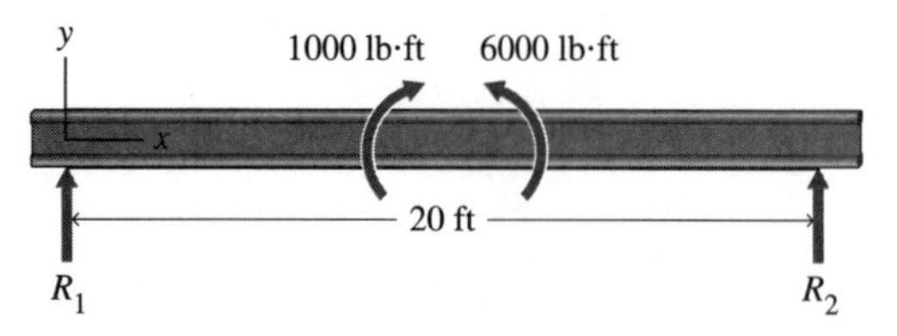

Figure E11.2b

Equations (a) and (b) yield

$$R_1 = +250 \text{ lb}$$
$$R_2 = -250 \text{ lb}$$

Since R_2 is negative, the right support pulls down on the beam (see Fig. E11.2c).

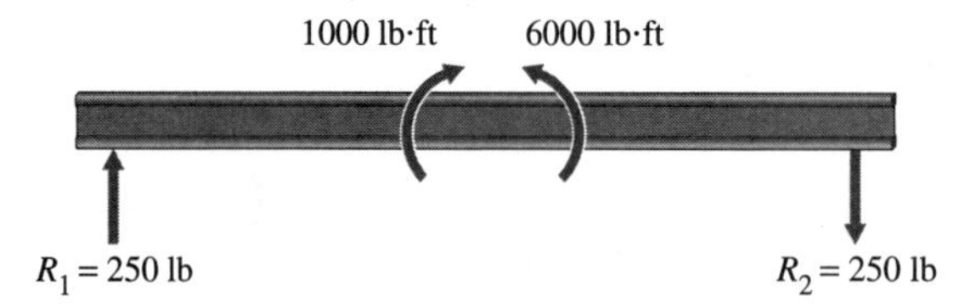

Figure E11.2c

Information Item Note that it was not necessary to specify the locations of the couples to determine the support reactions (external effects on the beam), since the moment of a couple that acts on a planar rigid body is independent of the location of the couple in the plane. However, the internal effects (shear forces and bending moments in the beam) do depend on the location of a couple (see Sec. 11.2). Also, in spite of its appearance, the roller support on the right end of the beam is assumed to resist forces acting either up or down. Hence, the beam will not lift off the roller support. ■

CHECKPOINT 

1. ***True or False:*** A simple beam is clamped at one end.
2. ***True or False:*** The forces and couples exerted on a beam by its supports are called support reactions.
3. ***True or False:*** A beam is usually loaded by forces that are transverse to the beam's axis.
4. ***True or False:*** A statically determinate beam is supported only by pin supports.

Answers: 1. False; 2. True; 3. True; 4. False

11.2 SHEAR FORCES AND BENDING MOMENTS IN BEAMS

After studying this section, you should be able to:

- Determine the internal shear force and bending moment at any cross section of a statically determinate beam loaded by concentrated forces and couples.

Key Concept Shear force and bending moment are internal forces in beams. They are caused by transverse loads and couples.

In the preceding section, we drew free-body diagrams of entire beams. Free-body diagrams of parts of beams are also used extensively in stress analysis. As an example, reconsider the beam of Example 11.1 (see Fig. 11.6a). Let's separate the beam into two parts by cutting it at section *a–a,* a distance x from the left support (x lies in the interval 2 m $<$ $x <$ 4 m). The free-body diagram of the part to the left of section *a–a* is shown in Fig. 11.6b. The support reaction of 1500 N was found in Example 11.1. In general, for equilibrium of this part, a force V and a couple whose moment is M must act at the cut. The force V and the moment M are the resultants of the forces distributed over the cross

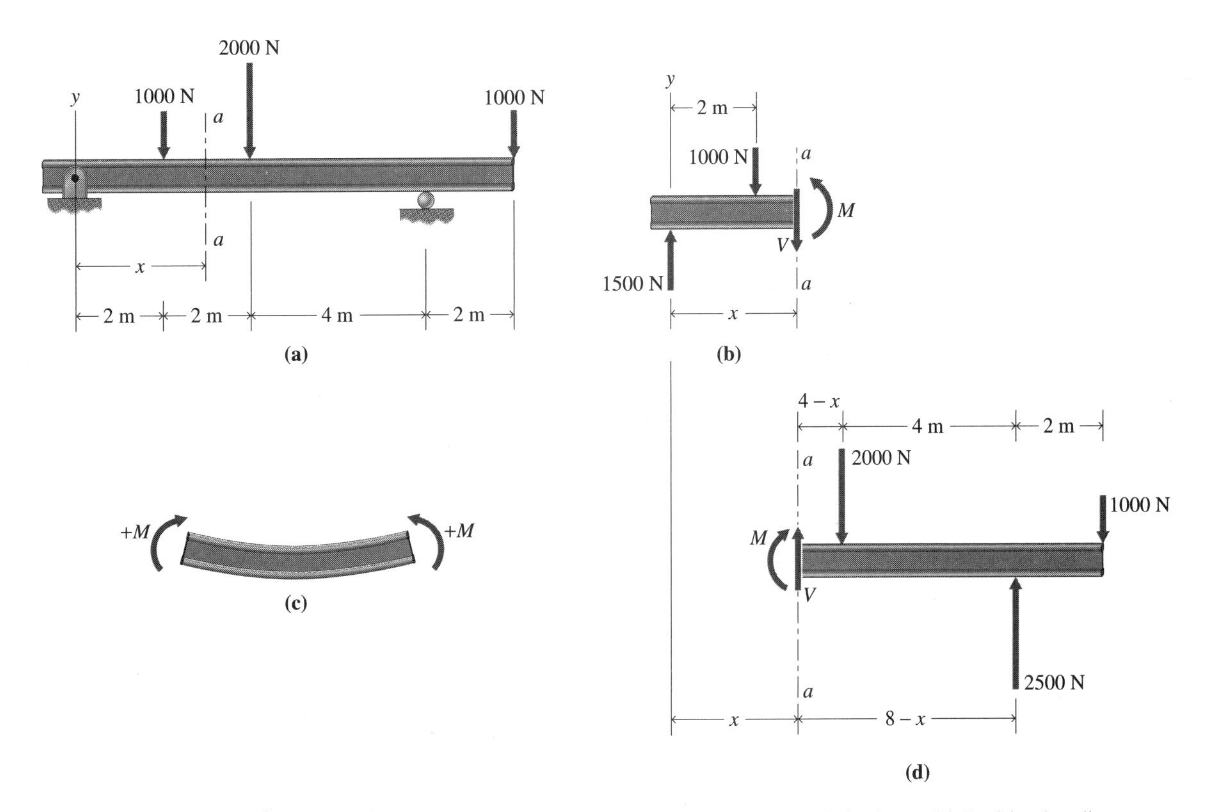

Figure 11.6 (**a**) Beam with overhang. (**b**) Part of beam to left of *a–a.* (**c**) Positive bending moment. (**d**) Part of beam to right of *a–a.*

external force **internal force** **shear force** **bending moment**

section.[1] In the free-body diagram of the isolated part of the beam (Fig. 11.6b), V and M are considered to be *external forces.* However, for the beam as a whole (Fig. 11.6a), they are *internal forces* at section *a–a.*

The force V is the resultant of the transverse forces that are distributed over the cross section at the cut. It is called the *shear force* (or simply the *shear*) acting on the beam at section *a–a.* Since V acts on section *a–a,* it is an internal shear. The couple with moment M is the resultant of the distributed axial force that acts on section *a–a.* It is called the *bending moment* acting on the beam at section *a–a.* Similar to V, it is an internal moment.

Figure 11.6b shows which senses of the shear force V and the couple M are usually considered to be positive. A positive bending moment tends to bend a beam element in such a way that it is concave-upward, with respect to the plane in which it lies (see Fig. 11.6c).

Equilibrium of the Left-Hand Part of the Beam The quantities V and M may be determined by the conditions of equilibrium for the isolated part of the beam (refer to Fig. 11.6b). The equation of equilibrium of forces is

$$+\uparrow \sum F_y = 1500 - 1000 - V = 0 \qquad \text{(a)}$$

or

$$V = 500 \text{ N}$$

The equation of equilibrium of moments about section *a–a* is

$$\circlearrowleft\!+ \sum M_{a\text{–}a} = -1500x + 1000(x - 2) + M = 0 \qquad \text{(b)}$$

or

$$M = 500x + 2000 \quad [\text{N·m}]$$

Equilibrium of the Right-Hand Part of the Beam Instead of determining the shear and the bending moment using the equilibrium conditions for the left part of the beam, we can derive them by considering the equilibrium of the right part, as follows: The free-body diagram on the right part of the beam is shown in Fig. 11.6d. By Newton's third law, the shear V and the bending moment M that act on the right part of the beam at section *a–a* are equal, respectively, to the shear and the bending moment that act on the left part of the beam at section *a–a,* but their senses are reversed. To be consistent with the sign conventions that were adopted for the left part of the beam, we specify that the shear and the bending moment on the right part are positive when they act as shown in Fig. 11.6d. Then, we can compute M and V by considering a free-body diagram of either the left part or the right part.

The equation of equilibrium of forces for the right part of the beam is (refer to Fig. 11.6d)

$$+\uparrow \sum F_y = V - 2000 + 2500 - 1000 = 0 \qquad \text{(c)}$$

where the support reaction of 2500 N was determined in Example 11.1. Consequently, $V = 500$ N, as before. The equation of equilibrium for moments about a point in section *a–a* is

$$\circlearrowleft\!+ \sum M_{a\text{–}a} = 2500(8 - x) - 1000(10 - x) - 2000(4 - x) - M = 0 \qquad \text{(d)}$$

Hence, $M = 500x + 2000$ [N·m], as before.

1. Distributed forces, like concentrated forces, may be reduced to a single resultant force and a couple. This reduction is analyzed in detail in Chapter 8.

Alternative Viewpoint Again considering the free-body diagram of the left part of the beam (Fig. 11.6b), we can write the equation of equilibrium of forces [Eq. (a)] as follows:

$$V = 1500 - 1000 = 500 \text{ N}$$

On the other hand, the equation of equilibrium of forces for the right part of the beam is [refer to Eq. (c)]

$$V = 2000 - 2500 + 1000 = 500 \text{ N}$$

These relations illustrate the following theorem.

Theorem 11.1

The shear V at any cross section of a horizontal beam is the algebraic sum of the vertical projections of all forces (including support reactions) to the left of the given cross section, where, in the sum, the projection of a force is positive if its sense is upward and negative if its sense is downward. Alternatively, the shear V is the algebraic sum of the vertical projections of all forces (including support reactions) to the right of the given cross section. In this case, the projection of a force is positive if its sense is downward and negative if its sense is upward.

A similar theorem applies for bending moments. For example, consider the free-body diagrams for the left and right parts of the beam (Figs. 11.6b and 11.6d, respectively). We can write the moment equation as (refer to Fig. 11.6b and Eq. (b))

$$M = 1500x - 1000(x - 2) = 500x + 2000 \quad [\text{N}\cdot\text{m}]$$

or (refer to Fig. 11.6d and Eq. (d))

$$M = 2500(8 - x) - 1000(10 - x) - 2000(4 - x) = 500x + 2000 \quad [\text{N}\cdot\text{m}]$$

These relations illustrate the following theorem.

Theorem 11.2

The bending moment M at any cross section of a horizontal beam is the sum of the moments about the given cross section of all the forces (including support reactions) to the left of the cross section, where, in the sum, a clockwise moment is positive and a counterclockwise moment is negative. Alternatively, the bending moment M is the sum of the moments about the given cross section of all the forces (including support reactions) to the right of the cross section. In this case, a counterclockwise moment is positive and a clockwise moment is negative.

These theorems are equivalent to the equations of equilibrium for each part of a beam. They may be used directly to determine the shear and the bending moment at any section of a beam subject to couples and transverse concentrated forces, or they may be used to check results obtained from considering the equilibrium of individual parts of the beam. Although these theorems have only been illustrated by a numerical example, they are true for any beam that is subject to couples and transverse concentrated forces.

PROBLEM-SOLVING TECHNIQUE

Shear Forces and Bending Moments in Beams Subjected to Concentrated Forces and Couples

To determine the shear forces and bending moments in a beam:

1. Determine the support reactions by the Problem-Solving Technique outlined in Sec. 11.1.
2. Pass a cutting plane through the cross section where the shear force and the bending moment are to be determined.
3. Draw a free-body diagram of the part of the beam to the left (or to the right) of the cut, and label the shear force at the cut with the letter V and the moment at the cut with the letter M. Select the positive senses of V and M, as described in the text.
4. Select xy axes, and, using the free-body diagram of Step 3, write the equations of equilibrium for the beam part.
5. Solve the equations of equilibrium for the shear force V and the bending moment M.
6. Repeat Steps 2 through 5 for each section of the beam for which V and M must be determined.
7. Alternatively, Steps 2 through 5 may be replaced by application of Theorems 11.1 and 11.2.

Example 11.3 Shear Force and Bending Moment in a Simple Beam

Problem Statement A simple steel beam is subjected to two transverse forces and a couple (see Fig. E11.3a).

a. Determine the support reactions.

b. Determine the shear force and the bending moment in each of the beam segments AB, BC, CD, and DE.

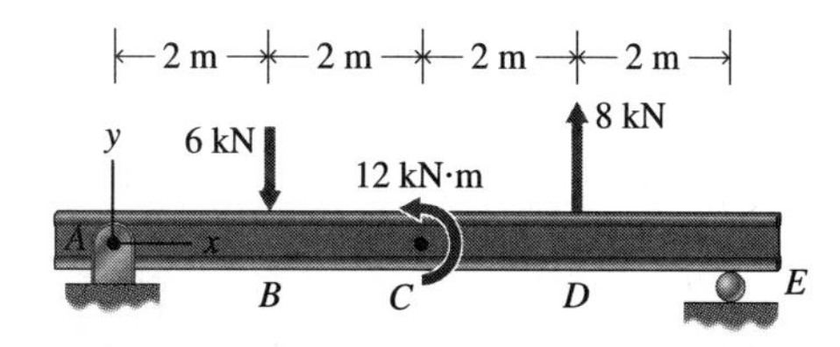

Figure E11.3a

Solution

This problem can be solved by either of the two methods discussed in this section—that is, either by considering the equations of equilibrium (with free-body diagrams) for individual parts of the beam or by applying Theorems 11.1 and 11.2. Each method has its merits. The approach utilizing free-body diagrams of beam parts and corresponding equations of equilibrium may appear to be more fundamental (and familiar). However, Theorems 11.1 and 11.2 are equally fundamental, as they are based on satisfaction of the equations of equilibrium. In this example, we use the theorems to solve the problem. You are encouraged to solve the example using equations of equilibrium for the individual parts of the beam and to compare that approach to the one used. The first step in either approach is to determine the support reactions on the beam.

a. We consider the free-body diagram of the entire beam and select xy axes (see Fig. E11.3b). The equations of equilibrium for the beam are

$$\sum F_x = A_x = 0$$

$$\sum F_y = A_y - 6 + 8 + E = 0 \tag{a}$$

$$\circlearrowleft^{+} \sum M_A = 8E + 6(8) + 12 - 2(6) = 0$$

Solution of Eqs. (a) gives

$$A_x = 0$$
$$A_y = 4 \text{ kN}$$
$$E = -6 \text{ kN}$$

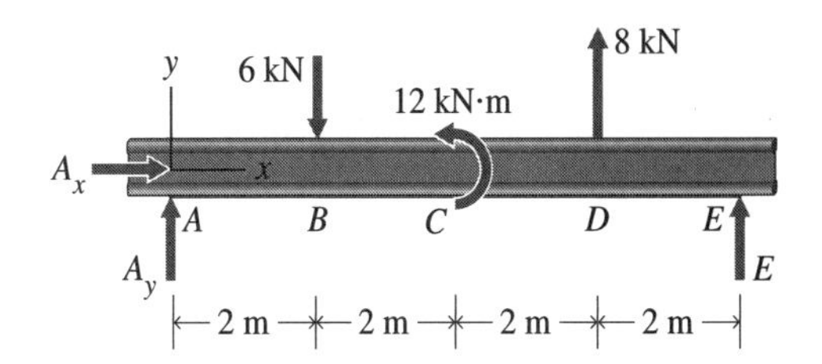

Figure E11.3b

Since E is negative, its sense is in the negative y direction. The reactions are shown with their correct senses in Fig. E11.3c.

b. To determine the shear force and the bending moment in each segment of the beam, we use Theorems 11.1 and 11.2. Hence, we consider cross sections a–a, b–b, c–c, and d–d in the beam (see Fig. E11.3c). Applying Theorem 11.1, we obtain the following:

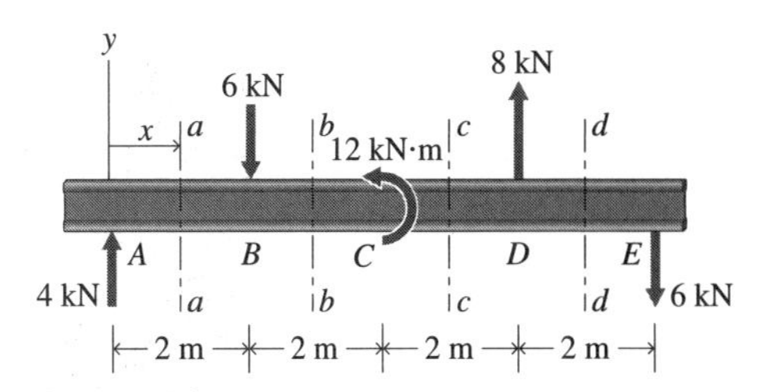

Figure E11.3c

Segment AB: $0 \text{ m} < x < 2 \text{ m}$ $V = +4 \text{ kN}$
Segment BC: $2 \text{ m} < x < 4 \text{ m}$ $V = -2 \text{ kN}$
Segment CD: $4 \text{ m} < x < 6 \text{ m}$ $V = -2 \text{ kN}$
Segment DE: $6 \text{ m} < x < 8 \text{ m}$ $V = +6 \text{ kN}$

Likewise, applying Theorem 11.2, we obtain the following:

Segment AB: $0 \text{ m} < x < 2 \text{ m}$ $M = 4x$
Segment BC: $2 \text{ m} < x < 4 \text{ m}$ $M = 4x - 6(x - 2) = -2x + 12$
Segment CD: $4 \text{ m} < x < 6 \text{ m}$ $M = 4x - 6(x - 2) - 12 = -2x$
Segment DE: $6 \text{ m} < x < 8 \text{ m}$ $M = 4x - 6(x - 2) - 12 + 8(x - 6) = 6x - 48$

11.3 BEAMS AND SHAFTS SUBJECTED TO AXIAL FORCES AND TWISTING MOMENTS

After studying this section, you should be able to:

- Determine the axial force, shear force, bending moment, and torque at any cross section of a statically determinate beam or shaft.

> ***Key Concept*** Axial forces and twisting moments (torques) are internal forces in beams and shafts. They can be caused by a wide variety of forces applied externally to a structural system.

axial force **twisting moment**

In addition to shear forces and bending moments, axial forces and twisting moments may act on long, slender structural members. An *axial force* acts normal to the cross section of the member. A *twisting moment* is a moment acting about the longitudinal axis of a member and tending to rotate the member's cross section about the member's longitudinal axis. When the primary internal force in a member is an axial force or a twisting moment, the member is referred to as a *shaft*. Use of the term *beam* generally suggests that the primary internal forces are shear forces and bending moments.

shaft

The horizontal drive shaft of this turbine generator transfers torque from one component to another. The large bearing on the left permits rotation of the shaft about its axis but prevents translation of the shaft.

AXIAL FORCES

To see an example of an axial force that acts on a beam's cross section, let's cut the L-shaped cantilever beam shown in Fig. 11.7a by cross-sectional plane *a–a*. The free-body diagram of the part of the beam to the right of section *a–a* is shown in Fig. 11.7b. The force system that acts on the cross section is represented by an axial component N, a transverse component V, and a bending moment M. The force N is the net axial force acting at section *a–a*. The force V is the shear on the cross section, and M is the moment of the couple on section *a–a*. The forces N, V, and M are the resultants of distributed forces (stresses) that act over the cross section of the beam.

It was shown in Sec. 4.5 that a force can be shifted laterally if a compensating couple is added. Accordingly, we may locate the line of action of the force N anywhere on the cross section. It is customarily located at the centroid of the cross section. (The definition of the centroid is given in Chapter 8.) For our purpose, we regard the centroid as the point where the axis of the beam (the dashed line in Figs. 11.7b and 11.7c) intersects the cross section of the beam. The bending moment M of the compensating couple depends on the specification of the line of action of the force N (see Sec. 4.5).

The quantities N, V, and M are determined by the conditions of equilibrium for the isolated part of the beam (see Fig. 11.7b):

$$\sum F_x = N - 400 = 0$$

$$\sum F_y = V + 200 - 100 = 0$$

$$\circlearrowleft\!\!+ \sum M_O = -M + 200(b - 12) - 100b - 400(8) = 0$$

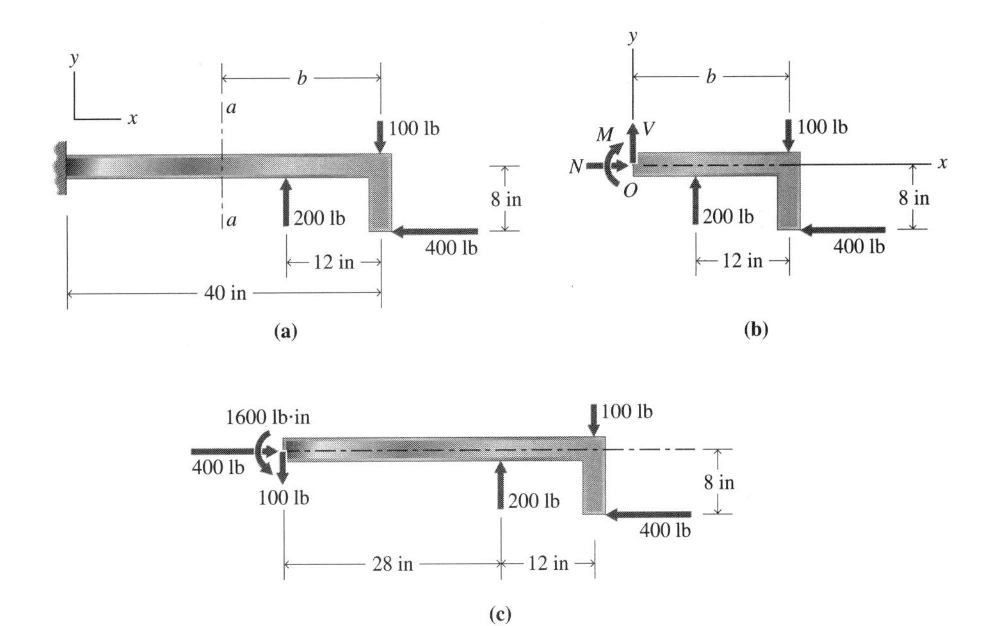

Figure 11.7

where M_O denotes the sum of the moments about point O (Fig. 11.7b). These equations yield

$$N = 400 \text{ lb}$$
$$V = -100 \text{ lb} \qquad \text{(a)}$$
$$M = -5600 + 100b \quad [\text{lb·in}]$$

Equations (a) may be used to determine the reaction of the support. For, if $b = 40$ in, Fig 11.7b becomes a free-body diagram of the entire beam. Then N, V, and M are the reactions of the supporting structure. From Eqs. (a), these reactions are

$$N = 400 \text{ lb}$$
$$V = -100 \text{ lb}$$
$$M = -1600 \text{ lb·in}$$

Figure 11.7c is a free-body diagram of the entire beam, showing the reactions with their correct senses.

TWISTING MOMENT, OR TORQUE

In mechanical systems, a shaft or a rod is often subjected to a moment about its longitudinal axis. This moment is called a twisting moment, or *torque.* Such moments occur in drive shafts, axles, and pulley shafts. As an example, let's consider a shaft that is twisted by a couple consisting of a pair of forces with magnitude F and lever arm a (see Fig. 11.8a). Let's cut the shaft by a cross-sectional plane A. Since the applied loads constitute a couple with moment Fa, the reaction at cross section A is a couple with moment $T = -Fa$ (see Fig. 11.8b). The resisting couple lies in the cross-sectional plane A. It is called the twisting moment or torque at the section or, more explicitly, the *resisting torque.* The couple that is applied to the transverse arm is called the *applied torque.* In general, a torque is a

torque

resisting torque

applied torque

couple that lies in a cross-sectional plane of a rod or a shaft on which it acts. Such a couple may be represented by a double-headed arrow that is perpendicular to the shaft's cross section, as in Fig. 11.8c. (The label T is generally used for a torque or twisting moment to distinguish it from a bending moment.) A shaft is twisted about its longitudinal axis by the action of a torque. The twisting action in a shaft is called *torsion*.

torsion

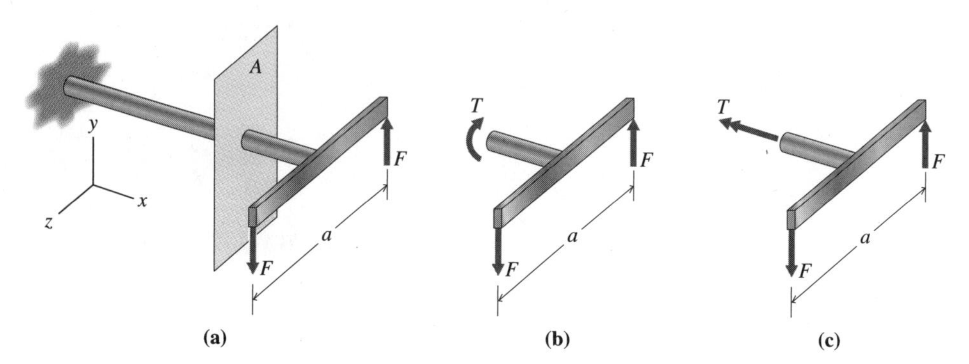

Figure 11.8 Torque applied to a shaft.

PROBLEM-SOLVING TECHNIQUE

Axial Force, Shear Force, Bending Moment, and Torque in a Statically Determinate Shaft

To determine the axial force, shear force, bending moment, and torque in a shaft:

1. Draw a free-body diagram of the entire shaft.
2. Select xy axes, and, from the free-body diagram, write the equations of equilibrium for the shaft.
3. Solve the equations of equilibrium for the support reactions.
4. Next, make a transverse cut of the shaft. Draw a free-body diagram of the part of the shaft, either to the left or to the right of the cut, and select the positive senses for the axial force, shear force, bending moment, and torque that act at the cut.
5. Choose xy axes, and, from the free-body diagram of the part, write the equations of equilibrium.
6. Solve the equations of equilibrium for the unknowns that act at the cut.
7. Repeat Steps 4, 5, and 6 for other sections in the shaft, as required.

Note: For a cantilever beam, it is often expedient to start at the free end, cut off a part from that end, draw a free-body diagram of the part, and go to Step 5 (see Figs. 11.7a and 11.7b).

Example 11.4

Internal Forces in Straight Beam

Problem Statement A beam is acted on by a force applied to a rigid extension arm (see Fig. E11.4a). Determine the axial force, shear force, and bending moment in segments AB and BC of the beam.

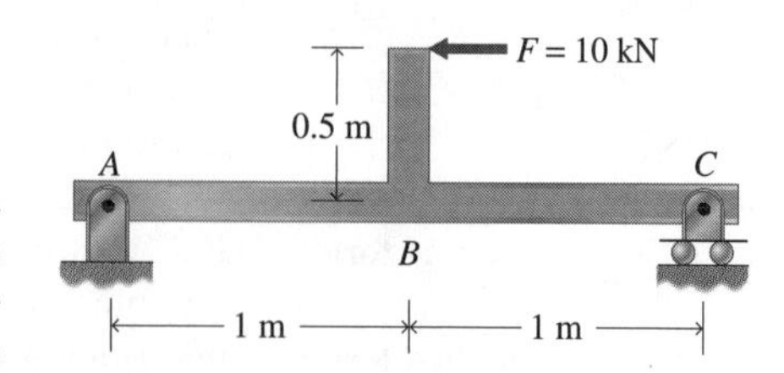

Figure E11.4a

Solution

To begin our solution, we draw a free-body diagram of the straight part of the beam removed from its supports and select xy axes (see Fig. E11.4b). To transfer the 10 kN force to the centroid of the beam's cross section at B, we must introduce a compensating couple with moment $M = 5$ kN·m (see Sec. 4.5).

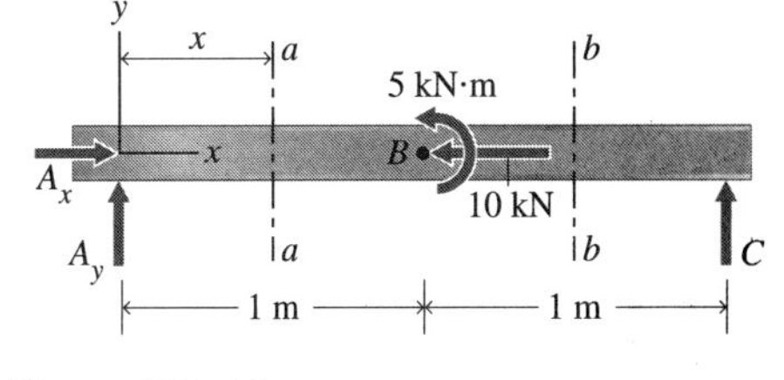

Figure E11.4b

The equations of equilibrium for the beam are

$$\sum F_x = A_x - 10 = 0$$
$$\sum F_y = A_y + C = 0$$
$$\circlearrowleft\!+ \sum M_A = 2C + 5 = 0$$

Solving these equations, we obtain

$$A_x = +10 \text{ kN}$$
$$A_y = +2.5 \text{ kN}$$
$$C = -2.5 \text{ kN}$$

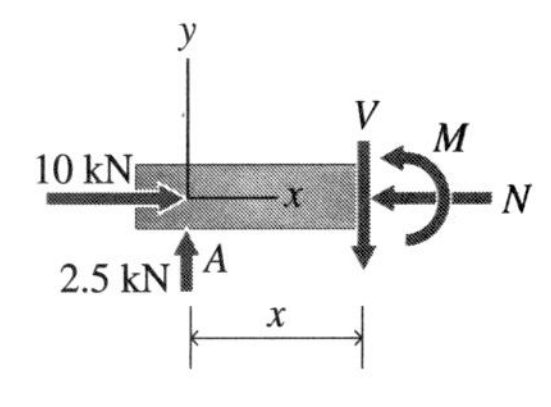

Figure E11.4c

Next, we consider the free-body diagram of the beam to the left of section a–a ($0 < x < 1$ m), shown in Fig. E11.4c. The equilibrium equations for this part are

$$\sum F_x = 10 - N = 0$$
$$\sum F_y = 2.5 - V = 0$$
$$\circlearrowleft\!+ \sum M_A = M - Vx = 0$$

Solving these equations, we obtain

$$N = 10 \text{ kN} \qquad 0 \text{ m} < x < 1 \text{ m}$$
$$V = 2.5 \text{ kN} \qquad 0 \text{ m} < x < 1 \text{ m} \qquad \text{(a)}$$
$$M = 2.5x \ [\text{kN·m}] \quad 0 \text{ m} < x < 1 \text{ m}$$

Since N, V, and M are all positive, their senses are correct as shown in Fig. E11.4c.

Finally, we consider the free-body diagram of the part of the beam to the left of section b–b ($1 \text{ m} < x < 2$ m), shown in Fig. E11.4d. (*Note:* Alternatively, we could consider the free-body diagram of the part of the beam to the right of section b–b. That is left for you to try.) The equations of equilibrium for this part of the beam are

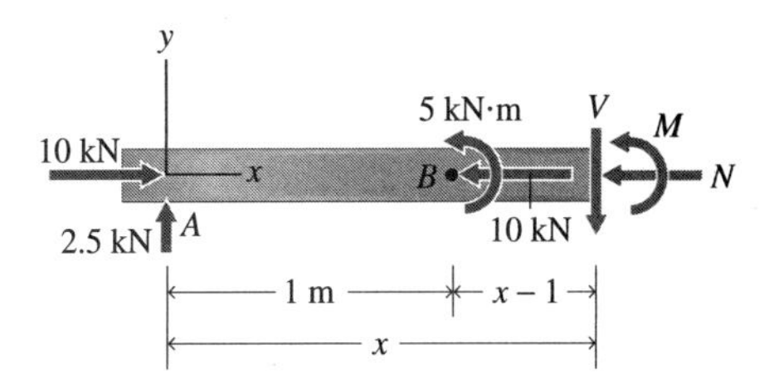

Figure E11.4d

$$\sum F_x = 10 - 10 - N = 0$$
$$\sum F_y = 2.5 - V = 0$$
$$\circlearrowleft\!+ \sum M_A = 5 + M - Vx = 0$$

Solution of these equations gives

$$
\begin{aligned}
N &= 0 && 1\text{ m} < x < 2\text{ m} \\
V &= 2.5\text{ kN} && 1\text{ m} < x < 2\text{ m} \qquad \text{(b)}\\
M &= 2.5x - 5 \quad [\text{kN·m}] && 1\text{ m} < x < 2\text{ m}
\end{aligned}
$$

Since V is positive, its sense is correct as shown in Fig. E11.4d. However, since M is negative in the range $1\text{ m} < x < 2\text{ m}$, the correct sense of M is *clockwise* (opposite to the sense shown in Fig. E11.4d). From the last of Eqs. (b), we see that $M = 0$ at $x = 2$ m, as it should be, since a roller support cannot resist a moment. Also, we note that the axial force N is constant (10 kN) in the segment AB and zero in segment BC. This result is consistent with the fact that the roller at C cannot resist a horizontal force. ■

Example 11.5

Curved Cantilever Beam

Problem Statement Figure E11.5a represents a cantilever beam that is curved in the form of a 90° circular arc. Determine the reactions of the wall support.

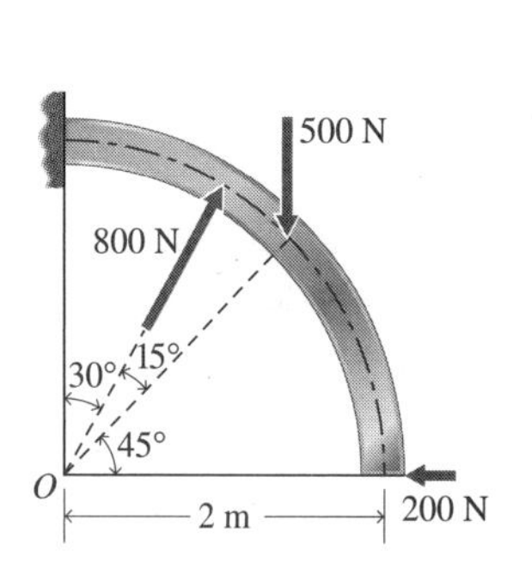

Figure E11.5a

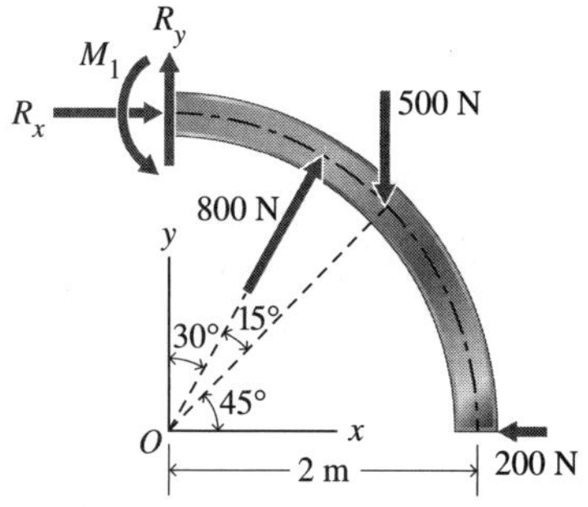

Figure E11.5b

Solution In the free-body diagram of the beam (Fig. E11.5b), the reactions of the support are the horizontal and vertical force components represented by R_x and R_y, respectively, and the couple with moment M_1. These reactions are due to the distributed forces exerted by the support (see Figs. 11.3b and 11.3c). Here, the support also exerts a horizontal reaction to balance the horizontal component of the 800 N load and the horizontal 200 N load. The horizontal and vertical reactions R_x and R_y are assumed to be positive in the x and y directions, respectively, and the moment M_1 is taken as positive in the counterclockwise sense.

In setting up the moment equation of equilibrium, we take moments about the origin O of the xy coordinates. This is a convenient moment center, since three forces (800 N, 200 N, and R_y) pass through O, and, therefore, the moments of these forces about O are zero (refer to Fig. E11.5b). The equilibrium equations are

$$
\begin{aligned}
\sum F_x &= R_x + 800\sin 30° - 200 = 0 \\
\sum F_y &= R_y + 800\cos 30° - 500 = 0 \\
\circlearrowleft^{+} \sum M_O &= M_1 - 500(2\cos 45°) - 2R_x = 0
\end{aligned}
$$

These equations yield

$$
\begin{aligned}
R_x &= -200.0\text{ N} \\
R_y &= -192.8\text{ N} \\
M_1 &= +307.1\text{ N·m}
\end{aligned}
$$

The quantities R_x, R_y, and M_1 are the axial force, the shear force, and the moment at the clamped end of the beam. Note that since R_x and R_y are negative, they actually act to the left and downward, respectively. Also, since R_x acts to the left, there is an axial tensile force at the clamped end.

Example 11.6

Pull of a Tractor

Problem Statement The tractor shown in Fig. E11.6a weighs 5500 lb. Each rear wheel weighs 350 lb. The driver tries to drag a rock up a 30° incline, but the rock does not move. If the front wheels just lift off the ground, determine the torque T_A that each rear axle transmits to its wheel and the tension R in the cable tied to the rock.

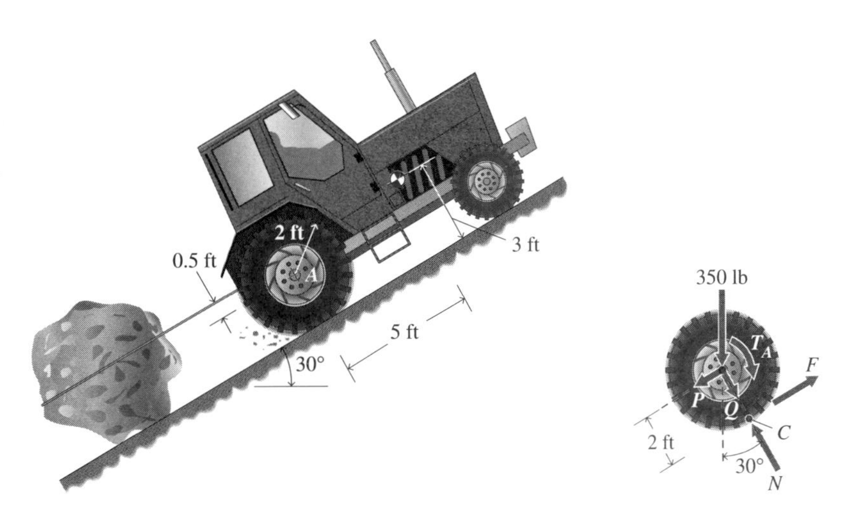

Figure E11.6a **Figure E11.6b**

Solution Figure E11.6b is the free-body diagram of a single rear wheel. The couple of magnitude T_A represents the torque transmitted to the wheel by the axle. The forces P and Q are also transmitted to the wheel by the axle. The equation of equilibrium of moments about C, the point of contact between the wheel and the ground, is

$$\circlearrowleft\!+ \quad \sum M_C = -T_A + 2P + 350(2\sin 30°) = 0$$

Hence,
$$2P = T_A - 350 \tag{a}$$

Figure E11.6c is the free-body diagram of the tractor, excluding the rear wheels. Since the front wheels lift off the ground, the ground exerts no force on them. The weight of this assembly is 5500 − 350 −350 = 4800 lb. This weight acts through the center of gravity of the assembly. The forces $2P$ and $2Q$ (with senses opposite to those of P and Q in Fig. E11.6b) represent the reactions of the two rear wheels on their axles. Note that the resultant of the forces that the wheels exert on their axles lies in the plane of symmetry of the tractor. The torque $2T_A$ (with sense

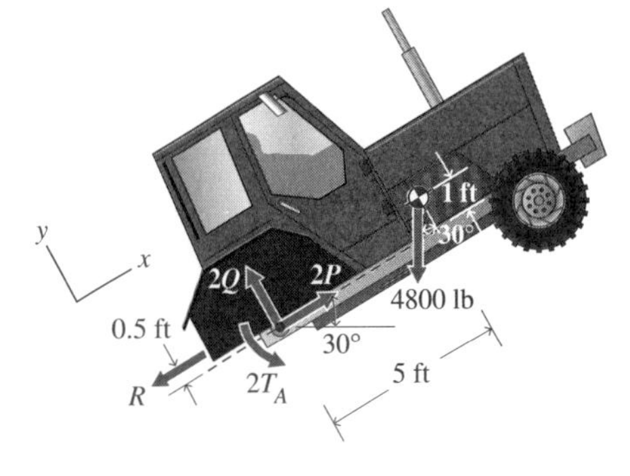

Figure E11.6c

opposite to that of T_A in Fig. E11.6b) represents the reaction to the two torques T_A that the two rear axles transmit to their wheels. The equation of equilibrium of forces in the x direction is

$$2P - R - 4800 \sin 30° = 0$$

Hence,

$$2P - R = 2400 \tag{b}$$

The equation of equilibrium of moments about the axis of the rear axles is

$$\circlearrowleft^{+} \sum M = 2T_A + 0.5R - 5(4800 \cos 30°) + 1(4800 \sin 30°) = 0$$

Hence,

$$4T_A + R = 36{,}769 \tag{c}$$

Substituting Eq. (a) into Eq. (b), we obtain

$$T_A - R = 2750 \tag{d}$$

Adding Eqs. (c) and (d), we obtain $5T_A = 39{,}519$. Hence, $T_A = 7904$ lb·ft. Equation (d) then yields the tension in the cable as $R = 5154$ lb. ■

Example 11.7 The Piston-Crank Mechanism

Problem Statement The piston-crank mechanism shown in Fig. E11.7a is in equilibrium. The pressure in the cylinder is $p = 2$ N/mm². Neglecting friction, calculate the torque and the force transmitted to the crankshaft.

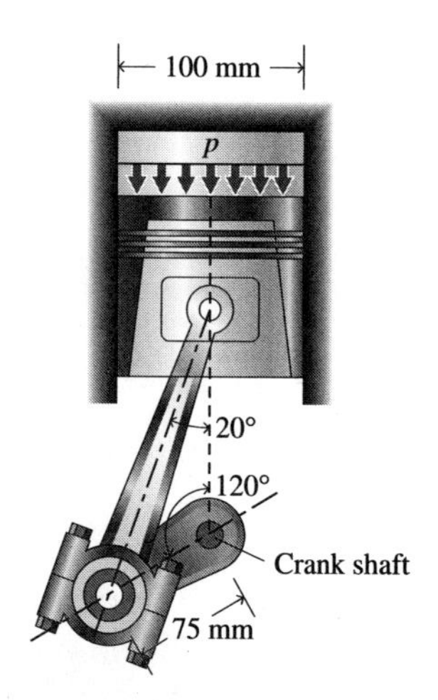

Figure E11.7a

Solution The connecting rod is a two-force member, since only two forces act on it and frictional couples of the bearings are neglected (see Fig. E11.7b). Accordingly, the connecting rod transmits a thrust P from the piston to the crank.

Figure E11.7c is the free-body diagram of the piston. The force caused by the pressure on the piston head is $\pi(50^2)2 = 15.71$ kN. For equilibrium of the vertical projections of the forces on the piston, we have

$$\sum F_y = P \cos 20° - 15.71 = 0$$

or

$$P = 16.72 \text{ kN} \tag{a}$$

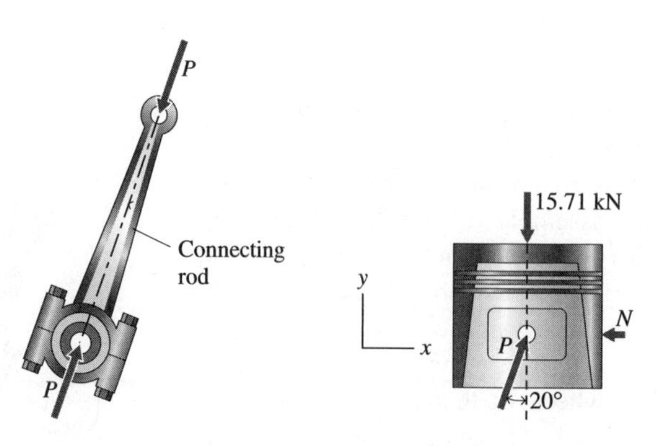

Figure E11.7b **Figure E11.7c**

Figure E11.7d is the free-body diagram of the crank. The force components R_x and R_y and the torque T_O are exerted on the crank by the crankshaft. For equilibrium of the crank, we have the equations

$$\sum F_x = R_x - P\sin 20^\circ = 0$$
$$\sum F_y = R_y - P\cos 20^\circ = 0 \qquad \text{(b)}$$
$$\circlearrowleft \!+\; \sum M_O = (P\cos 50^\circ)(75) - T_O = 0$$

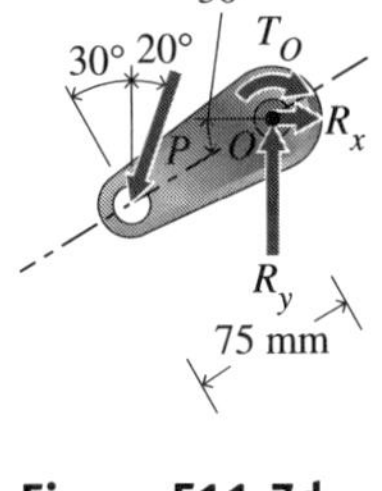

Figure E11.7d

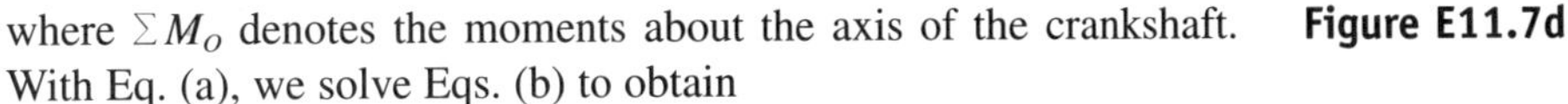

where ΣM_O denotes the moments about the axis of the crankshaft. With Eq. (a), we solve Eqs. (b) to obtain

$$R_x = P\sin 20^\circ = 5.72 \text{ kN}$$
$$R_y = P\cos 20^\circ = 15.71 \text{ kN}$$
$$T_O = 75(P\cos 50^\circ) = 806 \text{ N·mm}$$

The forces and the moment that act on the crankshaft are R_x, R_y, and T_O (see Fig. E11.7e). Note that the forces that act on the ends of the crank (refer to Fig. E11.7d) are not directed along the crank (i.e., they are not collinear), since a torque T_O is exerted on the crank by the crankshaft. In other words, the crank is not a two-force member.

T_O R_x O R_y

Figure E11.7e

CHECKPOINT

1. ***True or False:*** The axial component of a force that acts on a beam's cross section is usually assumed to act at the centroid of the cross section.
2. Is a torque the same as a bending moment? Explain.
3. ***True or False:*** A couple that lies in the cross section of a beam or a shaft is called a torque.

Answers: 1. True; 2. No—a torque is a moment about the longitudinal axis of a shaft, whereas a bending moment acts about an axis in the plane of the cross section; 3. True

11.4 INTERNAL FORCE DIAGRAMS FOR BEAMS AND SHAFTS SUBJECTED TO CONCENTRATED FORCES AND COUPLES

After studying this section, you should be able to:

- Draw internal force diagrams for a beam or a shaft subjected to concentrated forces and couples.

Key Concept Internal force diagrams are important tools used in the design of beams and shafts.

internal force diagrams

The axial force N, shear force V, bending moment M, and torque T at any section of a beam or a shaft may be represented graphically. Plots of these quantities, called *internal force diagrams,* are useful in the design of beams and shafts. They also provide a means of visualizing the results of computations. To construct such a plot, we determine expressions for N, V, M, and T as functions of the axial coordinate x of the member. (See Secs. 11.2 and 11.3 and Examples 11.3 and 11.4.) Then, we sketch each function and align it with a sketch of the beam or shaft. From these diagrams, we can easily read the values of N, V, M, and T for any value of x—that is, for any cross section of the beam or shaft. Important features of these plots are discussed in the following examples.

Example 11.8 Shear and Bending-Moment Diagrams of a Simple Beam Subjected to Concentrated Forces

Problem Statement Draw the shear and bending-moment diagrams for the beam of Example 11.1, shown again in Fig. E11.8a.

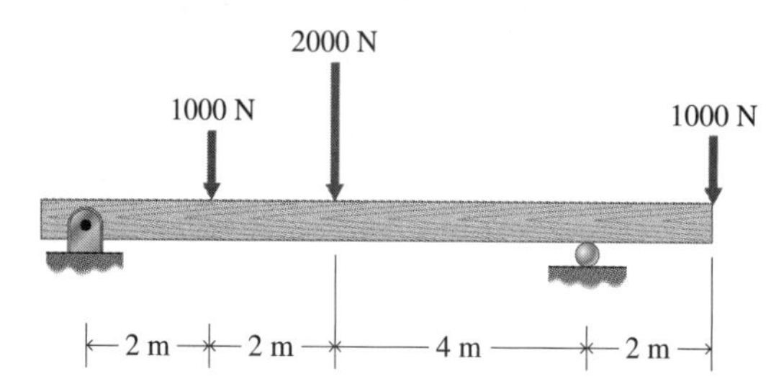

Figure E11.8a

Solution The shear force in the beam can be determined by applying Theorem 11.1. Using this method, we obtain the reactions $R_1 = 1500$ N and $R_2 = 2500$ N and the internal shears

$$\begin{aligned} V &= +1500 \text{ N} \quad 0 \text{ m} < x < 2 \text{ m} \\ V &= +500 \text{ N} \quad 2 \text{ m} < x < 4 \text{ m} \\ V &= -1500 \text{ N} \quad 4 \text{ m} < x < 8 \text{ m} \\ V &= +1000 \text{ N} \quad 8 \text{ m} < x < 10 \text{ m} \end{aligned} \tag{a}$$

In a similar manner, Theorem 11.2 can be used to find the bending moment in the beam. This approach gives

$$\begin{aligned} M &= +1500x \quad [\text{N·m}] \qquad 0 \text{ m} < x < 2 \text{ m} \\ M &= +500x + 2000 \quad [\text{N·m}] \qquad 2 \text{ m} < x < 4 \text{ m} \\ M &= -1500x + 10{,}000 \quad [\text{N·m}] \qquad 4 \text{ m} < x < 8 \text{ m} \\ M &= +1000x - 10{,}000 \quad [\text{N·m}] \qquad 8 \text{ m} < x < 10 \text{ m} \end{aligned} \tag{b}$$

The *shear* and *bending-moment diagrams* are graphs of the functions defined by Eqs. (a) and (b) and are plotted in Fig. E11.8b. For the case of a beam subject to concentrated vertical forces, the shear force V remains constant between adjacent forces. At a point where a concentrated force acts, the shear undergoes a change equal in magnitude to the force. Also, for a beam subject to concentrated forces, the bending moment M is a linear function of the axial coordinate x in each interval between forces; that is, the bending-moment diagram consists of straight line segments between forces. Therefore, only moments at the end points of the segments need be plotted and connected by straight lines. Also, at a concentrated load, the moment changes slope but not magnitude.

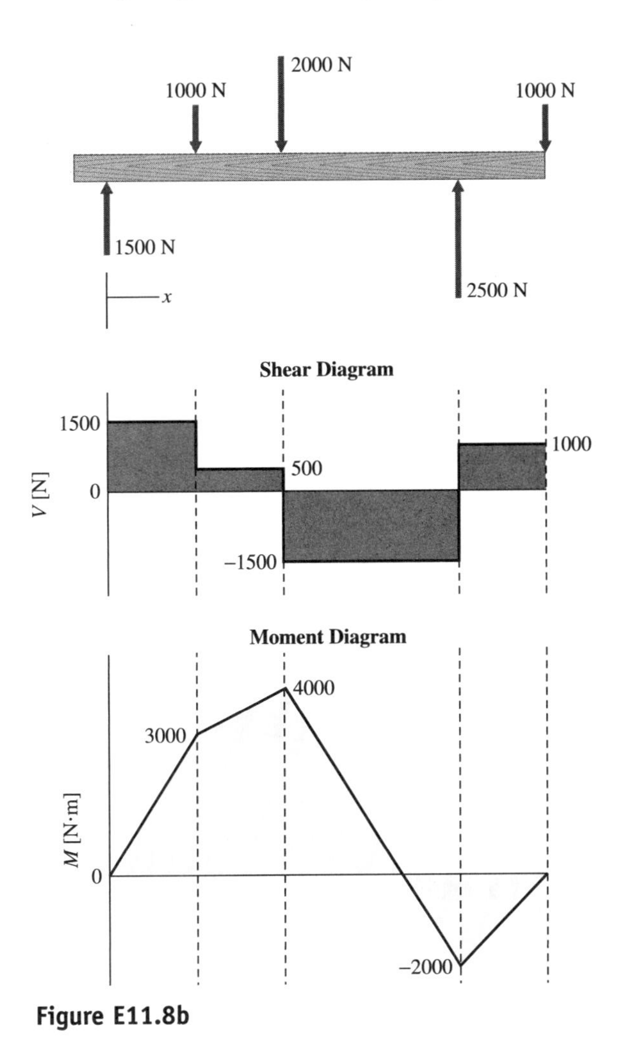

Figure E11.8b

Example 11.9

Axial Force, Shear Force, and Bending-Moment Diagrams

Problem Statement Draw the axial force, shear force, and bending-moment diagrams for the beam of Example 11.4 (see Fig. E11.4b).

Solution The *axial force diagrams, shear force diagrams,* and *bending-moment diagrams* are graphs of the functions defined by Eqs. (a) and (b) of Example 11.4. Like the shear force and the bending moment, the axial force can be plotted as a function of x (see Fig. E11.9). In Fig. E11.9, we have established an arbitrary sign convention making compressive axial force positive. Note that at a point where a couple acts, the shear remains constant, but the moment undergoes a change equal in magnitude to the moment of the couple.

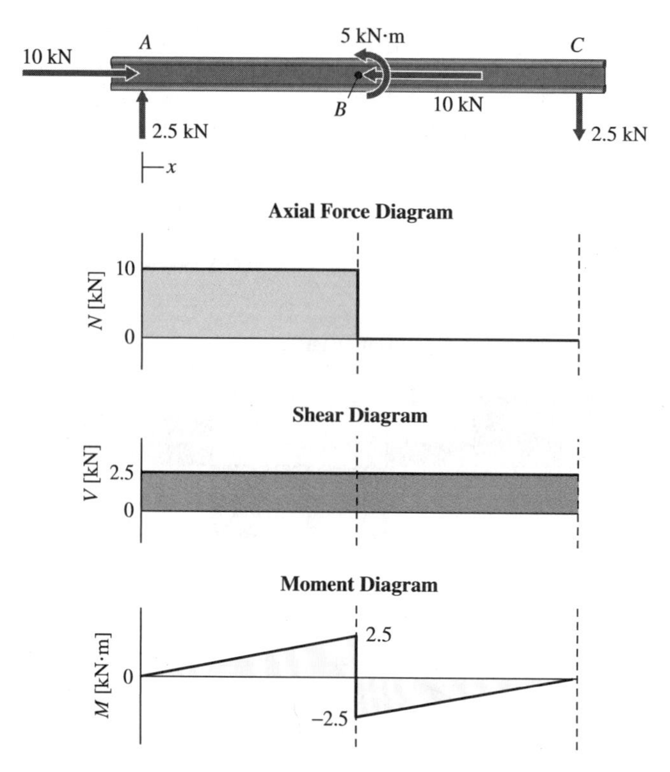

Figure E11.9

SHEAR/MOMENT RELATIONS

As illustrated in Example 11.8 (Fig. E11.8b), in the segment of a straight beam between two adjacent concentrated transverse loads, the shear force is constant and the bending moment varies linearly. This fact is indicative of a more general relation between the shear force and the bending moment in a straight beam. To determine this relationship, let's consider a beam subjected to a system of arbitrary concentrated transverse loads (see Fig. 11.9a). To relate the shear force V to the bending moment M, we first make a cut a–a at the location x between two of the concentrated loads and construct a free-body diagram of the part of the beam to the right of the cut (see Fig. 11.9b). At the cut a–a, the shear force V and the bending moment M are shown in their positive senses. Next, we make a second cut b–b, at the infinitesimal distance dx to the right of cut a–a, and draw a free-body diagram of the infinitesimal beam element of length dx between the cuts (see Fig. 11.9c). At cut b–b, the shear force $V + dV$ and the bending moment $M + dM$ are shown in their positive senses. The differential quantities dV and dM are the changes in V and M that occur over the interval dx.

The equation of equilibrium of forces in the y direction (Fig. 11.9c) is

$$\sum F_y = V - (V + dV) = 0$$

or

$$dV = 0 \tag{a}$$

Equation (a) indicates that the shear force V is constant in the interval between consecutive concentrated loads, since its change dV in the interval dx at any position x between the loads is zero.

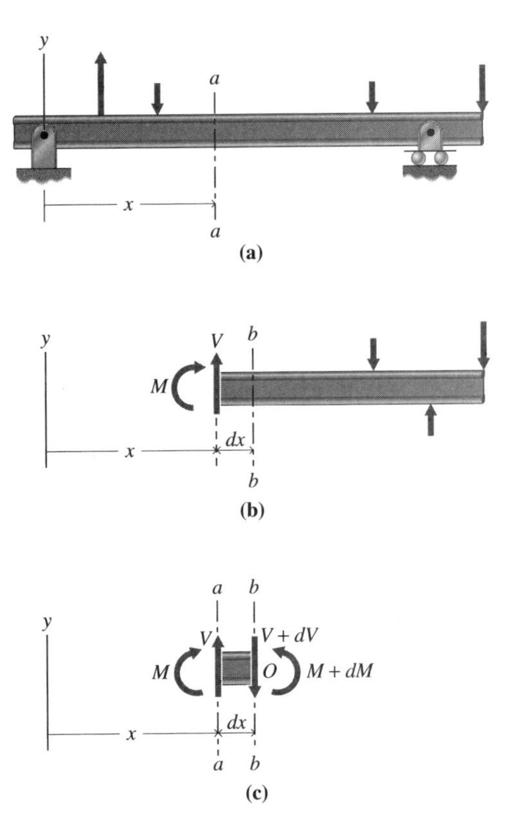

Figure 11.9 (a) Arbitrarily loaded beam. (b) Free-body diagram of beam segment to the right of section *a–a*. (c) Free-body diagram of infinitesimal segment of beam.

Similarly, the equation of equilibrium of moments for the element dx (Fig. 11.9c) is

$$\circlearrowleft\!+ \sum M_O = M + V dx - (M + dM) = 0 \qquad \text{(b)}$$

where ΣM_O denotes moments about point O at location $x + dx$. After cancellation of M, Eq. (b) yields the following relation between V, dM, and dx:

$$V = \frac{dM}{dx} \qquad (11.1)$$

Integration of Eq. (11.1) yields

$$M_2 - M_1 = \int_{x_1}^{x_2} V dx = V(x_2 - x_1) \qquad (11.2)$$

where $x_2 > x_1$ denote two positions between consecutive concentrated loads on the beam.

Equations (11.1) and (11.2) may be interpreted geometrically. Specifically, from Eq. (11.1), we derive the following rule:

RULE *Slope of the Bending Moment Diagram:* The rate of change dM/dx of the bending moment M at location x (the slope of the moment diagram at x) is equal to the shear force V at x. In other words, the rate of change of the bending moment at x is equal to the shear at x.

We note that a positive value of V denotes a positive slope of the bending moment diagram and a negative value of V denotes a negative slope.

Likewise, from Eq. (11.2), we derive the following rule:

RULE *Change in Bending Moment:* The change in the bending moment between x_1 and x_2 is equal to the area under the shear diagram from position x_1 to position x_2.

To further interpret this rule, we note that if V in Eq. (11.2) is positive, the change in M from x_1 to x_2 is positive; if V is negative, the change in M is negative.

Given the shear diagram, using these two rules, we can draw the bending-moment diagram for a beam subject to concentrated loads, without having to derive expressions for M in each load interval of the beam (see Example 11.10). In Sec. 11.5, these rules are extended to apply to beams subject to distributed loads.

Example 11.10

Construction of a Bending-Moment Diagram Using Shear/Moment Relations

Problem Statement Determine the support reactions and draw the shear and bending-moment diagrams for the beam shown in Fig. E11.10a, using Theorem 11.1 and the two rules presented above.

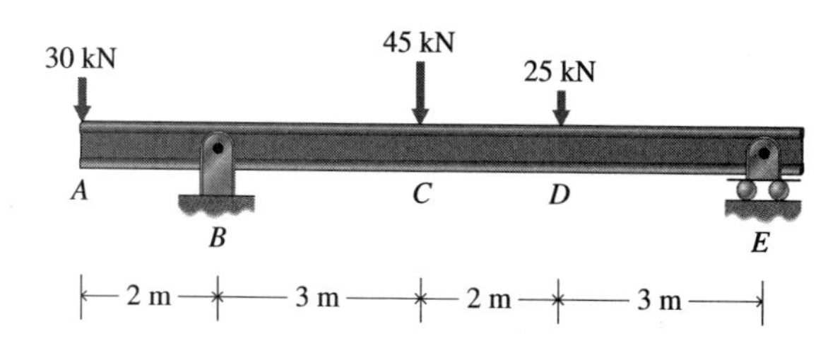

Figure E11.10a

Solution First, we draw the free-body diagram of the beam (Fig. E11.10b). Then, the equations of equilibrium for the beam are

$$\sum F_y = B + E - 30 - 45 - 25 = 0 \quad \text{(a)}$$

$$\circlearrowleft\!+ \sum M_B = 45(3) + 25(5) - 30(2) - 8E = 0 \quad \text{(b)}$$

Solving Eqs. (a) and (b) (see Fig. E11.10b), we obtain

$$E = 25 \text{ kN}$$

$$B = 75 \text{ kN}$$

To construct the shear diagram, we use Theorem 11.1. Starting at the left end of the beam, we note that the shear to the right of point A is $V = -30$ kN, and this shear remains constant to point B (see the shear diagram in Fig. E11.10b). To the right of point B, the shear is $V = 45$ kN (75 kN − 30 kN), and it remains constant to point C, where it changes to $V = 0$ (45 kN − 45 kN). That shear remains constant to point D, where it changes to $V = -25$ kN (0 − 25 kN). Again, from point D to point E, the shear remains constant. At point E, the shear returns to zero ($V = -25$ kN + 25 kN = 0).

The bending-moment diagram can now be constructed from the shear diagram using the two rules given above [see also Eqs. (11.1) and (11.2)]. To begin, we note that since no moment is applied at the left end of the beam (point A), $M = M_A = 0$. In the interval AB, by Eq. (11.1) or the first rule, the slope of the bending-moment diagram is given by $dM/dx = V = -30$ kN. By Eq. (11.2) or the second rule, the change in the bending moment from A to B is equal to the integral of V from A to B (the area "under" the shear diagram), or $M_B - M_A = (-30 \text{ kN})(2 \text{ m}) = -60$ kN·m. The moment at point B is given by $M_B = M_A - 60$ kN·m $= -60$ kN·m, since $M_A = 0$ (see the moment diagram

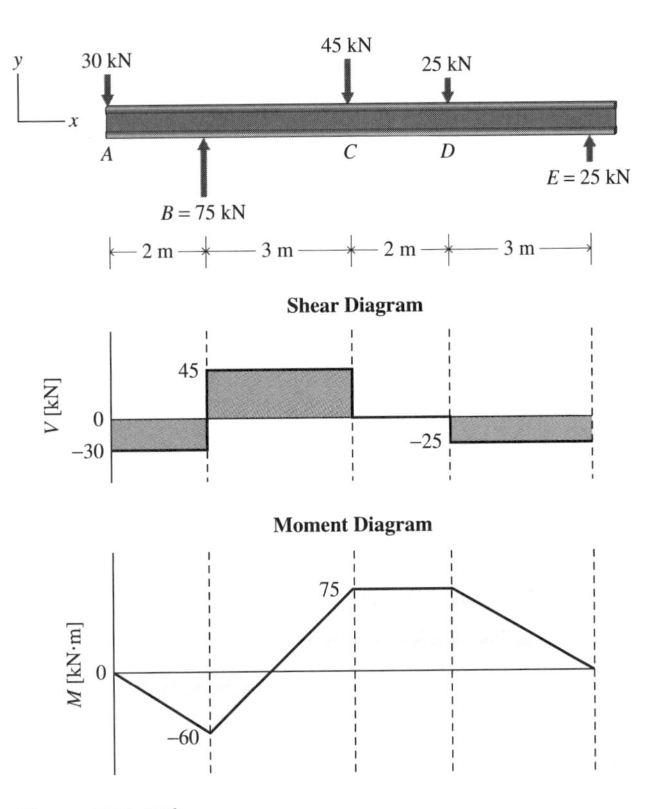

Figure E11.10b

in Fig. E11.10b). Likewise, in the interval *BC,* the slope of the bending-moment diagram is +45 kN (the value of V) and the change in moment from B to C is $M_C - M_B =$ (45 kN)(3 m) = 135 kN·m. Therefore, the moment at point C is $M_C = M_B + 135 = -60 + 135 = 75$ kN·m. Continuing in this manner, we construct the complete bending-moment diagram (see Fig. E11.10b). Again, note that although V is constant between concentrated loads, it undergoes a change at a concentrated load equal to the magnitude of the load. Also note that at a concentrated load, the slope of the bending-moment diagram changes, but the magnitude of the moment does not change. ∎

CHECKPOINT

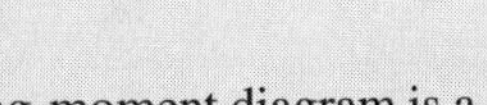

1. ***True or False:*** A bending-moment diagram is a graph of the bending moment M plotted as a function of the axial coordinate x of a beam.
2. ***True or False:*** The shear force in a beam changes abruptly at a concentrated transverse force.
3. ***True or False:*** The bending moment in a beam changes abruptly at a point where a concentrated transverse force acts.
4. ***True or False:*** If a beam is subjected to a couple at a point in the beam, the shear force does not change at that point.

Answers: 1. True; 2. True; 3. False; 4. True.

11.5 SHEAR AND BENDING-MOMENT DIAGRAMS FOR BEAMS SUBJECTED TO DISTRIBUTED LOADS

After studying this section, you should be able to:

- Draw shear and bending-moment diagrams for beams subjected to distributed loads.

Key Concept Mathematical relationships that relate distributed load w, shear force V, and bending moment M lead to geometric rules for constructing shear and bending-moment diagrams.

We will develop shear and bending-moment diagrams for beams subjected to distributed loads in this section, as we did for beams subjected to concentrated loads in Sec. 11.4. For a distributed load, the relationship between the shear force that acts at a section of a beam and the corresponding bending moment at the section has the same form as Eq. (11.1). In addition, two relationships between the distributed load that acts on a beam and the shear force that acts in the beam are useful for drawing shear and bending-moment diagrams.

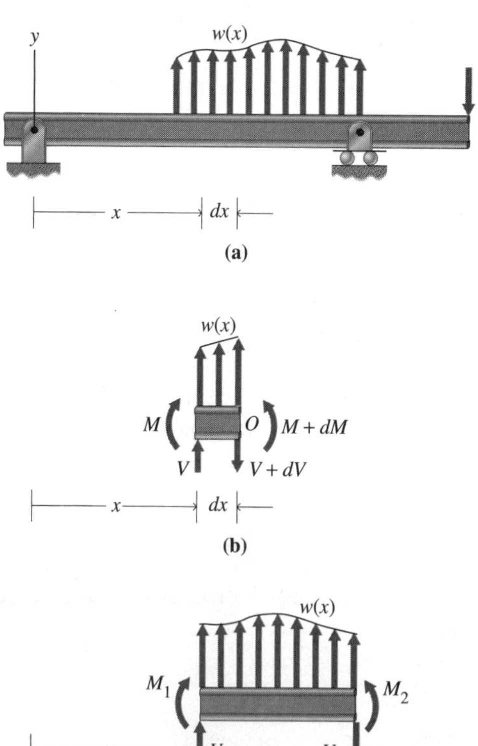

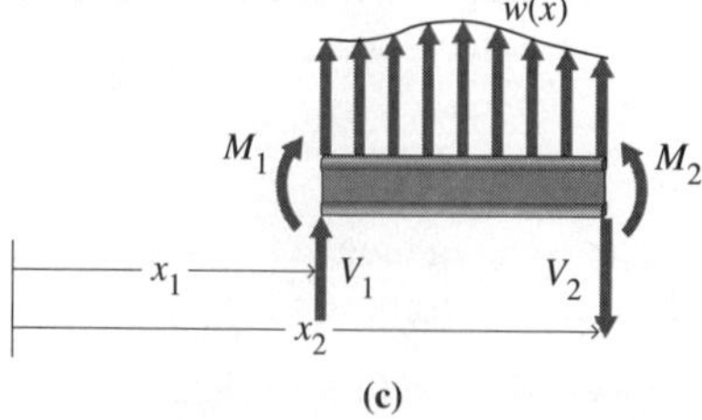

Figure 11.10

To develop the relationship between the distributed load and the shear force, let's consider a portion of a beam subjected to only a transversely distributed load that is a function of x—namely, $w(x)$. The load $w(x)$ is measured in force per unit length of the beam and is taken as *positive* when it acts *upward* (in the direction of the positive y axis in Fig. 11.10a). We cut a section at x and isolate an infinitesimal element of the beam of length dx to the right of the section at x. The free-body diagram of the element is shown in Fig. 11.10b. The positive senses of the shear forces and bending moments are taken as shown, with the shear force V and the moment M on the left side of the element and the shear force $V + dV$ and the bending moment $M + dM$ on the right side (refer to the discussion of shear and moment sign convention in Sec. 11.2).

The equilibrium equation for forces acting on the element dx in the y direction is

$$\sum F_y = V + w(x)\, dx - (V + dV) = 0$$

or

$$w(x)\, dx = dV \tag{a}$$

Dividing both sides of Eq. (a) by dx, we obtain

$$w(x) = \frac{dV}{dx} \tag{11.3}$$

Integration of Eq. (11.3) from $x = x_1$ to $x = x_2$ yields

$$\int_{V_1}^{V_2} dV = V_2 - V_1 = \int_{x_1}^{x_2} w(x)\, dx$$

or

$$V_2 = V_1 + \int_{x_1}^{x_2} w\, dx \tag{11.4}$$

where V_2 is the shear force at $x = x_2$ and V_1 is the shear force at $x = x_1$ (Fig. 11.10c).

Equations (11.3) and (11.4) can be interpreted geometrically in the form of two rules for drawing shear diagrams. The first rule relates to the slope of the shear diagram [see Eq. (11.3)]:

RULE *Slope of the Shear Diagram:* The rate of change dV/dx at x (the slope of the shear diagram at x) is equal to the value of the distributed load $w(x)$ at x. The slope of the shear diagram is positive if the sense of $w(x)$ is positive [if $w(x)$ is directed in the positive y direction (refer to Fig. 11.10a)], and it is negative if the sense of $w(x)$ is negative [if $w(x)$ is directed in the negative y direction].

The second rule relates a change in the shear force to the resultant of the distributed load [see Eq. (11.4)]:

RULE *Change in Shear Force:* The area under the distributed load curve between any two locations x_1 and x_2 on a beam (the integral of the distributed load between x_1 and x_2) is equal to the change $V_2 - V_1$ in the shear force V between x_1 and x_2.

If the integral in Eq. (11.4) is positive, the change $V_2 - V_1$ in the shear force V from x_1 to x_2 is positive; if the integral is negative, $V_2 - V_1$ is negative.

Once the support reactions of a beam have been determined, we can apply these two rules and draw the shear diagram for a beam subjected to distributed loads directly from the free-body diagram of the beam, without having to derive a formula for V. This procedure is illustrated in Examples 11.11, 11.12, and 11.13.

To develop the relationship between the shear force and the bending moment, we consider the equilibrium equation for moments that act on the infinitesimal beam element dx (see Fig. 11.10b). Summing moments about point O, we find

$$\sum M_O = M + V\,dx + [w(x)\,dx]\frac{dx}{2} - (M + dM) = 0 \tag{b}$$

Since the term $[w(x)\,dx](dx/2)$ in Eq. (b) is small compared to the terms $V\,dx$ and dM (we say it is a higher-order infinitesimal), it may be neglected. Then the equilibrium equation for moments reduces to

$$V = \frac{dM}{dx} \tag{11.5}$$

where V and dM/dx are the shear and the slope of the bending moment, respectively, at section x. Analogous to Eqs. (11.3) and (11.4), by integration, we can write Eq. (11.5) in the form

$$M_2 = M_1 + \int_{x_1}^{x_2} V\,dx \tag{11.6}$$

Equation (11.5) is equivalent to Eq. (11.1). However, in Eq. (11.6), V is a function of x. Except for this fact, the form of the relationship between the shear and the bending moment in a beam subject to distributed loads is the same as that for a beam subject to concentrated loads. *Therefore, the procedure described in Sec. 11.4 may be used to construct the bending-moment diagram of a beam subjected to distributed loads.*

Often, when designing a structure, an engineer must determine the magnitude of the largest moment in a beam to ensure that the beam is capable of carrying a given load. The largest moment may be the maximum positive moment or the maximum negative moment in the beam. Recall that a function of x attains a maximum (or minimum) value when the derivative of the function with respect to x equals zero. Therefore, referring to Eq. (11.5), we have the following rule to determine the maximum (or minimum) moment:

RULE *Location of the Maximum (Minimum) Moment:* The bending moment M attains a maximum (or minimum) value at a section in the beam where the shear force $V = 0$.

Example 11.11

Beam Subjected to a Uniformly Distributed Load

Problem Statement A beam carries a uniformly distributed load of 100 lb/ft over half its length (see Fig. E11.11a).

a. Calculate the support reactions.

b. Derive expressions for the shear force V and the bending moment M in the beam as a function of x.

c. Draw shear and bending-moment diagrams for the beam.

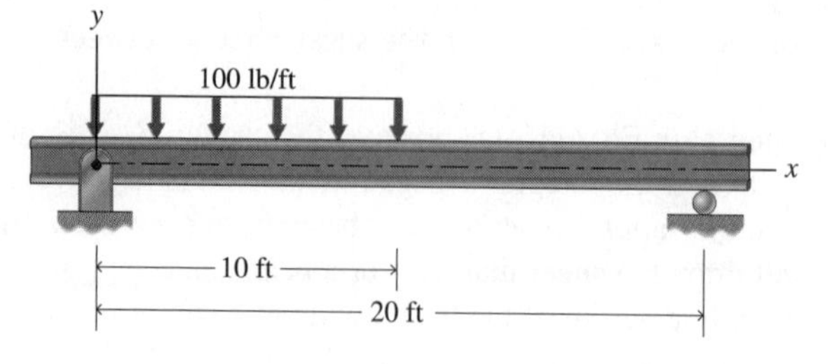

Figure E11.11a

Solution **a.** To calculate the support reactions (external forces on the beam), we can replace the distributed load by its resultant, which is located at the midpoint of the uniformly distributed load (see Fig. E11.11b). The magnitude of the resultant is equal to the magnitude of the total distributed load, or 100(10) = 1000 lb, and the line of action of the resultant lies at a distance of 5 ft from the left support. From Fig. E11.11b, the equations of equilibrium for the entire beam are

$$\oplus \sum M_1 = 20R_2 - 1000(5) = 0$$
$$\oplus \sum M_2 = 1000(15) - 20R_1 = 0 \tag{a}$$

where ΣM_1 and ΣM_2 indicate moments about the left support and the right support, respectively. The solutions of Eqs. (a) are

$$R_1 = 750 \text{ lb}$$
$$R_2 = 250 \text{ lb}$$

(Note that the equilibrium equation $\Sigma F_y = R_1 + R_2 - 1000 = 0$ serves as a check.)

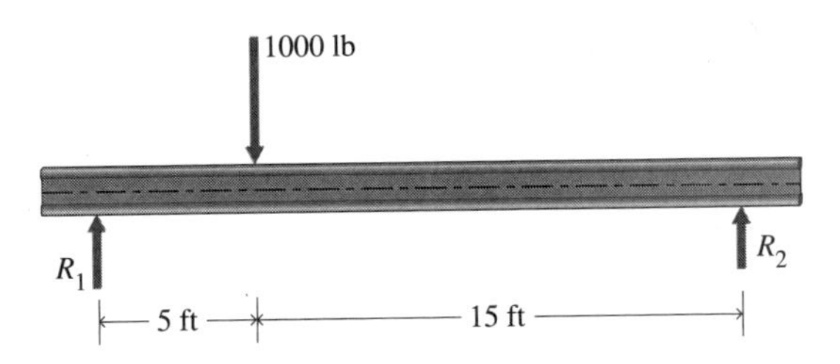

Figure E11.11b

b. To derive expressions for the shear and bending moments, we consider two free-body diagrams for different segments of the beam: one for $0 < x < 10$ ft and one for $10 \text{ ft} < x < 20$ ft, where x is a coordinate along the axis of the beam, measured from the left support. Figure E11.11c is a free-body diagram of the part of the beam from 0 to x for $x < 10$ ft. In this diagram, we consider the beam segment to be a rigid body, with V and M as external forces. In the beam itself, however, V and M are internal (refer to Fig. E11.11a). To determine V and M at x, we can replace the distributed load by its resultant. For the uniformly distributed load from 0 to x for $x < 10$ ft, the resultant force acts at $x/2$, and its magnitude is $100x$. The condition that the sum of all the vertical forces that act on the beam segment is zero is expressed by the equation

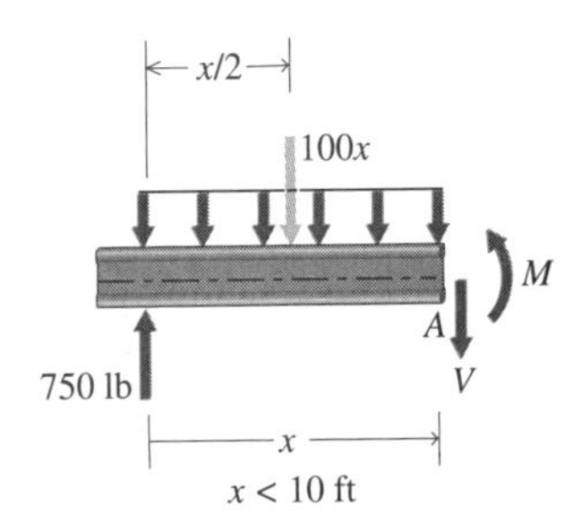

Figure E11.11c

$$\sum F_y = 750 - 100x - V = 0$$

where V is the shear force at section x (refer to Fig. E11.11c). Hence,

$$V = 750 - 100x \quad [\text{lb}], \qquad x < 10 \text{ ft} \tag{b}$$

The condition that the sum of the moments about A, the right end of the beam segment, is zero is expressed by the equation

$$\sum M_A = 750x - 100x\left(\frac{x}{2}\right) - M = 0$$

where M is the bending moment at section x. Hence,

$$M = 750x - 50x^2 \quad [\text{lb}\cdot\text{ft}], \qquad x < 10 \text{ ft} \tag{c}$$

Figure E11.11d is a free-body diagram of the left part of the beam from 0 to x, for 10 ft $< x <$ 20 ft. The condition of equilibrium of vertical forces is expressed by the equation

$$\sum F_y = 750 - 1000 - V = 0$$

or

$$V = -250 \text{ lb}, \qquad 10 \text{ ft} < x < 20 \text{ ft} \tag{d}$$

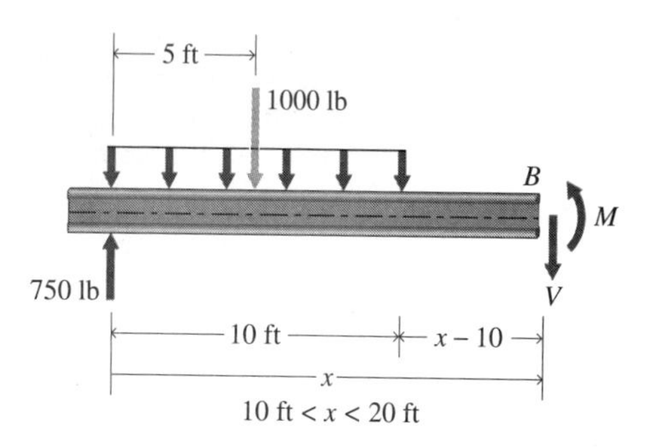

Figure E11.11d

The condition of equilibrium of moments about B, the right end of the beam segment, is given by

$$\sum M_B = 750x - 1000(x - 5) - M = 0$$

or

$$M = 5000 - 250x \quad [\text{lb}\cdot\text{ft}], \qquad 10 \text{ ft} < x < 20 \text{ ft} \tag{e}$$

Hence, the shear forces and bending moments in the two intervals of the beam are determined as functions of x.

c. The shear diagram is a graph of Eqs. (b) and (d) in the appropriate segments of the beam. Likewise, the bending-moment diagram is a graph of Eqs. (c) and (e) in the appropriate segments. For $x < 10$, the shear diagram is a linear function of x (a straight line), and the bending-moment diagram is a quadratic function of x (a parabolic arc). For 10 ft $< x <$ 20 ft, the shear is a constant (independent of x), and the bending-moment diagram is a straight line (a linear function of x). To plot the shear and bending-moment diagrams, we construct Table E11.11.

Table E11.11

	EQS. (b) AND (c)						EQS. (d) AND (e)	
x [ft]	0	2	4	6	8	10	10	20
V [lb]	750	550	350	150	−50	−250	−250	−250
M [lb·ft]	0	1300	2200	2700	2800	2500	2500	0

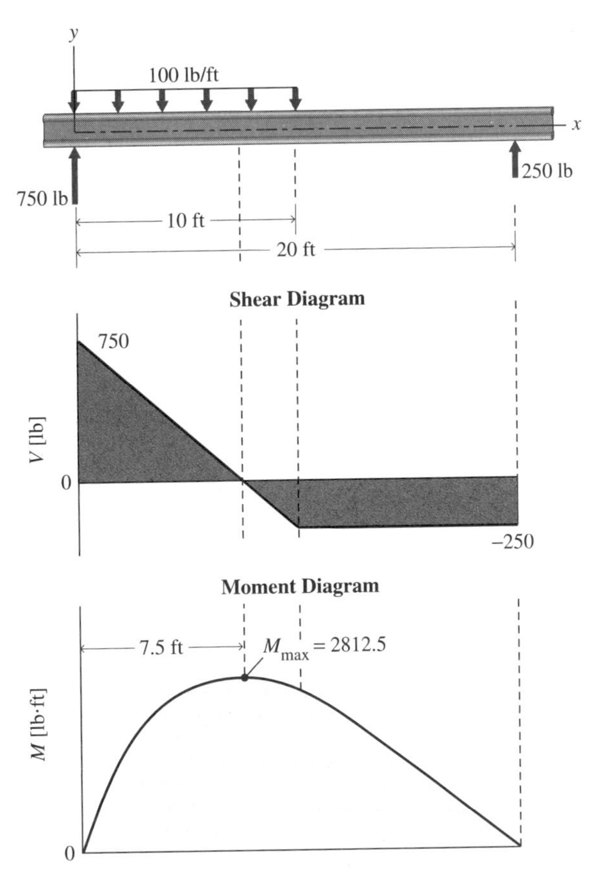

Figure E11.11e

Figure E11.11e shows the shear and the bending-moment diagrams, plotted by means of Table E11.11. The maximum value of the moment occurs in the range $x < 10$. The location of the maximum moment is obtained by setting the derivative of M with respect to x equal to zero—that is, by setting $V = 0$ [see Eq. (11.5)]. Thus, from Eq. (b), or Eq. (c),

$$V = \frac{dM}{dx} = 750 - 100x = 0$$

or

$$x = 7.5 \text{ ft}$$

Hence, the value of the maximum moment, by Eq. (c) is

$$M_{max} = (750)(7.5) - (50)(7.5)^2 = 2812.5 \text{ lb·ft}$$

Example 11.12

Beam Subjected to a Linearly Distributed Load

Problem Statement Construct and compare the shear and bending-moment diagrams for

a. a beam subjected to a linearly distributed load $w(x) = 0.2x$ [kips/ft] (directed downward in the free-body diagram in Fig. E11.12a)

b. the same beam as in part a but loaded by the resultant $F = 90$ lb of the distributed load (see the free-body diagram in Fig. E11.12b)

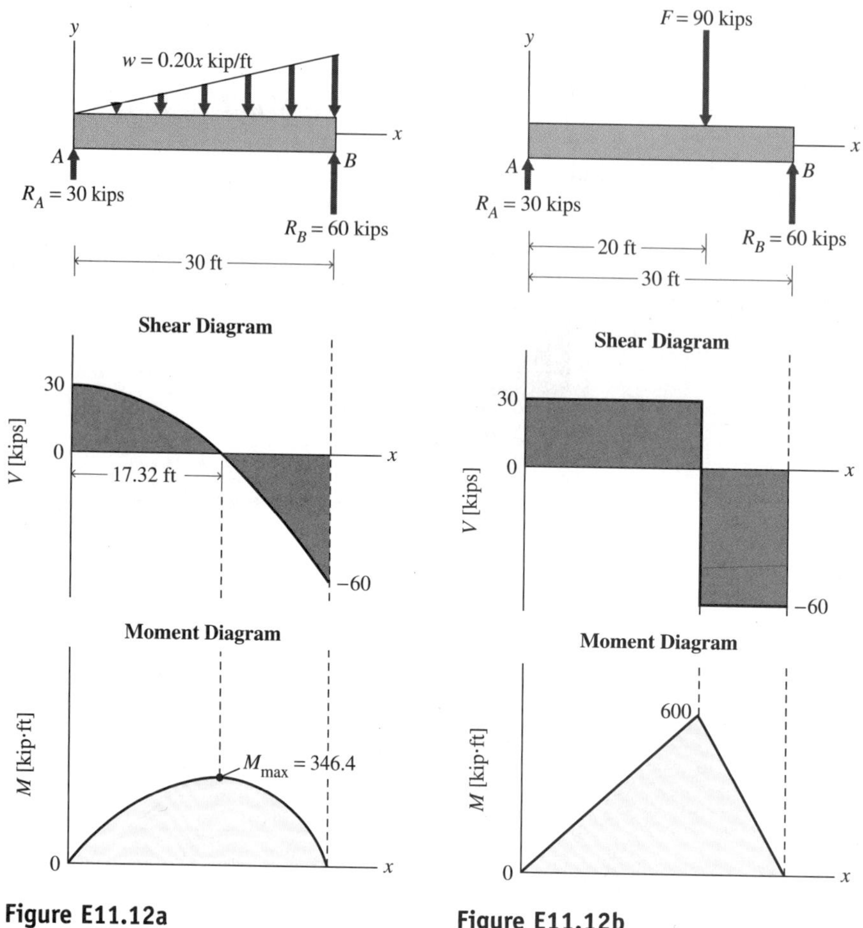

Figure E11.12a **Figure E11.12b**

Solution Since the loads of parts a and b are dynamically equivalent, they produce the same support reactions—namely, R_A =30 kips and R_B = 60 kips (see the free-body diagrams in Figs. E11.12a and E11.12b).

a. The shear and bending-moment diagrams are shown in Fig. E11.12a. The beam is subject to the linearly distributed load $w(x) = 0.2x$ [kips/ft] (directed downward, as shown in Fig. E11.12a). The shear force at support A ($x = 0$) is 30 kips. Since the slope of the shear diagram is $dV/dx = w(x)$ [see Eq. (11.3)], the slope changes from 0 at A [the value of $w(x)$ at $x = 0$] to -6 kips/ft at B [the value of $w(x)$ at $x = 30$ ft]. The change $V_B - V_A$ in the shear from A to B is -90 kips, equal to the total load on the beam (negative downward). Since $w(x) = 0.2x$ [kips/ft], $V_B - V_A$ can also be obtained from Eq. (11.4):

$$V_B - V_A = \int_0^{30} (-0.2x)\,dx = -\tfrac{1}{2}[0.2(30^2)] = -90 \text{ kips}$$

(the area of the triangular load; see the free-body diagram of the beam in Fig. E11.12a).

The maximum moment occurs at the section where $V = 0$, by the rule associated with Eq. (11.5). Let x_2 be the value of x at which $V_2 = 0$. Then, from Eq. (11.4), we have, integrating from $x_1 = 0$ to x_2, with $w(x) = 0.2x$, downward,

$$V_2 - V_1 = V_2 - V_A = 0 - 30 = -\tfrac{1}{2}(0.2)(x_2)^2$$

or $x_2 = 17.32$ ft (Fig. E11.12c).

The bending-moment diagram is drawn using the rules established in Sec. 11.4. Since the shear force varies parabolically, the area under the shear curve, between the point $x = 0$ (point A) and the point $x = x_2$ (where $V = V_2 = 0$), is

$$\frac{2}{3}V_A x_2 = \frac{2}{3}(30)(17.32) = 346.4 \text{ kip·ft}$$

Numerically, this area is equal to the change in the bending moment M from $x = 0$ to $x = 17.32$ ft [see Eq. (11.6)]. Hence, since $M = 0$ at $x = 0$, the maximum moment is $M_{max} = 346.4$ kip·ft.

Alternatively, we can determine the shear V and the bending moment M as functions of x by the method employed in Example 11.11. For example, consider the free-body diagram of a segment of the beam from 0 to x (see Fig. E11.12c). In this free-body diagram, V and M are *external* to the element (see also Example 11.11). However, in the actual beam, they are *internal* forces. From Fig. E11.12c, we find, by the sum of forces and moments, for $0 \leq x \leq 30$ ft:

$$V = 30 - \frac{1}{2}(0.2)x^2 \quad \text{[kips]} \tag{a}$$

$$M = 30x - \frac{1}{6}(0.2)x^3 \quad \text{[kip·ft]} \tag{b}$$

By Eq. (a), we find that $V = 0$ for $x = 17.32$ ft, and by Eq. (b), $M_{max} = 346.4$ kip·ft, as before. Also, Eqs. (a) and (b) show clearly that V varies parabolically (it is a quadratic function of x) and M varies with the cube of x.

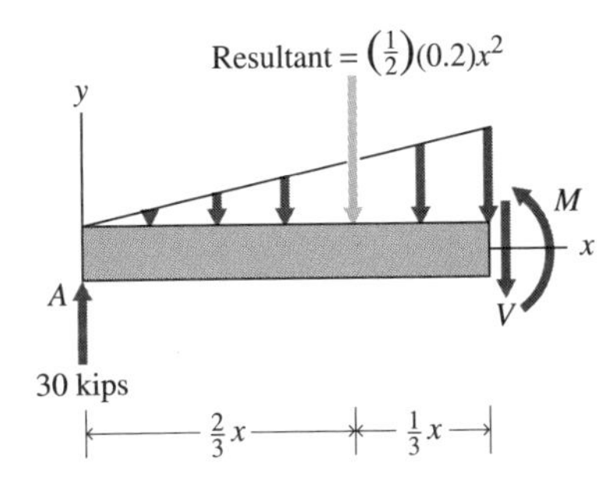

Figure E11.12c

b. The shear and bending-moment diagrams of Fig. E11.12b are drawn using the procedures outlined in Sec. 11.4. The differences in magnitude and shape that exist between the shear and moment diagrams of Figs. E11.12a and E11.12b are significant. In particular, the maximum moment for the case where the resultant acts on the beam is much larger than that for the distributed load. So, although the loads of parts a and b are dynamically equivalent and produce the same support reactions on the beam, their effects on internal forces in the beam (shear and bending moment) are very different. ■

Example 11.13

Beam Subjected to Distributed and Concentrated Loads

Problem Statement The beam shown in Fig. E11.13a carries a 1200 lb concentrated force at a distance of 4 ft from the left support. A distributed downward load $w(x) = 40x$,

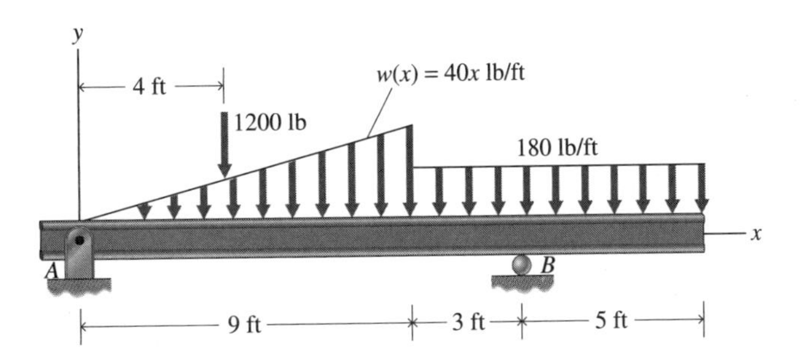

Figure E11.13a

which varies linearly from 0 at $x = 0$ to 360 lb/ft at $x = 9$ ft, acts on the left part of the beam. On the right part of the beam, there is a uniformly distributed load of 180 lb/ft.

a. Calculate the support reactions.

b. Derive expressions for the shear force and the bending moment in the beam as functions of x.

c. Draw shear and bending-moment diagrams for the beam.

Solution **a.** To calculate the support reactions, we replace the distributed loads by their resultants (see Fig. E11.13b). The total magnitude of the linearly distributed load is $9(360)/2 = 1620$ lb. The resultant axis of this load lies 6 ft from the left support. The total magnitude of the uniformly distributed load is $8(180) = 1440$ lb. The resultant axis of this load lies at the center of the uniformly loaded interval—that is, 4 ft from the right end.

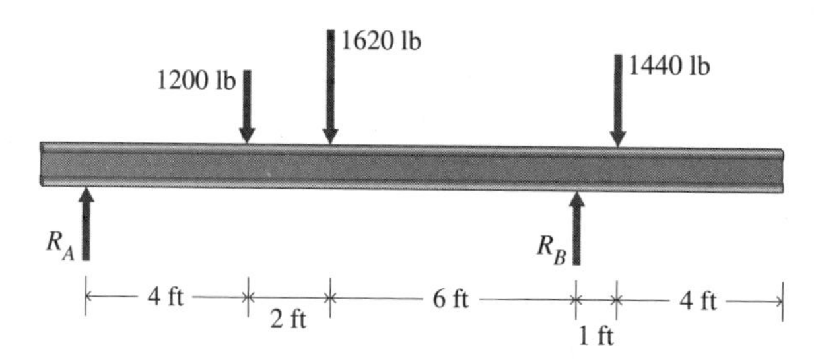

Figure E11.13b

Considering the resultant loads and taking moments about the right support, we obtain

$$\sum M_B = 12R_A - 8(1200) - 6(1620) + 1(1440) = 0$$

$$R_A = 1490 \text{ lb}$$

The total load on the beam is $1200 + 1620 + 1440 = 4260$ lb. Since the net vertical force on the beam is zero, $\Sigma F_y = R_A + R_B - 4260 = 0$. Consequently, $R_B = 2770$ lb.

b. To derive formulas for the shear and the bending moment in terms of x, we must consider four different segments of the beam and the associated free-body diagrams. We treat each segment as a rigid body subjected to external shear V and moment M. That is, we treat V and M as external forces relative to the beam segment. For example, Fig. E11.13c is a free-body diagram of a part of the beam in the range $0 < x < 4$ ft, where x is measured from the left support. The resultant of the distributed downward load in this interval is $x(40x)/2 = 20x^2$ [lb], since $w(x) = 40x$ [lb/ft] at x (see Fig. E11.13c). The coordinate of the resultant axis of this load is $2x/3$ [ft]. Equilibrium of vertical forces is expressed by the equation

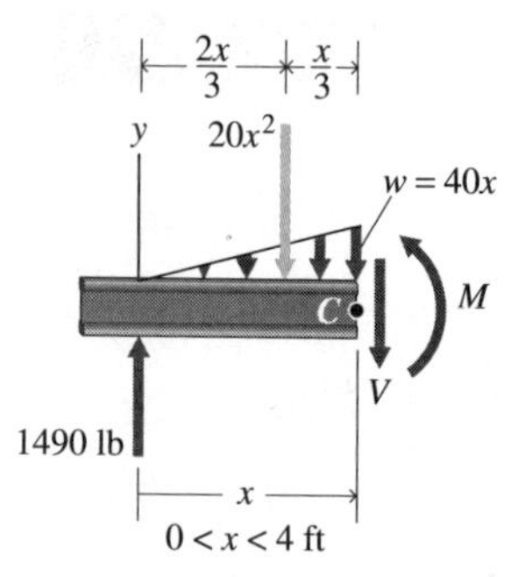

Figure E11.13c

$$\sum F_y = 1490 - 20x^2 - V = 0$$

where V, the shear at x, is external to the element. Hence,

$$V = 1490 - 20x^2 \quad \text{[lb]}, \qquad x < 4 \text{ ft} \tag{a}$$

Equilibrium of the moments about the right end of the segment is expressed by the equation

$$\circlearrowleft\!+\ \sum M_C = 1490x - 20x^2\left(\frac{x}{3}\right) - M = 0$$

where M, the bending moment at x, is external to the element. Hence,

$$M = 1490x - \frac{20x^3}{3} \quad [\text{lb·ft}], \qquad x < 4 \text{ ft} \tag{b}$$

Figure E11.13d is a free-body diagram of the left part of the beam, where 4 ft $< x <$ 9 ft. Equilibrium of vertical forces is expressed by the equation

$$\sum F_y = 1490 - 20x^2 - 1200 - V = 0$$

or

$$V = 290 - 20x^2 \quad [\text{lb}], \qquad 4 \text{ ft} < x < 9 \text{ ft} \tag{c}$$

Equilibrium of moments about the right end of the segment is

$$\circlearrowleft\!+\ \sum M_D = 1490x - 20x^2\left(\frac{x}{3}\right) - 1200(x - 4) - M = 0$$

or

$$M = 4800 + 290x - \frac{20x^3}{3} \quad [\text{lb·ft}], \qquad 4 \text{ ft} < x < 9 \text{ ft} \tag{d}$$

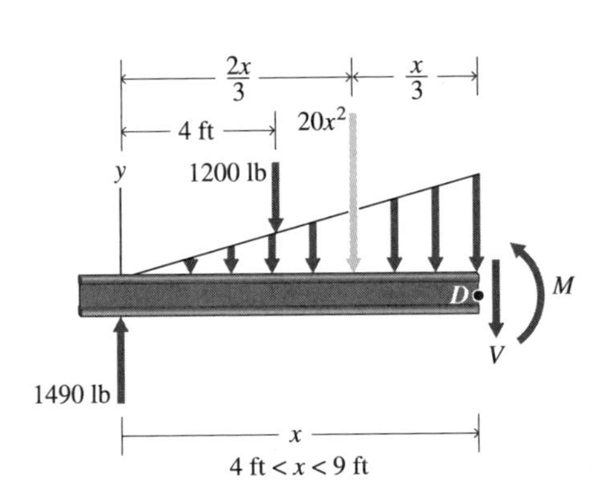

Figure E11.13d

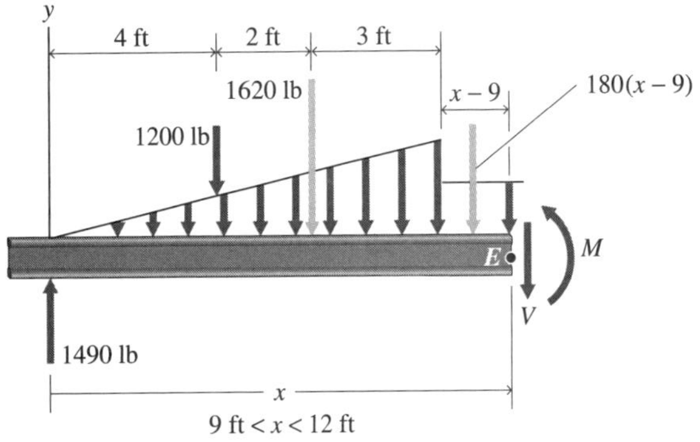

Figure E11.13e

Figure E11.13e is a free-body diagram of the left part of the beam, where 9 ft $< x <$ 12 ft. The condition of equilibrium of vertical forces is expressed by the equation

$$\sum F_y = 1490 - 1200 - 1620 - 180(x - 9) - V = 0$$

since the magnitude of the linearly distributed load is 1620 lb. Hence,

$$V = 290 - 180x \quad [\text{lb}], \qquad 9 \text{ ft} < x < 12 \text{ ft} \tag{e}$$

The condition of equilibrium of moments about the right end of the segment is expressed by the equation

$$\circlearrowleft\!+\ \sum M_E = 1490x - 1200(x - 4) - 1620(x - 6) - 180\frac{(x - 9)^2}{2} - M = 0$$

or

$$M = 7230 + 290x - 90x^2 \quad [\text{lb·ft}], \qquad 9 \text{ ft} < x < 12 \text{ ft} \tag{f}$$

Figure E11.13f is a free-body diagram of the beam in the range 12 ft $< x <$ 17 ft. The condition of equilibrium of vertical forces is expressed by the equation

$$\sum F_y = V - 180(17 - x) = 0$$

or

$$V = 3060 - 180x \quad \text{[lb]}, \qquad 12 \text{ ft} < x < 17 \text{ ft} \tag{g}$$

The condition of equilibrium of moments about the left end of the segment is expressed by the equation

$$\circlearrowleft\!+ \sum M_F = M + 180\,\frac{(17 - x)^2}{2} = 0$$

or

$$M = -90(17 - x)^2 \quad \text{[lb·ft]}, \qquad 12 \text{ ft} < x < 17 \text{ ft} \tag{h}$$

Thus, the shears and the bending moments in the various intervals of the beam have been determined in terms of x.

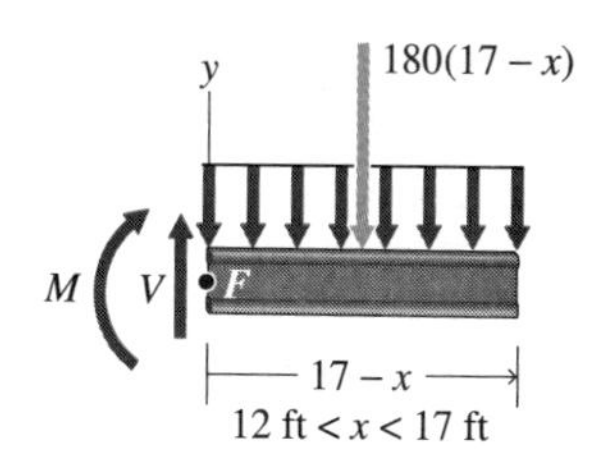

Figure E11.13f

c. The shear diagram is a graph of Eqs. (a), (c), (e), and (g) in the appropriate segments of the beam. Likewise, the bending-moment diagram is a graph of Eqs. (b), (d), (f), and (h) in the appropriate segments. Beneath the linearly distributed load, the shear diagram consists of arcs of parabolas, and the bending-moment diagram consists of arcs of cubic curves, as we see from Eqs. (a), (b), (c), and (d). Beneath the uniformly distributed load, the shear diagram consists of straight-line segments, and the bending-moment diagram consists of arcs of parabolas, as we see from Eqs. (e), (f), (g), and (h).

To plot the shear and bending-moment diagrams, we construct Table E11.13.

Table E11.13

EQS. (a) AND (b)						
x [ft]	0	1	2	3	4−	
V [lb]	1490	1470	1410	1310	1170	
M [lb·ft]	0	1483	2927	4290	5533	
EQS. (c) AND (d)						
x [ft]	4+	5	6	7	8	9−
V [lb]	−30	−210	−430	−690	−990	−1330
M [lb·ft]	5533	5417	5100	4543	3707	2550
EQS. (e) AND (f)						
x [ft]	9+	10	11	12−		
V [lb]	−1330	−1510	−1690	−1870		
M [lb·ft]	2550	1130	−470	−2250		
EQS. (g) AND (h)						
x [ft]	12+	13	14	15	16	17
V [lb]	900	720	540	360	180	0
M [lb·ft]	−2250	−1440	−810	−360	−90	0

In Table E11.13, the notations $x = 4-$ and $x = 4+$ denote values of x immediately to the left and to the right, respectively, of the point $x = 4$ ft. The notations $x = 9-$, $x = 9+$, $x = 12-$, and $x = 12+$ are interpreted similarly. The shear and bending-moment diagrams plotted by means of Table E11.13 are shown in Fig. E11.13g. Note that the shear force function is discontinuous at points of concentrated load. For example, the table shows that the values of V are different at the points $4-$ and $4+$. Although the bending moment M is a continuous function, cusps appear in the bending-moment diagram at points of concentrated load. The continuity of the bending moment at the points $x = 4$ ft, $x = 9$ ft, and $x = 12$ ft, and the continuity of the shear at the point $x = 9$ ft serve as checks on Eqs. (a) through (h) (see Table E11.13).

As an exercise, you may wish to construct the shear and bending-moment diagrams using the rules from Sec. 11.4 and those associated with Eqs. (11.3)–(11.6), rather than the values given in Table E11.13.

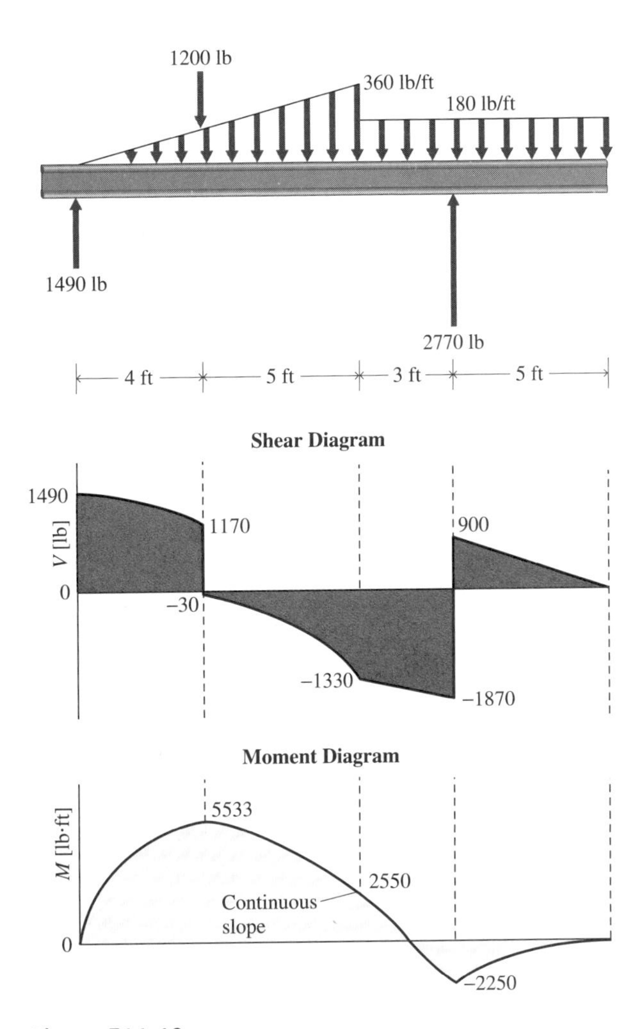

Figure E11.13g

CHECKPOINT 

1. ***True or False:*** The shear diagram for a beam is continuous at a point where a concentrated load is applied.
2. ***True or False:*** The bending-moment diagram for a beam is continuous at a point where a concentrated load is applied.
3. ***True or False:*** In segments where a beam is subjected to only distributed loads, the shear and bending-moment diagrams are continuous.

Answers: 1. False; 2. True; 3. True

Chapter Highlights

- A beam is a bar that is straight or curved, that is subjected primarily to lateral loads, and that is supported by hinges, rollers, or clamps.
- Beam supports exert forces on a beam. These forces are called support reactions.
- If the equations of statics are sufficient to determine the support reactions of a beam, the beam is statically determinate; if not, the beam is statically indeterminate.
- A cantilever beam is a straight beam that is loaded transversely and is clamped at one end and free at the other end. A cantilever beam is statically determinate.
- A simple beam is a straight beam subjected to transverse loads and supported at one end by a hinge and at the other by a roller. A simple beam is statically determinate.
- The resultant force V of the distributed transverse forces that act on a beam's cross section is called the shear force at the section. The shear force is internal to the beam.
- The resultant force N of the distributed axial forces that act on a beam's cross section is called the axial force at the section. It is internal to the beam. The force N is customarily located at the centroid of the beam's cross section.
- The resultant couple, with moment M, of the distributed axial forces that act on a beam's cross section is called the bending moment at the section. It is internal to the beam. The magnitude of M depends on the location of the axial force N. Ordinarily, N is located at the centroid of the beam's cross section, and M is calculated relative to the centroidal axis.
- A torque is a moment about the longitudinal axis of a shaft or a beam. It is also referred to as a twisting moment.
- Axial force, shear, bending-moment, and torque diagrams are graphs of the axial force N, the shear force V, the bending moment M, and the torque T in a beam or shaft, plotted as functions of the axial coordinate x of the beam.
- Complete axial force, shear, bending-moment, and torque diagrams of a beam or shaft should include a sketch of the beam or shaft aligned with the diagrams.
- The total distributed load on a beam and its resultant produce identical support reactions. However, the shear forces and bending moments in the beam change markedly if the distributed force is replaced by its resultant.

Problems

11.1 *Support Reactions: Beams Subjected to Concentrated Forces and Couples*

11.2 *Shear Forces and Bending Moments in Beams*

11.1 Determine the support reactions for each of the beams *AB* shown in Fig. P11.1.

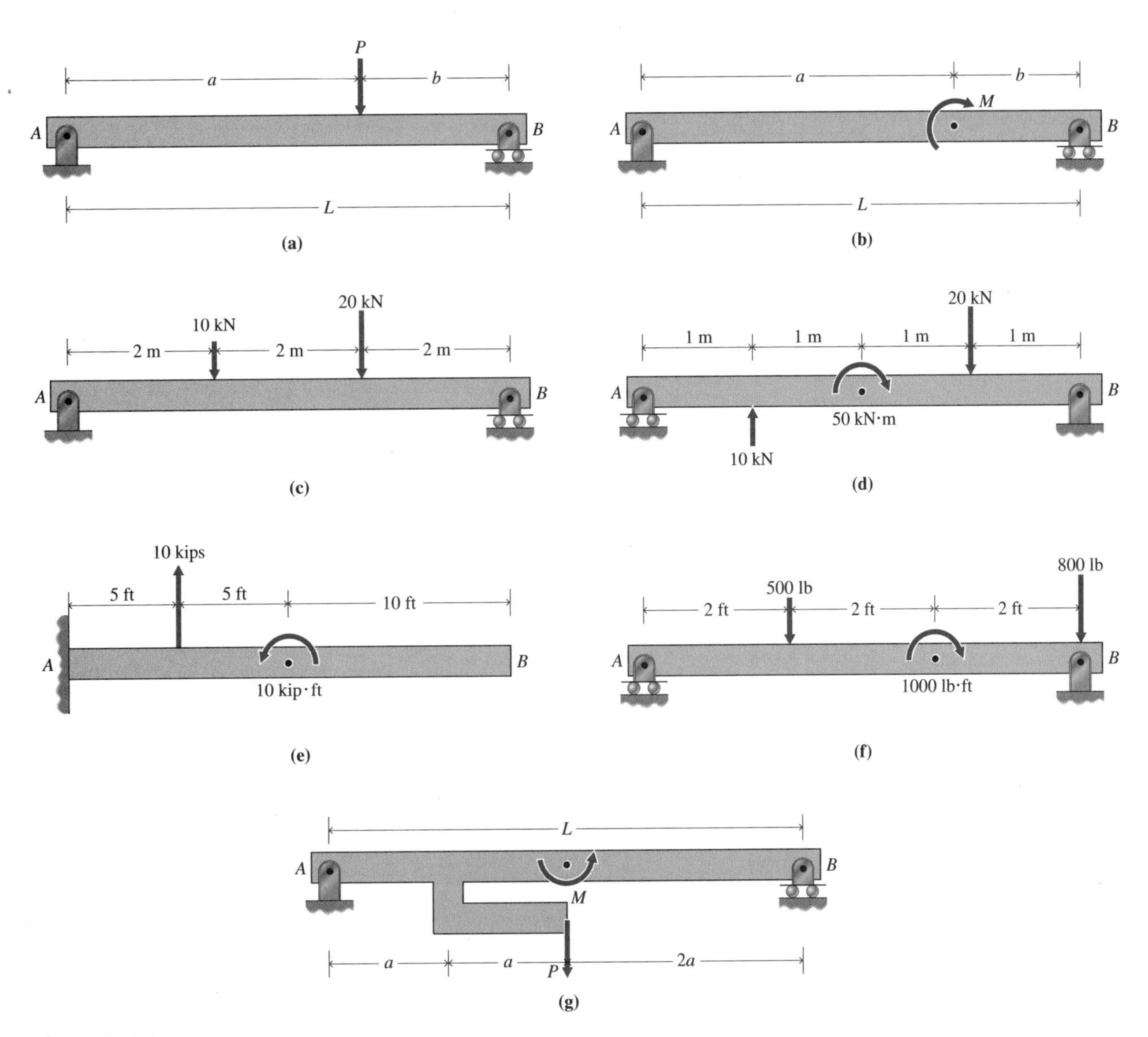

Figure P11.1

For Problems 11.2–11.6, compute the shear force and the bending moment at section *a–a* for each beam shown.

11.2

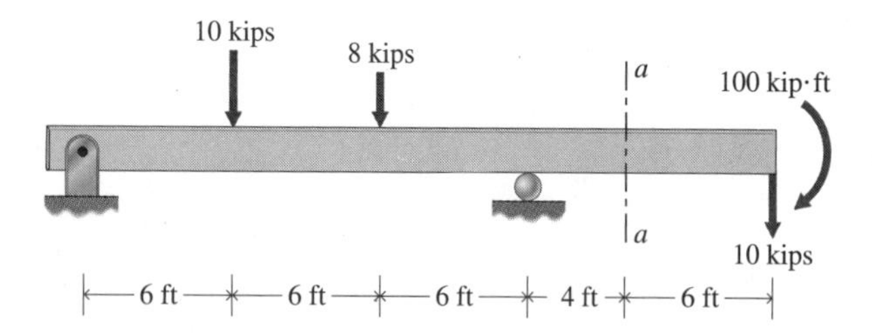

11.3

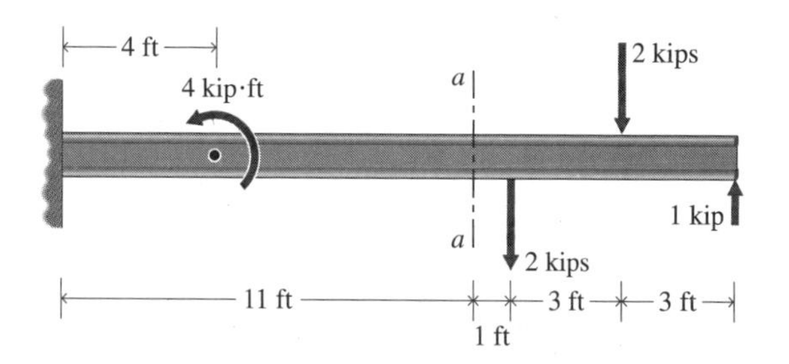

11.4

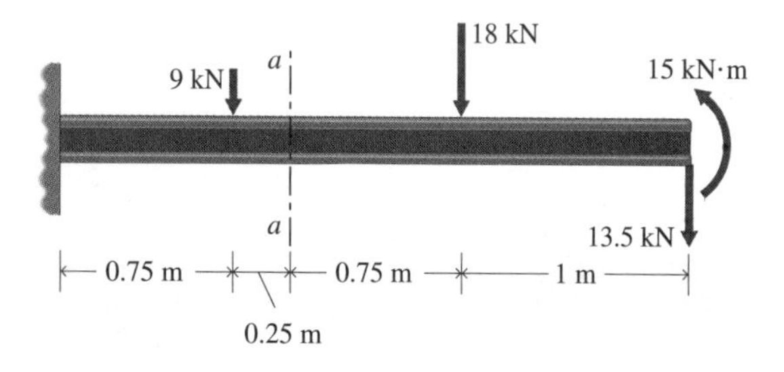

11.5

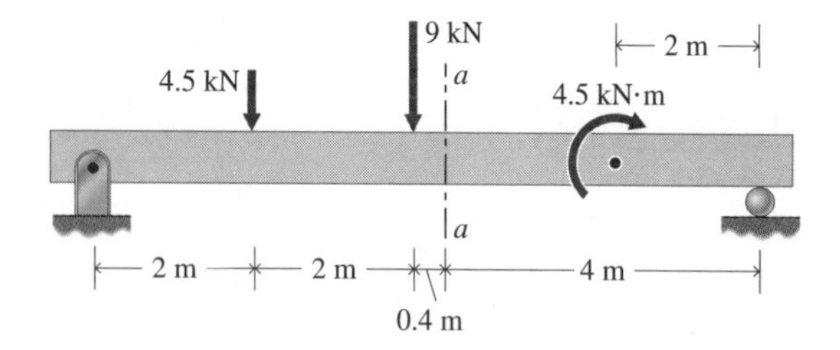

11.6

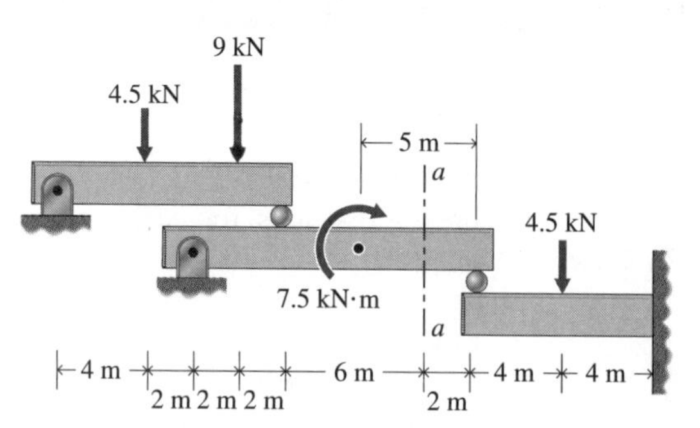

11.7 Determine the shear and the bending moment for each of the beams *AB* shown in Fig. P11.1, as a function of the distance x measured along the axis of the beam from the left end.

11.8 To get an automobile weighing 3000 lb over a wall 18 in high, a man lays two boards across the wall in front of the wheels, as shown in Fig. P11.8. He plans to drive the car onto the boards until they tip, allowing him to drive down the other side. If the bending moment in either board exceeds 40,000 lb·in, the boards will break. Determine whether the boards are strong enough to allow the man to drive the car slowly to the other side of the wall. Neglect the inclination of the board.

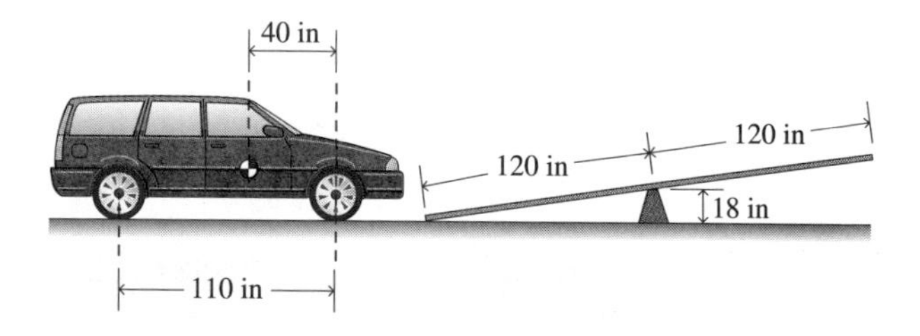

Figure P11.8

11.3 *Beams and Shafts Subjected to Axial Forces and Twisting Moments*

In the following problems, assume that the resultant axial force on a cross section acts at the center of the cross section.

11.9 To bend a bow into the shape shown in Fig. P11.9, an archer, holding the bow at A, must apply a 35 lb horizontal pull to the string at B. Calculate the shear force V, the axial force N, and the bending moment M at A. Ignore any forces that exist in the bow before the pull is applied.

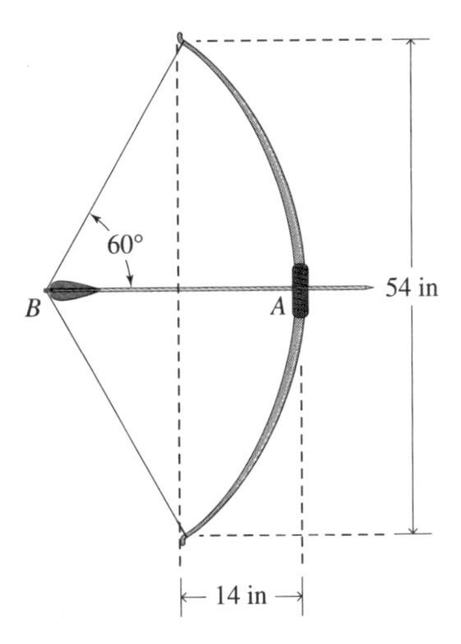

Figure P11.9

11.10 Neglecting friction, determine the torque T required in the camshaft A to hold the follower in the position designated in Fig. P11.10.

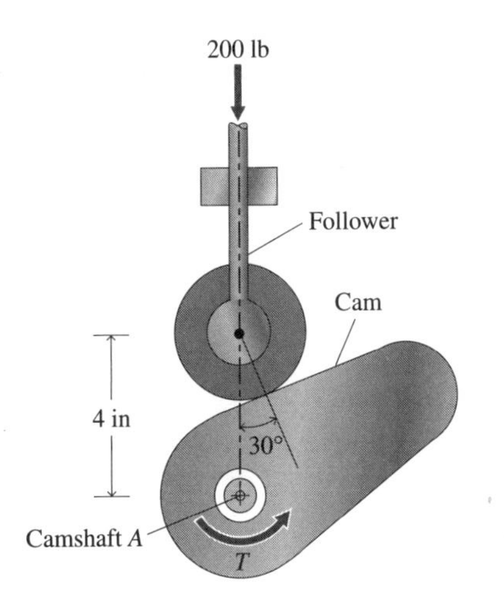

Figure P11.10

11.11 A bar magnet is hung in a horizontal position from a quartz fiber attached at the center of the magnet (see Fig. P11.11). If a uniform horizontal magnetic field H is applied perpendicular to the axis of the magnet, the magnet rotates through an angle θ. The force F that the field exerts on each pole of the magnet is proportional to H. The resisting torque in the fiber is proportional to θ. If $H = H_1$, then $\theta = 30°$. If $H = H_2$, then $\theta = 45°$. Compute H_2/H_1.

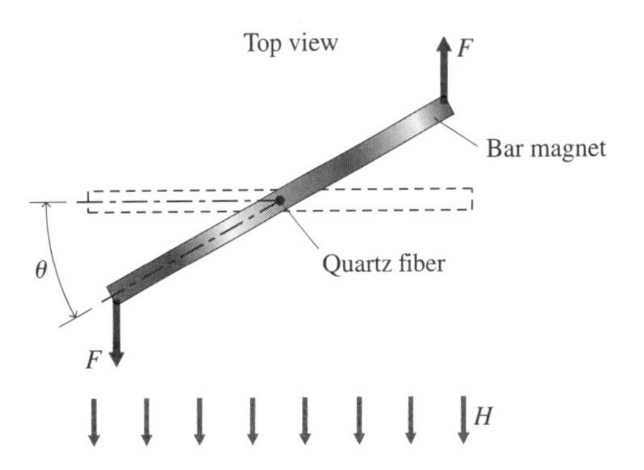

Figure P11.11

11.12 Two pulleys C and D are fixed to a shaft as shown in Fig. P11.12.

a. Calculate the torque T required to maintain the shaft and pulleys in equilibrium.

b. Calculate the reactions of the frictionless bearings at A and B, assuming that no transverse loads are applied outside the segment AB.

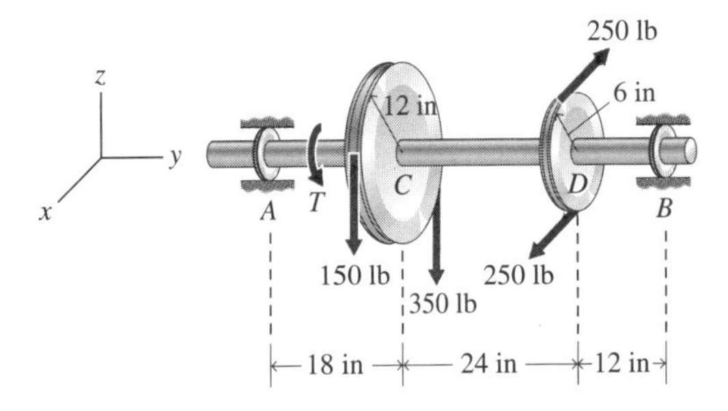

Figure P11.12

11.13 Figure P11.13 is a simplified representation of bevel gears (teeth not shown). A torque of 100 N·m is applied to shaft 1. Calculate the torque transmitted to shaft 2.

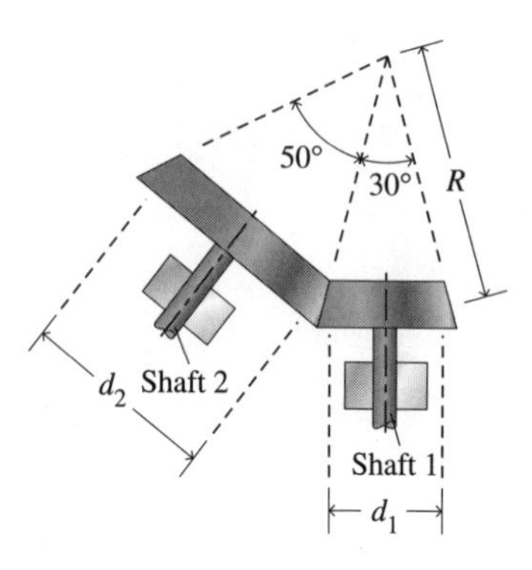

Figure P11.13

11.14 Assuming that equilibrium exists, determine the thrust F developed by the quick-return mechanism in Fig. P11.14 for the position shown. Neglect friction.

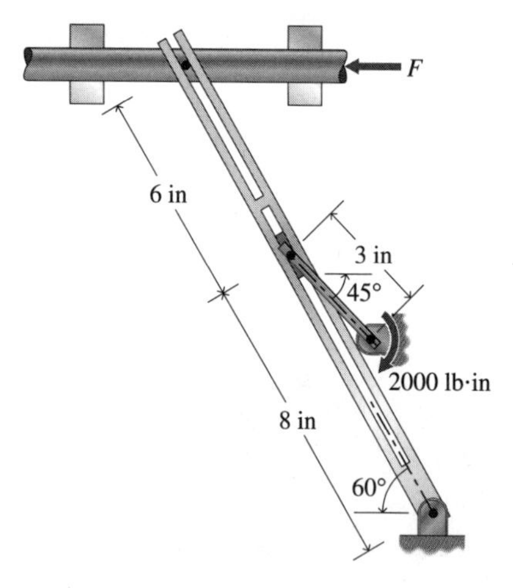

Figure P11.14

11.15 Figure P11.15 is a simplified representation of a train of gears (teeth not shown). The torque in the drive shaft A is 25 N·m. Shaft B is rigidly attached to its two gears.

a. Calculate the torque transmitted to shaft C, neglecting friction.

b. Calculate the force acting on a tooth of the smaller gear on shaft B, assuming that only one tooth makes contact with C.

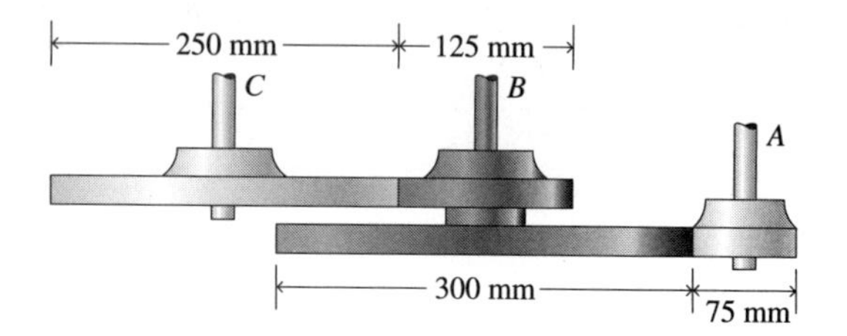

Figure P11.15

11.16 The linkage $ABCD$ in Fig. P11.16 is in equilibrium.

a. Neglecting friction, calculate the torque transmitted to shaft A.

b. Calculate the reactions R_x and R_y of the bearings on shaft A.

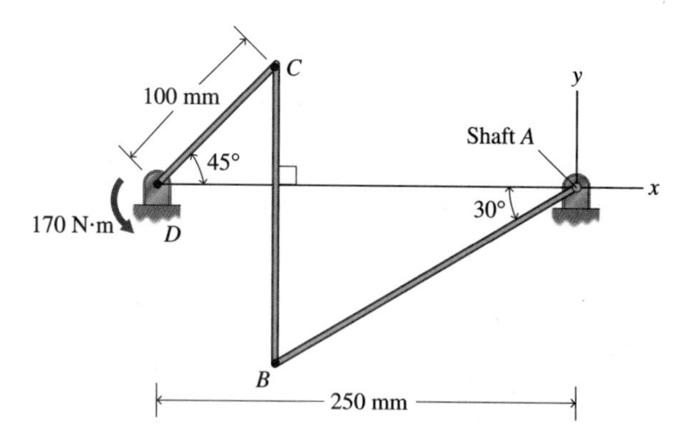

Figure P11.16

11.17 The uniform ring shown in Fig. P11.17 is subject to diametrically opposed forces P. The moment at section a–a due to these forces is $M_{a-a} = (\pi - 2)Pr/2\pi$.
a. Draw the free-body diagram of the upper half of the ring.
b. Write the expression for the moment M_{b-b} at the section b–b in terms of θ in the range $0 \le \theta \le \pi/2$.

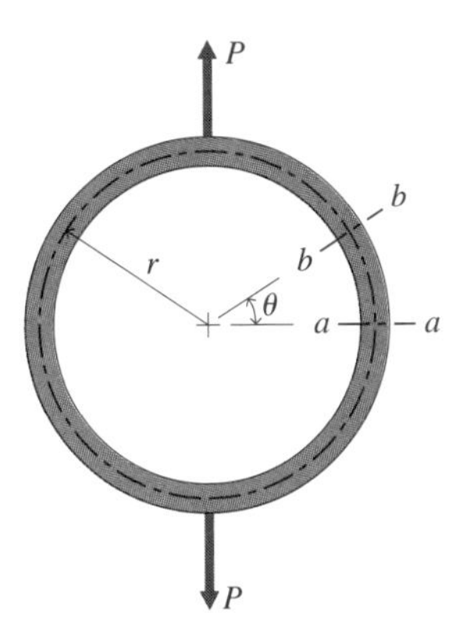

Figure P11.17

11.18 The structure shown in Fig. P11.18 is called a *bent.* The left leg is hinged at the bottom. The reactions of the support on the right leg are denoted by R_x, R_y, and M_O.
a. Express M_O in terms of P, R_y, a, and b.
b. Express the shear force V, the bending moment M, and the axial force N in each vertical leg in terms of P, R_x, R_y, a, b, and y. Let the bending moment be positive if it places the fibers on the outside face of the leg in compression.
c. Express the shear force V, the bending moment M, and the axial force N in the top member in terms of P, R_x, R_y, a, b, and x. Let the bending moment be positive if it places the outside fibers in compression.

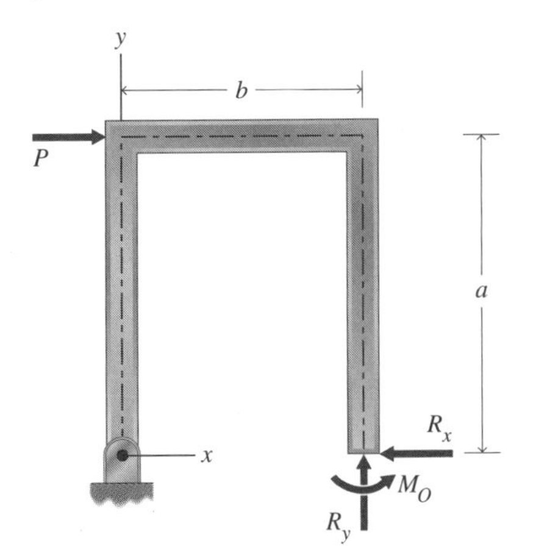

Figure P11.18

11.19 A torque T is applied to the shaft of a crank AC. The mechanism is held in equilibrium in the position shown in Fig. P11.19 by the force F. The L-shaped arm is pinned at B.
a. Neglecting friction, express F in terms of T and θ.
b. Determine the value of F for $T = 200$ N·m and $\theta = 30°$.

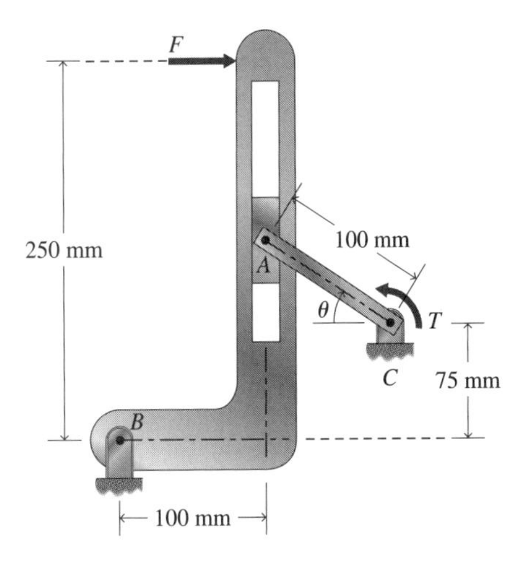

Figure P11.19

11.20 The torque in the drive shaft A of the four-bar linkage shown in Fig. P11.20 is 50 N·m.
a. Construct free-body diagrams of the members.
b. Calculate the torque transmitted to shaft B. Neglect friction.

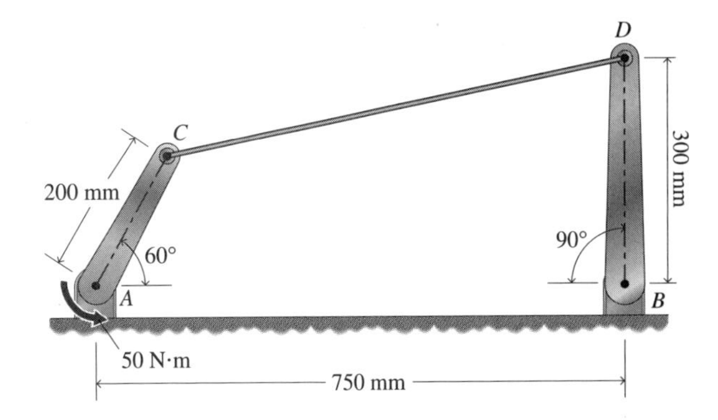

Figure P11.20

11.21 For the situation described in Problem 11.11, suppose the field direction lies at 60° counterclockwise from the original direction of the axis of the magnet. The force the field exerts on a pole of the magnet is in the direction of the field. Compute H_2/H_1.

11.22 Derive formulas for the bending moment, the shear force, and the net tension in the semicircular beam at the section located by the angle θ in Fig. P11.22.

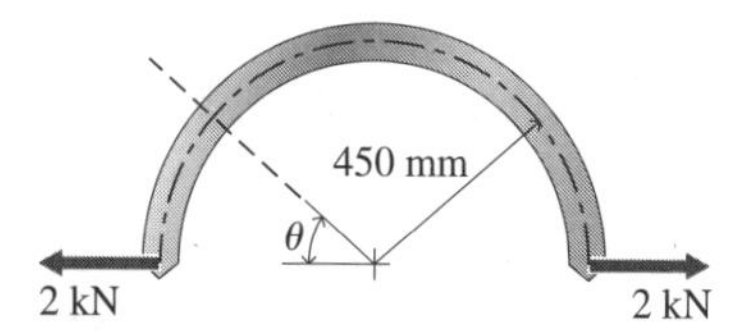

Figure P11.22

11.23 The shaft in Fig. P11.23 carries a pulley *A* (diameter of 150 mm) and a gear *B* (diameter of 100 mm). It is supported by frictionless bearings *D* and *E*. The gear *B* is subjected to the force $\mathbf{P} = 210\mathbf{i} - 75\mathbf{k}$, measured in newtons and lying in the *xz* plane. The point of application of the force is on the *z* axis. A torque of 20 N·m is applied to prevent rotation of the shaft.

a. Determine the magnitude *F* of the tension in the belt at point *G* and the reactions of the bearings *D* and *E*.

b. Determine the twisting moment in the shaft as a function of *y*.

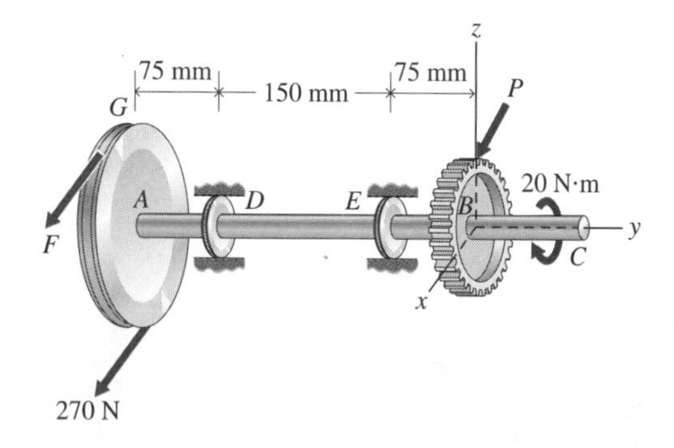

Figure P11.23

11.24 The shaft *O* in Fig. P11.24 transmits a torque of 2 kip·in to the crank arm *OA*. The force *F* acts on the piston *B*, which is in a smooth cylinder, and establishes equilibrium.

a. Calculate the force *P* in the connecting rod *AB*.

b. Calculate the force *F*.

c. Calculate the lateral force *L* that the cylinder exerts on piston *B*.

d. Calculate the horizontal and vertical reactions R_x and R_y of the crank bearing *O*.

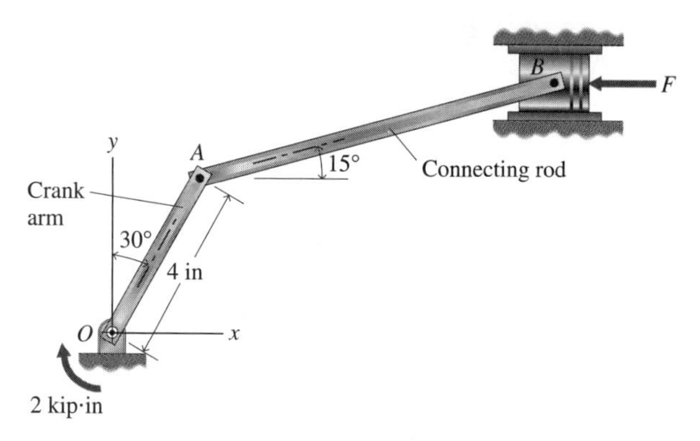

Figure P11.24

11.25 The torque of drive shaft *A* in Fig. P11.25 is 90 N·m. Construct appropriate free-body diagrams and compute the torque transmitted to shaft *B*.

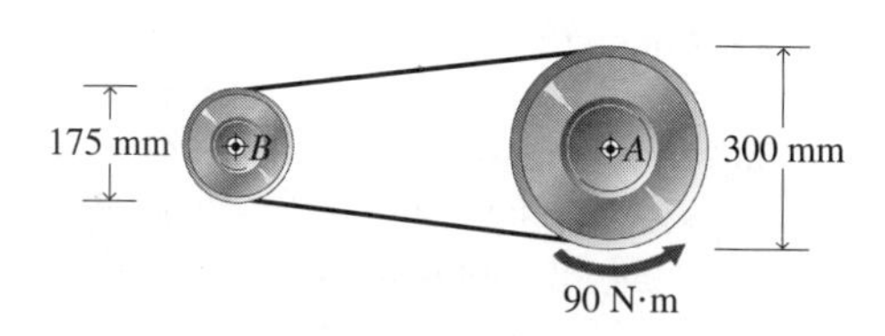

Figure P11.25

11.26 The equation of the center line of a parabolic arch (see Fig. P11.26) is $y = 120 - x^2/120$, where x and y are expressed in inches. Neglecting the weight of the arch, express the bending moment M, the shear force V, and the axial force N on the cross section with center at point P: (x, y) in terms of x for $x > 0$.

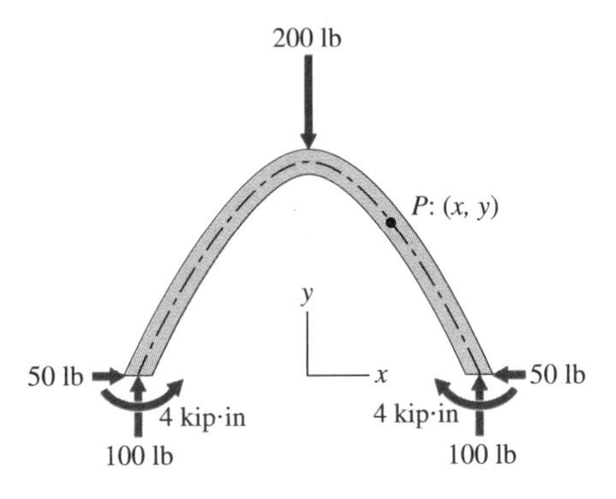

Figure P11.26

11.27 Derive formulas for the bending moment, the shear force, and the axial force in the arch in Fig. P11.27 at the section θ, for $\theta < \pi/2$ and for $\theta > \pi/2$. Express the results in terms of N_O, M_O, P, θ, and r.

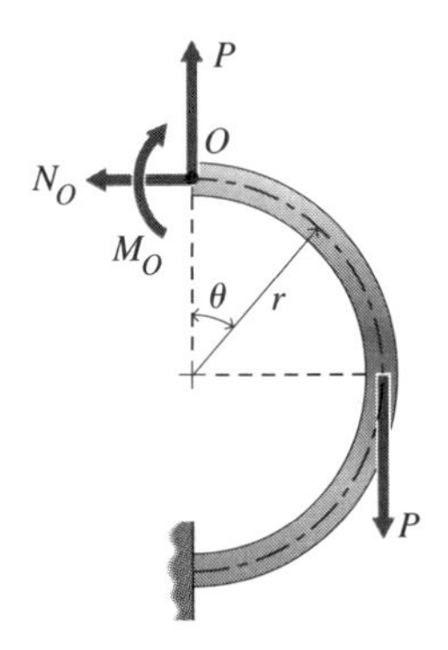

Figure P11.27

11.28 The surface on which the semicircular arch in Fig. P11.28 rests is frictionless. Express the bending moment, the shear force, and the axial force at section θ as functions of θ for the ranges $0 < \theta < \alpha$ and $\alpha < \theta < \pi$.

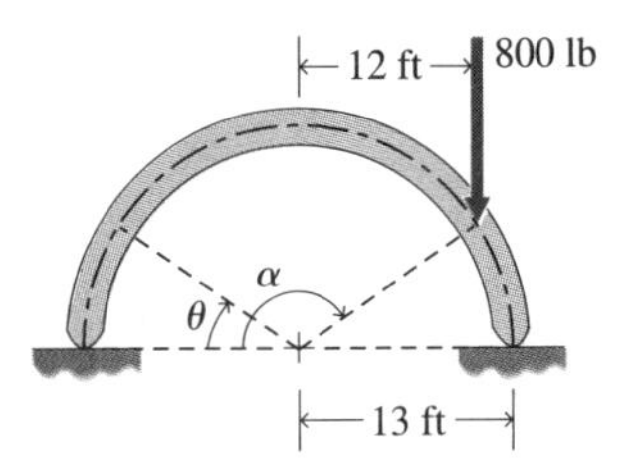

Figure P11.28

11.29 To maintain equilibrium of the bar in Fig. P11.29, a bearing O contains a spring that produces a resisting torque T about O to counteract the moment due to the force F. The resisting torque (in N·m) is given by the relationship $T = 500\theta$, where θ is expressed in radians.

a. Derive a formula for the magnitude F in terms of θ, and plot F as a function of θ for the ranges $0 \leq \theta < 180°$ and $180° < \theta < 360°$.

b. As a check, with the formula derived in part a, determine the magnitude F of the vertical force for $\theta = 120°$, and compare it to the plotted value.

c. How does the equilibrium load F behave as θ approaches 180°? As θ increases beyond 180° and approaches 360°?

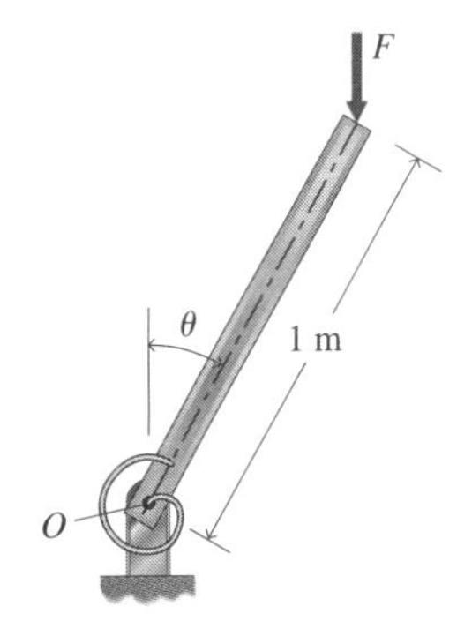

Figure P11.29

11.4 *Internal Force Diagrams for Beams and Shafts Subjected to Concentrated Forces and Couples*

Most of the problems for this section require you to draw shear and moment diagrams for beams and frame members. Unless otherwise specified in the problem statement or by your instructor, you may use either of the following approaches to draw the diagrams: (1) Develop expressions for shear V and moment M within each interval of the beam or member, and then simply plot those expressions. (2) Use the rules that relate shear and moment. In either case, you will want to begin by finding reactions on the member.

11.30 Determine the support reactions and construct the shear and bending-moment diagrams for the beam shown in Fig. P11.30. Neglect the weight of the beam.

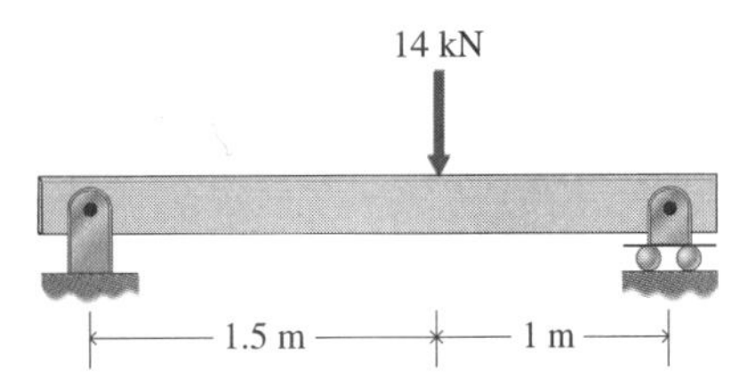

Figure P11.30

11.31 Construct the twisting-moment diagram for the shaft in Fig. P11.23.

11.32 Determine the support reactions and construct the shear and bending-moment diagrams for the beam shown in Fig. P11.32. Neglect the weight of the beam.

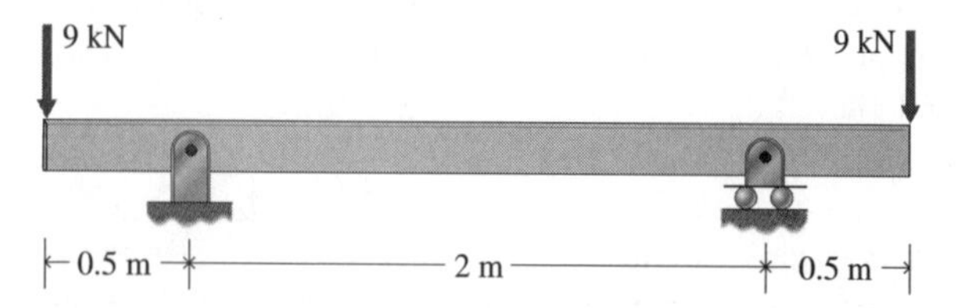

Figure P11.32

11.33 Determine the support reactions and construct the bending-moment and shear diagrams for the beam shown in Fig. P11.33. Neglect the weight of the beam.

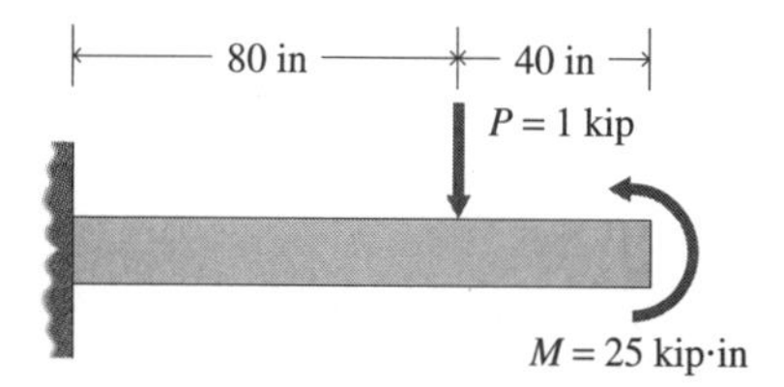

Figure P11.33

11.34 Determine the support reactions and construct the shear and bending-moment diagrams for the beam shown in Fig. P11.34. Neglect the weight of the beam.

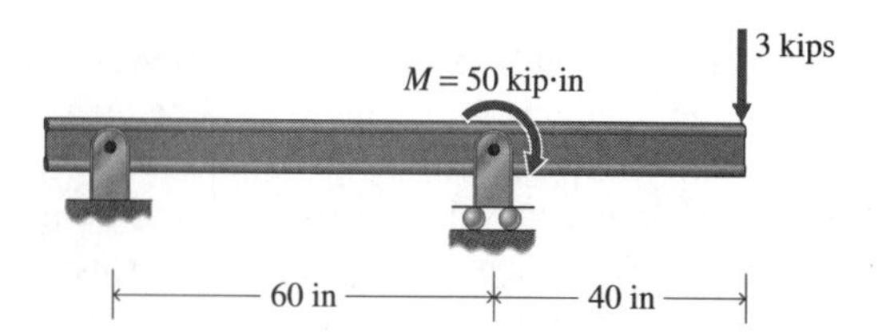

Figure P11.34

11.35 Draw the shear and bending-moment diagrams for each of the beams shown in Fig. P11.1.

11.36 Determine the support reactions and construct the shear and bending-moment diagrams for the beam shown in Fig. P11.36. Neglect the weight of the beam.

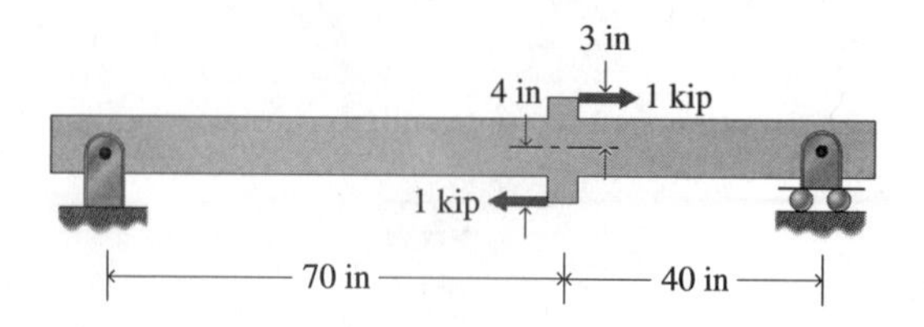

Figure P11.36

11.37 Construct shear and bending-moment diagrams for the beam in Fig. P11.2.

11.38 Construct shear and bending-moment diagrams for the beam in Fig. P11.3.

11.39 Construct shear and bending-moment diagrams for the beam in Fig. P11.4.

11.40 Construct shear and bending-moment diagrams for the U-shaped beam shown in Fig. P11.40. Neglect the weight of the beam.

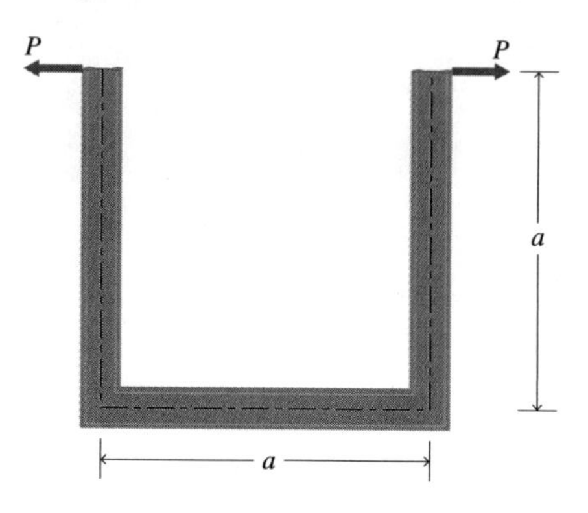

Figure P11.40

11.41 Construct shear and bending-moment diagrams for the beam in Fig. P11.5.

11.42 Construct shear and bending-moment diagrams for each of the beams in Fig. P11.6.

11.43 **a.** Derive formulas for the shear and the bending moment in the L-shaped cantilever beam *ABC* in Fig. P11.43.

b. Construct the shear and the bending-moment diagrams for the beam.

c. Construct the axial-force diagram for the beam.

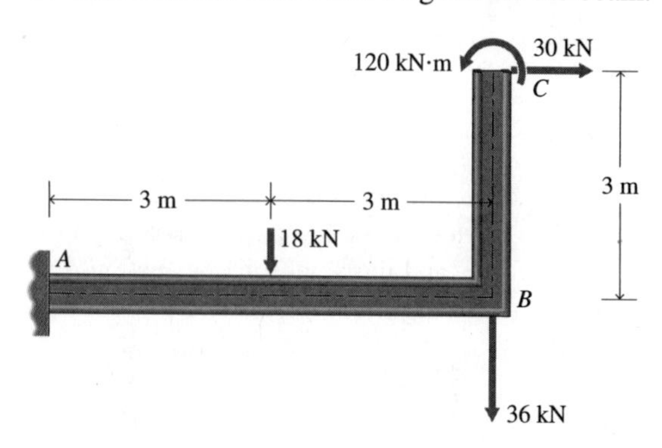

Figure P11.43

11.44 The layered system of two beams in Fig. P11.44 is subject to the loads shown. Draw the shear and bending-moment diagrams for the beam *ABCD*.

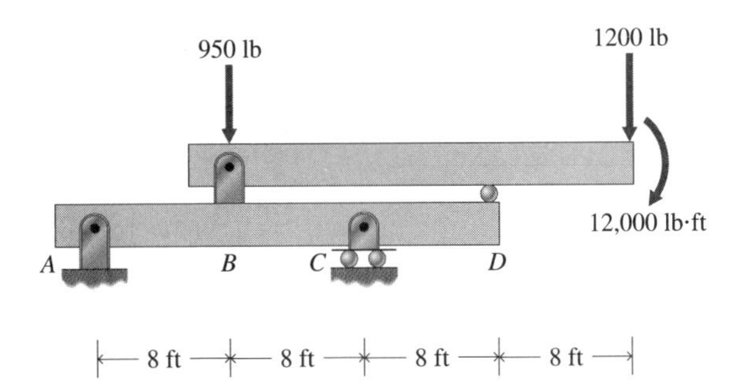

Figure P11.44

11.45 In Problem 11.18, let $P = 1$ kip, $R_y = 0.5$ kip, $R_x = 0.6$ kip, $a = 2b$ [ft]. Construct shear and bending-moment diagrams for each vertical leg and for the top member.

11.46 Construct shear and bending-moment diagrams for the split-ring beam shown in Fig. P11.46 in terms of P and a for $r = 5a$ and $0 \leq \theta \leq 180°$. Neglect the weight of the beam.

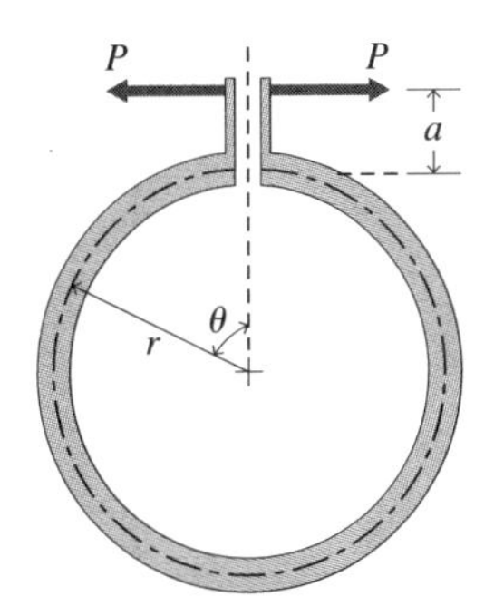

Figure P11.46

11.47 A truck travels over a single-span bridge (see Fig. P11.47). Its wheel span is s, and it weighs W. The length of the bridge is L. The front wheels of the truck carry two thirds of the weight W, and the rear wheels carry one third of the weight.

a. For $s = 15$ ft, $L = 115$ ft, and $W = 7500$ lb, plot diagrams showing the bridge's shear and bending moments due to the weight of the truck for $r = 42.5$ ft and for $r = 57.5$ ft, where r is the distance from the left support to the truck's rear wheels.

b. Plot the shear at the bridge's left support as a function of r for $0 \leq r \leq 100$ ft. For what value of r in this range is the shear a maximum?

c. Plot the bending moment at the bridge's midspan as a function of r for $0 \leq r \leq 42.5$ ft. For what value of r in this range is the bending moment a maximum?

d. Compare the values of the shear at the left support and the bending moment at midspan obtained in parts b and c to those obtained in part a for $r = 42.5$ ft.

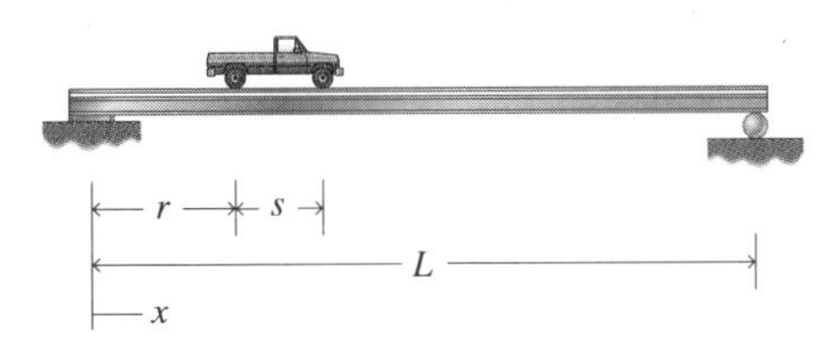

Figure P11.47

11.48 Consider the beam in Fig. P11.32. Determine a concentrated load or couple that, when applied between the supports, will reduce the midspan bending moment to zero.

Method: For the loads shown in Fig. P11.32, plot the bending-moment diagram. Then, consider a load and/or a couple placed between the supports. Locate the load and/or couple to reduce the bending moment at midspan to zero.

11.49 A straight beam with an overhang is subjected to two loads of magnitude P, as shown in Fig. P11.49. A new design requirement for the beam is that the magnitude of the maximum positive moment must be equal to the magnitude of the maximum negative moment. Move the support B to satisfy the new design requirement, and plot the bending-moment diagram for the new beam. *Method:* Determine the support reactions and the bending-moment diagram for the beam, as shown in Fig. P11.49. Decide which way support B must be moved (to the left or right of its initial position). Move the support a distance x, until the design requirement is satisfied.

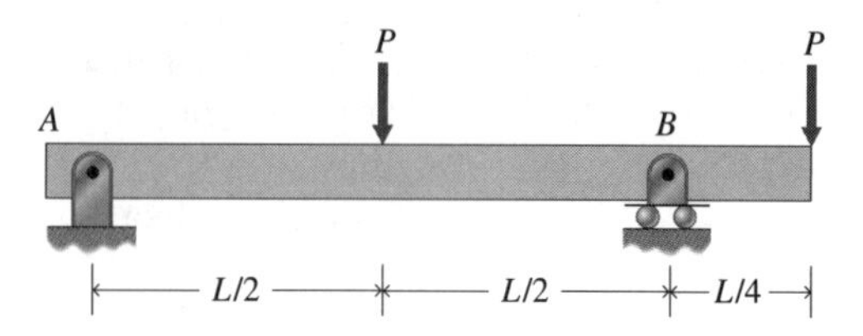

Figure P11.49

11.5 *Shear and Bending-Moment Diagrams for Beams Subjected to Distributed Loads*

Most of the problems for this section require you to draw shear and moment diagrams for beams and frame members. Unless otherwise specified in the problem statement or by your instructor, you may use either of the following approaches to draw the diagrams: (1) Develop expressions for shear V and moment M within each interval of the member, and then simply plot those expressions. (2) Use the rules that relate distributed load, shear, and moment. In either case, you will want to begin by finding reactions on the member.

11.50 Draw the shear and bending-moment diagrams for the beam shown in Fig. P11.50, and label numerical values at points A, B, C, and D.

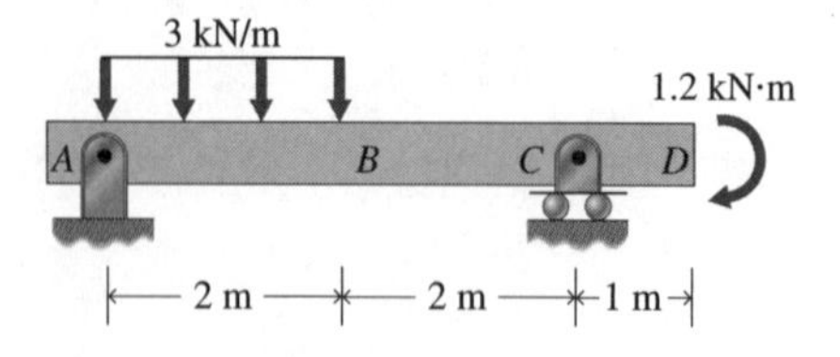

Figure P11.50

11.51 Draw the shear and bending-moment diagrams for the cantilever beam shown in Fig. P11.51, and label numerical values at points A, B, C, and D.

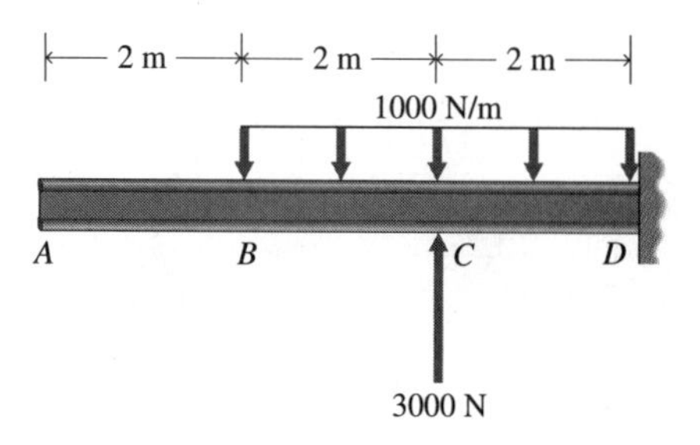

Figure P11.51

11.52 Draw the shear and bending-moment diagrams for the beam shown in Fig. P11.52, and label numerical values at points A, B, C, and D.

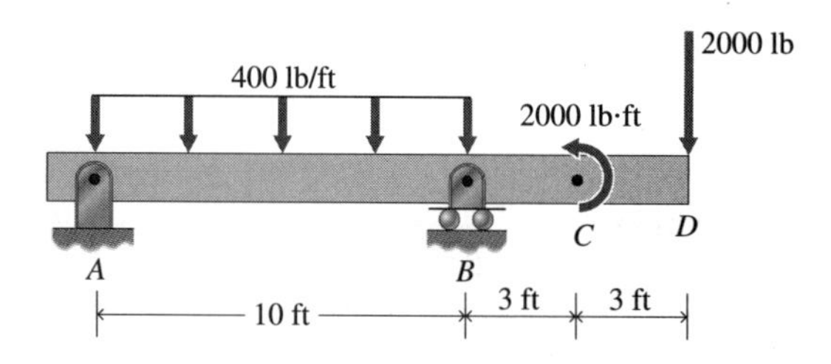

Figure P11.52

For each beam illustrated in Problems 11.53–11.62,

a. Draw free-body diagrams, and determine the support reactions.

b. Plot shear and bending-moment diagrams.

c. Locate the sections at which the shear force and bending moment attain their greatest absolute values.

d. Label points where the load undergoes a sudden change with numerical values of the shear force and bending moment at those points (see Example 11.13).

11.53

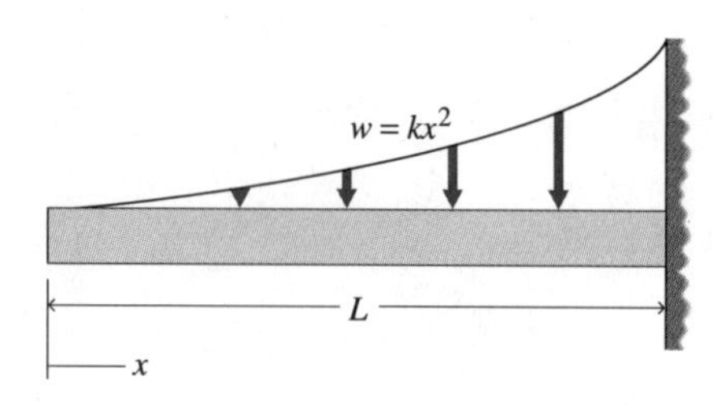

11.54

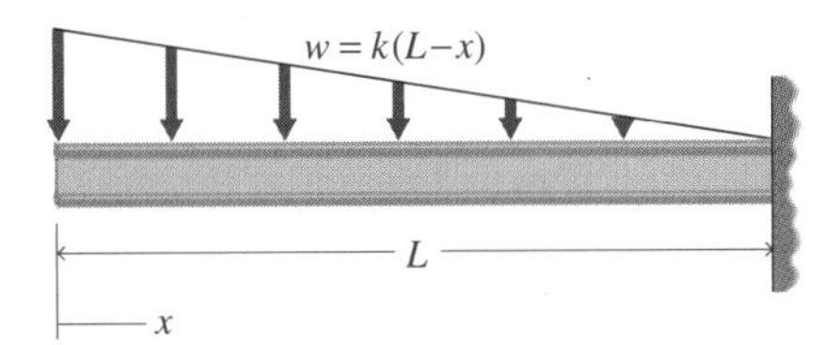

11.55

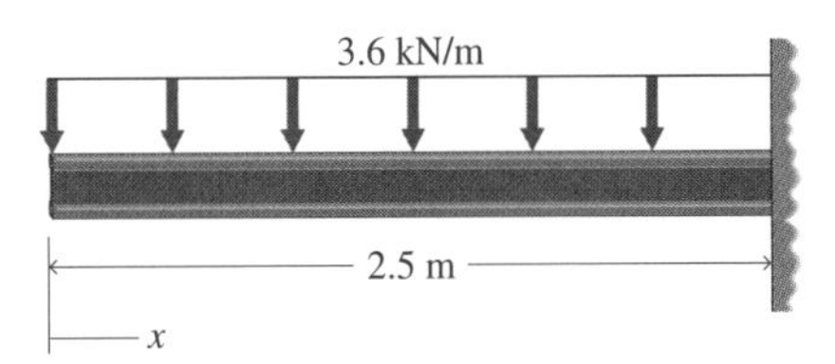

11.56

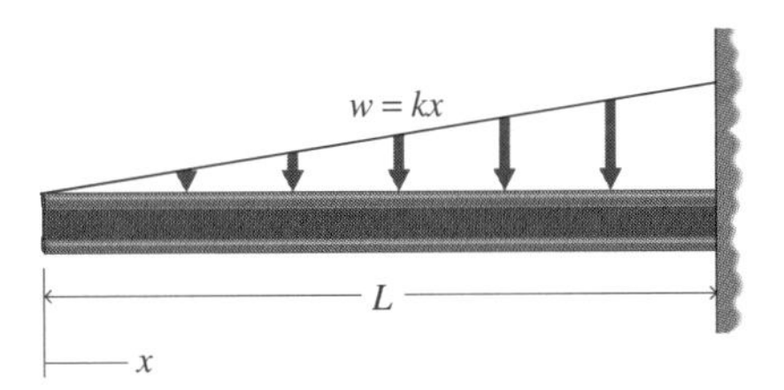

11.57

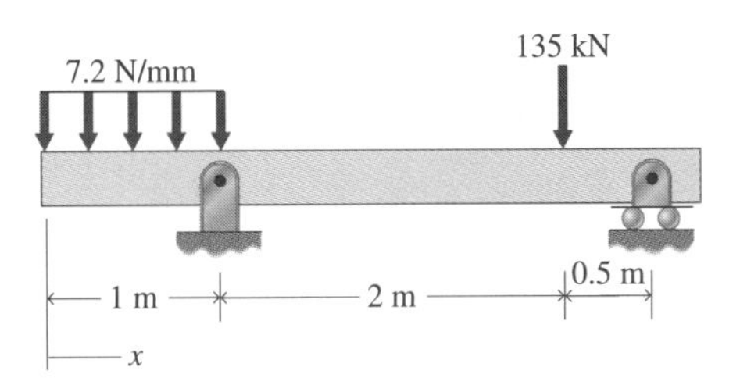

11.58

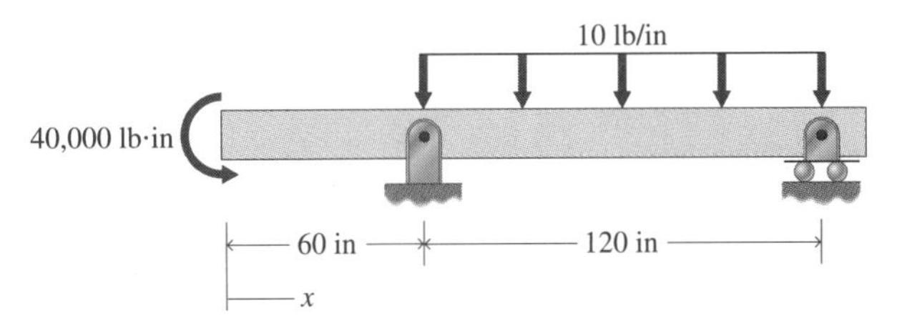

11.59

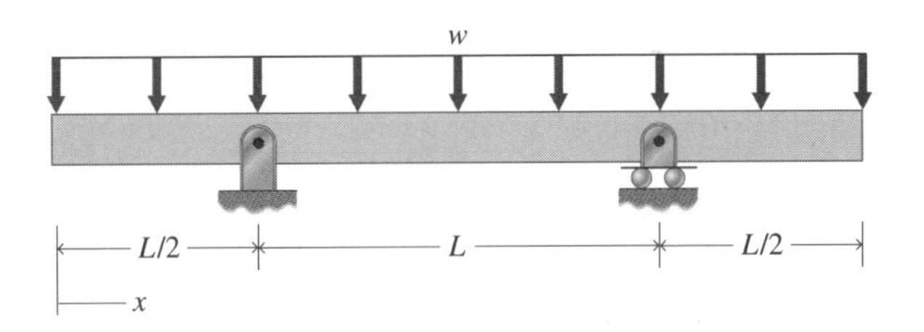

11.60

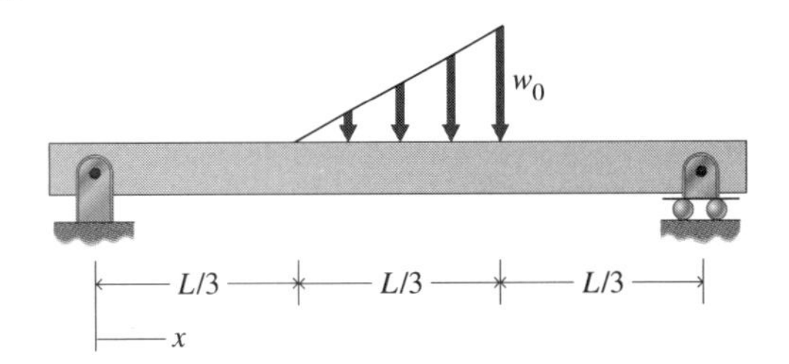

11.61

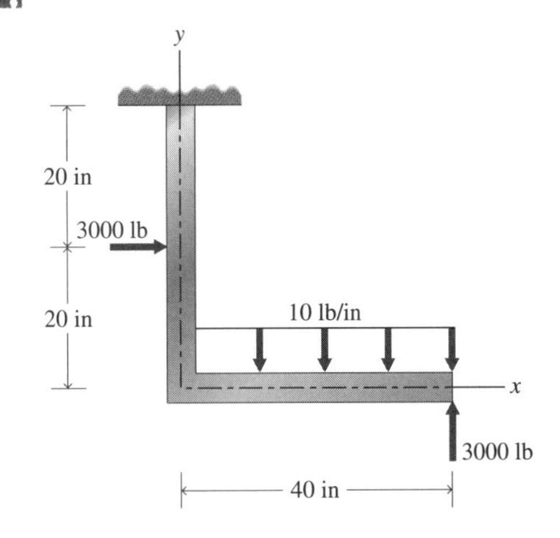

11.62

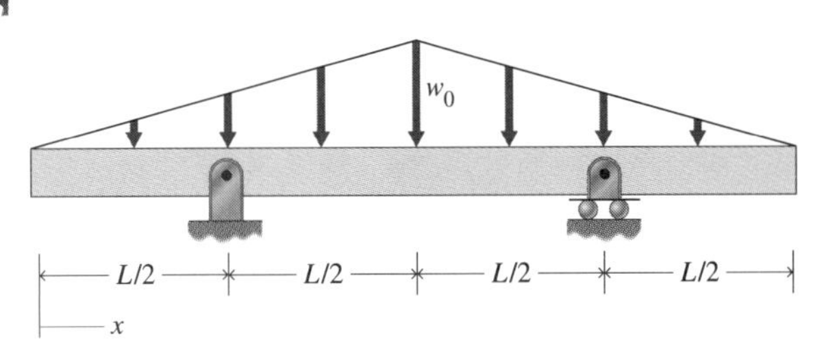

11.63 The top member of a wall frame (see Fig. P11.63) carries a uniformly distributed load of 3.6 kN/m.

a. Express the shear force, bending moment, and axial force in the top member as a function of x.

b. Plot the shear, bending-moment, and axial force diagrams for the top member.

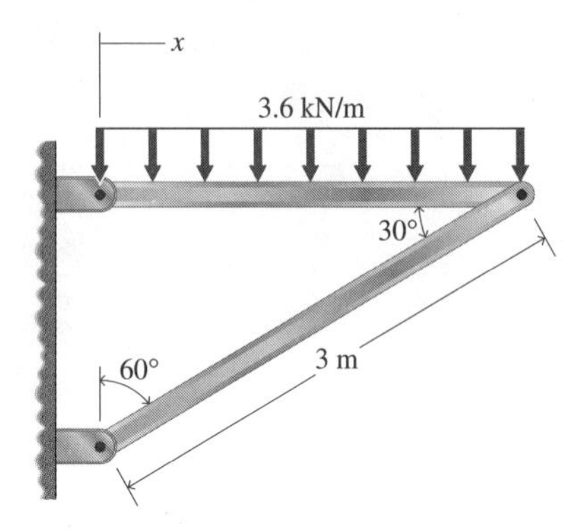

Figure P11.63

11.64 Figure P11.64 is a top view of a triangular cantilever beam used in an airplane structure. The specific weight of the beam is 180 lb/ft^3, and the beam is 2 in thick. Gravity acts in the $-z$ direction.

a. Derive expressions for the shear force V_z and bending moment M_y of the beam as functions of x.

b. Plot the shear and bending-moment diagrams for the beam.

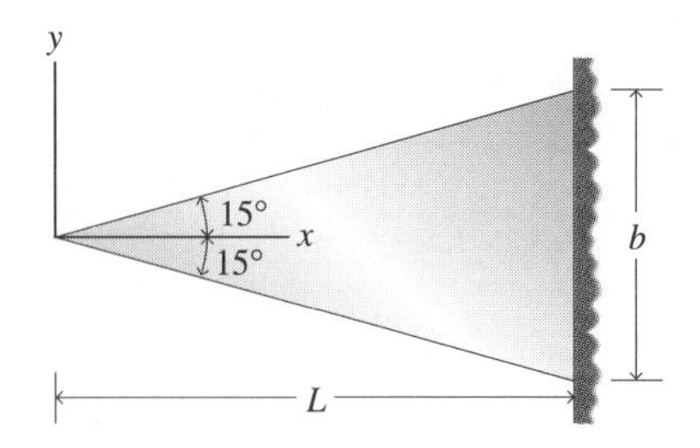

Figure P11.64

11.65 An airplane fuselage ring subtends an arc of 90° between radii OA and OB (see Fig. P11.65). It is subject to a constant pressure p distributed along its arc length. Draw the shear, bending-moment, and axial force diagrams for the fuselage ring. The thickness h of the ring is small compared to its radius a.

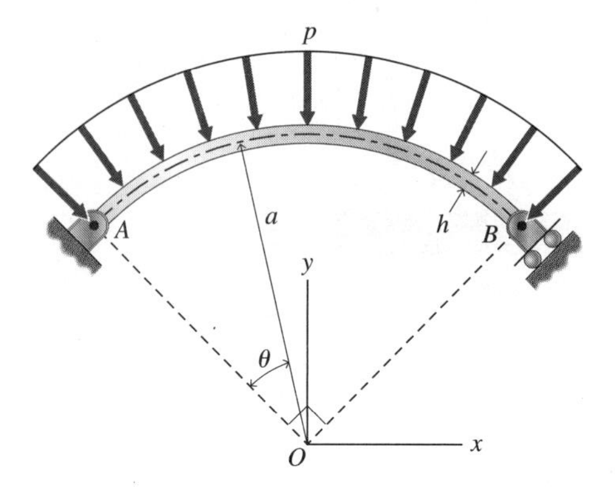

Figure P11.65

11.66 Draw the shear, bending-moment, and axial force diagrams for the L-beam shown in Fig. P11.66, and label numerical values at points A, B, C, and D.

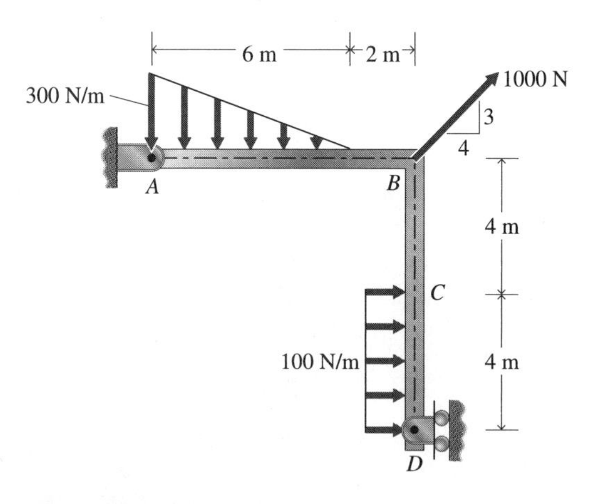

Figure P11.66

11.67 Figure P11.67 shows the approximate distribution of forces that act on an airplane wing, including the lift force and the weight of the wing.

a. Derive expressions for the shear force and the bending moment in the wing as a function of x.

b. Construct the shear and bending-moment diagrams for the wing.

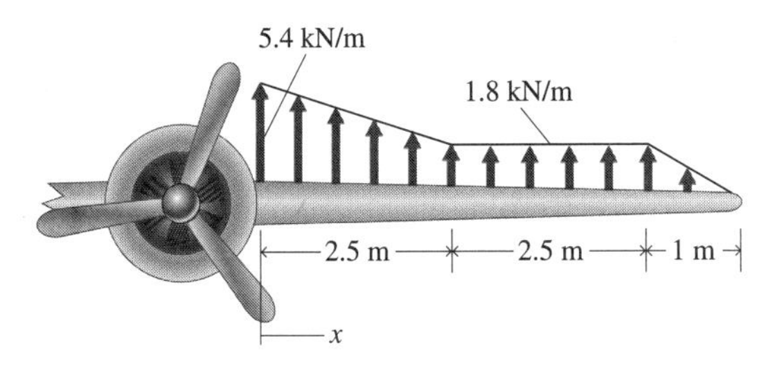

Figure P11.67

11.68 a. Express, in terms of x, the shear force, bending moment, and axial force in the wing of the small airplane shown in Fig. P11.68.

b. Construct the shear, bending-moment, and axial force diagrams for the wing.

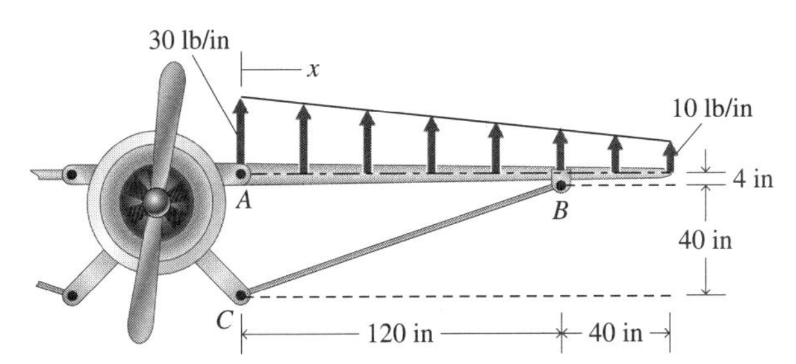

Figure P11.68

11.69 An army tank moves slowly across a symmetrical arch bridge in France (see Fig. P11.69). The tank weighs 20 metric tons, and its tracks are 5 m long. (A metric ton is the weight on earth of a 1000 kg mass.)

a. Derive formulas for the support reactions due to the tank's weight at A and C as functions of x, the position of the tank relative to support A.

b. Derive formulas for the shear force at A and the bending moment at B due to the tank's weight, as functions of x, for the range $0 \leq x \leq 85$ m.

c. Using the formulas derived in parts a and b, plot the support reactions at A and C, the shear force at A, and the bending moment at B as functions of x.

d. The bridge can support a maximum shear force of 200 kN at A and a maximum bending moment of 4 MN·m at B due to the tank's weight. Can the tank cross the bridge safely? Explain.

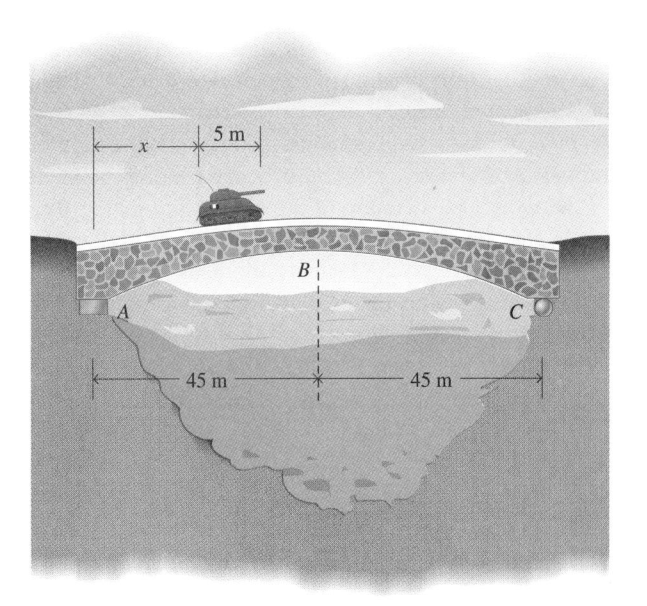

Figure P11.69

11.70 Consider the beam in Fig. P11.59. Move the left support to the left end of the beam, and locate the right support so that the magnitudes of the maximum positive bending moment and the maximum negative bending moment in the beam are equal.

Method: After moving the left support, with the right support located as shown in Fig. P11.59, determine the support reactions and plot the bending-moment diagram. If the maximum moments are not equal in magnitude, move the right support and repeat the process until the maximum moments are equal.

11.71 A design engineer finds that the bending moment at the root of the airplane wing shown in Fig. P11.67 is too large. He asks you to consider methods of reducing that moment by 50%. The wing itself cannot be rebuilt; nor can the lift force on the wing be changed, since it is due to aerodynamic pressures that occur in flight.

a. Consider methods of reducing the bending moment. For example, apply a weight that will counteract the distributed forces acting on the wing.

b. Assuming that you are successful in part a, how do you propose to implement your solution?

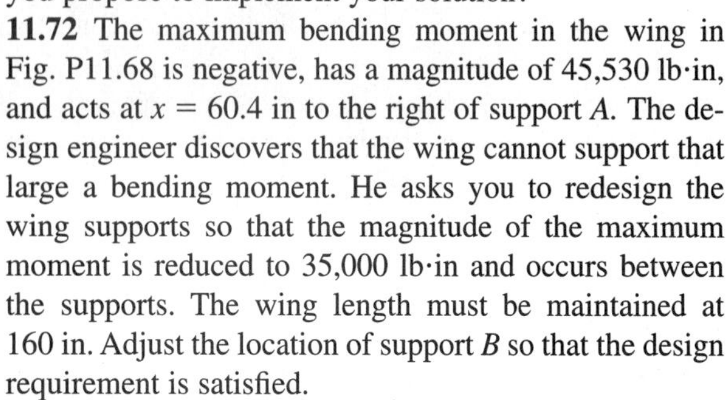

11.72 The maximum bending moment in the wing in Fig. P11.68 is negative, has a magnitude of 45,530 lb·in, and acts at $x = 60.4$ in to the right of support *A*. The design engineer discovers that the wing cannot support that large a bending moment. He asks you to redesign the wing supports so that the magnitude of the maximum moment is reduced to 35,000 lb·in and occurs between the supports. The wing length must be maintained at 160 in. Adjust the location of support *B* so that the design requirement is satisfied.

Method: Place the support *B* at a distance *d* to the right of support *A*, and derive a formula for the bending moment in the wing as a function of *x* and *d* in the region $0 \leq x \leq d$. Determine the condition that defines the maximum magnitude of the bending moment in terms of *x* and *d*. Then, search for the value of *d* that reduces the magnitude to 35,000 lb·in.

Caution: Make certain that the bending moment to the right of support *B* does not exceed 35,000 lb·in.

REVIEW QUESTIONS

11.73 Define *shear force at a section of a beam.*

11.74 Define *shear diagram.*

11.75 Define *bending moment at a section of a beam.*

11.76 What does the bending-moment diagram for a beam show?

11.77 How does the procedure for constructing shear and bending-moment diagrams for beams subjected to distributed loads differ from that for beams subjected to concentrated forces?

11.78 A pulley that has a diameter of 500 mm is tightly press-fit on a shaft. The pulley is rotated by a friction belt and transmits a torque of 90 N·m to the shaft. The tension T_2 in the slack side of the belt is 900 N. Determine the tension T_1 in the tight side of the belt.

11.79 A semicircular cantilever beam lies in the *xy* plane and is subjected to a couple T_O at its end (see Fig. P11.79). Consider a free-body diagram of the segment from point *O* to point *A*, the section subtended by angle θ. Assuming that the segment *OA* is in equilibrium, derive formulas for the couple directed along the radius *r* of the beam (the bending moment of the beam) and the couple tangent to the centerline of the beam (the torque, or twisting moment, of the beam), in terms of θ.

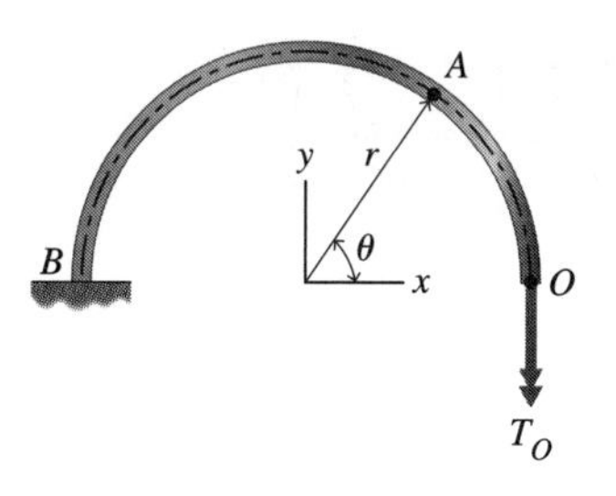

Figure P11.79

11.80 a. Draw a free-body diagram of the L-bar *ABC* shown in Fig. P11.80.

b. Draw shear, bending-moment, and axial force diagrams for the force components that act on the bar in the *yz* plane.

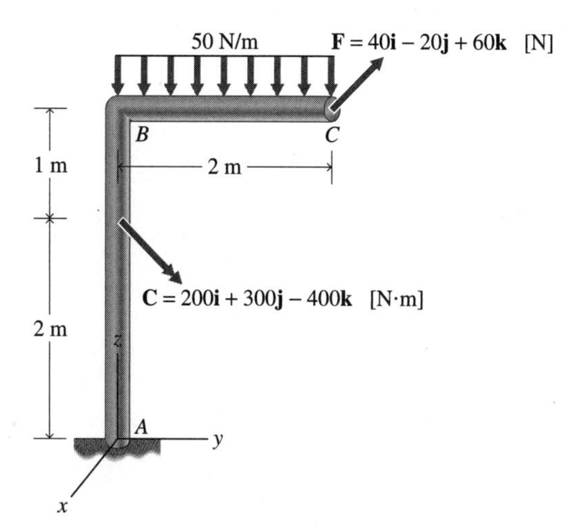

Figure P11.80

Chapter 12 WORK PRINCIPLES

In the 1976 Olympics, Vasili Alexeev lifted 2500 N (562 lb) from the floor to a height of 2 m. In doing so, he performed an astounding amount of work, 5000 N·m, and set a world record.

A Look Forward

IN PREVIOUS CHAPTERS, we studied applications of Newton's first and third laws to mechanical and structural systems. We used the concept of equilibrium to develop relationships among forces, couples, and physical quantities such as member dimensions, material densities, and so on. In this chapter, we will study principles of work. These principles provide another way of analyzing systems in equilibrium. For certain types of problems, applying work principles is more efficient than using the method of summation of forces and moments.

In this chapter, you will learn definitions of work used in physics and mechanics. You will also learn how to compute the work performed by gravity and by other forces and couples. You will be introduced to the principle of virtual work and its use to determine states of equilibrium.

Work principles are important in engineering because of their usefulness in analyzing and designing devices that perform work. All systems that do work use energy. For example, when a machine—such as a tractor—performs work, it uses energy (fuel); the amount of fuel depends on the work or task that the tractor performs. As another example, a pump used to lift water into an elevated storage tank uses energy in proportion to the amount of water pumped and the height to which the water is raised. In addition, work principles are used to determine whether a mechanical system is in a state of stable equilibrium—that is, whether the system tends to remain in its equilibrium position when subjected to a slight disturbance.

Survey Questions

- Is work a vector or a scalar quantity?
- What are the units of work?
- How does gravity do work?
- How do virtual displacements differ from other displacements?

12.1 WORK

After studying this section, you should be able to:

- Determine the work of a force acting on a body as the body moves along the line of action of the force.

Key Concept

Work is the product of a force and a distance; it is a scalar quantity.

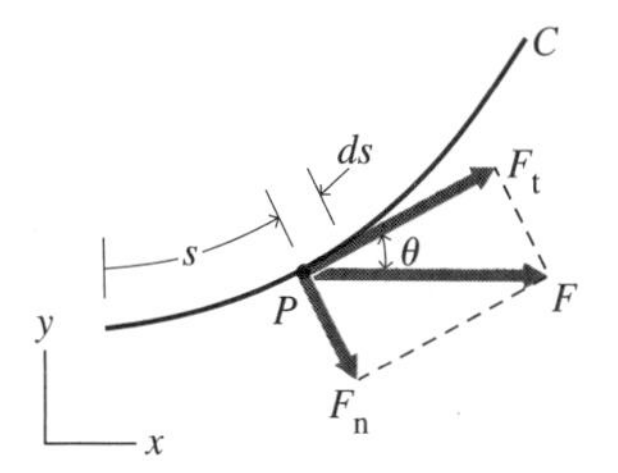

Figure 12.1

In the early days of steam engines, operators observed that the fuel consumption of such an engine when used to drive a water pump depended not only on the amount of water pumped but also on the height to which the water was raised. Evidently, fuel consumption is proportional to the quantity of water discharged. It is also proportional to the height to which the water is lifted. If water is hoisted in a bucket by means of a rope and a winch, the time the winch operates is proportional to the height the bucket is raised. In turn, fuel consumption is proportional to the duration of the operation. Accordingly, the fuel consumed by an engine that operates a hoist is proportional to the product of the weight of the body being lifted and the height to which that body is raised. More generally, if a constant force F translates a body in a straight line a distance d along the force's line of action, the fuel that is required is proportional to the product $U = Fd$. The product U is called the *work* that the force performs on the body.

work

Atlas does no work as he holds the earth on his shoulders, since there is no motion.

A slight generalization of this definition is required if the magnitude of the force varies with the displacement or if the line of action of the force does not coincide with the path along which the body is displaced. For example, consider a force **F** acting on a particle P that moves on a curved path C (see Fig. 12.1). The displacement of the point of application of the force is the same as the displacement of the particle. Hence, the increment of work dU that the force performs as the particle undergoes an infinitesimal displacement ds on C is defined by

$$dU = F_t\,ds \tag{12.1}$$

where F_t is the projection of the force **F** tangent to C (Fig. 12.1). Consequently, as particle P moves along curve C, the work that the force **F** performs on the particle is

$$U = \int_C F_t\,ds = \int_C F\cos\theta\,ds \tag{12.2}$$

where $F_t = F\cos\theta$, F is the magnitude of the force **F**, and θ is the angle between **F** and the tangent to C. Both F and θ are functions of s. The letter C below the integral sign designates the path of integration. The integrals in Eq. (12.2) are called line integrals. The work performed by **F** is a number with dimensions $[FL]$. Depending on the sign of $\cos\theta$, work may be a number that is positive, negative, or zero. In other words, work is a scalar quantity.

If the force **F** maintains a constant direction (say, the direction of the x axis, in Fig. 12.1), Eq. (12.2) may be expressed in the following form, since $dx = ds\cos\theta$:

$$U = \int_{x_1}^{x_2} F\,dx \tag{12.3}$$

Here, we regard F as a function of x. The limits x_1 and x_2 denote the initial and final values of x on the curve C.

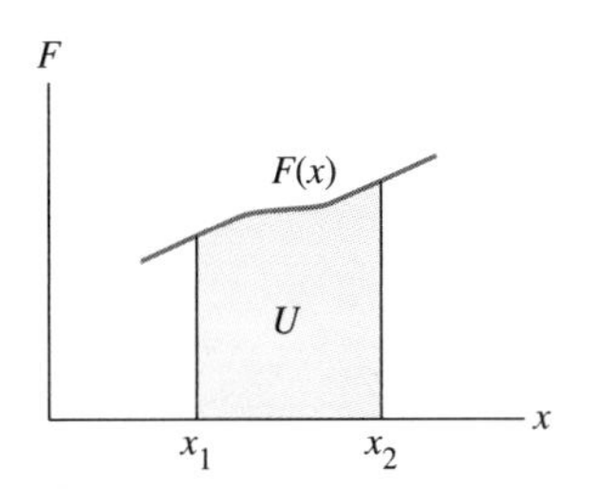

Figure 12.2 Graphical representation of work.

There is an appealing graphical interpretation of Eq. (12.3). The force F may be represented by a curve in a plane with abscissa x and ordinate F. Then, from Eq. (12.3), the work U is represented by the shaded area under the curve in the interval from x_1 to x_2 (see Fig. 12.2).

In the U.S. Customary System, the unit of work is the foot-pound (ft·lb). In the SI system, the unit of work is the newton-meter (N·m) or joule (J). The following conversions are useful (see Table 1.5):

$$1 \text{ ft·lb} = 1.356 \text{ J} = 1.356 \text{ N·m}$$

$$1 \text{ J} = 1 \text{ N·m} = 0.7376 \text{ ft·lb}$$

Example 12.1

Work Required to Fill a Water Tank

Problem Statement A cylindrical water tank with a diameter of 18 ft and a height of 27 ft is filled with water ($\gamma = 62.4$ lb/ft^3) from a cylindrical well with a diameter of 10 ft and a water level 30 ft below ground level. How much work does the pump do in filling the tank? Assume that there is no recharge (refilling) of the well during pumping.

Solution To fill the tank, the pump must lift $\pi(9)^2(27) = 6871$ ft^3 of water from the well. As the water is pumped, the water level in the well drops through a distance a (see Fig. E12.1). Therefore,

$$\pi(5)^2(a) = 6871 \text{ ft}^3$$

$$a = 87.48 \text{ ft}$$

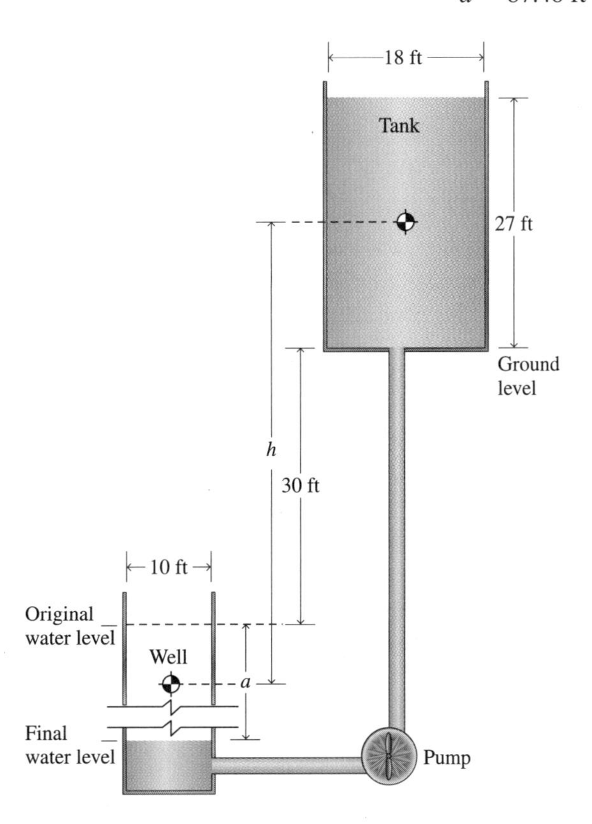

Figure E12.1

Effectively, 6871 ft^3 of water must be lifted a distance

$$h = \frac{a}{2} + \frac{27}{2} + 30$$

$$= 87.24 \text{ ft}$$

and the work done to lift the water is

$$U = (62.4)(6871)(87.24) = 37{,}404{,}000 \text{ ft·lb}$$

Example 12.2

Work Done by Gravity on an Airplane

Problem Statement An airplane has a mass of 12 000 kg. It takes off from sea level and attains a cruising altitude of 9 km.

a. What work is done on the plane by gravity from takeoff until it reaches cruising altitude?

b. How much work is required of the plane's engines to lift the plane to cruising altitude?

Solution **a.** Assuming that gravity is constant and noting that the gravitational force acts downward and the displacement of the plane is upward, we calculate the work done by gravity as

$$U_{\text{gravity}} = (-12\,000)(9.81)(9\,000) = -1.059 \text{ GN·m} = -1.059 \text{ GJ}$$

b. The lift (force) of the airplane's engines is directed vertically upward, so the work of the lift is the negative of the work of gravity; that is, $U_{\text{engines}} = 1.059$ GJ.

Information Item The plane's engines have to do much more work than that required simply to lift the plane 9 km. Consumption of additional fuel is necessary to overcome air resistance (drag), internal engine friction, and other mechanical inefficiencies.

Example 12.3

Work Done by Gravity on a Pendulum

Problem Statement A heavy steel ball of weight W is suspended from a light rope of length L to form a pendulum. The ball is raised so that the rope forms an angle ϕ with the vertical (see Fig. E12.3). From this position, it is released.

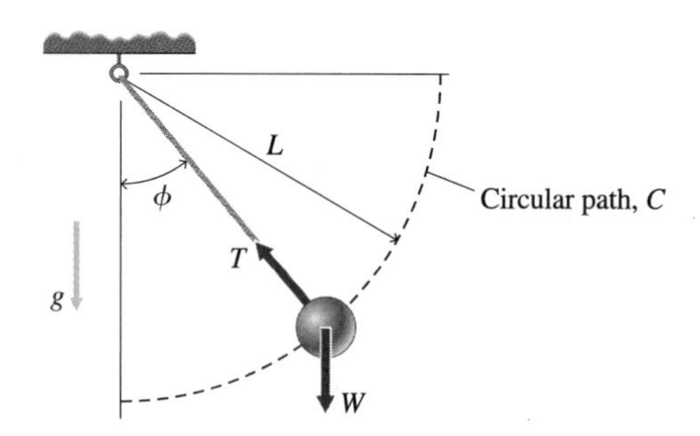

Figure E12.3

a. Determine the work done by gravity during the time it takes the ball to drop back to the vertical position.

b. Determine the work done by the tension T in the rope during this same time.

Solution **a.** As ϕ goes to zero, the ball follows path C, along a circle of radius L, and drops a vertical distance $L - L\cos\phi$. Therefore, the work done by gravity on the ball is

$$U_g = WL(1 - \cos\phi)$$

For example, if initially $\phi = 90°$, then $\cos\phi = 0$ and $U_g = WL$ as the ball drops a distance L.

b. As the ball drops, it travels along the circular path C (refer to Fig. E12.3). Since the tension T in the rope is perpendicular to C, it has no component tangent to C. Then, Eq. (12.2) yields, with $F = T$ and $\cos\theta = 0$, $U_T = 0$.

Example 12.4

Work Performed in Stretching a Spring

Problem Statement A helical spring is suspended from one end and stretched by an increasing load F (see Fig. E12.4a). The load-extension curve of the spring is a straight line (see Fig. E12.4b); that is, $F = ke$, where e is the extension of the spring and k is the *spring constant,* the slope of the curve.

a. Determine the work done by the load F in stretching the spring.

b. Determine the energy E_s stored in the spring.

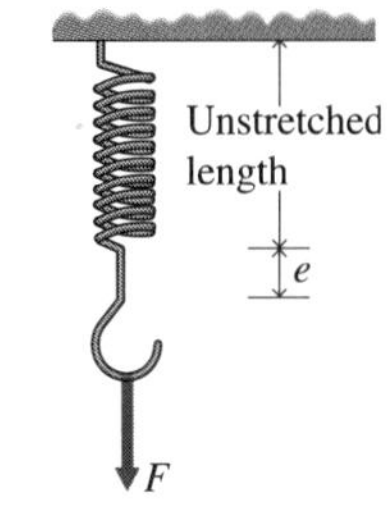

Figure E12.4a

Solution **a.** The work done by the load F in stretching the spring from extension e_1 to extension e_2 is

$$U_F = \int_{e_1}^{e_2} F\,de = \int_{e_1}^{e_2} ke\,de \tag{a}$$

Integration of Eq. (a) yields

$$U_F = \frac{k(e_2^2 - e_1^2)}{2} \tag{b}$$

This result is evident from Fig. E12.4b, since U_F is represented by the shaded area. Note that if $e_1 = 0$ and $e_2 = e$, Eq. (b) yields

$$U_F = \frac{ke^2}{2}$$

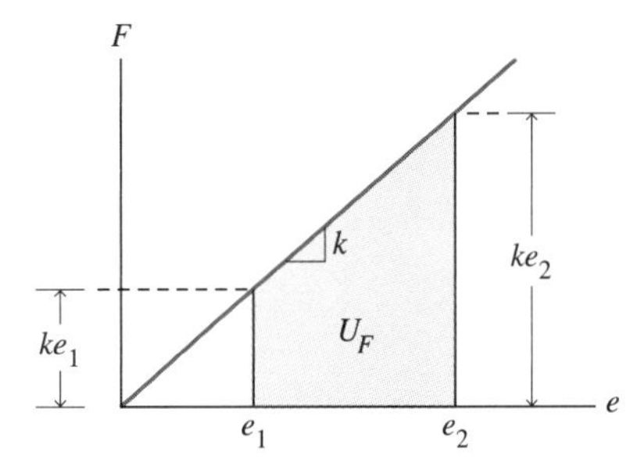

Figure E12.4b

b. The quantity $ke^2/2$ is also referred to as the *elastic energy* E_s stored in the spring. Therefore, the energy E_s stored in the spring is equal to the work done by the load F, as the spring is stretched from e_1 to e_2. Hence, from Eq. (b),

$$E_s = \frac{k(e_2^2 - e_1^2)}{2}$$

Information Item The force F in the spring is directly proportional to the extension of the spring (refer to Fig. E12.4b). The fact that the tension in a spring is directly proportional to the extension of the spring for extensions that are not too large was discovered by Robert Hooke (1635–1703) in 1678, and is known as *Hooke's law.* However, it is not a fundamental law of nature but rather a rule for a specific type of elastic behavior. Real springs do not always obey this rule precisely, but it is a practical model for many cases.

12.2 WORK AS A LINE INTEGRAL

After studying this section, you should be able to:

- Determine the work of a force acting on a particle as the particle moves along a given path.

Key Concept The work that a force performs on a particle may be expressed as a line integral.

If a force **F** that does work on a particle P varies in magnitude and direction, we use Eq. (12.1) to calculate work. The distance ds may be regarded as the magnitude of an infinitesimal displacement vector $d\mathbf{r}$ (see Fig. 12.3). Consequently, since $F_t = F\cos\theta$, where θ is the angle between the vectors **F** and $d\mathbf{r}$, Eq. (12.1) may be written

$$dU = F\,ds\cos\theta = \mathbf{F}\cdot d\mathbf{r}$$

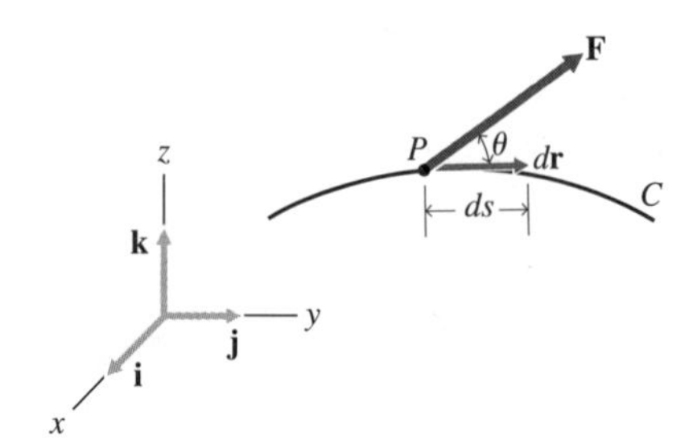

Figure 12.3 Particle moving on a curved path and subjected to a variable force.

By the definition of the scalar product (Sec. 5.1), this is equivalent to

$$dU = F_x\,dx + F_y\,dy + F_z\,dz = F_t\,ds \tag{12.4}$$

where dx, dy, dz and F_x, F_y, F_z are the projections of the vectors $d\mathbf{r}$ and **F** on rectangular Cartesian xyz axes. Equation (12.4) separates dU into a sum of three terms, the amounts of work performed by the force components $F_x\mathbf{i}$, $F_y\mathbf{j}$, and $F_z\mathbf{k}$.

Equations (12.2) and (12.4) yield

$$U = \int_C F_t\,ds = \int_C (F_x\,dx + F_y\,dy + F_z\,dz) = \int_C \mathbf{F}\cdot d\mathbf{r} \tag{12.5}$$

Equation (12.3) is a special case of Eq. (12.5) that results if $F_y = F_z = 0$. Note that since $\mathbf{F}\cdot d\mathbf{r}$ is a scalar, Eq. (12.5) demonstrates that work is a scalar quantity. Hence, if several forces act on a body, the total work performed on the body by the forces is the algebraic (signed) sum of the work done by each of the forces individually.

The curve C may be defined in terms of a parameter t by equations of the type

$$x = x(t) \qquad y = y(t) \qquad z = z(t) \tag{a}$$

For example, the parameter t may denote time. Then, Eqs. (a) tell the location of particle P at time t. Equations (a) yield

$$dx = \frac{dx}{dt}\,dt \qquad dy = \frac{dy}{dt}\,dt \qquad dz = \frac{dz}{dt}\,dt \tag{b}$$

Consequently, Eq. (12.5) may be written as

$$U = \int_{t_1}^{t_2}\left(F_x\frac{dx}{dt} + F_y\frac{dy}{dt} + F_z\frac{dz}{dt}\right)dt \tag{12.6}$$

Thus, the line integral [Eq. (12.5)] is reduced to an ordinary integral of a function of t.

If the particle P belongs to a mechanical system, the quantity U defined by Eq. (12.5) or (12.6) is the work that the force **F** performs on the system. This definition applies only if the force **F** acts continuously on the same particle P during the course of the motion. If the point of application of **F** jumps from one particle to another as the motion proceeds, the work **F** performs on the system is defined to be the sum of the amounts of work it performs on all the particles on which it acts. If several forces act on a moving system, the net work these forces perform on the system is defined to be the algebraic sum of their individual amounts of work.

Example 12.5

Work of a Constant Force on a Particle Traveling in a Straight Line

Problem Statement A particle P travels along a straight line that forms a 30° angle with respect to the x axis (see Fig. E12.5). It is acted on by a constant force

$$\mathbf{F} = 2\mathbf{i} + 3\mathbf{j} \quad [\text{N}] \tag{a}$$

Determine the work $\mathbf{F}$ performs on the particle in the time the particle travels 3 m along the line.

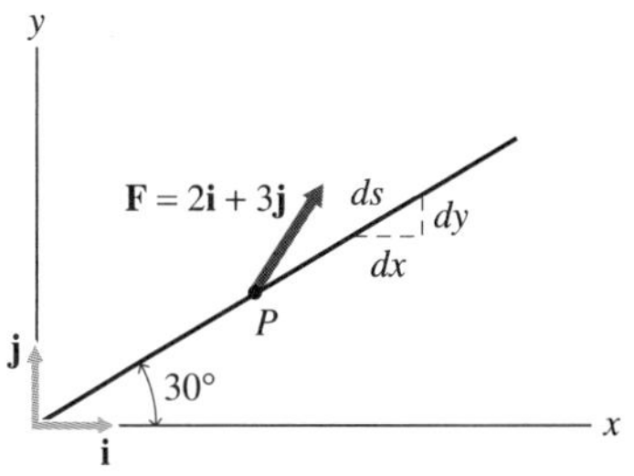

Figure E12.5

Solution Let the particle travel an infinitesimal distance ds along the straight line. The corresponding x and y increments are

$$dx = (\cos 30°)\, ds = \frac{\sqrt{3}}{2}\, ds$$
$$dy = (\sin 30°)\, ds = \frac{1}{2}\, ds \tag{b}$$

Therefore, the infinitesimal displacement vector along the straight line is

$$d\mathbf{r} = dx\,\mathbf{i} + dy\,\mathbf{j} = \frac{1}{2}\left(\sqrt{3}\,\mathbf{i} + \mathbf{j}\right) ds \quad [\text{m}] \tag{c}$$

Hence, by Eqs. (a), (c), and (12.5), the work done by the force $\mathbf{F}$ is

$$U = \int_C \mathbf{F}\cdot d\mathbf{r} = \frac{1}{2}\int_0^3 \left[(2\mathbf{i} + 3\mathbf{j})\cdot\left(\sqrt{3}\mathbf{i} + \mathbf{j}\right)\right] ds = \frac{1}{2}\int_0^3 \left(2\sqrt{3} + 3\right) ds \tag{d}$$

Integration of Eq. (d) yields $U = 9.70$ N·m $= 9.70$ J.

Alternatively, from Eqs. (a) and (b),

$$F_x = 2\text{ N} \qquad F_y = 3\text{ N} \qquad F_z = 0 \tag{e}$$

and

$$dx = \frac{\sqrt{3}}{2} ds \qquad dy = \frac{1}{2} ds \qquad dz = 0 \tag{f}$$

Then, by Eqs. (e), (f), and (12.5), we have

$$U = \int_0^3 \left(\sqrt{3} + \frac{3}{2}\right) ds = \left(\sqrt{3} + \frac{3}{2}\right)(3) = 9.70\text{ J}$$

Example 12.6

Work Performed on a Particle That Travels along a Curve

Problem Statement The position of a particle P in the xy plane is defined by $x = t^2$, $y = t^3$ (see Fig. E12.6), where x and y are measured in meters. The physical significance of the parameter t is irrelevant; it could be time or some other physical quantity. The particle is acted on by a force (in newtons) with x and y projections $F_x = t - 2$ and $F_y = 3/t$. Calculate the work that the force performs on the particle in the interval $1 \le t \le 2$.

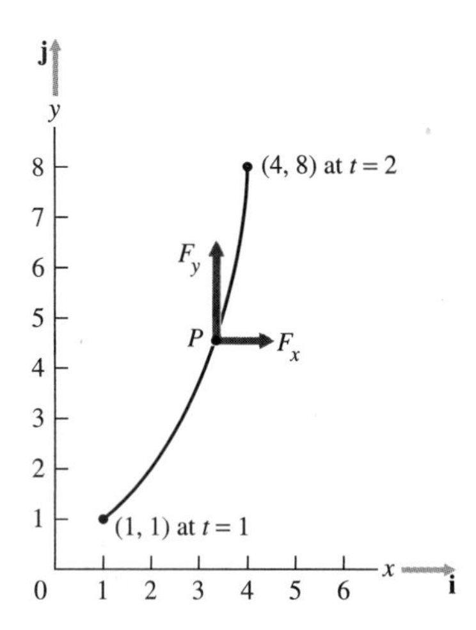

Figure E12.6

Solution Since $dx = 2t\,dt$ and $dy = 3t^2\,dt$, the infinitesimal displacement vector is

$$d\mathbf{r} = (2t\mathbf{i} + 3t^2\mathbf{j})\,dt \tag{a}$$

Also, the force **F** that acts on the particle P is

$$\mathbf{F} = (t - 2)\mathbf{i} + \left(\frac{3}{t}\right)\mathbf{j} \tag{b}$$

Hence, Eqs. (a), (b), and (12.5) yield

$$U = \int_1^2 \left[(t - 2)\mathbf{i} + \frac{3}{t}\mathbf{j}\right] \cdot \left(2t\mathbf{i} + 3t^2\mathbf{j}\right) dt = \int_1^2 \left(2t^2 + 5t\right) dt \tag{c}$$

Integration of Eq. (c) yields $U = 73/6 = 12.17$ J.

Alternatively, since $dx/dt = 2t$ and $dy/dt = 3t^2$, from Eq. (12.6), we have

$$U = \int_1^2 \left[(t - 2)(2t) + \left(\frac{3}{t}\right)(3t^2)\right] dt = 12.17 \text{ J}$$

Example 12.7

Particle Subjected to a Centrally Directed Force of Constant Magnitude

Problem Statement A particle P that moves in the xy plane is subject to a force **F** of constant magnitude F that is always directed toward a fixed point O [the origin of the xy axes shown in Fig. E12.7]. Such a force is called a *centrally directed force.* Determine the work done on the particle by the force **F** as the particle moves from point 1 to point 2.

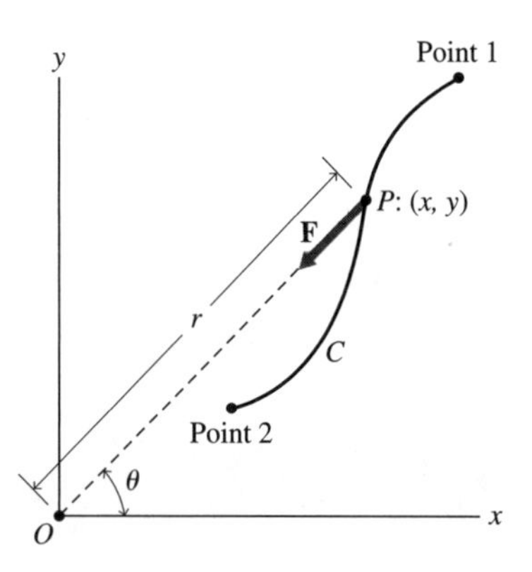

Figure E12.7

Solution From Fig. E12.7, we see that the (x, y) projections of **F** are

$$F_x = -F\cos\theta$$
$$F_y = -F\sin\theta \tag{a}$$

Also, from Fig. E12.7,

$$\cos\theta = \frac{x}{r}$$
$$\sin\theta = \frac{y}{r} \tag{b}$$

where r is the distance of the particle from the origin and (x, y) are the coordinates of the particle. Now, since $F_z = 0$, Eqs. (a), (b), and (12.5) yield

$$U = -F\int_C \frac{x\,dx + y\,dy}{r}$$

But $x\,dx + y\,dy = \frac{1}{2}d(x^2 + y^2) = \frac{1}{2}d(r^2) = r\,dr$, since $r^2 = x^2 + y^2$. Consequently,

$$U = -F\int_{r_1}^{r_2} dr = F(r_1 - r_2) \tag{c}$$

where r_1 and r_2 are the distances from points 1 and 2, respectively, to the origin (refer to Fig. E12.7).

Equation (c) means that the work done by a centrally directed force of constant magnitude F equals the product of F times the reduction in the distance of the particle from point O. If the particle moves away from point O, the work done by the force is negative.

It is easy to generalize this result to three dimensions. The requirement that the force be directed toward the origin is no restriction, since the origin can be located at any point. Consequently, we have

> If a particle that moves in space is subjected to a force of constant magnitude F that is always directed toward a fixed point O, the work that the force performs on the particle is the product of F with the reduction of the distance of the particle from point O. If the particle moves away from point O, the work is negative.

Information Item Note that, since the work performed on a particle by a constant centrally directed force depends only on the reduction in distance of the particle from point O, the particle may follow *any* path between points 1 and 2 without changing the work performed by the force. In particular, if the particle follows any closed path between points 1 and 2, the work done by the force is zero. Such a force is said to be a *conservative force.* ■

CHECKPOINT 

1. ***True or False:*** Work is the product of a force and a distance.
2. ***True or False:*** A force that acts on a particle in a direction perpendicular to the path on which the particle travels does no work on the particle.
3. ***True or False:*** The work of a force **F** that acts on a particle may be represented by a line integral of the scalar product of **F** and the infinitesimal displacement vector $d\mathbf{r}$ of the particle.
4. ***True or False:*** If several forces act on a moving system, the net work that these forces perform on the system is less than the algebraic sum of their individual amounts of work.
5. For a nonzero force that acts on a particle to do no work, must the particle be moving in a straight line? Explain.

Answers: 1. True; 2. True; 3. True; 4. False; 5. No—to do no work, the force need only act in a direction perpendicular to the path of the particle

12.3 WORK PERFORMED BY GRAVITY

After studying this section, you should be able to:

- Determine the work of gravity on a body as the body changes elevation.

Key Concept The work performed on a body by the force of gravity equals the weight of the body multiplied by the vertical distance through which its center of gravity descends.

Any mechanical system may be regarded as a system of particles. Let the particles be numbered 1, 2, 3, . . . ; let the weights of the respective particles be $w_1, w_2, w_3, \ldots$; and let their elevations above a fixed datum plane be $z_1, z_2, z_3, \ldots$. If the system is raised, the elevations $z_1, z_2, z_3, \ldots$ receive positive increments $\Delta z_1, \Delta z_2, \Delta z_3, \ldots$. Accordingly, if a constant gravitational force acts on the system while it is being raised, the work performed by gravity on the system is negative; that is,

$$U = -(w_1 \Delta z_1 + w_2 \Delta z_2 + w_3 \Delta z_3 + \cdots) \qquad \text{(a)}$$

If the system is lowered, $\Delta z_1, \Delta z_2, \Delta z_3, \ldots$ are negative. Then, the work of the force of gravity is positive.

The work performed by the diver in climbing to the top of the platform is the same as the work performed by gravity during his dive into the pool.

Since the weights $w_1, w_2, w_3, \ldots$ are constant, Eq. (a) may be written in the form

$$U = -\Delta(w_1 z_1 + w_2 z_2 + w_3 z_3 + \cdots) = -\Delta\left(\sum w_i z_i\right) \qquad \text{(b)}$$

where the operator Δ designates the increment of the quantity in parentheses and Σ denotes the sum of all the particles. By definition [see Eq. (8.1)], the elevation $\bar{z}$ of the center of gravity of the system is determined by the equation

$$W\bar{z} = \sum w_i z_i \qquad \text{(c)}$$

where W is the total weight of the system—that is, $W = \sum w_i$. Consequently, from Eqs. (b) and (c),

$$U = -\Delta(W\bar{z})$$

Since W is constant, this equation is equivalent to the relation

$$U = -W\Delta\bar{z} \tag{12.7}$$

This result may be stated as follows.

Theorem 12.1

If gravity is constant, the work that gravity performs on any mechanical system equals the total weight of the system multiplied by the vertical distance through which the center of gravity of the system descends.

Example 12.8

Work of the Force of Gravity on a Toppling Rod

Problem Statement A homogeneous rod of length L and weight W is pinned at one end (see Fig. E12.8). If the rod is pushed over, what work does the force of gravity perform as the rod turns through an angle θ from its vertical position?

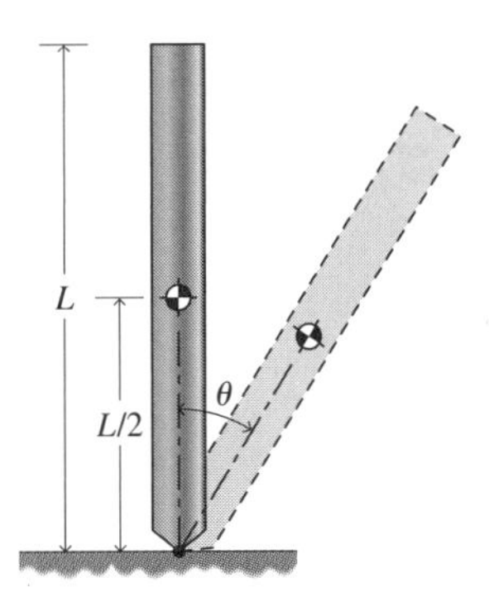

Figure E12.8

Solution The initial elevation of the center of gravity is $L/2$. The elevation of the center of gravity when the rod is inclined at the angle θ is $(L/2)(\cos\theta)$. Accordingly, the descent of the center of gravity is $(L/2)(1 - \cos\theta)$. Therefore, the work performed by the force of gravity is

$$U = \frac{WL(1 - \cos\theta)}{2}$$

Example 12.9

Work Done on a Block-Pulley System

Problem Statement A mechanical system consists of a block A of mass $m_A = 6$ slugs on an inclined plane of angle 30°. The coefficient of kinetic friction between the block and the plane is $\mu_k = 0.1$. Block A is connected by a cord that passes over a frictionless, massless pulley to a second block B of mass $m_B = 4$ slugs, hanging vertically (see Fig. E12.9a). Determine the total work done by the forces that act on the system during the time that block B drops 5 ft.

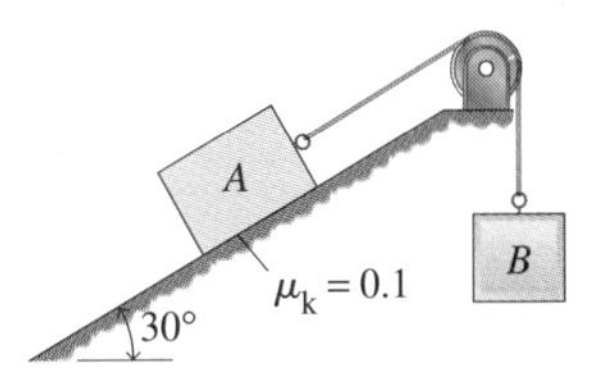

Figure E12.9a

To determine the forces that act on block A, we consider the free-body diagram of the block (Fig. E12.9b). The weight of block A is (6)(32.18) = 193.1 lb. So, summation of forces in the y direction yields

$$\sum F_y = N - (193.1)(\cos 30°) = 0$$

or

$$N = 167.2 \text{ lb}$$

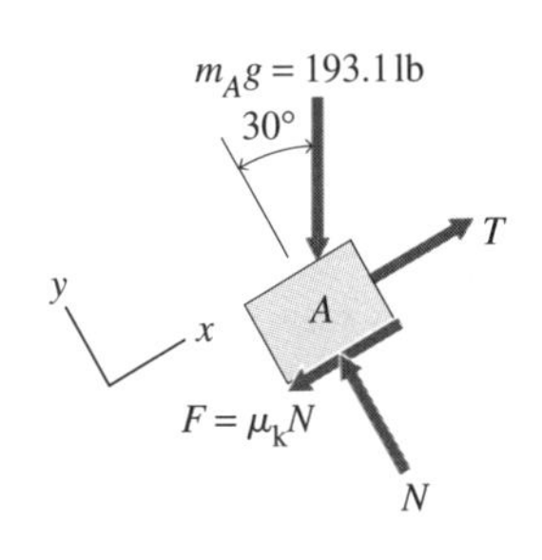

Figure E12.9b

Hence, the friction force is $F = \mu_k N = (0.1)(167.2) = 16.72$ lb.

Since the work done by the tension T on block A is the negative of the work done by T on block B, the net work performed by T on the system is zero. Also, the normal force N does no work on block A. Consequently, the total work done on the system as block B drops 5 ft is

$$U = 4(32.18)(5) - (193.1)(5 \sin 30°) - (16.72)(5) = 77.30 \text{ ft·lb}$$

Example 12.10

Work Performed on a Trunk

Problem Statement A man loads a trunk that weighs 900 N onto a truck, by dragging it 3.5 m up a loading ramp (see Fig. E12.10a). The coefficient of kinetic friction is 0.15. The man exerts just enough force to keep the trunk sliding at constant speed.

Figure E12.10a

a. Determine the total work done on the trunk by all the forces acting on it as it is dragged up the ramp.

b. Determine the work done on the trunk by the man.

Solution **a.** Since the man exerts just enough force to keep the trunk sliding at constant speed, equilibrium conditions apply to the trunk. Therefore, consider the free-body diagram of the trunk (Fig. E12.10b). Summation of forces in the x and y directions yields

$$\sum F_x = T \cos 30° - 900 \sin 20° - 0.15N = 0$$
$$\sum F_y = T \sin 30° - 900 \cos 20° + N = 0 \qquad \text{(a)}$$

Solving Eqs. (a), we obtain

$$T = 461.92 \text{ N} \qquad \text{(b)}$$

$$N = 614.76 \text{ N} \qquad \text{(c)}$$

Hence, the total work done on the trunk by all the forces as it is dragged up the ramp at constant speed is, with Eqs. (b) and (c),

$$U_{\text{total}} = (T \cos 30°)(3.5) - 900(3.5 \sin 20°) - (0.15N)(3.5) = 0$$

Alternatively, rather than computing the work done by each force and adding to get the total work, we can obtain the total work by computing the vector sum (resultant **F**) of the forces and then using the resultant **F** (or its components F_x, F_y, and F_z) in Eq. (12.5). In this example, $\mathbf{F} = 0$; hence, $F_x = F_y = F_z = 0$ [see Eqs. (a)]. Therefore, by Eq. (12.5), $U_{\text{total}} = 0$.

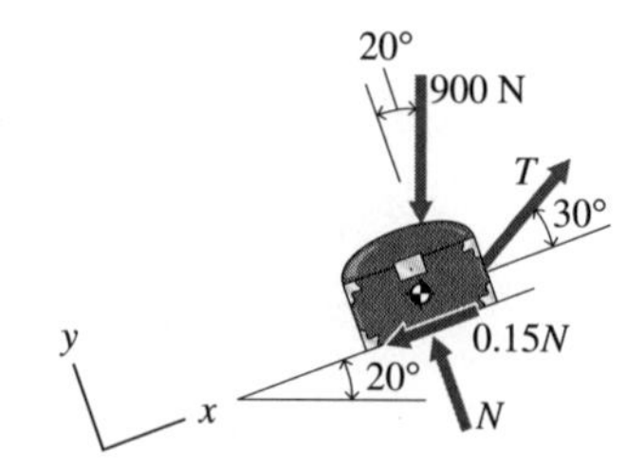

Figure E12.10b

b. To solve for the work done on the trunk by the man, we need to compute the work done by T. Since T is given by Eq. (b), the work done by the man is

$$U_{\text{man}} = (T \cos 30°)(3.5) = 1400 \text{ N·m}$$

12.4 WORK PERFORMED BY A COUPLE

After studying this section, you should be able to:

- Determine the work performed on a body by a couple as the body undergoes an angular displacement.

BODY DISPLACED IN A PLANE

Key Concept The work performed on a body by a couple is equal to the product of the moment of the couple and the angular displacement of the arm of the couple.

Consider a couple **M** that lies in the xy plane and acts on a body that moves in that plane (see Fig. 12.4). The couple consists of a force $-\mathbf{F}$ acting at point (x, y) and a force $\mathbf{F}$ acting at point (x', y'). If the rigid body on which the couple acts is displaced in a direction parallel to the xy plane, the coordinates (x, y) and (x', y') receive infinitesimal increments (dx, dy) and (dx', dy').[1] Then, by Eq. (12.4), the work performed by the forces of the couple is

$$dU = -F_x dx - F_y dy + F_x dx' + F_y dy' \tag{a}$$

where F_x and F_y are the projections of **F** on the x and y axes.

From Fig. 12.4, we have

$$x' = x + a\cos\theta$$

$$y' = y + a\sin\theta$$

where a is the arm of the couple. Hence, by differentiation,

$$\begin{aligned} dx' &= dx - a\sin\theta\, d\theta, \\ dy' &= dy + a\cos\theta\, d\theta \end{aligned} \tag{b}$$

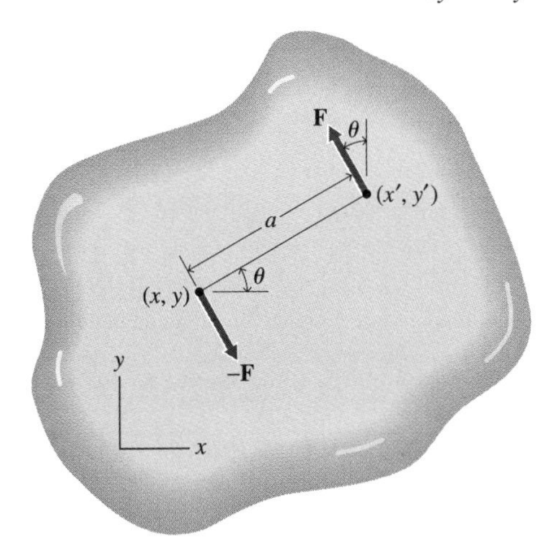

Figure 12.4 Work of a couple.

1. In general, a displacement is composed of a translation and a rotation. This book also refers to a rotation as an *angular displacement.*

Substituting Eqs. (b) into Eq. (a), we obtain

$$dU = -aF_x \sin\theta\, d\theta + aF_y \cos\theta\, d\theta \tag{c}$$

We see from Fig. 12.4 that

$$F_x = -F\sin\theta$$

$$F_y = F\cos\theta$$

Consequently, Eq. (c) yields

$$dU = aF\sin^2\theta\, d\theta + aF\cos^2\theta\, d\theta$$

Since $\sin^2\theta + \cos^2\theta = 1$, this yields

$$dU = aF\, d\theta \tag{d}$$

The moment of couple **M** is $M = aF$. Therefore, Eq. (d) yields

$$dU = M\, d\theta \tag{e}$$

Consequently, if M is a function of θ,

$$U = \int_{\theta_1}^{\theta_2} M\, d\theta \tag{12.8}$$

where θ_1 and θ_2 are the initial and final values, respectively, of the angle θ. Equation (12.8) is valid, regardless of the translation the body receives, provided only that the particles are all displaced parallel to the xy plane.

If M is constant, Eq. (12.8) yields $U = M(\theta_2 - \theta_1)$. Accordingly, if a rigid body that is subjected to a constant couple is displaced parallel to the plane of the couple, the work that the couple performs on the body is equal to the product of the moment of the couple and the angular displacement of the arm of the couple. Of course, the angle θ must be expressed in radians, otherwise Eqs. (b) are invalid.

BODY DISPLACED IN SPACE

Let the couple $\mathbf{M} = M_x\mathbf{i} + M_y\mathbf{j} + M_z\mathbf{k}$ act on a rigid body that undergoes a general angular displacement $d\mathbf{\Theta} = d\theta_x\mathbf{i} + d\theta_y\mathbf{j} + d\theta_z\mathbf{k}$, where **i**, **j**, and **k** are unit vectors along the positive x, y, and z axes, M_x, M_y, and M_z are moments of the couple **M** about the x, y, and z axes, and $d\theta_x$, $d\theta_y$, and $d\theta_z$ are angular displacements about the x, y, and z axes. The increment of work done by the couple under the angular displacement $d\mathbf{r}$ is, by Eq. (e),

$$dU = M_x\, d\theta_x + M_y\, d\theta_y + M_z\, d\theta_z = \mathbf{M}\cdot d\mathbf{\Theta}$$

Hence, if **M** is a function of $\mathbf{\Theta}$, the work done by **M** is

$$U = \int_{\mathbf{\Theta}_1}^{\mathbf{\Theta}_2} \mathbf{M}\cdot d\mathbf{\Theta} \tag{12.9}$$

where $\mathbf{\Theta}_1$ and $\mathbf{\Theta}_2$ denote the initial and final angular positions, respectively.

Example 12.11

Work of a Constant Torsional Couple

Problem Statement A constant torsional couple **T** of magnitude $T = 100$ N·m acts on a shaft (see Fig. E12.11).[2] While the couple acts, the shaft is allowed to rotate in the same sense as the couple through an angle of $\theta = 60°$. Then, the shaft reverses its rotation and

2. To distinguish a couple that twists (produces torsion in) a shaft about its longitudinal axis from a couple **M** that bends the shaft, we represent the former by the symbol **T** and call it a torsional couple (see Chapter 11).

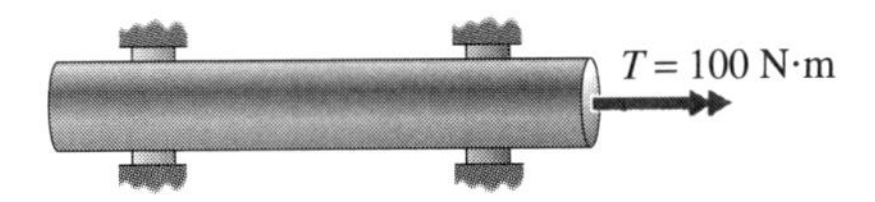

Figure E12.11

rotates in the sense opposite to that of the couple through an angle of 120°. Determine the work done on the shaft by the couple.

Solution During the first rotation of the shaft through the angle $\theta = 60°(\pi/3 \text{ rad})$, the couple performs work on the shaft equal to $T\theta = 100(\pi/3) = 104.7$ N·m. During the second rotation, the couple performs work $T\theta = (100)(-2\pi/3) = -209.4$ N·m. Therefore, the total work done by the couple is $104.7 - 209.4 = -104.7$ N·m.

Equivalently, the work done by the couple is

$$(100)(60° - 120°)(\pi/180°) = -104.7 \text{ N·m}$$

Example 12.12

Work of a Variable Torsional Couple

Problem Statement A variable torsional couple **T** acts on a rotating circular shaft (see Fig. E12.12). The magnitude of the couple (in newton-meters) is given by $T = 50 \sin 10t$, and the angular rotation of the shaft is given by the formula $\theta = 2 \sin 5t$, where t denotes time in seconds and θ is in radians. Determine the work done by the couple on the shaft from $t = 0$ to $t = \pi/2$.

Figure E12.12

Solution From Eq. (12.8), the work done by the couple is

$$U = \int_{\theta_1}^{\theta_2} T\, d\theta$$

where $\theta_1 = 0$ at $t = 0$, $\theta_2 = 2[\sin(5\pi/2)] = 2$ rad at $t = \pi/2$, and $d\theta = 10 \cos 5t\, dt$. Therefore,

$$U = \int_0^{\pi/2} (50 \sin 10t)(10 \cos 5t)\, dt \qquad \text{(a)}$$

Integration of Eq. (a) yields $U = 66.67$ N·m.

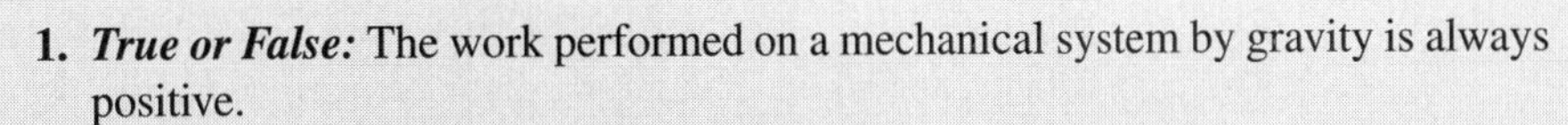

CHECKPOINT

1. ***True or False:*** The work performed on a mechanical system by gravity is always positive.
2. ***True or False:*** The work performed on a body by a couple is equal to the sum of the work performed by each of the forces of the couple.
3. ***True or False:*** The work performed by a couple of constant magnitude on a rigid body that is displaced parallel to the plane of the couple is equal to the product of the moment of the couple and the angular displacement of the arm of the couple.

Answers: 1. False; 2. True; 3. True

12.5 PRINCIPLE OF VIRTUAL WORK

After studying this section, you should be able to:

- Evaluate the equilibrium conditions for a system using the principle of virtual work.

Key Concepts

The virtual work of a real force is the work performed by the force during a virtual (imagined) displacement.

A motionless mechanical system is in equilibrium if the virtual work done on the system is negative or zero for all possible virtual displacements.

When a mechanical system begins to move from rest, the forces that act on the system perform positive work; since the system moves in the direction of the unbalanced force, the work is positive. This fact is a consequence of the law of kinetic energy, which will be introduced in the study of dynamics.

We may conclude, then, that a motionless mechanical system will not move by itself if there is no small displacement for which the forces perform positive work. To examine this conclusion, we place ourselves in the role of an imaginary experimentalist. We *imagine* that we give the system an arbitrary small displacement, and we determine the sign of the resulting work of the forces that act on the system, excluding the imaginary forces that we exert. Since this displacement is imagined and does not necessarily occur as an actual movement of the system, it is called a *virtual displacement,* and the work of the real forces that act during a virtual displacement is called *virtual work.*[3] During a virtual displacement, the real forces are taken to remain constant.

virtual displacement

virtual work

The French mathematician J. Fourier (1768–1830) summarized the foregoing reasoning, as follows.

Theorem 12.2

If the virtual work is negative or zero for every small virtual displacement of a motionless mechanical system, then the system is in equilibrium.

Fourier's inequality

This principle is known as *Fourier's inequality.* It expresses conditions that are *sufficient* for equilibrium, but not *necessary;* that is, in general, its converse is not true. Fourier's inequality is one formulation of a general law of equilibrium that has been employed in scientific investigations for more than 400 years under various names: "principle of virtual velocities," "principle of virtual displacements," and "principle of virtual work." This book designates it the *principle of virtual work.*

principle of virtual work

You should note that Theorem 12.2 places no restriction on the kinds of forces that act on the system; they may be friction forces, constraint forces (e.g., beam supports), applied loads, and so on. The virtual displacements must conform to the constraints.

As a simple application of the principle of virtual work, consider a brick that rests on the floor. If you lift the brick or tip it slightly, the force of gravity performs negative work. If you slide the brick, the frictional force performs negative work. Therefore, the brick is

3. Although virtual displacements are often considered to be infinitesimal, this restriction is not imposed here, where they need only be small. However, they must conform to the geometrical constraints placed on the body or system.

in equilibrium when it rests on the floor. This conclusion would remain valid if the floor were perfectly frictionless and horizontal, since the virtual work would then be zero for horizontal displacements and negative for vertical displacements.

A marble at rest at the bottom of a bowl serves as another example. If you give the marble any small virtual displacement, the force of gravity performs negative work. Friction also performs negative work. Therefore, the marble is in equilibrium when it is at rest at the bottom of the bowl. The marble may be given only *small* displacements, since larger ones could carry it over the rim of the bowl. Then, lowering it sufficiently could render the net work of the force of gravity positive. To avoid such situations, the virtual displacements must always be small.

A lever provides a more significant practical application of the principle of virtual work. If a horizontal lever is rotated through a virtual angle $\delta\theta$, as shown in Fig. 12.5, the work of the force of gravity is

$$\delta U = W_2 b\,\delta\theta - W_1 a\,\delta\theta \tag{a}$$

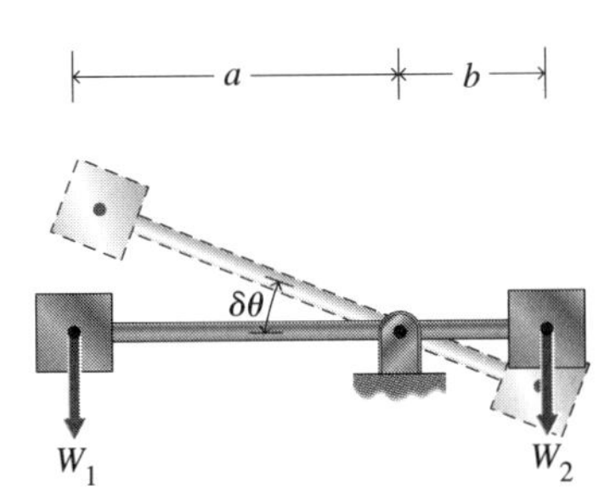

Figure 12.5 Horizontal lever.

Here, the symbols δU and $\delta\theta$ are used instead of dU and $d\theta$ to emphasize that the work and displacement are virtual (imagined). This notation is common in mechanics. By Eq. (a), if δU is negative for a positive angular displacement $\delta\theta$, it becomes positive when $\delta\theta$ is negative. Consequently, in this example, δU cannot be negative for all small virtual displacements. However, if $W_2 b - W_1 a = 0$, then δU is zero for all small virtual displacements. Accordingly, the lever is balanced if

$$W_1 a = W_2 b \tag{b}$$

Thus, we have derived the condition for moment equilibrium of the lever [Eq. (b)] by a type of reasoning that is quite different from that employed in the previous chapters. We can conclude that the principle of virtual work is independent of Newton's laws of motion.

Roberval's balance

The linkage shown in Fig. 12.6 is called *Roberval's balance.* In some respects, Roberval's balance is different from a lever. If the weight W_2 descends a distance δs, the

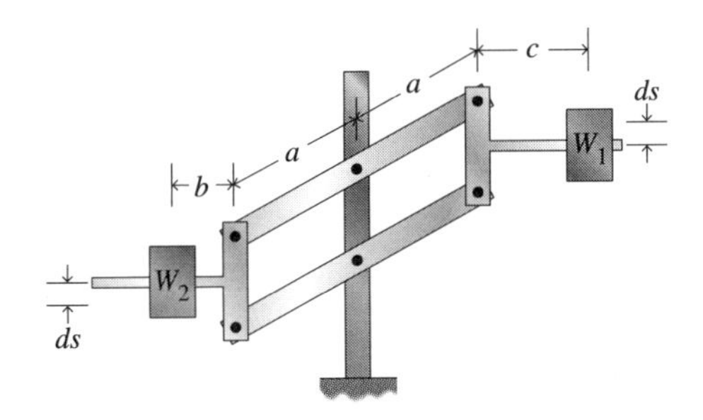

Figure 12.6 Roberval's balance.

weight W_1 ascends a distance δs. Consequently, if the joints are frictionless, the virtual work performed by gravity is $\delta U = W_2\,\delta s - W_1\,\delta s$. This expression cannot always be negative, since its sign is controlled by the sign of δs. However, δU vanishes if $W_1 = W_2$. By Theorem 12.2, this is an equilibrium condition; that is, equal weights placed on Roberval's balance are in equilibrium, irrespective of their locations b and c on the horizontal bars. The ease with which this conclusion has been derived indicates that applying the principle of virtual work is sometimes more convenient than using Newtonian methods.

PROBLEM-SOLVING TECHNIQUE

Applications of the Principle of Virtual Work

There are, basically, three types of equilibrium problems that are readily solved by the principle of virtual work. These are problems in which the unknowns are

a. the coordinates that describe an equilibrium position of a system

b. the external loads that maintain equilibrium of the system

c. the forces due to the support reactions

For cases a and b, the unknowns appear simply as variables in the virtual work expressions. For case c, the support is removed, and its reaction is applied to the system as a variable. For instance, if the reaction at a roller support is required, an unknown reaction force is applied to the system and the system is allowed to move as if the support did not exist.

To evaluate the conditions of equilibrium for a statically determinate system using the principle of virtual work:

1. Draw a diagram of the system, including all external loads that act on the system. If the problem involves determination of a support reaction, remove the support and replace it by its reaction. No more than one support can be removed and replaced by its reaction. If the objective of the problem is to find more than one support reaction, the principle of virtual work is applied once for each reaction.
2. Select a virtual displacement δu for the system, and draw a diagram of the system in its displaced position. The virtual displacement δu may be a translation of a point or a rotation of a part of the system. If a support reaction is to be determined (see Step 1), the virtual displacement must not violate any of the remaining support conditions (constraints).
3. From the diagram developed in Step 2, write expressions for the virtual work of the forces and couples in terms of the virtual displacement δu. The virtual work of a force is the magnitude of the force times the displacement of its point of application (written in terms of δu). The virtual work of a couple is the moment of the couple times the virtual rotation of its arm (also written in terms of δu). The virtual work of a force or a couple may be positive or negative, depending on the sense of the displacement or the rotation. The force and the couple remain constant during the virtual displacement.
4. Apply the principle of virtual work, and determine the values of the unknowns for which the virtual work of the system is zero or negative. Consider both positive and negative values of δu.

Example 12.13 Beam Reactions

Problem Statement A simply supported beam AB is subjected to a force P, as shown in Fig. E12.13a. By applying the principle of virtual work (Theorem 12.2), determine the support reaction at B.

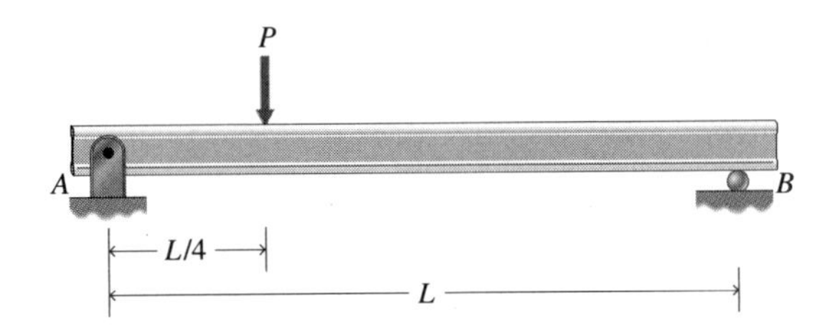

Figure E12.13a

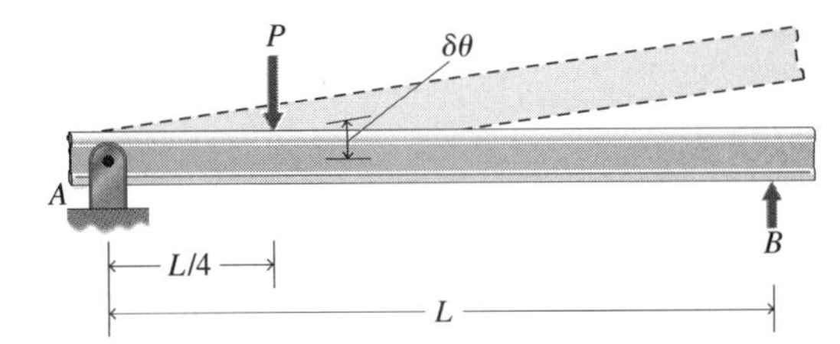

Figure E12.13b

Solution To solve this problem using the principle of virtual work, we remove the support at B and replace it by its reaction (see Fig. E12.13b). We choose a rotation θ of the beam about the pin at A as the displacement by which the positions of all points in the beam are described. Then, we let the beam undergo an imagined angular displacement (virtual rotation) $\delta\theta$ about A (refer to Fig. E12.13b). The virtual displacements of the points of application of forces B and P are $L\,\delta\theta$ and $L\,\delta\theta/4$, respectively. Under the virtual rotation, the frictionless pin at A must not move; that is, the reaction at A does no work. Hence, the virtual work done by the forces is

$$\delta U = B(L\,\delta\theta) - P\left(\frac{L}{4}\,\delta\theta\right) \tag{a}$$

If δU is negative for a positive angular displacement $\delta\theta$, it becomes positive when $\delta\theta$ is negative. Consequently, δU cannot be negative for all small virtual displacements. Therefore, since the beam must be in equilibrium for all $\delta\theta$, when $\delta U = 0$ (Theorem 12.2), Eq. (a) yields $B = P/4$.

Information Item The principle of virtual work provides only one equation ($\delta U \leq 0$). Hence, we can solve for only one unknown at a time. So, if we wanted to find the reaction at A using the principle of virtual work, we would perform the process again with the hinge at A replaced by its reaction force and the roller at B in place, as shown in Fig. E12.13a.

Example 12.14 Crate on a Ramp

Problem Statement A crate of weight W_1 is held in equilibrium on a ramp by a counterweight W_2, which is connected to the crate by a light rope that passes over a frictionless pulley (see Fig. E12.14a). The coefficient of static friction between the crate and the ramp is $\mu_s = 0.30$. By the principle of virtual work (Theorem 12.2), determine the range of the ratio W_2/W_1 for which the crate is in equilibrium.

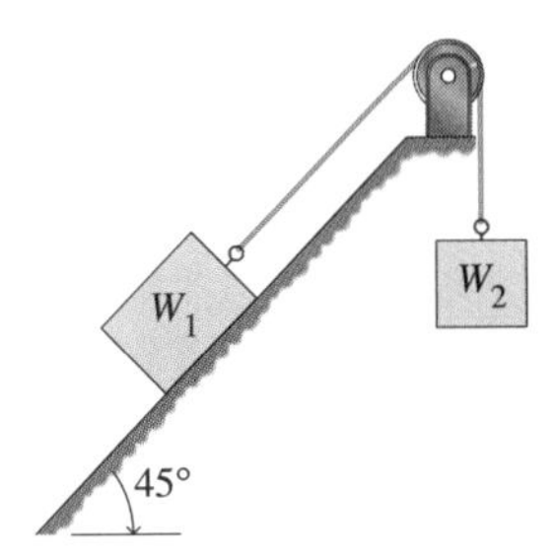

Figure E12.14a

Solution Since the pulley is frictionless and the rope is light, the tension $T = W_2$ in the rope is transmitted around the pulley to the crate (see Fig. E12.14b). Also, since the normal force acting on the crate is $N = W_1 \cos 45°$, the friction force at impending sliding of the crate is $\mu_s N = 0.30W_1(\cos 45°)$. Now, imagine that the crate is given a positive virtual displacement δs down the ramp (Fig. E12.14b). Since the sense of the frictional force is opposite to δU, regardless of whether δs is positive (down the ramp) or negative (up the ramp), the virtual work of friction is equal to $-0.30W_1(\cos 45°)|\delta s|$. Hence, the virtual work performed by the forces that act on the crate is, for δs positive,

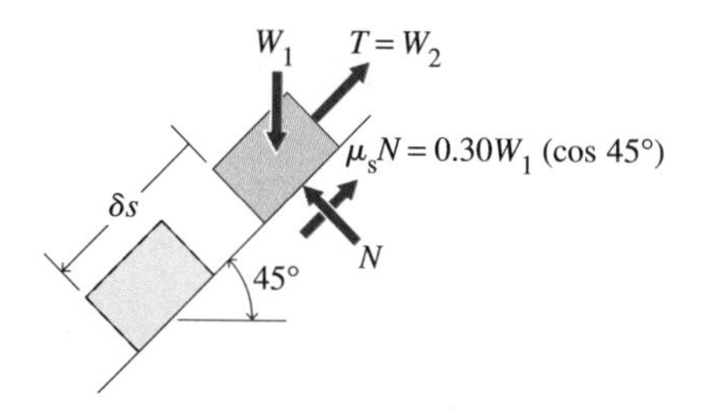

Figure E12.14b

$$\delta U = W_1 \sin 45°\, \delta s - W_2\, \delta s - 0.30W_1(\cos 45°)|\delta s| \tag{a}$$

By the principle of virtual work (Theorem 12.2), equilibrium exists provided that the virtual work $\delta U \leq 0$, whether δs is positive or negative. Hence, from Eq. (a),

$$(0.495\, W_1 - W_2) \leq 0, \text{ for } \delta s \text{ positive}$$
$$(W_2 - 0.919W_1) \leq 0, \text{ for } \delta s \text{ negative} \tag{b}$$

Dividing Eqs. (b) by W_1, we find that equilibrium exists if W_2/W_1 lies in the range

$$0.495 \leq \frac{W_2}{W_1} \leq 0.919$$

Example 12.15

Platform Scale

Problem Statement Figure E12.15 represents the mechanism of a platform scale that is used to measure a weight W. It is possible to adjust the length x so that the relationship between the balance weight P and the load W does not depend on the position of the weight W on the platform FD. By the principle of virtual work, determine this value of x and the corresponding relationship between P and W. Neglect the weights of the bars, since they are ordinarily compensated for by a counterweight (not shown in Fig. E12.15).

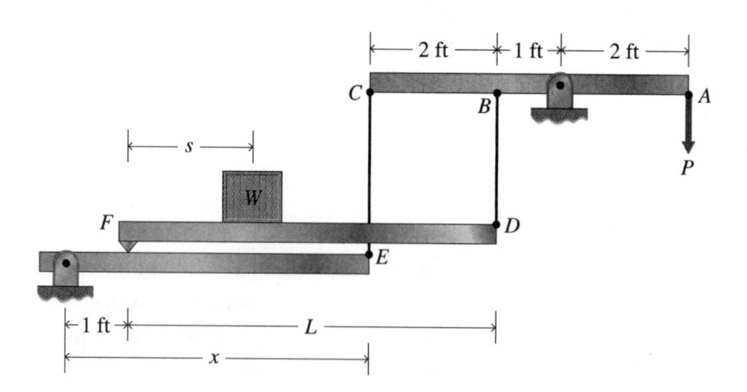

Figure E12.15

Solution Let the bar AC rotate through the virtual angle $\delta\theta$ (positive θ assumed clockwise). Then, point A descends the distance $\delta u_A = 2\,\delta\theta$. Points B and D ascend the distance $\delta u_D = \delta\theta$, and points C and E ascend the distance $\delta u_E = 3\,\delta\theta$. Consequently, point F ascends the distance $\delta u_F = (3/x)\,\delta\theta$. The ascension of weight W is $\delta u_W = \delta u_F + (s/L)(\delta u_D - \delta u_F)$. Therefore,

$$\delta u_W = \frac{3\,\delta\theta}{x} + \frac{s}{L}\left(\delta\theta - \frac{3\,\delta\theta}{x}\right)$$

The virtual work of the forces P and W is $\delta U = P\,\delta u_A - W\,\delta u_W$. Hence,

$$\delta U = \left[2P - \frac{3W}{x} - \frac{Ws}{L}\left(1 - \frac{3}{x}\right)\right]\delta\theta \tag{a}$$

The condition of equilibrium is that δU be negative or zero for all virtual displacements $\delta\theta$. Note that δU cannot be negative for all $\delta\theta$, since the sign of δU is controlled by the sign of $\delta\theta$. Therefore, the equilibrium condition is $\delta U = 0$. Since this condition must be satisfied for all values of $\delta\theta$, Eq. (a) yields

$$2P - \frac{3W}{x} - \frac{Ws}{L}\left(1 - \frac{3}{x}\right) = 0 \tag{b}$$

Since P must be independent of s, Eq. (b) shows that $(1 - 3/x) = 0$, or $x = 3$ ft. Then Eq. (b) yields $W = 2P$.

Information Item This example demonstrates the versatility of the principle of virtual work. Using Newtonian mechanics, we would have to consider three rigid bodies (the three bars) and six equations of equilibrium to find the relationship between P and W. With the principle of virtual work, the internal forces in the beams, the tensions in the cables, and the support reactions need not be considered.

Example 12.16

Chain Hoist with Bearing Friction

Problem Statement A differential chain hoist consists of a two-sprocket upper sheave and a single-sprocket lower sheave connected by a chain (see Fig. E12.16). The upper sheave consists of two connected sprockets, with radii $r_2 > r_1$, that rotate together when a pull P is applied to the chain. A weight W is carried by the lower sheave. The pressures in the bearings of the upper sheave are proportional to W. In turn, the frictional forces in those bearings are proportional to the pressures. Hence, the frictional torque exerted by the bearings when the upper sheave rotates is equal to μW, where μ is a constant that depends on the properties of the bearings and the relative sizes of the sprockets.

a. Determine the range of values of the ratio P/W for which the hoist is in equilibrium.

b. Determine the mechanical advantage $m = W/P$ of the hoist for the case where the bearings are frictionless.

Figure E12.16

Solution **a.** To analyze the differential chain hoist by the principle of virtual work, we give the upper sheave a clockwise virtual rotation $\delta\theta$. Then, the virtual work of the force P is $Pr_2\,\delta\theta$. By geometry, the center of the lower sheave is raised $(r_2 - r_1)\,\delta\theta/2$. Therefore, the virtual work done on the load W is $(r_2 - r_1)W\,\delta\theta/2$.

The virtual work done by the frictional torque is negative, irrespective of the sign of $\delta\theta$. Therefore, it is represented by $-\mu W|\delta\theta|$.

Summing the preceding expressions, we obtain the following equation for the total virtual work:

$$\delta U = Pr_2\,\delta\theta - \frac{(r_2 - r_1)W\,\delta\theta}{2} - \mu W|\delta\theta| \tag{a}$$

By the principle of virtual work, equilibrium exists provided that $\delta U \le 0$, whether $\delta\theta$ is positive or negative. Hence, by Eq. (a), the equilibrium conditions are

$$\begin{aligned} &Pr_2 - \frac{(r_2 - r_1)W}{2} - \mu W \le 0 \text{ if } \delta\theta > 0 \\ &\frac{(r_2 - r_1)W}{2} - \mu W - Pr_2 \le 0 \text{ if } \delta\theta < 0 \end{aligned} \tag{b}$$

Dividing Eq. (b) by Wr_2, we find that equilibrium exists if P/W lies in the following range:

$$\frac{1}{2}\left(1 - \frac{r_1}{r_2}\right) - \frac{\mu}{r_2} \le \frac{P}{W} \le \frac{1}{2}\left(1 - \frac{r_1}{r_2}\right) + \frac{\mu}{r_2} \tag{c}$$

b. The mechanical advantage m of the hoist with $\mu = 0$ is obtained from Eq. (b) as

$$m = \frac{W}{P} = \frac{2r_2}{r_2 - r_1}$$

CHECKPOINT

1. ***True or False:*** A virtual displacement of a system is an actual displacement of the system.
2. ***True or False:*** The work of the real forces that act on a system during a virtual displacement is called virtual work.
3. ***True or False:*** If the virtual work for a motionless mechanical system is zero for every small virtual displacement, then the system is in equilibrium.
4. Is the virtual work done by friction always negative? Explain.

Answers: 1. False; 2. True; 3. True; 4. Yes—the force of friction always opposes the direction of displacement

12.6 STABLE, UNSTABLE, AND NEUTRAL EQUILIBRIUM

After studying this section, you should be able to:

- Determine the state of equilibrium for a system, using the principle of virtual work.

Key Concept

A motionless mechanical system is in stable equilibrium if, and only if, the virtual work is negative for every small virtual displacement.

stable equilibrium

In a practical sense, a mechanical system is said to be in *stable equilibrium* if a slight disturbance causes only a slight departure of the system from its equilibrium position. In a mathematical sense, this means that an infinitesimal disturbance of the system causes only an infinitesimal change from the given equilibrium position. A definition of stability of a system based on the principle of virtual work is given below (see Theorem 12.3).

Near Petrified Forest National Park in Arizona, a large boulder rests in stable equilibrium on a rock formation. A slight disturbance, such as that due to wind, will not cause the boulder to fall. A larger disturbance, however, such as an earthquake, may bring the boulder down from its perch.

As noted in Sec 12.5, the principle of virtual work expresses conditions that are *sufficient* for equilibrium—namely, that the virtual work of the forces that act on a system in equilibrium is negative or zero for every small virtual displacement of the system (Theorem 12.2). However, these conditions are not *necessary* for equilibrium—there are states of *unstable equilibrium,* for which the virtual work is positive for small virtual displacements. Such a state is approximated if a marble is balanced on a dome (see Fig. 12.7a) or if a baseball bat is balanced on its end (see Fig. 12.7b). If the marble is displaced slightly or the bat is tipped slightly, gravity performs positive work, and the marble or the bat will move away from its equilibrium position. In other words, if a body is in a state of unstable equilibrium, a small displacement from its equilibrium position will allow the forces acting on the body to move the body further away from that position.

unstable equilibrium

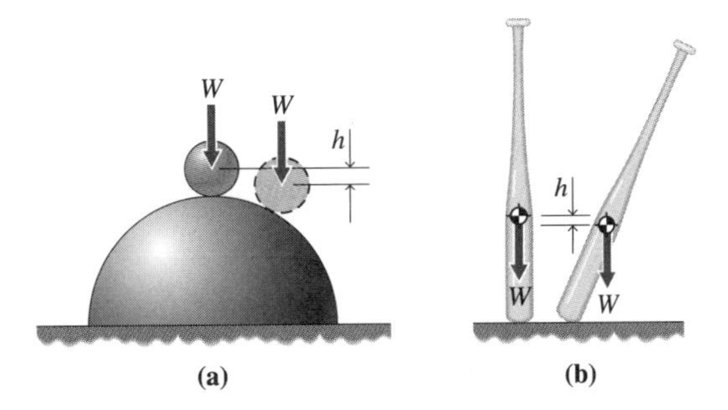

Figure 12.7 Unstable equilibrium: **(a)** marble on a smooth dome; **(b)** bat on a smooth plane.

Unstable states of equilibrium have little practical importance. But unless you study them, you cannot determine whether a body is stable or unstable in a given equilibrium configuration. For example, an engineering structure such as a truss may be stable under small loads, but may become unstable when the loads become large. Then, the structure collapses. This phenomenon is known as *buckling*. Buckling can be illustrated by placing one end of a yardstick against the floor and pressing on the other end until the yardstick bows out (buckles).

buckling

neutral equilibrium

If the virtual work on a body is zero, then the body is in *neutral equilibrium.* If a body is in a neutral equilibrium state, under a small displacement from this neutral position to a new position, the forces that act on the body will move the body neither away from nor toward the equilibrium position. Hence, the new position is an equilibrium position, and the equilibrium state is neutral. For example, if a marble on a horizontal plane is moved, it remains in the new equilibrium position without any tendency to return to its original equilibrium position (see Fig. 12.8). The lever (Fig. 12.5) and Roberval's balance (Fig. 12.6), discussed in Sec. 12.5, also illustrate neutral equilibrium.

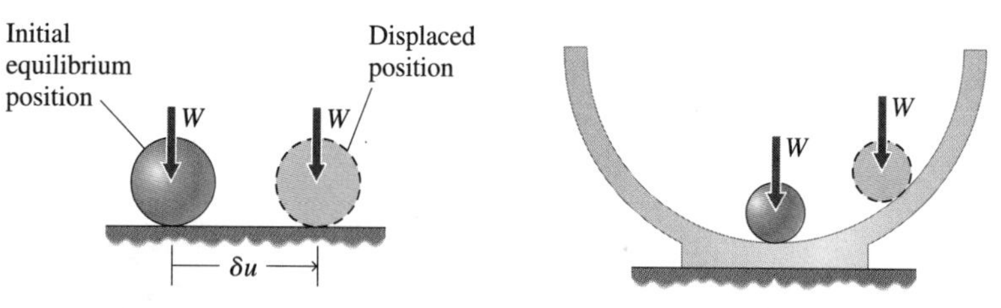

Figure 12.8 Neutral equilibrium.

Figure 12.9 Stable equilibrium.

Equilibrium is said to be stable if the virtual work is *always* negative for small displacements. Under a small displacement from a stable equilibrium position, the forces acting on a body will tend to move the body back to its original equilibrium position (see Fig. 12.9). Accordingly, the principle of virtual work (Theorem 12.2) has the following corollary.

Theorem 12.3

A motionless mechanical system is in stable equilibrium if, and only if, the virtual work is negative for every small virtual displacement.

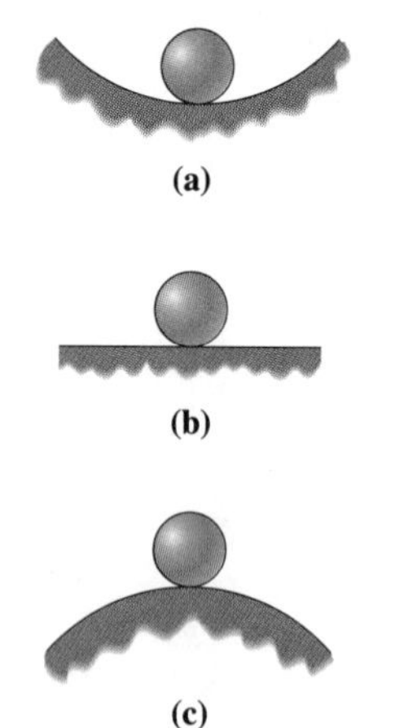

Figure 12.10 States of equilibrium of a marble: **(a)** stable; **(b)** neutral; **(c)** unstable.

This statement expresses a condition that is both *necessary* and *sufficient* for stable equilibrium. If stable equilibrium exists, the condition of the statement is fulfilled. And if the condition of the statement is fulfilled, stable equilibrium exists. For example, a brick that rests on a rough floor is in stable equilibrium. Likewise, a marble that rests at the bottom of a bowl is in stable equilibrium (refer to Fig. 12.9), because, under any small displacement, gravity does negative work and the marble tends to move back to the bottom of the bowl. Figure 12.10 illustrates stable, neutral, and unstable states of equilibrium of a marble.

Note that the above discussion uses relative terms such as "slight disturbances" and "small displacements" in the definition of stability. This is because stability is relative, in that one equilibrium state may be more or less stable than another equilibrium state. For example, consider a block of wood of width $2a$ and height $2h$ set on one end (see Fig. 12.11a). As the block is tipped about point A through a clockwise angle θ, its center

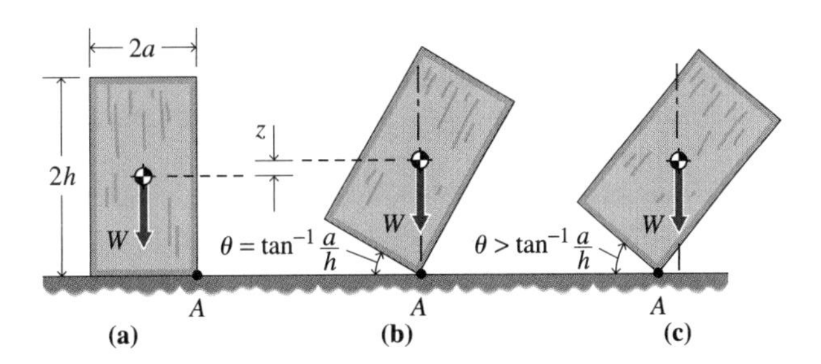

Figure 12.11 Degree of stability.

of gravity moves up vertically a distance $z = a\sin\theta + h\cos\theta - h$. Therefore, gravity performs *negative* work equal to $-Wz$, until $\tan\theta = a/h$; that is, until the line of action of W is directed through point A. Hence, for any disturbance that tips the block through an angle less that $\theta = \tan^{-1} a/h$, the block is stable and will return to its equilibrium position (see Fig. 12.11b). For larger values of θ, the center of gravity begins to descend and W does *positive* work. If the disturbance causes a rotation of the block through an angle greater than $\tan^{-1} a/h$, the block will continue to tip and then fall over (see Fig. 12.11c).

When $\tan\theta = a/h$, the center of gravity has been raised by an amount $z = \sqrt{a^2 + h^2} - h$, and the block is on the verge of falling over. The amount of work required to raise the center of gravity this distance is $U = W(\sqrt{a^2 + h^2} - h)$. This work is a measure of the *degree of stability* of the block. The degree of stability of the block can be increased by increasing U—that is, by increasing a (increasing the width of the block) or by decreasing h (decreasing the height of the block). Also, if the block has a variable density that makes its center of gravity nearer to one end, setting the block on the end nearest the center of gravity, as in Fig. 12.12a, increases the degree of stability. Setting the block on the end farthest from the center of gravity, as in Fig. 12.12b, decreases the degree of stability.

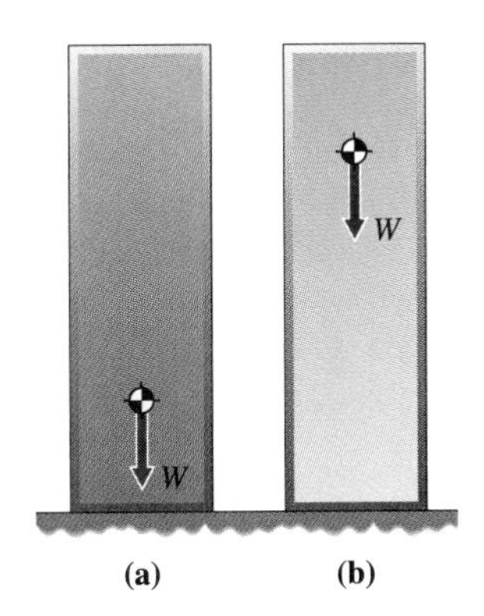

Figure 12.12 (a) More stable—low center of gravity. (b) Less stable—high center of gravity.

Example 12.17

States of Stability

Problem Statement A right circular cone is constrained to move on a smooth horizontal surface. Three possible equilibrium positions of the cone—(1) on its base, (2) on its side, and (3) on its apex—are shown in Fig. E12.17. Describe the states of equilibrium of the cone in these three positions; the cone is to maintain contact with the surface during any displacement.

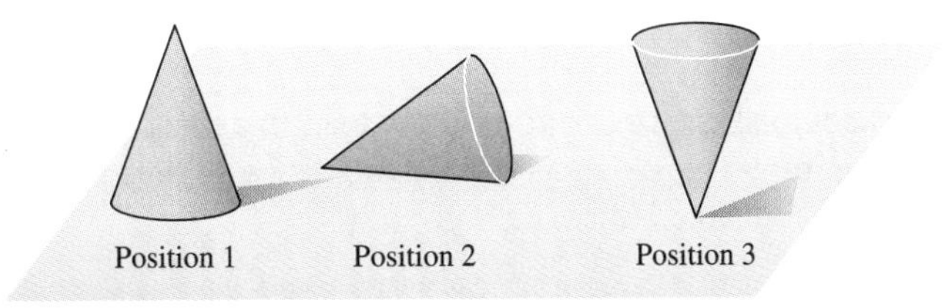

Figure E12.17

Solution In position 1, for any small virtual tipping displacement, gravity performs negative virtual work. Therefore, for tipping, the cone is in a state of stable equilibrium (Theorem 12.3). Also, the cone may undergo a rotation about its vertical axis or slide on its base. For these virtual displacements, in the absence of friction, the forces acting on the cone do no work. Therefore, in these types of displacements, the cone is in a state of neutral equilibrium.

In position 2, the cone rests on its side. Hence, it may roll or slide on the surface. For these displacements, the forces acting on the cone do no work, and the cone is in a state of neutral equilibrium. With regard to tipping of the cone about its base or tip, the cone is in a state of stable equilibrium, since gravity does negative work for any small virtual tipping displacement.

In position 3, the cone is in a state of neutral equilibrium under sliding or axial rotation (forces do no work), and it is unstable under tipping (gravity does positive work; see Theorem 12.3).

Information Item This example illustrates the fact that a system may be in stable equilibrium under certain virtual displacements and in unstable or neutral equilibrium under other virtual displacements. Thus, for tipping, the cone is stable in positions 1 and 2 in Fig. E12.17 and unstable in position 3. Since the surface is frictionless, the cone is in neutral equilibrium in all cases for sliding.

Generally, the equilibrium state of a mechanical system is determined with respect to an *arbitrary* virtual displacement. Then, of all possible equilibrium states, the *least* stable state of the system is defined as its state of stability. For instance, since the cone in position 1 in Fig. E12.17 is in stable equilibrium for tipping and neutral equilibrium for sliding, it is considered to be in neutral equilibrium, without additional qualifications. Likewise, in position 2, the cone is in neutral equilibrium, and in position 3, it is unstable.

Example 12.18

Stability of a Rigid Column with Spring Support

Problem Statement A rigid column of length L is hinged at the lower end and free at the upper end (see Fig. E12.18a). The hinge contains a rotational spring whose restoring moment for an angular rotation θ of the column is $M = k\theta$, where k is the spring constant. A vertical compressive load P is applied to the upper end of the column. Determine the maximum value of P, in terms of L and k, for which the column remains stable in the position shown.

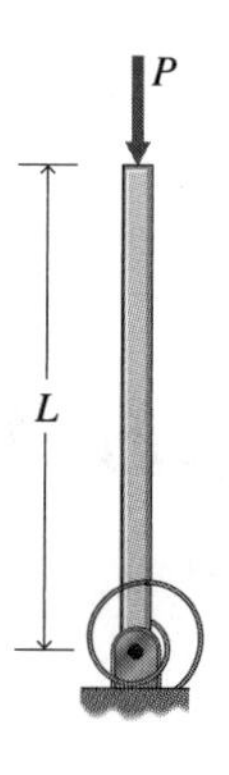

Figure E12.18a

Solution By definition, the column is stable if it returns to its initial position after being deflected a small amount from that position. Consider the deflected position shown in Fig. E12.18b. The load P undergoes a downward displacement of

$$y = L(1 - \cos\theta)$$

The virtual displacement δy of the load P is found by taking the variation of y. The variation of a function is analogous to the differential of the function. Hence, δy is given by

$$\delta y = L \sin\theta\, \delta\theta$$

and the virtual work of P is

$$\delta U_P = P\delta y = P(L \sin\theta\, \delta\theta) \tag{a}$$

The work of the spring on the column as it rotates to the deflected position is

$$U_s = -\tfrac{1}{2}k\theta^2 \tag{b}$$

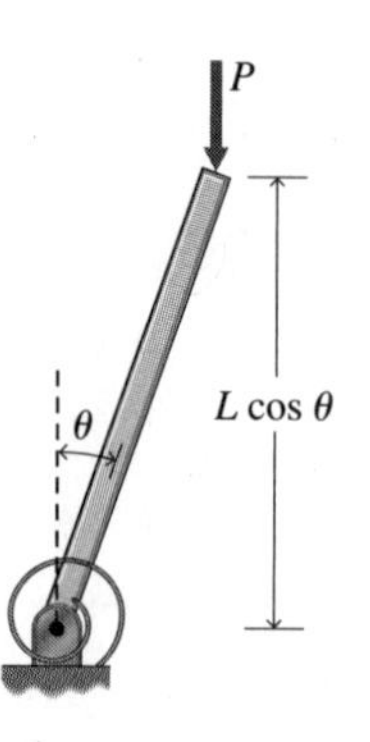

Figure E12.18b

Taking the variation of Eq. (b) with respect to θ, we obtain the virtual work of the spring as

$$\delta U_S = -k\theta\,\delta\theta \tag{c}$$

Hence, the total virtual work for the system is the sum of Eqs. (a) and (c):

$$\delta U = \delta U_P + \delta U_S = (PL\sin\theta - k\theta)\delta\theta \tag{d}$$

For stability of the column, $\delta U < 0$ (Theorem 12.3). So, by Eq. (d), for stability of the system, we have

$$PL\sin\theta - k\theta < 0$$

or

$$P < \frac{k\theta}{L\sin\theta} \tag{e}$$

Therefore, for the system to be stable in the vertical position, P must be less than the ratio $k\theta/(L\sin\theta)$ as $\theta \longrightarrow 0$. In the limit as $\theta \longrightarrow 0$, $\theta/(\sin\theta) \longrightarrow 1$. Then, from Eq. (e), we see that the column is stable at $\theta = 0$, provided that

$$P < \frac{k}{L}$$

The load $P = P_{cr} = k/L$ is known as the *critical load,* or the buckling load.

CHECKPOINT 

1. ***True or False:*** If a system is in stable equilibrium, the virtual work is negative for any virtual displacement of the system.
2. ***True or False:*** A system cannot be in equilibrium if the virtual work on the system is positive.

Answers: 1. True; 2. False

Chapter Highlights

- If a constant force moves a body along its line of action, the work it performs on the body is the product of the magnitude of the force and the distance the body is moved.
- The work U a force $\mathbf{F}$ performs on a particle that moves along a curved line C is given by the line integral $U = \int_C F\cos\theta\,ds$, where θ is the angle between the force $\mathbf{F}$ and the tangent to curve C and F is the magnitude of $\mathbf{F}$.
- The work performed by a constant gravitational force on any mechanical system equals the total weight of the system multiplied by the vertical distance through which the center of gravity of the system descends (positive work) or ascends (negative work).
- The work performed by a couple on a body that is displaced parallel to the plane of the couple is equal to the product of the moment of the couple and the angular displacement of the arm of the couple.
- Virtual displacements of a system are imagined displacements of the system.
- The work performed by the real forces that act on a system during a virtual displacement is called virtual work.
- The principle of virtual work states that a motionless mechanical system is in equilibrium if the virtual work is negative or zero for every small virtual displacement.
- A motionless mechanical system is in stable equilibrium if, and only if, the virtual work is negative for every small virtual displacement.

Problems

12.1 Work

12.1 The work required to lift 6 people 100 m on an elevator is 408 kJ. Determine the average mass per person and, hence, the weight per person.

12.2 The block shown in Fig. P12.2 weighs 150 N. The coefficient of sliding friction between the block and the surface is 0.40. Calculate the total work that all the forces perform on the block as it moves 2.5 m to the right.

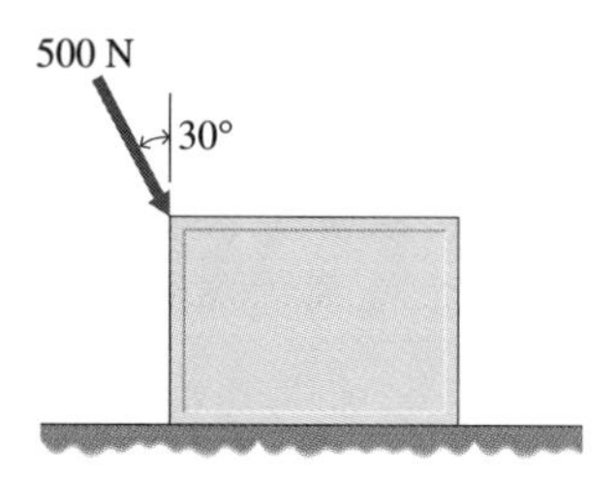

Figure P12.2

12.3 Two tugboats pull a barge on a river (see Fig. P12.3). The tensions in the lines AC and BC are $T_{AC} = 13.57$ kN and $T_{BC} = 10$ kN. Determine the work done on the barge by the tugboats as they move the barge 200 m.

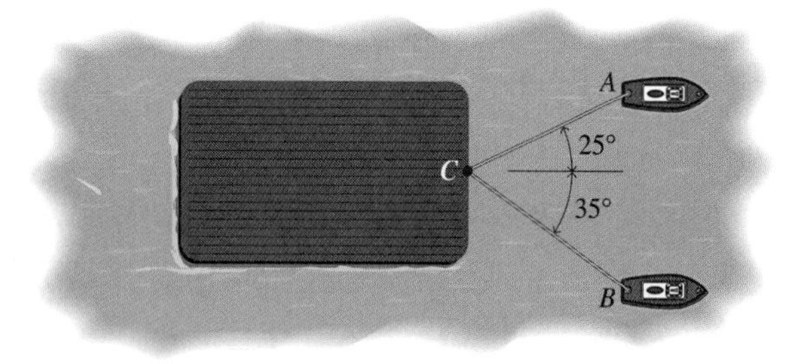

Figure P12.3

12.4 A block rests at O on a smooth, horizontal floor (see Fig. P12.4). At that position, it is connected to an unstretched linear spring with constant k.

a. Determine the work performed on the block by the spring as the block is moved from point O to point A, a distance x.

b. Determine the work performed on the block by the spring as the block moves from point O to point B, a distance x.

c. Determine the total work performed on the block by the spring as the block moves from O to A, then to B, then back to O.

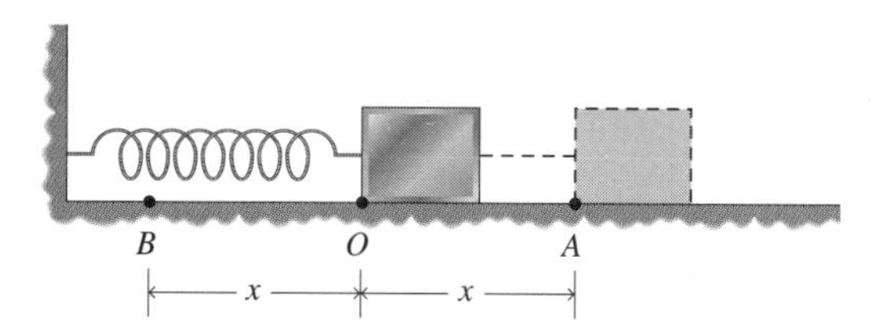

Figure P12.4

12.5 A linear spring is hung vertically in a gravity field (see Fig. P12.5). A body of mass m is attached to its lower end, causing the spring to stretch a distance d. Determine the spring constant k in terms of m, g, and d. (*Hint:* The work done by gravity on the mass is equal in magnitude to the energy stored in the spring as it is stretched a distance d.)

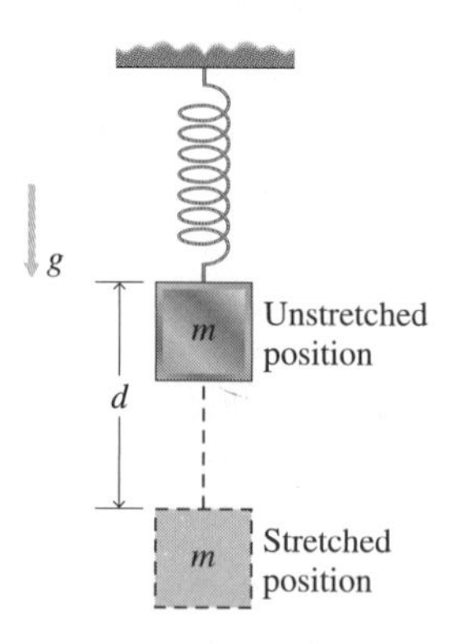

Figure P12.5

12.6 The graph in Fig. P12.6 represents the pull that the propellers of an airplane exert during the takeoff run. Calculate the work that the force F performs during takeoff for $0 \le x \le 2700$ ft.

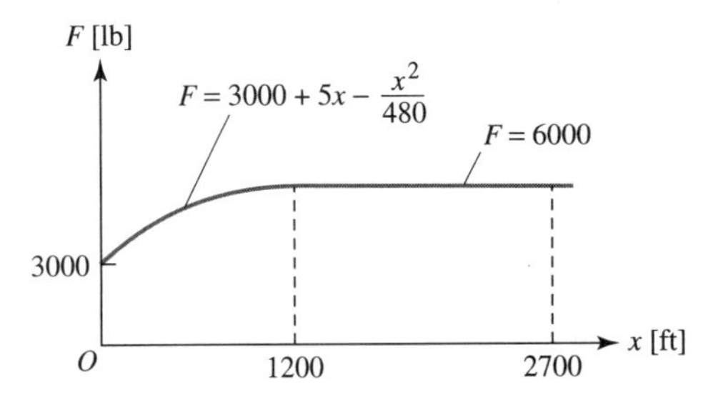

Figure P12.6

12.7 Determine the total work done on the crates A and B in Fig. P12.7 during the time that the force F moves the crates 3 m to the right. The coefficients of sliding friction are 0.20 for A and 0.30 for B.

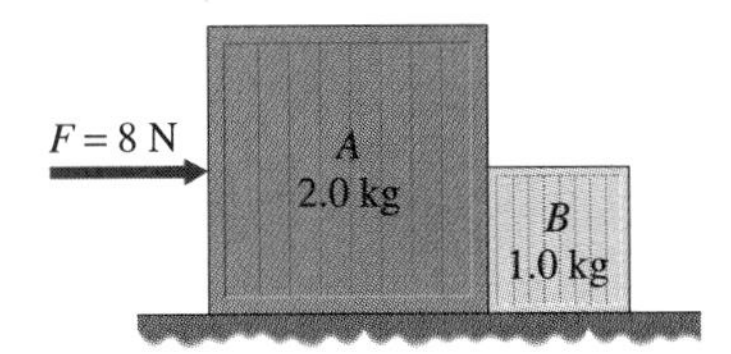

Figure P12.7

12.8 A wheelbarrow is filled with concrete (see Fig. P12.8). A force F is applied slowly at B to incline the wheelbarrow 15° from the horizontal. Determine the work done by each of the forces that act on the wheelbarrow in terms of F and W.

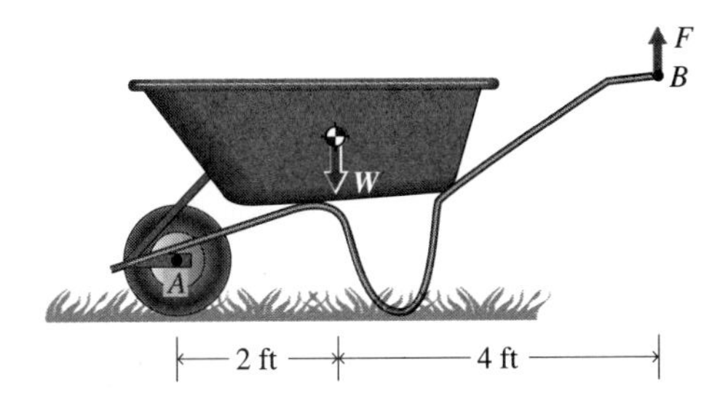

Figure P12.8

12.9 A force F pulls a body along a horizontal floor. The force varies with distance x, as shown in Fig. P12.9. Calculate the work done by the force in moving the body from $x = 0$ to $x = 8$ m.

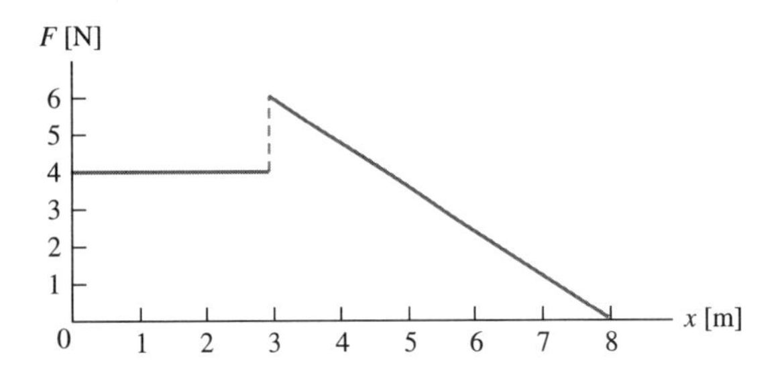

Figure P12.9

12.10 A crate that weighs 30 N slides down an inclined ramp that forms an angle of 50° with the horizontal (see Fig. P12.10). The coefficient of sliding friction is 0.50. Determine the work performed by the forces that act on the crate as it slides 3 m down the ramp.

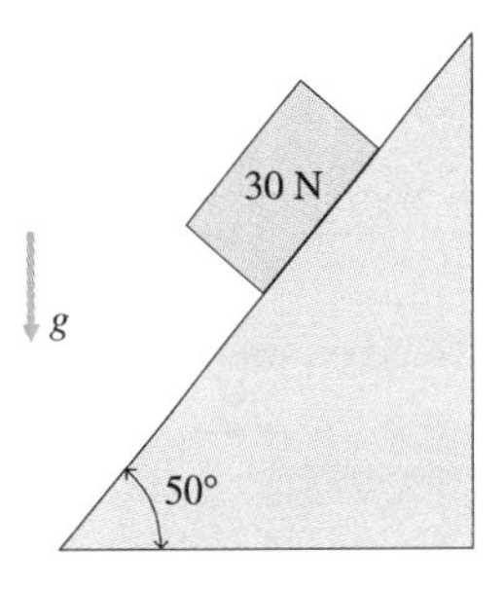

Figure P12.10

12.11 Figure P12.11 represents two identical bar magnets that rest on a wooden laboratory bench. The force of attraction (in newtons) between the poles is $F = 22\,500/x^2$, where x is the distance (in millimeters) between the poles. The poles are 6 mm from the ends of the magnets. Each magnet weighs 9 N. The coefficients of static and kinetic friction between magnet A and the bench are both 0.40. Magnet B is cemented to the bench. Magnet A is pushed slowly toward magnet B until it suddenly slides because of the force of attraction F. Calculate the work (in newton-millimeters) performed on magnet A by magnet B during sliding.

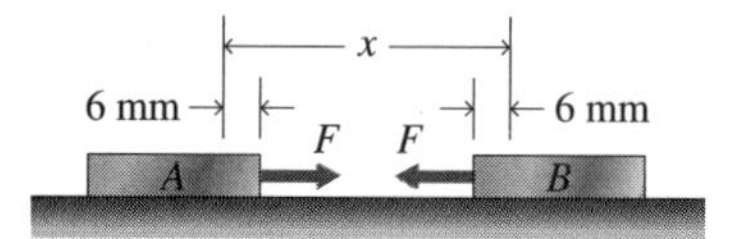

Figure P12.11

12.12 Three crates that rest on a horizontal floor are connected by light cords (see Fig. P12.12). They move slowly to the right under the action of a force $F = 18$ N. The coefficient of kinetic friction between the crates and the floor is 0.30.

a. Determine the work done on each of the crates as the force pulls the crates 4 m to the right.

b. Determine the total work performed on the crates.

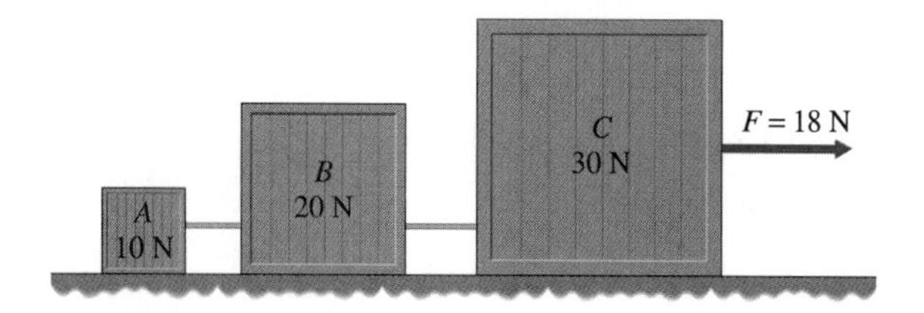

Figure P12.12

12.2 *Work as a Line Integral*

12.13 A constant force $\mathbf{F} = 3\mathbf{i} + 2\mathbf{j} + 3\mathbf{k}$ acts on a particle, where $\mathbf{i}$, $\mathbf{j}$, and $\mathbf{k}$ are unit vectors along the x, y, and z axes. Determine the work performed by the force (in pounds) as the particle undergoes a displacement (in feet) from the origin (0, 0, 0) of the xyz coordinate system to the point (0, 2, 4).

12.14 A force is given as $\mathbf{F} = -4t\mathbf{i} + 5\mathbf{j}$, where t denotes time in seconds and $\mathbf{i}$ and $\mathbf{j}$ are unit vectors in the x and y directions. The force acts on a particle that moves in the xy plane such that its coordinates are $x = 2.0 - 0.5t^3$ and $y = 1.6t^2$, in meters. Determine the work done by the force on the particle during the time interval $0 \leq t \leq 3$.

12.15 A particle P moves on the path $x = t$, $y = t^2$, $z = t^3$, where t is a parameter. The motion is resisted by a force $\mathbf{F}$, defined by its (x, y, z) projections $F_x = -3\,dx/dt$, $F_y = -3\,dy/dt$, $F_z = -3\,dz/dt$ (length in meters and force in newtons). Calculate the work that the resisting force performs on P during the interval from $t = 1$ to $t = 3$.

12.16 A block, whose mass is 6 kg, rests on a frictionless horizontal surface (see Fig. P12.16). In this equilibrium position, the block is attached to an unstretched nonlinear spring whose force-extension relation is $F = 400 \sin 10x$ [N], where x is the extension of the spring from its unstretched length. The block is moved a distance $x = 0.10$ m to the right and released.

a. Determine the work done by the spring on the block during the time it takes for the block to move back to its equilibrium position.

b. Determine the work done by the spring on the block as the block moves from $x = 0$ to $x = -0.10$ m.

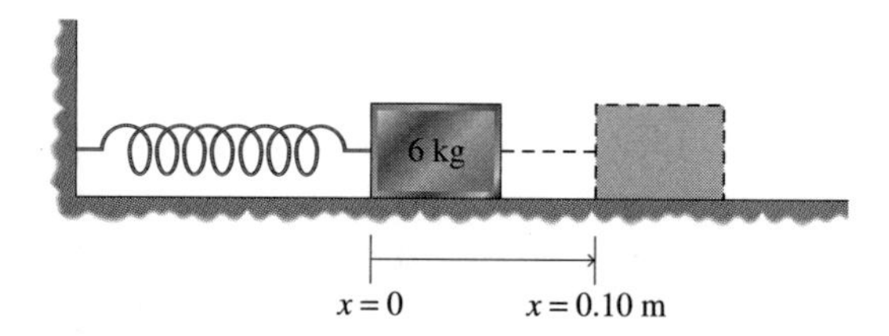

Figure P12.16

12.17 In a mechanism, a small part that weighs 50 N slides on a circular rod under the action of gravity (see Fig. P12.17). Its motion is resisted by a frictional force $F = 18\cos\theta$ (also in newtons). Determine the work of all the forces that act on the part as θ increases from 30° to 90°.

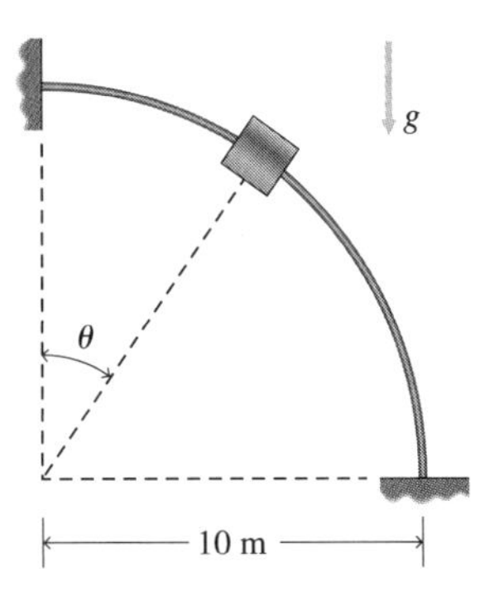

Figure P12.17

12.18 A particle moves on a circle under the action of a force **F**, defined by $F_x = -ky$, $F_y = kx$, where k is a constant and x and y are the coordinates of the particle. Show that, if the particle travels once around the circle in the counterclockwise sense, the work done by the force **F** is $U = 2kA$, where A is the area of the circle.

12.19 A particle P that moves in the xy plane is attracted to the origin by a force of magnitude $F = k/x$ (see Fig. P12.19).

a. Determine the work done by the force as the particle moves from point (1, 0) to point (2, 1) along the straight line from the point (1, 0) to the point (1, 1), and then along the straight line from the point (1, 1) to the point (2, 1).

b. Determine the work done by the force as the particle moves from point (1, 0) to point (2, 1) along the straight line from point (1, 0) to point (2, 0), and then along the straight line from point (2, 0) to point (2, 1).

c. Does the work done by the force depend on the path between points (1, 0) and (2, 1)?

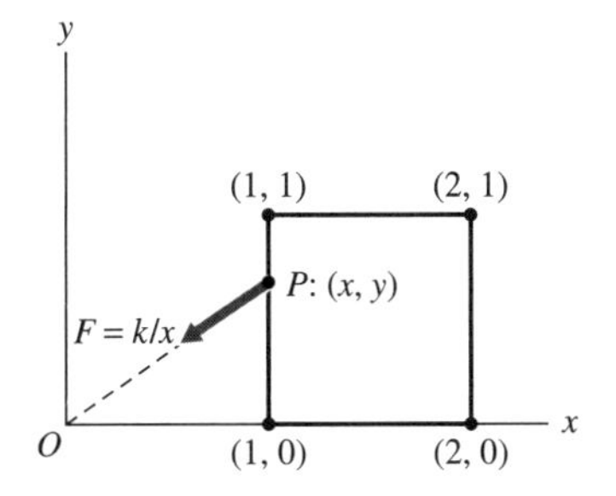

Figure P12.19

12.20 A particle P is constrained to move on a circular path (see Fig. P12.20). The particle is attracted toward point A by a force F proportional to its distance from A; that is, $F = kr$. At point B, the magnitude of the force is F_B. Determine the work done on the particle, in terms of F_B and R, by the force F as the particle moves from point A to point B.

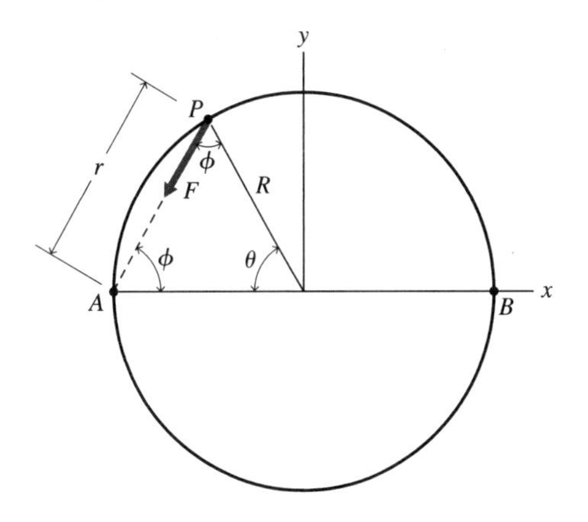

Figure P12.20

12.21 A small part P in a mechanism is constrained to move on a circular path of radius r (see Fig. P12.21). It is acted on by a driving force $F = as + b$, where a and b are constants and s is the arc length along the circular path. The force F remains tangent to the circular path at all times. The weight of the part is W. Determine the total work performed on the part, in terms of a, b, and W, as it moves from point A to point B.

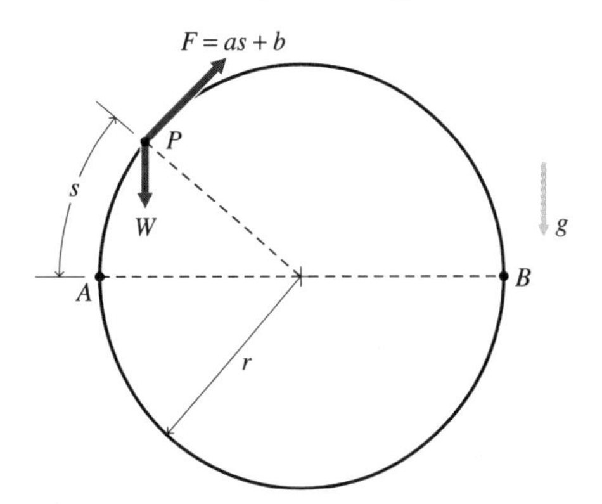

Figure P12.21

12.3 *Work Performed by Gravity*

12.22 The particle B shown in Fig. P12.22 weighs 10 N. It slides along the frictionless bar. Determine the work required to raise the particle from point O to point A.

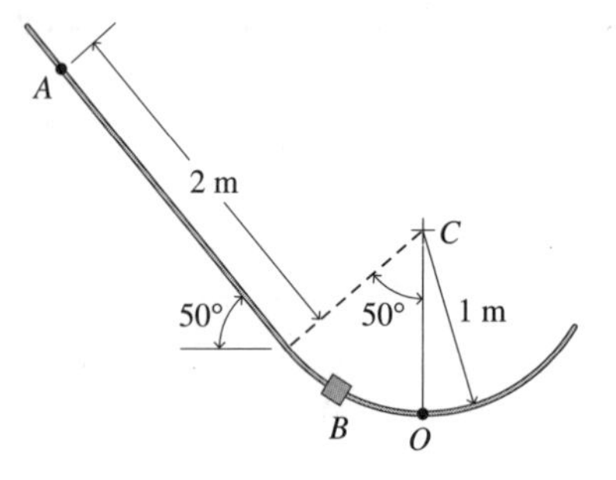

Figure P12.22

12.23 A cubical crate is set on edge at a 30° angle, as shown in Fig. P12.23, and is then released. The crate weighs 1260 N. Calculate the work performed by the force of gravity from the time the crate is released until its face strikes the floor.

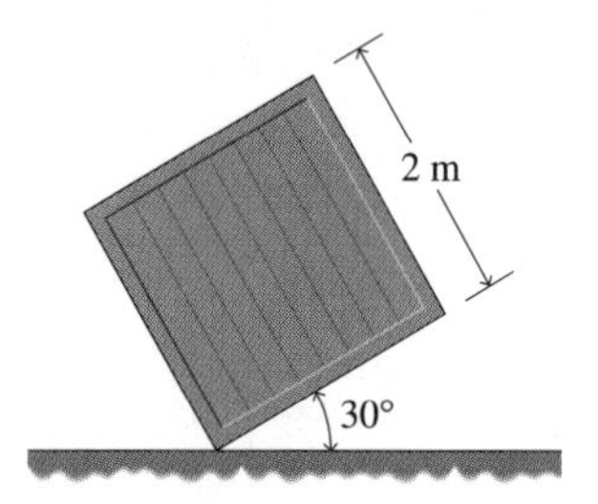

Figure P12.23

12.24 Two water tanks are connected by a pipe (see Fig. P12.24). Tank A has a horizontal cross section 4 m wide and 6 m long (the dimension perpendicular to the plane of the page). It is filled with water to a height of 4 m. Tank B is empty. It has a horizontal cross section 5 m wide and 6 m long. The valve C is opened to allow water to flow from tank A to tank B. Determine the work done by a gravity from the time the valve is opened to the time the water reaches the same level in both tanks. Neglect the small volume of water contained in the pipe.

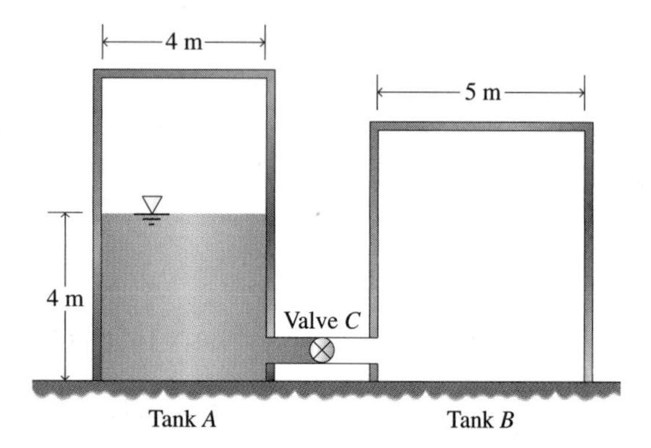

Figure P12.24

12.25 Tank B in Problem 12.24 is lowered 1 m. Solve the problem for this condition.

12.26 The uniform bar AB shown in Fig. P12.26 weighs 40 lb. It is hinged at A. The disk weighs 18 lb. Compute the work performed by the force of gravity as the bar and disk drop from the position shown to the position in which the bar is horizontal.

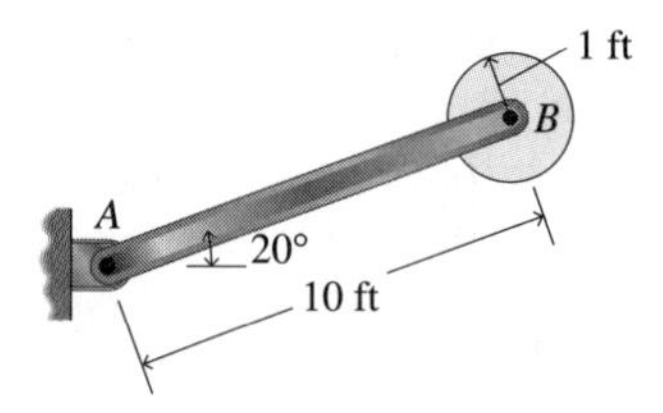

Figure P12.26

12.27 One end of a spring is attached to the body B in Fig. P12.22, and the other end is fixed at point C. The undeformed length of the spring is 900 mm. The spring constant is $k = 175$ N/m. Determine the work that must be done to lift the body from point O to point A.

12.28 A quantity of mercury weighing 270 N occupies a bent tube (see Fig. P12.28). The mercury is released from the position shown and begins to flow out the bottom of the tube. Calculate the work performed by the force of gravity from the time the mercury is released until the right end passes point B.

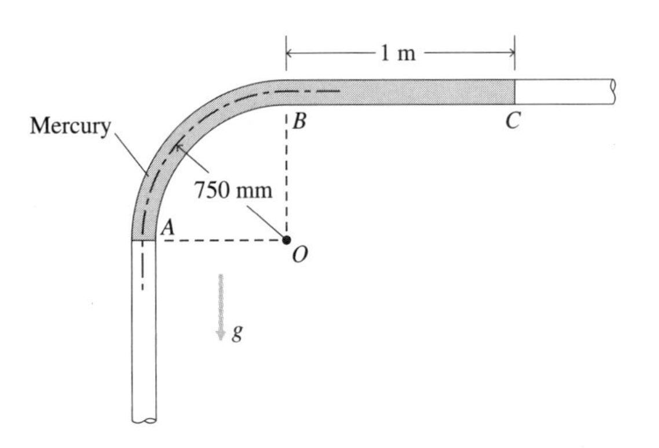

Figure P12.28

12.29 Water weighing 62.4 lb/ft^3 is pumped from a cylindrical well with a diameter of 20 ft into a cylindrical tank of diameter 30 ft. The bottom of the tank is originally 80 ft above the surface of the water in the well. Calculate the total work performed against gravity if 10,000 ft^3 of water are lifted. Assume that the well does not recharge (refill) during the pumping.

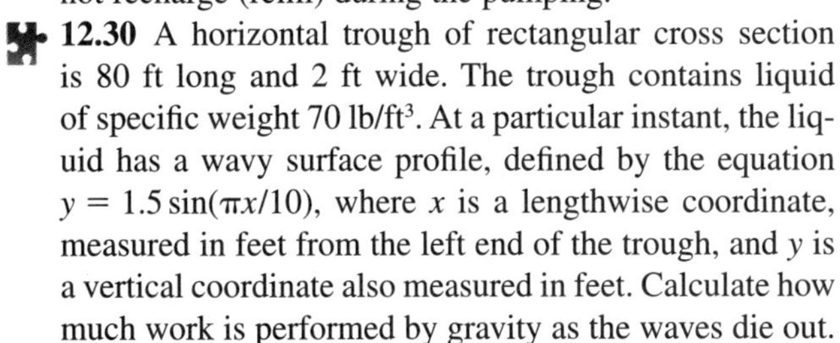

12.30 A horizontal trough of rectangular cross section is 80 ft long and 2 ft wide. The trough contains liquid of specific weight 70 lb/ft^3. At a particular instant, the liquid has a wavy surface profile, defined by the equation $y = 1.5 \sin(\pi x/10)$, where x is a lengthwise coordinate, measured in feet from the left end of the trough, and y is a vertical coordinate also measured in feet. Calculate how much work is performed by gravity as the waves die out.

12.4 *Work Performed by a Couple*

12.31 The cube shown in Fig. P12.31 rotates clockwise 45° in the xy plane when it is displaced from position A to position B. Calculate the work performed by the couple M during this displacement.

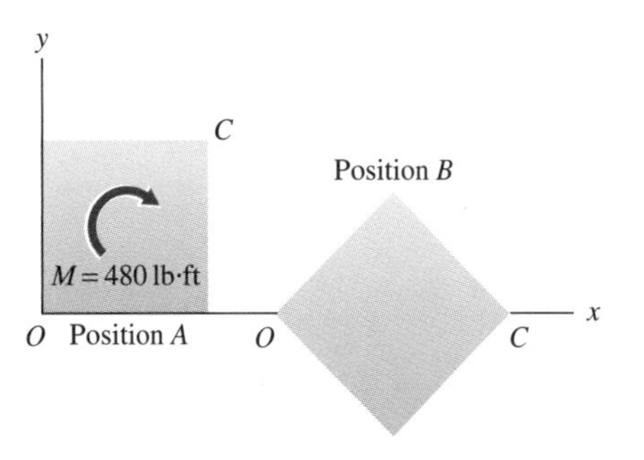

Figure P12.31

12.32 A torque with magnitude $M = a\theta^2 + b\theta$ is applied to a rotating shaft, where θ is the angular displacement of the shaft and a and b are constants. The sense of the torque is the same as the positive sense of θ. Express the work U performed by the torque as a function of the initial angle θ_0 and the final angle θ_1.

12.33 A couple is defined by the relation $\mathbf{M} = 5t\mathbf{i} + 16t^2\mathbf{j} + 12\mathbf{k}$, where t is a parameter. The couple acts on a rigid body that undergoes a rotation $\mathbf{\Theta} = e^{-t}\mathbf{i} + (1/t)\mathbf{j} + t^{1/2}\mathbf{k}$. Determine the work performed on the body by the couple over the range $0 \leq t \leq 10$.

12.5 *Principle of Virtual Work*

12.34 The crank ABC in Fig. P12.34 is free to rotate about the frictionless pin B. Determine the magnitude P of the horizontal force at C required to hold the crank in equilibrium.

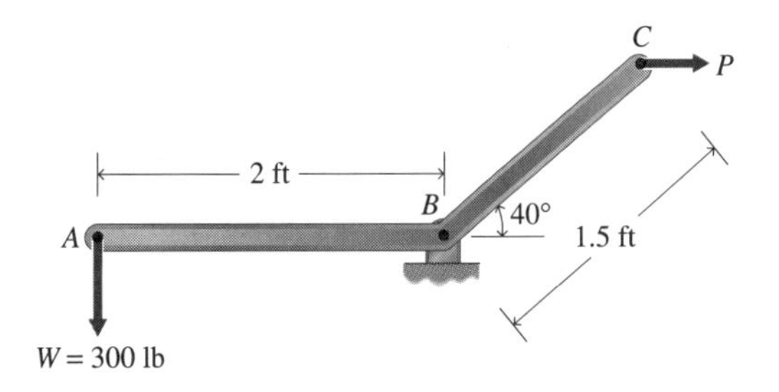

Figure P12.34

12.35 a. Apply the principle of virtual work to derive a formula in terms of *a, b,* and *W* for the magnitude *P* of the force required for equilibrium of the bell crank *ABC*, shown in Fig. P12.35. The pin at *B* is frictionless. Neglect the weight of the bell crank.
b. Check your result using the equation for equilibrium of moments.

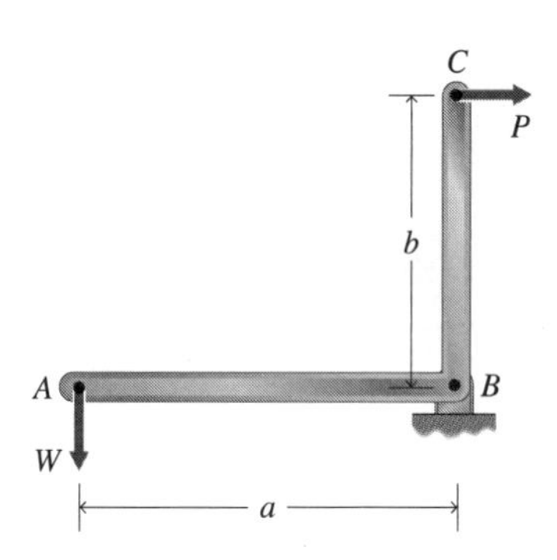

Figure P12.35

12.36 Figure P12.36 represents a pulley-crank mechanism that is used to raise the 400 lb weight. Using the principle of virtual work, determine the force *P*. Neglect the weight of the pulley.

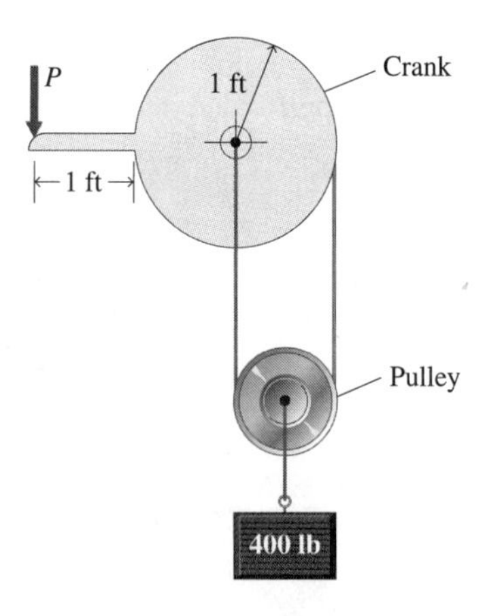

Figure P12.36

12.37 In Fig. P12.37, the weight of the block on the inclined plane is W_B. The system is held in equilibrium by the weight W_A. The system is frictionless. Determine the ratio W_A/W_B by the principle of virtual work. Check your result using the condition of equilibrium of forces.

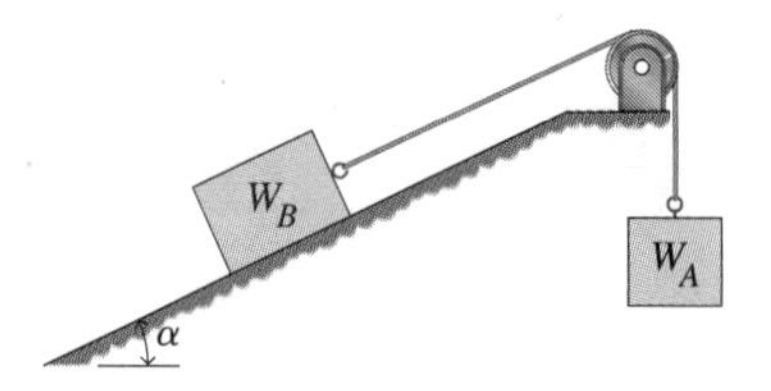

Figure P12.37

12.38 Using the principle of virtual work, determine the relationship between the forces *P* and *Q*, in terms of *a, b, c,* and *d,* for the frictionless mechanism that is in equilibrium in the position shown in Fig. P12.38. Neglect the weights of the bars.

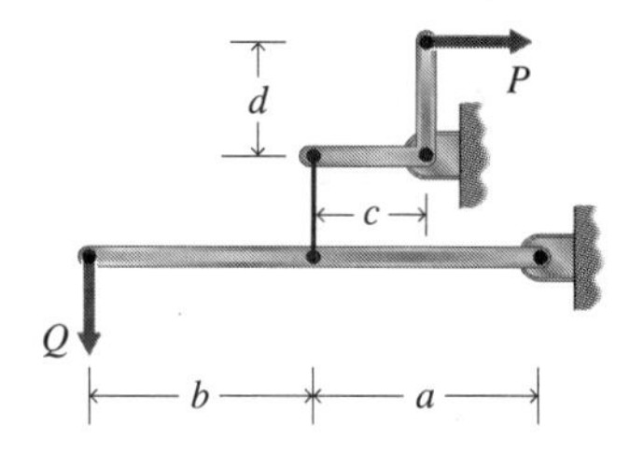

Figure P12.38

12.39 a. Use the principle of virtual work to determine the magnitude *P* of the force required to hold the linkage *ABCD* in equilibrium in the position shown in Fig. P12.39. The joints *A, B, C,* and *D* are frictionless. Neglect the weight of the bars of the frame.
b. Plot *P* versus θ, for the range $0 \leq \theta \leq \pi/2$.

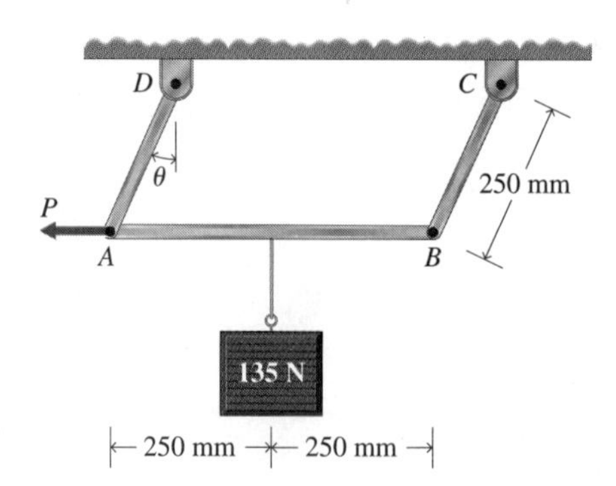

Figure P12.39

12.40 Rotational springs are inserted at joints D and C of the linkage in Fig. P12.39. These springs exert moments $M = -4.2\theta$ (in N·m) on bars AD and BC for any rotation θ. By the principle of virtual work, determine the magnitude P of the force required to hold the linkage at an angle $\theta = 30°$.

12.41 A rotational spring is inserted at joint B of the bell crank shown in Fig. P12.35. The moment on the bell crank due to the spring is $M = -k\theta$, where k is the rotational spring constant with dimensions $[FL]$, and θ is the rotation of the bell crank in radians.

a. Derive a formula, in terms of W, a, b, k, and θ, for the magnitude P of the force required for maintaining equilibrium. (*Hint:* Express U, the work done during the rotation, as a function of θ. Then, form δU by differentiation of U for a virtual rotation $\delta\theta$; that is, $\delta U = \delta U/\delta\theta \; \delta\theta$, where $\delta U/\delta\theta$ is the derivative of U with respect to θ.)

b. Check your formula using the condition of equilibrium of moments about B.

c. For $W = 500$ N, $a = 600$ mm, $b = 500$ mm, $k = 90$ N·m, calculate P for $\theta = 0°$, $\theta = 30°$, and $\theta = 45°$.

12.6 *Stable, Unstable, and Neutral Equilibrium*

12.42 The steel body in Fig. P12.42 consists of a solid rectangular prism welded to a solid semicylindrical roller. The roller rests on a horizontal surface. Friction prevents sliding. Prove that the upright position shown is unstable if $b/h < \sqrt{6}$. Let the prism undergo a clockwise rotation through an angle θ. See the hint in Problem 12.41.

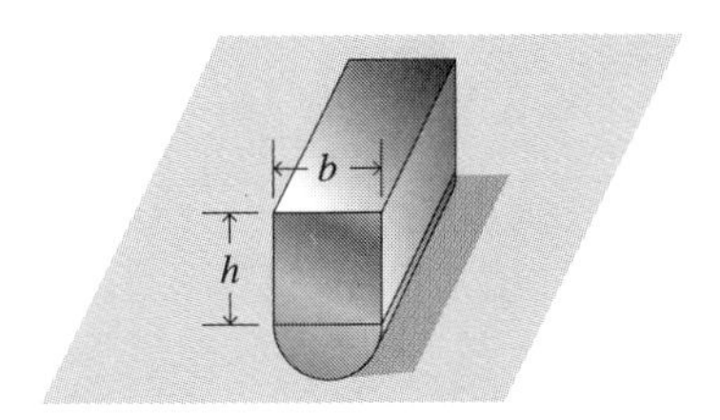

Figure P12.42

12.43 The solid homogeneous body in Fig. P12.43 consists of a cylinder of height h cemented to a hemispherical base of radius r. The body rests on a table in an upright position. Friction prevents sliding. Determine the range of the ratio h/r for which the configuration is stable. Let the prism undergo a clockwise rotation through an angle θ. See the hint in Problem 12.41.

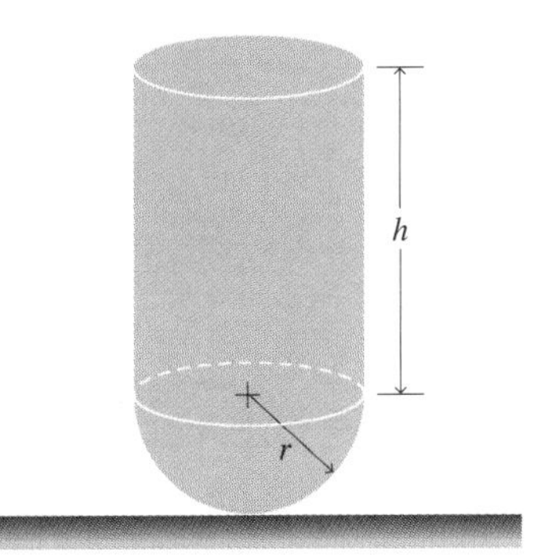

Figure P12.43

12.44 Two uniform slender rods of equal length are hinged together (see Fig. P12.44). They rest against a 45° inclined plane and are constrained to remain in a vertical plane (in the plane of the page).

a. Neglecting friction, determine the smallest nonzero angle θ for which equilibrium is possible.

b. Determine whether or not the equilibrium state is stable for this configuration. (See the hint in Problem 12.41.)

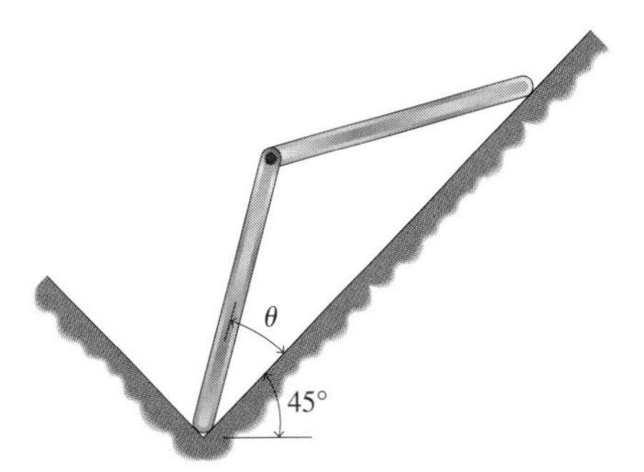

Figure P12.44

12.45 The rectangular linkage shown in Fig. P12.45 consists of rigid bars that are hinged at the ends. The hinges contain springs that produce a restoring moment of $k\theta$ when the two legs of a hinge experience a relative angular displacement θ. The forces of magnitude F remain directed along the diagonals of the parallelogram when the linkage is deformed. Show that the critical (buckling) load is $F_{cr} = 2kc^3a^{-2}b^{-2}$, where $c^2 = a^2 + b^2$. (See the hint in Problem 12.41.)

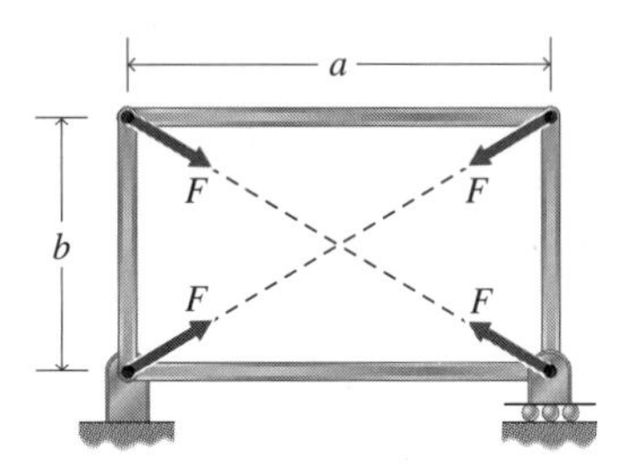

Figure P12.45

12.46 Show that, if the forces F in Problem 12.45 remain parallel to their original directions when the linkage buckles and if the bottom member is fixed, the buckling load is $F_{cr} = 2kcb^{-2}$. (See the hint in Problem 12.41.)

REVIEW QUESTIONS

12.47 Define the work of a force that maintains a constant direction. How is it represented on a graph of force versus displacement?

12.48 Express the work of a variable force as a line integral. Express the integrand in terms of the (x, y, z) projections of the force. Explain how the integral may be evaluated by expressing the integrand as a function of a parameter t.

12.49 A force that acts on a moving particle remains directed toward a fixed point. The magnitude of the force is constant. Give an algebraic formula for the work that the force performs on the particle.

12.50 State a theorem that determines the work performed by the force of gravity on a mechanical system.

12.51 State a theorem that determines the work performed by a variable couple acting on a rigid body that undergoes a displacement parallel to the plane of the couple.

12.52 State and derive the principle of virtual work in the form of Fourier's inequality. Explain the application of this principle to a book that rests on a table.

12.53 Derive the equilibrium equation for the lever by means of the principle of virtual work.

12.54 Derive the equilibrium conditions for Roberval's balance by means of the principle of virtual work.

12.55 State the criterion for stable equilibrium, in terms of virtual work. Hence, show that a book that rests on a table is in stable equilibrium.

Appendix A ALGEBRAIC EQUATIONS AND DETERMINANTS

A.1 ALGEBRAIC EQUATIONS

REDUNDANT ALGEBRAIC EQUATIONS

Several linear equations containing the same unknowns do not have a unique solution unless the equations are independent. For example, consider the equations

$$x - 2y = 4 \tag{a}$$

$$3x - 6y = 12 \tag{b}$$

redundant equation

Equation (b) is obtained by multiplying Eq. (a) by 3. Therefore, these really represent only one equation. Since Eq. (b) provides no new information, it is said to be *redundant.*

A less obvious example of redundancy is provided by the following three equations:

$$2x - y + z = 5 \tag{c}$$

$$x + 3y - 2z = 4 \tag{d}$$

$$7x - 14y + 11z = 13 \tag{e}$$

Equation (e) is redundant, but a cursory inspection does not reveal this fact. Equation (e) is obtained by multiplying Eq. (c) by 5 and Eq. (d) by 3 and taking the difference of the resulting equations. The general solution of the equations, obtained by simply ignoring Eq. (e), is

$$x = \frac{19}{7} - \frac{z}{7}$$

$$y = \frac{5z}{7} + \frac{3}{7}$$

The variable z may be assigned any arbitrary value.

In problems involving equilibrium of rigid bodies, redundant equations are obtained if moments are taken about too many points. For rigid bodies that are subject to coplanar forces, there are only three independent equations of equilibrium—for example, Eqs. (4.5). Although more than three equations can be derived (for example, by balancing moments about various points), any such equations are necessarily redundant.

CONSISTENCY OF EQUATIONS

Two or more linear algebraic equations can be inconsistent; that is, they may have no solution. For example, consider the equations

$$x + y + z = 2 \tag{f}$$

$$2x + 2y + 2z = 5 \tag{g}$$

If we multiply Eq. (f) by 2 and subtract it from Eq. (g), we obtain $0 = 1$. This absurd conclusion signifies that the equations are inconsistent.

A less obvious example of inconsistency is provided by the equations

$$3x + y + 2z = 5 \tag{h}$$

$$x + 2y + 4z = 2 \tag{i}$$

$$x - y - 2z = 0 \tag{j}$$

Adding Eq. (h) to Eq. (j), we obtain $x = \frac{5}{4}$. Substituting this result into Eqs. (h) and (i), we obtain $y + 2z = \frac{5}{4}$ and $2y + 4z = \frac{3}{4}$. Dividing the latter equation by 2 and subtracting the result from the first equation, we again obtain an absurd result: $0 = \frac{7}{8}$. Accordingly, Eqs. (h), (i), and (j) are inconsistent. There are systematic ways for testing equations for consistency, but these will not be covered here.

QUADRATIC EQUATIONS

The quadratic equation $ax^2 + bx + c = 0$ has two roots, x_1 and x_2, given by

$$x_1 = \frac{-b + \sqrt{b^2 - 4ac}}{2a}$$

$$x_2 = \frac{-b - \sqrt{b^2 - 4ac}}{2a}$$

The derivation of these roots may be found in any elementary algebra textbook. Depending on the values of the coefficients *a, b,* and *c,* the roots may be real or complex. In statics problems, complex roots are meaningless. Obtaining such results is usually an indication of an error in problem formulation or solution. In some cases, both roots are physically meaningful. However, this is not always the case. For some problems, even through both roots satisfy the mathematics of the equation, only one of the roots has a realistic relationship to the problem from which the equation was derived. The other root makes no sense.

A.2 DETERMINANTS

Determinants play important roles in several branches of applied mathematics, especially in the theory of linear algebra. This discussion is confined to two-rowed and three-rowed determinants, since this book does not use larger ones.

matrix

principal diagonal

square matrix

determinant

A group of numbers organized into rows and columns forming a rectangular array is called a *matrix.* A number in a matrix is denoted by a_{ij}, where i and j are the row and column locations, respectively. For instance, a_{23} occupies the position at row 2 and column 3 in the matrix. The *principal diagonal* of a matrix consists of the numbers that lie on the 45° line extending downward from the upper left corner. The principal diagonal contains the numbers $a_{11}, a_{22}, a_{33}, \ldots$ or, in general, a_{ii}, where $i = 1, 2, 3, \ldots$. A matrix is said to be *square* if the number of rows equals the number of columns.

A *determinant* Δ is a number that is derived from a square matrix by specific arithmetic processes. A determinant may be positive, negative, or zero. The determinant of a square matrix is represented by enclosing the numbers of the matrix between single vertical lines. For example, the determinant of a square matrix with two rows and two columns is defined as

$$\Delta = \begin{vmatrix} a_{11} & a_{12} \\ a_{21} & a_{22} \end{vmatrix} = a_{11}a_{22} - a_{12}a_{21}$$

Thus, the determinant of a square matrix with two rows and two columns is simply the product of the numbers on the principal diagonal minus the product of the other two numbers.

The determinant of a three-row-by-three-column matrix is defined as

$$\Delta = \begin{vmatrix} a_{11} & a_{12} & a_{13} \\ a_{21} & a_{22} & a_{23} \\ a_{31} & a_{32} & a_{33} \end{vmatrix} = a_{11}a_{22}a_{33} + a_{12}a_{23}a_{31} + a_{13}a_{21}a_{32} - a_{13}a_{22}a_{31} - a_{11}a_{23}a_{32} - a_{12}a_{21}a_{33}$$

The following technique is an aid for calculating the determinant Δ for a matrix with three rows and three columns. First, form the auxiliary matrix:

$$\begin{bmatrix} a_{11} & a_{12} & a_{13} & a_{11} & a_{12} \\ a_{21} & a_{22} & a_{23} & a_{21} & a_{22} \\ a_{31} & a_{32} & a_{33} & a_{31} & a_{32} \end{bmatrix}$$

The first three columns of the auxiliary matrix are identical to the corresponding columns of the original matrix. The fourth and fifth columns are duplications of the first and second columns. The determinant is formed by taking products of terms along diagonals of the matrix and then adding these products. The process is illustrated below.

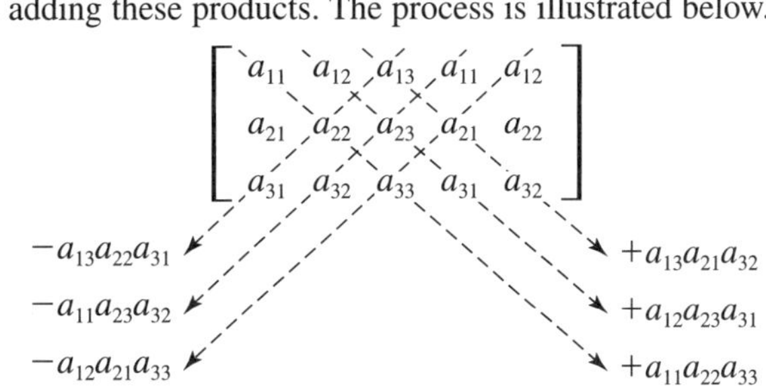

Note that products of terms that lie along diagonal lines parallel to the principal diagonal of the auxiliary matrix are *added* in forming the equation for Δ, whereas products of terms along diagonal lines transverse to the principal diagonal are *subtracted.* So, the determinant Δ of the original matrix is given by

$$\Delta = a_{11}a_{22}a_{33} + a_{12}a_{23}a_{31} + a_{13}a_{21}a_{32} - a_{13}a_{22}a_{31} - a_{11}a_{23}a_{32} - a_{12}a_{21}a_{33}$$

THEOREMS ON DETERMINANTS

A few theorems on determinants are stated below, without proofs. Proofs may be found in algebra texts.

1. The value of a determinant is unchanged if a row is interchanged with its corresponding column.
2. Only the sign of a determinant is changed if two rows (or two columns) are interchanged.
3. The value of a determinant is zero if all the numbers in a row (or column) are zero or if two rows (or two columns) are identical.
4. $\begin{vmatrix} a_{11} & a_{12} & a_{13} \\ a_{21} & a_{22} & a_{23} \\ a_{31} & a_{32} & a_{33} \end{vmatrix} \pm \begin{vmatrix} a_{11} & a_{12} & a_{13} \\ a_{21} & a_{22} & a_{23} \\ b_{31} & b_{32} & b_{33} \end{vmatrix} = \begin{vmatrix} a_{11} & a_{12} & a_{13} \\ a_{21} & a_{22} & a_{23} \\ a_{31} \pm b_{31} & a_{32} \pm b_{32} & a_{33} \pm b_{33} \end{vmatrix}$

The last theorem is not restricted to the case in which the differing entries occur in the last row; it is valid if they occur in any two corresponding rows or columns.

CRAMER'S RULE

Consider these simultaneous equations:

$$a_{11}x + a_{12}y + a_{13}z = b_1$$

$$a_{21}x + a_{22}y + a_{23}z = b_2$$

$$a_{31}x + a_{32}y + a_{33}z = b_3$$

We define the determinant Δ as follows:

$$\Delta = \begin{vmatrix} a_{11} & a_{12} & a_{13} \\ a_{21} & a_{22} & a_{23} \\ a_{31} & a_{32} & a_{33} \end{vmatrix}$$

If Δ is not zero, the equations have one, and only one, solution, given by

$$x = \frac{1}{\Delta}\begin{vmatrix} b_1 & a_{12} & a_{13} \\ b_2 & a_{22} & a_{23} \\ b_3 & a_{32} & a_{33} \end{vmatrix}, \qquad y = \frac{1}{\Delta}\begin{vmatrix} a_{11} & b_1 & a_{13} \\ a_{21} & b_2 & a_{23} \\ a_{31} & b_3 & a_{33} \end{vmatrix}, \qquad z = \frac{1}{\Delta}\begin{vmatrix} a_{11} & a_{12} & b_1 \\ a_{21} & a_{22} & b_2 \\ a_{31} & a_{32} & b_3 \end{vmatrix}$$

Cramer's rule

These relations are known as *Cramer's rule.* Cramer's rule is valid for n equations in n unknowns, although the foregoing definition of a determinant must be generalized if $n > 3$. As a practical device for solving equations, Cramer's rule is useful only for equations with two or three unknowns. For more unknowns, evaluation of the determinant is computationally inefficient (and possibly inaccurate) relative to alternative procedures for solution of simultaneous equations.

Appendix B

GEOMETRIC, TRIGONOMETRIC, AND HYPERBOLIC RELATIONS

As an aid for problem solving, the following geometric, trigonometric, and hyperbolic relations are provided without discussion. Their definitions and derivations are available in most trigonometry and calculus textbooks.

B.1 GEOMETRIC RELATIONS

With reference to the right triangle in Fig. B.1, the following relations hold:

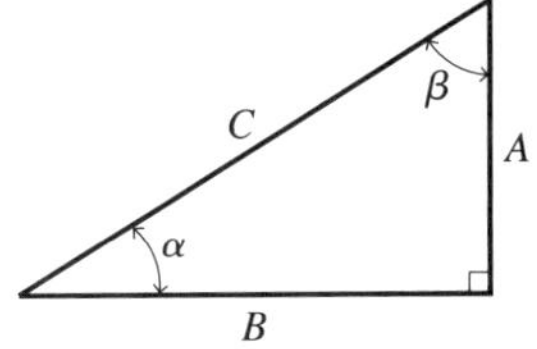

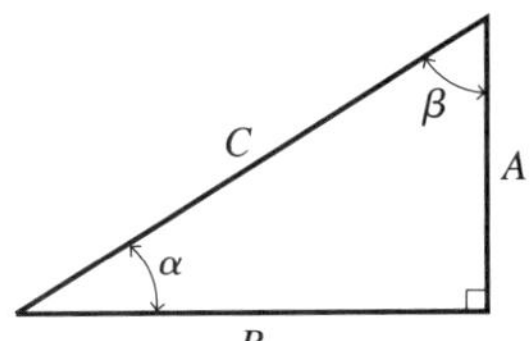

Figure B.1

$$A^2 + B^2 = C^2 \tag{B.1}$$

$$\sin\alpha = \frac{A}{C} \tag{B.2}$$

$$\cos\alpha = \frac{B}{C} \tag{B.3}$$

$$\tan\alpha = \frac{A}{B} = \frac{\sin\alpha}{\cos\alpha} \tag{B.4}$$

$$\sin^2\alpha + \cos^2\alpha = 1.0 \tag{B.5}$$

Similar relations hold for angle β.

With reference to the general plane triangle in Fig. B.2, the following relations hold:

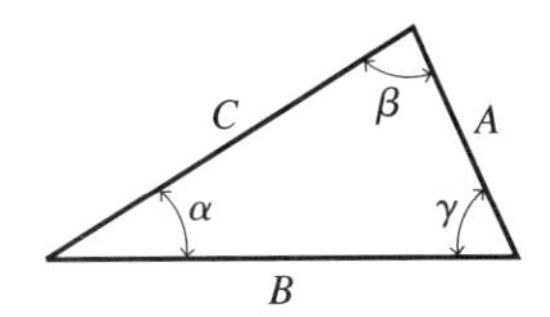

Figure B.2

$$\text{Law of sines: } \frac{\sin\alpha}{A} = \frac{\sin\beta}{B} = \frac{\sin\gamma}{C} \tag{B.6}$$

$$\text{Law of cosines: } A^2 + B^2 - 2AB\cos\gamma = C^2 \tag{B.7}$$

$$\alpha + \beta + \gamma = \pi \text{ [rad]} \tag{B.8}$$

B.2 TRIGONOMETRIC RELATIONS

The following fundamental trigonometric identities are useful in the solution of statics and dynamics problems.

$$\sin(-\alpha) = -\sin\alpha \tag{B.9}$$

$$\cos(-\alpha) = \cos\alpha \tag{B.10}$$

$$\sin\alpha = \pm\sqrt{1 - \cos^2\alpha} \tag{B.11}$$

$$\cos\alpha = \pm\sqrt{1 - \sin^2\alpha} \tag{B.12}$$

$$\sin(\alpha \pm \beta) = \sin\alpha\cos\beta \pm \cos\alpha\sin\beta \tag{B.13}$$

$$\sin(2\alpha) = 2\sin\alpha\cos\alpha \tag{B.14}$$

$$\sin\left(\pm\alpha + \frac{\pi}{2}\right) = \cos\alpha \tag{B.15}$$

$$\cos(\alpha \pm \beta) = \cos\alpha\cos\beta \mp \sin\alpha\sin\beta \tag{B.16}$$

$$\cos(2\alpha) = \cos^2\alpha - \sin^2\alpha = 2\cos^2\alpha - 1 = 1 - 2\sin^2\alpha \quad \text{(B.17)}$$

$$\cos\left(\pm\alpha + \frac{\pi}{2}\right) = \mp\sin\alpha \quad \text{(B.18)}$$

$$\tan(\alpha \pm \beta) = \frac{\tan\alpha \pm \tan\beta}{1 \mp \tan\alpha\tan\beta} \quad \text{(B.19)}$$

$$\tan(2\alpha) = \frac{2\tan\alpha}{1 - \tan^2\alpha} \quad \text{(B.20)}$$

$$\sin\left(\tfrac{1}{2}\alpha\right) = \pm\sqrt{\tfrac{1}{2}(1 - \cos\alpha)} \quad \text{(B.21)}$$

$$\cos\left(\tfrac{1}{2}\alpha\right) = \pm\sqrt{\tfrac{1}{2}(1 + \cos\alpha)} \quad \text{(B.22)}$$

$$\tan\left(\tfrac{1}{2}\alpha\right) = \frac{\sin\alpha}{1 + \cos\alpha} \quad \text{(B.23)}$$

$$\sin\alpha \pm \sin\beta = 2\sin\left[\tfrac{1}{2}(\alpha \pm \beta)\right]\cos\left[\tfrac{1}{2}(\alpha \mp \beta)\right] \quad \text{(B.24)}$$

$$\cos\alpha + \cos\beta = 2\cos\left[\tfrac{1}{2}(\alpha + \beta)\right]\cos\left[\tfrac{1}{2}(\alpha - \beta)\right] \quad \text{(B.25)}$$

$$\cos\alpha - \cos\beta = -2\sin\left[\tfrac{1}{2}(\alpha + \beta)\right]\sin\left[\tfrac{1}{2}(\alpha - \beta)\right] \quad \text{(B.26)}$$

$$\tan\alpha \pm \tan\beta = \frac{\sin(\alpha \pm \beta)}{(\cos\alpha)(\cos\beta)} \quad \text{(B.27)}$$

$$\sin^2\alpha - \sin^2\beta = \sin(\alpha + \beta)\sin(\alpha - \beta) \quad \text{(B.28)}$$

$$\cos^2\alpha - \cos^2\beta = -\sin(\alpha + \beta)\sin(\alpha - \beta) \quad \text{(B.29)}$$

$$\cos^2\alpha - \sin^2\beta = \cos(\alpha + \beta)\cos(\alpha - \beta) \quad \text{(B.30)}$$

$$\frac{d(\sin a\theta)}{d\theta} = a\cos a\theta \quad \text{(B.31)}$$

$$\frac{d(\cos a\theta)}{d\theta} = -a\sin a\theta \quad \text{(B.32)}$$

$$\int (\sin a\theta)\,d\theta = -\frac{\cos a\theta}{a} + C \quad \text{(B.33)}$$

$$\int (\cos a\theta)\,d\theta = \frac{\sin a\theta}{a} + C \quad \text{(B.34)}$$

B.3 HYPERBOLIC FUNCTIONS

hyperbolic functions

The functions $(e^x + e^{-x})/2$ and $(e^x - e^{-x})/2$ occur frequently in science and engineering. Since they can be described in terms of rectangular projections of points on a hyperbola, they are called *hyperbolic functions.* Hyperbolic functions share similarities with trigonometric relations and identities. The basic hyperbolic functions are the *hyperbolic cosine* (cosh) and the *hyperbolic sine* (sinh). Some hyperbolic functions, relations, and identities are listed here.

$$\cosh x = \frac{e^x + e^{-x}}{2} \qquad \sinh x = \frac{e^x - e^{-x}}{2} \tag{B.35}$$

$$e^x = \cosh x + \sinh x \qquad e^{-x} = \cosh x - \sinh x \tag{B.36}$$

$$\frac{d(\sinh x)}{dx} = \frac{e^x + e^{-x}}{2} = \cosh x \qquad \frac{d(\cosh x)}{dx} = \frac{e^x - e^{-x}}{2} = \sinh x \tag{B.37}$$

$$\int (\sinh x)\, dx = \cosh x \qquad \int (\cosh x)\, dx = \sinh x \tag{B.38}$$

$$\operatorname{sech} x = \frac{1}{\cosh x} = \frac{2}{e^x + e^{-x}} \qquad \operatorname{csch} x = \frac{1}{\sinh x} = \frac{2}{e^x - e^{-x}} \tag{B.39}$$

$$\tanh x = \frac{\sinh x}{\cosh x} = \frac{e^x - e^{-x}}{e^x + e^{-x}} \qquad \operatorname{ctnh} x = \frac{\cosh x}{\sinh x} = \frac{e^x + e^{-x}}{e^x - e^{-x}} \tag{B.40}$$

$$\cosh^2 x - \sinh^2 x = 1 \tag{B.41}$$

$$\cosh(-x) = \cosh x \tag{B.42}$$

$$\sinh(-x) = -\sinh x \tag{B.43}$$

$$\cosh(x \pm y) = \cosh x \cosh y \pm \sinh x \sinh y \tag{B.44}$$

$$\cosh 2x = \cosh^2 x + \sinh^2 x = 1 + 2\sinh^2 x = 2\cosh^2 x - 1 \tag{B.45}$$

$$\sinh(x \pm y) = \sinh x \cosh y \pm \cosh x \sinh y \tag{B.46}$$

$$\sinh 2x = 2 \sinh x \cosh x \tag{B.47}$$

$$(\cosh x + \sinh x)^n = \cosh nx + \sinh nx = e^{nx} \tag{B.48}$$

Appendix C AREA MOMENTS OF INERTIA

C.1 MOMENTS OF INERTIA BY INTEGRATION

Relationships between the internal force in a member, such as a beam or shaft, and the corresponding stress distribution over the member's cross section depend, in part, on certain geometric properties of the member's cross section. Consider a general cross-sectional area A of a member that lies in the xy plane (see Fig. C.1). The geometric properties referred to above include the *moments of inertia* of the cross-sectional area, defined as follows:

moments of inertia

$$I_x = \int y^2\, dA \tag{C.1}$$

$$I_y = \int x^2\, dA \tag{C.2}$$

$$J_O = \int r^2\, dA \tag{C.3}$$

$$I_{xy} = \int xy\, dA \tag{C.4}$$

where dA is an infinitesimal element of the cross section and the integral is evaluated over the area of the cross section. In Eq. (C.3), r is the radius from the origin O to the infinitesimal area (refer to Fig. C.1). Generally, r is written in terms of x and y as $r^2 = x^2 + y^2$. The terms I_x, I_y, J_O, and I_{xy} are often called the *second moments of area.* This name arises from the fact that the integrals involve second-order terms (x^2, y^2, r^2, and xy). Hence, these moments are analogous to the first moments Q_x and Q_y of area, which are integrals of y and x, respectively [see Eqs. (8.11)]. The term I_{xy} is also called the *product of inertia* of the area. The term J_O is called the *polar moment of inertia,* since it is defined in terms of r, a polar coordinate.

second moments of area

product of inertia

polar moment of inertia

Although it is physically more accurate to refer to I_x, I_y, J_O, and I_{xy} as second moments of area, they are commonly called moments of inertia (and are so called here), even though they are not inertial quantities. This practice has arisen because of the similarity of the area integrals in Eqs. (C.1) through (C.4) to the mass integrals that define mass moments of inertia in dynamics (to be discussed in Chapter 18).

The quantities I_x and I_y are used to determine bending stresses in beams. The quantity J_O is used in the torsional analysis of shafts. The product of inertia I_{xy} is also used in the analysis of beams subject to bending.

The moments of inertia defined above are used most commonly when the origin O of the xy coordinate axes coincides with the centroid of the member's cross section. Then the xy axes are centroidal axes of the area.

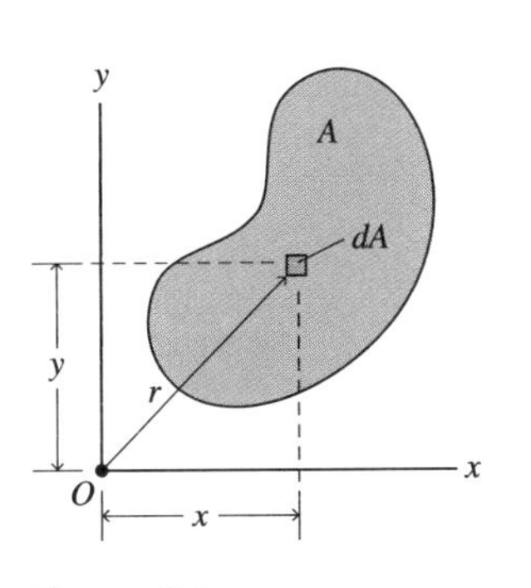

Figure C.1

Example C.1 Moments of Inertia of a Rectangle

Problem Statement Determine the moments of inertia I_x, I_y, J_O, and I_{xy} for the rectangular area in Fig. EC.1a, with the origin of the coordinate system at the centroid of the rectangle.

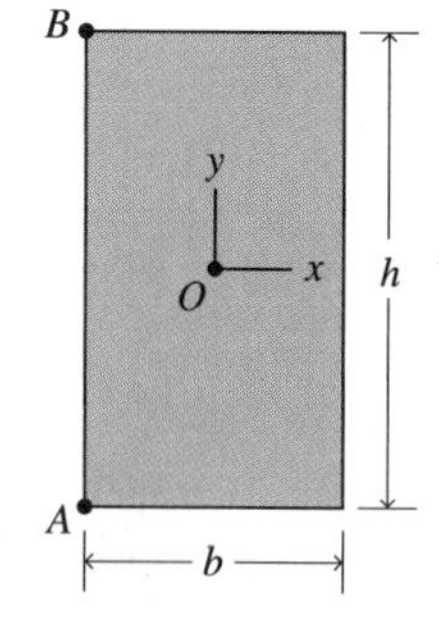

Figure EC.1a

Solution For I_x, we take $dA = b\,dy$ (see Fig. EC.1b). Then, from Eq. (C.1),

$$I_x = \int y^2\,dA = \int_{-h/2}^{h/2} y^2(b\,dy) = \frac{bh^3}{12}$$

To determine I_y, we take $dA = h\,dx$ (refer to Fig. EC.1b). Accordingly, from Eq. (C.2),

$$I_y = \int x^2\,dA = \int_{-b/2}^{b/2} x^2(h\,dx) = \frac{hb^3}{12}$$

For the polar moment of inertia J_O, we do not need to perform an additional integration, since we can write $r^2 = x^2 + y^2$. Then, from Eq. (C.3),

$$J_O = \int r^2\,dA = \int (x^2 + y^2)\,dA = \int x^2\,dA + \int y^2\,dA = I_y + I_x \qquad \text{(a)}$$

Hence, $J_O = hb^3/12 + bh^3/12 = (hb^3 + bh^3)/12$. Note that Eq. (a) applies for any area, not just a rectangle.

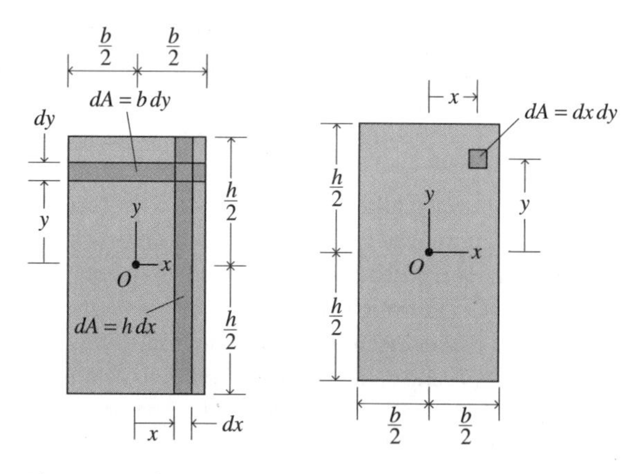

Figure EC.1b **Figure EC.1c**

Finally, for the product of inertia I_{xy}, we take $dA = dx\,dy$ (see Fig. EC.1c). Then, from Eq. (C.4),

$$I_{xy} = \int xy\,dA = \int_{-h/2}^{h/2} y\left(\int_{-b/2}^{b/2} x\,dx\right)dy$$

$$= \int_{-h/2}^{h/2} y\left(\frac{x^2}{2}\bigg|_{-b/2}^{b/2}\right)dy = 0$$

In this case, the xy axes were selected with the origin at the centroid of the rectangular area and directed along the axes of symmetry of the rectangle. For these axes, $I_{xy} = 0$. Axes for which the product of inertia I_{xy} is zero are known as *principal axes.* If the xy axes are not principal axes, I_{xy} will be nonzero. Also, for nonprincipal axes, I_{xy} may be either a positive or a negative number. For example, if the xy axes are taken parallel to the sides of the rectangle with origin at point A (refer to Fig. EC.1), then $I_{xy} = b^2h^2/4$. If the origin is taken at point B, then $I_{xy} = -b^2h^2/4$. The quantities I_x, I_y, and J_O are always positive.

In summary, for the rectangular area shown in Fig. EC.1a, the moments of inertia about the principal xy axes are

$$I_x = \frac{bh^3}{12}$$

$$I_y = \frac{hb^3}{12}$$

$$J_O = I_x + I_y = \frac{(hb^3 + bh^3)}{12}$$

$$I_{xy} = 0$$

C.2 PARALLEL AXIS THEOREM

Consider an area A for which the moments of inertia are known relative to centroidal xy axes with origin O. The moments of inertia relative to a second set of axes (x' and y') that are parallel to the xy axes and have origin O' are given by the *parallel axis theorem*.

parallel axis theorem

Theorem C.1

Parallel Axis Theorem

For a plane area A, the moment of inertia $I_{x'}$ about an axis x' that is parallel to the centroidal axis x of A equals the moment of inertia I_x plus the product $A\bar{y}^2$, where $\bar{y}$ is the coordinate distance from axis x to axis x'.

As stated, Theorem C.1 is for axes x and x'. However, it also applies for axes y and y'. Hence, it applies for $I_{y'}$, $J_{O'}$, and $I_{x'y'}$, as well as $I_{x'}$. The proof of Theorem C.1 follows.

Consider centroidal axes x and y and parallel axes x' and y' as shown in Fig. C.2. The $x'y'$ axes are located at distances $\bar{x}$ and $\bar{y}$ from the xy axes. Since $\bar{x}$ and $\bar{y}$ are measured relative to the xy axes, they are negative numbers in Fig. C.2. To prove Theorem C.1, we must show that $I_{x'} = I_x + A\bar{y}^2$. To begin, we use Eq. (C.1) to write an expression for $I_{x'}$ relative to $x'y'$ axes:

$$I_{x'} = \int (y')^2\, dA \qquad \text{(a)}$$

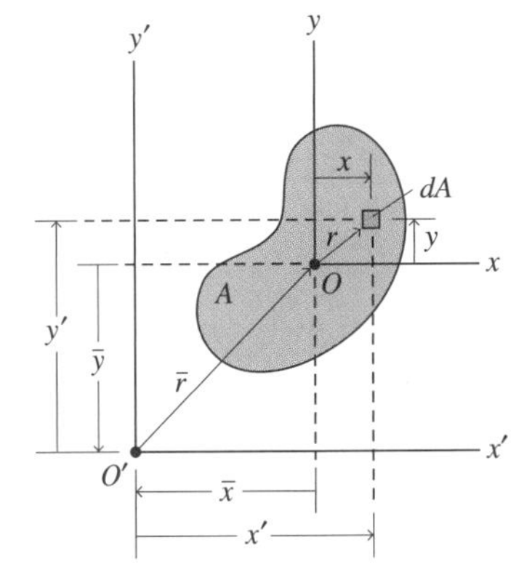

Figure C.2

From Fig. C.2, $y' = y - \bar{y}$. Substitution for y' into Eq. (a) yields

$$I_{x'} = \int (y - \bar{y})^2\, dA = \int y^2\, dA + \bar{y}^2 \int dA - 2\bar{y} \int y\, dA \qquad \text{(b)}$$

In Eq. (b), $\int y^2\, dA = I_x$. Also, $\bar{y}^2 \int dA = A\bar{y}^2$, since the integral of dA over the area is A. The y coordinate of the centroid of A is given by $y_O = \frac{1}{A}\int y\, dA$ [see Eq. (8.10b)]. Since the centroid of A is located at $x = 0$, $y = 0$, then $y_O = 0$. In other words, $\int y\,dA = 0$. Therefore, Eq. (b) reduces to

$$I_{x'} = \int (y')^2\,dA = I_x + A\bar{y}^2 \tag{C.5}$$

Steiner's theorem

Steiner component

Equation (C.5) verifies Theorem C.1. The parallel axis theorem is also known as *Steiner's theorem*, and the term $A\bar{y}^2$ is referred to as the *Steiner component.*

Similarly, it may be shown that

$$I_{y'} = \int (x')^2\,dA = I_y + A\bar{x}^2 \tag{C.6}$$

$$\begin{aligned} J_{O'} &= \int [(x')^2 + (y')^2]\,dA \\ &= I_y + I_x + A(\bar{x}^2 + \bar{y}^2) \\ &= J_O + A(\bar{r})^2 \end{aligned} \tag{C.7}$$

$$I_{x'y'} = \int (x'y')\,dA = I_{xy} + A\bar{x}\,\bar{y} \tag{C.8}$$

The parallel axis theorem allows you to express the moments of inertia of a plane area about $x'y'$ axes that are parallel to the centroidal xy axes of an area in terms of the moments of inertia relative to axes x and y. It does not permit you to relate moments of inertia about one set of noncentroidal axes to those about another. Nor does it allow you to relate moments of inertia about axes that are not parallel to each other.

The parallel axis theorem is particularly useful for finding the moments of inertia of a composite area (an area composed of two or more simple geometric shapes) based on the moments of inertia of its individual simple shapes. The following Problem-Solving Technique outlines the use of the theorem for this purpose.

PROBLEM-SOLVING TECHNIQUE

Moments of Inertia for a Composite Area

To determine the moments of inertia for a composite area:

1. Establish a convenient set of reference axes (X and Y) for the problem. Usually, the reference axes are located along the outer edges of the area.
2. Divide the area into simple component parts, and locate the centroid of each of the parts. Then, locate the centroid of the composite area. When possible, use Table D.2 to locate the centroid of the individual parts. Also, determine the area A_i of each individual part i.
3. For each individual part i, determine the moments of inertia I_{x_i}, I_{y_i}, I_{xy_i}, and J_{O_i} of the part relative to its own centroidal axes (x_i, y_i). Use the integrals presented in Sec. C.1 or the information in Table D.2 to perform this step.
4. For each individual part, determine the coordinate distances $\bar{x}_i$ and $\bar{y}_i$ from the centroid of the part to the centroid of the composite area. With these distances, the area of the part from Step 2, and the moments of inertia from Step 3, use Eqs. (C.5) through (C.8) to determine the contribution of the part to the moments of inertia of the composite area.
5. Sum the contributions of the individual parts to determine the moments of inertia for the composite area.

Example C.2

Moments of Inertia for a Composite Area

Problem Statement Determine the moments of inertia $I_{x'}$, $I_{y'}$, and $I_{x'y'}$, relative to the centroidal $x'y'$ axes for the L-shaped cross section of a beam shown in Fig. EC.2a. Take the $x'y'$ axes parallel to the horizontal and vertical edges of the cross section.

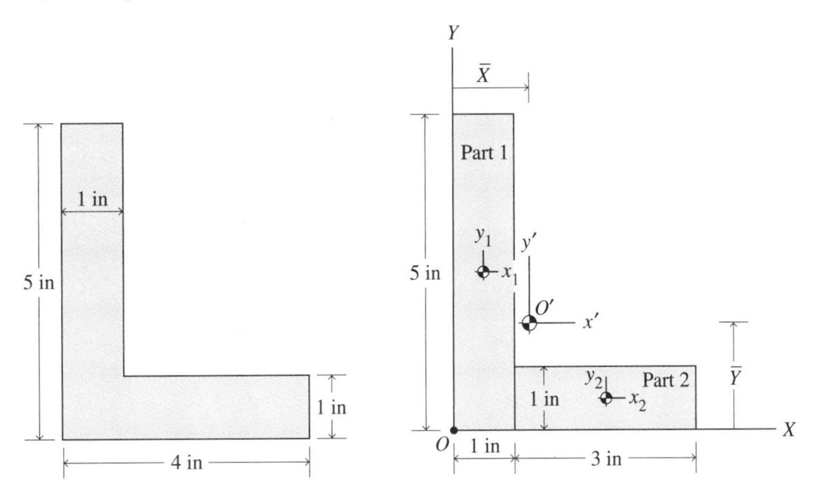

Figure EC.2a **Figure EC.2b**

Solution

Consider the L-shaped cross section to be composed of rectangular parts 1 and 2 (see Fig. EC.2b). Since the moments of inertia for a rectangle relative to its centroidal axes are known (see Example C.1), we can use the parallel axis theorem to find the moments of inertia of each rectangle about the centroidal axes $x'y'$ of the composite area. We must first locate the centroid O' of the composite area. To provide a convenient means of locating O', we establish reference axes (X and Y) with origin at the lower left corner of the area, as shown in Fig. EC.2b. The centroidal axes of part 1 are designated as x_1 and y_1; those for part 2 are x_2 and y_2.

The computations are organized in table form, as was done in Example 8.10. In Table EC.2a, X_i and Y_i locate the centroid of part i relative to the reference axes X and Y. From the entries in Table EC.2a and Eq. (8.19), the centroid O' is located at

$$\bar{X} = \frac{\sum (A_i X_i)}{\sum A_i} = \frac{10.0}{8.0} = 1.25 \text{ in}$$

$$\bar{Y} = \frac{\sum (A_i Y_i)}{\sum A_i} = \frac{14.0}{8.0} = 1.75 \text{ in}$$

Table EC.2a
Centroid of L-shaped area

i	A_i [in^2]	X_i [in]	Y_i [in]	$A_i X_i$ [in^3]	$A_i Y_i$ [in^3]
1	5.0	0.5	2.5	2.5	12.5
2	3.0	2.5	0.5	7.5	1.5
Sum	8.0	—	—	10.0	14.0

With the composite area's centroid known, we can calculate the moments of inertia of the cross section with the aid of the parallel axis theorem. Table EC.2b contains the various values used in the computations for moment of inertia. In Table EC.2b, $\bar{x}_i$ and $\bar{y}_i$ are the coordinates of O' relative to axes x_i and y_i (see Fig. EC.2b). With the sums in Table EC.2b and Eqs. (C.5), (C.6), and (C.8), we obtain the moments of inertia relative to the centroidal axes (x' and y'):

$$I_{x'} = \sum I_{x_i} + \sum A_i(\bar{y}_i)^2 = 10.667 + 7.50 = 18.167 \text{ in}^4$$

$$I_{y'} = \sum I_{y_i} + \sum A_i(\bar{x}_i)^2 = 2.667 + 7.50 = 10.167 \text{ in}^4$$

$$I_{x'y'} = \sum I_{x_i y_i} + \sum A_i \bar{x}_i \bar{y}_i = 0.0 + (-7.50) = -7.50 \text{ in}^4$$

Table EC.2b

Second moments of area of L-shaped area

PART	I_{x_i} [in^4]	I_{y_i} [in^4]	I_{xy_i} [in^4]	$\bar{x}_i = \bar{x} - x_i$ [in]	$\bar{y}_i = \bar{y} - y_i$ [in]	$A_i(\bar{x}_i)^2$ [in^4]	$A_i(\bar{y}_i)^2$ [in^4]	$A_i\bar{x}_i\bar{y}_i$ [in^4]
1	10.417	0.417	0.0	0.75	−0.75	2.8125	2.8125	−2.8125
2	0.25	2.25	0.0	−1.25	1.25	4.6875	4.6875	−4.6875
Sum	10.667	2.667	0.0	—	—	7.50	7.50	−7.50

The product of inertia $I_{x'y'}$ for the L-shaped cross section with respect to the centroidal axes x' and y' for the composite section is nonzero. This result indicates that the $x'y'$ axes are not principal axes for the cross section. Methods for finding the principal axes of a cross section are presented in the next section. ■

C.3 PRINCIPAL AXES OF AN AREA RELATIVE TO A POINT

principal axes

Consider the plane area A shown in Fig. C.3. Let xy axes be located at point O. Point O may lie inside the area or outside the area. Assume that the moments of inertia I_x, I_y, and I_{xy} are known and that I_{xy} is nonzero. To find the *principal axes* of the area at O, we must find the orientation θ of a pair of axes x' and y' such that $I_{x'y'} = 0$. To find such a pair of axes, we must first develop a set of equations that allows us to write $I_{x'}$, $I_{y'}$, and $I_{x'y'}$ in terms of I_x, I_y, I_{xy}, and θ.

The location of a fixed point P in the area (refer to Fig. C.3) can be given in terms of either (x, y) or (x', y') coordinates. The (x', y') coordinates are related to the (x, y) coordinates of P by the equations

$$\begin{aligned} x' &= x\cos\theta + y\sin\theta \\ y' &= y\cos\theta - x\sin\theta \end{aligned} \tag{C.9}$$

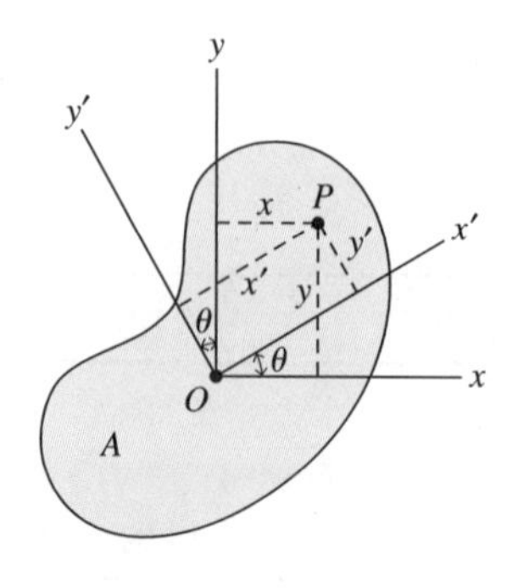

Figure C.3

where θ taken in the counterclockwise sense from the x axis is considered positive. Since we know I_x, I_y, and I_{xy}, with Eqs. (C.1), (C.2), (C.4) and (C.9), we can write the moments of inertia $I_{x'}$, $I_{y'}$, and $I_{x'y'}$ relative to the $x'y'$ axes:

$$\begin{aligned} I_{x'} &= \int (y\cos\theta - x\sin\theta)^2\, dA = I_x\cos^2\theta + I_y\sin^2\theta - 2I_{xy}\sin\theta\cos\theta \\ I_{y'} &= \int (x\cos\theta + y\sin\theta)^2\, dA = I_x\sin^2\theta + I_y\cos^2\theta + 2I_{xy}\sin\theta\cos\theta \\ I_{x'y'} &= \int (x\cos\theta + y\sin\theta)(y\cos\theta - x\sin\theta)\, dA \\ &= (I_x - I_y)\sin\theta\cos\theta + I_{xy}(\cos^2\theta - \sin^2\theta) \end{aligned} \tag{C.10}$$

With Eqs. (B.14) and (B.17), we can rewrite Eqs. (C.10) in the form

$$I_{x'} = \frac{I_x + I_y}{2} + \frac{I_x - I_y}{2}\cos 2\theta - I_{xy}\sin 2\theta \tag{C.11a}$$

$$I_{y'} = \frac{I_x + I_y}{2} - \frac{I_x - I_y}{2}\cos 2\theta + I_{xy}\sin 2\theta \tag{C.11b}$$

$$I_{x'y'} = \frac{I_x - I_y}{2}\sin 2\theta + I_{xy}\cos 2\theta \tag{C.11c}$$

By definition, $x'y'$ axes with origin O are principal axes for the area if $I_{x'y'} = 0$. Hence, setting $I_{x'y'} = 0$ in Eq. (C.11c), we find

$$\tan 2\theta = -\frac{2I_{xy}}{I_x - I_y} \tag{C.12}$$

Equation (C.12) defines the orientation angle θ for which the $x'y'$ axes are principal axes. There are two solutions for θ from Eq. (C.12); we designate them as θ_1 and θ_2. If we set $\theta_1 < \theta_2$, then θ_1 is the angle from the x axis to the x' axis, and θ_2 is the angle from the x axis to the y' axis. If we locate the xy axes such that $I_x > I_y$, then $I_{x'}$ is the maximum moment of inertia and $I_{y'}$ is the minimum moment of inertia. Since $I_x - I_y > 0$, if we substitute θ_1 from Eq. (C.12) into Eqs. (C.11a) and (C.11b), we find that

$$\begin{aligned} I_{x'} &= \frac{I_x + I_y}{2} + \sqrt{\left(\frac{I_x - I_y}{2}\right)^2 + I_{xy}^2} \\ I_{y'} &= \frac{I_x + I_y}{2} - \sqrt{\left(\frac{I_x - I_y}{2}\right)^2 + I_{xy}^2} \end{aligned} \tag{C.13}$$

principal moments of inertia

Then, $I_{x'}$ and $I_{y'}$ are the *principal moments of inertia* at point O for the area A.

If we add Eqs. (C.11a) and (C.11b), we obtain $I_{x'} + I_{y'} = I_x + I_y = J_O$. Hence, the polar moment of inertia J_O is independent of the orientation θ of the reference axes; it is a constant for the area A.

To show that $I_{x'}$ and $I_{y'}$ from Eq. (C.13) are indeed maximum and minimum values of moment of inertia for the area A, we can maximize (or minimize) $I_{x'}$ and $I_{y'}$ from Eq. (C.11). Differentiation of $I_{x'}$ with respect to θ gives

$$\frac{dI_{x'}}{d\theta} = -2\left(\frac{I_x - I_y}{2}\right)(\sin 2\theta) - 2I_{xy}\cos 2\theta = 0 \tag{a}$$

Solving for θ in Eq. (a), we obtain

$$\tan 2\theta = -\frac{2I_{xy}}{I_x - I_y}$$

This result is identical to Eq. (C.12), for which $I_{x'y'} = 0$. In beam theory, point O is usually assumed to coincide with the centroid of the area.

Example C.3

Principal Axes for a Z-Section

Problem Statement A purlin with a Z-shaped cross section is used to support roof panels of a building. The cross-sectional geometry is shown in Fig. EC.3a. Locate the principal axes, and determine the principal moments of inertia for the cross section relative to the centroidal axes xy.

Solution

Since the cross section has a point O of rotational symmetry, the centroid lies at that point, the origin of the xy coordinate axes. To determine the moments of inertia relative to the xy axes, we use the moments of inertia of the individual rectangles shown in Fig. EC.3b along with the parallel axis theorem. As in other examples involving composite areas, we organize the computations in table form. In Table EC.3, the term A_i is the area of the simple rectangle i; $\bar{x}_i$ and $\bar{y}_i$ are the coordinates of the centroid of rectangle i relative to the centroid of the composite cross section; I_{x_i}, I_{y_i}, and I_{xy_i} are the moments of inertia of rectangle i relative to its centroidal axes; $A_i\bar{x}_i^2$, $A_i\bar{y}_i^2$, and $A_i\bar{x}_i\bar{y}_i$ are the Steiner components of I_x, I_y, and I_{xy}, respectively.

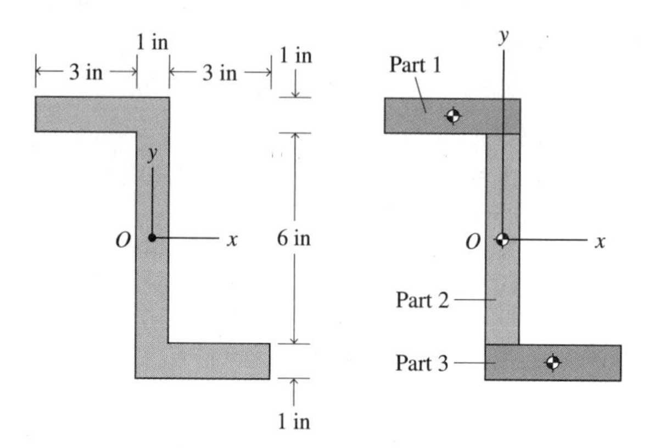

Figure EC.3a **Figure EC.3b**

Table EC.3

PART	A_i [in²]	$\bar{x}_i$ [in]	$\bar{y}_i$ [in]	I_{x_i} [in⁴]	I_{y_i} [in⁴]	I_{xy_i} [in⁴]	$A_i\bar{x}_i^2$ [in⁴]	$A_i\bar{y}_i^2$ [in⁴]	$A_i\bar{x}_i\bar{y}_i$ [in⁴]
1	4.0	−1.5	3.5	0.333	5.333	0	9.0	49.0	−21.0
2	6.0	0	0	18.0	0.5	0	0	0	0
3	4.0	1.5	−3.5	0.333	5.333	0	9.0	49.0	−21.0
Sums				18.67	11.17	0	18.0	98.0	−42.0

Using the sums in Table EC.3, we determine the moments of inertia as

$$I_x = \sum I_{x_i} + \Sigma A_i\bar{y}_i^2 = 18.67 + 98.0 = 116.7 \text{ in}^4$$

$$I_y = \sum I_{y_i} + \Sigma A_i\bar{x}_i^2 = 11.17 + 18.0 = 29.2 \text{ in}^4$$

$$I_{xy} = \sum I_{xy_i} + \Sigma A_i\bar{x}_i\bar{y}_i = 0 - 42 = -42 \text{ in}^4$$

Since I_{xy} is nonzero, the xy axes are not principal axes. The orientation of the principal axes (x' and y') is given by Eq. (C.12); see also Fig. EC.3c:

$$\tan 2\theta = -\frac{2(-42)}{116.7 - 29.2} = 0.960$$

$$\theta_1 = 21.92°, \ \theta_2 = 111.92°$$

Figure EC.3c

The principal moments of inertia for the cross section are obtained from Eq. (C.13):

$$I_{x'} = \frac{116.7 + 29.2}{2} + \sqrt{\left(\frac{116.7 - 29.2}{2}\right)^2 + (-42)^2} = 133.6 \text{ in}^4$$

$$I_{y'} = \frac{116.7 + 29.2}{2} - \sqrt{\left(\frac{116.7 - 29.2}{2}\right)^2 + (-42)^2} = 12.3 \text{ in}^4$$

As a check, we can find the polar moment of inertia by adding $I_{x'}$ and $I_{y'}$, or equivalently by adding I_x and I_y:

$$J_O = I_{x'} + I_{y'} = 133.6 + 12.3 = 145.9 \text{ in}^4$$

or

$$J_O = I_x + I_y = 116.7 + 29.2 = 145.9 \text{ in}^4$$

Problems

C.1 *Moments of Inertia by Integration*

C.1 Derive an expression for the moment of inertia with respect to the centroidal axis x of the right triangle in Fig. PC.1.

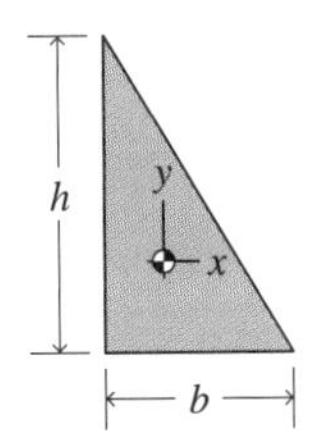

Figure PC.1

C.2 Derive an expression for the product of inertia with respect to the centroidal axes x and y of the right triangle in Fig. PC.1.

C.3 Derive an expression for the polar moment of inertia with respect to the center of a circular cross section having radius R. Use polar coordinates (r, θ) and an infinitesimal area $dA = (r\, d\theta)(dr)$.

C.4 Show that the polar moment of inertia with respect to the centroid of the cross section of a hollow circular tube with outer radius b and inner radius a is given by

$$J_O = \frac{\pi(b^4 - a^4)}{2}$$

C.5 For the semicircular cross section in Fig. PC.5, derive an expression for the moment of inertia I_y with respect to centroidal axis y.

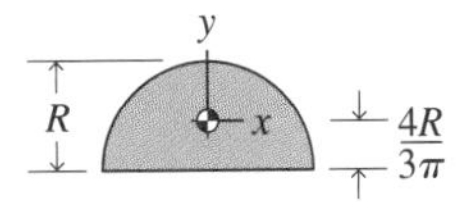

Figure PC.5

C.6 For the semielliptical cross section shown in Fig. PC.6 derive an expression for the moment of inertia I_x with respect to the diametral axis x.

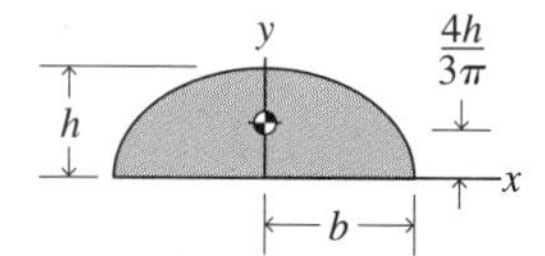

Figure PC.6

C.7 Refer to Problem C.4. For a thin-walled, hollow tube with a circular cross section, show that the polar moment of inertia is approximately $J_O = 2\pi R^3 t$, where R is the mean radius of the tube and t is the wall thickness ($t << R$). Determine the percent error in J_O as t increases from $0.001R$ to $0.2R$.

C.2 *Parallel Axis Theorem*

C.8 For the cross section of the symmetrical T-beam shown in Fig. PC.8, A–A is a centroidal axis. Calculate the moment of inertia with respect to A–A.

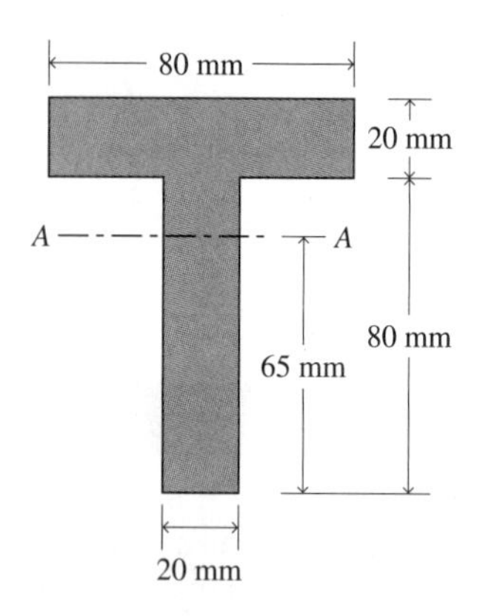

Figure PC.8

C.9 Determine the moments of inertia I_x, I_y, and J_O for the cross section of a regular hexagonal shaft, shown in Fig. PC.9.

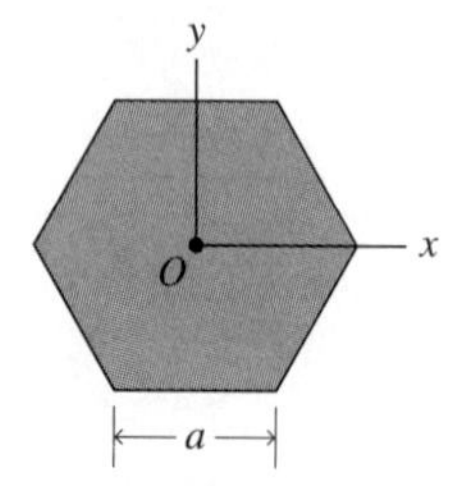

Figure PC.9

C.10 a. Determine the coordinate $\bar{y}$ of the centroid of the trapezoid in Fig. PC.10.

b. Determine the moment of inertia of the trapezoid with respect to the centroidal axis A–A.

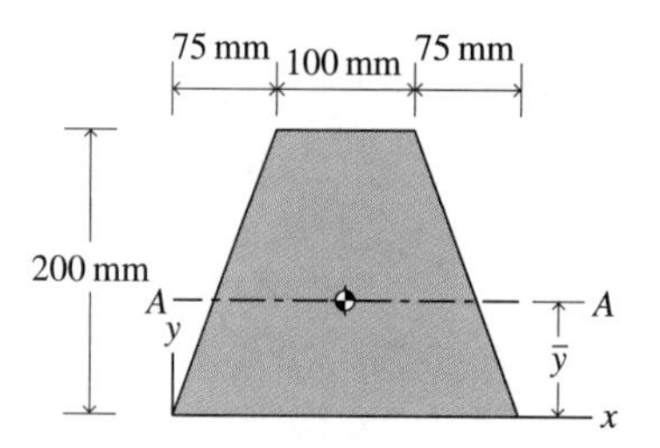

Figure PC.10

C.11 Determine the moment of inertia of the cross section of the slotted-beam shown in Fig. PC.11 with respect to the x axis.

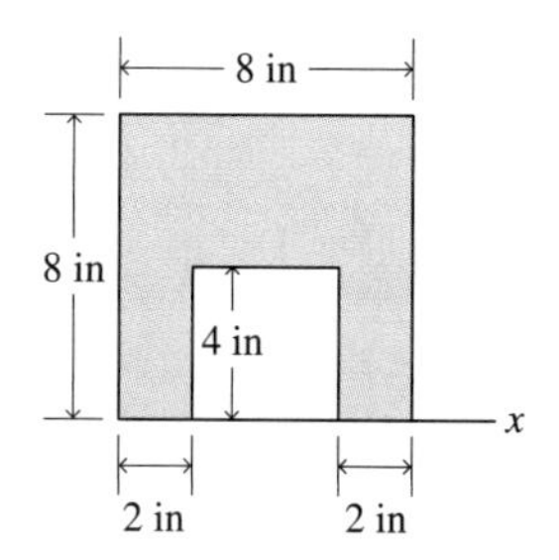

Figure PC.11

C.12 For the cross section of the bulb T-beam shown in Fig. PC.12,

a. Determine the y coordinate of the centroid.

b. Determine the moment of inertia of the cross section with respect to the y axis.

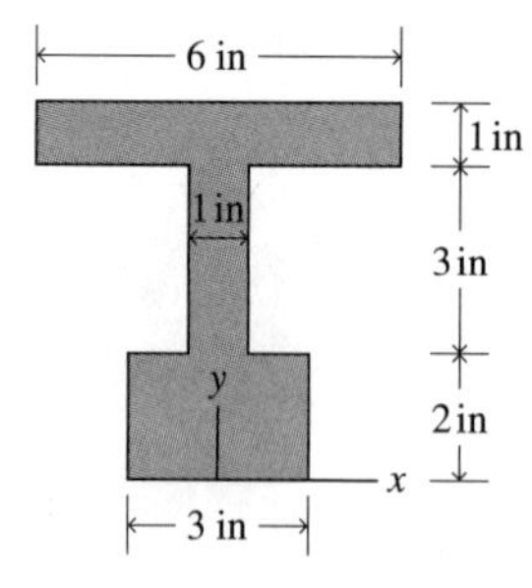

Figure PC.12

C.13 a. Determine the moment of inertia I_x of the area in Fig. PC.13 with respect to the x axis.
b. Determine the moment of inertia I_y of the area with respect to the y axis.
c. Locate the centroid of the area and then determine the moment of inertia $I_{y'}$ about a vertical axis through the centroid. What is the relationship between I_y from part b and $I_{y'}$?

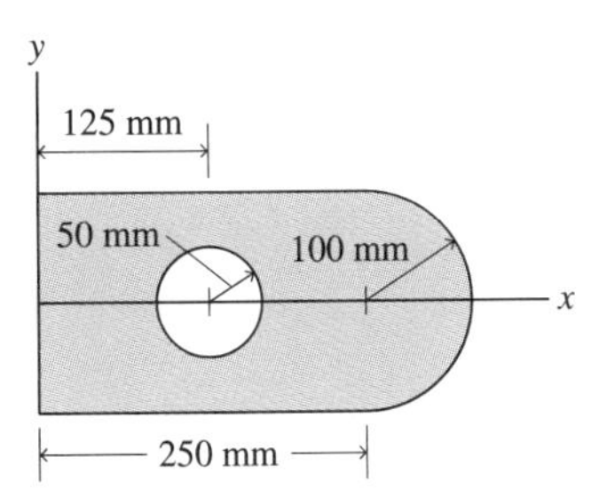

Figure PC.13

C.14 Locate the centroid for the L-shaped cross section in Fig. PC.14. Then, determine the moments of inertia I_x, I_y, and I_{xy} about centroidal xy axes that are parallel to the horizontal and vertical legs of the angle.

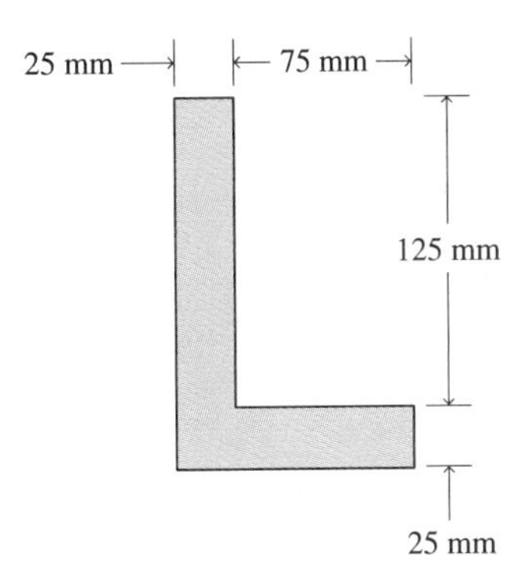

Figure PC.14

C.3 *Principal Axes of an Area*

C.15 Prove that for a square cross section, any centroidal axis is a principal axis.

C.16 Determine the orientation of the principal axes at the centroid and find the corresponding principal moments of inertia for an area composed of two semicircles (see Fig. PC.16). Each semicircle has radius r.

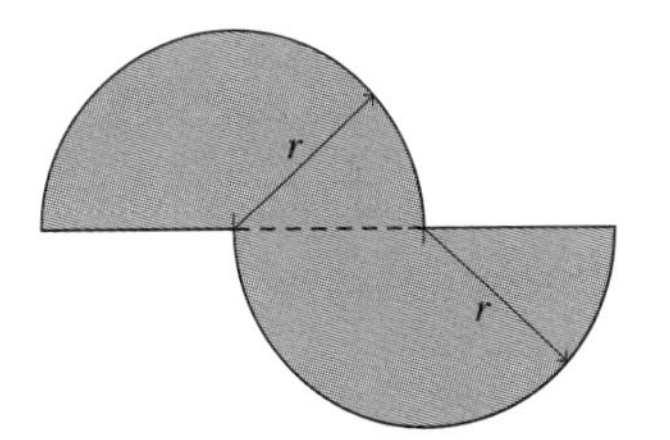

Figure PC.16

C.17 Determine the orientation of the principal axes at the centroid and find the corresponding principal moments of inertia for an area composed of two triangles (see Fig. PC.17). (*Hint:* Note that each individual triangle has a nonzero product of inertia relative to horizontal and vertical centroidal axes.)

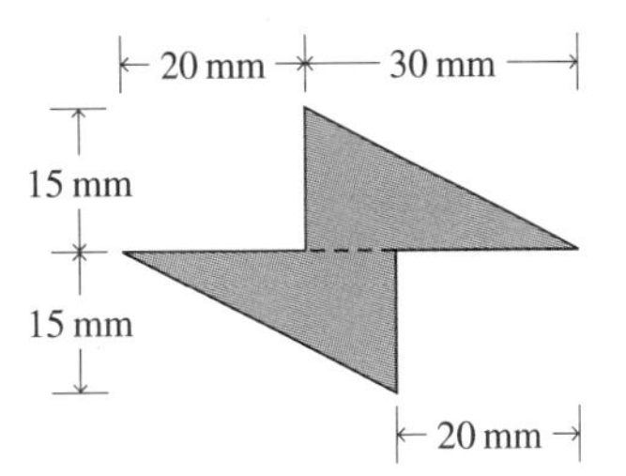

Figure PC.17

For each of the following problems:

a. Locate the centroid of the area indicated.

b. Determine the orientation of the principal axes at the centroid.

c. Determine the corresponding principal moments of inertia for the area.

C.18 The angle in Fig. PC.14

C.19 The square with semicircular cutouts in Fig. PC.19

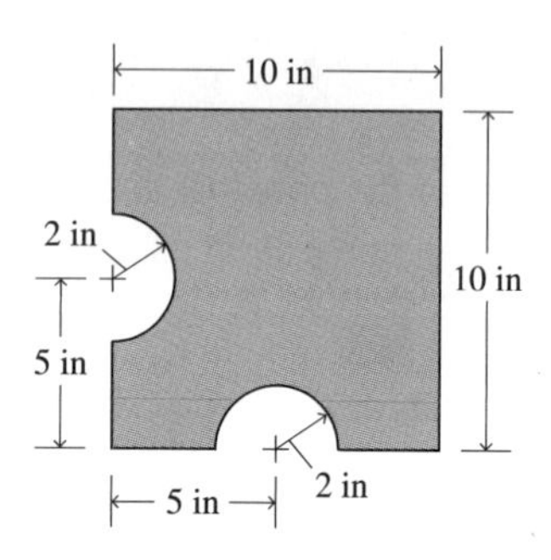

Figure PC.19

C.20 The J-shaped cross section in Fig. PC.20

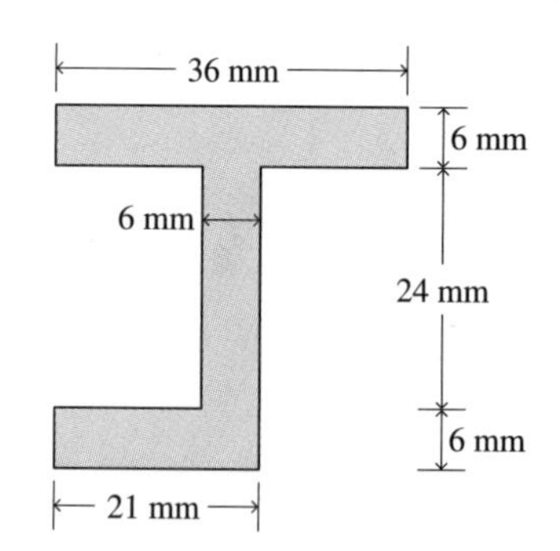

Figure PC.20

C.21 The skewed-trough cross section in Fig. PC.21 (*Hint:* Treat the inclined leg of the section as two isosceles right triangles, one with positive area and one with negative area.)

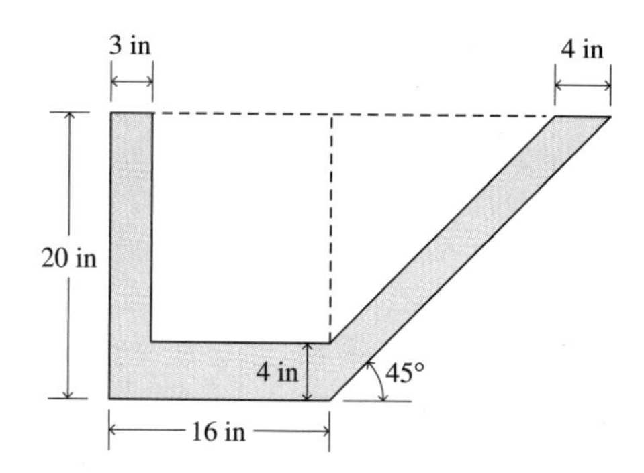

Figure PC.21

Appendix D

PROPERTIES OF LINES, AREAS, AND VOLUMES

As an aid for problem solving, the following geometric properties of lines, areas, and volumes are provided without discussion. The derivations of these formulas are available in various geometry, calculus, and mechanics textbooks.

Table D.1
Length and centroid of plane curves

Any plane curve

$L = \int ds$

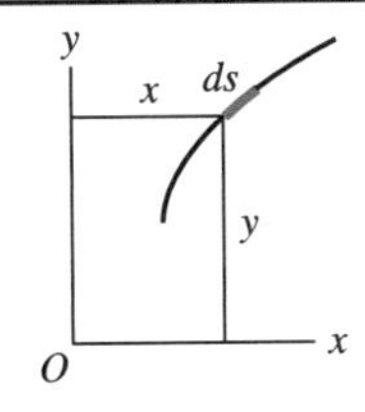

$$\bar{x} = \frac{\int x\,ds}{\int ds}, \qquad ds = \sqrt{1 + \left(\frac{dy}{dx}\right)^2}\,dx$$

$$\bar{y} = \frac{\int y\,ds}{\int ds}, \qquad ds = \sqrt{1 + \left(\frac{dx}{dy}\right)^2}\,dy$$

Circular arc

$L = 2r\theta$

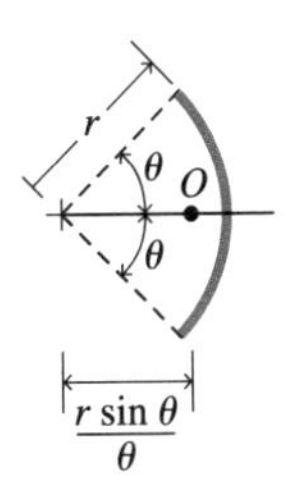

Quarter arc

$L = \frac{\pi r}{2}$

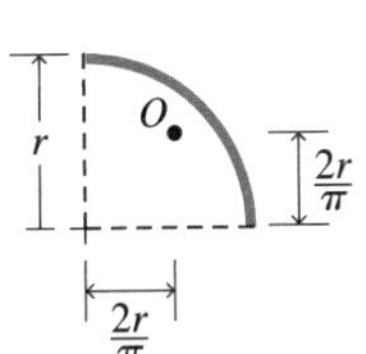

Table D.2
Area, centroid, and moments of inertia of plane areas

Rectangle
$A = bh$

$$I_x = \frac{bh^3}{12}$$

$$I_y = \frac{hb^3}{12}$$

$$I_{xy} = 0$$

$$J_O = \frac{(bh^3 + hb^3)}{12}$$

Right triangle
$A = bh/2$

$$I_x = \frac{bh^3}{36}$$

$$I_y = \frac{hb^3}{36}$$

$$I_{xy} = -\frac{b^2h^2}{72}$$

$$J_O = \frac{(bh^3 + hb^3)}{36}$$

Oblique triangle
$A = bh/2$

$$I_x = \frac{bh^3}{36}$$

$$I_y = \frac{bh(b^2 - ab + a^2)}{36}$$

$$I_{xy} = \frac{bh^2(b - 2a)}{72}$$

$$J_O = \frac{bh(h^2 + b^2 - ab + a^2)}{36}$$

Circle
$A = \pi R^2$

$= \dfrac{\pi D^2}{4}$

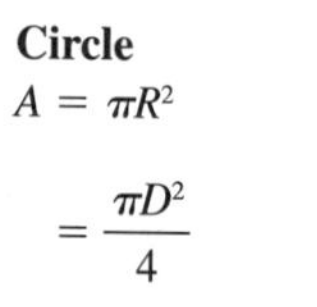

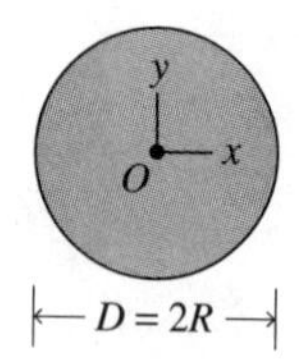

$$I_x = \frac{\pi D^4}{64} = \frac{\pi R^4}{4}$$

$$I_y = \frac{\pi D^4}{64} = \frac{\pi R^4}{4}$$

$$I_{xy} = 0$$

$$J_O = \frac{\pi D^4}{32} = \frac{\pi R^4}{2}$$

Ellipse

$A = \pi bh$

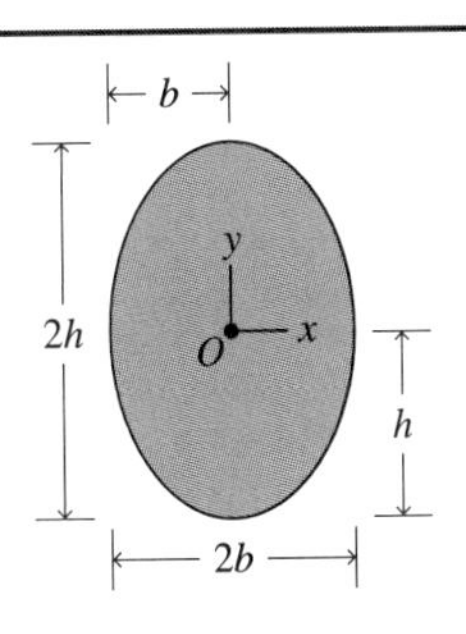

$$I_x = \frac{\pi bh^3}{4}$$

$$I_y = \frac{\pi hb^3}{4}$$

$$I_{xy} = 0$$

$$J_O = \frac{\pi(bh^3 + hb^3)}{4}$$

Semicircle

$$A = \frac{\pi R^2}{2}$$

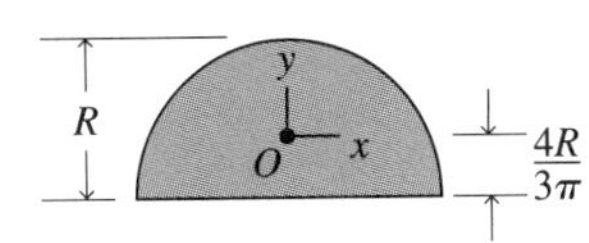

$$I_x = \pi R^4\left(\frac{1}{8} - \frac{8}{9\pi^2}\right)$$

$$I_y = \frac{\pi R^4}{8}$$

$$I_{xy} = 0$$

$$J_O = \pi R^4\left(\frac{1}{4} - \frac{8}{9\pi^2}\right)$$

Semiellipse

$$A = \frac{\pi bh}{2}$$

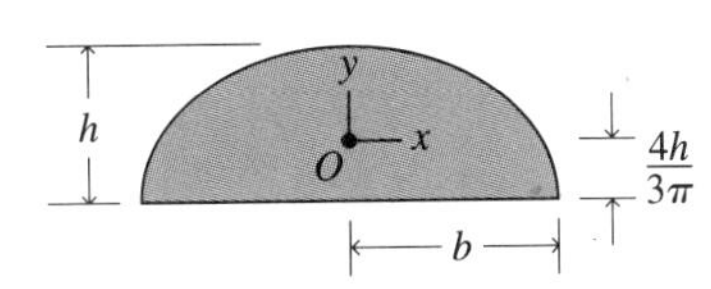

$$I_x = \pi bh^3\left(\frac{1}{8} - \frac{8}{9\pi^2}\right)$$

$$I_y = \frac{\pi hb^3}{8}$$

$$I_{xy} = 0$$

$$J_O = \pi bh\left(\frac{h^2}{8} - \frac{8h^2}{9\pi^2} + \frac{b^2}{8}\right)$$

Quarter circle

$$A = \frac{\pi R^2}{4}$$

$$I_x = R^4\left(\frac{\pi}{16} - \frac{4}{9\pi}\right)$$

$$I_y = R^4\left(\frac{\pi}{16} - \frac{4}{9\pi}\right)$$

$$I_{xy} = R^4\left(\frac{1}{8} - \frac{4}{9\pi}\right)$$

$$J_O = R^4\left(\frac{\pi}{8} - \frac{8}{9\pi}\right)$$

(continued)

Table D.2

Area, centroid, and moments of inertia of plane areas—(cont.)

Quarter ellipse

$$A = \frac{\pi b h}{4}$$

$$I_x = bh^3\left(\frac{\pi}{16} - \frac{4}{9\pi}\right)$$

$$I_y = hb^3\left(\frac{\pi}{16} - \frac{4}{9\pi}\right)$$

$$I_{xy} = b^2h^2\left(\frac{1}{8} - \frac{4}{9\pi}\right)$$

$$J_O = (bh^3 + hb^3)\left(\frac{\pi}{16} - \frac{4}{9\pi}\right)$$

Trapezoid

$$A = \frac{(a+b)h}{2}$$

$$I_x = h^3\left[\frac{a^2 + 4ab + b^2}{36(a+b)}\right]$$

$$I_y = \frac{h}{36(a+b)}[a^4 + b^4 + 2ab(a^2 + b^2) - c(a^3 + 3a^2b - 3ab^2 - b^3) + c^2(a^2 + 4ab + b^2)]$$

$$I_{xy} = \frac{h^2}{72(a+b)}[b(3a^2 - 3ab - b^2) + a^3 - c(2a^2 + 8ab + 2b^2)]$$

Trapezoid

$$A = ab$$

$$\bar{x} = \frac{a}{2}\sin\theta - \frac{b^2}{24a}(2 + \tan^2\theta)\sin\theta$$

$$\bar{y} = \frac{a}{2}\cos\theta + \frac{b^2}{24a}\sin\theta\tan\theta$$

Parabola

$$A = \frac{2bh}{3}$$

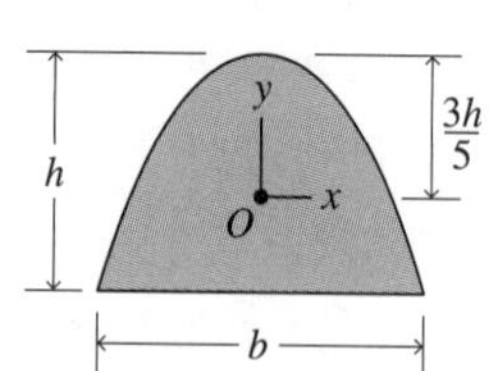

$$I_x = \frac{8bh^3}{175}$$

$$I_y = \frac{hb^3}{30}$$

$$I_{xy} = 0$$

$$J_O = bh\left(\frac{8h^2}{175} + \frac{b^2}{30}\right)$$

Half parabola

$A = \dfrac{2bh}{3}$

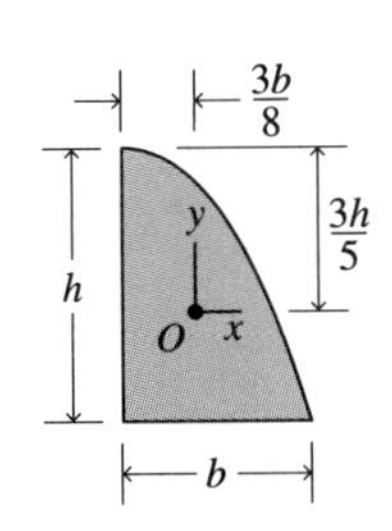

$$I_x = \frac{8bh^3}{175}$$

$$I_y = \frac{19b^3h}{480}$$

$$J_O = bh\left(\frac{8h^2}{175} + \frac{19b^2}{480}\right)$$

Circular sector

$A = r^2\theta$

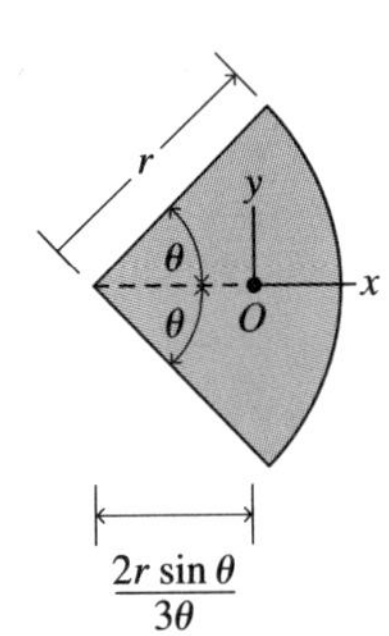

$$I_x = \frac{r^4}{4}(\theta - \sin\theta\cos\theta)$$

$$I_y = \frac{r^4}{4}\left(\theta + \sin\theta\cos\theta - \frac{16\sin^2\theta}{9\theta}\right)$$

$$I_{xy} = 0$$

$$J_O = \frac{r^4}{4}\left(2\theta - \frac{16\sin^2\theta}{9\theta}\right)$$

Circular segment

$A = \dfrac{r^2(2\theta - \sin 2\theta)}{2}$

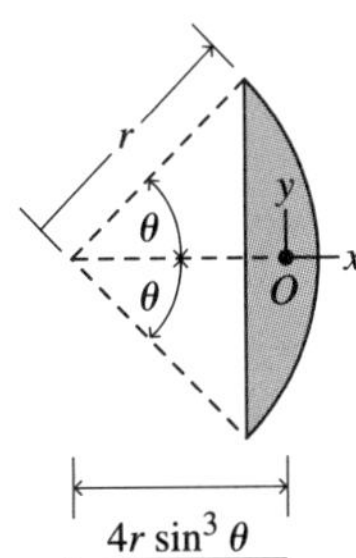

$$I_x = \frac{r^4}{48}(12\theta - 8\sin 2\theta + \sin 4\theta)$$

$$I_y = \frac{r^4}{16}(4\theta - \sin 4\theta) - \frac{8}{9}\frac{r^4\sin^6\theta}{(2\theta - \sin 2\theta)}$$

$$I_{xy} = 0$$

$$J_O = \frac{r^4}{72}\left[36\theta - 12\sin 2\theta - 3\sin 4\theta - \frac{64\sin^6\theta}{(2\theta - \sin 2\theta)}\right]$$

Table D.3
Surface area and centroid of hollow bodies

Right circular cone

$S = \pi r\sqrt{h^2 + r^2}$

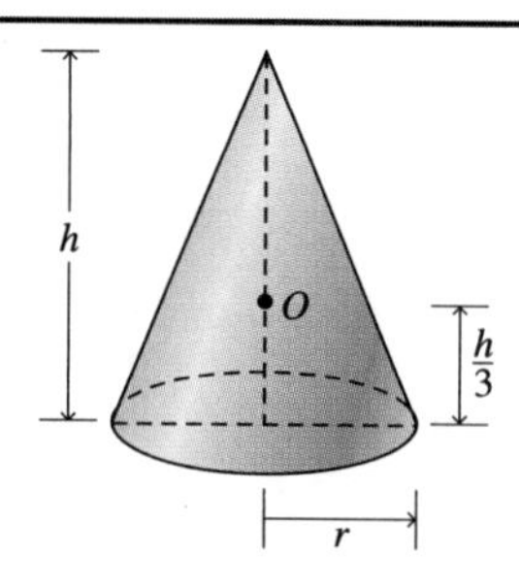

Hemisphere

$S = 2\pi r^2$

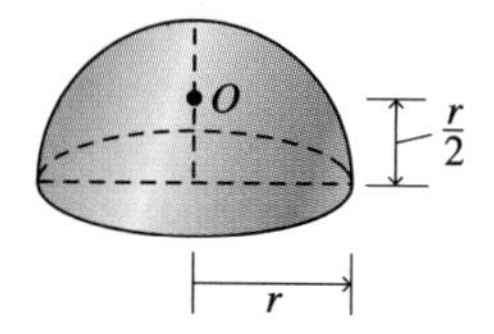

Table D.4
Volume and centroid of solid bodies

Right circular cone

$V = \dfrac{\pi r^2 h}{3}$

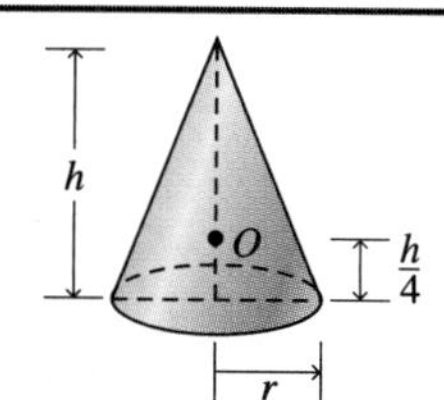

Hemisphere

$V = \dfrac{2\pi r^3}{3}$

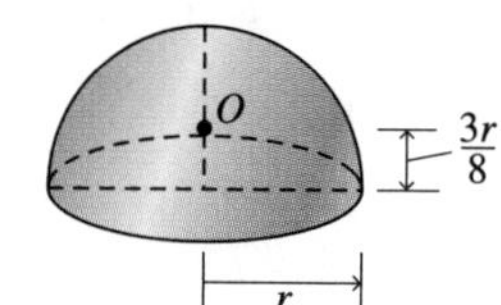

REFERENCES

AISC. (1994). *Manual of Steel Construction, Load & Resistance Factor Design, Vol. I,* 2nd ed. American Institute of Steel Construction, Inc., One East Wacker Drive, Suite 3100, Chicago, IL 60601-2001.

ASTM. (1992). *Standard Practice for Use of the International System of Units,* E380–92. American Society for Testing and Materials, 1916 Race Street, Philadelphia, PA 19103.

Bowden, F. P., and Tabor, D. (1986). *The Friction and Lubrication of Solids.* New York: Oxford University Press.

Hutchings, I. M. (1992). *Tribology: Friction and Wear of Engineering Materials.* Boca Raton, FL: CRC Press.

Jacobs, D., & Neville, A. E. (1968). *Bridges, Canals & Tunnels.* American Heritage Publishing and The Smithsonian Institution.

Lamb, H. (1924). *Statics: Including Hydrostatics and the Elements of the Theory of Elasticity,* 2nd ed. London: Cambridge University Press.

Langhaar, H. L. (1980). *Dimensional Analysis and Theory of Models.* Krieger Publishing, P.O. Box 9542, Melbourne, FL 32092-9542.

Orthwein, W. C. (1986). *Clutches and Brakes: Design and Selection.* New York: Marcel Dekker.

Orthwein, W. C. (1990). *Machine Component Design.* St. Paul, MN: West Publishing.

Osgood, W. F. (1937). *Mechanics.* New York: Dover Publications.

Peters, T. F. (1987). *Transitions in Engineering.* Boston: Birkhaeuser Verlag.

Robinson, F. P. (1970). *Effective Study,* 4th ed. New York: Harper & Row.

Shigley, J. E., and Mischke, C. R. (1989). *Mechanical Engineering Design.* New York: McGraw-Hill.

Shigley, J. E., and Vicker, J. J. (1995). *Theory of Machines and Mechanisms.* New York: McGraw-Hill.

Singer, C. J., Holmyard, E. J., Hall, A. R., and Williams, T. I. (Eds.). (1958). *A History of Technology, Vol. IV: The Industrial Revolution, 1750 to 1850.* Oxford, England: Clarendon Press.

Thomas, G. B., and Finney, R. L. (1995). *Calculus and Analytic Geometry.* Reading, MA: Addison-Wesley.

Tschann, C., and Modi, V. J. (1970). "A Comparative Study of Two Classical Damping Mechanisms for Gravity-Oriented Satellites." *Canadian Aeronautics and Space Institute Transactions,* 3(2): 135–146.

Turner, R., and Goulden, S. L. (Eds.). (1981). *Great Engineers and Pioneers in Technology, Vol. I: From Antiquity through the Industrial Revolution.* New York: St. Martin's Press.

Vigneron, F. R. (1970). "Effect of the Earth's Gravitational Forces on the Flexible Crossed-Dipole Satellite Configuration—Part 2: Attitude Stability." *Canadian Aeronautics and Space Institute Transactions,* 3(2): 127–134.

Vigneron, F. R., and Boresi, A. P. (1970). "Effect of the Earth's Gravitational Forces on the Flexible Crossed-Dipole Satellite Configuration—Part 1: Configuration Stability and Despin." *Canadian Aeronautics and Space Institute Transactions,* 3(2): 115–126.

Whipple, S. (1847). *A Work on Bridge Building.* Utica, NY: H. H. Curtiss, Printer.

Young, H. D. (1992). *University Physics,* 8th ed. Reading, MA: Addison-Wesley.

ANSWERS TO SELECTED EVEN-NUMBERED PROBLEMS

Chapter 1

1.2 **a.** (i) $\approx 0\%$, (ii) 0.005%, (iii) 0.504%; **b.** accurate; **c.** $m \longrightarrow \infty$
1.4 $F_x = 7.5$ N, $F_y = 9.375$ N
1.6 **a.** 5 teradollars; **b.** 18,939 times
1.8 meter, kilogram, second, newton (derived unit)
1.10 derived unit
1.12 If $W = 125$ lb, then $m = 3.88$ slugs $= 56.6$ kg.
1.14 **a.** Runner A is faster by 9.4%.
b. Runner A's distance is 3.56% longer than runner B's.
1.16 **a.** 1.941 slug/ft^3; **b.** 1.0 g/cm^3
1.18 5.735 L
1.20 **a.** 2.589; **b.** 1.308; **c.** 1296
1.22 \$512,000/mi^2, \$800/acre, \$0.20/m^2, \$0.02/ft^2
1.24 **a.** $[LT^{-2}]$, $[LT^{-4}]$; **b.** ft/s^2, ft/s^4; **c.** mi/h^2, mi/h^4; **d.** km/s^2, km/s^4
1.26 **a.** homogeneous; **b.** homogeneous; **c.** not homogeneous
1.28 $[FL^2]$
1.30 10,190
1.32 $\sigma = 1{,}723$ lb/in^2

Chapter 2

2.2 **a.** **A** and **B** are collinear. **b.** **B** is a zero vector. **c.** **A** and **B** are perpendicular.
2.10 **a.** $R = (A^2 + B^2 + 2AB\cos\theta)^{1/2}$, $\phi = \sin^{-1}[(B\sin\theta)/R]$
b. For $\theta = 0°$: $R = A + B$, $\phi = 0$
For $\theta = 90°$: $R = (A^2 + B^2)^{1/2}$, $\phi = \sin^{-1}(B/R)$
For $\theta = 180°$: $R = A - B$, $\phi = 0°$ if $A > B$, $\phi = 180°$ if $A < B$
2.12 **a.** $R = \sqrt{10 - 6\cos\theta}$
b. $\phi = \sin^{-1}\left(\dfrac{\sin\theta}{\sqrt{10 - 6\cos\theta}}\right)$
c. For $\theta = 0°$: $R = 2$ kN, $\phi = 0$
For $\theta = 90°$: $R = 3.16$ kN, $\phi = 18.4°$, cw
For $\theta = 180°$: $R = 4$ kN, $\phi = 0°$
2.14 **a.** $R = \sqrt{10 - 6\sin\theta}$
b. $\phi = \sin^{-1}\left(\dfrac{\cos\theta}{\sqrt{10 - 6\sin\theta}}\right)$
c. For $\theta = 0°$: $R = 3.16$ kN, $\phi = 18.43°$, ccw
For $\theta = 90°$: $R = 2$ kN, $\phi = 0$
For $\theta = 180°$: $R = 3.16$ kN, $\phi = 18.43°$, cw
2.16 $A = 0.897$ kN, $B = 0.732$ N
2.18 $\theta = 90°$, $B = 1414$ lb
2.20 $R = 20.07$ kips, $\theta = 88.78°$ cw
2.22 component along $AP = 1932$ lb, component along $BP = 2732$ lb
2.24 $R = 6.19$ kips, $\theta = 14.3°$ ccw
2.26 $R_A = 30.6$ lb, $\theta = 22.5°$ cw; $R_B = 73.9$ lb, $\theta = 112.5°$ cw
2.28 $R = 73$ N, $\theta = 55°$ ccw
2.30 $R = 1.5$ kips, $\theta = 8°$ ccw
2.32 $R = 157$ lb, $\theta = 0.24$ rad
2.34 $F = 20$ lb, $T = 34.6$ lb
2.36 $R = 353.6$ lb, $F = 70.71$ lb
2.38 41.4°, 55.8°, 82.8°
2.42 **a.** $6\mathbf{A} + 4\mathbf{B} - 25\mathbf{C} = 0$
b. $\mathbf{F} = 9\mathbf{A} - 24\mathbf{B} + 21\mathbf{C}$
c. $31\mathbf{A} - 17\mathbf{B} = 0$
2.44 **a.** yes; **b.** no; **c.** yes;
d. yes, provided their polygon is a closed triangle
2.46 **a.** $R = 76.54$, $\theta = 22.5°$ cw;
b. $R = 124.5$, $\theta = 145.4°$ ccw;
c. $D = 76.54$, $\theta = 157.5°$ ccw;
d. $D = 124.5$, $\theta = 145.4°$ ccw
2.48 **a.** $R_x = A + B\cos\theta$, $R_y = B\sin\theta$; $R^2 = R_x^2 + R_y^2$; $\phi = \tan^{-1}(R_y/R_x)$
b. For $\theta = 0°$: $R_x = A + B$, $R_y = 0$, $R = A + B$, $\phi = 0$
For $\theta = 90°$: $R_x = A$, $R_y = B$, $R = (A^2 + B^2)^{1/2}$, $\phi = \tan^{-1}(B/A)$
For $\theta = 180°$: $R_x = A - B$, $R_y = 0$, $R = A - B$, $\phi = 0$
2.50 **a.** $R_x = 3 - \cos\theta$, $R_y = -\sin\theta$, $R = (R_x^2 + R_y^2)^{1/2} = (10 - 6\cos\theta)^{1/2}$
b. $\phi = \tan^{-1}(R_y/R_x) = \tan^{-1}[(-\sin\theta)/(3 - \cos\theta)]$
c. For $\theta = 0°$: $R = 2$ kN, $\phi = 0$
For $\theta = 90°$: $R = 3.16$ kN, $\phi = 18.43°$ cw relative to 3 kN
For $\theta = 180°$: $R = 4$ kN, $\phi = 0°$
2.52 **a.** $R_x = -\cos\theta$, $R_y = 3 - \sin\theta$, $R = (R_x^2 + R_y^2)^{1/2} = (10 - 6\sin\theta)^{1/2}$
b. $\phi = \tan^{-1}[|R_x|/R_y] = \tan^{-1}[|\cos\theta|/(3 - \sin\theta)]$
c. For $\theta = 0°$: $R = 3.16$ kN, $\phi = 18.43°$ ccw relative to 3 kN
For $\theta = 90°$: $R = 2$ kN, $\phi = 0$

For $\theta = 180°$: $R = 3.16$ kN, $\phi = 18.43°$ cw relative to 3 kN
2.54 $A = 0.897$ kN, $B = 0.732$ kN
2.56 $B = 1414$ lb, $\theta = 90°$
2.58 $R = 20.06$ kips, $\theta = 88.77°$ cw
2.60 $A = 453.7$ lb, $B = 673.0$ lb
2.64 $\mathbf{R} = (2, 4, 4)$
2.66 **a.** $(-126.50, 94.87, 81.37)$;
b. $\theta_x = 135.3°$, $\theta_y = 57.76°$, $\theta_z = 62.77°$
2.70 **a.** $\mathbf{F} = 70.71\mathbf{i} + 70.71\mathbf{j}$ [N];
b. $\mathbf{F} = 141.4\mathbf{i} + 100\mathbf{n}$ [N]
2.72 $\mathbf{A} + \mathbf{B} = 18\mathbf{i} + 14\mathbf{j}$
$\mathbf{A} - \mathbf{B} = 2\mathbf{i} + 26\mathbf{j}$
$\mathbf{A} + 3\mathbf{B} = 34\mathbf{i} + 2\mathbf{j}$
$3\mathbf{A} - \mathbf{B} = 22\mathbf{i} - 66\mathbf{j}$
2.74 $\mathbf{R} = 100\mathbf{i} = 138.6\mathbf{j} - 120\mathbf{k}$ [N]
2.76 $\mathbf{R} = 2.69\mathbf{i} - 3.59\mathbf{j} + 2.01\mathbf{k}$ [kN]
2.78 $\mathbf{R} = 3\mathbf{i} - 35.1\mathbf{j} - 21.2\mathbf{k}$ [lb]

Chapter 3

3.4 0.732 kip $< F <$ 2 kips
3.8 **a.** $T = 50/\sin\theta$
3.10 $F = 25.98$ N, $T_{AB} = 39.69$ N, $T_{BC} = 30$ N, $\theta = 40.9°$
3.12 $T_1 = 772.6$ N, $T_2 = 451.8$ N, $T_3 = 606.2$ N, $W = 571.5$ N
3.14 $\theta = 77.8°$
3.16 **a.** $\mathbf{E} = 126.5\mathbf{i} - 94.87\mathbf{j} - 81.37\mathbf{k}$ [N];
b. $\theta_x = 44.7°$
3.18 **a.** $x = 113.2$ mm; **b.** $T = 1491$ N
3.22 $\mathbf{F}_{AB} = \mathbf{i} - \mathbf{j}$ [kips], $\mathbf{F}_{BC} = -4.464\mathbf{i}$ [kips]
3.24 $\mathbf{F}_{AB} = -\mathbf{i} + \mathbf{j}$ [kN], $\mathbf{F}_{BC} = 0$
3.26 $F_{AB} = 8.624$ kN (tension), $F_{BC} = 8.196$ kN (compression)
3.28 $T = 35.36$ N
3.30 $F_{AD} = 1.155W$ (tension), $F_{AB} = F_{AC} = 0.577W$ (compression)
3.34 $F_x = 0.5R_1 - 0.866P$
$F_y = 0.866R_1 + R_2 + 0.5P - W$
3.36 $\theta = 120°$, $F = 866$ lb, $P = 500$ lb
3.38 $A = 26.6$ kN, $B = 34.2$ kN
3.40 **c.** $m_1 = m_2/\sin 30°$; **d.** $T = m_1 g \sin 30°$;
e. $R = m_1 g \cos 30°$
3.42 **b.** $\theta = \cos^{-1}(r_1/r_2)$
3.44 Top roller: $\mathbf{R}_1 = -44.72\mathbf{i}$ [N],
$\mathbf{R}_2 = 44.72\mathbf{i} + 50\mathbf{j}$ [N]
Bottom roller: $\mathbf{R}_2 = -44.72\mathbf{i} - 50\mathbf{j}$ [N],
$\mathbf{R}_3 = 44.72\mathbf{i}$ [N], $\mathbf{R}_4 = 100\mathbf{j}$ [N]
3.46 $A_x = -5.726$ kN, $A_y = 28.63$ kN, $P = 22.12$ kN
3.48 $\alpha = 53.1°$, $\beta = 36.9°$
3.50 $P = 41.67$ lb, $R = 108.3$ lb
3.52 $T_1 = 3200$ lb, $T_2 = 1971$ lb, $T_3 = 2263$ lb
3.56 Maximum value of F_{PA} is 75.4 kN at $\theta = 166°$ and $\theta = -14°$. F_{PA} is never negative; that is, member PA is never in compression.

Chapter 4

4.2 $M = 893.5$ N·m
4.4 $M_{(0,0)} = -480$ lb·ft, $M_{(3,7)} = 320$ lb·ft, $M_{(-2,5)} = 50$ lb·ft
4.6 $M_O = -2.843$ N·m
4.8 $M = 164.4$ N·m
4.10 **a.** $F = 1941.6$ lb, $\theta = -34.51°$, $b = 58.13$ in;
c. $y = -0.688x + 58.13$
4.12 **a.** $F = 140.8$ N, $\theta = 83.88°$, $b = 1886.7$ mm;
c. $y = 9.33x + 1886.7$
4.14 **a.** $F = 600$ N, $\theta = 0°$, $b = -14.67$ m;
c. $y = -14.67$ m
4.16 **a.** $\mathbf{F} = -10\mathbf{i} + 166.7\mathbf{j}$ [lb]; **b.** $x = 7.44$ in from A
4.18 **a.** $\mathbf{F} = -1.716\mathbf{i} + 15.60\mathbf{j}$ [N];
b. $\mathbf{F} = -11.72\mathbf{i} - 1.716\mathbf{j}$ [N]
4.20 **a.** $\mathbf{F} = 4.10\mathbf{i} + 3.83\mathbf{j}$ [kN]; **b.** $s = 0.535$ m
4.22 $\mathbf{F} = -1.25\mathbf{i} + 6.40\mathbf{j}$ [kips], $x = 142.9$ ft
4.24 $\mathbf{F} = -60\mathbf{j}$ [kips]
4.26 $\mathbf{F} = 80\mathbf{j}$ [N] to the right of the left 160 N force
4.28 **a.** $\mathbf{F}_1 = -100\mathbf{i} + 240\mathbf{j}$ [N], $\mathbf{F}_2 = 100\mathbf{i} + 160\mathbf{j}$ [N];
b. and **c.** $\mathbf{R} = 400\mathbf{j}$ [N], $r = 90$ mm to the right of the 240 N force
4.30 **a.** no; **b.** no
4.32 $\mathbf{F} = -8\mathbf{j}$ [kips]
4.34 **a.** $\mathbf{R} = -80\mathbf{j}$ [N]; **b.** $\mathbf{R} = 60\mathbf{i}$ [N];
c. $\mathbf{R} = 30\mathbf{i} - 40\mathbf{j}$ [N]; **d.** $\mathbf{R} = 40\mathbf{i} - 30\mathbf{j}$ [N];
e. $\mathbf{R} = -60\mathbf{j}$ [N]; **f.** $M = -40$ N·m;
g. $\mathbf{R} = 20\mathbf{j}$ [N]; **h.** $M = 0$ [N·m];
i. $\mathbf{R} = -40\mathbf{i}$ [N]; **j.** $\mathbf{R} = 0$ [N]; **k.** $\mathbf{R} = 20\mathbf{i}$ [N];
l. $\mathbf{R} = 20\mathbf{j}$ [N]; **m.** $\mathbf{R} = -10\mathbf{i} + 10\mathbf{j}$ [N];
n. $\mathbf{R} = -20\mathbf{j}$ [N]; **o.** $\mathbf{R} = 30\mathbf{j}$ [N]
4.36 $\mathbf{R} = -10\mathbf{j}$ [N]
4.38 $\mathbf{A} = 200$ lb $\uparrow$, $\mathbf{B} = 200$ lb $\downarrow$
4.40 $M = 0$ lb·ft
4.42 $\mathbf{R} = 0.2\mathbf{j}$ [kN]
4.46 **a.** $M = 10{,}900$ lb·in
4.48 **a.** $\mathbf{R} = -130.7\mathbf{i} + 30.71\mathbf{j}$ [lb], $M_O = -74.31$ lb·ft
4.50 **a.** $\mathbf{R} = 100\mathbf{i} - 50\mathbf{j}$ [lb], $M_O = 220$ lb·ft
4.52 $\mathbf{R} = 236.6\mathbf{i} - 36.60\mathbf{j}$ [lb], $M_O = -846.4$ lb·ft
4.54 $\mathbf{F} = -200\mathbf{j}$ [N], $r = 0.66$ m to the right of O
4.56 $\mathbf{A} = 28.87\mathbf{i}$ [lb], $\mathbf{B} = 100\mathbf{j}$ [lb]
4.58 $F = 100$ lb, $M = 10{,}500$ lb·in
4.60 $\mathbf{A} = 5.02\mathbf{j}$ [kN], $\mathbf{B} = 10.32\mathbf{j}$ [kN]
4.62 $T = 66$ lb, $R_x = 76.67$ lb, $R_y = 53.33$ lb

4.64 $\mathbf{A} = 6\mathbf{i} + 10.2\mathbf{j}$ [kN], $\mathbf{E} = 10.2\mathbf{j}$ [kN]
4.66 $\mathbf{A} = -2000\mathbf{i} + 857.1\mathbf{j}$ [lb], $\mathbf{B} = 3142.9\mathbf{j}$ [lb]
4.68 $\mathbf{E} = 16.31\mathbf{i} + 16.31\mathbf{j}$ [kN], $\mathbf{F} = -11.51\mathbf{i} - 8.31\mathbf{j}$ [kN]
4.70 $R_1 = (W/12)(7\cos\theta - 3.5\sin\theta)$
$R_2 = (W/12)(5\cos\theta + 3.5\sin\theta)$
$F = W\sin\theta$
4.72 **b.** $O_x = -36.26$ lb, $O_y = -42.8$ lb
4.74 $\mathbf{A} = -431.25\mathbf{i} + 2325\mathbf{j}$ [lb], $\mathbf{B} = -543.75\mathbf{i} + 2775\mathbf{j}$ [lb], $F = 975$ lb
4.78 $T = 3.33x + 2.25$ [kN]
4.84 **a.** $R = 0$ kip; **b.** $M_x = -6.5$ kip·ft, $M_y = 15$ kip·ft
4.86 **a.** $R = 0$ N; **b.** $M_x = -2625$ N·m, $M_y = -4675$ N·m
4.88 $\mathbf{B} = -500\mathbf{k}$ [N], $\mathbf{C} = 500\mathbf{k}$ [N], $\mathbf{D} = -500\mathbf{k}$ [N]
4.90 A carries $\frac{21}{44}W$. B carries $\frac{15}{44}W$. C carries $\frac{2}{11}W$.
4.110 $M = 70$ N·m
4.112 $C = 7500$ N·mm, cw
4.114 $C = 5000$ N·mm, cw

Chapter 5

5.2 **a.** 5.196 N; **b.** 1.73 N
5.4 −8.02
5.6 **a.** 54.03°; **b.** 5.874 N; **c.** 5.874 N
5.8 **a.** 10, 12, 24, −20; **b.** 47.6°, 36.9°, 133.6°; **c.** 2.71
5.10 −54.55 N
5.12 **a.** 5.2; **b.** $4.16\mathbf{i} + 3.12\mathbf{k}$
5.14 $-0.397\mathbf{i} - 0.330\mathbf{j} + 0.857\mathbf{k}$
5.18 **b.** $r = (x^2 + y^2 + z^2 + 2xy\cos\gamma + 2xz\cos\beta + 2yz\cos\alpha)^{1/2}$
5.24 $0.396\mathbf{i} + 0.330\mathbf{j} - 0.857\mathbf{k}$
5.28 **a.** $6\mathbf{i} - 3\mathbf{j} + 21\mathbf{k}$; **b.** $-16\mathbf{i} - 4\mathbf{j} + 6\mathbf{k}$; **c.** $34\mathbf{i} - 33\mathbf{j} - 4\mathbf{k}$; **d.** $-3\mathbf{k}$
5.30 164
5.32 **a.** $-7\mathbf{i} + 26\mathbf{j} + 10\mathbf{k}$; **b.** $7\mathbf{i} - 26\mathbf{j} - 10\mathbf{k}$; **c.** $-30\mathbf{i} - 24\mathbf{j} + 9\mathbf{k}$; **d.** $\mathbf{j} + 2\mathbf{k}$; **e.** $\mathbf{i} + 4\mathbf{j}$
5.34 **b.** 13.42 m^2
5.36 114.2 ft
5.38 75 in
5.40 **a.** $7x + 4y + 13z - 54 = 0$
5.42 **a.** $5\mathbf{j}$ lb·ft; **b.** (0, 1, 0)
5.44 $-67\mathbf{i} - 4\mathbf{j} + 15\mathbf{k}$
5.46 **a.** $-2000\mathbf{i} + 2000\mathbf{j} - 1000\mathbf{k}$ [lb]; **b.** $-4000\mathbf{j} - 8000\mathbf{k}$ [lb·in]
5.52 $34\mathbf{i} + 12\mathbf{j} - 19\mathbf{k}$ [kN·m]
5.54 692.8 lb·in
5.56 3.625 kN·m
5.58 8.018 kN·m
5.62 **a.** −0.163 kN·m; **b.** 0
5.64 **b.** $-0.919\mathbf{i} - 0.379\mathbf{j} - 0.108\mathbf{k}$ or $0.919\mathbf{i} + 0.379\mathbf{j} + 0.108\mathbf{k}$
5.66 $-28\mathbf{i} + 20\mathbf{j} - 4\mathbf{k}$ [kN·m]
5.68 −12.35 kN·m
5.70 **a.** 0; **b.** $11\mathbf{i} - 2\mathbf{j} - \mathbf{k}$
5.72 $-80\mathbf{i} - 60\mathbf{j} - 40\mathbf{k}$ [lb·in]
5.74 $146\mathbf{i} + 17\mathbf{j} + 32\mathbf{k}$ [kN·m]
5.76 $7\mathbf{i} - 9\mathbf{j} - \mathbf{k}$ [kN·m]
5.78 $\mathbf{F} = 4\mathbf{i} + 5\mathbf{j} + \mathbf{k}$ [kN], $\mathbf{M}_O = 25\mathbf{i} + 14\mathbf{j} + 16\mathbf{k}$ [kN·m]
5.80 $\mathbf{F} = -100\mathbf{i} - 1040\mathbf{j} + 220\mathbf{k}$ [lb], $\mathbf{M} = 80{,}570\mathbf{i} + 11{,}470\mathbf{j} - 64{,}830\mathbf{k}$ [lb·ft]
5.82 $\mathbf{F} = -16\mathbf{i} + 45.21\mathbf{j} + 14.79\mathbf{k}$ [N], $\mathbf{M} = -27.63\mathbf{i} + 36.84\mathbf{j} + 228.84\mathbf{k}$ [N·m]
5.84 **a.** $\mathbf{F} = 3\mathbf{i} + \mathbf{j} - 2\mathbf{k}$ [kip], $\mathbf{M}_1 = 10\mathbf{i} - 5\mathbf{j} + 4\mathbf{k}$ [kip·ft]; **b.** 15.33 kip·ft
5.86 $\mathbf{P} = -16\mathbf{i} + 45.21\mathbf{j} + 14.79\mathbf{k}$ [lb], $\mathbf{M} = -27.63\mathbf{i} + 96\mathbf{j} + 48\mathbf{k}$ [lb·ft]
5.88 $\mathbf{F} = 1000\mathbf{i} + 185\mathbf{j} - 80\mathbf{k}$ [N], $\mathbf{M} = 448.3\mathbf{i} - 136.3\mathbf{j} - 107.8\mathbf{k}$ [N·m]
5.90 $\mathbf{F} = 40\mathbf{j} + 20\mathbf{k}$ [lb], $\mathbf{M} = -40\mathbf{i} - 190\mathbf{j} + 320\mathbf{k}$ [lb·in]
5.92 $\mathbf{F} = 120\mathbf{i} + 150\mathbf{k}$ [N], $\mathbf{M}_A = 200\mathbf{i} + 360\mathbf{j}$ [N·m]
5.94 $\mathbf{F} = -1.943\mathbf{i} + 3.238\mathbf{j} - 3.202\mathbf{k}$ [N], $\mathbf{M} = 9.72\mathbf{i} + 5.83\mathbf{j}$ [N·m]
5.96 **a.** $\mathbf{F} = 8500\mathbf{j}$ [lb], $\mathbf{M} = 61{,}064\mathbf{i} + 72{,}777\mathbf{k}$ [lb·ft] **b.** $M_x = 61{,}064$ lb·ft, $M_z = 72{,}777$ lb·ft
5.98 $R = 17.74$ kN, (0.512, 0.477, 0.714)
5.100 **a.** $\mathbf{R} = -2\mathbf{i} + 2\mathbf{j} + 2\mathbf{k}$ [kN], $\mathbf{M} = (7 - 2b)\mathbf{i} + (-6 + 2a)\mathbf{j} + (15 - 2a - 2b)\mathbf{k}$ [kN·m] [$\mathbf{R}$ acts at $(a, b, 0)$]; **b.** $a = 3.333$ m, $b = 3.833$ m, with $\alpha = 125.8°$, $\beta = \gamma = 54.7°$
5.102 **a.** $\mathbf{F} = -10\mathbf{i} + 5\mathbf{j} - 8\mathbf{k}$ [kips], $\mathbf{M} = (129 + 8b)\mathbf{i} + (134 - 8a)\mathbf{j} + (-45 - 5a - 10b)\mathbf{k}$ [kip·ft] **b.** $(-1985 + 80a - 40b)\mathbf{k} = 0$ **c.** a = 17.61 ft, b = −14.41 ft
5.104 **a.** $\mathbf{F} = 3\mathbf{i} - 2\mathbf{j} + \mathbf{k}$ [lb], $\mathbf{M} = (12 - b)\mathbf{i} + (-28 + a)\mathbf{j} + (18 + 2a + 3b)\mathbf{k}$ [lb·ft] **b.** $(-8 - 5a - 6b)\mathbf{i} + (-42 - 6a - 10b)\mathbf{j} + (-60 + 3a - 2b)\mathbf{k} = 0$, (0.8018, −0.5345, 0.2673) **c.** $a = 12.29$ ft, $b = -11.57$ ft
5.106 $\mathbf{F} = -100\mathbf{k}$ [lb] at (20, 12, 0), $\mathbf{M} = 0$
5.108 $\mathbf{F} = 1000\mathbf{i} + 185\mathbf{j} - 80\mathbf{k}$ [N] in the direction $\mathbf{n} = 0.980\mathbf{i} + 0.181\mathbf{j} - 0.078\mathbf{k}$ at point (−40.33, 0, 1.92) [cm], $\mathbf{M} = 41.485\mathbf{i} + 7675\mathbf{j} - 3319\mathbf{k}$ [N·cm]

5.110 $\mathbf{F} = -40\mathbf{i} - 30\mathbf{j} - 60\mathbf{k}$ [N] in the direction $\mathbf{n} = -0.512\mathbf{i} - 0.384\mathbf{j} - 0.768\mathbf{k}$ at point $(-0.803, 0.538, 0)$ [m], $\mathbf{M} = 32.27\mathbf{i} + 24.20\mathbf{j} + 48.39\mathbf{k}$ [N·m]

5.112 $\mathbf{F} = -400\mathbf{k}$ in the direction $\mathbf{n} = -\mathbf{k}$ at point $(0.5, 0.3, 0)$ [m], $\mathbf{M} = 0$

5.126 $A_x = 450$ N, $A_y = 900$ N, $B_x = -450$ N

5.128 **a.** $\mathbf{F} = -40\mathbf{i} + 60\mathbf{j} + 30\mathbf{k}$ [N], $\mathbf{M} = 45\mathbf{i} + 60\mathbf{k}$ [N·m]
b. $A_x = 40$ N, $A_y = -60$ N, $A_z = -30$ N, $M_x = -45$ N·m, $M_y = 30$ N·m, $M_z = -120$ N·m

5.130 $\mathbf{F}_A = 300\mathbf{i} + 400\mathbf{k}$ [lb], $\mathbf{M}_A = -1600\mathbf{i} - 2400\mathbf{j} + 1200\mathbf{k}$ [lb·ft]

5.132 130.2 kN

5.134 $\mathbf{R}_O = 3.759\mathbf{i} + 1.368\mathbf{k}$ [kN], $\mathbf{M}_O = 12.31\mathbf{i}$ [kN·m]

5.136 $\mathbf{F} = 346.4\mathbf{i}$ [lb], $\mathbf{C} = -346.4\mathbf{i} + 100\mathbf{j}$ [lb]

5.138 $\mathbf{R}_O = -1.414\mathbf{i} + 1.152\mathbf{j}$ [kN], $\mathbf{R}_B = 1.363\mathbf{j}$ [kN]

5.140 $A = 20.00$ lb, $B_x = -10.00$ lb, $B_y = 22.68$ lb, $C_x = -10.00$ lb, $C_y = 62.68$ lb, $M_C = -170.72$ lb·ft

5.142 $BD = -114.9$ kN (compression), $BE = 154.7$ kN (tension)

5.144 $\mathbf{R}_O = 137.9\mathbf{i} - 311.6\mathbf{k}$ [N], $\mathbf{R}_C = -34.5\mathbf{i} + 187.3\mathbf{k}$ [N], $\mathbf{F}_{DE} = -103.4\mathbf{i} + 299.3\mathbf{k}$ [N]

5.146 **b.** $d \geq 4.26$ ft

5.148 **a.** $A_x = -P\sin(30° - \phi)$, $A_y = -P\cos(30° - \phi)$, $B_x = P\sin(30° - \phi)$, $B_y = P\cos(30° - \phi)$, $B_z = 0$

5.162 $\theta = 90°$

5.164 $(4, -4, 13)$

Chapter 6

6.2 $F_{AB} = -3.59$ kN (compression), $F_{BC} = -0.0717$ kN (compression)

6.4 $T_{AB} = T_{BC} = F/(2\sin\alpha)$ (tension), $T_{CD} = T_{AD} = F/(2\sin\beta)$ (tension), $T_{AC} = (-F/2)(1/\tan\alpha + 1/\tan\beta)$ (compression)

6.6 $T_{AB} = T_{BD} = -100$ N (compression), $T_{AC} = T_{CD} = 86.6$ N (tension), $T_{BC} = 100$ N (tension)

6.8 $T_{AB} = T_{DE} = -115.5$ kips (compression), $T_{AG} = T_{EF} = 57.7$ kips (tension), $T_{BC} = T_{CD} = -115.5$ kips (compression), $T_{BG} = T_{DF} = 115.5$ kips (tension), $T_{CG} = T_{CF} = 0$, $T_{FG} = 115.5$ kips (tension)

6.10 $A_x = -P$, $A_y = -P$, $D_y = P$, $T_{AB} = 0$, $T_{AC} = P/(\cos 45°)$ (tension), $T_{BC} = -P$ (compression), $T_{BD} = 0$, $T_{CD} = -P$ (compression)

6.12 $T_{AC} = 3$ kips (tension), $T_{AD} = 5$ kips (tension), $T_{BD} = -9$ kips (compression), $T_{CD} = 0$

6.14 $T_{AB} = 62.5$ kips (tension), $T_{BC} = 125.0$ kips (tension), $T_{BD} = -108.3$ kips (compression), $T_{BE} = 0$, $T_{CD} = -217$ kips (compression), $T_{DE} = -187.5$ kips (compression)

6.16 $T_{BD} = -3.82$ kN (compression), $T_{CD} = 2.60$ kN (tension), $T_{CE} = 0.577$ kN (tension)

6.18 $T_{AB} = 0$, $T_{AD} = -293$ N (compression), $T_{AE} = 800$ N (tension), $T_{BC} = -400$ N (compression), $T_{CE} = 800$ N (tension), $T_{CD} = -693$ N (compression), $T_{ED} = 414$ N (tension)

6.20 $\theta = -12.09°$, $T_{AB} = 13.18$ kN (tension), $T_{BC} = 6.74$ kN (tension), $T_{CD} = 9.32$ kN (tension)

6.22 $T_{AB} = -10$ kN (compression), $T_{BD} = 7.07$ kN (tension), $T_{DE} = -3$ kN (compression)

6.24 $T_{BD} = 2$ kN (tension), $T_{CD} = 1.155$ kN (tension), $T_{CE} = 1.155$ kN (tension)

6.26 $T_{CE} = -55$ kN (compression), $T_{DF} = -15$ kN (compression), $T_{EF} = -20$ kN (compression)

6.28 $T_{BC} = 0.5$ kN (tension), $T_{BE} = 0$, $T_{EF} = -1.25$ kN (compression)

6.30 $T_{BC} = 1$ kN (tension), $T_{BD} = -3.46$ kN (compression), $T_{BE} = 1$ kN (tension)

6.32 $T_{top} = -177.5$ kN (compression), $T_{diagonal} = 31.8$ kN (tension), $T_{bottom} = 157.7$ kN (tension)

6.34 $T_{BC} = 36.8$ kN (tension), $T_{CD} = -26.0$ kN (compression), $T_{DF} = -92.0$ kN (compression), $T_{CE} = 66.0$ kN (tension)

6.36 $T_{BD} = 4.8$ kN (tension), $T_{CD} = 0$, $T_{CE} = 0$, $T_{BC} = 0$

6.38 $T_{BC} = -11.55$ kN (compression), $T_{CD} = 0$, $T_{DE} = 0$, $T_{CE} = -28.9$ kN (compression)

6.40 $A_x = 0$, $A_y = 0$, $B_y = 60$ kN, $T_{AB} = T_{AD} = 0$, $T_{BC} = T_{CE} = 0$, $T_{BD} = -50$ kN (compression), $T_{BE} = -50$ kN (compression), $T_{DE} = 40$ kN (tension)

6.44 Move the horizontal member that connects the supports to the top of the truss. There are three different solutions.

6.46 **a.** $T_{AB} = 17.32$ N (tension), $T_{AE} = T_{CD} = T_{DE} = 0$, $T_{BE} = T_{CE} = -20$ N (compression)
b. $T_{AB} = T_{BE} = T_{CE} = 0$, $T_{AE} = T_{DE} = 20$ N (tension), $T_{CD} = -17.32$ N (compression)

6.48 $T_{AB} = 1.56P$ (tension), $T_{AC} = -3.16P$ (compression), $T_{AD} = 2.2P$ (tension), $T_{BC} = T_{BD} = 0$, $T_{CD} = P$ (tension), $B_x = -1.299P$, $B_y = -0.866P$, $B_z = 0$, $C_x = 3P$, $D_x = 2.2P$, $D_z = P$

6.50 $T = -8.00$ kN (compression)

6.58 $T_{BD} = T_{BE} = 1$ kN (tension)

Chapter 7

7.2 $d = 220$ mm

7.4 $T_A = 1.256$ kN, $T_D = 3.84$ kN

7.6 $L = 4.29 \times 10^{19}$ km

7.8 $W = 8.462$ kN

7.14 $W/P = 8$
7.16 $F_{AB} = 0$, $F_{BE} = 2$ kN (tension), $F_{DE} = -2$ kN (compression)
7.18 $W = 9.6$ kips
7.20 In Fig. P7.20a, $P = 1000$ lb; in Fig. P7.20b, $P = 800$ lb
7.22 In Fig. P7.22a, $P = 150$ lb; in Fig. P7.22b, $P = 150$ lb; in Fig P7.22c, $P = 33.3$ lb
7.26 $T_{AB} = -8$ kips (compression), $T_{CE} = 0$, $T_{DE} = -2$ kips (compression)
7.28 Choices b and c are both correct.
7.30 $\mathbf{A} = 0.5\mathbf{i} + \mathbf{j}$ [kips], $\mathbf{B} = -0.5\mathbf{i}$ [kips], $\mathbf{D} = -\mathbf{j}$ [kips]
7.32 **b.** $B_x = -3.85$ kN, $B_y = 2.1$ kN, $D = 3.85$ kN
7.34 **b.** $A = 525$ lb, $C_x = 225$ lb, $C_y = 400$ lb
7.36 **b.** $\mathbf{A} = -13.5\mathbf{i} - 2.5\mathbf{j}$ [kN], $\mathbf{B} = 14.5\mathbf{i} + 3.5\mathbf{j}$ [kN], $\mathbf{C} = -\mathbf{i} - \mathbf{j}$ [kN]
7.38 **a.** $T_{BD} = -63.5$ kips (compression), $T_{CD} = 0$, $T_{CE} =$ 39.5 kips (tension), $D_x = 0$, $D_y = 63.5$ kips, $E_x = 0$, $E_y = -39.5$ kips
b. $T_{BD} = -58.5$ kips (compression), $T_{CD} = 0$, T_{CE} $= 34.5$ kips (tension), $D_x = 0$, $D_y + 58.5$ kips, $E_x = 0$, $E_y = -34.5$ kips
7.40 $A_x = 3600$ N, $A_y = 2825$ N, $B_x = -3600$ N, $B_y = 2575$ N, $C_x = 3600$ N, $C_y = 2125$ N
7.42 Member AD: $\mathbf{A} = -0.333\mathbf{j}$ [kN], $\mathbf{D} = 1.333\mathbf{j}$ [kN], $\mathbf{P} = -\mathbf{j}$ [kN]
Member ABC: $\mathbf{A} = 0.333\mathbf{j}$ [kN], $\mathbf{B} = -0.667\mathbf{j}$ [kN], $\mathbf{C} = 0.333\mathbf{j}$ [kN]
Member BDE: $\mathbf{B} = 0.667\mathbf{j}$ [kN], $\mathbf{D} = -1.333\mathbf{j}$ [kN], $\mathbf{E} = 0.667\mathbf{j}$ [kN]
7.44 Member BCD: $\mathbf{B} = 3.12\mathbf{i} + 0.96\mathbf{j}$ [kips], $\mathbf{C} = -1.92\mathbf{i} - 1.92\mathbf{j}$ [kips], $\mathbf{D} = -1.2\mathbf{i} + 0.96\mathbf{j}$ [kips]
Member CE: $\mathbf{C} = 1.92\mathbf{i} + 1.92\mathbf{j}$ [kips], $\mathbf{E} = -1.92\mathbf{i} - 1.92\mathbf{j}$ [kips]
Member $ABEF$: $\mathbf{A} = 1.2\mathbf{i}$ [kips], $\mathbf{B} = -3.12\mathbf{i} - 0.96\mathbf{j}$ [kips], $\mathbf{E} = 1.92\mathbf{i} + 1.92\mathbf{j}$ [kips], $\mathbf{F} = -0.96\mathbf{j}$ [kips]
7.46 $A_x = 40$ kN, $A_y = 0$, $E_x = 40$ kN, $E_y = 10$ kN, $T_{BC} = -23.09$ kN (compression), $T_{BD} = 46.2$ kN (tension)
7.48 **a.** $A_x = F(\cos\theta - 1.25\sin\theta)$, $A_y = -(2/3)F\sin\theta$, $P = 2.083F\sin\theta$
7.50 **a.** $A_x = 2.08$ kN, $A_y = 0$, $B_x = 2.08$ kN, $B_y = 0$, $C_x = 2.08$ kN, $C_y = 2$ kN
7.52 $D_x = 300$ lb, $D_y = 0$, $G = 400$ lb
Pulley: $\mathbf{B} = -100\mathbf{i} + 100\mathbf{j}$ [lb]
Member BCE: $\mathbf{B} = 100\mathbf{i} - 100\mathbf{j}$ [lb], $\mathbf{C} = -70\mathbf{i} + 60\mathbf{j}$ [lb], $\mathbf{E} = -30\mathbf{i} + 40\mathbf{j}$ [lb]
Member $ACDFG$: $\mathbf{A} = 100\mathbf{j}$ [lb], $\mathbf{C} = 70\mathbf{i} - 60\mathbf{j}$ [lb], $\mathbf{D} = 300\mathbf{i}$ [lb], $\mathbf{G} = -400\mathbf{i}$ [lb]
Member EF: $\mathbf{E} = 30\mathbf{i} - 40\mathbf{j}$ [lb], $\mathbf{F} = -30\mathbf{i} + 40\mathbf{j}$ [lb]
7.54 $T_{AB} = 1500$ lb (tension), $T_{AD} = 0$, $T_{BD} = 0$, $T_{CD} =$ -2121 lb (compression), $T_{DE} = -2121$ lb (compression)
7.56 **a.** $F = 3$ kN; **b.** yes
7.58 $A_x = 8.62$ kN, $A_y = 16$ kN, $B_x = 8.62$ kN, $T_{AB} =$ 1.072 kN (tension), $T_{AC} = 17.24$ kN (tension)
7.60 **a.** $T_{CE} = 9769$ N; **b.** no
7.68 $T_{\max} = 4723$ kN
7.70 **a.** $T_{\max} = 101.8$ kN; **b.** $L = 40.0$ m
7.72 **b.** $W = 2.00$ kN/m; **c.** $L = 109.8$ m
7.74 **b.** $W = 2.10$ kN/m; **c.** $L = 112.74$ m
7.76 **b.** $T_0 = 1939$ kN, $T_1 = 2099$ kN
7.78 $d/a = 0.3947$
7.82 **a.** $H = 2.998$ kN; **b.** $\theta_{\max} = 60°$
7.84 14.79 m
7.86 **c.** $K/H = -4$, $w/H = 0.1743\ \text{m}^{-1}$, $C = -2.07$
7.88 $L = 2580$ ft
7.90 **a.** $y_0 = 15.11$ ft, $y_1 = 170.2$ ft; **b.** $L = 182.8$ ft, $W = 365.6$ lb; **c.** $H = 16$ lb, $T_0 = 30.2$ lb, $T_1 = 340$ lb
7.96 Fig. P7.96a, first-class lever; P7.96b second-class lever; P7.96c, second-class lever; P7.96d, third-class lever; P7.96e, third-class lever
7.98 $O_x = 1.2$ N, $O_y = 13.4$ N
7.100 $T_{CD} = 1.5$ kN
7.102 Fig. P7.102a, mechanism, P7.102b, mechanism; P7.102c, truss; P7.102d, frame; P7.102e, frame
7.106 $T_{\max} = 83.9$ lb

Chapter 8

8.2 $(\bar{y}, \bar{z}) = (0.485, 0.0462)$ [m]
8.4 $(\bar{y}, \bar{z}) = (4.83, 4.70)$ [in]
8.6 $(\bar{x}, \bar{y}) = (-1.273, 0.73)$
8.8 $\bar{y} = -0.222$ ft
8.10 The gravity axis moves 21.5 mm toward the nose of the plane.
8.12 no
8.14 $\bar{y} = -166.4$ mm
8.18 $(\bar{x}, \bar{y}, \bar{z}) = (3.00, 4.83, 4.70)$ [in]
8.20 $(\bar{x}, \bar{y}, \bar{z}) = (0.385, 0.485, 0.0462)$ [m]
8.22 **a.** $(\bar{x}, \bar{y}, \bar{z}) = (36.7, -166.4, 0)$ [mm]; **b.** 249 mm
8.24 $(\bar{x}, \bar{y}) = (3.84, 1.832)$ [m]
8.26 $(\bar{x}, \bar{y}) = (3.02, 4.47)$ [ft]
8.28 $(\bar{x}, \bar{y}) = (-0.427, 5.70)$ [m]

8.30 $0.375a$

8.32 $\frac{2b}{3}$

8.36 $(\bar{x}, \bar{y}) = (2.4, 0.857)$

8.38 $(\bar{x}, \bar{y}) =$

$$\left(\frac{-h^2\sqrt{r^2 - h^2} + r^2\sqrt{r^2 - h^2} - (r^2 - h^2)^{3/2}/3}{r^2 \sin^{-1}(\sqrt{r^2 - h^2}/r) - h\sqrt{r^2 - h^2}}, 0\right)$$

8.40 $(\bar{x}, \bar{y}) = \left(\frac{1}{a\sqrt{\pi}}, \frac{1}{\sqrt{2}}\right)$

8.42 $\frac{4a(\sin^3 \theta)}{3(2\theta - \sin 2\theta)}$

8.44 $(\bar{x}, \bar{y}) = (0, 0)$

8.46 1.886 m

8.48 $(\bar{x}, \bar{y}, \bar{z}) = (0.385, 0.485, 0.0462)$ [m]

8.50 $(\bar{x}, \bar{y}) = (-0.838, 3.77)$ [in]

8.52 **a.** 17.25 in; **b.** 0.640 [in]

8.54 $(\bar{x}, \bar{y}) = (-0.354, -0.354)$ [in]

8.56 $(\bar{x}, \bar{y}) = (0, 74.1)$ [mm]

8.58 3.46 m

8.60 $(\bar{x}, \bar{y}) = (3.79, 2.98)$ [in]

8.62 $(\bar{x}, \bar{y}) = \left(\frac{4(3r^2 - 3rh - h^2)}{3[2(4 + \pi)r + \pi h]}, 0\right)$

8.68 $2\pi a^2$

8.70 $2R/\pi$

8.72 $4b/3\pi$

8.74 **a.** $4R/3\pi$; **b.** $4R/3\pi$

8.76 231 ft^3

8.78 **a.** $V = \frac{\pi^2 D^2 R}{2}$; **b.** $A = 2\pi^2 DR$;
c. $V = 5.30 \times 10^6$ m^3, $A = 4.24 \times 10^6$ m^2;
d. excavation: $53,000,000, lining: $3,820,000, total cost: $56,820,000

8.80 $h/2$

8.82 **a.** Triangular load: $R = 900$ N, $\bar{x} = 2$ m, constant load: $R = 400$ N, $\bar{y} = 2$ m
b. $A_x = -525$ N, $A_y = 300$ N, $D_x = 675$ N

8.84 **a.** $R = 300$ N, $M = -300$ N·m;
b. $A_y = 420$ N, $B_x = 300$ N, $B_y = 80$ N

8.86 **a.** Constant load: $R = 9$ kN, $\bar{x} = 0.6$ m, triangular load: $R = 8.1$ kN, $\bar{y} = 0.6$ m
b. $A_x = 8.1$ kN, $A_y = -3.6$ kN, $B_y = 17.1$ kN

8.88 **a.** 37.8 kN; **b.** 77.8 kN·m;
c. $A_x = 0$, $A_y = 16.2$ kN, $B_y = 21.6$ kN

8.92 **a.** $A_y = B_y = 2400 \sin(\phi/2)$

Chapter 9

9.2 7040 psi

9.4 4.74 psi

9.6 14.324 psi

9.8 **a.** 40.72 in Hg; **b.** 46.2 ft H_2O

9.10 $h = 25.74$ mm

9.12 $h = 1.843$ m

9.14 766 lb/ft^2

9.16 $h = 72.1$ mm

9.18 $p_B - p_C = -896$ lb/ft^2, $p_C - p_A = -1792$ lb/ft^2, $p_A - p_B = 2688$ lb/ft^2

9.20 $\theta = 1.841°$

9.22 $y = 14.75$ m

9.24 $\gamma_3 = \frac{\gamma_2 h_2 + \gamma_1 h_1}{h_2 + h_1}$

9.26 **a.** $h = \frac{aH}{2A}$;
b. $\Delta p = \gamma H\left(\frac{a}{A} + 0.5\right)$

9.28 $k = 133 + 123.22\left(\frac{d^2}{D^2}\right)$

9.30 $p_B - p_C = 171{,}300$ N/m^2, $p_C - p_A = -236{,}500$ N/m^2, $p_A - p_B = 65{,}130$ N/m^2

9.32 45.2 kN

9.34 $F = \pi$, $(x_p, y_p) = \left(0, \frac{3}{2\pi}\right)$

9.36 $b = \left(\frac{3\mu_v Q a^2}{W}\right)^{1/3}$

9.38 **a.** 1200 kN;
b. $y_p = 10.13$ m below the water surface

9.40 **a.** 118.5 kN;
b. $y_p = 3.07$ m below the water surface

9.42 4564 lb·ft

9.44 $F = -24.05$ kN, $y_p = 0.659$ m

9.46 **a.** 132,030 kN; **b.** $y_p = 12.5$ m

9.48 172.6 kN

9.50 **a.** $b = \left\{\left(\frac{12\mu_v Q a^2}{W}\right)\left(\frac{1}{4}\right) - \left(\frac{k^2}{8}\right)\left[\ln\left(\frac{2}{k}\right) + \frac{1}{2}\right]\right\}^{1/3}$

9.52 **a.** $A_x = -624h^2$ [lb], $A_y = 13.87h^3 - 624h^2 + 75{,}000$ [lb], $C_y = 150{,}000 + 624h^2 - 13.87h^3$ [lb]

9.54 1944 lb/ft

9.56 4241 lb

9.58 **a.** 28.27 kN; **b.** $\sigma_\theta = 4.00$ MPa, $\sigma_A = 1.905$ MPa

9.60 $R_{min} = 50.38$ kN

9.62 **a.** 125.6 kN at $\theta = 33.0°$; **b.** 525.0 kN·m

9.64 1.236 kN

9.66 3.32 mm

9.68 **a.** 300 kips; **b.** 86 tanks

9.70 $M = 0.490h^4$

9.72 **a.** $\theta = 0°$; **b.** 44%

Chapter 10

10.2 **b.** $A_x = 320$ lb, $F = 20$ lb, $N = 480$ lb

10.4 $P = 900$ N

10.6 **a.** The blocks are in equilibrium. **b.** $\mu_s = 0.1155$
10.8 **a.** Block A will slide. **b.** Block B will slide.
10.10 $\phi = 78.7°$
10.12 $M = 104.0$ lb·ft
10.14 $P = 186.5$ N, $\mu_s = 0.467$
10.16 **b.** $P = 84.5$ N
10.18 $\mu_s = \dfrac{\sin\phi}{1 + \cos\phi}$
10.20 **c.** $P = 300$ N
10.22 $W = 3.47$ kN
10.24 **a.** $\theta = \tan^{-1}\mu_s$;
b. $P = W[\sin(\phi + \phi_s)]$,
$N = W\cos\phi_s$
$\cos(\phi_s + \phi)$
10.26 **b.** $P = \dfrac{W(\mu_{sB})(1 + \mu_{sA})}{1 + 2\mu_{sB}\mu_{sA}}$;
c. $P = 93.6$ N
10.28 $\mu_s = 1 - \dfrac{2b}{a}$
10.32 **a.** 0.268; **b.** the bar's surface
10.36 $\theta = 57.46°$ or 1.003 rad
10.40 $\dfrac{P}{W} = \dfrac{\sin\phi + \mu_s\cos\phi}{\cos\theta + \mu_s\sin\theta}$
$\dfrac{N}{W} = \dfrac{\cos(\theta + \phi)}{\cos\theta + \mu_s\sin\theta}$
10.42 **a.** $\mu_{sA} = 0.3972$, $\mu_{sB} = 0.1924$; **b.** $W_B/W_A = 1.532$
10.44 The upper crate will slide.
10.46 **a.** $\sum F_x = R\cos\phi - N - F\sin\phi = 0$,
$\sum F_y = R\sin\phi - P/2 + F\cos\phi = 0$
10.48 $0.2222Q \le P \le 1.401Q$
10.50 **b.** $P = W\left[\dfrac{\sin(\phi - \phi_s)}{\cos(\phi_s + \theta)}\right]$,
$N = W\left[\dfrac{\cos\phi_s\cos(\theta + \phi)}{\cos(\phi_s + \theta)}\right]$
10.54 **a.** the larger wheel; **b.** $M = -18{,}000x$ [lb·ft];
c. yes
10.56 **a.** $P = \mu_s W\left(\dfrac{3\sin\theta - 2\mu_s\cos\theta}{\sin\theta - \mu_s\cos\theta}\right)$, 130 lb;
b. $\theta \le 11.31°$
10.62 7.107 lb·in
10.64 **a.** 71.91 lb; **b.** 28.51 lb
10.66 31.4 mm
10.68 **a.** $P = 8793$ lb; **b.** $R = 15.83$ lb
10.70 1393 lb·in
10.72 **a.** $\mu_s = 0.07074$; **b.** 49.75%
10.76 **a.** 72.1; **b.** 42.4 N·m, ccw
10.78 **a.** $\eta_a = \dfrac{\tan\phi}{\tan(5.711° + \phi)}$
b. $\eta_b = \dfrac{\tan\phi}{\tan(\phi - 5.711°)}$
c. $\dfrac{\eta_b}{\eta_a} = \dfrac{\tan(\phi + 5.711°)}{\tan(\phi - 5.711°)}$
10.80 $P = 77.5$ lb
10.82 0.075
10.84 $T_{\text{left}} = 1006$ lb, $T_{\text{right}} = 320$ lb,
$M = 286$ lb·ft, ccw
10.86 **a.** 313 N·m; **b.** 313 N·m
10.88 **a.** 416.5 N·m
10.90 **a.** $F = 14.50$ kN, $T_B = 1790$ N, $T_C = 11{,}790$ N
b. $F = 13.58$ N, at 373.6 mm to the right of A
10.96 **a.** $p = \dfrac{W}{\pi(r_C^2 - r_S^2)}$; **b.** $T = \dfrac{2\mu_s W(r^3 - r_S^3)}{3(r_C^2 - r_S^2)}$
10.98 **a.** $T = 1556$ lb·ft
10.100 759 lb·ft
10.104 **b.** $T_k = \left(\dfrac{2\mu_k L}{3}\right)\left(\dfrac{r_2^3 - r_1^3}{r_2^2 - r_1^2}\right)$
10.106 168,750 lb·in
10.108 **a.** $p = \dfrac{3W(r_C - r)}{\pi(r_C - r_S)^2(r_C + 2r_S)}$
b. $T = \dfrac{\mu_s W(r_C^2 + 2r_C r_S + 3r_S^2)}{2(r_C + 2r_S)}$
10.110 $M_d = 2\mu_k p_A b r^2\tan(\phi/2)$

Chapter 11

11.2 $V = 10$ kips, $M = -160$ kip·ft
11.4 $V = 31.5$ kN, $M = -22.1$ kN·m
11.6 $V = -3.63$ kN, $M = 7.25$ kN·m.
11.8 The boards are strong enough.
11.10 $T = 462$ lb·in
11.12 $T = 5400$ lb·in; **b.** $A_z = 333$ lb, $B_z = 166.7$ lb
11.14 $F = 342$ lb
11.16 **a.** 431 N·m; **b.** $R_x = 0$, $R_y = -2404$ N
11.18 **a.** $M_O = Pa - R_y b$
b. Right leg: $V = R_x, N = R_y, M = Pa - R_y b - R_x y$
Left leg: $V = P - R_x, N = R_y, M = y(P - R_x)$
c. $V = R_y, N = R_x, M = Pa - R_y x - R_x a$
11.20 **b.** 97.58 N·m, ccw
11.22 $M = 0.90\sin\theta$ [kN·m],
$V = -2\cos\theta$ [kN], $N = -2\sin\theta$ [kN]
11.24 **a.** $P = 0.7071$ kips;
b. $F = 0.683$ kips;
c. $L = 0.183$ kips;
d. $R_x = 0.683$ kips, $R_y = 0.183$ kips
11.26 $M = -2000 + 100x - 5x^2/12$ [lb·in]
$V = \dfrac{50x - 6000}{\sqrt{x^2 + 3600}}$ [lb]
$N = \dfrac{-(3000 + 100x)}{\sqrt{x^2 + 3600}}$ [lb]

11.28 For $0 < \theta < \alpha$: $M = 400(1 - \cos\theta)$ [lb·ft], $V = 30.77 \sin\theta$ [lb], $N = -30.77 \cos\theta$ [lb]
For $\alpha < \theta < \pi$: $M = 10{,}000(1 + \cos\theta)$ [lb·ft], $V = -769.23 \sin\theta$ [lb], $N = 769.23 \cos\theta$ [lb]
11.30 $A_x = 0$, $A_y = 5.6$ kN, $B_y = 8.4$ kN, $|V_{max}| = 8.4$ kN, $M_{max} = 8.4$ kN·m
11.32 $A_x = 0$, $A_y = 9$ kN, $B_y = 9$ kN, $V_{max} = 9$ kN, $|M_{max}| = 4.5$ kN·m
11.34 $A_x = 0$, $A_y = -2.833$ kips, $B_y = 5.833$ kips, $V_{max} = 3$ kips, $|M_{max}| = 170$ kip·in
11.36 $A_x = 0$, $A_y = -63.64$ lb, $B_y = 63.64$ lb, $|V_{max}| = 63.64$ lb, $|M_{max}| = 4455$ lb·in
11.38 $V_{max} = 3$ kips, $|M_{max}| = 32$ kip·ft
11.40 $V_{max} = P$, $M_{max} = Pa$
11.42 Top beam: $|V_{max}| = 9$ kN, $M_{max} = 18$ kN·m
Middle beam: $V_{max} = 5.375$ kN, $M_{max} = 21.5$ kN·m
Bottom beam: $|V_{max}| = 8.125$ kN, $|M_{max}| = 47$ kN·m
11.44 $V_{max} = 2550$ lb, $|M_{max}| = 20{,}400$ lb·ft
11.46 $|V_{max}| = P$, $M_{max} = 11Pa$
11.50 $V_{max} = 4.2$ kN at A, $M_{max} = 2.94$ kN·m at 1.4 m from A
11.52 $|V_{max}| = 3000$ lb at B, C, and D, $|M_{max}| = 10{,}000$ lb·ft at B
11.54 **c.** $|V_{max}| = (kL^2)/2$ at support, $|M_{max}| = (kL^3)/3$ at support
11.56 **c.** $|V_{max}| = (kL^2)/2$ at support, $|M_{max}| = (kL^3)/6$ at support
11.58 **c.** $V_{max} = 933.33$ lb at left support, $|M_{max}| = 40{,}000$ lb·in from left end to left support
11.60 **c.** $|V_{max}| = (5w_0L)/54$ at $x = 2L/3$, $M_{max} = (26w_0L^2)/729$ at $x = 5L/9$
11.62 **c.** $V_{max} = (3w_0L)/8$ at $x = L/2$ and $x = 3L/2$, $M_{max} = (w_0L^2)/12$ at $x = L/2$ and $x = 3L/2$
11.64 **a.** $V_z = -8.038x^2$ [lb], $M_y = 2.679x^3$ [lb·ft]
11.66 Vertical leg: $V_{max} = 675$ N, $M_{max} = 3000$ N·m
Horizontal leg: $|V_{max}| = 600$ N, $|M_{max}| = 3000$ N·m
11.68 **a.** For $0 \le x < 120$:
$V = -0.0625x^2 + 30x - 1583.8$ [lb],
$N = 4848.6$ lb,
$M = -0.02083x^3 + 15x^2 - 1583.8x$ [lb·in]
For $120 < x \le 160$:
$V = -(20 - 0.0625x)(160 - x)$ [lb], $N = 0$,
$M = (8.333 - 0.02083x)(160 - x)^2$ [lb·in]
b. $|V_{max}| = 1583.8$ lb, $|M_{max}| = 45{,}529$ lb·in, $N_{max} = 4848.6$ lb
11.78 $T_1 = 1260$ N
11.80 **b.** Vertical leg: $|V_{max}| = 20$ N, $M_{max} = 280$ N·m, $N_{max} = 40$ N
Horizontal leg: $|V_{max}| = 60$ N, $M_{max} = 36$ N·m, $N_{max} = 20$ N

Chapter 12

12.2 $U = 42$ J
12.4 **a.** $U_{OA} = -{}^1/_2kx^2$; **b.** $U_{OB} = -{}^1/_2kx^2$; **c.** $U_{O\text{-}O} = 0$
12.6 $U = 1.5 \times 10^7$ ft·lb
12.8 $U_W = -0.518W$, $U_F = 1.553F$
12.10 $U = 40.0$ J
12.12 **a.** $U_A = U_B = U_C = 0$; **b.** $U_{total} = 0$
12.14 $U = 193.5$ J
12.16 **a.** $U = 18.39$ J; **b.** $U = 18.39$ J
12.20 $U_A = -F_BR$
12.22 $U_O = 18.89$ N·m
12.24 $U = 1.046$ MJ
12.26 $U = 129.97$ ft·lb
12.28 $U = 154.95$ N·m
12.30 $U = 6300$ ft·lb
12.32 $U = (a/3)(\theta_1^3 - \theta_0^3) + (b/2)(\theta_1^2 + \theta_0^2)$
12.34 $P = 622.29$ lb
12.36 $P = 100$ lb
12.38 $\dfrac{P}{Q} = \left(\dfrac{c}{d}\right)\left(1 + \dfrac{b}{a}\right)$
12.40 $P = 98.26$ N
12.44 **a.** $\theta = 26.56°$; **b.** unstable

Appendix C

C.2 $I_{xy} = -(b^2h^2)/72$
C.4 $J_O = (\pi/2)(b^2 - a^2)$
C.6 $I_x = (\pi/8)bh^3$
C.8 $I_{A\text{-}A} = 2.907 \times 10^6$ mm^4
C.10 **a.** $\bar{y} = 85.714$ mm; **b.** $I_{A\text{-}A} = 9.286 \times 10^{-5}$ m^4
C.12 **a.** $\bar{y} = 3.3$ in; **b.** $I_y = 22.75$ in^4
C.14 $\bar{x} = 29.17$ mm, $\bar{y} = 54.17$ mm, $I_x = 1.201 \times 10^7$ mm^4, $I_y = 4.20 \times 10^6$ mm^4, $I_{xy} = -3.906 \times 10^6$ mm^4
C.16 $\theta = -29.75°$ and $60.25°$, $I_{x'} = 1.952r^4$, $I_{y'} = 0.404r^4$, $I_{x'y'} = 0$
C.18 **a.** $\bar{x} = 29.17$ mm, $\bar{y} = 54.17$ mm;
b. $\theta = 22.5°$ and $112.5°$;
c. $I_{x'} = 1.363 \times 10^7$ mm^4, $I_{y'} = 2.582 \times 10^6$ mm^4
C.20 **a.** $\bar{x} = 16.056$ mm, $\bar{y} = 20.78$ mm;
b. $\theta = -17.64°$ and $72.36°$;
c. $I_{x'} = 86\,479$ mm^4, $I_{y'} = 28\,298$ mm^4

INDEX

CREDITS

This page constitutes an extension of the copyright page. We have made every effort to trace the ownership of all copyrighted material and to secure permission from copyright holders. In the event of any question arising as to the use of any material, we will be pleased to make the necessary corrections in future printings. Thanks are due to the following photographers, publishers, and agents for permission to use the photos indicated.

Preface **p. xv (clockwise from upper left),** NASA; Jon Eisberg/FPG International LLC; Lester Lefkowitz/FPG International LLC; Nick Wheeler/Corbis; Lester Lefkowitz/FPG International LLC,

Chapter 1 **p. 1,** University of Toronto Photographic Services; **4 (top),** Chuck Muhlstock/SportsLight, 1994; **4 (bottom),** Copyright 1998—Saab. Used with permission of GM Media Archives. All rights reserved; **5 (top),** Smithsonian Institution; **5 (bottom),** NASA; **8,** Photo by H. Mark Helfer, courtesy of NIST; **12,** Chuck Muhlstock/SportsLight, 1994; **23,** UPI/Corbis–Bettmann; **24,** Richard J. Schmidt

Chapter 2 **p. 33,** Paul Sakuma/Wide World Photos; **36,** Richard J. Schmidt; **43,** Age Fotostock 1994/FPG International LLC

Chapter 3 **p. 85,** Camerique, Inc./The Picture Cube; **88,** Richard J. Schmidt; **89,** Camerique, Inc./The Picture Cube; **97,** Tom Pantages; **107,** Bob Daemmrich, Stock Boston

Chapter 4 **p. 127,** Neil Rabinowitz/Corbis; **135,** Nicolas Russell/The Image Bank; **143, 154,** Richard J. Schmidt; **159,** Courtesy of Lee L. Lowery, Jr., Ph.D., P.E.

Chapter 5 **p. 197,** Michael Busselle/Corbis; **212,** Michael S. Yamashita/Corbis; **234,** Courtesy of Case Corporation

Chapter 6 **p. 267,** Michael Dalton/Fundamental Photographs, NYC; **269,** Courtesy of Vulcraft Division of Nucor Corporation; **269, 270, 275, 296 (left),** Richard J. Schmidt; **296 (right),** PhotoDisc

Chapter 7 **p. 311,** Dean Siracusa/FPG International LLC; **313 (left),** Arthur P. Boresi; **313 (right),** Courtesy of Case Corporation; **317,** Richard J. Schmidt; **321,** Timber frame by Riverbend Timber Framing™, structural insulated sandwich panels by Insulspan™; **334,** American Institute of Steel Construction; **343,** Lester Lefkowitz/FPG International LLC; **345,** Richard J. Schmidt

Chapter 8 **p. 373,** Vince Streano/Corbis; **375,** Richard J. Schmidt; **396,** Kindra Clineff/The Picture Cube; **416,** Dan Netzel

Chapter 9 **p. 439,** Peter B. Kaplan/Photo Researchers, Inc.; **444,** Bob Daemmrich, Stock Boston; **446,** Jeff Greenberg, Stock Boston; **470,** Richard J. Schmidt

Chapter 10 **p. 489,** Dean Abramson, Stock Boston; **491,** David Gnizak/Phototake NYC; **491,** Neal Preston/Corbis; **494,** Bob Daemmrich, Stock Boston; **511, 525, 527,** Tom Pantages

Chapter 11 **p. 553,** TempSport/Corbis; **555,** American Institute of Steel Construction; **564,** Tim Barnwell, Stock Boston

Chapter 12 **p. 605,** Neil Leifer; **614,** Tony Duffy/Allsport Photography (USA), Inc. All rights reserved; **606,** Corbis; **627,** Fred Hirschman